Acute Phase Proteins

Molecular Biology, Biochemistry, and Clinical Applications

Acute Phase Proteins

Molecular Biology, Biochemistry, and Clinical Applications

Edited by

Andrzej Mackiewicz, M.D., Ph.D.
Department of Cancer Immunology
Academy of Medicine
at GreatPoland Cancer Center
Poznań, Poland

Irving Kushner, M.D.
Departments of Medicine and Pathology
Case Western Reserve University
MetroHealth Medical Center
Cleveland, Ohio

Heinz Baumann, Ph.D.
Department of Molecular and Cellular Biology
Roswell Park Cancer Institute
Buffalo, New York

CRC Press
Boca Raton Ann Arbor London Tokyo

Library of Congress Cataloging-in-Publication Data

Acute phase proteins: molecular biology, biochemistry, and clinical applications/edited by Andrzej Mackiewicz, Irving Kushner, Heinz Baumann.

p. cm.

Includes bibliographical references and index.

ISBN 0-8493-6913-4

1. Acute phase proteins. 2. Acute phase reaction. I. Mackiewicz, Andrzej. II. Kushner, Irving, 1929- III. Baumann, Heinz, 1947-

[DNLM: 1. Acute Phase Proteins—therapeutic use. QU 55 A189 1933]

RB131.A275 1993

612.1'1--dc20

DNLM/DLC

for Library of Congress

93-128

CIP

Direct all inquiries to CRC Press, Inc., 2000 Corporate Blvd., N.W., Boca Raton, Florida 33431.

International Standard Book Number 0-8493-6913-4

Library of Congress Card Number 93-128

Printed in the United States of America 1 2 3 4 5 6 7 8 9 0

Printed on acid-free paper

THE EDITORS

Andrzej Mackiewicz, M.D., Ph.D., is Head of the Department of Cancer Immunology, Chair of Oncology, Academy of Medicine at the GreatPoland Cancer Center, Poznań, Poland.

He received his M.D. and Ph.D. degrees from the Academy of Medicine at Poznań in 1978 and 1984, respectively. From 1978 to 1985 he served as Senior Instructor of Medicine in the Departments of Pathology and Infectious Diseases at the Academy of Medicine. From 1985 to 1986 and from 1988 to 1990, he was Assistant Professor of Medicine and Head of the Laboratory of Immunochemistry of the Department of Immunology and Rheumatology at the Academy of Medicine. From 1986 to 1987 he served as a Senior Instructor of Medicine in the Department of Medicine, Case Western Reserve University, Metropolitan General Hospital, Cleveland, Ohio. In 1990 he became Associate Professor of Immunology and Head of the Department of Cancer Immunology. In 1991 he was visiting Professor in the Department of Biochemistry, Faculty of Medicine, RWTH Aachen, Germany. During his career he trained for from one to four months at the Institute Pasteur de Lyon (France), at the University of Ghent (Belgium), at the University of Copenhagen (Denmark), and at Jonkoping Central Hospital (Sweden).

Dr. Mackiewicz is a member of the Polish Immunological Society (currently President of the Poznań Region), the Polish Oncological Society, the Polish Society for Pathology, the Polish Society for Internal Medicine, and the International Society for Oncodevelopmental Biology and Medicine. He has received the SOWA Award from the Students' Scientific Society, the Fellows Award from the American Rheumatism Association (Central Region), the L. Hirszfeld Award from the Polish Immunological Society, and a number of awards from the Ministry of Health and Social Welfare for scientific achievements.

He is the author or co-author of more than 80 research papers. His current major research interests relate to molecular mechanisms regulating glycosylation of plasma proteins in diseases and to gene therapy of cancer.

Irving Kushner, M.D., is Professor of Medicine and Pathology at Case Western Reserve University, MetroHealth Medical Center, Cleveland, Ohio.

Dr. Kushner received his B.A. degree with honors from Columbia College of Columbia University, New York City, in 1950, and obtained his M.D. degree cum laude from Washington University School of Medicine, St. Louis, Missouri, in 1954. He trained in Internal Medicine at the Grace New Haven Community Hospital in New Haven, Connecticut, and on the Harvard Service at the Boston City Hospital, and served as a clinical associate in the Laboratory of Clinical Investigation at the National Institute of Allergy and Infectious Diseases of the National Institutes of Health (NIH) in Bethesda, Maryland. Subsequent training was as a Research Fellow of the Helen Hay Whitney Foundation at Cleveland Metropolitan General Hospital. He has been on the faculty of Case Western Reserve University since 1958, except for the year 1973–74 when he served as Professor of Medicine and Chairman of the Division of Rheumatology at West Virginia University. During the period 1974 through 1985, he served as the Director of the Division of Rheumatology at Cleveland Metropolitan General Hospital, and was Medical Director of Highland View Rehabilitation Hospital from 1985 to 1991.

Dr. Kushner is a member of Phi Beta Kappa, Alpha Omega Alpha, the American Association for the Advancement of Science, the American Association of Immunologists, the American College of Physicians, the American Federation for Clinical Research, the American College of Rheumatology, the Central Society for Clinical Research, the Society for Experimental Biology in Medicine, and the New York Academy of Sciences. He has served as chairman of special study section site visit teams for the NIH on numerous occasions. He has been a senior international fellow of the Fogarty International Center of the NIH.

Dr. Kushner is the author of more than 125 papers and of 19 book chapters. He has previously been the editor or co-editor of two books. His current major research interest is in the regulation and function of the acute phase response.

Heinz Baumann, Ph.D., is a Cancer Research Scientist at the Department of Molecular and Cellular Biology, Roswell Park Cancer Institute, Buffalo, New York, and Associate Research Professor, Department of Cellular and Molecular Biology, State University of New York at Buffalo, Roswell Park Division.

He received his Ph.D. in Zoology from the University of Zurich in 1974. After completing postdoctoral fellowships in enzymology, biochemistry, and cell biology at the Institute of Biochemistry, University of Zurich, and at the Roswell Park Cancer Institute Department of Molecular Biology, he assumed the position of Cancer Research Scientist at Roswell Park Cancer Institute in 1978. From 1983 to 1990, he was Assistant Professor, and since 1990 Associate Research Professor in the Department of Cellular and Molecular Biology, State University of New York at Buffalo, Roswell Park Division.

Dr. Baumann received an Established Investigatorship Award from the American Heart Association from 1985 to 1990. His research is supported by grants from the National Institutes of Health (NIH). His memberships include the Swiss Society for Cell and Molecular Biology, the American Society for Cell Biology, the American Society for Biological Chemists, and the New York Academy of Science. From 1987 to 1991, he served on a NIH Study Section for Biological Sciences and is currently a member of the Editorial Board for the *Journal of Biological Chemistry*.

He is the author or co-author of over 80 research papers as well as over 21 reviews and chapters. His long-standing major research interests relate to the genetics, cell biology, and molecular biology of acute phase protein gene regulation.

CONTRIBUTORS

Alok Agrawal, Ph.D.
Department of Medicine
University of Alabama at Birmingham
Birmingham, Alabama

Angela R. Aldred, Ph.D.
Department of Biochemistry
University of Melbourne
Parkville, Victoria, Australia

Rosamonde E. Banks, Ph.D.
Institute for Cancer Studies
St. James's University Hospital
Leeds, United Kingdom

Heinz Baumann, Ph.D.
Department of Molecular and Cellular Biology
Roswell Park Cancer Institute
Buffalo, New York

Pierre Baumann, Ph.D.
Department of Biochemistry and Psychopharmacology
University Department of Adult Psychiatry
Prilly-Lausanne, Switzerland

Earl P. Benditt, M.D.
Department of Pathology
University of Washington
Seattle, Washington

Jonathan Betts, Ph.D.
Department of Internal Medicine
Howard Hughes Medical Institute
University of Michigan Medical Center
Ann Arbor, Michigan

Peter G. Brouckaert, M.D.
Laboratory of Molecular Biology
University of Gent
Gent, Belgium

Susana P. Campos
Department of Pediatrics
Children's Hospital of Buffalo
Buffalo, New York

Gennaro Ciliberto
Department of Virology
Research Institute of Molecular Biology
Rome, Italy

Timothy J. Cole, Ph.D.
Institute of Cell and Tumor Biology
German Cancer Research Center
Heidelberg, Germany

Harvey R. Colten, M.D.
Department of Pediatrics
St. Louis Children's Hospital
Washington University School of Medicine
St. Louis, Missouri

Riccardo Cortese
Department of Scientific Direction
Research Institute of Molecular Biology
Rome, Italy

Mei-Zhen Cui, Ph.D.
Department of Immunology
The Scripps Research Institute
La Jolla, California

Gretchen J. Darlington, Ph.D.
Department of Pathology
Baylor College of Medicine
Houston, Texas

Frederick C. de Beer, M.D.
Department of Medicine
University of Kentucky
Lexington, Kentucky

Wanda Dobryszycka, Pharm. Sci. D., Ph.D.
Department of Biochemistry
Academy of Medicine
Wroclaw, Poland

Chin B. Eap, Ph.D.
Department of Biochemistry and Clinical Psychopharmacology
University Department of Adult Psychiatry
Prilly-Lausanne, Switzerland

Mark R. Edbrooke, Ph.D.
Department of Molecular Science
Glaxo Group Research
Middlesex, United Kingdom

Jan J. Enghild, Ph.D.
Department of Pathology
Duke University Medical Center
Durham, North Carolina

Nils Eriksen, Ph.D.
Department of Pathology
University of Washington
Seattle, Washington

Stuart W. Evans, Ph.D.
Department of Clinical Medicine
University of Leeds
Leeds, United Kingdom

Georg H. Fey, Ph.D.
Department of Genetics
University of Erlangen
Erlangen, Germany

Philippa Francis, Ph.D.
Department of Developmental Biology and Anatomy
Middlesex Hospital
University College
London, England

Erik Fries, Ph.D.
Department of Medical and Physiological Chemistry
University of Uppsala
Uppsala, Sweden

Gerald M. Fuller, Ph.D.
Department of Cell Biology
University of Alabama at Birmingham
Birmingham, Alabama

Mahrukh K. Ganapathi, Ph.D.
Departments of Medicine and Biochemistry
Case Western Reserve University
Cleveland, Ohio

Jack Gauldie, Ph.D.
Department of Pathology
McMaster University
Hamilton, Ontario, Canada

Aleksander Górny, M.D.
Department of Diagnostics and Immunology of Cancer
GreatPoland Cancer Center
Poznań, Poland

Peter C. Heinrich, Ph.D.
Department of Biochemistry
RWTH Aachen
Aachen, Germany

David Heney, M.B., Ch.B.
Department of Pediatrics and Child Health
St. James's University Hospital
Leeds, United Kingdom

Gertrud M. Hocke, Ph.D.
Department of Genetics
University of Erlangen
Erlangen, Germany

Friedemann Horn, Ph.D.
Department of Biochemistry
RWTH Aachen
Aachen, Germany

Todd S.-C. Juan, B.S.
Institute for Molecular Genetics
Baylor College of Medicine
Houston, Texas

Patricia A. Kalonick, M.S.
Department of Medicine
Case Western Reserve University
MetroHealth Medical Center
Cleveland, Ohio

John M. Kilpatrick, Ph.D.
Department of Medicine
University of Alabama at Birmingham
Birmingham, Alabama

Aleksander Koj, M.D., Ph.D.
Institute of Molecular Biology
Jagiellonian University
Krakow, Poland

Irving Kushner, M.D.
Departments of Medicine and Pathology
Case Western Reserve University
MetroHealth Medical Center
Cleveland, Ohio

Claude R. F. Libert, Ph.D.
Laboratory of Molecular Biology
University of Gent
Gent, Belgium

Stephen S. Macintyre, M.D., Ph.D.
Department of Medicine
Case Western Reserve University
MetroHealth Medical Center
Cleveland, Ohio

Andrzej Mackiewicz, M.D., Ph.D.
Department of Cancer Immunology
Academy of Medicine
GreatPoland Cancer Center
Poznań, Poland

Sanja Marinkovic-Pajovic, Ph.D.
Department of Molecular and Cellular Biology
Roswell Park Cancer Institute
Buffalo, New York

Rick L. Meek, Ph.D.
Department of Pathology
University of Washington
Seattle, Washington

David H. Perlmutter, M.D.
Departments of Pediatrics, Cell Biology, and Physiology
Washington University School of Medicine
St. Louis, Missouri

Karen R. Prowse, Ph.D.
Geron Corp.
Menlo Park, California

Dipak Ramji
Department of Genetics
Research Institute of Molecular Biology
Rome, Italy

Carl D. Richards, Ph.D.
Department of Pathology
McMaster University
Hamilton, Ontario, Canada

Jürgen A. Ripperger, M.S.
Department of Genetics
University of Erlangen
Erlangen, Germany

Ronald C. Roberts, Ph.D.
Marshfield Medical Research Foundation
Marshfield, Wisconsin

Hanna Rokita, Ph.D.
Department of Biochemistry
Jagiellonian University
Krakow, Poland

Stefan Rose-John, Ph.D.
Department of Biochemistry
RWTH Aachen
Aachen, Germany

Guy Salvesen, Ph.D.
Department of Pathology
Duke University Medical Center
Durham, North Carolina

Gerhard Schreiber, M.D.
Department of Biochemistry
Melbourne University
Parkville, Victoria, Australia

Pravin B. Sehgal, M.D., Ph.D.
Department of Microbiology and Immunology
New York Medical College
Valhalla, New York

Mohammed Shoyab, Ph.D.
Growth Regulators Department
Bristol-Myers Squibb
Pharmaceutical Research Institute
Seattle, Washington

Jean D. Sipe, Ph.D.
Department of Biochemistry
Boston University School of Medicine
Boston, Massachusetts

E. Mathilda Sjöberg, Ph.D.
Department of Molecular Biology
Karolinska Institute
Huddinge, Sweden

Douglas Thompson, B.S.
Department of Chemical Pathology
The General Infirmary at Leeds
Leeds, United Kingdom

Willem van Dijk, Ph.D.
Department of Medical Chemistry
Faculty of Medicine
Vrije Universiteit
Amsterdam, The Netherlands

John E. Volanakis, M.D.
Department of Medicine
University of Alabama at Birmingham
Birmingham, Alabama

Glenda Watson, Ph.D.
Department of Molecular Rheumatology
Clinical Research Center
Harrow, Middlesex, United Kingdom

Ursula M. Wegenka, M.S.
Department of Biochemistry
RWTH Aachen
Aachen, Germany

John T. Whicher, M.B., B.Chir.
Department of Molecular Pathology and
Experimental Cancer Research
Institute for Cancer Studies
St. James's University Hospital
Leeds, United Kingdom

Deborah R. Wilson, Ph.D.
Department of Pathology
Baylor College of Medicine
Houston, Texas

Kwang-Ai Won, Ph.D.
Department of Molecular Biology
The Scripps Research Institute
La Jolla, California

Patricia Woo, M.D., Ph.D.
Department of Molecular Rheumatology
MRC Clinical Research Center
Harrow, Middlesex, United Kingdom

Yu-Chung Yang, Ph.D.
Department of Medicine
Indiana University School of Medicine
Indianapolis, Indiana

TABLE OF CONTENTS

A. INTRODUCTION TO THE ACUTE PHASE RESPONSE

B. THE MAJOR ACUTE PHASE PROTEINS: THEIR STRUCTURE AND POSSIBLE FUNCTION IN ACUTE PHASE RESPONSE

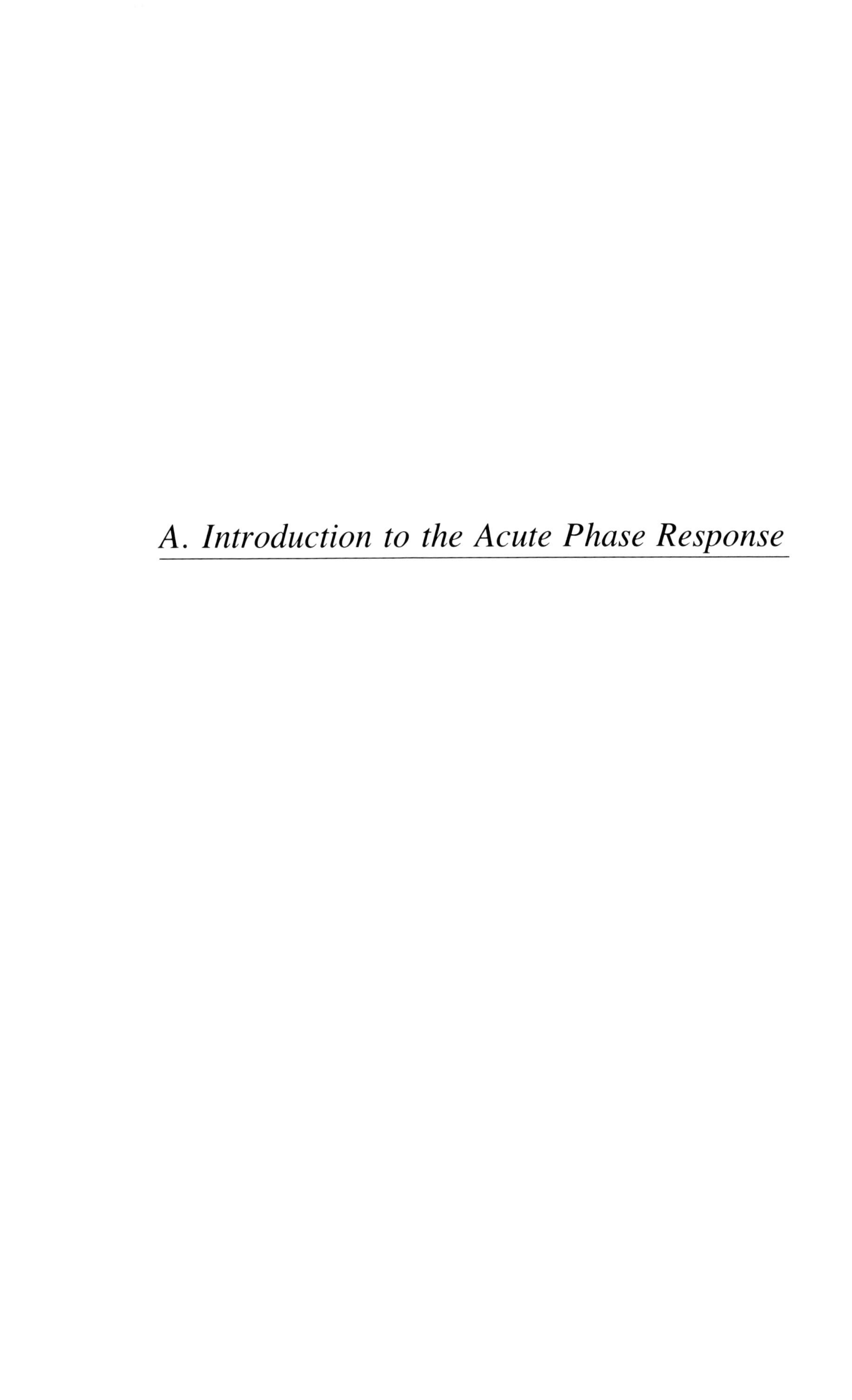

A. Introduction to the Acute Phase Response

Chapter 1

THE ACUTE PHASE RESPONSE: AN OVERVIEW

Irving Kushner and Andrzej Mackiewicz

TABLE OF CONTENTS

0-8493-6913-4/93/$0.00 + $.50

I. ACUTE PHASE RESPONSE

During good health, our homeostatic mechanisms maintain an optimal internal environment in the face of a constantly changing external environment. The adaptive mechanisms by which living organisms respond to external threats range from the increased intracellular synthesis of heat-shock proteins manifested by prokaryotes to the broad array of more complex responses which occur in vertebrates. Following a wide variety of inflammatory stimuli, the localized response, inflammation, is accompanied in vertebrates by a large number of systemic and metabolic changes which are referred to collectively as the acute phase response.[1] Stimuli which commonly give rise to the acute phase response include bacterial infection, surgical or other trauma, bone fracture, neoplasms, burn injury, tissue infarction, various immunologically mediated and crystal-induced inflammatory states, and child birth.[1] The acute phase response represents the substitution of new "set points" and a setting aside of many of the homeostatic processes that normally maintain stability in the face of an inconstant external environment. During significant tissue injury and infection, defense mechanisms must take priority over optimal homeostatic states.

The initial recognition of the acute phase response can probably be attributed to the ancient Greeks, who observed that the blood of healthy individuals formed a homogeneous clot upon standing, while the blood of sick persons rapidly sedimented into four relatively distinct layers before clotting. They hypothesized that the failure of these four blood elements, or humors, to blend was responsible for disease. This belief governed medical thought unchallenged for over 2000 years, until it was gradually appreciated that rapid sedimentation of the red blood cells was the result rather than the cause of disease. The potential clinical usefulness of determining the erythrocyte sedimentation rate was recognized by Fahraeus, who, in 1921, quantitated the rate of blood sedimentation in normal individuals during pregnancy and in various disease states. Subsequent studies have revealed that a rapid erythrocyte sedimentation rate reflects elevated concentrations of several acute phase plasma proteins, particularly fibrinogen.

It is probable that only a small fraction of the systemic, metabolic, humoral, nutritional, and physiologic changes that occur during the acute phase response have as yet been delineated. Among the changes in homeostatic settings described during the acute phase response are fever, somnolence, anorexia, increased synthesis of a number of endocrine hormones, decreased erythropoiesis, thrombocytopenia, alterations in plasma cation concentrations, inhibition of bone formation, negative nitrogen balance (largely resulting from proteolysis and decreased protein synthesis in skeletal muscle) with consequent gluconeogenesis, and alterations in lipid metabolism.[1,2] The acute phase response may be relatively transient, reverting to normal with recovery, or can be persistent in chronic disease.

One of the major components of the acute phase response is the alteration in concentrations of a large number of plasma proteins referred to as acute phase proteins (APP),[3] the focus of this volume. Both increases and decreases in concentration are seen and changes in different proteins occur at differing rates and to different degrees, constituting, all in all, a reorchestration of the pattern of gene expression of secretory proteins in hepatocytes.[4] In addition to these quantitative changes, most of the acute phase glycoproteins undergo qualitative alterations manifested by changes in the arrangement and composition of the heteroglycan side chains.

II. ACUTE PHASE PROTEINS

A. DEFINITION

APP have been empirically defined as those whose plasma concentration changes by 25% or more following inflammatory stimulus.[5] Those proteins whose concentrations in-

TABLE 1
Best-Studied Positive Human Acute Phase Proteins

Protein	Normal plasma concentration (mg/l)
Group I (concentration may increase by 50%)	
Ceruloplasmin	150–600
Complement C3	800–1700
Complement C4	150–650
Group II (concentration may increase two- to fivefold)	
α_1-acid glycoprotein (AGP)	550–1400
α_1-protease inhibitor (PI)	2000–4000
α_1-antichymotrypsin (ACT)	300–1600
Haptoglobin (Hp)	400–1800
Fibrinogen	2000–4500
Group III (concentration may increase up to 1000-fold)	
C-reactive protein (CRP)	<5.0
Serum amyloid A (SAA)	<10.0

crease are referred to as positive APP, while those whose levels decline are termed negative APP. In man, positive APP are represented by a large number of proteins which may be divided into three classes based on their increase in concentration (Table 1): Class I, those whose concentrations rise by about 50%; Class II, those whose concentrations increase about two- to fivefold; and Class III, those whose levels may increase up to 1000-fold. The best-studied human APP are tabulated in Table 1. From the clinical point of view, it is important to note that these quantitative changes are paralleled by kinetic changes. C-reactive protein (CRP), serum amyloid A (SAA), and α_1-antichymotrypsin (ACT) are characterized by a rise in concentration as early as 4 h after inflammatory stimulus, attainment of maximum levels within 24 to 72 h, and very rapid decline. In contrast, the concentrations of most class II APP begin to increase 24 to 48 h after stimulus, reach a maximum in about 7 to 10 d, and need about 2 weeks to return to normal. Other reported APP include several complement components, mannan-binding protein, lipopolysaccharide-binding protein, angiotensinogen, kininogen, kininogenase, ferritin, plasminogen activation inhibitor type I, pancreatic secretory trypsin inhibitor, and phospholipase A2.[2] The negative human APP include albumin, prealbumin, transthyretin, α_2-HS-glycoprotein, and α-fetoprotein, an oncofetal protein which is present in the plasma of newborns and in some cancer patients and behaves both *in vivo* and *in vitro* as a negative APP.

B. INTERSPECIES AND SEX DIFFERENCES

Great variablity in the APP response between different species is observed (Table 2).[5] In all mammalian species studied, there is increased synthesis of haptoglobin, α_1-acid glycoprotein, and fibrinogen, and reduced synthesis of albumin. In contrast, significant interspecies differences in the behavior of CRP, SAA, serum amyloid P (SAP), and α_2-macroglobulin (α_2-M) are found.

TABLE 2
Interspecies Differences in Acute Phase Proteins

	CRP	SAA	SAP	α_2-M	AGP	Tf
Man	+ + +	+ + +	0	0	+ +	−
Rabbit	+ + +	+ + +	?	+ +	?	+
Mouse	+	+ + +	+ +	?	+ +	−
Rat	+	ND	0	+ + +	+ +	(−)(+)

Note:

0	no significant change
+	increase by about twofold
+ +	increase about two- to tenfold
+ + +	increase by 100-fold or more
−	decrease
ND	not detected in plasma
?	not known to the authors
(−)(+)	conflicting reports
SAP	Serum amyloid P
α_2-M	α_2-macroglobulin
Tf	transferrin

In man, both CRP and SAA are major APP, while SAP and α_2-M do not behave as APP. Similarly, in the rabbit, CRP and SAA show the greatest response to injury. In contrast to man, however, serum levels of both α_2-M and transferrin increase during the acute phase response in the rabbit.[6] In the mouse, the major APP is SAA; only minor changes in CRP concentration are seen following stimulus, and SAP is a significant APP. In the rat, α_2-M is the major APP, and SAA is not detectable in plasma. In this species, the normal serum level of CRP is high and no more than a doubling of its concentration usually occurs during the acute phase response. Fibronectin is apparently an APP in the rat and the mouse, although this is not the case in man.[7]

Finally, the pattern of the APP response often depends on the sex of the animal. Considerable differences in changes of plasma α_1-M and α_2-M have been reported in male and female rats after injection with cortisol and turpentine.[8] Sexual dimorphism has also been shown in Syrian hamsters. A protein named "female protein", belonging to the pentraxin family (together with CRP and SAP), is present in high concentration in the plasma of female hamsters and in relatively low concentration in males. During the acute phase response, its concentration may increase threefold in males, while in females it decreases by about 50%.[9]

C. BIOLOGICAL FUNCTION

The known biological functions of APP can be divided into three major categories: participation in host adaptation or defense, inhibition of serine proteinases, and transport of proteins with antioxidant activity, as recently reveiwed by Volanakis.[10] His review serves as the basis for portions of this section. These categories, however, do not include all of the APP, since biological functions of some others, notably SAA, are not known.

1. Host Defense Acute Phase Proteins

CRP is the prototypic APP. A major function seems to relate to its ability to bind foreign pathogens and damaged cells of the host and to initiate their elimination by interacting with humoral and cellular effector systems.[11] The functions of CRP are discussed in detail in Chapter 4 of this volume. In brief, CRP displays a number of biologically significant binding specificities,[12-16] including the ability to bind to galactans,[16a] and displays two major effector functions: complement activation,[17] with consequent opsonizing activity,[18] and modulation of phagocytic cell function. In addition, CRP apparently is able to inhibit platelet-activating factor.[19]

Mannan-binding protein (MBP) is an acute phase lectin with specificity for terminal nonreducing *N*-acetylglucosamine, mannose, fucose, and glucose residues,[20,21] which are present in a number of pathogens. MBP has structural similarities to Clq and conglutinin,[22] and activates the complement system through either the classical[23] or the alternative pathway,[24] thereby serving as an opsonin. *Complement* is a major effector system consisting of more than 30 proteins. The complement components, many of which are acute phase reactants, can, on activation, affect chemotaxis, opsonization, vascular permeability, and vascular dilation, and can lead to cytotoxicity. *Fibrinogen* plays a major role in hemostasis, tissue repair, and wound healing.[25] Fibrinogen binds to activated platelets, forming interplatelet bridges which restore the structural integrity of injured blood vessels.[26] Fibrinogen and fibrin interact with endothelial cells, promoting the adhesion, motility, and cytoskeletal organization of these cells.[27]

2. Inhibitors of Serine Proteinases

The acute phase serine proteinase inhibitors (serpins) control extracellular matrix turnover, fibrinolysis, and complement activation.[28] Since inflammation leads to activation of a number of serine proteinases and the release of others from phagocytic cells, serpins play a critical role in limiting the activity of these enzymes, thus protecting the integrity of host tissues.

Among these APP are *α_1-proteinase inhibitor* (PI) (α_1-antitrypsin) which inhibits neutrophil elastase, *α_1-antichymotrypsin* (ACT) which inhibits chymotrypsin-like serine proteinases, neutrophil cathepsin G, and mast cell chymase, and *C1 inhibitor* (C1 Inh), a key control protein of the complement classical activation pathway which inactivates factors XIIa and XIIf.[29]

3. Transport Proteins with Antioxidant Activity

The APP which belong to this group play an important role in protecting host tissues from toxic oxygen metabolites released from phagocytic cells during inflammatory states,[30] an important function since reactive oxygen metabolites, when not controlled, can cause injury of host cells. *Ceruloplasmin* is involved in copper transport and antioxidant defense,[31] the latter by inhibiting copper ion-stimulated formation of reactive oxidants and the scavengers H_2O_2 and superoxide. *Hemopexin* binds heme released from damaged heme-containing proteins.[32] *Haptoglobin* (Hp) binds hemoglobin released during hemolysis. Recently, it has also been shown to stimulate angiogenesis.[33] Clearance of free hemoglobin[34] is very important since hemoglobin can accelerate lipid peroxidation, leading to production of toxic molecules.[35]

D. REGULATION OF ACUTE PHASE PROTEIN BIOSYNTHESIS

1. Role of Cytokines and Cytokine Receptors

The observation that local tissue injury, at distant sites, leads to the APP response implied the existence of circulating messengers which lead to hepatocyte stimulation.[3] *In vivo* studies have been of limited value because injected inflammatory mediators, such as cytokines, were able to induce synthesis of biologically active secondary endogenous molecules.[36] Studies in primary hepatocyte cultures, hepatoma-derived cell lines, and cells transfected with either plasma protein genes or indicator gene constructs have been carried out to clarify the roles of cytokines and corticosteroids in inducing APP changes. Although each of these model systems has its limitations and findings cannot be interpreted with

TABLE 3
Complexity of the Cytokine Signaling Language

Cytokines may induce themselves.
Cytokines up- or downregulate other cytokines.
Cytokines up- or downregulate receptors for other cytokines.
Cytokine combinations can be additive, inhibitory, synergistic, or cooperative.
Cytokine effects may be influenced by other extracellular messengers.
Cytokine effects are influenced by inhibitors, soluble receptors, autoantibodies, and binding to plasma proteins.

absolute confidence as reflecting *in vivo* events, meaningful conclusions can be drawn if data are interpreted with restraint.[37]

It is clear that cytokines, the secreted products of activated cells, play a central role in the induction of APP synthesis, but it is equally evident that the role of cytokines in the induction of APP is complex. Recent studies have indicated that there are concurrent, overlapping pathways of APP induction, and that a number of cytokines, alone or in a network, may influence the synthesis of various APP. In addition, it appears likely that certain cofactors such as corticosteroids may participate in APP regulation.

Thus far, eight cytokines, including interleukin-6 (IL-6), IL-11, IL-1, tumor necrosis factor-α (TNFα), leukemia inhibitory factor (LIF), transforming growth factor-β (TGFβ), oncostatin M, and interferon-γ (INFγ), either alone or in combination, have been reported to be capable of affecting human plasma protein synthesis in liver cell cultures.[38-46] However, it is possible that still more cytokines will demonstrate such capabilities. In general, these cytokines, where employed alone, each have their own unique spectra of activities. While it is not yet clear which of these cytokines participate in the human acute phase response *in vivo,* there is little doubt that IL-6 serves as a major mediator. This view is based on (1) the large number of plasma proteins whose synthesis is affected and the magnitude of change induced in *in vitro* model systems and (2) the observation of elevated serum IL-6 levels which correlate with acute phase changes in a number of inflammatory states (see below). The spectrum of changes induced by IL-11, LIF, and oncostatin M is similar to that of IL-6, probably reflecting a structural relationship[47] and shared receptor-transducing subunits.[48] IL-1 and TNFα induce more restricted patterns of APP changes which roughly resemble one another. INFγ has been found to downregulate the synthesis of human ACT, PI, and Hp, and to enhance the synthesis of α_2-M and several complement components. TGFβ has been found to influence the synthesis of two antiproteinases (ACT and PI), plasminogen activator inhibitor type I, and some negative APP.[42,44,49,50]

The cytokine-signaling language (Table 3) is very complex. It should be pointed out that the biological effects of circulating cytokines may be inhibited by circulating receptor antagonists or receptors,[51,52] by autoantibodies,[53] or by binding to plasma proteins.[54]

Recently, soluble cytokine receptors have been found in biological fluids.[52] They represent extracellular fragments of their membrane-bound counterparts lacking cytoplasmic and transmembrane domains.[55] Generally, they are believed to compete with membrane-bound receptors for the ligand, which leads to a decrease of cytokine activities. Interestingly, soluble IL-6 receptor (sIL-6-R) was found to display an unique activity resulting in enhancement of IL-6's effect of inducing the synthesis of APP by human hepatoma cells. Moreover, hepatocytes continuously exposed to IL-6 became desensitized to IL-6, and sIL-6-R reconstituted their responsiveness to IL-6.[56]

It is likely that hepatocytes are usually not exposed to individual factors, but, rather, to complex mixtures of extracellular messenger molecules. Cytokines and their soluble receptors appear to function as part of a complex regulatory network, a signaling language in which informational content resides in the combinations, and perhaps sequence, of cytokines and

TABLE 4
Effects of Cytokines on Four Positive Acute Phase Proteins in Hep G2 Cells

	IL-6	TGFβ	TNFα	IL-1α
α_1-Antichymotrypsin	+	+	+	+
α_1-Protease inhibitor	+	+	0	0
Haptoglobin	+	0	↓	*
Fibrinogen	+	↓	↓	↓

Note:
+ induction alone or in combination
0 no effect alone or in combination
↓ inhibition of induction by IL-6
* Inhibition only at low IL-6 concentrations

other extracellular messenger molecules received by a cell.[57] The effects of combinations of cytokines are complex and have often been found to be unpredictable. Additive, inhibitory, and synergistic effects have all been seen, as have cooperative interactions in which cytokines which have no apparent effect when employed alone produce marked changes in combination.

Recent studies of the effects of cytokine combinations on plasma protein production by human hepatoma cell lines lead to three major conclusions. (1) The different plasma proteins respond differently to different combinations of cytokines, each protein exhibiting a somewhat specific response to different combinations of cytokines. (2) The effect of a complex combination of cytokines is not always predictable based on prior knowledge of the effects of the cytokines employed alone, or even on the effects of their binary combinations. (3) Different concentrations of a cytokine and different sequences of delivery of cytokines may produce different effects.

In our studies, we have evaluated the effects of simple binary combinations of four cytokines — IL-6, IL-1α, TNFα, and $TGF\beta_1$ — on production of the six positive APP — PI, ACT, Hp, fibrinogen, CRP, and SAA — and on the negative APP albumin and α-fetoprotein in human hepatoma cells. Each protein exhibited a specific pattern of response to these cytokines, alone or in combination. The effects of ACT, PI, Hp, and fibrinogen[44,58] are shown in Table 4. In addition, induction of CRP and SAA in Hep 3B cells was found to require the cooperative interaction of IL-6 and IL-1, neither cytokine alone having a significant effect on the synthesis of these proteins.[59] However, the combination of IL-6 and TNFα was also effective in inducing SAA, while this combination had no effect on CRP induction.[60] Finally, while each of these four cytokines downregulated production of the negative APP α-fetoprotein and albumin, binary combinations of the cytokines were simply additive, for the most part, in inhibiting α-fetoprotein production, while the inhibitory effects of combinations of cytokines on albumin production differed significantly from simple additive effects.[58]

We recently extended our studies to evaluate the effects of two additional cytokines, LIF and INFγ, in binary combinations with IL-6 and $TGF\beta_1$, on the synthesis of the antiproteinases PI and ACT in Hep G2 cells.[61] In addition, we investigated the effects of even more complex combinations of three or four of the above cytokines on the synthesis of these proteins. Binary combinations of LIF with TGFβ and of IL-6 with TGFβ had simple additive effects, while the combination of LIF and IL-6 was less than additive. INFγ significantly inhibited the synthesis of both proteins when used alone; when combined with each of the other three cytokines, it significantly decreased their inducing effects. Interestingly, INFγ alone, in high concentrations, which caused less inhibition than lower, optimal concentrations, had more pronounced inhibitory effects when employed in binary combinations than the lower optimal dose.

The unpredictable effects of cytokine combinations are underscored by recent studies of the induction of human C1 Inh and CRP. Both IL-6 and INFγ induced C1 Inh, while IL-1 alone had only a marginal effect on the production of this protein.[62] However, in combination, IL-1 inhibited the upregulating effect of INFγ, while it was synergistic with IL-6 in inducing C1 Inh. The effects of different concentrations or sequences of delivery of cytokines are demonstrated in studies in which TGFβ in concentrations of less than 0.1 pg/ml enhanced CRP induction by simultaneously administered IL-6 or IL-1 in NPLC cells, while concentrations in the range of 10 pg/ml or more inhibited CRP induction.[63] In contrast, TGFβ at concentrations of 3 pg/ml administered 6 to 10 h prior to induction with IL-6 enhanced CRP synthesis. Taken together, these observations suggest that the potential exists for highly specific regulation of the individual plasma proteins by particular combinations, sequences, or concentrations of cytokines.

2. Cofactors

It is now clear that factors other than cytokines, cytokine inhibitors, and cytokine receptors are constituents of the network controlling the synthesis of APP. Corticosteroids are the best-studied cofactors. Their role is different in different species and varies for different APP. Dexamethasone, with a few possible exceptions, has no direct effect on the synthesis of human APP, while it has been shown to potentiate the effect of cytokines.[64]

In one study using primary hepatocyte cultures, dexamethasone alone was found to cause moderate induction of SAA and to synergize with IL-6 in inducing this APP, while it had no effect on CRP induction, either alone or in combination with IL-6.[41] Our own data from studies in Hep 3B cells contrasted with these findings: dexamethasone alone had no detectable effect on the synthesis of either SAA or CRP, but led to a dose-dependent potentiation of the induction of both APP by cytokines.[60] In our recent studies of the effect of complex cytokine combinations on the synthesis of PI and ACT by Hep G2 cells, we found that dexamethasone very significantly enhanced IL-6-inducing and INFγ-inhibiting activities, but had only moderately enhancing effects on LIF and TGFβ activity. Accordingly, when dexamethasone was added to binary combinations of these four cytokines, it promoted IL-6 and INFγ effects. When both IL-6 and INFγ were employed together or were constituents of complex cytokine combinations, dexamethasone potentiated IL-6 effects.[61] The mechanisms underlying this potentiation are not clear, although dexamethasone has been shown to induce the mRNAs of both components of the hepatic IL-6 receptor.[65,66] Other possible mechanisms include effects on the initiation of transcription mediated by glucocorticoid-responsive elements[67] or effects on mRNA stability.

Other cofactors have been studied less extensively. Insulin has been found to modulate APP production, interacting with cytokines and dexamethasone in a complex fashion.[68,69] Thrombin had a synergistic effect on the induction of plasminogen-activator inhibitor type 1 by TGFβ,[49] a finding of considerable interest since thrombin receptors have been identified on eukaryotic cells.[70] Finally, histamine was found to potentiate the induction of C3 by IL-6 in mouse hepatocytes,[71] a somewhat surprising finding in view of the downregulatory effect of histamine on IL-6 binding in Hep G2 cells.[72]

3. Intracellular Events

The specific signal transduction mechanisms which mediate APP induction have not yet been definitively evaluated. Well-recognized pathways, including cAMP, Ca^{2+}, phosphoinositol turnover, and activation of protein kinases A and C, do not directly mediate cytokine-induced changes;[73-75] rather, noval signal transduction mechanisms may be involved. It is likely that interactions between downstream signaling elements or branching and cross-talk between pathways may occur.[76] Multiple pathways appear to participate in the induction of different APP, and these effects may be different for different proteins, cytokines, or cell systems. Thus, phorbol esters, which activate protein kinase C, led to increased mRNA levels of fibrinogen in both a rat and a human hepatoma cell line,[73,77] but inhibited CRP and SAA induction by the combination of IL-6 and IL-1 in Hep 3B cells.[78]

Similarly, the processes by which gene transcription is regulated appear to be very complex; the binding of a number of different *trans*-acting nuclear factors to multiple *cis*-acting regulatory components, in specific modular arrangements, seems to be required to achieve full expression of acute phase changes.[79] The effects of cytokines are not mediated by cytokine-specific nuclear factors; rather, multiple cytokines may utilize the same or similar nuclear factors. Overlapping effects and overlapping binding specificities of the different transcription factors are commonly seen,[80,81] and both inhibitory and activating effects of different nuclear factors have been demonstrated.[80,82] As suggested by Muegge and Durum,[83] the apparent specific effect of a single cytokine may be related to the production of particular combinations of transcription factors rather than to a single unique factor.

In addition to transcriptional control, posttranscriptional events participate in APP regulation. Both *in vivo* and *in vitro* studies have suggested that processing or stabilization of mRNAs for various APP may play a role in inducing some APP changes.[84-87] In addition, translational regulation has been implicated.[63,88]

The process of CRP secretion, as distinct from synthesis, appears to be separately regulated during the course of the acute phase response, and may represent another cytokine-mediated phenomenon. Intracellular transport of newly synthesized rabbit CRP is altered during the acute phase response. Intracellular transit time decreases dramatically in hepatocytes from inflamed rabbits.[89] Studies employing the technique of gene transfection[90] and subcellular fractionation[91] have indicated that pentameric CRP is retained within the endoplastic reticulum of the resting hepatocyte, probably via interaction with a specific high-affinity binding site identified by radioligand-binding studies.

E. GLYCOSYLATION ALTERATIONS OF ACUTE PHASE PROTEINS

Most of the APP except CRP, SAA, and albumin are glycoproteins. They almost exclusively bear N-glucosidically linked complex-type glycans which may contain two to four *N*-acetyllactosamine residues — branches (biantennary, triantennary, and tetraantennary structures) arising from the 1,3- and 1,6-α-linked mannose (Man) residues of the pentasaccharide inner core structure [Man$\alpha_{1,3}$(Man$\alpha_{1,6}$)Man$\beta_{1,4}$GlcNAc$\beta_{1,4}$GlcNAc]. *N*-acetyllactosamine residues can in turn bear sialic acid, fucose, or other sugars in a number of different configuratons. Moreover, in these antennary structures, a bisecting GlcNAc in a 1,4 linkage on the β-linked core Man may be present. Variations in the glycan structure present at a given glycosylation site have been referred to as microheterogeneity. Two types of microheterogeneity have been distinguished: major microheterogeneity, which reflects differences in the number of branches on the antennary structures, and minor microheterogeneity, referring to variations in sialic acid or fucose content.[92]

Major micorheterogeneity of APP may be conveniently determined in biological fluids, including patients' sera, by crossed-affinity immunoelectrophoresis (CAIE) employing the lectin concanavalin A (Con A) as a ligand.[92-96] Con A binds the unsubstituted groups of α-linked 2-O-substituted Man residues at carbons 3, 4, and 6, with at least two interacting Man molecules being required for the binding. Thus, this lectin binds with bi-, but not tri- or tetraantennary structures. The presence of an additional bisecting GlcNAc residue on the Manβ of the biantennary units inhibits the binding. In the case of multiheteroglycan proteins, the degree of reactivity with Con A depends on the number of biantennary structures present on the molecule.[92]

1. Acute and Chronic Types of Glycosylation Changes

Changes in serum concentrations of APP are often accompanied by alterations of the relative amounts of various types of the antennary structures on the polypeptide backbone.[92] While an increase of positive and decrease of negative APP concentration is observed in both acute and chronic inflammatory states, changes in the glycosylation patterns of APP in patients' sera during acute inflammation differ from those seen in chronic inflammation.[93,94] Two types of glycosylation alterations have been distinguished. Type I changes have been found in acute inflammation and are charcterized by an increase of the relative amounts of biantennary vs. more complex antennary structures — increased Con A reactivity. In contrast, type II changes have been seen in some chronic inflammatory states and are characterized by a relative increase of tetra- and/or triantennary compared to biantennary structures — decreased Con A reactivity. There are some exceptions to the above classification in which changes in APP concentration are not accompanied by any change in glycosylation pattern.[92-96] It should also be noted that both positive and negative acute phase glycoproteins undergo changes of each type.

2. Influence of Glycosylation Changes on the Function of Acute Phase Proteins

Significant biological roles are played by N-linked glycans of membrane and secretory proteins; different glycoforms of serum APP have been found to have various biological effects. For example, desialylated α_1-acid glycoprotein is associated with increased expression of activity inhibitory to platelet aggregation[97] or with loss of the ability to inhibit mitogen-induced lymphocyte proliferation.[98] The Con A-nonreactive form of serum α_1-acid glycoprotein is more effective in the modulation of lymphocyte proliferation[99] and induction of release of IL-1 inhibitory activity by monocytes[100] than are the Con A-reactive forms. The Con A-nonreactive form of PI is more effective in the inhibition of natural killer cell activity than are Con A-reactive forms.[101] More fucosylated antennary structures of α_1-acid glycoprotein inhibit the adherence of leukocytes to endothelial cells more effectively.[102] Accordingly, changes of the profile of the glycosylation of APP seen during the acute phase response may be associated with altered specific biological activity which affects immune response or clotting.

3. Regulation of Glycosylation Alterations

It has been suggested that changes of the APP glycosylation pattern observed in patients' sera may result from selective clearance of particular glycoforms from the circulation through lectin-like receptor systems such as that on the hepatocyte membrane.[103] It has also been postulated that qualitative alterations of APP might be secondary to quantiative changes, reasoning that the increased rate of APP synthesis is not compensated for by an increase of synthesis or activity of particular glycosylating enzymes.[104] In the light of recent studies, neither hypothetical mechanism seems to be the major one.[105-106a] Using human hepatoma cell lines as a model, we have demonstrated that glycosylation alterations (major microheterogeneity) of a number of APP, both positive and negative, seen in patients' sera occur within hepatocytes at the biosynthetic level following cytokine induction.[107] These results were later confirmed in human and rat primary hepatocyte cultures as well as in mouse primary hepatocyte cultures from rat α_1-acid glycoprotein transgenic mice, and could be induced by a number of cytokines and glucocorticoids, some of them acting directly and some modulating the effects of the others.[108] IL-6, LIF, TGFβ, INFγ, TNFα, IL-1, and dexamethasone were found to be capable of affecting the glycosylation profile of APP secreted by human hepatoma cell lines.[61,108,109] Moreover, sIL-6-R was able to enhance the IL-6 effect on induction of these changes.[56] The final glycosylation pattern, either type I or II, depended on the composition and amounts of cytokines and glucocorticoids affecting the hepatocyte. Finally, dissociation between alterations of the level of gene expression and changes in the pattern of glycosylation of APP was observed.[108]

Posttranslational modification of oligosaccharide side chains of glycoproteins is a multistep enzymatic process. A series of highly specific glycosidases and glycosyltransferases sequentially process an oligosaccharide precursor to yield various types of N-linked glycans.[110] The branching that occurs on complex-type glycans is initiated by the incorporation of a GlcNA residue catalyzed by a group of enzymes, the GlcNAc transferases (GnT). GnT III catalyzes the formation of biantennary structures with bisecting GlcNAc, and GnT IV and GnT V catalyze the formation of tri- and tetraantenary units. Control of the level of relative activity of different GnT is one of the mechanisms which regulates branching during the synthesis of complex-type oligosaccharides. Recently, IL-6 has been demonstrated to increase GnT IV and GnT V activity and to decrease GnT III activity, accompanied by an increase of tri- and tetraantenary glycans on glycoproteins.[111] It should be noted that reduced GnT III activity might also contribute to formation of more branched structures, since GnT III and GnT V compete for the same substrate.[112] Consequently, changes of the glycosylation profile of APP are most probably due to the regulation of GnT gene expression by a network of cytokines, cytokine receptors, and glucocorticoids.

III. ROLE OF CYTOKINES IN OTHER ACUTE PHASE PHENOMENA

Fever, somnolence, and anorexia are the most readily observed acute phase phenomena. Fever, the readjustment of the thermoregulatory set point, has been the subject of scientific study for the longest period of time. It is now appreciated that fever is regulated by factors referred to as endogenous pyrogens. Thus far, IL-1, IL-6, TNFα, INFγ, and macrophage inflammatory protein 1 have all been found to fulfill at least some of the criteria for characterization as endogenous pyrogens.[113] The precise roles played by these cytokines and their interactions in the pathogenesis of fever are as yet unclear.

Recent studies of the role of cytokines in the pathogenesis of somnolence[114] have shown that IL-1α, IL-1β, TNFα, and $INF\alpha_2$, when injected into the lateral ventricle of rabbits, enhanced non-REM sleep while inhibiting REM sleep, presumably directly affecting the anterior hypothalamic preoptic area.

Several cytokines have been implicated in the pathogenesis of anorexia. IL-1α, but not IL-1β,[115] when administered into the cerebral ventricles, suppressed food intake by rats. TNFα caused marked anorexia in rats in experiments in which Chinese hamster ovary (CHO) cells which continuously secreted human TNFα were injected intracerebrally.[116] The same cells injected intramuscularly did not induce anorexia, although comparable serum TNFα levels were achieved, suggesting that local TNFα production in the brain may be critical in this process.

Effects of cytokines on the hypothalamic-pituitary-adrenal axis have been shown in both *in vivo* and *in vitro* studies. In general, there is evidence that at least IL-1 and IL-6 can activate this axis at all three levels.[117] Both IL-1β and IL-6 activate this axis at all three levels.[117] Both IL-1β and IL-6 directly stimulated release of corticotropin-releasing hormone (CRH) from rat hypothalamus explant cultures;[118] no effect of TNFα, INFγ, $INF\alpha_2$, IL-2, or IL-8 was found. Neither IL-1β nor IL-6 had a direct effect on ACTH release, suggesting secondary stimulation by CRH.

Both lipid and protein metabolism are altered during the acute phase response. Recently, TNF has been found to cause skeletal muscle protein breakdown, while IL-1 had no effect.[119] In contrast, IL-1 caused a marked decrease of protein synthesis in skeletal muscle,[120] in studies which suggested that corticosteroids may play a role in, but are not completely responsible for, muscle proteolysis.

IV. RELATIONSHIP OF CYTOKINE LEVELS TO THE ACUTE PHASE RESPONSE AND POSSIBLE CLINICAL USEFULNESS

One would expect the complexity of cytokine regulation of the acute phase response observed in *in vitro* studies to be reflected at the clinical level, leading to different patterns of APP changes. Indeed, it is well recognized clinically that not all acute phase phenomena occur in all sick people; e.g., in many afebrile patients, CRP levels are elevated, and vice versa. The various APP do not always respond in unison in disease states; CRP levels are often not elevated in patients with systemic lupus erythematosus who do have elevated erythrocyte sedimentation rates,[121] and renal transplant rejection is often accompanied by SAA, but not CRP elevation.[122] In addition, in several instances, no correlation was found between APP concentrations and levels of cytokines felt to be responsible for induction of these proteins.

Recent studies of patients with severe sepsis or major trauma (in whom the full-blown acute phase response is usually seen) suggest that IL-6 plays a critical role in the acute phase response in such patients and that determination of its serum concentration may be clinically useful. Serum IL-6 levels are regularly found to be elevated in postoperative patients, while elevation of other cytokines is usually not seen. In these and similar patients, correlations between IL-6 and CRP responses have been found. Finally, striking reductions in serum CRP levels have been seen in patients receiving anti-IL-6 antibodies for therapeutic purposes.[123]

ACKNOWLEDGMENTS

Portions of the work described here were supported by NIH Grant AG 02467, The Marie Sklodowska-Curie Fund II MZ/HHS-92-104, and KBN Grants 40791, 41121, and 41076.

REFERENCES

1. **Kushner, I.,** The phenomenon of the acute phase response, *Ann. N.Y. Acad. Sci.*, 389, 39, 1982.
2. **Kushner, I.,** Regulation of the acute phase response by cytokines, in *Clinical Applications of Cytokines: Role in Pathogenesis, Diagnosis and Therapy,* Gearing, A., Rossio, J., and Oppenheim, J., Eds., Oxford University Press, New York, 1993.
3. **Koj, A.,** Acute phase reactants, in *Structure and Function of Plasma Proteins,* Allison, A. C., Ed., Plenum Press, New York, 1974, 73.
4. **Fey, G. H. and Gauldie, J.,** The acute phase response of the liver in inflammation, *Prog. Liver Dis.*, 9, 89, 1990.
5. **Kushner, I. and Mackiewicz, A.,** Acute phase proteins as disease markers, *Dis. Markers,* 5, 1, 1987.
6. **Mackiewicz, A., Ganapathi, M. K., Schultz, D., Samols, D., Reese, J., and Kushner, I.,** Regulation of rabbit acute phase protein biosynthesis by monokines, *Biochem. J.*, 253, 851, 1988.
7. **Scott, D. L., Robinson, M. W., and Yoshino, W.,** Fibronectin in chronic inflammation: studies using the rat pouch model of chronic allergic inflammation, *Br. J. Exp. Pathol.*, 66, 519, 1985.
8. **Bosanquet, A. G., Chandler, A. M., and Gordon, H.,** Effects of injury on the concentration of alpha-1-macroglobulin and alpha-2-macroglobulin in the plasmas of male and female rats, *Experientia,* 32, 1348, 1976.
9. **Coe, J. E., Margossian, S. S., Slayter, H. S., and Sogn, J. A.,** Hamster female protein, a new pentraxin structurally and functionally similar to C-reactive protein and amyloid P component, *J. Exp. Med.*, 153, 977, 1981.
10. **Volanakis, J. E.,** Acute phase proteins, in *Arthritis and Allied Conditions. A Textbook of Rheumatology,* 12th ed., McCarty, D. J. and Koopman, W. J., Eds., Lea & Febiger, Philadelphia, 1993.
11. **Volanakis, J. E.,** Complement activation by C-reactive protein complexes, *Ann. N.Y. Acad. Sci.*, 389, 235, 1982.

12. **Volanakis, J. E. and Kaplan, M. H.,** Specificity of C-reactive protein for choline phosphate residues of pneumococcal C-polysaccharide, *Proc. Soc. Exp. Biol. Med.*, 136, 612, 1971.
13. **Salonen, E. M., Vartio, T., Hedman, K., and Vahert, A.,** Binding of fibronectin by the acute phase reactant C-reactive protein, *J. Biol. Chem.*, 259, 1496, 1984.
14. **Robey, F. A., Jones, K. D., Tanaka, T., and Liu, T. Y.,** Binding of C-reactive protein to chromatin and nucleosome core particles. Possible physiological role of C-reactive protein, *J. Biol. Chem.*, 259, 7311, 1984.
15. **DuClos, T. W., Zlock, L. T., and Rubin, R. L.,** Analysis of the binding of C-reactive protein to histones and chromatin, *J. Immunol.*, 141, 4266, 1988.
16. **DuClos, T. W.,** C-reactive protein reacts with the U1 and small nuclear ribonucleoprotein, *J. Immunol.*, 143, 2553, 1989.

16a. **Kottgen, E., Hell, B., Kage, A., and Tauber, R.,** Lectin specificity and binding charcteristics of human C-reactive protein, *J. Immunol.*, 149, 445, 1992.

17. **Kaplan, M. H. and Volanakis, J. E.,** Interaction of C-reactive protein complexes with the complement system. I. Consumption of human complement associated with the reaction of C-reactive protein with penumococcal C-polysaccharide and with choline phosphatides, Lecithin and sphingomyelin, *J. Immunol.*, 112, 2135, 1974.
18. **Mortensen, R. F., Osmand, A. P., Lint, T. F., and Gewurz, H.,** Interaction of C-reactive protein with lymphocytes and monocytes: complement dependent adherence and phagocytosis, *J. Immunol.*, 117, 774, 1976.
19. **Kilpatrick, J. M. and Virella, G.,** Inhibition of platelet activating factor by rabbit C-reactive protein, *Clin. Immunol. Immunopathol.*, 37, 276, 1985.
20. **Kawasaki, T., Etoh, R., and Yamashina, I.,** Isolation and charcterization of a mannan-binding protein from rabbit liver, *Biochem. Biophys. Res. Commun.*, 81, 1018, 1978.
21. **Ezekowitz, R. A. B., Day, L. E., and Herman, G. A.,** A human mannose-binding protein is an acute phase reactant that shares sequence homology with other vertebrate lectins, *J. Exp. Med.*, 167, 1034, 1988.
22. **Thiel, S. and Reid, K. B. M.,** Structures and functions associated with the group of mammalian lectins containing collagen-like sequences, *FEBS Lett.*, 250, 78, 1989.
23. **Ohta, M., Okada, M., Yamashina, I., and Kawasaki, T.,** The mechanism of carbohydrate-mediated complement activation by the serum mannan-binding protein, *J. Biiol. Chem.*, 265, 1980, 1990.
24. **Schweinle, J. E., Ezekowitz, R. A., Tenner, A. J., Kuhlman, M., and Joiner, K. A.,** Human mannose-binding protein activates the alternative complement pathway and enhances serum bactericidal activity on a mannose-rich isolate of Salmonella, *J. Clin. Invest.*, 84, 1821, 1989.
25. **Doolittle, R. F.,** Fibrinogen and fibrin, *Annu. Rev. Biochem.*, 53, 195, 1984.
26. **Marguerie, G. A., Ginsberg, M. H., and Plow, E. F.,** Fibrinogen and platelet function, *Adv. Exp. Med. Biol.*, 192, 41, 1985.
27. **Dejana, E., Zanetti, A., and Conforti, G.,** Biochemical and functional characteristics of fibrinogen interaction with endothelial cells, *Haemostasis*, 18, 262, 1988.
28. **Travis, J. and Salvesen, G. S.,** Human plasma proteinase inhibitors, *Annu. Rev. Biochem.*, 52, 655, 1983.
29. **Davis, A. E., III,** C1 inhibitor and hereditary angioneurotic edema, *Annu. Rev. Immunol.*, 6, 595, 1988.
30. **Halliwell, B. and Gutteridge, J. M. C.,** The antioxidants of human extracellular fluids, *Arch. Biochem. Biophys.*, 280, 1, 1990.
31. **Samokyszyn, V. M., Miller, D. M., Reif, D. W., and Aust, S. D.,** Inhibition of superoxide and ferritin-dependent lipid peroxidation by ceruloplasmin, *J. Biol. Chem.*, 264, 21, 1989.
32. **Gutterrdge, J. M. and Smith, A.,** Antioxidant protection by haemopexin of haem-stimulated lipid peroxidation, *Biochem. J.*, 256, 861, 1988.
33. **Cid, M. C., Grant, D. S., Hoffman, G. S., Auerbach, R., Fauci, A. S., and Klinman, H. K.,** Identification of haptoglobin as an antigen factor in sera from patients with systemic vasculitides, *J. Clin. Invest.*, 91, 977, 1993.
34. **McCormick, D. J. and Atassi, M. Z.,** Hemoglobin binding with haptoglobin: delineation of the haptoglobin bindig site on the alpha chain of human hemoglobin, *J. Protein Chem.*, 9, 735, 1990.
35. **Oshiro, S. and Nakajima, H.,** Intrahepatocellular site of the catabolism of heme and globin moiety of hemoglobin-haptoglobin after intravenous administration to rats, *J. Biol. Chem.*, 263, 16032, 1988.
36. **Wahl, S. M., McCartney, N., and Mergenhagen, S. E.,** *Immunol. Today*, 258, 261, 1989.
37. **Kushner, I.,** Post-modernism, the acute phase response, and interpretation of data, *Ann. N.Y. Acad. Sci.*, 389, 39, 1989.
38. **Ramadori, G., Sipe, J. D., Dinarello, C. A., Mizel, S. B., and Colten, H. R.,** Pretranslational modulation of acute phase hepatic protein synthesis by murine recombinant interleukin 1 (IL-1) and purified human IL-1, *J. Exp. Med.*, 162, 930, 1985.
39. **Perlmutter, D. H., Dinarello, C. A., Punsal, P. I., and Colten, H. R.,** Cachectin/tumor necrosis factor regulates hepatic acute phase gene expression, *J. Clin. Invest.*, 78, 1349, 1986.

40. **Gauldie, J., Richards, C., Harnish, D., and Baumann, H.**, Interferon β2/B-cell stimulatory factor type 2 shares identity with monocyte-derived hepatocyte-stimulating factor and regulates the major acute phase protein response in liver cells, *Proc. Natl. Acad. Sci. U.S.A.*, 84, 7251, 1987.
41. **Castell, J. V., Gomez-Lechon, M. J., David, M., Hirano, T., Kishimoto, T., and Heinrich, P. C.**, Recombinant human interleukin-6 (IL-6/BSF-2/HSF) regulates the synthesis of acute phase proteins in human hepatocytes, *FEBS Lett.*, 232, 347, 1988.
42. **Magielska-Zero, D., Bereta, J., Czuba-Pelech, B., Pajdak, W., Gauldie, J., and Koj, A.**, Inhibitory effect of human recombinant interferon gamma on synthesis of acute phase proteins in human hepatoma cells stimulated by leukocyte cytokines, TNFα and INFβ2/BSF-2/IL-6, *Biochem. Int.*, 17, 17, 1988.
43. **Baumann, H. and Wong, G. G.**, Hepatocyte-stimulating factor III shares structural and functional identity with leukemia-inhibitory factor, *J. Immunol.*, 143, 1163, 1989.
44. **Mackiewicz, A., Ganapathi, M. K., Schultz, D., Brabenec, A., Weinstain, J., Kelley, M. F., and Kushner, I.**, Transforming growth factor β1 regulates synthesis of acute phase proteins, *Proc. Natl. Acad. Sci. U.S.A.*, 87, 1491, 1990.
45. **Baumann, H. and Schendel, P.**, Interleukin-11 regulates the hepatic expression of the same plasma protein genes as interleukin 6, *J. Biol. Chem.*, 266, 20424, 1991.
46. **Richards, C. D., Brown, T. J., Shoyab, M., Baumann, H., and Gauldie, J.**, Recombinant oncostatin M stimulates the production of acute phase proteins in Hep G2 cells and rat primary hepatocytes in vitro, *J. Immunol.*, 148, 1731, 1992.
47. **Rose, T. M. and Bruce, A. G.**, Oncostatin M is a member of a cytokine family that incudes leukemia-inhibitory factor, granulocyte colony-stimulating factor, and interleukin 6, *Proc. Natl. Acad. Sci. U.S.A.*, 88, 8641, 1991.
48. **Gearing, D. P., Comeau, M. R., Friend, D. J., Gimpel, S. D., Thut, C. J., McGourty,J., Brasher, K. K., King, J. A., Gillis, S., Mosley, B., Ziegler, S. F., and Cosman, D.**, The IL-6 signal transducer, gp130: an oncostatin M receptor and affinity converter for the LIF receptor, *Science*, 255, 1434, 1992.
49. **Hopkins, W. E., Fujii, S., and Sobel, B. E.**, Synergistic induction of plasminogen activator inhibitor type-1 in Hep G2 cells by thrombin and transforming growth factor-β, *Blood*, 79, 75, 1992.
50. **Nako, K., Nakata, K., Mitsuoka, S., Ohtsuru, A., Ido, A., Hatano, M., Sato, Y., Nakayama, T., Shima, M., Kusumoto, Y., Koji, T., Tamaoki, T., and Nagataki, S.**, Transforming growth factor β1 differentially regulates α-fetoprotein and albumin in HuH-7 human hepatoma cells, *Biochem. Biophys. Res. Commun.*, 174, 1294, 1991.
51. **Arend, W. P. and Dayer, J.-D.**, Cytokines and cytokine inhibitors or antagonists in rheumatoic arthritis, *Arthritis Rheum.*, 33, 305, 1990.
52. **Fernandez-Botran, R.**, Soluble cytokine receptors: their role in immunoregulation, *FASEB J.*, 5, 2567, 1991.
53. **Hansen, M. D., Svenson, M., Diamant, M., and Bendtzen, K.**, Anti-interleukin-6 antibodies in normal human serum, *Scand. J. Immunol.*, 33, 777, 1991.
54. **Taylor, A. W. and Mortensen, R. F.**, Effect of alpha-2-macroglobulin on cytokine-mediated human C-reactive protein production, *Inflammation*, 15, 61, 1991.
55. **Taga, T., Hibi, M., Hirata, Y., Yamasaki, K., Yasukawa, K., Matsuda, T., Hirano, T., and Kishimoto, T.**, Interleukin-6 triggers the association of its receptor with a possible signal transducer, gp130, *Cell*, 58, 573, 1989.
56. **Mackiewicz, A., Schooltink, H., Heinrich, P. C., and Rose-John, S.**, Complex of soluble human interleukin-6-receptor/interleukin-6 up-regulates expression of acute-phase proteins, *J. Immunol.*, 149, 2021, 1992.
57. **Balkwill, F. R. and Burke, F.**, The cytokine network, *Immunol. Today*, 10, 299, 1989.
58. **Mackiewicz, A., Speroff, T., Ganapathi, M. K., and Kushner, I.**, Effects of cytokine combinations on acute phase protein production in two human hepatoma cell lines, *J. Immunol.*, 146, 3032, 1991.
59. **Ganapathi, M. K., May, L. T., Schultz, D., Brabebec, A., Weinstain, J., Sehgal, P. B., and Kushner, I.**, Role of interleukin-6 in regulating synthesis of C-reactive protein and serum amyloid A in human hepatoma cell lines, *Biochim. Biophys. Res. Commun.*, 157, 271, 1989.
60. **Ganapathi, M. K., Rzewnicki, D., Samols, D., Jiang, S.-L., and Kushner, I.**, Effect of combinations of cytokines and hormones on synthesis of serum amyloid A and C-reactive protein in Hep 3B cells, *J. Immunol.*, 147, 1261, 1991.
61. **Mackiewicz, A., Laciak, M., Gorny, M., and Baumann, H.**, Leukemia inhibitory factor, interferon γ and dexamethasone regulate N-glycosylation of α1-protease inhibitor in Hep G2 cells, *Eur. J. Cell. Biol.*, 1993, in press.
62. **Zuraw, B. L. and Lotz, M.**, Regulation of the hepatic synthesis of C1 inhibitor by the hepatocyte stimulating factors interleukin-6 and interferon γ, *J. Biol. Chem.*, 265, 12664, 1990.
63. **Taylor, A. W., Ku, N.-O., and Mortensen, R. F.**, Regulation of cytokine-induced human C-reactive protein production by transforming growth factor β1, *J. Immunol.*, 145, 2507, 1990.

64. **Baumann, H., Richards, C., and Gauldie, J.,** Interaction among hepatocyte-stimulating factors, interleukin 1, and glucocorticoides for regulation of acute phase plasma proteins in human hepatoma (Hep G2) cells, *J. Immunol.*, 139, 4122, 1987.
65. **Rose-John, S., Schooltink, H., Lenz, D., Hipp, G., Dufhues, G., Schmitz, H., Schiel, X., Hirano, T., Kishimoto, T., and Heinrich, P. C.,** Studies on the structure and regulation of the human hepatic interleukin-6 receptor, *Eur. J. Biochem.*, 190, 79, 1990.
66. **Schooltink, H., Schmitz-Van de Leur, H., Heinrich, P. C., and Rose-John, S.,** Up-regulation of the interleukin-6-signal transducing protein (gp130) by interleukin-6 and dexamethasone in Hep G2 cells, *FEBS Lett.*, 297, 263, 1992.
67. **Baumann, H., Jahreis, G. P., and Morella, K. K.,** Interaction of cytokine- and glucocorticoid-response elements of acute phase plasma protein genes, *J. Biol. Chem.*, 265, 22275, 1990.
68. **Thompson, D., Harrison, S. P., Evans, S. W., and Whicher, J. T.,** Insulin modulation of acute phase protein production in a human hepatoma cell line, *Cytokine,* 3, 619, 1991.
69. **Campos, S. P. and Baumann, H.,** Insulin is a prominent modulator of cytokine-stimulated expression of acute phase plasma protein genes, *Mol. Cell. Biol.,* 12, 1789, 1992.
70. **Lefkowitz, R. J.,** Thrombin receptor. Variations on a theme, *Nature (London),* 351, 353, 1991.
71. **Falus, A., Rokita, H., Walcz, E., Brozik, M., Hidvegi, T., and Meretey, K.,** Hormonal regulation of complement biosynthesis in human cell lines. II. Upregulation of the biosynthesis of complement components C3, factor B and C1 inhibitor by interleukin-6 and interleukin-1 in human hepatoma cell line, *Mol. Immunol.*, 27, 197, 1990.
72. **Meretey, K., Falus, A., Taga, T., and Kishimoto, T.,** Histamine influences the expression of the interleukin-6 receptor on human lymphoid, monocytoid and hepatoma cell lines, *Agents Actions,* 33, 189, 1991.
73. **Baumann, H., Isseroff, H., Latimer, J. J., and Jahreis, G. P.,** Phorbol ester modulates interleukin-6 and interleukin-1 regulated expression of acute phase plasma proteins in hepatoma cells, *J. Biol. Chem.*, 263, 17390, 1988.
74. **Ganapathi, M. K., Mackiewicz, A., Samols, D., Brabenec, A., Kushner,I., Hu, S.-I., and Schultz, D.,** Induction of C-reactive protein by cytokines in human hepatoma cell lines is potentiated by caffeine, *Biochem. J.*, 269, 41, 1990.
75. **Heinrich, P. C., Castell, J. V., and Andus, T.,** Interleukin-6 and the acute phase response, *Biochem. J.*, 265, 621, 1990.
76. **Robertson, M.,** Membrane proteins and minimalism, *Nature (London),* 354, 183, 1991.
77. **Kurdowska, A., Bereta, J., and Koj, A.,** Comparison of the action of interleukin-6, phorbol myristate acetate and glucagon on the acute phase protein production and amino acid uptake by fat cultured hepatocytes, *Ann. N.Y. Acad. Sci.*, 557, 506, 1989.
78. **Ganapathi, M. K., Brabenec, A., Schultz, D., and Kushner, I.,** Differential modulation by caffine and phorbal esters of the induction of acute phase proteins by cytokines, *Ann. N. Y. Acad. Sci.*, 557, 497, 1989.
79. **Ganapathi, M. K.,** Okadaic acid, an inhibitor of protein phosphatase 1 and 2A inhibits induction of acute phase proteins by interleukin 6 alone or in combination with interleukin 1 in human hepatoma cell lines, *Biochem. J.*, 284, 645, 1992.
80. **Won, K.-A. and Baumann, H.,** The cytokine response element of the rat α1-acid glycoprotein gene is a complex of several interacting regulatory sequences, *Mol. Cell. Biol.*, 10, 3965, 1990.
81. **Falvey, E. and Schibler, U.,** How are the regulators regulated?, *FASEB J.*, 5, 309, 1991.
82. **Edbrooke, M. R., Foldi, J., Chesire, J. K., Li, F., Faulkes, D. J., and Woo, P.,** Constitutive and NF-kB-like proteins in the regulation of the serum amyloid A gene by interleukin 1, *Cytokine,* 3, 380, 1991.
83. **Muegge, K. and Durum, S. K.,** Cytokines and transcription factors, *Cytokine,* 2, 1, 1990.
84. **Gitlin, J. D.,** Transcriptional regulation of ceruloplasmin gene expression during inflammation, *J. Biol. Chem.*, 263, 6281, 1988.
85. **Morrone, G., Cortese, R., and Sorrentino, V.,** Posttranscriptional control of negative acute phase genes by transforming growth factor beta, *EMBO J.*, 8, 3767, 1989.
86. **Westerhausen, D. R., Hopkins, W. E., and Billadello, J. J.,** Multiple transforming growth factor-β-inducible elements regulate expression of the plasminogen activator inhibitor type-1 gene in Hep G2 cells, *J. Biol. Chem.*, 266, 1092, 1991.
87. **Lappin, D. F., Guc, D., Hill, A., McShane, T., and Whaley, K.,** Effect of interferon-γ on complement gene expression in different cell types, *Biochem. J.*, 281, 437, 1992.
88. **Rogers, J. T., Bridges, K. R., and Durmowicz, G. P., Glass, J., Auron, P. E., and Munro, H. N.,** Translational control during the acute phase response. Ferritin synthesis in response to interleukin 1, *J. Biol. Chem.*, 265, 14572, 1990.
89. **Macintyre, S. S., Kushner, I., and Samols, D.,** Secretion of C-reactive protein becomes more efficient during the course of the acute phase response, *J. Biol. Chem.*, 260, 4169, 1985.

90. **Hu, S.-I., Macintyre, S. S., Schultz, D., Kushner, I., and Samols, D.**, Secretion of rabbit C-reactive protein by transfected human cell lines is more rapid than by cultured rabbit hepatocytes, *J. Biol. Chem.*, 263, 1500, 1988.
91. **Macintyre, S. S.**, Regulated export of a secretory protein from the ER of the hepatocyte: a specific binding site retaining C-reactive protein within the ER is down-regulated during the acute phase response, *J. Cell Biol.*, 118, 253, 1992.
92. **Bręborowicz, J. and Mackiewicz, A.**, Affinity electrophoresis for diagnosis of cancer and inflammatory conditions, *Electrophoresis*, 10, 568, 1989.
93. **Mackiewicz, A., Pawlowski, T., Gorny, A., and Kushner, I.**, Glycoforms of α_1-acid glycoprotein in the management of rheumatic patients, in *Affinity Electrophoresis: Principles and Application*, Bręborowicz, J. and Mackiewicz, A., Eds., CRC Press, Boca Raton, FL, 1992, 229.
94. **Bręborowicz, J., Gorny, A., Drews, K., and Mackiewicz, A.**, Glycoforms of alpha$_1$-acid glycoprotein in cancer, in *Affinity Electrophoresis: Principles and Application*, Bręborowicz, J. and Mackiewicz, A., Eds., CRC Press, Boca Raton, FL, 1992, 191.
95. **Mackiewicz, A., Marcinkowska-Pieta, R., Ballou, S., Mackiewicz, S., and Kushner, I.**, Microheterogeneity of alpha-1-acid glycoprotein in the detection of intercurrent infection in systemic lupus erythematosus, *Arthritis Rheum.*, 30, 513, 1987.
96. **Fassbender, K., Zimmerli, W., Kissling, R., Sobieska, M., Aeschlimann, A., Kellner, M., and Muller, W.**, Glycosylation of α_1-acid glycoprotein in relation to duration of disease in acute and chronic infection and inflammation, *Clin. Chim. Acta*, 203, 315, 1991.
97. **Costello, M., Fiedel, B. A., and Gewurz, H.**, Inhibition of platelet aggregation by native and desialised α1-acid glycoprotein, *Nature (London)*, 281, 677, 1979.
98. **Bennett, M. and Schmidt, K.**, Immunosuppression by human plasma α1-acid glycoprotein: importance of the carbohydrate moiety, *Proc. Natl. Acad. Sci. U.S.A.*, 77, 6109, 1980.
99. **Pos, O., Moshage, H. J., Yap, S. H., Schnieders, J. P. M., Aarden, L. A., Van Gool, J., Boers, W., Brugman, A. M., and Van Dijk, W.**, Effects of monocytic products, recombinant interleukin-1, and recombinant interleukin-6 on glycosylation of α_1-acid glycoprotein: studies with primary human hepatocyte cultures and rats, *Inflammation*, 13, 415, 1989.
100. **Durand, G.**, Glycan variants of human α1-acid glycoprotein modulate the biology of macrophages, *Prog. Clin. Biol. Res.*, 300, 247, 1989.
101. **Le Jeune, P.-J., Mallet, B., Farnarier, C., and Kaplanski, S.**, Changes in serum levels and affinity for concanavalin A of human α1-protease inhibitor in severe burn patients: relationship to natural killer cell activity, *Biochim. Biophys. Acta*, 990, 122, 1989.
102. **De Graaf, T. W., Van der Stelt, M. E., Anbergen, W. G., and Van Dijk, W.**, Inflammation induced increase in sialyl-Lewis^x-bearing glycans on α1-acid glycoprotein in human sera, submitted.
103. **Ashwell, G. and Harford, N.**, Carbohydrate-specific receptors of the liver, *Annu. Rev. Biochem.*, 51, 531, 1982.
104. **Raynes, J.**, Variations in the relative proportions of microheterogeneous forms of plasma glycoproteins in pregnancy and disease, *Biomedicine*, 36, 77, 1982.
105. **Bierhuizen, M., De Wit, M., Govers, C., Ferwerda, W., Koeleman, C., Pos, O., and Van Dijk, W.**, Glycosylation of three molecular forms of human α_1-acid glycoprotein having different interactions with concanavalin A. Variations in the occurrance of di-, tri-, and tetraantenary glycans and the degree of sialylation, *Eur. J. Biochem.*, 175, 387, 1988.
106. **Pawlowski, T. and Mackiewicz, A.**, Minor microheterogeneity of α1-acid glycoprotein in rheumatoid arthritis, *Prog. Clin. Biol. Res.*, 300, 223, 1989.
106a. **Parivar, K., Tolentino, L., Taylor, G., and Oie, S.**, Elimination of non-reactive and weakly reactive human α_1-acid glycoprotein after induction of the acute phase response in rats, *J. Pharm. Pharmacol.*, 44, 447, 1992.
107. **Mackiewicz, A., Ganapathi, M. K., Schultz, D., and Kushner, I.**, Monokines regulate glycosylation of acute phase proteins, *J. Exp. Med.*, 166, 253, 1987.
108. **Mackiewicz, A., Pos, O., Van der Stelt, M. E., Yap, S.-H., Kapcinska, M., Laciak, M., Dewey, M. J., Berger, F. G., Baumann, H., Kushner, I., and Van Dijk, W.**, Regulation of glycosylation of acute phase proteins by cytokines *in vitro*, in *Affinity Electrophoresis: Principles and Application*, Bręborowicz, J. and Mackiewicz, A., Eds., CRC Press, Boca Raton, FL, 1992, 135.
109. **Mackiewicz, A. and Kushner, I.**, Affinity electrophoresis for studies of mechanisms regulating glycosylation of plasma proteins, *Electrophoresis*, 10, 830, 1990.
110. **Schachter, H., Narasimhan, S., Gleeson, P., and Vella, G.**, Control of branching during the biosynthesis of asparagine-linked oligosaccharides, *Can. J. Biochem. Cell Biol.*, 16, 1049, 1983.
111. **Nakao, H., Nishikawa, A., Karasuno, T., Nishiura, T., Iida, M., Kanayama, Y., Yonezawa, T., Tarui, S., and Tanigushi, N.**, Modulation of N-acetyl-glucosaminyltransferase III, IV, and V activates an alteration of the surface oligosaccharide structure of a myeloma cell line by interleukin 6, *Biochem. Biophys. Res. Commun.*, 172, 1260, 1990.

112. **Brockenhausen, I., Hull, E., Hindsgaul, O., Schachter, H., Shah, R. N., Michnick, S. W., and Carver, J. P.,** Control of glycoprotein synthesis. Detection and characterization of a novel branching enzyme from hen oviduct, UDP-N-acetylglucosamine: GlcNAcβ1-6(GlcNAcβ1-2)Manα-R (GlcNAc to Man). Beta-4-N-acetylglucosaminyltransferase VI, *J. Biol. Chem.*, 264, 1121, 1989.
113. **Kluger, M. J.,** Fever: role of pyrogens and cryogens, *Physiol. Rev.*, 71, 93, 1991.
114. **Krueger, J. M., Ferenc, O., Jr. Opp, M., Toth, L., Johannsen, L., and Cady, A. B.,** Somnogenic cytokines and models concerning their effects on sleep, *Yale J. Biol. Med.*, 63, 157, 1990.
115. **Uehara, Y., Shimizu, H., Shimomura, Y., Negishi, M., Kobayashi, I., and Kobayashi, S.,** Effects of recombinant human interleukins on food intake of previously food-deprived rats, *Proc. Soc. Exp. Biol. Med.*, 195, 197, 1990.
116. **Tracey, K. J., Morgello, S., Koplin, B., Fahey, T. J., III, Fox, J., Aledo, A., Manogue, K. R., and Cerami, A.,** Metabolic effects of cachectin/tumor necrosis factor are modified by site of production, *J. Clin. Invest.*, 86, 2014, 1990.
117. **Busbridge, N. J. and Grossman, A. B.,** Stress and the single cytokine: interleukin modulation of the pituitary-adrenal axis, *Mol. Cell. Endocrinol.*, 82, C209, 1991.
118. **Navarra, P., Tsagarakis, S., Faria, M. S., Rees, L. H., Besser, G. M., and Grossman, A. B.,** Interleukin 1 and 6 stimulate the release of corticotropin-releasing hormone-41 from rat hypothalamus in vitro via the eicosanoid cyclooxygenase pathway, *Endocrinology*, 128, 37, 1991.
119. **Goodman, M. N.,** Tumor necrosis factor induces skeletal muscle protein breakdown in rats, *Endocrinol. Metab.*, 23, E727, 1991.
120. **Ballmer, P. E., McNurlan, M. A., Southorn, B. G., Grant, I., and Garlick, P. J.,** Effects of human recombinant interleukin-1β and protein synthesis in rat tissues compared with a classical acute phase reaction induced by turpentine, *Biochem. J.*, 279, 683, 1991.
121. **Pereira Da Silva, J. A., Elkon, K. B., Hughes, G. R. V., Dyck, R. F., and Pepys, M. B.,** C-reactive protein levels in systemic lupus erythematosus: a classification criterion, *Arthritis Rheum.*, 23, 770, 1980.
122. **Maury, C. P. J.,** Comparative study of serum amyloid A protein and C-reactive protein in disease, *Clin. Sci.*, 68, 233, 1985.
123. **Klein, B., Wijdenes, J., Zhang, X.-G., Jourdan, M., Boiron, J. M., Brochier, J., Liautard, J., Merlin, M., Clement, C., Morel-Fournier, B., et al.,** Murine anti-interleukin-6 monoclonal antibody therapy for a patient with plasma cell leukemia, *Blood*, 78, 1198, 1991.

Chapter 2

THE NEGATIVE ACUTE PHASE PROTEINS

Angela R. Aldred and Gerhard Schreiber

TABLE OF CONTENTS

0-8493-6913-4/93/$0.00 + $.50

I. INTRODUCTION

Simultaneous with the increase in concentrations of many plasma proteins in the blood during the acute phase response to tissue injury, a decrease in concentration is observed for several proteins, such as albumin,[1,2] α_2-HS-glycoprotein,[3] transthyretin,[4,5] and retinol-binding protein.[5] The latter proteins have become known as the negative acute phase proteins. The fall in plasma concentration of albumin after injury was originally thought to be partly due to its transfer from the vascular to the extravascular space,[6] or due to increased catabolism[6] (reviewed in Reference 7). However, more recent *in vivo* investigations,[8,9] some of which are presented in this chapter, have shown that cytokine-mediated reduction in the transcriptional activity of the genes for negative acute phase proteins in the liver is responsible for their decreased plasma concentrations after tissue injury.

II. MATERIALS AND METHODS

Details of all materials and methods are presented in the cited publications.

III. RESULTS AND DISCUSSION

A. MECHANISM OF THE NEGATIVE ACUTE PHASE RESPONSE

1. Reduced Concentrations of Negative Acute Phase Proteins in Blood Plasma

An acute phase response can be readily and reproducibly induced in experimental rats by subcutaneous injection of a small amount of mineral turpentine to produce a transient acute inflammation. After induction of an acute inflammation in rats, the concentration of negative acute phase proteins in the blood serum or plasma is reduced considerably, simultaneously with an increase in the concentration of positive acute phase proteins (Figure 1). The serum concentration of albumin decreased from 35 to 23 mg/ml[10] (Figure 1A, top right panel) and the plasma concentration of transthyretin decreased to about one third of the normal value[4] (Figure 1B) 2 d after inducing inflammation. The plasma concentration of α_1-inhibitor III, another strongly downregulated negative acute phase protein in the rat,[11] falls to 36% of its normal concentration 2 d after induction of inflammation by turpentine injection. An average threefold decrease of serum corticosterone binding has been observed in rats 40 h after inducing inflammation by turpentine injection, indicating the negative acute phase nature of transcortin.[12]

The simultaneous decrease in the plasma concentration of negative acute phase proteins, while the concentration of positive acute phase proteins is rapidly increasing, prevents the concentration of total plasma protein from rising more than moderately during the acute phase response (Figure 1A, top left panel).

2. Reduced Concentrations in Plasma Are Due to Decreased Synthesis Rates

The reduced plasma concentrations of negative acute phase proteins, observed during the acute phase response, could result from altered distribution between extravascular and intravascular compartments, an increased rate of degradation, or reduced rate of synthesis of the proteins. The reduced concentration of the negative acute phase proteins in the blood during the acute phase response cannot be explained by changes in equilibration between the vascular system and extravascular spaces. In fact, the relative changes in the total body pools of transthyretin[4] (Figure 2) and albumin[10] (see Figure 6 of Chapter 13, this volume) during an acute phase response are similar to the relative changes in the concentrations of the proteins in the bloodstream (Figure 1). This suggests rapid equilibration of plasma proteins between intravascular and extravascular compartments. The rates of degradation of

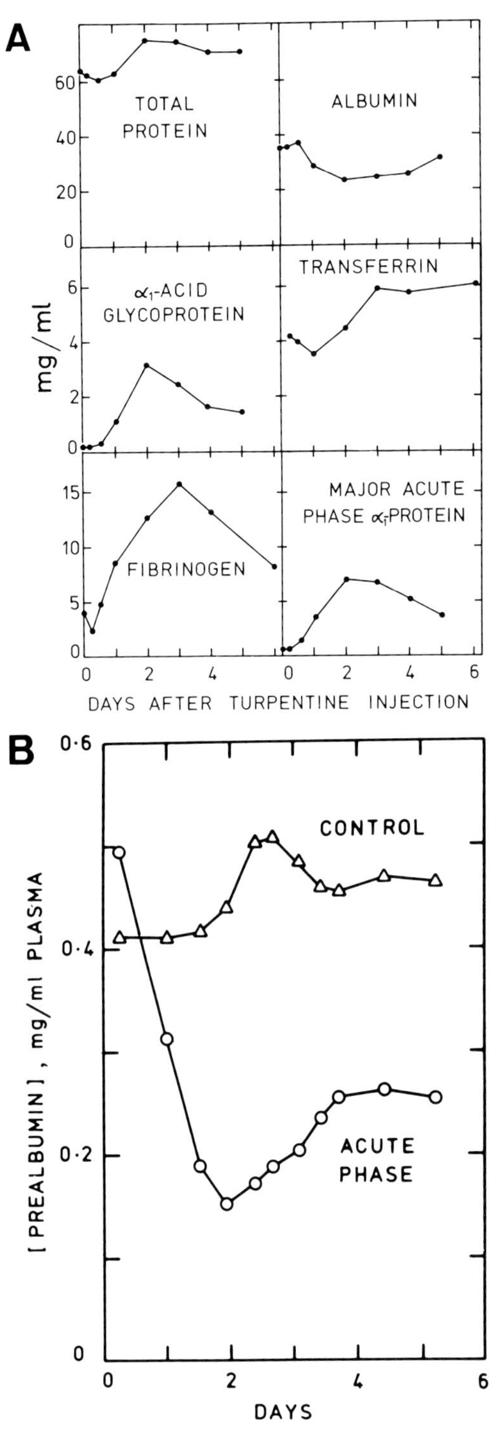

FIGURE 1. (A) Concentrations of proteins in the serum (total protein, albumin, α_1-acid glycoprotein, fibrinogen, major acute phase α_1-protein) or plasma (transferrin) of rats at various times after induction of inflammation by subcutaneous injection of turpentine. (From Schreiber, G., Howlett, G., Nagashima, M., Millership, A., Martin, H., Urban, J., and Kotler, L., *J. Biol. Chem.*, 257, 10271, 1982. With permission.) (B) Concentration of transthyretin (prealbumin) in plasma of healthy rats (control) and in rats suffering an acute inflammation (acute phase) induced by turpentine. (From Dickson, P. W., Howlett, G. J., and Schreiber, G., *Eur. J. Biochem.*, 129, 289, 1982. With permission.)

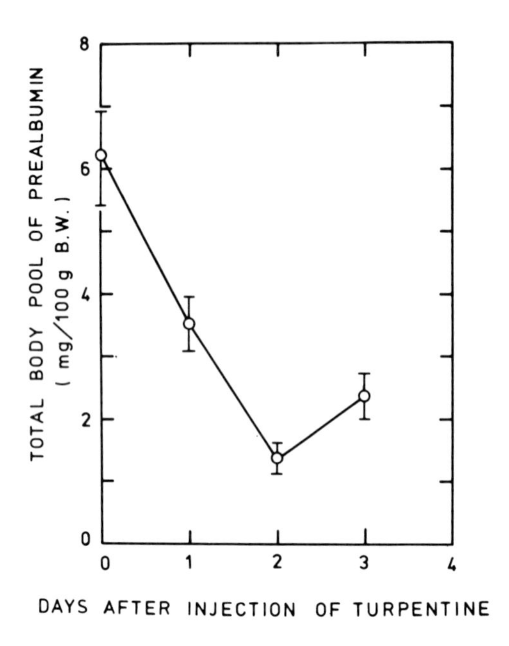

FIGURE 2. Effect of acute inflammation on the total body pool of transthyretin (prealbumin) in rats. (From Dickson, P. W., Howlett, G. J., and Schreiber, G., *Eur. J. Biochem.*, 129, 289, 1982. With permission.)

transthyretin[4] (Figure 3A) and transferrin[10] (Figure 3B) during an acute phase response were not different from the rates of degradation in healthy animals. Thus, increased rates of degradation could not explain the reduced plasma concentration of transthyretin (and probably other plasma proteins) during the acute phase response.

Distinct changes can be observed in the rates of incorporation of [^{14}C]leucine into proteins during an acute inflammation[10,13] (Figure 4). The changes in the rates of incorporation occur in the same direction as the changes in the concentration of the plasma proteins in the bloodstream. The reduction in the rate of incorporation of radioactive amino acid into albumin precedes the reduction in its plasma concentration, similar to the observation for positive acute phase proteins.

Schreiber et al.[10] showed that several plasma proteins, including albumin, are synthesized from a common precursor pool of amino acids in the liver. They calculated the rates of synthesis of plasma proteins in the livers of healthy rats from the total body pools and the turnover of injected radioactive proteins. Using the constant rate of synthesis of transferrin during the first 24 h of the acute phase response as a reference, the net synthesis rate of albumin in rats 24 h after induction of inflammation decreased from 91 to 32 mg/100 g of body weight per day (see Table 4 of Chapter 13, this volume). Jamieson et al.[14] observed a similar decrease in the rate of synthesis of albumin in liver slices obtained from rats suffering an acute inflammation induced by turpentine injection. The rate of synthesis of transthyretin in healthy rats, 4 mg/100 g of body weight per day, was estimated to decrease to near zero during the first 2 d of an acute phase response.[4]

3. Reduced Synthesis Rates Are Due to Decreased Transcription

Reduced synthesis rates of a negative acute phase protein can be the result of reduction in the rate of transcription of its gene or translation of its mRNA. The availability of specific cDNA probes has allowed measurement of intracellular levels of mRNA for specific plasma proteins by hybridization. Changes in the relative concentrations of mRNAs for negative acute phase plasma proteins observed in liver after induction of an inflammation[15] are

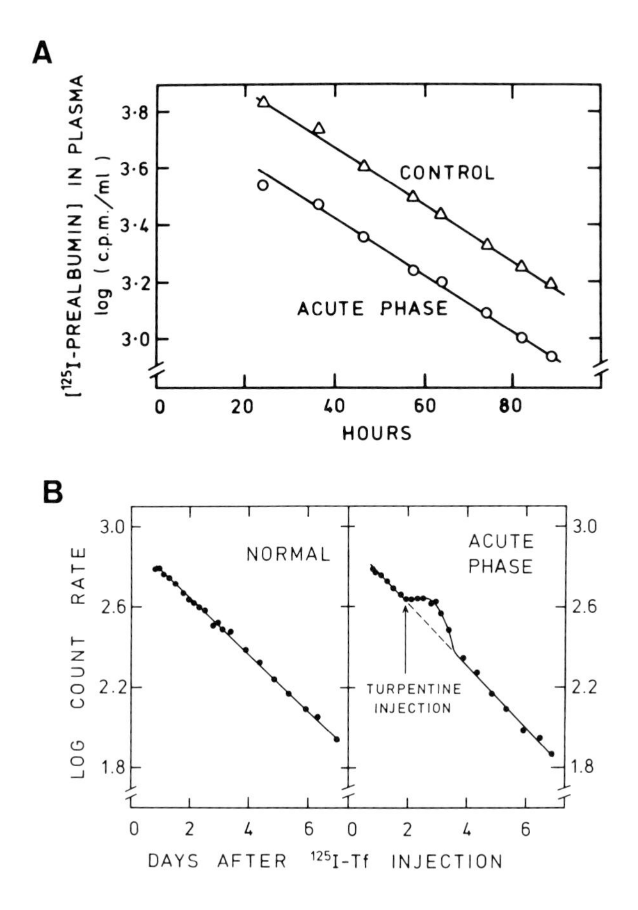

FIGURE 3. (A) Rate of removal of [^{125}I]transthyretin ([^{125}I]prealbumin) from the bloodstream of healthy rats (control) and rats suffering from an acute inflammation (acute phase) induced by subcutaneous injection of turpentine. (From Dickson, P. W., Howlett, G. J., and Schreiber, G., *Eur. J. Biochem.*, 129, 289, 1982. With permission.) (B) Disappearance of [^{125}I]transferrin ([^{125}I]Tf) from a healthy rat (normal) and a rat injected subcutaneously with turpentine (acute phase) 2 d after injection of [^{125}I]transferrin. Radiation was detected by two NaI detectors on opposite sides of the animal. Deviation from a straight line in the semilogarithmic plot for the acute phase rat is caused by the transient accumulation of fluid containing plasma proteins in the inflamed area. This increases the efficiency with which [^{125}I]-radiation is detected because the half-thickness for ^{125}I-radiation in the rat tissue was only 2 cm. (From Schreiber, G., Howlett, G., Nagashima, M., Millership, A., Martin, H., Urban, J., and Kotler, L., *J. Biol. Chem.*, 257, 10271, 1982. With permission.)

presented in Figure 5A. A rapid decrease was observed in the levels of mRNA for albumin,[16] transthyretin,[17] α_{2u}-globulin,[15] retinol-binding protein,[15] α_2-HS-glycoprotein[18] (see Section III.C below), and inter-α-trypsin inhibitor I[18] (see Section III.C below) after inducing inflammation. The minimum in the mRNA level after induction of inflammation was reached first for retinol-binding protein (24 h), then for albumin and transthyretin (36 h). The level of α_{2u}-globulin mRNA continued to decrease for several days. A similar decrease in the level of mRNA for α_2-HS-glycoprotein after turpentine injection has been reported.[19] Similar decreases in the level of albumin mRNA have been reported after induction of inflammation by injection of turpentine[20,21] or bacterial lipopolysaccharide.[22] A less rapid decrease in the mRNA levels for albumin and α_{2u}-globulin was observed by Baumann et al.[23] after injecting a smaller dose of turpentine. A rapid decrease in the level of mRNA for α_1-inhibitor III was observed after intraperitoneal injection of complete Freund's adjuvant,[24] whereas a delayed decrease (after 6 h) was observed after subcutaneous injection of turpentine.[25] In contrast

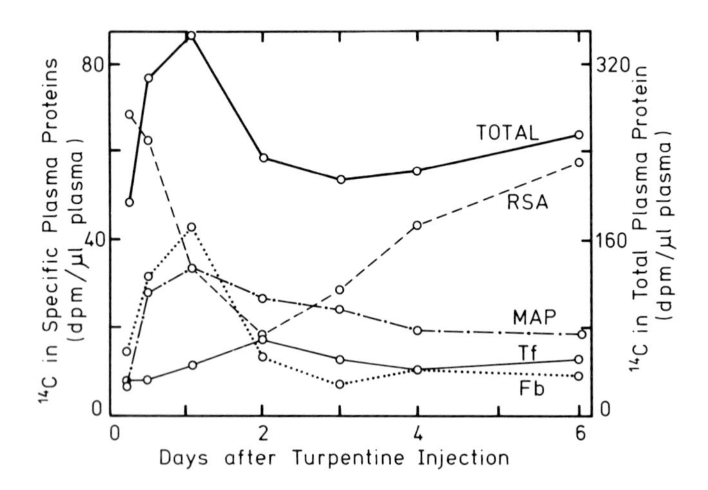

FIGURE 4. Incorporation of L-[1-^{14}C]leucine into plasma proteins during acute experimental inflammation. Incorporation of [^{14}C]leucine during a 2-h period was measured for total protein (TOTAL), albumin (RSA), and thiostatin (MAP) in serum and for transferrin (Tf), and fibrinogen (Fb) in plasma of rats at various times after induction of inflammation by subcutaneous injection of turpentine. (From Schreiber, G. and Howlett, G., in *Plasma Protein Secretion by the Liver,* Glaumann, H., Peters, T., Jr., and Redman, C., Eds., Academic Press, London, 1983, 423. With permission.)

to observations for the other negative acute phase proteins, mRNA levels in liver for the two negative acute phase apolipoproteins, apolipoprotein AIV and apolipoprotein E, remained unchanged for the first 18 h after induction of inflammation[26] (Figure 5A). However, after this time, a dramatic decrease in the level of apolipoprotein AIV mRNA and a more moderate decrease in the level of apolipoprotein E mRNA were observed.

Levels of mRNA in the liver for some intracellular proteins are also influenced by the acute phase response[15] (Figure 5B). Levels of mRNA for ornithine transcarbamoylase decreased dramatically in rat liver immediately after induction of inflammation. Alcohol dehydrogenase and phosphoenolpyruvate carboxykinase mRNA levels decreased moderately before increasing to above-normal levels 40 h after induction of inflammation.

Changes in the mRNA levels for acute phase proteins closely corresponded to changes in the rate of incorporation of radiolabeled amino acid into the proteins. The ratio of the rate of incorporation of amino acid into protein in plasma to the relative mRNA level in liver remained fairly constant during the acute phase response, irrespective of the dramatic changes in either the incorporation rate or mRNA level[16] (Figure 6). This indicates that changes in the synthesis of plasma proteins during the acute phase response are not caused by regulation at the level of translation.

The reduced level of mRNA in the liver for negative acute phase proteins observed during the acute phase response could be brought about by reduced transcriptional activity of the genes involved, or by decreased stability of the specific mRNAs. The transcriptional activities of genes for several negative and positive acute phase proteins in rat liver were measured after induction of an acute inflammation[27] (Figure 7). For all three negative acute phase proteins tested — α_{2u}-globulin, transthyretin, and albumin — transcriptional activity decreased dramatically immediately after induction of inflammation. The early decrease in transcriptional activity preceded the reduction in mRNA levels. About 24 h after induction of inflammation, the transcriptional activities of the genes for transthyretin and albumin began to increase, followed by an increase in mRNA levels. Thus, the negative acute phase response of α_{2u}-globulin, transthyretin, and albumin is achieved by changes at the level of gene transcription, with little, if any, change in the rate of degradation of mRNA. Similar

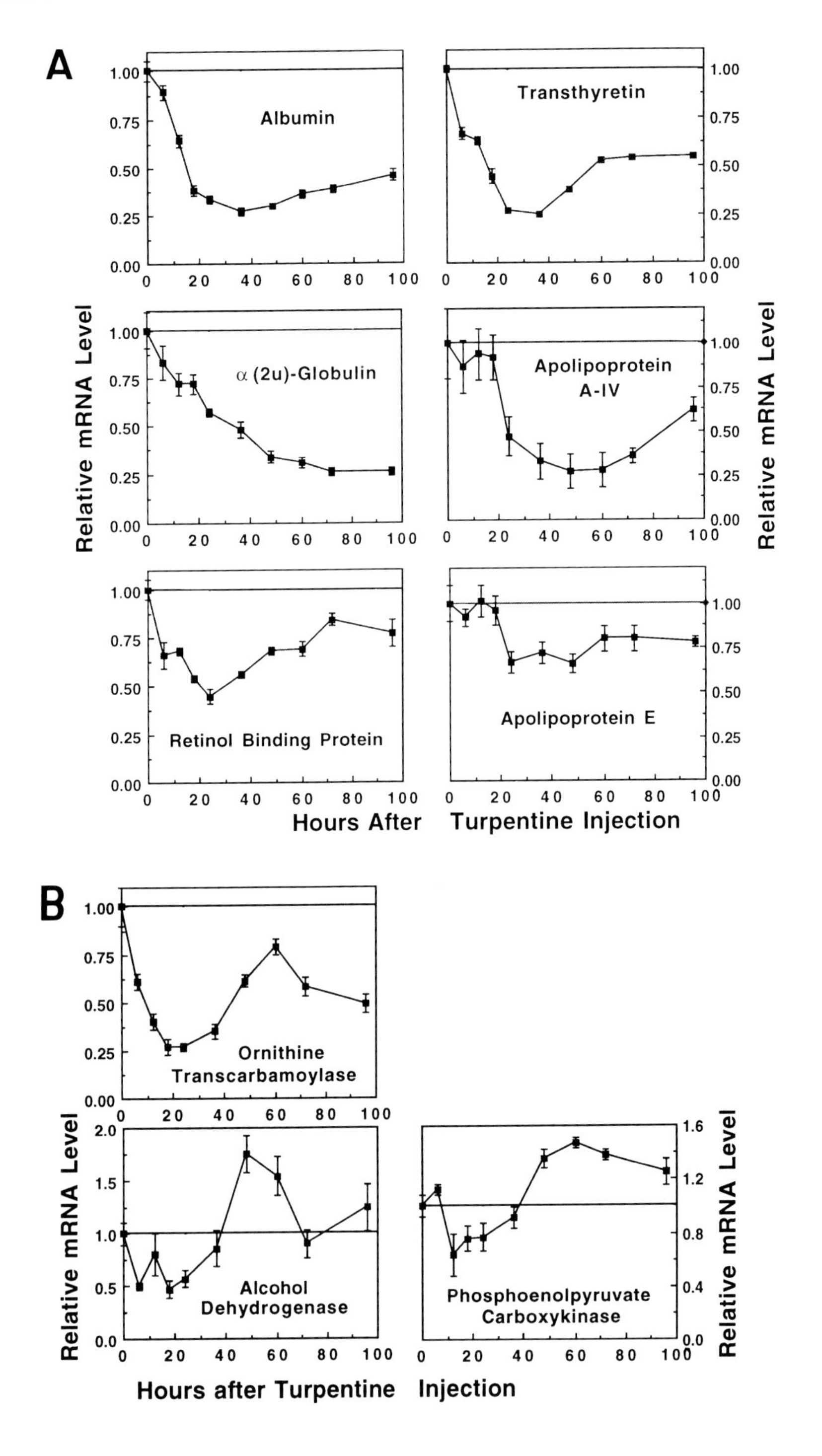

FIGURE 5. Relative mRNA levels for (A) negative acute phase plasma proteins and (B) intracellular proteins in the liver of rats suffering from various lengths of time from an acute inflammation induced by subcutaneous injection of turpentine. Levels of mRNA were determined by hybridization of specific cDNA probes to mRNA in cytoplasmic extracts from liver. Each point is the mean ± standard error for eight rats. The level in healthy rats is indicated by a horizontal line. (Modified from Schreiber, G., Tsykin, A., Aldred, A. R., Thomas, T., Fung, W.-P., Dickson, P. W., Cole, T., Birch, H., de Jong, F. A., and Milland, J., *Ann. N.Y. Acad. Sci.*, 557, 61, 1989. With permission.)

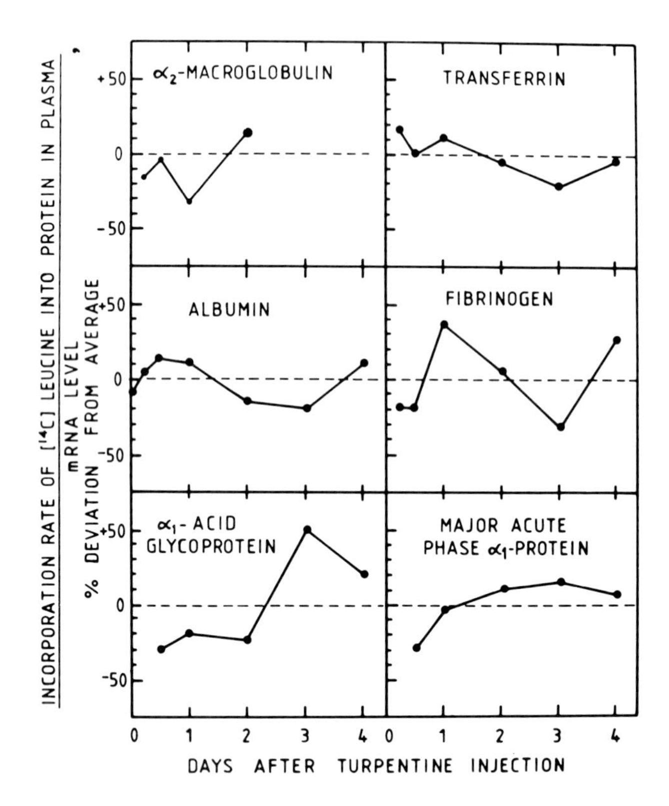

FIGURE 6. Relationship between rates of incorporation of radioactive leucine into plasma proteins and levels of corresponding messenger ribonucleic acids in the liver of rats suffering an acute inflammation. Rates of incorporation of L-[1-^{14}C]leucine into plasma proteins[10] (see Figure 4) were divided by mRNA levels[16] (see Figure 5) at various times after induction of inflammation. The average of these ratios for a particular protein was considered to be 100% (indicated by a broken line) for comparison with the ratios for individual time points. (From Schreiber, G., Aldred, A. R., Thomas, T., Birch, H. E., Dickson, P. W., Guo-Fen, T., Heinrich, P. C., Northemann, W., Howlett, G. J., de Jong, F. A., and Mitchell, A., *Inflammation,* 10, 59, 1986. With permission.)

reductions in the transcriptional activity of the albumin and α_{2u}-globulin genes during turpentine-induced inflammation,[28] and the α_1-inhibitor III gene during adjuvant-induced inflammation,[29] have been reported. In contrast, for some positive acute phase proteins, e.g., α_1-acid glycoprotein and α_2-macroglobulin, changes in mRNA stability have great importance in determining the acute phase levels of mRNA.

In cases where the transcriptional activity is low (α_{2u}-globulin, transthyretin, and albumin), or where changes in mRNA levels are large (retinol-binding protein and apolipoprotein AIV), it is possible to calculate an upper threshold for the half-life of specific mRNAs during the acute phase response. Upper threshold values for the half-life of mRNAs for several negative acute phase proteins,[15] calculated from information from Figure 5, are presented in Table 1.

4. Cytokines Involved

The liver is the major site of synthesis of acute phase proteins (reviewed in Chapter 3 of this volume). Since injury to another part of the body results in altered synthesis rates of acute phase proteins in the liver, the existence of hormone-like mediators of the acute phase response was suggested and has been intensively investigated. Much of this work initially concentrated on interleukin-1, also previously known as lymphocyte-activating factor, endogenous pyrogen, or leukocyte endogenous mediator (extensively reviewed in Reference

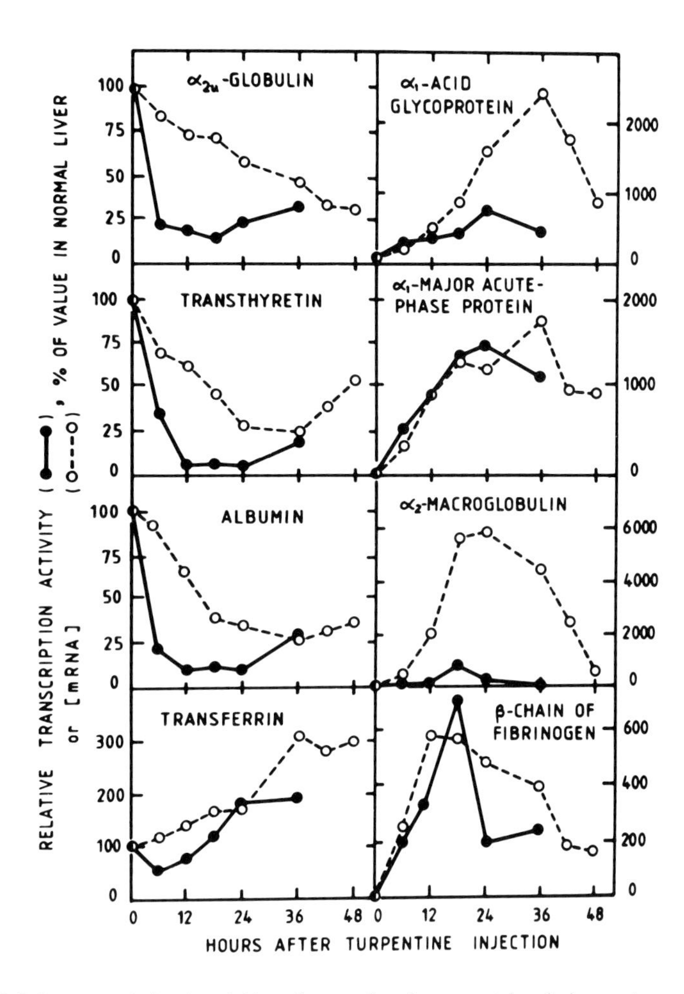

FIGURE 7. Relative transcriptional activities of genes for plasma proteins during acute experimental inflammation. "Nuclear run-off" transcription assays were performed *in vitro* using nuclei isolated from livers of rats removed at various times after induction of acute inflammation by turpentine injection. Transcriptional activity relative to normal (●——●) and mRNA levels (○– – –○). (From Birch, H. E. and Schreiber, G., *J. Biol. Chem.*, 261, 8077, 1986. With permission.)

30). When recombinant human interleukin-1α was intraperitoneally injected into rats at 2-h intervals, hepatic mRNA levels reached a maximal response after 6 to 10 h[31] (Figure 8). The changes in mRNA levels for the negative acute phase proteins albumin, transthyretin, and apolipoprotein E were similar to the changes observed after turpentine-induced inflammation (see Figure 5). However, changes in mRNA levels for positive acute phase proteins did not exhibit typical acute phase behavior after injection of interleukin-1α, indicating that factors other than interleukin-1α are required for a comprehensive acute phase response.

In recent years, it has been recognized that several polypeptide cytokines are responsible for eliciting an acute phase response. The synthesis of albumin[8,9] and α_{2u}-globulin[9] is decreased after intraperitoneal administration of interleukin-6 *in vivo. In vitro,* the synthesis of albumin in human hepatocytes and hepatoma lines is decreased by interleukin-6, interleukin-1, or tumor necrosis factor, and the synthesis of α_2-HS-glycoprotein in Hep G2 cells is decreased by interleukin-6 or interleukin-1 (reviewed in Reference 32). Differences in the rate and extent of changes observed in the acute phase response in various model systems have demonstrated the heterogeneous nature of the mechanism responsible for acute phase

TABLE 1
Upper Threshold Values for the Half-Lives of mRNAs for Negative Acute Phase Proteins in Liver During the Acute Phase Response

mRNA	Upper threshold for half-life (h)[a]
α_{2u}-globulin	32
Retinol-binding protein	20
Transthyretin	9
Ornithine transcarbamoylase	9
Albumin	7
Apolipoprotein AIV	6

[a] To estimate the upper threshold values for mRNAs, the data presented in Figure 5 were plotted in semilogarithmic plots of mRNA level against time, and the half-lives were obtained from the slopes of sections of the curves approximating a straight line.

Adapted from Schreiber, G., Tsykin, A., Aldred, A. R., Thomas, T., Fung, W.-P., Dickson, P. W., Cole, T., Birch, H., de Jong, F. A., and Milland, J., *Ann. N.Y. Acad. Sci.*, 557, 61, 1989. With permission.

regulation (reviewed in Reference 33). In the presence of sufficient levels of glucocorticoids, the major regulator of acute phase protein gene expression is interleukin-6, with modulation of that regulation by interleukin-1.[34] Other cytokines such as tumor necrosis factor-α and hepatocyte-stimulating factors identified in the human squamous carcinoma cell line COLO 16 may also be involved (reviewed in Reference 35). The cytokines and hormones implicated in the regulation of acute phase protein synthesis are the subject of Section D of this volume.

B. NEGATIVE ACUTE PHASE RESPONSE IN THE FETUS AND IN THE NEONATE

The concentration of many plasma proteins differs in the bloodstream of adult and neonatal rats. Nevertheless, the neonatal liver is capable of a typical acute phase response.[36] After induction of an acute inflammation in 4-d-old rats, a coordinated change in the pattern of concentrations of plasma proteins in the bloodstream was observed (Figure 9). The concentration of albumin and transthyretin decreased while the concentration of the positive acute phase proteins increased, similar to observations for adult rats (See Figure 1).

Subcutaneous injection of turpentine into 18-d-old fetal rats *in utero* also elicits a fetal plasma protein acute phase response.[37] Two days after induction of experimental inflammation, increases in the plasma concentration of positive acute phase proteins were accompanied by decreased concentrations of albumin, transthyretin, α_1-fetoprotein, and transcortin.

C. NEGATIVE ACUTE PHASE RESPONSE AND PARTIAL HEPATECTOMY

Regeneration of the liver after partial hepatectomy, with its accompanying acute phase response, provides a model to study the integration of growth-related and acute phase-related regulatory stimuli for the synthesis rates of individual proteins. Previously, the level of albumin mRNA in regenerating liver has been reported to decrease,[38] decrease with a delay,[39] or show no significant change.[40-42] Hepatic mRNA levels for negative acute phase proteins were determined after partial hepatectomy and compared with acute phase levels measured after induction of an acute inflammation[18] (Figure 10). The pattern of the decrease in hepatic mRNA levels for albumin, transthyretin, and α_2-HS-glycoprotein was similar during the first 12 to 18 h after partial hepatectomy or induction of inflammation (Figure 10A). After the initial decrease, mRNA levels in regenerating liver began to increase toward normal levels, whereas mRNA levels during the acute phase response to inflammation continued

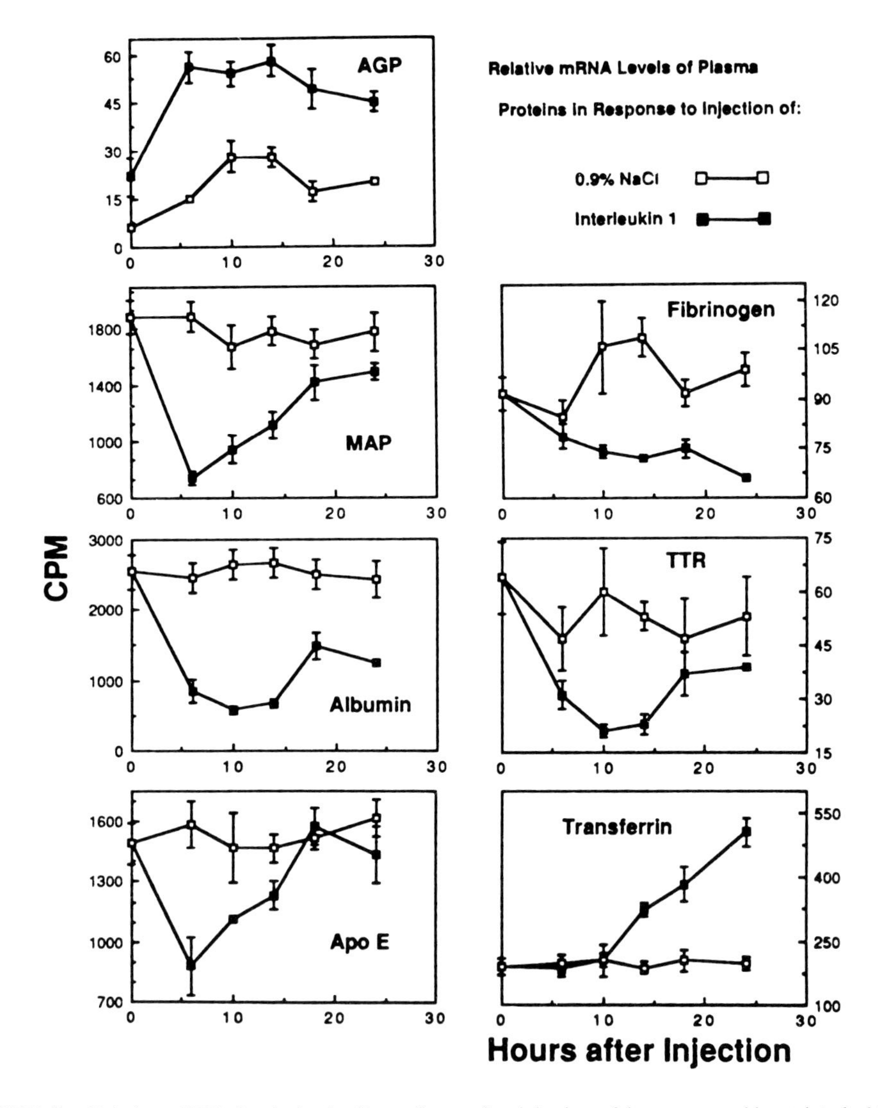

FIGURE 8. Relative mRNA levels in the liver of rats after injection of human recombinant interleukin-1α (■——■) or 0.9% NaCl (□——□). Rats were injected intraperitoneally with 4500 units of interleukin-1 in 1 ml of 0.9% NaCl, or 1 ml of 0.9% NaCl, at 0, 2, and 4 h and sacrificed at 6, 10, 14, 18, and 24 h. Values are expressed as cpm of [^{32}P]cDNA bound to RNA in cytoplasmic extracts from liver. Each point represents the mean ± standard error for six animals injected with interleukin-1 or three animals injected with 0.9% NaCl. AGP, α_1-acid glycoprotein; MAP, thiostatin; TTR, transthyretin; ApoE, apolipoprotein E. (From de Jong, F. A., Birch, H. E., and Schreiber, G., *Inflammation,* 12, 613, 1988. With permission.)

to decrease. A dramatic decrease in the level of α_2-HS-glycoprotein mRNA 24 h after partial hepatectomy[19] and a five- to sixfold decline in mRNA levels for α_1-inhibitor III 24 h after partial hepatectomy[25] have also been reported. Changes in the transcriptional activity of the albumin and transthyretin genes were sufficient to explain the changes in their mRNA levels.[18] Attenuation of the acute phase component of the response to partial hepatectomy, observed at the time of onset of liver growth (Figure 10A, 18 h after partial hepatectomy), was similar to that observed for positive acute phase proteins.[18] The consequence of the attenuation of the negative acute phase response of albumin mRNA levels in regenerating liver is that during both acute inflammation and liver regeneration, comparable total amounts of albumin mRNA are available to the animal.[18]

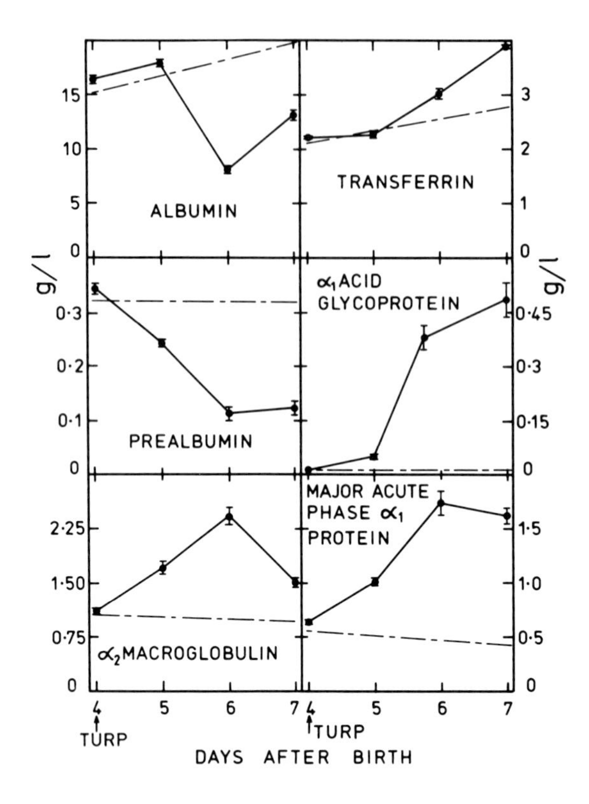

FIGURE 9. Changes in the concentration of plasma proteins in neonatal rats after induction of acute inflammation by turpentine injection. Each point is the mean ± standard error of values for six rats from two or three litters. The concentrations which would have been measured in healthy rats are indicated by a broken line. (From Thomas, T. and Schreiber, G., *Inflammation,* 9, 1, 1985. With permission.)

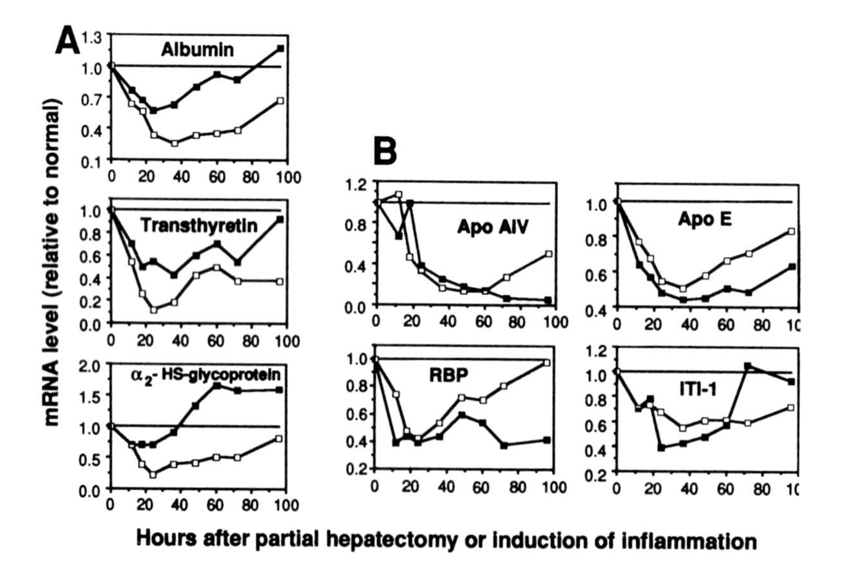

FIGURE 10. Changes in hepatic mRNA levels for (A) albumin, transthyretin, and α_2-HS-glycoprotein, and (B) apolipoproteins AIV (Apo AIV) and E (Apo E), retinol-binding protein (RBP) and inter-α-trypsin inhibitor I (ITI-1) at various times after partial hepatectomy (■——■) or induction of inflammation (□——□). Levels of mRNA were determined by Northern analysis. Samples were pooled from four to eight animals for each time point. The mRNA level in healthy rats is indicated by a horizontal line. (From Milland, J., Tsykin, A., Thomas, T., Aldred, A. R., Cole, T., and Schreiber, G., *Am. J. Physiol.,* 259, G340, 1990. With permission.)

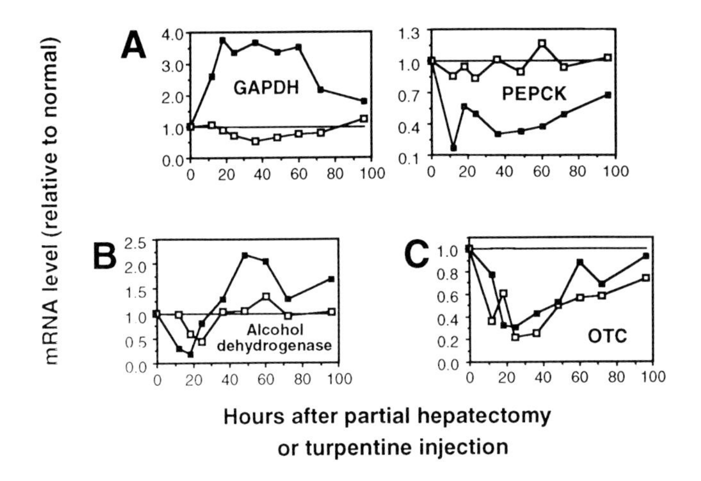

FIGURE 11. Changes in hepatic mRNA levels for (A) glyceraldehyde-3-phosphate dehydrogenase (GAPDH) and phosphoenolpyruvate carboxykinase (PEPCK), (B) alcohol dehydrogenase, and (C) ornithine transcarbamoylase (OTC) at various times after partial hepatectomy (■——■) or induction of inflammation (□——□). Levels of mRNA were determined by Northern analysis. Samples were pooled from four to eight animals for each time point. The mRNA level in healthy rats is indicated by a horizontal line. (Modified from Milland, J., Tsykin, A., Thomas, T., Aldred, A. R., Cole, T., and Schreiber, G., *Am. J. Physiol.*, 259, G340, 1990. With permission.)

Another pattern of similar changes in hepatic mRNA levels during liver regeneration and the acute phase response to inflammation was observed for apolipoprotein AIV, retinol-binding protein, apolipoprotein E, and inter-α-trypsin inhibitor I[18] (Figure 10B). For these negative acute phase proteins, the pattern of changes in mRNA levels are similar for about the first 60 h after partial hepatectomy or induction of inflammation. After this time, the mRNA levels for apolipoprotein AIV, retinol-binding protein, and apolipoprotein E (but not inter-α-trypsin inhibitor I) during liver regeneration remained low, whereas mRNA levels during the acute phase response to inflammation began to increase. An attenuation of the negative acute phase component of the response to partial hepatectomy was observed only for inter-α-trypsin inhibitor I within this group of proteins.

The changes in mRNA levels after partial hepatectomy for the glycolytic enzyme glyceraldehyde-3-phosphate dehydrogenase (increased to 3.5 times the normal level) and the gluconeogenic enzyme phosphoenolpyruvate carboxykinase (reduced to one third of the normal level) were very large compared with those observed after induction of inflammation (Figure 11A). Changes in the transcriptional activity of the phosphoenolpyruvate carboxykinase gene in regenerating liver were sufficient to account for the changes observed in the mRNA level.[43] The directions of change in the mRNA levels of glyceraldehyde-3-phosphate dehydrogenase and phosphoenolpyruvate carboxykinase were unexpected, consistent with favoring glucose degradation over gluconeogenesis. However, the regenerating liver is predominantly a gluconeogenic organ.[44,45] Increased phosphoenolpyruvate carboxykinase mRNA levels 1 to 4 h after partial hepatectomy have been reported.[46] Phosphoenolpyruvate carboxykinase enzyme activity is increased during liver regeneration,[43,45] maintaining the gluconeogenic nature of regenerating liver, despite decreased expression of the phosphoenolypyruvate carboxykinase gene.

For alcohol dehydrogenase, the pattern of changes in the mRNA level was similar after partial hepatectomy or after induction of inflammation, but the amplitude of changes was greater during liver regeneration (Figure 11B). Changes in hepatic mRNA levels for ornithine transcarbamoylase were similar during liver regeneration and the acute phase response to inflammation (Figure 11C). This is surprising since both are conditions with negative nitrogen balance, although ornithine transcarbamoylase is not a rate-limiting enzyme in the urea cycle.

IV. CONCLUDING REMARKS

The functional significance of the reduced plasma concentrations of the negative acute phase proteins during the acute phase response is not clear. It has been proposed that albumin may act as a "metabolic adapter" during the acute phase response.[13] The decrease in the rate of albumin synthesis may contribute considerably to reduce the simultaneous increased demand for ATP and amino acyl-tRNA, caused by the increased synthesis of acute phase proteins during the acute phase response. Albumin is well suited to the role of a metabolic adapter during sudden increases in the synthesis rates of acute phase proteins, as it has a high rate of synthesis,[10] very large body pool, slow turnover, and is not indispensible for healthy life (as indicated by the existence of healthy albumin-deficient individuals in both man[47] and rat[48]).

This role of metabolic adaptation during the acute phase response may be shared, to a lesser extent, by other negative acute phase proteins such as transthyretin.[4] However, the acute phase behavior of other negative acute phase proteins may occur merely as a consequence of regulatory features shared with albumin, with no specific negative acute phase function for each of these other proteins. The component of the negative acute phase response due to fasting during experimental inflammation probably varies among proteins. Following a 48-h period of fasting, mRNA levels in rat liver for transthyretin decreased by 30%, whereas those for albumin decreased by only 10 to 20%.[49]

During liver regeneration, the onset of liver growth would lead to increased demand for amino acyl tRNA and ATP in addition to that caused by the inflammation-related increased synthesis rates of positive acute phase proteins. Thus, one might expect more pronounced decreases in mRNA levels for negative acute phase proteins in regenerating liver than in liver during an acute phase response alone, i.e., a greater metabolic adaptation role. However, the opposite of this expectation was observed (see Figure 10A). Liver regeneration in analbuminemic rats is inhibited, suggesting that albumin may be necessary for normal liver regeneration.[50]

The negative-acute phase regulation of plasma protein synthesis has evolved as a liver-specific feature. The transthyretin gene is expressed at a higher level (per gram of tissue) in the choroid plexus of the brain than in the liver (See Chapter 3). All transthyretin synthesized by the choroid plexus is secreted toward the brain, into the cerebrospinal fluid.[51] During evolution, synthesis of transthyretin in the choroid plexus (in reptiles[52]) occurred much earlier than in the liver (in marsupials[53]). However, transthyretin synthesis in the choroid plexus is not under acute phase control,[54] even though acute phase regulation in the liver is present in evolutionary older animals (the fishes[55,56]) than those first synthesizing transthyretin in the choroid plexus. Thus, the structure and function (binding of thyroxine) of transthyretin and the negative acute phase regulation of its gene developed separately, the first in the brain, the second in the liver. It is an open question whether functional selective pressure led to the evolution of negative acute phase regulation of the transthyretin gene in liver, or whether this regulation appeared as a "chance accident" in evolution.

The function of albumin and transthyretin in the bloodstream is transport of thyroid hormones, and in the case of transthyretin, also the indirect transport of retinol by binding of retinol-binding protein. The decreased serum concentrations of these carrier proteins during the acute phase response may have consequences for the transport and distribution of their ligands. After surgical trauma,[57] or during turpentine-induced acute inflammation[58] in rats, the serum concentration of total thyroxine is decreased, while the concentration of free thyroxine is increased above the normal level. Free thyroxine, although amounting to only 0.03% of total thyroxine in blood plasma, is thought to be the "biologically relevant" form of thyroxine (the "free hormone hypothesis", reviewed in Reference 59). The increased concentration of free thyroxine in blood would lead to an increase of free thyroxine in tissues and, possibly, increased production of triiodothyronine, the more potent form of thyroid hormone, found predominantly within cells and interacting with nuclear receptors. It has been proposed that the decreased serum concentration of carrier proteins during an acute phase response, with a concomitant increase in the concentration of the unbound form of their ligands, increases the availability of the ligands as an antiinflammatory self-defense mechanism.[58]

Elucidation of the specific mechanisms regulating synthesis of negative acute phase proteins during an acute phase response will provide insight about the extent of common mechanisms in the coordinated regulation of these proteins. Although the function of the decreased serum concentration of these proteins during an acute phase response is not yet clear, they represent an important group of proteins involved in the complex homeostatic processes by which the body responds to trauma and injury.

ACKNOWLEDGMENT

The work on which some of this review is based was supported by grants from the National Health and Medical Research Council of Australia and the Australian Research Council.

REFERENCES

1. **Werner, M. and Cohnen, G.,** Changes in serum proteins in the immediate postoperative period, *Clin. Sci.*, 36, 173, 1969.
2. **Cuthbertson, D. P. and Tompsett, S. L.,** Note on the effect of injury on the level of the plasma proteins, *Br. J. Exp. Pathol.*, 16, 471, 1935.
3. **van Oss, C. J., Bronson, P. M., and Border, J. R.,** Changes in the serum alpha glycoprotein distribution in trauma patients, *J. Trauma*, 15, 451, 1975.
4. **Dickson, P. W., Howlett, G. J., and Schreiber, G.,** Metabolism of prealbumin in rats and changes induced by acute inflammation, *Eur. J. Biochem.*, 129, 289, 1982.
5. **Moody, B. J.,** Changes in the serum concentrations of thyroxine-binding prealbumin and retinol-binding protein following burn injury, *Clin. Chim. Acta*, 118, 87, 1982.
6. **Davies, J. W. L., Ricketts, C. R., and Bull, J. P.,** Studies of plasma protein metabolism. I. Albumin in burned and injured patients, *Clin. Sci.*, 23, 411, 1962.
7. **Owen, J. A.,** Effect of injury on plasma proteins, *Adv. Clin. Chem.*, 9, 1, 1967.
8. **Geiger, T., Andus, T., Klapproth, J., Hirano, T., Kishimoto, T., and Heinrich, P.,** Induction of rat acute phase proteins by interleukin 6 *in vivo*, *Eur. J. Immunol.*, 18, 717, 1988.
9. **Marinkovic, S., Jahreis, G. P., Wong, G. G., and Baumann, H.,** IL-6 modulates the synthesis of a specific set of acute phase plasma proteins in vivo, *J. Immunol.*, 142, 808, 1989.
10. **Schreiber, G., Howlett, G., Nagashima, M., Millership, A., Martin, H., Urban, J., and Kotler, L.,** The acute phase response of plasma protein synthesis during experimental inflammation, *J. Biol. Chem.*, 257, 10271, 1982.
11. **Gauthier, F. and Ohlsson, K.,** Isolation and some properties of a new enzyme-binding protein in rat plasma, *Hoppe-Seyler's Z. Physiol. Chem.*, 359, 987, 1978.
12. **Savu, L., Lombart, C., and Nunez, E. A.,** Corticosterone binding globulin: an acute phase "negative" protein in the rat, *FEBS Lett.*, 113, 102, 1980.
13. **Schreiber, G. and Howlett, G.,** Synthesis and secretion of acute phase proteins, in *Plasma Protein Secretion by the Liver*, Glaumann, H., Peters, T., Jr., and Redman, C., Eds., Academic Press, London, 1983, 423.
14. **Jamieson, J. C., Morrison, K. E., Molasky, D., and Turchen, B.,** Studies on acute phase proteins of rat serum. V. Effect of induced inflammation on the synthesis of albumin and α_1-acid glycoprotein by liver slices, *Can. J. Biochem.*, 53, 401, 1975.
15. **Schreiber, G., Tsykin, A., Aldred, A. R., Thomas, T., Fung, W.-P., Dickson, P. W., Cole, T., Birch, H., de Jong, F. A., and Milland, J.,** The acute phase response in the rodent, *Ann. N.Y. Acad. Sci.*, 557, 61, 1989.

16. **Schreiber, G., Aldred, A. R., Thomas, T., Birch, H. E., Dickson, P. W., Guo-Fen, T., Heinrich, P. C., Northemann, W., Howlett, G. J., de Jong, F. A., and Mitchell, A.,** Levels of messenger ribonucleic acids for plasma protein in rat liver during acute experimental inflammation, *Inflammation,* 10, 59, 1986.
17. **Dickson, P. W., Howlett, G. J., and Schreiber, G.,** Rat transthyretin (prealbumin): molecular cloning, nucleotide sequence and gene expression in liver and brain, *J. Biol. Chem.,* 260, 8214, 1985.
18. **Milland, J., Tsykin, A., Thomas, T., Aldred, A. R., Cole, T., and Schreiber, G.,** Gene expression in regenerating and acute phase rat liver, *Am. J. Physiol.,* 259, G340, 1990.
19. **Daveau, M., Davrinche, C., Djelassi, N., Lemetayer, J., Julen, N., Hiron, M., Arnaud, P., and Lebreton, J. P.,** Partial hepatectomy and mediators of inflammation decrease the expression of liver α_2-HS-glycoprotein gene in rats, *FEBS Lett.,* 273, 79, 1990.
20. **Princen, H. M. G., Selten, G. C. M., Nieuwenhuizen, W., and Yap, S. H.,** Changes of synthesis of fibrinogen polypeptide and albumin mRNAs in rat liver after turpentine injection and after partial hepatectomy or laparotomy, in *Fibrinogen — Structure, Functional Aspects and Metabolism,* Vol. 2, Haverkate, F., Henschen, A., Nieuwenhuizen, W., and Straub, P. W., Eds., Walter de Gruyter, Berlin, 1983, 263.
21. **Hayashida, K., Okubo, H., Noguchi, M., Yoshida, H., Kangawa, K., Matsuo, H., and Sakaki, Y.,** Molecular cloning of DNA complementary to rat α_2-macroglobulin mRNA, *J. Biol. Chem.,* 260, 14224, 1985.
22. **Morrow, J. F., Stearman, R. S., Peltzman, C. G., and Potter, D. A.,** Induction of hepatic synthesis of serum amyloid A protein and actin, *Proc. Natl. Acad. Sci. U.S.A.,* 78, 4718, 1981.
23. **Baumann, H., Firestone, G.L., Burgess, T. L., Gross, K. W., Yamamoto, K. R., and Held, W. A.,** Dexamethasone regulation of α_1-acid glycoprotein and other acute phase reactants in rat liver and hepatoma cells, *J. Biol. Chem.,* 258, 563, 1983.
24. **Braciak,T. A., Northemann, W., Hudson, G. O., Shiels, B. R., Gehring, M. R., and Fey, G. H.,** Sequence and acute phase regulation of rat α_1-inhibitor III messenger RNA, *J. Biol. Chem.,* 263, 3999, 1988.
25. **Aiello, L. P., Shia, M. A., Robinson, G. S., Pilch, P. F., and Farmer, S. R.,** Characterization and hepatic expression of rat α_1-inhibitor III mRNA, *J. Biol. Chem.,* 263, 4013, 1988.
26. **Tu, G.-F., de Jong, F., Apostolopoulos, J., Nagashima, M., Fidge, N., Schreiber, G., and Howeltt, G.,** Effect of acute inflammation on rat apolipoprotein mRNA levels, *Inflammation,* 11, 241, 1987.
27. **Birch, H. E. and Schreiber, G.,** Transcriptional regulation of plasma protein synthesis during inflammation, *J. Biol. Chem.,* 261, 8077, 1986.
28. **Kulkarni, A. B., Reinke, R., and Feigelson, P.,** Acute phase mediators elevate α_1-acid glycoprotein gene transcription, *J. Biol. Chem.,* 260, 15386, 1985.
29. **Northemann, W., Shiels, B. R., Braciak, T. A., and Fey, G. H.,** Structure and negative transcriptional regulation by glucocorticoids of the acute phase rat α_1-inhibitor III gene, *Biochemistry,* 28, 84, 1989.
30. **Gordon, A. H. and Koj, A., Eds.,** *The Acute Phase Response to Injury and Infection,* Elsevier, Amsterdam, 1985.
31. **de Jong, F. A., Birch, H. E., and Schreiber, G.,** Effect of recombinant interleukin-1 on mRNA levels in rat liver, *Inflammation,* 12, 613, 1988.
32. **Heinrich, P. C., Castell, J. V., and Andus, T.,** Interleukin-6 and the acute phase response, *Biochem. J.,* 265, 621, 1990.
33. **Baumann, H.,** Hepatic acute phase reaction in vivo and in vitro, *In Vitro Cell. Dev. Biol.,* 25, 115, 1989.
34. **Gauldie, J., Richards, C., Northemann, W., Fey, G., and Baumann, H.,** IFNβ2/BSF2/IL-6 is the monocyte-derived HSF that regulates receptor-specific acute phase gene regulation in hepatocytes, *Ann. N.Y. Acad. Sci.,* 557, 46, 1989.
35. **Fey, G. H. and Gauldie, J.,** The acute phase response of the liver in inflammation, in *Progress in Liver Diseases,* Popper, H. and Schaffner, F., Eds., W. B. Saunders, Philadelphia, 1990, 89.
36. **Thomas, T. and Schreiber, G.,** Acute phase response of plasma protein synthesis during experimental inflammation in neonatal rats, *Inflammation,* 9, 1, 1985.
37. **Vranckx, R., Savu, L., Cohen, A., Maya, M., and Nunez, E.,** Inflammatory competence of fetal rat: acute phase plasma protein response of the fetus treated by turpentine in utero, *Inflammation,* 13, 79, 1989.
38. **Princen, H. M. G., Selten, G. C. M., Selten-Versteegen, A.-M. E., Mol-Backx, G. P. B. M., Nieuwenhuizen, W., and Yap, S. H.,** Distribution of mRNAs of fibrinogen polypeptides and albumin in free and membrane-bound polyribosomes and induction of α-fetoprotein mRNA synthesis during liver regeneration after partial hepatectomy, *Biochim. Biophys. Acta,* 699, 121, 1982.
39. **Krieg, L., Alonso, A., Winter, H., and Volm, M.,** Albumin messenger RNA after partial hepatectomy and sham operation, *Biochim. Biophys. Acta,* 610, 311, 1980.
40. **Petropoulos, C., Andrews, G., Tamaoki, T., and Fausto, N.,** α-Fetoprotein and albumin mRNA levels in liver regeneration and carcinogenesis, *J. Biol. Chem.,* 258, 4901, 1983.

41. **Fausto, N.,** Messenger RNA in regenerating liver: implications for the understanding of regulated growth, *Mol. Cell. Biochem.*, 59, 131, 1984.
42. **Friedman, J. M., Chung, E. Y., and Darnell, J. E., Jr.,** Gene expression during liver regeneration, *J. Mol. Biol.*, 179, 37, 1984.
43. **Milland, J. and Schreiber, G.,** Transcriptional activity of the phosphoenolpyruvate carboxykinase gene decreases in regenerating rat liver, *FEBS Lett.*, 279, 184, 1991.
44. **Brinkmann, A., Katz, N., Sasse, D., and Jungermann, K.,** Increase of the gluconeogenic and decrease of the glycolytic capacity of rat liver with a change of the metabolic zonation after partial hepatectomy, *Hoppe-Seyler's Z. Physiol. Chem.*, 359, 1561, 1978.
45. **Katz, N.,** Correlation between rates and enzyme levels of increased gluconeogenesis in rat liver and kidney after partial hepatectomy, *Eur. J. Biochem.*, 98, 535, 1979.
46. **Mohn, K. L., Laz, T. M., Melby, A. E., and Taub, R.,** Immediate-early gene expression differs between regenerating liver, insulin-stimulated H-35 cells, and mitogen-stimulated Balb/c 3T3 cells. Liver-specific induction patterns of gene 33, phosphoenolpyruvate carboxykinase, and the *Jun, Fos* and *Egr* families, *J. Biol. Chem.*, 265, 21914, 1990.
47. **Boman, H., Hermodson, M., Hammond, C. A., and Motulsky, A. G.,** Analbuminemia in an American Indian girl, *Clin. Genet.*, 9, 513, 1976.
48. **Nagase, S., Shimamune, K., and Shumiya, S.,** Albumin-deficient rat mutant, *Science*, 205, 590, 1979.
49. **de Jong, F. A., Howeltt, G. J., and Schreiber, G.,** Messenger RNA levels of plasma proteins following fasting, *Br. J. Nutr.*, 59, 81, 1988.
50. **Hicks, B. A., Drougas, J., Arnaout, W., Felcher, A., Moscioni, A. D., Levenson, S. M., and Demetriou, A. A.,** Impaired liver regeneration in the analbuminemic rat, *J. Surg. Res.*, 46, 427, 1989.
51. **Schreiber, G., Aldred, A. R., Jaworowski, A., Nilsson, C., Achen, M. G., and Segal, M. B.,** Thyroxine transport from blood to brain via transthyretin synthesis in choroid plexus, *Am. J. Physiol.*, 258, R338, 1990.
52. **Harms, P. J., Tu, G.-F., Richardson, S. J., Aldred, A. R., Jaworowski, A., and Schreiber, G.,** Transthyretin (prealbumin) gene expression in choroid plexus is strongly conserved during evolution of vertebrates, *Comp. Biochem. Physiol.*, 99B, 239, 1991.
53. **Aldred, A. R., Pettersson, T. M., Harms, P. J., Richardson, S. J., Duan, W., Tu, G.-F., Achen, M. G., Nichol, S., and Schreiber, G.,** The position of the echidna in the evolution of thyroid hormone-binding plasma proteins, in *"Platypus and Echidnas"*, Augee, M. L., Ed., The Royal Zoological Society of New South Wales, Sydney, 1992, 44.
54. **Dickson, P. W., Aldred, A. R., Marley, P. D., Bannister, D., and Schreiber, G.,** Rat choroid plexus specializes in the synthesis and the secretion of transthyretin (prealbumin) — regulation of transthyretin synthesis in choroid plexus in independent from that in liver, *J. Biol. Chem.*, 261, 3475, 1986.
55. **White, A. and Fletcher, T. C.,** The effects of adrenal hormones, endotoxin and turpentine on serum components of the plaice (*Pleuronectes platessa* L.), *Comp. Biochem. Physiol.*, 73C, 195, 1982.
56. **White, A. and Fletcher, T. C.,** The influence of hormones and inflammatory agents on C-reactive protein, cortisol and alanine aminotransferase in the plaice (*Pleuronectes platessa* L.), *Comp. Biochem. Physiol.*, 80C, 99, 1985.
57. **Haibach, H. and McKenzie, J. M.,** Increased free thyroxine postoperatively in the rat, *Endocrinology*, 81, 435, 1967.
58. **Savu, L., Zouaghi, H., Vranckx, R., and Nunez, E. A.,** Total and free thyroid hormones and thyroxine binding prealbumin in sera of post-natal developing rats undergoing acute inflammation, *Exp. Clin. Endocrinol.*, 6, 151, 1987.
59. **Mendel, C. M.,** The free hormone hypothesis: a physiologically based mathematical model, *Endocr. Rev.*, 10, 232, 1989.

Chapter 3

EXTRAHEPATIC SYNTHESIS OF ACUTE PHASE PROTEINS

Gerhard Schreiber and Angela R. Aldred

TABLE OF CONTENTS

0-8493-6913-4/93/$0.00 + $.50

I. INTRODUCTION: CONSTANCY OF THE INTERNAL MILIEU AND PROTEIN HOMEOSTASIS

Cells in higher animals can only survive and function properly in an environment of appropriate composition (''la fixité du milieu intérieur est la condition de la vie libre'').[1] Various regulatory mechanisms ensure that the composition of this environment is kept relatively constant. The achieved state of equilibrium has been called homeostasis.[2,3] Proteins are essential components of the fluids filling the extracellular environment. The ultimate function of the acute phase proteins and the purpose of the acute phase response is the maintenance of extracellular homeostasis. The site for this function is extravascular, i.e., in the immediate vicinity of cells or at the barrier structures delineating body compartments.

II. COMPARTMENTATION AND DISTRIBUTION OF PLASMA PROTEINS

The composition of the extracellular environment shows tissue-specific variations, in particular, different patterns of protein concentrations. Various extracellular compartments can be distinguished within the body. Various forms of transport with quite different mechanisms and with different kinetics provide communication between the compartments. For example, intravenously injected radioactive albumin is distributed throughout the intravascular compartment within a few minutes. In a much slower process, the injected albumin equilibrates with albumin in the interstitial compartment.[4] Equilibrium between the intravascular and the extravascular compartments is reached after about 36 h (Figure 1). Trauma and acute inflammation, such as those occurring after partial hepatectomy, lead to greater changes in the total and extravascular plasma protein pools than in the intravascular pool (Figure 2). Special barrier systems (e.g., blood-brain barrier, blood-cerebrospinal fluid barrier, blood-testis barrier, placenta, and fetal membranes) separate the extracellular com-

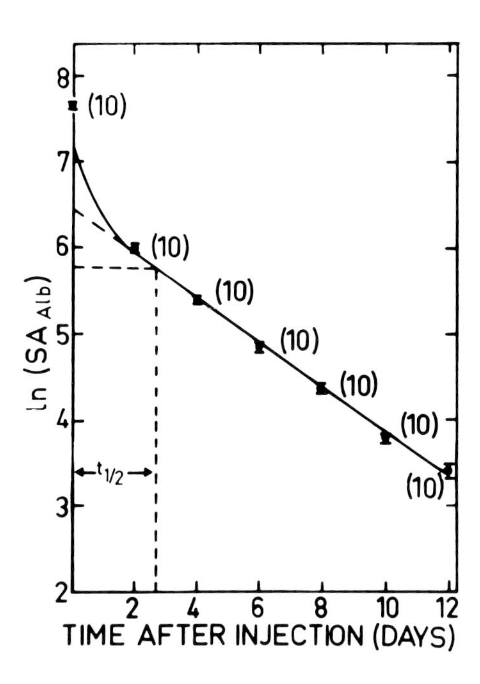

FIGURE 1. Turnover of albumin in the serum of Buffalo rats. Ordinate, natural logarithm of the specific radioactivity of albumin in the bloodstream; abscissa, time after intravenous injection of [^{14}C] albumin. Per 100 g of body weight, 8.77 mg of [^{14}C] albumin, specific radioactivity 27,100 dpm/mg, dissolved in 0.37 ml of 0.9% NaCl solution, was injected into the caval vein at zero time. After 10 min, 0.7 ml of blood was taken from the caval vein, whereas all other blood samples were taken from the femoral veins at the indicated times. After coagulation, albumin concentration and protein radioactivity were determined in the serum and the specific radioactivity was calculated. Linearity of the semilogarithmic decay curve, i.e., equilibrium between extravascular and intravascular pools, was reached in parentheses. The mean of the natural logarithm of the specific radioactivity of albumin for each group, $\ln(SA_{Alb})$, and its standard error are given. $t_{1/2}$, half-life. (From Schreiber, G., Urban, J., Zähringer, J., Reutter, W., and Frosch, U., *J. Biol. Chem.*, 246, 4531, 1971. With permission.)

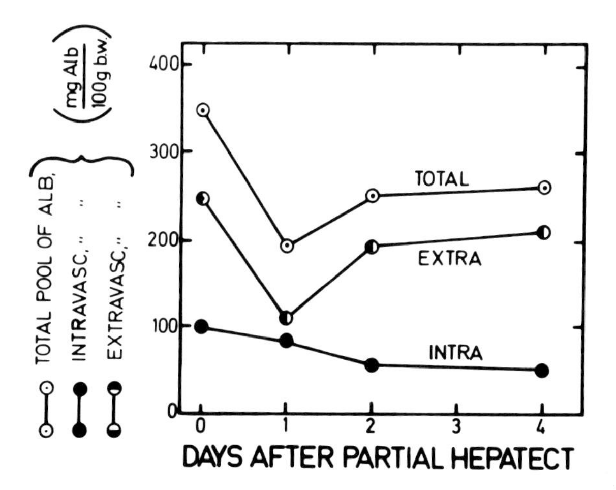

FIGURE 2. Intravascular (INTRA), extravascular (EXTRA), and total body pool (TOTAL) of albumin (Alb) per 100g of body weight (b.w.), in partially hepatectomized rats. Pools were determined by measuring equilibration and dilution of injected [^{14}C] albumin, as described in detail elsewhere.[4] (From Schreiber, G., Urban, J., Zähringer, J., Reutter, W., and Frosch, U., *J. Biol. Chem.*, 246, 4531, 1971. With permission.)

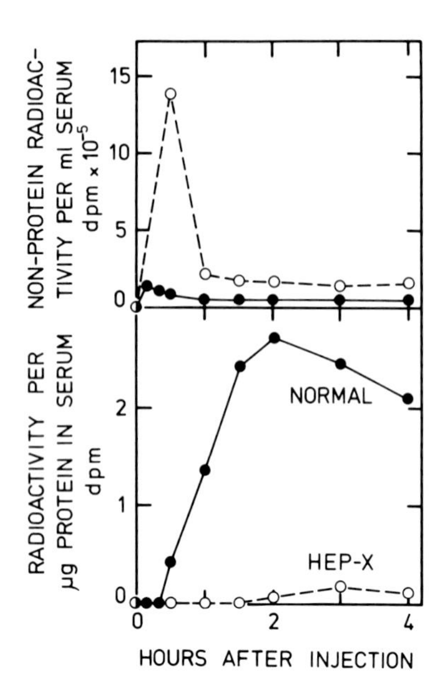

FIGURE 3. Nonprotein (upper half of figure) and protein radioactivity (lower half) in serum from healthy, intact (NORMAL, ●——●) and hepatectomized (HEP-X, [○– – –○) rats at different times after the intraperitoneal injection of a mixture of uniformly ^{14}C-labeled leucine, histidine, and lysine. Despite the much higher labeling pulse in the serum of the hepatectomized rats (upper panel), protein in the serum of the hepatectomized animals incorporated only a very small amount of radioactivity after 3 h (lower panel). (Drawn from the data of Schreiber, G., Boutwell, R. K., Potter, V. R., and Morris, H. P., *Cancer Res.*, 26, 2357, 1966. With permission.)

partments within special organs or tissues (e.g., brain, testis, and those of the growing embryo) from the general interstitial and vascular compartment of the body. However, homeostatic control, keeping protein concentrations in an appropriate range, is required in *all* extracellular compartments. Details of this control differ among tissues and are related to the particular function of organs.

III. SITES OF PLASMA PROTEIN SYNTHESIS

The organization of the extracellular environment of the body into different compartments, which contain plasma proteins in different concentrations, leads to the question of the origin of these proteins. Two types of studies have convincingly shown that the liver is the dominant source for plasma proteins in the vascular compartment. In a study of the first type, Miller et al.[5] investigated albumin synthesis in the isolated perfused rat liver. Goldsworthy et al.[6] analyzed the synthesis of albumin, transferrin, and fibrinogen in the perfused bovine liver. The amounts of albumin produced per hour per gram of isolated liver in these and other similar studies (e.g., Gordon and Humphrey[7] and Katz et al.[8]) were found to be in the same range as the turnover rates obtained for albumin *in vivo* with biosynthetically labeled [^{14}C]albumin injected into living rats.[4] In studies of the second type, the incorporation of radioactive amino acids into plasma proteins was compared for healthy and hepatectomized animals. Tarver and Reinhardt[9] reported that removal of the liver from dogs reduced the incorporation of methionine into albumin at least 20-fold. This was confirmed by Kukral et al.[10] Schreiber et al.[11] observed that in rats, hepatectomy almost completely abolished incorporation of intraperitoneally injected radioactive amino acids into plasma proteins (Figure 3). Similar observations were made with hepatectomized rhesus monkeys.[12]

A precursor-product relationship was indicated by structural[13-15] and kinetic[16-19] studies for an albumin-like protein synthesized in liver and albumin secreted into the medium of

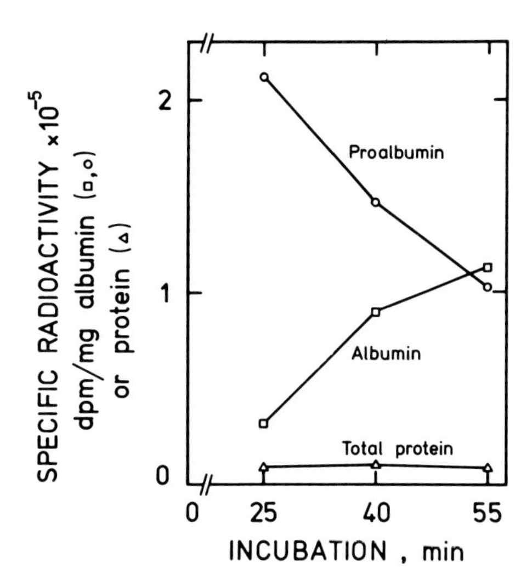

FIGURE 4. Specific radioactivities of total protein, albumin, and albumin-like protein from rat liver cells incubated for 25 min with L-[1-^{14}C]leucine, followed by a chase with excess nonlabeled leucine. The specific radioactivity of total protein (△) was measured in the homogenate. Albumin (□) and albumin-like protein (proalbumin, ○) were isolated to constant specific radioactivity. (From Edwards, K., Schreiber, G., Dryburgh, H., Urban, J., and Inglis, A. S., *Eur. J. Biochem.*, 63, 303, 1976. With permission.)

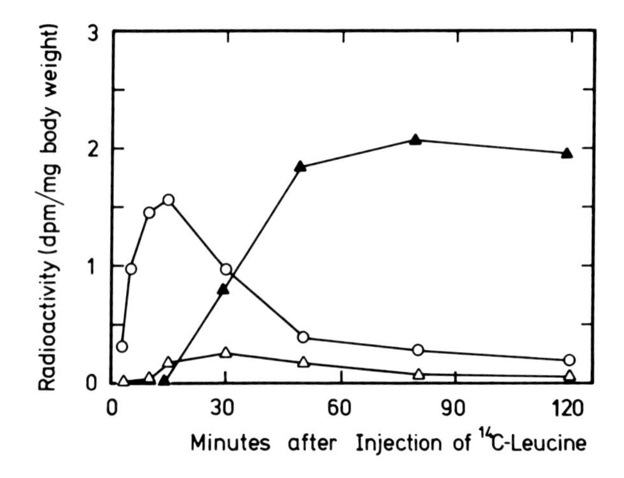

FIGURE 5. Radioactivity in albumin-like protein, extravascular albumin in liver, and albumin in plasma after intraportal injection of [^{14}C]leucine *in vivo* into Buffalo rats. The radioactivity in albumin and albumin-like protein is given as disintegrations per minute, normalized for body weight, as explained elsewhere.[18] Radioactivity in albumin-like protein in liver (○), extravascular albumin in liver (△), and albumin in plasma (▲). (From Urban, J., Chelladurai, M., Millership, A., and Schreiber, G., *Eur. J. Biochem.*, 67, 477, 1976. With permission.)

cell cultures (Figure 4) or into the bloodstream of living animals (Figure 5). From the bloodstream, plasma proteins migrate into the extracellular compartments of the body. For most plasma proteins, the amount found in the extravascular space is larger than in the intravascular space (see, e.g., References 4 and 20). Usually, the rates of migration of proteins from the vascular to the extravascular compartment are in the same range (see Reference 20 for rates of migration of albumin, transferrin, thiostatin, and α_1-acid glycoprotein in healthy rats and in rats suffering from an acute inflammation).

The name ''plasma protein'' has been coined and retained because this type of protein is most easily obtained from the bloodstream, even though its major proportion and function

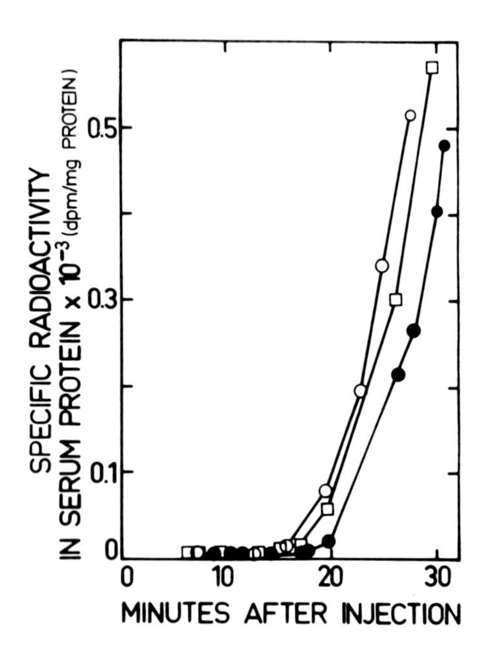

FIGURE 6. Specific radioactivity of protein in the serum of normal rats after intracaval injection of [^{14}C]leucine. □——□, ○——○, and ●——●: values from one animal each. (From Schreiber, G., Urban, J., Zähringer, J., Reutter, W., and Frosch, U., *J. Biol. Chem.*, 246, 4531, 1971. With permission.)

might be located outside the bloodstream. The name "acute phase" protein is based on the observation of a changed concentration in the blood plasma or serum, although such a change might not occur for the particular protein in other body compartments (or animal species). For example, the pattern of concentrations of proteins in cerebrospinal fluid varies greatly from that found in serum or plasma (see Reference 21). Selective transport[22] of individual proteins between the compartments and selective degradation of transported proteins[23] have been proposed to explain the differences in the concentration of particular proteins in cerebrospinal fluid and serum. Another mechanism leading to different patterns of concentration of proteins in plasma and in the fluid in other extracellular compartments would be the extrahepatic synthesis and secretion of plasma proteins.

IV. MEASUREMENT OF EXTRAHEPATIC SYNTHESIS OF PLASMA PROTEINS

A. INCORPORATION OF RADIOACTIVE AMINO ACIDS INTO PROTEINS

1. Necessity for Sampling Within the Secretion Time ("Minimum Transit Time")

The study of extrahepatic protein synthesis by measuring incorporation of radioactive amino acids into proteins *in vivo* is hampered by the fact that all tissues contain plasma proteins, derived from the liver, in their vascular and interstitial compartments. However, newly synthesized plasma proteins are secreted from the liver into medium or the bloodstream with a delay (Figure 6). Values reported for this delay ranged from 8 min[24] to about 1 h.[5] The lag period is the sum of the times required for the injected amino acid to travel to the liver, its uptake into the hepatocyte and binding to tRNA, the synthesis and posttranslational modification of the protein to be exported, the transport of the protein from the rough endoplasmic reticulum via the Golgi apparatus to the cell membrane, and its release into the bloodstream. More precise measurements of the length of this delay gave a value of about 15 min for albumin.[4,18,25-27] It has been called the "secretion time"[4] or "minimum transit time''.[27] The secretion time is shorter after partial hepatectomy, when the endoplasmic reticulum is rearranged in the liver cell.[4] For the demonstration and investigation of extrahepatic plasma protein synthesis by incorporation of injected radioactive amino acids *in vivo,* it is necessary to remove tissues within the secretion time, i.e., before radioactive plasma proteins are secreted by the liver and ''contaminate'' the extrahepatically synthesized proteins.

2. Proof that the Incorporation of Radioactive Amino Acids into Protein is Due to Protein Biosynthesis

In most cases, the rates of extrahepatic synthesis of plasma proteins per gram of tissue will be lower than those in liver. The measurement of small amounts of radioactivity ensuing from low rates of incorporation can present various difficulties. Obviously, for a 98% radiochemically pure amino acid preparation and incorporation rates lower than 2%, it is necessary to show that the incorporated radioactivity represents amino acid (98% of offered radioactivity) and not contamination (2% of offered radioactivity). Nonspecific binding, binding of amino acid to specific sites on the protein, and enzyme-catalyzed amino acid transfer can lead to nonbiosynthetic labeling of proteins. Such labeling can be ATP dependent,[28] similar to proper protein biosynthesis. This labeling might even be susceptible to inhibitors of amino acid transfer from amino acyl-tRNA, such as puromycin.[29] The incorporation of radioactive amino acids into proteins by nonprotein biosynthetic mechanisms has been reviewed in detail elsewhere.[30]

For quantitation of the incorporation of a radioactive amino acid into a specific protein, the protein should be radiochemically pure, i.e., it should have been purified to constant specific radioactivity. When this was attempted for albumin from liver[28,31] or hepatoma cells,[28,31] albumin and an albumin-like protein were obtained. For more than 2 h after injection of radioactive leucine *in vivo* into rats, the albumin-like protein was more highly labeled than albumin.[18] For the first 30 min after injection, the total radioactivity in the albumin-like protein in liver was greater than that in total albumin in the bloodstream.[18] Analysis of the amino acid sequence showed that the albumin-like protein differed from albumin by an oligopeptide extension at the N terminus.[13-15] The albumin-like protein was found to be an intracellular precursor form of albumin,[16-19] and the name ''proalbumin'' was suggested for this precursor.[32] The existence of precursor forms can considerably complicate investigations of the site of synthesis of plasma proteins.

In view of the difficulties associated with the demonstration of extrahepatic plasma protein synthesis by incorporation of radioactive amino acids, it is not surprising that conflicting reports about the extrahepatic synthesis of plasma proteins, and in particular of albumin, exist in the literature. Miller et al.[33] found no radioactivity in albumin when the lower half of the rat carcass was perfused for 5 to 6 h with [^{14}C]lysine. Albumin with low radioactivity could be isolated from cultures of chick embryo and HeLa cells,[34,35] and lymphocytes[36] incubated with radioactive leucine for several days.

The best method for identification of a protein is analysis of its primary structure. Demonstration of incorporation of offered radioactive amino acid into the right positions along the polypeptide chain is the best method to prove that protein biosynthesis is the mechanism of the incorporation of the radioactive amino acid.[37-40] ''Thyroid albumin'', synthesized by human thyroid tissue,[41] differed from albumin in its amino acid composition,[42] and hence could not be albumin. None of the reports of extrahepatic albumin synthesis has been confirmed by measurement of albumin mRNA concentrations in the various extrahepatic tissues (see discussion below).

B. ESTIMATION OF mRNA LEVELS

Messenger RNA for proteins is only found within cells. Messenger RNAs can be detected and measured in a very sensitive and specific way by hybridization to cDNA probes. In recent years, a large number of cDNAs for plasma proteins have become available (for reviews, see References 21 and 43). Using such hybridization to cDNAs, extrahepatic synthesis could be unambiguously demonstrated for various plasma proteins in various locations. In most cases, the cells or tissues synthesizing plasma proteins were found to be associated with special extracellular compartments, such as the ventricles and subarachnoidal space of the brain (choroid plexus), male reproductive tract (Sertoli cells), or fetus (yolk sac, placenta). The blood-brain or blood-cerebrospinal fluid barrier, blood-testis barrier, and fetal membranes separate these compartments from the main extracellular compartment of the body. The paracompartmental location of these cells, synthesizing plasma proteins, and the unidirectional secretion of the synthesized proteins into the associated extracellular compartments provide for effective participation of these cells in the control of protein homeostasis in the respective compartments. The synthesis and secretion of transthyretin by the choroid plexus seems to be of particular importance for extracellular homeostasis in the brain, and is discussed in greater detail below.

Four major methods of hybridization have been used for the estimation of mRNA: Northern analysis, dot blot hybridization, absolute quantitation of mRNA by hybridization in solution followed by ribonuclease protection assay, and *in situ* hybridization.

1. Northern Analysis

Expression of the transthyretin gene was studied by Northern analysis of RNA from rat liver, testis, and brain with full-length transthyretin cDNA[44] as a probe (Figure 7A). A distinct autoradiographic band indicating the presence of transthyretin mRNA was obtained in the analysis of total RNA from liver and brain. No transthyretin mRNA was detectable in RNA from testis. Analysis of a dilution series of RNA (results not shown) suggested that 1 g of rat brain tissue contained about 7% of the amount of transthyretin mRNA found in 1 g of rat liver.

The Northern analysis was repeated using various fragments of transthyretin cDNA as probes to see whether the overall structural organization of transthyretin mRNA in rats was similar in liver RNA and brain RNA. The results obtained indicated that transthyretin mRNA from liver is very similar, if not identical, to transthyretin mRNA from brain (Figure 7B–D). This conclusion was later confirmed by cloning and sequence analysis of transthyretin cDNA from brain, for both rats[45] and sheep.[46] The nucleotide sequences for hepatic and cerebral transthyretin cDNAs were found to be completely identical.[45] An analysis of rat genomic DNA and the cloned transthyretin gene showed that the transthyretin gene was present as a single copy per haploid genome in the rat,[47] similar to the situation in humans[48-50] and mice.[50,51]

2. Dot-Blot Hybridization

The structure of the brain is not homogeneous. It seemed unlikely that the expression of the transthyretin gene would be distributed evenly throughout the brain. The choroid plexus is known to produce most of the cerebrospinal fluid (for a review, see Reference 52). The most likely site for the expression of a protein in the brain for secretion into the cerebrospinal fluid would therefore seem to be the choroid plexus.

RNA was prepared from choroid plexus, the ''rest of brain'' (i.e., brain from which choroid plexus had been removed), and the liver. This RNA was analyzed by dot-blot hybridization with cDNA probes for various rat plasma proteins, as shown in Figure 8. One gram of choroid plexus contained far more transthyretin mRNA than 1 g of liver. Transferrin

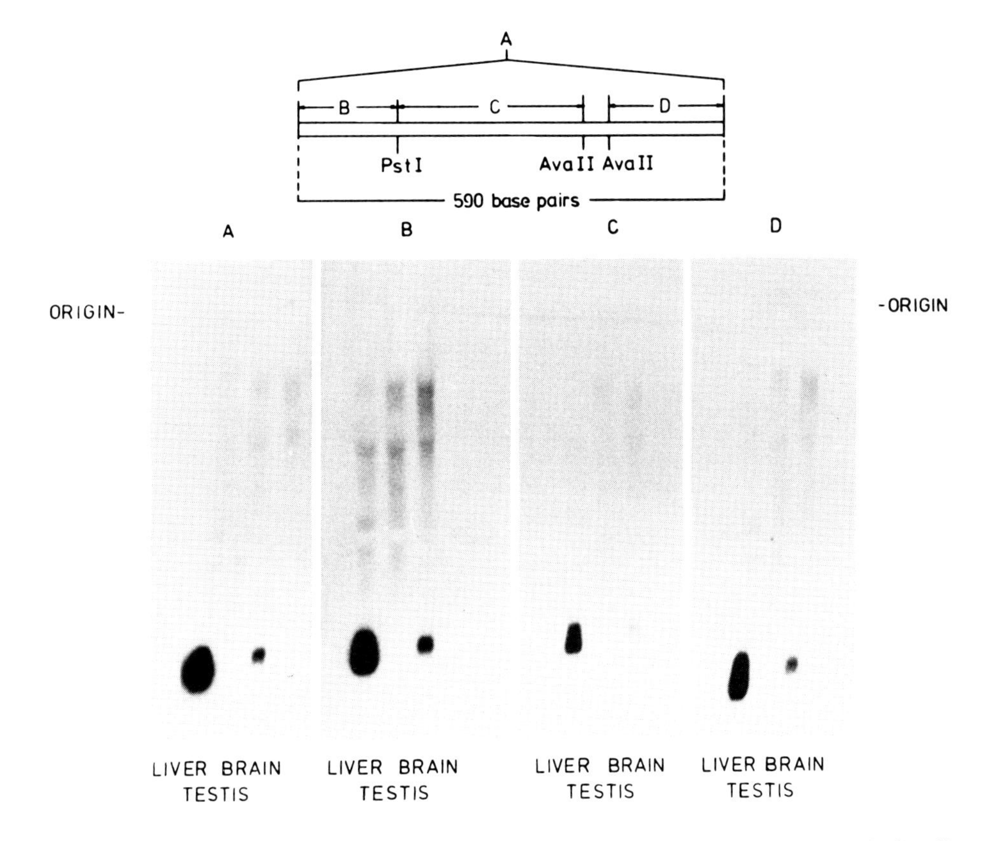

FIGURE 7. Northern analysis of transthyretin mRNA from rat liver and brain. Polyadenylated RNA from liver, brain, and testes, 7 μg per track, was separated by electrophoresis in 1.4% agarose gel containing formaldehyde. (A) Autoradiograph obtained when ^{32}P-labeled transthyretin cDNA containing the whole structural region plus some untranslated flanking regions was used for hybridization; (B–D) results obtained when subfragments of transthyretin cDNA, representing different sections of transthyretin cDNA, as indicated at the top of the figure, were used for hybridization. The box represents the transthyretin cDNA clone used as a probe; the sites of cleavage by restriction enzymes are also shown. (From Dickson, P. W., Howlett, G. J., and Schreiber, G., *J. Biol. Chem.*, 260, 8214, 1985. With permission.)

mRNA was present in choroid plexus and liver in roughly similar amounts per gram of tissue. Messenger RNAs for albumin, the β-subunit of fibrinogen, thiostatin, and α_1-acid glycoprotein were detectable in liver only.[53]

3. Absolute Quantitation of mRNA by Hybridization in Solution Followed by Ribonuclease Protection Assay

Northern analysis and dot-blot hybridization only allow an approximate, relative quantitation of RNA. Molecular titration of specific mRNA is possible if hybridization is carried out in solution and followed by ribonuclease protection assay.[54] The antisense RNA used for hybridization can be synthesized *in vitro* with SP6 RNA polymerase to a given specific radioactivity.[55] The principle of this assay is illustrated in Figure 9. It is possible to include an unrelated, tritiated RNA of size similar to the specific RNA as an internal standard for determination of yield. In this way, it has been possible to obtain precise values for the amount of transthyretin mRNA in micrograms or micromoles per gram of wet weight of tissue for both liver and choroid plexus in rats[56] and chickens.[57] The values obtained for both liver and choroid plexus are summarized in Table 1.

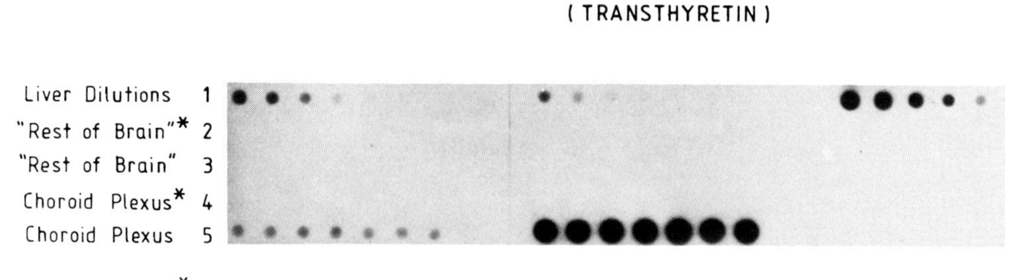

FIGURE 8. Determination of mRNA for transferrin (A), prealbumin (B), and albumin (C) in rat tissue extracts by hybridization to specific cDNA. Row 1, cytoplasmic extracts corresponding to 2500, 100, 500, 250, 100, 50, 25, and 10 μg of liver, wet weight, per spot. The extracts used in rows 2 to 5 were prepared from the tissue of seven individual animals and processed separately Rows 2 and 3, cytoplasmic extracts corresponding to 500 μg of choroid plexus per spot, excluding choroid plexus; rows 4 and 5, cytoplasmic extracts corresponding to 500 μg of choroid plexus per spot; rows 2 and 4, cytoplasmic extracts incubated with ribonuclease prior to processing for hybridization. (From Dickson, P. W., Aldred, A. R., Marley, P. D., Guo-Fen, T., Howlett, G. J., and Schreiber, G., *Biochem. Biophys. Res. Commun.*, 127, 890, 1985. With permission.)

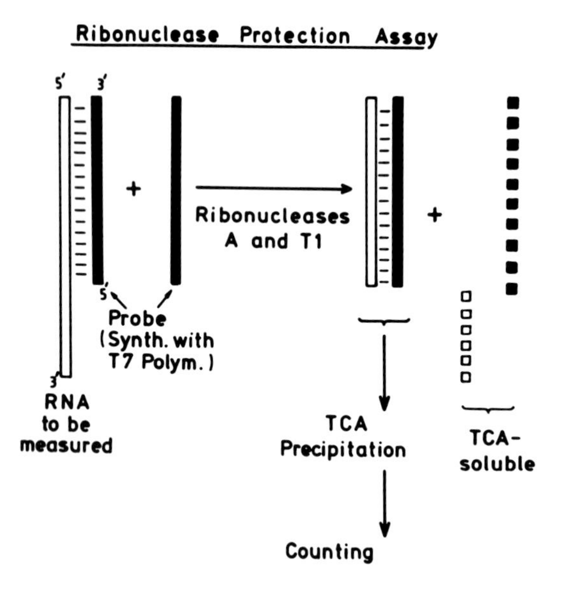

FIGURE 9. Principle of mRNA quantitation by hybridization in solution, followed by ribonuclease protection assay. Nonradioactive RNA is indicated by open bars, radioactive RNA-probe by solid bars. Ribonucleases A and T1 degrade single-stranded RNA. Degraded RNA is indicated by squares. TCA, trichloroacetic acid.

4. *In Situ* Hybridization

Localization of gene expression at the cellular level is possible by *in situ* hybridization of RNA in tissue sections on microscope slides. The choroid plexus is composed of cells of quite different structure and function. Transthyretin mRNA in histological sections was hybridized to a cDNA probe and the expression of the transthyretin gene in rat choroid plexus was found to occur exclusively in the epithelial cell layer (ependyma) on the brain side of the choroid plexus[58] (Figure 10). These epithelial cells are linked by tight junctions and form the blood-cerebrospinal fluid barrier.

TABLE 1
Comparison of the Absolute Concentrations of Transthyretin mRNA in Various Tissues and Species

Species	Tissue	Concentration of transthyretin mRNA	
		μg/g wet weight tissue	nmol/kg wet weight tissue
Rat	Liver	0.39	1.76
	Choroid plexus (4th vent.)	4.38	19.91
Chicken	Liver	0.30	1.31
	Choroid plexus (lat. vent.)	6.48	27.55

C. SYNTHESIS AND SECRETION OF PLASMA PROTEINS BY *IN VITRO* INCUBATION OF ISOLATED TISSUE OR CELLS OTHER THAN LIVER

The synthesis and secretion of plasma proteins can be conveniently studied by analyzing the incorporation of radioactive amino acids into proteins in suspensions of cells[59-65] (for reviews, see References 66 and 67) or in thin layers of tissue, such as the choroid plexus,[68-71] in which cells are easily accessible to the surrounding fluid. An example of such a study is the analysis of the synthesis and secretion of transthyretin and transferrin by rat choroid plexus,[68] as shown in Figure 11. Secretion of the two proteins occurred with the typical time delay necessary for intracellular synthesis and transport of the proteins. About 12% of the newly synthesized total protein in cells was transthyretin, and about 43% of the secreted protein was transthyretin. Thus, the choroid plexus is greatly specialized for the synthesis and secretion of transthyretin.

D. SYNTHESIS AND SECRETION OF PROTEINS BY PERFUSED ORGANS

An extracorporeal perfusion system was developed for sheep choroid plexus by Pollay et al.[72] This system allows the introduction of radioactive amino acids into the perfusion medium on the body side of the organ perfusion system and continuous sampling of secreted protein from both the body and the brain side of the choroid plexus. The synthesis and secretion of transthyretin by the choroid plexus was studied using this method. Figure 12 shows a diagram of the perfusion system and the results of fluorography following SDS gel-electrophoresis of proteins synthesized and secreted by perfused sheep choroid plexus.[56] Newly synthesized transthyretin was secreted in increasing amounts exclusively toward the brain side of the perfused tissue system. No newly synthesized transthyretin could be isolated from the perfusion medium on the body side, even after extensive concentration of the medium to which unlabeled transthyretin had been added as a carrier.[56]

V. FUNCTIONAL SIGNIFICANCE OF THE EXTRAHEPATIC SYNTHESIS OF PLASMA PROTEINS

A. TRANSPORT OR BINDING OF VARIOUS COMPOUNDS

1. Transport of Compounds Insoluble or Sparely Soluble in Water

a. Transferrin

All cells require iron as a cofactor for proteins involved in the transport of oxygen or electrons. However, iron is virtually insoluble in water. Therefore, in extracellular fluids, iron is transported bound to a protein, transferrin. Transferrin is synthesized in the liver via a precursor protein and secreted into the bloodstream.[73,74] Transferrin has a molecular weight of about 76,500[69,73] (reviewed in Reference 75) and cannot easily penetrate barriers between the bloodstream and extravascular compartments separated from the bloodstream by layers

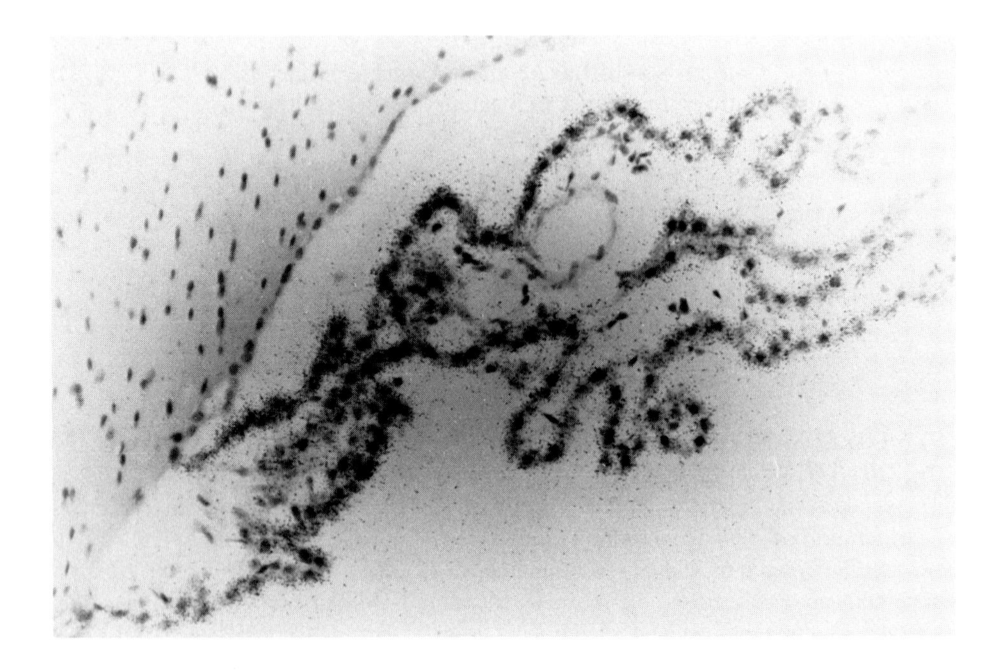

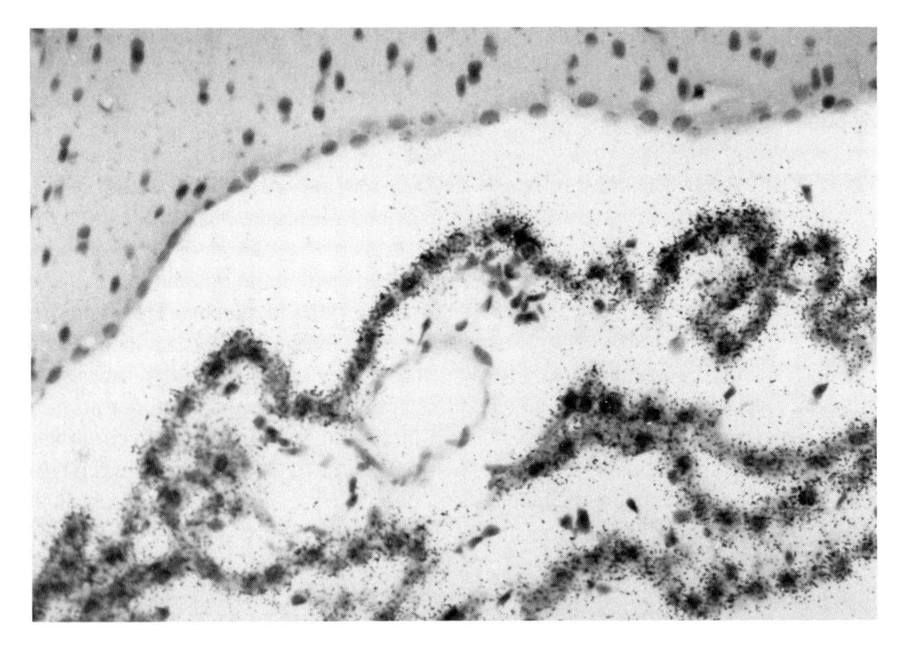

FIGURE 10. *In situ* hybridization followed by autoradiography of a section of rat choroid plexus in which mRNA was hybridized *in situ* to transthyretin cDNA labeled with ^{35}S. The lower picture shows a central region of the upper picture in higher magnification. Silver grains are seen only above the epithelial cells. Ependyma and stroma of the choroid villi, in particular vascular endothelial cells, are free of silver grains. (From Schreiber, G., in *The Plasma Proteins,* Putnam, F. W., Ed., Academic Press, New York, 1987, 293. With permission.)

of cells linked with tight junctions. The presence of transferrin receptors on the endothelial surface of the cells forming the blood-brain barrier has been reported.[76] However, the occurrence of such receptors does not necessarily indicate the presence of a transcytotic transport system for transferrin. As a rule, transferrin receptors at the surface of cell mem-

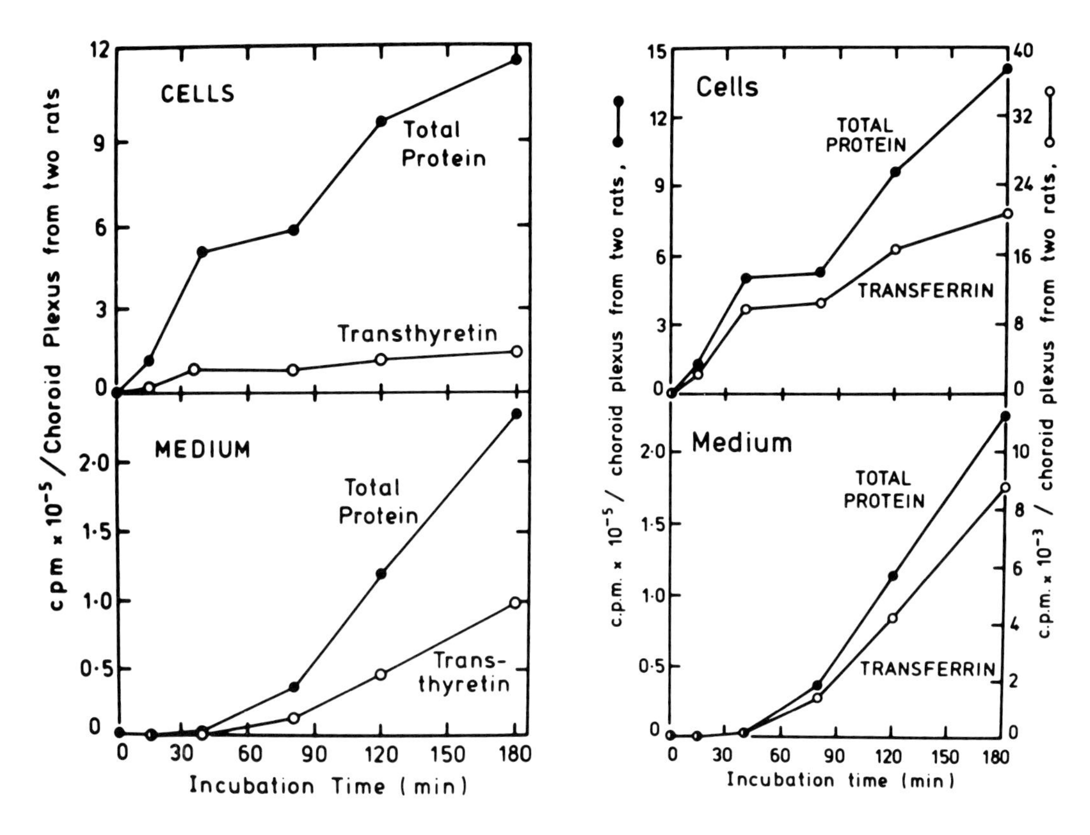

FIGURE 11. Incorporation of [^{14}C]leucine into total protein, transthyretin, and transferrin synthesized and secreted by isolated choroid plexus. Abscissa, incubation time; ordinate, [^{14}C]leucine incorporated into protein. Choroid plexus was dissected from rats weighing between 300 and 400g (pooled tissue from two rats per time point) and incubated for various times in the presence of [^{14}C]leucine. Tissue pieces and medium were separated by centrifugation. Transthyretin and transferrin were immunoprecipitated from samples of tissue homogenates and medium with transthyretin antiserum and transferrin antiserum, respectively. Radioactivity in solubilized immunoprecipitates and in trichloroacetic acid precipitates of total protein was determined by liquid scintillation spectrometry. Cycloheximide (20 μM) added to control incubations inhibited incorporation into total protein by 97% (curve not shown). (Adapted from Dickson, P. W., Aldred, A. R., Marley, P. D., Bannister, D., and Schreiber, G., *J. Biol. Chem.*, 261, 3475, 1986; Aldred, A. R., Dickson, P. W., Marley, P. D., and Schreiber, G., *J. Biol. Chem.*, 262, 5293, 1987. With permission.)

branes are a component of the iron uptake system of cells for their own use of iron (reviewed in References 77 and 78). The question arises as to where the transferrin, found in compartments not freely accessible to proteins from the bloodstream, originates from. The extrahepatic occurrence of transferrin mRNA was demonstrated by hybridization to cloned transferrin cDNA.[69,79-83] In particular, the transferrin gene was found to be expressed in the cells forming the barrier between the bloodstream and the cerebrospinal fluid,[21,53,69,83] in cells of the blood-testis barrier,[84-86] in the visceral yolk sac of the rat,[82] and also in cells not linked in a barrier by tight junctions, such as oligodendrocytes.[80,81] Figure 13 shows the tissue distribution of the expression of the transferrin gene in various rat tissues as determined by Northern analysis.[69] Recently, it was reported that the distribution of expression of the transferrin gene in the brain shows great differences between species[83] (Figure 14, Table 2). This might not be surprising, since the size of the brain, and therefore probably also the iron demand, varies greatly between species. Extrapolation of observations with rat brain to the human[87] does not seem to be appropriate in attempts to understand the effects of serotonin on the metabolism of transferrin in human brain.

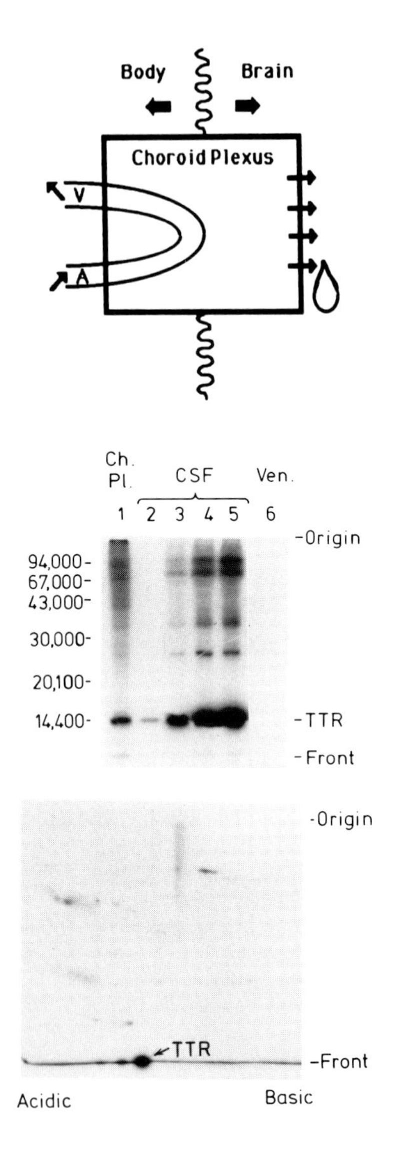

FIGURE 12. Secretion of newly synthesized protein by perfused sheep choroid plexus. Top panel, principle of the experiment. The position of the blood-brain barrier is indicated schematically by a wavy line, that of the choroid plexus by a box. A, anterior choroidal artery; V, great vein of Galen. The drop indicates newly formed cerebrospinal fluid, which was collected from the surface of the exposed choroid plexus. Middle panel, fluorograph of ^{14}C-labeled proteins synthesized by the perfused choroid plexus and secreted into newly formed cerebrospinal fluid, separated by polyacrylamide gel electrophoresis under denaturing conditions. Lane 1, proteins in choroid plexus homogenate (Ch. Pl); lanes 2–5, proteins in newly formed cerebrospinal fluid, collected 35 min (lane 2), 50 min (lane 3), 80 min (lane 4), and 90 min (lane 5) after addition of [^{14}C]leucine to perfusion medium,; lane 6, proteins in venous effluent (Ven.) after 90 min of perfusion. The position of authentic transthyretin from sheep serum is indicated. Molecular weight markers were bovine α-lactalbumin (14,400), soybean trypsin inhibitor (20,100), bovine carbonic anhydrase (30,000), ovalbumin (43,000), bovine serum albumin (67,000), and rabbit muscle phosphorylase b (94,000). Bottom panel, two-dimensional isolectric focusing/electrophoretic analysis of proteins in newly secreted cerebrospinal fluid collected after 90 to 120 min of perfusion with medium containing [^{14}C]leucine. The position of transthyretin subunits is indicated with an arrow. Isoelectric focusing (pH 3.5 to 10) was carried out in the first dimension and polyacrylamide gel electrophoresis (12.5% polyacrylamide gel) in the second dimension. Radioactive proteins were detected by fluorography. (From Schreiber, G., Aldred, A. R., Jaworowski, A., Nilsson, C., Achen, M. G., and Segal, M. B., *Am. J. Physiol.*, 258, R338, 1990. With permission.)

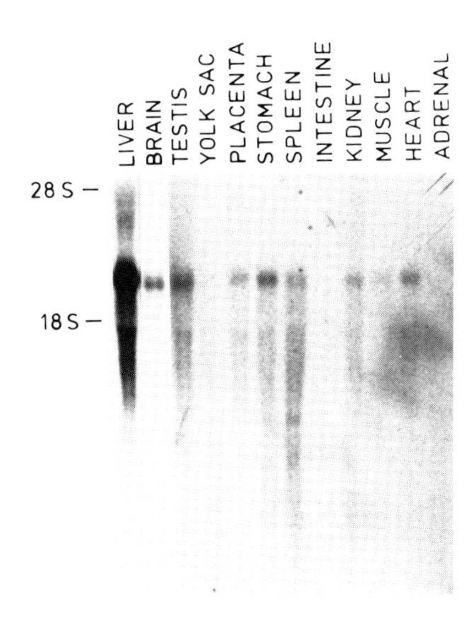

FIGURE 13. Tissue distribution of transferrin mRNA studied by Northern analysis. Total RNA was extracted from liver, brain, testes, stomach, spleen, small intestine, kidney, muscle, heart, and adrenal glands of three male rats, weighing between 250 and 350 g, and from yolk sac and placenta from a female rat at day 17 of gestation. RNA, 20 μg per track, was subjected to electrophoresis in 1% agarose gel containing formaldehyde, transferred to a nitrocellulose membrane, and hybridized with ^{32}P-labeled transferrin cDNA. Autoradiography at −70°C using Kodak X-Omat AR film with an intensifying screen was for 24 h (liver and brain tracks) or 18 d (all other tracks). The positions of 28 S and 18 S ribosomal RNA bands are indicated. (From Aldred, A. R., Dickson, P. W., Marley, P. D., and Schreiber, G., *J. Biol. Chem.*, 262, 5293, 1987. With permission.)

b. Ceruloplasmin

The best-known function of ceruloplasmin is the transport of copper, but ceruloplasmin also possesses various enzymatic activities, such as those of a ferroxidase, an amine oxidase, a superoxide dismutase, and a deaminase.[88-90] The levels of ceruloplasmin in the bloodstream[91] and of its mRNA[92,93] in liver increase during the acute phase response of the rat to inflammation. After cloning of ceruloplasmin cDNA from rat liver RNA, the tissue distribution of the expression of the ceruloplasmin gene was studied in the rat.[93] Extrahepatic synthesis of ceruloplasmin mRNA was observed in choroid plexus, yolk sac, placenta, and testis of the rat (Figure 15). High levels of ceruloplasmin mRNA were measured in the uterus of both pregnant and nonpregnant rats[94] (Figure 16). Ceruloplasmin synthesis by the uterus could be part of a system transporting copper to the fetus.

c. Retinol-Binding Protein

Retinol is transported in the blood stream bound to retinol-binding protein. Retinol-binding protein is, itself, bound to transthyretin. Using a genomic DNA fragment containing the exons coding for retinol-binding protein as a probe, extrahepatic expression of the gene for retinol-binding protein was demonstrated[95] for choroid plexus, other areas of the brain, and liver by Northern analysis of RNA extracts, as illustrated in Figure 17.

2. Transport Between Compartments

a. Transthyretin

The epithelial cells of the choroid plexus are among the most highly specialized cells for the synthesis and secretion of one particular protein, transthyretin.[68] All of the transthyretin synthesized by the choroid plexus is secreted toward the brain.[56] The brain is one of the main target organs for thyroid hormones.[96,97] The question arises as to whether trans-

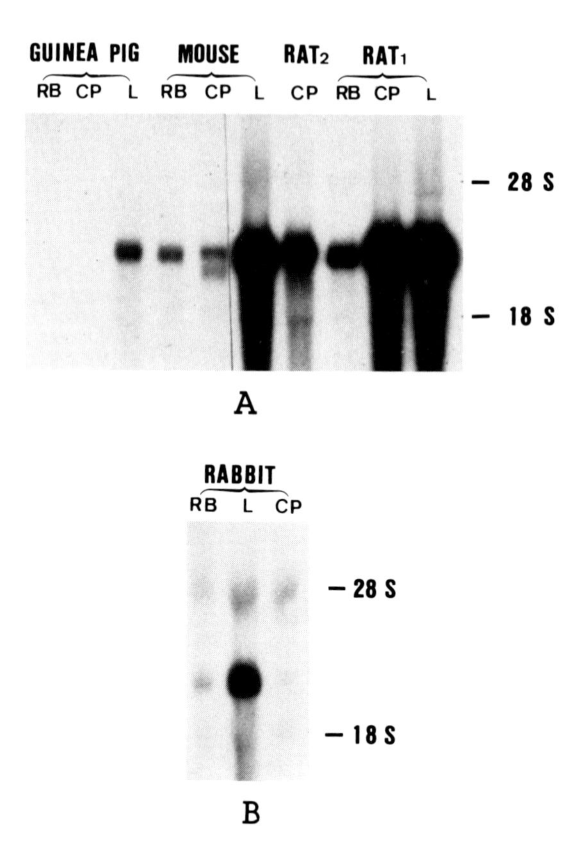

FIGURE 14. Northern analysis of transferrin mRNA in choroid plexus (CP), brain without choroid plexus (RB), and liver (L) in various species. RAT_1, Buffalo rats; RAT_2, brown Norway rats; MOUSE, C57 black × CBA mice. Ten μg of total RNA per track. (A, B) Hybridization with a rat transferrin cDNA probe; (C, D) hybridization with a pig transferrin cDNA probe. The positions of 28 S and 18 S ribosomal RNA bands are indicated at the right margins of the figures. (From Tu, G.-F., Achen, M. G., Aldred, A. R., Southwell, B. R., and Schreiber, G., *J. Biol. Chem.*, 266, 6201, 1991. With permission.)

thyretin synthesis and secretion by the choroid plexus may be linked to the transport of thyroid hormones from the bloodstream to the brain. Intravenously injected [^{125}I]thyroxine, but not [^{125}I]triiodothyronine, was found to accumulate specifically in the choroid plexus.[98] Within a few hours after intravenous injection, [^{125}I]thyroxine, but not [^{125}I]triiodothyronine, accumulated first in choroid plexus,[98] thereafter in the cerebrospinal fluid,[56] and finally in more peripheral parts of the brain, such as the striatum, cortex, and cerebellum.[98] The kinetics of this accumulation are shown in Figures 18 to 20. Thyroid hormones were found to partition into liposomes, reaching their highest concentration in the middle of the lipid membrane.[98] In choroid plexus pieces incubated *in vitro,* both thyroxine and triiodothyronine were taken up in a nonsaturable process.[98]

These observations were used to propose the hypothesis summarized in Figure 21 for the mechanism of the transport of thyroid hormones from the bloodstream to the brain. According to this hypothesis, the transfer of thyroid hormones from the bloodstream to the brain is mediated by transthyretin synthesis and secretion by the epithelial cells of the choroid plexus, which form the blood-cerebrospinal fluid barrier. It is proposed that both free and protein-bound thyroid hormones reach the stroma of the choroid plexus freely from the fenestrated capillaries of the choroid plexus. From the interstitial space in the stroma, free thyroid hormones partition into the membrane system of the choroid plexus epithelial cells.

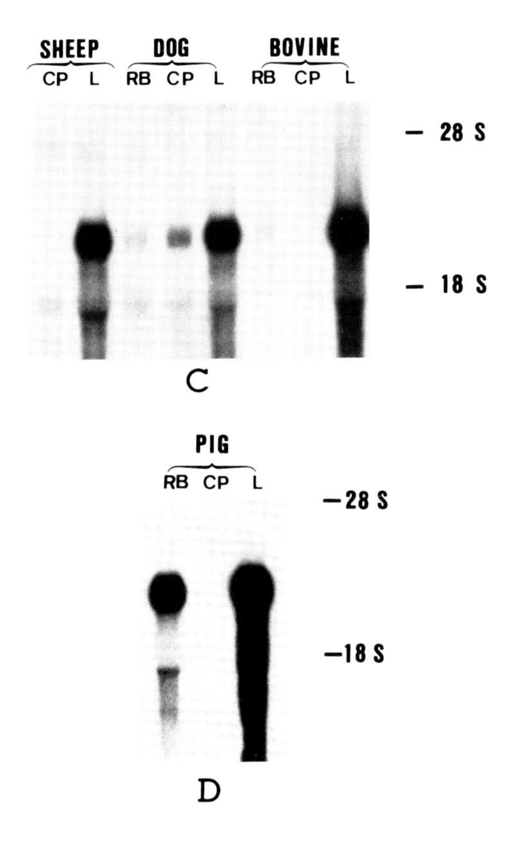

FIGURE 14C and D.

The newly synthesized transthyretin is secreted unidirectionally from the epithelial cells of the choroid plexus into the cerebrospinal fluid. This transthyretin binds thyroxine. The precise site for this binding (intracellular, near membranes, or extracellular) remains to be identified.

The observation that thyroxine, but not triiodothyronine, is the predominant form of thyroid hormone transported through the choroid plexus to the brain can be explained as follows: (1) the molar concentration of free thyroxine in plasma is much higher than that of free triiodothyronine (about sevenfold higher in both the rat[99] and humans[100]) and (2) the affinity of transthyretin for thyroxine is much greater than that for triiodothyronine (by a factor of 12.5 in the human;[101] the affinity for rat transthyretin is not known).

After transport to the brain, thyroxine can be deiodinated near the synapses and taken up into synaptosomes.[102] The fact that the transport of thyroid hormones from the bloodstream to the brain occurs in the form of thyroxine, but that the effects of thyroid hormones within the brain are mainly due to triiodothyronine (see Reference 96), would create the possibility of independent regulation of the transport and action of thyroid hormone within the brain.

b. Other Proteins

The expression of the genes for transferrin, ceruloplasmin, and retinol-binding protein in cells linked by tight junctions for the formation of barriers between compartments raises the question of whether these proteins might be involved in the transcytotic transport of iron, copper, and retinol between compartments.

TABLE 2
Amounts of Transferrin mRNA in Various Tissues and Species

Tissue	ng transferrin mRNA per gram of tissue	fmol transferrin mRNA[a] per gram of tissue
	Rat	
Liver	3700	4300
Brain[b]	15	17
Choroid plexus[c]	470	550
	Sheep	
Liver	570	660
Brain[b]	6.8	7.9
Choroid plexus[c,d]	<1.2	<1.4
	Pig	
Liver	510	590
Brain[b]	10	12
Choroid plexus[c,d]	<1.5	<1.7

Note: RNA was extracted from tissues as described in Reference 83. Recovery of RNA was monitored by including a radioactive RNA, of a size similar to that of transferrin mRNA, as an internal standard. Transferrin mRNA was quantitated by hybridization in solution with transferrin mRNA antisense RNA probes followed by a ribonuclease protection assay.[54]

[a] Molecular weight of transferrin mRNA was assumed to be 8.6×10^5 (based on length of mRNA and average molecular weight of ribonucleotides).
[b] Brains from which choroid plexus had been removed.
[c] Lateral ventricle choroid plexus.
[d] The lowest level of detection is given.

Modified from Tu, G.-F., Achen, M. G., Aldred, A. R., Southwell, B. R., and Schreiber, G., *J. Biol. Chem.*, 266, 6201, 1991. With permission.

3. Creation of Circulating Intracompartmental Pools for Control of the Distribution of Ligands

a. Transthyretin

It is a common textbook error (see, e.g., Reference 103) to assume that thyroid hormone-binding proteins are necessary to make thyroid hormones "soluble" for transport in the aqueous phase of blood plasma. The concentration of thyroxine in serum, about 24 p*M* (see Reference 100), is well within the limits of its physical solubility. Values of 2 μ*M* at pH 7.0 and 2.8 μ*M* at pH 8.0 have been reported for the solubility of thyroxine in an aqueous medium.[104] Mendel et al.[105] have shown that the addition of thyroid hormone-binding proteins to perfusion medium is necessary to achieve even distribution of thyroid hormones in the cells of the lobules of the isolated perfused rat liver. Without addition of thyroid hormone-binding proteins, thyroid hormones are taken up by the hepatocytes at the periportal pole of the perfused lobule. Due to the rapid partitioning of the highly lipid-soluble thyroid

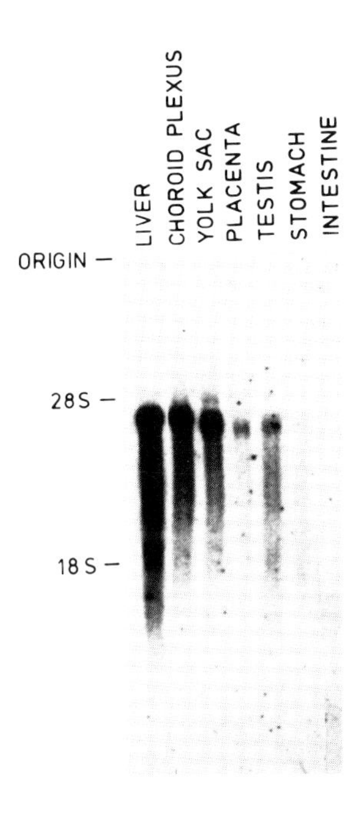

FIGURE 15. Determination of the tissue distribution of ceruloplasmin mRNA by Northern analysis. Polyadenylated RNA from choroid plexus, approximately 2 μg, and other tissues, approximately 7 μg (liver, yolk sac, placenta, testis, stomach, and small intestine), was analyzed by electrophoresis in a 1% agarose gel containing formaldehyde and transferred to nitrocellulose. The filter-bound RNA was hybridized with ^{32}P-labeled ceruloplasmin cDNA. The single species of ceruloplasmin mRNA detected in liver, choroid plexus, yolk sac, placenta, and testis showed evidence of degradation. The slightly greater mobility of the mRNA in RNA from yolk sac, placenta, and testis was due to uneven migration in the gel. In other experiments (not shown), we found that the size of the ceruloplasmin mRNA was similar in all these tissues. No ceruloplasmin mRNA was detected in RNA from stomach and small intestine. (From Aldred, A. R., Grimes, A., Schreiber, G., and Mercer, J. F. B., *J. Biol. Chem.*, 262, 2875, 1987. With permission.)

hormones into the membranes of the periportal hepatocytes, the perfusion medium is depleted of thyroid hormones before the medium can reach more central areas of the liver lobule. Apparently, thyroid hormone-binding proteins create circulating intracompartmental pools for ensuring appropriate distribution of hormones[106] (Figure 22), and such a system of binding proteins has evolved concurrently with the increase in membrane systems seen in endothermic animals (see below).

b. Transferrin

Virtually all circulating iron in the extracellular fluid is bound to transferrin (only small amounts are bound to extracellularly located ferritin). It has been shown recently that the intracerebral distribution of the transferrin gene shows great differences between species.[83] It is conceivable that the regional differences in transferrin synthesis in the brain could create regional differences in the extracellular concentration of iron and that such differences might have physiological implications.

B. PROTECTION OF THE INTEGRITY OF COMPARTMENTS, CELLS, AND SURFACE STRUCTURES

1. Proteinase Inhibitors

Proteinase inhibitors are an important group of acute phase proteins (see Chapter 13). Extrahepatic expression has been observed for the proteinase inhibitors cystatin C and α_2-macroglobulin.

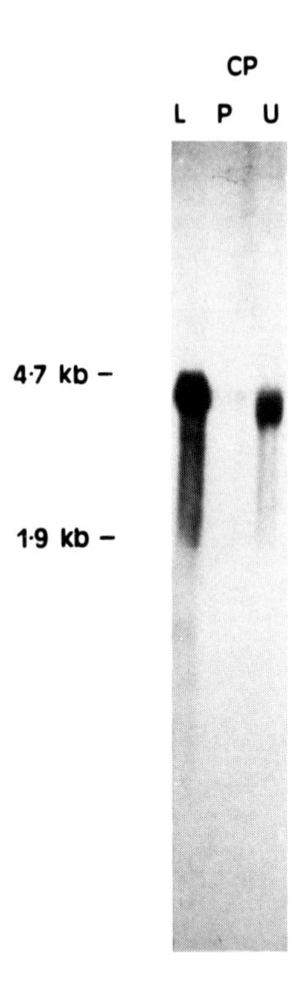

FIGURE 16. Northern analysis of the ceruloplasmin mRNA levels in rat uterus during pregnancy. Hybridization of a rat ceruloplasmin cDNA probe[93] to ceruloplasmin mRNA in total RNA (15 μg) from maternal liver (L), placenta (P), and uterus (U). (From Thomas, T. and Schreiber, G., *FEBS Lett.*, 243, 381, 1989. With permission.)

a. Cystatin C

Cystatin C is the extracellular brain protein with the highest ratio of the concentration in cerebrospinal fluid over that in blood plasma (for data on the human, see Reference 107). The primary structure of the cDNA has been reported for human cystatin C.[108] Oligonucleotide probes were synthesized based on the sequence of human cystatin C cDNA and found to cross-hybridize with the corresponding mRNA in the rat.[109,110] In the rat, the tissue with the highest cystatin C mRNA level was found to be the choroid plexus.[109] Cystatin C mRNA was also detected in lower levels in other areas of the brain, in testis, epididymis, seminal vesicles, prostate, ovary, submandibular gland, and, in trace amounts, in liver.[109] Choroid plexus pieces incubated with radioactive leucine secreted radioactive cystatin C.[109] Possibly, the function of cystatin C is the inhibition of extracellular proteinases, such as those released from destroyed or dying cells. Thus, cystatin C could be important in maintaining the integrity of cell-surface proteins in the brain. The presence of large numbers of lysosomes, cell organelles rich in proteinases, has been described for neuronal cells.[111]

b. α_2-Macroglobulin

α_2-Macroglobulin is the positive acute phase protein with the greatest relative change in plasma concentration during the acute phase response in the rat (for reviews, see References 21 and 112). It inhibits proteinases by binding them to a particular section of α_2-macroglobulin, the so-called "bait" region. Development and growth involves extensive tissue remodeling, with local increases in the intensity of both protein synthesis and protein degradation. Such tissue remodeling occurs, for example, in the rat placenta when the decidua capsularis begins to degenerate.[113] For orderly growth, synthesis and degradation of proteins

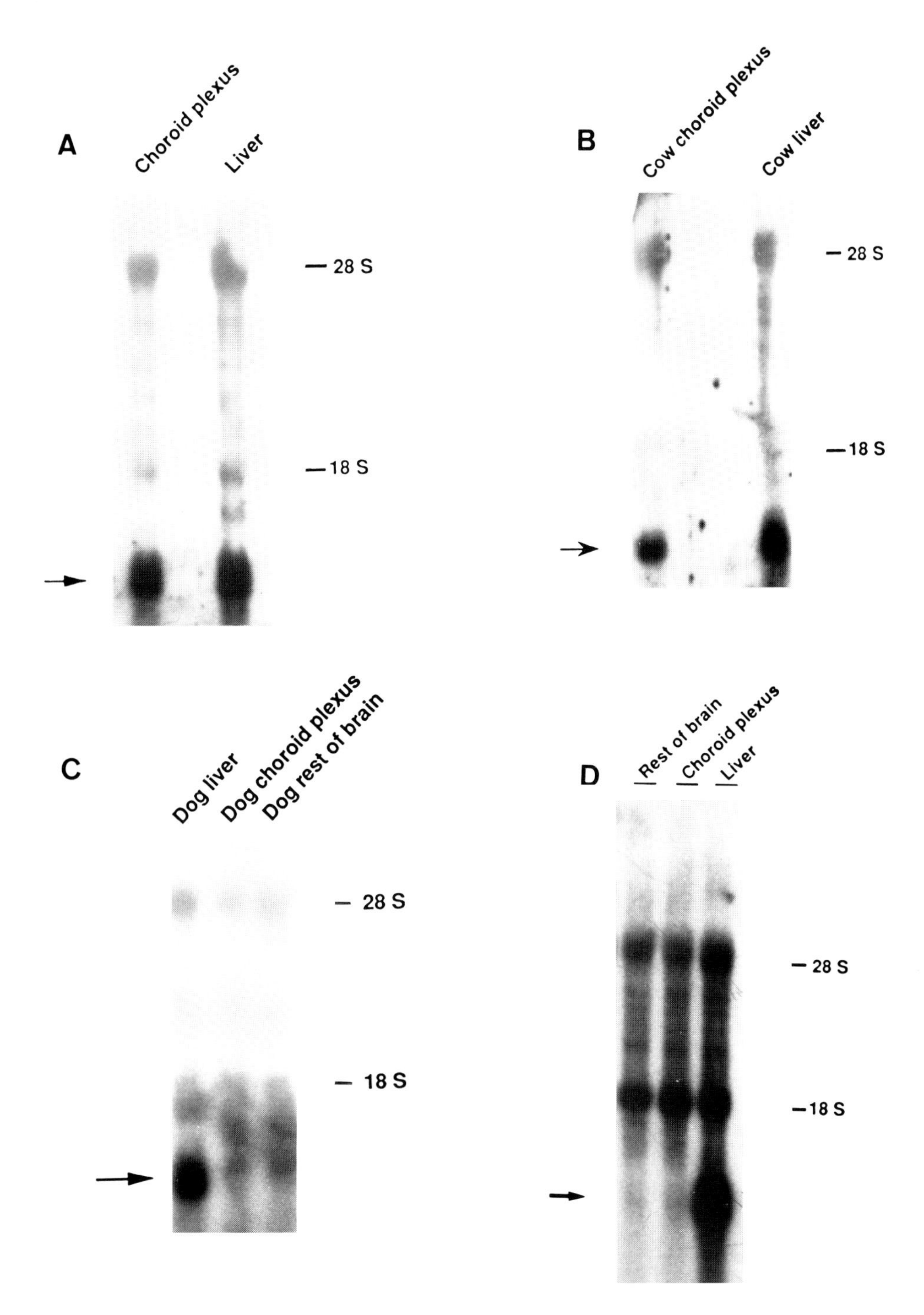

FIGURE 17. Northern analysis of retinol-binding protein mRNA in total RNA (20 μg per track) with a ^{32}P-labeled human retinol-binding protein genomic DNA fragment as a probe. The positions of 28 S and 18 S ribosomal RNA bands are shown to the right of the autogradiographs. The position of retinol-binding protein mRNA is indicated by the arrow on the left. (A, B, C, D) RNA from tissues from sheep, cattle, dogs, and mice, respectively. Choroid plexus from several animals were pooled for RNA preparation. (From Duan, W. and Schreiber, G., *Comp. Biochem. Physiol.* B., 101 B, 391, 1992. With permission.)

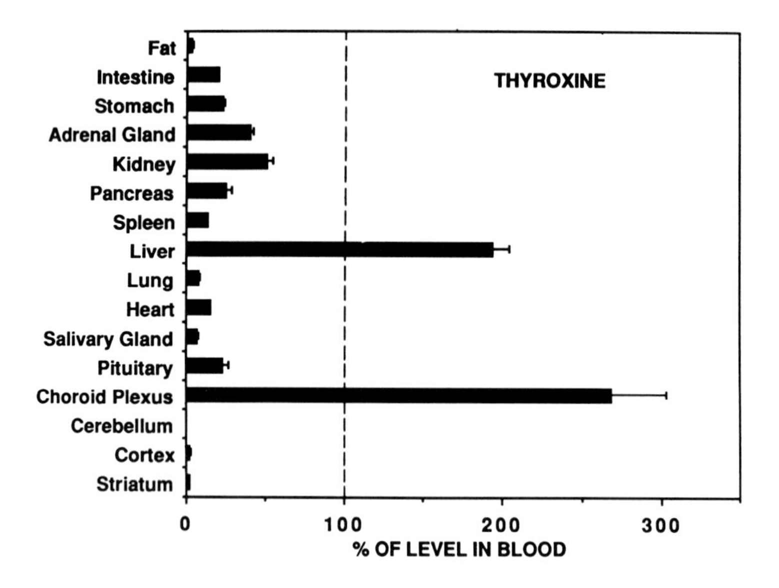

FIGURE 18. Tissue distribution of injected [^{125}I]thyroxine *in vivo,* 10 min after injection of the labeled hormone into the caval vein. The results are given as radioactivity in 1 mg of tissue as a percent of the radioactivity in 1 μl of blood. The radioactivity in each tissue has been corrected for blood contamination. Results are given as the mean for three animals ± SE. (From Dickson, P. W., Aldred, A. R., Menting, J. G. T., Marley, P. D., Sawyer, W. H., and Schreiber, G., *J. Biol. Chem.,* 262, 13907, 1987. With permission.)

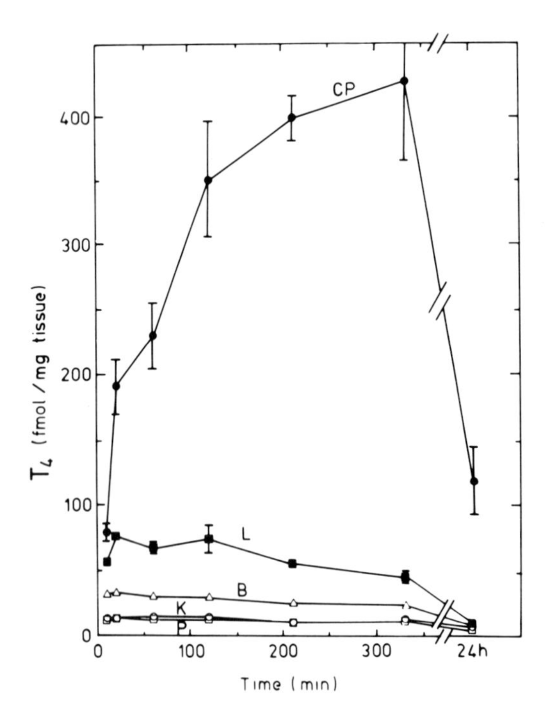

FIGURE 19. Kinetics of uptake of intravenously injected [^{125}I]thyroxine into choroid plexus, liver, kidney, and pituitary at various times after an intravenous injection of [^{125}I]thyroxine. CP, choroid plexus; L, liver; B, blood; K, kidney, P, pituitary. Each point is the mean for three animals ± SE. (From Dickson, P. W., Aldred, A. R., Menting, J. G. T., Marley, P. D., Sawyer, W. H., and Schreiber, G., *J. Biol. Chem.,* 262, 13907, 1987. With permission.)

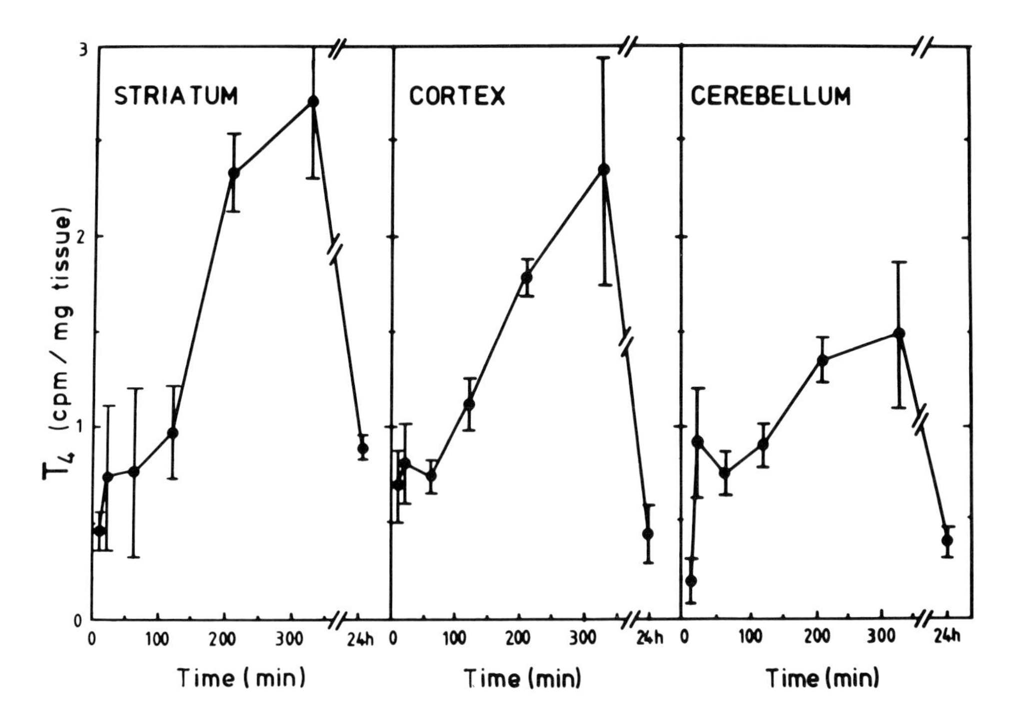

FIGURE 20. Kinetics of uptake of intravenously injected [^{125}I]thyroxine into striatum, cortex, and cerebellum at various times after an intravenous injection of [^{125}I]thyroxine. Each point is the mean for three animals ± SE. (From Dickson, P. W., Aldred, A. R., Menting, J. G. T., Marley, P. D., Sawyer, W. H., and Schreiber, G., *J. Biol. Chem.*, 262, 13907, 1987. With permission.)

have to be regulated. In this regard, it is of interest that the decidua is a site of α_2-macroglobulin gene expression which shows a distinct maximum during pregnancy in the rat around day 11.5, i.e., in midpregnancy (Figure 23).

C. CASES IN WHICH THE FUNCTIONAL SIGNIFICANCE IS NOT WELL UNDERSTOOD

1. α_1-Acid Glycoprotein

The structure, cDNA cloning, and suggestions for possible functions of α_1-acid glycoprotein have been reviewed elsewhere.[21] The largest amount of α_1-acid glycoprotein in the body is synthesized in the liver. This synthesis occurs via a precursor protein.[74,114,115] The rate of synthesis of α_1-acid glycoprotein increases strongly during the acute phase response.[20] However, high levels of α_1-acid glycoprotein mRNA are also found during pregnancy in the decidua of the rat, with a maximum at about day 11.5[115] (Figure 24). The maximum in the curve describing α_1-acid glycoprotein mRNA levels in the decidua as a function of time occurs about 1 d later than the mild transient increase in the α_1-acid glycoprotein mRNA level seen in the maternal liver.[116] A discussion of the functional relevance of an increase in the local synthesis of α_1-acid glycoprotein in the border region between maternal and fetal tissue during pregnancy can be found elsewhere.[116]

2. Other Proteins

Two other plasma proteins that are synthesized in the liver and at extrahepatic sites are β_2-microglobulin and βA4-amyloid precursor protein. β_2-Microglobulin is the protein with the second-highest ratio of the concentration in cerebrospinal fluid to that in blood plasma (see Reference 107). It is the light chain of the type I histocompatibility antigen. The highest

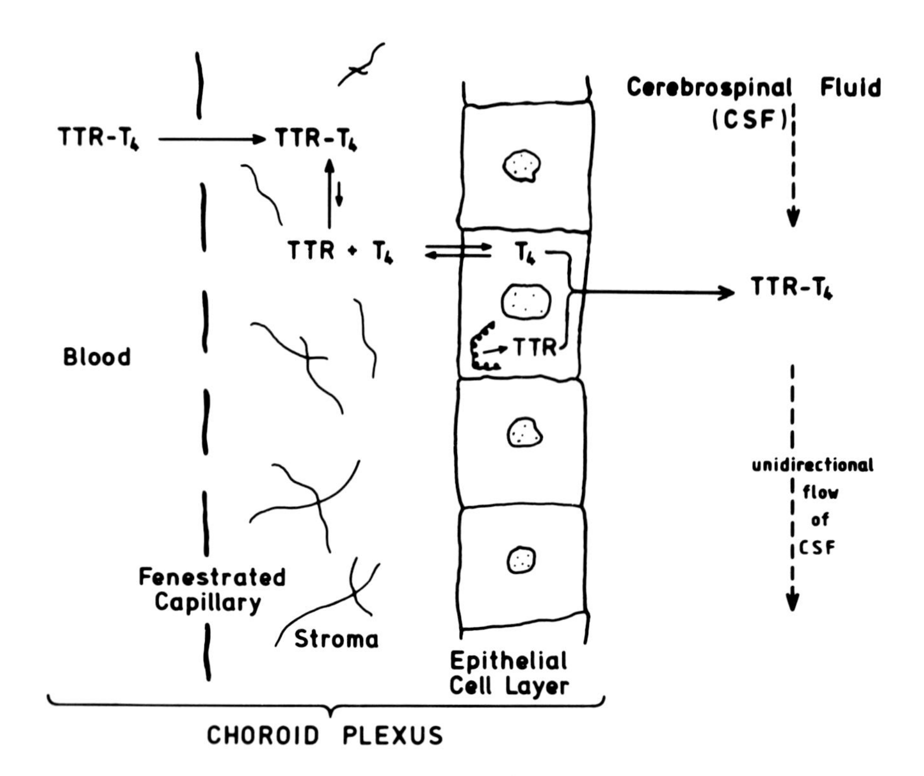

FIGURE 21. Hypothesis for the transport of thyroxine from blood to cerebrospinal fluid via the choroid plexus. The direction of migration of thyroxine is from the bloodstream at the left side of the figure to the central nervous system at the right side of the figure. TTR, transthyretin; T_4, thyroxine; CSF, cerebrospinal fluid. (From Dickson, P. W., Aldred, A. R., Menting, J. G. T., Marley, P. D., Sawyer, W. H., and Schreiber, G., *J. Biol. Chem.*, 262, 13907, 1987. With permission.)

levels of β_2-microglobulin mRNA were observed in the choroid plexus and liver of rats.[109] Other parts of the brain and testis contained lower levels of β_2-microglobulin mRNA.

βA4-Amyloid precursor protein mRNA is absent from, or present only in low levels in, the liver of mammals. Therefore, since it is not synthesized by the liver, βA4-amyloid precursor protein does not qualify as a plasma protein in this class of animals. However, in birds, βA4-amyloid precursor protein mRNA was found to be present in high levels in the liver, in addition to the extrahepatic expression of the gene.[117]

VI. REGULATION OF EXTRAHEPATIC PLASMA PROTEIN SYNTHESIS

Regulatory changes in the rates of synthesis of extracellular proteins are observed when protein homeostasis is challenged. Changes in the rates of incorporation of radioactive amino acids into plasma proteins usually reflect changes in the levels of mRNAs, indicating that translational or posttranslational control of protein synthesis is of no, or minor, importance.[118] Alteration of the transcriptional activity of the genes is the major mechanism leading to changes in mRNA levels, but in some cases, changes in the stability of mRNAs also play a role.[119] The challenges to extracellular homeostasis differ in the various extracellular compartments of the body. Thus, it is not surprising that regulation of the expression of the transthyretin gene in choroid plexus is independent of that in liver[68] (Table 3).

The levels of mRNAs for plasma proteins in different parts of the female reproductive tract are usually lower than those in liver, with the exception of α_2-macroglobulin[94] (Table

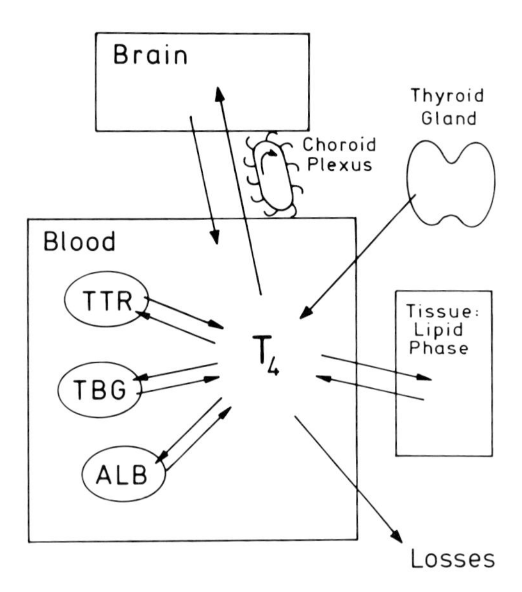

FIGURE 22. Thyroxine (T_4) pools in the body. TTR, transthyretin; TBG, thyroxine-binding globulin; ALB, albumin; TTR-T_4, transthyretin with bound thyroxine; TTR + T_4, free transthyretin and free thyroxine. (From Aldred, A. R., Pettersson, T. M., Harms, P. J., Richardson, S. J., Duan, W., Tu, G.-F., Achen, M. G., Nichol, S., and Schreiber, G., in *"Platypus and Echidnas"*, M. L. Augee, Ed., The Royal Zoological Society of New South Wales, Sydney, Australia, 1992, 44. With permission.)

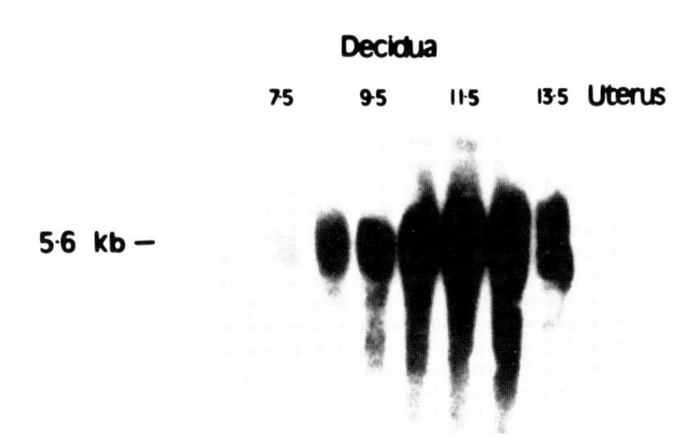

FIGURE 23. Northern analysis of the expression of α_2-macroglobulin in rat decidua RNA and uterus RNA, 7.5 to 13.5 d after conception. Fifteen micrograms of RNA per track. (From Thomas, T. and Schreiber, G., *FEBS Lett.*, 243, 381, 1989. With permission.)

4). Again, regulation of gene expression in the liver is independent of that in extrahepatic tissues (see Figure 24 for α_1-acid glycoprotein mRNA in the decidua and in the maternal liver during pregnancy). In some cases, e.g., for the synthesis of transthyretin in the fetus, mRNA levels in tissues are regulated in a reciprocal way. Thus, in the first part of pregnancy, transthyretin mRNA levels are high in the yolk sac, but absent from, or low in, fetal liver. In the second half of pregnancy, transthyretin mRNA levels steadily increase in the liver, while they decrease in the yolk sac[47] (Figure 25).

The pattern of the levels of mRNA for extracellular proteins in choroid plexus shows stage-specific changes during early development of the brain (Figure 26). In the rat, tran-

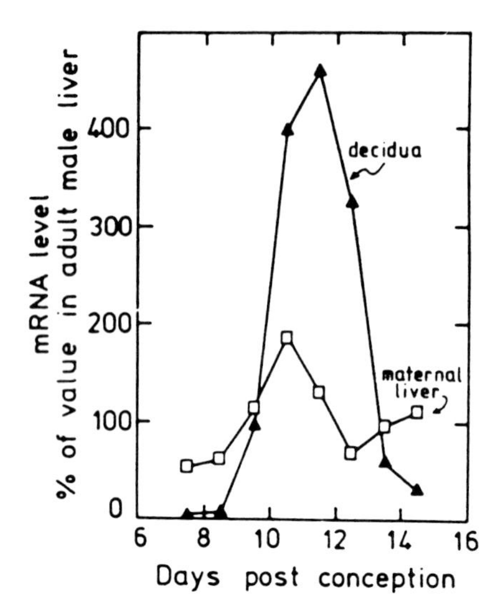

FIGURE 24. Levels of the mRNA for α_1-acid glycoprotein in the decidua and maternal liver of rats. Dot hybridization of ^{32}P-labeled α_1-acid glycoprotein cDNA to total RNA purified from the decidua and maternal liver between 7.5 and 14.5 d after conception. Between 8 and 14 animals were used per time point. The α_1-acid glycoprotein mRNA levels are expressed as a percentage of the level found in adult male rat liver (tissue pooled from ten rats). (From Thomas, T., Fletcher, S., Yeoh, G. C. T., and Schreiber, G., *J. Biol. Chem.*, 264, 5784, 1989. With permission.)

scription of the transthyretin gene is activated first, followed by that of the genes for cystatin C and β_2-microglobulin. The transferrin gene is expressed only later, beginning at the time of weaning. The rat is an altrical species, i.e., a species with relatively slow brain development. Patterns of expression of the genes for extracellular proteins in the choroid plexus of precocial animals, such as sheep, differ from the altrical pattern by a shift in the onset of expression of genes to earlier times (Figure 26).[70,110,117]

VII. EVOLUTION OF EXTRAHEPATIC PLASMA PROTEIN SYNTHESIS

Primary structures, regulation of synthesis rates, and tissue specificity of expression of plasma proteins differ among different species (for reviews, see Reference 21 and 112). Some proteins (e.g., α_2-macroglobulin) are strong acute phase proteins in the rat, but are constitutively expressed in other species. C-reactive protein is an acute phase protein in humans and rabbits, but not in rats. Generally, immunochemical cross-reactivity between proteins from different species and the similarity of amino acid sequences are less for acute phase proteins than for other plasma proteins. Therefore, it is surprising to find very strong conservation of both primary and three-dimensional structures of transthyretin for a wide range of species (Figures 27 and 28, Table 5).

Strong conservation of the structure of a protein, or parts thereof, is usually interpreted as an indication of the functional importance of the conserved molecule or regions thereof. In the bloodstream of larger mammals, transthyretin shares with albumin and thyroxine-binding globulin in the transport of thyroid hormones.[131] However, transthyretin is the only thyroid hormone-binding protein which is strongly expressed in the brain. Analysis of the phylogeny of expression of transthyretin by the choroid plexus showed that this function is

TABLE 3
Levels of mRNAs for Transthyretin, Albumin, α_1-Acid Glycoprotein, and Major Acute Phase α_1-Protein in Liver and Choroid Plexus of Rats During the Acute Phase Response to Inflammation

	Liver		Choroid plexus	
	cpm/g	%	cpm/g	%
	Transthyretin			
Healthy rats	295 ± 45	100	8,926 ± 605	100
Turpentine-injected rats	80 ± 20	27	7,660 ± 277	86
Talc-injected rats	170 ± 6	58	9,203 ± 742	103
	Albumin			
Healthy rats	2,921 ± 119	100	Not detected	
Turpentine-injected rats	1,662 ± 128	57	Not detected	
Talc-injected rats	1,628 ± 146	56	Not detected	
	α_1-Acid Glycoprotein			
Healthy rats	681 ± 93	100	Not detected	
Turpentine-injected rats	21,770±851	3,196	Not detected	
Talc-injected rats	15,266±775	2,241	Not detected	
	Major Acute Phase α_1-Protein			
Healthy rats	72 ± 17	100	Not detected	
Turpentine-injected rats	2,364 ± 81	3,297	Not detected	
Talc-injected rats	1,759 ± 122	2,454	Not detected	

Note: Levels of mRNA were measured by dot hybridization of specific [^{32}P]cDNA probes to mRNA in tissue extracts from five rats per group. The extract for each dot was derived from 0.25 mg of liver tissue or 0.05 mg of choroid plexus. Backgrounds were subtracted before calculation. Means ± 1 SE are given. Acute experimental inflammation was produced by subcutaneous injection of a small amount of mineral turpentine or talc in the dorsolumbar region.

Modified from Dickson, P. W., Aldred, A. R., Marley, P. D., Bannister, D., and Schreiber, G., *J. Biol. Chem.*, 261, 3475, 1986. With permission.

first found at the reptile stage (Figures 29 and 30). The importance of thyroid hormones for brain development may be the explanation for the strong conservation of the features of transthyretin involved in the binding of thyroxine. In the bloodstream, large amounts of transthyretin, synthesized and secreted by the liver, are first observed at the marsupial stage. This change in the tissue distribution of transthyretin gene expression changed the system for the transport of thyroid hormones in the blood. However, the system of thyroxine transport by transthyretin evolved much earlier in the brain, at the stage of the stem reptiles, about 350 million years ago. Thus, transthyretin evolved and existed for more than 200 million years as a protein synthesized exclusively outside the liver. It participated in maintaining homeostasis in the cerebral extracellular comparment. When transthyretin was expressed in the liver, it became a plasma protein, with a distribution of 22% in the intravascular and

TABLE 4
Summary of Plasma Protein Gene Expression in the Reproductive Tract of the Female Rat

mRNA	Tissue	Level (maximum detected)
Ceruloplasmin	Uterus	50% of normal liver level
Transferrin	Uterus	1% of normal liver level
α_2-Macroglobulin	Decidua	165% of maximum level observed in acute phase liver
	Uterus	Not detected
α_1-Acid glycoprotein	Decidua	30% of maximum level observed in acute phase liver
	Uterus	Not detected

Note: Maximum levels of plasma protein mRNA in RNA from decidua and from uterus are shown as a percentage of the value for normal or acute phase liver RNA.

Adapted from Thomas, T. and Schreiber, G., *FEBS Lett.*, 243, 381, 1989. With permission.

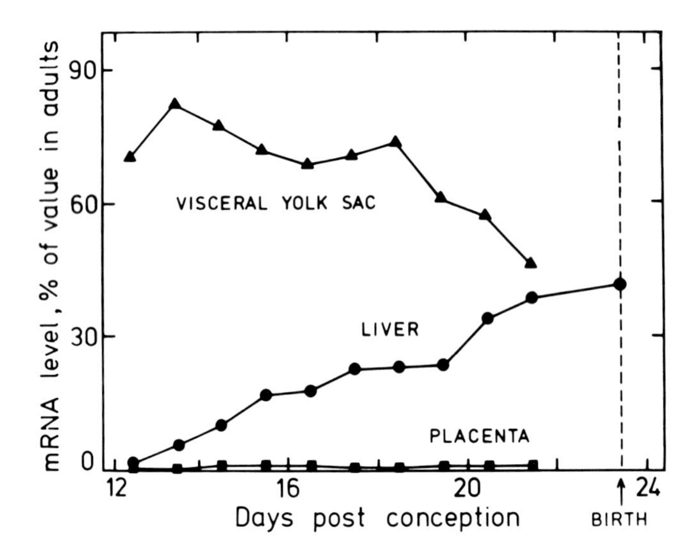

FIGURE 25. Levels of transthyretin mRNA in fetal rat tissues during development. The levels are expressed as a percentage of the amount of cDNA probe hybridized to the same quantity of total RNA purified from adult male rat liver. (From Fung, W.-P., Thomas, T., Dickson, P. W., Aldred, A. R., Milland, J., Dziadek, M., Power, B., Hudson, P., and Schreiber, G., *J. Biol. Chem.*, 263, 480, 1988.

78% in the extravascular compartment (calculated for rats from data in References 20 and 133). Expression of the transthyretin gene in the choroid plexus is not infuenced by trauma and inflammation. It is not known when, during evolution, transthyretin gene expression in the liver came under negative acute phase regulation.

ACKNOWLEDGEMENTS

We are grateful to Samantha J. Richardson for the critical reading and discussion of the manuscript and to Judy Guest for her dedicated word processing. Our work, on part of which this review is based, was supported by the Australian Research Council and the National Health and Medical Research Council of Australia.

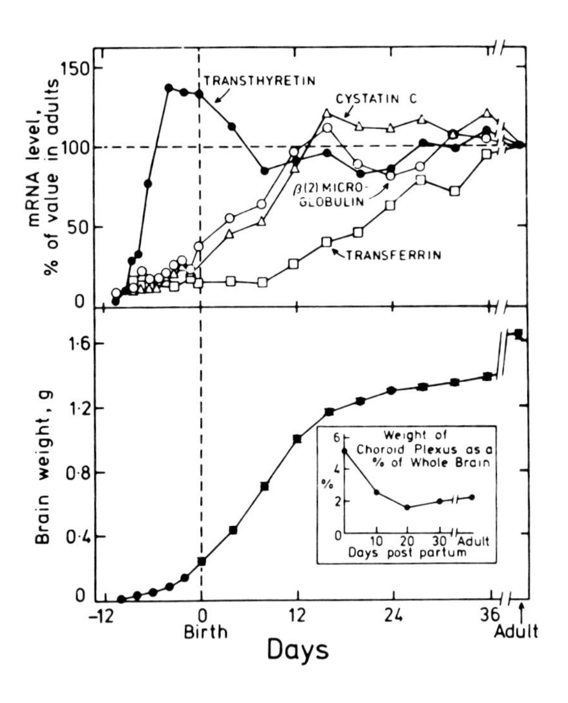

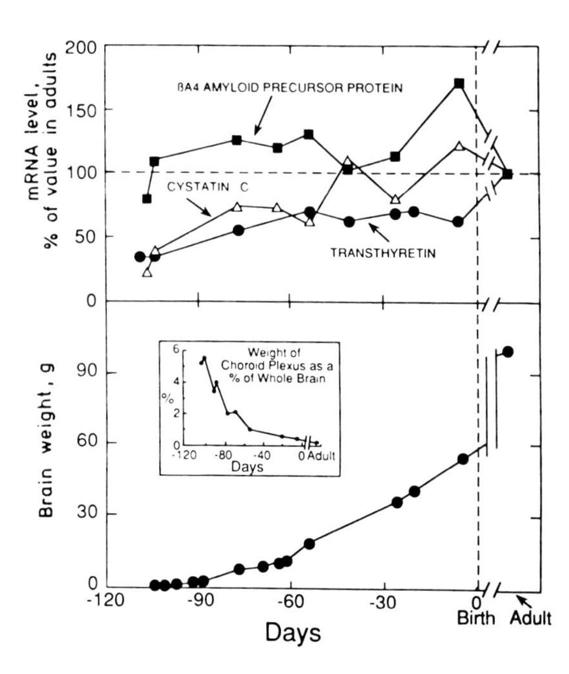

FIGURE 26. Growth of brain and choroid plexus and levels of mRNAs for transthyretin, transferrin, cystatin C, β_2-microglobulin, and βA4-amyloid precursor protein during development. After Northern analysis with ^{32}P-labeled cDNAs, radioactive bands were cut out and the radioactivity in the bands measured by liquid scintillation spectrometry. Results are expressed as a percentage of the value obtained for choroid plexus from adult animals. (From Schreiber, G. and Aldred, A. R., in *Barrier Concepts and CSF Analysis,* Proc. Int. Quincke Symp. Göttingen, Felgenhauer, K., Ed., Springer-Verlag, Heidelberg, in press. With permission.)

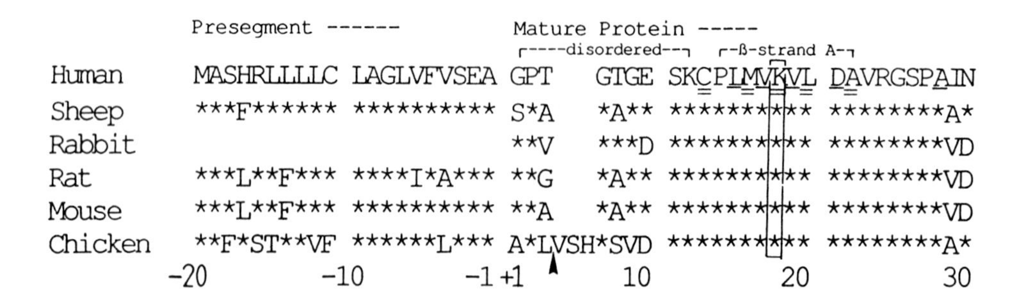

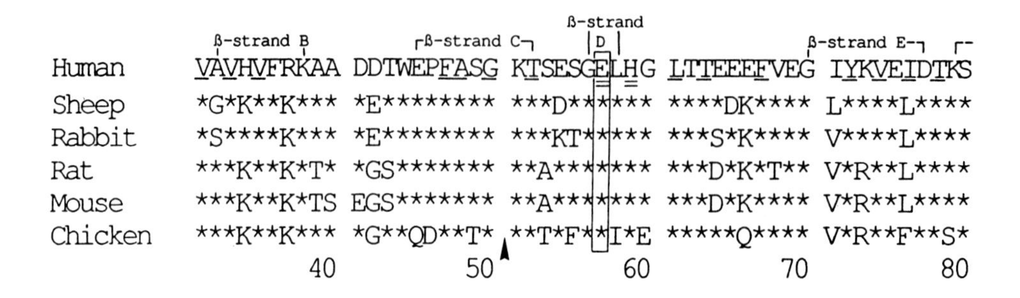

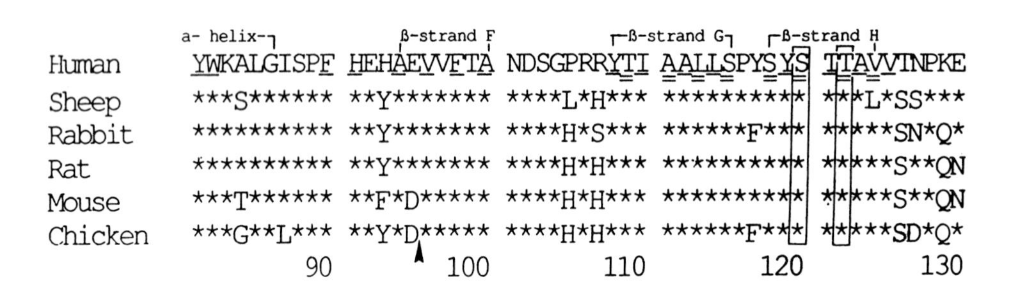

FIGURE 27. Comparison of the primary structures of six transthyretins. The amino acid sequence of rabbit transthyretin[121] and those deduced from cDNA sequences for sheep,[46] rat,[44,45] mouse,[50] and chicken[57] transthyretins are aligned with the deduced amino acid sequence of human transthyretin.[122] The N-terminal residue of mature transthyretin (designated +1) has been identified for the human,[123] rabbit,[121] rat,[124] and chicken,[57] but has been inferred for mouse and sheep from this alignment. Numbers below the sequences refer to the amino acids of chicken transthyretin. Those residues in sheep, rabbit, rat, mouse, and chicken transthyretins which are identical to those in human transthyretin are presented by asterisks. Presegments contain the residues numbered from −20 to −1. Features of secondary structure in human transthyretin[125,126] are indicated above the sequence for human transthyretin. Residues located in the core of the transthyretin subunit are highly underlined, those located in the central channel of the tetramer are doubly underlined, and those which participate in the binding of thyroxine[127] are enclosed in rectangles. The positions of the three arrows below the amino acid sequences correspond to the positions of the intron-exon boundaries in the human[49] and mouse[128] transthyretin genes. (From Duan, W., Achen, M. G., Richardson, S. J., Lawrence, M. C., Wettenhall, R. E., Jaworowski, A., and Schreiber, G., *Eur. J. Biochem.*, 200, 670, 1991. With permission.)

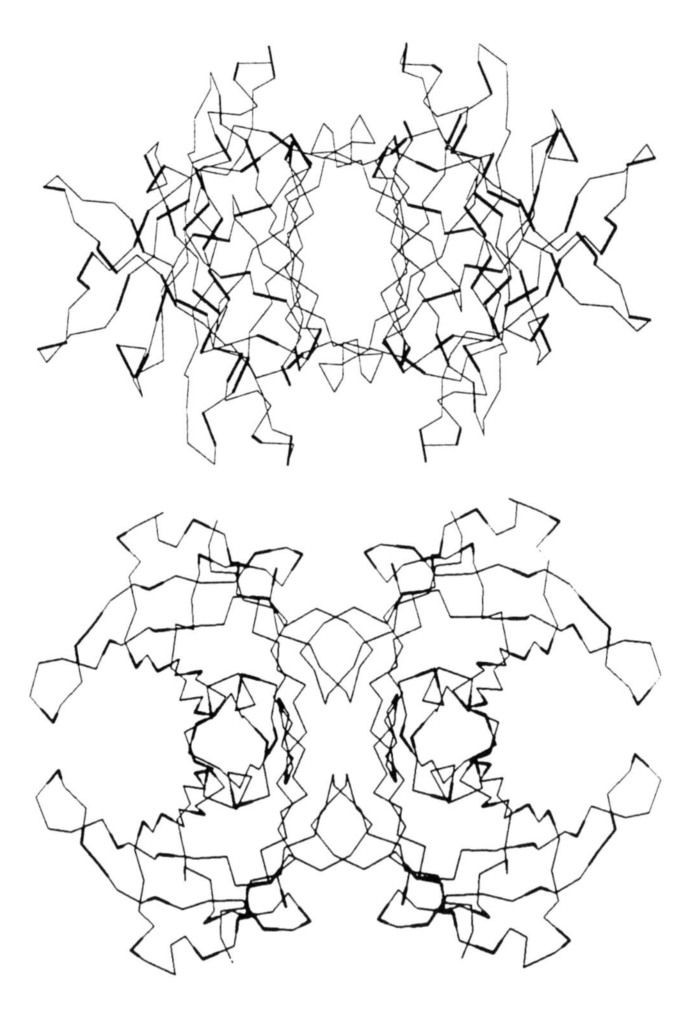

FIGURE 28. Position of the differences in amino acid sequence between human and chicken transthyretin. A trace of the α-carbon backbone of the human transthyretin tetramer is shown in which the positions of amino acid differences in the two species are shown as bold segments. To obtain the diagram, a model for the three-dimensional structure of chicken transthyretin, based on the X-ray structure of human transthyretin,[125] was built using the interactive molecular-graphics program QUANTA (Polygen Corporation). The coordinates for human transthyretin were obtained from the Brookhaven Protein Data Base[129] and, using the "mutate" facility in QUANTA, were replaced on a residue-by-residue basis to obtain the chicken transthyretin sequence. In the X-ray structure study of human transthyretin, residues 1 to 9 and 124 to 127 are not defined; hence, sequence differences between the chicken and human species in these regions were not considered. The plausibility of the model-building was investigated further by subjecting both the model and the human structure to 40 cycles of conjugate gradient energy minimization using the program XPLOR.[130] To avoid spurious conformational changes during the minimization, the coordinates of the backbone atoms of both structures were restrained in a harmonic fashion to the X-ray coordinates of the human structure. The structure is viewed down the crystallographic z-axis (upper panel) and x-axis (lower panel). The absence of changes in the central channel containing the thyroxine-binding site is apparent. (From Duan, W., Achen, M. G., Richardson, S. J., Lawrence, M. C., Wettenhall, R. E., Jaworowski, A., and Schreiber, G., *Eur. J. Biochem.*, 200, 670, 1991. With permission.)

TABLE 5
Rates of Evolutionary Change for Different Sections of the Transthyretin Molecule

Species compared	Part of transthyretin molecule	Accepted point mutations per 100 residues in 10^8 years	Ref.
Human/chick	Whole molecule	10	57, 122
	Surface residues	17	
	Central channel	0	
Human/sheep	Whole molecule	18	46, 122
	Surface residues	34	
	Central channel	8[a]	
Human/rabbit	Whole molecule	21	121, 122
	Surface residues	37	
	Central channel	0	

[a] Derived from only one substitution per subunit.

From Duan, W., Achen, M. G., Richardson, S. J., Lawrence, M. C., Wettenhall, R. E., Jaworowski, A., and Schreiber, G., *Eur. J. Biochem.*, 200, 670, 1991. With permission.

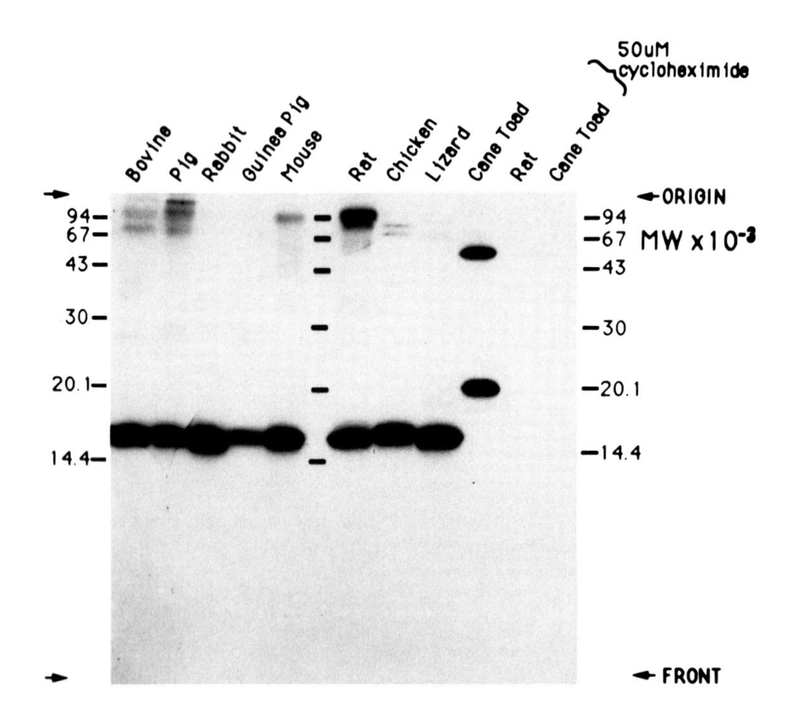

FIGURE 29. Comparison of proteins synthesized and secreted by choroid plexus of various animal species by sodium dodecyl sulfate-polyacrylamide gel electrophoresis (SDS-PAGE) and fluorography. Choroid plexus, dissected from the species indicated, were incubated with [^{14}C]leucine and analyzed by electrophoresis in SDS polyacrylamide gel, followed by fluorography. The protein standards used and subunit weights assumed (in brackets) were phosphorylase b (94,000), bovine serum albumin (67,000), ovalbumin (43,000), carbonic anhydrase (30,000), soybean trypsin inhibitor (20,100), and α-lactalbumin (14,400). Control lanes, medium from rat and cane toad choroid plexus incubated with [^{14}C]leucine in the presence of 50 μ*M* cycloheximide. (From Harms, P. J., Tu, G.-F., Richardson, S. J., Aldred, A. R., Jaworowski, A., and Schreiber, G., *Comp. Biochem. Physiol.*, 99B, 239, 1991. With permission.)

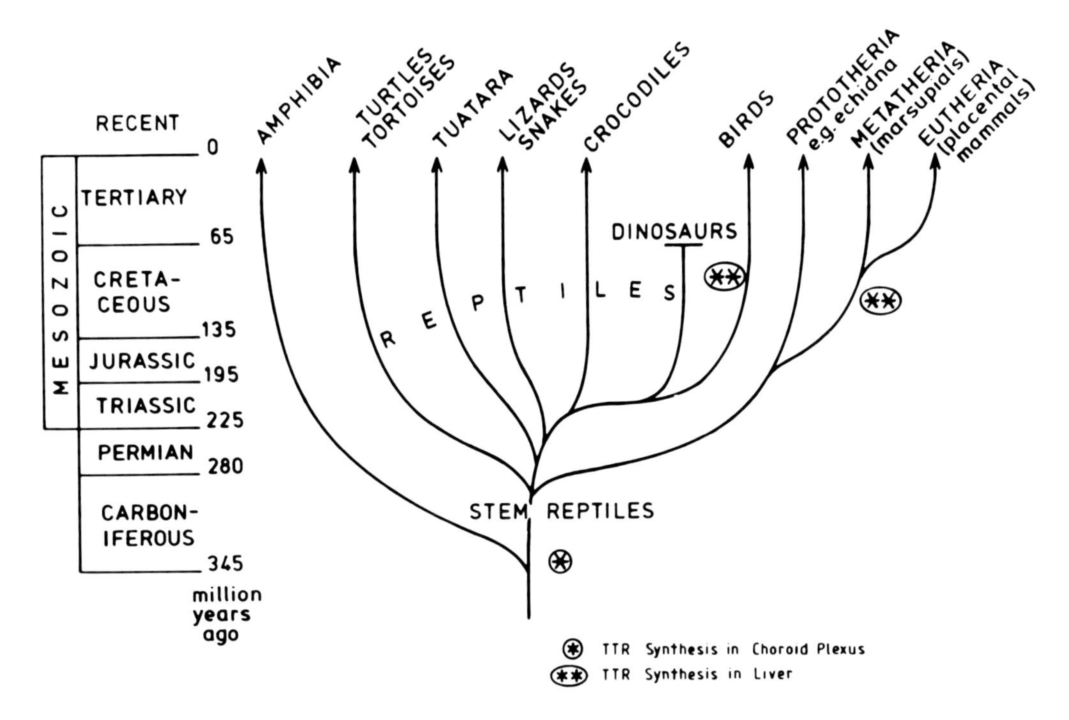

FIGURE 30. Evolutionary tree indicating the development of transthyretin gene expression in choroid plexus (*) and liver (**). Geological periods adapted from Young,[132] not drawn to scale.

REFERENCES

1. **Bernard, C.,** *Leçcons sur les Phénomènes de La Vie Communs aux Animaux et aux Vegetaux,* Baillière et Fils, Paris, 1878, 113.
2. **Cannon, W. B.,** Organization for physiological homeostasis, *Physiol. Rev.,* 9, 399, 1929.
3. **Cannon, W. B.,** *The Wisdom of the Body,* W. W. Norton, New York, 1932.
4. **Schreiber, G., Urban, J., Zähringer, J., Reutter, W., and Frosch, U.,** The secretion of serum protein and the synthesis of albumin and total protein in regenerating rat liver, *J. Biol. Chem.,* 246, 4531, 1971.
5. **Miller, L. L., Bly, C. G., Watson, M. L., and Bale, W. F.,** The dominant role of the liver in plasma protein synthesis. A direct study of the isolated perfused rat liver with the aid of lysine-ε-C^{14}, *J. Exp. Med.,* 94, 431, 1951.
6. **Goldsworthy, P. D., McCartor, H. R., McGuigan, J. E., Peppers, G. F., and Volwiler, W.,** Relative albumin, transferrin, and fibrinogen synthesis rates in perfused bovine liver, *Am. J. Physiol.,* 218, 1428, 1970.
7. **Gordon, A. H. and Humphrey, J. H.,** Methods for measuring rates of synthesis of albumin by the isolated perfused rat liver, *Biochem. J.,* 75, 240, 1960.
8. **Katz, J., Bonorris, G., Okuyama, S., and Sellers, A. L.,** Albumin synthesis in perfused liver of normal and nephrotic rats, *Am. J. Physiol.,* 212, 1255, 1967.
9. **Tarver, H. and Reinhardt, W. O.,** Methionine labeled with radioactive sulfur as an indicator of protein formation in the hepatectomized dog, *J. Biol. Chem.,* 167, 395, 1947.
10. **Kukral, J. C., Kerth, J. D., Pancner, R. J., Cromer, D. W., and Henegar, G. C.,** Plasma protein synthesis in the normal dog and after total hepatectomy, *Surg. Gynecol. Obstet.,* 113, 360, 1961.
11. **Schreiber, G., Boutwell, R. K., Potter, V. R., and Morris, H. P.,** Lack of secretion of serum protein by transplanted rat hepatomas, *Cancer Res.,* 26, 2357, 1966.
12. **Mullins, F., Weissman, S. M., and Konen, J. A.,** Extraabdominal serum protein synthesis in the rhesus monkey, *J. Surg. Res.,* 6, 315, 1966.

13. **Urban, J., Inglis, A. S., Edwards, K., and Schreiber, G.,** Chemical evidence for the difference between albumins from microsomes and serum and a possible precursor-product relationship, *Biochem. Biophys. Res. Commun.*, 61, 494, 1974.
14. **Russell, J. H. and Geller, D. M.,** The structure of rat proalbumin, *J. Biol. Chem.*, 250, 3409, 1975.
15. **Millership, A., Edwards, K., Chelladurai, M., Dryburgh, H., Inglis, A. S., Urban, J., and Schreiber, G.,** N-terminal amino acid sequence of proalbumin from inbred Buffalo rats, *Int. J. Pept. Protein Res.*, 15, 248, 1980.
16. **Urban, J. and Schreiber, G.,** Biological evidence for a precursor protein of serum albumin, *Biochem. Biophys. Res. Commun.*, 64, 778, 1975.
17. **Edwards, K., Schreiber, G., Dryburgh, H., Urban, J., and Inglis, A. S.,** Synthesis of albumin via a precursor protein in cell suspensions from rat liver, *Eur. J. Biochem.*, 63, 303, 1976.
18. **Urban, J., Chelladurai, M., Millership, A., and Schreiber, G.,** The kinetics in vivo of the synthesis of albumin-like protein and albumin in rats, *Eur. J. Biochem.*, 67, 477, 1976.
19. **Edwards, K., Fleischer, B., Dryburgh, H., Fleischer, S., and Schreiber, G.,** The distribution of albumin precursor protein and albumin in liver, *Biochem. Biophys. Res. Commun.*, 72, 310, 1976.
20. **Schreiber, G., Howlett, G., Nagashima, M., Millership, A., Martin, H., Urban, J., and Kotler, L.,** The acute phase response of plasma protein synthesis during experimental inflammation, *J. Biol. Chem.*, 257, 10271, 1982.
21. **Schreiber, G.,** Synthesis, processing, and secretion of plasma proteins by the liver and other organs and their regulation, in *The Plasma Proteins*, Putnam, F. W., Ed., Academic Press, New York, 1987, 293.
22. **Brightman, M. W.,** Ultrastructural characteristics of adult choroid plexus: relation to the blood-cerebrospinal fluid barrier to proteins, in *The Choroid Plexus in Health and Disease*, Netsky, M. G. and Shuangshoti, S., Eds., John Wright and Sons, Bristol, 1975, 86.
23. **Hurley, J. V., Anderson, R. McD., and Sexton, P. T.,** The fate of plasma protein which escapes from blood vessels of the choroid plexus of the rat — an electron microscope study, *J. Pathol.*, 134, 57, 1981.
24. **Jungblut, P. W.,** Biosynthese von Ratten-Serumalbumin. II. Untersuchung der Synthese und Sekretion von Serumalbumin mit der isoliert durchströmten Rattenleber, *Biochem. Z.*, 337, 285, 1963.
25. **Peters, T., Jr.,** The biosynthesis of albumin. II. Intracellular phenomena in the secretion of newly formed albumin, *J. Biol. Chem.*, 237, 1186, 1962.
26. **Morgan, E. H. and Peters, T., Jr.,** The biosynthesis of rat serum albumin. V. Effect of protein depletion and refeeding on albumin and transferrin synthesis, *J. Biol. Chem.*, 246, 3500, 1971.
27. **Peters, T., Jr. and Peters, J. C.,** The biosynthesis of rat serum albumin. VI. Intracellular transport of albumin and rates of albumin and liver protein synthesis *in vivo* under various physiological conditions, *J. Biol. Chem.*, 247, 3858, 1972.
28. **Rotermund, H.-M., Schreiber, G., Maeno, H., Weinssen, U., and Weigand, K.,** The ratio of albumin synthesis to total protein synthesis in normal rat liver, in host liver, and in Morris hepatoma 9121, *Cancer Res.*, 30, 2139, 1970.
29. **Leibowitz, M. J. and Soffer, R. L.,** Enzymatic modification of proteins. III. Purification and properties of a leucyl, phenylalanyl transfer ribonucleic acid-protein transferase from *Escherichia coli*, *J. Biol. Chem.*, 245, 2066, 1970.
30. **Schreiber, G. and Urban, J.,** The synthesis and secretion of albumin, in *Reviews of Physiology, Biochemistry and Pharmacology*, Springer-Verlag, New York, 1978, 27.
31. **Schreiber, G., Rotermund, H.-M., Maeno, H., Weigand, K., and Lesch, R.,** The proportion of the incorporation of leucine into albumin to that into total protein in rat liver and hepatoma Morris 5123TC, *Eur. J. Biochem.*, 10, 355, 1969.
32. **Judah, J. D., Gamble, M., and Steadman, J. H.,** Biosynthesis of serum albumin in rat liver. Evidence for the existence of 'proalbumin', *Biochem. J.*, 134, 1083, 1973.
33. **Miller, L. L., Bly, C. G., and Bale, W. F.,** Plasma and tissue proteins produced by non-hepatic rat organs as studied with lysine-ϵ-C^{14}, *J. Exp. Med.*, 99, 133, 1954.
34. **Abdel-Samie, Y., Broda, E., Kellner, G., and Zischka, W.,** Production of serum albumin and of globulins by chick mesenchymal tissue in culture, *Nature (London)*, 184, 361, 1959.
35. **Abdel-Samie, Y. M., Broda, E., and Kellner, G.,** The autonomous production of individual serum proteins by tissue in culture, *Biochem. J.*, 75, 209, 1960.
36. **Goussault, Y., Sharif, A., and Bourrilon, R.,** Serum albumin biosynthesis and secretion by resting and lectin stimulated human lymphocytes, *Biochem. Biophys. Res. Commun.*, 73, 1030, 1976.
37. **Bishop, J., Leahy, J., and Schweet, R.,** Formation of the peptide chain of hemoglobin, *Proc. Natl. Acad. Sci. U.S.A.*, 46, 1030, 1960.
38. **Dintzis, H. M.,** Assembly of the peptide chains of hemoglobin, *Proc. Natl. Acad. Sci. U.S.A.*, 47, 247, 1961.
39. **Naughton, M. A. and Dintzis, H. M.,** Sequential biosynthesis of the peptide chains of hemoglobin, *Proc. Natl. Acad. Sci. U.S.A.*, 48, 1822, 1962.

40. **Sargent, J. R. and Campbell, P. N.,** The sequential synthesis of the polypeptide chain of serum albumin by the microsome fraction of rat liver, *Biochem. J.*, 96, 134, 1965.
41. **Otten, J., Jonckheer, M., and Dumont, J. E.,** Thyroid albumin. II. *In vitro* synthesis of a thyroid albumin by normal human thyroid tissue, *J. Clin. Endocrinol. Metab.*, 32, 18, 1971.
42. **Jonckheer, M. H. and Karcher, D. M.,** Thyroid albumin. I. Isolation and characterization, *J. Clin. Endocrinol. Metab.*, 32, 7, 1971.
43. **Schreiber, G., Tsykin, A., Aldred, A. R., Thomas, T., Fung, W.-P., Dickson, P. W., Cole, T., Birch, H., de Jong, F. A., and Milland, J.,** The acute phase response in the rodent, *Ann. N.Y. Acad. Sci.*, 557, 61, 1989.
44. **Dickson, P. W., Howlett, G. J., and Schreiber, G.,** Rat transthyretin (prealbumin): molecular cloning, nucleotide sequence, and gene expression in liver and brain, *J. Biol. Chem.*, 260, 8214, 1985.
45. **Duan, W., Cole, T., and Schreiber, G.,** Cloning and nucleotide sequencing of transthyretin (prealbumin) cDNA from rat choroid plexus and liver, *Nucleic Acids Res.*, 17, 3979, 1989.
46. **Tu, G.-F., Cole, T., Duan, W., and Schreiber, G.,** The nucleotide sequence of transthyretin cDNA isolated from a sheep choroid plexus cDNA library, *Nucleic Acids Res.*, 17, 6384, 1989.
47. **Fung, W.-P., Thomas, T., Dickson, P. W., Aldred, A. R., Milland, J., Dziadek, M., Power, B., Hudson, P., and Schreiber, G.,** Structure and expression of the rat transthyretin (prealbumin) gene, *J. Biol. Chem.*, 263, 480, 1988.
48. **Sasaki, H., Yoshioka, N., Takagi, Y., and Sakaki, Y.,** Structure of the chromosomal gene for human serum prealbumin, *Gene*, 37, 191, 1985.
49. **Tsuzuki, T., Mita, S., Maeda, S., Araki, S., and Shimada, K.,** Structure of the human prealbumin gene, *J. Biol. Chem.*, 260, 12224, 1985.
50. **Wakasugi, S., Maeda, S., Shimada, K., Nakashima, H., and Migita, S.,** Structural comparisons between mouse and human prealbumin, *J. Biochem. Tokyo*, 98, 1707, 1985.
51. **Costa, R. H., Lai, E., and Darnell, J. E., Jr.,** Transcriptional control of the mouse prealbumin (transthyretin) gene: both promoter sequences and a distinct enhancer are cell specific, *Mol. Cell. Biol.*, 6, 4697, 1986.
52. **Cserr, J. F.,** Physiology of the choroid plexus, *Physiol. Rev.*, 51, 273, 1971.
53. **Dickson, P. W., Aldred, A. R., Marley, P. D., Guo-Fen, T., Howlett, G. J., and Schreiber, G.,** High prealbumin and transferrin mRNA levels in the choroid plexus of rat brain, *Biochem. Biophys. Res. Commun.*, 127, 890, 1985.
54. **Lee, J. J. and Costlow, N. A.,** A molecular titration assay to measure transcript prevalence levels, *Methods Enzymol.*, 152, 633, 1987.
55. **Krieg, P. A. and Melton, D. A.,** *In vitro* RNA synthesis with SP6 RNA polymerase, *Methods Enzymol.*, 155, 397, 1987.
56. **Schreiber, G., Aldred, A. R., Jaworowski, A., Nilsson, C., Achen, M. G., and Segal, M. B.,** Thyroxine transport from blood to brain via transthyretin synthesis in choroid plexus, *Am. J. Physiol.*, 258, R338, 1990.
57. **Duan, W., Achen, M. G., Richardson, S. J., Lawrence, M. C., Wettenhall, R. E., Jaworowski, A., and Schreiber, G.,** Isolation, characterization, cDNA cloning and gene expression of chicken transthyretin: implications for the evolution of structure and functions of transthyretin in vertebrates, *Eur. J. Biochem.*, 200, 670, 1991.
58. **Stauder, A. J., Dickson, P. W., Aldred, A. R., Schreiber, G., Mendelsohn, F. A. O., and Hudson, P.,** Synthesis of transthyretin (prealbumin) mRNA in choroid plexus epithelial cells, localized by *in situ* hybridization in the rat brain, *J. Histochem. Cytochem.*, 34, 949, 1986.
59. **Weigand, K., Müller, M., Urban, J., and Schreiber, G.,** Intact endoplasmic reticulum and albumin synthesis in rat liver cell suspensions, *Exp. Cell Res.*, 67, 27, 1971.
60. **Schreiber, G. and Schreiber, M.,** Protein synthesis in single cell suspensions from rat liver. I. General properties of the system and permeability of the cells for leucine and methionine, *J. Biol. Chem.*, 247, 6340, 1972.
61. **Müller, M., Schreiber, M., Kartenbeck, J., and Schreiber, G.,** Preparation of single cell suspensions from normal liver, regenerating liver, and Morris hepatomas 9121 and 5123 TC, *Cancer Res.*, 32, 2568, 1972.
62. **Schreiber, M., Schreiber, G., and Kartenbeck, J.,** Protein and ribonucleic acid metabolism in single-cell suspensions from Morris hepatoma 5123 TC and from normal rat liver, *Cancer Res.*, 34, 2143, 1974.
63. **Edwards, K., Schreiber, G., Dryburgh, H., Millership, A., and Urban, J.,** Biosynthesis of albumin via a precursor protein in Morris hepatoma 5123 TC, *Cancer Res.*, 36, 3113, 1976.
64. **Edwards, K., Nagashima, M., Dryburgh, H., Wykes, A., and Schreiber, G.,** Secretion of proteins from liver cells is suppressed by the proteinase inhibitor N-α-tosyl-L-lysyl chloromethane, but not tunicamycin, an inhibitor of glycosylation, *FEBS Lett.*, 100, 269, 1979.

65. **Howeltt, G., Nagashima, M., and Schreiber, G.,** The acute phase pattern of plasma protein synthesis and secretion persists in isolated hepatocytes, *Biochem. Int.*, 3, 93, 1981.
66. **Schreiber, G. and Schreiber, M.,** The preparation of single cell suspensions from liver and their use for the study of protein synthesis, *Sub-Cell. Biochem.*, 2, 1973, 307 (review).
67. **Schreiber, G. and Schreiber, M.,** Protein synthesis and excretion in single cell suspensions from liver and Morris hepatoma 5123 TC, in *Gene Expression and Carcinogenesis in Cultured Liver,* Gerschenson, L. E. and Thompson, E. B., Eds., Academic Press, New York, 1975, 46.
68. **Dickson, P. W., Aldred, A. R., Marley, P. D., Bannister, D., and Schreiber, G.,** Rat choroid plexus specializes in the synthesis and the secretion of transthyretin (prealbumin) — regulation of transthyretin synthesis in choroid plexus is independent from that in liver, *J. Biol. Chem.*, 261, 3475, 1986.
69. **Aldred, A. R., Dickson, P. W., Marley, P. D., and Schreiber, G.,** Distribution of transferrin synthesis in brain and other tissues in the rat, *J. Biol. Chem.*, 262, 5293, 1987.
70. **Thomas, T., Schreiber, G., and Jaworowski, A.,** Developmental patterns of gene expression of secreted proteins in brain and choroid plexus, *Dev. Biol.*, 134, 38, 1989.
71. **Harms, P. J., Tu, G.-F., Richardson, S. J., Aldred, A. R., Jaworowski, A., and Schreiber, G.,** Transthyretin (prealbumin) gene expression in choroid plexus is strongly conserved during evolution of vertebrates, *Comp. Biochem. Physiol.*, 99B, 239, 1991.
72. **Pollay, M., Stevens, A., Estrada, E., and Kaplan, R.,** Extracorporeal perfusion of choroid plexus, *J. Appl. Physiol.*, 32, 612, 1972.
73. **Schreiber, G., Dryburgh, H., Millership, A., Matsuda, Y., Inglis, A., Phillips, J., Edwards, K., and Maggs, J.,** The synthesis and secretion of rat transferrin, *J. Biol. Chem.*, 254, 12013, 1979.
74. **Schreiber, G., Dryburgh, H., Weigand, K., Schreiber, M., Witt, I., Seydewitz, H., and Howlett, G.,** Intracellular precursor forms of transferrin, α_1-acid glycoprotein and α_1-antitrypsin in human liver, *Arch. Biochem. Biophys.*, 212, 319, 1981.
75. **Putnam, F. W.,** Transferrin, in *The Plasma Proteins: Structure, Function, and Genetic Control,* 2nd ed., Putnam, F. W., Ed., Academic Press, New York, 1975, 265.
76. **Jefferies, W. A., Brandon, M. R., Hunt, S. V., Williams, A. F., Gatter, K. C., and Mason, D. Y.,** Transferrin receptor on endothelium of brain capillaries, *Nature (London),* 312, 162, 1984.
77. **Huebers, H. A. and Finch, C. A.,** The physiology of transferrin and transferrin receptors, *Physiol. Rev.*, 67, 520, 1987.
78. **Thorstensen, K. and Romslow, I.** The role of transferrin in the mechanism of cellular iron uptake, *Biochem. J.*, 271, 1, 1990.
79. **Levin, M. J., Tuil, D., Uzan, G., Dreyfus, J.-C., and Kahn, A.,** Expression of the transferrin gene during development of non-hepatic tissues: high level of transferrin mRNA in fetal muscle and adult brain, *Biochem. Biophys. Res. Commun.*, 122, 212, 1984.
80. **Bloch, B., Popovici, T., Levin, M. J., Tuil, D., and Kahn, A.,** Transferrin gene expression visualized in oligodendrocytes of the rat brain by using *in situ* hybridization and immunohistochemistry, *Proc. Natl. Acad. Sci. U.S.A.*, 82, 6706, 1985.
81. **Bloch, B., Popovici, T., Chouham, S., Levin, M. J., Tuil, D., and Kahn, A.,** Transferrin gene expression in choroid plexus of the adult rat brain, *Brain Res. Bull.*, 18, 573, 1987.
82. **Thomas, T., Southwell, B. R., Schreiber, G., and Jaworowski, A.,** Plasma protein synthesis and secretion in the rat visceral yolk sac of fetal rat: gene expression, protein synthesis and secretion, *Placenta,* 11, 413, 1990.
83. **Tu, G.-F., Achen, M. G., Aldred, A. R., Southwell, B. R., and Schreiber, G.,** The distribution of cerebral expression of the transferrin gene is species-specific, *J. Biol. Chem.*, 266, 6201, 1991.
84. **Skinner, M. K. and Griswold, M. D.,** Sertoli cells synthesize and secrete transferrin-like protein, *J. Biol. Chem.*, 255, 9523, 1980.
85. **Wright, W. W., Musto, N. A., Mather, J. P., and Bardin, C. W.,** Sertoli cells secrete both testis-specific and serum proteins, *Proc. Natl. Acad. Sci.U.S.A.*, 78, 7565, 1981.
86. **Skinner, M. K., Cosand, W. L., and Griswold, M. D.,** Purification and characterization of testicular transferrin secreted by rat Sertoli cells, *Biochem. J.*, 218, 313, 1984.
87. **Tsutsumi, M., Skinner, M. K., and Sanders-Bush, E.,** Transferrin gene expression and synthesis by cultured choroid plexus epithelial cells — regulation by serotonin and cyclic adenosine 3′,5′-monophosphate, *J. Biol. Chem.*, 264, 9629, 1989.
88. **Frieden, E.,** Ceruloplasmin: the serum copper transport protein with oxidase activity, in *Copper in the Environment,* Part II, Nriagu, J.O., Ed., John Wiley & Sons, Chichester, 1979, 241.
89. **Gutteridge, J. M. C. and Stocks, J.,** Caeruloplasmin: physiological and pathological perspectives, *CRC Crit. Rev. Clin. Lab. Sci.*, 14, 257, 1981.
90. **Cousins, R. J.,** Absorption, transport, and hepatic metabolism of copper and zinc: special reference to metallothionein and ceruloplasmin, *Physiol. Rev.*, 65, 238, 1985.
91. **Meyer, B. J., Meyer, A. C., and Horwitt, M. K.,** Factors influencing serum copper and ceruloplasmin oxidative activity in the rat, *Am. J. Physiol.*, 194, 581, 1958.

92. **Milland, J., Tsykin, A., Thomas, T., Aldred, A. R., Cole, T., and Schreiber, G.,** Gene expression in regenerating and acute phase rat liver, *Am. J. Physiol.,* 259, G340, 1990.
93. **Aldred, A. R., Grimes, A., Schreiber, G., and Mercer, J. F. B.,** Rat ceruloplasmin: molecular cloning and gene expression in liver, choroid plexus, yolk sac, placenta and testis, *J. Biol. Chem.,* 262, 2875, 1987.
94. **Thomas, T. and Schreiber, G.,** The expression of genes coding for positive acute phase proteins in the reproductive tract of the female rat: high levels of ceruloplasmin mRNA in the uterus, *FEBS Lett.,* 243, 381, 1989.
95. **Duan, W. and Schreiber, G.,** Expression of retinol-binding protein mRNA in mammalian choroid plexus, *Comp. Biochem. Physiol. B,* 101 B, 399, 1992.
96. **Shambaugh, G. E., III,** Thyroid hormone action-biologic and cellular effects, in *Werner's The Thyroid,* 5th ed., Ingbar, S.H. and Braverman, L. E., Eds., Lippincott, Philadelphia, 1986, 201.
97. **Morreale de Escobar, G.,** Thyroid hormones and the developing brain, in *Frontiers in Thyroidology,* Vol. 1, Medeiros-Neto, G. and Gaitan, E., Eds., Plenum Press, New York, 1986, 5.
98. **Dickson, P. W., Aldred, A. R., Menting, J. G. T., Marley, P. D., Sawyer, W. H., and Schreiber, G.,** Thyroxine transport in choroid plexus, *J. Biol. Chem.,* 262, 13907, 1987.
99. **Young, Y. A., Rajatanavin, R., Moring, A. F., and Braverman, L. E.,** Fasting induces the generation of serum thyroxine-binding globulin in Zucker rats, *Endocrinology,* 116, 1248, 1985.
100. **Larsen, P. R.,** Thyroid hormone concentrations, in *Werner's The Thyroid,* 5th ed., Ingbar, S. H. and Braverman, L. E., Eds., Lippincott, Philadelphia, 1986, 479.
101. **Robbins, J. and Edelhoch, H.,** Thyroid hormone transport proteins: their nature, biosynthesis, and metabolism, in *Werner's the Thyroid,* 5th ed., Ingbar, S. H. and Braverman, L. E., Eds., Lippincott, Philadelphia, 1986, chap. 6.
102. **Dratman, M. B. and Crutchfield, F. L.,** Synaptosomal [^{125}I]triiodothyronine after intravenous [^{125}I]thyroxine, *Am. J. Physiol.,* 235, E638, 1978.
103. **Alberts, B., Bray, D., Lewis, J., Raff, M., Roberts, K., and Watson, J. D.,** *Molecular Biology of the Cell,* 2nd ed., Garland, New York, 1989, 688.
104. **Rotzsch, W., Köhler, H., and Martin, H.,** Zur Löslichkeit von Thyroxin, *Hoppe-Seyler's Z. Physiol. Chem.,* 348, 939, 1967.
105. **Mendel, C. M., Weisiger, R. A., Jones, A. L., and Cavalieri, R. R.,** Thyroid hormone-binding proteins in plasma facilitate uniform distribution of thyroxine within tissues: a perfused rat liver study, *Endocrinology,* 120, 1742, 1987.
106. **Aldred, A. R., Pettersson, T. M., Harms, P. J., Richardson, S. J., Duan, W., Tu, G.-F., Achen, M. G., Nichol, S., and Schreiber, G.,** The position of the echidna in the evolution of thyroid hormone-binding plasma proteins, *Aust. Zool. Rev.,* in *''Platypus and Echidnas''* M. L. Augee, Ed., The Royal Zoological Society of New South Wales, Sydney, Australia, 1992, 44.
107. **Schreiber, G., Aldred, A. R., Dickson, P. W., Thomas, T., and Cole, T.,** Extracellular proteins in the brain, *Aust. J. Ageing,* 43, 1989.
108. **Abrahamson, M., Grubb, A., Olafson, I., and Lundwall, Å.,** Molecular cloning and sequence analysis of cDNA coding for the precursor of the human cysteine proteinase inhibitor cystatin C, *FEBS Lett.,* 216, 229, 1987.
109. **Cole, T., Dickson, P. W., Esnard, F., Averill, S., Risbridger, G., Gauthier, F., and Schreiber, G.,** The cDNA structure and expression analysis of the genes for the cysteine proteinase inhibitor cystatin C and for β(2)-microglobulin in rat brain, *Eur. J. Biochem.,* 186, 35, 1989.
110. **Tu, G.-F., Cole, T., Southwell, B. R., and Schreiber, G.,** Expression of the genes for transthyretin, cystatin C and βA4 amyloid precursor protein in sheep choroid plexus during development, *Dev. Brain Res.,* 55, 203, 1990.
111. **Wolff, J. R., Leutgeb, U., Holzgraefe, M., and Teuchert, G.,** Synaptive remodelling during primary and reactive synaptogenesis, *Prog. Zool.,* 37, 68, 1989.
112. **Schreiber, G. and Howlett, G.,** Synthesis and secretion of acute phase proteins, in *Plasma Protein Secretion by the Liver,* Glaumann, H., Peters, T., Jr., and Redman, C., Eds., Academic Press, New York, 1983, 423.
113. **Welsh, A. O. and Enders, A. C .,** Light and electron microscopic examination of the mature dedicual cells of the rat with emphasis on the antimesometrial decidua and its degeneration, *Am. J. Anat.,* 172, 1, 1985.
114. **Nagashima, M., Urban, J., and Schreiber, G.,** Intrahepatic precursor form of rat α_1-acid glycoprotein, *J. Biol. Chem.,* 255, 4951, 1980.
115. **Weigand, K., Dryburgh, H., and Schreiber, G.,** Der Nachweis einer intrazellulären Vorstufe von α_1-Antitrypsin (AT) in menschlicher Leber, *Verh. Dtsch. Ges. Inn. Med.,* 87, 888, 1981.
116. **Thomas, T., Fletcher, S., Yeoh, G. C. T., and Schreiber, G.,** The expression of α_1-acid glycoprotein mRNA during rat development: high levels of expression in the decidua, *J. Biol. Chem.,* 264, 5784, 1989.

117. **Tu, G.-F., Southwell, B. R., and Schreiber, G.,** Species specificity and developmental patterns of the β amyloid precursor protein (APP) gene in brain, liver and choroid plexus in birds, *Comp. Biochem. Physiol. B,* 101 B, 391, 1992.
118. **Schreiber, G., Aldred, A. R., Thomas, T., Birch, H. E., Dickson, P. W., Tu, G.-F., Heinrich, P. C., Northemann, W., Howlett, G. J., de Jong, F. A., and Mitchell, A.,** Levels of messenger ribonucleic acids for plasma proteins in rat liver during acute experimental inflammation, *Inflammation,* 10, 59, 1986.
119. **Birch, H. E. and Schreiber, G.,** Transcriptional regulation of plasma protein synthesis during inflammation, *J. Biol. Chem.,* 261, 8077, 1986.
120. **Schreiber, G. and Aldred, A. R.,** Origin and function of proteins in the cerebrospinal fluid, in *Barrier Concepts and CSF Analysis,* Proc. Int. Quincke Symp. Göttingen, Felgenhauer, K., Ed., Springer-Verlag, Heidelberg, in press.
121. **Sundelin, J., Melhus, H., Das, S., Eriksson, U., Lind, P., Trägärdh, L., Peterson, P. A., and Rask, L.,** The primary structure of rabbit and rat prealbumin and a comparison with the tertiary structure of human prealbumin, *J. Biol. Chem.,* 260, 6481, 1985.
122. **Mita, S., Maeda, S., Shimada, K., and Araki, S.,** Cloning and sequence analysis of cDNA for human prealbumin, *Biochem. Biophys. Res. Commun.,* 124, 558, 1984.
123. **Kanda, Y., Goodman, D. S., Canfield, R. E., and Morgan, F. J.,** The amino acid sequence of human plasma prealbumin, *J. Biol. Chem.,* 249, 6796, 1974.
124. **Navab, M., Mallia, A. K., Kanda, Y., and Goodman, D. S.,** Rat plasma prealbumin — isolation and partial characterization, *J. Biol. Chem.,* 252, 5100, 1977.
125. **Blake, C. C. F., Geisow, M. J., Oatley, S. J., Rérat, B., and Rérat, C.,** Structure of prealbumin: secondary, tertiary and quaternary interactions determined by Fourier refinement at 1.8 Å, *J.Mol. Biol.,* 121, 339, 1978.
126. **Blake, C. C. F., Geisow, J. M., Swan,I. D. A., Rérat, C., and Rérat, B.,** Structure of human plasma prealbumin in 2.5 A resolution — a preliminary report on the polypeptide chain conformation, quaternary structure and thyroxine binding, *J. Mol. Biol.,* 88, 1, 1974.
127. **Blake, C. C. F. and Oatley, S. J.,** Protein — DNA and protein-hormone interactions in prealbumin: a model of the thyroid hormone nuclear receptor?, *Nature (London),* 268, 115, 1977.
128. **Wakasugi, S., Maeda, S., and Shimada, K.,** Structure and expression of the mouse prealbumin gene, *J. Biochem.,* 100, 49, 1986.
129. **Bernstein, F. C., Koetzle, T. F., Williams, G. J. B., Meyer, E. F., Jr., Brice, M. D., Rodgers, J. R., Kennard, O., Shimanouchi, T., and Tasumi, M.,** The protein data bank: a computer-based archival file for macromolecular structures, *J. Mol. Biol.,* 112, 535, 1977.
130. **Brünger, A. T.,** Crystallographic refinements by simulated annealing — application to a 2.8 Å resolution structure of aspartate aminotransferase, *J. Mol. Biol.,* 203, 803, 1988.
131. **Larsson, J., Pettersson, T., and Carlström, A.,** Thyroid hormone binding in serum of 15 vertebrate species: isolation of thyroxine-binding globulin and prealbumin analogs, *Gen. Comp. Endocrinol.,* 58, 360, 1985.
132. **Young, J. Z.,** *The Life of Vertebrates,* 3rd ed., Clarendon Press, Oxford, 1981, 294.
133. **Dickson, P. W., Howlett, G. J., and Schreiber, G.,** Metabolism of prealbumin in rats and changes induced by acute inflammation, *Eur. J. Biochem.,* 129, 289, 1982.

B. The Major Acute Phase Proteins: Their Structure and Possible Function in Acute Phase Response

Chapter 4

STRUCTURE AND FUNCTION OF HUMAN C-REACTIVE PROTEIN

Alok Agrawal, John M. Kilpatrick, and John E. Volanakis

TABLE OF CONTENTS

0-8493-6913-4/93/$0.00 + $.50

I. INTRODUCTION

C-reactive protein (CRP) is the most characteristic human acute phase protein, since its plasma concentration rises by several hundredfold within 24 to 48 h from tissue injury.[1] These high levels persist for the duration of the acute phase response, returning to the normal low concentrations with restoration of tissue structure and function. Based on its structure and calcium-dependent binding specificities, CRP is classified as a pentraxin.[2] Members of the pentraxin family of proteins exhibit a remarkable conservation of structure and binding reactivities. All pentraxins consist of single polypeptide chain subunits, arranged in pentagonal or, rarely, hexagonal cyclic symmetry and discernible by electron microscopy. All pentraxins bind Ca^{2+} ions, which are necessary for the expression of ligand-binding activity. In vertebrates, there are two main branches of pentraxins. CRP-like members bind phosphocholine (PCh) and SAP (serum amyloid P)-like members bind carbohydrate moieties. Not all pentraxins are acute phase proteins. For example, in humans, CRP is an acute phase protein and SAP is not, while in mice the reverse is true.[3]

Comparison of the primary structure of mammalian pentraxins from several species indicates an identity range from 50 to 75% (Figure 1). The stable conservation of structure and binding specificities suggests an important biologic function for pentraxins. In this review, the structure and function of human CRP are described.

II. STRUCTURE OF HUMAN CRP

CRP is made up of five identical, noncovalently bound subunits[4] exhibiting a planar pentagonal appearance in electron micrographs.[2] A molecular weight of 118,000 has been calculated for the native pentameric molecule from sedimentation equilibrium studies.[5] The primary structure of CRP has been determined by protein sequencing,[6] and has been subsequently confirmed and corrected by sequencing cDNA and genomic clones.[7,8] Each subunit of CRP consists of 206 amino acid residues with a calculated molecular weight of 23,017.[7,8] A single disulfide bond links the two half-cystines at positions 36 and 97.[6] There are no potential N-glycosylation sites in the amino acid sequence and no carbohydrate is present in human CRP.[4] Several regions of the amino acid sequence of CRP have been found to be homologous to regions of known function in other proteins, particularly PCh-binding myeloma proteins, calmodulin, tuftsin, and histones (Table 1).

Circular dichroism studies have shown that CRP contains 34% α-helix and 45% β-structure.[9] Two different crystal forms of human CRP have been grown from solutions of 2-methyl-2,4-pentanediol. Both forms are tetragonal, and their unit cell parameters have been reported.[17] Rotation function studies of the two crystal forms of human CRP have confirmed the pentameric structure of the molecule.[18]

III. FUNCTIONS OF HUMAN CRP

A. LIGAND-BINDING PROPERTIES

1. Binding to Calcium and Phosphocholine

The first described reactivity of CRP, which led to its discovery and naming, was for the C-polysaccharide of the cell wall of pneumococci (PnC).[19] The precipitation reaction of CRP and PnC requires calcium ions.[20] Equilibrium gel filtration studies indicated that CRP binds Ca^{2+} with a binding constant of 1.3×10^{-4} *M*.[21] More recently, equilibrium dialysis experiments indicated a K_d value of 6×10^{-5} *M* for the binding of Ca^{2+} to CRP and a valence of two per CRP subunit.[22] Binding of Ca^{2+} induces an allosteric conformational change in the CRP molecule, as demonstrated by changes in the aromatic region of the CD spectrum of the molecule[9] and by differential reactivity with monoclonal antibodies.[23] A highly conserved region in the amino acid sequence of all pentraxins corresponding to residues 138 to 147 of human CRP (Figure 1) has been proposed to bind calcium.[11] This region is homologous to the Ca^{2+} binding motifs of calmodulin, fibrinogen, and parvalbumin, and has been recently shown to be involved in the binding of Ca^{2+} by human CRP.[22]

Studies of the structure of PnC indicating its ribitol teichoic acid nature led to the demonstration that phosphate groups on PnC constitute the main determinants recognized by CRP.[21,24] A number of phosphate monoesters were shown to bind to CRP with binding constants of 2 to 3 $\times$ 10^{-4} M and a valence of one per subunit.[21] It was subsequently shown that PCh is a far more potent inhibitor of the CRP-PnC precipitation reactions than other phosphate monoesters, leading to the proposal that PCh residues on PnC provide the major determinant group for the binding of CRP.[25] The specificity of CRP for PCh was further supported by results indicating binding of the protein to emulsions containing phosphatidylcholine and to other cell wall and capsular polysaccharides containing PCh.[26-28] Equilibrium dialysis experiments indicated that in the presence of Ca^{2+}, CRP binds PCh with an association constant of 1.6 $\times$ 10^{-5} M and a valence of one per noncovalent subunit.[29]

Available information on the structure of the PCh binding site of CRP is indirect and inconclusive. On the basis of quantitative precipitation tests using PCh and phosphorylethanolamine-substituted BSA and hapten inhibition assays, it was proposed that the binding site in CRP consists of a primary locus for binding of the anionic phosphoryl group and a secondary locus for the binding of the cationic choline group.[30] The zwitterionic nature of the PCh binding site was subsequently confirmed by using phosphate monoesters, choline derivatives, and dipeptides as inhibitors of the binding of CRP to PCh-substituted KLH.[31] Electron spin resonance studies indicated a shallow PCh binding site which does not exceed 5 Å in depth.[32]

Immunoelectron microscopy indicated that all the PCh binding sites are on the same face of the molecule and are nearly perpendicular to the plane of the molecule.[33] It was also shown that CRP bears antigenic determinant(s) cross-reacting with T15 idiotypic determinants on PCh-binding myeloma proteins.[34,35] These experiments were interpreted to suggest that the CRP tetrapeptide 39-FYTE is part of the PCh binding site of CRP, because the homologous sequence 32-FYME, which is unique to PCh-binding myeloma proteins, contains two antigen-binding residues (Y_{33} and E_{35}), as determined by crystallographic studies.[9,36] However, site-directed mutagenesis studies indicated that this tetrapeptide does not play a major role in the formation of the PCh binding site of CRP.[37] Three mutant recombinant CRP (rCRP) — Y40F, E42Q, and Y40F,E42Q — were constructed and expressed in COS cells. Wild-type and all mutant rCRP bound to solid-phase PnC and to PCh-BSA with similar avidities. Inhibition experiments using PCh and dAMP as inhibitors indicated that the relative inhibitory power of the haptens for Y40F; E42Q, and the double mutant Y40F,E42Q was similar to that for the wild-type CRP.[38]

Based on the conservation of amino acids in CRPs from several species (Figure 1), it was proposed that the PCh binding region is composed of the highly conserved residues 51 to 66; the positively charged residues Lys_{57} Arg_{58} were thought to bind to the phosphate group and the negatively charged residues Asp_{60} and Glu_{62} to the choline moiety of PCh.[11] A synthetic peptide corresponding to residues 47 to 63 of CRP was found to mimic the PCh binding characteristics of the intact CRP molecule.[39] One of the several monoclonal antibodies against this synthetic peptide recognized an epitope on the PCh-binding myeloma protein TEPC-15 that is also recognized by an anti-T15 anti-idiotypic monoclonal antibody.[40] We tested the role of Lys_{57} and Arg_{58} residues in the formation of PCh-binding sites by site-specific mutagenesis. Three mutant cDNAs — K57Q, R58G, and K57Q,R58G — were constructed and expressed in a eukaryotic system. Two more mutant cDNAs, W67K and

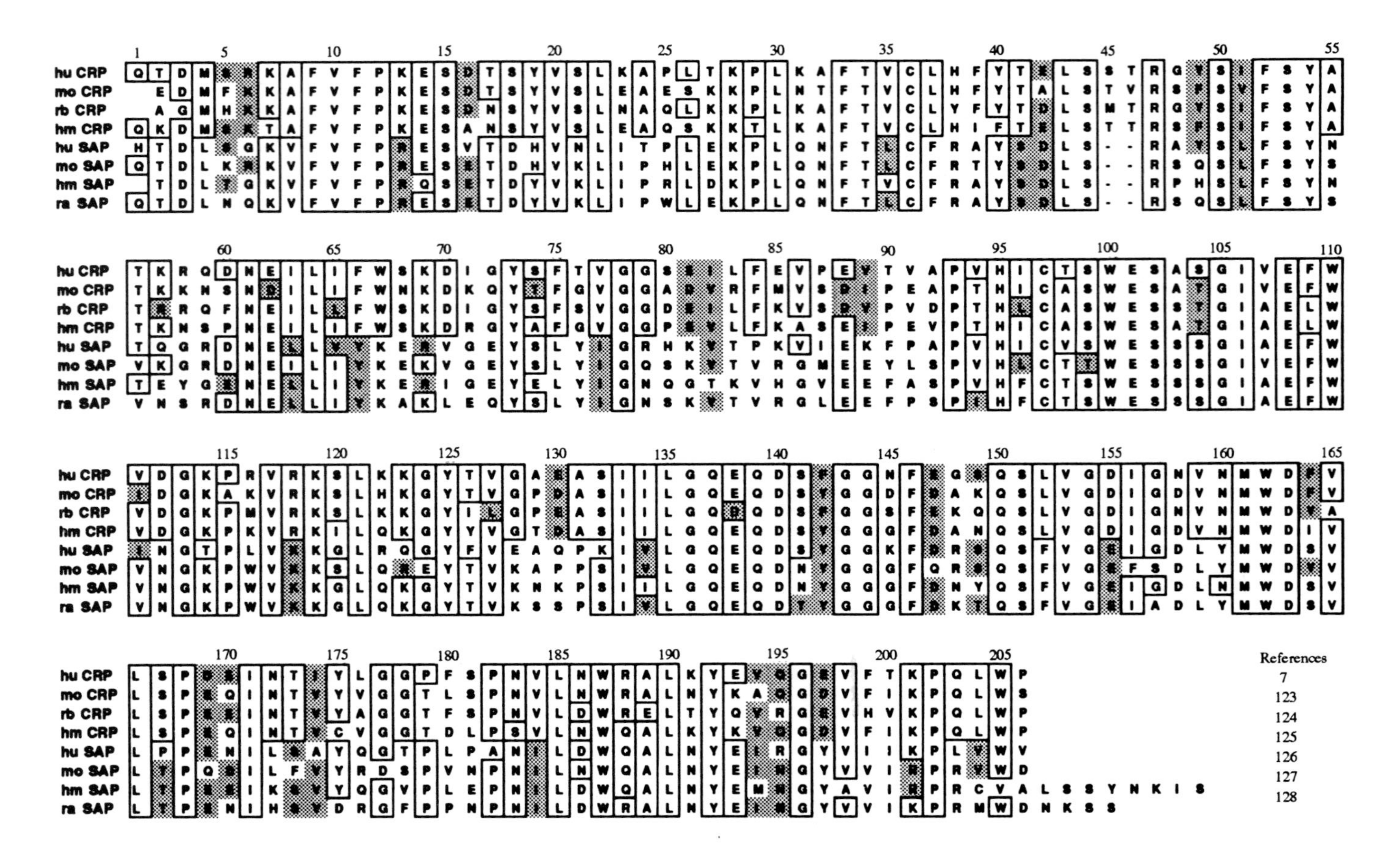

Figure 1. Comparison of the amino acid sequence of eight mammalian pentraxins. Gaps, indicated by dashes, have been introduced to maximize homologies. The numbering system refers to human CRP. hu, human; mo, mouse; rb, rabbit; hm, hamster; ra, rat. Boxes indicate indentities to human CRP residues of more than four pentraxins. Shading indicates chemical similarity to human CRP residues.

TABLE 1
Homology in the Amino Acid Sequence of CRP and Other Proteins

CRP	Homologous region in other proteins	Significance	Ref.
39-FYTE	32-FYME of myeloma protein MoPC 603	PCh binding?	9, 10
138-EQDSFGGNFE	Consensus sequence for Ca^{2+} binding in calmodulin and related molecules	Ca^{2+} binding	11, 12
27-TKPL 113-GKPR 200-TKPQ	TKPR, the bioactive tetrapeptide, tuftsin	Stimulation of phagocytes	13, 14
118-RKSLKK	GKKRSK in histone H2B, PKKKRK in SV40 large T-antigen, VTKKRK in lamin A, AAKRVK in human c-Myc	Nuclear localization signal	15, 16

Note: Single-letter abbreviatins of amino acids are shown.

K57Q,R58G,W67K, were also constructed and expressed. The results of the binding of these mutant rCRP to PCh-BSA indicated that W67 is the critical residue for the formation of the PCh binding site of CRP. Inhibition experiments indicated that K57 and R58 also contribute to the structure of the PCh binding site.[38]

2. Binding to Nuclear Material

The first evidence for the interaction of CRP with nuclear material was provided by the observation that CRP was associated with nuclei of synoviocytes and histiocytes in synovial biopsies of patients with rheumatoid arthritis.[41] Robey et al. later showed that CRP binds chromatin and chromatin fragments in a Ca^{2+}-dependent, PCh-inhibitable manner.[42] CRP pentamers bound to chromatin with a K_d of 8×10^{-7} *M* and a stoichiometry of approximately one CRP binding site for every 160 base pairs of DNA in chromatin.[42] CRP-chromatin complexes were shown to activate complement, and it was suggested that one of the functions of CRP is to mediate the removal of exposed nuclear DNA by complement-dependent solubilization and subsequent removal of chromatin by phagocytes.[43,44] The binding of CRP to nuclei enhanced micrococcal nuclease digestion, suggesting that CRP binding alters chromatin structure by increasing linker DNA exposure. In addition, saturation binding of CRP to chromatin suppressed DNA transcription. It was hypothesized that this suppression could limit aberrant transcription of damaged DNA.[44] It was shown subsequently by using solid-phase assays that, among the histones, CRP binds to H1, H2A, and H2B, but not to H3 and H4.[45] CRP was shown to bind only to the C-terminal fragment of H1, the site which is responsible for condensation of chromatin. The binding determinant for CRP on histone H2A was localized to the N-terminal 15 amino acids.[46] DuClos et al.[47] suggested that the interaction of chromatin with individual core histones does not appear to be responsible for the binding of CRP to native chromatin. However, binding to core particles could be mediated by differentially exposed determinants on H2A and H2B.[47] Transport of CRP into nuclei was reported to be mediated through a nuclear localization signal present in the primary structure of CRP[15] (Table 1). Recently, CRP was also found to bind the 70-KDa polypeptide of U1 small nuclear ribonucleoprotein in a Ca^{2+}-dependent and PCh-inhibitable manner.[48,49] CRP has also been shown to bind to nuclear envelope proteins in a Ca^{2+}-dependent manner.[49]

3. Binding to Other Ligands

Higginbotham et al.[50] reported that CRP reacts with depyruvylated pneumococcal type IV capsular polysaccharide which does not contain PCh, although this reaction can be inhibited by PCh.[51] Subsequently, CRP was found to react with several snail galactans, perhaps through trace phosphate groups that are minor constituents of those galactans.[52,53] CRP was also reported to react with agarose,[54] leishmanial galactans,[55] and fungal materials.[56] Certain polycations, including protamine sulfate, polymers of L-lysine, and leukocyte cationic proteins, were also found to react with CRP.[57,58] DiCamelli et al. showed direct binding of CRP to a variety of arginine-rich and lysine-rich cationic molecules in a Ca^{2+}-independent manner and suggested that the polycation binding site on the CRP molecule is probably distinct from, but perhaps proximal to, the PCh binding site.[59] Interestingly, the precipitin reactivity of CRP with cationic molecules is inhibited by physiological concentrations of calcium and promoted by PCh in the presence of calcium.[60]

Initial suggestive evidence for the interaction of CRP and lipids was presented by Aho, who showed that CRP-positive sera gave flocculation reactions with emulsions of phosphatidylcholine (PC), cholesterol, and the detergent, Span 60.[61] This report was confirmed by Kaplan and Volanakis.[26] Using liposomes of various compositions, binding of CRP to the PCh polar head group of phospholipids has also been demonstrated.[62] Binding of CRP to multilamellar liposomes or unilamellar vesicles of egg-PC also required the presence of lysophosphatidylcholine (LPC) in the bilayer. The binding was found to be Ca^{2+} dependent and could be inhibited by PCh.[54] It was suggested that a disturbance of the normal architecture of the lipid bilayer is required for the formation of a complex between CRP and the polar head group of PC.[62] Furthermore, the presence of galactosyl residues on the surface of the bilayer was found to enhance the binding of CRP, perhaps through interaction with a secondary binding site on the protein.[54] Studies by others suggested that CRP interactions with liposomes might involve binding to sites other than the PCh group.[63-65] Studies on the binding of CRP to positively charged liposomes indicated that it was mediated through the putative polycation binding site of the protein.[64,65] DeBeer et al. reported that when CRP is aggregated on a solid phase at a sufficient density, it selectively binds low density lipoprotein (LDL) and traces of very low density lipoprotein from whole human serum. This phenomenon has been suggested to have significance in the clearance and deposition of LDL.[66]

Salonen et al.[67] reported that CRP binds to the extracellular matrix protein, fibronectin (Fn). Each CRP molecule binds nine Fn molecules with a K_d of 1.47×10^{-7} M when Fn is on the solid phase,[68] or with a K_d of 1.5×10^{-8} M when CRP is on the solid phase.[67] No binding could be demonstrated between soluble CRP and Fn. Ca^{2+} at a concentration of more than 1 mM was reported to be inhibitory for the CRP-Fn interaction, which suggests that the polycation binding site on CRP is the binding target for Fn.[67] We have recently observed that the CRP mutant E42Q binds Fn approximately 40 times more avidly than wt rCRP.[38] However, the E42Q CRP mutant and wt rCRP have similar binding avidities for PCh-containing ligands. This suggests that residue E42 may be involved in Fn binding but not in PCh binding, and that the binding requirements for PCh and Fn are not identical. The C-terminal domain of Fn, including the cell-binding and the heparin-binding regions, is involved in CRP binding.[67,68] Binding of CRP to Fn was reported to interfere with the cell attachment-promoting activity of Fn.[69] The CRP-Fn interaction may explain in part the selective deposition of CRP at sites of tissue injury and may play a role in the formation of the extracellular matrix needed for tissue repair.[67,68] CRP also binds to the basement membrane protein laminin in a Ca^{2+}-dependent manner.[70] The binding was saturable at a molar ratio of 4 CRP/laminin. The CRP-laminin interaction may serve as a means of concentrating CRP at sites of tissue damage so that CRP might function as a ligand for leukocytes, an event that will result in removal of necrotic tissue and cell debris.[69,70]

B. CRP AND COMPLEMENT ACTIVATION

Activation of the complement (C) system by human CRP was first demonstrated for CRP-PnC complexes and CRP-lipid emulsions consisting of cholesterol and either PC or sphingomyelin.[26] Analysis of C-component depletion indicated that the activation proceeded through the classical pathway. Interestingly, CRP complexes failed to activate guinea pig C, and this defect could be corrected by human Clq.[71] This report was confirmed later by Claus et al.,[72] who showed that human Clq binds to and agglutinates CRP-coated particles. It was subsequently reported that complexes of CRP with a variety of other ligands, including chromatin, several synthetic and naturally occurring polycations, as well as polycation-polyanion complexes, can also activate complement.[43,57,58,73] CRP can also activate C upon binding to liposomes with a strong positive charge in the bilayer structure, or to the PC liposomes containing LPC.[54,63,64,74] Like immune aggregates, insoluble CRP-PnC precipitates can be solubilized by C, and fragments of C3 bind covalently to CRP and to PnC during the solubilization reaction.[75] In addition, formation of covalent complexes between CRP and the α'-chain of C4 was also demonstrated.[76] Thus, CRP-initiated activation of the classical pathway of C leads to the assembly of an effective C3-convertase, and it seems reasonable to assume that it results in the generation of host defense-related C fragments, including the anaphylotoxins C3a and C4a and the opsonins C4b, C3b, and iC3b. However, CRP complexes do not appear able to lead to the formation of an efficient C5-convertase.[77] Thus, CRP-initiated C activation may not result in the generation of the C chemotactic factor C5a and the cell membrane lytic complex C5b-9.

C. INTERACTIONS OF CRP WITH EFFECTOR CELLS

A number of studies conducted over the last 50 years have shown that CRP interacts with and affects the function of phagocytic cells. Initial experimental evidence indicated that CRP induces agglutination and capsular swelling of certain types of *Streptococcus pneumoniae* and "gives a slight temporary protection to mice against infection with pneumococci types 27 and 28".[28,78] Further studies indicated that certain bacterial species were phagocytosed more rapidly and in greater numbers by leukocytes of normal human blood after incubation with CRP.[79] Using purified CRP and washed unfractionated blood leukocytes as a source of polymorphonuclear leukocytes, it was clearly shown that CRP enhances the phagocytosis of a variety of Gram-positive and Gram-negative pathogens.[80,81] Mortensen et al.[82] showed that when both CRP and complement fragments were attached to membranes of sheep erythrocytes, these cells were ingested by human monocytes. More recently, isolated human neutrophils were used in a chemiluminescence assay to investigate opsonization of *S. penumoniae* by CRP. The opsonic properties of CRP were found to depend on its ability to activate complement and on the presence of PCh in the bacterial capsule.[83,84] These studies indicated that both CRP and C fragments are necessary for the generation of a phagocytic signal. We have demonstrated that CRP alone can elicit a phagocytic signal, provided the neutrophils are stimulated by either phorbol esters or a less than 10-KDa product(s) of stimulated blood mononuclear cells.[85,86] In addition, solid-phase complexes of CRP and PnC stimulated degranulation of neutrophils. Degranulation could be potentiated by treatment of the neutrophils with the low molecular weight factor(s) from stimulated mononuclear cells.[87]

The relevance of the aforementioned studies on the *in vitro* opsonic properties of CRP to the *in vivo* function of the protein has also been demonstrated. It has been shown that CRP can affect the uptake of PnC-conjugated erythrocytes by the spleen and liver.[88] By using a murine pneumococcal infection model, human CRP was shown to passively protect mice against fatal infection with type 3 and 4 *S. pneumoniae*.[89,90] This protective effect was also observed in mice depleted of C3 by administration of cobra venom factor.[91] However, optimal protective activity of CRP requires a functioning complement system, and it can be demonstrated even in the *xid* mouse, which has virtually no naturally occurring anti-pneumococcal antibody.[92,93] In addition to its opsonic properties, CRP has been reported to induce macrophage tumoricidal activity.[94-96] Treatment of C57BL/6J mice bearing the syngeneic tumors T241 fibrosarcoma or MAC-38 colon carcinoma with human CRP delivered in liposomes or erythrocyte ghosts inhibited lung or liver metastases, respectively.[97-99] Tumor inhibitory activity of peritoneal exudate cells (PEC) stimulated by CRP-multilamellar vesicle (MLV) was comparable to that of PEC activated by another macrophage-activating agent, liposomal muramyl tripeptide (MTP-MLV).[94,100] Antitumor effects of CRP-MLV may also involve other cytotoxic cells bearing T- and/or natural killer (NK)-cell markers.[101] Human peripheral blood monocytes and alveolar macrophages can also be activated *in vitro* by CRP to generate tumoricidal activity a wide variety of human tumors.[96,102]

The aforementioned studies are consistent with CRP-receptor interactions mediating the observed effects. Biochemical and structural data on CRP receptors were obtained recently. Several laboratories have examined the direct binding of radiolabeled CRP to neutrophils and monocytes.[103-108] These studies have documented specific, saturable, and relatively low-affinity binding of CRP to phagocytic cells. CRP binding to many of these cells is inhibited by prior exposure to IgG;[58,103-105,107] however, CRP does not inhibit or modulate IgG binding sites, suggesting that CRP may bind to phagocytic cells via a unique receptor.[85,104,108,109] Recent studies have reported distinct CRP binding proteins of 38 to 40 KDa in the human promonocyte cell line U-937 and 57 to 60 KDa in the mouse macrophage cell line PU5 1.8.[104,108] Scatchard analysis of the binding data indicated both high- and low-affinity binding sites for CRP on murine and human cell lines.[104,108] In addition, Tebo and Mortensen have shown that CRP receptor complexes are internalized in an endosomal compartment.[110] Crowell et al.[111] analyzed the binding of CRP to peripheral blood leukocytes and U-937 cells by immunofluorescence, immunoprecipitation, and cross-linking, and reported that CRP binds to several membrane proteins on U-937 cells, including a protein of 43 to 45 KDa slightly larger than the previously reported 38 to 40 KDa protein and a protein that appears to be identical to the IgG receptor, FcγRI, thus explaining the presence of both high- and low-affinity binding sites for CRP on U-937 cells. The finding that CRP also binds to FCγRI could explain, in part, the earlier findings of inhibition of CRP by IgG-containing aggregates.

D. CRP AND PLATELET-ACTIVATING FACTOR

CRP has also been postulated to play a protective role through the inhibition of platelet-activating factor (PAF), 1-0-alkyl-2-0-acetyl-sn-glycero-3-phosphocholine.[112-116] Hokama et al.[112] first reported that CRP inhibited platelet aggregation by PAF. Vigo demonstrated that CRP inhibits the release of arachidonic acid from both PC and phosphatidylinositol of PAF-stimulated platelet membranes.[113] CRP has also been shown to inhibit PAF-induced degranulation and superoxide production, and hence chemiluminescence by human neutrophils.[115,116] The effect of CRP was attributed to inhibition of the binding of PAF to neutrophils.[114] It was also shown in these studies that CRP protected neutrophils and platelets from the lytic effects of lysolecithin.[113,115] The *in vivo* significance of the CRP-PAF interaction has not been determined, but these *in vitro* data support the hypothesis that, in addition to its role in host defense and recognition and elimination of damaged cells of the host, CRP also palys a role in the control of the acute inflammatory response.

E. FUNCTIONS OF PEPTIDES GENERATED FROM CRP

Buchta et al.[117] showed that the primary structure of CRP contained three regions with peptide sequences closely resembling the amino acid sequence of the phagocytic stimulator tuftsin (Table 1). Using chemically synthesized peptides based on the tuftsin-like CRP sequences, superoxide production, chemotaxis, and degranulation of neutrophils were found to be inhibited,[117] while phagocytosis, chemotaxis, and superoxide and interleukin production by monocytes was stimulated.[118] The likely explanation for this seeming discrepancy was proposed to be the slight differences in the amino acid sequences of the peptides used by the two groups.[119] It was proposed that at sites of inflammation, phagocyte-derived enzymes cleave CRP and generate bioactive peptides.[117,120] Shephard et al. showed that during incubation of CRP with neutrophils, a membrane-derived enzyme can also degrade CRP to small, soluble peptides which inhibit many of the proinflammatory functions of activated neutrophils.[121] Recently, the amino acid sequences of the small CRP peptides were determined, and the ability of synthetic CRP peptides, modeled on these identified sequences, to influence neutrophil functions was investigated. It was found that these peptides act synergistically, and their action involves the signal transduction pathways for neutrophil activation.[119]

IV. CONCLUSIONS

The principal binding specificity of CRP is for PCh and polycations. Binding of CRP to PCh in Ca^{2+} dependent, while binding to polycations is Ca^{2+} independent. Other known ligands bind to CRP through either the PCh binding site or the polycation binding site. Residues Lys_{57}, Arg_{58}, and Trp_{67} are important determinants of the PCh binding site of CRP. The structure of the binding site for other ligands of CRP is not known. The biological significance of the PCh binding site is underscored by the wide distribution of this group in eukaryotic cell membranes and prokaryotic cell walls. In damaged tissues, CRP interacts with nuclear material and probably aids in their removal through interactions with the complement system and cells of the phagocytic system. Thus, the functions of CRP relate to its ability to specifically recognize foreign pathogens and damaged cells of the host and to initiate their elimination by interacting with humoral and cellular effector systems in the blood. CRP binds to phagocytic cells in a specific and reversible manner, and upon binding, a biological response in the form of a phagocytic or tumoricidal signal is elicited. In addition, CRP may modulate the early phases of the acute inflammatory response.

REFERENCES

1. **Kushner, I.,** The phenomenon of the acute phase response, *Ann. N.Y. Acad. Sci.*, 389, 39, 1982.
2. **Osmand, A. P., Friedenson, B., Gewurz, H., Painter, R. H., Hofmann, T., and Shelton, E.,** Characterization of C-reactive protein and the complement subcomponent Clt as homologous proteins displaying cyclic pentameric symmetry (pentraxins), *Proc. Natl. Acad. Sci. U.S.A.*, 74, 739, 1977.
3. **Pepys, M. B., Baltz, M. L., Gomer, K., Davies, A. J. S., and Doenhoff, M.,** Serum amyloid P component is an acute phase reactant in the mouse, *Nature (London)*, 278, 259, 1979.
4. **Gotschlich, E. C. and Edelman, G. M.,** C-reactive protein: a molecule composed of subunits, *Proc. Natl. Acad. Sci. U.S.A.*, 54, 558, 1965.
5. **Volanakis, J. E., Clements, W. L., and Schrohenloher, R. E.,** C-reactive protein: purification by affinity chromatography and physicochemical characterization, *J. Immunol. Methods*, 23, 285, 1978.
6. **Oliveira, E. B., Gotschlich, E. C., and Liu, T.-Y.,** Primary structure of human C-reactive protein, *J. Biol. Chem.*, 254, 489, 1979.
7. **Lei, K.-J., Liu, T., Zon, G., Soravia, E., Liu, T.-Y., and Goldman, N. D.,** Genomic DNA sequence for human C-reactive protein, *J. Biol. Chem.*, 260, 13377, 1985.
8. **Woo, P., Korenberg, J. R., and Whitehead, A. S.,** Characterization of genomic and complementary DNA sequence of human C-reactive protein, and comparison with the complementary DNA sequence of serum amyloid P component, *J. Biol. Chem.*, 260, 13384, 1985.
9. **Young, N. M. and Williams, R. E.,** Comparison of the secondary structures and binding sites of C-reactive protein and the phosphorylcholine-binding murine myeloma proteins, *J. Immunol.*, 121, 1893, 1978.
10. **Kabat, E. A., Wu, T. T., and Bilofsky, H.,** Attempts to locate residues in complementarity-determining regions of antibody combining sites that make contact with antigen, *Proc. Natl. Acad. Sci. U.S.A.*, 73, 617, 1976.

11. **Liu, T.-Y., Syin, C., Nguyen, N. Y., Suzuki, A., Boykins, R. A., Lei, K.-J., and Goldman, N.**, Comparison of protein structure and genomic structure of human, rabbit and *Limulus* C-reactive proteins: possible implications for function and evolution, *J. Prot. Chem.*, 6, 263, 1987.
12. **Dang, C. V., Ebert, R. F., and Bell, W. R.**, Localization of a fibrinogen calcium binding site between γ-subunit positions 311 and 336 by terbium fluorescence, *J. Biol. Chem.*, 260, 9713, 1985.
13. **Fiedel, B. A.**, Influence of tuftsin-like synthetic peptides derived from C-reactive protein (CRP) on platelet behavior, *Immunology,* 64, 487, 1988.
14. **Najjar, V. A. and Nishioka, K.**, Tuftsin: a natural phagocytosis stimulating peptide, *Nature (London),* 228, 672, 1970.
15. **DuClos, T. W., Mold, C., and Stump, R. F.**, Identification of a polypeptide sequence that mediates nuclear localization of the acute phase protein C-reactive protein, *J. Immunol.*, 145, 3869, 1990.
16. **Roberts, B.**, Nuclear location signal-mediated protein transport, *Biochim. Biophys. Acta,* 1008, 263, 1989.
17. **DeLucas, L. J., Greenhough, T. J., Rule, S. A., Myles, D. A. A., Babu, Y. S., Volanakis, J. E., and Bugg, C. E.**, Preliminary X-ray study of crystals of human C-reactive protein, *J. Mol. Biol.*, 196, 741, 1987.
18. **Myles, D. A. A., Rule, S. A., DeLucas, L. J., Babu, Y. S., Xu, Y., Volanakis, J. E., Bugg, C. E., Bailey, S., and Greenhough, T. J.**, Rotation function studies of human C-reactive protein, *J.Mol. Biol.*, 216, 491, 1990.
19. **Tillett, W. S. and Francis, T., Jr.**, Serological reactions in pneumonia with a non-protein somatic fraction of pneumococcus, *J. Exp. Med.*, 52, 561, 1930.
20. **Abernethy, T. J. and Avery, O. T.**, The occurrence during acute infections of a protein not normally present in the blood. I. Distribution of the reactive protein in patients' sera and the effect of calcium on the flocculation reaction with C-polysaccharide of pneumococcus, *J. Exp. Med.*, 73, 173, 1941.
21. **Gotschlich, E. C. and Edelman, G. M.**, Binding properties and specificity of C-reactive protein, *Proc. Natl. Acad. Sci. U.S.A.*, 57, 706, 1967.
22. **Kinoshita, C. M., Ying, S.-C., Hugli, T. E., Siegel, J. N., Potempa, L. A., Jiang, H., Houghten, R. A., and Gewurz, H.**, Elucidation of a protease-sensitive site involved in the binding of calcium to C-reactive protein, *Biochemistry,* 28, 9840, 1989.
23. **Kilpatrick, J. M., Kearney, J. F., and Volanakis, J. E.**, Demonstration of calcium-induced conformational change(s) in C-reactive protein by using monoclonal antibodies, *Mol. Immunol.*, 19, 1159, 1982.
24. **Tomasz, A.**, Choline in the cell wall of a bacterium: novel type of polymer-linked choline in pneumococcus, *Science,* 157, 694, 1967.
25. **Volanakis, J. E. and Kaplan, M. H.**, Specificity of C-reactive protein for choline phosphate residues of pneumococcal C-polysaccharide, *Proc. Soc. Exp. Biiol. Med.*, 136, 612, 1971.
26. **Kaplan, M. H. and Volanakis, J. E.**, Interaction of C-reactive protein complexes with the complement system. I. Consumption of human complement associated with the reaction of C-reactive protein with penumococcal C-polysaccharide and with the choline phosphatides, lecithin and sphingomyelin, *J. Immunol.*, 112, 2135, 1974.
27. **Baldo, B. A., Fletcher, T. C., and Pepys, J.**, Isolation of a peptido-polysaccharide from the dermatophyte *Epidermophyton floccosum* and a study of its reaction with human C-reactive protein and a mouse anti-phosphorylcholine myeloma serum, *Immunology,* 32, 831, 1977.
28. **Löfström, G.**, Comparison between the reactions of acute phase serum with pneumococcus C-polysaccharide and with pneumococcus type 27, *Br. J. Exp. Pathol.*, 25, 21, 1944.
29. **Anderson, J. K., Stroud, R. M., and Volanakis, J. E.**, Studies on the binding specificity of human C-reactive protein for phosphorylcholine, *Fed. Proc.*, 37, 1495, 1978.
30. **Oliveira, E. B., Gotschlich, E. C., and Liu, T.-Y.**, Comparative studies on the binding properties of human and rabbit C-reactive proteins, *J. Immunol.*, 124, 1396, 1980.
31. **Barnum, S., Narkates, A. J., Suddath, F. L., and Volanakis, J. E.**, Comparative studies on the binding specificities of C-reactive protein (CRP) and HOPC 8, *Ann. N.Y. Acad. Sci.*, 389, 431, 1982.
32. **Robey, F. A. and Liu, T.-Y.**, Synthesis and use of new spin labeled derivatives of phosphorylcholine in a comparative study of human, dogfish and *Limulus* C-reactive proteins, *J. Biol. Chem.*, 258, 3895, 1983.
33. **Roux, K. H., Kilpatrick, J. M., Volanakis, J. E., and Kearney, J. F.**, Localization of the phosphocholine-binding sites on C-reactive protein by immunoelectron microscopy, *J. Immunol.*, 131, 2411, 1983.
34. **Volanakis, J. E. and Kearney, J. F.**, Cross-reactivity between C-reactive protein and idiotypic determinants on a phosphocholine-binding murine myeloma protein, *J. Exp. Med.*, 153, 1604, 1981.
35. **Vasta, G. R., Marchalonis, J. J., and Kohler, H.**, Invertebrate recognition protein cross-reacts with an immunoglobulin idiotype, *J. Exp. Med.*, 159, 1270, 1984.
36. **Padlan, E. A., Davies, D. R., Rudikoff, S., and Potter, M.**, Structural basis for the specificity of phosphorylcholine binding immunoglobulins, *Immunochemistry,* 13, 945, 1976.
37. **Volanakis, J. E., Xu, Y., and Macon, K. J.**, Human C-reactive protein and host defense, in *Defense Molecules,* Marchalonis, J. J. and Reinish, C. L., Eds., Alan R. Liss, New York, 1990, 161.

38. **Agrawal, A., Xu, Y., Ansardi, D., Macon, K. J., and Volanakis, J. E.,** Probing the phosphocholine-binding site of human C-reactive protein by site-directed mutagenesis, submitted. *J. Biol. Chem.*, 267, 25352, 1992.
39. **Swanson, S. J. and Mortensen, R. F.,** Binding and immunological properties of a synthetic peptide corresponding to the phosphorylcholine-binding region of C-reactive protein, *Mol. Immunol.*, 27, 679, 1990.
40. **Swanson, S. J., Lin, B.-F., Mullenix, M. C., and Mortensen, R. F.,** A synthetic peptide corresponding to the phosphorylcholine (PC)-binding region of human C-reactive protein possesses the TEPC-15 myeloma PC-idiotype, *J. Immunol.*, 146, 1596, 1991.
41. **Gitlin, J. D., Gitlin, J. I., and Gitlin, D.,** Localization of C-reactive protein in synovium of patients with rheumatoid arthritis, *Arthritis Rheum.*, 20, 1491, 1977.
42. **Robey, F. A., Jones, K. D., Tanaka, T., and Liu, T.-Y.,** Binding of C-reactive protein to chromatin and nucleosome core particles: a possible physiological role of C-reactive protein, *J. Biol. Chem.*, 259, 7311, 1984.
43. **Robey, F. A., Jones, K. D., and Steinberg, A. D.,** C-reactive protein mediates the solubilization of nuclear DNA by complement *in vitro, J. Exp. Med.*, 161, 1344, 1985.
44. **Shephard, E. G., VanHelden, P. D., Strauss, M., Böhm, L., and DeBeer, F. C.,** Functional effects of CRP binding to nuclei, *Immunology,* 58, 489, 1986.
45. **DuClos, T. W., Zlock, L. T., and Rubin, R. L.,** Analysis of the binding of C-reactive protein to histones and chromatin, *J. Immunol.*, 141, 4266, 1988.
46. **DuClos, T. W., Zlock, L. T., and Marnell, L.,** Definition of a C-reactive binding determinant on histones, *J. Biol. Chem.*, 266, 2167, 1991.
47. **DuClos, T. W., Marnell, L., Zlock, L. R., and Burlingame, R. W.,** Analysis of the binding of C-reactive protein to chromatin subunits, *J. Immunol.*, 146, 1220, 1991.
48. **DuClos, T. W.,** C-reactive protein reacts with the U1 small nuclear ribonucleoprotein, *J. Immunol.*, 143, 2553, 1989.
49. **Shephard, E. G., Smith, P. J., Coetzee, S., Strachan, A. F., and DeBeer, F. C.,** Pentraxin binding to isolated rat liver nuclei, *Biochem. J.*, 279, 257, 1991.
50. **Higginbotham, J. D., Heidelberger, M., and Gotschlich, E. C.,** Degradation of a penumococcal type-specific polysaccharide with exposure of group-specificity, *Proc. Natl. Acad. Sci. U.S.A.*, 67, 138, 1970.
51. **Heidelberger, M., Gotschlich, E. C., and Higginbotham, J. D.,** Inhibition experiments with penumococcal C and depyruvylated type-IV polysaccharides, *Carbohydr. Res.*, 22, 1, 1972.
52. **Uhlenbruck, G., Karduck, D., Haupt, H., and Schwick, H. G.,** C-reactive protein (CRP), 9.5Sα_1-glycoprotein and Clq: serum proteins with lectin properties?, *Z. Immun.-Forsch.*, 155, 262, 1979.
53. **Soelter, J. and Uhlenbruck, G.,** The role of phosphate groups in the interaction of human C-reactive protein with galactan polysaccharides, *Immunology,* 58, 139, 1986.
54. **Volanakis, J. E. and Narkates, A. J.,** Interaction of C-reactive protein with artificial phosphatidylcholine bilayers and complement, *J. Immunol.*, 126, 1820, 1981.
55. **Pritchard, D. G., Volanakis, J. E., Slutsky, G. M., and Greenblatt, C. L.,** C-reactive protein binds leishmanial excreted factors, *Proc. Soc. Exp. Biol. Med.*, 178, 500, 1985.
56. **Jensen, T. D. B., Schonheyder, H., Anderson, P., and Stenderup, A.,** Binding of C-reactive protein to *Aspergillus fumigatus* fractions, *J. Med. Microbiol.*, 21, 173, 1986.
57. **Siegel, J., Rent, R., and Gewurz, H.,** Interactions of C-reactive protein with the complement system. I. Protamine-induced consumption of complement in acute phase sera, *J. Exp. Med.*, 140, 631, 1974.
58. **Siegel, J., Osmand, A. P., Wilson, M. F., and Gewurz, H.,** Interaction of C-reactive protein with the complement system. II. C-reactive protein-mediated consumption of complement by poly-L-lysine polymers and other polycations, *J. Exp. Med.*, 142, 709, 1975.
59. **DiCamelli, R., Potempa, L. A., Siegel, J., Suyehira, L., Petras, K., and Gewurz, H.,** Binding reactivity of C-reactive protein for polycations, *J. Immunol.*, 125, 1933, 1980.
60. **Potempa, L. A., Siegel, J. N., and Gewurz, H.,** Binding reactivity of C-reactive protein for polycations. II. Modulatory effects of calcium and phosphocholine, *J. Immunol.*, 127, 1509, 1981.
61. **Aho, K.,** Studies of syphilitic antibodies. IV. Evidence of reactant partner common for C-reactive protein and certain anti-lipoidal antibodies, *Br. J. Vener. Dis.*, 45, 13, 1969.
62. **Volanakis, J. E. and Wirtz, K. W. A.,** Interaction of C-reactive protein with artificial phosphatidylcholine bilayers, *Nature (London),* 281, 155, 1979.
63. **Richards, R. L., Gewurz, H., Osmand, A. P., and Alving, C. R.,** Interactions of C-reactive protein and complement with liposomes, *Proc. Natl. Acad. Sci. U.S.A.*, 74, 5672, 1977.
64. **Richards, R. L., Gewurz, H., Siegel, J., and Alving, C. R.,** Interactions of C-reactive protein and complement with liposomes. II. Influence of membrane composition, *J. Immunol.*, 122, 1185, 1979.
65. **Mold, C., Rodgers, C. P., Richards, R. L., Alving, C. R., and Gewurz, H.,** Interaction of C-reactive protein with liposomes. III. Membrane requirements for binding, *J. Immunol.*, 126, 856, 1981.

66. **DeBeer, F. C., Soutar, A. K., Baltz, M. L., Trayner, I. M., Feinstein, A., and Pepys, M. B.,** Low density lipoprotein and very low density lipoprotein are selectively bound by aggregated C-reactive protein, *J. Exp. Med.*, 156, 230, 1982.
67. **Salonen, E.-M., Vartio, T., Hedman, K., and Vaheri, A.,** Binding of fibronectin by the acute phase reactant C-reactive protein, *J. Biol. Chem.*, 259, 1496, 1984.
68. **Tseng, J. and Mortensen, R. F.,** Binding of human C-reactive protein (CRP) to plasma fibronectin occurs via the phosphorylcholine-binding site, *Mol. Immunol.*, 25, 679, 1988.
69. **Tseng, J. and Mortensen, R. F.,** The effect of human C-reactive protein on the cell-attachment activity of fibronectin and laminin, *Exp. Cell Res.*, 180, 303, 1989.
70. **Swanson, S. J., McPeek, M. M., and Mortensen, R. F.,** Characteristics of the binding of human C-reactive protein (CRP) to laminin, *J. Cell. Biochem.*, 40, 121, 1989.
71. **Volanakis, J. E. and Kaplan, M. H.,** Interaction of C-reactive protein complexes with the complement system. II. Consumption of guinea pig complement by CRP complexes: requirement for human Clq, *J. Immunol.*, 113, 9, 1974.
72. **Claus, D. R., Siegel, J., Petras, K., Osmand, A. P., and Gewurz, H.,** Interactions of C-reactive protein with the first component of human complement, *J. Immunol.*, 119, 187, 1977.
73. **Claus, D. R., Siegel, J., Petras, K., Skor, D., Osmand, A. P., and Gewurz, H.,** Complement activation by interaction of polyanions and polycations. III. Complement activation by interaction of multiple polyanions and polycations in the presence of C-reactive protein, *J. Immunol.*, 118, 83, 1977.
74. **Volanakis, J. E.,** Complement activation by C-reactive protein complexes, *Ann. N.Y. Acad. Sci.*, 389, 235, 1982.
75. **Volanakis, J. E.,** Complement-induced solubilization of C-reactive protein-pneumococcal C-polysaccharide precipitates: evidence for covalent binding of complement proteins to C-reactive protein and to pneumococcal C-polysaccharide, *J. Immunol.*, 128, 2745, 1982.
76. **Volanakis, J. E. and Narkates, A. J.,** Binding of human C4 to C-reactive protein-pneumococcal C-polysaccharide complexes during activation of the classical complement pathway, *Mol. Immunol.*, 20, 1201, 1983.
77. **Berman, S., Gewurz, H., and Mold, C.,** Binding of C-reactive protein to nucleated cells leads to complement activation without cytolysis, *J. Immunol.*, 136, 1354, 1986.
78. **Löfström, G.,** Nonspecific capsular swelling in pneumococci: a serologic and clinical study, *Acta Med. Scand.*, 141, 1, 1943.
79. **Hokama, Y., Coleman, M. K., and Riley, R. F.,** *In vitro* effects of C-reactive protein on phagocytosis, *J. Bacteriol.*, 83, 1017, 1962.
80. **Ganrot, P. O. and Kindmark, C.-O.,** C-reactive protein: a phagocytosis-promoting factor, *Scand. J. Clin. Lab. Invest.*, 24, 215, 1969.
81. **Kindmark, C. O.,** Stimulating effect of C-reactive protein on phagocytosis of various species of pathogenic bacteria, *Clin. Exp. Immunol.*, 8, 941, 1971.
82. **Mortensen, R. F., Osmand, A. P., Lint, T. F., and Gewurz, H.,** Interaction of C-reactive protein with lymphocytes and monocytes: complement-dependent adherence and phagocytosis, *J. Immunol.*, 117, 774, 1976.
83. **Edwards, K. M., Gewurz, H., Lint, T. F., and Mold, C.,** A role for C-reactive protein in the complement-mediated stimulation of human neutrophils by type 27 *Streptococcus pneumoniae, J. Immunol.*, 128, 2493, 1982.
84. **Mold, C., Edwards, K. M., and Gewurz, H.,** Effect of C-reactive protein on the complement-mediated stimulation of human neutrophils by *Streptococcus pneumoniae* serotypes 3 and 6, *Infect. Immun.*, 37, 987, 1982.
85. **Kilpatrick, J. M. and Volanakis, J. E.,** Opsonic properties of C-reactive protein: stimulation by phorbol myristate acetate enables human neutrophils to phagocytize C-reactive protein-coated cells, *J. Immunol.*, 134, 3364, 1985.
86. **Kilpatrick, J. M., Gresham, H. D., Griffin, F. M., Jr., and Volanakis, J. E.,** Peripheral blood mononuclear leukocytes release a mediator(s) that induces phagocytosis of C-reactive protein-coated cells by polymorphonuclear leukocytes, *J. Leuk. Biol.*, 41, 150, 1987.
87. **Poston, R., Hyman, B., and Kilpatrick, J. M.,** Effects of a low molecular weight lymphokine on neutrophil lysosomal release stimulated by C-reactive protein and IgG, *FASEB J.*, 2, A1459, 1988.
88. **Nakayama, S., Mold, C., Gewurz, H., and DuClos, T. W.,** Opsonic properties of C-reactive protein *in vivo, J. Immunol.*, 128, 2435, 1982.
89. **Mold, C., Nakayama, S., Holzer, T. J., Gewurz, H., and DuClos, T. W.,** C-reactive protein is protective against *Streptococcus penumoniae* infection in mice, *J. Exp. Med.*, 154, 1703, 1981.
90. **Yother, J., Volanakis, J. E., and Briles, D. E.,** Human C-rective protein is protective against fatal *Streptococcus pneumoniae* infection in mice, *J. Immunol.*, 128, 2374, 1982.

91. **Nakayama, S., Gewurz, H., Holzer, T., DuClos, T. W., and Mold, C.,** The role of the spleen in the protective effect of C-reactive protein in *Streptococcus pneumoniae* infection, *Clin. Exp. Immunol.*, 54, 319, 1983.
92. **Horowitz, J., Volanakis, J. E., and Briles, D. E.,** Blood clearance of *Streptococcus pneumoniae* by C-reactive protein, *J. Immunol.*, 138, 2598, 1987.
93. **Briles, D. E., Forman, C., Horowitz, J. C., Volanakis, J. E., Benjamin, W. H., Jr., McDaniel, L. S., Eldridge, J., and Brooks, J.,** Antipneumococcal effects of C-reactive protein and monoclonal antibodies to pneumococcal cell wall and capsular antigens, *Infect. Immun.*, 57, 1457, 1989.
94. **Barna, B. P., Deodhar, S. D., Gautam, S., Yen-Lieberman, B., and Roberts, D.,** Macrophage activation and generation of tumoricidal activity by liposome associated human C-reactive protein, *Cancer Res.*, 44, 305, 1984.
95. **Zahedi, K. and Mortensen, R. F.,** Macrophage tumoricidal activity induced by human C-reactive protein, *Cancer Res.*, 46, 5077, 1986.
96. **Barna, B. P., James, K., and Deodhar, S. D.,** Activation of human monocyte tumoricidal activity by C-reactive protein, *Cancer Res.*, 47, 3959, 1987.
97. **Deodhar, S. D., James, K., Chiang, T., Edinger, M., and Barna, B. P.,** Inhibition of lung metastases in mice bearing a malignant fibrosarcoma by treatment with liposomes containing human C-reactive protein, *Cancer Res.*, 42, 5084, 1982.
98. **Gautam, S., Barna, B., Chiang, T., Pettay, J., and Deodhar, S.,** Use of resealed erythrocytes as delivery system for C-reactive protein (CRP) to generate macrophage-mediated tumoricidal activity, *J. Biol. Response Mod.*, 6, 346, 1987.
99. **Thombre, P. S. and Deodhar, S. D.,** Inhibition of liver metastases in murine colon adenocarcinoma by liposomes containing human C-reactive protein or crude lymphokine, *Cancer Immunol. Immunother.*, 16, 145, 1984.
100. **Gautam, S., James, K., and Deodhar, S. D.,** Macrophage-mediated tumoricidal activity generated by human C-reactive protein (CRP) encapsulated in liposomes is complement-dependent, *Cleveland Clin. Q.*, 53, 235, 1986.
101. **Gautam, S. and Deodhar, S.,** Generation of tumoricidal effector cells by human C-reactive protein and muramyl tripeptide: a comparative study, *J. Biol. Response Mod.*, 8, 560, 1989.
102. **Barna, B. P., Thomassen, M. J., Wiedemann, H. P., Ahmad, M., and Deodhar, S. D.,** Modulation of human alveolar macrophage tumoricidal activity by C-reactive protein, *J. Biol. Response Mod.*, 7, 483, 1988.
103. **Müller, H. and Fehr, J.,** Binding of C-reactive protein to human polymorphonuclear leukocytes: evidence for association of binding sites with Fc receptors, *J. Immunol.*, 136, 2202, 1986.
104. **Tebo, J. M. and Mortensen, R. F.,** Characterization and isolation of a C-reactive protein receptor from the human monocytic cell line U-937, *J. Immunol.*, 144, 231, 1990.
105. **Buchta, R., Pontet, M., and Fridkin, M.,** Binding of C-reactive protein to human neutrophils, *FEBS Lett.*, 211, 165, 1987.
106. **Dobrinich, R. and Spagnuolo, P. J.,** Specific C-reactive protein binding to human neutrophils, *Clin Res.*, 35, 897A, 1987.
107. **Ballou, S. P., Buniel, J., and Macintyre, S. S.,** Specific binding of human C-reactive protein to human monocytes *in vitro*, *J. Immunol.*, 142, 2708, 1989.
108. **Zahedi, K., Tebo, J. M., Siripont, J., Klimo, G. F., and Mortensen, R. F.,** Binding of human C-reactive protein to mouse macrophages is mediated by distinct receptors, *J. Immunol.*, 142, 2384, 1989.
109. **Zeller, M., Kubak, B. M., and Gewurz, H.,** Binding sites for C-reactive protein on human monocytes are distinct from IgG Fc receptors, *Immunology*, 67, 51, 1989.
110. **Tebo, J. M. and Mortensen, R. F.,** Internalization and degradation of receptor bound C-reactive protein by U937 cells: induction of H_2O_2 production and tumoricidal activity, *Biochim. Biophys. Acta*, 1095, 210, 1991.
111. **Crowell, R. E., DuClos, T. W., Montoya, G., Heaphy, E., and Mold, C.,** C-reactive protein receptors on the human monocytic cell line U-937: evidence for additional binding to FcγRI, *J. Immunol.*, 147, 3445, 1991.
112. **Hokama, Y., Garcia, D. A., Bonilla, M. R., Lam, M. P., and Shimizu, R. M.,** Inhibition of platelet aggregation by C-reactive protein (CRP) in platelets stimulated by 1-0-alkyl-2-0-acetyl-sn-glycero-3-phosphocholine, (PAF) and related analogs, *Fed. Proc.*, 43, 461, 1984.
113. **Vigo, C.,** Effect of C-reactive protein on platelet-activating factor-induced platelet aggregation and membrane stabilization, *J. Biol. Chem.*, 260, 3418, 1985.
114. **Kilpatrick, J. M. and Virella, G.,** Inhibition of platelet activating factor by rabbit C-reactive protein, *Clin. Immunol. Immunopathol.*, 37, 276, 1985.
115. **Tatsumi, N., Hashimoto, K., Okuda, K., and Kyougoku, T.,** Neutrophil chemiluminescence induced by platelet activating factor and suppressed by C-reactive protein, *Clin. Chim. Acta*, 172, 85, 1988.

116. **Filep, J. and Földes-Filep, E.,** Effects of C-reactive protein on human neutrophil granulocytes challenged with N-formyl-methionyl-leucyl-phenylalanine and platelet-activating factor, *Life Sci.,* 44, 517, 1989.
117. **Buchta, R., Fridkin, M., Pontet, M., and Romeo, D.,** Synthetic peptides from C-reactive protein containing tuftsin-related sequences, *Peptides,* 7, 961, 1986.
118. **Robey, F. A., Ohura, K., Futaki, S., Fujii, N., Yajima, H., Goldman, N., Jones, K. D., and Wahl, S.,** Proteolysis of human C-reactive protein produce peptides with potent immunomodulating activity, *J. Biol. Chem.,* 262, 7053, 1987.
119. **Shephard, E. G., Anderson, R., Rosen, O., Myer, M. S., Fridkin, M., Strachan, A. F., and DeBeer, F. C.,** Peptides generated from C-reactive protein by a neutrophil membrane protease: amino acid sequence and effects of peptides on neutrophil oxidative metabolism and chemotaxis, *J. Immunol.,* 145, 1469, 1990.
120. **Shephard, E. G., Anderson, R., Beer, S. M., Van Rensburg, C. E. J., and DeBeer, F. C.,** Neutrophil lysosomal degradation of human CRP: CRP-derived peptides modulate neutrophil function, *Clin. Exp. Immunol.,* 73, 139, 1988.
121. **Shephard, E. G., Beer, S. M., Anderson, R., Strachan, A. F., Nel, A. E., and DeBeer, F. C.,** Generation of biologically active C-reactive protein peptides by a neutral protease on the membrane of phorbol myristate acetate-stimulated neutrophils, *J. Immunol.,* 143, 2974, 1989.
122. **Ohnishi, S., Maeda, S., Nishiguchi, S., Arao, T., and Shimada, K.,** Structure of the mouse C-reactive protein gene, *Biochem. Biophys. Res. Commun.,* 156, 814, 1988.
123. **Syin, C., Gotschlich, E. C., and Liu, T.-Y.,** Rabbit C-reactive protein: biosynthesis and characterization of cDNA clones, *J. Biol. Chem.,* 261, 5473, 1986.
124. **Dowton, S. B. and Holden, S. N.,** C-reactive protein (CRP) of the Syrian hamster, *Biochemistry,* 30, 9531, 1991.
125. **Mantzouranis, E. C., Dowton, S. B., Whitehead, A. S., Edge, M. D., Bruns, G. A. P., and Colten, H. R.,** Human serum amyloid P component: cDNA isolation, complete sequence of pre-serum amyloid P component, and localization of the gene to chromosome 1, *J. Biol. Chem.,* 260, 7752, 1985.
126. **Nishiguchi, S., Maeda, S., Araki, S., and Shimada, K.,** Structure of mouse serum amyloid P component gene, *Biochem. Biophys. Res. Commun.,* 155, 1366, 1988.
127. **Dowton, S. B., Woods, D. E., Mantzouranis, E. C., and Colten, H. R.,** Syrian hamster female protein: analysis of female protein primary structure and gene expression, *Science,* 228, 1206, 1985.
128. **Dowton, S. B. and McGrew, S. D.,** Rat serum amyloid P component: analysis of cDNA sequence and gene expression, *Biochem. J.,* 270, 553, 1990.

Chapter 5

THE SAA LIPOPROTEIN FAMILY

Nils Eriksen, Rick L. Meek, and Earl P. Benditt

TABLE OF CONTENTS

0-8493-6913-4/93/$0.00 + $.50

I. INTRODUCTION

The history of serum amyloid A (SAA) protein, a major acute phase reactant, had its beginning in the isolation and characterization of a unique protein found in tissues obtained at autopsy from patients with amyloidosis secondary to chronic inflammatory diseases.[1-3] Amyloidosis is a tissue abnormality recognized by the presence of extracellular proteinaceous deposits (amyloid) with a distinctive fibrillar ultrastructure and an affinity for the dye Congo red. Specimens stained with the dye exhibit orange-yellow to blue-green dichroism when they are examined by polarized light as the plane of polarization is being rotated.[32,36] These properties of amyloid are essentially invariant, irrespective of the tissue involved or the underlying pathology; nevertheless, in the last 2 decades, wide differences have been observed in the composition of the protein component of amyloid. Amino acid composition analyses of the inflammation-associated amyloid protein (now called amyloid A, or AA) clearly distinguished it from amyloid proteins of immunoglobulin origin, leading to the concept of chemical classes of amyloid substance.[4] In comparative work, a protein isolated from the liver of a monkey with a chronic granulomatous disease was shown by N-terminal amino acid sequence analysis to be closely related to human AA.[5] Since then, the list of animals found to be susceptible to amyloidosis of the AA type has been extended to include at least ten species, experimental or domesticated.

Using antibodies to AA protein, two independent groups discovered an antigenically related protein in human serum. In almost simultaneous publications, Levin et al.[6] and Husby et al.[7] described double immunodiffusion tests of such antibodies against samples of serum from normal individuals and from patients representing a variety of disease categories. Positive results in over 50% of the cases were obtained in most disease categories, whereas less than 10% of the control samples gave positive results; a high incidence of positive results in association with amyloidosis secondary to a predisposing disease was apparent in both studies. The mass of the serum particle with the AA immunoreactivity was reported by the first group to be slightly in excess of that of serum albumin and by the second group, approximately 100 kDa.

Early in 1975, Linke et al.[8] described the recovery of a 12.5-kDa protein with AA immunoreactivity from a guanidine-dissociated sample of serum that, in the native state, showed a 200-kDa component with AA immunoreactivity. Very shortly thereafter, Anders et al.[9] published similar results, based on the use of formic acid as the dissociating agent; in an addendum, these authors reported a 20-residue N-terminal amino acid sequence identity between the low-molecular-mass serum component (SAA) and the antigenically related tissue component (AA), which generally has a mass about 30% less than that of SAA. Confirmation of these later findings came quickly from Franklin and co-workers.[10]

As its story unfolded, SAA gained recognition not only as the probable precursor of the tissue protein AA, but also as an indicator of acute inflammatory states and thus as an acute phase protein; early insights in the latter regard were provided by McAdam and Anders,[11] McAdam et al.,[12] and Franklin et al.[10]

The essential character of the high-molecular-mass circulating components with which SAA is associated in subjects with elevated SAA levels was revealed in our analysis of fractions separated by sequential density-increment centrifugation of serum obtained from a human adult 20 h after an injection for immunization against typhoid fever[13] and of pooled plasma obtained from mice 24 h after intraperitoneal injection of bacterial lipopolysaccharide (LPS).[14] In each experiment, the largest amount of the SAA was found in the flotation layer at density 1.21, after centrifugal removal of material floating at density 1.125. Chromatographic analysis of the flotation fraction confined to the density interval 1.125 to 1.21 yielded particle-mass estimates of 160 kDa for the human and 220 kDa for the mouse SAA-carrying components; a similar component, at a lower concentration, was found in the density 1.063 to 1.125 fraction of the mouse plasma. In both species, therefore, the major SAA-carrying components displayed properties of high density lipoproteins (HDL); minor or trace amounts of SAA were found in the infranatant fraction after centrifugation at density 1.21 (lower three quarters of tube contents) and, particularly in the mouse, in the very low and low density lipoprotein fractions (VLDL and LDL). Further work in our laboratory[15] and investigations elsewhere[16-18] have established that generally, as a consequence of severe tissue damage, SAA is demonstrable in the LDL and VLDL fractions, although consistently the peak concentration of SAA is in the HDL_3 fraction.

The heterogeneous nature of SAA derived from human HDL was disclosed by ion-exchange chromatography.[19-21] At least six and possibly eight isoforms were separated by this technique. A complete sequence analysis of one of the major isoforms showed it to be a 104-residue protein identical to tissue AA in its first 76 residues.[22,23]

The primary structure of SAA has now been determined partially or completely, by amino acid sequence analysis or by deduction from nucleotide sequences of SAA genes, for nine species (man, cow, dog, hamster, horse, mink, mouse, rabbit, and sheep). Sequence variations have been noted within each species, additional differences exist between species, and in several species SAA contains short insertions between amino acid residues 69 and 70 (human numbering system), but the sequence DKYFHARGNYDAA (residues 33 to 45) occurs intact, with only rare instances of substitution, in the known SAA and AA isoforms of a total of 12 species, including monkey, cat, and duck.[24-29] The highly conserved stretch continues as QRGPGG (positions 46 to 51), with the exception of several Q → K or Q → R substitutions at position 46. This conservation of sequence suggests that SAA gene products have an important biological function.

Recent investigations have focused attention on SAA gene families, the cloning of individual member genes, and their expression in various tissues and cell types, notably from humans and rodents. At present, a minimum of four human SAA genes are recognized.[30,31] Three and two allelic variants have been detected in, respectively, the human SAA_1 and SAA_2 genes;[31] SAA proteins corresponding to four of these variants have been found in acute phase HDL.[22,32] A protein corresponding to the SAA_4 gene has recently been found in non-acute phase human HDL.[32a,32b] A protein corresponding to the human SAA_3 gene has not yet been identified. The amino acid sequence predicted by this gene, also known as GSAA1,[33] is unusual in having V instead of the almost invariably occurring A at position 45; also, the deduced sequence shows 80% similarity to that of a collagenase-inducing SAA protein produced by rabbit synovial fibroblasts,[34] but only 71% similarity to that of rabbit acute phase SAA lipoprotein[28] and 75 to 76% similarity to the human SAA_1 and SAA_2 protein sequences.

In our laboratory, aspects of SAA induction and expression related to the acute phase response have been studied in the mouse, hamster, and rat. In the mouse, the known SAA gene family comprises three genes, SAA_1, SAA_2, and SAA_3, and a pseudo gene.[35,36] The mRNA for each of the three genes is elevated 500- to 2000-fold in mice after LPS injection. SAA_1 and SAA_2 proteins were found in association with HDL in mice challenged with LPS;[37] only SAA_2, derived from circulating HDL, proved to be a precursor of AA protein in mouse amyloid deposits.[38,39] The mouse SAA_3 protein has recently been identified; its expression differs among various tissues (see Section III). Also, a fifth SAA gene product (SAA_5) has recently been discovered in HDL from LPS-injected BALB/c mice.[29] The SAA_5 protein is slightly more massive than mouse SAA_1 and SAA_2, the difference being attributed to an insertion in an unsequenced portion of the SAA_5 molecule. The mouse SAA_5 sequences shows only 58, 56, and 61% similarity, respectively, to the SAA_1, SAA_2, and SAA_3 sequences in the 89 positions for which identifications were available for comparison. The SAA_5 protein is unusual in that it contains no methionine residues insofar as it has been sequenced and shows four amino acid substitutions in the commonly conserved region (positions 33 to 45). A study of SAA gene structure and expression in the Syrian hamster has indicated three SAA genes; the corresponding deduced amino acid sequences were designated $hSAA_1$, $hSAA_2$, and $hSAA_3$.[40] There is a 96.7% sequence similarity between $hSAA_1$ and $hSAA_2$ and an 88.5% sequence similarity between $hSAA_1$ and $hSAA_3$. The three sequences are quite similar (84.5 to 88.3%) to the deduced sequence of mouse SAA_3. The rat presents a special case: although genes related to mouse SAA mRNAs are expressed in some of its tissues in response to turpentine or LPS stimulation,[41] the rat does not develop AA amyloidosis nor show evidence thus far of an AA-related plasma protein.

The dramatic increases (up to 1000-fold) in levels of circulating SAA observed in experimental animals within 24 h after administration of an inflammatory agent[42-44] are the result of highly elevated SAA mRNA levels controlled by transcriptional and posttranscriptional mechanisms.[45,46]

Early investigations of SAA regulation demonstrated that a circulating factor produced by macrophages in response to LPS stimulated the synthesis of SAA in mice.[47] A variety of systems have been used to identify the cytokines that may be involved in the regulation of human, rabbit, and mouse SAA synthesis.[34,48-58] Although not complete, available evidence indicates that interleukin-1 (IL-1), IL-6, and tumor necrosis factor (TNF), singly or in various combinations, stimulate SAA gene expression. *Cis*-acting regulatory sequences, identified in human, mouse, and rat SAA genes, have been found to participate in specific gene expression via NFκB-like and C/EBP-like transcription factors.[59-63] By all indications, the molecular mechanisms controlling SAA gene expression will prove to be intricate.

II. MATERIALS AND METHODS

A. EXPERIMENTS IN MICE

An acute phase response was elicited in male CBA mice by a subcutaneous injection of casein and in male BALB/c mice by an intraperitoneal injection of *Salmonella typhosa* LPS. The next day, blood was collected by cardiac puncture of the injected mice and uninjected (control) mice after ether anesthetization. Tissues were then removed and processed for RNA isolation, immunocytochemistry (Carnoy's fixative), and *in situ* hybridization (paraformaldehyde fixative) with an SAA_1/SAA_2-specific [^{3}H]cRNA probe; Northern-blot hybridizations and dot-blot analyses were done with a nick-translated ^{32}P-labeled SAA cDNA probe.[64]

Expression of the SAA_3 gene in BALB/c mice was studied in tissues obtained 18 hours after LPS injection and in control tissue;[65] details of tissue fixation and *in situ* hybridization have been described.[64] In addition to the aforementioned SAA_1/SAA_2 specific probe, a cRNA probe designed to detect only SAA_3 RNA was used.[66]

B. EXPERIMENTS IN RATS

An acute phase response was elicited in male Sprague-Dawley rats by subcutaneous injection of gum turpentine or by intravenous injection (via tail vein) of *Escherichia coli* LPS. RNA was prepared from tissues removed 24 h after injection and from tissues removed from unstimulated (control) rats. Details of RNA preparation, mouse SAA cDNA and cRNA probe preparation, hybrid selection of mRNA, and cell-free translation have been presented.[66]

C. EXPERIMENTS IN HAMSTERS

An acute phase response was elicited in female golden Syrian hamsters by injection of *E. coli* LPS intraperitoneally or gum turpentine subcutaneously. Lipoprotein fractions isolated by sequential density-increment centrifugation of serum samples obtained from injected hamsters 20 h poststimulus and from uninjected (control) hamsters were examined by urea/sodium dodecyl sulfate-polyacrylamide gel electrophoresis (SDS-PAGE) followed by electroimmunoblotting. Details of the procedures and descriptions of the primary antibodies used in the blotting have been presented;[67] whole antiserums were used at a 200-fold dilution and affinity-purified antibody preparations were used at a 50-fold dilution. Additional antibody preparations (whole rabbit antiserum to hamster AA and affinity-purified rabbit anti-bovine AA) were gifts from Dr. T. A. Niewold; both were used at a 500-fold dilution.

III. RESULTS

A. STUDIES IN MICE

Within 5 h after injection of mice with casein, the level of SAA mRNA in the liver had begun to increase (SAA_1 + SAA_2 mRNA, measured with a common probe). The level reached a maximum at about 16 h, remained constant for about 10 h, and then slowly decreased; the level of circulating SAA protein showed a similar time course, but lagged behind the mRNA by several hours, Although expression of SAA_1/SAA_2 mRNA after casein injection was not detected in extrahepatic tissues (kidney, spleen, lung, brain, testis, heart, adrenal gland, and skeletal muscle), immunocytochemical staining with antibody which reacts with SAA_1 and SAA_2, but not SAA_3, showed the presence of SAA protein in the liver, kidney, testis, spleen, and adrenal gland. In recent experiments, we have found immunostaining of macrophages fixed in plasma. Except for hepatocytes, the data indicate that these tissues must take up SAA or SAA-HDL from the circulation.

A time-course study of SAA accumulation in liver, kidney, testis, and spleen revealed two general patterns. Staining intensity in hepatocytes and the epithelia of proximal convoluted tubules, slight at 4 h after injection, reached a plateau by 9 to 16 h. A delayed response was observed in Leydig cells of the testis and in perifollicular cells of the spleen, where in each case staining was negative at 9 h after injection, became evident by 16 h, and reached a maximum intensity 20 h after casein injection, when circulating SAA was at the highest level. Adrenal cortical cells displayed high levels of SAA at 20 h; earlier time points were not investigated. Whether the temporally different patterns of SAA uptake are determined by local SAA concentrations or by specific uptake mechanisms is not clear. The question remains to be explored.

The cellular location of SAA_1 and/or SAA_2 mRNA in the liver, kidney, and intestine of mice injected with LPS was determined by *in situ* hybridization of ^{3}H-labeled SAA_1/SAA_2-cRNA to tissue sections. Hepatocytes were heavily labeled, but other cell types of the liver, such as endothelial cells, Kupffer cells, and epithelial cells of bile ducts, were negative. In kidney tissue, hybridization was confined to epithelial cells of the proximal and distal convoluted tubules; the probe did not hybridize to other cells of the cortex or to any cells of the medulla. In the ileum and large intestine of LPS-injected mice, hybridization was confined primarily to epithelial cells lining the mucosal membrane; corresponding tissues taken from control mice were negative.

The expression of SAA_3 mRNA in response to the administration of LPS differs strikingly from its expression in response to the administration of casein. Hepatic and extrahepatic SAA_3 mRNA are expressed only at low or undetectable levels after casein administration, whereas after LPS administration SAA_3 mRNA is expressed in a wide variety of tissues at levels covering more than a 100-fold overall range of difference, the highest levels having been found in liver and in peritoneal macrophages.[68] The widespread distribution and varied levels of SAA_3 mRNA expression in tissues (Table 1) prompted consideration of the possible existence of a single dispersed cell system for this activity. The possibility was explored by

TABLE 1
LPS Stimulation of SAA Gene Family Expression in Mouse Tissues

	SAA_1	SAA_2	SAA_3
Liver	+ + + +	+ + + +	+ + + +
Kidney	+	+	+
Large intestine	+	—	+
Ileum	+	—	+
Lung	—	—	+ + +
Spleen	—	—	+ +
Heart	—	—	+ +
Testis	—	—	+
Adrenal	—	—	+
Skeletal muscle	—	—	+
Stomach	—	—	+ +
Peritoneal macrophages	—	—	+ + + +

Note: Ten micrograms of total RNA was examined by Northern-blot hybridization with specific ^{32}P-labeled oligonucleotides.[68] Autoradiographic exposures were quantitated by densitometry and compared to liver RNA standards, arbitrarily set at 100% and indicating ~6000 molecules of each SAA mRNA per cell.[45] + + + +, 100%; + + +, 50 to 75%; + +, 25 to 50%; +, 5 to 25% of liver mRNA; —, no detectable hybridization.

examination of sections of various tissues from control and LPS-stimulated mice by *in situ* hybridization with a cRNA probe specific for SAA_3 mRNA.[65] The study revealed that SAA_3 mRNA is expressed by adipocytes of fat tissues in or adherent to organs such as adrenal gland, aorta, lung, ileum, and large intestine. In addition to this system of dispersed cells, specialized cells in the testis (Leydig cells), spleen (most likely macrophages), and liver, as well as hepatocytes themselves, were found to express SAA_3 mRNA.

Furthermore, it is clear that mouse monocytes express SAA_3: a cultured mouse monocyte line (J-774), upon stimulation with LPS, expressed SAA_3 mRNA. SAA_3 protein was found in the culture medium by electroimmunoblotting with rabbit antiserum made against a biosynthetic fragment (C-terminal 55 residues) of mouse SAA_3. Immunohistochemical application of the antiserum revealed SAA_3 immunoreactivity in mononuclear cells of blood clots, spleen, lung, and other tissues from mice injected with LPS.[69]

B. STUDIES IN RATS

RNA from normal and injected (LPS or turpentine) rats was examined by RNA blot hybridization with SAA cDNA and cRNA probes for the three mouse SAA genes. A ~400-nucleotide SAA mRNA related to mouse SAA_1 and SAA_2 mRNAs reached a high level in liver removed 24 h after injection of LPS. No extrahepatic tissues were found to express the SAA_1/SAA_2-related mRNA. Turpentine induced two hepatic SAA_1/SAA_2-related mRNAs, ~400 and ~500 nucleotides in length. Whether they represent mRNA from separate genes or differently processed mRNA is not known. Liver SAA_1/SAA_2-related mRNA from LPS- and turpentine-injected rats was hybrid selected in the presence of [^{35}S]methionine. The translation product was a single protein of estimated (urea/SDS-PAGE) molecular mass ~8 kDa, which is approximately 6 kDa less than that of mouse SAA preproteins. Further hybridization experiments with an appropriate probe indicated that rat liver SAA mRNA most likely lacks a region that codes for a sequence encompassing the highly conserved stretch of amino acid residues (33 to 51) in SAA. In a brief report, Liao et al.[70] described sequence analyses of several rat liver SAA cDNAs, all of which lacked a sequence coding for residues 12 to 58. The lack of this coding sequence is consistent with the production of the low-molecular-mass protein coded by the mouse SAA_1/SAA_2-related mRNA from the livers of LPS- and turpentine-injected rats. A ~600-nucleotide mRNA related to mouse SAA_3 mRNA was expressed at a high level in rat lung after LPS, but not after turpentine, injection. This mRNA was also expressed at high levels in the ileum and large intestine of control rats, where further elevation did not occur after injection of the rats with LPS or turpentine; the mouse SAA_3-related mRNA was not found in the liver of control or challenged rats.

C. STUDIES IN HAMSTERS

A prominent doublet in the 12-kDa region was revealed by urea/SDS-PAGE of the HDL_3 fraction (d 1.125 to 1.21 g/ml) isolated from serum of hamsters injected subcutaneously with turpentine.[67] The doublet was present but less intense in the HDL_2 fraction (d 1.063 to 1.125 g/ml), fainter in the lower density lipoprotein fractions (d $<$1.063 g/ml), and hardly detectable in any of the lipoprotein fractions derived from noninjected hamsters. The response to a subcutaneous injection of 0.5 ml of turpentine was greater than that evoked by the intraperitoneal injection of 50 μg of LPS; a more toxic dose of LPS (250 μg) diminished the response.

Antibodies to hamster AA reacted strongly with the leading member of the 12-kDa doublet; the reaction extended over the whole region of the doublet, with a suggestion, but not a clear indication, of resolution into two bands. Resolution was not improved by reductions in the amount of sample analyzed. Antibodies to human, monkey, and bovine AA reacted preferentially with the leading member of the doublet, whereas antibodies to mouse AA reacted preferentially with the trailing member. Antibodies to human and mouse SAA and to duck AA, which is very similar to known SAA proteins in size and sequence,[71] reacted with both members. In addition to their reactions in the 12-kDa region, antibodies to hamster and mouse AA and to mouse SAA showed a band (also a doublet) of moderate intensity at ~40 kDa. Weak reactions occasionally observed in the 25-kDa region of control as well as acute phase HDL_3 fractions are believed to be mainly nonspecific interactions involving the major apolipoprotein apoA-I, although a distinct doublet revealed in this region by anti-hamster AA may indicate AA relatedness. A diagrammatic summary of the various immunoreactions described above is shown in Figure 1.

IV. DISCUSSION

In the complex series of events that constitute the acute phase response, the functions of several of the participating reactants (e.g., fibrinogen and the protease inhibitors) are readily understood.[72] Still remaining largely unexplained is the role played by one of the major participants, the SAA family, which in itself presents a complexity because of the multiplicity of genes and the existence of posttranslational modifications.[32]

Our recent experiments have yielded information regarding sites of expression of the three mouse SAA genes, the identity of the cells wherein the expression takes place, the anatomic distribution of the encoded proteins, and the differential response of mice to the inflammation-inducing agents casein and LPS.[64,65,73,74] In addition, these experiments have revealed the existence of SAA genes in an amyloidosis-resistant species, the rat.[66] One of these genes, related to mouse SAA_1 and SAA_2, is expressed as an mRNA of shortened form, lacking a part that includes the coding region for the highly conserved domain in the N-terminal portion of SAA in other species. The other gene, related to mouse SAA_3, is expressed as an SAA mRNA of expected length.

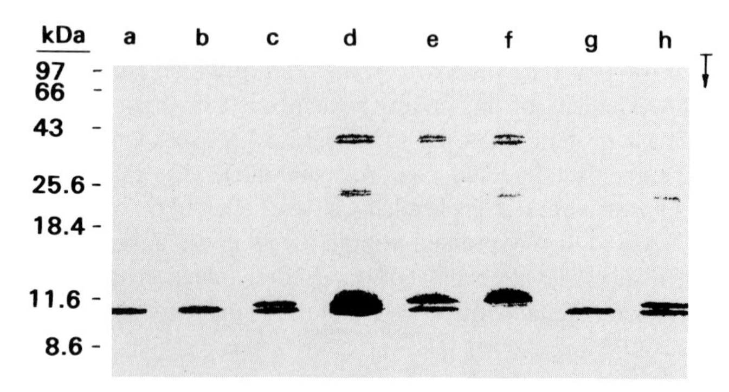

FIGURE 1. Diagrammatic representation of immunoreactions of electroblotted samples of hamster acute phase HDL_3 after separation by urea/SDS-PAGE. The diagram is a composite of analyses done in the same way, but at different times and with different batches of reagents. Primary antibodies: (a) anti-monkey AA, whole antiserum; (b) anti-human AA, whole antiserum; (c) anti-human $apoSAA_2$, affinity purified; (d) anti-hamster AA, whole antiserum; (e) anti-mouse SAA, sodium sulfate ppt. absorbed with normal mouse serum; (f) anti-mouse AA, affinity purified; (g) anti-bovine AA, affinity purified; (h) anti-duck AA, whole antiserum. The intensities of the principal bands in the 10- to 12-kDa region have been rendered approximately equal; the original intensities were determined largely by the amount of antibody bound and the time allowed for color development (alkaline-phosphatase/nitro blue tetrazolium/5-bromo-4-chloro-3-indolyl-phosphate system). Molecular-mass standards are, in descending order, phosphorylase b, bovine serum albumin, α-chymotrypsinogen, ovalbumin, β-lactoglobulin, duck AA, and monkey AA.

With respect to SAA mRNA similarity, there is a closer resemblance between the mouse and hamster than between the mouse and rat. In the hamster, as in the mouse, three SAA genes have been recognized (see Section I). Two of the corresponding deduced hamster SAA amino acid sequences are almost identical (two differences in a total of 103 residues) and the third differs only slightly from the others (11 and 13 differences in the same total). All three of the hamster SAA amino acid sequences show a close resemblance to mouse SAA_3 (at least 83% similarity), but it is not known whether, as in the mouse, only one of the three represents an AA protein precursor, a sequence analysis of hamster AA having not yet been reported.

Also not known is how many of the hamster SAA proteins are included in the acute phase HDL apolipoproteins. The 12-kDa doublet that became evident on urea/SDS-PAGE of the acute phase HDL could represent separate gene products or posttranslational modifications of a single gene product. The differences among the immunoreactions of members of the doublet against antibodies to hamster, human, cow, monkey, mouse, and duck AA and/or SAA must depend upon reactions with different epitopes and are compatible with the existence of more than one gene product. Four SAA isoforms have been recognized by isoelectric focusing of acute phase hamster HDL;[75] these isoforms were characterized only by second-dimension SDS-PAGE, so their genetic individuality remains in question. The higher-molecular-mass components (~40-kDa doublet) detected in hamster acute phase HDL by antibodies to hamster and mouse AA and to mouse SAA could be SAA aggregates, complexes of SAA with carriers, or novel proteins embodying SAA epitopes.

Several functions of an inhibitory nature have been proposed for SAA. Early evidence suggested suppression of the immune response of murine spleen cells to sheep red blood cells by SAA.[76,77] In 1984, it was reported that the natural-killer-cell activity of cells from casein-treated CBA/J mice was suppressed by mouse serum with an elevated SAA level, as well as by partially purified AA protein;[78] a few years later, inhibition of lymphocyte proliferation and growth of HeLa and MRC_5 cell cultures,[79] and inhibition of plasma cholesterol esterification[80] were described. A 1990 paper[82] and a group of reports at the 6th International Symposium on Amyloidosis described additional effects of SAA or fragments thereof: inhibition of platelet aggregation,[82-84] inhibition of thromboxane B_2 generation and serotonin release by thrombin-activated platelets,[83] modulation of activated-human-neutrophil superoxide anion production,[85] and inhibition of growth and potentiation of differentiation of myeloid leukemia cells.[86] Closely following the symposium reports came a description of the inhibition by SAA of the pyrogenic activity of recombinant human IL-1β and TNFα.[87] That SAA has measurable *in vitro* activities thus seems well established; whether any of them may be considered an actual biological function of SAA is not so certain.

The weight of evidence linking SAA to lipoproteins, HDL in particular, favors a role in lipid transport and metabolism as the major function of SAA. As possibilities within this framework, we have suggested that the SAAs may be part of an early responding network of defensive molecules acting in ways analogous to those of other apolipoproteins, but specialized to operate in response to specific requirements of tissue injury and inflammation. SAA could act by influencing the rates of intra- and extracellular transport of cholesterol and other lipids, and the activities of enzymes involved in steroid biosynthesis, and also could aid in the removal of lipid debris resulting from tissue-destructive processes.[74] Others have pointed out that the changes in lipoprotein metabolism observed during an acute phase response are compatible with the idea that the delivery of cholesterol to the liver for excretion is altered to allow cholesterol to remain in the tissues where it is needed for repair and regeneration of damaged membranes.[88] Adding detail to this theme, Kisilevsky[89] has postulated that the function of SAA is to direct HDL to sites of tissue destruction where cholesterol is being collected by macrophages; there, the HDL particles would take up cholesterol from the lipid-laden macrophages (foam cells) for delivery to and reutilization by proliferating cells in areas of repair, or to sites of ultimate cholesterol excretion.

That each SAA may have a specific function is a distinct possibility, particularly in the mouse, in which SAA_3 is regulated differently than are SAA_1 and SAA_2. The injection of silver nitrate, like the injection of casein, induces maximal expression of SAA_1 and SAA_2 mRNA, but does not stimulate expression of SAA_3 mRNA; the injection of amyloid-enhancing factor, like the injection of LPS, induces expression of all three mRNAs.[90] Moreover, SAA_3 mRNA is expressed in a variety of cell types, whereas SAA_1 and SAA_2 mRNA expression is primarily in the liver, by hepatocytes. Macrophages, major effector cells of the acute phase response,[91] are found at sites of tissue injury, destruction, and repair. In the mouse, these cells express only SAA_3. Similar involvement of the SAA lipoproteins may occur in other vertebrate species.

The discovery of the dramatic increases in SAA levels following tissue injury generated clinical interest in possible correlations between circulating SAA levels and the extent of injury or severity of inflammation. Random or serial SAA levels have been measured in such diverse conditions as myocardial infarction, accidental or surgical trauma, bacterial or viral infection, neoplastic disease, organ transplantation and rejection, rheumatoid arthritis, inflammatory bowel disease, and others.[92-96] Additional references are cited in the reports just mentioned. Although SAA levels are not diagnostic, their measurements have been deemed useful for monitoring disease activity and response to therapy, in some cases for prognosis, and in general for the most sensitive indication of the acute phase.[96]

Despite years of effort devoted to the task of assaying SAA, a completely reliable method has yet to be achieved. Some of the remaining problems were discussed and some recent developments were presented at the 6th International Symposium on Amyloidosis.[97-100] The existence of several SAA isoforms probably contributes to the difficulties. As more is learned about the characteristics of these isoforms, direct and accurate procedures for their individual quantification are likely to be devised. If successful, such efforts may add diagnostic or prognostic value to the determination of SAA levels.

Finally, from the foregoing it is clear that the SAA family of proteins must represent important ingredients in the acute phase response to various forms of tissue damage. The studies recorded thus far suggest that proteins of this family may help modulate several facets of the response to injury. The effects may be immediate and local or may be widespread.

REFERENCES

1. **Benditt, E. P., Lagunoff, D., Eriksen, N., and Iseri, O. A.,** Amyloid. Extraction and preliminary charcterization of some proteins, *Arch. Pathol.*, 74, 323, 1962.
2. **Benditt, E. P. and Eriksen, N.,** Amyloid. II. Starch gel electrophoretic analysis of some proteins extracted from amyloid, *Arch. Pathol.*, 78, 325, 1964.
3. **Benditt, E. P. and Eriksen, N.,** Amyloid. III. A protein related to the subunit structure of human amyloid fibrils, *Proc. Natl. Acad. Sci. U.S.A.*, 55, 308, 1966.

3a. **Benditt, E. P., Eriksen, N., and Berglund, C.,** Congo red dichroism with dispersed amyloid fibrils, an extrinsic cotton effect, *Proc. Natl. Acad. Sci., U.S.A.*, 66, 1044, 1970.

3b. **Taylor, D. L., Allen, R. D., and Benditt, E. P.,** Determination of the polarization optical properties of the amyloid-Congo red complex by phase modulation microspectrophotometry, *J. Histochem. Cytochem.*, 22, 1105, 1974.

4. **Benditt, E. P. and Eriksen, N.,** Chemical classes of amyloid substance, *Am. J. Pathol.*, 65, 231, 1971.
5. **Benditt, E. P., Eriksen, N., Hermodson, M. A., and Ericsson, L. H.,** The major proteins of human and monkey amyloid substance: common properties including unusual N-terminal amino acid sequences, *FEBS Lett.*, 19, 169, 1971.
6. **Levin, M., Pras, M., and Franklin, E. C.,** Immunologic studies of the major nonimmunoglobulin protein of amyloid. I. Identification and partial characterization of a related serum component, *J. Exp. Med.*, 138, 373, 1973.
7. **Husby, G., Natvig, J. B., Michaelsen, T. E., Sletten, K., and Höst, H.,** Unique amyloid protein subunit common to different types of amyloid fibril, *Nature (London)*, 244, 362, 1973.
8. **Linke, R. P., Sipe, J. D., Pollock, P. S., Ignaczak, T. F., and Glenner, G. G.,** Isolation of a low-molecular-weight serum component antigenically related to an amyloid fibril protein of unknown origin, *Proc. Natl. Acad. Sci. U.S.A.*, 72, 1473, 1975.
9. **Anders, R. F., Natvig, J. B., Michaelsen, T. E., and Husby, G.,** Isolation and characterization of amyloid-related serum protein SAA as a low molecular weight protein, *Scand. J. Immunol.*, 4, 397, 1975.
10. **Franklin, E. C., Rosenthal, C. J., and Pras, M.,** Studies on the amyloid A protein (AA protein) and a related serum component — purification — partial charcterization and tissue origin and distribution in different types of amyloidosis, *Adv. Nephrol.*, 5, 89, 1975.
11. **McAdam, K. P. W. J. and Anders, R. F.,** Association of amyloidosis with lepromatous reactions, in *Amyloidosis*, Wegelius, O. and Pasternack, A., Eds., Academic Press, London, 1976, 425.
12. **McAdam, K. P. W. J., Anders, R. F., Smith, S. R., Russell, D. A., and Price, M. A.,** Association of amyloidosis with erythema nodosum leprosum reactions and recurrent neutrophil leucocytosis in leprosy, *Lancet*, II, 572, 1975.
13. **Benditt, E. P. and Eriksen, N.,** Amyloid protein SAA is associated with high density lipoprotein from human serum, *Proc. Natl. Acad. Sci. U.S.A.*, 74, 4025, 1977.
14. **Benditt, E. P., Eriksen, N., and Hanson, R. H.,** Amyloid protein SAA is an apoprotein of mouse plasma high density lipoprotein, *Proc. Natl. Acad. Sci. U.S.A.*, 76, 4092, 1979.
15. **Eriksen, N. and Benditt, E. P.,** Trauma, high density lipoproteins, and serum amyloid protein A, *Clin. Chim. Acta*, 140, 139, 1984.
16. **Marhaug, G., Sletten, K., and Husby, G.,** Characterization of amyloid related protein SAA complexed with serum lipoproteins (apoSAA), *Clin. Exp. Immunol.*, 50, 382, 1982.
17. **Feussner, G. and Ziegler, R.,** Detection of human serum amyloid A protein in very low density and high density lipoproteins of patients after acute myocardial infarction, *Electrophoresis*, 10, 776, 1989.
18. **Saïle, R., Kabbaj, O., Visvikis, S., Steinmetz, J., Steinmetz, A., Férard, G., Fruchart, J. C., and Métais, P.,** Variations in apolipoproteins serum amyloid A, A-I, A-II, and C-III in severely head-injured patients, *J. Clin. Chem. Clin. Biochem.*, 28, 519, 1990.
19. **Bausserman, L. L., Herbert, P. N., and McAdam, K. P. W. J.,** Heterogeneity of human serum amyloid A proteins, *J. Exp. Med.*, 152, 641, 1980.
20. **Eriksen, N. and Benditt, E. P.,** Isolation and charcterization of the amyloid-related apoprotein (SAA) from human high density lipoprotein, *Proc. Natl. Acad. Sci. U.S.A.*, 77, 6860, 1980.

21. **Skogen, B., Sletten, K., Lea, T., and Natvig, J. B.,** Heterogenity of human amyloid protein AA and its related serum protein, SAA, *Scand. J. Immunol.*, 17, 83, 1983.
22. **Parmelee, D. C., Titani, K., Ericsson, L. H., Eriksen, N., Benditt, E. P., and Walsh, K. A.,** Amino acid sequence of amyloid-related apoprotein (apoSAA_1) from human high-density lipoprotein, *Biochemistry,* 21, 3298, 1982.
23. **Sletten, K., Marhaug, G., and Husby, G.,** The covalent structure of amyloid-related serum protein SAA from two patients with inflammatory disease, *Hoppe-Seyler's Z. Physiol. Chem.*, 364, 1039, 1983.
24. **Sletten, K., Husebekk, A., and Husby, G.,** The amino acid sequence of an amyloid fibril protein AA isolated from the horse, *Scand. J. Immunol.*, 26, 79, 1987.
25. **Marhaug, G., Sletten, K., and Husby, G.,** Characterization of serum amyloid A (SAA) protein in the sheep, in *Amyloid and Amyloidosis,* Isobe, T., Araki, S., Uchino, F., Kito, S., and Tsubura, E., Eds., Plenum Press, New York, 1988, 217.
26. **Johnson, K. H., Sletten, K., Werdin, R. E., Westermark, G. T., O'Brien, T. D., and Westermark, P.,** Amino acid sequence variations in protein AA of cats with high and low incidences of AA amyloidosis, *Comp. Biochem. Physiol.*, 94B, 765, 1989.
27. **Syversen, P. V., Sletten, K., and Husby, G.,** Evolutionary aspects of protein SAA, in *Amyloid and Amyloidosis 1990,* Natvig, J. B., Førre, O., Husby, G., Husebekk, A., Skogen, B., Sletten, K., and Westermark, P., Eds., Kluwer, Dordrecht, 1991, 111.
28. **Liepnieks, J. J., Dwulet, F. E., Benson, M. D., Kluve-Beckerman, B., and Kushner, I.,** The primary structure of serum amyloid A protein in the rabbit: comparison with serum amyloid A proteins in other species, *J. Lab. Clin. Med.*, 118, 570, 1991.
29. **De Beer, M. C., Beach, C. M., Shedlofsky, S. I., and de Beer, F. C.,** Identification of a novel serum amyloid A protein in BALB/c mice, *Biochem. J.*, 280, 45, 1991.
30. **Steinkasserer, A., Weiss, E. H., Schwaeble, W., and Linke, R. P.,** Heterogeneity of human serum amyloid A protein. Five different variants from one individual demonstrated by cDNA sequence analysis, *Biochem. J.*, 268, 187, 1990.
31. **Betts, J. C., Edbrooke, M. R., Thakker, R. V., and Woo, P.,** The human acute phase serum amyloid A gene family: structure, evolution and expression in hepatoma cells, *Scand. J. Immunol.*, 34, 471, 1991.
32. **Dwulet, F. E., Wallace, D. K., and Benson, M. D.,** Amino acid structures of multiple forms of amyloid-related serum protein SAA from a single individual, *Biochemistry,* 27, 1677, 1988.
33. **Sack, G. H., Jr. and Talbot, C. C., Jr.,** The human serum amyloid A (SAA)-encoding gene GSAA1: nucleotide sequence and possible autocrine-collagenase-inducer function, *Gene,* 84, 509, 1989.
32a. **Whitehead, A. S., de Beer, M. C., Steel, D. M., Rits, M., Lelias, J. M., Lane, W. S., and de Beer, F. C.,** Identification of novel members of the serum amyloid A protein superfamily as constitutive apolipoproteins of high density lipoproteins, *J. Biol. Chem.*, 267, 3862, 1992.
32b. **Watson, G., Coade, S., and Woo, P.,** Analysis of the genomic and derived protein structure of a novel human serum amyloid A gene, SAA4, *Scand. J. Immunol.*, 36, 703, 1992.
34. **Mitchell, T. I., Coon, C. I., and Brinckerhoff, C. E.,** Serum amyloid A (SAA3) produced by rabbit synovial fibroblasts treated with phorbol esters or interleukin 1 induces synthesis of collagenase and is neutralized with specific antiserum, *J. Clin. Invest.*, 87, 1177, 1991.
35. **Lowell, C. A., Potter, D. A., Stearman, R. S., and Morrow, J. F.,** Structure of the murine serum amyloid A gene family. Gene conversion, *J. Biol. Chem.*, 261, 8442, 1986.
36. **Yamamoto, K. and Migita, S.,** Complete primary structures of two major murine serum amyloid A proteins deduced from cDNA sequences, *Proc. Natl. Acad. Sci. U.S.A.*, 82, 2915, 1985.
37. **Hoffman, J. S. and Benditt, E. P.,** Changes in high density lipoprotein content following endotoxin administration in the mouse: formation of serum amyloid protein-rich subfractions, *J. Biol. Chem.*, 257, 10510, 1982.
38. **Hoffman, J. S., Ericsson, L. H., Eriksen, N., Walsh, K. A., and Benditt, E. P.,** Murine tissue amyloid protein AA. NH_2-terminal sequence identity with only one of two serum amyloid protein (apoSAA) gene products, *J. Exp. Med.*, 159, 641, 1984.
39. **Meek, R. L., Hoffman, J. S., and Benditt, E. P.,** Amyloidogenesis. One serum amyloid A isotype is selectively removed from the circulation, *J. Exp. Med.*, 163, 499, 1986.
40. **Webb, C. F., Tucker, P. W., and Dowton, S. B.,** Expression and sequence analyses of serum amyloid A in the Syrian hamster, *Biochemistry,* 28, 4785, 1989.
41. **Meek, R. L. and Benditt, E. P.,** Rat liver and lung express serum amyloid A related mRNAs, in *Amyloid and Amyloidosis,* Isobe, T., Araki, S., Uchino, F., Kito, S., and Tsubura, E., Eds., Plenum Press, New York, 1988, 283.
42. **Kushner, I.,** The phenomenon of the acute phase response, *Ann. N.Y. Acad. Sci.*, 389, 39, 1982.
43. **Benson, M. D., Scheinberg, M. A., Shirahama, T., Cathcart, E. S., and Skinner, M.,** Kinetics of serum amyloid protein A in casein-induced murine amyloidosis, *J. Clin. Invest.*, 59, 412, 1977.

44. **McAdam, K. P. W. J. and Sipe, J. D.,** Murine model for human secondary amyloidosis: genetic variability of the acute phase serum protein SAA response to endotoxin and casein, *J. Exp. Med.*, 144, 1121, 1976.
45. **Lowell, C. A., Stearman, R. S., and Morrow, J. F.,** Transcriptional regulation of serum amyloid A gene expression, *J. Biol. Chem.*, 261, 8453, 1986.
46. **Rienhoff, H. Y., Jr. and Groudine, M.,** Regulation of amyloid A gene expression in cultured cells, *Mol. Cell. Biol.*, 8, 3710, 1988.
47. **Sipe, J. D., Vogel, S. N., Ryan, J. L., McAdam, K. P. W. J., and Rosenstreich, D. L.,** Detection of a mediator derived from endotoxin-stimulated macrophages that induces the acute phase serum amyloid A response in mice, *J. Exp. Med.*, 150, 597, 1979.
48. **Ganapathi, M. K., Rzewnicki, D., Samols, D., Jiang, S. L., and Kushner, I.,** Effect of combinations of cytokines and hormones on synthesis of serum amyloid A and C-reactive protein in Hep 3B cells, *J. Immunol.*, 147, 1261, 1991.
49. **Raynes, J. G., Eagling, S., and McAdam, K. P. W. J.,** Acute phase protein synthesis in human hepatoma cells: differential regulation of serum amyloid A (SAA) and haptoglobin by interleukin-1 and interleukin-6, *Clin. Exp. Immunol.*, 83, 488, 1991.
50. **Limburg, P. C., Aarden, L. A., and van Rijswijk, M. H.,** Tumor necrosis factor (TNF) inhibits interleukin (IL)-1 and/or IL-6 stimulated synthesis of C-reactive protein (CRP) and serum amyloid A (SAA) in primary cultures of human hepatocytes, *Biochim. Biophys. Acta,* 1091, 405, 1991.
51. **Schultz, D. R. and Arnold, P. I.,** Properties of four acute phase proteins: C-reactive protein, serum amyloid A protein, alpha 1-acid glycoprotein, and fibrinogen, *Semin. Arthritis Rheum.*, 20, 129, 1990.
52. **Castell, J. V., G'omez-Lech'on, M. J., David, M., Andus, T., Geiger, T., Trullenque, R., Fabra, R., and Heinrich, P. C.,** Interleukin-6 is the major regulator of acute phase protein synthesis in adult human hepatocytes, *FEBS Lett.*, 242, 237, 1989.
53. **Johns, M. A., Sipe, J. D., Melton, L. B., Strom, T. B., and McCabe, W. R.,** Endotoxin-associated protein: interleukin-1-like activity on serum amyloid A synthesis and T-lymphocyte activation, *Infect. Immun.*, 56, 1593, 1988.
54. **Moshage, H. J., Roelofs, H. M. J., van Pelt, J. F., Hazenberg, B. P. C., van Leeuwen, M. A., Limburg, P. C., Aarden, L. A., and Yap, S. H.,** The effect of interleukin-1, interleukin-6 and its interrelationship on the synthesis of serum amyloid A and C-reactive protein in primary cultures of adult human hepatocytes, *Biochem. Biophys. Res. Commun.*, 155, 112, 1988.
55. **Ghezzi, P. and Sipe, J. D.,** Dexaméthasone modulation of LPS, IL-1, and TNF stimulated serum amyloid A synthesis in mice, *Lymphokine Res.*, 7, 157, 1988.
56. **Ganapathi, M. K., Schultz, D., Mackiewicz, A., Samols, D., Hu, S.-I., Brabenec, A., Macintyre, S. S., and Kushner, I.,** Heterogeneous nature of the acute phase response. Differential regulation of human serum amyloid A, C-reactive protein, and other acute phase proteins by cytokines in Hep 3B cells, *J. Immunol.*, 141, 564, 1988.
57. **Sipe, J. D., Vogel, S. N., Douches, S., and Neta, R.,** Tumor necrosis factor/cachectin is a less potent inducer of serum amyloid A synthesis than interleukin 1, *Lymphokine Res.*, 6, 93, 1987.
58. **Woo, P., Sipe, J., Dinarello, C. A., and Colten, H. R.,** Stucture of a human serum amyloid A gene and modulation of its expression in transfected L cells, *J. Biol. Chem.*, 262, 15790, 1987.
59. **Rienhoff, H. Y., Jr.,** Identification of a transcriptional enhancer in a mouse amyloid gene, *J. Biol. Chem.*, 264, 419, 1989.
60. **Li, X., Huang, J. H., Rienhoff, H. Y., Jr., and Liao, W. S.-L.,** Two adjacent C/EBP-binding sequences that participate in the cell-specific expression of the mouse serum amyloid A3 gene, *Mol. Cell. Biol.*, 10, 6624, 1990.
61. **Edbrooke, M. R., Burt, D. W., Cheshire, J. K., and Woo, P.,** Identification of *cis*-acting sequences responsible for phorbol ester induction of human serum amyloid A gene expression via a nuclear factor κB-like transcription factor, *Mol. Cell. Biol.*, 9, 1908, 1989.
62. **Li, X. and Liao, W. S.-L.,** Expression of rat serum amyloid A1 gene involves both C/EBP-like and NFκB-like transcription factors, *J. Biol. Chem.*, 266, 15192, 1991.
63. **Rienhoff, H. Y., Jr., Huang, J. H., Li, X., and Liao, W. S.-L.,** Molecular and cellular biology of serum amyloid A, *Mol. Biol. Med.*, 7, 287, 1990.
64. **Meek, R. I., Eriksen, N., and Benditt, E. P.,** Serum amyloid A in the mouse. Sites of uptake and mRNA expression, *Am. J. Pathol.*, 135, 411, 1989.
65. **Benditt, E. P. and Meek, R. L.,** Expression of the third member of the serum amyloid A gene family in mouse adipocytes, *J. Exp. Med.*, 169, 1841, 1989.
66. **Meek, R. L. and Benditt, E. P.,** Rat tissues express serum amyloid A protein-related mRNAs, *Proc. Natl. Acad. Sci. U.S.A.*, 86, 1890, 1989.
67. **Eriksen, N., Meek, R. L., and Benditt, E. P.,** Serum amyloid A (SAA) induction in the serum high density lipoproteins of the Syrian hamster, in *Amyloid and Amyloidosis 1990,* Natvig, J., Førre, O., Husby, G., Husebekk, A., Skogen, B., Sletten, K., and Westermark, P., Eds., Kluwer, Dordrecht, 1991, 99.

68. **Meek, R. L. and Benditt, E. P.,** Amyloid A gene family expression in different mouse tissues, *J. Exp. Med.,* 164, 2006, 1986.
69. **Meek, R. L. and Benditt, E. P.,** Mouse macrophages express and secrete serum amyloid SAA3, *J. Cell Biol.,* 115, (Abstr.), 452a, 1991.
70. **Liao, W. S.-L., Li, X., and Caldwell, C. K.,** Expression and structure analysis of rat serum amyloid A gene, in *Regulation of Liver Gene Expression,* Cold Spring Harbor Laboratory, Cold Spring Harbor, NY, 1987, 182 (Abstr.).
71. **Ericsson, L. H., Eriksen, N., Walsh, K. A., and Benditt, E. P.,** Primary structure of duck amyloid protein A. The form deposited in tissues may be identical to its serum precursor, *FEBS Lett.,* 218, 11, 1987.
72. **Ciliberto, G.,** Transcriptional regulation of acute phase response genes with emphasis on the human C-reactive protein gene, in *Acute Phase Proteins in the Acute Phase Response,* Pepys, M. B., Ed., Springer-Verlag, London, 1989, 29.
73. **Rokita, H., Shirahama, T., Cohen, A. S., Meek, R. L., Benditt, E. P., and Sipe, J. D.,** Differential expression of the amyloid SAA 3 gene in liver and peritoneal macrophages of mice undergoing dissimilar inflammatory episodes, *J. Immunol.,* 139, 3849, 1987.
74. **Benditt, E. P., Meek, R. L., and Eriksen, N.,** ApoSAA: structure, tissue expression and possible functions, in *Acute Phase Proteins in the Acute Phase Response,* Pepys, M. B., Ed., Springer-Verlag, London, 1989, 59.
75. **Niewold, T. A. and Tooten, P. C. J.,** Purification and characterization of hamster serum amyloid-A protein (SAA) by cholesteryl hemisuccinate affinity chromatography, *Scand. J. Immunol.,* 31, 389, 1990.
76. **Benson, M. D. and Aldo-Benson, M.,** Effect of purified protein SAA on immune response *in vitro:* mechanisms of suppression, *J. Immunol.,* 122, 2077, 1979.
77. **Aldo-Benson, M. A. and Benson, M. D.,** SAA suppression of immune response *in vitro:* evidence for an effect on T cell-macrophage interaction, *J. Immunol.,* 128, 2390, 1982.
78. **Kimura, K.,** Changes in natural killer activities in experimental secondary amyloidosis, *Shikoku Acta Med.,* 40, 377, 1984.
79. **Peristeris, P., Gaspar, A., Gros, P., Laurent, P., Bernon, H., and Bienvenu, J.,** Effects of serum amyloid A protein on lymphocytes, HeLa, and MRC_5 cells in culture, *Biochem. Cell Biol.,* 67, 365, 1989.
80. **Steinmetz, A., Hocke, G., Saïle, R., Puchois, P., and Fruchart, J.-C.,** Influence of serum amyloid A on cholesterol esterification in human plasma, *Biochim. Biophys. Acta,* 1006, 173, 1989.
81. **Zimlichman, S., Danon, A., Nathan, I., Mozes, G., and Shainkin-Kestenbaum, R.,** Serum amyloid A, an acute phase protein, inhibits platelet activation, *J. Lab. Clin. Med.,* 116, 180, 1990.
82. **Levartowsky, D. and Pras, M.,** The effect of SAA-derived fragment — SAA_{2-82} — on platelet aggregation, in *Amyloid and Amyloidosis 1990,* Natvig, J. B., Førre, O., Husby, G., Husebekk, A., Skogen, B., Sletten, K., and Westermark, P., Eds., Kluwer, Dordrecht, 1991, 129.
83. **Zimlichman, S., Danon, A., Nathan, I., Mozes, G., and Shainkin-Kestenbaum, R.,** Serum amyloid A, an acute phase protein, inhibits platelet activation, in *Amyloid and Amyloidosis 1990,* Natvig, J. B., Førre, O., Husby, G., Husebekk, A., Skogen, B., Sletten, K., and Westermark, P., Eds., Kluwer, Dordrecht, 1991, 133.
84. **Shainkin-Kestenbaum, R., Levartowsky, D., Zimlichman, S., Fridkin, M., and Pras, M.,** Antiplatelet aggregation activity of serum amyloid A (SAA) related peptides, in *Amyloid and Amyloidosis 1990,* Natvig, J. B., Førre, O., Husby, G., Husebekk, A., Skogen, B., Sletten, K., and Westermark, P., Eds., Kluwer, Dordrecht, 1991, 139.
85. **Levartowsky, D., Pras, M., Shephard, E., Rosen, O., and Fridkin, M.,** Serum amyloid A (SAA)-related peptide isolated from synovial fluid modulates superoxide production by human neutrophils, in *Amyloid and Amyloidosis 1990,* Natvig, J. B., Førre, O., Husby, G., Husebekk, A., Skogen, B., Sletten, K., and Westermark, P., Eds., Kluwer, Dordrecht, 1991, 135.
86. **Nathan, I., Goldfarb, D., Dvilansky, A., Zolotov, Z., and Shainkin-Kestenbaum, R.,** Effect of purified serum amyloid A on growth and differentiation of transformed cells, in *Amyloid and Amyloidosis 1990,* Natvig, J. B., Førre, O., Husby, G., Husebekk, A., Skogen, B., Sletten, K., and Westermark, P., Eds., Kluwer, Dordrecht, 1991, 143.
87. **Shainkin-Kestenbaum, R., Berlyne, G., Zimlichman, S., Sorin, H. R., Nyska, M., and Danon, A.,** Acute phase protein, serum amyloid A, inhibits IL-1- and TNF-induced fever and hypothalamic PGE_2 in mice, *Scand. J. Immunol.,* 34, 179, 1991.
88. **Cabana, V. G., Siegel, J. N., and Sabesin, S. M.,** Effects of the acute phase response on the concentration and density distribution of plasma lipids and apolipoproteins, *J. Lipid Res.,* 30, 39, 1989.
89. **Kisilevsky, R.,** Serum amyloid A (SAA), a protein without a function: some suggestions with reference to cholesterol metabolism, *Med. Hypotheses,* 35, 337, 1991.

90. **Brissette, L., Young, I., Narindrasorasak, S., Kisilevsky, R., and Deeley, R.,** Differential induction of the serum amyloid A gene family in response to an inflammatory agent and to amyloid-enhancing factor, *J. Biol. Chem.,* 264, 19327, 1989.
91. **Smith, L. L.,** Acute inflammation: the underlying mechanism in delayed onset muscle soreness?, *Med. Sci. Sports Exerc.,* 23, 542, 1991.
92. **Maury, C. P. J.,** Comparative study of serum amyloid A protein and C-reactive protein in diseae, *Clin. Sci.,* 68, 233, 1985.
93. **Hazenberg, B. P. C., Limburg, P. C., Bijzet, J., and van Rijswijk, M. H.,** SAA versus CRP serum levels in different inflammatory conditions, studied by ELISA using polyclonal anti-AA and monoclonal anti-SAA antibodies, in *Amyloid and Amyloidosis,* Isobe, T., Araki, S., Uchino, F., Kito, S., and Tsubura, E., Eds., Plenum Press, New York, 1988, 229.
94. **Maury, C. P. J., Tötterman, K.J., Gref, C.-G., and Ehnholm, C.,** Serum amyloid A protein, apolipoprotein A-I, and apolipoprotein B during the course of acute myocardial infarction, *J. Clin. Pathol.,* 41, 1263, 1988.
95. **Bausserman, L. L., Sadaniantz, A., Saritelli, A. L., Martin, V. L., Nugent, A. M., Sady, S. P., and Herbert, P. N.,** Time course of serum amyloid A response in myocardial infarction, *Clin. Chim. Acta,* 184, 297, 1989.
96. **Mozes, G., Friedman, N., and Shainkin-Kestenbaum, R.,** Serum amyloid A: an extremely sensitive marker for intensity of tissue damage in trauma patients and indicator of acute response in various diseases, *J. Trauma,* 29, 71, 1989.
97. **Sipe, J. D., de Beer, F. C., Pepys, M., Husebekk, A., Skogen, B., Kisilevsky, R., Selkoe, D., Buxbaum, J., Linke, R. P., and Gertz, M. A.,** Report of special session on bioassays and standardization of amyloid proteins and precursors, in *Amyloid and Amyloidosis 1990,* Natvig, J. B., Førre, O., Husby, G., Husebekk, A., Skogen, B., Sletten, K., and Westermark, P., Eds., Kluwer, Dordrecht, 1991, 883.
98. **De Beer, F. C., de Beer, M. C., and Sipe, J. D.,** Identification of apo-SAA isoforms in man and mouse, in *Amyloid and Amyloidosis 1990,* Natvig, J. B., Førre, O., Husby, G., Husebekk, A., Skogen, B., Sletten, K., and Westermark, P., Eds., Kluwer, Dordrecht, 1991, 890.
99. **Sipe, J. D., Gonnerman, W. A., Knapschaefer, G., and Xie, W.-J.,** Normal lipoproteins inhibit binding of SAA-rich lipoproteins to polyvinylchloride surfaces, in *Amyloid and Amyloidosis 1990,* Natvig, J. B., Førre, O., Husby, G., Husebekk, A., Skogen, B., Sletten, K., and Westermark, P., Eds., Kluwer, Dordrecht, 1991, 894.
100. **Hazenberg, B. P. C., Limburg, P. C., Bijzet, J., and van Rijswijk, M. H.,** Monoclonal antibody based ELISA for human SAA, in *Amyloid and Amyloidosis 1990,* Natvig, J. B., Førre, O., Husby, G., Husebekk, A., Skogen, B., Sletten, K., and Westermark, P., Eds., Kluwer, Dordrecht, 1991, 898.

Chapter 6

THE α_1-ACID GLYCOPROTEIN: STRUCTURE AND POSSIBLE FUNCTIONS IN THE ACUTE PHASE RESPONSE

Chin B. Eap and Pierre Baumann

TABLE OF CONTENTS

0-8493-6913-4/93/$0.00 + $.50

I. INTRODUCTION

α_1-Acid glycoprotein (α_1-AGP) was isolated in pure form 40 years ago,[1] and extensive research on its biochemistry performed in the 1960s and 1970s led to the determination of its complete amino acid sequence as well as its carbohydrate composition (for a review, see Reference 2). Although several studies have investigated its possible physiological role, no clear *in vivo* function has yet been demonstrated. This chapter reviews the biochemistry, molecular biology, physiology, and possible functions of α_1-acid glycoprotein.

II. BIOCHEMISTRY OF α_1-ACID GLYCOPROTEIN

α_1-AGP, with an isoelectric point of 2.7 and an isoionic point of 3.5, is one of the most acidic proteins in the blood.[2] It has a molecular weight of 40 kDa, with a high (45%) proportion of carbohydrates, which form five oligosaccharide chains. The five chains show some heterogeneity, with bi-, tri-, or tetraantennary structures, allowing the differentiation of three types of α_1-AGP:[3]

1. α_1-AGP not bound to concanavalin A (α_1-AGP-A), which represents about 46% of the total α_1-AGP in nonpathological conditions
2. α_1-AGP weakly bound to concanavalin A (α_1-AGP-B), which represents about 39% of the total α_1-AGP
3. α_1-AGP strongly bound to concanavalin A (α_1-AGP-C), which represents about 15% of the total α_1-AGP.

The proportion of these three types of α_1-AGP may vary considerably in acute phase conditions.[4,5]

α_1-AGP consits of a single polypeptide chain containing 181 amino acids, of which 21 can be substituted.[2] It is synthesized mainly in the liver,[6] but leukocytes (lymphocytes, granulocytes, and monocytes) also synthesize a membranal proform with a molecular weight of 52 kDa which, after cleavage at residue 181, yields the 40-kDa circulating form.[7,8] It has been suggested that leukocytes may contribute significantly to the synthesis of α_1-AGP during the acute phase response.[7] Some synthesis of α_1-AGP by human-breast epithelial cells has also been shown.[9] Degradation of α_1-AGP is mediated by the liver. The plasma half-life of native α_1-AGP is around 3 d[10] compared with the plasma half-life of asialo-α_1-AGP (a form without terminal sialic acid on the polysaccharide chains) of only a few minutes.[11] This shorter plasma half-life is due to the presence of receptors for asialoglycoproteins on hepatocytes; these receptors bind and internalize glycoproteins from which the terminal sialic acid has been removed (for a review, see Reference 12).

Two genetic polymorphisms of α_1-AGP have been described for α_1-AGP (for a review, see Reference 13). The first is explained by differential linkage of the terminal sialic acid residue (on the carbohydrate chains) to the carbon atom of the galactose residue (C2, C3, C4, or C6), which leads to differences in the pK values of the carboxylic groups of the sialic acid residues.[2] By electrophoresis of the native protein at pH 2.9 (i.e., near its isoelectric point), patterns of five to eight bands can be observed. The genetic transmission of these patterns has been shown in family, including twin, studies.[2] The second genetic polymorphism of α_1-AGP is revealed when its desialylated form is analyzed by electrophoresis at pH 5.1, i.e., near its pI. The phenotype is determined by the relative intensity of two main bands called the F- and S-bands (fast- and slow-migrating bands). During the last 5 years, improved analytical methods such as isoelectric focusing with ampholine or immobiline gels have shown that the various forms of desialylated α_1-AGP are encoded by two loci (ORM1 and ORM2), of which several variants have been described in various populations.[13]

III. MOLECULAR BIOLOGY OF α_1-ACID GLYCOPROTEIN

The α_1-AGP locus has been mapped on chromosome 9[14] and is associated with adenylate kinase-1, ABO, and red cell δ-amino-levulinate dehydrase in the distal portion (band q34) of the long arm of chromosome 9.[15] α_1-AGP is encoded by three genes (α_1-AGP-A, α_1-AGP-B, and α_1-AGP-B′) in a cluster[16,17] (for a review, see Reference 18), and each gene has a similar internal organization (six exons and five introns). α_1-AGP-B and α_1-AGP-B′ genes are identical, but the proteins deduced from the nucleotide sequence of α_1-AGP-A differ from those of α_1-AGP-B/B′ by 22 amino acid substitutions.[17] The existence of two different genes, revealed by molecular biology techniques, accords with the existence of two loci postulated by population polymorphism studies.[13] A study using transgenic mice has shown that the ORM1 F1 variant (one of the two main variants of the first locus) is encoded by the α_1-AGP-A gene, and ORM2 A (the main variant of the second locus) by the α_1-AGP-B/B′ genes.[19] In the rat, an increase in the α_1-AGP plasma concentrations is correlated with an increase in hepatic mRNA,[20] a finding that confirms earlier observations on other acute phase proteins.[18] It is of interest that induction by inflammatory stimuli of human hepatoma cells or of liver cells of transgenic mice (carrying human α_1-AGP genes) results in a severalfold increase of mRNA derived mainly from the α_1-AGP-A gene.[21] Moreover, studies analyzing the relative proportions of the variants derived from the α_1-AGP-A and α_1-AGP-B/B′ genes in patients with burn injury,[22] and before and after various orthopedic operations,[23] showed increased concentrations of proteins derived from both genes. Altogether, these findings[21-23] strongly suggest that both transcriptional and post-transcriptional factors are involved in the regulation of α_1-AGP expression. Post-transcriptional events include stabilization of the primary transcript[24] by deadenylation of α_1-AGP mRNA,[25,26] transport of the mature mRNA out of the nucleus,[27] and post-translational processing during externalization of α_1-AGP after its synthesis in the rough endoplasmic reticulum.[28,29]

IV. PLASMA LEVELS OF α_1-ACID GLYCOPROTEIN

Plasma levels of α_1-AGP are highly variable even among healthy subjects: 0.36 to 1.46 g/l, with a mean value of about 0.77 g/l.[30] Moreover, diurnal variations around individual means can be as high as 49%, as measured in a study with 13 healthy volunteers.[31] Family studies show that environmental factors contribute more than do genetic factors to the variance of α_1-AGP plasma levels.[30]

α_1-AGP concentrations are very low in the fetus, and the protein is hardly detectable before 16 weeks. It then increases constantly during gestation, but the ratio of the fetal α_1-AGP concentration to the maternal α_1-AGP concentration is still around 0.37 at birth.[32] After a rapid increase during the first week of life, it increases slowly, reaching the adult concentration at about 1 year.[33]

α_1-AGP plasma levels are decreased in liver cirrhosis[34] and after intake of oral contraceptives containing estrogens.[35] The intake of oral contraceptive drugs by a part of the female population may explain why in one study slightly higher concentrations of α_1-AGP were found in men than in women.[35]

Plasma α_1-AGP concentrations are elevated in obesity,[36] old age,[37,38] malnutrition,[39] and chronic renal failure.[40] α_1-AGP is an acute phase protein reactant, and its concentration may rise to three to four times its mean value in response to a wide variety of inflammatory stimuli such as infection or inflammation, tissue injury, burn injury, bone fracture, trauma, surgical operation, malignancy,or tissue infarction.[41] Peak levels are reached 3 to 6 days after stimulation.[42] This increased synthesis is the response of hepatocytes to cytokines such

TABLE 1
Physiological Conditions or Disease States in which Varying α_1-AGP Levels Have Been Measured

Acidosis, age, alcohol use, allergy (ventricular), arrhythmia, and arthritis
Bacterial infection in neonatal period, burn
Cancer (breast, colorectal, lung, and ovaries), chest pain, chronic inactive pyelonephritis, chronic hemodialysis patients, chronic pain, chronic renal failure, chronic ulcerative colitis, and Crohn's disease
Depression, drug treatment
Epilepsy
Genetic factor, gliomas
Hepatitis, hormonal contraceptive use, hyperlipoproteinemia, hyperlipidemia, and hypertension
Inflammation
Liver cirrhosis, liver carcinoma
Multiple sclerosis, myocardial infarction
Nephrotic disease
Obesity
Pregnancy
Renal disease
Sex, smoking, stress, and surgery
Trauma
Uremic diseae
Wound healing

Adpated from Kremer, J. M. H., Wilting, J., and Janssen, L. M. H., *Pharmacol. Rev.*, 40, 1, 1988. With permission.

as interleukin-1 and interleukin-6, tumor necrosis factor-α, hepatocyte-stimulating factor III, and to corticosteroids released by monocytes and other cells during the early phase of the acute phase reaction.[43] Table 1 lists some physiological conditions or disease states in which varying α_1-AGP concentrations have been measured.

V. PHYSIOLOGICAL FUNCTIONS OF α_1-ACID GLYCOPROTEIN

It must first be emphasized that almost all studies on the function of α_1-AGP have been performed *in vitro* and, as mentioned in Section I, it is the present author's view that the physiological role of this protein remains to be demonstrated. The most interesting results concern the immunomodulating effects of α_1-AGP.

Thus, α_1-AGP inhibits markedly the proliferative response of human peripheral blood lymphocytes to phytohemagglutinin, and the blastogenesis induced by concanavalin A and porkeweed mitogen.[45] It also suppresses the lymphocyte mitogenic responses to concanavalin A, lipopolysaccharides, and alloantigens, the antibody responses to sheep erythrocytes, and the induction of cell-mediated lympholysis against allogenic target cells.[46] Interaction of α_1-AGP with the lymphoid cell surface has been demonstrated,[47] and it has been hypothesized that α_1-AGP acts as a nonspecific immunosuppressive agent which, during the acute phase response, may contribute to limiting, modulating, or directing the host immune response.[46] This nonspecific blocking activity of α_1-AGP and other acute phase proteins may be of particular relevance to cancer in that, as has been suggested, it may contribute to the immune escape of the tumor;[48,49] however, α_1-AGP has also been found to inhibit the growth of tumor cells *in vitro* at a concentration normally observed in inflammation.[50] Also the carbohydrate moiety of α_1-AGP is an important determinant of its biological activity, as the concanavalin A-bound fraction (α_1-AGP-C; see Section II) is the strongest inhibitor of lymphocyte proliferation.[51] As the relative proportions of the different types of glycanic variants of α_1-AGP (α_1-AGP-A, -B, and -C) vary greatly in acute phase conditions,[4,5] it has been suggested that the cellular immunity of the host may be partially controlled by changes in the proportions of these variants. It is of interest that another variant, the concanavalin A unbound fraction (α_1-AGP-A), inhibits the thymocyte proliferative activity of interleukin-1, while α_1-AGP-C has little effect.[52] In one of the few experiments performed *in vivo,* injections of monosodium urate crystals, with and without α_1-AGP, into the footpads of rats showed, by measuring the diameter of the inflamed foot, an antiinflammatory action of α_1-AGP.[53] However, this effect was found only at high doses (10 g/l) and for a short time.

Several other inhibitory activities of α_1-AGP have been described. In an *in vitro* study and at concentrations reached in malaria, α_1-AGP was found to inhibit by 80% the multiplication of the malaria parasite *Plasmodium falciparum.* This inhibitory activity, caused by nonspecific competition for the cell surfaces of parasites, may depend on the protein sialic acid moiety.[54] However, the results of this study have not been reproduced.[55] The filamentous form of α_1-AGP polymers, produced by heating the sodium salt of the α_1-AGP monomer in water, inhibits strongly the hemagglutination of some sensitive strains of influenza virus;[56] and α_1-AGP, at a concentration of 1 g/l, significantly depressed the phagocytosis of *Escherichia coli* and *Staphylococcus aureus* by human neutrophils *in vitro.*[57] It also inhibits the activity of natural killer cells on "fast target" K562 cells, probably by interacting with a cytotoxic factor secreted from natural killer cells after effector-target interaction.[58] Some experiments suggest that both the soluble and the membranal form of α_1-AGP present on the surface of lymphocytes may have a physiological function, as they may be involved in the T3-Ti antigen-specific pathway of T-cell activation.[59]

Several stimulating effects of α_1-AGP have been demonstrated. It influences the formation of fibrous, long-spacing fibers of collagen,[60] and its administration to rats with chronic liver injury accelerates collagen synthesis and hepatic fibrosis.[61] It has a growth-stimulating or growth-inhibiting effect on cells in culture[62] (HeLa epithelial, HEL fibroblast, and lymphoblastoid cells), depending on the concentrations used: stimulating effects are observed between 0.001 and 0.03 g/l (i.e., at far lower than normal physiological concentrations) and inhibiting effects at higher concentractions. *In vitro* stimulation of mitogenesis of human peripheral blood lymphocytes has been observed at a very low concentration[63] (0.05 g/l). At a concentration of 0.1 g/l or more, α_1-AGP promotes the passage of erythrocytes through micropores and reduces the degree of hemolysis, and it has been hypothesized that α_1-AGP facilitates the microvascular passage of plasma cells.[64]

α_1-AGP increases the number (as shown by measurements of the binding of (−)-dihydroalprenolol) and the function (by measurements of the L-isoproterenol-induced cyclic AMP elevation) of β-adrenoceptors on human mononuclear leukocytes *in vitro.*[65] It also promotes nerve growth, but is approximately 1500 times less potent than nerve-growth factor, with maximum activity obtained at a lower-than-physiological concentration[66] (0.1 g/l). It has been hypothesized that α_1-AGP produced during the acute phase response accumulates in the diseased tissue because of increased vascular permeability and is involved in various aspects of wound healing,[67] including regulation of cell growth and nerve regeneration.[66]

One study has shown an effect of α_1-AGP on lipid metabolism: it seems to be a cofactor in the lipoprotein lipase reaction, therefore increasing lipolysis by a factor of two in the presence of C-II apolipoprotein in a lipoprotein lipase assay system.[68] Severe hyperlipidemia is a common disorder in patients or experimental animals with nephrotic syndrome. The injection of α_1-AGP into rats with induced nephrotic syndrome showing a decrease in triglyceride clearance restores the lipid clearance to normal. Therefore, it has been proposed that the hyperlipidemia observed in the nephrotic syndrome may be caused by a relative deficiency, by urinary loss, of α_1-AGP in plasma.[68]

Several studies on the interaction of α_1-AGP with the blood coagulation system have shown α_1-AGP to be a potent inhibitor of platelet activation,[69-72] but it also promotes some coagulation activity due to its antiheparin effect.[72-74]

A good overview of the possible physiological functions of α_1-AGP should refer to studies that try to find an homology between α_1-AGP and other proteins. The amino-terminal 43-residue segment of the CNBr I fragment shows some homology with the amino terminal of the variable region of the K-type L-chain of human IgG,[75] and the amino-terminal 22-residue segment of the CNBr II fragment with the α-chain of haptoglobin;[76] also, α_1-AGP seems to present an homology with the epidermal growth-factor receptor.[77] These results have been contested by some very convincing studies which show, instead, an homology of α_1-AGP with serum retinol-binding protein and other proteins such as haptoglobin or bilin-binding protein.[78-80] It has been proposed that these proteins be called lipocalins, as they share a common feature — the binding and transport of lipophilic molecules.[79] This is consistent with the very-well-known ability of α_1-AGP to bind and transport a wide variety of drugs, which is of particular relevance to clinical pharmacology (for a review, see Reference 44). It is of interest that α_1-AGP also binds endogenous compounds such as steroids[81] and autocoids (lipophilic endogenous amines with or without biological activities), in particular a factor called platelet-activating factor (PAF), which is a potent bioactive phospholipid released from platelets, neutrophils, basophils, and macrophages, and which can provoke platelet and neutrophil activation, hypotension, and bronchoconstriction.[82] Thus, the hypothesis may be formulated that α_1-AGP has no physiological role in itself, but, rather, plays an indirect role by binding and carrying other molecules. This may explain the wide heterogeneity of the effects attributed to this protein, such as described here, and why some results seem to be contradictory.

VI. CONCLUSIONS

α_1-AGP has been extensively studied since it was isolated 40 years ago. Its structure, biochemistry, and molecular biology are thus fairly well known. However, its function, particularly during the acute phase response, has yet to be demonstrated despite the numerous reports published on the subject. As this protein presents an homology with the family of proteins called lipocalins, the function of which is to bind and transport various lipophilic molecules, it may be hypothesized that this could also be the primary function of α_1-AGP. Further studies are needed to test this hypothesis.

ACKNOWLEDGMENTS

We gratefully acknowledge the editorial assistance of Mrs. C. Bertschi and the bibliographic work of Mrs. J. Bourquin, Mrs. M. Gobin, and Mrs. T. Bocquet. Our studies cited in this chapter were partly supported by the Swiss National Research Foundation (project no. 3.962-0.85).

REFERENCES

1. **Weimer, H. E., Mehl, J. W., and Winzler, R. J.,** Studies on the mucoproteins of human plasma. V. Isolation and characterization of a homogeneous mucoprotein, *J. Biol. Chem.*, 185, 561, 1950.
2. **Schmid, K.,** α1-Acid glycoprotein, in *The Plasma Proteins: Structure, Function, and Genetic Control*, Putman, F. W., Ed., Academic Press, New York, 1975, 183.

3. **Bierhuizen, M. F. A., De Wit, M., Govers, C. A. R. L., Ferwerda, W., Koeleman, C., Pos, O., and Van Dijk, W.,** Glycosylation of three molecular forms of human α1-acid glycoprotein having different interactions with concanavalin A. Variations in the occurrence of di-, tri-, and tetraantennary glycans and the degree of sialylation, *Eur. J. Biochem.*, 175, 387, 1988.
4. **Hansen, J. E. S., Jensen, S. P., Nørgaard-Pedersen, B., and Bøg-Hansen, T. C.,** Electrophoretic analysis of the glycan microheterogeneity of orosomucoid in cancer and inflammation, *Electrophoresis*, 7, 180, 1986.
5. **Jezequel, M., Seta, N. S., Corbic, M. M., Feger, J. M., and Durand, G. M.,** Modifications of concanavalin A patterns of α1-acid glycoprotein and α2-HS glycoprotein in alcoholic liver disease, *Clin. Chim. Acta*, 176, 49, 1988.
6. **Sarcione, E. J.,** Synthesis of α1-acid glycoprotein by the isolated perfused rat liver, *Arch. Biochem. Biophys.*, 100, 516, 1963.
7. **Gahmberg, C. G. and Andersson, L. C.,** Leukocyte surface origin of human α1-acid glycoprotein (orosomucoid), *J. Exp. Med.*, 148, 507, 1978.
8. **Eap, C. B., Baumann, P., and Moretta, A.,** Synthesis of alphal-acid glycoprotein by human T lymphocytes, *Experientia*, 45, A52, 1989.
9. **Gendler, S. J., Dermer, G. B., Silverman, L. M., and Tökés, Z. A.,** Synthesis of α1-antichymotrypsin and α1-acid glycoprotein by human breast epithelial cells, *Cancer Res.*, 42, 4567, 1982.
10. **Bré, F., Houin, G., Barré, J., Moretti, J. L., Wirquin, V., and Tillement, J. P.,** Pharmacokinetics of intravenously administered ^{125}I-labelled human α1-acid glycoprotein, *Clin. Pharmacokinet.*, 11, 336, 1986.
11. **Morell, A. G., Gregoriadis, G., Scheinberg, I. H., Hickman, J., and Ashwell, G.,** The role of sialic acid in determining the survival of glycoproteins in the circulation, *J. Biol. Chem.*, 246, 1461, 1971.
12. **Weiss, P. and Ashwell, G.,** The asialoglycoprotein receptor: properties and modulation by ligand, in *Alphal-Acid Glycoprotein: Genetics, Biochemistry, Physiological Functions, and Pharmacology*, Baumann, P., Eap, C. B., Müller, W. E., and Tillement, J. P., Eds., Alan R. Liss, New York, 1988, 169.
13. **Eap, C. B. and Baumann, P.,** The genetic polymorphism of human alpha 1-acid glycoprotein, in *Alphal-Acid Glycoprotein: Genetics, Biochemistry, Physiological Functions, and Pharmacology*, Baumann, P., Eap, C. B., Müller, W. E., and Tillement, J. P., Eds., Alan R. Liss, New York, 1989, 111.
14. **Cox, D. W. and Francke, U.,** Direct assignment of orosomucoid to human chromosome 9 and α2HS-glycoproteinn to chromosome 3 using human fetal liver × rat hepatoma hybrids, *Hum. Genet.*, 70, 109, 1985.
15. **Eiberg, H., Mohr, J., and Nielsen, L. S.,** Linkage of orosomucoid (ORM) to ABO and AK1, *Cytogenet. Cell Genet.*, 32, 272, 1982.
16. **Dente, L., Ciliberto, G., and Cortese, R.,** Structure of the human α1-acid glycoprotein gene: sequence homology with other human acute phase protein genes, *Nucleic Acids Res.*, 13, 3941, 1985.
17. **Dente, L., Pizza, M. G., Metspalu, A., and Cortese, R.,** Structure and expression of the genes coding for human α1-acid glycoprotein, *EMBO J.*, 6, 2289, 1987.
18. **Dente, L.,** Human α-1-acid glycoprotein genes, in *Alphal-Acid Glycoprotein: Genetics, Biochemistry, Physiological Functions, and Pharmacology*, Baumann, P., Eap, C. B., Müller, W. E., and Tillement, J. P., Eds., Alan R. Liss, New York, 1988, 85.
19. **Tomei, L., Eap, C. B., Baumann, P., and Dente, L.,** Use of transgenic mice for the characterization of human α1-acid glycoprotein (orosomucoid) variants, *Hum. Genet.*, 84, 89, 1989.
20. **Ricca, G. A., Hamilton, R. W., McLean, J. W., Conn, A., Kalinyak, J. E., and Taylor, J. M.,** Rat α1-acid glycoprotein mRNA. Cloning of double-stranded cDNA and kinetics of induction of mRNA levels following acute inflammation, *J. Biol. Chem.*, 256, 10362, 1981.
21. **Dente, L., Rüther, U., Tripodi, M., Wagner, E. F., and Cortese, R.,** Expression of human α1-acid glycoprotein genes in cultured cells and in transgenic mice, *Genes Dev.*, 2, 259, 1988.
22. **Van Dijk, W., Pos, O., Van Der Stelt, M. E., Moshage, H. J., Yap, S. H., Dente, L., Baumann, P., and Eap, C. B.,** Inflammation-induced changes in expression and glycosylation of genetic variants of α1-acid glycoprotein — studies with human sera, primary cultures of human hepatocytes and transgenic mice, *Biochem. J.*, 276, 343, 1991.
23. **Eap, C. B., Fischer, J. F., and Baumann, P.,** Variations in relative concentrations of variants of human α1-acid glycoprotein after acute phase conditions, *Clin. Chim. Acta*, 203, 379, 1991.
24. **Vannice, J. L., Taylor, J. M., and Ringold, G. M.,** Glucocorticoid-mediated induction of α1-acid glycoprotein: evidence for hormone-regulated RNA processing, *Proc. Natl. Acad. Sci. U.S.A.*, 81, 4241, 1984.
25. **Shiels, B. R., Northemann, W., Gehring, M. R., and Fey, G. H.,** Modified nuclear processing of α1-acid glycoprotein RNA during inflammation, *J. Biol. Chem.*, 262, 12826, 1987.
26. **Carter, K. C., Bryan, S., Gadson, P., and Papaconstantinou, J.,** Deadenylation of α1-acid glycoprotein mRNA in cultured hepatic cells during stimulation by dexamethasone, *J. Biol. Chem.*, 264, 4112, 1989.

27. **Clawson, G. A., Button, J., Woo, C. H., Liao, Y. C., and Smuckler, E. A.**, In vitro release of α1-acid glycoprotein RNA sequences shows fidelity with the acute phase response in vivo, *Mol. Biol. Rep.*, 11, 163, 1986.
28. **Haffar, O. K., Edwards, C. P., and Firestone, G. L.**, Regulation of α1-acid glycoprotein externalization and intracellular accumulation in glucocorticoid-induced rat hepatoma cells, *Arch. Biochem. Biophys.*, 246, 449, 1986.
29. **Drechou, A., Perez-Gonzalez, N., Agneray, J., Féger, J., and Durand, G.**, Increased affinity to concanavalin A and enhanced secretion of α1-acid glycoprotein by hepatocytes isolated from turpentine-treated rats, *Eur. J. Cell Biol.*, 50, 111, 1989.
30. **Blain, P. G., Mucklow, J. C., Rawlins, M. D., Roberts, D. F., Routledge, P. A., and Shand, D. G.**, Determinants of plasma α1-acid glycoprotein (AAG) concentration in health, *Br. J. Clin. Pharmacol.*, 20, 500, 1985.
31. **Yost, R. L. and DeVane, C. L.**, Diurnal variation of α1-acid glycoprotein concentration in normal volunteers, *J. Pharm. Sci.*, 74, 777, 1985.
32. **Krauer, B., Dayer, P., and Anner, R.**, Changes in serum albumin and α1-acid glycoprotein concentrations during pregnancy: an analysis of fetal-maternal pairs, *Br. J. Obstet. Gynaecol.*, 91, 875, 1984.
33. **Bienvenu, J., Sann, L., Bienvenu, F., Lahet, C., Divry, P., Cotte, J., and Bethenod, M.**, Laser nephelometry of orosomucoid in serum of newborns: reference intervals and relation to bacterial infections, *Clin. Chem.*, 27, 721, 1981.
34. **Barré, J., Houin, G., Rosenbaum, J., Zini, R., Dhumeaux, D., and Tillement, J. P.**, Decreased α1-acid glycoprotein in liver cirrhosis: consequences for drug protein binding, *Br. J. Clin. Pharmacol.*, 18, 652, 1984.
35. **Routledge, P. A., Stargel, W. W., Kitchell, B. B., Barchowsky, A., and Shand, D. G.**, Sex-related differences in the plasma protein binding of lignocaine and diazepam, *Br. J. Clin. Pharmacol.*, 11, 245, 1981.
36. **Benedek, I. H., Blouin, R. A., and McNamara, P. J.**, Serum protein binding and the role of increased α1-acid glycoprotein in moderately obese male subjects, *Br. J. Clin. Pharmacol.*, 18, 941, 1984.
37. **Abernethy, D. R. and Kerzner, L.**, Age effects on α1-acid glycoprotein concentration and imipramine plasma protein binding, *J. Am. Geriatr. Soc.*, 32, 705, 1984.
38. **Verbeeck, R. K., Cardinal, J. A., and Wallace, S. M.**, Effect of age and sex on the plasma binding of acidic and basic drugs, *Eur. J. Clin. Pharmacol.*, 27, 91, 1984.
39. **Jagadeesan, V., and Krishnaswamy, K.**, Drug binding in the undernourished: a study of the binding of propranolol to α1-acid glycoprotein, *Eur. J. Clin. Pharmacol.*, 27, 657, 1985.
40. **Docci, D., Bilancioni, R., Pistocchi, E., Mosconi, G., Turci, F., Salvi, G., Baldrati, L., and Orsi, C.**, Serum α1-acid glycoprotein in chronic renal failure, *Nephron*, 39, 160, 1985.
41. **Kushner, I. and Mackiewicz, A.**, Acute phase proteins as disease markers, *Dis. Markers*, 5, 1, 1987.
42. **Werner, M.**, Serum protein changes during the acute phase reaction, *Clin. Chim. Acta*, 25, 299, 1969.
43. **Baumann, H. and Gauldie, J.**, Regulation of hepatic acute phase plasma protein genes by hepatocyte stimulating factors and other mediators of inflammation, *Mol. Biol. Med.*, 7, 147, 1990.
44. **Kremer, J. M. H., Wilting, J., and Janssen, L. M. H.**, Drug binding to human α1-acid glycoprotein in health and disease, *Pharmacol. Rev.*, 40, 1, 1988.
45. **Chiu, K. M., Mortensen, R. F., Osmand, A. P., and Gewurz, H.**, Interactions of α1-acid glycoprotein with the immune system. I. Purification and effects upon lymphocyte responsiveness, *Immunology*, 32, 997, 1977.
46. **Bennett, M. and Schmid, K.**, Immunosuppression by human plasma α1-acid glycoprotein: importance of the carbohydrate moiety, *Proc. Natl. Acad. Sci. U.S.A.*, 77, 6109, 1980.
47. **Cheresh, D. A., Haynes, D. H., and Distasio, J. A.**, Interaction of an acute phase reactant α1-acid glycoprotein (orosomucoid) with the lymphoid cell surface: a model for non-specific immune suppression, *Immunology*, 51, 541, 1984.
48. **Samak, R., Edelstein, R., and Israel, L.**, Immunosuppressive effect of acute phase reactant proteins in vitro and its relevance to cancer, *Cancer Immunol. Immunother.*, 13, 38, 1982.
49. **Tamura, K., Shibata, Y., Matsuda, Y., and Ishida, N.**, Isolation and charcterization of an immunosuppressive acidic protein from ascitic fluids of cancer patients, *Cancer Res.*, 41, 3244, 1981.
50. **Watanabe, M., Iwai, K., Shibata, S., Takahashi, K., Narui, T., and Tashiro, T.**, Purification and charcterization of mouse α1-acid glycoprotein and its possible role in the antitumor activity of some lichen polysaccharides, *Chem. Pharm. Bull.*, 34, 2532, 1986.
51. **Fujii, M., Takahashi, N., Hayashi, H., Furusho, T., Matsunaga, K., and Yoshikumi, C.**, Comparative study of α1-acid glycoprotein molecular variants in ascitic fluid of cancer and non-cancer patients, *Anticancer Res.*, 8, 303, 1988.
52. **Bories, P. N., Guenounou, M., Féger, J., Kodari, E., Agneray, J., and Durand, G.**, Human α1-acid glycoprotein-exposed macrophages release interleukin 1 inhibitory activity, *Biochem. Biophys. Res. Commun.*, 147, 710, 1987.

53. **Denko, C. W., and Wanek, K.,** Anti-inflammatory action of α1-acid glycoprotein in urate crystal inflammation, *Agents Actions,* 15, 5, 1984.
54. **Friedman, M. J.,** Control of malaria virulence by α1-acid glycoprotein (orosomucoid), an acute phase (inflammatory) reactant, *Proc. Natl. Acad. Sci. U.S.A.,* 80, 5421, 1983.
55. **Gupta, S. K., Oppenheim, J. D., Glick, J., Schulman, S., and Vanderberg, J. P.,** Lack of inhibitory effects of α1-acid glycoprotein (orosomucoid) on Plasmodium falciparum invasion of human erythrocytes, *Am. J. Trop. Med. Hyg.,* 34, 841, 1985.
56. **Barclay, G. R., Flewett, T. H., Keller, E., Halsall, H. B., and Spragg, S. P.,** Effect of polymerized orosomucoid on some strains of influenza virus, *Biochem. J.,* 111, 353, 1969.
57. **Van Oss, C. J., Gillman, C. F., Bronson, P. M., and Border, J. R.,** Phagocytosis-inhibiting properties of human serum α1-acid glycoprotein, *Immunol. Commun.,* 3, 321, 1974.
58. **Okumura, Y., Kudo, J., Ikuta, T., Kurokawa, S., Ishibashi, H., and Okubo, H.,** Influence of acute phase proteins on the activity of natural killer cells, *Inflammation,* 9, 211, 1985.
59. **Stefanini, G. F., Dirienzo, W., Arnaud, P., Nel, A., Canonica, G. W., and Fudenberg, H. H.,** Inhibitory effect of an antibody against alphal-acid glycoprotein (α1-AGP) on autologous mixed lymphocyte reaction and anti-T3 T-lymphocyte activation, *Cell. Immunol.,* 103, 65, 1986.
60. **Franzblau, C., Schmid, K., Faris, B., Beldekas, J., Garvin, P., Kagan, H. M., and Baum, B. J.,** The interction of collagen with α1-acid glycoprotein, *Biochem. Biophys. Acta,* 427, 302, 1976.
61. **Ozeka, T., Kan, M., Iwaki, K., and Ohuchi, K.,** Orosomucoid as the accelerator of hepatic fibrosis, *Br. J. Exp. Pathol.,* 67, 731, 1986.
62. **Maeda, H., Murakami, O., Kann, M., and Yamane, I.,** The growth-stimulating effect of α1-acid glycoprotein in cells in culture (40751), *Proc. Soc. Exp. Biol. Med.,* 163, 223, 1980.
63. **Singh, V. K. and Fudenberg, H. H.,** Lymphocyte stimulation in vitro by orosomucoid glycoprotein, *Immunol. Lett.,* 14, 9, 1986.
64. **Maeda, H., Morinaga, T., Mori, I., and Nishi, K.,** Further characterization of the effects of α1-acid glycoprotein on the passage of human erythrocytes through micropores, *Cell Struct. Funct.,* 9, 279, 1984.
65. **Sager, G., Sandnes, D., Aakesson, I., and Jacobsen, S.,** Effect of serum, α1-acid glycoprotein, lipoproteins and albumin on human mononuclear leucocyte β-adrenoceptors, *Acta Pharmacol. Toxicol.,* 58, 193, 1986.
66. **Liu, H. M., Takagaki, K., and Schmid, K.,** In vitro nerve-growth-promoting activity of human plasma α1-acid glycoprotein, *J. Neurosci. Res.,* 20, 64, 1988.
67. **Powanda, M. C.,** Plasma proteins and wound healing, *Surg. Gynecol. Obstet.,* 153, 749, 1981.
68. **Staprans, I. and Felts, M. J.,** The effect of α1-acid glycoprotein (orosomucoid) on triglyceride metabolism in the nephrotic syndrome, *Biochem. Biophys. Res. Commun.,* 79, 1272, 1977.
69. **Snyder, S. and Coodley, E. L.,** Inhibition of platelet aggregation by α1-acid glycoprotein, *Arch. Intern. Med.,* 136, 778, 1976.
70. **Costello, M., Fiedel, B. A., and Gewurz, H.,** Inhibition of platelet aggregation by native and desialised α1-acid glycoprotein, *Nature (London),* 281, 677, 1979.
71. **Andersen, P. and Eika, C.,** Thrombin-, epinephrine- and collagen-induced platelet aggregation inhibited by α1-acid glycoprotein, *Scand. J. Haematol.,* 24, 365, 1980.
72. **Andersen, P. and Godal, H. C.,** The antiheparin effect of α1-acid glycoprotein probably due to steric hindrance of the heparin-thrombin interaction, *Thromb. Res.,* 15, 857, 1979.
73. **Klatzow, D. J. and Vos, G. H.,** The effect of seromucoid on coagulation, *S. Afr. Med. J.,* 60, 424, 1981.
74. **Fiedel, B. A.,** Interaction of the acute phase reactants α1-acid glycoprotein, C-reactive protein and serum amyloid P-component with platelets and the coagulation system, in *Marker Proteins in Inflammation,* De Gruyter, W., Ed., Walter de Gruyter, New York, 1984, 99.
75. **Schmid, K., Kaufmann, H., Isemura, S., Bauer, F., Emura, J., Motoyama, T., Ishiguro, M., and Nanno, S.,** Structure of α1-acid glycoprotein. The complete amino acid sequence, multiple amino acid substitutions, and homology with the immunoglobulins, *Biochemistry,* 12, 2711, 1973.
76. **Ikenaka, T., Ishiguro, M., Emura, J., Kaufmann, H., Isemura, S., Bauer, W., and Schmid, K.,** Isolation and partial characterization of the cyanogen bromide fragments of α1-acid glycoprotein and the elucidation of the amino acid sequence of the carboxyl-terminal cyanogen bromide fragment, *Biochemistry,* 11, 3817, 1972.
77. **Toh, H., Hayashida, H., Kikuno, R., Yasunaga, T., and Miyata, T.,** Sequence similarity between EGF receptor and α1-acid glycoprotein, *Nature (London),* 314, 199, 1985.
78. **Marchalonis, J. J., Vasta, G. R., Warr, G. W., and Barker, W. C.,** Probing the boundaries of the extended immunoglobulin family of recognition molecules: jumping domains, convergence and minigenes, *Immunol. Today,* 5, 133, 1984.
79. **Pervaiz, S. and Brew, K.,** Homology and structure-function correlations between α1-acid glycoprotein and serum retinol-binding protein and its relatives, *FASEB. J.,* 1, 209, 1987.

80. **Godovac-Zimmermann, J.,** The structural motif of beta-lactoglobulin and retinol-binding protein: a basic framework for binding and trnsport of small hydrophobic molecules?, *TIBS,* 13, 64, 1988.
81. **Kute, T. and Westphal, U.,** Steroid-protein interactions. XXXIV. Chemical modification of α1-acid glycoprotein for characterization of the progesterone binding site, *Biochim. Biophys. Acta,* 420, 195, 1976.
82. **McNamara, P. J., Brouwer, K. R., and Gillespie, M. N.,** Autocoid binding to serum proteins. Interaction of platelet activating factor (PAF) with human serum alpha-1-acid glycoprotein (AAG), *Biochem. Pharmacol.,* 35, 621, 1986.

Chapter 7

PROTEINASE INHIBITORS: AN OVERVIEW OF THEIR STRUCTURE AND POSSIBLE FUNCTION IN THE ACUTE PHASE

Guy Salvesen and Jan J. Enghild

TABLE OF CONTENTS

0-8493-6913-4/93/$0.00 + $.50

I. INTRODUCTION

Proteinase inhibitors constitute a major category of proteins whose synthesis is elevated during the acute phase response in mammals. The simplest explanation is that the organism requires increased synthesis of these proteins to gain control of proteinases released by invading pathogens or by the host during its response to injury and infection. Proteinase inhibitors are usually selective for certain proteinases, and the goal of proteinase inhibitor research in this context is to elucidate the specificity of the interactions. Thus has developed the concept that certain proteinases are targeted for inhibition by certain inhibitors.[1,2]

The two mammals whose response to inflammation has been studied in most detail possess major acute phase response proteinase inhibitors with quite different specificities for target proteinases, α_1-antichymotrypsin in humans and α_2-macroglobulin and T-kininogen (thiostatin) in rats. From this observation, one may reason that humans and rats have quite different acute phase responses. However, some proteinase inhibitors, or proteinase/inhibitor complexes (human α_1-proteinase inhibitor, for example), exhibit cytokine-like effects on cells *in vitro*. In this context, one could think of proteinase inhibitors as nature's best way of signaling the presence of proteinases released by invading pathogens or responsive inflammatory cells. Consequently, biological effects of proteinase inhibitors that are not directly related to suppressing proteolysis must be considered as part of the acute phase response.

The major acute phase proteinase inhibitors are described in detail elsewhere in this volume, so our goal is to set the proteinase inhibitors in the context of the groups to which they belong. Our review concentrates on descriptions of the primary sequence and three-dimensional structures of mammalian proteinase inhibitors (when it is known), and we have broadened the review to incude descriptions of proteinase inhibitors that are not covered in other chapters of this volume. Readers interested in other general reviews on proteinase inhibitors are directed to references.[1-3] The following reviews give more specific information on individual proteinase inhibitor superfamilies: kunins,[4] kazals,[5] serpins,[6-8] α-macroglobulins,[9] and cystatins.[10,11] In this chapter, we discuss inhibitory specificities, mechanisms, and structural motifs that are important in delineating the biological role of proteinase inhibitors, and their possible role in generation and control of the acute phase response. To do this, we classify the inhibitors according to structural relationships, since the role of a given inhibitor is likely to be analogous to the role of related proteins.

A. DEFINITIONS

Superfamily — Proteins that are suspected to be derived from a common ancestral gene are members of the same superfamily. This definition is very arbitrary, but usually applies to proteins with identities of more than 30%. Exon shuffling during the course of evolution of a protein can blur the superfamily distinction, since domains from one ancestor have sometimes combined with domains from an unrelated gene. Nevertheless, the superfamily distinction is usually relatively obvious with mammalian proteinase inhibitors. Members of a superfamily often possess comparable activities, so that if a protein is only 25% identical to another one, but the two share a common activity, they are probably descended from a common ancestor. The genomic equivalent of a protein superfamily is the supergene family.

Family — Proteins that share at least 50% primary sequence identity are considered to be members of the same family.[12]

Orthologous — This term describes the direct line of descent in the evolution of a particular protein shared by several species. Orthologous proteins are the equivalent of the same protein in different organisms, and they usually perform the same function. Examples of orthologous proteinase inhibitors are human and bovine antithrombin III.

Paralogous — Paralogous proteins are species-specific proteins whose genes diverged from the main lineage during recent speciation events. Examples of paralogous proteinase inhibitors are the multiple α_1-antichymotrypsins of the mouse and the T-kininogens of the rat which are not found in humans. These gene loci probably multiplied during rodent speciation, although we do not know why, to give a number of proteins with apparently distinct functions.

Proteinase — Many authors use the terms protease and proteinase interchangeably, but they do have separate definitions. Proteolytic enzymes are divided into exopeptidases and endopeptidases, depending upon where in a peptide or protein chain they cut. The term proteinase is synonymous with endopeptidases, whereas the term protease refers to both exo- and endopeptidases. Recent nomenclature revisions by the International Union of Biochemistry recommend replacing protease by peptidase and proteinase by endopeptidase, but we stay with the established terms in this chapter.

Proteinase inhibitor — In principle, any compound that decreases the measured rate of hydrolysis of a given substrate could be called an inhibitor. This definition could appear to apply to a competing substrate in a reaction where one can decrease the apparent rate of hydrolysis of the other, particularly when the competitor is at a concentration close to its K_m. Many of the proteins considered in this chapter act in a substrate-like manner, but the usual property that distinguishes an inhibitor from a substrate is that the inhibitor will form extremely tight complexes with its target enzymes and that hydrolysis, when it does occur, proceeds very slowly. Substrates bind more loosely, but are hydrolyzed much more rapidly.

Target enzyme — This term has previously been used in a general sense to describe the proteinase component of a proteinase/proteinase inhibitor interaction,[1] and in a more specific sense to describe the likely physiological target for a particular inhibitor.[2] We use it in the second, more specific sense in this chapter.

Reactive site loop (RSL) — This is a contiguous stretch of amino acids on an inhibitor that form primary contacts with the substrate binding sites of a proteinase. It does not include inhibitor/enzyme contacts distant from the enzyme's substrate binding cleft.

Virgin and modified inhibitors — These terms are used in the sense introduced by Laskowski and Kato,[1] where "virgin" describes an inhibitor that has not interacted with a

TABLE 1
Classification of Mammalian Proteinase Inhibitor Superfamilies

Inhibitor superfamily	Proteinase class inhibited	Common examples (human, unless otherwise indicated)	Distinguishing charcteristics
Kunins	Serine	Aprotinin (bovine) Trypstatin (rat) Inter-α- and pre-α-inhibitor Proteinase nexin-2/APP_{751} LACI collagen type VI *β-bungarotoxin B-subunit*	Inhibitor domain of 58 residues often embedded in larger protein, sometimes multiple chain, often multiheaded
Kazals	Serine	PSTI Seminal plasma inhibitor PEC-60 (pig) Submandibular inhibitor (cat) agrin	Inhibitor domain of 56 residues, sometimes multiheaded
ALPs	Serine	ALP Elafin	
Serpins	Serine	α_1-Proteinase inhibitor **α_1-Antichymotrypsin** Antithrombin III α_2-Antiplasmin PAI-1, PAI-2 Proteinase nexin-1 C1 inhibitor Protein C inhibitor *Angiotensinogen*	Single-chain, large folding unit (350–400 residues); most common type of inhibitor in mammals.
α-Macroglobulins	Serine Cysteine Metalloaspartic	α_2-Macroglobulin Pregnancy zone protein α_1-Inhibitor$_3$ (rat) **α_2-Macroglobulin** (rat) *Complement component C3*	Composed of one, two, or four subunits of ~1500 residues each; most contain intrachain thiolester
Cystatins	Cysteine	Cystatin A (family 1) Cystatin B (family 1) Cystatin C (family 2) Cystatin S (family 2) Kininogens (family 3) **T-kininogen** (rat) *Histidine-rich glycoprotein*	Single chain, sometimes multiheaded (kininogens); some can inhibit an exopeptidase
TIMPs	Metallo-	TIMP-1 TIMP-2	Specific for matrix metalloproteinases

Note: Representative members of each superfamily are listed, together with some characteristics that they share. Some superfamilies contain proteins that are probably not inhibitors of proteinases (italicized), and some contain proteins whose inhibitory capacity has not yet been confirmed (underlined). Major acute phase members of each family are bolded.

proteinase, and "modified" describes one that has been proteolyzed somewhere in the RSL as a result of interaction with a target enzyme or another, noninhibited, proteinase.

B. PROTEINASE INHIBITOR SUPERFAMILIES

A classification system for proteinase inhibitors based on specificities for target enzyme presents problems due to the overlapping specificity of individual inhibitors. Consequently, we classify proteinase inhibitors by superfamily relationships (see Table 1). This may actually aid in determining functional and evolutionary relationships. In mammals, we currently recognize seven protein superfamilies, each containing at least two individuals, whose members are often proteinase inhibitors: the serpins, the kunins, the kazals, the ALPs, the α-macroglobulins, the cystatins, and the TIMPs.* There also exists an inhibitor, calpastatin, that seems to be in a superfamily of one member. This protein is selective for the intracellular calpains (distant papain relatives). Since this review concentrates on extracellular proteolysis, we do not consider the calpain/calpastatin system further.

C. VARIATION OF PROTEINASE BINDING REGIONS

Although the general specificity of inhibitors of serine proteinases is often dictated by the P_1 residue of the RSL that fits into the S_1 pocket of the proteinase, other adjacent RSL sidechains have equally important roles to play. When the primary sequences of members of the superfamilies that inhibit serine proteinases are compared, the most striking variations are in the RSLs. These are the regions that dictate selectivity of the inhibitors for individual target enzymes. This hypervariability of RSL sequences was first documented by Laskowski and colleagues in their classic study of the sequences and activities of 100 kazal domains from avian ovomucoids.[5] Similar hypervariability in serpin[13] RSLs led to the conclusion that the specificity determinants of serine proteinase inhibitors are under positive Darwinian selection.

Kunin RSLs do not show such obvious hypervariability, but the sketchy data on α-macroglobulin bait regions suggest that they also are hypervariable. The nature of the selective pressure is open to speculation, but it is assumed that proteinases are the key players. In the case of avian ovomucoids, the variation may be an attempt to fit the specificity of rapidly varying pathogenic proteinases,[5,14] although there are few data on what these pathogens may be. In the case of the serpins, it is possible that selection of RSL sequences is not driven by target enzymes, but is an attempt to evade proteolytic inactivation by pathogenic proteinases, since serpins are very susceptible to cleavage in their RSLs.[15,16] Cystatins exhibit far less variability in proteinase contact sites, so the reasons for their apparent selectivities between cysteine proteinases are still unknown. The proteinase contact sites in TIMPs have yet to be defined.

D. PROTEINASE INHIBITORS AS CYTOKINE-LIKE FACTORS

One of the perplexing functions of proteinase inhibitors, but of important potential to the acute phase response, is their apparent activity as growth factors for a number of cells. For example, human bikunin and PSTI are able to stimulate the growth of human endothelial cells,[17] and TIMP-1 is a growth factor for a number of different cell types.[18] Such activities have no consensus biological explanations, but it is likely that the effects are due to the inhibition of proteolytic activity, and a common scenario that may explain the effects is suggested below. Proteinase zymogens are secreted by many cells and are also present in animal sera used in cell culture. Upon activation, these proteinases can degrade cell-surface adhesion molecules necessary for cell movement and growth. Inhibition of these proteinases would allow maintenance of the adhesion molecules, and therefore appear to be growth stimulatory *in vitro*. Whether these effects occur *in vivo* is technically difficult to assess.

One of the acronyms of the serpin proteinase nexin-1 (PN-1) is ''glial-derived neurite promoting factor'',[19] which describes the ability of this serpin to promote the outgrowth of neurites in culture. This property was shown to be due to inhibition of thrombin present in neurite outgrowth media.[20] Since thrombin inhibits neurite outgrowth, thrombin inhibitors will appear to stimulate it over the background, and it is not clear whether the outgrowth activity is real or a cell-culture artifact.

* Abbreviations are described in the relevant sections.

Other serpins show cytokine-like activity upon a number of cells. Most studies have been performed on human α_1-PI since it is the easiest one to obtain, and have shown proteinase complexes and/or proteolytically modified forms of this serpin to be able to upregulate its own synthesis in monocytes and act as a neutrophil chemoattractant. The bases of these claims are discussed at the end of Section V.

II. KUNINS

Kunins are a superfamily of proteins homologous with aprotinin, also known as pancreatic trypsin inhibitor (Kunitz), or under the common trade name Trasylol®. The term originated from "bikunin", a contraction of *bi*s *Kun*itz *in*hibitor,[21] which is the tandemly repeated two-domain Kunitz-type proteinase inhibitor that comprises one of the chains of human inter-α-inhibitor. Since single-chain polypeptides containing one, two, or three Kunitz-type domains are now known, we introduced the term "kunin" to encompass these types of proteins.[22]

Contrary to a recent report,[23] kunins are currently known only in animals. The soybean trypsin inhibitor and the aprotinin superfamily have both been designated "Kunitz" inhibitors after their discoverer,[1,24] although the two families are not related. The origin of the common nomenclature was described by Laskowski and Kato,[1] and in this chapter we use the term kunin to describe homologs of aprotinin. Possible target enzymes and biological roles of the kunins are described in the individual entires.

A. STRUCTURE AND MECHANISM

A kunin domain, exemplified by aprotinin, is composed of 58 amino acid residues (Figure 1). Since we now recognize over 30 vertebrate kunins, the degree of protein sequence identity between them is rather low and members of the family are recognized mainly by the position of six cysteine residues that comprise the three disulfide bridges of aprotinin (Figure 2). The inhibitory activity of aprotinin is dependent on a compact structure that relies on the correct formation of the three disulfide bonds,[25] and it is probable that other kunins have the same requirement. Based on sequence identity relationships, we recognize some kunins that are not thought to be proteinase inhibitors, the most well known being the β-bungarotoxin B-subunits. Presumably, inhibitory activity has been lost as a result of deletion of the P_1 residue.[26] Mammalian kunins are often embedded in other proteins, which often undergo complex posttranslational processing events. We currently recognize the mammalian kunins listed in Figure 3. At the moment, there are no definite roles assigned to these inhibitors, although it is possible that two of them, trypstatin and LACI, may be involved in regulating blood coagulation.

Kunins operate by the "standard mechanism" of inhibition.[1] Inhibition results from interactions between the substrate-like RSL and the substrate binding sites on the proteinase. The conformation of the RSL is highly complementary to the surface of the enzyme. One can think of the formation of kunin/proteinase complexes as a classic "lock-and-key" interaction wherein the RSL is fixed in a conformation that is preformed to fit the substrate binding mode of the enzyme.[27,28] Small adjustments occur during complex formation and the enzyme's active site becomes blocked. The conformation of the RSL is maintained by a disulfide bridge on the N-terminal side and a two-stranded β-sheet on the C-terminal side (Figure 1). The loop sits on top of a protein scaffold that helps maintain its conformation, and the scaffold may be thought of as the infrastructure that maintains the loop's rigidity. Disruptions of the scaffold cause increased RSL flexibility, which leads to decreased inhibitory strength and a more substrate-like loop.

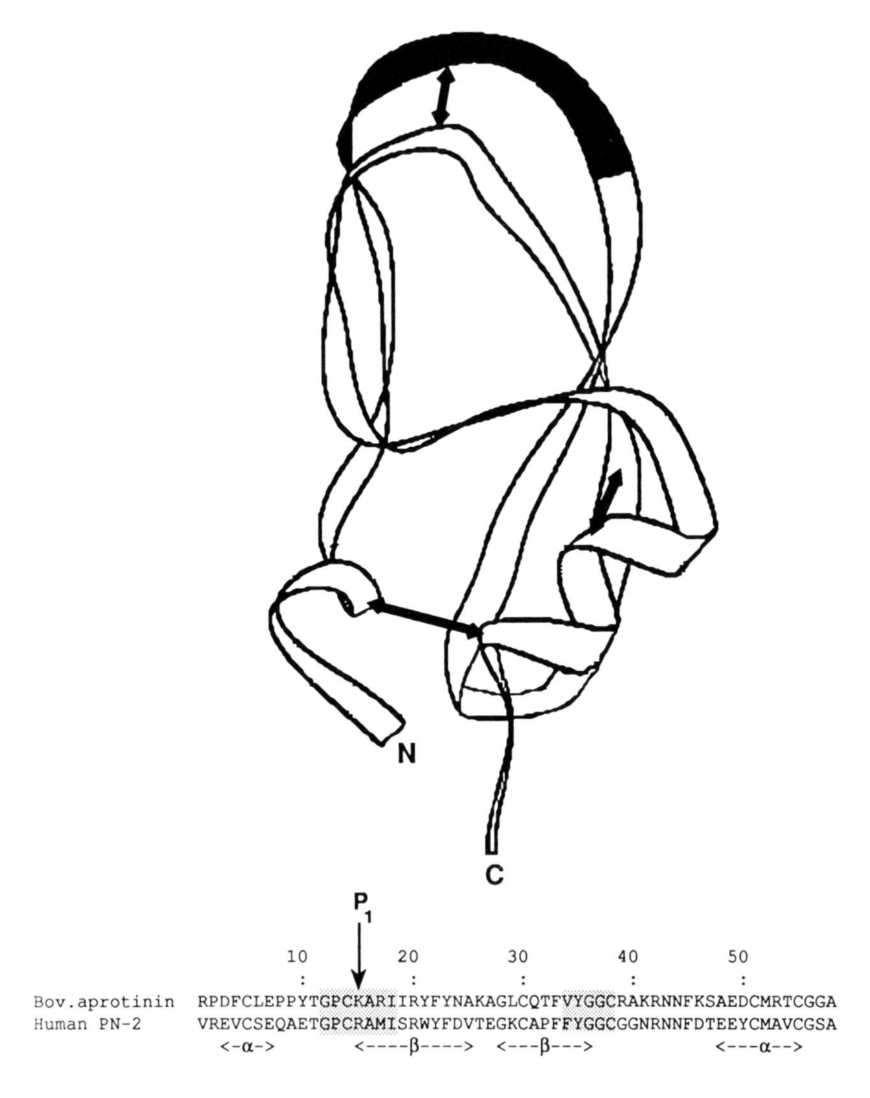

FIGURE 1. Peptide chain fold of a kunin domain. The ribbon follows the amino acid α-carbons of the peptide chain and shows the main units of secondary structure that comprise the domain. The ribbon was constructed from the molecular coordinates for bovine aprotinin,[145] using output from the program CHAOS.[146] The chain fold of the kunin domain of protease nexin-2 is very similar,[147] and the fold is probably shared by all kunins. The location of the three disulfide bonds is shown by thick black lines with their arrows pointing to the respective cysteine α-carbons. The RSL is shown by the filled-in region of the ribbon. Note that the RSL is bracketed on the N-terminal side by a disulfide to an underlying strand, and on the C-terminal side by a two-stranded β-sheet. The structure maintains the rigid, extended conformation of the RSL, so that the main interactions with target proteinases are to the peptide backbone and side chains of the RSL. Linear sequences of representative domains are shown below the ribbon, as an aid to identifying the main regions of secondary structure (α-helices and β-sheet strands). Shaded regions of the linear sequence correspond to segments whose main chain or side chains form contacts with target enzymes.

1. Trypstatin

Trypstatin is found in rat mast cells,[29] the same cellular location as aprotinin in cows.[30] Indeed, aprotinin may be the bovine ortholog of trypstatin. The inhibitor is located in the same intracellular granules as tryptase, with which it may exist as a preformed complex. Tryptase and coagulation Xa are both efficient activators of prothrombin, and are both

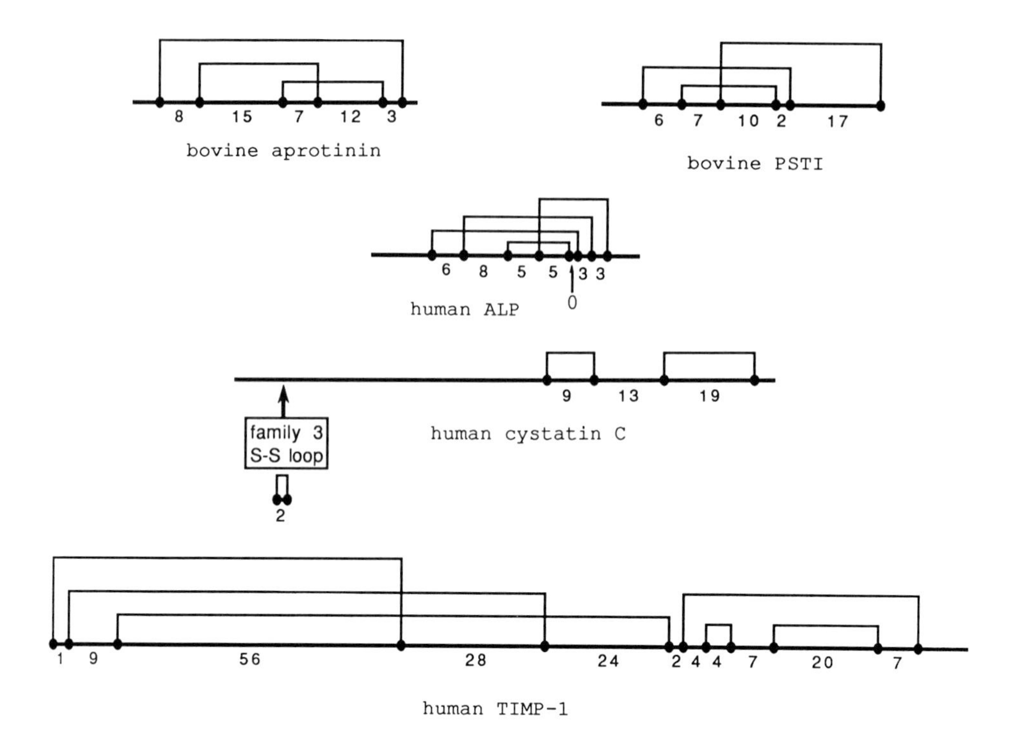

FIGURE 2. Disulfide connectivities of members of the proteinase inhibitor superfamilies. The spacing and topology of disulfides are characteristic of superfamily members, and are often used as a basis for their categorization.[144] This figure presents the relative spacing of cysteine residues (shown as beads on the chains) by giving the number of amino acid residues between each cysteine. The number of residues between cysteines is not absolute; sometimes one or two residues are deleted or inserted when divergent superfamily members are compared. However, the connectivities are thought to be absolute, since their topology is constrained by the chain folds of the respective molecules. In cystatins, the small disulfide loop is present near the N terminus of domains 2 and 3 of family 3 members, but is not present in the noninhibitory first domain. Serpins are omitted from the scheme since they do not have a highly conserved disulfide topology; indeed, some of them do not have any disulfides. α-Macroglobulins are omitted because they are too large to fit on the figure, but the disulfide connectivity of human α_2-M is shown in Figure 6.

efficiently inhibited by trypstatin[29] (K_is in the 10^{-10} *M* range). Although the role of the protein is unclear, Katanuma and colleagues have postulated an intriguing possibility for it as a regulator of coagulation in the environs of degranulating mast cells.[29,31] Such a proposal may only apply to rodents, since there is no evidence that human mast cells contain a trypstatin ortholog.

2. Pre-α-Inhibitor and Inter-α-Inhibitor (PαI and IαI)

Unusually stable proteinase inhibitors are present in human blood. These inhibitors can be visualized if a small sample of blood is run in sodium dodecyl sulfate-polyacrylamide gel (SDS-PAGE) followed by a trypsin inhibitory counterstain. The proteins migrate with molecular weights of 225 and 125 kDa. One of these proteins, IαI, was first described by Steinbuch and Loeb.[32] Another one, PαI, has recently been purified and characterized by us.[33] IαI is composed of two distinct noninhibitory heavy chains and a two-domain kunin chain known as "bikunin" (Figure 3). PαI is composed of bikunin linked to another heavy chain.[33] The protein chains are assembled by a chondroitin-4-sulfate chain that originates from Ser_{10} of bikunin.[34]

We recently analyzed the structure of the cross-link that assembles the two chains of PαI and found it to comprise a previously undescribed structure that we christened a *protein-*

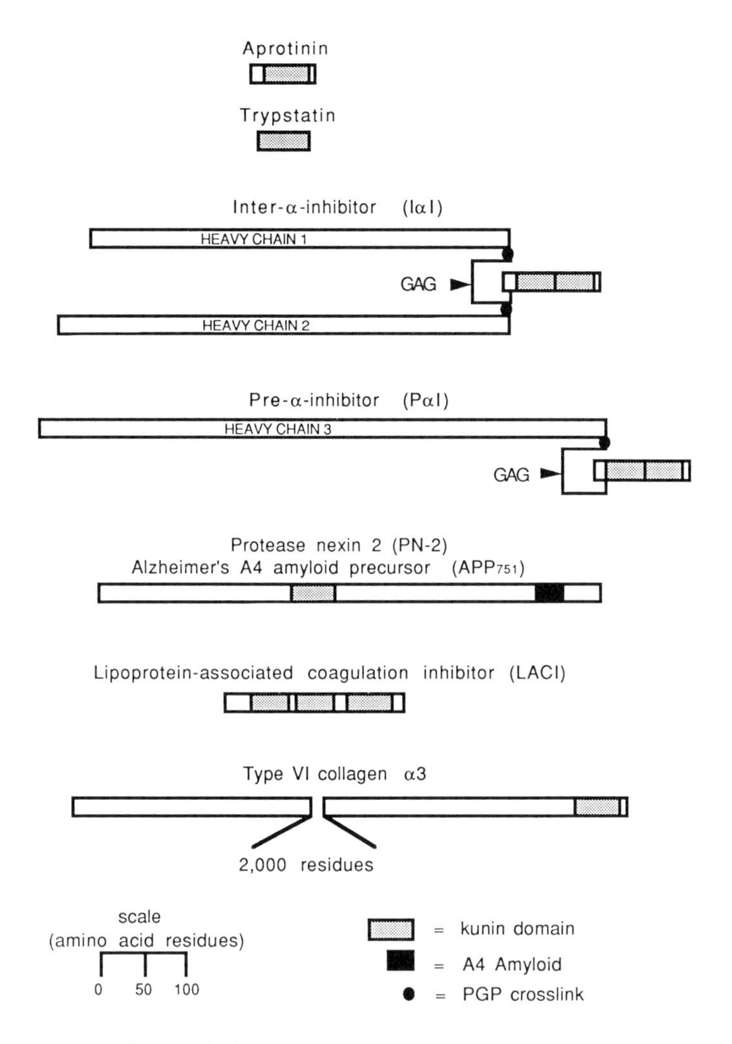

FIGURE 3. Proteins containing kunin domains. The prototypic kunin, bovine aprotinin, contains a single domain with short extensions at each end.[148] Homologous domains in different proteins are indicated by the stippled boxes. The diagram is to scale (but note the 2000-residue abbreviation of collagen IV α_3) and shows examples of the kunins so far encountered in mammals.

*g*lycosaminoglycan-*p*rotein (PGP) cross-link.[34] Specifically, a chondroitin-4-sulfate chain that originates from Ser_{10} of bikunin is covalently bound to HC3 via an ester bond between the α-carbon of the C-terminal Asp and carbon 6 of an internal *N*-acetylgalactosamine of the chondroitin-4-sulfate chain. We presume that IαI is assembled in the same way, except that two heavy chains must be linked to the chondroitin-4-sulfate chain. We consistently observe HC2-bikunin during our purification of PαI and IαI, and this protein probably represent a separate protein.[33]

The function of these proteins *in vivo* is not understood and we cannot explain the reason for the unusual PGP cross-link. It is interesting, however, that these proteins are found in all mammalian species examined, including the North American opossum, a marsupial.[35] Several nonmammalian species were examined, however, and no SDS-stable proteinase inhibitors with a molecular weight similar to the protein found in mammals was detected. Furthermore, the SDS-stable proteins identified in these animals were not sensitive to chondroitin-degrading enzymes. In the chicken, for example, we found that the SDS-stable proteinase inhibitor in the blood was ovoinhibitor (45 kDa), a member of the kazal superfamily.[35]

3. Lipoprotein-Associated Coagulation Inhibitor (LACI)

LACI is composed of three kunin domains and is therefore a trikunin. The protein circulates in plasma at a rather low concentration (2.5 n*M*) and is found both as free LACI and associated with lipoproteins. This heterogeneity is at least partly reversed after reduction of the disulfides, suggesting that the high molecular weight LACI is the result of disulfide formations between LACI and other proteins.[36]

LACI inhibits blood coagulation factor Xa and trypsin with K_i in the nonomolar range; the inhibition of factor Xa is enhanced in the presence of heparin.[23] LACI does not significantly decrease the activity of human neutrophil elastase, urokinase, protein C, tissue plasminogen activator, thrombin, or kallikrein. The activity of plasmin and chymotrypsin are slightly decreased.[23] It has been suggested that LACI inhibits the extrinsic pathway of coagulation by decreasing the activity of the calcium-dependent factor VIIa/tissue factor (TF or thromboplastin) activator complex.[36,37] The factor VIIa/TF activator complex is not inhibited by LACI alone, but only in the presence of factor Xa. Factor Xa binds to the second kunin domain, and this binary complex is then able to bind to factor VIIa/TF via the first kunin domain. This hypothesis is supported by expressiion of mutant LACI where the P_1 amino acid residue of each of the kunin domains had been changed from Lys or Arg to Ile or Leu by site-directed mutagenesis.[38] No function has been assigned to the third kunin domain of LACI nor to the heparin-dependent factor Xa inhibition.

The concentration of LACI in plasma is below the K_i of the LACI/factor Xa complex formation. This means that LACI will not be an effective inhibitor of factor Xa in the fluid phase of blood. The initial interaction between factor Xa and LACI is therefore unlikely. Since this complex is necessary for the binding to TF and for the subsequent inhibition of the extrinsic pathway, more information regarding the proposed inhibitory mechanism of LACI is imperative in order to firmly establish the function of this protein.

4. Proteinase Nexin-2 (PN-2)

PN-2 is a proteinase inhibitor that was originally identified by Cunningham and colleagues through its ability to bind to thrombin.[39] When the N-terminal sequence of PN-2 became available, it was clear that it was identical to one of the forms of the Alzheimer's peptide precursor (APP),[40] a protein that had been shown to contain a kunin domain with no known function.[41] The forms of this protein that contain the kunin are APP_{751} and APP_{770}, the subscript denoting the total number of residues in the unprocessed primary translate. The form lacking the kunin is designated APP_{695}.

APPs are synthesized as proteins of 90 to 110 kDa which undergo glycosylation to become 100- to 130-kDa species.[42] Secretion of APPs, to liberate what may be termed PN-2, is thought to take place after the C-terminal trans-membrane domain, which includes part of the A4 amyloid region, is removed.[43] Under normal circumstances, the release of APP is mediated by a cleavage inside the amyloid domain[44] and is thought to destroy the ability of A4 to aggregate. Thus, the deposition of the A4 peptide is the result of incorrect proteolytic processing of the APPs.

The discovery that PN-2 and APPs share the same polypeptide precursor has led to several hypotheses regarding the role played by this kunin. PN-2 has been implicated in the regulation of coagulation via inhibition of the intrinsic pathway proteinase factor XIa, which is inhibited with a K_i in the range of 300 to 650 p*M*.[45,46] Presumably, two molecules of PN-2 will interact with factor XIa in order to inhibit this proteinase because factor XIa is a dimer composed of two identical disulfide-linked kallikrein-like proteinases. This hypothesis will unfortunately have to share the same fate as the LACI hypothesis, as the concentration of PN-2 in plasma is well below the K_i of the interaction, making it unlikely that any inhibition of factor XIa will occur in circulation. Heparin will, however, enhance the inhibition severalfold, and it is possible that the inhibitory efficiency may be enhanced in local environments.[45,46]

5. Collagen Type VI (α_3)

This protein is one of the building blocks of the microfibrils found in tissue and cell cultures.[47] It is composed of three chains called α_1, α_2, and α_3. A partial cDNA sequence of chicken collagen type VI chain α_3 and the complete sequence of a human homolog were recently published.[48,49] The human mRNA is ≈10 kilobases, encoding one of the biggest single-chain proteins so far encountered, composed of 2943 amino acid residues. Collagen type VI α_3 is composed of several different domains previously seen in von Willebrand factor, fibronectin, actin, and salivary proteins. A single kunin domain is found near the C terminal of this gigantic protein. No data regarding proteinase inhibition are available and the function of this kunin remain unknown.

B. KUNINS IN THE ACUTE PHASE

The deposition of A4 amyloid as a central component of the senile plaques that charcterize Alzheimer's disease implicate APPs in its development. Several theories[50,51] have been raised to account for this. One attractive possibility regards the cause of the disease as a mild inflammatory condition that leads to enhanced expression of proteinases and proteinase inhibitors in local brain areas. From this perspective, it would be of interest to examine the acute phase response of the various APPs. The other kunin likely to be involved in the acute phase is bikunin. Its primary site of synthesis is liver, and elevated levels accumulate in urine following the course of certain carcinomas.[52] The origin of the urinary bikunin is unknown, although it could be from degradation products of IαI or PαI,[53] or from discordinate synthesis of bikunin and heavy chains leading to aberrant assembly.[54] No acute phase response has been proposed for the other kunins described in this chapter.

III. KAZALS

Four kazal families are known in mammals: the PSTI, submandibular inhibitor, seminal plasma inhibitor, and PEC-60 families. A recent report[55] implicates the rat protein agrin as a multiheaded kazal. This protein, which causes aggregation of acetylcholine receptors on cultured muscle fibers, contains nine homologous repeats that have cysteine spacings close to that of a typical kazal domain[55] (Figure 4). There is no evidence yet that the agrin segments have the kazal disulfide topology, nor whether some of them are proteinase inhibitors. Other mammalian kazals are not normally multiheaded (with the exception of the two-headed submandibular inhibitors), but nonmammalian vertebrates contain proteins that can have as many as seven tandem domains.[1] Kazals are particularly rich in birds and reptiles, especially in their eggs, where they can account for a substantial portion of total protein.

A. STRUCTURE AND MECHANISM

A kazal domain is in some ways similar to a kunin domain. It is small, compact, contains three disulfides, and the conformation of the RSL is almost identical to that of kunins. However, the chain fold and disulfide topology are quite different (Figures 2 and 4), which probably means that nature has found another type of scaffold upon which to place a RSL. It is now apparent that several superfamilies of small proteinase inhibitors have found discrete scaffolds upon which to place RSLs that are very similar in conformation, and that the

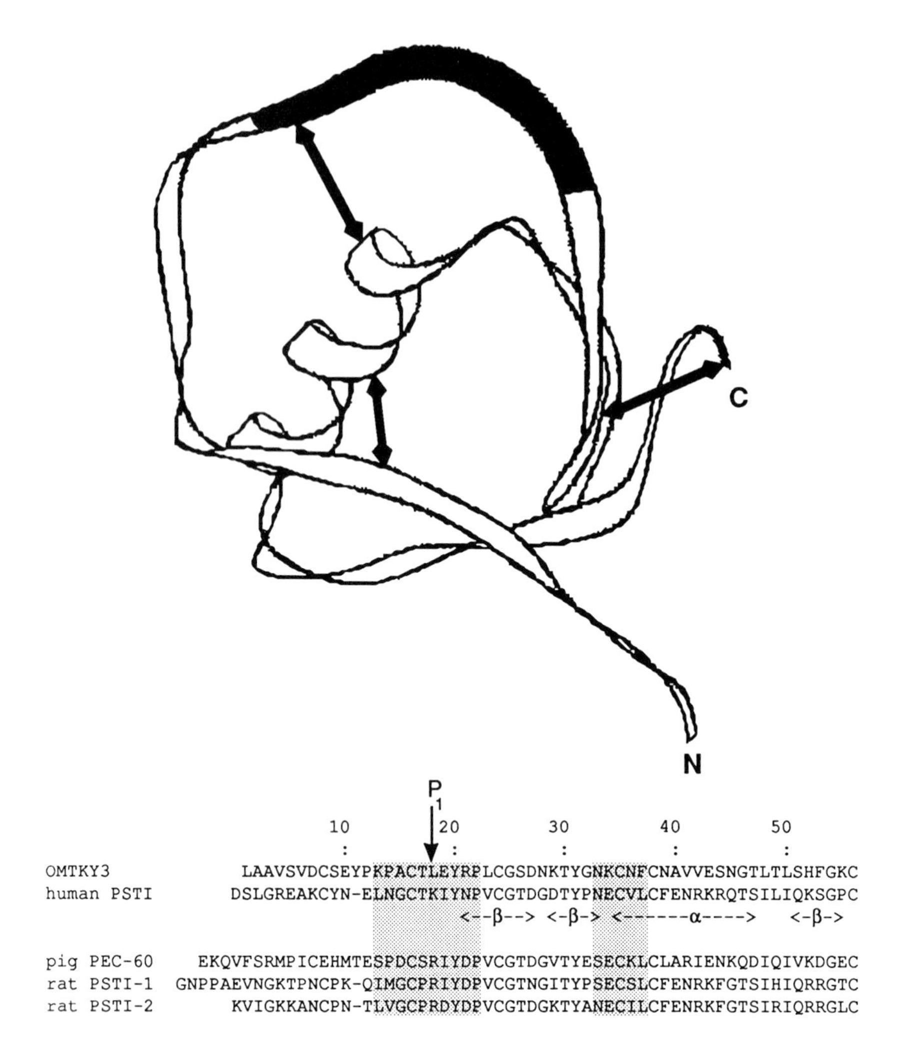

FIGURE 4. Peptide chain fold of a kazal domain. The kazal ribbon was constructed from coordinates for turkey ovomucoid third domain,[149] and is very similar to the PSTI fold. The location of the three disulfide bonds and the RSL is as in Figure 1. Note that the RSL is bracketed on the N-terminal side by a disulfide to an underlying helix for the kazal, and on the C-terminal side by a two-stranded β-sheet in each type of domain. Linear sequences of representative domains are shown below each ribbon, with units of secondary structure and proteinase contact sites identified as described in Figure 1.

kunins and the kazals are just two examples of these. Other examples are the eglin superfamily (from medicinal leeches and plants, but so far not found in mammals), soybean trypsin inhibitor family, and ALP family. In each case, the scaffolds are topologically distinct, implying separate evolutions, but the RSLs are superimposable.[27,28] The conformation of the RSL in such inhibitors, most of which employ the standard mechanism of inhibition, has recently been called the "canonical" loop because of its high degree of conformational conservation.[56] Although slightly larger compensations in RSL geometry are formed when some kazals bind to their target enzymes than when kunins bind,[27] the interaction can still be thought of as lock and key. Consequently, the mechanism of inhibition is the same as for kunins.

1. Pancreatic Secretory Trypsin Inhibitor (PSTI)

Rats contain two PSTI genes, one of which (PSTI-1) is expressed mainly in the pancreas and the other of which is expressed in pancreas and liver (PSTI-2).[57] The two genes are very closely related in primary sequence, indicating that they diverged during rodent speciation. Only one ortholog is found in cows and humans. The role of PSTI is thought to be the prevention of adventitious proteolysis in the pancreas due to premature activation of trypsin, although the expression of human PSTI and rat PSTI-2 in liver raises the possibility of an extrapancreatic role.

2. PEC-60

PEC-60 is the acronym describing a ''peptide with N-terminal glutamic acid, C-terminal cysteine, and a total of 60 residues'' introduced to describe a kazal domain, isolated by its ability to stimulate cholecystokinin release, from pig intestine.[58] The original report concluded that PEC-60, unlike other mammalian kazals, does not possess a strong trypsin inhibitory capacity. Interestingly, PEC-60 shares the ability to release cholecystokinin with rat PSTI-1.[59]

3. Seminal Plasma Inhibitor

In humans, kazal seminal plasma inhibitors are sometimes called HUSI-2 or acrosin inhibitor.[60] The function of this protein is not known.

4. Submandibular Inhibitors

The best-characterized members of this family are from carnivorous mammals, since these tend to have large submandibular glands, enabling isolation of large quantities of starting material. These inhibitors are the only double-headed kazals known in mammals, each containing Arg or Lys in P_1 of the first domain and Met in P_1 of the second domain.[61] Again, the function of these proteins is not known.

B. KAZALS IN THE ACUTE PHASE

There is no evidence that PEC-60, seminal, or submandibular inhibitors respond to inflammation; indeed, their restricted locations and site of synthesis argue against this. PSTI, however, has been reported to increase during inflammation[62] and is found elevated in the urine of patients suffering from certain neoplasias,[63] which explains the name ''*t*umor-*a*ssociated *t*rypsin *i*nhibitor'' (TATI) sometimes associated with it. The demonstration that IL-6 stimulates the production of PSTI in cultured hepatoma cells[64] firmly implicates this kazal as an inhibitor to be considered as an acute phase responder. Presumably, its function in the acute phase is distinct from its normal pancreatic function, although there is little idea of what this is yet.

IV. ANTILEUKOPROTEASES (ALPs)

We do not recommend the term ''ALPs'' to define this superfamily. At the moment, no other suggestions have been made, but we hope someone invents a better name. Currently, the superfamily is composed of two members, ALP itself, and elafin.

A. STRUCTURE

The crystal structure of recombinant human ALP indicates a compact, disulfide-rich shape for the inhibitory domain, similar to kunins and kazals in that the N-terminal side of the RSL is held rigid by a disulfide, while the C-terminal side is organized by a two-stranded β-sheet.[65] However, the disulfide topology and polypeptide fold describe a previously unobserved structural motif distinct from other inhibitor superfamilies.

1. ALP

This protein, originally identified in bronchial secretions and given the name bronchial mucous inhibitor,[66] is now known to be present in many human secretory fluids,[67] including seminal plasma (where it is given the name HUSI-1).[65] Since its role is thought to involve regulation of proteinases present from dead and degranulated leukocytes, it is sometimes given the name "secretory leukocyte proteinase inhibitor" (SLPI). The sequence of the protein shows it to comprise 107 residues that are organized into two domains.[68] The second domain contains the inhibitory capacity of the molecule against all enzymes tested, which include elastases, chymotrypsin, cathepsin G, and trypsin.[69,70] The function of the first domain is unknown. Although the function of ALP is unknown, it is assumed to act as an inhibitor of neutrophil elastase in humans, in those parts of the body α_1-PI is not normally found.[71]

2. Elafin

Elafin, isolated from horny layers of human skin, comprises 57 residues with 38% identity to the second domain of ALP.[72] The inhibitor is effective against human neutrophil elastase, pig pancreatic elastase,[72] and human neutrophil proteinase 3,[73] but shows no activity against cathepsin G, chymotrypsin, or trypsin (K_i calculated to be above 10^{-7} *M* from published data[73]). Its function is unknown, although its location and selectivity would suggest a role in protecting skin layers from neutrophil elastolytic enzymes.

V. SERPINS

Serpins encompass a superfamily of proteins homologous with human α_1-proteinase inhibitor ($\alpha - $PI), which serves as the archetype of the group. The term serpin was coined by Carrell and Travis,[6] and is a contraction of *ser*ine *p*roteinase *in*hibitor, yet it should not be assumed that all members of the superfamily are inhibitors of serine proteinases. For example, angiotensinogen[74] and hen ovalbumin[75] are both homologs of α_1-PI and are classified as serpins, yet it is almost certain that they possess no proteinase inhibitory activity. Moreover, human blood contains two other serpins, corticosteroid-binding globulin and thyroxin-binding globulin, that are hormone transporters and not, apparently, proteinase inhibitors. Nevertheless, giving the superfamily an easily remembered name has helped to establish the relatedness of proteins from disparate sources, and we now recognize more than 60 serpins from vertebrates, insects, plants, and viruses, although there is yet no evidence for their presence in prokaryotes or lower eukaryotes such as fungi.

At present, we have good evidence for the function of only a handful of the known serpins, confined mainly to those that regulate coagulation, fibrinolysis, and neutrophil proteinases in humans. How many of the serpins currently identified only by nucleic sequence analyses will turn out to be proteinase inhibitors is anybody's guess, but it is certain that some of them will have other functions.

A. STRUCTURE AND MECHANISM

α_1-PI, the serpin archetype, is a glycoprotein of 394 residues, with carbohydrate accounting for about 20% of its 52-kDa mass, and until 1990, a proteolytically modified form of this inhibitor was the only serpin structure resolved at the molecular level.[76] Despite the low level of primary sequence identity between members of the superfamily, it has been suggested that α_1-PI serves as a template upon which the structure of other serpins could be modeled.[8] This suggestion has been borne out with the recent elucidations of the structures of virgin ovalbumin,[77] modified ovalbumin,[78] modified human α_1-antichymotrypsin (α_1-ACT),[79] modified human antithrombin III (ATIII), and latent human plasminogen-activator inhibitor 1 (PAI-1). Although we now have these six serpin structures available for analysis, we still lack a clear picture of the native inhibitory conformation, since the structures are either of proteolytically modified forms (Figure 5), a noninhibitor (ovalbumin), or a non-inhibitory latent form of PAI-1.

Inhibitory serpins obey at least part of the standard mechanism, since they rapidly form tight equimolar complexes with their target proteinases, but they deviate in two significant aspects: (1) modified inhibitors are inactive and cannot recombine with proteinases and (2) complexes appear to be covalently stabilized when visualized under denaturing conditions. The reason for the first deviation is apparent when the structure of proteolytically modified α_1–PI, α_1–ACT, or ATIII is examined (Figure 5). Although Met_{358} and Ser_{359} must be joined in the virgin molecule, in the modified protein they are separated by the length of the molecule. The virgin inhibitor can be thought of as analogous to an unstable intermediate in a protein-folding pathway to the thermodynamically stable, modified form.[7] Proteolysis in the RSL allows the formation of a fully formed antiparallel β-sheet, thereby placing Met_{358} and Ser_{359} very far from each other. The second deviation from the standard mechanism, the apparent covalent stabilization of serpin-proteinase complexes visualized under denaturing conditions, led some workers to consider that the complex exists in a state equivalent to the acyl intermediate of the hydrolysis of substrate.[2,80,81] This would mean that peptide bond cleavage of the inhibitor had occurred, which is unlikely since we know that RSL-cleaved serpins are no longer inhibitors. Data showing binding of catalytically inactive anhydroproteinases to serpins,[82,83] and the reversibility of certain serpin-proteinase complexes,[84,85] argue against stabilization at an acyl level. The recent demonstration, based on noninvasive nuclear magnetic resonance (NMR) data,[86] that at least a portion of the α_1-PI pig pancreatic elastase complex exists at the level of the preacyl, tetrahedral intermediate also argues aginst peptide bond cleavage as a mechanism for inhibition. Possibly the covalent bond formed between serpins and the catalytic serine of target proteinases is an artifact and forms as a result of the denaturing conditions required for its detection.[2]

This lack of molecular structures of a native inhibitory serpin or a serpin-proteinase complex has led to a number of theories of the possible structure of the active inhibitory conformation, based on kinetic, thermodynamic, and modeling approaches. The simplest model regards serpins as standard mechanism inhibitors that have evolved a new solution to the problem of maintaining a rigid RSL.[87] This solution relates to the way in which a region upstream from the RSL interacts with the dominant unit of secondary structure, the A-sheet (Figure 5). This interaction is not well understood, but likely results in tying down the RSL in a manner analogous to the disulfides on the RSLs of kunins and kazals (Figures 1 and 4). The region downstream from the RSL is held in place as strands of a β-sheet, much as in kunins and kazals. Thus,the RSL may be held in the canonical structure,[56] although the RSL of serpins probably has more freedom of movement (at least in the virgin state) than other standard mechanism inhibitors.[88]

B. α_1-PROTEINASE INHIBITOR AND α_1-ANTICHYMOTRYPSIN FAMILIES

In humans, the animal whose serpins have been studied in most detail, the inhibitory members can be divided into those that regulate coagulation (ATIII, PN-1, and protein C inhibitor), fibrinolysis (α_2-antiplasmin, PAIs), complement (C1 inhibitor), and inflammatory proteinases (α_1-PI, α_1-ACT). It is highly significant that almost all of the inhibitors that regulate these systems are serpins, and we presume that the superfamily has diverged to fulfill required regulatory roles during the evolution of vertebrate hemostatic systems. In keeping with the scope of this chapter, we concentrate on α_1-PI and α_1-ACT, since they are the main acute phase serpins. Although related genes, possible pseudogenes, have been identified,[89] it is currently thought that humans express only one α_1-PI and one α_1-ACT.

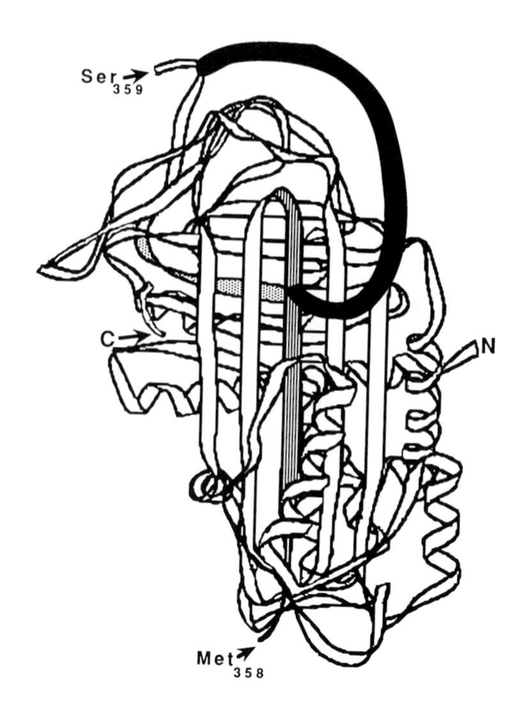

FIGURE 5. Peptide chain fold of a serpin domain. The ribbon follows the chain fold of α_1-PI that has been modified at the Met_{358}-SER_{359} reactive-site peptide bond.[76] The sequence is about six times longer than a kunin or kazal domain, and therefore contains far more secondary structure. Despite its large size, it should probably be thought of as a single domain, since all units are closely packed and deletion of any large regions would probably interfere drastically with molecule folding. The view looks down onto the dominant unit of secondary structure, the six-stranded A-β-sheet that runs the length of the molecule. One helix lies above the plane of the A-sheet, and the other eight lie below it. The two other β-sheets of the molecule lie under the top of the A-sheet. Since no active virgin serpins have been crystallized, we can only estimate the conformation of the RSL. In this structure, the RSL has been cut and the P_1 and P_1' residues at opposite ends of the molecule, and most of the RSL is now the central strand of the A-sheet (vertical shading). To recreate the virgin conformation, it is necessary to pull the RSL out of the A-sheet and connect it to the top of the molecule. The simplest solution, compatible with most current data, would join the cut ends of the loop as shown by the solid black shading, leaving the first three or four residues of the strand inserted into the sheet.[56,88,150] The sheet would compact to take up the space vacated by removing its central strand. The rigid, extended conformation typical of kunin and kazal RSLs could be achieved by the tendency of the loop to insert into the A-sheet being opposed by the C-sheet. The C-sheet is equivalent to the β-sheets on the C-terminal sides of the small inhibitor RSLs, which would make the tendency to move into the A-sheet the structural analog of the disulfides on the N-terminal sides of the RSLs of the smaller inhibitors (α_1-PI contains no disulfides). We stress that this topology is hypothetical and that our views may change dramatically when the molecular structure of an active virgin serpin is solved. The stippled region, part of the B-β-sheet, is the location of the segment reported to be the receptor recognition site for serpin/proteinase complexes.[151] It is in the hydrophobic core of the molecule and, in the modified structure at least, its associated side chains are not available to interact with external molecules. Thus, the molecular structure and biological data on receptor-recognized determinants are in apparent conflict.

The function of α_1-PI is to regulate the extracellular activity of neutrophil elastase, while the function of α_1-ACT appears to be to regulate neutrophil cathepsin G and, possibly, mast cell chymases.[2,90]

The situation is far from clear in rodents, since multiple α_1-PI and α_1-ACT family members have been identified, mainly at the genomic level,[91,92] but also by protein identification.[93] The origin of this diversity is obscure, but a comparison of serpin sequences reveals some interesting facts about conservation of primary structures. The overall degree of identity between human α_1-ACT and the various rodent homologs varies from 50 to 64%.

However, conservation of residues in the RSL region (residues 352 to 365) is much lower, ranging from 16 to 20%. Similar variations in RSL residues are found when human and rodent α_1-PI family members are compared. On the other hand, human and rodent homologs of ATIII, PN-1, and PAI-1 are 83 to 86% identical throughout, with levels of identity in their RSLs of 80 to 100%. Since the RSL dictates specificity, it is likely that α_1-ACT and α_1-PI homologs serve different functions in different mammals. The high degree of conservation of the specificity regions of ATIII, PN-1, and PAI-1 argues that they probably serve the same function in different mammals. Significantly, those serpins that show the greatest variation are the ones that regulate the proteolytic systems of leukocytes, whereas those that show the least are the ones that regulate coagulation and fibrinolysis. We presume that the fit between coagulation and fibrinolytic proteinases and the serpins that regulate them was fixed at an early stage of mammalian evolution; the fit between the leukocyte serine proteinases and α_1-PI and α_1-ACT family members is still evolving.

C. SERPINS IN THE ACUTE PHASE

The elevation of human α_1-PI and α_1-ACT during the acute phase appears to make sense, since they would be needed to guard against the damaging effects of proteinases released from neutrophils responding to inflammatory signals. Although an excellent case may be made for α_1-PI, since human neutrophil elastase is a very powerful degradative endopeptidase, it is not quite clear why α_1-ACT must rise even more dramatically, since its target enzymes have quite low activity on most substrates. Perhaps we have not found the substrate against which adventitious proteolysis by cathepsin G or chymase must be prevented, or perhaps α_1-ACT has another function.

Serpins are extremely sensitive to proteolytic inactivation by proteinases they do not inhibit,[16] and it is apparent that proteolytically modified serpins are able to act as cytokines. Thus, proteolytically modified α_1-PI was found to be chemotactic for neutrophils[94] and able to stimulate its own synthesis in monocytes.[95] Neither activity was found in the virgin molecule, although elastase/α_1-PI complexes were found to have the same effect on monocytes. The receptor thought to be responsible for the binding of α_1-PI/proteinase complexes to monocytes and cultured hepatoma cells[96] may not be the same as that responsible for the removal of α_1-PI/proteinase from circulation,[97] since the former recognizes proteolytically modified serpins[98] whereas the latter does not.[87] The ability of modified α_1-PI to act as a cytokine is interesting, but the possibility that other serpins may act similarly following proteolytic inactivation by proteinases that they do not inhibit is potentially very significant. For example, α_1-ACT is inactivated more readily by proteolysis than α_1-PI,[16] and this is more likely to be of relevance in a physiologic medium, provided that the modified α_1-ACT is a cytokine.

VI. α-MACROGLOBULINS (α-Ms)

The α-macroglobulin (α-M) subunit chain is about 1500 residues long, giving a size of about 180 kDa. Subunits have multimerized during the course of evolution of the superfamily, so that dimers or tetramers of identical subunits charcterize some of the best-known members. For example, humans contain the tetrameric α_2-M and the dimeric pregnancy zone protein (PZP), both of which are proteinase inhibitors, as well as complement components C3, C4, and C5, which are monomeric α-Ms with no known inhibitory capacity. Rats contain the tetrameric α_2-M and α_1-M and, in addition to the usual monomeric complement components, they contain the monomeric proteinase inhibitor known as α_1-inhibitor$_3$ (α_1-I$_3$). More precisely, rats contain at least two α_1-I$_3$ paralogs, whose orthologs in the mouse are known as murinoglobulins. Rat α-Ms constitute an interesting acute phase paradigm, since α_1-M varies

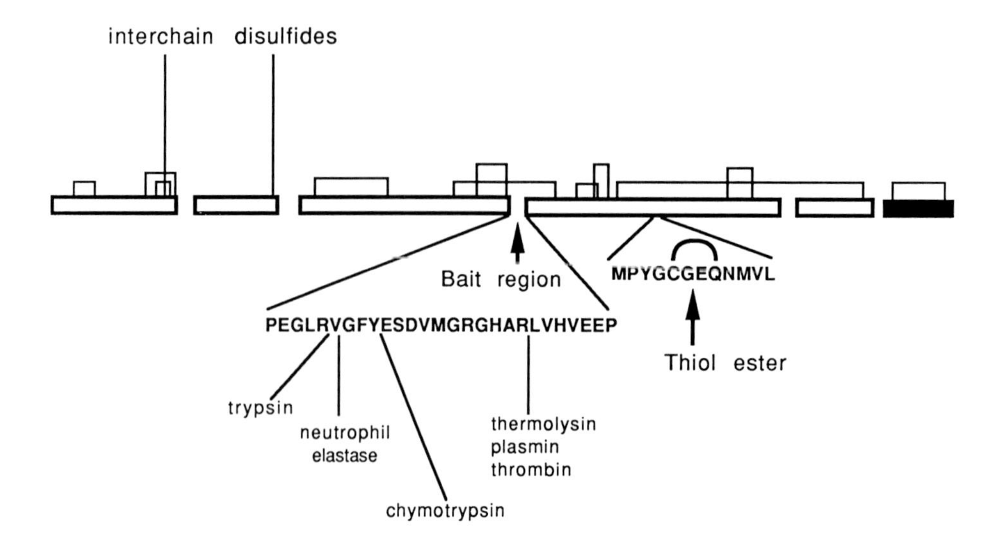

FIGURE 6. Features of the human α_2-M subunit. The disulfide connectivity of the subunit is indicated by the thin joined lines. The two interchain disulfides enable the subunit to link to another identical subunit to form a disulfide dimer.[102] The domain structure of the human subunit has not been determined, and the structure shown above is based on the major fragmentation pattern of the homologous rat α_1-I_3.[103] The C-terminal domain (filled-in segment) contains the receptor-recognized determinants responsible for removing proteinase-reacted α_2-M from the circulation. The amino acid sequence of the bait region includes sites of primary cleavage by a number of proteinases that are trapped by α_2-M. The neutrophil elastase site supersedes the one previously identified,[152] which we now suspect to be a secondary site of cleavage.[35]

little with the health of the animals, whereas α_2-M is a major acute phase responder and α_1-I_3, present at high concentrations in healthy animals, declines dramatically during the acute phase of inflammation. α_1-I_3, therefore, is known as *negative* acute phase responder.

On a functional basis, we can divide the α-Ms into two groups: the inhibitory ones and the noninhibitory complement components. Which came first, and why nature used related proteins to accomplish apparently unrelated tasks are interesting questions that we are beginning to address. There are no known examples of mammalian α-Ms that possess both proteinase-inhibitory and complement capacities. The ancient phylum in which α-Ms first appeared is unknown, although we have isolated a proteinase inhibitor with the characteristic properties of α-Ms from the mollusc *Octopus vulgaris*.[99] Since molluscs do not have a complement system, one could presume that the proteinase-inhibitory activity predated complement activity. However, this is not a valid presumption, since the properties of primitive complement components can only be guessed at. For example, the horseshoe crab (*Limulus polyphemus*) contains a protein, probably an α-M, that is a proteinase inhibitor[100] and able to participate in a lytic system following binding to foreign cells.[101] We stress that we do not yet consider the limulus protein to be a complement factor, since a mammalian-like complement system may not be present in this primitive species, but we do consider that it may be functionally related to a primitive progenitor of the mammalian complement factors.

Aside from primary sequence considerations, there are three properties of most α-Ms that aid in characterizing them: (1) they all have subunits of about the same size, (2) they all contain a region, about the middle of the chain, that is susceptible to proteolysis (called the "bait region"), and (3) all except chicken egg white ovostatin contain an interchain thiolester located about two thirds of the way along the chain (Figure 6).

A. STRUCTURE AND MECHANISM

α-Ms are huge proteins, so no molecular structures are yet available. The disulfide pattern of human α_2-M has been determined[102] (Figure 6), and most of the cysteines are conserved between proteinase-inhibitory mammalian α-Ms, but not necessarily between complement components. α-Ms are almost certainly composed of multidomain subunits, and the results of limited proteolysis of rat α_1-I_3 have allowed us to designate tentative locations for the major independent folding units of the proteinase-inhibitory α-Ms.[103]

Cleavage of any of the peptide bonds in the bait region initiates a series of changes that often result in compacting of the subunits. Since the bait region probably acts as a true substrate for a number of proteinases, members of all four catalytic classes of endopeptidases are inhibited. The inhibitory capacity of the α-Ms results from a conformational change following a proteinase reaction that physically entraps the reacting proteinase molecule(s).[104] α-Ms are thus mechanistically distinct from the active site-directed proteinase inhibitors that directly block the active sites of proteinases. α-M-bound proteinases retain activity against small substrates such as peptides, but are almost completely inactive against large ones (proteins larger than 30 kDa). This unusual reactivity results from steric hindrance of proteinases by the large α-M molecule. The retention of activity by α-M-proteinase complexes can comlicate the analysis of proteolytic activities. For example, Kuehn et al.[105] realized that a high molecular weight proteinase reported by several authors to be present in rat muscle was really a complex of rat α_1-M with small lysosomal cysteine proteinases, possibly resulting from isolation artifacts.

Unique to α-Ms is an intrachain thiolester[106] located two thirds of the distance from the N terminus of each subunit (Figure 6). This group is usually reactive with small amines in native α-Ms, and becomes highly reactive with a broader range of nucleophiles during the conformational change caused by reaction with proteinases. Often, proteinase molecules become covalently linked to the thiolester, although this seems to be essential only for inhibition by monomeric α-Ms such as rat α_1-I_3[107] and the dimeric human PZP.[108] The thiol ester's function in tetrameric α-Ms is unknown, since its ability to covalently link to proteinases is not required for inhibition; moreover, tetrameric chicken ovostatin does not possess a thiolester, yet is still able to trap and inhibit proteinases.[109]

B. α-MACROGLOBULINS IN THE ACUTE PHASE

The biological role of the complement α-Ms is fairly certain, but the role of proteinase-inhibitory α-Ms has been debated, without resolution, for many years. Human α_2-M is a good inhibitor of a number of proteinases, but in almost every case better and more specific inhibitors are present in the body.[2] This has led to the idea that α_2-M is a "backup" inhibitor that exists to prevent adventitious proteolysis only when the specific inhibitors reach saturation levels by their target enzymes.[2] Alternatively, human α_2-M may help prevent invasion by parasites to whose proteinases the body has not had time to evolve selective inhibitors.[110] Whatever the role, the lack of α_2-M is apparently incompatible with life, since no complete deficiencies have been detected.

Although human α_2-M increases slightly during inflammation, it is in the rat that the most dramatic changes in α-M levels occur. While it can be assumed that the dramatic increase in rat α_2-M occurs to aid in proteinase inhibition, the reason for the fall of α_1-I_3 levels is not easy to explain. We have shown that α_1-I_3 is a proteinase inhibitor, but it is far less efficient than tetrameric α-Ms. Sometimes, several molecules of α_1-I_3 are turned over by some proteinases (particularly those deficient in lysine) before an inhibition event occurs.[107] To us, this suggests that α_1-I_3 must be downregulated to protect it from turnover by the many exogenous and endogenous proteinases released during the acute phase. The downregulation may not be so much for the protection of α_1-I_3 itself as for the protection from saturation of the general receptor pathway for the clearance of α-M-proteinase complexes.[111]

VII. CYSTATINS

The name cystatin was used by Barrett (reviewed in Reference 10) to describe a protein from chicken egg white that inhibits members of the papain superfamily of cysteine proteinases. The name was extended to describe the superfamily of related proteins and later adopted by many groups working the field.[10] Four branches of the cystatin superfamily are currently recognized. c-Ha-*ras* oncogene protein p21 and a domain of human fibronectin have been concluded by some authors[112,113] to be related to cystatins, although Rawlings and Barrett[124] have recently excluded these from the superfamily based on protein sequence similarity comparisons with all other known members. Family 4 is loosely defined and includes the two noninhibitory cystatins, histidine-rich glycoprotein and α_2-HS-glycoprotein (the human ortholog of bovine fetuin), that share identity with other members of the superfamily, but little with each other.

A. STRUCTURE AND MECHANISM

Molecular structures are available for the family 1 member cystatin B (in complex with carboxymethyl-papain)[114] and the family 2 member chicken cystatin.[115] The key units of the inhibitory site of cystatins are seen to be the N-terminal segment of the molecule and two hairpin loops, which together form a hydrophobic wedge that is complementary to the substrate-binding cleft of papain (Figure 7). In the case of the cystatins, it is not appropriate to think in terms of a RSL, since the portions of the molecule required for inhibition are composed of noncontiguous segments.

The mode of binding of cystatins by their target enzymes, described as competitive and tight binding,[116] is expected to be similar to that of papain, since proteins of the papain superfamily have similar V-shaped substrate-binding crevices.[117] In contrast, the cystatin wedge would not fit into the substrate binding sites of serine proteinases, which explains why cystatins do not inhibit serine proteinases.[115] Indeed, human cystatin C is a substrate for the serine proteinase neutrophil elastase, which cuts between Val_8 and Gly_9 to produce a truncated inhibitor with greatly decreased inhibitory potential.[118] Since it is the N-terminal "trunk" of cystatins that most closely approaches the primary specificity pockets of the proteinases, and since N-terminal truncated cystatin forms possess greatly altered binding constants for individual proteinases,[118-120] it is likely that this region dictates the inhibitory selectivity of the superfamily members. Presumably, the other two sites of interaction, the hairpin loops, provide orientation and increase binding energy in a less selective way.

1. Family 1. The Stefins

Family 1 members lack intrachain disulfides, carbohydrate and signal peptides, and probably reside mainly in the cytoplasm, although small amounts (0.3 to 0.4 μM) of human cystatin A are found in saliva and amniotic fluids.[121] The best-characterized members are human cystatin A and rat cystatin-α, and human cystatin B and rat cystatin-β. The rat cystatins are probably orthologs of the human proteins. There is currently no consensus for biological roles of family 1 members, so our ideas of possible functions are based mainly upon circumstantial evidence. Family 1 cystatins are found in most cells of the body, although the distribution of A and α differs from that of B and β. A and α are found predominantly in epidermal layers of the skin, epithelial cell types that line the gastrointestinal tract, and neutrophils, whereas B and β have a much more even distribution throughout cells and tissues. The restricted distribution of A and α, in cell types thought to be involved in "front-line defense" against pathogens, led Barrett and colleagues[10] to speculate a function in the inhibition of pathogenic proteinases. The much broader distribution of B and β, reminiscent of the distribution of lysosomes, led the same authors to suggest a protective role for these inhibitors in regulating the activity of lysosomal cysteine proteinases.

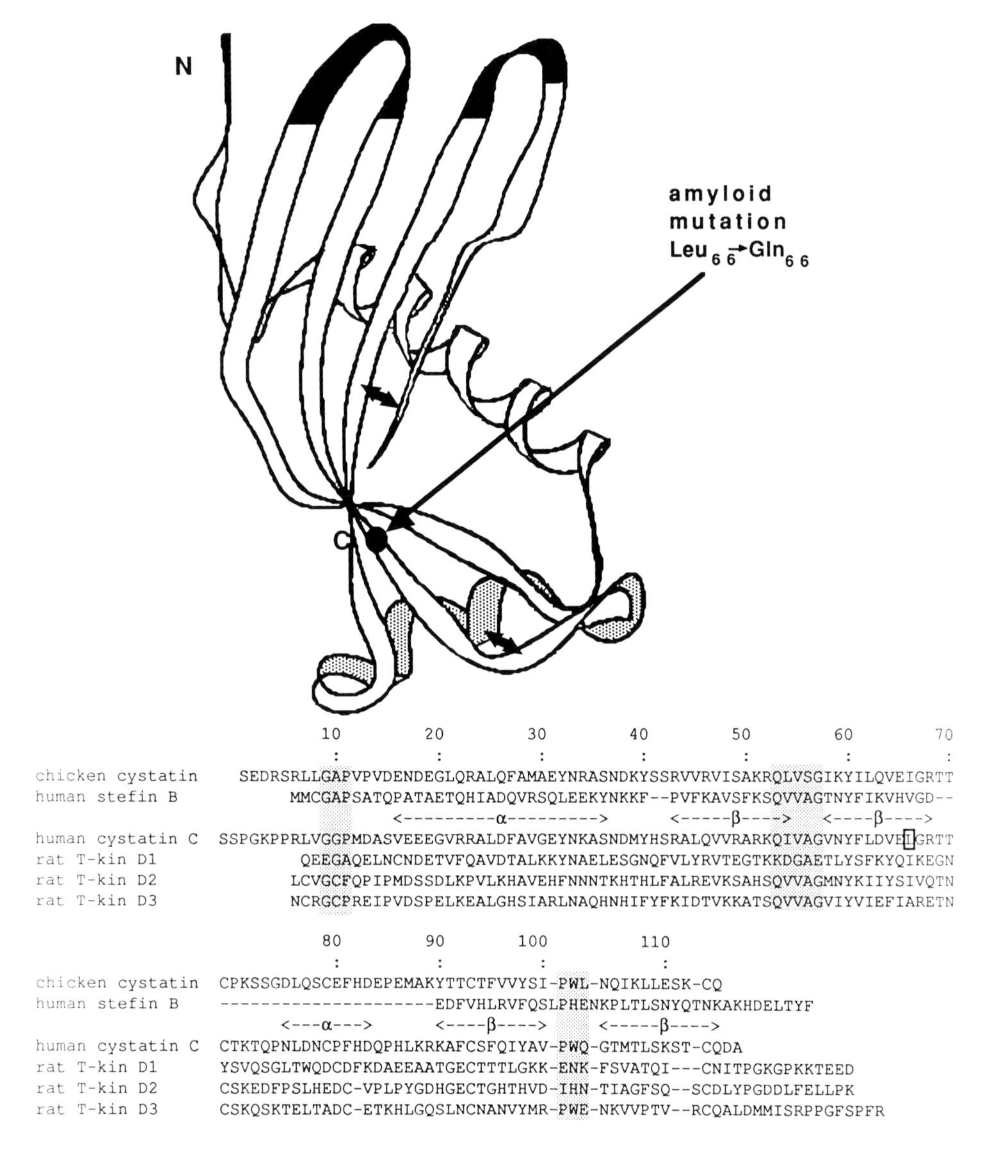

FIGURE 7. Peptide fold of a cystatin domain. The ribbon follows the peptide fold of chicken egg white cystatin,[115] a member of family 2. The two disulfides are shown as black lines whose arrowheads point at their respective α-carbons. The position of the single-residue substitution that causes a hereditary form of cerebral amyloidosis,[153] presumably by deposition of mutated cystatin C, is shown by the arrow in the folidng diagram, and by a box around the residue in the aligned sequences. The cystatins do not have a RSL in the sense of the serine proteinase inhibitors, but the regions of the molecule whose side chains are thought to form contacts with the substrate-binding groove of papain are shown as filled-in segments of the ribbon. These comprise the N-terminal segment and two hairpin loops connecting β-sheet strands. The molecular structure of the family 1 cystatin stefin B, in complex with papain, showed a highly superimposable structure and confirmed the proposed binding mode.[114] The shaded helix at the bottom of the chicken cystatin structure is absent in stefin B, as are the disulfides. The discovery of this deletion required a revised alignment of cystatin superfamily primary sequences, since previous alignment strategies had not allowed such a large gap. The revised alignment[114] is shown under the structure, with units of secondary structure identified as α for helix and β for β-sheet strands, and papain-contact regions shaded. Family 3 cystatins align well, but some problems in alignment around the cysteines of the second disulfide will be clarified when structures become available. Interestingly, the folding topology of cystatins is equivalent to ribonuclease A, hinting at a possible common ancestor. However, the central helix adopts a different location with respect to the main sheet, and the segments of ribonuclease equivalent to the hairpin loops of cystatins are much more extended, which eliminates ribonuclease A as a potential cysteine proteinase inhibitor.

2. Family 2. The Cystatins

Family 2 contains the archetypal cystatin from chicken egg white and mammalian C-type and S-type cystatins (reviewed in Reference 10), and a mRNA transcript expressed in parotid glands, designated cystatin D, whose protein product has yet to be identified.[122] Since the family 2 cystatins contain signal peptides, they are normally found in extracellular locations. In humans, cystatin C is found throughout the body, although it is at its highest concentration (4 μM) in seminal plasma.[121] Human type S cystatins are found almost exclusively in seminal plasma, tears, and salivary secretions.[121,123] As with family 1 cystatins, little is known about the physiological role of the members of this family. Calculations based on inhibitor concentrations and kinetic inhibition constants[121] implicate cystatin C in the regulation of cathepsin B in human extracellular fluids, although this interaction is only likely to occur when the compartmentation between the enzyme (lysosomal) and the inhibitor breaks down. Such an occurrence is still hypothetical, although it is easy to speculate how this may happen as a result of cell death or secretion of lysosomal enzymes, events associated with the inflammatory stimuli of certain cells.

3. Family 3. The Kininogens

An analysis of how kininogens may participate in the acute phase response is complicated by their reported multiple roles. The human kinogens are apparently (1) able to participate in the intrinsic coagulation system, (2) serve as precursors for the generation of kinins, and (3) inhibit cysteine proteinases of the papain superfamily. In all species studied, H-kininogen and L-kininogen are products of the same gene;[124] they contain identical cystatin domains and kinin segments, but diverge in their C-terminal regions as a result of alternate exon splicing. H-kininogen participates in the intrinsic coagulation pathway by modifying the events that lead to activation of prokallikrein and factors XI and XII.[125] L-kininogen cannot substitute for H-kininogen in this function, although both are equivalent substrates for the excision of kinins by tissue or plasma kallikreins.

Rat T-kininogens appear not to be substrates for kinin excision by kallikreins due to the replacement of residues flanking the kinin segment. Although it is possible that the kinin sequence (which is identical to that of H- and L-kininogens) may be released by other proteinases, it is not clear whether this is physiologic. More likely, T-kininogens have lost their ability to act as kinin substrates and are now just proteinase inhibitors. The two rat T-kininogens are encoded by distinct genes[126] that separated recently, though at distinct times, from the rodent H/L-kininogen gene[124] and are therefore paralogs of Rat H- and L-kininogen. T-kininogens, therefore, probably do not exist in mammals other than rodents.

The first cystatin domain of the kininogens has apparently lost its inhibitory function, but the second and third domains retain inhibitory activity against a number of cysteine proteinases. One unusual feature of kininogens is the ability of the second domain of human kininogen to inhibit calpain,[127] a cytosolic cysteine proteinase that is only distantly related to other members of the papain superfamily. Whether this is an accident of evolution or whether some significance may be attached to it is unknown, since other cystatins, including other species kininogens, do not share this property.

B. TARGET PROTEINASES

The chief characteristic of cystatins, as implied by their name, is the ability to inhibit cysteine proteinases. Cystatins do not inhibit proteinases with other catalytic mechanisms, and they are usually thought to be selective for cysteine proteinases of the papain superfamily, which include the lysosomal proteinases cathepsin B, H, and L, and the cytosolic calpains. Some evidence suggests that other types of cysteine proteinases, including clostripain and polioviral proteinases, may be inhibited, although the interactions have not been studied in detail. Members of this superfamily are unique among the inhibitors considered in this chapter, since some of the members of families 1 and 2 are able to inhibit the exopeptidase known as dipeptidyl peptidase I, an enzyme that sequentially removes dipeptides from the N terminus of proteins.

If cystatins exist to inhibit proteinases, then the number of possible targets is quite small, and restricted to environments in or around cells where the lysosomal proteinases are likely to be active. As with all proteinase inhibitors, one must also consider that the cystatins serve to inhibit pathogenic proteinases, including those encoded by viral genomes. Currently, the target enzymes for individual cystatins, if that concept is appropriate in the context of cysteine proteinases, are unknown.

C. CYSTATINS IN THE ACUTE PHASE

Although the family 3 member rat T-kininogen is the most well-known major acute phase responder of the superfamily, it is apparent that members of families 1 and 2 also show an increase in levels during inflammation. For instance, the plasma concentration of cystatins A and C is elevated in inflammatory and autoimmune diseases.[10] However, it is not known whether these increases are due to specific responses to inflammatory signals or just to leakage from damaged cells. Further work is required to clarify this point. This leaves us with T-kininogen, also known as the major acute phase protein of rat plasma. There can be little doubt that the enormous increases in the concentration of T-kininogens in response to inflammation must endow rats with some type of defense function. Since it is likely that the function of T-kininogens is to inhibit cysteine proteinases, and not to serve as physiological substrates for kinin release, the protection must be against cysteine proteinases. But what are the targets, are they endogenous or exogenous?

VIII. TISSUE INHIBITORS OF METALLOPROTEINASES (TIMPs)

Class-specific inhibitors of metalloproteinases seem to be rarer than inhibitors of serine or cysteine proteinases. Currently, two TIMPs are known, and designated TIMP-1 and TIMP-2. Both are about 185 residues long, giving proteins of 20 to 29 kDa (depending upon the extent of glycosylation), with a degree of identity of 38% for the human proteins.[128,129] Orthologs of each are found in a number of mammals.[130] All TIMPs so far sequenced contain 12 cysteine residues whose positions are conserved, and an analysis of human TIMP-1 indicates that these are arranged to form six disulfides, which organize the molecule into two domains.[131] The inhibitory activity of TIMP-1 is confined to the N-terminal domain,[132] but finer mapping of the inhibitory site has not yet been reported.

A. STRUCTURE AND MECHANISM

No data are yet available on the molecular structure of TIMPs. However, since the N-terminal disulfide-limited segment of human TIMP-1 can be expressed free of the C-terminal disulfide segment,[132] we presume that there are at least two topologically distinct chain folds in the molecule. Mechanistic studies of the TIMPs have been hindered by the small amounts of inhibitors and target enzymes available for study and by the unavailability of efficient substrates. It is apparent that TIMP-1 forms equimolar complexes with K_d in the 10^{-10} *M* range with some target proteinases.[133] With the application of new synthetic substrates to the investigation of TIMP/proteinase interactions, we expect that by the time this volume is published, ongoing studies will have led to much clearer conclusions.

B. TARGET ENZYMES

TIMPs seem to be selective for enzymes of the matrix metalloproteinase (MMP) superfamily found exclusively (at present) in animals.[130] They do not inhibit metalloproteinases from bacteria, nor members of the astacin superfamily of metalloproteinases.[134] The MMPs currently number about 11 members whose functions are thought to be to degrade connective tissue proteins during development (following apoptosis, for example), inflammation, and wound healing.[130] TIMP-1 is found in association with proMMP-9 (92-kDa gelatinase/type IV collagenase) and TIMP-2 with proMMP-2 (72-kDa gelatinase/type IV collagenase) during the production of these two proteinase zymogens in cell culture,[135,136] in associations that are apparently not dependent on the inhibitory site of the inhibitors.[137] Since all known MMPs are secreted as inactive precursors (whether or not they have TIMPs attached), the significance of TIMP association is obscure. Cross-linked higher-order complexes consisting of two molecules of MMP and one molecule of TIMP have been observed in cell culture,[138] but the significance of these observations is also obscure. It is presumed that the function of TIMPs is to regulate the activity of MMPs, but since they do not seem to be highly selective, it is difficult to decide upon which interactions are likely to take place *in vivo,* although Howard et al.[139] have made a good case for some specificity in the interaction of TIMP-2 and MMP-2 and MMP-9.

C. TIMPs IN THE ACUTE PHASE

MMPs are implicated in some models of tumor progression. Thus, a potential role for TIMPs exists as antimetastatic factors. For example, Swiss 3T3 cells take on an oncogenic phenotype when transformed with an antisense TIMP-1 RNA,[140] and transformation of metastatic rat cells with TIMP-2 decreased their invasive potential.[141] These data indicate that TIMPs may normally act to suppress MMP-mediated invasion *in vivo.* The relationship between MMPs, TIMPs, and oncogenesis has led to many investigations of the response of these proteins to growth regulation and transforming substances.[130] Thus, recent studies[142,143] have shown increased synthesis and secretion of TIMP-1, but not its target enzymes, in response to IL-6.

IX. CONCLUSIONS

There is little doubt that a close link exists between proteolysis and the acute phase response to inflammation. Inflammation is usually produced by infectious pathogens, which use proteinases to help invade. The host cells primarily involved in the early stages of controlling infection (neutrophils, macrophages, and mast cells) contain huge quantities of proteinases. Consequently, the body is proteolytically assaulted from without and from within. It is reasonable to conclude that the assault is regulated by increasing the production of selective and nonspecific proteinase inhibitors, and that this is why many acute phase reactants are proteinase inhibitors. This is about as far as our knowledge goes, for we have not yet been able to answer the crucial question, "what are the target enzymes of the major acute phase-responding proteinase inhibitors?" Future studies will attempt to answer this question.

We will also need to increase our understanding of the role of proteinase inhibitors as potential signals during the acute phase. Do kunins, serpins, and α-macroglobulins act as growth factors and/or cytokines *in vivo,* or are these symptoms of cultured cells? Are the elevations in the levels of some kunins, kazals, and cystatins real acute phase responses? As is often the case, elucidations of biological roles have lagged behind structural understanding, but this is acceptable when we consider that if we do not know the structure of a protein, there is little chance that we will be able to accurately predict or understand all its functions. Now that several structures are available, and the familial relationships clearer, we can concentrate on functions.

ACKNOWLEDGMENTS

We thank Wolfram Bode for the molecular coordinates of chicken cystatin, Jane Richardson and David Richardson for helpful discussions and for pointing out the similarity between cystatin and ribonuclease chain folds, Magnus Abrahamson, Tomoko Komiyama, Michael Laskowski, Jr., and Hideaki Nagase for helpful suggestions, and Louise Taft for help with the art work. Molecular graphics were generated with the aid of facilities of the Duke Comprehensive Cancer Center using the computer program CHAOS written by David Richardson and Michael Zalis.

REFERENCES

1. **Laskowski, M. and Kato, I.,** Protein inhibitors of proteinases, *Annu. Rev. Biochem.,* 49, 593, 1980.
2. **Travis, J. and Salvesen, G. S.,** Human plasma proteinase inhibitors, *Annu. Rev. Biochem.,* 52, 655, 1983.
3. **Barrett, A. J. and Salvesen, G.,** *Proteinase Inhibitors,* Elsevier, Amsterdam, 1986.
4. **Gebhard, W., Tscheche, H., and Fritz, H.,** Biochemistry of aprotinin and aprotinin-like inhibitors, in *Proteinase Inhibitors,* Barrett, A. J. and Salvesen, G. S., Eds., Elsevier, Amsterdam, 1986, 375.
5. **Laskowski, M., Kato, I., Ardelt, W., Cook, J., Denton, A., Empie, M. W., Kohr, W. J., Park, S. J., Parks, K., Schatzley, B. L., Schoenberger, O. L., Tashiro, M., Vichot, G., Whatley, H. E., Wieczorek, A., and Wieczorek, M.,** Ovomucoid third domain from 100 avian species: isolation, sequences, and hypervariability of enzyme-inhibitor contact residues, *Biochemistry,* 26, 202, 1987.
6. **Carrell, R. W. and Travis, J.,** α_1-Antitrypsin and the serpins: variation and countervariation, *Trends Biochem. Sci.,* 10, 20, 1985.
7. **Carrell, R. W. and Boswell, D. R.,** Serpins: the superfamily of plasma serine proteinase inhibitors, in *Proteinase Inhibitors,* Barrett, A. J. and Salvesen, G., Eds., Elsevier, Amsterdam, 1986, 403.
8. **Huber, R. and Carrell, R. W.,** Implications of the three-dimensional structure of α_1-antitrypsin for structure and function of serpins, *Biochemistry,* 28, 8966, 1989.
9. **Sottrup-Jensen, L.,** α_2-Macroglobulin and related thiol ester plasma proteins, in *Plasma Proteins,* Putnam, F. W., Ed., Academic Press, New York, 1987, 191.
10. **Barrett, A. J., Rawlings, N. D., Davies, M. E., Machledit, W., Salvesen, G., and Turk, V.,** Cysteine proteinase inhibitors of the cystatin superfamily, in *Proteinase Inhibitors,* Barrett, A. J. and Salvesen, G., Eds., Elsevier, Amsterdam, 1986, 515.
11. **Barrett, A. J.,** The cystatins: a new class of peptidase inhibitors, *Trends Biochem. Sci.,* 12, 193, 1987.
12. **Dayhoff, M. O., Barker, W. C., and Hunt, L. T.,** Protein superfamilies, in *Atlas of Protein Sequence and Structure,* Dayhoff, M. O., Ed., National Biomedical Research Foundtion, Washington, D.C., 1978, 9.
13. **Hill, R. E. and Hastie, N. D.,** Accelerated evolution in the reactive centre regions of serine protease inhibitors, *Nature (London),* 326, 96, 1987.
14. **Laskowski, M. J., Kato, I., Kohr, W. J., Park, S. J., Tashiro, M., and Whatley, H. E.,** Positive Darwinian selection in evolution of protein inhibitors of serine proteinases, *Cold Spring Harbor Symp. Quant. Biol.,* 52, 545, 1987.
15. **Potempa, J., Watorwek, W., and Travis, J.,** The inactivation of human plasma α_1-proteinase inhibitor by proteinases from *Staphylococcus aureus, J. Biol. Chem.,* 261, 14330, 1986.
16. **Mast, A. E., Enghild, J. J., Nagase, H., Suzuki, K., Pizzo, S. V., and Salvesen, G.,** Kinetics and physiologic relevance of the inactivation of α_1-proteinase inhibitor, α_1-antichymotrypsin, and antithrombin III by matrix metalloproteinases-1 (tissue collagenase), -2 (72-kDa gelatinase/type IV collagenase), and -3 (stromelysin), *J. Biol. Chem.,* 266, 15810, 1991.
17. **McKeehan, W. L., Sakagami, Y., Hoshi, H., and McKeehan, K. A.,** Two apparent human endothelial cell growth factors from human hepatoma cells are tumor-associated proteinase inhibitors, *J. Biol. Chem.,* 261, 5378, 1986.
18. **Hayakawa, T., Yamashita, K., Tanzawa, K., Uchijima, E., and Iwata,K.,** Growth-promoting activity of tissue inhibitor of metalloproteinases-1 (TIMP-1) for a wide range of cells. A possible new growth factor in serum, *FEBS Lett.,* 298, 29, 1992.
19. **Sommer, J., Gloor, S. M., Rovelli, G. F., Hofsteenge, J., Nick, H., Meier, R., and Monard, D.,** cDNA sequence coding for a rat glia-derived nexin and its homology to members of the serpin superfamily, *Biochemistry,* 26, 6407, 1987.

20. **Gurwitz, D. and Cunningham, P. D.**, Thrombin modulates and reverses neuroblastoma neurite outgrowth, *Proc. Natl. Acad. Sci. U.S.A.*, 85, 3440, 1988.
21. **Gebhard, W., Schreitmüller, T., Hochstrasser, K., and Wachter, E.**, Two out of the three kinds of subunits of inter-α-trypsin inhibitor are structurally related, *Eur. J. Biochem.*, 181, 571, 1989.
22. **Gebhard, W., Hochstrasser, K., Fritz, H., Enghild, J. J., Pizzo, S. V., and Salvesen, G.**, Structure of inter-α-inhibitor (inter-α-trypsin inhibitor) and pre-α-inhibitor: current state and proposition of a new terminology, *Biol. Chem. Hoppe-Seyler*, 371 (Suppl.), 13, 1990.
23. **Broze, G. J., Girard, T. J., and Novotny, W. F.**, Regulation of coagulation by a multivalent Kunitz-type inhibitor, *Biochemistry*, 29, 7539, 1990.
24. **Kunitz, M. and Northrop, J. H.**, Isolation from beef pancreas of crystalline trypsinogen, trypsin, a trypsin inhibitor, and an inhibitor-trypsin compound, *J. Gen. Physiol.*, 19, 991, 1936.
25. **Creighton, T. E.**, Kinetic study of protein unfolding and refolding using urea gradient electrophoresis, *J. Mol. Biol.*, 137, 61, 1980.
26. **Dufton, M. J.**, Proteinase inhibitors and dendrotoxins. Sequence classification, structural prediction and structure/activity, *Eur. J. Biochem.*, 153, 647, 1985.
27. **Read, R. J. and James, M. N. G.**, Introduction to the protein inhibitors: X-ray crystallography, in *Proteinase Inhibitors*, Barrett, A. J. and Salvesen, G. S., Eds., Elsevier, Amsterdam, 1986, 301.
28. **Hubbard, S. J., Campbell, S. F., and Thornton, J. M.**, Molecular recognition. Conformational analysis of limited proteolytic sites and serine proteinase protein inhibitors, *J. Mol. Biol.*, 220, 507, 1991.
29. **Kido, H., Hokogoshi, Y., and Katanuma, N.**, Kunitz-type proteinase inhibitor found in rat mast cells, *J. Biol. Chem.*, 263, 18104, 1988.
30. **Fritz, H. and Kruck, J.**, Immunofluorescence studies indicate that the basic trypsin-kallikrein-inhibitor of bovine organs (trasylol) originates from mast cells, *Hoppe-Seyler's Z. Physiol. Chem.*, 360, 437, 1979.
31. **Kido, H., Fukusen, N., and Katunuma, N.**, Chymotrypsin- and trypsin-type serine proteases in rat mast cells: properties and functions, *Arch. Biochem. Biophys.*, 239, 436, 1985.
32. **Steinbuch, M. and Loeb, J.**, Isolation of an α_2-globulin from human plasma, *Nature (London)*, 182, 1196, 1961.
33. **Enghild, J., Thøgersen, I., Pizzo, S., and Salvesen, G.**, Analysis of inter-α-trypsin inhibitor and a novel trypsin inhibitor, pre-α-trypsin inhibitor, from human plasma: polypeptide chain stoichiometry and assembly by glycan, *J. Biol. Chem.*, 264, 15975, 1989.
34. **Enghild, J. J., Salvesen, G., Hefta, S. A., Thogersen, I. B., Rutherfurd, S., and Pizzo, S. V.**, Chondroitin 4-sulfate covalently cross-links the chains of the human blood protein pre-α-inhibitor, *J. Biol. Chem.*, 266, 747, 1991.
35. **Enghild, J. J. and Salvesen, G.**, unpublished data, 1990.
36. **Novotny, W. F., Girard, T. J., Miletich, J. P., and Broze, G. J.**, Purification and characterization of the lipoprotein-associated coagulation inhibitor from human plasma, *J. Biol. Chem.*, 264, 18832, 1989.
37. **Rapaport, S. I.**, The extrinsic pathway inhibitor: a regulator of tissue factor-dependent blood coagulation, *Thromb. Haemost.*, 66, 6, 1991.
38. **Girard, T. J., Warren, L. A., Novotny, W. F., Likert, K. M., Brown, S. G., Miletich, J. P., and Broze, G. J.**, Functional significance of the Kunitz-type inhibitory domains of lipoprotein-associated coagulation inhibitor, *Nature (London)*, 338, 518, 1989.
39. **Baker, J. B., Low, D. A., Simmer, R. L., and Cunningham, D. D.**, Protease-nexin: a cellular component that links thrombin and plasminogen activator and mediates their binding to cells, *Cell*, 21, 37, 1980.
40. **Oltersdorf, T., Fritz, L. C., Schenk, D. B., Lieberburg, I., Johnson-Wood, K. L., Beattie, E. C., Ward, P. J., Blacher, R. W., Dovey, H. F., and Sinha, S.**, The secreted form of the Alzheimer's amyloid precursor protein with the Kunitz domain is protease nexin II, *Nature (London)*, 341, 144, 1989.
41. **Tanzi, R. E., McClatchey, A. I., Lamperti, E. D., Villa-Komaroff, L., Gusella, J. F., and Neve, R. L.**, Protease inhibitor domain encoded by an amyloid protein precursor mRNA associated with Alzheimer's disease, *Nature (London)*, 331, 571, 1988.
42. **Weidemann, A., Konig, G., Bunke, D., Fischer, P., Salbaum, J. M., Masters, C. L., and Beyreuther, K.**, Identification, biogenesis and localization of precursors of Alzheimer's disease A4 amyloid protein, *Cell*, 57, 115, 1989.
43. **Palmert, M. R., Podlisny, M. B., Witker, D. S., Olterdorf, T., Younkin, L. H., Selkoe, D. J., and Younkin, S. G.**, The β-amyloid protein precursor of Alzheimer's diseae has soluble derivatives found in human brain and cerebrospinal fluid, *Proc. Natl. Acad. Sci. U.S.A.*, 86, 6338, 1989.
44. **Sisodia, S. S., Koo, E. H., Beyreuther, K., Unterbeck, A., and Price, D. L.**, Evidence that β-amyloid protein in Alzheimer's disease is not derived by normal processing, *Science*, 248, 492, 1990.
45. **Van Nostrand, W. E., Wagner, S. L., Farrow, J. S., and Cunningham, D. D.**, Immunopurification and protease inhibitory properties of protease nexin-2/amyloid beta-protein precursor, *J. Biol. Chem.*, 265, 9591, 1990.

46. **Smith, R. P., Higuchi, D. A., and Broze, G. J.,** Platelet coagulation factor XIa-inhibitor, a form of Alzheimer amyloid precursor protein, *Science,* 248, 1126, 1990.
47. **Mayne, R. and Burgeson, R. E.,** *Structure and Function of Collagen Types,* Academic Press, Orlando, 1987.
48. **Doliana, R., Bonaldo, P., and Columbatti, A.,** Multiple forms of chicken alpha 3(VI) collagen generated by alternative splicing in type A repeated domains, *J. Cell Biol.,* 111, 2197, 1990.
49. **Chu, M. L., Mann, K., Deutzmann, R., Pribula-Conway, D., Hsu-Chen, C. C., Bernard, M. P., and Timpl, R.,** Characterization of three constituent chains of collagen type VI by peptide sequences and cDNA clones, *Eur. J. Biochem.,* 168, 309, 1987.
50. **Abraham, C. R., Selkoe, D. J., and Potter, H.,** Immunochemical identification of the protease inhibitor alpha 1-antichymotrypsin in the brain amyloid deposits of Alzheimer's disease, *Cell,* 52, 487, 1988.
51. **Selkoe, D. J.,** Aging, amyloid, and Alzheimer's disease, *N. Engl. J. Med.,* 320, 1484, 1989.
52. **Chawla, R. K., Rausch, D. J., Miller, F. W., Vogler, W. R., and Lawson, D. H.,** Abnormal profile of serum proteinase inhibitors in cancer patients, *Cancer Res.,* 44, 2718, 1984.
53. **Pratt, C. W. and Pizzo, S. V.,** Mechanism of action of inter-α-trypsin inhibitor, *Biochemistry,* 26, 2855, 1987.
54. **Salier, J.-P.,** Inter-α-trypsin inhibitor: emergence of a family within the Kunitz-type protease inhibitor superfamily, *Trends Biochem. Sci.,* 15, 435, 1990.
55. **Rupp, F., Payan, D. G., Magill-Solc, C., Cowan, D. M., and Scheller, R. H.,** Structure and expression of rat agrin, *Neuron,* 6, 811, 1991.
56. **Bode, W. and Huber, R.,** Ligand binding: proteinase-proteinase inhibitor interactions, *Curr. Opin. Struct. Biol.,* 1, 45, 1991.
57. **Horii, A., Tomita, N., Yokouchi, H., Doi, S., Uda, K., Ogawa, M., Mori, T., and Matsubara, K.,** On the cDNAs for two types of rat pancreatic secretory trypsin inhibitor, *Biochem. Biophys. Res. Commun.,* 162, 151, 1989.
58. **Agerbeth, B., Soderling-Barros, J., Jornvall, H., Chen, Z. W., Ostenson, C. G., Efendic, S., and Mutt, V.,** Isolation and characterization of a 60-residue intestinal peptide structurally related to the pancreatic secretory type of trypsin inhibitor: influence on insulin secretion, *Proc. Natl. Acad. Sci. U.S.A.,* 86, 8590, 1989.
59. **Lin, Y.-Z., Isaac, D. D., and Tam, J. P.,** Synthesis and properties of cholecystokinin-releasing peptide (monitor peptide), a 61-residue trypsin inhibitor, *Int. J. Pept. Protein Res.,* 36, 433, 1990.
60. **Moritz, A., Lilja, H., and Fink, E.,** Molecular cloning and sequence analysis of the cDNA encoding the human acrosin-trypsin inhibitor (HUSI-II), *FEBS Lett.,* 278, 127, 1991.
61. **Reisinger, P. W. M., Hochstrasser, K., Goettlicher, I., Eulitz, M., and Wachter, E.,** The amino acid sequence of the double-headed proteinase inhibitors from cat, lion and dog submandibular glands, *Biol. Chem. Hoppe-Seyler,* 368, 717, 1987.
62. **Ogawa, M.,** Pancreatic secretory trypsin inhibitor as an acute phase reactant, *Clin. Biochem.,* 21, 19, 1988.
63. **Turpeinen, U., Koivunen, E., and Stenman, U.-H.,** Reaction of tumor-associated trypsin inhibitor with serine proteinases associated with coagulation and tumor invasion, *Biochem. J.,* 254, 911, 1988.
64. **Yasuda, T., Ogawa, M., Murata, A., Oka, Y., Uda, K., and Mori, T.,** Response to IL-6 stimulation of human hepatoblastoma cells: production of pancreatic secretory trypsin inhibitor, *Biol. Chem. Hoppe-Seyler,* 371, 95, 1990.
65. **Grütter, G. M., Fendrich, G., Huber, R., and Bode, W.,** The 2.5 Å X-ray structure of the acid-stable proteinase inhibitor from human mucous secretions analyzed in its complex with bovine α-chymotrypsin, *EMBO J.,* 7, 345, 1988.
66. **Hochstrasser, K., Albrecht, G. J., Schonberger, O. L., Rasche, B., and Lempart, K.,** An elastase-specific inhibitor from human bronchial mucus. Isolation and characterization, *Hoppe-Seyler's Z. Physiol. Chem.,* 362, 1369, 1981.
67. **Salvesen, G. and Travis, J.,** properties of naturally occurring elastase inhibitors, in *Elastin and Elastases,* Robert, L. and Hornebeck, W., Eds., CRC Press, Boca Raton, FL, 1988, 95.
68. **Thompson, R. C. and Ohlsson, K.,** Isolation, properties, and complete amino acid sequence of human secretory leukocyte protease inhibitor, a potent inhibitor of leukocyte elastase, *Proc. Natl. Acad. Sci. U.S.A.,* 83, 6692, 1986.
69. **Meckelein, B., Nikiforov, T., Clemen, A., and Appelhans, H.,** The location of inhibitory specificities in human mucus proteinase inhibitor (MPI): separate expression of the COOH-terminal domain yields an active inhibitor, *Protein Eng.,* 3, 215, 1990.
70. **Van-Seuningen, I. and Davril, M.,** Separation of the two domains of human mucus proteinase inhibitor, *Biochem. Biophys. Res. Commun.,* 179, 1587, 1991.

71. **Vogelmeier, C., Hubbard, R. C., Fells, G. A., Schnebli, H. P., Thompson, R. C., Fritz, H., and Crystal, R. G.,** Anti-neutrophil elastase defense of the normal human respiratory epithelial surface provided by the secretory leukoprotese inhibitor, *J. Clin. Invest.*, 87, 482, 1991.
72. **Wiedow, O., Schroder, J.-M., Greogry, H., Young, J. A., and Christophers, E.,** Elafin: an elastase-specific inhibitor of human skin, *J. Biol. Chem.*, 265, 14791, 1990.
73. **Wiedow, O., Lüdemann, J., and Utecht, B.,** Elafin is a potent inhibitor of proteinase 3, *Biochem. Biophys. Res. Commun.*, 174, 6, 1991.
74. **Doolittle, R. F.,** Angiotensinogen is related to the antitrypsin-antithrombin-ovalbumin family, *Science*, 222, 417, 1983.
75. **Hunt, L. T. and Dayhoff, M. O.,** A surprising new protein superfamily containing ovalbumin, antithrombin-III, and α_1-proteinase inhibitor, *Biochem. Biophys. Res. Commun.*, 95, 864, 1980.
76. **Loebermann, H., Tokuoka, R., Deisenhofer, J., and Huber, R.,** Human α_1-proteinase inhibitor: crystal structure analysis of two modifications, molecular model and preliminary analysis of the implications for function, *J. Mol. Biol.*, 177, 531, 1984.
77. **Stein, P. E., Leslie, A. G. W., Finch, J. T., Turnall, W. G., McLaughlin, P. J., and Carrell, R. W.,** Crystal structure of ovalbumin as a model for the reactive center of serpins, *Nature (London)*, 347, 99, 1990.
78. **Wright, H. T., Qian, H. X., and Huber, R.,** Crystal structure of plakalbumin, a proteolytically nicked form of ovalbumin: its relationship to the structure of cleaved α_1-proteinase inhibitor, *J. Mol. Biol.*, 213, 513, 1990.
79. **Baumann, U., Huber, R., Bode, W., Grosse, D., Lesjak, M., and Laurell, C. B.,** Crystal structure of cleaved human α_1-antichymotrypsin at 2.7 Å resolution and its comparison with other serpins, *J. Mol. Biol.*, 218, 595, 1991.
80. **Longas, M. O. and Finley, T. H.,** The covalent nature of the human antithrombin III-thrombin bond, *Biochem. J.*, 189, 481, 1980.
81. **Mahoney, W. C., Kurachi, K., and Hermodson, M. A.,** Formation and dissociation of the covalent complexes between trypsin and two homologous inhibitors, α_1-antitrypsin and antithrombin III, *Eur. J. Biochem.*, 105, 545, 1980.
82. **Moroi, M. and Aoki, N.,** On the interaction of α_2-plasmin inhibitor and proteases: evidence for the formation of a covalent crosslinkage and non-covalent weak bondings between the inhibitor and proteases, *Biochim. Biophys. Acta*, 482, 412, 1977.
83. **Tomono, T. and Sawada, E.,** Preparation of anhydro-thrombin and its interaction with plasma antithrombin III, *Acta Haematol. Jpn.*, 49, 163, 1986.
84. **Shieh, B.-H., Potempa, J., and Travis, J.,** The use of α_2-antiplasmin as a model for the demonstration of complex reversibility in serpins, *J. Biol. Chem.*, 264, 13420, 1989.
85. **Longstaff, C. and Gaffney, P. J.,** Serpin-serine protese binding kinetics: α_2-antiplasmin as a model inhibitor, *Biochemistry*, 30, 979, 1991.
86. **Matheson, N. R., Van-Halbeek, H., and Travis, J.,** Evidence for a tetrahedral intermediate complex during serpin-proteinase interactions, *J. Biol. Chem.*, 266, 13489, 1991.
87. **Mast, A. E., Enghild, J. J., Pizzo, S. V., and Salvesen, G.,** Analysis of the plasma elimination kinetics and conformational stabilities of native, proteinase-complexed, and reactive site cleaved serpins: comparison of α_1-proteinase inhibitor, α_1-antichymotrypsin, antithrombin III, α_2-antiplasmin, angiotensinogen, and ovalbumin, *Biochemistry*, 30, 1723, 1991.
88. **Mast, A. E., Enghild, J. J., and Salvesen, G.,** Conformation of the reactive site loop of human α_1-proteinase inhibitor probed by limited proteolysis, *Biochemistry*, 31, 2720, 1992.
89. **Kidd, V. J. and Woo, S. L. C.,** Molecular analysis of the serine proteinase inhibitor gene family, in *Proteinase Inhibitors*, Barrett, A. J. and Salvesen, G., Eds., Elsevier, Amsterdam, 1986, 421.
90. **Beatty, K., Bieth, J., and Travis, J.,** Kinetics of association of serine proteinases with native oxidized α-1 proteinase inhibitor and α-1-antichymotrypsin, *J. Biol. Chem.*, 255, 3931, 1980.
91. **Hill, R. E., Shaw, P. H., Boyd, P. A., Baumann, H., and Hastie, N. D.,** Plasma protease inhibitors in mouse and man: divergence within the reactive centre regions, *Nature (London)*, 311, 175, 1984.
92. **Inglis, J. D. and Hill, R. E.,** The murine Spi-2 proteinase inhibitor locus: a multigene family with a hypervariable reactive site domain, *EMBO J.*, 10, 255, 1991.
93. **Pirie-Shepherd, S. R., Miller, H. R. P., and Ryle, A.,** Differential inhibition of rat mast cell proteinase I and II by members of the alpha-1-proteinase inhibitor family of serine proteinase inhibitors, *J. Biol. Chem.*, 266, 17314, 1991.
94. **Banda, M. J., Rice, A. G., Griffin, G. L., and Senior, R. M.,** α_1-proteinase inhibitor is a neutrophil chemoattractant after proteolytic inactivation by a macrophage elastase, *J. Biol. Chem.*, 263, 4481, 1988.
95. **Perlmutter, D. H., Travis, J., and Punsal, P. I.,** Elastase regulates the synthesis of its inhibitor, α_1-proteinase inhibitor, and exaggerates the defect in homozygous PiZZ α_1PI deficiency, *J. Clin. Invest.*, 81, 1774, 1988.

96. **Perlmutter, D. H., Joslin, G., Nelson, P. C. S., Adams, S. P., and Fallon, R. J.,** Endocytosis and degradation of α_1-antitrypsin-protease complexes is mediated by the serpin-enzyme complex (SEC) receptor, *J. Biol. Chem.*, 265, 16713, 1990.
97. **Pizzo, S. V., Mast, A. E., Feldman, S. R., and Salvesen, G.,** In vivo catabolism of α_1-antichymotrypsin is mediated by the serpin receptor which binds α_1-proteinase inhibitor, antithrombin III and heparin cofactor II, *Biochim. Biophys. Acta,* 967, 158, 1988.
98. **Barbey-Morel, C. and Perlmutter, D. H.,** Effect of pseudomonas elastase on human mononuclear phagocyte α_1-antitrypsin expression, *Pediatr. Res.*, 29, 133, 1991.
99. **Thogersen, I. B., Salvesen, G., Brucato, F. H., Pizzo, S. V., and Enghild, J. J.,** Purification and characterization of an α-macroglobulin from the mollusc *Octopus vulgaris, Biochem. J.*, in press.
100. **Quigley, J. P. and Armstrong, P. B.,** An endopeptidase inhibitor, similar to mammalian α_2-macroglobulin, detected in the hemolymph of an invertebrate, *Limulus polyphemus, J. Biol. Chem.*, 258, 7903, 1983.
101. **Enghild, J. J., Thogersen, I. B., Salvesen, G., Figler, N. L., Gonias, S. L., and Pizzo, S. V.,** The α-macroglobulin from *Limulus polyphemus* exhibits proteinase-inhibitory activity and participates in a hemolytic system, *Biochemistry,* 29, 10070, 1990.
102. **Jensen, P. E. H. and Sottrup-Jensen, L.,** Primary structure of human α2-macroglobulin: complete disulfide bridge assignment and localization of two interchain bridges in the dimeric proteinase binding unit, *J. Biol. Chem.*, 261, 15863, 1986.
103. **Rubenstein, D. S., Enghild, J. J., and Pizzo, S. V.,** Limited proteolysis of the alpha-macroglobulin rat alpha$_1$-inhibitor-3, *J. Biol. Chem.*, 266, 11252, 1991.
104. **Barrett, A. J. and Starkey, P. M.,** The interaction of α_2-macroglobulin with proteinases: characteristics and specificity of the reaction, and a hypothesis concerning its molecular mechanism, *Biochem. J.*, 133, 709, 1973.
105. **Kuehn, L., Dahlmann, B., Gauthier, F., and Neubauer, H. P.,** High-molecular-mass proteinases in rabbit reticulocytes: the multicatalytic proteinase is an ATP-independent enzyme and ATP-activated proteolysis is in part associated with a cysteine proteinase complexed to alpha 1-macroglobulin, *Biochim. Biophys. Acta,* 991, 263, 1989.
106. **Howard, J. B.,** Reactive site in human alpha 2-macroglobulin: circumstantial evidence for a thiolester, *Proc. Natl. Acad. Sci. U.S.A.*, 78, 2235, 1981.
107. **Enghild, J. J., Salvesen, G. S., Thørgensen, I. B., and Pizzo, S. V.,** Proteinase binding and inhibition by the monomeric α-macroglobulin rat α_1-inhibitor-3, *J. Biol. Chem.*, 264, 11428, 1989.
108. **Christensen, U., Simonsen, M., Harrit, N., and Sottrup-Jensen, L.,** Pregnancy zone protein, a proteinase-binding macroglobulin. Interactions with proteinases and methylamine, *Biochemistry,* 28, 9324, 1989.
109. **Nagase, H., Harris, E. D. J., Woessner, J. F. J., and Brew, K.,** Ovostatin: a novel proteinase inhibitor from chicken egg white, *J. Biol. Chem.*, 258, 7481, 1983.
110. **Barrett, A. J., Brown, M. A., and Sayers, C. A.,** The electrophoretically 'slow' and 'fast' forms of the α_2-macroglobulin molecule, *Biochem. J.*, 181, 401, 1979.
111. **Fuchs, H. E., Shifman, M. A., and Pizzo, S. V.,** In vivo catabolism of α_1-proteinase inhibitor-trypsin, antithrombin III-thrombin and α_2-macroglobulin-methylamine, *Biochim. Biophys. Acta,* 716, 151, 1982.
112. **Hiwasa, T., Yokoyama, S., Ha, J.-M., Nuguchi, S., and Sakiyama, S.,** c-Ha-*ras* gene products are potent inhibitors of cathepsins B and L, *FEBS Lett.*, 211, 23, 1987.
113. **Keil-Dlouha, V. and Turk, V.,** Structural analogies between adhesive proteins and cysteine proteinase inhibitors, *Biol. Chem. Hoppe-Seyler,* 369, 199, 1988.
114. **Stubbs, M. T., Laber, B., Bode, W., Huber, R., Jerala, R., Lenarcic, B., and Turk, V.,** The refined 2.4 Å X-ray crystal structure of recombinant human stefin B in complex with the cysteine proteinase papain: a novel type of proteinase inhibitor interaction, *EMBO J.*, 9, 1939, 1990.
115. **Bode, W., Engh, R., Musil, D., Thiele, U., Huber, R., Karshikov, A., Brzin, J., Kos, J., and Turk, V.,** The 2.0 Å X-ray crystal structure of chicken egg white cystatin and its possible mode of interaction with cysteine proteinases, *EMBO J.*, 7, 2593, 1988.
116. **Nicklin, M. J. H. and Barrett, A. J.,** Inhibition of cysteine proteinases and dipeptidyl peptidase I by egg-white cystatin, *Biochem. J.*, 223, 245, 1984.
117. **Musil, D., Zucic, D., Turk, D., Engh, R. A., Mayr, I., Huber, R., Popovic, T., Turk, V., Towatari, T., Katunuma, N., and Bode, W.,** The refined 2.15 Å X-ray crystal structure of human liver cathepsin B: the structural basis for its specificity, *EMBO J.*, 10, 2321, 1991.
118. **Abrahamson, M., Mason, R. W., Hansson, H., Buttle, D. J., Grubb, A., and Ohlsson, K.,** Human cystatin C. Role of the N-terminal segment in the inhibition of human cysteine proteinases and in its inactivation by leucocyte elastase, *Biochem. J.*, 273, 621, 1991.

119. **Machleidt, W., Thiele, U., Laber, B., Assfalg-Machleidt, I., Esterl, A., Wiegand, G., Kos, J., Turk, V., and Bode, W.,** Mechanism of inhibition of papain by chicken egg white cystatin. Inhibition constants of N-terminally truncated forms and cyanogen bromide fragments of the inhibitor, *FEBS Lett.*, 243, 234, 1989.
120. **Abrahamson, M., Ritonja, A., Brown, M. A., Grubb, A., Machleidt, W., and Barrett, A. J.,** Identification of the probable inhibitory reactive site of the cysteine proteinase inhibitors human cystatin C and chicken cystatin, *J. Biol. Chem.*, 262, 9688, 1987.
121. **Abrahamson, M., Barrett, A. J., Salvesen, G., and Grubb, A.,** Isolation of six cysteine proteinase inhibitors from human urine. Their physiochemical and enzyme kinetic properties and concentrations in biological fluids, *J. Biol. Chem.*, 261, 11282, 1986.
122. **Freije, J. P., Abrahamson, M., Olafsson, I., Velasco, G., Grubb, A., and López-Otin, C.,** Structure and expression of the gene encoding cystatin D, a novel human cysteine proteinase inhibitor, *J. Biol. Chem.*, 266, 20538, 1991.
123. **Isemura, S., Saitoh, E., and Sanada, K.,** Charcterization and amino acid sequence of a new acidic cysteine proteinase inhibitor (cystatin SA) structurally closely related to cystatins, *J. Biochem.*, 102, 693, 1987.
124. **Rawlings, N. D. and Barrett, A. J.,** Evolution of proteins of the cystatin superfamily, *J. Mol. Evol.*, 30, 60, 1990.
125. **Jackson, C. M. and Nemerson, Y.,** Blood coagulation, *Annu. Rev. Biochem.*, 49, 765, 1980.
126. **Kitamura, N., Takagaki, Y., Furuto, S., Tanaka, T., Nawa, H., and Nakanishi, S.,** A single gene for bovine high molecular weight and low molecular weight kininogens, *Nature (London)*, 305, 545, 1983.
127. **Salvesen, G., Parkes, C., Abrahamson, M., Grubb, A., and Barrett, A. J.,** Human low-Mr kininogen contains three copies of a cystatin sequence that are divergent in structure and in inhibitory activity for cysteine proteinases, *Biochem. J.*, 234, 429, 1986.
128. **Docherty, A. J. P., Lyons, A., Smith, B. J., Wright, E. M., Stephens, P. E., Harris, T. J. R., Murphy, G., and Reynolds, J. J.,** Sequence of human tissue inhibitor of metalloproteinases and its identity to erythroid-potentiating activity, *Nature (London)*, 318, 66, 1985.
129. **Boone, T. C., Johnson, M. J., De Clerck, Y. A., and Langley, K. E.,** cDNA cloning and expression of a metalloproteinase inhibitor related to tissue inhibitor of metalloproteinases, *Proc. Natl. Acad. Sci. U.S.A.*, 87, 2800, 1990.
130. **Woessner, J. F. J.,** Matrix metalloproteinases and their inhibitors in connective tissue remodeling, *FASEB J.*, 5, 2145, 1991.
131. **Williamson, R. A., Marston, F. A. O., Angal, S., Koklitis, P., Panico, M., Morris, H. R., Carne, A. F., Smith, B. J., Harris, T. J. R., and Freedman, R. B.,** Disulphide bond assignment in human tissue inhibitor of metalloproteinases (TIMP), *Biochem. J.*, 268, 267, 1990.
132. **Murphy, G., Houbrechts, A., Cockett, M. I., Williamson, R. A., O'Shea, M., and Docherty, A. J. P.,** The N-terminal domain of tissue inhibitor of metalloproteinases retains metalloproteinase inhibitory activity, *Biochemistry*, 30, 8097, 1991.
133. **Cawston, T. E.,** Protein inhibitors of metallo-proteinases, in *Proteinase Inhibitors*, Barrett, A. J. and Salvesen, G., Eds., Elsevier, Amsterdam, 1986, 589.
134. **Dumermuth, E., Sterchi, E. E., Jiang, W., Wolz, R. L., Bond, J. S., Flannery, A. V., and Beynon, R. J.,** The astacin family of metalloendopeptidases, *J. Biol. Chem.*, 266, 21381, 1991.
135. **Stetler-Stevenson, W. G., Krutzsch, H. C., and Liotta, L. A.,** Tissue inhibitor of metalloproteinase (TIMP-2). A new member of the metalloproteinase inhibitor family, *J. Biol. Chem.*, 264, 17374, 1989.
136. **Goldberg, G. I., Mermer, B. L., Grant, G. A., Eisen, A. Z., Wilhelm, S., and He, C. S.,** Human 72-kilodalton type IV collagenase forms a complex with a tissue inhibitor of metalloproteases designated TIMP-2, *Proc. Natl. Acad. Sci. U.S.A.*, 86, 8207, 1989.
137. **Howard, E. W. and Banda, M. J.,** Binding of tissue inhibitor of metalloproteinases 2 to two distinct sites on human 72-kDa gelatinase. Identification of a stabilization site, *J. Biol. Chem.*, 266, 17972, 1991.
138. **Kleiner, D. E. J., Unsworth, E. J., Krutsch, H. C., and Stetler-Stevenson, W. G.,** Higher order complex formation between the 72-kilodalton type IV collagenase and tissue inhibitor of metalloproteinases-2, *Biochemistry*, 31, 1665, 1992.
139. **Howard, E. W., Bullen, E. C., and Banda, M. J.,** Preferential inhibition of 72- and 92-kDa gelatinases by tissue inhibitor of metalloproteinases-2, *J. Biol. Chem.*, 266, 13070, 1991.
140. **Khokha, R., Waterhouse, P., Yagel, S., Lala, P. K., Overall, C. M., Norton, G., and Denhart, D. T.,** Antisense RNA-induced reduction in murine TIMP levels confers oncogenicity on Swiss 3T3 cells, *Science*, 243, 947, 1989.
141. **DeClerck, Y. A., Perez, N., Shimada, H., Boone, T. C., Langley, K. E., and Taylor, S. M.,** Inhibition of invasion and metastasis in cells transfected with an inhibitor of metalloproteinases, *Cancer Res.*, 52, 701, 1992.

142. **Sato, T., Ito, A., and Mori, Y.,** Interleukin 6 enhances the production of tissue inhibitor of metalloproteinases (TIMP) but not that of matrix metalloproteinases by human fibroblasts, *Biochem. Biophys. Res. Commun.,* 170, 824, 1990.
143. **Lotz, M. and Guerne, P. A.,** Interleukin-6 induces the synthesis of tissue inhibitor of metalloproteinases-1/erythroid potentiating activity (TIMP-1/EPA), *J. Biol. Chem.,* 266, 2017, 1991.
144. **Warne, N. W. and Laskowski, M. J.,** All fifteen arrangements of three disulfide bridges in proteins are known, *Biochem. Biophys. Res. Commun.,* 172, 1364, 1990.
145. **Marquart, M., Walter, J., Diesenhofer, J., Bode, W., and Huber, R.,** The geometry of the reactive site and of the peptide groups in trypsin, trypsinogen and its complexes with inhibitors, *Acta Crystallogr.,* B39, 480, 1983.
146. The computer program "CHAOS" was written by David Richardson and Michael Zalis.
147. **Hynes, T. R., Randal, M., Kennedy, L. A., Eigenbrot, C., and Kossiakoff, A. A.,** X-ray crystal structure of the protease inhibitor domain of Alzheimer's amyloid β-protein precursor, *Biochemistry,* 29, 10018, 1990.
148. **Creighton, T. E. and Charles, I. G.,** Sequence of the genes and polypeptide precursors for two bovine protease inhibitors, *J. Mol. Biol.,* 194, 11, 1987.
149. **Read, R., Fujinaga, M., Sielecki, A., Ardelt, W., Laskowski, M., Jr., and James, M.,** Conformational flexibility in the third domain of the turkey ovomucoid inhibitor bound to SGPB and α-chymotrypsin, *Acta Crystallogr.,* A40, C50, 1984.
150. **Carrell, R. W., Evans, D. L., and Stein, P. E.,** Mobile reactive centre of serpins and the control of thrombosis, *Nature (London),* 353, 576, 1991.
151. **Joslin, G., Fallon, R. J., Bullock, J., Adams, S. P., and Perlmutter, D. H.,** The SEC receptor recognizes a pentapeptide neodomain of alpha1-antitrypsin-protease complexes, *J. Biol. Chem.,* 266, 11282, 1991.
152. **Virca, G. D., Salvesen, G., and Travis, J.,** Human neutrophil elastase and cathepsin G cleavage sites in the bait region of α_2-macroglobulin, *Hoppe-Seyler's Z. Physiol. Chem.,* 364, 1297, 1983.
153. **Palsdottir, A., Abrahamson, M., Thorsteinsson, L., Arnason, A., Olafsson, I., Grubb, A., and Jensson, O.,** Mutation in cystatin C gene causes hereditary brain haemorrhage, *Lancet,* 8611, 603, 1988.

Chapter 8

α_1-ANTITRYPSIN: STRUCTURE, FUNCTION, PHYSIOLOGY

David H. Perlmutter

TABLE OF CONTENTS

0-8493-6913-4/93/$0.00 + $.50

I. INTRODUCTION

α_1-Antitrypsin (α_1-AT) is the principle inhibitor of neutrophil elastase in human plasma and body fluids. Because neutrophil elastase is capable of degrading many components of the extracellular connective tissue matrix, its inhibitor, α_1-AT, is thought to play a critical role in maintenance of the "elastase-anti-elastase balance" and regulation of connective tissue turnover. Unopposed neutrophil elastase activity is probably responsible for destructive lung disease in individuals with inherited deficiencies of α_1-AT and contributes to lung injury in disease states (adult respiratory distress syndrome, septic shock, and cystic fibrosis) associated with proteolytic and oxidative inactivation of α_1-AT (reviewed in References 1 to 4). α_1-AT also inhibits neutrophil cathepsin G and may inhibit several serine proteases derived from mast cells and cytolytic T-lymphocytes. Since this array of proteolytic enzymes, as well as similar enzymes released by invading microorganisms, is likely to be present at sites of inflammation or tissue injury, α_1-AT is also thought to be involved in preventing incidental destruction to surrounding tissues and in allowing orderly initiation of tissue repair at inflammatory foci.

α_1-AT is the archetype of a family of proteins, serpins, which form covalently stabilized complexes with their cognate serine proteases.[5,6] These proteins bear ~30% primary sequence homology, higher degrees of homology within functional domains, have similar mechanisms of action, and are thought to play important roles in the host response to tissue injury/inflammation.

α_1-AT is also referred to as a "positive hepatic acute phase reactant" because it is predominantly derived from the liver[7,8] and because its plasma concentration increases three- to fourfold during the inflammatory response.[9,10] Work in several different laboratories has indicated that the acute phase mediator interleukin-6 (IL-6) mediates an increase in synthesis of α_1-AT in liver cells[11-14] and, hence, has made it possible to implicate IL-6 in the acute phase response of α_1-AT. IL-6 also regulates synthesis of α_1-AT in extrahepatic cell types, including human monocytes, macrophages,[12] and epithelial cells derived from the intestine.[110] Synthesis of α_1-AT in extrahepatic cell types is also regulated by cell type-specific mechanisms, e.g., bacterial endotoxin mediates a five- to tenfold increase in the synthesis of α_1-AT in human mononuclear phagocytes, but has no effect on the synthesis of α_1-AT in liver cells or intestinal epithelial cells.[15] It is not yet known to what extent the extrahepatic sites of synthesis, or tissue-specific regulation of extrahepatic α_1-AT synthesis, contributes to the overall acute phase response of this protein. However, a recently identified, novel regulatory mechanism in which the synthesis of α_1-AT is linked to its functional activity, inhibition of neutrophil elastase, may be an even more important determinant of the acute phase response of α_1-AT than are cytokines or endotoxin. In this mechanism, α_1-AT which has been structurally rearranged during the formation of an inhibitory complex with neutrophil elastase presents a pentapeptide neodomain to a specific cell-surface receptor, the serpin-enzyme complex or SEC receptor, and, in so doing, activates a signal transduction pathway for increased synthesis of α_1-AT.[16-22] In this chapter, I review these mechanisms of regulation of α_1-AT gene expression as well as the current literature on the structure and function of this inhibitor molecule.

II. STRUCTURE

α_1-AT is a single-chain ~52- to 55-kDa polypeptide with 394 amino acids and three asparagine-linked complex carbohydrate side chains.[23] There are two major isoforms in serum, depending on the presence of a bi- or triantennary configuration for the carbohydrate side chains.[24] In X-ray crystallography studies,it has been shown that α_1-AT has a globular shape and a highly ordered internal domain predominantly composed of α-helices and β-pleated sheets.[5,25]

α_1-AT is the archetype of the serpin supergene family, members of which incude antithrombin III, α_1-antichymotrypsin (α_1-ACT), C1 inhibitor, α_2-antiplasmin, protein C inhibitor, heparin cofactor II, plasminogen-activator inhibitors I and II, protease nexin-1, ovalbumin, angiotensinogen, corticosteroid-binding globulin, and thyroid-binding globulin.[5] These proteins share ~25 to 40% primary structural homology with higher degrees of regional homology in functional domains. Most serpins function as suicide inhibitors, forming equimolar complexes with a specific target protease. Other serpins are not inhibitory. For instance, corticosteroid- and thyroid hormone-binding globulins form complexes, but do not activate their hormone ligands, presumably subserving a transport function for these ligands. Neither ovalbumin nor angiotensinogen inactivate cognate serine protease, but do serve as substrates for serine proteases *in vitro*.

Comparison of α_1-AT to other members of the serpin family has generated several important concepts of the structure and function of α_1-AT. For instance, the reactive site P1 residue of α_1-AT is localized to a highly stressed "reactive loop" that protrudes from the serpin molecule. This loop may provide a certain degree of flexibility to the functional activity of the inhibitor. The reactive loop conformation of serpins is also thought to make them susceptible to proteolytic cleavage by thiolenzymes and metalloenzymes. The P1 residue itself is the most important determinant of functional specificity for each serpin molecule. This concept was dramatically confirmed by the discovery of α_1-AT Pittsburgh, a variant in which the P1 residue of α_1-AT, Met_{358}, is replaced by Arg_{358}, the amino acid which is present in the P1 position of ATIII. As might have been predicted, α_1-AT functioned as a thrombin inhibitor in the child with α_1-AT Pittsburgh, and severe bleeding diathesis resulted.[26]

The amino-terminal head of α_1-AT and the other serpins is also considered an important domain on the basis of structure-function relationships: it is variable in length in individual serpins, relatively lacking in higher-order structure, exteriorly located, and accessible for cleavage. Angiotensin I and II are cleaved from this domain of angiotensinogen, and the heparin binding site occupies this region of antithrombin III.

The carboxyl-terminal fragment of α_1-AT and the other serpins also bears important structural and functional chracteristics. There is a much higher degree of sequence homology among serpins in the carboxyl terminus. A small fragment at this terminus is cleaved during formation of the inhibitory complex with serine protease. This carboxyl-terminal fragment possesses chemotactic activity.[27,28] Moreover, this fragment bears the receptor binding domain for cell-surface binding, internalization of α_1-AT-elastase and other serpin-enzyme complexes, and for activating a signal transduction pathway for upregulation of α_1-AT gene expression.[16-22]

Structural variants of α_1-AT in humans are classified according to the protease inhibitor (Pi) phenotype system as defined by agarose electrophoresis or isoelectric focusing of plasma. The Pi classification assigns a letter to variants according to the position of migration of α_1-AT in these gel systems, using alphabetical order from low to high isoelectric point. For example, the most common normal variant migrates to an intermediate isoelectric point designated M. Individuals with the most common severe deficiency have an α_1-AT allelic variant that migrates to a high isoelectric point designated Z. In these individuals, there is an ~85 to 90% reduction in plasma concentrations of α_1-AT, premature development of emphysema, chronic liver disease, and hepatocellular carcinoma. A single nucleotide substitution results in a single amino acid substitution, Glu_{342} to Lys_{342}, and in an abnormal protein which accumulates within the endoplasmic reticulum of liver cells. Emphysema is presumably caused by deficient serum and lower respiratory tract α_1-AT concentrations,

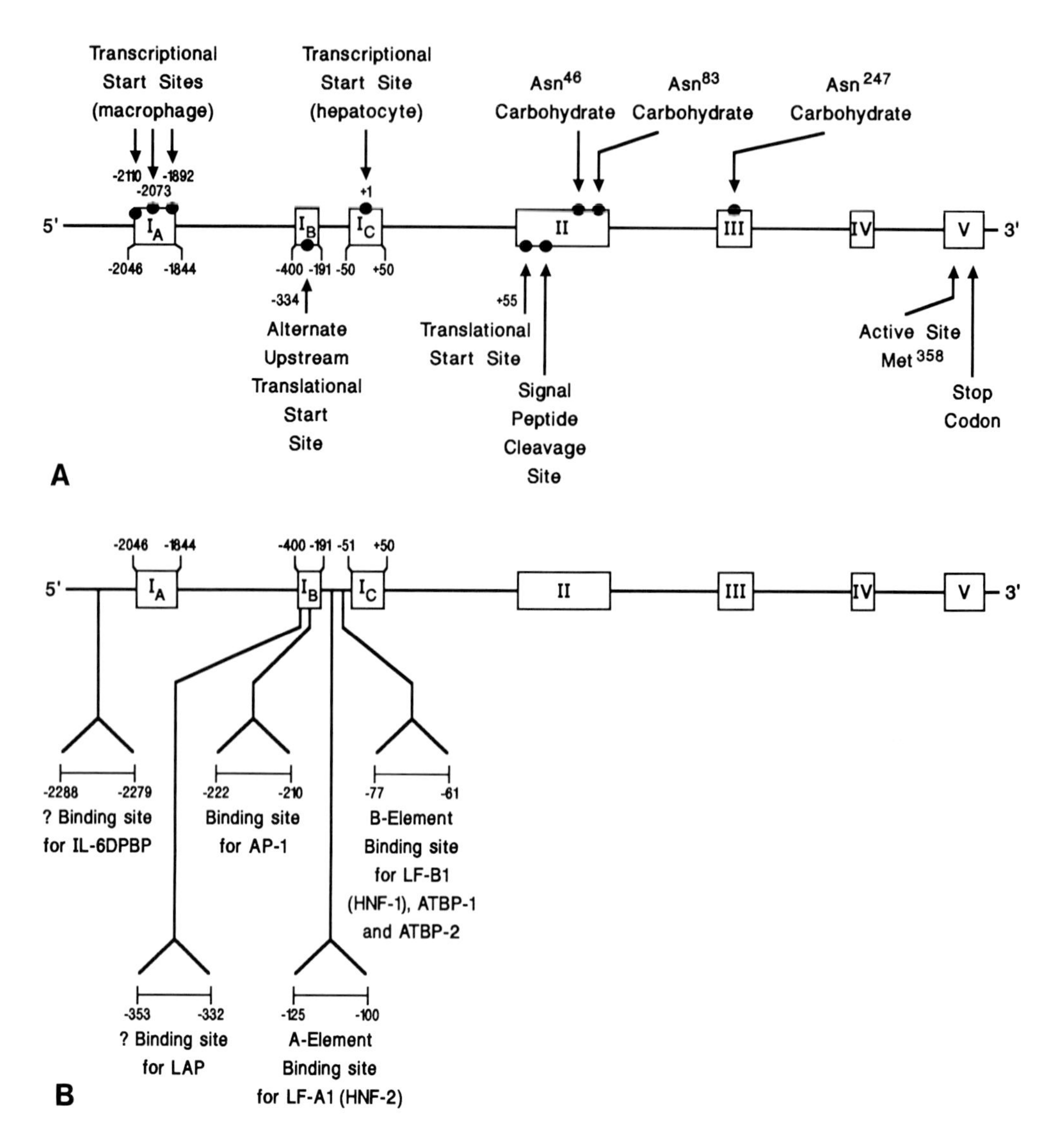

FIGURE 1. Map of the human α_1-antitrypsin gene. (A) Transcriptional and translational initiation sites; (B) regulatory elements. (Adapted from Perlmutter, D. H., *Hepatology*, 13, 172, 1991. With permission.)

allowing unregulated elastolytic attack on the lungs. Liver disease is thought to be a consequence of intracellular accumulation of an abnormally folded protein (reviewed in Reference 6).

α_1-AT is encoded by an ~12.2-kb gene (Figure 1A) located on human chromosome 14q31–32.2.[29-31] It is organized in seven exons and six introns.[32,33] The first two exons and a short 5′ segment of the third exon code for 5′ untranslated regions of the α_1-AT mRNA. Most of the fourth exon and the remaining three exons encode the protein sequence of α_1-AT. A 72-base sequence constitutes the 24-amino acid amino-terminal signal sequence. The three sites for asparagine-linked carbohydrate attachment are located at residues 46, 83, and 247. The active site — the so-called P_1 residue, Met_{358} — is encoded in the seventh exon. There is a "sequence-related gene" ~12 kb downstream from the α_1-AT gene.[29,34-36] Because there is no evidence that the sequence-related gene is expressed, it is considered a pseudogene. The genes for two other serpins, α_1-antichymotrypsin and corticosteroid-binding globulin, are also closely linked on chromosome 14.[31,37]

III. FUNCTION

Several lines of evidence suggest that inhibition of neutrophil elastase is the major physiological function of α_1-AT. First, the kinetics of association of α_1-AT and neutrophil elastase are more favorable, by several orders of magnitude, than those of α_1-AT and any other serine proteinase.[4] Second, α_1-AT constitutes >90% of the neutrophil elastase inhibitory activity in the one body fluid that has been examined, pulmonary alveolar lavage fluid.[38] Third, individuals with α_1-AT deficiency are susceptible to premature development of emphysema, a lesion that may be induced in experimental animals by intratracheal instillation of neutrophil elastase.[39] α_1-AT inhibits two other destructive human neutrophil proteases, cathepsin G and proteinase 3,[40] as well as the mast cell proteinase II.[41] It may also inhibit several cytolytic serine esterases of T-lymphocytes which are structurally related to neutrophil cathepsin G.[42] However, its capacity to inhibit other serine proteases, including trypsin, is much slower, and thus these inhibitory reactions may not be physiologically significant.

α_1-AT acts competitively by allowing its target enzyme to attack a substrate-like sequence in the carboxyl-terminal region of the inhibitor molecule.[43] This reaction between enzyme and inhibitor is second order, and the resulting complex contains one molecule of each of the reactants. A peptide bond in the inhibitor may be hydrolyzed during formation of the enzyme-inhibitor complex. However, hydrolysis of this reactive-site peptide bond does not proceed to completion. An equilibrium, near unity, is established between complexes in which the reactive-site peptide bond of α_1-AT is intact (native inhibitor) and those in which this peptide bond is cleaved (modified inhibitor). The complex of α_1-AT and serine proteinase is a covalently stabilized structure, resistant to dissociation by denaturing compounds, including sodium dodecyl sulfate and urea. During complex formation and hydrolysis of the reactive-site peptide bond, a ~4-kDa carboxy-terminal fragment of the inhibitor may be generated. This peptide fragment probably remains attached to modified α_1-AT by hydrophobic associations at the extreme carboxyl terminus.[44,45]

The net functional activity of α_1-AT in complex biological fluids may be modified by several factors. First, the reactive-site methionine of α_1-AT may be oxidized and thereby rendered inactive as an elastase inhibitor. The relationship of oxidation to the net biological activity of α_1-AT *in vivo* is not fully understood. However, α_1-AT is oxidatively inactivated *in vitro* by activated neutrophils[46,47] and by oxidants released by alveolar macrophages of cigarette smokers.[48] α_1-AT purified from the bronchoalveolar lavage fluid of smokers has a reduced association rate constant for neutrophil elastase.[49] Second, the functional activity of α_1-AT may be modified by proteolytic inactivation. A metalloproteinase secreted by mouse macrophages,[50] neutrophil collagenase,[51,52] neutrophil gelatinase,[52-54] interstitial collagenase,[55] stromelysin,[56] thiol proteinase cathepsin L,[57] and pseudomonas elastase[58] represent examples of proteinases shown to cleave and inactivate α_1-AT. Moreover, secreted products of rabbit alveolar macrophages have been shown to modify α_1-AT functional activity by proteolytic inactivation.[59] Interaction of elastase and other serine proteases with components of extracellular matrix, specifically the glycosaminoglycans heparin and vitronectin, may affect the rate of inactivation of these proteases by α_1-AT.[60]

IV. SYNTHESIS

The predominant site of synthesis of plasma α_1-AT is liver. This is most clearly shown by conversion of plasma α_1-AT to donor phenotype after orthotopic liver transplantation.[7,8] It is synthesized in human hepatoma cells as a 52-kDa precursor, undergoes posttranslational dolichol phosphate-linked glycosylation at three asparagine residues, and also undergoes tyrosine sulfation.[61,62] It is secreted as a 55-kDa native single-chain glycoprotein with a half-time for secretion of 35 to 40 min.

α_1-AT is also synthesized and secreted in primary cultures of human blood monocytes and bronchoalveolar and breast milk macrophages.[63] The cellular defect in homozygous PiZZ α_1-AT deficiency, a selective defect in the secretion of α_1-AT, is expressed in monocytes and macrophages from deficient individuals.[64] Expression of α_1-AT in monocytes and macrophages is profoundly influenced by products generated during inflammation.

In addition to hepatocytes and mononuclear phagocytes, α_1-AT mRNA has been isolated from multiple tissues in transgenic mice,[65-67] but it has not been possible to distinguish whether such α_1-AT mRNA is in ubiquitious tissue macrophages or other cell types. We have recently demonstrated the synthesis and secretion of α_1-AT in a human colonic adenocarcinoma cell line (Caco2) that differentiates into a villous enterocyte.[68] α_1-AT mRNA is also present in human jejunal epithelium by ribonuclease protection assay[69] and is localized to villous enterocytes and subjacent macrophages by *in situ* hybridization.[111] Expression of α_1-AT increases markedly in Caco2 cells as they differentiate into villous-type enterocytes.

Liver-specific expression of α_1-AT is predominantly directed by two *cis*-acting structural elements (Figure 1B) within a 125-nucleotide region upstream from the hepatocyte-specific transcriptional initiation site, the so-called A element at residues -125 to -100 and B element at residues -77 to -61.[70] In liver-derived nuclear extracts, there are at least five different proteins which bind to one of these two elements.[71] LF-B1, also called HNF-1, is a homeodomain-containing protein[72,73] which, together with a dimerizing cofactor (DCoH),[74] binds as homo- and heterodimers to the B element. Although it has been considered the *trans*-acting nuclear factor which predominantly determines liver-specific expression of α_1-AT and other hepatic proteins, LF-B1 is also expressed in kidney, intestine, stomach, and, to a lesser extent, lung and ovary.[74,75] Thus, other mechanisms such as the relative abundance of LF-B1, DCoH, and other *trans*-acting nuclear factors, must determine the relatively high levels of expression of α_1-AT in liver compared to that in extrahepatic cell types. The other four *trans*-acting nuclear proteins include LF-A1, or HNF-2, which binds to the A element,[76] and LF-B2, ATBP-1, and ATBP-2, which may be negative regulators at the B element.[71,77]

There are at least four transcriptional initiation sites (Figure 1A) in the 5′ flanking region of the human α_1-AT gene.[33,69] Three of these (-2110, -2073, and -1892) are termed "macrophage specific" because these initiation sites are only used for constitutive α_1-AT gene expression in cells of mononuclear phagocyte origin.[69] The downstream transcriptional initiation site ($+1$) is termed "hepatocyte specific" because it is used by hepatocytes for constitutive α_1-AT gene expression.[33,69] This downstream transcriptional initiation site is also used for constitutive α_1-AT gene expression by cells of enterocytic origin, whether crypt-like or having undergone differentiation to villous-like enterocytes.[69] It is not yet known whether an alternative upstream translational initiation site at -354 is used in mononuclear phagocytes. This translational initiation site is followed by a short open-reading frame and termination codon.[33] Furthermore, this translational codon is encoded within a context which is favorable for initiation of translation according to the Kozak consensus sequence principles.[78,79] Other factors, such as a surrounding upstream flanking region which is heavily encumbered by secondary structure, may prevent efficient initiation at this site but still allow initiation of translation at the downstream site, which is known to precede the amino-terminal amino acid for the α_1-AT protein produced in these cell types. There is another potential translational initiation codon at -204, but according to the Kozak principles,[78,79] it may be encumbered by the presence of intronic sequences within a short distance downstream.

V. CATABOLISM

The half-time of α_1-AT in plasma is ~5 d.[80,81] There is a slight increase in the rate of clearance of radiolabeled PiZ α_1-AT compared with PiM α_1-AT when infused into PiMM individuals, but this difference does not account for the decrease in plasma levels of α_1-AT in deficient individuals.[82,83]

Some data exist on the clearance of α_1-AT-proteinase complexes in mice.[84-86] In these studies, α_1-AT-trypsin complexes were cleared with a half-time of 20 min. More than 70% of these complexes were distributed to the liver. Clearance of α_1-AT-trypsin complexes was blocked by antithrombin III-thrombin complexes, α_1-ACT-cathepsin G complexes, and heparin cofactor II-thrombin complexes. In each case, clearance of complexes was much more rapid than that of the native proteins, and was also significantly more rapid than that of proteolytically modified serpins.[86,87] Thus, clearance of α_1-AT-proteinase complexes may involve a pathway common to at least several serpin-enzyme complexes.

α_1-AT diffuses into most tissues and is found in most body fluids.[88] The concentration of α_1-AT in lavage fluid from the lower respiratory tract is approximately equivalent to that in serum.[38] α_1-AT is also found in feces, and increased fecal concentrations of α_1-AT correlated with inflammatory lesions of the bowel.[89] In each case, it has been assumed that the α_1-AT is derived from serum. Local sites of synthesis, such as macrophages and epithelial cells as noted above, may also make important contributions to the α_1-AT pool in these tissues and body fluids. In fact, it has been reported that fecal α_1-AT clearance is higher in individuals with homozygous PiZZ α_1-AT deficiency than in normal individuals.[90] Since the former individuals have only 10 to 15% of the normal serum concentrations of α_1-AT, a local intestinal source for fecal α_1-AT is implicated. It is therefore possible that the bulk of α_1-AT in feces is derived from sloughed enterocytes. Increased fecal α_1-AT in individuals with homozygous PiZZ α_1-AT deficiency would result from intracellular accumulation of the abnormal α_1-AT molecule, and in normal PiMM individuals with inflammatory-related protein-losing enteropathy, from increased sloughing of enterocytes.

VI. REGULATION

Plasma concentrations of α_1-AT increase three- to fourfold during the acute phase response.[9,10] This phenomenon can be recapitulated in a number of types of cell culture systems from a number of species using recombinant human IL-6.[11-14] For instance,IL-6 mediates an ~threefold increase in steady-state levels of α_1-AT mRNA and in *de novo* synthesis of α_1-AT in human hepatoma Hep G2 cells.[12] Il-6 also mediates an increase in the synthesis of α_1-AT in human peripheral blood monocytes, human bronchoalveolar macrophages,[12] and the human intestinal epithelial cell lines Caco2 and T84.[110] Ribonuclease protection assays have shown that IL-6 mediates a modest increase in α_1-AT mRNA transcripts initiated at the downstream hepatocyte-specific transcriptional initiation site in Hep G2 and Caco2 cells, but also induces α_1-AT mRNA transcripts initiated at the three upstream macrophage-specific transcriptional initiation sites in these cell lines.[69] In blood monocytes and tissue macrophages, IL-6 mediates an increase only in transcripts initiated at the upstream transcriptional initiation sites. It is not yet known whether these effects are mediated by two sequences, similar to the proximal portion of the IL-6 response element[91] and approximately 200 nucleotides upstream from the hepatocyte transcriptional initiation site, and/or by one sequence, even more homologous with the IL-6 response element and approximately 200 nucleotides upstream from the most remote macrophage transcriptional initiation site. It is also not yet known if these effects involve the IL-6-inducible *trans*-acting DNA binding protein called IL-6DBP, or LAP.[92,93] This *trans*-acting protein is characterized by leucine

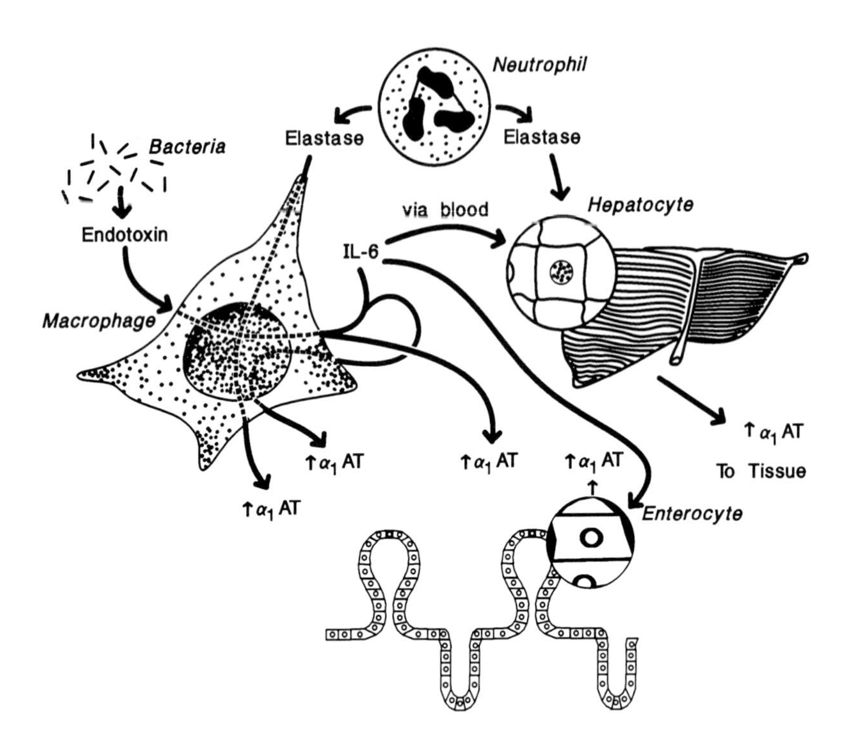

FIGURE 2. Physiological factors which regulate α_1-antitrypsin synthesis. (Adapted from Perlmutter, D. H. and Pierce, J. A., *Am. J. Physiol.*, 257, L147, 1989. With permission.)

zipper domains similar to that of C/EBP, a transcription factor implicated in terminal differentiation. A recent study has shown that use of a downstream translational initiation site within the IL-6DBP/LAP mRNA leads to a shorter protein which binds to the IL-6 response element with high affinity but cannot activate transcription.[94] It is, in effect, a transcriptional repressor. Relative ratios of the two translational products and their relative affinities for *cis*-acting structural elements upstream of IL-6-resposive genes may therefore determine the magnitude of the effect of IL-6 on that particular gene. Finally, it is not yet known whether the alternative upstream translational initiation site within the IL-6-inducible, longer α_1-AT mRNA is actually used. There is no evidence that α_1-AT gene expression is significantly modulated by other acute phase mediators, including IL-1β, TNFα, and IFNγ.[12,95-97]

Plasma concentrations of α_1-AT also increase during oral contraceptive therapy and pregnancy. Nominal changes in plasma α_1-AT levels follow administration of synthetic androgen danazol.[98] A recent study suggests that these effects can also be recapitulated in cell culture using the human breast cancer cell line, MCF-7.[99] Estradiol and dihydrotestosterone independently mediated a three- to fourfold increase in α_1-AT mRNA levels and in *de novo* synthesis in MCF-7 cells. There has not yet been any convincing evidence that sex steroid hormones mediate changes in expression of the α_1-AT gene in liver cell culture systems.

Cell-specific regulation of α_1-AT synthesis in mononuclear phagocytes is now well established (Figure 2). Bacterial endotoxin mediates a marked increase in α_1-AT synthesis in human blood monocytes and bronchoalveolar macrophages,[15] but has no effect on Hep G2, Caco2, or T84 cells.[110] The effect of endotoxin can be distinguished from the completely independent direct effect of IL-6 on mononuclear phagocyte α_1-AT synthesis.[12] Regulation of α_1-AT synthesis by endotoxin is also distinctive in that there are increases in both specific mRNA levels and the specific translational efficiency of α_1-AT mRNA.[15,17] In a recent study employing ribonuclease protection assays and primer extension analysis, the increase in translational efficiency of α_1-mRNA mediated by endotoxin could not be attributed to a change in transcriptional initiation to the downstream hepatocyte-specific promoter.[69] Endotoxin mediated increases in three different α_1-AT mRNA species, all initiated at the upstream macrophage-specific transcriptional initiation sites. By exclusion of the latter mechanism, the translational effect of endotoxin on α_1-AT gene expression must therefore involve a change in specific RNA-protein interactions or a change in specific RNA folding.

We have recently spent a considerable amount of time characterizing the mechanism by which α_1-AT-elastase complexes regulate α_1-AT synthesis.[16-22] Because this type of regulatory mechanism would potentially allow the integration of α_1-AT production with its functional activity, we believed it would be important in the overall physiological fate of α_1-AT and that it could represent an example of mechanisms by which the expression of other serpins, and other types of proteinase inhibitors, was controlled.

Evidence for the existence of this regulatory mechanism arose during studies of the expression of the α_1-AT gene in human mononuclear phagocytes. A decrease in steady-state levels of α_1-AT mRNA and a corresponding decrease in α_1-AT synthesis was found to accompany the maturation of monocytes in tissue culture.[62] This decrease in expression of α_1-AT was not caused by a change in viability or total metabolic activity because total mRNA content, steady-state levels of specific mRNAs, total protein synthesis, and synthesis of other specific secretory proteins was increasing during the same interval. This observation, therefore, suggested that expression of α_1-AT in monocytes was downregulated by a factor elaborated in tissue culture or by removal from an *in vivo* upregulating factor. We examined the latter possibility first and found that addition of exogenous elastase to primary cultures of peripheral blood monocytes prevented the decrease in α_1-AT synthesis during the first week in culture.[16] These results suggested that expression of α_1-AT decreased during *in vitro* maturation of monocytes because the cells were removed from elastase present *in vivo,* or because there was a decrease in the spontaneous release of the small amount of neutrophilic elastase present in monocytes early in primary culture.[100,101]

Subsequent experiments had the following results: (1) neutrophil elastase, in nanomolar concentrations, mediated concentration- and time-dependent increases in steady-state levels of α_1-AT mRNA and in the synthesis of α_1-AT in human monocytes and bronchoalveolar macrophages, (2) the regulatory effect was specific for the effector, elastase, and for the responder, α_1-AT, and required enzymatically active elastase, and (3) the effect required the formation of a complex of exogenous elastase with endogenous α_1-AT or an exogenous, preformed elastase-α_1-AT complex.[16] The last observation suggested that structural rearrangement of the α_1-AT or elastase molecules, during formation of an α_1-AT-elastase complex, exposed a domain that could be recognized by a specific cell-surface receptor, or receptors. In order to test this hypothesis, we used synthetic peptides based on the sequence of the carboxy-terminal fragment of α_1-AT as candidate mediators for regulation of α_1-AT synthesis and as candidate ligands for cell-surface binding.[18] This region was selected because it had been previously implicated in the chemoattractant properties of α_1-AT-elastase complexes[27,28] and because crystal structure analysis predicted that a domain within this region was exteriorly exposed after formation of a complex.[5,25] The results indicated that a synthetic peptide (peptide 105Y) based on amino acids 359 to 374 of α_1-AT (Figure 3A) mediated a selective increase in α_1-AT synthesis in human monocytes and in human hepatoma HepG2 cells. Radioiodinated peptide 105Y bound specifically and saturably to HepG2 cells, defining a single class of receptors with a K_d of $\sim$40 nM at a density of $\sim 4.5 \times 10^5$ plasma membrane receptor molecules per cell.[18] Binding of [^{125}I] peptide 105Y was blocked by unlabeled elastase-α_1-AT complexes, and unlabeled peptide 105Y blocked binding of [^{125}I]α_1-AT-elastase complexes, [^{125}I]elastase-α_1-AT complexes, and [^{125}I]trypsin-α_1-AT com-

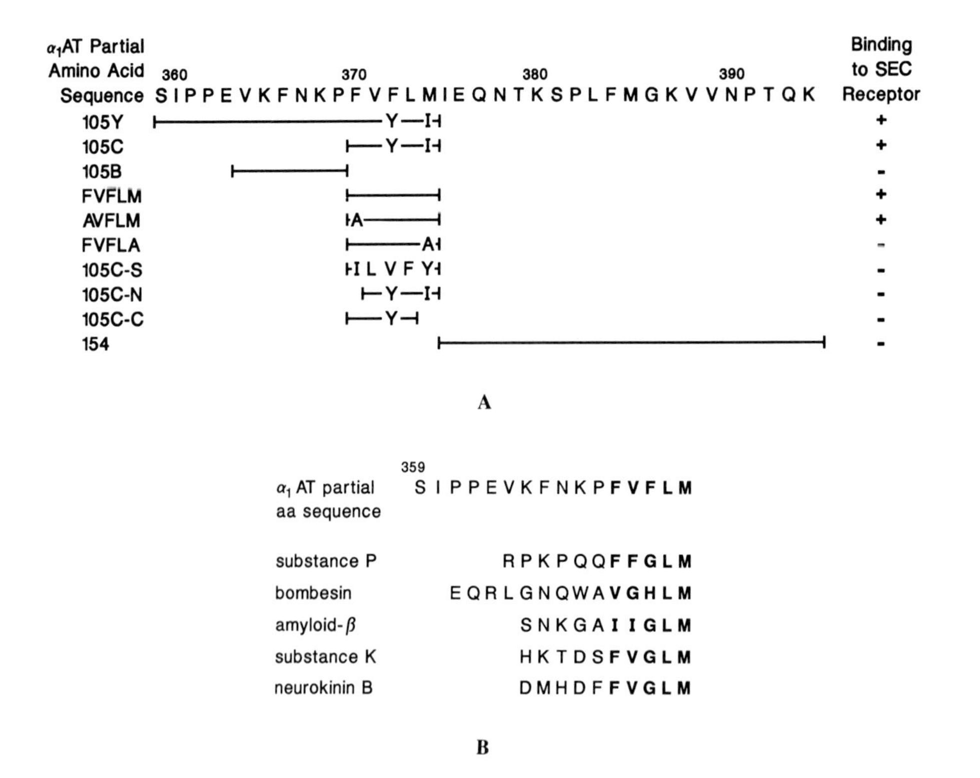

FIGURE 3. Map of synthetic peptides used to define the SEC receptor. (A) α_1-Antitrypsin peptides; (B) other bioactive peptides. (Adapted from Joslin, G., Krause, J. E., Hershey, A. D., Adams, S. P., Fallon, R. J., and Perlmutter, D. H., *J. Biol. Chem.*, 266, 21897, 1991. With permission.)

plexes.[18,19] Antisera to keyhole-limpet hemocyanin-coupled peptide 105Y blocked the binding of $[^{125}I]\alpha_1$-AT-elastase complexes and the increase in synthesis of α_1-AT mediated by α_1-AT-elastase complexes.[112] These results provided confirmatory evidence that at least part of the region corresponding to peptide 105Y represented the receptor-binding domain of α_1-AT-elastase complexes and was capable of transducing a signal to increase synthesis of α_1-AT.

Next, we examined the significance of the high degree of primary sequence homology within this receptor-binding domain of α_1-AT and in the corresponding regions of serpin ATIII, α_1-ACT, and C1 inhibitor (Figure 4). In competitive binding assays, we found that the binding of $[^{125}I]$peptide 105Y was displaced by ATIII-thrombin, α_1-ACT-cathepsin G, and, to a lesser extent, C1 inhibitor-C1s complexes, but not by the corresponding proteins in their native forms.[18] These data indicated that the receptor that recognizes peptide 105Y and α_1-AT-elastase complexes also recognizes these other serpin-enzyme complexes, so we have called it the serpin-enzyme complex, or SEC receptor. These data also showed that the SEC receptor only recognizes the serpin after it has undergone the structural rearrangement that accompanies formation of a complex with its cognate enzyme. Other experimental results showed that the SEC receptor also recognizes α_1-AT after it has undergone proteolytic modification at its reactive site by the action of the metalloelastase of *Pseudomonas aeruginosa*.[20] In each case, the SEC receptor recognizes a domain within the carboxyl-terminal fragment of α_1-AT which has remained associated with the rest of the α_1-AT molecule by tenacious hydrophobic interactions at the extreme carboxyl terminus, and thereby carries to the cell-surface receptor-binding site the larger amino-terminal portion of α_1-AT. The SEC

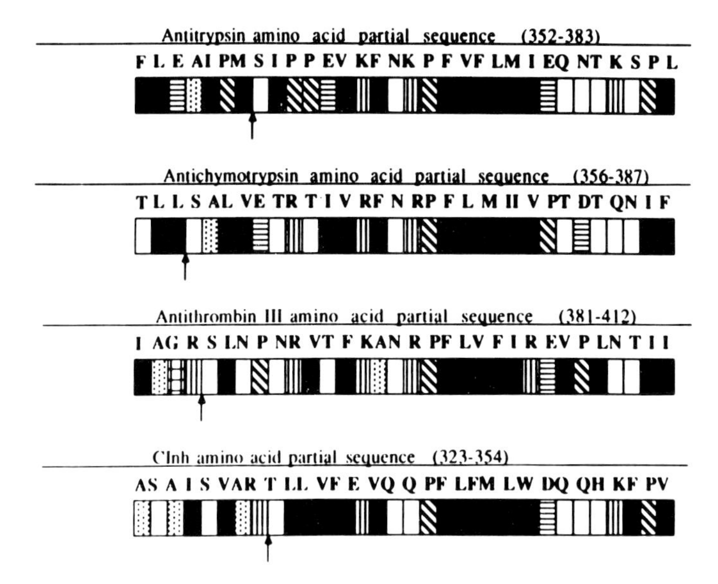

FIGURE 4. Functional homology within the carboxyl-terminal fragments of serpins (kindly provided by C. Schasteen, St. Louis, MO). Amino acids with similar functional units are indicated by open bars, solid bars, diagonal hatched bars, vertical hatched bars, horizontal hatched bars, or bars with dotted lines.

receptor has subsequently been found on a number of different cell types, including hepatoma cells, mononuclear phagocytes, neutrophils, human intestinal epithelial cell line Caco2, mouse fibroblast L cells, Cos cells, and PC12 cells, but is not present on chinese hamster ovary cells or Hela cells.

We have also examined the possibility that the SEC receptor is involved in clearance/catabolism of serpin-enzyme complexes *in vivo* by determining whether it mediated internalization and/or degradation of serpin-enzyme complexes in tissue culture.[19] As mentioned above α_1-AT-protease complexes are subject to rapid *invivo* clearance and are predominantly catabolized in the liver. The pathway for clearance/catabolism is shared by other serpin-enzyme complexes, including ATIII-thrombin and α_1-ACT-cathepsin G. Our studies showed that α_1-AT-elastase and α_1-AT-trypsin complexes were internalized in Hep G2 cells by SEC receptor-mediated endocytosis and delivered to an acidic compartment, either late endosome or lysosome, for degradation. Thus, these results provide evidence that the characteristics of the SEC receptor in cell culture are similar to those that would be expected for the receptor responsible for *in vivo* clearance/catabolism of at least several serpin-enzyme complexes.

In more recent studies, we have used synthetic peptides to determine the minimal structural requirements for the binding of α_1-AT-elastase complexes by the SEC receptor.[21] These studies have shown that a pentapeptide domain within the carboxyl-terminal fragment of α_1-AT (amino acids 370 to 374, FVFLM) is sufficient for binding to the SEC receptor (Figure 3A). A synthetic analog of this pentapeptide (peptide 105C, FVYLI) blocked binding and internalization of [^{125}I]trypsin-α_1-AT complexes by HepG2 cells. [^{125}I]peptide 105C bound specifically and saturably to Hep G2 cells, and its binding was blocked by unlabeled trypsin-α_1-AT or elastase-α_1-AT complexes. Alterations of the sequence of the pentapeptide introduced into synthetic peptides (mutations, deletions, or scrambling) demonstrated that recognition by the SEC receptor was sequence specific. Synthetic pentapeptides were also capable of mediating an increase in the synthesis of α_1-AT. As might have been predicted from competitive binding of other serpin-enzyme complexes to the SEC receptor, the SEC receptor-binding pentapeptide neodomain of α_1-AT is highly conserved in the corresponding regions of these other serpins.

The SEC receptor-binding pentapeptide of α_1-AT was also found to be remarkably similar to sequences in substance P, several other tachykinins, bombesin, and the amyloid-β peptide (Figure 3B). These peptides have a number of different biological activities. In many cases, these biological activities are mediated by specific cell-surface receptors, including tachykinin receptors NK-1 (substance P), NK-2 (substance K), NK-3 (neurokinin B), and several bombesin receptors.[102,103] Because these cell-surface receptors have only recently been described and because there are only a few highly selective, high-affinity receptor antagonists, it has not yet been possible to attribute all of the biological activities of these peptides to the known receptors. Furthermore, recent data have suggested that the amyloid-β peptide, the major proteinaceous component of the extracellular deposits found in Alzheimer's disease and Down's syndrome (reviewed in References 104 and 105), has neurotrophic/neurotoxic effects that could be blocked by substance P,[106] but a specific cell-surface receptor had not yet been identified. With these considerations in mind, we examined the possibility that the tachykinins, bombesin, and amyloid-β peptide bind to the SEC receptor.[22] The results indicated that substance P, several other tachykinins, bombesin, and amyloid-β peptide compete for binding to, and cross-linking of, the SEC receptor. These other ligands also mediated an increase in the synthesis of α_1-AT in monocytes and Hep G2 cells. These results were not surprising in that the two residues within the receptor-binding pentapeptide of α_1-AT that were most affected by mutations, the carboxyl-terminal leucine and methionine residues,[21] are the ones most highly conserved among the tachykinins, bombesin, and amyloid-β peptide.

The SEC receptor was found to be distinct from the substance P receptor by several criteria. There was no substance P receptor mRNA in Hep G2 cells or human liver as assessed by ribonuclease protection assays with human substance P receptor cRNA as probe. The SEC receptor recognized synthetic peptide ligands with carboxyl-terminal, carboxy-acid, or carboxy-amide moieties with equivalent affinity, whereas the substance P receptor recognized substance P carboxy-amide with an affinity several orders of magnitude higher than that for substance P carboxy-acid. The SEC receptor was present in much higher density on receptor-bearing cells and bound its ligands at lower affinity than the substance P receptor. The SEC receptor was much less restricted in the specificity with which it recognized ligands, i.e., ligands for the SEC receptor, including peptide 105Y (based on α_1-AT sequence 359 to 374), α_1-AT-protease complexes, bombesin, and amyloid-β peptide did not compete for binding of substance P to a stable transfected cell line expressing the substance P receptor. Several of these criteria also make it highly likely that the SEC receptor is distinct from the substance K, neurokinin B, and bombesin receptors. Partial structural characterization of the SEC receptor also suggests that the SEC receptor is distinct from the tachykinin and bombesin receptors. The ligand-binding subunit of the SEC receptor in Hep G2 cells is a single-chain polypeptide of ~78 kDa, as determined by covalent photoaffinity cross-linking with a radioiodinated photoreactive derivative of peptide 105Y (based on α_1-AT sequence 359 to 374).[22] Cross-linking of the SEC receptor with [^{125}I]ASA-peptide 105Y is highly specific in that it is blocked by unlabeled peptide 105Y, pentapeptide 105C, substance P, bombesin, amyloid-β peptide, and α_1-AT-protease complexes, but not by negative-control peptides, mutant pentapeptides, deleted pentapeptides, a substance P receptor antagonist, or native α_1-AT. The SEC receptor has also been purified to homogeneity as a ~80-kDa polypeptide from Hep G2 cell membranes by ligand-affinity chromatography with α_1-AT-elastase complexes. Amino-terminal amino acid sequence analysis clearly demonstrates that it is distinct from the tachykinin and bombesin receptors.[113]

A sequence similar to α_1-AT 368 to 374, the receptor-binding region, has also been found in the carboxyl-terminal propeptide of mouse β-glucuronidase[107] and the collagen-binding protein gp46.[108] β-Glucuronidase is ordinarily a lysosomal enzyme, but a significant proportion of newly synthesized β-glucuronidase is retained within the endoplasmic reticulum of murine hepatocytes in a serpin-like complex with the esterase-active site of the protein egasyn. Similarly, gp46, a collagen-binding protein found in human fibroblasts, is also retained in the endoplasmic reticulum.[108] It is homologous with a rat skeletal myoblast protein, expression of which is modulated by heat shock, with the J6 protein from mouse F9 embryonal carcinoma cells, expression of which is modulated by retinoic acid, and with hsp 47 from chick embryo fibroblasts. The β-glucuronidase-egasyn complex and collagen-gp46 complex are presumably retained in the endoplasmic reticulum by a KDEL-like sequence in egasyn and gp46, respectively, but it will be interesting to determine whether these interact with the SEC receptor and whether these potential interactions shed some light on the mechanism by which the mutant Z α_1-AT protein is retained in the endoplasmic reticulum in individuals with homozygous PiZZ α_1-AT deficiency.

We have recently examined the possibility that the SEC receptor mediates the neutrophil chemotactic effects of α_1-AT-elastase complexes.[114] Previous studies have shown that α_1-AT-elastase complexes and proteolytically modified α_1-AT are chemotactic for neutrophils and that the carboxyl-terminal fragment of α_1-AT possessed all of this biological activity.[27,28]

First, receptor-binding studies with [^{125}I] peptide 105Y showed that there was a single class of receptors with a K_d (~43 n*M*) almost identical to that previously reported for Hep G2 cells.[18] There were, however, only ~13,000 plasma membrane SEC receptor molecules on each human neutrophil compared to ~450,000 plasma membrane receptors on each human hepatoma Hep G2 cell. Second, chemotactic studies showed that peptide 105Y and pentapeptide 105C mediated neutrophil chemotaxis with maximal stimulation at 10^{-9} to 10^{-8} *M*. The magnitude of the effect was comparable to that of the chemotactic peptide fMLP of 10^{-8} *M*. The specificity of the effect was consistent with its being mediated by a SEC receptor, as shown by negative control peptides. Most importantly, the neutrophil chemotactic effect of α_1-AT-elastase complexes was completely blocked by antiserum to keyhole-limpet hemocyanin-coupled peptide 105Y and antiserum to purified SEC receptor, but not by a control antiserum. Other ligands for the SEC receptor, including the amyloid-β peptide, mediated neutrophil chemotaxis. Finally, preincubation of neutrophils with peptide 105Y completely abrogated the chemotactic effect of amyloid-β peptide by inducing homologous desensitization of the SEC receptor. Thus, the SEC receptor mediates the previously recognized chemotactic effect of α_1-AT-elastase complexes and the previously unrecognized chemotactic effect of amyloid-β peptide. It is also likely to mediate the recently described chemotactic effect of α_1-ACT-cathepsin G complexes.[87] One might also predict that it mediates any chemotactic effect of HCII-thrombin complexes, although a structurally distinct region in the amino-terminal domain of HCII has been shown to possess neutrophil chemotactic activity.[109] Further studies will be necessary to determine whether two regions of HCII can mediate neutrophil chemotactic effects through two distinct receptors. Although it has not been completely excluded, there is no current evidence to suggest that other regions within the α_1-AT molecule, or other serpin molecules, contribute to binding to the SEC receptor.

Taken together, these studies define the cellular biochemistry of an interesting and physiologically relevant network for the regulation of α_1-AT activity, and hence, extracellular proteolytic activity (Figure 5). Formation of covalently stabilized inhibitory complexes with neutrophil elastase, or proteolytic modification by metalloelastases such as that of *P. aeruginosa,* induces structural rearrangement of α_1-AT and, in so doing, exposes a pentapeptide receptor-binding domain in the carboxyl-terminal frgment of α_1-AT. This domain can, in turn, be recognized by a single class of receptors with a K_d of ~40 n*M* and a ligand-binding subunit of ~78 kDa. The receptor-binding domain is highly conserved among the serpin family and several serpin-enzyme complexes can be recognized by the same receptor, the

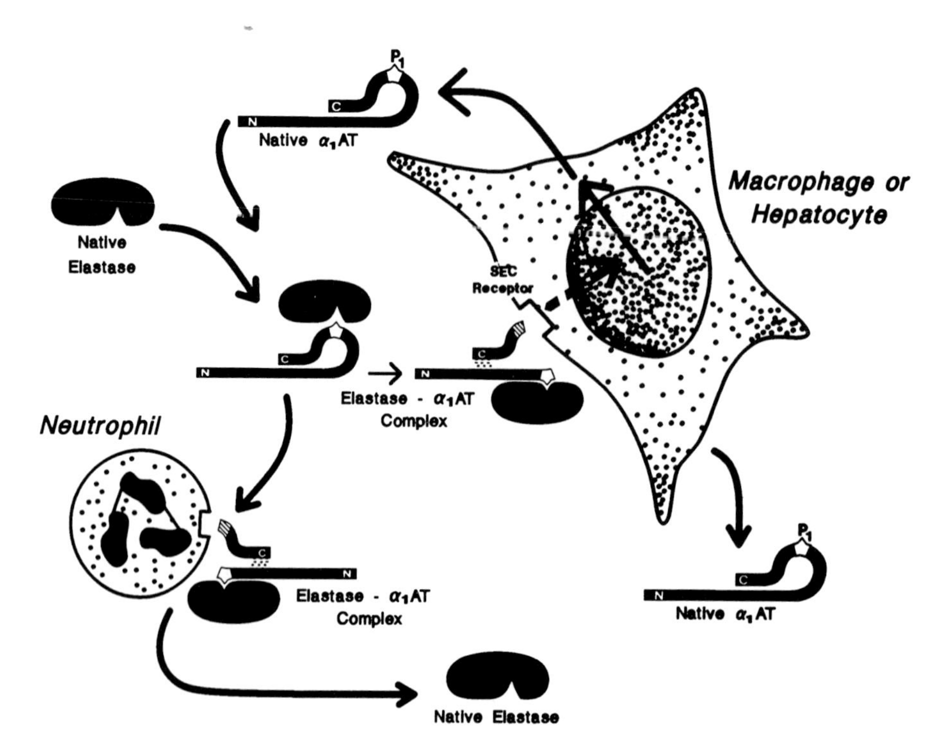

FIGURE 5. Regulation of α_1-antitrypsin synthesis by the SEC receptor. (Adapted from Perlmutter, D. H., *Hepatology*, 13, 172, 1991. With permission.)

SEC receptor. The receptor-binding domain is also conserved among several tachykinins, bombesin, and the amyloid-β peptide. These other ligands are also recognized by the SEC receptor. Once engaged, the SEC receptor is capable of activating a single transduction pathway for increased synthesis of α_1-AT. Thus, the regulatory effect maintains an excess of inhibitor in the extracellular milieu, an effect which is likely to be important for the control of limited proteolytic cascade pathways at sites of inflammation, orderly initiation of tissue repair, or prevention of excessive connective tissue destruction around migrating cells and sprouting cell processes. The SEC receptor is also capable of internalizing its ligands and delivering them to an acidic compartment, either late endosome or lysosome, for degradation. Based on this property and the similarity of its specificity for recognition of ligands to the specificity of pathways for *in vivo* clearance/catabolism of serpin-enzyme complexes, the SEC receptor probably mediates *in vivo* clearance/catabolism of certain serpin-enzyme complexes. The SEC receptor also mediates neutrophil chemotactic activities. Finally, identification of amyloid-β peptide as a ligand for the SEC receptor and identification of the SEC receptor in cells of neuronal origin[22] raise the possibility that the SEC receptor is involved in neutrophic/neurotoxic effects which may characterize the neutropathology of Alzheimer's disease and presenile dementia in Down's syndrome.

ACKNOWLEDGMENTS

The studies described in this review were supported in part by an Established Investigator Award from the American Heart Association, by grants from the Arthritis Foundation, the Monsanto-Washington University Biomedical Science Program, and the United States Public Health Service HL37784.

REFERENCES

1. **Carrell, R. W.,** Alpha-1-antitrypsin: molecular pathology, leukocytes and tissue damage, *J. Clin. Invest.,* 77, 1427, 1986.
2. **Crystal, R. G.,** Alpha-1-antitrypsin deficiency, emphysema and liver disease: genetic basis and strategies for therapy, *J. Clin. Invest.,* 95, 1343, 1990.
3. **Perlmutter, D. H. and Pierce, J. A.,** The alpha-1-antitrypsin gene and emphysema, *Am. J. Physiol.,* 257, L147, 1989.
4. **Travis, J. and Salvesen, G.,** Human plasma proteinase inhibitors, *Annu. Rev. Biochem.,* 52, 655, 1983.
5. **Huber, R. and Carrell, R. W.,** Implications of the three-dimensional structure of alpha-1-antitrypsin for structure and function of serpins, *Biochemistry,* 28, 8951, 1990.
6. **Perlmutter, D. H.,** The cellular basis for liver injury in alpha-1-antitrypsin deficiency, *Hepatology,* 12, 172, 1991.
7. **Alper, C. A., Raum, D., Awdeh, Z. I., Petersen, R. H., Taylor, P. D., and Starzl, T. E.,** Studies of hepatic synthesis in vivo of plasma proteins including orosomucoid, transferrin, alpha-1-antitrypsin, C8 and factor B, *Clin. Immunol. Immunopathol.,* 16, 84, 1990.
8. **Hood, J. M., Koep, L., Peters, R. F., Schroter, G. P. J., Well, R., Redeker, A. G., and Starzl, T. E.,** Liver transplantation for advanced liver disease with alpha-1-antitrypsin deficiency, *N. Engl. J. Med.,* 302, 272, 1989.
9. **Aronsen, K.-F., Ekelund, G., Kindmark, C.-O., and Laurell, C.-B.,** Sequential changes of plasma proteins after myocardial infarction, *Scand. J. Clin. Lab. Invest.,* 29 (Suppl. 24), 127, 1972.
10. **Dickson, I. and Alper, C. A.,** Changes in serum proteinase inhibitor levels following bone surgery, *Clin. Chim. Acta,* 54, 381, 1974.
11. **Gauldie, J., Richards, C., Harnish, D., Landsdorp, P., and Baumann, H.,** Interferonβ2/B-cell stimulatory factor type 2 shares identity with monocyte-derived hepatocyte-stimulating factor and regulates the major acute phase protein response in liver cells, *Proc. Natl. Acad. Sci. U.S.A.,* 84, 7251, 1987.
12. **Perlmutter, D. H., May, L. T., and Sehgal, P. G.,** Interferonβ2/interleukin-6 modulates synthesis of alpha-1-antitrypsin in human mononuclear phagocytes and in human hepatoma cells, *J. Clin. Invest.,* 84, 138, 1989.
13. **Andus, T., Geiger, T., Hirano, T., Kishimoto, T., and Heinrich, P. C.,** Action of recombinant human interleukin-6, interleukin-1β and tumor necrosis factor on the mRNA induction of acute phase proteins, *Eur. J. Immunol.,* 18, 739, 1988.
14. **Ganapathi, M. K., Schultz, D., Mackiewicz, A., Samols, D., Mu, S.-I., Brabeneck, A., MacIntyre, S. S., and Kushner, I.,** Heterogeneous nature of the acute phase response. Differential regulation of human serum amyloid A, C-reactive protein and other acute phase proteins by cytokines in Hep3B cells, *J. Immunol.,* 141, 564, 1988.
15. **Barbey-Morel, C., Pierce, J. A., Campbell, E. J., and Perlmutter, D. H.,** Lipopolysaccharide modulates the expression of alpha-1-proteinase inhibitor and other serine proteinase inhibitors in human monocytes and macrophages, *J. Exp. Med.,* 166, 1041, 1987.
16. **Perlmutter, D. H., Travis, J., and Punsal, P. I.,** Elastase regulates the synthesis of its inhibitor alpha-1-proteinase inhibitor, and exaggerates the defect in homozygous PiZZ alpha-1-proteinase inhibitor deficiency, *J. Clin. Invest.,* 81, 1744, 1988.
17. **Perlmutter, D. H. and Punsal, P. I.,** Distinct and additive effects of elastase and endotoxin on alpha-1-proteinase inhibitor expression in macrophage, *J. Biol. Chem.,* 263, 16499, 1988.
18. **Perlmutter, D. H., Glover, G. I., Rivetna, M., Schasteen, C. S., and Fallon, R. J.,** Identification of a serpin-enzyme complex (SEC) receptor on human hepatoma cells and human monocytes, *Proc. Natl. Acad. Sci. U.S.A.,* 87, 3753, 1990.
19. **Perlmutter, D. H., Joslin, G., Nelson, P., Schasteen, C. S., Adams, S. P., and Fallon, R. J.,** Endocytosis and degradation of alpha-1-antitrypsin-proteinase complexes is mediated by the SEC receptor, *J. Biol. Chem.,* 265, 16713, 1990.
20. **Barbey-Morel, C. and Perlmutter, D. H.,** Effect of pseudomonas elastase on human mononuclear phagocyte alpha-1-antitrypsin expression, *Pediatr. Res.,* 29, 133, 1991.
21. **Joslin, G., Fallon, R. J., Bullock, J., Adams, S. P., and Perlmutter, D. H.,** The SEC receptor recognizes a pentapeptide neodomain of alpha-1-antitrypsin-protease complexes, *J. Biol. Chem.,* 266, 11282, 1991.
22. **Joslin, G., Krause, J. E., Hershey, A. D., Adams, S. P., Fallon, R. J., and Perlmutter, D. H.,** Amyloid-β peptide, substance P, and bombesin bind to the serpin-enzyme complex receptor, *J. Biol. Chem.,* 266, 21897, 1991.
23. **Carrell, R. W., Jeppsson, J.-O., Laurell, C.-B., Brennan, S. O., Owen, M. C., Vaughn, L., and Boswell, D. R.,** Structure and variation of human alpha-1-antitrypsin, *Nature (London),* 298, 329, 1982.
24. **Vaughn, L., Lorier, M. A., and Carrell, R. W.,** Alpha-1-antitrypsin microheterogeneity: isolation and physiological significance of isoforms, *Biochim. Biophys. Acta,* 701, 339, 1982.

25. **Loebermann, H., Tokuoka, R., Deisenhofer, J., and Huber, R.,** Human alpha-1-proteinase inhibitor: crystal structure analysis of two crystal modifications, molecular model and preliminary analysis of the implications for function, *J. Mol. Biol.,* 177, 531, 1984.
26. **Owen, M. C., Brennan, S. O., Lewis, J. H., and Carrell, R. W.,** Mutation of antitrypsin to antithrombin: alpha-1-antitrypsin Pittsburgh (358 Met-Arg), a fatal bleeding disorder, *N. Engl. J. Med.,* 309, 694, 1983.
27. **Banda, M. J., Rice, A. G., Griffin, G. L., and Senior, R. M.,** Alpha-1-proteinase inhibitor is a neutrophil chemoattractant after proteolytic inactivation by macrophage elastase, *J. Biol. Chem.,* 263, 4481, 1988.
28. **Banda, M. J., Rice, A. G., Griffin, G. L., and Senior, R. M.,** The inhibitory complex of human alpha-1-proteinase inhibitor and human leukocyte elastase is a neutrophil chemoattractant, *J. Exp. Med.,* 168, 1608, 1988.
29. **Lai, E. C., Kao, F.-F., Law, M. L., and Woo, S. L. C.,** Assignment of the alpha-1-antitrypsin gene and sequence-related gene to human chromosome 14 by molecular hybridization, *Am. J. Hum. Genet.,* 35, 385, 1983.
30. **Pearson, S. J., Tetri, P., George, D. L., and Francke, U.,** Activation of human alpha-1-antitrypsin gene in rat hepatoma × human fetal liver cell hybrids depends on presence of human chromosome 14, *Somat. Cell. Mol. Genet.,* 9, 567, 1983.
31. **Rabin, M., Watson, M., Kidd, V., Woo, S. L. C., Breg, W. R., and Ruddle, F. H.,** Regional location of alpha-1-antichymotrypsin and alpha-1-antitrypsin genes on human chromosome 14, *Somat. Cell. Mol. Genet.,* 12, 209, 1988.
32. **Long, G. L., Chandra, T., Woo, S. L. C., Davie, E. W., and Kurachi, K.,** Complete nucleotide sequence of the cDNA for human alpha-1-antitrypsin and the gene for the S variant, *Biochemistry,* 23, 4828, 1984.
33. **Perlino, E., Cortese, R., and Ciliberto, G.,** The human alpha-1-antitrypsin gene is transcribed from two different promoters in macrophages and hepatocytes, *EMBO J.,* 6, 2767, 1987.
34. **Hofker, M. H., Nelen, M., Klasen, E. C., Nukiwa, T., Curiel, D., Crystal, R. G., and Frants, R. R.,** Cloning and charcterization of an alpha-1-antitrypsin-like gene 12 kb downstream of the genuine alpha-1-antitrypsin gene, *Biochem. Biophys. Res. Commun.,* 155, 634, 1988.
35. **Kelsey, G. D., Parker, M., and Povey, S.,** The human alpha-1-antitrypsin-related sequence gene: isolation and investigation of its sequence, *Ann. Hum. Genet.,* 52, 151, 1988.
36. **Sefton, L., Kelsey, G., Kearney, P., Povey, S., and Wolfe, J.,** A physical map of human PI and AACT genes, *Genomics,* 7, 382, 1990.
37. **Seralini, G.-E., Berube, D., Gagne, R., and Hammond, G. L.,** The human corticosteroid binding globulin gene is located on chromosome 14q31-q32.1 near two other serine protease inhibitor genes, *Hum. Genet.,* 80, 75, 1990.
38. **Gadek, J. E., Fells, G. A., Zimmerman, R. L., Renard, S. I., and Crystal, R. G.,** Antielastases of the human alveolar structure: implications for the protease-antiprotease theory of emphysema, *J. Clin. Invest.,* 68, 889, 1981.
39. **Janoff, A.,** Elastase and emphysema: current assessment of the protese-antiprotease hypothesis, *Am. Rev. Respir. Dis.,* 132, 417, 1985.
40. **Roa, N. V., Wehner, N. G., Marshall, B. C., Gray, W. R., Gray, B. H., and Hoidal, J. R.,** Characterization of proteinase-3, a neutrophil serine proteinase, *J. Biol. Chem.,* 266, 9540, 1991.
41. **Pirie-Shepherd, S. R., Miller, H. R. P., and Ryle, A.,** Differential inhibition of rat mast cell proteinase I and II by members of the alpha-1-proteinase inhibitor family of serine proteinase inhibitors, *J. Biol. Chem.,* 266, 17314, 1991.
42. **Hudig, D., Allison, N. J., Pickett, T. M., Winkler, U., Chin-Min, K., and Powers, J. C.,** The function of lymphocyte proteases: inhibition and restoration of granule-mediated lysis with isocoumarin serine protease inhibitors, *J. Immunol.,* 147, 1360, 1991.
43. **Matheson, N. R., van Halbeek, H., and Travis, J.,** Evidence for a tetrahedral intermediate complex during serpin-proteinase interactions, *J. Biol. Chem.,* 266, 13489, 1991.
44. **Carrell, R. W., Owen, M., Brennan, S., and Vaughn, L.,** Carboxy terminal fragment of alpha-1-antitrypsin from hydroxylamine cleavage: homology with antithrombin III, *Biochem. Biophys. Res. Commun.,* 91, 1032, 1979.
45. **Morii, M., Odani, S., and Ikenaka, T.,** Characterization of a peptide released during the reaction of human alpha-1-antitrypsin and bovine chymotrypsin, *J. Biochem.,* 86, 915, 1979.
46. **Carp, H. and Janoff, A.,** In vitro suppression of serum elastase inhibitor capacity by reactive oxygen species generated by phagocytosing polymorphonuclear leukocytes, *J. Clin. Invest.,* 63, 793, 1989.
47. **Ossanna, P. J., Test, S. T., Matheson, N. R., Regiani, S., and Weiss, S. J.,** Oxidative regulation of neutrophil elastase-alpha-1-proteinase inhibitor interactions, *J. Clin. Invest.,* 89, 1366, 1987.

48. **Hubbard, R. C., Ogushi, F., Fells, G. A., Cantin, A. M., Jallat, S., Courtney, M., and Crystal, R. G.,** Oxidants spontaneously released by alveolar macrophages of cigarette smokers can inactivate the active site of alpha-1-antitrypsin rendering it ineffective as an inhibitor of neutrophil elastase, *J. Clin. Invest.*, 80, 1289, 1987.
49. **Ogushi, F., Hubbard, R. C., Vogelmeier, C., Fells, G. A., and Crystal, R. G.,** Risk factors for emphysema: cigarette smoking is associated with a reduction in the association rate constant of lung alpha-1-antitripsin for neutrophil elastase, *J. Clin. Invest.*, 87, 1060, 1991.
50. **Banda, M. J., Clark, E. J., and Werb, Z.,** Limited proteolysis by macrophage elastase inactivates human alpha-1-protease inhibitor, *J. Exp. Med.*, 152, 1563, 1980.
51. **Desrochers, P. E. and Weiss, S. J.,** Proteolytic inactivation of alpha-1-proteinase inhibitor by neutrophil metallo-proteinase, *J. Clin. Invest.*, 81, 1646, 1988.
52. **Michaelis, J., Vissers, M. C. M., and Winterbourn, C. C.,** Human neutrophil collagenase cleaves alpha-1-antitrypsin, *Biochem. J.*, 270, 809, 1990.
53. **Senior, R. M., Griffin, G. L., Fliszar, C. J., Shapiro, S. D., Goldberg, G. I., and Welgus, H. G.,** Human 92- and 72-kilodalton type IV collagenases are elastases, *J. Biol. Chem.*, 266, 7870, 1991.
54. **Mast, A. E., Enghild, J. J., Nagase, H., Suzuki, K., Pizzo, S. V., and Salvesen, G.,** Kinetics and physiologic relevance of the inactivation of alpha-1-proteinase inhibitor, alpha-1-antichymotrypsin, and antithrombin III and matrix metalloproteinases-1 (tissue collagenase), -2 (72-kDa gelatinase/type IV collagenase), and -3 (stromelysin), *J. Biol. Chem.*, 266, 15810, 1991.
55. **Desrochers, P. E., Jeffrey, J. J., and Weiss, S. J.,** Interstitial collagenase (matrix metalloproteinase-1) expresses serpinase activity, *J. Clin. Invest.*, 87, 2258, 1991.
56. **Winyard, P. G., Zhang, Z., Chidwick, K., Blake, D. R., Carrell, R. W., and Murphy, G.,** Proteolytic inactivation of human alpha-1-antitrypsin by human stromelysin, *FEBS Lett.*, 279, 91, 1991.
57. **Johnson, D. A., Barrett, A. J., and Mason, R. W.,** Cathepsin L inactivates alpha-1-proteinase inhibitor by cleavage in the reactive site region, *J. Biol. Chem.*, 261, 14748, 1986.
58. **Morihara, K., Tsuzuki, H., and Oda,K.,** Protease and elastase of Pseudomonas aeruginosa: inactivation of human plasma alpha-1-proteinase inhibitor, *Infect. Immun.*, 24, 188, 1989.
59. **Banda, M. J., Clark, E. M., and Werb, Z.,** Regulation of alpha-1-proteinase inhibitor function by rabbit alveolar macrophages: evidence for proteolytic rather than oxidative inactivation, *J. Clin. Invest.*, 75, 1758, 1985.
60. **Frommherz, K. M., Faller, B., and Bieth, J. G.,** Heparin strongly decrases the rate of inhibition of neutrophil elastase by alpha-1-proteinase inhibitor, *J. Biol. Chem.*, 266, 15356, 1991.
61. **Lodish, H. F., Kong, N., Hirani, S., and Rasmussen, J.,** A vesicular intermediate in the transport of hepatoma secretory proteins from the rough endoplasmic reticulum to Golgi complex, *J. Cell Biol.*, 104, 221, 1987.
62. **Liu, M.-C., Yu, S., Sy, J., Redman, C. M., and Lipmann, F.,** Tyrosine sulfation of proteins from human hepatoma cell line HepG2, *Proc. Natl. Acad. Sci. U.S.A.*, 82, 7160, 1985.
63. **Perlmutter, D. H., Cole, F. S., Kilbridge, P., Rossing, T. H., and Colten, H. R.,** Expression of alpha-1-proteinase inhibitor gene in human monocytes and macrophages, *Proc. Natl. Acad. Sci. U.S.A.*, 82, 795, 1985.
64. **Perlmutter, D. H., Kay, R. M., Cole, F. S., Rossing, T. H., Van Thiel, D. H., and Colten, H. R.,** The cellular defect in alpha-1-proteinase inhibitor deficiency is expressed in human monocytes and in xenopus oocytes injected with human liver mRNA, *Proc. Natl. Acad. Sci. U.S.A.*, 82, 6918, 1985.
65. **Kelsey, G. D., Povey, S., Bygrave, A. E., and Lovell-Badge, R. H.,** Species- and tissue-specific expression of human alpha-1-antitrypsin in transgenic mice, *Genes Dev.*, 1, 161, 1987.
66. **Koopman, P., Povey, S., and Lovell-Badge, R. H.,** Widespread expression of human alpha-1-antitrypsin in transgenic mice revealed by in situ hybridization, *Genes Dev.*, 3, 16, 1989.
67. **Carlson, J. A., Rogers, B. B., Sifers, R. N., Hawkins, H. K., Finegold, M. J., and Woo, S. L. C.,** Multiple tissues express alpha-1-antitrypsin in transgenic mice and man, *J. Clin. Invest.*, 82, 26, 1988.
68. **Perlmutter, D. H., Daniels, J. D., Auerbach, H. S., De Schryver-Kecskemeti, K., Winter, H. S., and Alpers, D. A.,** The alpha-1-antitrypsin gene is expressed in a human intestinal epithelial cell line, *J. Biol. Chem.*, 264, 9485, 1989.
69. **Hafeez, W., Ciliberto, G., and Perlmutter, D. H.,** Constitutive and modulated expression of the human alpha-1-antitrypsin gene: different transcriptional initiation sites used in three different cell types, *J. Clin. Invest.*, 89, 1214, 1992.
70. **DeSimone, V., Ciliberto, G., Hardon, E., Paonessa, G., Palla, F., Lundberg, L., and Cortese, R.,** Cis- and trans-acting elements responsible for the cell specific expression of the human alpha-1-antitrypsin gene, *EMBO J.*, 6, 2759, 1987.
71. **Monaci, P., Nicosia, A., and Cortese, R.,** Two different liver-specific factors stimulate in vitro transcription from the human alpha-1 antitrypsin promoters, *EMBO J.*, 7, 2075, 1988.

72. **Frain, M., Swart, G., Monaci, P., Nicosia, A., Stampfli, S., Frank, R., and Cortese, R.,** The liver-specific transcription factor LF-B1 contains a highly diverged homebox DNA binding domain, *Cell,* 59, 145, 1989.
73. **Mendel, D. B. and Crabtree, G. R.,** HNF-1, a member of a novel class of dimerizing homeodomain proteins, *J. Biol. Chem.,* 266, 677, 1991.
74. **Mendel, D. B., Khavari, P. A., Conley, P. B., Graves, M. K., Hansen, L. P., Admon, A., and Crabtree, G. R.,** Charcterization of a cofactor that regulates dimerization of a mammalian homeodomain protein, *Science,* 254, 1762, 1991.
75. **Bamhueter, S., Mendel, D. B., Conley, P. B., Kuo, C. J., Turk, C., Graves, M. K., Edwards, C. A., Courtois, G., and Crabtree, G. R.,** HNF-1 shares three sequence motifs with the pou domain proteins and is identical to LF-B1 and APF, *Genes Dev.,* 4, 372, 1990.
76. **Rangan, V. S. and Das, C. G.,** Purification and biochemical characterization of hepatocyte nuclear factor 2 involved in liver-specific transcription of the human alpha-1-antitrypsin gene, *J. Biol. Chem.,* 265, 8874, 1990.
77. **Mitchelmore, C., Trabotti, C., and Cortese, R.,** Isolation of two cDNAs encoding zinc finger proteins which bind to the alpha-1-antitrypsin promoter and to the major histocompatibility complex class I enhancer, *Nucleic Acids Res.,* 19, 141, 1990.
78. **Kozak, M.,** Structural features in eukaryotic mRNAs that modulate the initiation of translation, *J. Biol. Chem.,* 266, 19867, 1991.
79. **Kozak, M.,** An analysis of vertebrate mRNA sequence: intimations of translational control, *J. Cell Biol.,* 115, 887, 1991.
80. **Jones, E. A., Vergalla, J., Steer, C. J., Bradley-Moore, P. R., and Vierling, J. M.,** Metabolism of intact and desialylated alpha-1-antitrypsin, *Clin. Sci. Mol. Med.,* 55, 139, 1978.
81. **Makino, S. and Reed, C. E.,** Distribution and elimination of exogenous alpha-1-antitrypsin, *J. Lab. Clin. Med.,* 75, 742, 1970.
82. **Glaser, C. B., Karic, L., Fallat, R. J., and Stockert, R.,** Plasma survival studies in rat of the normal and homozygote-deficient forms of alpha-1-antitrypsin, *Biochem. Biophys. Acta,* 495, 87, 1977.
83. **Laurell, C.-B., Nosslin, B., and Jeppsson, J.-O.,** Catabolic rate of alpha-1-antitrypsin of Pi type M and Z in man, *Clin. Sci. Mol. Med.,* 52, 457, 1977.
84. **Fuchs, H. E., Shifman, M. A., and Pizzo, S. V.,** In vivo catabolism of alpha-1-proteinase inhibitor-trypsin, antithrombin III-thrombin and alpha-2-macroglobulin-methylamine, *Biochim. Biophys. Acta,* 716, 151, 1982.
85. **Pizzo, S. V., Mast, A. W., Feldman, S. R., and Salvesen, G.,** In vivo catabolism of alpha-1-antichymotrypsin is mediated by the serpin receptor which binds alpha-1-proteinase inhibitor, antighrombin III, and heparin co-factor II, *Biochim. Biophys. Acta,* 967, 158, 1988.
86. **Mast, A. E., Enghild, J. J., Pizzo, S. V., and Salvesen, G.,** Analysis of plasma elimination kinetics and conformational stabilities of native, proteinase-complexes and reactive site cleaved serpins: comparison of alpha-1-proteinase inhibitor, alpha-antichymotrypsin, antithrombin III, alpha-2-antiplasmin, angiotensinogen, and ovalbumin, *Biochemistry,* 30, 1723, 1991.
87. **Potempa, J., Fedak, D., Duhn, A., Mast, A., and Travis, J.,** Proteolytic inactivation of alpha-1-antichymotrypsin: site of cleavage and generation of chemotactic activity, *J. Biol. Chem.,* 266, 21482, 1991.
88. **Gadek, J. E. and Crystal, R. G.,** Alpha-1-antitrypsin deficiency, in *The Metabolic Basis of Inherited Disease,* 5th ed., Stanbury, J. B., Wyngaarden, J. B., Fredrickson, D. S., Goldstein, J. L., and Brown, M. S., Eds., McGraw-Hill, New York, 1983, 1450.
89. **Strygler, B., Nicas, M. J., Santangelo, W. C., Porter, J. L., and Fordtran, J. S.,** Alpha-1-antitrypsin excretion in stool in normal subjects and in patients with gastrointestinal disorders, *Gastroenterology,* 99, 1380, 1990.
90. **Grill, B., Tinghitella, T., Hillemeier, C., and Gryboski, J.,** Increased intestinal clearance of alpha-1-antitrypsin in patients with alpha-1-antitrypsin deficiency, *J. Pediatr. Gastroenterol. Nutr.,* 2, 95, 1983.
91. **Poli, V. and Cortese, R.,** Interleukin 6 induces a liver-specific nuclear protein that binds to the promoter of acute phase genes, *Proc. Natl. Acad. Sci. U.S.A.,* 86, 8202, 1989.
92. **Poli, V., Mancini, F. P., and Cortese, R.,** IL-6DBP, a nuclear protein involved in interleukin-6 signal transduction, defines a new family of leucine zipper protein related to C/EBP, *Cell,* 63, 643, 1990.
93. **Descombes, P., Chojker, M., Lichsteiner, S., Falvey, E., and Schibler, U.,** LAP, a novel member of the C/EBP gene family, encodes a liver enriched transcriptional activator protein, *Genes Dev.,* 4, 1541, 1990.
94. **Descombes, P. and Schibler, U.,** A liver-enriched transcriptional activator protein, LAP, and a transcriptional inhibitor protein, LIP, are translated from the same mRNA, *Cell,* 67, 569, 1991.
95. **Perlmutter, D. H., Goldberger, G., Dinarello, C. A., Mizel, S. B., and Colten, H. R.,** Regulation of class III major histocompatibility complex gene products by interleukin-1, *Science,* 232, 850, 1986.

96. **Perlmutter, D. H., Dinarello, C. A., Punsal, P. I., and Colten, H. R.,** Cachectin/tumor necrosis factor regulates hepatic acute phase gene expression, *J. Clin. Invest.*, 78, 1349, 1986.
97. **Perlmutter, D. H.,** Cytokines and the hepatic acute phase response, in *Multiple Systems Organ Failure: Hepatic Regulation of Systemic Host Defence,* Matuschak, G. M., Ed., Marcel Dekker, New York, in press.
98. **Laurell, C.-B. and Rannevik, G.,** A comparison of plasma protein changes induced by danazol, pregnancy and estrogens, *J. Clin. Endocrinol. Metab.*, 49, 719, 1979.
99. **Tamir, S., Kadner, S. S., Katz, J., and Finley, T. H.,** Regulation of antitrypsin and antichymotrypsin synthesis by MCF-7 breast cancer cell sublines, *Endocrinology,* 127, 1319, 1990.
100. **Sandhaus, R. A., MacCarthy, K., Masson, R., and Henson, P.,** Elastolytic proteinases of the human macrophage, *Chest,* 83, 60, 1983.
101. **Welgus, H. G., Connolly, N. L., and Senior, R. M.,** 12-O-tetradecanoly-phorbol-13-acetate-differentiated U937 cells express a macrophage-like profile of neutral proteinases. High levels of secreted collagenase and collagenase inhibitor accompany low levels of intracellular elastase and cathepsin G, *J. Clin. Invest.*, 77, 1675, 1986.
102. **Helke, C. J., Krause, J. E., Mantyh, P. W., Couture, R., and Bannon, M. J.,** Diversity in mammalian tachykinin peptidergic neurons: multiple peptides, receptors and regulatory mechanisms, *FASEB J.*, 4, 1605, 1990.
103. **Battey, J. and Wada, E.,** Two distinct receptor subtypes for mammalian bombesin-like peptides, *Trends Neurol. Sci.*, 14, 524, 1991.
104. **Selkoe, D. J.,** Amyloid protein precursor and the pathogenesis of Alzheimer's disease, *Cell,* 58, 611, 1989.
105. **Katzman, R. and Saitoh, T.,** Advances in Alzheimer's disease, *FASEB J.*, 5, 278, 1991.
106. **Yankner, B. A., Duffy, L. K., and Kirschner, D. A.,** Neurotrophic and neurotoxic effects of amyloid protein: reversal by tachykinin neutropeptides, *Science,* 250, 279, 1990.
107. **Li, H., Takeuchi, K. H., Manly, K., Chapman, V., and Swank, R. T.,** The propeptide of β-glucuronidase. Further evidence of its involvement in compartmentalization of β-glucuronidase and sequence similarity with portions of the reactive site region of the serpin superfamily, *J. Biol. Chem.*, 265, 14732, 1990.
108. **Clarke, E. P., Cates, G. A., Ball, E. H., and Sanwal, B. D.,** A collagen-binding protein in the endoplasmic reticulum of myoblasts exhibits relationship with serine proteinase inhibitors, *J. Biol. Chem.*, 266, 17230, 1991.
109. **Church, F. C., Pratt, C. W., and Hoffman, M.,** Leukocyte chemoattractant peptides from the serpin heparin cofactor II, *J. Biol. Chem.*, 266, 704, 1991.
110. **Molmenti, E., Ziambaras, T., and Perlmutter, D. M.,** Evidence for an acute phase response in human intestinal epithelial cells, *J. Biol. Chem.*, in press.
111. **Molmenti, E., Perlmutter, D. H. and Rubin, D.,** Cell-specific expression of the α_1 antitrypsin gene in human intestinal epithelium, *J. Clin. Invest.*, in press.
112. **Perlmutter, D. H.,** unpublished data.
113. **Fallon, R. J., Joslin, G., and Perlmutter, D. H.,** unpublished data.
114. **Joslin, G., Griffin, G. L., August, A. M., Adams, S., Fallon, R. J., Senior, R. M., Perlmutter, D. H.,** The serpin-enzyme complex (SEC) receptor mediates the neutrophil chemotactic effect of α_1 antitrypsin-elastase complexes and amyloid-β peptide, *J. Clin. Invest.*, 90, 1150, 1992.

Chapter 9

FIBRINOGEN: A MULTIFUNCTIONAL ACUTE PHASE PROTEIN

Gerald M. Fuller

TABLE OF CONTENTS

0-8493-6913-4/93/$0.00 + $.50

I. INTRODUCTION

The acute phase responses of an organism are the mechanisms by which it both defends itself and recovers from noxious insults. The two major systems that carry out these responses are the immune and coagulation systems. The immune response involves both humoral and cellular events which utilize dozens of different chemical responses (e.g., antibody bindings, secretion of "alarm" response molecules, opsinization, phagocytosis, etc.). The coagulation system is equally complex, involving a cascade of enzymatic amplifications with the end result of transforming a soluble protein (fibrinogen) to an insoluble fibrin polymer that, together with the aggregation of cells, forms a proteinaceous plug at the site of vessel damage.

Within the last 4 years, there has been a quantum leap in information on the molecular details of the acute phase component in the inflammatory response (for reviews of cytokines in the inflammatory response, see References 1 to 4). In particular, it is now known that the appearance and action of three cytokines orchestrate most of the cellular responses during the early period of the inflammatory reaction. These cytokines — interleukin-1 (IL-1), interleukin-6 (IL-6), and tumor necrosis factor (TNF) — initiate a very large number of cellular responses. All three molecules are quite different from one another and have different signaling systems. Two of them, IL-1 and TNF, have functions that significantly overlap; for example, both IL-1 and TNF stimulate IL-6 production, induce fever and hypotension, initiate neutrophil diapedesis, and stimulate the synthesis of prostaglandin E_2. Recent studies have shown that these two cytokines stimulate some of the same intracellular signaling pathways and transcription factors.[5,6] The cellular responses controlled by IL-6 are, for the most part, different from IL-1 and TNF, and include inducing the final maturation of B-lymphocytes to form high-producing IgG plasma cells[7] and controlling the production of many hepatic proteins involved in the inflammatory response.[8] Two features of most cytokines that are important in the context of this discussion are (1) cytokines are not expressed constitutively, and therefore must be induced (usually as a result of tissue/cell insult) and (2) the half-life of the cytokine is brief.

It should be emphasized that the acute phase response serves two critically important functions for the organism: damage control and the immediate initiation of repair processes. In certain instances, these are carried out by the same molecule(s). The hepatic component of the acute phase response is a major alteration in the expression of a subset of plasma proteins. These proteins belong to one or more classes of molecules that are involved in the defense, repair, or salvage of important cellular and bodily components. Fibrinogen is one of these acute phase-responding proteins. Its synthesis increases four- to tenfold during the early hours of an acute inflammatory response.

The central focus of this chapter is to review the basic molecular design of the clotting protein fibrinogen and to describe some of its functions in addition to forming a clot. Fibrinogen has been the subject of intensive investigation from many laboratories for more than 5 decades. Several recent reviews focus on the details of the structure of the fibrinogen molecule.[9-11] In the present discussion, salient features of the protein are emphasized as a way to point out how it functions in different physiological roles.

II. STRUCTURE AND FUNCTION OF FIBRINOGEN

A. BASIC DESIGN OF THE FIBRINOGEN MOLECULE

Fibrinogen is a rod-like (240 × 50 Å) dimeric glycoprotein (mol wt, 340 kDa) that is produced by hepatocytes and circulates in the plasma at a concentration of 200 to 400 mg/dl. The half-molecule (monomer) is comprised of three different polypeptides — Aα, Bβ,

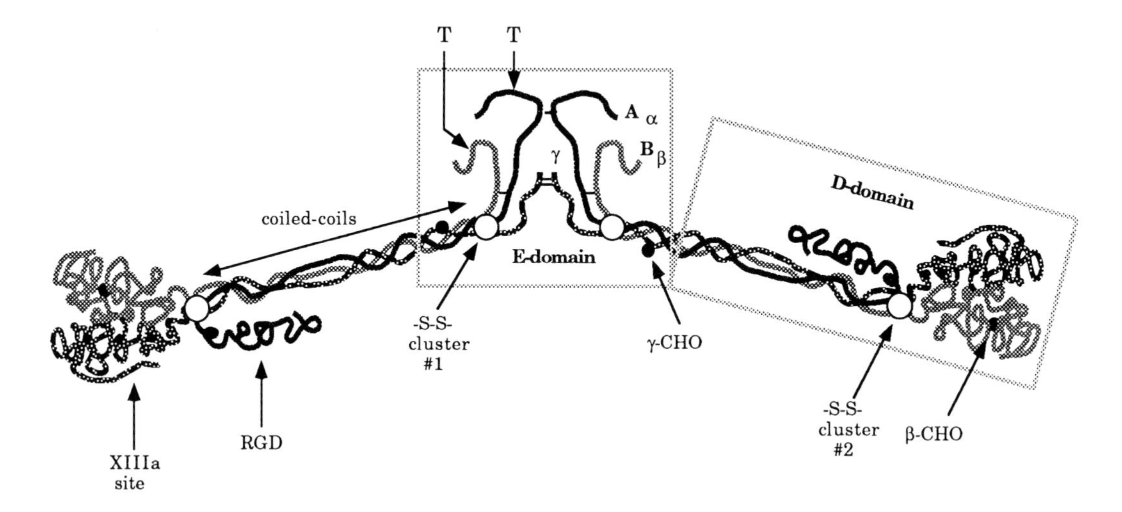

FIGURE 1. Schematic model of fibrinogen. The schematic drawing shown is a rendition of the Doolittle model.[9] Shown here are the thrombin cleavage site (T) in the Aα- and Bβ-chains. Removal of the A-fibrinopeptides unmasks the two polymerizing knobs within the E domain. The circles represent regions in the molecule that contain at least two cysteines on each chain involved in interchain disulfide bonding. There are two disulfide clusters in the E domain and one each in the D domains. The location of the carbohydrates on the γ- and β-subunits are noted. The tightly wound supercoiled α-helix regions are shown as coiled coils under the arrows. The C-terminal region of the Bβ- and γ-chains form "random coils". In this region of the γ-chain reside the Gln-Lys cross-linking sites catalyzed by activated factor XIIIa. The C-terminal portion of the Aα-chain carries the RGD binding motif for the platelet receptor.

and γ, (63,000, 55,000, and 47,500 Da, respectively) — that are linked to one another through 13 pairs of disulfide bonds. The two monomers are bridged together through three additional disulfide bonds located near the NH_2 terminals, which enable the monomers to face each other, such that the amino terminals of the six chains occur near the center of the rod (Figure 1).

The amino-terminal region of the molecule (located in the middle of the rod) provides the thrombin-binding site,[12] the polymerizing domains,[13-15] and the relevant cytokine-signaling fragment (discussed in a following section). The carboxy-terminal regions contain the principle cross-linking sites for factor XIII,[16,17] the platelet-binding sites,[18] and the fibronection-binding domains.[19] Two of the three chains, Bβ and γ, have single N-linked carbohydrate clusters which are first added to the different chains during the translation/translocational event.[20]

Physicochemical data suggest that the overall shape of the fibrinogen molecule is a prolate ellipsoid, somewhat reminiscent of the shape of a cigar. Electron microscopic examination of fibrinogen[21,22] showed that the protein possesses three nodules, one at each end and one in the middle of the rod. The central nodule of the protein is smaller than those at the ends and appears as an open globular motif which allows for the binding of thrombin and cleavage of the fibrinopeptides.[12] If one examines the stylized model of fibrinogen (Figure 1) beginning at the central domain and proceeding laterally along its length, one encounters first a disulfide cluster stitching the three chains together. Immediately following this cluster is the first carbohydrate group linked to Asn_{53} on the γ-chain. A less flexible region comprised of "supercoiled" portions of the three chains constitutes the next part of the protein. This configuration is somewhat reminiscent of the coiled-coil portion of a collagen superhelix. The coiled coils of fibrinogen end, however, at another cluster of interchain disulfide linkers. From this distal disulfide knot, the β- and γ-chains each form another open globular domain[23] from which a carbohydrate protrudes from the β-subunit

(Asn_{364}). The cross-linking regions of the γ-chain also located within this distal domain (Gln_{39};9 and Lys_{407}) are available for isopeptide bond formation with another fibrin molecule, a process catalyzed by activated factor XIII (factor XIIIa).[24] Also proceeding from the distal disulfide girdle is the α-chain C terminal that appears to double back toward the middle of the protein and is not a part of the globular domains of the β- and γ-chains which constitute the terminal knob. This region of the α-chain possesses an RGD binding motif (see below) as well as several proteolytic cleavage sites of plasmin.

It has been difficult to obtain protein crystals of intact fibrinogen. Some very large fragments of bacterially proteolyzed fibrinogen, however, have been examined in detail by X-ray diffraction analyses.[23] Results from these studies have provided additional information on the overall structure of the protein. Although the newer findings add important "fine structure" to the basic molecular pattern first shown by Hall and Slater some 30 years ago, the so-called "trinodular" form of fibrinogen remains as the accepted molecular design.

Studies of the step by step degradation of fibrinogen/fibrin by plasmin (described in more detail below) led to the identification of major proteolytic domains of the molecule.[25] Following exhaustive plasminolysis, a large amino-terminal domain arises from plasmin cleavage sites located just distal to the first disulfide girdles. This region is referred to as the E domain.[26] The remaining structures, including the coiled coils and the distal disulfide connectors, form what is termed the D fragments or domain, as shown in Figure 1.

In summary, fibrinogen is a disulfide-linked heterodimer composed of three pairs of nonidentical polypeptide chains which are held together by an extensive array of inter- and intradisulfide bonds. At the macromolecular level, it is an elongated protein with a rigid coiled-coil portion in the middle of each monomer. The two end regions and the central portion of the protein contain a number of specialized sites for proteolysis, cross-linking, polymerization, and binding to other macromolecular structures, including matrix molecules and cells. Insofar as it has been studied, the fundamental molecular arrangement of fibrinogen has been preserved from hagfishes to mammals,[27] an evolutionary period of some 450 million years, attesting to the physiological importance of its molecular design.

B. ASSEMBLY OF FIBRINOGEN

All three of the fibrinogen genes are located in a tight cluster on the long arm of chromosome 4 in man[28] and chromosome 2 in rats.[29] Each fibrinogen subunit is transcribed and processed to form a separate mRNA species.[30] The mRNAs are delivered to the cytoplasm, where each is individually translated and the resulting polypeptide translocated into the endoplasmic reticulum.[31] Assembly of fibrinogen is a complex process and most of the existing details of the steps involved have been worked out by Redman and his colleagues.[32-37] Hepatocytes of all species examined have unequal amounts of the three fibringen chains within the lumen of the endoplasmic reticulum. All species examined have an excess of γ-subunits, but different species have different amount of Aα and Bβ-chains. The reasons for the differences in the amounts of the chains are not understood. Unequal rates of synthesis and degradation likely rise to the differences.[38] The assembly pattern for human fibrinogen is described below.

Within the lumen of the endoplasmic reticulum of the human liver cell are surplus pools of Aα- and γ-chains which exist both as individual chains and as Aα-γ complexes. There is no indiction that Aα or γ dimerizes with itself. As the Bβ-subunit is translated and translocated into the ER lumen, the Aα-γ dimer attaches to the elongating Bβ-polypeptide. Once the translation/translocation of the Bβ-subunit has occurred, chain rearrangements undoubtedly occur and a competent monomeric unit of fibrinogen forms, which then dimerizes to finalize the assembly process. These steps are shown in Figure 2.

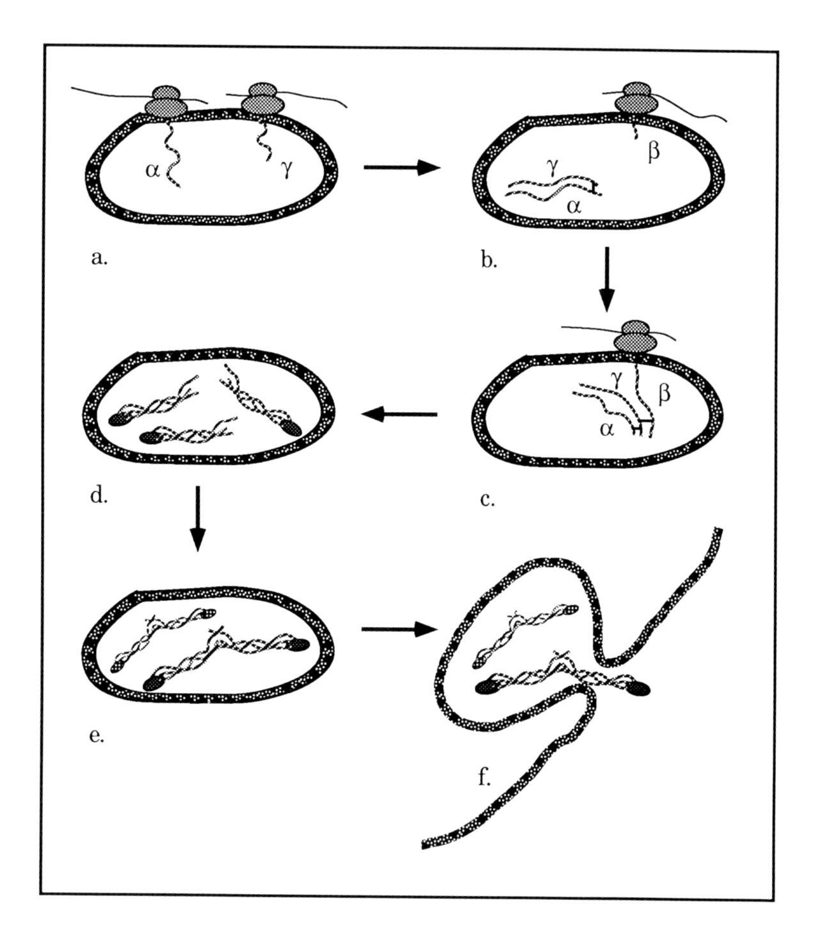

FIGURE 2. The assembly of human fibrinogen. (a) Cross-section of hepatic endoplasmic reticulum with ribosomes translating the Aα- and γ-mRNAs. Translocation of nascent Aα- and γ-chains into the ER are also indicated. (b) Aα-γ dimer in the lumen of the ER and the beginning of translation/translocation of the Bβ-chain; (c) attachment of the Aα-γ complex to the enlongating Bβ-chain. Note that this occurs prior to completion of the Bβ-translation/translocation. (d) Fibrinogen monomers consisting of an Aα-, Bβ- and γ-chain complex. Rearrangement and additional folding likely occurs to prepare the monomers for dimerization. (e) Completed fibrinogen molecule prior to secretion. It is not known if this is accomplished in ER or in Golgi (f) Secretion vesicle. The fibrinogen molecule is being secreted into the sinusoids of the liver. Only the intact heterodimer of fibrinogen is secreted.

Recently, it has been demonstrated that transfection of a human hepatocarcinoma cell line, HepG2, with the Bβ-chain cDNA led to a threefold increase in the production of fibrinogen.[36] Not only did the transfected cell increase its synthesis of Bβ, due to the exogenous Bβ-cDNA, but also Aα- and γ-chain synthesis increased, suggesting that the Bβ-chain may have an effect on the synthesis of the other chains. This finding also implied that maintenance of unequal and surplus amounts of Aα and γ are essential in the production of fibrinogen.[39] The role of excess chains in fibrinogen assembly is not understood, but one possibility is that surplus chains may direct or drive the assembly by a mass action process.

While the steps of assembly described above appear to be the general paradigm, important details of the process are still lacking. It is clear that in the species examined (human, chicken, and rat), the Bβ-chain is the ''nucleating'' subunit in the assembly of this multichain molecule. In order to gain information on which portion of the polypeptide is essential for assembly, Redman and colleagues cotransfected COS cells with truncated forms of the Bβ-cDNA together with intact Aα- and γ-cDNAs.[35] Polypeptides containing only the first 207 amino acids of this chain could assemble with completed Aα- and γ-subunits and be secreted by the cell. However, when the COS cells were similarly cotransfected with Bβ-cDNAs lacking amino acids 1 to 207 or 1 to 93, the molecule did not assemble, nor were any of the subunits secreted. The domain considered essential for assembly contains part of the coiled-coils region, indicating that proper alignment of the chains occurs by interchain α-helical foldings. These findings suggest that the coiled regions are essential in the alignment and folding of the protein to allow for correct disulfide bond formation and not vice versa.

C. FIBRINOGEN ACTIVATION — FIBRIN FORMATION

The end result of either the intrinsic or extrinsic clotting cascade is the conversion of prothrombin to thrombin. The thrombin molecule binds to the N-terminal region of fibrinogen and initiates two steps of proteolysis. First is a rapid removal of fibrinopeptide A (aa 1 to 16), followed by a slower hydrolysis of fibrinopeptide B (aa 1 to 14). Excision of these peptides removes a cluster of negatively charged amino acids in the middle of the protein, allowing for closer molecular packing of the freshly cleaved fibrinogens. The first step of polymerization appears to be an adherence of fibrinogen molecules in a half-overlapped fashion brought about following the removal of the charged clusters of the fibrinopeptides. An essential feature of fibrin formation in locking the polymer together in a "leggo-like" formation is the unmasking of "polymerizing knobs" following the release of fibrinopeptide A.[13,14] Figure 3 schematically shows the appearance of the "knobs" formed by tertiary arrangements of amino acids 337 to 379 in the γ-chain.[40,41] Although the B-peptide is removed more slowly from fibrinogen, its polymerizing knobs, Gly-Pro-His, also participate in stabilizing the fibrin polymer.[15] The pattern of polymer formation is driven in part by the shape of the molecules and the way in which each associates with one another. A second stage of fibrin polymer stabilization occurs by the formation of isopeptide bonds between lysyl and glutaminyl residues located at the C-terminal regions of the γ-chains that have aligned end to end.[42,43] This antiparallel arrangement leads to the formation of two isopeptide cross-links at each end of a fibrin monomer. Factor XIIIa, the transglutaminase that catalyzes the formation of these bonds, is activated by thrombin. Additional cross-linking occurs through the α-chains, but these are formed later in the clot-stabilizing process.

This brief account of the polymerization process leaves undiscussed many of the subtle and remarkable steps involved in the construction of the fibrin lattice (for additional details, see Reference 44). Molecular rearrangements leading to the formation of new binding domains for cells and for plasminogen, fibronectin, α_2-plasmin inhibitor, thrombin, and factor XIII point to the importance and versatility of the fibrin polymer.

D. FIBRINOLYSIS

One of the special features of fibrin is its transient nature. As described above, the formation of fibrin is a rapid process and together with platelet aggregation provides the structural basis for hemostasis. In physiologic terms, however, the demand of the organism is to keep the vessels patent and to allow blood flow to continue. An elegant system has evolved for the regulated dissolution of the fibrin/platelet matrix following wound stabilization, and the fibrinogen/fibrin molecule is involved in the regulatory processes.

The process of physiologic fibrin digestion is called fibrinolysis and is accomplished by the enzyme plasmin, also a liver-derived protein. Plasmin is a trypsin-like serine protease with a fairly broad specificity capable of hydrolyzing many proteins at lysine and arginine residues. It does exert strong substrate specificity for fibrin and is the protease primarily involved in fibrin digestion. Plasmin circulates in the blood as an inactive precursor plasminogen.

Two different strategies have evolved to ensure a regulated removal of the clot or thrombus. First is the appearance of specific plasminogen-binding domains on the fibrin

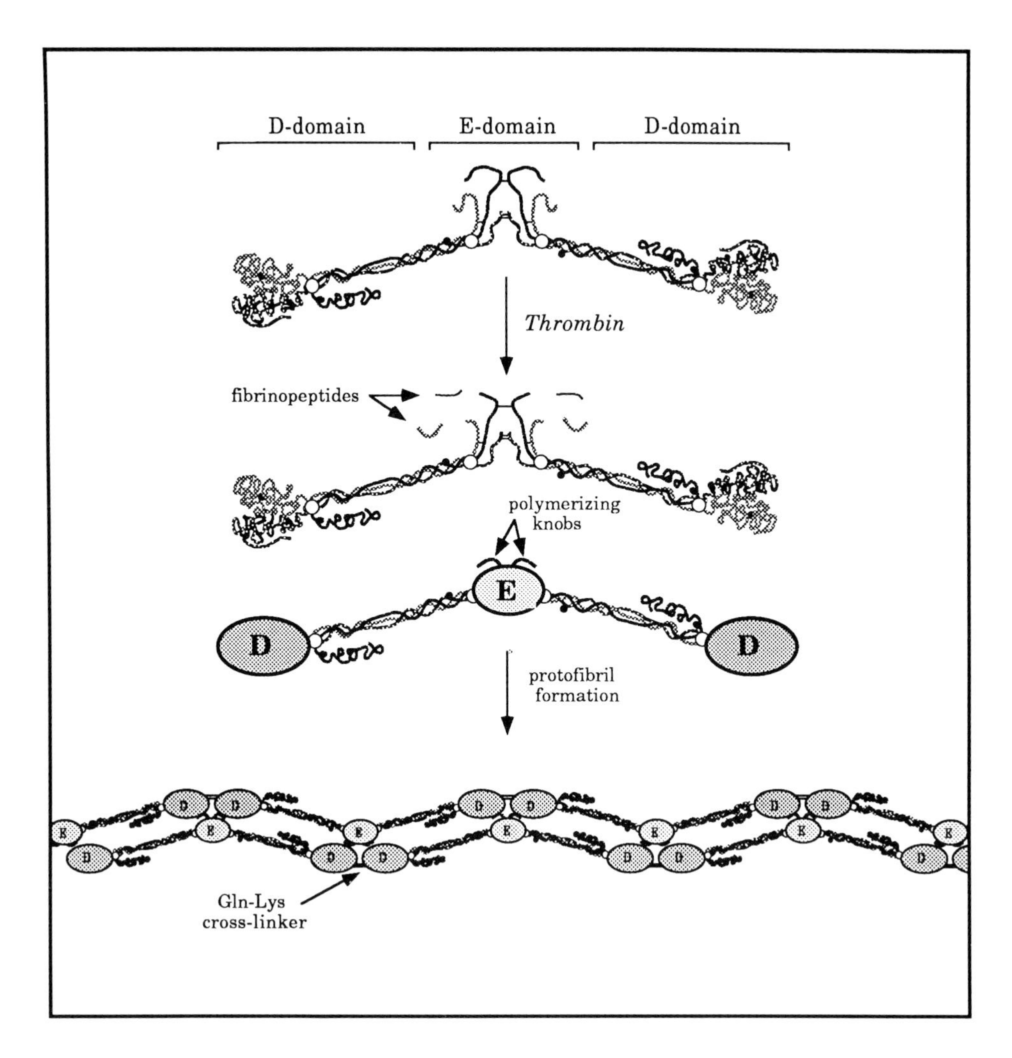

FIGURE 3. Schematic events of the clotting reaction. Intact fibrinogen is depicted in the top of the figure. Following the activation of prothrombin to thrombin, the enzyme binds to fibrinogen and then cleaves out the fibrinopeptides, thus generating the polymerizing knobs as shown in the diagram. The "activated" fibrinogen forms a half-staggered, overlapped arrangement and lines up end to end, leading to the formation of a protofilament. Thrombin also "activates" factor XIII in the presence of Ca^{2+}, which then carries out a transamidination reaction leading to the formation of two isopeptide bonds between γ-chains of different fibrin molecules.

molecule during the clotting event.[45] Once available, these domains permit plasminogen attachment to fibrin during polymer formation, resulting in a sequestered proteolytic zymogen throughout the fibrin meshwork. The second control point is the gradual release of enzyme(s) required to activate plasminogen to plasmin. Again, the fibrin molecule is part of the process. The plasminogen activators (mostly of the tissue-secreted type, tissue plasminogen activator, tPA) exhibit a binding affinity to fibrin, placing it in close proximity to plasminogen. Thus, both the dissolving enzyme and its activator become part of the fibrin clot. It is apparent that the fibrinogen/fibrin molecule has been specifically tailored to allow for the binding, activation, and degradation of this extracellular matrix component. Not all of the lysines and arginines in the molecule are cleavage points of the enzyme, however; rather, there are preferential sites, such that predictable fragments are systematically excised from the polymer. Figure 4 depicts the generation of the core fragments of fibrin. These fragments are

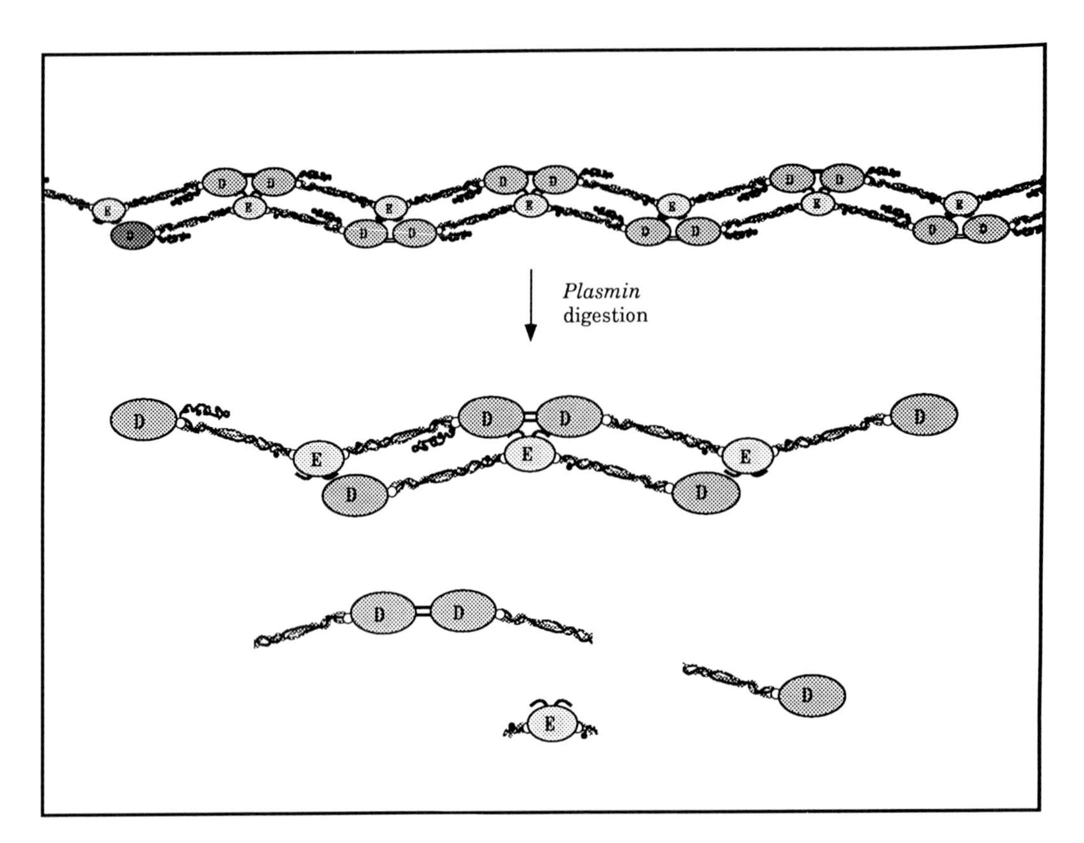

FIGURE 4. The events of fibrinolysis. This figure depicts events opposite of coagulation. Plasminogen-binding sites are expressed on fibrin as it is polymerizing. Plasminogen binds to these sites and is incorporated into the clot. Following stabilization at the wound site, the surrounding cells secrete a plasminogen activator (tissue plasminogen activator, tPA) that converts plasminogen into plasmin. Plasmin, a trypsin-like protease, begins a systematic cleavage of the fibrin matrix. Three basic core peptides resulting from exhaustive plasminolysis have been found in the circulation: fragment D, fragment E, and cross-linked fragment D known as the D-D dimer. These peptides are derived from the regions of the molecule shown in the diagram.

the constituents of circulating fibrin degradative products (FDPs) and in certain instances are diagnostic for a serious clotting disorder known as desseminated intravascular coagulation (DIC).[46]

In summary, the process of clot or thrombus removal is vitally important. The dissolution of this matrix, however, must be carried out in a orderly fashion. Fibrin, the matrix to be dissolved, participates in the process by forming specific binding sites for enzymes ultimately involved in its removal.

III. RECEPTORS FOR FIBRINOGEN/FIBRIN

A. PLATELETS

Disruption of the endothelial cell attachment to its basement membrane exposes a collagen surface. Platelets have a collagen receptor (integrin receptor, $\alpha_2\beta_1$) which binds to this collagen surface. It is believed that binding of the platelet to collagen almost instantaneously initiates a rearrangement of a fibrinogen receptor ($\alpha_{IIb}\beta_3$) on the surface of the platelet, and fibrinogen quickly attaches to this surface.[18] The attachment of fibrinogen to its receptor recruits and activates other receptors, leading to a rapid aggregation of platelets. Other agents that are present at a wound site, e.g., thrombin, adenosine diphosphate, epinephrine, and prostaglandins, also activate the fibrinogen receptors, and within seconds there is the formation of a meshwork of platelets-fibrinogen-collagen and other extracellular matrix components that continues to build and entrap blood cells, including other platelets, leading to formation of a clot. A common feature of the agents that initiate the expression of functional fibrinogen receptors is that all are released at a vascular injury. Thus, the hemostatic property of fibrinogen is in part related to its ability to bind platelets together. Additionally, the "activated" platelet secretes a lectin-like protein, called thrombospondin (TSP),[47] that also binds to the platelet surface. Once TSP is bound to the platelet surface, fibrinogen binds to it, leading to a more stabilized platelet aggregate.[48,49] In many instances, the clot is extensive enough to halt the catastrophic loss of body fluids. Fibrinogen/receptor complexes usually cluster with themselves and with the collagen receptor complex. The clustering events lead to the formation of a strong cell-cell interaction.

In the truest sense, fibrinogen functions as a ligand in cell-to-cell attachment complexes. It has been shown that the sequence motif of Arg-Gly-Asp (RGD) located in the C-terminal region of the α-chain constitutes the binding region of fibrinogen for its receptor,[50,51] as does another sequence (a dodecamer) located near the C-terminal region of the γ-chain.[52,53] Further inspection of the fibrinogen molecule also reveals an RGD sequence near the central domain. Thus, there appears to be three ligand-binding regions of the fibrinogen monomer (six per molecule). The location of two of these binding domains is at opposite ends of the protein, which allows for the potential cross-linking of two receptors through a single fibrinogen molecule.

B. LEUKOCYTES

Fibrinogen association with monocytes or neutrophils has been recognized as a component of delayed-type hypersensitivity, transplant rejection, and the part of the pathophysiology of vascular obstruction and atherogenesis.[54-56] Recent evidence has shown that fibrinogen has specific binding domains to two additional specific membrane receptors on the surface of neutrophils, monocytes, and macrophages. These receptors (integrins) are CD11b/CD18 (also known as Mac-1[57]) and CD11c/CD18.[58] Interestingly, the binding specificity of fibrinogen to CD11b/CD18 occurs through a site which does not include the RGD binding region of the Aα-chain nor the dodecapeptide of the γ-chain. The binding of fibrinogen to the integrim CD11c/CD18 occurs through a domain located in the NH_2 region of the Aα-chain; this binding region has been identified more precisely and includes a domain that contains amino acids 17 to 19 (Gly-Pro-Arg),[58] the polymerizing knobs.

To date, at least four distinct eukaryotic cell-binding receptors have been identified for the fibrinogen/fibrin molecule: those that bind to (1) IIb/IIIa platelet receptors ($\alpha_{IIb}\beta_3$) (2) $\alpha_2\beta_3$-integrin on endothelial cells,[59] (3) CD11b/CD18 on macrophages and neutrophils,[60] and (4) CD11c/CD18 on macrophages.[58] In another context, it has been shown that certain strains of *Staphylococcus* bind preferentially to fibrinogen, which led to the development of a sensitive agglutination test for fibrinogen. The specific regions of the protein which bind to the bacterial cell wall have been shown to involve a 15-amino acid stretch near the C-terminal domain of the γ-chain.[61] Table 1 lists the cell types and functions of fibrinogen receptors.

C. FIBRINOGEN/FIBRIN SPECIFICALLY ASSOCIATES WITH EXTRACELLULAR MATRIX MOLECULES

Information presented in the previous section demonstrates that fibrinogen is a specific ligand to several different types of surface receptors. It has also become apparent that fibrinogen/fibrin has specific binding interactions with a variety of other proteins. Fibronectin, a major external cell-surface protein, has specific sequences to which fibrinogen

TABLE 1
Cellular Binding Domains of Fibrinogen

Receptor	Cell type with receptor	Molecular binding domain: Subunit	Molecular binding domain: Sequence	Function
$\alpha_{IIb}\beta_3$	Platelet	α	R-G-D	Aggregation Adhesion
$\alpha_v\beta_3$	Endothelial cell		K-Q-A-G-D-V	Cell-cell attachment
CD11b/CD18	Neutrophils Macrophages	D domain	K-Q-A-G-D-V	Inflammation PMN aggregate
CD11c/CD18	Neutrophils Macrophages Fibroblasts	E domain	G-P-R	IL-6 induction
Cell wall	*Staphyloccus*	γ	Decapenta-peptide	Agglutination

binds.[62,63] Fibronectin has, as part of its molecular structure, specific binding regions for fibrin, collagen, heparin, and cell-surface integrins, making it a critically important substrate for the adhesion and migration of cells repairing the damaged tissue.[64] Fibronectin is also involved in phagocytosis, and thus may also play a role in the removal of debris of various types. The extensive interaction between fibrin and fibronectin at a wound and the numerous interactions in which they participate during the remodeling and repair process underscore the importance of these two extracellular matrix molecules. A detailed description of these interactions has recently been published.[65]

It has also been observed that there is a specific interaction between fibrinogen and hyaluronic acid (HA).[66] HA is a major glycosaminoglycan component of external coats of proliferating and migrating cells that are active in developing, regenerating, and repairing tissues. HA concentrations increase in the extravascular spaces, particularly in regions that have suffered damage and a fibrin meshwork has been laid down. This observation has led to the prediction that the specific interaction between HA and fibrinogen/fibrin participates in the formation and organization of the early wound matrix.[67]

IV. FIBRINOGEN BIOSYNTHESIS AND THE IL-6 LOOP

Numerous studies on the biosynthesis of fibrinogen using a variety of experimental systems from animals undergoing an acute phase response to primary heptocyte cultures and hepatocarcinoma cells all reveal that interleukin-6 (IL-6) is the cytokine that controls the increase in fibrinogen gene expression. It has also been shown that glucocorticoids synergize with IL-6 and amplify the cytokine signal.[68-70]

The kinetics of the appearance and disappearance of the fibrinogen mRNAs showed that their expression is stringently coordinated. For example, increases in cytoplasmic levels of Aα-,Bβ-, and γ-mRNA's occur within 30 min following exposure to IL-6 and a glucocorticoid analog, dexamethasone.[71] Additionally the half-life for each fibrinogen mRNA has been shown to be identical. While the precise molecular details controlling the expression of fibrinogen have not been fully determined, it is remarkable how tightly coupled the expression of these genes is with one another. Currently, the precise second-messenger pathways involved are unknown, although it has been shown that IL-6 increases the binding of specific nuclear proteins to IL-6-responsive elements in the promoter regions of the three fibrinogen genes.[72-74]

Approximately 25 years ago, Barnhart and colleagues reported that they could increase the hepatic production of fibrinogen in animals by injecting plasmin digestion products of fibrinogen/fibrin (the core fragments employed were fragments D and E).[75] This observation suggested that there were specific receptors for fibrinogen fragments on the hepatocyte surface and that binding of the fragments somehow increased fibrinogen biosynthesis. When the isolated fragments were added to a primary hepatocyte culture, no increase in fibrinogen biosynthesis was detected.[76] If, however, the fragments were added to cultures of peripheral blood monocytes, and this conditioned medium then added to the hepatocytes, fibrinogen biosynthesis increased more than fivefold. These findings indicated that the monocytes produced a factor(s) whose expression was upregulated by fragments of fibrin. Since this initial observation was made, it has been shown that the stimulating factor produced by activated monocytes is IL-6.[8]

In a more detailed analysis of the plasmin-digested fragments of fibrinogen and fibrin, it has been recently shown that only the E fragment of fibrin is capable of initiating the production of IL-6 in macrophages.[77] Furthermore, the binding domain of the E fragment to the cell involves a portion of the polymerizing knobs uncovered by the action of thrombin. As shown in the diagram in Figures 3 and 4, these polymerizing regions are located on the α- and β-chains and are cryptic until fibrinopeptide removal.

Is the fact that one of the plasmin-digested fragments of fibrin is capable of inducing the expression of IL-6 of physiologic relevance, or is it simply one of the quirks of biology? Several recent reports[78,79] implicate IL-6 as an important participant in tissue repair. One can envision a situation in which following the initial formation of the fibrin scaffolding, repairative cells enter this wound site and initiate the dissolving of fibrin by secretion of plasminogen activator. As the matrix dissolves, a continual production of a signaling molecule (fragment E) allows for a continual production of IL-6. It has recently been reported, however, that IL-6 upregulates the expression of an inhibitor of enzymes that dissolve collagen and other components of the extracellular matrix.[78] The fact that circulating levels of IL-6 are low suggests that levels produced by cells within the repairing site are using the cytokine in an autocrine/paracrine loop. Neither all of the genes under IL-6 control nor all that necessarily participate in the repair process have been identified; however, it seems reasonable that IL-6 is important in the regenerative process. That one can now demonstrate the presence of an endogenous regulator for IL-6 expression within a wound site implies that IL-6 is important in cell renewal and tissue reconstruction. These findings also imply that fibrinogen can participate in regulating its own synthesis through an indirect feedback loop that involves a specific fragment of fibrin and certain IL-6-producing cells such as macrophages (Figure 5).

V. THE ROLE OF FIBRIN IN WOUND HEALING

Wound healing is a complicated multistep process which is crucial to the survival of the organism and which parallels many of the complex events that occur during embryonic development. During the repair process, the extracellular matrix is sequentially remodeled and rebuilt by the concerted action of different cell types. The wound itself becomes a transitory organ or structure whose function is to remodel successively a series of increasingly complex and ordered extracellular matrices. The matrix develops sequentially from a lesion (wound) to a platelet fibrin plug, to a relatively loose matrix of glycosaminoglycan and collagen, to a denser granulation tissue, and then to the final repaired tissue.

Thus, fibrin's role in the wound-healing process is to provide a scaffolding for the trapping of blood cells, platelets, neutrophils, monocytes, and lymphocytes. These cells carry with them arsenals of molecules involved in the inactivation of bacteria, secretion of proteolytic enzymes for digesting damaged cells, and secretion of cytokines for stimulating cells in the adjacent areas to release substances for either the killing of foreign organism or

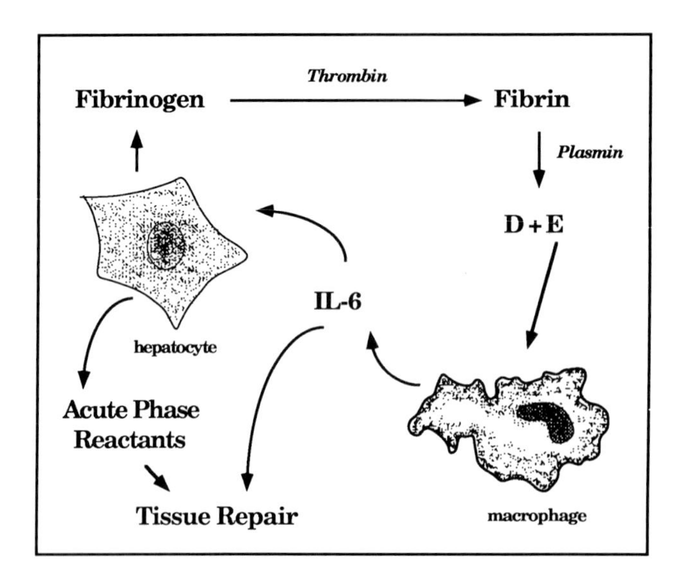

FIGURE 5. A feedback loop showing how fibrinogen can be involved in regulating its own synthesis via the cytokine IL-6. The E fragment of fibrin has been shown to bind to the surface of macrophages and induce the expression of IL-6, which increases the synthesis of fibrinogen in hepatocytes. This diagram also shows a possible linkage between fibrinolysis during wound healing and the generation of IL-6, a potentially potent regulator of tissue repair processes.

the chemotaxis of other cells. It is essential to have a rich, structurally insoluble matrix to which cells attach almost immediately. Fibrin polymers form a somewhat porous gel-like meshwork containing other proteins important for the remodeling process. Fibronectin is incorporated into the clot and distributed along the fibrin strands.[80] It has been suggested that HA enters the wound and that its concentration increases significantly in this environment. The presence of HA has been postulated to play an important part in the healing events, including angiogenesis.[67] The precise signals involved in increasing HA synthesis by repariative cells is unknown; however, it is conceivable that IL-6 could play an important role.

VI. CONCLUSION AND SUMMARY

In this chapter, I have briefly described the structure, biosynthesis, assembly, and activation of fibrinogen as a way to emphasize the many remarkable features of this protein. Closer examination of fibrinogen reveals that it is involved in a number of processes related to, but distinct from, its hemostatic role. That it constitutes a major ligand for platelets and their activation, that it has specific receptors on the surface of neutrophils, monocytes, and endothelials cells, and that its proteolytic peptides are involved in the induction of a critically important cytokine all point to the fact that fibrinogen is involved in a number of important homeostatic processes.

ACKNOWLEDGMENTS

I thank my graduate student, James E. Nesbitt, for the creative computer drawings presented in the chapter. I also acknowledge Heart and Lung and Blood Institute Grant No. HL 43155 for support of my research.

REFERENCES

1. **Dinarello, C. A.,** Biology of interleukin-1, *FASEB J.,* 2, 108, 1988.
2. **Gauldie, J. and Baumann, H.,** Cytokines and acute phase protein expression, in *Cytokines and Inflammation,* Kimball, E. S., Ed., CRC Press, Boca Raton, FL, 1991, 275.
3. **Koj, A.,** Biological functions of acute phase proteins, in *The Acute Phase Response to Injury and Infection,* Gordon, A. H. and Koj, A., Eds., Elsevier, Amsterdam, 1985, 145.
4. **Kishimoto, T.,** The biology of interleukin-6, *Blood,* 74, 1, 1989.
5. **Osborn, L., Kunkel, S., and Nabel, G. J.,** Tumor necrosis factor alpha and interleukin-1 stimulate immunodeficiency virus enhancer by activation of nuclear factor kappa B, *Proc. Natl. Acad. Sci. U.S.A.,* 86, 2336, 1989.
6. **Zhang, Y., Lin, J.-X., and Vilcek, J.,** Interleukin-6 induction by tumor necrosis factor alpha and interleukin-1 in human fibroblasts involves activation of a nuclear factor to a kappa B like sequence, *Mol. Cell. Biol.,* 10, 3818, 1990.
7. **Hirano, T. et al.,** Complementary DNA for a novel human interleukin (BSF-2) that induces B lymphocytes to produce immunoglobulin, *Nature (London),* 324, 73, 1986.
8. **Gauldie, J. et al.,** Interferon β2/B-cell stimulatory factor type 2 shares identity with monocyte derived hepatocyte stimulating factor and regulates the major acute phase protein response in liver cells, *Proc. Natl. Acad. Sci. U.S.A.,* 84, 7251, 1987.
9. **Doolittle, R. F.,** Fibrinogen and fibrin, *Annu. Rev. Biochem.,* 53, 196, 1984.
10. **Crabtree, G. R.,** The molecular biology of fibrinogen, in *The Molecular Basis of Blood Diseases,* Stamatoyannopoulos, G., Neinhuis, A. W., and Majerus, P. W., Eds., W. B. Saunders, Philadelphia, 1987, 631.
11. **Chung, D. W., Harris, J. E., and Davies, E. W.,** Nucleotide sequences of the three genes coding for human fibrinogen, *Adv. Exp. Med. Biol.,* 281, 39, 1990.
12. **Mann, K. G. and Lundblad, R. L.,** Biochemistry of thrombin, in *Hemostasis and Thrombosis,* Coleman, R. W., Hirsh, J., Marder, V. J., and Salzman, E. W., Eds., J.B. Lippincott, Philadelphia, 1982, 112.
13. **Laudano, A. P. and Doolittle, R. F.,** Synthetic derivatives which bind to fibrinogen and prevent the polymerization of fibrin monomers, *Proc. Natl. Acad. Sci. U.S.A.,* 75, 3085, 1978.
14. **Laudano, A. P. and Doolittle, R. F.,** Studies on synthetic peptides that bind to fibrinogen and prevent fibrin polymerization. Structural requirement, numbers of binding sites and species differences, *Biochemistry,* 19, 1013, 1980.
15. **Furlan, M. et al.,** Effects of calcium and synthetic peptides on fibrin polymerization, *Thromb. Haemost.,* 47, 118, 1982.
16. **Chen, R. and Doolittle, R. F.,** γ-γCross-linking sites in human and bovine fibrin, *Biochemistry,* 10, 4486, 1971.
17. **Olexa, S. A. and Budzynske, A. Z.,** Evidence for four different polymerization sites involved in human fibrin formation, *Proc. Natl. Acad. Sci. U.S.A.,* 77, 1374, 1980.
18. **Marguerie, G. A. et al.,** The binding of fibrinogen to its platelet receptor, *J. Biol. Chem.,* 257, 11872, 1982.
19. **Grinnell, F., Feid, M., and Minter, D.,** Fibroblast adhesion to fibrinogen and fibrin substrata: requirement for cold-insoluble globulin (plasma fibronectin), *Cell,* 19, 517, 1980.
20. **Nickerson, J. M. and Fuller, G. M.,** Modification of fibrinogen chains during synthesis: glycosylation of Bβ and γ chains, *Biochemistry,* 20, 2818, 1981.
21. **Hall, C. E. and Slater, H. S.,** The fibrinogen molecule: its size, shape and mode of polymerization, *J. Biophys. Cytol.,* 5, 11, 1959.
22. **Fowler, W. E. and Erickson, H. P.,** Trinodular structure of fibrinogen, *J. Mol. Biol.,* 134, 241, 1979.
23. **Weisel, J. W. et al.,** A model for fibrinogen: domains and sequence, *Science,* 230, 1388, 1985.
24. **Folk, J. E. and Chung, S. I.,** Molecular and catalytic properties of transglutaminases, *Adv. Enxymol.,* 38, 109, 1973.
25. **Marder, V. J., Shulman, N. R., and Carroll, W. R.,** The importance of intermediate degradation products of fibrinogen in fibrinolytic hemorrhage, *Trans. Assoc. Am. Phys.,* 53, 156, 1967.
26. **Marder, V. J., Shulman, N. R., and Carroll, W. R.,** High molecular weight derivatives of human fibrinogen produced by plasmin. I. Physicochemical and immunological characterization, *J. Biol. Chem.,* 244, 451, 1969.
27. **Doolittle, R. F.,** The structure and evolution of vertebrate fibrinogen, *Ann. N.Y. Acad. Sci.,* 408, 28, 1983.
28. **Kant, J. et al.,** Organization and evolution of the human fibrinogen gene locus on chromosome 4, *Proc. Natl. Acad. Sci. U.S.A.,* 82, 2344, 1985.
29. **Marino, M. W., Fuller, G. M., and Elder, F. F. B.,** Chromosomal localization of human and rat Aα, Bβ and γ fibrinogen genes by in situ hybridization, *Cytogenet. Cell Genet.,* 42, 36, 1986.

30. **Bouma, H., III, Kwan, S.-W., and Fuller, G. M.,** Radioimmunological identification of polysomes synthesizing fibrinogen polypeptide chains, *Biochemistry,* 14, 4787, 1975.
31. **Nickerson, J. M. and Fuller, G. M.,** In vitro synthesis of rat fibrinogen: identification of pre-Aα, pre-Bβ and pre-γ polypeptides, *Proc. Natl. Acad. Sci. U.S.A.,* 78, 303, 1981.
32. **Yu, S. et al.,** Intracellular assembly of human fibrinogen, *J. Biol. Chem.,* 258, 13407, 1983.
33. **Yu, S. et al.,** Fibrinogen precursors, *J. Biol. Chem.,* 259, 10574, 1984.
34. **Yu, S., Kudryk, B., and Redman, C. M.,** A scheme for the intracellular assembly of human fibrinogen, in *Fibrinogen-Fibrin Formation and Fibrinolysis,* Lane, D. A., Henschen, A., and Jasani, M. K., Ed., Walter de Gruyter, New York, 1986, 3.
35. **Zhang, J.-Z., Roy, S., and Redman, C. M.,** Identification of Bβ chain domains involved in human fibrinogen assembly, *FASEB J.,* 6, A226, 1992.
36. **Roy, S. N., Mukhopadhyay, G., and Redman, C. M.,** Regulation of fibrinogen assembly, *J. Biol. Chem.,* 265, 6389, 1990.
37. **Roy, S. N. Et al.,** Assembly and secretion of recombinant human fibrinogen, *J. Biol. Chem.,* 266, 4758, 1991.
38. **Grieninger, G., Plant, P. W., and Chiasson, M. A.,** Selective intracellular degradation of fibrinogen and its reversal in cultured hepatocytes, *J. Biol. Chem.,* 259, 14973, 1984.
39. **Roy, S. N. et al.,** Assembly and secretion of recombinant human fibrinogen, *J. Biol. Chem.,* 266, 4758, 1991.
40. **Shimizu, S., Nagel, G. M., and Doolittle, R. F.,** Photoaffinity labeling of the primary polymerizing site: isolation and characterization of a labeled cyanogen bromide fragment corresponding to γ-chain residues 337–379, *Proc. Natl. Acad. Sci. U.S.A.,* 89, 2888, 1992.
41. **Yamazumi, K. and Doolittle, R. F.,** Photoaffinity labeling of the primary fibrin polymerization site: localization of the label to γ-chain Tyr-363, *Proc. Natl. Acad. Sci. U.S.A.,* 89, 2893, 1992.
42. **Pisano, J. J., Finlayson, J. S., and Peyton, M. P.,** Cross-link fibrin polymerized by factor XIII: ε-(γ-glutamyl) lysine, *Science,* 160, 892, 1968.
43. **Folk, J. E. and Finlayson, J. S.,** The ε-(γ-glutamyl) lysine cross-link and the catalytic role of transglutaminases, *Adv. Protein Chem.,* 31, 1, 1977.
44. **Marder, V. J., Francis, C. W., and Doolittle, R. F.,** Fibrinogen structure and physiology, in *Hemostasis and Thrombosis,* Coleman, R. W., Hirsh, J., Marder, V. J., and Salzman, E. W., Eds., J. B. Lippincott, Philadelphia, 1982, 145.
45. **Wiman, B. and Collen, D.,** Molecular mechanism of physiological fibrinolysis, *Nature (London),* 272, 549, 1978.
46. **Coleman, R. W. and Marder, V. J.,** Desseminated intravascular coagulation (DIC): pathogenesis, pathophysiology and laboratory abnormalities, in *Hemostasis and Thrombosis,* Coleman, R. W., Hirsh, J., Marder, V. J., and Salzman, E. W., Eds., J.B. Lippincott, Philadelphia, 1982, 654.
47. **Lawler, J.,** The structural and functional properties of thrombospondin, *Blood,* 67, 1197, 1986.
48. **Bale, M. D., Westrick, L. G., and Mosher, D. F.,** Incorporation of thrombospondin into fibrin clots, *J. Biol. Chem.,* 260, 7502, 1985.
49. **Bacon-Baguley, T., Kudryk, B. J., and Walz, D. A.,** Thrombospondin interaction with fibrinogen, *J. Biol. Chem.,* 262, 1927, 1987.
50. **Ruoslahti, E. and Vaheri, A.,** Interaction of soluble fibroblast surface antigen with fibrinogen and fibrin, *J. Exp. Med.,* 141, 497, 1975.
51. **Pytela, R. et al.,** Platelet membrane glycoprotein IIb/IIIa: member of a family of Arg-Gly-Asp specific adhesion receptor, *Science,* 231, 1559, 1986.
52. **Kloczewiak, M., Timmons, S., and Hawiger, J.,** Recognition sites for the platelet receptor is present on the 15 residue carboxy-terminal of the gamma chain of human fibrinogen and is not involved in the fibrin polymerization reaction, *Thromb. Res.,* 29, 249, 1983.
53. **Plow, E. F. et al.,** Evidence that three adhesive proteins interact with a common recognition site on activated platelets, *J. Biol. Chem.,* 259, 5388, 1984.
54. **Geczy, C. L. et al.,** Macrophage procoagulant activity as a measure of cell-mediated immunity in the mouse, *J. Immunol.,* 130, 2743, 1983.
55. **Gerrity, R. G.,** The role of the monocyte in atherogenesis. II. Migration of foam cells from the atherosclerotic lesion, *Am. J. Pathol.,* 103, 191, 1981.
56. **Hattler, B. G. et al.,** Functional features of lymphocytes recovered from human renal allograft, *Cell Immunol.,* 9, 289, 1973.
57. **Hynes, R. O.,** Integrins: a family of cell surface receptors, *Cell,* 48, 549, 1987.
58. **Loike, J. D., et al.,** CD11c/CD18 on neutrophils recognize a domain at the N terminus of the Aα chain of fibrinogen, *Proc. Natl. Acad. Sci. U.S.A.,* 88, 1044, 1991.
59. **Ruoslahti, E. and Pierschbacher, M. D.,** New perspectives in cell adhesion: RGD and integrins, *Science,* 238, 491, 1987.

60. **Altier, D. C. et al.,** A unique recognition site mediates the interaction of fibrinogen with the leukocyte integrin Mac-1 (CD11b/CD18), *J. Biol. Chem.,* 265, 12119, 1990.
61. **Hawiger, J. et al.,** Identification of a region of human fibrinogen with staphylococcal clumping factor, *Biochemistry,* 21, 1407, 1982.
62. **Pytcha, R., Pierschbacher, M., and Rouslahti, E.,** Identification and isolation of a 140 kd cell surface glycoprotein with properties expected of a fibronectin receptor, *Cell,* 40, 191, 1985.
63. **Ruoslahti, E. and Pierschbacher, M. D.,** Arg-Gly-Asp: a versatile cell recognition signal, *Cell,* 44, 517, 1986.
64. **Sekiguchi, K., Fududa, M., and Hakomori, S.,** Domain structures of human plasma fibronectin, *J. Biol. Chem.,* 256, 6452, 1981.
65. **Hynes, R. O.,** *Fibronectins,* Springer-Verlag, New York, 1990, 593.
66. **LeBoeuf, R. D. et al.,** Human fibrinogen specifically binds hyaluronic acid, *Biochemistry,* 261, 12586, 1986.
67. **Weigel, P. H., Fuller, G. M., and LeBouef, R. D.,** A model for the role of hyaluronic acid and fibrin in the early events during the inflammatory response and wound healing, *J. Theor. Biol.,* 119, 219, 1986.
68. **Otto, J. M., Grenett, H. E., and Fuller, G. M.,** The coordinated regulation of fibrinogen gene transcription by hepatocyte-stimulating factor and dexamethasone, *J. Cell Biol.,* 105, 1067, 1987.
69. **Baumann, H., Jahreis, G. P., and Morella, K. K.,** Interaction of cytokine and glucocorticoid-response elements of acute phase plasma protein genes, *J. Biol.,* 265, 22275, 1990.
70. **Grieninger, G. et al.,** Regulation of plasma protein synthesis in cultured hepatocytes: effects of hormones and serum factors, in *The Liver: Quantitative Aspects of Structure and Function,* Preisig, R. and Bircher, J., Eds., Aulendorf, W. Germany, 1979, 118.
71. **Nesbitt, J. E. and Fuller, G. M.,** Transcription and translation are required for fibrinogen mRNA degradation in hepatocytes, *Biochim. Biophys. Acta,* 1089, 88, 1991.
72. **Fey, G. H. et al.,** Regulation of rat liver acute phase genes by interleukin-6 and production of hepatocyte-stimulating factors by hepatoma cells, *Ann. N.Y. Acad. Sci.,* 557, 1989, 317.
73. **Fowlkes, D. M. et al.,** Potential basis for regulation of the coordinately expressed fibrinogen genes: homology in the 5′ flanking regions, *Proc. Natl. Acad. Sci. U.S.A.,* 81, 2313, 1984.
74. **Tsuchiya, Y. et al.,** Sequence analysis of the putative regulatory region of rat α-2 macroglobulin gene, *Gene,* 57, 73, 1987.
75. **Barnhart, M. I. et al.,** Influence of fibrinolytic products on hepatic release and synthesis of fibrinogen, *Thromb. Diath. Haematol.,* (Suppl.), 139, 143, 1970.
76. **Ritchie, D. G. et al.,** Regulation of fibrinogen synthesis by plasmin derived fragments of fibrinogen and fibrin, *Proc. Natl. Acad. Sci. U.S.A.,* 79, 1530, 1981.
77. **Grenett, H. E. et al.,** An endogenous regulator of IL-1 and IL-6: A new functional role for a fibrin peptide, *J. Cell. Biol.,* 111, 511, 1991.
78. **Lotz, M. and Guerne, P.-A.,** Interleukin-6 induces the synthesis of tissue inhibitor of metalloproteinases-1/erythroid potentiating activity (TIMP-1/EPA), *J. Biol. Chem.,* 266, 2017, 1991.
79. **Matrisian, L. M.,** Metalloproteinases and their inhibitors in matrix remodeling, *Trends Genet.,* 6, 121, 1990.
80. **Grinnell, F. and Feld, M. K.,** Distribution of fibronectin on peripheral blood cells in freshly clotted blood, *Thromb. Res.,* 24, 397, 1981.

Chapter 10

HAPTOGLOBIN: RETROSPECTIVES AND PERSPECTIVES

Wanda Dobryszycka

TABLE OF CONTENTS

0-8493-6913-4/93/$0.00 + $.50

I. INTRODUCTION

A. HAPTOGLOBIN STRUCTURE — MAJOR PHENOTYPES

Plasma haptoglobin is a genetically determined α_2-acidic glycoprotein which combines specifically with hemoglobin, showing an activity of the "true" peroxidase type (for a review, see Reference 1). The binding of haptoglobin to hemoglobin is one of the strongest known noncovalent interactions in biology, the association constant being greater than 10^{-15} *M*.[2] Hemoglobin that has been released into the bloodstream by red cell hemolysis is effectively bound by haptoglobin. This mechanism prevents passage of hemoglobin through the glomerules and/or its deposition in the renal tubules; hence, iron remains available for further metabolic use. However, the complex is catabolized in the reticuloendothelial system in the liver; thus, haptoglobin cannot be considered as a regular transport protein (e.g., transferrin for iron, hemopexin for heme, etc.), but is regarded as a suicidal one.

Essentially, three major phenotypic forms of human haptoglobin, designated Hp 1-1, Hp 2-2, and Hp 2-1, have been isolated from human sera. In its simplest form, Hp 1-1 is a tetrachain structure composed of two α^1 (light, 9.1 kDa, each 83 residues)- and two β (heavy, 40.0 kDa, each 245 amino acids)-chains linked by disulfide bridges. The amino acid sequence and distribution of inter- and intradisulfide bridges are known.[1] Haptoglobin type 1-1 has the shape of a barbell with two spherical head groups, the β-chains. These are connected by a thin filament with a central knob, which corresponds to the α-chains. The overall length of the molecule is about 124 Å, and the interhead distance is 87 Å.[3]

Haptoglobin 2-2 and 2-1 are polymerized forms of higher molecular mass, showing multiple bands in polyacrylamide gel electrophoresis. The polymorphism of haptoglobin was found to be related to α-subunits (the α^2-chain, comprising 142 residues, is almost a duplicate of the α^1-chain, although in a cross-over event, some amino acid residues have been deleted). (Figure 1).

The structural gene controlling the synthesis of the α-chain is located on chromosome 16q22.105-16q22.108.[4,5] In humans, three common alleles (Hp^{1F}, Hp^{1S}, and Hp^2) control the synthesis of haptoglobins. The genes Hp^{1F} and Hp^{1S} code for the α-chain polypeptides α^{1F} and α^{1S}, and for the β-chain. The gene Hp^2 codes for α_2 and for the β-chains. The β-chains coded by the three genes are identical.

All the carbohydrate content of haptoglobin, which constitutes approximately 20%, is found on the β-chains. Oligosaccharide moieties are attached to asparaginyl residues 23, 46, 50, and 80, two of these being biantennary and two triantennary branches of the complex type with the terminal sialic acid. The ratio of 2-6-linked sialic acid to 2-3-linked sialic acid is about 4:1.[8,9] Fucose may be present in both a core $\alpha_{1\text{-}6}$-position and in an $\alpha_{1\text{-}3}$-position on an external *N*-acetylglucosamine residue.[1,10]

Haptoglobin β-chains were found to play a special role in the biological properties of the haptoglobin molecule. Binding of one half molecule of hemoglobin is reported to occur on the β-chain.[3] Residues in the 130 to 137 region of the β-chain are implicated in the contact with hemoglobin. When the dimers of hemoglobin are binding to the β-chains, the interhead group distance for the haptoglobin-hemoglobin complex is 117 Å, i.e., 30 Å greater than for haptoglobin.

Most antigenic determinants are located on the β-chain.[11] The lectin concanavalin A (Con A) is bound to the mannobiosyl-N-acetyl-glucosamine core of the carbohydrate on the β-chains.[12] Structural similarity of the β-chain to the family of serine proteases might suggest that this region has been preserved intact during evolution longer than other areas of the haptoglobin molecule.[13]

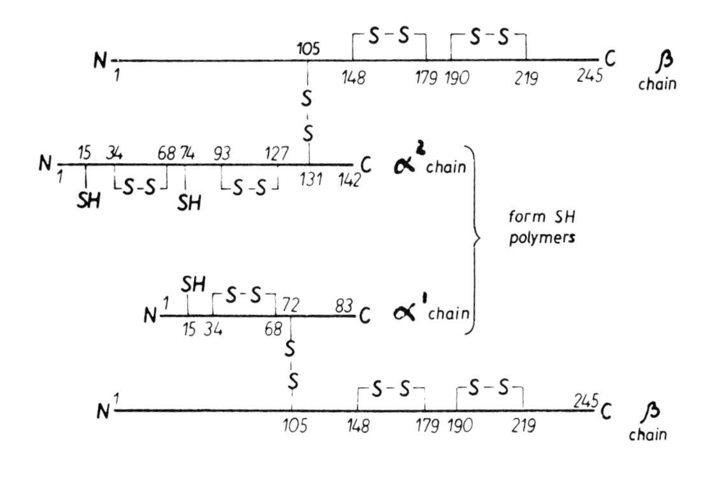

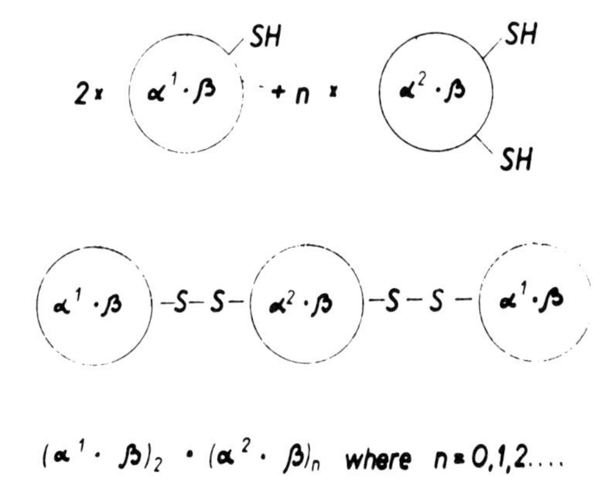

FIGURE 1. Structural representation and scheme of the subunit arrangements in haptoglobin 2-1.[1] Sulfhydryl groups at Cys_{15} and Cys_{74} are postulated to be involved in polymer formation. (From Bowman, B. H. and Kurosky, A., *Adv. Hum. Genet.*, 12, 189, 1982. With permission.)

B. RELATIONSHIPS OF HAPTOGLOBIN WITH OTHER PROTEINS

Occasionally, homologous sequences of α-chains of haptoglobin with either κ- and λ-chains of immunoglobulin[14] or a regulator protein of complement cascade, the so-called "decay accelerating factor",[15] were reported. Kurosky et al.[13] presented a detailed account of the homology of the haptoglobin β-chain with the chymotrypsinogen family of serine proteases.

The haptoglobin gene diverged from the ancestral protease gene about 500 million years ago.[16] Relationships within this family of functionally unrelated proteins are shown in Figure 2. A new branch was identified thanks to the finding that the exon-intron structure of the human complement C1s gene displays a striking similarity to the gene encoding haptoglobin.[17] Figure 2 includes Con A (which is surprisingly similar to the haptoglobin β-chain) and the glycophorin variant St^2 (the nucleotide sequence surrounding the putative cross-over point is homologous to the cross-over point proposed for haptoglobin genes).[19]

C. BIOLOGICAL ACTIVITIES OF HAPTOGLOBIN

Despite years of investigations, the physiological functions of haptoglobin have remained enigmatic. It has been generally accepted that the role of haptoglobin in hemoglobin me-

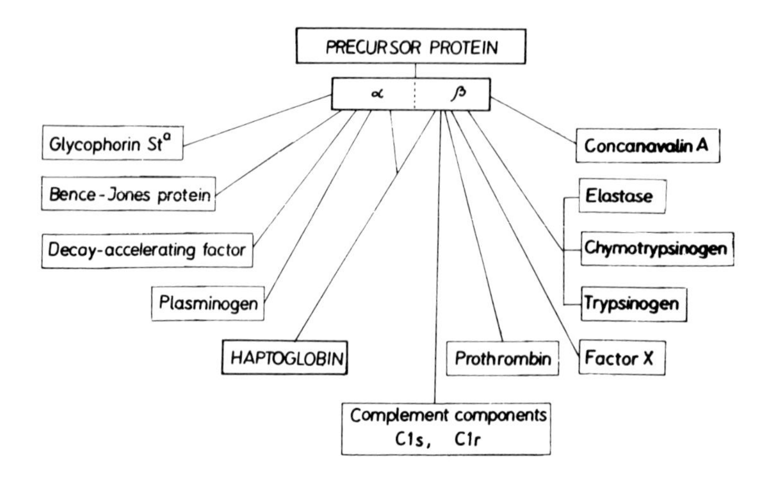

FIGURE 2. Structural relationships of haptoglobin with other proteins. Data taken from References 13 to 19.

tabolism is to protect the kidneys from tissue destruction by binding free hemoglobin following hemolysis. In 1963, it was suggested that heme α-methenyl oxygenase would convert pyridine-hemichromogen and hemoglobin-haptoglobin complex into a precursor of biliverdin.[20] On the other hand, hemoglobin binding is apparently not an essential function, since patients with ahaptoglobinemia are clinically silent. Moreover, under normal conditions, there is more than enough circulating haptoglobin to bind the intravascular free hemoglobin that occurs daily.

Other data suggest that haptoglobin may have a temporary but specific physiological role other than its role in hemoglobin metabolism. One of the likely roles seems to be in the host defense responses to infection and inflammation. Native haptoglobin blocks the human neutrophil response to a variety of agonists with defined plasma membrane receptors. Neutrophils possess specific binding sites for haptoglobin with an affinity similar to that of Con A binding. This suggests that haptoglobin may act as a natural antagonist for receptor-ligand activation of the immune system. Of the various functional parameters assessed, neutrophil respiratory burst activity (superoxide production) and rise in intracellular calcium were inhibited by haptoglobin.[21]

Other biological activities of haptoglobin include the inhibition of cathepsin B[22] (?), viral hemagglutination,[23] and prostaglandin H synthase,[24] suppression of lectin- and lipopolysaccharide-produced proliferation of lymphocytes and B-cell mitogenesis,[25] modulation of macrophage function, agglutination of T4 streptococci,[26] and binding of *Streptococcus pyogenes*.[27] Haptoglobin stimulates the formation of prostaglandin E in isolated osteoblasts and synergistically potentiates the effect of bradykinin and thrombin.[28] Formation of the haptoglobin-hemoglobin complex *in vivo* can play a role in prevention of the hemoglobin-driven hydroxyl radical and lipid peroxide generation in areas of inflammation.[29]

D. HAPTOGLOBIN-RELATED PROTEIN, SER-HAPTOGLOBIN, FETAL HAPTOGLOBIN. AN ONCOFETAL SPECIES, WHETHER NEOPLASTIC AND/OR FETAL HAPTOGLOBINS?

Analysis of human genomic clones led to the isolation of an haptoglobin-related protein (Hpr) sequence and to a demonstration of its structural homology to the Hp gene. The two genes are closely linked, with Hpr being 2.2 kb downstream of Hp. The sequences of Hp and Hpr are highly homologous, although the colinearity is interrupted by an Alu sequence inserted in the 5′ flanking region and by a retrovirus-like sequence inserted in the first intron of Hpr.[30] The events that have shaped the Hp and Hpr genes in primates were very complicated, including duplication (nonhomologous chromosomal breakage and reunion), triplication, insertions of two independent retrovirus-like sequences, deletion, fusion (homologous unequal cross-over), and several other recombinational events.[31] In the black population, chromosomes with multiple copies of Hpr genes were formed, always connected with phenotypes Hp 2-2 and Hp 2-1 (never Hp 1-1).[16,32] Multiplication of the Hpr gene automatically increases a number of retrovirus-like elements in the Hp gene cluster, although the significance of this fact has remained undefined.

Originally, the Hpr gene was thought to be nonfunctional, or a pseudogene, since several studies failed to detect Hpr mRNA in either fetal or adult liver.[30,32,33] However, transfection experiments *in vivo* indicate that the Hpr promoter is active and cell specific.[34] Recent studies have proved that the Hpr gene is still able to produce a protein. Hpr was detected in normal human plasma and characterized in the plasma of pregnant women, and the expression of Hpr in human breast cancers was demonstrated.[35,36] Clinically, these properties manifested as a dramatically worsened prognosis. Thus, detection and quantitation of Hpr might prove to be a useful diagnostic procedure in cancer, and possibly, pregnancy.

These data suggest that Hpr synthesis might not be restricted to the liver; an alternative site of synthesis besides breast cancer cells may be the placenta or decidua.[32,33]

The Hpr sequence codes for a protein whose α (light, 16.5 kDa)- and β (heavy, 40 kDa)-chains are distinct from, but highly homologous to, haptoglobin type 1-1.[35] A major difference in the Hpr sequence is the presence of a retrovirus-like element. The α-chain of Hpr differs from the α-chain of normal haptoglobin in their apparent molecular mass. One of the minor changes in the β-chain is the substitution of the histidine resiude (no. 22) for a serine residue close to one of the glycosylation sites. As this change switches off the glycosylation site, Hpr has only three potential sites of glycosylation compared with the normal four. Therefore, the lectin-binding properties of Hpr and haptoglobin will probably be different. Moreover, it is suggested that mutation in the β-chain resulted in abolition of the characteristic activity of haptoglobin, i.e., hemoglobin binding.[33]

An immune suppressive factor found in ascitic fluids of ovarian cancer patients was identified as a polymeric form of haptoglobin (850 to 900 kDa).[37] This SER-haptoglobin (SER = E-receptor-like suppressor factor) was 100 to 1000 times more potent an immunosuppressor than normal plasma haptoglobin and had a larger molecular mass than any of the normal adult-type haptoglobin.

Some macromolecular forms of fetal haptoglobin have been described;[38,39] therefore, it was suggested that SER-haptoglobin would have been identified as such a variant, since it was immunologically analogous to the fetal or neonatal haptoglobin found in cord serum.[40] However, the latter disclosed an activity to bind hemoglobin, whereas SER-haptoglobin did not. This could result from modifications of β-chains (hemoglobin-binding site) by oxidation, polymerization, or the action of proteolytic enzymes. Because of the lack of this characteristic property, SER-haptoglobin cannot be considered as an ordinary polymer of haptoglobin type 2-2 or 2-1.

It is suggested that SER-haptoglobin represents an oxidized form of plasma haptoglobin generated from macrophages activated during an inflammatory response and might be a negative feedback regulator of immune response produced by activated macrophages. SER-haptoglobin interferes with the functions of interleukin-1 on T-cell activation and inhibits phagocytic functions of normal macrophages. The sequential levels of SER-haptoglobin measured in the plasma of tumor-bearing patients were diagnostic of their clinical response to immunotherapy. Its production might be responsible for dysfunction of the immunological system in cancer patients (T-cell-mediated phagocytosis of macrophages, cytotoxic cells, synthesis of immunoglobulins, etc.).[37] Another hypothesis has assumed that SER-haptoglobin might be a product of gene transcription occurring during cancer development, mutation of a normal gene of haptoglobin, or of a haptoglobin gene modified by the effect of malignant growth.[7]

The carbohydrate moiety of haptoglobin (located exclusively on β-chains) derived from ascitic fluids of patients with ovarian carcinoma was apparently changed compared to normal β-chains. This was demonstrated by means of the reaction with the plant lectin, Con A.[41] Also, abnormally glycosylated forms of haptoglobin in sera from women with ovarian or breast cancer, using fucose-specific lectin from *Lotus tetragonolobus,* were described.[10] The expression of such a ''neoplastic'' haptoglobin appears to be associated with the development of malignancy, and thus has been used in monitoring the effects of therapy in ovarian cancer, in the differential diagnosis of malignant vs. nonmalignant tumors.[42,43] Haptoglobin is normally synthesized by the liver, but it may also be synthesized by some tumor cells.[44,45] On the other hand, the tumor may release soluble factors, e.g., interleukin-6, which promote the change in the carbohydrate moiety of haptoglobin in the liver.[46] Since the molecular structure of ''neoplastic'' is still unclear, its relationship (if any) with Hpr or SER-haptoglobin has not been identified.

Fetal (neonatal) haptoglobin was isolated from human fetuses (850 kDa in the complex with hemoglobin)[47] or from cord blood (900 to 1000 kDa).[39] Attempts have been made to verify the possibility that this haptoglobin form could be an oncofetal protein, as neonatal haptoglobin isolated from cord serum exhibited an immunosuppressive potency similar to that of SER-haptoglobin obtained from malignant effusions.

Table 1 summarizes some properties of normal human haptoglobin, Hpr, fetal haptoglobin, and SER-haptoglobin. Aside from normal haptoglobin, hemoglobin binding has revealed only fetal haptoglobin.

Biochemical-morphological analogies between neoplastic states and pregnancy are often drawn. The identification of Hpr from pregnancy plasma and its localization in breast carcinoma, the presence of SER-haptoglobin in ascitic fluids from ovarian cancer patients, and the isolation of fetal forms of haptoglobin broaden the spectrum of pathophysiologic roles of haptoglobin.[35,37,39] Clinical data raise the possibility that abnormal haptoglobins (Hpr, SER-haptoglobin) are synthesized by neoplastic cells, and that such haptoglobins have a biologically active role in the phenotypic expression and potential of malignancy. Continued expression occurs in phenotypically aggressive neoplasia, and metastases from initially negative tumors express Hpr epitopes, suggesting the participation of Hpr as a mediator in the malignant proliferation of cells. On the other hand, the detection of Hpr in maternal serum and the immunological similarity of SER-haptoglobin to a fetal species indicate similar functions in pregnancy and neoplasia, aiding development of the placenta and fetus on the one hand while facilitating tumor invasion and metastasis on the other (modulating the host response to a malignancy so as to enable it to survive and expand clonally).

E. HAPTOGLOBIN SYNTHESIS AND CATABOLISM

The site of haptoglobin biosynthesis is the liver.[44] Induced synthesis of acute phase proteins in the liver occurs after the stimulation of hepatocytes by cytokines produced in macrophages and other cells at the site of injury. The transcription rate of the haptoglobin gene in hepatic cells from rats, mice, and rabbits is stimulated by interleukin-1 (IL-1), tumor necrosis factor-α, IL-6, leukemia inhibitory factor (LIF), and glucocorticoids, while in human liver cells, the transcription rate is primarily stimulated by IL-6 and LIF, and this action is synergistically enhanced by glucocorticoids.[51-56]

Induction of haptoglobin mRNA accumulation corresponds to transcriptional activation of the chloramphenicol acetyl-transferase fusions. The effect of IL-6 is exerted at the level of transcription and short segments of the 5′ flanking sequences of the inducible gene contain IL-6-responsive elements.[52] Analysis of the haptoglobin promoter by site-directed mutage-

TABLE 1
Comparison of Some Properties of Normal and Pathophysiological Variants of Human Haptoglobin

Properties	Normal human haptoglobin[a]	Haptoglobin-related protein (Hpr)[b]	"Fetal" haptoglobin[c]	"Neoplastic" SER-haptoglobin[d]
Hemoglobin binding	+	–	+	–
Molecular mass	100–400 kDa, depending on phenotype	Three kinds of subunits: 40, 20, and 16.5 kDa $\times$ 2 = 153 kDa	900–1000 kDa	850–900 kDa
Occurrence	Serum ~1.2 g/l	Plasma of pregnant women, breast cancer cells	Human fetus, cord serum; normal serum ~1 mg/l	Ascitic fluid from patients with ovarian carcinoma; normal serum ~1 mg/l
Reaction with antibodies	Polyclonal antibodies do not differentiate among major phenotypes; monoclonal antibodies recognize α- and β-chains	Antibodies do not react with α^1- and α^2-chains of normal haptoglobin	Difference in immunoprecipitation with α-chains	Reaction of polyclonal antibodies with α^1- and α^2-chains of normal haptoglobin, while monoclonals react only with α^2-chains or with SER-haptoglobin

[a] Data from References 1, 11, 48, and 49.
[b] Data from References 30, 35, 36, and 50.
[c] Data from References 38–40 and 47.
[d] Data from References 10, 13, and 37.

nesis has led to identification of *cis*-acting elements that are responsible for the activation of transcription and the response to IL-6.[53]

Regulation of the expression of the haptoglobin gene occurs on at least three levels: (1) control associated with ontogenesis, responsible for the lack of minimal expression in fetal liver, (2) control of tissue specificity, and (3) modulation of expression during the acute phase reaction. The mechanisms of regulation of haptoglobin gene expression are described in Chapters 21 and 23, references 57 and 58, respectively.

The primary product of haptoglobin mRNA is a single polypeptide chain containing a signal peptide with α- and β-sequences. This preprohaptoglobin is cotranslationally glycated in the β-region, followed by posttranslational proteolytic cleavage of prohaptoglobin into α- and β-dimers connected by disulfide bonds. The last step of "maturation" is the formation of tetramers (haptoglobin 1-1) or polymers (types 2-2 and 2-1).[59]

Catabolic experiments with haptoglobin have been carried out by administration of radioactively labeled haptoglobin preparations to humans, dogs, rabbits, chickens, etc. The half-life of human haptoglobin labeled *in vitro* with iodine administered to normal humans appeared to be 4.5 d. Human haptoglobin labeled *in vivo* with either ^{14}C-glucosamine or ^{35}S-methionine gave similar results in the state of health, whereas in patients with acute burns, the half-life was shortened to 2.8 d, analogous to the accelerated turnover rate. The haptoglobin-hemoglobin complex was cleared rapidly from the circulation, with a half-life of 1 to 4 h, similar to that of haptoglobin from which terminal sialic acid was removed by the action of neuraminidase.[60-64]

Development of the concept of exposure of specific carbohydrate moieties as a biological strategy for steering circulating glycoproteins for hepatic uptake led to the current explanation of mechanisms of protein degradation in the liver. Hepatic plasma membrane receptors mediate the specific binding and uptake (receptor-mediated endocytosis) of partially deglycosylated glycoproteins. The role of the exposure of the penultimate galactose residues of several plasma glycoproteins (asialoglycoproteins) in their rapid clearance from the mammalian circulation and binding by liver receptors (referred to as hepatic lectins), followed by degradation in the lysosomes, was described. These receptors have been isolated from mammalian and avian species. The mammalian hepatocyte system recognizes fucose, galactose, *N*-acetylgalactosamine, and glucose, whereas the reticuloendothelial system recognizes mannose, *N*-acetylglucosamine, and glucose. The cellular mechanism of this process involves the binding of desialylated glycoprotein to a Ca-dependent, galactose-specific receptor, endocytosis of the receptor-ligand complex, dissociation of the ligand from the receptor in an acidified endosomal compartment, and transport of the ligand to lysosomes, where it is degraded. The receptor does not enter the lysosomes and is not degraded, but recycled back to the cell surface to participate in future rounds of endocytosis.[65-72]

A specific receptor for asialohaptoglobin in the liver was identified and characterized.[73,74] On the other hand, it was shown that haptoglobin complex with hemoglobin was not recognized by normal receptors for asialoglycoproteins.[75] According to Kino et al.,[76] specific receptors for the haptoglobin-hemoglobin complex in hepatocytes are able to recognize changes in the conformation of hemoglobin-bound haptoglobin. The complex and its degradation products are initially bound by the Golgi subfractions in rat liver, and then by the lysosomal fraction of the hepatocytes. It was found that the receptor for the haptoglobin-hemoglobin complex is distinct from several previously described binding proteins that recognize terminal carbohydrate residues. In contradistinction to these receptors, no role for carbohydrate was detected in haptoglobin-hemoglobin binding to the receptor.[77,78]

F. CLINICAL APPLICATIONS

Haptoglobin is present in most body fluids (serum, urine, saliva, cerebrospinal fluid, amniotic fluid, and ascites). The mean values of haptoglobin concentrations in normal human serum obtained from various laboratories differ significantly (0.5 to 1.5, 0.75 to 1.75, and 0.7 to 1.3 g/l). These variations are explained in part by the differences in the levels of the three common haptoglobin types (1.36, 1.08, and 0.82 g/l for haptoglobins 1-1, 2-1, and 2-2, respectively) as well as by the dependence on sex (1.13 and 0.94 g/l for men and women, respectively).[48]

Ahaptoglobinemia is common in the newborn, and is present in 80 to 90% of infants up to 3 months of age. This has been attributed to both the hemolysis of fetal red cells and the immaturity of the parenchymal cells and reticuloendothelial system of the liver, for the biosynthesis and catabolism of haptoglobin, respectively. Some cases of ahaptoglobinemia in Nigerian populations have been described.[1,25,79] A significant increase of ahaptoglobinemia was found among leukemia patients.[80]

Profound changes in serum haptoglobin level occur in a great variety of disease states. Haptoglobin should be considered as a protein undergoing two pathophysiological phenomena: an increase in production during an inflammatory reaction and a decrease (as primary phenomenon) in production during severe chronic hepatocellular deficiency (a failure in biosynthesis or in secretion by hepatocytes), or a decrease (secondary phenomenon) under hemolytic conditions.

Haptoglobin is an acute phase protein. Its serum level is increased up to three- to fivefold in response to injury agents (surgery and burns, bacterial or parasitic infections, ischemic necrosis, connective tissue diseases, chemical irritants, and malignancy).[44] Haptoglobin measurements may help to assess the disease status of cancer patients, particularly during chemotherapy, radiation therapy, or following surgery, providing valuable information on the duration of remission, and eventually on the recurrence of malignancy, which could have been unnoticed by means of other tests commonly used and under apparently good subjective conditions and clinical status of patients. Figure 3 shows examples of serial analyses of haptoglobin and lipid-bound sialic acid in patients with ovarian carcinoma undergoing chemotherapy. As can be seen, the haptoglobin levels indicated satisfactorily progression of the disease, verified subsequently by other clinical and biochemical data.[81] The introduction of affinity chromatography and crossed-affinity electrophoresis of haptoglobin with lectins (Con A and wheat germ agglutinin), leading to the appearance of several microheterogeneous forms, was shown to be of value in the differential diagnosis of ovarian disorders, incuding benign and malignant tumors and inflammation.[84]

Low levels of haptoglobin are observed in intra- or extravascular hemolysis, ineffective erythropoiesis, severe liver disease, genetic factors in seemingly normal individuals, and in a number of abnormal hemoglobin diseases such as sickle cell anemia, hemoglobin C disease, and thalassemia, as well as after adverse transfusion reactions and autoimmune hemolysis, as in acquired hemolytic anemia. In an acute hemolytic crisis, all of the haptoglobin is usually depleted because of the one-way transit of the complex with hemoglobin to the liver, where it is catabolized. The use of serum haptoglobin determinations in the differential diagnosis of a variety of primary and secondary hemolytic processes vs. anemias without hemolysis has been described.[85] It was generally believed that haptoglobin values, although decreased in hemolysis, were of little practical value. This attitude was in part due to the arduous, insensitive techniques previously available for haptoglobin measurements. Highly sensitive methods are now in use, and the present evidence suggests that haptoglobin may be used in diagnostic tests and procedures not only as a traditional acute phase reactant but also in clinical cases associated with decreased values, particularly in patients exhibiting hemolysis and concurrent acute phase reaction, where the decline in serum haptoglobin concentration occurs significantly faster than serum haptoglobin increases in the acute phase.[86] Enzyme-linked immunosorbent assay (ELISA) systems based on specific interactions of haptoglobin with hemoglobin, Con A, or *S. pyogenes,* developed in our laboratory, are of

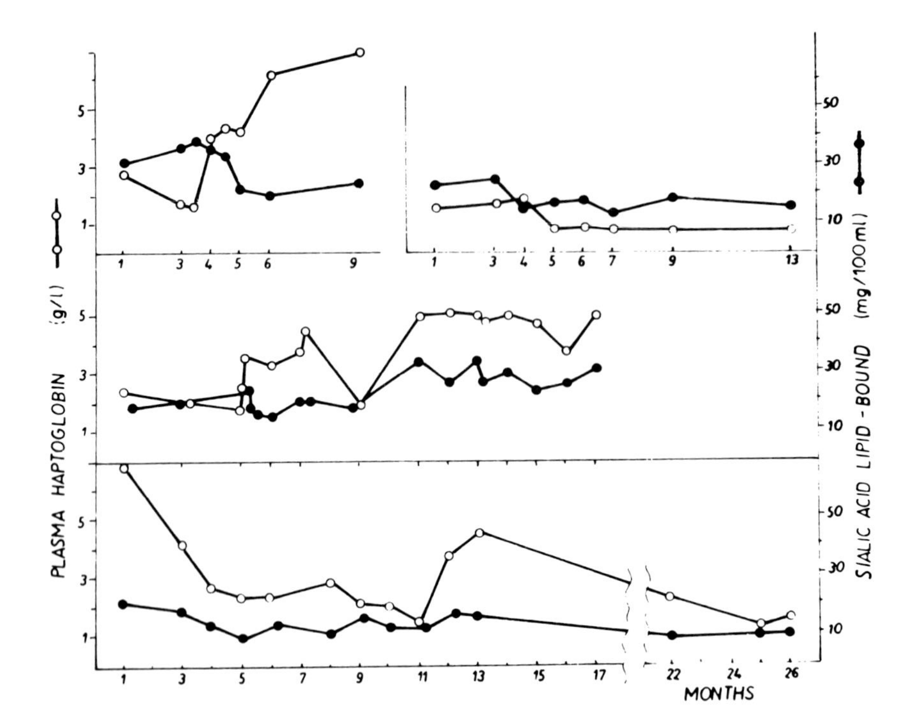

FIGURE 3. Examples of serial mesurements of haptoglobin and lipid-bound sialic acid. Haptoglobin concentration was determined by the peroxidase method,[82] and lipid-bound sialic acid according to Dnistrian et al.[83] (From Dobryszycka, W., Gerber, J., Zuwala-Jagiello, J., and Ujec, M., *Arch. Immunol. Ther. Exp.*, 39, 41, 1991. With permission.)

special significance in measurements of haptoglobin levels in biological fluids of low haptoglobin content i.e., in cord sera and amniotic fluids (diagnosis of intrauterine fetal infections), effusions of different etiology (discrimination of transudative vs. exudative ascites), and in sera from hemolytic anemias, etc.[41,87,88]

Haptoglobin polymorphism has been used in paternity cases in forensic science for several years. Now, thanks to modern laboratory techniques, haptoglobin subtyping can also be carried out in dried blood stains.[85]

A possible therapeutic application of human haptoglobin consists of the selective removal of free hemoglobin from the blood of patients suffering from severe traumatic shock. Such an application would require an abundant and safe source of material accessible through recombinant DNA technology. Human prohaptoglobin (not cleaved into subunits, but glycosylated), as synthesized in insect cells, was able to bind hemoglobin *in vitro,* although less efficiently than plasma-derived haptoglobin.[90]

Sclerotherapy is a method for controlling acute bleeding from esophageal varices, but may cause acute renal damage, as excess free hemoglobin (exceeding 22 to 35 mg/100 ml) in the circulating blood is excreted through renal glomeruli. Administration of haptoglobin protected against renal damage from ethanolamine sclerosant.[91]

There is still no generally accepted satisfactory explanation for the mechanism of maintaining genetic polymorphisms in human populations. Interactions between single- or multilocus genetic systems have been extensively examined in the search for evidence of selection acting on man. The occurrence of polymorphism at the haptoglobin locus has motivated many investigations directed at the determination of possible associations between haptoglobin phenotype distribution and different disorders. Identification of persons at high risk for a disease on the basis of haptoglobin phenotype could have been a rather stimulating goal of preventive medicine. Since a report on significant increases of haptoglobin type 1-1 among leukemia patients compared to other cancer patients as well as to normal controls, numerous reports have appeared. These include investigations of bladder cancer, ovarian tumors, renal cell carcioma, multiple myeloma, leukemia, multiple endocrine neoplasia type 2A, motor neutron disease, hemoglobinopathies, Marner's cataract, essential hypertension, Batten disease, acute myocardial infarction, and others.[92-104] Some of these investigations clearly show a significant association, while others show contradictory results, i.e., sometimes an association is shown and occasionally this association cannot be demonstrated. For the time being, it seems that these inconclusive reports have limited any practical use of haptoglobin phenotype for clinical science.

G. EFFECTS OF MODIFICATIONS ON BIOLOGICAL PROPERTIES OF HAPTOGLOBIN

Studies on the reactivity of amino acid residues in protein as carried out by specific chemical modifications are a very efficient tool for recognizing the participation of particular amino acid residues in the stabilization of secondary and tertiary structures of the protein as well as their role in biological properties. One can use several modifying reagents of various molecular size and different physicochemical character. Taking into consideration the character of the group introduced on an enzyme, one may define a catalytic active center; by modifying the hydrophobic area on complementary sites of antigen and antibody, it is possible to gain insight into the immunological reactivity, etc. Limited proteolysis or stepwise degradation of the carbohydrate moiety in glycoproteins may supply information concerning the dependence of biological properties on the native conformation of a protein. Site-directed mutagenesis is a potent tool in the structural analysis of proteins. Protein modifications can also be used in determining the uptake of proteins by receptor-mediated endocytosis and their targeting to specific intracellular locations.

For several years, the studies in our laboratory have dealt with "active sites" on the surface of the haptoglobin molecule: hemoglobin-, antibody-, and plant, hepatic, and bacterial lectin-binding sites. The molecular basis of the complex formation of haptoglobin with these ligands has been investigated by means of various modifications of the haptoglobin structure. Three types of modifications have been applied: (1) chemical modifications of some amino acid residues,[105] (2) enzymatic or chemical deglycosylation, and (3) reduction of disulfide bonds, and trypsin digestion.[106,107]

Tyrosine and histidine residues of human and rabbit haptoglobins were subjected to the reaction with diazotized sulfanilic acid.[108] From haptoglobin, either sialic acid or the carbohydrate moiety was removed; disulfide bonds were reduced, followed by isolation of α^2-, α^1-, and β-subunits. Formation of an active complex with hemoglobin, the binding of polyclonal and monoclonal antibodies, and the reaction with plant lectin (Con A) *in vitro* and with specific hepatic receptors (lectins) *in vivo* by clearance of modified preparations from the rabbit circulation were studied.

II. MATERIALS AND METHODS

Samples of ascitic fluids from patients with ovarian carcinoma were obtained from the Second Clinic of Gynecology and Obstetrics of the Medical Academy in Wroclaw, Poland.

A. PREPARATION OF HAPTOGLOBINS AND ISOLATED SUBUNITS

Human haptoglobins of types 1-1, 2-1, and 2-2 were prepared from ascitic fluids of patients with ovarian carcinoma, as previously reported.[109] Further purification was done on Sephacryl S-300 (Pharmacia, Sweden) in 0.1 *M* Tris-HCl buffer, pH 8.0 (containing 0.5 *M* NaCl). Rabbit haptoglobin was prepared essentially by the same method.

The haptoglobin α^2-, α^1-, and β-chains were isolated by the reductive cleavage of purified haptoglobin. The purity of the preparation was checked by polyacrylamide gel electrophoresis and immunoelectrophoresis. The haptoglobin concentration was determined by the method of Jayle,[82] or by enzyme immunoassay.[87]

B. POLYCLONAL AND MONOCLONAL ANTIBODIES

Antiserum directed against haptoglobin type 2-1 or isolated β-chain was produced in a goat.[111] Quantitative precipitin analysis was carried out by the method of Zschocke and Bezkorovainy.[112] Antigenic valency was calculated by plotting the antibody/antigen molar ratio against the antigen concentration. The vertical intercept of this plot i.e., at zero antigen concentration, gives the maximum number of antibody molecules bound to one antigen molecule.[113]

Monoclonal antibodies were produced in BALB/c/J/I i W mice with haptoglobin type 2-1. Spleen cells were fused with SP 2/O-Ag 14 hybrid plasmacytoma cells. Selected clones were expanded *in vitro* and injected into BALB/c mice. Monoclonal antibodies 7.60.66.55 and 18.4.40.80 (both of the IgG class) were used in further experiments.[11]

C. MODIFICATIONS OF THE HAPTOGLOBIN MOLECULE

Preparation of sulfanilazo derivatives of haptoglobin and the number of monoazotyrosines. C2-monoazohistidines, and C4-monoazohistidines was carried out according to Pielak et al.[114] Trypsin digestion and chromatography on the DEAE-Sephadex A-50 (Pharmacia, Sweden) column were carried out as previously described.[115,116] Titration of sulfhydryl groups following 2-mercaptoethanol reduction was carried out by the use of ^{14}C-iodoacetamide and sodium dodecyl sulfate-polyacrylamide gel electrophoresis.[106,117]

The removal of sialic acid from haptoglobin was done with neuraminidase from *Clostridium perfringens* (Behringwerke, Germany); 95% of the sialic acid was removed. Chemical deglycosylation was performed by the use of trifluoromethane-sulfonic acid (TFMS).[118] The effects of deglycosylation were examined by quantitative estimation of the remaining sugar compositions by gas liquid chromatography, using inositol as an internal standard.[119] A Varian Aerograph (U.S.) was used.

D. CATABOLIC EXPERIMENTS

These were carried out by the administration of ^{125}I-labeled haptoglobin preparations to rabbits. The drinking water of the animals was supplemented for 3 d prior to the experiments with 0.1 g/l KI. Radioiodination was effected by the chloramine-T method,[120] with carrier-free ^{125}Na (Nuclear Institute, Poland), and measured in a TESLA (Czechoslovakia) scintillation counter. Female rabbits, weighing about 3 kg, were injected in the ear vein with 0.5 mg of labeled preparations. Blood samples drawn from the ear vein were placed in heparinized tubes and centrifuged. Protein-bound plasma radioactivity was expressed as a percentage of the radioactivity at 5 min after injection.

III. RESULTS

A. PROPERTIES OF SULFANILAZO DERIVATIVES OF HAPTOGLOBIN[62,108]

1. Peroxidase Activity of the Complex with Hemoglobin

Sulfanilazo derivatives of haptoglobin were prepared by using diazotized sulfanilic acid/haptoglobin molar ratios of 10 to 100. At the highest reagent/haptoglobin ratio, approximately 28 tyrosine, two C-4, and three C-2 histidine residues were modified.

Peroxidase activities of the modified haptoglobin preparations with hemoglobin were found to decrease gradually as the number of sulfanilazo-Tyr and sulfanilazo-His increased

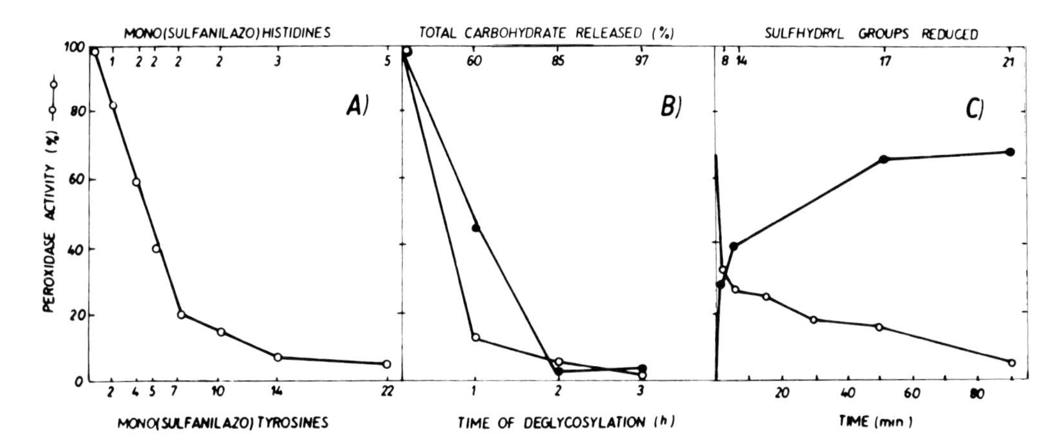

FIGURE 4. Effects of sulfanilazo modification (A),[108] deglycosylation (B), and reduction of disulfide bonds (C)[106] of haptoglobin on peroxidase activity of the complex hemoglobin. (From Dobryszycka, W. and Guszczynski, T., *Int. J. Biochem.*, 20, 321, 1988; Dobryszycka, W. and Guszczynski, T., *Biochim. Biophys. Acta*, 829, 13, 1985. With permission.)

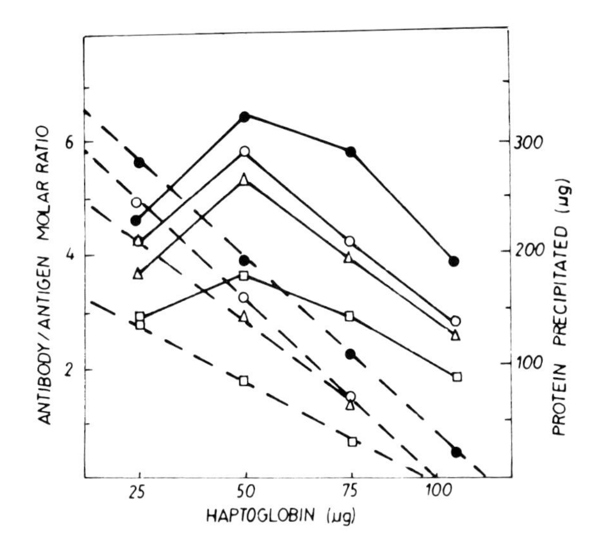

FIGURE 5. Quantitative immunoprecipitation of sulfanilazo derivatives of haptoglobin with antihaptoglobin serum. (From Dobryszycka, W. and Guszczynski, T., *Int. J. Biochem.*, 829, 13, 1985. With permission.)

(Figure 4). However, 6.8% of the original activity persisted even at the maximum reagent/haptoglobin ratio.

2. Reaction with Antibodies

Quantitative immunoprecipitation with anti-haptoglobin serum is shown in Figure 5. Sulfanilazo modification of one histidine and ten tyrosine residues diminished the antigen-antibody reaction by approximately 20%, while the blockage of an additional histidine and four tyrosine residues diminished the reaction by 50%. Calculations of antigenic valency indicate that in the latter case, out of six sites of antibody binding per mole of native haptoglobin, 3 mol of the antigen became inaccessible to the ligand. Haptoglobin with 22 azo-Tyr and 5 azo-His did not form any precipitate with the antiserum.

TABLE 2
Catabolism of Native and Modified Haptoglobin Derivatives in the Rabbit — Kinetic Analysis

^{125}I-labeled preparations	Fractional catabolic rate (h^{-1})	Kinetic data — biexponential model[a] Slopes (h^{-1}) b_1	b_2	Intercepts c_1	c_2	Half-life (h)	Disappearance of 90% of radioactivity (h)
Rabbit haptoglobin	0.030	0.007	0.054	0.11	0.89	96.0	48
Human haptoglobin	0.051	0.010	0.079	0.08	0.92	72.0	36
Haptoglobin β-chain	0.065	0.011	0.090	0.05	0.95	67.2	22
Haptoglobin α^2-chain	0.138	0.029	0.340	0.21	0.79	27.0	24
Haptoglobin α^1-chain	0.101	0.029	1.410	0.27	0.73	19.0	36
Sulfanilazo-Hp (10 azoTyr, 2 azoHis)	0.108	0.031	1.560	0.27	0.73	20.8	40
Sulfanilazo-Hp (22 azoTyr, 5 azoHis)	0.088	0.058	0.530	0.62	0.38	11.4	16
Asialohaptoglobin	0.277	0.130	0.443	0.25	0.75	5.5	7
Asialo β-chain	0.342	0.138	1.730	0.35	0.65	5.0	8
Asialo-sulfanilazo-Hp (14 azoTyr, 4 azoHis)	0.336	0.170	0.947	0.40	0.60	4.2	7

[a] Kinetic analysis was carried out according to the biexponential model $b_1 = c_1 \exp(-b_1 t) + c_2 \exp(-b_2 t)$; the fractional catabolic rate — $(c_1/b_1 + c_2/b_2)^{-1}$.

From Dobryszycka, W., Guszczynski, T., and Kubicz, Z., *Int. J. Biochem.*, 20, 325, 1988. With permission.

3. Reaction with Plant and Hepatic Lectins

The reaction of Con A (Serva, Heidelberg, Germany) was only slightly affected by sulfanilazo modification; even maximum modification of the preparation resulted in a 20% decrease of the haptoglobin-Con A precipitate.

Results of the clearance of sulfanilazo derivatives of haptoglobin from the rabbit circulation are summarized in Table 2. The half-lives (T/2) of the sulfanilazohaptoglobins were 20.8 and 11.4 h vs. 72 h for human haptoglobin.

B. EFFECTS OF CHANGES IN CARBOHYDRATE MOIETY

1. Enzymatic Removal of Sialic Acid

Asialohaptoglobin obtained by the action of neuraminidase from *C. perfringens* showed 81% of the reaction with hemoglobin and practically unchanged reactions with antihaptoglobin serum and Con A. However, the clearance of asialohaptoglobin from the rabbit circulation proceeded 13 times faster than that of native haptoglobin. Asialohaptoglobin revealed the shortest T/2 (4.2 h) and highest fractional catabolic rate (0.336 h) of all the preparations shown in Table 2.

2. Chemical Deglycosylation

Gradual removal of particular sugars from the carbohydrate moiety of haptoglobin by the use of trifluoromethane sulfonic acid (TFMS) was examined.[118] The action of TFMS for 1 h resulted in the removal of 100% of the sialic acid, 92.4% of the fucose, and approximately 57% of the mannose, *N*-acetylglucosamine, and galactose. After 3 h of deglycosylation, practically all the sugars were removed. The effect of such modified preparations on the formation of an active complex with hemoglobin is shown in Figure 4. The carbohydrate of haptoglobin does not participate in the reaction with hemoglobin, but the action of TFMS for 1 h resulted in a drastic, 85% decrease of peroxidase activity. The preparation obtained following 1 h of deglycosylation retained 39% of the antigenic reactivity with antihaptoglobin serum and 10% of the precipitate with Con A. Further action of TFMS abolished completely the three reactivities examined. Deglycosylated preparations were rapidly cleared from the rabbit circulation with a T/2 of 3.6 to 4.1 h.

C. DISINTEGRATION OF THE NATIVE STRUCTURE

1. Reduction of Disulfide Bonds

Gradual reduction of disulfide bonds in haptoglobin resulted in the formation of a number of intermediates ($\beta.\alpha^1.\alpha^2.\beta$, $\beta.\alpha^1.\beta$, $\beta.\alpha^2.\alpha^2$, $\beta.\alpha^2$, and $\beta.\alpha^1$), followed by the complete dissociation into α^1-, α^2-, and β-subunits.[106]

The peroxidase activity of the complex with hemoglobin decreased almost immediately by 50% after addition of the reducing agent and the appearance of free SH groups, followed by a further decrease to approximately 10% when all of the disulfide bonds were reduced (Figure 4). Of the isolated chains of haptoglobin, only the β-chain exhibited 10 to 15% of the activity of native haptoglobin. However, an additional purification of the β-chain on Con A-Sepharose resulted in threefold enhanced reactivities with both hemoglobin and Con A.[121] The half-lives of the α^1-, α^2-, and β-chains were 19.0, 27.0, and 67.2 h, respectively (Table 2).

The immunological reactivity with anti-haptoglobin serum was unchanged throughout the course of the reduction. However, in the immunodiffusion test, the β-chain was found to be deficient in antigenic determinants as related to haptoglobin type 2-1 and other reduction products.[106] The reaction of isolated chains with polyclonal and monoclonal antibodies directed against haptoglobin 2-1 was examined.[11] The reaction of particular chains was weaker than with native haptoglobin. The β-chain reacted more strongly than the α^2-chain, while the rection with the α^1-chain was very weak. Two monoclonal antibodies formed complexes only with the β-chain, whereas one of them reacted with both the β- and α^2-chains.

2. Trypsin Digestion[116]

Five glycopeptides were obtained by limited trypsin digestion of the β-chain. None of them revealed hemoglobin-binding capacity, one showed dominant antigenic determinants in relation to native haptoglobin and the β-chain, one inhibited strongly the reaction of haptoglobin and the β-chain with polyclonal antibodies, and the reaction of one glycopeptide with Con A was almost twice the corresponding reaction of haptoglobin.

IV. DISCUSSION

Our earlier supposition concerning the existence of an asymmetric distribution of tyrosine and histidine residues in two separate active sites on the haptoglobin molecule — for hemoglobin and antibody binding[122,123] — has been confirmed. Moderately modified derivatives of haptoglobin (10 azo-Tyr and 2 azo-His) reduced the peroxidase activity of the complex with hemoglobin to 20% (Figure 4), whereas the reaction with antibody amounted to about 80% (Figure 5). Three of the six antigenic determinants of haptoglobin were blocked by the action of diazotized sulfanilic acid. The presence of six antigenic determinants in haptoglobin was also observed in the profile of hydrophilicity.[124] Thus, if the active sites for hemoglobin and antibody binding overlap, the common surface would be relatively small.

Although TFMS has proved useful as a reagent for the deglycosylation of glycoproteins, the amino acid composition of a glycoprotein is unaffected by TFMS treatment; enzymatic or hormonal activities may be changed or lost.[125] Both hemoglobin- and antibody-binding capacities were annulled following deglycosylation of haptoglobin by TFMS. A question arose as to whether TFMS affected the polypeptide portion of haptoglobin. However, removal of 40% of the carbohydrate from haptoglobin by the use of exoglycosidases totally inhibited both activities.[126] Also, independently of the method of deglycosylation, lack of sugars diminished the solubility of haptoglobin and resulted in precipitation of the protein. These data indicate that the carbohydrate moiety of haptoglobin, although hardly having any direct role in the studied activities, is essential for the functionally active form of haptoglobin. Since haptoglobins 2-2 and 2-1 occur in dense clusters with the β-chains on the surface, the removal of oligosaccharides induces obvious changes in the conformation of the molecule; thus, the active sites are no longer accessible.

As a rule, the biological properties of proteins depend on their uniquely folded conformation. The effect of the reduction of disulfide bonds on biological activities may vary, as sometimes a native set of -S-S- groups is not essential for activity.[127] Also, cleavage of a glycoprotein into (glyco)peptides does not necessarily mean the loss of biological properties. Gradual reduction of the disulfide bonds in haptoglobin resulted in a rapid decrease of hemoglobin-binding capacity. The residual capacity left (Figure 4) may be attributed to the presence of the β-chain, and dimers and trimers with its content. Thus, disulfide linkages in haptoglobin are essential for maintenance of the proper conformation enabling hemoglobin binding. Almost all peroxidase activity was renatured following oxidative reassociation when a native set of disulfide bonds was formed. This was demonstrated for haptoglobin type 1-1.[117] Although the β-chain is the site of hemoglobin binding, certain parts of the α-chains, linked by disulfide bonds, are probably necessary to ensure a proper hydrophobic cavity environment for hemoglobin.

Experiments on disintegration of the haptoglobin structure with either reduction or limited proteolysis have proved that the active site for hemoglobin binding was almost completely destroyed, while these modifications did not affect epitopes on fragments of the haptoglobin molecules (isolated chains and some glycopeptides reacted with either polyclonal or monoclonal antibodies).

The aim of catabolic experiments has been among other things, to learn whether modifications of the haptoglobin polypeptide moiety, which is not directly involved in lectin binding, might affect haptoglobin-lectin interactions. Deglycosylated and asialo derivatives of haptoglobin disappeared rapidly from the rabbit circulation, whereas α and α-chains (no carbohydrate present) were cleared approximately threefold faster than native haptoglobin or the β-chain, but four to five times slower than asialo derivatives. Similar results were obtained in the internalization and degradation of asialoorosomucoid and its cyanogen bromide fragment in cultured rat hepatocytes.

In hepatocytes, there are probably two kinetically distinct pathways by which galactosyl receptors can mediate ligand uptake, dissociation and degradation.[129] If this is the case, one could suppose that in the model experiments with asialoglycoproteins, one subpopulation of cell-surface receptors would participate in asialoglycoprotein binding, whereas another subpopulation would be prepared for the acceptance of slowly desialylated "full" glycoproteins. One could postulate that under physiological conditions there exists a population of receptors able to identify native glycoprotein molecules. Such receptors may be species specific, allowing even recognition of subtle differences in the primary structure of heterologous proteins. In reality, human haptoglobin displayed a shorter T/2 in the rabbit circulation than the homologous rabbit haptoglobin (Table 2). The clearance rate of human-porcine haptoglobin hybrid in the pig circulation was significantly shorter than that of native porcine haptoglobin.[130] Chicken haptoglobin under homologous conditions revealed a T/2 twofold longer than heterologous mammalian haptoglobin.[131] The above observations imply that different pathways of haptoglobin catabolism exist *in vivo,* with polypeptide determinants in addition to carbohydrate being involved.

Sulfanilazo derivatives of haptoglobin with accelerated catabolism (at the unchanged carbohydrate moiety) are probably taken up by scavenger receptors present on the plasma membranes of Kupffer cells or endothelial cells, analogous to the glycoproteins carrying high mannose-type oligosaccharide side chains,[132] and albumin modified with formaldehyde or by reduction and acetylation of disulfide bonds.[133,134] Endocytosis of aldehyde-modified albumin[134] followed by intralysosomal proteolysis was also demonstrated in renal plasma membranes, and nonenzymatically glycosylated albumin in rat peritoneal or splenic macrophages.[135] This suggests that besides the liver, other organs may be involved in the clearance of denatured or chemically modified glycoproteins. For instance, liver endothelium, not hepatocytes or Kupffer cells, have transferrin receptors, which emphasizes the generally unappreciated role of the endothelium in transport across the tissue-blood barrier.[136] Moreover, studies on the mechanism of protein degradation in primary cultured hepatocytes indicated that short- and long-lived proteins are catabolized via different pathways.[137]

There have been retrospectives. And what about perspectives? The real metabolic role of haptoglobin should be solved first, as the formation of a complex with hemoglobin, which reveals *in vitro* (at pH 4.5) peroxidase activity, obviously cannot constitute such a role, even if we take into consideration the binding of intravascularly released hemoglobin.

Finding a purpose for haptoglobin is not simply a teleological problem, because different aspects of quantitative and qualitative changes in haptoglobin in health and diseases are involved.

Are people without any detectable haptoglobin truly ahaptoglobinemic? Previous methods could detect haptoglobin in concentrations over 0.01 g/l. Novel ELISA systems, based on specific interactions of haptoglobin with hemoglobin, Con A, or *S. pyogenes,* developed in our laboratory with the use of polyclonal and monoclonal antibodies, are 1000-fold more sensitive, and thus might help to answer this question.

Hpr and neoplastic SER-haptoglobin do not form the characteristic complex with hemoglobin. Could one of these variants exist in apparently ahaptoglobinemic subjects? The significance (if any) of the occurrence of Hpr (with multiple retrovirus-like elements), fetal, or neoplastic haptoglobins in pathophysiological states requires future attention.

Independently of future applications of these haptoglobin forms in clinical diagnostics, differences in haptoglobin glycosylation, as studied by means of two ELISA systems, have given interesting results in the laboratory diagnosis of cancer, inflammations, and liver diseases.[138]

Considering its inhibition of prostaglandin cyclase *in vitro,* does haptoglobin have any curative significance *in vivo* (with other acute phase reactants) in inflammations?

The multiple aspects of haptoglobin structure and function are readily apparent in research gaps and inconsistencies. During almost 50 years of study, scientists have gathered a great array of facts in order to establish the contribution of haptoglobin to processes in living organisms. Our knowledge of this complex area continues to evolve, but we are still arranging the many pieces of the puzzle, trying to uncover the mysteries of haptoglobin.

REFERENCES

1. **Bowman, B. H. and Kurosky, A.,** Haptoglobin: the evolutionary product of duplication, unequal crossing over, and point mutation, *Adv. Hum. Genet.,* 12, 189, 1982.
2. **Hwang, P. K. and Greer, J.,** Interaction between hemoglobin subunits in the hemoglobin-haptoglobin complex, *J. Biol. Chem.,* 255, 3038, 1980.
3. **Wejman, J. C., Hovsepian, D., Wall, J. S., Hainfeld, J. F., and Greer, J.,** Structure of haptoglobin and the haptoglobin-hemoglobin complex by electron microscopy, *J. Mol. Biol.,* 174, 319, 1984.

4. **Bashir, A., Dawson, S., Vincent, J., Powell, J., Humphries, S., and Henney, A.,** An Ava II polymorphism in the haptoglobin α gene (HPA), *Nucleic Acids Res.,* 17, 4906, 1989.
5. **Hyland, V. J.,** A TaqI RFLP detected by the human haptoglobin (HP) cDNA probe, pULB1148, *Nucleic Acids Res.,* 16, 8203, 1988.
6. **Smithies, O., Connell, G. E., and Dixon, G. H.,** Inheritance of haptoglobin subtypes, *Am. J. Hum. Genet.,* 14, 14, 1962.
7. **Raugei, G., Bensi, G., Colantuoni, V., Romano, V., Santoro, C., Constanzo, F., and Cortese, R.,** Sequence of human haptoglobin cDNA; evidence tht the alpha and beta subunits are encoded by the same mRAN, *Nucelic Acids Res.,* 11, 5811, 1983.
8. **Baenziger, J. U.,** The oligosaccharides of plasma glycoproteins. Synthesis, structure and function, in *The Plasma Proteins. Structure, Function and Genetic Control,* Vol. 4, 2nd ed., Putnam, F. W., Ed., Academic Press, New York, 1984, 272.
9. **Nilsson, B., Lowe, M., Osada, J., Ashwell, G., and Zopf, D.,** The carbohydrate structure of human haptoglobin 1-1, paper presented at 6th Int. Symp. Glycoconjugates, Tokyo, September 1981.
10. **Thompson, S., Cantwell, B. M. J., Cornell, C., and Turner, G. A.,** Abnormally-fucosylated haptoglobin: a cancer marker for tumour burden but not gross liver metastasis, *Br. J. Cancer.,* 64, 386, 1991.
11. **Katnik, I., Steuden, I., Pupek, M., Wiedlocha, A., and Dobryszycka, W.,** Monoclonal antibodies against human haptoglobin, *Hybridoma,* 8, 551, 1989.
12. **Dobryszycka, W. and Katnik, I.,** Interaction of haptoglobin with lectin, in *Lectins — Biology, Biochemistry, Clinical Biochemistry,* Vol. 2, Walter de Gruyter, New York, 1982, 381.
13. **Kurosky, A., Barnett, D. R., Lee, T.-H., Touchstone, B., Hay, R. E., Arnott, M. S., Bowman, B. M., and Fitch, W. M.,** Covalent structure of human haptoglobin: a serine protease homology, *Proc. Natl. Acad. Sci. U.S.A.,* 77, 3388, 1980.
14. **Black, J. A. and Dixon, W.,** Amino-acid sequence of alpha chains of human haptoglobins, *Nature (London),* 218, 736, 1968.
15. **Nicholson-Weller, A., Zaia, J., Raum, M. G., and Coligan, J. E.,** Decay accelerating factor (DAF) peptide sequences share homology with a consensus sequence found in the superfamily of structurally related complement proteins and other proteins including haptoglobin, factor XIII, β-glycoprotein I and IL-2 receptor, *Immunology Lett.,* 14, 307, 1986/87.
16. **Maeda, N. and Smithies, O.,** The evolution of multigene families: human haptoglobin genes, *Annu. Rev. Genet.,* 20, 81, 1986.
17. **Tosi, M., Duponchel, C., Meo, T., and Couture-Tosi, E.,** Complement genes C1r and C1s feature an intronless serine protease domain closely related to haptoglobin, *J. Mol. Biol.,* 208, 709, 1989.
18. **Dobryszycka, W. and Przysiecki, B.,** Structural similarities among concanavalin A, haptoglobin and trypsin, *FEBS Lett.,* 171, 85, 1984.
19. **Rearden, A., Phan, H., Dubnicoff, T., Kudo, S., and Fukuda, M.,** Identification of the crossing-over point of a hybrid gene encoding human glycophorin variant St. Similarity to the crossing-over point in haptoglobin-related genes, *J. Biol. Chem.,* 265, 9259, 1990.
20. **Nakajima, H., Takemura, T., Nakajima, O., and Yamaoka, K.,** Studies on heme α-methenyl oxygenase. I. The enzymatic conversion of pyridine-hemichromogen and hemoglobin-haptoglobin into a possible precursor of biliverdin, *J. Biol. Chem.,* 238, 3784, 1963.
21. **Oh, S.-K., Pavlotsky, N., and Tauber, A. J.,** Specific binding of haptoglobin to human neutrophils and its functional consequences, *J. Leuk. Biol.,* 47, 142, 1990.
22. **Pagano, M., Engler, R., Gelin, M., and Jayle, M. F.,** Kinetic study of the interaction between rat haptoglobin and rat liver cathepsin B, *Can. J. Biochem.,* 58, 410, 1979.
23. **Lisowska, E. and Dobryszycka, W.,** Effect of degradation on properties of haptoglobin. II. Cleavage of disulphide bonds, *Biochim. Biophys. Acta,* 133, 338, 1967.
24. **Beisembaeva, R. U., Mursaglieva, A. T., Dzhumalieva, L. M., Shaikenov, T. E., and Mevkh, A. T.,** Identification of haptoglobin as an endogenous inhibitor of prostaglandin H synthase in the cytosol fraction of primary cells from sheep vesicular glands, *FEBS Lett.,* 269, 125, 1990.
25. **Koj, A.,** Biological functions of acute phase proteins and the cytokines involved in their induced synthesis, in *Modulation of Liver Cell Expression,* Proc. 43rd Falk Symp., Basel, MTP Press, Boston, 1986, 331.
26. **Kohler, W. and Prokop, O.,** Relationship between haptoglobin and Streptomyces pyogenes T4 antigens, *Nature (London),* 271, 373, 1978.
27. **Lammler, C., Chatwal, G. S., and Blobel, H.,** Binding of α2-macroglobulin and haptoglobin to Actinomyces pyogenes, *Can. J. Microbiol.,* 31, 657, 1985.
28. **Frohlander, N., Ljunggren, O., and Lerner, U. H.,** Haptoglobin synergistically potentiates bradykinin and thrombin induced prostaglandin biosynthesis in isolated osteoblasts, *Biochim. Biophys. Res. Commun.,* 178, 343, 1991.
29. **Gutteridge, J. M. C.,** The antioxidant activity of haptoglobin towards haemoglobin-stimulated lipid peroxidation, *Biochim. Biophys. Acta,* 917, 219, 1987.

30. **Maeda, N.,** Nucleotide sequence of the haptoglobin and haptoglobin-related gene pair. The haptoglobin-related gene contains a retrovirus-like element, *J. Biol. Chem.,* 260, 6698, 1985.
31. **McEvoy, S. M. and Maeda, N.,** Complex events in the evolution of the haptoglobin gene cluster in primates, *J. Biol. Chem.,* 263, 15740, 1988.
32. **Maeda, N., McEvoy, S. M., Harris, H. F., Huisman, T. H. J., and Smithies, O.,** Polymorphisms in the human haptoglobin gene cluster: chromosomes with multiple haptoglobin-related (Hpr) genes, *Proc. Natl. Acad. Sci. U.S.A.,* 83, 7395, 1986.
33. **Bensi, G., Raugei, G., Klefenz, H., and Cortese, R.,** Structure and expression of the human haptoglobin locus, *EMBO J.,* 4, 119, 1985.
34. **Oliviero, S., Morrone, G., and Cortese, R.,** The human haptoglobin gene: transcriptional regulation during development and acute phase induction, *EMBO J.,* 6, 1905, 1987.
35. **Kuhajda, F. P., Katumuluwa, A. I., and Pasternack, G. R.,** Expression of haptoglobin-related protein and its potential role as tumor antigen, *Proc. Natl. Acad. Sci. U.S.A.,* 86, 1188, 1989.
36. **Kuhajda, F. P., Piantadosi, G., and Pasternack, G. R.,** Haptoglobin-related protein (Hpr) epitopes in breast cancer as a predictor of recurrence of the disease, *N. Engl. J. Med.,* 321, 636, 1989.
37. **Oh, S. K., Very, D. L., Ettinger, R., Walker, J., Giampaolo, C., and Bernardo, J.,** Monoclonal antibody to SER immune suppressor detects polymeric forms of haptoglobin, *Hybridoma,* 8, 449, 1989.
38. **Mucchielli, A. and Masseyeff, R.,** Purification and characterization of a haptoglobin-hemoglobin complex isolated from human fetuses, *Oncodev. Biol. Med.,* 2, 371, 1981.
39. **Raam, S., Lewis, G. P., and Cohen, J. L.,** Immunological characteristics of peroxidase protein in human neonatal sera, *Clin. Chim. Acta,* 79, 533, 1977.
40. **Oh, S.-K., Very, D. L., Walker, J., Raam, S., and Ju, S.-T.,** An analogy between fetal haptoglobin and a potent immunosuppressor in cancer, *Cancer Res.,* 47, 5120, 1987.
41. **Katnik, I. and Dobryszycka, W.,** Development of concanavalin A-enzyme immunosorbent assay for glycated haptoglobin using polyclonal and monoclonal antibodies, *J. Immunoassay,* 13, 145, 1992.
42. **Warwas, M., Dobryszycka, W., Gerber, J., and Pietkiewicz, A.,** Clinical usefulness of serum acute phase reactants in patients with ovarian tumors, *Neoplasma,* 28, 485, 1981.
43. **Warwas, M., Gerber, J., and Pietkiewicz, A.,** Haptoglobin and proteinase inhibitors in the blood serum of women with inflammatory, benign and neoplastic lesions of the ovary, *Neoplasma,* 33, 79, 1986.
44. **Koj, A.,** Acute phase reactants. Their synthesis, turnover and biological significance, in *Structure and Function of Plasma Proteins,* Vol. 1, Allison, A. C., Ed., Plenum Press, London, 1974, 73.
45. **Yoshimura, S., Tamaoki, N., Ueyama, Y., and Hata, J.-J.,** Plasma protein production by human tumors xenotransplanted in nude mice, *Cancer Res.,* 38, 3474, 1978.
46. **Mackiewicz, A., Ganapathi, M. K., Schultz, D., and Kushner, I.,** Effect of cytokines on glycosylation of acute phase proteins in human hepatoma cell lines, *Clin. Exp. Immunol.,* 75, 70, 1989.
47. **Mucchielli, A.,** Estrogen binding properties of a haptoglobin-hemoglobin complex isolated from human fetus, *J. Steroid Biochem.,* 16, 713, 1982.
48. **Dobryszycka, W.,** Clinical significance of haptoglobin measurements, *Diagn. Lab.,* 23, 117, 1987 (in Polish).
49. **Dobryszycka, W.,** Clinical significance of haptoglobin measurements in Abstr. 6th Eur. Congr. Clinical Chemistry, Jerusalem, 1985, 61.
50. **Fawcett, H. A. C., Al-Hawi, Z., and Brzeski, H.,** Identification of the products of the haptoglobin-related gene, *Biochim. Biophys. Acta,* 1048, 187, 1990.
51. **Baumann, H. and Wong, G. G.,** Hepatocyte-stimulating factor III shares structural and functional identity with leukemia inhibitory factor, *J. Immunol.,* 143, 1163, 1989.
52. **Morrone, G., Ciliberto, G., Oliviero, R., Arcone, L., Dente, L., Content, J., and Cortese, R.,** Recombinant interleukin 6 regulates transcriptional activation of a set of human acute phase genes, *J. Biol. Chem.,* 263, 12554, 1988.
53. **Oliviero, S. and Cortese, R.,** The human haptoglobin gene promoter: interleukin-6 responsive elements interact with a DNA-binding protein by interleukin-6, *EMBO J.,* 8, 1145, 1989.
54. **Baumann, H., Morella, K. K., Jahreis, G. P., and Marinkovic, S.,** Distinct regulation of the interleukin-1 and interleukin-6 response elements of the rat haptoglobin gene in rat and human hepatoma cells, *Mol. Cell. Biol.,* 10, 5967, 1990.
55. **Mackiewicz, A., Ganapathi, M. K., Schultz, D., Samols, D., Reese, J., and Kushner, I.,** Regulation of rabbit acute phase protein biosynthesis by monokines, *Biochem. J.,* 253, 851, 1988.
56. **Kordula, T., Rokita, H., Koj, A., Fiers, W., Gauldie, J., and Baumann, H.,** Effects of interleukin-6 and leukemia inhibitory factor in the acute phase response and DNA synthesis in cultured rat hepatocytes, *Lymphokine Cytokine Res.,* 10, 23, 1991.
57. **Romji, D. P., Cortese, R., and Ciliberto, G.,** Regulation of C-reactive protein, haptoglobin, and hemopexin gene expression in *Acute Phase Proteins: Molecular Biology, Biochemistry, Clinical Applications,* Mackiewicz, A., Kushner, I., and Baumann, H., Eds., CRC Press, Boca Raton, FL, 1993, chap. 21.

58. **Baumann, H., Won, K-A., and Marinkovic-Pajovic, S.**, Regulation of α_1-acid glycoprotein genes and relationship to other types of acute phase plasma proteins, in *Acute Phase Proteins: Molecular Biology, Biochemistry, Clinical Applications*, Mackiewicz, A., Kushner, I., and Baumann, H., Eds., CRC Press, Boca Raton, FL, 1993, chap. 23.
59. **Haugen, T. H., Hanley, J. M., and Heath, E. C.**, Haptoglobin: a novel mode of biosynthesis of a liver secretory glycoprotein, *J. Biol. Chem.*, 256, 1055, 1981.
60. **Moretti, J., Borel, J., Dobryszycka, W., and Jayle, M. F.**, Determination de la demi-vie de li haptoglobine plasmatique humaine, *Biochim. Biophys. Acta*, 69, 205, 1963.
61. **Dobryszycka, W., Zeineh, R., Ebroon, E., and Kukral, J. C.**, Metabolism of plasma proteins in injury states. II. Incorporation of ^{14}C-glucosamine and ^{35}S-methionine into human and canine haptoglobin, *Clin. Sci.*, 36, 231, 1969.
62. **Dobryszycka, W., Guszczynski, T., and Kubicz, Z.**, Clearance of certain modified haptoglobins from the rabbit circulation, *Int. J. Biochem.*, 20, 325, 1988.
63. **Wozniak, M.**, Effect of inhibition of carbohydrate-mediated endocytosis on catabolism of equine haptoglobin and its complex with haemoglobin in hen, *Acta Biochim. Pol.*, 33, 65, 1986.
64. **Engler, R., Moretti, J., and Jayle, M. F.**, Catabolisme du complexe haptoglobine-hemoglobine, *Bull. Soc. Chim. Biol.*, 49, 263, 1967.
65. **Ashwell, G. and Morell, A. G.**, The role of surface carbohydates in the hepatic recognition and transport of circulating glycoproteins, *Adv. Enzymol.*, 41, 99, 1974.
66. **Ashwell, G. and Harford, J.**, Carbohydrate-specific receptors of the liver, *Annu. Rev. Biochem.*, 51, 531, 1982.
67. **Drickamer, K., Mamon, J. F., Binns, G., and Leung, J. O.**, Primary structure of the rat liver asialoglycoprotein receptor: structural evidence for multiple polypeptide species, *J. Biol. Chem.*, 259, 770, 1984.
68. **Schwartz, A. L. and Rup, D.**, Biosynthesis of the human asialoglycoprotein receptor, *J. Biol. Chem.*, 258, 11249, 1983.
69. **Kawasaki, T. and Ashwell, G.**, Isolation and characterization of an avian hepatic binding protein specific for N-acetyl-glucosamine terminated glycoproteins, *J. Biol. Chem.*, 252, 6536, 1977.
70. **Neufeld, E. F. and Ashwell, G.**, Carbohydrate recognition systems for receptor-mediated pinocytosis, in *The Biochemistry of Glycoproteins and Proteoglycans*, Lennarz, W. J., Ed., Plenum Press, New York, 1980, 261.
71. **Schwartz, A. L.**, The hepatic asialoglycoprotein receptor, *CRC Crit. Rev. Biochem.*, 16, 207, 1984.
72. **Stockert, R. J. and Morell, A. G.**, Hepatic binding protein: the galactose-specific receptor of mammalian hepatocytes, *Hepatology*, 3, 750, 1983.
73. **Hudgin, R. L., Pricer, W. E., Jr., Ashwell, G., Stockert, R. J., and Morell, A. G.**, The isolation and properties of a rabbit liver binding protein specific for asialoglycoproteins, *J. Biol. Chem.*, 249, 5536, 1974.
74. **Kawasaki, T. and Ashwell, G.**, Chemical and physical properties of an hepatic membrane protein that specifically binds asialoglycoproteins, *J. Biol. Chem.*, 251, 1296, 1976.
75. **Tsunoo, H., Higa, Y., Kino, K., Nakajima, H., and Hamaguchi, H.**, Studies on hemoglobin metabolism. III. Identification of rat liver cell membrane receptor specific for hemoglobin-haptoglobin complex in vivo, *Proc. Jpn. Acad.*, 53, 22, 1977.
76. **Kino, K., Tsunoo, H., Higa, Y., Takamai, M., Hamaguchi, H., and Nakajima, H.**, Hemoglobin-haptoglobin receptor in rat liver plasma membrane, *J. Biol. Chem.*, 255, 9616, 1980.
77. **Higa, Y., Oshiro, S., Kino, K., Tsunoo, H., and Nakajima, H.**, Catabolism of globin-haptoglobin complex to rats, *J. Biol. Chem.*, 256, 12322, 1981.
78. **Lowe, M. E. and Ashwell, G.**, Solubilization and assay of an hepatic receptor for the haptoglobin-hemoglobin complex, *Arch. Biochem. Biophys.*, 216, 704, 1982.
79. **Gbenle, G. O.**, Haptoglobin distribution among a group of Nigerian blood bank donors, *IRCS Med. Sci.*, 9, 639, 1981.
80. **Frohlander, N.**, Haptoglobin groups and leukemia, *Hum. Hered.*, 34, 311, 1984.
81. **Dobryszycka, W., Gerber, J., Zuwala-Jagiello, J., and Ujec, M.**, Acute phase reactants and circulating immune complexes in patients with ovarian carcinoma, *Arch. Immun. Ther. Exp.*, 39, 41, 1991.
82. **Jayle, M. F.**, Methode de dosage de l'haptoglobine serique, *Bull. Soc. Chim. Biol.*, 33, 876, 1951.
83. **Dnistrian, A. M., Schwartz, M. K., Katopodis, N., Fracchia, A. A., and Stock, C. C.**, Serum lipid-bound sialic acid as a marker in breast cancer, *Cancer*, 50, 1815, 1982.
84. **Dobryszycka, W. and Katnik, I.**, Interaction of haptoglobin with concanavalin A and wheat germ agglutinin. Basic research and clinical applications, in *Affinity Electrophoresis: Principles and Application*, Bręborowicz, J. and Mackiewicz, A., Eds., CRC press, Boca Raton, FL, 1992, 211.
85. **Marchand, A., Galen, R. S., and Van Lente, F.**, The predictive value of serum haptoglobin in hemolytic disease, *JAMA*, 243, 1909, 1980.

86. **Warkentin, D. L., Marchand, A., and Van Lente, F.**, Serum haptoglobin concentrations in concurrent hemolysis and acute phase reaction, *Clin. Chem.*, 33, 1265, 1987.
87. **Katnik, I. and Dobryszycka, W.**, Enzyme immunoassay to measure low levels of haptoglobin in biological fluids, *J. Immunoassay*, 11, 503, 1990.
88. **Katnick, I., Lammler, C., Guszczynski, T., and Dobryszycka, W.**, Quantitative measurement of human haptoglobin by the use of Streptococcus pyogenes cells as solid phase, *Arch. Immun. Ther. Exp.* in press.
89. **Teige, B., Olaisen, B., Pedersen, L., and Teisberg, P.**, Forensic aspects of haptoglobin: electrophoretic patterns of haptoglobin allotype products and an evaluation of typing procedure, *Electrophoresis*, 9, 384, 1988.
90. **Heinderyckx, M., Jacobs, P., and Bollen, A.**, Secretion of glycosylated human recombinant haptoglobin in baculovirus-infected insect cells, *Mol. Biol. Resp.*, 13, 225, 1989.
91. **Hashizumi, M., Kitano, S., Yamaga, H., and Sugimachi, K.**, Haptoglobin to protect against renal damage from ethanolamine oleate sclerosant, *Lancet*, 6, 340, 1988.
92. **Peacock, A. C.**, Serum haptoglobin type and leukemia: an association with possible etiological significance, *J. Natl. Cancer Inst.*, 36, 631, 1966.
93. **Benkman,, H.-G., Hansen, H.-P., Ovenbeck, R., and Goedde, H. W.**, Distribution of alpha-1-antitrypsin and haptoglobin phenotypes in bladder cancer patients, *Hum. Hered.*, 37, 290, 1987.
94. **Dobryszycka, W. and Warwas, M.**, Haptoglobin type in ovarian carcinoma, *Neoplasma*, 30, 169, 1983.
95. **Frohlander, N., Ljunberg, B., and Roos, G.**, Genetic variation of haptoglobin and transferrin in relation to DNA content and stage in renal cell carcinoma, *Cancer*, 63, 1138, 1989.
96. **Mitchell, R. J. and Carzino, R.**, Haptoglobin groups and transferrin subtypes in multiple myeloma, *Hum. Hered.*, 38, 117, 1988.
97. **Mitchell, R. J., Carzino, R., and Janardhana, V.**, Association between the two serum proteins haptoglobin and transferrin and leukemia, *Hum. Hered.*, 38, 144, 1988.
98. **Kidd, K. K., Kidd, J. R., Castiglione, C. M., Pakstis, A. J., and Sparkes, R. S.**, Progress toward resolving the possible linkage of multiple endocrine neoplasia type 2A to haptoglobin and group-specific loci: use of restriction fragment length polymorphisms extends exclusion region, *Genet. Epidemiol.*, 3, 195, 1986.
99. **Frohlander, N. and Forsgren, L.**, Haptoglobin groups in motor neuron disease, *J. Neurol Neurosurg. Psychiatry*, 51, 440, 1988.
100. **Moreira, H. W. and Naoum, P. C.**, Serum haptoglobin types in patients with hemoglobinopathies, *Hereditas*, 113, 227, 1990.
101. **Eiberg, M., Marner, E., Rosenberg, T., and Mohr, J.**, Marner's cataract (CAM) assigned to chromosome 16; linkage to haptoglobin, *Clin. Genet.*, 34, 272, 1988.
102. **Surya Pradha, P., Padma, T., and Ramaswamy, M.**, Haptoglobin patterns in essential hypertension and associated conditions — increased risk for Hp 2-2, *Hum. Hered.*, 37, 345, 1987.
103. **Eiberg, H., Gardiner, R. M., and Mohr, J.**, Batten disease (Spielmayer-Sjogren disease) and haptoglobins (HP): indication of linkage and assignment to chr. 16, *Clin. Genet.*, 36, 217, 1989.
104. **Frohlander, N. and Johnson, O.**, Haptoglobin groups in acute myocardial infarction, *Hum. Hered.*, 39, 345, 1989.
105. **Dobryszycka, W. and Bec-Katnik, I.**, Effect of modification on physico-chemical and biological properties of haptoglobin. Acetylation, iodination, nitration, *Acta Biochim. Pol.*, 22, 143, 1975.
106. **Dobryszycka, W. and Guszczynski, T.**, Reduction of disulphide bonds in human haptoglobin 2-1, *Biochim. Biophys. Acta*, 829, 13, 1985.
107. **Dobryszycka, W. and Guszczynski, T.**, Products of trypsin digestion of haptoglobin β (heavy) chain, *Int. J. Biochem.*, 17, 917, 1985.
108. **Dobryszycka, W. and Guszczynski, T.**, Properties of sulfanilazo-haptoglobin, *Int. J. Biochem.*, 20, 321, 1988.
109. **Dobryszycka, W. and Lisowska, E.**, Effect of degradation on the chemical and biological properties of haptoglobin. I. Products of trypsin digestion, *Biochim. Biophys. Acta*, 121, 42, 1966.
110. **Bernini, L. F. and Borri-Voltattorni, C.**, Studies on the structure of human haptoglobins. I. Spontaneous refolding after extensive reduction and dissociation, *Biochim. Biophys. Acta*, 200, 203, 1970.
111. **Katnik, I., Podgorska, M., and Dobryszycka, W.**, Polyethylene glycol enzyme immunoassay for screening anti-haptoglobin monoclonal antibodies, *J. Immunol. Methods*, 102, 279, 1987.
112. **Zschocke, R. H. and Bezkorovainy, A.**, Some immunochemical properties of succinylated human transferrin and related proteins, *Biochim. Biophys. Acta*, 200, 241, 1970.
113. **Hunneyball, J. M. and Stanworth, O. R.**, The effects of chemical modification on the antigenicity of human and rabbit immunoglobulin G, *Immunology*, 30, 881, 1976.
114. **Pielak, G. J., Ordea, M. S., and Legg, J. I.**, Preparation and characterization of sulfanilazo and arsenilazoproteins, *Biochemistry*, 23, 596, 1984.

115. **Katnik, I. and Dobryszycka, W.**, The effects of limited proteolysis by trypsin on human haptoglobin, *Biochim. Biophys. Acta*, 670, 17, 1981.
116. **Katnik, I., Guszczynski, T., and Dobryszycka, W.**, Immunological comparison of glycopeptides obtained from haptoglobin, *Arch. Immun. Ther. Exp.*, 32, 111, 1984.
117. **Dobryszycka, W. and Krawczyk, E.**, Intermediates in the reassociation of reduced haptoglobin, *Acta Biochim. Pol.*, 30, 203, 1983.
118. **Edge, A. S. B., Flatynek, C. R., Hof, L., Reichert, L. E., and Weber, P.**, Deglycosylation of glycoprotein by tri-fluoromethane-sulfonic acid, *Anal. Biochem.*, 118, 131, 1981.
119. **Sawardeker, J. S., Slokar, J. M., and Jeaves, A.**, Quantitative determination of monosaccharides and their acetates by gas liquid chromatography, *Anal. Chem.*, 37, 1062, 1965.
120. **McConahay, P. J. and Dixon, F. J.**, A method of trace iodination of proteins for immunological studies, *Int. Arch. Allergy*, 29, 185, 1966.
121. **Katnik, I., Guszczynski, T., and Dobryszycka, W.**, Only the con A-binding fraction of the β chain of haptoglobin is biologically active, in *Lectins — Biology, Biochemistry, Clinical Biochemistry*, Vol. 4, Walter de Gruyter, New York, 1985, 253.
122. **Katnik, I. and Dobryszycka, W.**, Influence of modification on physicochemical and biological properties of haptoglobin. VII. Immunochemical reactivity of haptoglobin with modified tyrosine residues, *Arch. Immunol. Ther. Exp.*, 25, 541, 1977.
123. **Osada, J., Sawaryn, A., and Dobryszycka, W.**, Studies on the structure of haptoglobin and the haptoglobin-haemoglobin complex by spin and fluorescence labelling, *Acta Biochim. Pol.*, 25, 333, 1978.
124. **Dobryszycka, W., Kubisa, W., and Kubicz, Z.**, Prediction of secondary structure of human haptoglobin, *Acta Biochim. Pol.*, 34, 377, 1987.
125. **Ranta, T., Chen, H.-C., Jalkanen, J., Nikula, H., Shimohigashi, Y., and Huhtaniemi, J.**, Receptor binding properties and biologic action of deglycosylated human chorionic gonadotropin in human ovary and testis, *Obstet. Gynecol.*, 70, 171, 1987.
126. **Kaartinen, V. and Mononen, I.**, Hemoglobin binding to deglycosylated haptoglobin, *Biochim. Biophys. Acta*, 953, 345, 1988.
127. **Massague, J. and Czech, M. P.**, Role of disulfides in the subunit structure of the insulin receptor, *J. Biol. Chem.*, 267, 7629, 1982.
128. **Chang, T., Chakraborti, P., and Chang, C. H. L.**, The cyanogen bromide fragment I of asialoorosomucoid is transported more efficiently than asialoorosomucoid in rat hepatocytes, *Biochim. Biophys. Acta*, 1010, 166, 1989.
129. **Weigel, P. H., Clarke, B. L., and Oka, J. A.**, The hepatic galactosyl receptor system: two different ligand dissociation pathways are mediated by distinct receptor population, *Biochem. Biophys. Res. Commun.*, 140, 43, 1986.
130. **Dobryszycka, W., Osada, J., and Wozniak, M.**, Metabolic studies on hybrid haptoglobins, *Int. J. Biochem.*, 10, 75, 1974.
131. **Wozniak, M.**, Catabolism of avian and mammalian haptoglobins and their complexes with hemoglobin in chicken, *Comp. Biochem. Physiol.*, 79, 413, 1984.
132. **Steube, K., Gross, V., Haussinger, D., Tran-Thi, T.-A., Decker, K., Gerok, W., and Heinrich, P. C.**, Clearance of acute phase plasma proteins with no, high-mannose-, hybrid- or complex type oligosaccharide side chains by the isolated perfused rat liver, *Biochem. Biophys. Res. Commun.*, 141, 949, 1986.
133. **Horiuchi, S., Murakami, M., Takata, K., and Morino, V.**, Scavenger receptor for aldehyde-modified proteins, *J. Biol. Chem.*, 261, 4962, 1986.
134. **Ranganathan, P. N. and Mego, J. L.**, Renal plasma membrane receptors for certain modified serum albumins, *Biochem. J.*, 239, 537, 1986.
135. **Takata, K., Horiuchi, S., Araki, N., Shiga, M., Saitoh, M., and Morino, Y.**, Endocytic uptake of nonenzymatically glycosylated proteins is mediated by a scavenger receptor for aldehyde-modified proteins, *J. Biol. Chem.*, 263, 14819, 1988.
136. **Soda, R. and Tavassoli, M.**, Liver endothelium and not hepatocytes or Kupffer cells have transferrin receptors, *Blood*, 63, 270, 1984.
137. **Kato, H., Shin-Ichiro, T., Takenaka, A., Funabiki, R., Noguchi, T., and Naito, H.**, Degradation of endogenous proteins and internalized asialofetuin in primary cultured heptocytes of rats, *Int. J. Biochem.*, 21, 483, 1989.
138. **Katnik, I.**, personal communication.

Chapter 11

THE ACUTE PHASE COMPLEMENT PROTEINS

Harvey R. Colten

TABLE OF CONTENTS

0-8493-6913-4/93/$0.00 + $.50

I. INTRODUCTION

It has long been recognized that the acute phase response is complex and involves many diverse metabolic changes. Fever, leukocytosis, and changes in fat, carbohydrate, and trace metal metabolism are features of the acute phase response. Moreover, the concentrations of many plasma proteins change following an acute phase stimulus, although the kinetics, magnitude, and direction of the response is characteristic for each. Among these proteins are some for which a role in host defense is well known (e.g., fibrinogen, complement proteins, and α_1-antitrypsin), while the precise physiological function of others is less clear (e.g., C-reactive protein and serum amyloid A).

Although the acute phase serum complement proteins are quantitatively not among the most prominent, the increases in these proteins are generally sustained for several days. Moreover, among the complement proteins not recognized as ''acute phase'' by serum measurements, several are ''extrahepatic acute phase reactants''; that is, tissue injury and inflammation results in local extrahepatic upregulation of complement gene expression which is often manyfold greater than the hepatic (serum) response and includes proteins such as the second component (C2), the serum concentration of which does not change following an acute phase stimulus.

The acute phase complement components are representative of positive acute phase proteins responsive to the broadest array of extracellular signals, incuding the full gamut of ''acute phase cytokines''. It should be noted, however, that the mechanisms by which changes in acute phase protein gene expression are accomplished vary from one protein to another, are tissue and species specific, and developmentally regulated.

II. COMPLEMENT PROTEINS

The complement system consists of a set of more than 20 effector and regulatory proteins that mediate inflammation initiated by tissue injury, microbial infection, and/or interaction between specific antibodies and corresponding antigens. Other nonhost defense-related functions of the complement system have also been recognized. The complement cascade is activated by limited proteolysis of the classical (antibody-dependent) pathway and alternative (antibody-independent) pathway proteins. The interaction of antigens with antibodies of immunoglobulin class IgM and several IgG subclasses initiates binding and activation of the classical complement pathway. The specificity of this reaction is imposed by the antibody. Under some conditions, however, viral agents,[1-4] DNA, C-reactive protein,[5,6] and mitochondrial membranes[7,8] can activate the classical pathway in the absence of antibody. Activation of the alternative complement pathway always involves antibody-independent recognition of structures that are highly represented among pathogenic microorganisms. This activation therefore involves relatively nonspecific binding. Localization of the alternative pathway reaction is accomplished by permissive propagation of the cascade, i.e., conditions limiting control protein inactivation of the effector proteins. Amplification is an important feature of both pathways, since deposition of one or a few molecules results in enzymatic cleavage of thousands of later components in the cascade and the appearance of complement-dependent biological activities. Further amplification results from the positive activation loop involving the proteins of the alternative pathway. The proteins of the classical pathway include C1 (a macromolecular complex comprised of the products of five genes and the fourth [C4] and second [C2] components). The latter two are encoded by genes within the major histocompatibility complex along with one of the alternative pathway proteins, factor B. The other alternative pathway constituents include factor D and C3. Two distinct unstable enzymes capable of cleaving the third component of complement (C3) are generated by complexing active fragments of the proteins of the two pathways.

The C3 protein is a principle source of biologically active cleavage products that mediate inflammation, solubilize and clear immune complexes, and further propogate the complement cascade, resulting in assembly of the terminal complement protein complex (the membrane attack complex, MAC) and cytolysis via the generation of discrete membrane channels. The complement effector proteins are controlled by an elaborate network of regulatory proteins, some of which also serve as cell-surface receptors for complement activation products. A detailed account of the complement effector and regulatory genes and gene products is beyond the scope of this chapter. The reader should consult relatively recent reviews.[9-11]

The genes encoding the effector and regulatory proteins of the complement cascade are expressed in liver and in many extrahepatic tissues.[12] Liver (almost certainly the hepatocyte) is the source of most of the complement proteins circulating in plasma.[13-15] Interest in the liver as a site of complement production was first suggested at the turn of the century by Ehrlich and Morgenroth,[14] who demonstrated that complement, but not antibody, was depressed in sera of animals with experimental liver injury. Liver perfusion studies showed an increase in complement activity in the effluate, lending further support that the liver was a site of complement synthesis.[16] Much later, Alper and co-workers[13,17] established the liver as a major site of serum complement synthesis in studies of a patient following liver transplantation by a shift from the recipient allotype to the donor allotype of several complement proteins exhibiting genetic polymorphism. It is now known that the liver is a site of synthesis of all the serum complement effector proteins, with the exception of C1q, factor D, and properdin. Experiments using immunofluorescent staining for complement proteins in liver tissue and study of the synthesis of proteins in primary hepatocyte cultures[18] and in well-differentiated hepatoma-derived cell lines[19] have indicated that in liver, the cell of origin of the complement proteins is the hepatocyte. Studies by Thorbecke and colleagues[20] suggested that cell types of endodermal, mesodermal, and ectodermal origin in several organs were capable of synthesizing proteins of the complement system. This concept has been confirmed and extended with the use of modern molecular biological methods. Studies of complement metabolism and mRNA expression *in vivo*[21-23] and biosynthesis in tissue culture indicate a role for complement produced at extrahepatic sites, especially in the immunopathogenesis of autoimmune injury.

Complement proteins of the classical and alternative pathways and C5 are synthesized in several cell types. Thus, activation of the cascade through C5 can be accomplished by proteins synthesized at the extrahepatic sites without a requirement for liver-derived proteins. The widespread synthesis of the prominent acute phase complement proteins, C3 and factor B, which have been identified in virtually every cell type studied, suggests that this extrahepatic synthesis provides a source of alternative complement proteins for local host defense before the influx of serum proteins. The synthesis of two other alternative complement proteins, factor D and properdin, at extrahepatic sites is consistent with this conept and because of their distribution with novel potential functions unrelated to host defenses. Mononuclear phagocytes provide the single richest extrahepatic source of complement proteins, with synthesis of all the activator and inhibitor proteins in both activation pathways, with the exception of factor H. Fibroblasts are also a source of complement proteins, synthesizing at least seven effectors and inhibitors of the system. The functional importance of this extrahepatic complement production is supported by the elaborate tissue-specific and developmentally regulated controls of complement gene expression now recognized in several species.

Factor B, C3, and C4 are the quantitatively most prominent of the acute phase complement proteins in plasma, increasing in concentration two- to threefold with reproducible kinetics following an acute phase stimulus. These changes result from an increase in hepatic complement gene expression. Clinical[23,24] and experimental[21,22] observations established that inflammation and tissue injury also modulate complement gene expression at extrahepatic sites. As noted above, for some, the magnitude of acute phase-induced changes in extrahepatic complement expression exceed the effects in liver. Expression of the classical activation pathway protein C2 (not an acute phase *plasma* protein) is highly regulated at extrahepatic sites of inflammation.

The acute phase complement gene response is also affected by genetic background and development. In the past few years, the molecular bases of this phenomenon have come under scrutiny. The proximate soluble mediators and their respective receptors, the cell biological events, and the *cis*- and *trans*-acting elements that govern regulated complement gene expression have been examined. In addition, other less-well-studied modifiers of complement expression such as sex hormones, corticosteroids, and nutrition offer interesting models for the elucidation of mechanisms by which these influence inflammation in host defenses and autoimmune injury, thereby modifying the acute phase response.

A. THE FOURTH (C4) AND SECOND (C2) COMPONENTS

The constituents of the classical pathway C3 cleaving enzyme are synthesized in similar cell types, but the co- and postsynthetic processing of each differs considerably. C2 and C4 are synthesized in hepatocytes,[19] fibroblasts[25] (only C2),[26] and mononuclear phagocytes.[27,28] Expression of C2 and C4 in mononuclear phagocytes is a function of cellular maturation, i.e., in marrow, only about 10% of the cells produce C4 (no C2 production is detected), whereas 45 to 50% of tissue macrophages synthesize C4.[29] The proportion of tissue macrophages that synthesize and secrete C2 varies from ~2% in bronchoalveolar cells to ~45% in breast milk[30] and splenic and peritoneal cells.[29] These represent interesting and as yet unexplored developmental and tissue-specific differences in the constitutive expression of complement proteins.

C4 is a heterotrimeric glycoprotein of approximately 200 kDa.[31] The protein is synthesized as a single-chain precurosr (prepro-C4)[32] programmed by a 5.5-kb mature mRNA.[33] In humans, two genes, C4A and C4B, polymorphic in size and fine structure, encode the C4 transcripts.[34-36] Processing prepro-C4 requires proteolytic cleavage by a signal peptidase, excision of two interchain linking peptides, generation of a thiolester bridge, glycosylation of α- and β-chain residues via a dolichol phosphate intermediate, sulfation, and cleavage of a carboxy terminal fragment of the α-chain by an extracellular metalloproteinase.[37-39] Processing and secretion of C4 occurs with a half-time of about 60 to 90 min,[40] a rate similar to the kinetics of the synthesis, processing, and secretion of C3 and C5 in hepatoma Hep G2 cells.[19,41] The mature C4 protein is activated by cleavage of C4a (a weak anaphylotoxin), a 9-kDa NH_2 terminal fragment from the α-chain, leaving the remainder, C4b. The latter serves as a subunit of the C3 cleaving enzyme of the classical pathway.

Constitutive expression of C4 is under genetic control and is tissue specific. This phenomenon has been most thoroughly studied in inbred mouse strains, but similar observations have been made in studies of human and guinea pig C4 expression. The 20-fold difference in plasma C4 concentration among inbred mouse strains of different H-2 haplotypes[42] is primarily a reflection of differences in transcriptional regulation of C4 expression.[43-45] The tissue specificity of this phenomenon is evidenced by the finding of equal C4 expression (steady-state mRNA and protein) in extrahepatic macrophages from C4 high and C4 low strains.[45,46]

C2 is a single-chain glycoprotein (M_r ~100,000) that binds to C4b in the presence of Mg^{2+} and is cleaved into an amino terminal, 223-amino acid polypeptide C2b and carboxy terminal C2a (509-amino acid) fragment. The latter bears the active enzymatic site and, like factor B (see below), is unusual among serine proteinases, since the active polypeptide is the larger of the two fragments generated during activation/cleavage.[47] The C3 cleaving enzyme, C4b2a, is unstable, i.e., provides an intrinsic control mechanism, but its decay is accelerated by cell-surface regulatory proteins.

C2 is encoded by a gene (15 kb) that is located ~400 base pairs 5′ to the factor B gene. Several C2 primary translation products (84, 79, and 70 kDa) have been recognized.[48] The largest of these is glycosylated and secreted within about 60 min; the others also are glycosylated, but remain cell associated. At least two murine C2 transcripts have been identified,[49] one of which exhibits a deletion of a seven-amino acid peptide within the serine proteinase binding pocket. This transcript is expressed at low levels in all tissues thus far examined except in heart, where about equal amounts of the two C2 transcripts have been detected.

Volanakis and colleagues[50] suggest that translational control of C2 is due to differences in transcriptional initiation. These generate differences in the 5′ untranslated region, the sequence of which governs the rate of C2 translation. Further analysis of this interesting finding in the context of tissue specificity of C2 expression is warranted.

B. FACTOR B

Factor B is synthesized as an ~80 kDa propolypeptide which undergoes signal peptide cleavage, glycosylation, and secretion in about 60 to 90 min. Two factor B transcripts generated from alternative transcriptional initiation sites have been recognized in studies of murine Bf expression in kidney and intestine.[21,51] The large of the two is expressed in amounts equal to the shorter message only in those two tissues; in liver, the long transcript represents <5% of Bf mRNA.[49] No obvious promoter sequence has been recognized at the upstream initiation sites, but mutations near this site which parallel differences in expression of the Bf long transcript in different mouse strains should help in the analysis of the specific sequences essential for upstream transcriptional initiation.[52]

Constitutive and regulated expression of C2 and Bf has been most extensively examined in mice. Conservation of overall gene structure, including ~65% identity in nucleotide sequence between humans and mice for C2 and Bf,[49,53] suggests that conclusions from these studies will pertain to humans as well. Among inbred mouse strains, marked differences in C2 and B serum concentrations are paralleled by differences in hepatic mRNA content for each.[21] Expression of C2 and Bf in these strains vary independently. Constitutive control is tissue specific, as indicated by the observation that the relative content of specific C2 or B mRNA in macrophages is independent of the relative hepatic expression of each among the murine strains examined. Recent studies by Garnier et al.[52] have begun to identify the *cis*- and *trans*-acting elements responsible for these phenomena.

C. COMPLEMENT PROTEIN C3

Prepro C3 is programmed by a ~5.2-kb mature mRNA derived from a >40-kb gene on chromosome 19[54] in humans and 17 (outside of the histocompatibility complex) in mice.[55] Co- and postsynthetic modification involves cleavage of a signal peptide, excision of a single interchain linking peptide, and glycosylation. Generation of the functionally critical thiolester bridge of C3 probably proceeds via isomerization of a lactam to a thiolactone,[56] perhaps involving a specific enzyme.[57] The hepatic biosynthetic rate estimated *in vivo* from fractional catabolic rates of ^{125}I-labeled C3 is 0.45 to 2.7 mg/kg/h,[58] a rate comparable to that observed in cell culture.[19]

Constitutive expression of C3[59] also varies among inbred mouse strains, but effects of age and sex[55] complicate analysis of the *cis*- and *trans*-acting elements accounting for these differences. Nevertheless, a detailed structural and functional study of the sequences 5′ to the murine and human C3 genes[60,61] and the nucleoproteins that interact with these regions should soon elucidate the pertinent molecular details accounting for constitutive and regulated control of C3 expression.

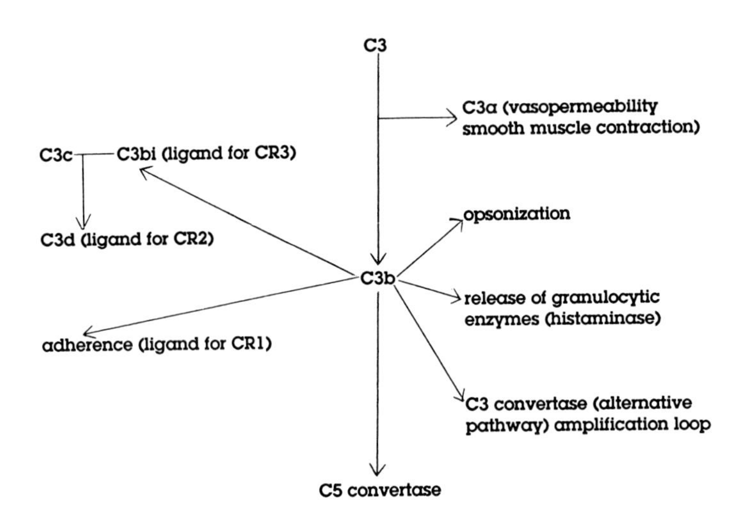

FIGURE 1. Biological function of C3. Some of the functions of C3 cleavage products are summarized. Cr1, Cr2, and Cr3 are complement receptors that mediate many C3-dependent cell biological effects. (Modified from Colten, H. R. and Gitlin, J. D., in *Blood: Principles and Practice of Hematology,* Handin, R. I., Lux, S. E., and Stossel, T. P., Eds., J.B. Lippincott, Philadelphia, in press. With permission.)

The C3 protein (M_r ~185 kDa) is a disulfide-linked heterodimer of α (~110 kDa)- and β(75 kDa)-chains.[62] Cleavage of C3 at Arg_{77}-Ser_{78} of the α-chain liberates the C3a fragment, a potent anaphylatoxin which elicits histamine release, smooth muscle contraction, and noncytotoxic liberation of arachadonate metabolites.[63] These effects are mediated via specific receptors for C3a on mast cells, smooth muscles cells and leukocytes. Covalent binding of C3b to a carboxyl or amino acceptor site is accomplished via its labile thiolester.[64,65] Deposition of C3 on a cell is relatively inefficient (<5 to 10% bound under optimal conditions),[66] so that much of the metastable C3b reacts with water and cannot bind to the target cell or immune complex. Nevertheless, C3b, whether surface bound or hydrolyzed, can bind C5, factors B, H, and properdin, and the C3b receptor, CR1.

Binding and activation of factor B to C3b generates the alternative pathway amplification loop, the flux of which is controlled by intrinsic decay of the enzyme C3bBb, control/cofactor protein dissociation of the complex, and cleavage of the C3b.[67] Fluid phase and cell-surface proteins participate in the regulation of the complement cascade at this point in the sequence. Interestingly, some microogranisms have evolved strategies for control of C3b activity that mimic these physiological regulatory mechanisms. A second C3b bound to the C3bBb complex results in generation of a C5 cleaving enzyme and assembly of the membrane attack complex.[9,68] However, this is but one of the several pathways for further generation of active complement proteins, since many of the products of C3b cleavage have important biological activities (Figure 1).

Cleavage of C3b by one of the natural control proteins, factor I, is facilitated by the fluid-phase component factor H and on erythrocytes, lymphocytes, granulocytes, mononuclear phagocytes, and other cells by the C3b receptor CR1. The C3b α-chain is cleaved by factor I at two sites to generate C3bi, consisting of disulfide-linked α-chain fragments of 62.5 and 29.5 kDa, bound to the β-chain, and a 2-kDa fragment (C3f) of uncertain function. A member of the integrin family of proteins found on phagocytes, CR3 is a specific receptor for C3bi. The C3bi undergoes further cleavage by factor I, dissociating a 62.5-kDa amino-terminal fragment to generate C3 dg, a 40-kDa C3 product that remains covalently bound to the target of complement activation. Further cleavage of a 5-kDa fragment, C3g, from the amino terminus by plasmin generated C3d (35 kDa), for which a specific receptor,

CR2, is found on B-lymphocytes. The balance of the C3 molecule, C3c, is released from the site. The sequential reactions in which C3 serves as substrate are reviewed in Reference 65.

III. REGULATED COMPLEMENT GENE EXPRESSION

A. ENDOTOXIN

Recent studies have exploited modern molecular biological tools to investigate the acute phase response of the complement system. From several of these, abundant evidence established that endotoxin (lipopolysaccharide, LPS), a constituent of the cell wall of Gram-negative bacteria, has a potent direct and indirect effect on regulation of complement gene expression. The analysis of this endotoxin effect is complicated by the many genes induced by endotoxin, products of which also regulate complement gene expression. Moreover, the biochemical mechanisms for endotoxin recognition on cell surfaces by LPS binding sites and the signal transduction pathways mediating its effects are poorly characterized. These constitute problems even in studies of homogeneous cell populations in culture, but the problem of ascertaining which among the LPS-induced effects are of most importance *in vivo* is more complex. Nevertheless, LPS administration *in vivo* or to most cells in culture results in a brisk increase in the net synthesis of several of the complement proteins up to 20+-fold over constitutive synthesis rates. The LPS effect is clearly due to the lipid A moiety.[69] Certain cells that express complement genes such as monocytes, type II lung cells, and fibroblasts are highly LPS responsive, whereas others, the hepatocyte (Hep G2 hepatoma) and intestinal epithelium (CaCo2 cells), are not.

The LPS effect on C3 and factor B expression serves as a useful model to evaluate this phenomenon, since these have been most well studied. The quantitative effect of LPS on these two genes is substantial, and both gene products are of considerable importance in host defenses against Gram-negative bacteria. In human peripheral blood monocytes and macrophages and in murine macrophages, LPS at nanogram concentrations induces a five- to 30-fold increase in C3 and Bf synthesis and secretion.[69-71] Most of this effect is pretranslational (probably transcriptional), as indicated by corresponding increases in C3 and Bf mRNA. However, two lines of evidence suggest an important translational regulatory effect as well. For example, in adult human fibroblasts, the increased Bf protein synthesis rates following LPS stimulation far exceed the increase in Bf mRNA. This effect is similar to that observed for LPS induction of TNF synthesis[72] and others.

The second line of evidence is derived from studies of the developmental biology of C3 and Bf gene expression. St. John Sutton et al.[70] noted that lipid A failed to upregulate C3 and Bf synthesis in newborn (umbilical cord) blood monocytes in contrast to its well-recognized effect on adult monocytes. This difference was specific for the complement genes, since other LPS-induced phenomena were equally demonstrable in neonatal and adult cells. Further exploration of this phenomenon in fibroblasts permitted evaluation of transcriptional regulation and extension to fetal cells. These data show LPS induction of specific C3 and Bf mRNA in newborn fibroblasts, but no increase in C3 or Bf protein synthesis,[73] and in the fetal cells, no induction of mRNA.[101] Thus, the response to LPS involves a sequential maturation of transcriptional, then translational, competence from fetal to newborn and adult.

Translational control is also important in C2[50] and perhaps factor B expression as well. In most inbred mice, two Bf mRNA transcripts are observed in about equal amounts in kidney and intestine.[51,74] The longer transcript has multiple out-of-frame AUG initiation codons that could alter translation rates by a so-called "stutter" mechanism, i.e., effete translation products are generated, decreasing the efficiency of productive translation reactions. The 5′ sequences flanking the human and murine C3 genes contain many TATA boxes extending well upstream of the box essential for hepatocellular transcription. This sequence raises the possibility of multiple C3 transcripts with 5′ extensions. Multiple human Bf and C3 transcripts have not yet been identified.

B. CYTOKINES

As alluded to above, LPS induces many of the cytokines (IL-1, IL-6, and TNF) that have a direct effect on complement gene expression. Accordingly, an extensive evaluation of cytokine-mediated complement gene regulation has been undertaken.

1. Interleukin-1 (IL-1)

IL-1 is a polypeptide of ~17 kDa, first recognized by its manifold biological activities.[75] Mononuclear phagocytes constitute a major source of this cytokine, although other cells produce IL-1 as well in response to several stimuli, including LPS.

Incubation of Hep G2 hepatoma cells or primary murine hepatocyte cultures in IL-1 induces a dose-dependent increase in C3 and Bf synthesis and secretion, but has no effect on C2 or C4 synthesis.[76] This increase in C3 and Bf is due to an increase in transcription, requires specific nucleotide sequences 5′ to the transcriptional initiation site of each of the responsive genes, and is mediated by a signal transduction pathway separable from those for other cytokines.[77,78] IL-1 also regulates complement genes at extrahepatic sites,[21,79] but in these cells it is more difficult to evaluate whether the effect of IL-1 is direct or on another gene such as IL-6; in the latter case, IL-6 would be the proximate mediator of complement gene expression. For instance, IL-1 induces a dose-dependent increase in murine C3 and Bf synthesis in macrophages[79] and human fibroblasts, but macrophages and fibroblasts also express IL-6 and/or TNF, both of which could, via a paracrine or autocrine mechanism, induce C3 and Bf. It is, in fact, probable that at extrahepatic sites, multiple cytokines are induced by a single stimulus (e.g., IL-1) and that this amplifies the primary event. For example,[80] IL-1 increases Bf and C3 expression by ten- to 100-fold in fibroblasts, but in Hep G2, which does *not* express IL-6 even when stimulated with IL-1 or TNF, the increase in C3 and Bf is only two- to threefold. The difference in magnitude of the observed induction cannot be directly compared because of differences in constitutive expression in the two cell types, but the principle is almost certainly valid. Moreover, the kinetics of induction of IL-6 by IL-1 (2 h) and of Bf (4 h) is consistent with an effect of more than one cytokine. In any case, the IL-1 effect is modified by dexamethasone (inhibited) and interferon-γ (enhanced). Species differences in the modulation of cytokine induction have been noted. For instance, a substantial number of studies suggest that corticosteroids enhance cytokine-mediated upregulation of acute phase genes in rat.[81] Early experiments showed that hydrocortisone at a concentration of approximately 10^{-6} *M* stimulated synthesis of C3 and transferrin as judged by semiquantitative methods.[26] In experiments designed to quantitate the magnitude of this effect, incubation of a well-differentiated rat hepatoma in hydrocortisone succinate (4×10^{-7} *M*) increased rates of C3 production up to ninefold.[82] However, these findings are not characteristic of the effect of corticosteroid on the same gene(s) in other species, including mice and man.

Within a single species, genetic control of IL-1-regulated C3 and Bf expression has been observed. For example, Falus et al.[21] found that in B10.PL ($H\text{-}2^u$) mice,IL-1 induced a slight increase in hepatic and a marked increase in splenic, kidney, and pulmonary C3 mRNA. In contrast, C3 mRNA decreased in liver and spleen of B10.AKM ($H\text{-}2^k$) mice, although the response to IL-1 in kidney and lung was similar to that observed in the B10.PL. Examination of C3 expression in peritoneal macrophages showed similar strain differences, but ruled out an effect closely linked to H-2 since the response of C3 to IL-1 was opposite in two H-2^k strains. Recent work by Kawamura et al.[60] and Darlington et al.,[61] elucidating the *cis* elements required for the C3 response to IL-1 and IL-6, should be extended. The strain- and tissue-specific differences observed in inbred mice will facilitate further identification of important *trans*-acting regulatory elements.

Specific sequences 5′ to the human and mouse Bf gene[78,83] extending into the 3′ untranslated region of the C2 gene confer IL-1 and IL-6 responsiveness to a chimeric gene construct (Bf 5′ flanking sequence — chloramphenicol acetyltransferase gene) when transfected into Hep G2 hepatoma or L-cells (fibroblasts). This intergenic region displays an extraordinary phylogenetic conservation (65% identity between human and mouse) and within the specific IL-1/IL-6 response elements, even greater conservation.[41] Similar elements are recognized in C3 as well, but deletion analysis indicates that only some of these structural elements are of functional importance in hepatocytes or fibroblasts.

2. Tumor Necrosis Factor (TNF)

Like Il-1, TNF is a cytokine expressed in mononuclear phagocytes and other cell types.[72] TNF is capable of upregulating the expression of several of the complement genes.[80,84] For some, such as C3 and Bf, the kinetics, magnitude, and direction of the response to TNF and IL-1 are similar. Recent data[102] show a tissue specificity of the response to TNF of Bf long and short transcripts, providing a model for elucidating potential branch points in the IL-1/TNF signal transduction pathway. Factor H is an example of another complement gene differentially regulated by TNF, and IL-1/IL-6.[80] Little attention has been given to the molecular mechanisms involved in TNF regulation of complement gene expression, but several obvious areas need further work. For example, the mechanisms involved in the marked effect of starvation on factor D expression and the known effects of TNF on fat cell metabolism[85] certainly should stimulate additional studies of these phenomena.

3. Interleukin-6

Some investigators have suggested that the regulation of all acute phase genes can be explained by converging pathways with IL-6 as the proximate mediator.[86,87] Some genes, such as α_1-antitrypsin and fibrinogen, are clearly only responsive to IL-6 and not at all responsive to IL-1,[84] but thus far it seems that each of the IL-6-responsive complement genes is also IL-1 responsive. Some, such as C2 and C4, are responsive to neither IL-1 nor IL-6, but are only regulated by interferon (see below).

At least four lines of evidence indicate that IL-6 is an important mediator of complement gene expression, but that other cytokines also exert direct effects on complement genes. First, IL-1 upregulates C3 and Bf in cells (Hep G2) that do not express detectable IL-6 mRNA or protein.[77,88] Some[89] have reported IL-6 mRNA in Hep G2. These discrepancies may be a result of genetic drift in Hep G2 cell lines and are not likely due to technical difficulties. Second, the use of monoclonal antibodies against IL-1 and IL-6 in Hep G2 cultures incubated with IL-1 or IL-6 demonstrated independent increases in Bf expression with each cytokine and specific inhibition of the effects by the monoclonal antibodies.[77] A subclone of Hep G2 provides the third line of evidence. Factor B expression in this cell, which bears IL-1 receptors similar in number and affinity to the parent cell line, is not increased by IL-1, but is by IL-6.[77] Nevertheless, addition of both IL-1 and IL-6 augments the IL-6 response, suggesting either an effect on IL-6 receptors or common as well as parallel signal transduction pathways and/or nucleoproteins for IL-1 and IL-6. Finally, Baumann et al.[90] have shown that phorbol ester modulates IL-6 induction of an acute phase promoter-CAT gene construct by increasing its expression, but the phorbol counterregulates the Il-1 induction. The complexity of this effect is also revealed by the fact that IL-1 and IL-6 together had at least an additive and perhaps synergistic effect on the chimeric gene in Hep G2.

4. Growth Factor Counterregulation

Selective counterregulation of cytokine-mediated induction of complement genes has been observed[91,92] in studies of the effect of growth factors on complement genes. IL-4 (a B-lymphocyte growth factor), platelet-derived growth factor (PDGF), and epidermal growth factor (EGF) abrogate the stimulatory effect of IL-1, IL-6, and TNF on factor B expression in human fibroblasts, but IL-4, in fact, augments the effect of TNF on C3 expression. That the counter regulation is also cytokine specific was demonstrated in experiments showing no effect of IL-4 on interferon-dependent upregulation of Bf expression. These counterregulatory effects of growth factors are independent of, and preceed their effect on, cellular proliferation. In fact, this may be an important mechanism by which the complement-dependent inflammatory response is decreased during tissue repair, i.e., the end of an acute phase response may not simply be a relaxation of stimulatory mechanisms, but, rather, an active downregulation. Exploration of the molecular mechanisms involved should therefore prove quite fruitful and important.

5. Interferon

Among the well-characterized cytokines that modulate complement gene expression, interferon-γ (IFNγ) is the most pluripotent, i.e., even those complement genes unresponsive to IL-1, IL-6, and TNF, such as C2 and C4, are upregulated by IFN. In addition, C3 (in some species) and Bf are responsive to IFN. Moreover, several inhibitors and cofactors of inhibitors of complement activation, such as C1 inhibitor and factor H, are also regulated by IFN, thus indirectly affecting net biological expression of the acute phase reactants. Regulated complement gene expression by IFN has been recognized in many different cell types (hepatocytes,[93] mononuclear phagocytes,[94] fibroblasts,[71,95,96] and endothelial cells[97]).

Each of the human class III MHC genes — C2, Bf, C4A, and C4B — is upregulated by IFNγ, but the magnitude and kinetics of the response varies considerably. For example, the response of Bf to IFN is more rapid (<4 h) and of greater magnitude than the response of C2 (>10 h). Likewise, the response of C4A is greater and more prolonged than for C4B.[93] In human monocytes, IFNγ prevents the rapid extinction of C4A or C4B expression with time in culture. Surprisingly, this effect of IFN on C4 expression in human monocytes is counterregulated by picogram concentrations of LPS. The LPS inhibitory effect on C4 is not mediated by IL-1, IL-6, or TNF.[98] In contrast,[71] LPS synergistically enhances the response of factor B to IFNγ.

These observations and the detailed studies of *cis*-acting elements essential for the IFN response raise several interesting general questions about regulated gene expression. An IFN response element has been identified in sequence analyses of human and murine DNA in the intergenic region between C2 and factor B. Moreover, this element serves as a true enhancer (orientation and position independent) in functional studies of a chimeric gene construct with CAT as the reporter element.[41,78,97] Uncertainty remains about the 5′ boundary of the C2 gene, but thus far an IFN response element has not been recognized 5′ to the C2 gene within a cosmid containing ~2 kb of sequences upstream of the C2 coding region. L-cells transfected with this cosmid express C2, and its expression is increased by IFN. Taken together, these data suggest the possibility that the same IFN response element is functional for regulating C2 and Bf expression. IL-1/IL-6 response elements are within the 3′ untranslated region of the C2 gene, perhaps accounting for the effect of these cytokines on Bf, not on C2.

The response of the C3 gene to IFN is less easily understood in this context, however, since the magnitude and even the direction of the IFN effect are species specific. The response of human C3 to IFN is uncertain. IFN has either no effect or a small negative or positive regulatory effect on C3 expression in several different human cell types.[94,99] In contrast, the positive effect of IFN on murine (at least some strains) C3 expression is up to 25-fold. This effect on murine C3 is, in part, transcriptional and is augmented by inhibition of protein synthesis with cyclohexamide.[100] The *cis* elements responsible for this difference are not obvious, even though detailed comparisons of the murine and human C3 genes have been reported.[60]

IV. CONCLUSION

The acute phase complement proteins represent a first-line response to tissue injury and microbial infection. The complement proteins in plasma are produced in liver, but extrahepatic complement gene expression has been detected in many other tissues. The acute phase complement genes are upregulated by bacterial products (endotoxin) and by a complex array of cytokines. Recently, much has been learned about the structure and function of the complement proteins, the complement genes, and the mechanisms controlling their expression. The impact of these studies extends beyond an appreciation of the acute phase response. Hence, in addition to observations pertinent to basic molecular and cellular biology, one can now begin to conceive of strategies for the control of human diseases in which complement plays a role.

REFERENCES

1. **Welsh, R. M., Cooper, N. R., Jensen, F. C., and Oldstone, M. B. A.,** Human serum lyses RNA tumor viruses, *Nature (London)*, 257, 612, 1975.
2. **Cooper, N. R., Jensen, F. C., Welsh, R. M., and Oldstone, M. B. A.,** Lysis of RNA tumor viruses by human serum: direct antibody-independent triggering of the classical complement pathway, *J. Exp. Med.*, 144, 970, 1976.
3. **Bartholomew, R. M. and Esser, A. F.,** Mechanism of antibody-independent activation of the first component of complement (C1) on retrovirus membranes, *Biochemistry*, 19, 2847, 1980.
4. **Bartholomew, R. M., Esser, A. F., and Muller-Eberhard, H. J.,** Lysis of oncocornaviruses by human serum. Isolation of the viral complement (C1) receptor and identification as p15E, *J. Exp. Med.*, 147, 844, 1978.
5. **Kaplan, M. H. and Volanakis, J. E.,** Interaction of C-reactive protein complexes with the complement system. I. Consumption of human complement associated with the reaction of C-reactive protein with pneumococcal C-polysaccharide and with choline phosphatides, lecithin and sphingomyelin, *J. Immunol.*, 112, 2135, 1974.
6. **Richards, R. L., Gewurz, H., Osmand, A. P., and Alving, C. R.,** Interactions of C-reactive protein and complement with liposomes, *Proc. Natl. Acad. Sci. U.S.A.*, 74, 5672, 1977.
7. **Pinckard, R. N., O'Rourke, R. A., and Crawford, M. H.,** Complement localization and mediation of ischemic injury in baboon myocardium, *J. Clin. Invest.*, 66, 1050, 1980.
8. **Storrs, S. B., Kolb, W. P., Pinckard, R. N., and Olson, M. S.,** Characterization of the binding of purified human C1q to heart mitochondrial membranes, *J. Biol. Chem.*, 256, 10924, 1981.
9. **Muller-Eberhard, H. J.,** Molecular organization and function of the complement system, *Annu. Rev. Biochem.*, 57, 321, 1988.
10. **Frank, M. M.,** Complement: a brief review, *J. Allergy Clin. Immunol.*, 84, 411, 1989.
11. **Colten, H. R. and Gitlin, J. D.,** Immunoproteins, in *Blood: Principles and Practice of Hematology*, Handin, R. I., Lux, S. E., and Stossel, T. P., Eds., J.B. Lippincott, Philadelphia, in press.
12. **Colten, H. R.,** The biosynthesis of complement components, in *New Comprehensive Biochemistry*, Harrison, R. A., Ed., Elsevier, Amsterdam, in press.
13. **Alper C. A., Raum, D., Awdeh, Z., Petersen, B. H., Taylor, P. D., and Starzl, T. E.,** Studies of hepatic synthesis in vivo of plasma proteins including orosomucoid, transferrin, alpha-1-antitrypsin, C8 and factor B, *Clin. Immunol. Immunopathol.*, 16, 84, 1980.
14. **Ehrlich, P. and Morgenroth, J.,** Ueber Haemolysine, *Berl. Klin. Wochenschr.*, 37, 453, 1900.

15. **Perlmutter, D. H. and Colten, H. R.,** Complement: molecular genetics, in *Inflammation: Basic Principles and Clinical Correlates,* 2nd ed., Gallin, J. I., Goldstein, I. M., and Snyderman, R., Eds., Raven Press, New York, 1988, 75.
16. **Muller, L.,** *Zentralbl. Bakteriol. Parasitenkd. Infektionskr. Abt. 1,* 57, 577, 1911.
17. **Alper, C. A., Johnson, A. M., Birtch, A. G., and Moore, R. D.,** Human C3: evidence for the liver as the primary site of synthesis, *Science,* 163, 286, 1969.
18. **Ramadori, G., Tedesco, F., Bitter-Suermann, D., and Meyer zum Buschenfelde, K. H.,** Biosynthesis of the third (C3), eighth (C8), and ninth (C9) complement components by guinea pig hepatocyte primary cultures, *Immunobiology,* 170, 203, 1985.
19. **Morris, K. M., Aden, D. P., Knowles, B. B., and Colten, H. R.,** Complement biosynthesis by the human hepatoma derived cell line HepG2, *J. Clin. Invest.,* 70, 906, 1982.
20. **Thorbecke, G. J., Hochwald, B. M., Van Furth, L. R., Muller-Eberhard, H. J., and Jacobson, E. B.,** Problems in determining the sites of synthesis of complement components, in *CIBA Symposium: Complement,* Wolstenholme, G. E. W. and Knight, J., Eds., Churchill Livingstone, London, 1965, 99.
21. **Falus, A., Beuscher, H. U., Auerbach, H. S., and Colten, H. R.,** Constitutive and IL-1 regulated murine complement gene expression is strain and tissue specific, *J. Immunol.,* 138, 856, 1987.
22. **Passwell, J., Schreiner, G. F., Nonaka, M., Beuscher, H. U., and Colten, H. R.,** Local extrahepatic expression of complement genes C3, factor B, C2 and C4 is increased in murine lupus nephritis, *J. Clin. Invest.,* 82, 1676, 1988.
23. **Ruddy, S. and Colten, H. R.,** Rheumatoid arthritis: biosynthesis of complement proteins by synovial tissues, *N. Engl. J. Med.,* 290, 1284, 1974.
24. **Ahrenstedt, O., Knutson, L., Nilsson, B., Nilsson-Ekdahl, K., Odlind, B., and Hallgren, R.,** Enhanced local production of complement components in the small intestines of patients with Crohn's disease, *N. Engl. J. Med.,* 322, 1345, 1990.
25. **Katz, Y. and Strunk, R. C.,** Synovial fibroblast-like cells synthesize seven proteins of the complement system, *Arthritis Rheum.,* 31, 1365, 1988.
26. **Stecher, V. J. and Thorbecke, G. J.,** Sites of synthesis of serum proteins. III. Production of beta-1-C, beta-1-E and transferrin by primate and rodent cell lines, *J. Immunol.,* 99, 660, 1967.
27. **Einstein, L. P., Schneeberger, E. E., and Colten, H. R.,** Synthesis of the second component of complement by long-term primary cultures of human monocytes, *J. Exp. Med.,* 143, 114, 1976.
28. **Whaley, K.,** Biosynthesis of the complement components and the regulatory proteins of the alternative complement pathway by human peripheral blood monocytes, *J. Exp. Med.,* 151, 501, 1980.
29. **Alpert, S. E., Auerbach, H. S., Cole, F. S., and Colten, H. R.,** Macrophage maturation: differences in complement secretion by marrow, monocyte and tissue macrophages detected with an improved hemolytic plaque assay, *J. Immunol.,* 130, 102, 1983.
30. **Cole, F. S., Auerbach, H. S., Goldberger, G., and Colten, H. R.,** Tissue specific pretanslational regulation of complement production in human mononuclear phagocytes, *J. Immunol.,* 134, 2610, 1985.
31. **Schreiber, R. D. and Muller-Eberhard, H. J.,** Fourth component of human complement: description of a three polypeptide chain structure, *J. Exp. Med.,* 140, 1324, 1974.
32. **Hall, R. E. and Colten, H. R.,** Cell-free synthesis of the fourth component of guinea pig complement (C4): identification of a precursor of serum C4 (pro-C4), *Proc. Natl. Acad. Sci. U.S.A.,* 75, 1707, 1977.
33. **Whitehead, A. S., Goldberger, G., Woods, D. E., Markham, A. F., and Colten, H. R.,** Use of a cDNA clone for the fourth component of human complement for analysis of a genetic deficiency of C4 in guinea pig, *Proc. Natl. Acad. Sci. U.S.A.,* 80, 5387, 1983.
34. **Belt, K. T., Carroll, M. C., and Porter, R. R.,** The structural basis of the multiple forms of human complement component C4, *Cell,* 36, 907, 1984.
35. **Carroll, M. C., Belt, K. T., Palsdottir, A., and Porter, R. R.,** Structure and organization of C4 genes, *Philos. Trans. R. Soc. London,* 306, 379, 1984.
36. **Schneider, P. M., Carroll, M. C., Alper, C. A., Rittner, C., Whitehead, A. S., Yunis, E. J., and Colten, H. R.,** Polymorphism of the human complement C4 and steroid 21-hydroxylase genes: restriction fragment length polymorphisms revealing structural deletions, homoduplications and size variants, *J. Clin. Invest.,* 78, 650, 1986.
37. **Goldberger, G. and Colten, H. R.,** Precursor complement protein (pro-C4) is converted in vitro to native C4 in plasmin, *Nature (London),* 286, 514, 1980.
38. **Karp, D. R.,** Post-translational modification of the fourth component of complement. Sulfation of the alpha chain, *J. Biol. Chem.,* 258, 12745, 1983.
39. **Karp, D. R.,** Post-translational modification of the fourth component of complement: effect of tunicamycin and amino acid analogs on the formation of the internal thiol ester and disulfide bonds, *J. Biol. Chem.,* 258, 14490, 1983.
40. **Roos, M. H., Kornfeld, S., and Shreffler, D. C.,** Characterization of the oligosaccharide units of the fourth component of complement (Ss protein) syntehsized by murine macrophages, *J. Immunol.,* 124, 2860, 1980.

41. **Morris, K. M., Goldberger, G., Colten, H. R., Aden, D. P., and Knowles, B. B.,** Biosynthesis and processing of a human precursor complement protein, pro-C3, in a hepatoma-derived cell line, *Science,* 215, 399, 1982.
42. **Shreffler, D. C.,** The S region of the mouse major histocompatibility complex (H-2): genetic variation and functional role in complement system, *Transplant. Rev.,* 32, 140, 1976.
43. **Newell, S. L. and Atkinson, J. P.,** Biosynthesis of C4 by mouse peritoneal macrophages. II. Comparison of C4 synthesis by resident and elicited cell populations, *J. Immunol.,* 130, 834, 1983.
44. **Rosa, P. A. and Shreffler, D. C.,** Cultured hepatocytes from mouse strains expressing high and low levels of the fourth component of complement differ in rate of synthesis of the protein, *Proc. Natl. Acad. Sci. U.S.A.,* 80, 2332, 1983.
45. **Sackstein, R. and Colten, H. R.,** Molecular regulation of MHC class III (C4 and factor B) gene expression in mouse peritoneal macrophages, *J. Immunol.,* 133, 1618, 1984.
46. **Cox, B. J. and Robins, D. M.,** Tissue-specific variation in C4 and Slp gene regulation, *Nucleic Acids Res.,* 16, 6857, 1988.
47. **Cooper, N. R.,** Enzymatic activity of the second component of complement, *Biochemistry,* 14, 4245, 1975.
48. **Perlmutter, D. H., Cole, F. S., Goldberger, G., and Colten, H. R.,** Distinct primary translation products from human liver mRNA give rise to secreted and cell-associated forms of complement protein C2, *J. Biol. Chem.,* 259, 10380, 1984.
49. **Ishikawa, N., Nonaka, M., Westsel, R. A., and Colten, H. R.,** Murine complement C2 and factor B genomic and cDNA cloning reveals different mechanisms for multiple transcripts of C2 and B, *J. Biol. Chem.,* 265, 19040, 1990.
50. **Horiuchi, T., Macon, K. J., Kidd, V. J., and Volanakis, J. E.,** Translational regulation of complement protein C2 expression by differential utilization of the 5′-untranslated region of mRNA, *J. Biol. Chem.,* 265, 6521, 1990.
51. **Nonaka, M., Ishikawa, N., Passwell, J., Natsuume-Sakai, S., and Colten, H. R.,** Tissue specific initiation of murine complement factor B mRNA transcription, *J. Immunol.,* 142, 1377, 1989.
52. **Garnier, G., Ault, B., Kramer, M., and Colten, H. R.,** Cis and trans elements differ among mouse strains with high and low extrahepatic complement factor B gene expression, *J. Exp. Med.,* 175, 471, 1992.
53. **Bentley, D. R.,** Primary structure of human complement component C2: homology to two unrelated protein families, *Biochem. J.,* 239, 339, 1986.
54. **Whitehead, A. S., Solomon, E., Chambers, S., Bodmer, W. F., Povey, S., and Fey, G.,** Assignment of the structural gene for the third component of human complement of chromosome 19, *Proc. Natl. Acad. Sci. U.S.A.,* 79, 5021, 1982.
55. **DaSilva, F. P., Hoecker, G. E., Day, N. K., Vienne, D., and Rubinstein, P.,** Murine complement component 3: genetic variation and linkage to H-2, *Proc. Natl. Acad. Sci. U.S.A.,* 75, 963, 1978.
56. **Khan, S. A. and Erickson, B. W.,** An equilibrium model of the metastable binding sites of alpha 2 macroglobulin and complement proteins C3 and C4, *J. Biol. Chem.,* 257, 11864, 1982.
57. **Iijima, M., Tobe, T., Sakamoto, T., and Tomita, M.,** Biosynthesis of the internal thioester bond of the third component of complement, *J. Biochem.,* 96, 1539, 1984.
58. **Alper, C. A., Abramson, N., Johnston, R. B., Jandl, J. H., and Rosen, F. S.,** Increased susceptibility to infection associated with abnormalities of complement-mediated functions and of the third component of complement (C3), *N. Engl. J. Med.,* 282, 349, 1970.
59. **Dieli, F., Lio, D., Sereci, G., and Salerno, A.,** Genetic control of C3 production by the S region of the mouse MHC, *J. Immunogenet.,* 15, 339, 1988.
60. **Kawamura, N., Singer, L., Wetsel, R. A., and Colten, H. R.,** Cis- and transacting elements required for constitutive and cytokine (IL-1/IL-6) regulated expression of the murine complement C3 gene, *Biochem. J.,* 283, 705, 1992.
61. **Wilson, D. R., Juan, T. S. C., Wilde, M. D., Fey, G. H., and Darlington, G. J.,** A 58-base pair region of the human C3 gene confers synergistic inducibility by interleukin-1 and interleukin-6, *Mol. Cell. Biol.,* 10, 6181, 1990.
62. **Tack, B. F., Janatova, J., and Thomas, M. L.,** The third, fourth, and fifth components of human complement: isolation and biochemical properties, *Methods Enzymol.,* 80, 64, 1981.
63. **Hugli, T. E.,** Biochemistry and biology of anaphylotoxins, *Complement,* 3, 111, 1986.
64. **Thomas, M. L., Janatova, J., Gray, W. R., and Tack, B. F.,** Third component of human complement: localization of the internal thiolester bond, *Proc. Natl. Acad. Sci. U.S.A.,* 79, 1054, 1982.
65. **Wetsel, R. A. and Barnum, S. R.,** Molecular biology and biochemistry of the third (C3) and fifth (C5) complement components, in *Biochemistry and Molecular Biology of Complement,* Sim, R. B., Ed., MTP Press, Lancaster, in press.
66. **Colten, H. R. and Alper, C. A.,** Hemolytic efficiencies of genetic variants of human C3, *J. Immunol.,* 108, 1184, 1972.

67. **Pangburn, M. K.,** The alternative pathway, in *Immunobiology of the Complement System,* Ross, G. D., Ed., Academic Press, New York, 1986, 45.
68. **Muller-Eberhard, H. J.,** The membrane attack complex of complement, *Annu. Rev. Immunol.,* 4, 503, 1986.
69. **Strunk, R. C., Whitehead, A. S., and Cole, F. S.,** Pretranslational regulation of the synthesis of the third component of complement in human mononuclear phagocytes by the lipid A portion of lipopolysaccharide, *J. Clin. Invest.,* 76, 985, 1985.
70. **St. John Sutton, M. B., Strunk, R. C., and Cole, F. S.,** Regulation of the synthesis of the third component of complement and factor B in cord blood monocytes by lipopolysaccharide, *J. Immunol.,* 136, 1366, 1986.
71. **Katz, Y., Cole, F. S., and Strunk, R. C.,** Synergism between gamma interferon and lipopolysaccharide for synthesis of factor B, but not C2, in human fibroblasts, *J. Exp. Med.,* 167, 1, 1988.
72. **Beutler, B., Krochin, N., Milsark, I. W., Luedke, C., and Cerami, A.,** Control of cachectin (tumor necrosis factor) synthesis: mechanisms of endotoxin resistance, *Science,* 232, 977, 1986.
73. **Schien, J.,** unpublished data.
74. **Falus, A., Wakeland, E. K., McConnell, T. J., Gitlin, J. D., Whitehead, A. S., and Colten, H. R.,** DNA polymorphism of MHC III genes in inbred and wild mouse strains, *Immunogenetics,* 25, 290, 1987.
75. **Dinarello, C. A.,** Interleukin-1, *Rev. Infect. Dis.,* 6, 51, 1984.
76. **Perlmutter, D. H., Goldberger, G., Dinarello, C. A., Mizel, S. B., and Colten, H. R.,** Interleukin-1 regulates class III major histocompatibility complex (MHC) gene products, *Science,* 232, 850, 1986.
77. **Perlmutter, D. H., Colten, H. R., Adams, S. P., May, L. T., Sehgal, P. B., and Fallon, R. J.,** A cytokine selective defect in IL-1β mediated acute phase gene expression in a subclone of the human hepatoma cell line (HepG2), *J. Biol. Chem.,* 264, 7669, 1989.
78. **Nonaka, M., Gitlin, J. D., and Colten, H. R.,** Regulation of human and murine complement: comparison of 5′ structural and functional elements regulating human and murine complement factor B gene expression, *Mol. Cell. Biochem.,* 89, 1, 1989.
79. **Beuscher, H. U., Fallon, R. J., and Colten, H. R.,** Macrophage membrane IL-1 regulates the expression of acute phase proteins in human hepatoma Hep3B cells, *J. Immunol.,* 139, 1896, 1987.
80. **Katz, Y. and Strunk, R. C.,** Interleukin-1 and tumor necrosis factor: similarities and differences in stimulation of expression of alternative pathway of complement and interferon-beta 2/interleukin-6 genes in human fibroblasts, *J. Immunol.,* 142, 3862, 1989.
81. **Baumann, H., Onorato, V., Gauldie, J., and Jahreis, G. P.,** Distinct sets of acute phase plasma proteins are stimulated by separate human hepatocyte stimulatory factors and monokines in rat hepatoma cells, *J. Biol. Chem.,* 262, 9756, 1987.
82. **Strunk, R. S., Tashjian, A. H., and Colten, H. R.,** Complement biosynthesis in vitro by rat hepatoma cell strains, *J. Immunol.,* 114, 331, 1975.
83. **Wu, L.-C., Morley, B. J., and Campbell, R. D.,** Cell specific expression of the human complement protein factor B gene: evidence for the role of two distinct 5′ flanking elements, *Cell,* 48, 331, 1987.
84. **Perlmutter, D. H., Dinarello, C. A., Punsal, P. I., and Colten, H. R.,** Cachectin/tumor necrosis factor regulates hepatic acute phase gene expression, *J. Clin. Invest.,* 78, 1349, 1986.
85. **Beutler, B. and Cerami, A.,** The biology of cachectin/tumor necrosis factor — a primary mediatory of the host response, *Annu. Rev. Immunol.,* 7, 625, 1989.
86. **Gauldie, J., Richards, C., Harnish, D., Landsdorp, P., and Baumann, H.,** Interferon beta 2/B-cell stimulatory factor type 2 shares identity with monocyte-derived hepatocyte-stimulating factor and regulates the major acute phase protein response in liver cells, *Proc. Natl. Acad. Sci. U.S.A.,* 84, 7251, 1987.
87. **Kishimoto, T. K., Larson, R. S., Corbi, A. L., Dustin, M. L., Stauton, D. E., and Springer, T. A.,** The leukocyte integrins, *Adv. Immunol.,* 46, 149, 1989.
88. **Perlmutter, D. H.,** IFNβ2/IL-6 is one of the several cytokines that modulate acute phase gene expression in human hepatocytes and human macrophages, *Ann. N.Y. Acad. Sci.,* 557, 332, 1989.
89. **Akira, S., Hirano, T., Taga, T., and Kishimoto, T.,** Biology of multifunctional cytokines: IL-6 and related molecules (IL-1 and TNF), *FASEB J.,* 4, 2860, 1990.
90. **Baumann, H., Isseroff, H., Latimer, J. J., and Jahreis, G. P.,** Phorbol ester modules IL-6 and IL-1 regulated expression of acute phase plasma proteins in hepatoma cells, *J. Biol. Chem.,* 263, 17390, 1988.
91. **Katz, Y., Revel, M., and Strunk, R. C.,** IL-6 stimulates synthesis of complement proteins factor B and C3 in human skin fibroblasts, *Eur. J. Immunol.,* 19, 983, 1989.
92. **Circolo, A., Pierce, G. F., Katz, Y., and Strunk, R. C.,** Antiinflammatory effects of polypeptide growth factors: PDGF, EGF and FGF inhibit the cytokine induced expression of the alternative complement pathway activator factor B in human fibroblasts, *J. Biol. Chem.,* 265, 5066, 1990.
93. **Miura, N., Prentice, H. L., Schneider, P. M., and Perlmutter, D. H.,** Synthesis and regulation of the two human complement C4 genes in stable transfected mouse fibroblasts, *J. Biol. Chem.,* 262, 7298, 1987.

94. **Strunk, R. C., Cole, F. S., Perlmutter, D. H., and Colten, H. R.,** Gamma interferon increases expression of class III complement genes C2 and factor B in human monocytes and in murine fibroblasts transfected with human C2 and factor B genes, *J. Biol. Chem.,* 260, 15280, 1985.
95. **Katz, Y. and Strunk, R. C.,** Synthesis and regulation of complement protein factor H in human skin fibroblasts, *J. Immunol.,* 141, 559, 1988.
96. **Katz, Y. and Strunk, R. C.,** Synthesis and regulation of C1 inhibitor in human skin fibroblasts, *J. Immunol.,* 142, 2041, 1989.
97. **Ripoche, J., Mitchell, A., Erdei, A., Madin, C., Moffatt, B., Mokoena, T., Gordon, S., and Sim, R. B.,** Interferon-gamma induces synthesis of complement alternative pathway proteins by human endothelial cells in culture, *J. Exp. Med.,* 168, 1917, 1988.
98. **Kulics, J., Colten, H. R., and Perlmutter, D. H.,** Counter-regulatory effects of interferon gamma and endotoxin on expression of the human C4 genes, *J. Clin. Invest.,* 85, 943, 1990.
99. **Strunk, R. C. and Colten, H. R.,** Regulation of complement synthesis in mononuclear phagocytes, in *The Reticulo-Endothelial System: A Comprehensive Treatise,* Filkins, J. P. and Reichard, S. M., Eds., Plenum Press, New York, 1985, 25.
100. **Celada, A., Klemsz, M. J., and Maki, R. A.,** Interferon-gamma activates multiple pathways to regulate the expression of the genes for MHC II I-A beta, tumor necrosis factor and C3 in mouse macrophages, *Eur. J. Immunol.,* 19, 1103, 1989.
101. **Strunk, R. C. et al.,** unpublished data.
102. **Garnier, G. and Colten, H. R.,** unpublished data.

Chapter 12

RAT α_2-MACROGLOBULIN AND RELATED α-MACROGLOBULINS IN THE ACUTE PHASE RESPONSE

Ronald C. Roberts

TABLE OF CONTENTS

0-8493-6913-4/93/$0.00 + $.50

I. INTRODUCTION

The inclusion of a chapter on the rat α-macroglobulin family of proteins in a book devoted to the acute phase reaction is most appropriate. This group of three proteins, all of which apparently have the same general function as endoproteinase inhibitors, undergo markedly different quantitative changes in blood plasma during the acute phase response. This phenomenon, insofar it involves three different α-macroglobulin family members, is currently only known to occur in the rat. Although extensive studies in other species have only been done in humans. We are certain that the analogous situation does not occur in man, where the major α-macroglobulin, α_2-macroglobulin (α_2-M), is a constitutive protein whose concentration does not undergo major changes during the acute phase reaction. In the rat, however, the protein now designated as α_2-M has been recognized since the mid-1960s as the major positive acute phase reactant in plasma.[1-3] By the original methods, this protein rose from undetectable levels in the normal adult rat to a prominent electrophoretic or immunoelectrophoretic band within 24 h of injury. At the same time, the concentration of a protein of similar electrophoretic and immunological properties, later known as α_1-macroglobulin (α_1-M), was noted to change very little during the acute phase reaction. The proteinase-inhibiting properties of these two proteins were rapidly demonstrated after human α_2-M was found to be a plasmin inhibitor[4] and able to form complexes with trypsin.[5] The third member of the α-macroglobulin family in the rat was not recognized to be such until the middle 1980s. This protein, named α_1-inhibitor$_3$ (α_1-I_3) by Gauthier and Ohlsson[6] or murinoglobulin by Saito and Sinohara,[7] was recognized by Lonberg-Holm et al.[8] to be a major negative acute phase reactant in rat plasma. Rat α_1-I_3 is the major globulin of normal rat plasma at about 7.5 mg/ml, but during the acute phase reaction its concentration declines about 0.5 mg/ml at a rate of decline nearly reciprocal to the rise in the α_2-M concentration.

This chapter briefly reviews the history of the discovery of these proteins and discusses the *in vivo* kinetics of the acute phase reactions for these proteins, their structure, and the relationship of various structural elements to the protease inhibition function of these proteins. The α-macroglobulins are multifunctional proteins which also bind divalent metal ions such as zinc and nickel, bind many cytokines, and modulate the immune response and the inflammatory reaction. The functions of these interesting proteins other than proteinase inhibition are also briefly reviewed, with emphasis on what has specifically been found for the rat proteins. Attention is also directed to other comprehensive reviews of α_2-M[9] and the α-macroglobulin family of proteins.[10] Geiger et al.[11] have written a concise review of the rat α-macroglobulins.

II. BRIEF HISTORY OF THE RAT α-MACROGLOBULINS

Electrophoresis on solid supports — first paper, then agarose, starch, and polyacrylamide — progressively improved the separation of plasma proteins, leading to many studies seeking changes in the electrophoretogram patterns of serum or plasma proteins in various disease states. Serum proteins were initially labeled according to their mobility at alkaline pH in the moving boundary electrophoresis apparatus of Tselius, with the designation of albumin for the largest and most anionic peak, then α-, β-, and γ-globulins in order of decreasing mobility toward the anode. As the resolution improved the α-peak was split into α_1- and α_2-peaks using paper, agarose, or cellulose acetate for the support media. There was no molecular sieving in these media, so separation was primarily on the basis of net charge. The introduction of starch gel and then polyacrylamide gels as supports resulted in separation by size as well as charge and resolved the blood proteins into many more bands. The molecular sieving became progressively more pronounced at molecular weights over about

70 kDa in these nondenaturing gel systems, with proteins of around 1 million barely moving into the gel. References to α-, β-, and γ-mobility to new proteins were then established by determining their mobility on one of the nonmolecular sieving media. With this brief summary, perhaps the original logic of the nomenclature of these proteins may be better appreciated.

The first identification of the protein which would eventually be recognized as rat α_2-M was made by Darcy,[12] who identified a precipitin line in Ouchterlony double-diffusion plates that was present in sera from rats with malignant tumors, but not present in normal rats. The antisera was made by immunizing rabbits with serum from rats bearing transplanted Walker tumors. Further work by Darcy[13] established that this protein was not cancer specific, but also occurred in partially hepatectomized rats during liver regeneration, during skin regeneration in injured rats, and in pregnant and newborn rats. The mobility of this protein was found to be α to fast-β by immunoelectrophoresis. Darcy speculated that the protein was associated with tissue growth in general. Heim[14] and Beaton et al.[15] used the higher-resolving vertical starch gel electrophoretic technique to demonstrate a distinct protein component, called slow α_2-globulin, in the sera of pregnant and neonatal rats independently. This protein was soon recognized to be the same as the one described by Darcy, but was not yet associated with the acute phase reaction. Others studying the relationship of the appearance of this protein with other physiologic and pathologic events included Boffa et al.[16] Heim and Lane[17] and Weimer and Benjamin[18] were the first to establish that rat α_2-M was an acute phase reactant. Weimer and Benjamin introduced a new name for the protein, α_2-acute phase globulin, which has been used in a number of subsequent publications.

The similarity of slow α_2-globulin with the α_2-M of human and other species was first pointed out by Heim.[19] This homology subsequently became well established by a variety of studies. The existence of two rat α-macroglobulins, the normal constitutive protein α_1-M and the acute phase reactant α_2-M, was also established by a number of studies during this time period (the late 1960s). Rat α_2-M showed greater immunological identity with human α_2-M than rat α_1-M. Once the relationship of the two rat proteins to human α_2-M was established, further studies on their isolation and characterization as well as functional studies on their protese-inhibiting properties followed, and paralleled similar studies on the human α_2-M.

The realization that there was a third member of the rat α-macroglobulin family and that it was a major negative acute phase reactant did not occur until the mid-1980s. The relationship of α_1-I_3 to the other two α-macroglobulins was first pointed out by Lonberg-Holm et al.,[8,20] by similarities in N-terminal sequence, the presence of a thiol ester bond, and its protease-binding properties. This relationship was further substantiated by the degree of homology found in the sequence of the messenger RNA[21] and further studies on the mode of protease binding.[22]

III. BEHAVIOR DURING THE ACUTE PHASE REACTION

A number of studies have measured the change in concentration of α_2-M and α_1-M in the blood of rats using a variety of methods to induce an inflammatory response. In fact, measurement of α_2-M concentration was promoted as a method of estimating the intensity of the inflammatory response (see, e.g., Reference 23). The strong correlation between the amount of α_2-M synthesis and degree of trauma, as well as the obligatory requirement for glucocorticords to obtain increased synthesis of α_2-M, was shown in this early work. Interleuken-6 is now known to be the principle trigger initiating α_2-M synthesis in the rat.[24]

The concentration of α_2-M in blood plasma can be increased over 100-fold within 48 h in cases of severe injury. Over the same time period, the concentration of α_1-I_3 falls about

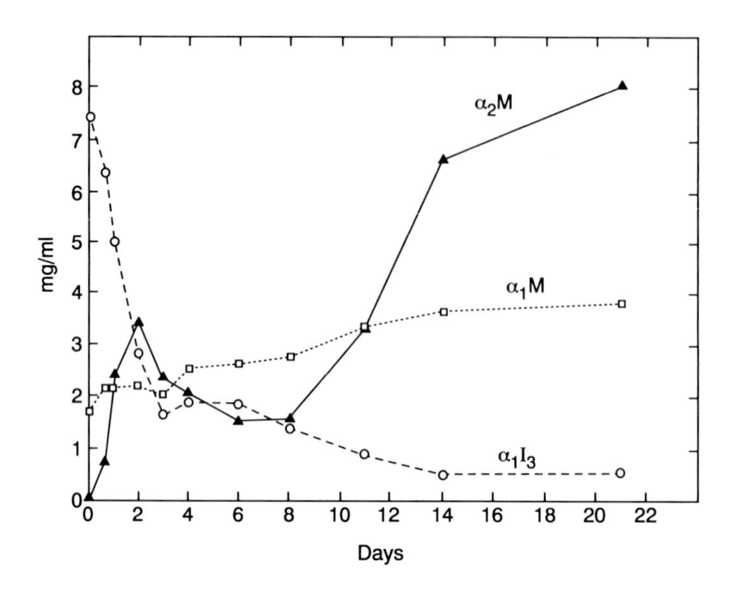

FIGURE 1. Immunochemical determination of changes in the three rat α-macroglobulins following induction of adjuvant arthritis in the rat. (From Lonberg-Holm, K., Reed, D. L., Roberts, R. C., Hebert, R. R., Hillman, M. C., and Kutney, R. M., *J. Biol. Chem.*, 262, 440, 1987. With permission.)

the same amount, while the concentration of α_1-M changes very little. A representative study showing the simultaneous change in all three of these proteins after induction of adjuvant arthritis is shown in Figure 1.[8] The proteins were quantitated by rocket immunoelectrophoresis. The changes were followed over a period of 3 weeks and reveal that after the peak of 3.5 mg/ml at 48 h, the α_2-M concentration declined about 50%, then rose to about 8 mg/ml at 21 d. The total increase was from less than 40 to 8000 μg/ml, an increase of 200-fold. During the same period, α_1-I_3 decreased from about 7.5 to 0.5 mg/ml. The α_1-M concentration, on the other hand, increased less than twofold. The second rise in α_2-M paralleled the development of polyarthritis, as evidenced by the swelling of joints in all extremities. Bacterial endotoxin (25 μg/kg intravenously or 250 μg/kg intraperitoneally) produced a maximum response at day 2 of about 3.5 mg/ml and declined to starting values in about 1 week. Thus, the development of a chronic inflammatory state such as adjuvant-induced arthritis produces a secondary α_2-M response of greater magnitude than the primary response.

IV. PHYSICOCHEMICAL PROPERTIES

A. THE PRIMARY STRUCTURE

All rat α-macroglobulins are assembled from single-subunit glycosylated polypeptide chains with relative molecular weights of about 180,000. The polypeptide chains of each of these proteins have regions shared in common which are important for the various activities identified for these proteins. These regions include the sequence giving rise to a hallmark feature of this family of proteins, the β-cysteinyl-γ-glutamyl thiol ester bond, which is directly involved in the proteinase inhibitor functions of these proteins, as discussed below. Other important regions of the polypeptide chain include the "bait" region, which is involved in the proteinase inhibitor activity and the cell receptor recognition site involved in the specific clearance of the proteins after they have complexed with proteinases. The interrelationships of the positions of these features on the linear subunit chains are shown diagrammatically in Figure 2. The bait region is located approximately in the center of the

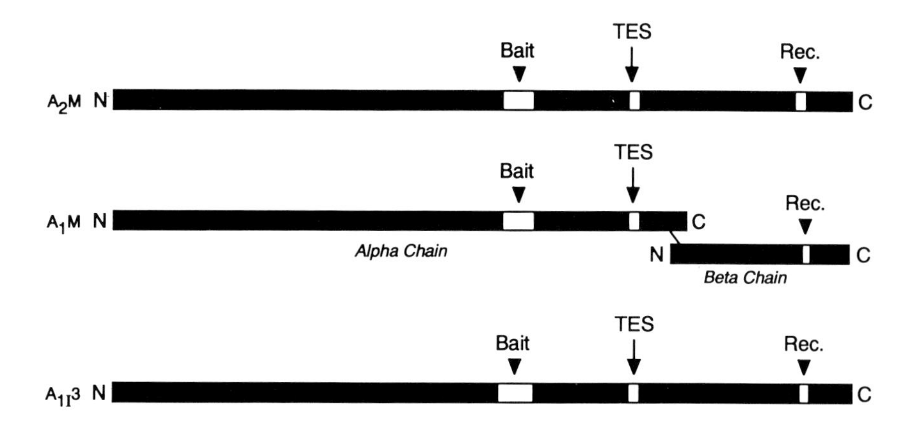

FIGURE 2. Schematic illustration of the similarity in the location of the bait region, thiol ester bond,and cell receptor recognition site in subunits of each of the three rat α-macroglobulins. The α_1-M subunit is shown after the posttranslational proteolytic cleavage as a disulfide-linked α- and β-chain.

TABLE 1
Primary Structure Comparisons of the Rat α-Macroglobulins

Property	α_2-M	Pro-α_1-M	α_1-I_3
Number of AA res. (mature protein)	1445	1476	1453
Polypeptide Chain mol wt	160,670	164,368	161,053
% overall homology	57.2	100	53.0
% similarity	72.3	100	68.3
Number of ½ Cys.	25	21	23

chain in each case, the thiol ester site about two thirds of the distance from the N terminal to the C terminal, and the receptor recognition site is near the C terminal. The α_1-M subunit chain undergoes an additional posttranslational proteolytic cleavage which occurs sometime during assembly of the final mature protein; therefore, the intact α_1-M subunit is properly referred to as the pro-α_1-M subunit. The relative locations of the bait region, thiol ester bond, and cell receptor recognition site are the same on the pro-α_1-M subunit chain, but change after the cleavage to the mature α- and β-chains. The cell receptor recognition site ends up on the β-chain, and the bait regions and thiol ester bond sites are shifted toward the carboxyl terminal end of the α-chain.

The amino acid sequences of these subunit chains have all been deduced from their cDNA sequences. The sequence for rat α_2-M was published by Gehring et al.,[25] that for α_1-I_3 by Braciak et al.,[21] and that for α_1-M by Eggertsen et al.[26] The sequences show greater than 50% overall homology to each other and to human α_2-M. This homology is conserved most rigidly in regions identified as contributing to specific functions of the protein — with one important exception, the bait region. The conserved areas include the location of the disulfide bridges, the sequence giving rise to the β-cysteinyl-γ-glutamyl thiol ester bond, and a cell receptor recognition sequence. The bait regions, while all in the center of the subunit chains, show the greatest sequence diversity. A number of the statistical features of the amino acid sequences for the three proteins are compared in Table 1. These proteins are all glycosylated; however, the details of the carbohydrate chain composition and the number of chains attached per subunit have not yet been determined. Jamieson et al.[27] reported that rat α_2-M has a carbohydrate content of 15.9%. Gauthier and Mouray[28] reported 4.25% hexose, 3.4% glucosamine, 2% sialic acid, and 0.2% fucose for α_2-M. Gordon[29] compared

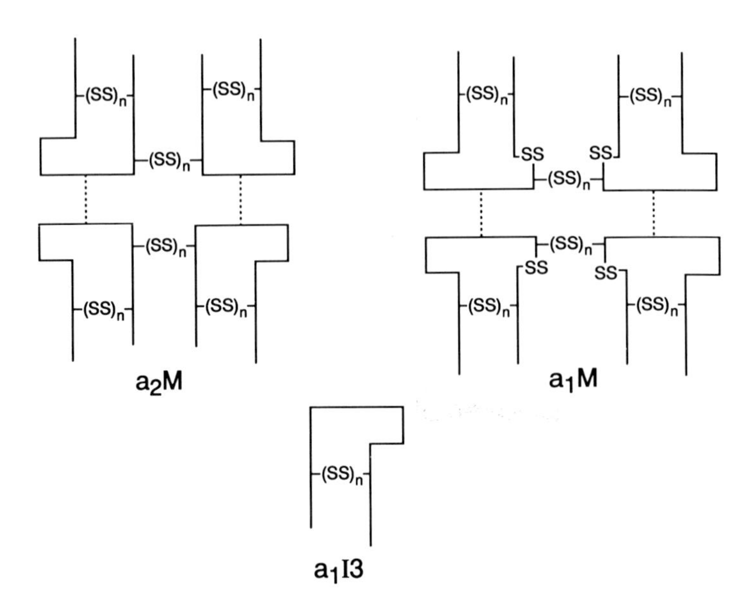

FIGURE 3. Schematic comparison of the subunit organization of the three rat α-macroglobulins. α_2-M, upper left; α_1-M, upper right; α_1-I_3, center bottom. The arrangement of the subunits and the disulfide linkages are shown only to indicate which components are thus linked, and do not reflect their actual positions in the molecule. The dashed lines in the α_2-M and α_1-M diagrams indicate nocovalent salt linkages.

the carbohydrate content of rat α_1-M and α_2-M, and found 5.44 and 6.26% mannose plus galactose, respectively, and 2.60 and 2.61% glucosamine, respectively. Fucose could not be detected.

B. QUATERNARY STRUCTURE

The organization of the subunit polypeptide chain into the mature circulating active protein differs somewhat for each of the rat α-macroglobulins. This final organization for each of the proteins is illustrated in the line diagrams of Figure 3. In α_2-M, two of the basic subunit chains are combined by disulfide bonds into a dimer, and two of these dimers are combined through noncovalent linkages into a protein with a relative molecular weight of about 720,000.[29-33] The rat α_2-M quaternary structure organization is identical to that of human α_2-M. Rat α_1-M also is a tetramer of the basic subunit chain, organized as two disulfide-linked dimers held together by noncovalent forces. However, since the pro-α_1-M subunit chain undergoes at least one posttranslational proteolytic cleavage between amino acid residues 1221 and 1222,[26] reduced SDS-polyacrylamide gels show bands at approximate M_rs of 156,000 and 42,000 instead of the usual 180-kDa band for α_1-M. These two fragments, referred to as α- and β-chains, respectively, are held together by a disulfide bridge.[32,33] This disulfide bridge has been predicted to be between Cys_{1320} (β-chain, using the residue-numbering system based on the pro-α_1-M chain) and either Cys_{911} or Cys_{253} of the α-chain.[26] Finally, α_1-I_3 exists as the monomer,[8] the only member of the α-macroglobulin family known to do so. This remarkable difference from the other α-macroglobulins leads to differences in its interaction with proteinases and speculation concerning its true physiological function, as discussed further below.

C. THE THIOL ESTER BOND

The presence of an internal thiol ester bond in these proteins is a unique characteristic of this family of proteins. Also included in this family are the complement proteins C3, C4, and C5. C5 is the only member of this structurally and functionally related family that does not contain the thiol ester group. (see Reference 10 for a comprehensive review of the evidence for the existence of this novel posttranslational modification and its chemical properties.) The increased reactivity of the thiol ester bond toward nucleophilic attack by ϵ-amino groups of proteinases after conformation changes precipitated by the attack of the proteinase on the bait regions of the molecules is strong circumstantial evidence for proteinase inhibition being an important physiological function of α_1-M and α_2-M. The possibility that α_1-I_3 may also covalently link with other proteins besides proteinases in a similar manner has been suggested by Gehring et al.[25]

The thiol ester bond is a β-cysteinyl-γ-glutamyl bond between a cysteinyl residue and a glutamyl residue separated by glycinyl and glutamyl residues. The sequence, CGEQ, which forms the thiol ester bond is found in both rat and human α_2-M at positions 970 to 973.[25] In α_1-I_3, the sequence is located at positions 951 to 954.[21] The thiol ester sequence in α_1-M is at positions 962 to 965.

The chemical properties resulting from the presence of the thiol ester bond in these proteins include (1) the incorporation of 1 mol of small nucleophiles per thiol ester bond, (2) heat-induced autolytic cleavage of the peptide chain at the site of the glutaminyl residue, and (3) after proper activation, the ability to form γ-glutamyl peptide bonds through amino groups (generally the ϵ-amino group of lysyl residues) of other proteins. The latter property leads to the frequent formation of covalent linkages between α-macroglobulins and the proteinases which have been complexed with them.

D. THE PROTEASE INHIBITOR FUNCTION

1. The Trap Mechanism of Proteinase Inhibition

The interaction of proteinases with the α-macroglobulins has been the subject of intense investigation since the ability of human α_2-M to form complexes with proteinase, so that its active site is still available to low molecular weight substrates, was discovered.[5,34] Most proteinase inhibitors function by forming a stable complex between the active site of the protease and the binding site on the inhibitor. This requirement for multiple specific interactions limits these inhibitors to a narrow range of proteinases that they can inhibit. Human α_2-M and its homologs in other species, on the other hand, are able to inhibit a wide range of endoproteinases of all mechanistic classes by entrapping them. This entrapment leaves the active site open to low molecular weight substrates or inhibitors specific for the proteinase involved. In the process of forming the α_2M-proteinase complex, the basic 180-kDa subunit chain of α_2M is cleaved approximately in the middle, giving rise to two 90-kDa chains. In most cases, two of the four subunits are cleaved per molecule of proteinase bound, suggesting that there is one active proteinase site for each dimer of subunit chains or two per intact α_2-M tetramer. In addition, for each molecule of proteinase bound, two thiol ester bonds are broken, leading to the generation of two new thiol groups and, frequently, to covalent linkage of the proteinase via the formation of γ-glutamyl-ϵ-lysyl bonds. Cleavage of the subunit chains and thiol ester bonds of α_2M initiates a rapid change in conformation to a more compact structure, the "fast" electrophoretic form. This conformation change leads to the exposure of an endothelial cell receptor recognition site on α_2-M complexed with proteinase, which allows the rapid endocytosis of the complexes and their clearance from the circulation. This receptor recognition site is apparently located in a papain-generated, 140-residue, C-terminal peptide fragment for both rat and human α_2-M and α_1-I_3,[26] and presumably in the α_1-M β-chain as well. All three of the rat α-macroglobulins are recognized by the same cell receptor via this cell receptor recognition site,[35] confirming the conservation of this feature of the proteins.

The overall hypothesis tying all these observations together to explain the mechanism of proteinase inhibition for this class of inhibitors is that the α-macroglobulins possess "bait" regions. These "bait" regions are probably hydrophilic, flexible loops on the surface of the protein which are highly susceptible to cleavage by a variety of endopeptidases. Nuclear magnetic resonance studies by Gettins and Cunningham[36] provide physical evidence supporting the flexibility of the bait region. Cleavage of this bait region leds to conformational changes somehow activating the thiol ester bond, whose cleavage further destabilizes the structure and covalently links the attacking proteinase while the α-macroglobulin folds around the proteinase, engulfing and entrapping it. This trap hypothesis was originally proposed by Barrett and Starkey[37] in 1973, and for the most part has been supported and further refined by many subsequent studies. The basic entrapping unit for the tetrameric forms of the α-macroglobulins is believed to be the noncovalently linked dimers, thus giving the tetrameric forms the potential of binding 2 mol of proteinase. The actual binding capacity of the tetrameric α-macroglobulins for proteinases varies between 1 and 2 mol/mol, depending on the initial rate of reaction of the protease with the α-macroglobulin, the concentrations of the inhibitor and enzyme, and the relative size of the proteinase.[38]

Whether a proteinase is inhibited by an α-macroglobulin appears to depend primarily upon the bait region having the appropriate residues that meet the specificity requirements of the proteinase. To a lesser degree, proteinases of a molecular size on the order of 100 kDa or more may not be inhibited, even though they efficiently cleave the bait region, because they are too large to be effectively entrapped. Regarding this limitation on the specificity of the inhibitors, it is of interest that the least conserved regions of the peptide chains among the α-macroglobulin homologs appear to be the bait regions.[11] This variation has led to speculation that despite the wide range of proteinases that are inhibited by these proteins, the different members of the family do have different proteinase-inhibiting spectra.

With this general background about the proteinase-inhibiting properties of the α-macroglobulins in general, we now consider more specifically what is known about the different forms of the rat α-macroglobulins regarding their interaction with proteinases.

2. Comparisons of the Bait Regions and Proteinase-Inhibiting Spectra

The sequence diversity and identification of the specific cleavage sites in the rat α-macroglobulins (along with human α_2-M and human pregnancy zone protein) for a variety of proteinases was explored in depth by Sottrup-Jensen et al.[39] Sottrup-Jensen and Birkedal-Hansen[40] extended these studies to include the bait region cleavage sites for human fibroblast collagenase for the same α-macroglobulins. The positions for the cleavage sites for papain and human neutrophil elastase in the bait region of α_1-I_3 were determined by Enghild et al.[22] The proteinases for which bait region cleavage sites for each of the α-macroglobulins have been established are listed in Table 2, and a diagram of the bait region sequence showing the different cleavage sites is presented in Figure 4.

Even though human neutrophil elastase cleaves the bait region of α_1-I_3, Enghild et al.[22] found that this monomeric form of α-macroglobulin did not inhibit this proteinase, apparently because it does not contain a lysine residue to form a covalent cross-link with the glutamyl residue of the thiol ester bond. Thermolysin, on the other hand, did not cleave the bait region of α_1-I_3 because of the lack of a suitable cleavage site.

Even though there are suitable cleavage sites for the various proteinases in the bait regions of the different α-macroglobulins, the rates of cleavage and consequently the rates of inhibition can vary considerably. As pointed out by Travis and Salvesen,[41] a proteinase inhibitor can only be considered an effective inhibitor for a given proteinase if it inhibits it at a rate comparable to that at which it hydrolyzes its natural substrates. While the rates of inhibition of human fibroblast collagenase by rat α_1-M and α_2-M have been estimated to be greater than $10^6\ M^{-1}\ s^{-1}$, which is sufficiently fast to be a good inhibitors,[40] there is very little information about the rates of inhibition for other proteinases by the rat α-macroglob-

TABLE 2
Proteinases Whose Bait Region Cleavage Sites Have Been Determined for the Rat α-Macroglobulins

Proteinase	Abbrev.[a]	Class
Collagenase, *Clostridium histolyticum*	cc	Metallo-
Collagenase, human fibroblast	fc	Metallo-
Trypsin, bovine	tr	Serine
Thrombin, bovine	th	Serine
Trypsin, *Streptomyces griseus*	st	Serine
Proteinase B, *S. griseus*	sb	?
Elastase, porcine pancreatic	pe	Elastase
Cathepsin G, human	cg	Serine
Chymosin, bovine	cs	Aspartic
Chymotrypsin, bovine	ct	Serine
V-8 proteinase, *Staphylococcus aureus*	sp	Serine
Subtilisin	su	Serine
Elastase, human leukocyte	he	Serine
Thermolysin	tl	Metallo-
Papain	pa	Cystiene

[a] Abbreviations are those used in Figure 4.

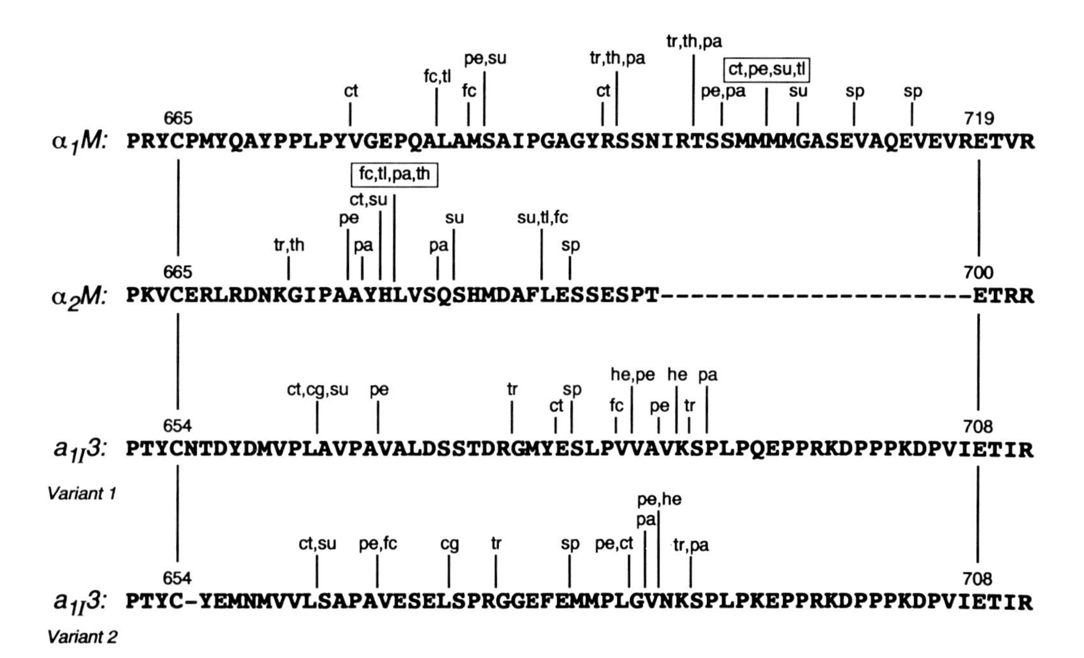

FIGURE 4. Comparison of the cleavage sites for proteinases in the bait regions of the rat α_1-M, α_2-M,and two main variants of α_1-I_3. The bold capital letters represent the one-letter-code amino acid sequences for each of the bait regions. The cleavage points for the various proteinases are indicated by the short vertical lines leading to the lower-case abbreviations of the proteinases. The proteinases associated with the abbreviations may be found in Table 2. The sequence for α_2-M is from Gehring et al.,[25] the α_1-M sequence from Eggertsen et al.,[26] and the α_1-I_3 sequence from Braciak et al.[21] The numbering of the sequences and the gaps introduced in the α_2-M and α_1-I_3 variants are as presented by Eggertsen et al.[26] to align with the α_1-M sequence. (Adapted from Sottrup Jensen, L., et al., *J. Biol. Chem.*, 264, 15781, 1989; Eggertsen, G., et al., *Mol. Biol. Med.*, 8, 287, 1991.)

ulins. In fact, there have not been any studies on the ability of these proteins to inhibit specific rat proteinases, particularly those released during the inflammatory process. However, Horne et al.[42] concluded that only rat α_1-M, and not α_2-M, inhibited plasmin *in vivo*. Streptokinase and human or rat plasmin infusion into rats, or streptokinase infusion into rats with immune complex nephritis, produced decreases in α_1-M (slow α_1-globulin), but not α_2-M. The decrease in α_1-M was attributed to the rapid clearance of α_1-M-plasmin complexes. The interaction of plasmin with the rat α-macroglobulins in purified protein systems apparently has not been studied yet. A systematic comparative study of the interaction of rat endogenous extracellular proteinases is needed to provide further information on why the rat has maintained three different α-macroglobulins.

3. The Thiol Ester Bond and Conformational Changes

The conformational change resulting in the trapping mechanism for human α_2-M may be initiated by either proteolytic cleavage of the bait region or cleavage of the thiol ester bond by a low molecular weight nucleophile such as methylamine. This conformational change can be readily observed as a change from a slower to a faster electrophoretic form on nondenaturing polyacrylamide electrophoresis (for references, see Reference 9). After nucleophilic cleavage of the thiol ester bond, α_2-M no longer binds and inhibits proteinases. However, the bait region remains highly susceptible to proteolytic cleavage. Rat α_1-M behaves in an identical manner. However, methylamine treatment of rat α_2-M does not lead to the conformational change from a slow to a fast form, nor does it completely abolish the inhibition of trypsin by the trapping mechanism.[20,43] Consistent with the observation that methylamine does not cause a conformational change in rat α_2-M is the observation of Gliemann et al.[44] that methylamine-treated α_2-M is also not cleared by rat cells. This indicates that the receptor recognition site has not been exposed.

Bovine α_2-M is similar to rat α_2-M with regard to its resistance to conformational change following methylamine treatment. Bjork and Jornvall[45] have attributed this greater stability to changes in the amino acid residues three and four residues upstream from the Glu residue of the thioester bond. However, the rat α_2-M sequence in these positions is the same as human and rat α_1-M, suggesting that the causes of the differences in conformational stability upon cleavage of the thiol ester bond in these α-macroglobulins resides elsewhere in the molecule.

Rat α_1-I_3 electrophoretic mobility on nondenaturing polyacrylamide gels slows upon treatment with trypsin, but does not change after methylamine treatment. These differences are considered to be more related to the lower native molecular weight of this inhibitor and therefore less molecular sieving. Methylamine treatment abolishes its inhibiting activity.

V. PHYSIOLOGICAL FUNCTION

A. PROTEINASE INHIBITION

The fact that the α-macroglobulins have an elaborately developed mechanism for entrapping a wide range of proteinases and clearing them rapidly from the circulation through a common cell receptor found primarily on reticuloendothelial cells strongly suggests that proteinase inhibition is an integral part of the physiological function of these proteins. However, there also is ever-increasing evidence that these proteins are multifunctional, as discussed below.

Proteinase inhibitors make up a major fraction of the plasma proteins in vertebrate species. All of these inhibitors except the α-macroglobulins have a limited specificity indicating the proteinase they most likely control. Each of the proteinases of the coagulation pathway, complement cascade, and granulocytic proteinases released by white blood cells during an inflammatory reaction seem to have a specific proteinase inhibitor available to block it and limit its destructive capacity. The broad specificity of the α-macroglobulins allows them to act as backup inhibitors available when the more specific inhibitors are consumed during traumas which lead to major activation of the organism's proteolytic cascades. This may be particularly true in the case of the coagulation and fibrinolytic cascades which occur within the blood circulatory system. The presence of three different α-macroglobulins in the rat and other rodents suggests that these proteins play a greater role than simply backup inhibitors and relics of evolution.

The necessity of an organism having large quantities of proteinase inhibitors present during the acute phase reaction, to prevent widespread tissue destruction, has been appreciated for the last decade or more (see, e.g., Reference 46 for a review of the biological functions of acute phase proteins). Since α_2-M in the rat is, quantitatively, the major positive acute phase reactant in the plasma of the rat, and even α_1-M rises modestly it appears that these two proteins play a major role in protecting this species from uncontrolled proteolysis during the acute phase reaction.

A more subtle modulation of the inflammatory response due to proteinase inhibition by the α-macroglobulins (and other plasma proteinase inhibitors) is the reduction in the bioactive peptides which are also released by the limited proteolysis steps of the coagulation and complement cascades.

B. IMMUNOMODULATORY ACTIVITY

α_2-M has been shown to influence a wide variety of phenomena associated with the inflammatory reaction and the immune response in both *in vivo* assays in animals (usually rats or mice) and *in vitro* assays utilizing various cells of the immune system. The majority of these studies have been carried out on human α_2-M rather than the rat proteins currently under discussion. However, the high level of homology in both structure and other functions between these proteins in the two species suggests that rat α_2-M may behave similarly, particularly when the studies frequently used rat cells (for reviews of early studies on the influence of α_2-M on the immune system, see References 9 and 47). The effect of α_2-M may be either inhibitory or stimulatory, depending on the phenomenon being measured and, in some cases, the total concentration of α_2-M available.

Both homologous and heterologous α_2-M were found to promote leukopoiesis and erythropoiesis in lethally irradiated mice and significantly reduced the incidence of mortality.[48] Graham et al.[49] have suggested that this effect may be on the differentiation of lymphoreticular cells rather than on their proliferation. Human α_2-M was also found to promote restoration of humoral responsiveness in sublethally irradiated mice.[50] These *in vivo* studies strongly suggest that α_2-M plays a key role in facilitating the maintenance of the lymphoreticular cell systems, but the molecular mechanisms of these activities remain unknown. This could be an organism-wide manifestation of specific proteinase inhibitor, or be due to the preservation of a small amount of an essential proteinase activity by protecting it from inactivation by other proteinases. The latter was suggested by Teodorescu and co-workers[51] to be the source of polyclonal B-cell activator activity. In these studies, the polyclonal B-cell activator activity appeared to reside in α_2-M complexed with a proteinase (showing serine proteinase esterase activity) produced by T-cells in close proximity to the B-cells.[52]

Macrophages from a variety of sites synthesize α_2-M and have cell receptors for the fast (proteinase-complexed) form of the α-macroglobulins. Hoffman et al.[53] have shown that the fast form of α_2-M is also capable of modulating effector functions of mouse peritoneal macrophages. For example, the fast α_2-M antagonized the expression of interferon-γ-induced histocompatibility antigens (Ia). This effect was due to the interaction with the α_2-M receptors and not to the residual activity of the bound proteinases. In another study,[54] antigens conjugated to α_2-M were effectively endocytosed by murine macrophages via the α_2-M receptor. The stimulation of the T-cell proliferation response by macrophages fed the same amount of unconjugated antigen was less. The fact that a subpopulation of murine macrophages as well as liver cells also synthesize α_2-M, thereby ensuring an immediate local supply of the protein at inflammatory sites, supports a regulatory role for this protein.[55] Thus, it appears that the α_2-M interaction with its specific receptor, at least on macrophages, promotes some effects on the immune response.

α_2-M has been reported to be a potent inhibitor of antigen- and mitogen-induced T-cell proliferation,[56-60] the mixed lymphocyte reaction,[56] and cell-mediated cytotoxicity.[61,62] These immunosuppressive activities may result from either polypeptide cytokines (see below) associated with α_2-M or the residual activity of proteinases bound to α_2-M. However, Petersen et al.[63] have presented evidence which strongly suggests that these immunosuppressive activities arise from α_2-M inhibiting membrane-associated proteases necessary for the activation of these immune responses by the effector cells.

Rat α_2-M has been shown by Ufkes et al.[64] to be a selective inhibitor of antigen-induced leukotrienes in lung perfusion models. This activity explains the ability of α_2-M to inhibit the increase in pulmonary resistance during antigen-induced bronchoconstriction in rats *in vivo*. The mechanism of this inhibitory activity has not yet been clarified.

Rat α_2-M has been shown to inhibit complement activity in a complement-induced immune hemolysis test by Bellot et al.[65] The complement-inhibiting activity was not affected by the reaction of α_2-M with trypsin or methylamine to convert it to the fast form. The authors suggest that this activity may be due to inhibition of the regulatory activity of the complement component, factor H.

C. CYTOKINE BINDING

α_2-M have been found to specifically bind a number of growth factors and cytokines in recent years. The cytokines that thus far have been found to bind to α_2-M include growth hormone,[66] platelet-derived growth factor (PDGF),[67] transforming growth factors (TGFs) β_1 and β_2,[68] fibroblast growth factor (FGF),[69] interleuken-6 (IL-6),[70] interleukin-1β (IL-1β),[71] tumor necrosis factor-α,[72] and nerve growth factor (NGF).[73] The study of the nature of the binding sites, the form of α-macroglobulin involved, and the physiological consequences of this binding are actively being pursued in several laboratories. A thorough review of the status of current work on cytokine binding by α-macroglobulins has recently been prepared by Bonner and Brody.[74] Most of the studies have used human or bovine α_2-M. However, a rat α-macroglobulin produced by alveolar macrophages was found to bind PDGF.[75] Immunochemical methods have tentatively identified this protein as α_1-M.[80] Thus far, the physiological consequences of the cytokine binding has not been clarified. However, these findings have further strengthened the suspicions that α-macroglobulins have multiple physiological functions. Comparative studies of cytokine binding by the three different rat α-macroglobulins may be useful in clarifying the physiological roles of these proteins.

D. OTHER FUNCTIONS

1. Divalent Ion Binding

Human α_2-M is one of the major zinc-binding proteins of human plasma. The α-macroglobulins of other species retain this affinity for zinc, which is a useful property to utilize when isolating the proteins. Zinc chelate affinity chromatography columns have been used to isolate and separate rat α_1-M and α_2-M,[32] demonstrating that both also have zinc-binding sites. Numerous studies on human α_2-M have shown that there is no interaction between the Zinc-binding sites and the functioning of the protein as a proteinase inhibitor. However, Borth and Luger[71] found that the presence of Zn^{2+} increased the binding of IL-

1β to the fast form of human α_2-M. The ability of α_1-I_3 to bind Zn has apparently not been studied.

2. Antiviral Activity

Rat α_1-M and α_1-I_3 (murinoglobulin) have been identified as the proteins responsible for inhibition of infuenza C virus hemagglutination.[76,77] This activity was destroyed by treatment of either protein with sodium hydroxide or methylamine, but not by oxidation with sodium metaperiodate. The activity was also destroyed by treatment with neuraminidase from *Arthrobacter ureafaciens*. These results suggest that the native slow forms of the proteins have a specific type of sialic acid residue at the binding site which is no longer exposed following the conformational changes brought about by cleavage of the thioester bond or alkaline denaturation. Evidence was presented that this binding site was very similar to the specific influenza C virus binding sites erythrocytes. Since only normal rat serum was used, it is not clear whether rat α_2-M also shares this property. Recently, guinea pig and horse α_2-M have also been shown to be potent infuenza virus hemagglutination inhibitors because of the presence of 4-O-acetyl-*N*-acetylneuraminic acid residues in their N-linked carbohydate side chains.[78]

3. Phosphorylation of α_1-I_3

Rat α_1-I_3[79] has recently been identified as the 195-kDa protein in rat muscle and liver homogenates that is tyrosine phosphorylated by insulin receptor in the presence of insulin and polylysine. Its presence in the homogenates was shown to be due to plasma contamination. This phosphorylation was observed in an *in vitro* system, and it is not known as yet whether it occurs *in vivo*. Also of interest, the level of α_2-I_3 was found to be decreased to about 25% of normal levels in streptozotocin-diabetic rats, suggesting an acute phase reaction-like change for this protein upon the induction of diabetes.

REFERENCES

1. **Heim, W. G. and Lane, P. H.,** Appearance of slow α_2-globulin during the inflammatory response of the rat, *Nature (London),* 203, 1077, 1964.
2. **Darcy, D. A.,** Response of a serum glycoprotein to tissue injury and necrosis, *Br. J. Exp. Pathol.,* 45, 281, 1964.
3. **Weimer, H. E. and Benjamin, D. C.,** Immunochemical detection of an acute phase protein in rat serum, *Am. J. Physiol.,* 209, 736, 1965.
4. **Schultze, H. E., Heimburger, N. H., Heide, K., Haupt, H., Storiko, K., and Schwick, H. G.,** Preparation and chracterization of α_1-trypsin inhibitor and α_2-plasmin inhibitor of human serum, in *Proceedings of the 9th Congress of the European Society of Haematologists,* S. Karger, Basel, 1963, 1315.
5. **Haverback, B. J., Dyce, H. F., Bundy, H. F., et al.,** Protein binding of pancreatic proteolytic enzymes, *J. Clin. Invest.,* 41, 972, 1962.
6. **Gauthier, F. and Ohlsson, K.,** Isolation and some properties of a new enzyme binding protein in rat plasma, *Hoppe-Seyler's Z. Physiol. Chem.,* 359, 987, 1978.
7. **Saito, A. and Sinohara, H.,** Rat plasma murinoglobulin: isolation, characterization, and comparison with rat α-1- and α-2-macroglobulins, *J. Biochem. (Tokyo),* 98, 501, 1985.
8. **Lonberg-Holm, K., Reed, D. L., Roberts, R. C., Hebert, R. R., Hillman, M. C., and Kutney, R. M.,** Three high molecular weight protease inhibitors of rat plasma. Isolation, characterization, and acute phase changes, *J. Biol. Chem.,* 262, 438, 1987.
9. **Roberts, R. C.,** Alpha$_2$-macroglobulin, in *Reviews of Hematology,* Vol. 2, *Protease Inhibitors of Human Plasma, Biochemistry and Pathophysiology,* Murano, G., Ed., PJD Publications, Westbury, NY, 1986, 129.

10. **Sottrup-Jensen, L.**, α_2-Macroglobulin and related thiol ester plasma proteins, in *The Plasma Proteins: Structure, Function, and Genetic Control,* Vol. 5, 2nd ed., Putnam, F. W., Ed., Academic Press, Orlando, 1987, 191.
11. **Geiger, T., Andus, T., Kunz, D., Heiseg, M., Bauer, J., Nothoff, N., Gauthier, F., Tran-Thi, T. A., and Heinrich, P. C.**, Regulation of proteinase activity by high molecular weight inhibitors: biosynthesis of rat α-macroglobulins, *Adv. Exp. Med. Biol.*, 240, 183, 1988.
12. **Darcy, D. A.**, Immunological discrimination between blood of normal and tumour bearing rats, *Nature (London),* 176, 643, 1955.
13. **Darcy, D. A.**, Immunological demonstration of a substance in rat blood associated with tissue growth, *Br. J. Cancer,* 11, 137, 1957.
14. **Heim, W. G.**, A reproduction associated protein in the rat, *Am. Zool.*, 1, 359, 1961.
15. **Beaton, G. A., Selby, A. E., Veen, M. J., and Wright, A. M.**, Starch gel electrophoresis of rat serum proteins. II. Slow α_2-globulin and other serum proteins in pregnant, tumor-bearing and young rats, *J. Biol. Chem.*, 236, 2005, 1961.
16. **Boffa, G. A., Nadal, C., Zajedla, F., and Fine, J. M.**, Slow alpha$_2$-globulin of rat serum, *Nature (London),* 203, 1182, 1964.
17. **Heim, W. G. and Lane, P. H.**, Appearance of slow α_2-globulin during the inflammatory response of the rat, *Nature (London),* 203, 1077, 1964.
18. **Weimer, H. E. and Benjamin, D. C.**, Immunochemical detection of an acute phase protein in rat serum, *Am. J. Physiol.*, 209, 736, 1965.
19. **Heim, W. G.**, Relation between rat slow α_2-globulin and α_2-macroglobulin of other mammals, *Nature (London),* 217, 1057, 1968.
20. **Lonberg-Holm, K., Reed, D. L., Roberts, R. C., and Damato-McCabe, D.**, Three high molecular weight protese inhibitors of rat plasma: reactions with trypsin, *J. Biol. Chem.*, 262, 4844, 1987.
21. **Braciak, T. A., Northemann, W., Hudson, G. D., Shills, B. R., Gehring, M. R., and Fey, G. H.**, Sequence and acute phase regulation of rat α_1-inhibitor III messenger RNA, *J. Biol. Chem.*, 263, 3999, 1988.
22. **Enghild, J. J., Salvesen, G., Thøgersen, and Pizzo, S. V.**, Proteinase binding and inhibition by the monomeric α-macroglobulin rat α_1-inhibitor-3, *J. Biol. Chem.*, 264, 11428, 1989.
23. **Bogden, A. E. and Gray, J. H.**, Glycoprotein synthesis and steroids. I. Relationship of trauma, cortisol administration and α-2-GP synthesis, *Endocrinology,* 82, 1077, 1968.
24. **Fey, G. H., Hattori, M., Hocke, G., Bechner, T., Baffet, G., and Baumann, M.**, Gene regulation by interleuken-6, *Biochimie,* 73, 47, 1991.
25. **Gehring, M. R., Shiels, B. R., Northemann, W., de Bruijn, M. H. L., Kan, C.-C., Chain, A. C., Noonan, D., and Fey, G. H.**, Sequence of rat liver α_2-macroglobulin and acute phase control of its messenger RNA, *J. Biol. Chem.*, 262, 4973, 1987.
26. **Eggertsen, G., Hudson, G., Shiels, B., Reed, D., and Fey, G. F.**, Sequence of rat $\alpha_1$macroglobulin, a broad-range proteinase inhibitor from the α-macroglobulin-complement family, *Mol. Biol. Med.*, 8, 287, 1991.
27. **Jamieson, J. C., Friesen, A. D., Ashton, F. E., and Chou, B.**, Studies on acute phase proteins of rat serum. I. Isolation and partial characterization of an α_1-acid glycoprotein and an α_2-macroglobulin, *Can. J. Biochem.*, 50, 856, 1972.
28. **Gauthier, F. and Mouray, H.**, Rat α_2 acute phase macroglobulin, *Biochem. J.*, 159, 661, 1976.
29. **Gordon, A. H.**, The α macroglobulins of rat serum, *Biochem. J.*, 159, 643, 1976.
30. **Nieuwenhuizen, W., Emeis, J. J., and Hemmink, J.**, Purification and properties of rat α_2-acute phase macroglobulin, *Biochem. Biophys. Acta,* 580, 129, 1979.
31. **Okuba, H., Miyanago, O., Nagano, M., Ishibashi, H., Kudo, J., Ikuta, T., and Shibata, K.**, Purification and immunological determination of α_2-macroglobulin in serum from injured rats, *Biochem. Biophys. Acta,* 668, 257, 1981.
32. **Nelles, L. P. and Schnebli, H. P.**, Subunit structure of the rat α-macroglobulin proteinase inhibitors, *Hoppe-Seyler's Z. Physiol. Chem.*, 363, 677, 1982.
33. **Schaeufele, J. T. and Koo, P. H.**, Structural comparison of rat α_1- and α_2-macroglobulins, *Biochem. Biophys. Res. Commun.*, 108, 1, 1982.
34. **Mehl, J. W., O'Connell, W., and Degroot, J.**, Macroglobulin from human plasma which forms an enzymatically active compound with trypsin, *Science,* 145, 821, 1964.
35. **Gliemann, J. and Sottrup-Jensen, L.**, Rat plasma α_1-inhibitor 3 binds to receptors for α_2-macroglobulin, *FEBS Lett.*, 221, 55, 1987.
36. **Gettins, P. and Cunningham, L. W.**, Identification of ^{1}H resonances from the bait region of human α_2-macroglobulin and effects of proteases and methylamine, *Biochemistry,* 25, 5011, 1986.
37. **Barrett, A. J. and Starkey, P. M.**, The interaction of α_2-macroglobulin with proteinases, *Biochem. J.*, 133, 709, 1973.

38. **Strickland, D. K., Larsson, L. J., Neuenschwander, D. E., and Bjork, I.,** Reaction of proteinases with alpha-2-macroglobulin. Rapid kinetic evidence for a conformational rearrangement of the initial alpha-2-macroglobulin-trypsin complex, *Biochemistry,* 30, 2797, 1991.
39. **Sottrup-Jensen, L., Sand, O., Kristensen, L., and Fey, G. F.,** The α-macroglobulin bait region. Sequence diversity and localization of cleavage sites for proteinases in five mammalian α-macroglobulins, *J. Biol. Chem.,* 264, 15781, 1989.
40. **Sottrup-Jensen, L. and Birkedal-Hansen, H.,** Human fibroblast collagenase-α-macroglobulin interactions. Localization of cleavage sites in the bait regions of five α-macroglobulins, *J. Biol. Chem.,* 264, 393, 1989.
41. **Travis, J. and Salvesen, G. S.,** Human plasma proteinase inhibitors, *Annu. Rev. Biochem.,* 52, 655, 1983.
42. **Horne, C. H. W., Forbes, C. D., and Prentice, C. R. M.,** Antiplasmin activity of rat serum slow α-globulins, *Br. J. Haematol.,* 24, 115, 1972.
43. **Gonias, S. L., Balber, A. E., Hubbard, W. J., and Pizzo, S. V.,** Ligand binding, conformational change and plasma elimination of human, mouse, and rat alpha-macroglobulin proteinase inhibitors, *Biochem. J.,* 209, 99, 1983.
44. **Gliemann, J., Davidson, O., Sottrup-Jensen, L., and Sonne, O.,** Uptake of rat and human α2macroglobulin-trypsin complexes into rat and human cells, *FEBS Lett.,* 188, 352, 1985.
45. **Bjork, I. and Jornvall, H.,** The structure around the thioester bond in bovine alpha-2-macroglobulin, *FEBS Lett.,* 205, 87, 1986.
46. **Koj, A.,** Biological functions of acute phase proteins, in *The Acute Phase Response to Injury and Infection,* Gordon, A. H. and Koj, A., Eds., Elsevier, Amsterdam 1985, chap. 13.
47. **Jones, K.,** Alpha$_2$ macroglobulin and its possible importance in immune systems, *Trends Biol. Sci.,* 5, 43, 1980.
48. **Hanna, M. G., Nettesheim, P., Fisher, W. D., Peters, L. C., and Francis, M.,** Serum alpha-globulin fraction: survival and recovery effect in irradiatedmice, *Science,* 157, 1458, 1967.
49. **Graham, J. D., Earney, W. W., and Hilton, P. K.,** Serum macroglobulin stimulation of the proliferation and differentiation of granulocytic precursors, *Trans. Am. Microsc. Soc.,* 94, 375, 1975.
50. **Tunstall, A. M. and James, K.,** The effect of human $\alpha_2$macroglobulin on the restoration of humoral responsiveness in X-irradiated mice, *Clin. Exp. Immunol.,* 21, 173, 1975.
51. **Teodorescu, M., Chang, J.-L., and Skosey, J. L.,** Polyclonal B cell activator associated with alpha-2-macroglobulin in the serum of patients with rheumatoid arthritis, *Int. Arch. Allergy Appl. Immunol.,* 66, 1, 1981.
52. **Chang, J.-L., Ganea, D., Dray, S., and Teodorescu, M.,** An $\alpha_2$macroglobulin associated factor produced by T lymphocytes which provides polyclonal stimulation of B lymphocytes to maintain the turnover of their surface Ig, *Immunology,* 44, 745, 1981.
53. **Hoffman, M. R., Pizzo, S. V., and Weinberg, J. B.,** Modulation of mouse peritoneal macrophage Ia and human peritoneal macrophage HLA-DR expression by α_2-macroglobulin "fast" forms, *J. Immunol.,* 139, 1885, 1987.
54. **Osada, T., Noro, N., Kurjoda, Y., and Ikai, A.,** Murine T cell proliferation can be specifically augmented by macrophages fed with specific antigen:α_2-macroglobulin conjugate, *Biochem. Biophys. Res. Commun.,* 146, 26, 1987.
55. **Godfrey, H. P., Malorny, U., Michels, E., Habicht, G. S., Atlas, A., Randazzo, B., and Sorg, C.,** Murine alpha-2-macroglobulin: localization on a subpopulation of macrophages, *Immunobiology,* 175, 183, 1987.
56. **Hubbard, W. J., Hess, A. D., Hsia, S., and Amos, D. B.,** The effects of electrophoretically "slow" and "fast" $\alpha_2$macroglobulin on mixed lymphocyte cultures, *J. Immunol.,* 126, 292, 1981.
57. **Alomran, A., Shenton, B. K., Proud, G., Francis, D. M. A., Donnelly, K., Hubbard, W. J., and Taylor, R. M. R.,** Possible mechanism of immune regulation produced by $\alpha_2$macroglobulin, *Lancet,* 2, 1168, 1982.
58. **Hubbard, W. J., Anderson, B. D., and Balber, A. E.,** Immunosuppression by human $\alpha_2$macroglobulin, a 3800 Mr peptide, and other derivatives, *Ann. N.Y. Acad. Sci.,* 421, 332, 1983.
59. **Rastoga, S. C. and Clausen, J.,** Kinetics of inhibition of mitogen-induced proliferation of human lymphocytes by α_2-macroglobulin in serum-free medium, *Immunobiology,* 169, 37, 1985.
60. **Goutner, A., Simmler, M. C., Tapon, J., and Rosenfeld, C.,** Modulation by α_2-macroglobulin of human lymphocyte proliferation in response to mitogens and antigen, *Differentiation,* 5, 171, 1976.
61. **Ades, E. W., Hinson, C., Chapius-Cellier, C., and Arnaud, P.,** Modulation of the immune response by plasma protease inhibitors. I. Alpha1-macroglobulin and alpha1-antitrypsin inhibit natural killing and antibody-dependent cell-mediated cytotoxicity, *Scand. J. Immunol.,* 15, 109, 1982.
62. **Gravgna, P., Gianazza, E., Arnoud, P., Neels, M., and Ades, E. W.,** Modulation of the immune response by plasma protease inhibitors. II. Alpha2-macroglobulin subunits inhibit natural killer cell cytotoxicity and antibody-dependent cell-mediated cytotoxicity, *Scand. J. Immunol.,* 15, 115, 1982.

63. **Petersen, C. M., Ejlersen, E., Moestrup, S. K., Jensen, P. H., Sand, O., and Sottrup-Jensen, L.,** Immunosuppressive properties of electrophoretically "slow" and "fast" form α_2-macroglobulin. Effects on cell-mediated cytotoxicity and (allo-)antigen induced T cell proliferation, *J. Immunol.*, 142, 629, 1989.
64. **Ufkes, J. G. R., Ottenhof, M., Von Rooij, J. J., and Van Gool, J.,** Rat α_2-macroglobulin is a selective inhibitor of antigen-induced leucotrienes in rat isolated lungs, *Br. J. Exp. Pathol.*, 69, 457, 1988.
65. **Bellot, R., Bon, A., Lestage, J., Giroud, J.-P., and Chateaureynaud, P.,** Evidence for an α_2-macroglobulin with complement-inhibiting activity in rat serum, *Int. J. Exp. Pathol.*, 72, 151, 1991.
66. **Adham, N.F., Chakmakjian, Z. H., Wehl, J. W., and Bethune, J. E.,** Human growth hormone and α_2-macroglobulin: a study of binding, *Arch. Biochem. Biophys.*, 132, 175, 1969.
67. **Huang, J. S., Huang, S. S., and Deuel, T. S.,** Human platelet-derived growth factor: radioimmunoassay and discovery of a specific binding protein, *J. Cell Biol.*, 97, 383, 1983.
68. **Huang, S. S., O'Grady, P., and Huang, J. S.,** Human transforming growth factor β/α_2-macroglobulin complex is a latent form of transforming growth factor, *J. Biol. Chem.*, 263, 1535, 1988.
69. **Dennis, P. A., Saksela, O., Harpel, P., and Rifkin, D. B.,** α_2Macroglobulin is a binding protein for basic fibroblast growth factor, *J. Biol. Chem.*, 264, 7210, 1989.
70. **Matsuda, T., Hirano, T., Nagasawa, S., and Kishimoto, T.,** Identification of $\alpha_2$macroglobulin as a carrier protein for interleukin 6, *J. Immunol.*, 142, 148, 1989.
71. **Borth, W. and Luger, T. A.,** Identification of α_2-macroglobulin as a cytokine binding protein: binding of interleukin 1β to "F" α_2-macroglobulin, *J. Biol. Chem.*, 264, 5818, 1989.
72. **Wollenberg, G. K., LaMarre, J., Rosendal, S., Gonias, S. L., and Hayes, M. A.,** Binding of tumor necrosis factor alpha to activated forms of human plasma alpha$_2$-macroglobulin, *Am. J. Pathol.*, 138, 265, 1991.
73. **Koo, P. H. and Stach, R. W.,** Interaction of nerve growth factor with murine α-macroglobulin, *J. Neurosci. Res.*, 22, 247, 1989.
74. **Bonner, J. A. and Brody, A. R.,** Cytokine-binding proteins, in *Lung Biology in Health and Disease V,* Kelley, J., Ed., Marcel Dekker, New York, 1992, chap. 15.
75. **Bonner, J. C., Hoffman, M., and Brody, A. R.,** Alpha macroglobulin secreted by alveolar macrophages serves as a binding protein for a macrophage-derived homologue of platelet-derived growth factor, *Am. J. Respir. Cell. Mol. Biol.*, 1, 171, 1989.
76. **Herrler, G., Geyer, R., Muller, H.-P., Stirm, S., and Klenk, H.-D.,** Rat α_1-macroglobulin inhibits hemagglutination by influenza C virus, *Virus Res.*, 2, 183, 1985.
77. **Kitame, F., Nakamura, K., Saito, A., Sinohara, H., and Homma, M.,** Isolation and characterization of influenza C virus inhibitor in rat serum, *Virus Res.*, 3, 231, 1985.
78. **Pritchett, T. J. and Paulson, J. C.,** Basis for the potent inhibition of influenza virus infection by equine and guinea pig α_2-macroglobulin, *J. Biol. Chem.*, 264, 9850, 1989.
79. **Komori, K., Robinson, K. A., Block, N. E., Roberts, R. C., and Buse, M. G.,** Phosphorylation of the rodent negative acute phase protein α_1-inhibitor III, by the insulin receptor tyrosine kinase, *Endocrinology,* 131, 1288, 1992.
80. **Bonner, J. A.,** personal communication.

Chapter 13

RAT THIOSTATIN: STRUCTURE AND POSSIBLE FUNCTION IN THE ACUTE PHASE RESPONSE

Gerhard Schreiber and Timothy J. Cole

TABLE OF CONTENTS

0-8493-6913-4/93/$0.00 + $.50

INTRODUCTION

It has long been known that the properties of plasma or serum change during diseases, particularly in response to trauma and inflammation (for reviews, see References 1 to 3). Measurement of the erythrocyte sedimentation rate and the precipitation of proteins from serum by salts, organic solvents, or heating were used as early diagnostic methods to monitor the inflammatory response. After the introduction of electrophoretic separation of serum proteins, it was recognized that the basis for changes from the normal values which were observed with the earlier diagnostic methods was an altered pattern of concentrations of plasma proteins. The most prominent changes were an increase in the concentration of protein in the globulin region and a decrease in the albumin region (for reviews, see References 4 and 5). In the rat, an increase in the concentration of an α_1-globulin serum was very prominent during the acute phase response.[6,7] The protein causing the increase in concentration of serum protein in the α_1-globulin region was isolated to homogeneity and its physicochemical properties were characterized in detail.[8] More recently, cDNA[9,10] and genomic[11] clones for the protein have been isolated and sequenced. During inflammation, the protein can account for at least 10%, or more, of total serum protein, and was therefore tentatively termed the major acute phase α_1-protein[8] (MAP). It was immunochemically identical to the α_1-glycoprotein originally observed by Darcy[6,12] as shown by double immunodiffusion.[13] After demonstration of its immunochemical identity with the α_1-cysteine proteinase inhibitor isolated by Esnard and Gauthier,[14] the name thiostatin was suggested for the protein.[9] Thiostatin is synthesized in the liver via a precursor protein,[13] which is posttranslationally modified into the mature protein. This protein is then secreted into the serum.

II. MATERIALS AND METHODS

All materials and methods used are described in the quoted original publications.

III. RESULTS AND DISCUSSION

A. IMMUNOCHEMICAL IDENTIFICATION IN SERUM AND PHYSICOCHEMICAL PROPERTIES OF THIOSTATIN

The relationship between the changes in concentration of thiostatin and those of other proteins in serum during the acute phase response can best be demonstrated by crossed immunoelectrophoresis. Patterns for crossed immunoelectrophoresis of normal serum (Figure 1A) and of serum obtained 48 h after the induction of an acute phase response in rats by subcutaneous injection of a small amount of mineral turpentine (Figure 1B) are shown. In Figure 1B, it can be clearly seen that two protein peaks in the α_1-globulin region and a peak in the prealbumin region (the protein formerly known as prealbumin, which is now called transthyretin) are decreased compared with those for normal serum, whereas at least ten peaks are increased, five of them in the α_1-globulin region. In the sera analyzed in Figures 1A and B, the albumin concentration, measured by single radial immunodiffusion with purified serum albumin as a standard,[15,16] decreased from (34 $\pm$ 1) g/l in normal serum to (23 $\pm$ 2) g/l in acute phase serum (obtained 48 h after induction of inflammation). The most conspicuous peak in the α_1-globulin region, apart from albumin, was that caused by thiostatin. Its concentration rose from 0.46 to 7.2 g/l 48 h after induction of inflammation. The total protein concentration remained virtually unchanged, with (67 $\pm$ 1) g/l in normal serum and (72 $\pm$ 2) g/l in the acute phase serum.

Thiostatin was isolated by ammonium sulfate fractionation, followed by ion-exchange chromatography, gel filtration, affinity chromatography, and preparative electrophoresis in

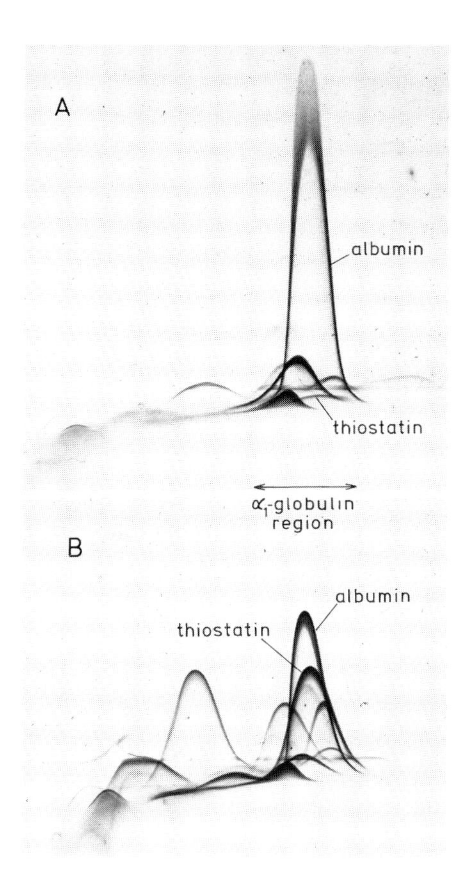

FIGURE 1. Crossed immunoelectrophoresis of normal rat serum (A) and acute phase rat serum (B). The acute phase serum was obtained 48 h after turpentine-induced injury. The cathode for electrophoresis in the first and second dimension was, respectively, to the left and to the bottom of the plates. Two microliters of normal serum (A) and of acute phase serum (B) were applied to the gels. Electrophoresis in the first dimension was in 1% agarose at pH 8.6. In the second dimension, electrophoresis was carried out in agarose containing antiserum to rat plasma proteins. (From Urban, J., Chan, D., and Schreiber, G., *J. Biol. Chem.*, 254, 10565, 1979. With permission.)

polyacrylamide gel.[8] The purified protein was found to be homogeneous in crossed immunoelectrophoresis, immunoelectrophoresis, and double immunodiffusion, using antiserum to total acute phase serum.[8] Thiostatin, when analyzed by polyacrylamide gel electrophoresis in the presence of sodium dodecyl sulfate, under reducing conditions, migrated as a single band corresponding to a molecular weight of 68,000. However, sedimentation equilibrium analysis of thiostatin in concentrations of 1.2, 0.9, and 0.6 g of protein per liter in 0.1 *M* Tris-HCl, pH 7.8, gave a molecular weight of 56,000 ± 1000, using 0.71 as the partial specific volume for calculation. Both the log displacement vs. r^2 plot[17] and similarity of the molecular weight values obtained with the two methods indicated a protein composed of a single polypeptide chain. The amino acid and carbohydrate compositions of the purified thiostatin are summarized in Table 1. The circular dichroic spectrum of thiostatin was measured between 190 and 240 nm, and the content of the α-helix structure was found to be about 26.5%,[9] calculated according to Chen and Yang.[22]

The crossed immunoelectrophoresis pattern presented clearly showed one homogeneous peak for thiostatin in the serum of Buffalo rats. However, the cysteine proteinase inhibitor preparation from the plasma of Wistar rats by Esnard and Gauthier,[14] which was immunochemically identical to thiostatin, was electrophoretically resolved into two distinct pro-

TABLE 1
Amino Acid and Carbohydrate Compositions of Thiostatin

Residue	Mol/100 mol of total amino acids	Percentage of dry weight	Mol/56,000 g
Lys	8.10	6.58	29
His	3.48	3.03	12
Arg	3.39	3.36	12
Asp	10.45	7.62	37
Thr[a]	7.43	4.77	26
Ser[a]	5.94	3.29	21
Glu	12.98	10.61	46
Pro + Cys[b]	5.29	3.26	19
Gly	6.70	2.43	24
Ala	7.47	3.37	27
Half-cystine	0.94	0.60	3
Val[c]	7.08	4.46	25
Met	1.38	1.15	5
Ile	3.81	2.73	14
Leu[c]	7.77	6.38	32
Tyr	2.96	3.06	11
Phe	4.20	3.92	15
Trp[d]	0.64	0.71	2
Total amino acids	100.00	71.33	360
N-Acetylglucosamine		4.62	12
N-Acetylneuraminic acid		7.24	13
Neutral hexoses		7.30	23
Total carbohydrate		19.16	48
Total Residues		90.49	408

Note: Amino acids and amino sugars were determined with a Beckman 121 MB amino acid analyzer (AA-10 resins). Values for amino acids, except residues footnoted a, c, and d, are averaged from duplicate determinations on 24-, 48-, and 72-h 6 *N* HCl hydrolysates (110°C) of 100 μg of protein per milliliter. Values for carbohydrate are averaged from two determinations in duplicate. Separate determinations differed by not more than 5%. *N*-Acetylglucosamine was analyzed after hydrolysis of 412 μg of protein in 1 ml of 3 *N* CH_3SO_3H for 24 h at 100°C, a modification of a published method.[18] *N*-Acetylneuraminic acid and neutral hexoses were determined, respectively, by the thiobarbituric acid method[19] with *N*-acetylneuraminic acid as a reference standard and by the phenol/water method[20] with mannose/galactose (1:1) as a reference standard.

[a] Corrected by extrapolation to zero time.
[b] Comparison of $A_{440\ nm}/A_{570\ nm}$ to proline and cysteine reference standards suggests that this peak contained about 25% cysteine.
[c] Value after hydrolysis for 72 h (Val) or 24 h (Leu).
[d] Average of two determinations in duplicate from 24-h 4 *N* CH_3SO_3H hydrolysates (115°C) of 412 μg of protein per milliliter in the presence of 2% (w/v) 3-(2-aminoethyl) indole.[21]

From Urban, J., Chan, D., and Schreiber, G., *J. Biol. Chem.*, 254, 10565, 1979. With permission.

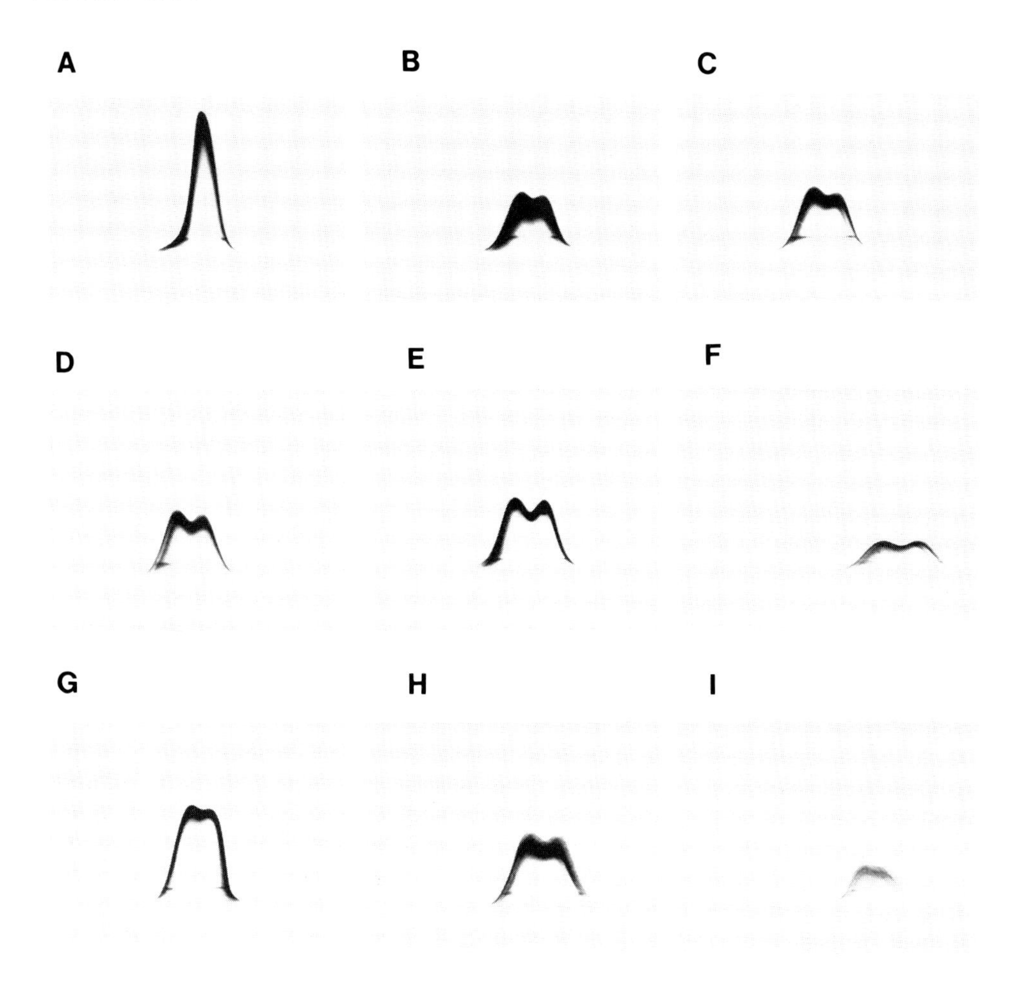

FIGURE 2. Analysis of plasma samples from nine different strains of *R. norvegicus* by crossed immunoelectrophoresis using antiserum against purified thiostatin. Two microliters of plasma, isolated from rats suffering an acute inflammation, were analyzed by electrophoresis in 0.9% agarose gels in the first dimension. Protein was then moved by electrophoresis into a second gel composed of 0.9% agarose containing 3.0% (v/v) of antiserum against purified thiostatin (Buffalo rat). Plates were stained with Coomassie Blue-R250. Plasma was analyzed from the following rat strains: (A) Buffalo, (B) Wistar, (C) Sprague Dawley, (D) Hooded Wistar, (E) Brown Norway, Katholiek, (F) Porton-albino, (G) Long Evans, (H) ACI, and (I) Lewis. (From Cole, T. and Schreiber, G., *Comp. Biochem. Physiol.*, 93B, 813, 1989. With permission.)

teins. To clarify this apparent contradiction, a systematic study of thiostatin in nine different strains of *Rattus norvegicus* was performed. The result is shown in Figure 2. The detection of a single peak is characteristic of the electrophoretic analysis of thiostatins from the Buffalo strain only. Analysis of the thiostatin genes at the genomic level[11] showed the presence of two different thiostatin gene copies in rats. These two genes are transcribed into two slightly different mRNAs. These are translated into two slightly different polypeptides, clearly distinguishable from each other by electrophoretic analysis, except in the plasma of Buffalo rats.

B. CLONING AND STRUCTURE OF THIOSTATIN cDNA

The cDNAs for thiostatins have been cloned and sequenced from Buffalo rat liver,[9,10] Sprague-Dawley rat liver,[24] and Wistar rat liver.[25,26] Details of the structure of thiostatin genomic DNAs and implications for protein structures are discussed in Chapter 9 of this

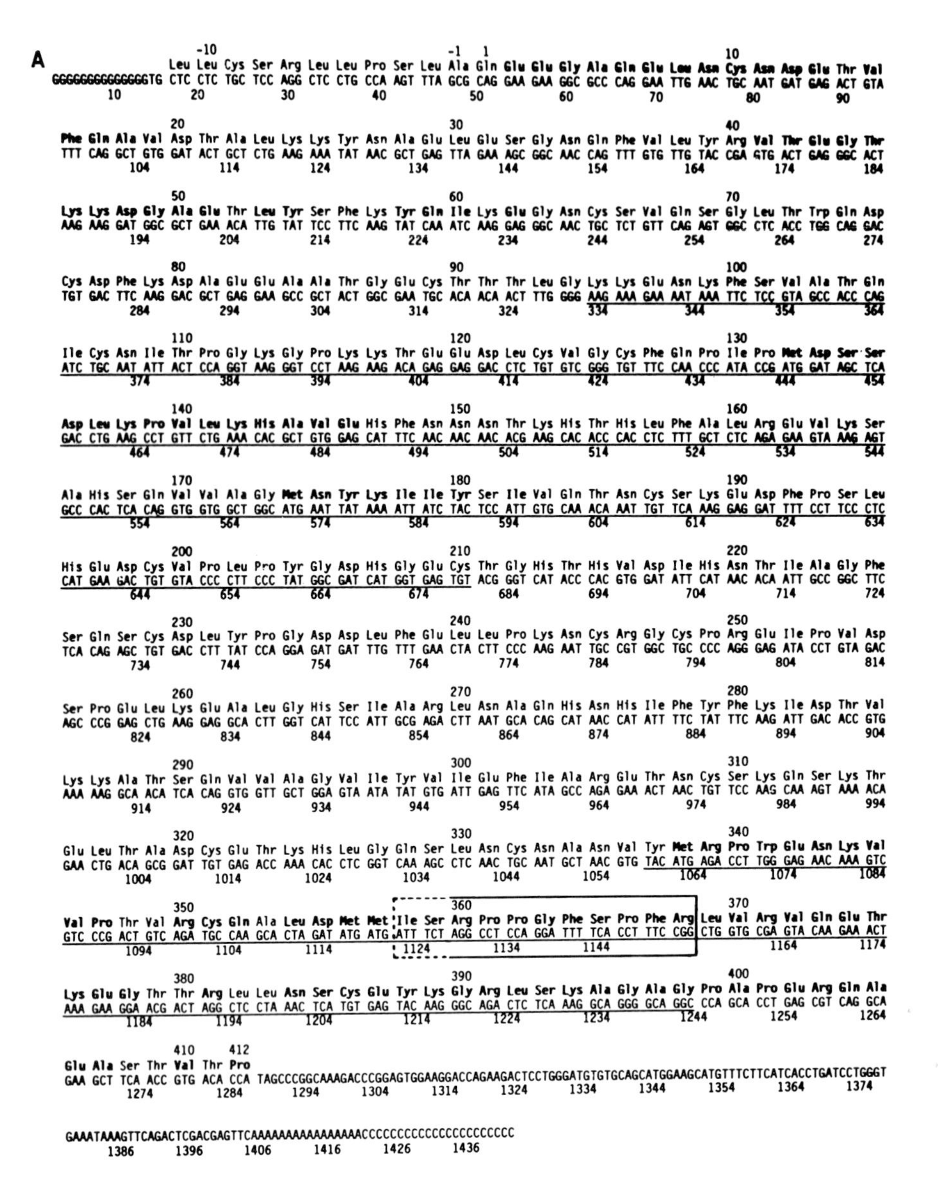

FIGURE 3. Nucleotide sequence of thiostatin cDNA and derived amino acid sequence. The nucleotide sequence derived from independent sequencing of both strands is indicated by underlining. Amino acid sequences deduced from both nucleotide sequencing and direct analysis of thiostatin fragments are indicated by bold letters. Numbering of amino acids begins at the N terminus of the mature protein. The sequence of bradykinin is indicated by a solid box and that of T-kinin by a dotted extension of the box. (Adapted from Cole, T., Inglis, A. S., Roxburgh, C. M., Howlett, G. J., and Schreiber, G., *FEBS Lett.*, 182, 57, 1985. With permission.)

volume. The nucleotide sequence for thiostatin cDNA from Buffalo rat liver and the derived amino acid sequence are shown in Figure 3. The amino acid sequence deduced for thiostatin from the nucleotide sequence of its cDNA showed two surprising features. The first is the presence of the sequence for bradykinin between amino acids 360 and 368 from the N terminus of the mature protein. The second is the high similarity of the amino acid sequences of rat thiostatin to those of bovine and human kininogens,[9,10,27-29] as illustrated in Figure 4. This structural relationship between thiostatin and the kininogens raises interesting aspects concerning the possible function of thiostatin (see below).

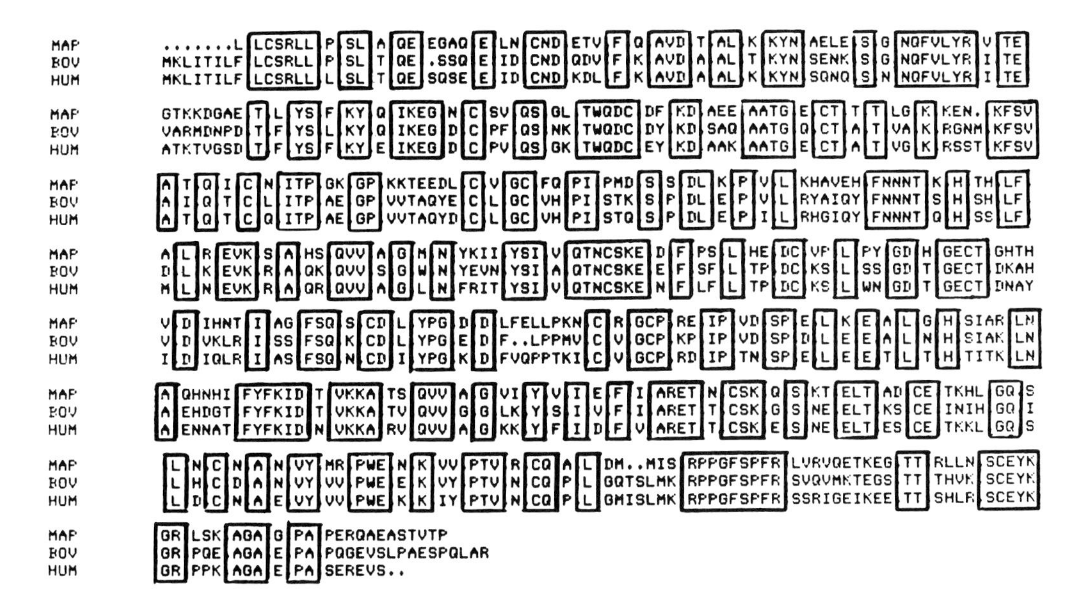

FIGURE 4. Comparison of the amino acid sequences for bovine kininogen, human kininogen, and rat thiostatin. Based on sequence data from references 9, 10, and 27 to 29. Conserved sequences are indicated by boxes. The fully conserved sequence for the nonapeptide bradykinin is located in the right half of the second-to-last row. (From Schreiber, G., in *The Plasma Proteins,* Vol. 5, Putnam, F. W., Ed., Academic Press, New York, 1987, 293. With permission.)

C. MECHANISM OF SYNTHESIS AND SECRETION OF THIOSTATIN

All proteins synthesized and secreted by liver are synthesized via precursor forms (for a review, see Reference 5). After removal of the presegment, the intracellular precursors of the plasma proteins are modified further. This modification concerns the polypeptide chain or, in the case of glycoproteins, the carbohydrate moieties. In many cases, intracellular precursors of plasma proteins can be isolated (see, e.g., References 30 to 42). Also, in the case of thiostatin from Buffalo rats, it is possible to clearly separate a form of thiostatin found only in the liver from one that occurs in the plasma. Figure 5 shows the separation by anion-exchange chromatography of the two forms of thiostatin from Buffalo rat liver. The first peak corresponds to the liver form of thiostatin, the second peak to serum thiostatin. Table 2 gives a summary of the carbohydrate analysis of both liver and serum forms of Buffalo rat thiostatin. The liver thiostatin does not possess any *N*-acetylneuraminic acid, in contrast to the 12 mol of *N*-acetylneuraminic acid per 56,000 g of serum thiostatin. Also, the *N*-acetylglucosamine content of serum thiostatin is higher than that of liver thiostatin.

D. SYNTHESIS RATES AND BODY POOLS OF THIOSTATIN DURING THE ACUTE PHASE RESPONSE

Thiostatin in the bloodstream is at equilibrium with thiostatin in the interstitial space of the body. Therefore, not only do the concentrations of thiostatin in the bloodstream change drastically during the acute phase response, but also the total body pools of thiostatin increase considerably (Figure 6). The half-life of thiostatin in the bloodstream of male Buffalo rats is (1.5 $\pm$ 0.27) d.[43] Thus, under steady-state conditions in healthy, normal Buffalo rats, the synthesis rate of thiostatin can be calculated from the turnover and total body pool, and was found to be (2.28 $\pm$ 0.42) mg/100 g of body weight per day (Table 3).

The rates of incorporation of radioactive amino acids into plasma proteins are proportional to the rates of their synthesis.[45] Such incorporation rates can be easily measured if mono-

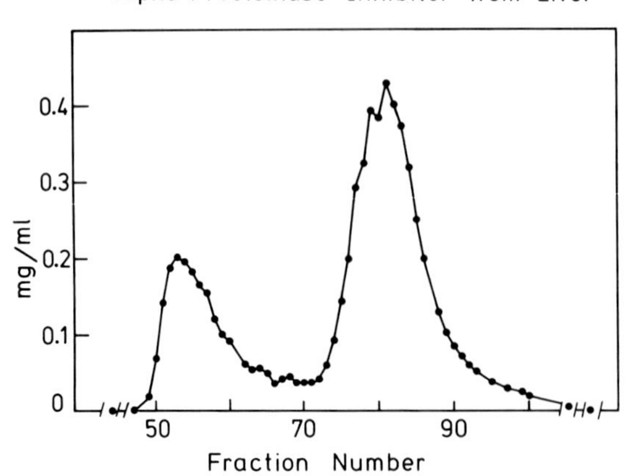

FIGURE 5. Ion-exchange chromatography of an extract of acetone-dried powder from rat livers. Livers were taken from rats 48 h after induction of inflammation. An extract of acetone-dried powder from three livers containing a total of 6 g of protein in 20 m*M* Tris-HCl, pH 7.7, was chromatographed on a column of DEAE-cellulose (2.5 × 90 cm) at 96 ml/h with 1.9 l of a linear gradient of 0 to 240 m*M* NaCl in the same buffer. Fractions of 16.2 ml were collected. The concentration of thiostatin was measured by single radial immunodiffusion. The first peak is liver thiostatin, the second peak is serum thiostatin. (From Urban, J., Nagashima, M., and Schreiber, G., *Biochem. Int.*, 4, 75, 1982. With permission.)

TABLE 2
Carbohydrate Composition of Thiostatin Isolated from Liver and Serum of Buffalo Rats

Carbohydrate	Liver thiostatin (mol/51,600 g)[a]	Thiostatin from serum (mol/56,000 g)
N-Acetylglucosamine	7	12
Neutral hexoses	27	25
N-Acetylneuraminic acid	0	12

Note: *N*-Acetylglucosamine was determined as described earlier.[8,18] Neutral hexoses and *N*-acetylneuraminic acid were measured according to References 19 and 20, using the reference standards described in Reference 8. The value for *N*-acetylglucosamine in liver thiostatin was obtained from a single protein preparation. Each of the other values represents the average from two independently analyzed protein preparations. Deviations between independent analyses did not exceed 10%.

[a] This value was calculated from the M_r of thiostatin from serum based on the assumption that the two species of thiostatin differ only in their carbohydrate, and not peptide, content.

From Urban, J., Nagashima, M., and Schreiber, G., *Biochem. Int.*, 4, 75, 1982. With permission.

specific antisera are available for isolation of proteins by immunoprecipitation (Figure 7). For some proteins, such as transferrin, the rate of synthesis and pools in the body do not change for the first 24 h after inducing an acute phase response.[43] Assuming that plasma proteins are made from the same intracellular pool of amino acids, the rates of protein synthesis, which change during the acute phase response, can be calculated from the in-

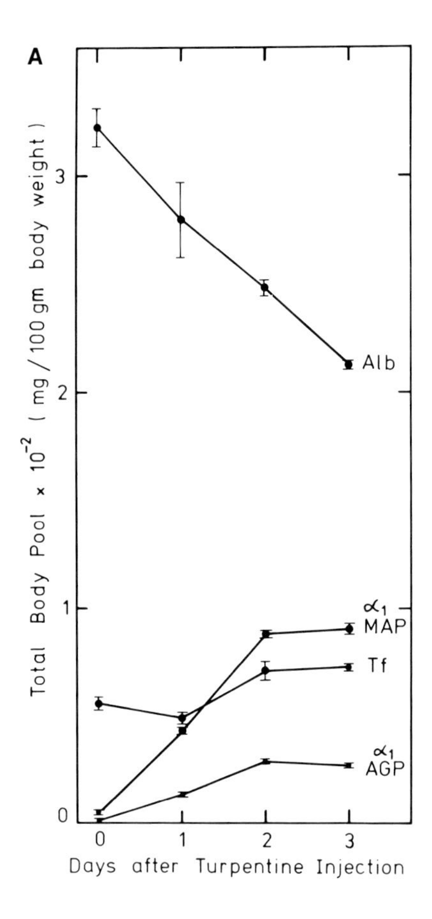

FIGURE 6. Total body pool of albumin (Alb), thiostatin (α_1-MAP), transferrin (Tf), and α_1-acid glycoprotein (α_1-AGP) during acute inflammation in Buffalo rats. (From Schreiber, G., Howlett, G., Nagashima, J., Millership, A., Martin, H., Urban, J., and Kotler, L., *J. Biol. Chem.*, 257, 10271, 1982. With permission.)

TABLE 3
Total Body Pools, Half-Lives, and Synthesis Rates (Mean ± S.E.) for Four Plasma Proteins in Male Buffalo Rats on a Diet Containing 20% Protein

	Albumin	Transferrin	Thiostatin	α_1-Acid glycoprotein
Half-life, (d, n = 8)	2.45[a]	2.06 ± 0.04	1.5 ± 0.27	1.3 ± 0.27
Total body pool (mg/100 g of body weight, n = 5)	323 ± 9	55.8 ± 2.7	4.94 ± 0.09	1.89 ± 0.41
Synthesis rate calculated from turnover and total body pool (mg/100 g of body weight per day)	91.4	18.77 ± 0.97	2.28 ± 0.42	1.01 ± 0.33

Note: For the determination of half-lives, proteins were labeled with ^{125}I and "screened" for 2 d in other rats before injection into experimental animals.

[a] Mean value from literature data.[44]

From Schreiber, G., Howlett, G., Nagashima, M., Millership, A., Martin, H., Urban, J., and Kotler, L., *J. Biol. Chem.*, 257, 10271, 1982. With permission.

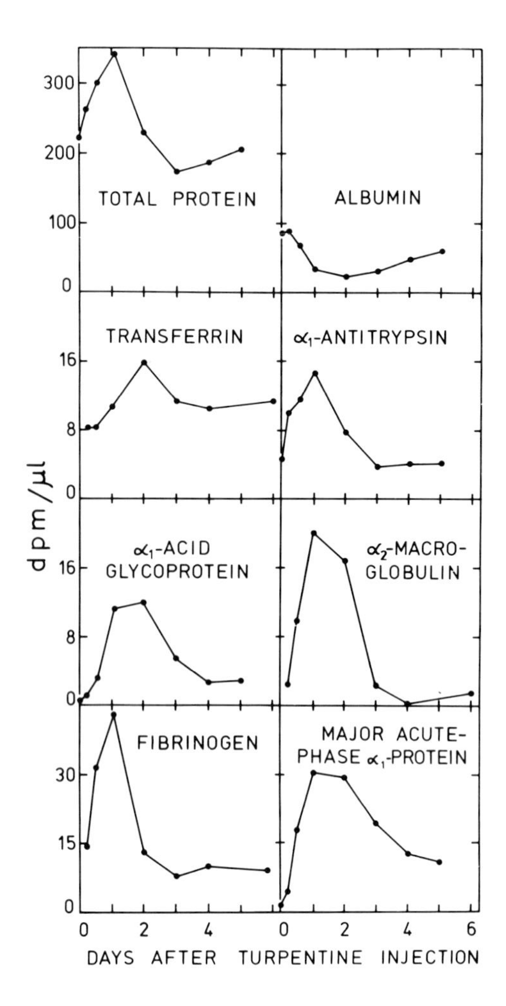

FIGURE 7. Incorporation of intravenously injected L-[1-^{14}C]leucine into plasma proteins in the bloodstream during acute experimental inflammation in Buffalo rats. Proteins were isolated from plasma obtained at the indicated times after induction of inflammation by immunoprecipitation with the appropriate monospecific antisera. (From Schreiber, G., Tsykin, A., Aldred, A. R., Thomas, T., Fung, W.-P., Dickson, P. W., Cole, T., Birch, H., de Jong, F. A., and Milland, J., *Ann. N.Y. Acad. Sci.*, 557, 61, 1989. With permission.)

corporation rates of [^{14}C]leucine into proteins, using a nonchanging protein as a standard, as outlined in detail in Reference 43. In this way, a 64-fold increase in the rate of thiostatin synthesis in the liver, 24 h after inducing inflammation, has been obtained (Table 4). Inducing acute inflammation by subcutaneous injection of turpentine also leads to a change of the ratio of the intravascular pool of thiostatin to the extravascular pool of thiostatin.[43] After correcting for this change, 24 h after inducing acute inflammation, a 97-fold increase in the rate of thiostatin synthesis was observed.[43]

E. POSSIBLE FUNCTIONS OF THIOSTATIN

Two possible physiological functions have to be considered for thiostatin. The first is related to the presence of the sequence for bradykinin in the polypeptide chain of thiostatin. Bradykinin is released from kininogen by the action of kallikrein. One of the recognition sites for this proteinase is at the N-terminal end of the bradykinin segment. In thiostatin, this recognition site differs from that in kininogen by the presence of the residues isoleucine and serine in positions 358 and 359 of the thiostatin polypeptide chain, immediately preceding

TABLE 4
Relative Rates of Incorporation of L-[1-^{14}C[Leucine into Plasma Proteins and Their Rates of Synthesis in Healthy Rats on a Diet Containing 20% Protein (H), and in Rats 24 H after Including Inflammation plus Fasting (I)

	State of rat	Albumin	Transferrin	Thiostatin	α_1-Acid glycoprotein
Rate of incorporation in percent of that into plasma total protein	H	41	5.1	0.53	0.24
	I	9.2	3.1	9.0	3.2
Factor by which the rate of synthesis changes 24 h after inducing inflammation[a]		0.37	1.0	28 42[b]	22
Rate of synthesis in healthy rats (from row 3 in Table 3) (mg/100 g of body weight per day)		91	19	2.3	1.0
Rates of synthesis 24 h after inducing inflammation (mg/100 g of body weight per day)		32	19	64 97[b]	22

Note: Each value is the mean from four animals.

[a] Calculated using the assumption that the rate of synthesis of transferrin did not change 24 h after inducing inflammation.
[b] Value corrected for increase in space of distribution of thiostatin during inflammation.[43]

From Schreiber, G., Howlett, G., Nagashima, M., Millership, A., Martin, H., Urban, J., and Kotler, L., *J. Biol. Chem.*, 257, 10271, 1982. With permission.

the N-terminal arginine residue of the bradykinin segment. The evolution of the structure of the thiostatin gene leading to this particular structural feature at the N-terminal end of the bradykinin segment is discussed in Chapter 9 of this volume. The bradykinin moiety of thiostatin, unlike that in the kininogens, cannot be released from the protein by digestion with kallikreins or with low concentrations of trypsin.[47] Also, the concentration of kininogens in the blood does not increase during the acute phase response.[48] It is therefore unlikely that thiostatin acts as a kininogen. A strong hypotensive effect, such as produced by bradykinin, would be inappropriate during the acute phase response.

The second alternative for the physiological function of thiostatin is that of an inhibitor of cysteine proteinases. Such proteinases are released from lysosomes during inflammation.[14] Inhibition of these proteinases becomes important during the acute phase response. Not only cysteine proteinases released from lysosomes, such as cathepsins, but also papain, whose amino acid sequence is partially homologous[49] to that of cathepsin H and B, are inhibited by thiostatin. In summary, the function of thiostatin seems more likely to be that of a proteinase inhibitor rather than that of a kinin precursor, suggesting that the name ''thiostatin'' is more appropriate for the protein than the name ''T-kininogen''.

ACKNOWLEDGMENTS

We thank S. Richardson for critical discussion and J. Guest and M. Moritz for word processing. The work was supported by grants from the National Health and Medical Research Council of Australia and the Australian Research Council.

REFERENCES

1. **Aronson, K.-F., Ekelund, G., Kindmark, C.-O., and Laurell, C.-B.,** Sequential changes of plasma proteins after surgical trauma, *Scand. J. Clin. Lab. Invest.*, 29 (Suppl. 124), 127, 1971.
2. **Koj, A.,** Acute phase reactants, their synthesis, turnover and biological significance, in *Structure and Function of Plasma Proteins,* Vol. 1, Allison, A. C., Ed., Plenum Press, London, 1974, 73.
3. **Gordon, A. H.,** The acute phase plasma proteins, in *Plasma Protein Turnover,* Bianchi, R., Mariani, G., and McFarlane, A. S., Eds., University Park Press, Baltimore, 1976, 381.
4. **Schreiber, G. and Howlett, G.,** Synthesis and secretion of acute phase proteins, in *Plasma Protein Secretion by the Liver,* Glaumann, H., Peters, T., Jr., and Redman, C., Eds., Academic Press, New York, 1983, 423.
5. **Schreiber, G.,** Synthesis, processing, and secretion of plasma proteins by the liver and other organs and their regulation, in *The Plasma Proteins,* Vol. 5, Putnam, F. W., Ed., Academic Press, New York, 1987, 293.
6. **Darcy, D. A.,** Immunological demonstration of a substance in rat blood associated with tissue growth, *Br. J. Cancer,* 11, 137, 1957.
7. **Gordon, A. H. and Louis, L. N.,** α_1 Acute phase globulins of rats, isolation and some properties, *Biochem. J.,* 113, 481, 1969.
8. **Urban, J., Chan, D., and Schreiber, G.,** A rat serum glycoprotein whose synthesis rate increases greatly during inflammation, *J. Biol. Chem.,* 254, 10565, 1979.
9. **Cole, T., Inglis, A., Nagashima, M., and Schreiber, G.,** Major acute phase alpha (1) protein in the rat: structure, molecular cloning, and regulation of mRNA levels, *Biochem. Biophys. Res. Commun.,* 126, 719, 1985.
10. **Cole, T., Inglis, A. S., Roxburgh, C. M., Howlett, G. J., and Schreiber, G.,** Major acute phase α_1-protein of the rat is homologous to bovine kininogen and contains the sequence for bradykinin. Its synthesis is regulated at the mRNA level, *FEBS Lett.,* 182, 57, 1985.
11. **Fung, W.-P. and Schreiber, G.,** Structure and expression of the genes for major acute phase α_1-protein (thiostatin) and kininogen in the rat, *J. Biol. Chem.,* 262, 9293, 1987.
12. **Darcy, D. A.,** A rat α_1-globulin associated with growth, in *Protides of the Biological Fluids,* Vol. 10, Peeters, H., Ed., Pergamon Press, Oxford, 1962, 131.
13. **Urban, J., Nagashima, M., and Schreiber, G.,** A hepatic precursor form of the major acute phase α_1-protein in the rat, *Biochem. Int.,* 4, 75, 1982.
14. **Esnard, F. and Gauthier, F.,** Rat α_1-cysteine proteinase inhibitor — an acute phase reactant identical with α_1-acute phase globulin, *J. Biol. Chem.,* 258, 12443, 1983.
15. **Urban, J., Chelladurai, M., Millership, A., and Schreiber, G.,** The kinetics *in vivo* of the synthesis of albumin-like protein and albumin in rats, *Eur. J. Biochem.,* 67, 477, 1976.
16. **Mancini, G., Carbonara, A. O., and Heremans, J. F.,** Immunochemical quantitation of antigens by single radial immunodiffusion, *Immunochemistry,* 2, 235, 1965.
17. **Yphantis, D. A.,** Equilibrium ultracentrifugation of dilute solutions, *Biochemistry,* 3, 297, 1964.
18. **Allen, A. K. and Neuberger, A.,** The quantitation of glucosamine and galactosamine in glycoproteins after hydrolysis in p-toluenesulphonic acid, *FEBS Lett.,* 60, 76, 1975.
19. **Warren, L.,** The thiobarbituric acid assay of sialic acids, *J. Biol. Chem.,* 234, 1971, 1959.
20. **Dubois, M., Gilles, K. A., Hamilton, J. K., Rebers, P. A., and Smith, F.,** Colorimetric method for determination of sugars and related substances, *Anal. Chem.,* 28, 350, 1956.
21. **Simpson, R. J., Neuberger, M. R., and Liu, T.-Y.,** Complete amino acid analysis from a single hydrolysate, *J. Biol. Chem.,* 251, 1936, 1976.
22. **Chen, Y. H. and Yang, J. T.,** A new approach to the calculation of secondary structures of globular proteins by optical rotatory dispersion and circular dichroism, *Biochem. Biophys. Res. Commun.,* 44, 1285, 1971.
23. **Cole, T. and Schreiber, G.,** Synthesis of thiostatins (major acute phase α_1 proteins) in different strains of *Rattus norvegicus, Comp. Biochem. Physiol.,* 93B, 813, 1989.
24. **Anderson, K. P. and Heath, E. C.,** The relationship between rat major acute phase protein and the kininogens, *J. Biol. Chem.,* 260, 12065, 1985.
25. **Furuto-Kato, S., Matsumoto, A., Kitamura, N., and Nakanishi, S.,** Primary structures of the mRNAs encoding the rat precursors for bradykinin and T-kinin, *J. Biol. Chem.,* 260, 12054, 1985.
26. **Kageyama, R., Kitamura, N., Ohkubo, H., and Nakanishi, S.,** Differential expression of the multiple forms of rat prekininogen mRNAs after acute inflammation, *J. Biol. Chem.,* 260, 12060, 1985.
27. **Anderson, K. P., Martin, A. D., and Heath, E. C.,** Rat major acute phase protein: biosynthesis and characterization of a cDNA clone, *Arch. Biochem. Biophys.,* 233, 624, 1984.

28. **Ohkubo, I., Kurachi, K., Takasawa, T., Shiokawa, H., and Sasaki, M.,** Isolation of a human cDNA for α_2-thiol proteinase inhibitor and its identity with low molecular weight kininogen, *Biochemistry,* 23, 5691, 1984.
29. **Nawa, H., Kitamura, N., Hirose, T., Asai, M., Inayama, S., and Nakanishi, S.,** Primary structures of bovine liver low molecular weight kininogen precursors and their two mRNAs, *Proc. Natl. Acad. Sci. U.S.A.,* 80, 90, 1983.
30. **Urban, J., Inglis, A. S., Edwards, K., and Schreiber, G.,** Chemical evidence for the difference between albumins from microsomes and serum and a possible precursor-product relationship, *Biochem. Biophys. Res. Commun.,* 61, 494, 1974.
31. **Urban, J. and Schreiber, G.,** Biological evidence for a precursor protein of serum albumin, *Biochem. Biophys. Res. Commun.,* 64, 778, 1975.
32. **Edwards, K., Schreiber, G., Dryburgh, H., Urban, J., and Inglis, A. S.,** Synthesis of albumin via a precursor protein in cell suspensions from rat liver, *Eur. J. Biochem.,* 63, 303, 1976.
33. **Schreiber, G., Urban, J., and Edwards, K.,** Possible functions of the oligopeptide extension in the albumin precursor, *J. Theor. Biol.,* 60, 41, 1976.
34. **Edwards, K., Schreiber, G., Dryburgh, H., Millership, A., and Urban, J.,** Biosynthesis of albumin via a precursor protein in Morris hepatoma 5123 TC, *Cancer Res.,* 36, 3113, 1976.
35. **Urban, J., Chelladurai, M., Millership, A., and Schreiber, G.,** The kinetics *in vivo* of the synthesis of albumin-like protein and albumin in rats, *Eur. J. Biochem.,* 67, 477, 1976.
36. **Schreiber, G., Dryburgh, H., Millership, A., Matsuda, Y., Inglis, A., Phillips, J., Edwards, K., and Maggs, J.,** The synthesis and secretion of rat transferrin, *J. Biol. Chem.,* 254, 12013, 1979.
37. **Millership, A., Edwards, K., Chelladurai, M., Dryburgh, H., Inglis, A. S., Urban, J., and Schreiber, G.,** N-terminal amino acid sequence of proalbumin from inbred Buffalo rats, *Int. J. Pept. Protein Res.,* 15, 248, 1980.
38. **Nagashima, M., Urban, J., and Schreiber, G.,** Intrahepatic precursor form of rat α_1-acid glycoprotein, *J. Biol. Chem.,* 255, 4951, 1980.
39. **Weigand, K., Dryburgh, H., and Schreiber, G.,** Der Nachweis einer intrazellulären Vorstufe von α_1-Antitrypsin (AT) in menschlicher Leber, *Verh. Dtsch. Ges. Inn. Med.,* 87, 888, 1981.
40. **Nagashima, M., Urban, J., and Schreiber, G.,** Identical NH_2-terminal amino acid sequence of the intrahepatic and the secreted form of rat α_1-acid glycoprotein, *J. Biol. Chem.,* 256, 2091, 1981.
41. **Edwards, K., Fleischer, B., Dryburgh, H., Fleischer, S., and Schreiber, G.,** The distribution of albumin precursor protein and albumin in liver, *Biochem. Biophys. Res. Commun.,* 72, 310, 1976.
42. **Schreiber, G., Dryburgh, H., Weigand, K., Schreiber, M., Witt, I., Seydewitz, H., and Howlett, G.,** Intracellular precursor forms of transferrin, α_1-acid glycoprotein, and α_1-antitrypsin in human liver, *Arch. Biochem. Biophys.,* 212, 319, 1981.
43. **Schreiber, G., Howlett, G., Nagashima, M., Millership, A., Martin, H., Urban, J., and Kotler, L.,** The acute phase response of plasma protein synthesis during experimental inflammation, *J. Biol. Chem.,* 257, 10271, 1982.
44. **Schreiber, G. and Urban, J.,** The synthesis and secretion of albumin, in *Reviews of Physiology, Biochemistry and Pharmacology,* Vol. 82, Springer-Verlag, New York, 1978, 27.
45. **Schreiber, G., Aldred, A. R., Thomas, T., Birch, H. E., Dickson, P. W., Tu, G.-F., Heinrich, P. C., Northemann, W., Howlett, G. J., de Jong, F. A., and Mitchell, A.,** Levels of messenger ribonucleic acids for plasma proteins in rat liver during acute experimental inflammation, *Inflammation,* 10, 59, 1986.
46. **Schreiber, G., Tsykin, A., Aldred, A. R., Thomas, T., Fung, W.-P., Dickson, P. W., Cole, T., Birch, H., de Jong, F. A., and Milland, J.,** The acute phase response in the rodent. *Ann. N.Y. Acad. Sci.,* 557, 61, 1989.
47. **Okamoto, H. and Greenbaum, L. M.,** Kininogen substrates for trypsin and cathepsin D in human, rabbit and rat plasmas, *Life Sci.,* 32, 2007, 1983.
48. **Pagano, M., Engler, R., Esnard, F., and Gauthier, F.,** On the interaction between human liver cathepsin L and the two cysteine proteinase inhibitors present in human serum, in *Marker Proteins in Inflammation,* Vol. 2, Arnaud, P., Ed., Walter de Gruyter, Berlin, 1984, 203.
49. **Takio, K., Towatari, T., Katunuma, N., Teller, D. C., and Titani, K.,** Homology of amino acid sequences of rat liver cathepsins B and H with that of papain, *Proc. Natl. Acad. Sci. U.S.A.,* 80, 3666, 1983.

C. Systems in which to Study Regulation of Acute Phase Proteins

Chapter 14

EXPERIMENTAL SYSTEMS FOR STUDYING HEPATIC ACUTE PHASE RESPONSE

Kwang-Ai Won, Susana P. Campos, and Heinz Baumann

TABLE OF CONTENTS

0-8493-6913-4/93/$0.00 + $.50

I. DEFINITION OF THE HEPATIC ACUTE PHASE RESPONSE

The hepatic acute phase response is characterized by a coordinate change in the production of a subset of plasma proteins (the acute phase plasma proteins, APP).[1] Although there is no universally accepted definition of an APP, it has been proposed that an APP is a plasma protein whose concentration increases by as much as 25% during acute inflammation or injury.[2] However, since individual variations in the circulating concentration of a given plasma protein can exceed 25%, an inflammation-dependent increase of at least 100% above basal level has also been proposed as a definition of a positive APP.[3] The qualitative and quantitative patterns as well as the kinetics of APP expression established *in vivo* serve as standards for the reproduction of the hepatic response in tissue culture systems. These features guide the search for potential inflammatory mediators and the characterization of cellular and molecular mechanisms of APP gene regulation. Given the variability in response patterns among tissue culture systems, the relevance of information obtained from *in vitro* studies has to be verified *in vivo*. Such follow-up experiments in animals are greatly facilitated because specific regulatory processes are predictable.

II. *IN VIVO* STUDIES

The nature and magnitude of the *in vivo* APP response strongly depends on the conditions used to elicit the hepatic acute phase reaction and the time points at which the response is monitored. The hepatic acute phase response in man has been studied in limited conditions, such as during elective surgery, traumatic and thermal injuries, parturition, and sepsis.[4,5] The obvious restrictions on the use of human subjects for acute phase research has forced investigators to create animal models. Mouse, rat, rabbit, chicken, and fish have all been employed as experimental animals. Most of the current studies utilize rats and mice, mainly because of the wealth of information accumulated for the rodent systems and the accessibility of these animals.

Scalding,[11] injection of turpentine, endotoxin, lipopolysaccharides, Freund's adjuvant,[6-10] or potential mediators[12-18] are the experimental means to induce an acute phase response. Although the qualitative pattern of the APP response does not seem to be significantly influenced by the type of injury sustained,[19,20] there is a limited linear relationship between the severity of the injury and the degree of plasma protein expression.[21,22] The final result is quite similar for every type of injury, but not every type of injury utilizes the same systemic signaling pathway. One striking example is that the endotoxemia-induced acute phase is paralleled by an activation of hepatic expression of interleukin-6 (IL-6),[23] whereas sterile tissue damage causes the same type of acute phase response, but without hepatic expression of IL-6.[24] The type and dosage of mediator administered, emergence of mediator in the tissue, and presence of other inflammation-controlling cytokines should be considered[27] when the role of a potential mediator is being investigated, particularly when antibodies which block the biological activity of suspected mediators are used.[25,26]

The most frequently measured acute phase response is the increase in the plasma concentration of APP. The plasma composition has been analyzed by two-dimensional polyacrylamide gel electrophoresis[6,28] or crossed immunoelectrophoresis using polyspecific antibodies against total serum proteins.[29] A specific protein is quantitated by immunoprecipitation[30] using Western-blot analysis,[31] rocket immunoelectrophoresis,[29,32] enzyme-limited immunosorbant assay (ELISA),[33] or radioimmunoassay.[34] Representative APPs include C-reactive protein (CRP) and serum amyloid A (SAA) in humans,[2,19,35] α_1-acid glycoprotein (AGP), thiostatin, and α_2-macroglobulin in rats,[36] and serum amyloid A (SAA) and serum amyloid P (SAP) in mouse,[12,33] fibrinogen and fibronectin in chicken;[8] and CRP, SAA, haptoglobin, ceruloplasmin, α_2-macroglobulin, and transferrin in rabbit.[10]

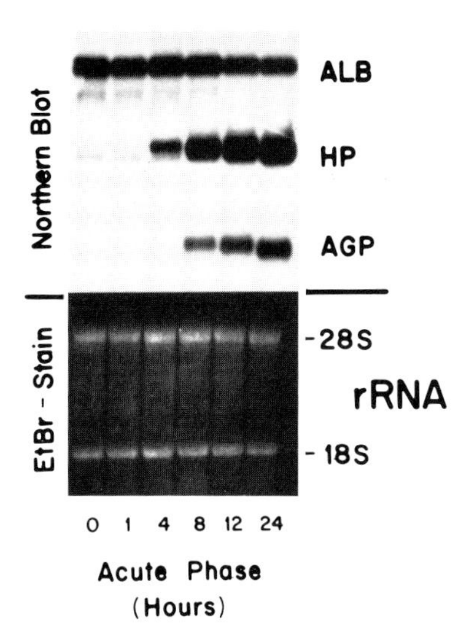

FIGURE 1. Hepatic acute phase in the mouse. Three-month-old males of *M. domesticus* (strain C57BL/6J) received a subcutaneous injection of 25 μl of turpentine. At the indicated time, the animals were killed and RNA was extracted from the liver. An untreated animal served as control (0 h acute phase). Equal amounts (15 μg) of RNA were separated on a formamide/formaldehyde-containing agarose gel and stained with ethidium bromide to verify equal loading (lower panel: pattern of 18 and 28S rRNA bands). The RNAs were transferred to nitrocellulose and hybridized to ^{32}P-labeled cDNA encoding mouse albumin, haptoglobin, and AGP. The autoradiograms were exposed for 6 h. The size reduction of haptoglobin and AGP mRNA during the 24-h period is probably due to trimming of the polyadenylate tail.[46]

The change in APP levels is generally only an increase of a few fold, although human CRP, human and mouse SAA, and rat α_2-macroglobulin can be stimulated several hundredfold.[1,36] Such an impressive magnitude of change is mainly due to the extremely low basal levels of these proteins in healthy individuals. Maximal levels of acute phase stimulation in small animals are generally recorded between 24 and 48 h,[37-39] whereas in humans, peak values are maintained up to several days following tissue injury.[40]

An alternative approach for examining the acute phase response involves the analysis of APP gene expression in the liver. This provides a direct measure of the hepatic response and circumvents the problems that are associated with plasma composition analysis, such as consideration of pool size and the dynamics of the vascular and extravascular compartments. The animals are treated with an optimal dose of inflammation-inducing agents or irritants and sacrificed at various times after treatment. The total composition of liver mRNAs from treated animals is analyzed and compared to that from untreated, healthy control animals (Figure 1). The change in liver mRNA can be visualized by *in vitro* translation of mRNA followed by one- or two-dimensional electrophoresis of protein products.[6,22,41] mRNAs specific for APPs are quantitated by Northern-blot analysis using either cloned genomic sequences or cDNAs (Figure 1),[9,15,22,38,39,42-44] or RNA protection analysis.[45] A comparison of the changes in APP liver mRNAs and the plasma concentrations of these APPs revealed that the increase in hepatic mRNA level was accompanied by a concomitant increase in the plasma level.[38,39] However, when transcription rates of APP genes were measured in liver by nuclear run-on reactions, the change in transcription rates and the corresponding mRNA levels was not always proportional.[38,46] This discrepancy suggests the existence of posttranscriptional regulatory mechanisms which may affect the processing of the primary transcript[46,47] or mRNA stability.[48] The concentration of proteins produced by hepatocytes may also be influenced by various posttranslational control mechanisms involving the glycosylation, transport, and secretion of these proteins.[49,50]

The phenomenon of the acute phase response is evolutionarily conserved in vertebrates,[1,19] but there are significant variations in APP expression between different species. For example, α_2-macroglobulin and thiostatin are characteristic of rats,[9,51] and CRP of humans and rabbits.[35] APPs also differ in their kinetics of expression. Whereas mRNA and protein expression of most rodent APPs reach a peak in 18 to 24 h,[22,39,52] mouse SAA,[48] human CRP in transgenic mouse,[53] and rat fibrinogen[42] genes are maximally expressed at an earlier time point.

Since injury at a distant site induced the synthesis of proteins from the liver, it was assumed that humoral factors must mediate the signal from the site of injury to the liver.[1] Analysis of the acute phase response in adrenalectomized rats revealed that glucocorticoids and noncorticosteroid mediators separately contributed to the acute phase phenotype.[52,54-56] IL-1 was the first cytokine identified as a modulator of the acute phase response.[13,57,58] The APP-regulating activity of other cytokines observed in tissue culture cells subsequently suggested the involvement of tumor necrosis factor-α, IL-6, IL-11, leukemia inhibitory factor, and oncostatin M in the *in vivo* acute phase response. Moreover, a modulating activity on the level of APP gene expression has been ascribed to interferon-γ,[59] transforming growth factor-β,[60,61] and insulin.[62,63] A suggested criterion for the involvement of humoral mediators in the acute phase response *in vivo* is that the circulating concentration of humoral mediators should correlate with the changes in APP gene expression, and that the increase of the mediator should temporally precede the rise in APP concentration. The correlation between the level of the mediators and changes in APP expression is, however, not always perfect. For example, the peak concentration of circulating IL-6 was detected between 4 and 6 h after surgery/trauma, and the concentrations of CRP and α_1-antichymotrypsin were still rising at 24 h.[5,64]

In vivo studies showed that it was difficult to evaluate the direct and indirect effects of potential mediators on the regulated pattern of APP expression. Indeed, studies can even fail to find the expected correlations between the concentration of a suspected humoral mediator, e.g., LIF, and changes in APP expression.[65-67] The intricate network of cytokine interactions,[68] along with the induction of glucocorticoids by IL-1 and IL-6,[69,70] interferes with the evaluation of the direct effect of cytokines in controlling the hepatic acute phase response.

III. TISSUE CULTURE SYSTEMS

A. PRIMARY CULTURES OF HEPATOCYTES

A variety of hepatic tissue culture systems have demonstrated a direct interaction of cytokines and steroids with liver cells which result in activation of the plasma protein genes. Short-term cultures of primary adult hepatocytes from rat, mouse, rabbit, and human exhibit regulation of most of the APP genes in response to hormone mediators.[10,33,71-78] However, the unstable phenotypes, as manifested by the loss of liver-specific properties in long-term cultures[79] and potential contamination with other cell types,[80] necessitated a cautious interpretation of results obtained and thus limited their utility.

Despite difficulties maintaining the phenotype of adult hepatocytes *in vitro,* conditions for culturing these cells have been established which have allowed their use not only for defining hormone response in long-term cultures, but also for testing regulatory properties of cloned plasma protein gene elements[81] (and citations in Reference 81). Although the acute phase response is considered to be most prominently manifested in adult liver, fetal and neonatal hepatocytes have also been used,[82] since these cells are easy to isolate[83] and have the advantage of growing well in tissue culture. A disadvantage, however, is that several major acute phase genes in fetal hepatocytes are still developmentally inactive or minimally activated.[84]

B. HEPATOMA CELLS

Simian virus-40 (SV40) immortalized hepatocytes have been proposed as a more suitable alternative to primary hepatocyte culture.[85] These cells provided the ease of being passaged as immortalized cells, and retained their responsiveness to most of the inflammatory mediators and their ability to regulate a broad range of APP genes characteristic of "normal" liver cells.[86] However, the general use of SV40-immortalized cells never materialized because at the same time period, the search for regulatable hepatic cells identified hepatoma cell lines which also displayed a comparable regulated expression of APP genes. Several human and rat hepatoma cell lines have since become preferred experimental systems in numerous laboratories.

Hepatoma cell lines have greatly facilitated analyses of the effects of potential mediators at the level of receptor-signal transduction and transcriptional activation of APP genes. Several lines have emerged as specifically useful for acute phase analysis, including the rat hepatoma lines Reuber H-35,[87,88] Fao,[89] FAZA,[73] Fto2B,[90] and the human hepatoma lines Hep G2,[91] Hep 3B2,[92] and PLC/PRF/5.[93] An attractive feature of hepatoma cells over primary hepatocytes is the convenience of maintaining these in continuous culture and the stable clonal phenotypes. The pattern of expression of APP genes and their regulation, however, are not always identical to primary hepatocytes.[125] In addition, variation seems to exist within subclones of a given cell line.[94] For example, CRP and SAA in PLC are only sensitive to IL-6,[93] whereas in Hep 3B these require both IL-1 and IL-6.[95,96] Hep G2 cells do not coregulate CRP with other APPs.[92] Therefore, the properties of each cell line with regard to the hormone-specific responses and the spectrum of APP gene regulation should be clearly defined before general conclusions are made.[3,97] Availability of subclones of the hepatoma cell lines deficient in specific cytokine responses can be useful in further delineation of signal transduction mechanisms.[3,94]

Although hepatoma cells can be maintained in continuous culture for extended periods of time without appreciable loss of specific function, periodic clonal analysis of both rat and human hepatoma cell lines has, nevertheless, revealed phenotypic drifting and the appearance of variant clones.[3] H-35 cells provide a striking example. The clone T-7 of H-35 cells[88] has been subjected to 18 rounds of successive single-cell cloning, yielding the clonal line T-7-18. This line is exceptional in that all major positive APP genes are inducible to a level approximating that in normal liver.[88] Since in these cells the basal expression of most APP genes is low to nondetectable, the numerical values for the magnitude of stimulation of APP genes are extremely high and can exceed by several hundredfold those measured in rat liver.[98] Clonal analysis of T-7-18 revealed that one specific regulatory phenotype arises with a frequency of 1×10^{-4} to 1×10^{-5}. Clone 237-7 phenotype is recognized by the extremely reduced cytokine response in comparison with H-35 (T-7-18) (Figure 2). The frequent occurrence of this pleiotropic phenotype suggests the loss of a key factor required for cytokine signaling or coordinate gene regulation.

IV. HORMONE REGULATION OF APP GENES IN TISSUE CULTURE SYSTEM

The initial search for factors which are active on liver cells and which modulate APP gene expression revealed numerous potential cellular sources. Analysis of conditioned media from lipopolysaccharide-stimulated peripheral blood monocytes or macrophages,[71,72,99,100] activated T-cells,[93] fibroblasts,[99,101] monocytic leukemia cells,[102] and squamous carcinoma cells[91] indicated the presence of factors with specific liver-regulating activities. Recognizing the complex interaction of potential factors, it has been concluded that optimal expression of a given APP in a given hepatic culture system requires a specific combination of hor-

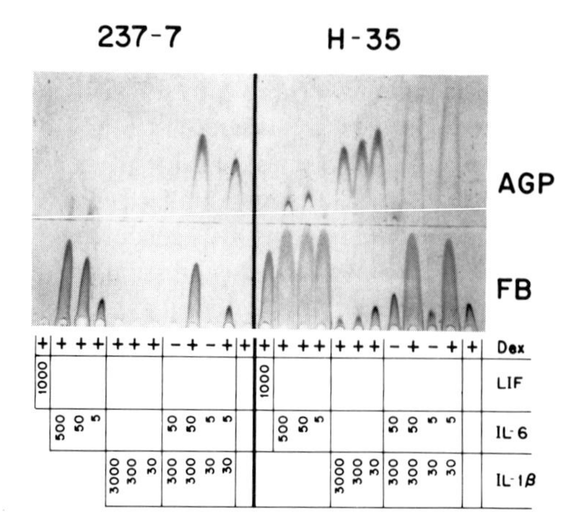

FIGURE 2. Loss of cytokine response in rat hepatoma cells. H-35 (T-7-18) cells[88] were subjected to single-cell cloning by the limited dilution technique. One variant clone, 237-7, and one normal clone were expanded and passaged into 24-well cluster plates. When cells reached confluency (for both ~3 × 10^5 cells per well), each well was treated for 24 h with 300 μl of serum-free medium containing the indicated human recombinant cytokines at the concentrations marked in units per milliliter. The concentration of dexamethasone (Dex) was 1 μ*M*. The amounts of AGP and fibrinogen secreted in the culture medium during the treatment period were determined by rocket immunoelectrophoresis on a double-section agarose gel. The upper section contained antibodies against rat AGP and the lower section against rat fibrinogen. The Coomassie blue-stained pattern is shown.

mones.[3,97,103] IL-1, TNFα, IL-6, LIF, and glucocorticoids act as general mediators of the hepatic acute phase response, whereas TGFβ, IFNγ, glucagon, insulin, epidermal growth factor (EGF), catecholamines, and prostaglandins all have the potential to influence the expression of specific APPs.[60,104]

Acute inflammation *in vivo* and cytokines *in vitro* cause changes in the transcription of target genes.[38,43,46,56,105] The transcriptional activation of most APP genes by cytokines and dexamethasone in tissue culture cells is immediate and is not dependent on *de novo* protein synthesis[73,105] (Figure 3). Regulation of the APP genes differs from that of the immediate early transcription factor genes in that cotreatment with cycloheximide and cytokines does not result in superinduction. Although there is a temporally coordinated stimulation of the "early" APP genes, such as fibrinogens, hemopexin, thiostatin, and haptoglobin in rat hepatic cells, the activation of α_2-macroglobulin and AGP genes in some cells such as H-35 is delayed.[105] The fact that there is a temporal difference in the activation of APP genes suggests that not all genes are controlled by the same regulatory mechanism.

With the aid of hepatoma cells as assay systems and the stimulation of specific APPs as indicators, the present list of APP-regulating factors has been assembled. Hepatoma cells proved their value not only for identifying liver-regulating factors, but also for dissecting the precise contribution of each factor to the overall cell response (representative example in Figure 4). Utilizing rat hepatoma cells, IL-1 and TNFα have been similarly shown to stimulate type 1 APP, including AGP, complement C3, haptoglobin, hemopexin, and angiotensinogen.[87-89,105-107] IL-6, IL-11, LIF, and oncostatin M share similar activity in that they stimulate all type 2 APP genes, including α_2-macroglobulin, α_1-antichymotrypsin (contrapsin), thiostatin, α_1-antitrypsin, and fibrinogen, and enhance IL-1 action on type 1 APP genes.[73,87,90,99] Glucocorticoids, such as dexamethasone, exert a strong synergistic action with cytokines that is especially prominent for AGP and α_2-macroglobulin.[90,108]

Although there appears to be a redundance between the action of IL-6, IL-11, LIF, and oncostatin M, the comparison of the cell response nevertheless indicates quantitative dif-

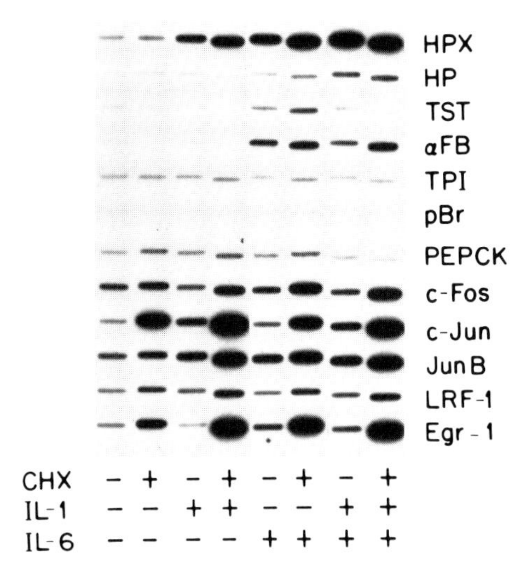

FIGURE 3. Transcription activation of APP genes. Duplicate cultures of confluent H-35 cells (clone T-7-18) were treated for 30 min with serum-free medium containing 100 U/ml human IL-1β or 100 U/ml human IL-6. One culture from each duplicate was pretreated for 5 min with 10 μg/ml cycloheximide (CHX) and maintained in the presence of the inhibitor. Nuclei were prepared and subjected to the run-on reaction.[98] The ^{32}P-labeled RNA (35×10^6 cpm each) were hybridized to nitrocellulose filter strips carrying slot-blotted cDNA encoding the following proteins: hemopexin (HPX), haptoglobin (HP), thiostatin (TST), α-fibrinogen (αFB), triosephosphate isomerase (TPI), phosphoenolpyruvate carboxykinase (PEPCK), c-*Fos*, c-*Jun*, *Jun*B, liver-regenerating factor 1 (LRF-1), and early growth-regulated factor 1 (Egr-1). The autoradiogram was exposed for 24 h.

ferences. LIF, IL-11, and oncostatin M are less effective than IL-6 on type 2 APPs and fail to produce a prominent synergism with dexamethasone. IL-11, but not LIF or oncostatin M, is equal to IL-6 in synergizing with IL-1 on type 1 APP.[65,109,110] In rat hepatic cells, the response produced by IL-1, IL-6, and dexamethasone from type 1 APP and by IL-6 and dexamethasone from type 2 APPs is never exceeded by any other combination of factors.[88,109-111]

The gene specificity, kinetics, and concentration dependence of cytokine action seem to be generally conserved in rodent and human hepatic cells, whether tested in primary cultures or established cell lines.[78,91,97] Species-specific differences in the APP patterns have to be taken into consideration; for instance, SAA responds as a type 1 APP in mouse[58,112] but not in rabbit hepatocytes,[10] and human haptoglobin has the characteristics of a type 2 APP.[43,113] Comparison of the cytokine response achieved in separate hepatoma cell lines from the same species revealed distinct differences which have been ascribed to the abnormal phenotypes of these cells.[95] Some possible mechanisms involved include loss of specific signaling components or loss of receptor function. The latter is a primary cause for some of the differences observed between cell lines: Hep 3B cells lack LIF and oncostatin M receptors and have greatly reduced IL-11 receptor activity, while Hep G2 cells lack IL-11 receptor activity.[98,110] By introducing expression vectors for the appropriate receptor components, the receptor-deficient cell lines will serve as a target for receptor function reconstitution. The approach will be essentially as described for reconstituting the rat IL-6 receptor function in Hep 3B cells.[114]

The treatment of primary hepatocyte and hepatoma cells has defined the specific activity for various cytokines. In each case, the predicted role of the cytokine *in vivo* has to be confirmed. Obviously, this task is not a simple one. Previous *in vivo* studies have shown that injection of IL-1,[12] TNFα,[14] IL-6,[15] or LIF[66] always results in a change in the expression of APP genes. However, an assessment of how much is due to direct or indirect action of the cytokines on liver remains to be worked out.

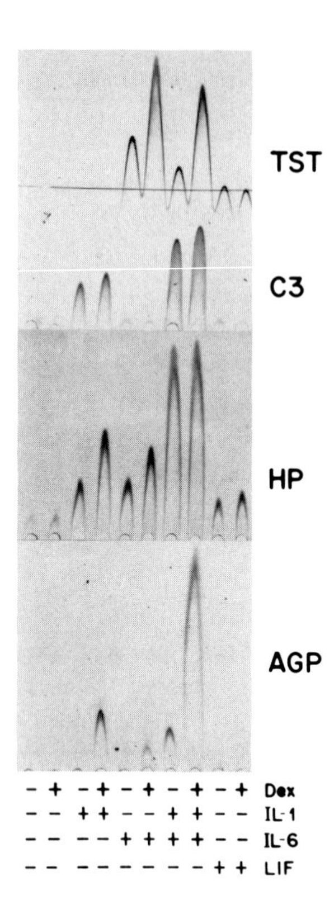

FIGURE 4. Cytokine-specific regulation of APPs. Confluent monolayers of H-35 cells (clone T-7-18) were treated for 24 h with serum-free medium containing 1 μM dexamethasone, 100 U/ml human recombinant IL-1β, 100 U/ml human IL-6, and/or 10 U/ml LIF. The effect on the production of thiostatin (TST), complement C3, haptoglobin (HP), and AGP was determined by rocket immunoelectrophoresis.

V. SYSTEMS USED TO STUDY MOLECULAR MECHANISMS OF APP GENE REGULATION

A. MOLECULAR ANALYSIS IN TISSUE CULTURE

The availability of cloned APP genes from various species has stimulated the search for *cis*-acting elements controlling their hormonal inducibility.[96,108,113,115-120] Since hormone-specific regulation of some APP genes has been achieved by introducing entire structural genes into nonhepatic cell types such as mouse fibroblasts (L-cells) and HeLa cells,[115,121-124] these cells have also been used for functional dissection of potential *cis*-acting regulatory sequences in APP genes.

Numerous independent studies confirmed that Hep 3B and Hep G2 cell lines[125] are preferred as a test system for the function of cytokine response elements. These cells have not only retained liver-specific gene expression and a broad hormone responsiveness, but also exhibit a high DNA transfection efficiency.[126] Using Hep G2 cells, response elements (REs) were localized in rat and mouse AGP genes,[76,119] rat haptoglobin gene (GRE, IL-1, and IL-6 REs),[127] rat angiotensinogen gene (GRE and IL-1 REs),[107,119a] rat fibrinogen gene (GRE and IL-6 REs);[128] and mouse complement factor B gene (IL-1 REs).[129] Hep 3B cells were used to define the IL-6 REs of the genes for the human haptoglobin[113,130] and hemopexin.[116] The regulation of CRP gene elements W as studied in Hep 3B[96,131] and in PLC/PRF/5.[132]

Not only human, but also rat hepatoma cells were used for characterization of the regulatory elements of APP genes. The glucocorticoid response element of the rat AGP gene was determined in HTC cells,[133] and the cytokine RE of rat haptoglobin, AGP, and β-fibrinogen genes was identified in H-35 cells.[128] In most instances, the cytokine-specific changes in the expression of the transgene construct followed that seen with the endogenous gene. However, the magnitude of the cytokine response was not always comparable. The difference has been ascribed in part to the loss or modified function of specific regulatory sequences[132] and improper presentation of regulatory sequences in the test plasmid.[98]

The described hormone REs have been used for identifying *trans*-acting factors which may mediate the hormone signal.[106,107,134] NF-κB has been proposed to be involved in transcriptional control of human and mouse SAA, rat angiotensinogen, and mouse complement factor B in HeLa, Hep 3B, and Hep G2 cells,[119a,124,129,135] as shown by *in vitro* DNA binding assays. IL-6 DBP was identified in rat liver[134] by cloning the cDNA encoding the protein binding to the IL-6 RE of the human haptoglobin, hemopexin, and CRP genes. The mouse homolog of IL-6 DBP, called AGP/EBP, was independently cloned as the protein binding to the promoter proximal sequence of the AGP gene.[136] These factors proved to be members of the C/EBP gene family which share sequence-similar DNA binding domains and therefore interact with identical, or near identical sequences.[137-139]

With cloning of the DNA-binding proteins, their function as transcription-controlling factors for APP genes could be established by utilizing hepatoma cells as the primary test system.[105,134,140] By cotransfecting a mixture of a high-expression plasmid for one of the C/EBP isoforms and a cytokine-response reporter gene construct, the positive effect of C/EBP isoforms on basal and cytokine-induced expression has been observed. Surprisingly, the spectrum of genetic targets of the C/EBP isoform is diverse and includes REs for IL-6, IL-1, and glucocorticoids.[105] Although C/EBP binding sites have been found in various APP genes, their functional involvement in the cytokine response still remains to be better defined.[141-143] Transient and stable transfection of transcription factor expression vectors and reporter gene constructs into hepatoma cells is a major tool for establishing a role for C/EBP or any other transcription factor in APP gene expression.

The observation that C/EBP isoforms are capable of increasing the expression of plasmids containing the C/EBP binding site in the absence of cytokine treatment[105,134] suggests that these transcription factors might also function as determinants for the basal-level activity of plasma protein genes. Indeed, many hepatoma cells which show low to nondetectable levels of APPs also have low expression of C/EBP isoforms.[140] The contribution of C/EBP to the basal expression of APP was illustrated, for instance, by stably introducing a mouse C/EBPα expression plasmid into H-35 cells. Cell lines with elevated C/EBPα were obtained which showed a corresponding increase in several, but not all, APP genes which have known C/EBP-sensitive elements[105] (Figure 5). A major value of hepatoma cell lines with constant, but different levels of C/EBP expression is that the immediate effects of cytokines on the transcription of both endogenous and transiently transfected reporter gene constructs can be measured as a function of the C/EBP concentrations. Analysis of C/EBP function at later stages (after ~30 min) of APP gene regulation is complicated by the fact that the treatment of hepatic cells with IL-1, IL-6, LIF, or dexamethasone is invariably followed by an increase of the endogenous C/EBPβ (or IL-6 DBP,[134] NF-IL-6,[142] or AGP/EBP).[136] Stimulation of the endogenous C/EBPβ gene is not immediate, and mRNA and protein activity reach a maximum level after 4 to 18 h.[105,144] Therefore, the stimulation of APP genes in cytokine-treated cells, and probably in liver cells as well, is brought about by two events: (1) an immediate activation step that is independent of protein synthesis and (2) a subsequent change in transcription factor concentration which contributes to sustained high transcription rates for the APP gene. Obviously, the latter event is dependent upon protein synthesis and

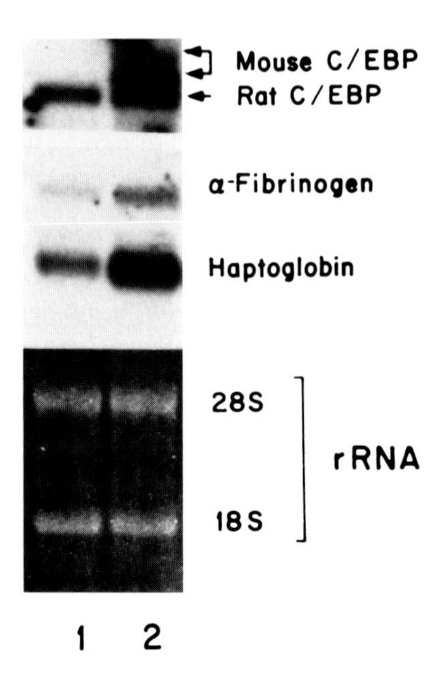

FIGURE 5. Effect of C/EBPα on basal expression of plasma proteins. H-35 cells (clone T-7-18) were stably transformed with a mixture of pCD-mC/EBP, SV40 promoter-containing expression plasmid containing the cDNA to C/EBPα of *M. caroli*, and pSV2neo.[105] Clone 2 was expanded and RNA was extracted from these cells as well as from nontransformed H-35 cells. The mRNAs for haptoglobin, α-fibrinogen, and the endogenous rat and transgene-derived mouse C/EBPα were analyzed by Northern blotting. The latter mRNA appear as two bands larger than the rat C/EBP mRNA due to transcribed 3′ untranslated SV40 sequences.[14]

explains why in some cell systems, stimulation of APP genes is severely reduced or prevented in the presence of cycloheximide.[88]

B. TRANSGENIC MICE

Although *in vitro* transfection studies provide a rapid means of identifying regulatory gene elements, the question remains whether these function *in vivo* as they do in tissue culture. The whole gene or selected gene sequences can be introduced into the genome and exposed to the normal liver environment of the animal in transgenic mice.[145] The regulation and tissue specificity of the gene expression have been studied in transgenic mice and compared to those shown in the natural environment from which they have been isolated.

CRP is highly inducible in man, but not in mice. In transgenic mice, human CRP behaves as a major APP, as it does in man, suggesting that the species specificity is due to the interaction of the individual gene regulatory regions with evolutionarily conserved *trans*-acting factors.[53]

Expression of rat AGP in transgenic mice verifies the observation from transfection experiments that a minimum of two upstream sequences are responsible for the inflammatory induction of rat AGP.[146] Although AGP is not normally expressed in the heart and spleen, low levels of transgene expression are detectable in these organs, suggesting that a *cis*-acting sequence responsible for silencing AGP in these tissues exists and that these sequences are missing or nonfunctional in the gene fragment tested.

In contrast to the "generic" activity of a cloned promoter of a human AGP-A gene, when introduced into hepatic and nonhepatic human cell lines, the identical promoter sequences in transgenic mice are expressed only in liver and stimulated only during inflammation.[147] Therefore, mechanisms other than simple cell-specific, *trans*-acting factors might be involved in differential gene expression. The observation that the liver specificity of human AGP gene expression cannot be reproduced in cultured cells emphasizes the importance of *in vivo* studies.

VI. FUTURE DIRECTION

Hepatic tissue culture cells have been instrumental in defining potential mediators of hepatic acute phase response, identifying the genetic targets of these mediators, and pinpointing components acting at the APP gene level. These cell systems will continue to be of crucial importance in the current quest to learn the pathways that link the mediators with the regulated APP genes. Identification of the possible *trans*-acting factors as the link between signaling pathways provides the investigator with a starting point for back-tracking on the underlying intracellular signal transduction. Several approaches are being pursued, each of which will depend on hepatic tissue culture cells, especially hepatoma cells. Receptor cDNAs and antibodies to the receptors will provide a way to analyze the effect of a specific hormone, predict possible signal pathways, and dissect the cross-talk between different signal transduction pathways.[68,148]

Currently addressed issues are the mechanisms of *trans*-membrane signaling by cytokine receptors and the nature of the intracellular messengers generated. Since many different hepatic and nonhepatic cell types are responsive to the same cytokines, one would expect that both general and cell type-specific regulatory elements are involved in signaling. A useful asset for defining the complexity of cytokine signaling in hepatic cells will undoubtedly be the hepatoma cells with variant response patterns (e.g., Figure 2). Because deviations from the normal phenotype are likely in each tissue culture system, only the combined results of several *in vitro* systems will yield answers to the true cellular and molecular mechanisms of acute phase gene regulation.

ACKNOWLEDGMENTS

Research in the authors' laboratory was supported by NIH grants CA26122 and DK33886. We thank Karen K. Morella and Gerald P. Jahreis for technical assistance, and Marcia Held for secretarial support.

REFERENCES

1. **Koj, A.,** Acute phase reactants, in *Structure and Function of Plasma Proteins,* Vol. 1, Allison, A. C., Ed., Plenum Press, New York, 1974, 73.
2. **Kushner, I.,** Acute phase proteins as disease markers, *Ann. N.Y. Acad. Sci.,* 389, 39, 1982.
3. **Baumann, H.,** Hepatic acute phase reaction in vivo and in vitro, *In Vitro,* 25, 115, 1989.
4. **Myers, M. A., Fleck, A., Sampson, B., Colley, C. M., Bent, J., and Hall, G.,** Early plasma protein and mineral changes after surgery: a two stage process, *J. Clin. Pathol.,* 37, 862, 1984.
5. **Fong, Y., Moldawer, L. L., Marano, M., Wei, H., Tatter, S. B., Clarick, R. H., Santhanam, U., Sherris, D., May, L. T., Sehgal, P. P., and Lowry, S. F.,** Endotoxemia elicits increased circulating β2-IFN/IL-6 in man, *J. Immunol.,* 142, 2321, 1989.
6. **Baumann, H. and Held, W. A.,** Biosynthesis and hormone-regulated expression of secretory glycoproteins in rat liver and hepatoma cells. Effect of glucocorticoids and inflammation, *J. Biol. Chem.,* 256, 10145, 1981.
7. **White, A. and Fletcher, T. C.,** The effects of adrenal hormones, endotoxin and turpentine on serum components of the plaice (*Pleuronectes platessa* L.), *Comp. Biochem. Physiol.,* 73C, 195, 1982.
8. **Amrani, D. L., Manzy-Melitz, D., and Mosesson, M. W.,** Effect of hepatocyte-stimulating factor and glucocorticoids on plasma fibronectin levels, *Biochem. J.,* 238, 365, 1986.

9. **Gehring, M. R., Shield, B. R., Northemann, W., de Bruijn, M. H. L., Kan, C.-C., Chain, A. C., Noonan, D. J., and Fey, G. H.,** Sequence of rat liver α_2-macroglobulin and acute phase control of its messenger RNA, *J. Biol. Chem.*, 262, 446, 1987.
10. **Mackiewicz, A., Ganapathi, M. K., Schultz, D., Samols, D., Rees, J., and Kushner, I.,** Regulation of rabbit acute phase protein biosynthesis by monokines, *Biochem. J.*, 253, 851, 1988.
11. **Sevalievic, L., Glibetic, M., Poznanovic, G., et al.,** Thermal injury-induced expression of acute phase proteins in rat liver, *Burns*, 14, 280, 1988.
12. **Sipe, J. D., Vogel, S. N., Ryan, J. L., McAdam, K. P. W. J., and Rosenstreich, D. L.,** Detection of a mediator derived from endotoxin-stimulated macrophages that induces the acute phase amyloid A response in mice, *J. Exp. Med.*, 150, 597, 1979.
13. **Kampschmidt, R. F. and Mesecher, M.,** Interleukin-1 from P388D: effects upon neutrophils, plasma iron, and fibrinogen in rats, mice and rabbits, *Proc. Soc. Exp. Biol. Med.*, 179, 197, 1985.
14. **Doucher, S. and Neta, R.,** Interaction of recombinant IL-1 and recombinant tumor necrosis factor in the induction of mouse acute phase proteins, *J. Immunol.*, 140, 2260, 1988.
14a. **Gresser, I., Delers, F., Guangs, N. T., Marion, S., Engler, R., Maruey, C., Soria, J., Fiers, W., and Travernier, J.,** Tumor necrosis factor induces acute phase proteins in rats, *J. Biol. Regul. Homeost. Agents*, 1, 173, 1987.
15. **Geiger, T., Andus, T., Klapproth, J., Hirano, T., Kishimoto, T., and Heinrich, P. C.,** Induction of rat acute phase proteins by interleukin 6 in vivo, *Eur. J. Immunol.*, 18, 717, 1988.
16. **Marinkovic, S., Jahreis, G. P., Won, G. G., and Baumann, H.,** IL-6 modulates the synthesis of a specific set of acute phase plasma proteins in vivo, *J. Immunol.*, 142, 808, 1989.
17. **Cornell, R. P.,** Acute phase responses after acute liver injury by partial hepatectomy in rats as indicators of cytokine release, *Hepatology*, 11, 923, 1990.
18. **Neta, R., Vogel, S. N., Sipe, J. P., Wong, G. G., and Nordan, R.,** Comparison of in vivo effects of human recombinant IL-1 and human recombinant IL-6 in mice, *Lymphokines Res.*, 7, 403, 1988.
19. **Kushner, I. and Mackiewicz, A.,** Acute phase proteins as disease markers, *Dis. Markers*, 5, 1, 1987.
20. **Glibetic, M. D. and Baumann, H.,** Influence of chronic inflammation on the level of mRNAs for acute phase reactants in the mouse liver, *J. Immunol.*, 137, 1616, 1986.
21. **Kageyama, R., Kitamura, N., Ohkubo, H., and Nakanishi, S.,** Differential expression of the multiple forms of rat prekininogen mRNA after acute inflammation, *J. Biol. Chem.*, 260, 12060, 1985.
22. **Goldberger, G., Bing, D. H., Sipe, J. D., Rits, M., and Colten, H. R.,** Transcriptional regulation of genes encoding the acute phase proteins CRP, SAA and C3, *J. Immunol.*, 138, 3967, 1987.
23. **Dei, S. K., McMaster, M. T., and Andrews, G. K.,** Endotoxin induction of murine metallothionine gene expression, *J. Biol. Chem.*, 265, 15267, 1990.
24. **Gauldie, J., Northemann, W., and Fey, G. H.,** IL-6 functions as an exocrine hormone in inflammation. Hepatocytes undergoing acute phase responses require exogenous IL-6, *J. Immunol.*, 144, 3804, 1990.
25. **Tracey, K. J., Fong, Y., Herse, D. G., Manogue, K. R., Lee, A. T., Kuo, G. C., Coury, S. F., and Cerami, A.,** Anti-cachectin/TNF monoclonal antibodies prevent septic shock during lethal bacteraemia, *Nature (London)*, 330, 662, 1987.
26. **Fong, Y., Tracey, K. J., Moldawer, L. L., Hesse, D. G., Manogue, K. R., Kenney, J. S., Lee, A. T., Kuo, G. C., Allison, A. C., Lowry, S. F., and Cerami, A.,** Antibodies to cachectin/tumor necrosis factor reduce interleukin 1β and interleukin-6 appearance during lethal bacteremia, *J. Exp. Med.*, 170, 1627, 1989.
27. **Heremans, H. and Billiau, A.,** The potential role of interferons and interferon antagonists in inflammatory disease, *Drugs*, 38, 957, 1989.
28. **Lonberg-Holm, K., Reed, D. L., Roberts, R. C., Hebert, R. R., Hillman, M. C., and Kutney, R. M.,** Three high molecular weight protease inhibitors of rat plasma. Isolation, characterization and acute phase changes, *J. Biol. Chem.*, 262, 438, 1987.
29. **Baumann, H.,** Electrophoretic analysis of acute phase plasma proteins, *Methods Enzymol.*, 163, 566, 1988.
30. **Macintyre, S. S., Schultz, D., and Kushner, I.,** Synthesis and secretion of C-reactive protein by rabbit primary hepatocyte cultures, *Biochem. J.*, 210, 707, 1983.
31. **Dente, L., Cilberto, G., and Cortese, R.,** Structure of the human α_1-acid glycoprotein gene: sequence homology with other human acute phase protein genes, *Nucleic Acids Res.*, 13, 3941, 1989.
32. **Magielska-Zero, D., Rokita, H., Cieszka, K., Kurdowska, A., Koj, A., Sipe, J. D., and Gauldie, J.,** Comparison of the acute phase response of cultured Morris hepatoma 7777 cells and of rat hepatocytes, *Br. J. Exp. Pathol.*, 68, 485, 1987.
33. **Le, P. T. and Mortensen, R. F.,** Induction and regulation by monokines of hepatic synthesis of mouse serum amyloid P component (SAP), *J. Immunol.*, 136, 2526, 1986.
34. **Darlington, G. J., Kelly, J. H., and Buffone, G. J.,** Growth and hepatospecific gene expression of human hepatoma cells in defined medium, *In Vitro*, 23, 349, 1987.

35. **Pepys, M. B. and Baltz, M. L.,** Acute phase proteins with special reference to C-reactive protein and related proteins (pentaxins) and serum amyloid A protein, *Adv. Immunol.*, 34, 141, 1983.
36. **Gordon, A. H. and Koj, A.,** The acute phase response to injury and infection, in *Research Monographs in Cell and Tissue Physiology,* Vol. 10, Elsevier/North-Holland, New York, 1985.
37. **Schreiber, G., Alfred, A. R., Thomas, T., Birch, H. E., Dickson, P. W., Tu, G.-F., Heinrich, P. C., Northemann, W., Howlett, G. J., De Jong, F. A., and Mitchell, A.,** Levels of messenger ribonucleic acids for plasma proteins in rat liver during acute experimental inflammation, *Inflammation,* 10, 59, 1986.
38. **Birch, H. E. and Schreiber, G.,** Transcriptional regulation of plasma protein synthesis during inflammation, *J. Biol. Chem.*, 261, 8077, 1986.
39. **Ricca, G. A., Hamilton, R. W., McLean, J. W., Conn, A., Kallinyak, J. E., and Taylor, J. M.,** Rat α_1-acid glycoprotein messenger RNA, *J. Biol. Chem.*, 256, 10362, 1981.
40. **Killingsworth, L. M.,** Plasma proteins implicated in the inflammatory response, in *Marker Proteins in Inflammation,* Allen, R. C., Bienvenu, J., Laurent, P., and Suskind, R. M., Eds., Walter de Gruyter, Berlin, 1982, 21.
41. **Baumann, H. and Berger, F. G.,** Genetics and evolution of the acute phase proteins in mice, *Mol. Gen. Genet.*, 201, 505, 1985.
42. **Crabtree, G. R. and Kant, J. A.,** Coordinate accumulation of the mRNAs for the α, β and γ chains of rat fibrinogen following defibrination, *J. Biol. Chem.*, 257, 7277, 1982.
43. **Morrone, G., Giliberto, G., Oliviero, S., Arcone, R., Dente, L., Content, J., and Cortese, R.,** Recombinant interleukin 6 regulates the transcriptional activation of a set of human acute phase genes, *J. Biol. Chem.*, 263, 12554, 1988.
44. **Aiello, A. P., Shia, M. A., Robinson, G. S., Pilch, P. F., and Farmer, S. R.,** Characterization and hepatic expression of the rat α_1-inhibitor III mRNA, *J. Biol. Chem.*, 263, 4013, 1988.
45. **Evans, E., Courtois, G. M., Kilian, P. L., Fuller, G. M., and Crabtree, G. R.,** Induction of fibrinogen and a subset of acute phase response genes involves a novel monokine which is mimicked by phorbol esters, *J. Biol. Chem.*, 262, 10850, 1987.
46. **Shiels, B. R., Northeman, W., Gehring, M. R., and Fey, G. H.,** Modified nuclear processing of α_1-acid glycoprotein RNA during inflammation, *J. Biol. Chem.*, 262, 12826, 1987.
47. **Vannice, J. L., Taylor, J. M., and Ringold, G. M.,** Glucocorticoid-mediated induction of α_1-acid glycoprotein evidence for hormone-regulated RNA processing, *Proc. Natl. Acad. Sci. U.S.A.*, 81, 4241, 1984.
48. **Lowell, C. A., Stearman, R. S., and Morrow, J. F.,** Transcriptional regulation of serum amyloid A gene expression, *J. Biol. Chem.*, 261, 8453, 1986.
49. **Drechou, A., Perez-Gonzalez, N., Agneray, J., Fdeger, J., and Durand, G.,** Increased affinity to concanavalin A and enhanced secretion of alpha 1-acid glycoprotein by hepatocytes isolated from turpentine-treated rats, *Eur. J. Cell. Biol.*, 50, 111, 1989.
50. **Pos, O., Drechou, A., Durand, G., Bierhuizen, M. F., van der Stelt, M. E., and van Dijk, W.,** Con A affinity of rat alpha 1-acid glycoprotein (rAGP): changes during inflammation, dexamethasone or phenobarbital treatment as detected by crossed affino-immunoelectrophoresis (CAIE) are not only a reflection of biantennary glycan content, *Clin. Chim. Acta,* 184(2), 121, 1989.
51. **Fung, W.-P. and Schreiber, G.,** Structure and expression of the genes for major acute phase α-protein (thiostatin) and kininogen in the rat, *J. Biol. Chem.*, 262, 9298, 1987.
52. **Baumann, H., Firestone, G. L., Burgess, T. L., Gross, K. W., Yamamoto, K. R., and Held, W. A.,** Dexamethasone regulation of alpha-1-acid glycoprotein and other acute phase reactants in rat liver and hepatoma cells, *J. Biol. Chem.*, 258, 563, 1983.
53. **Ciliberto, G., Arcone, R., Ruther, U., and Wagner, E.,** Inducible and tissue-specific expression of human C-reactive protein in transgenic mice, *EMBO J.*, 6, 4017, 1987.
54. **Heim, W. G. and Ellenson, S. R.,** Adrenal cortical control of the appearance of rat slow alpha$_2$-globulin, *Nature (London),* 213, 1260, 1967.
55. **Gordon, A. H. and Limaos, E. A.,** Effects of bacterial endotoxin and corticosteroids on plasma concentration of α_2-macroglobulin, haptoglobin and fibrinogen in rats, *Br. J. Exp. Pathol.*, 60, 434, 1979.
56. **Kulkarni, A. B., Reinke, R., and Feigelson, P.,** Acute phase mediators and glucocorticoids elevate α_1-acid glycoprotein gene transcription, *J. Biol. Chem.*, 260, 15386, 1985.
57. **Selinger, M. J., McAdam, K. P. W. J., Kaplan, M. M., Sipe, J. D., Vogel, S. N., and Rosenstreich, D. L.,** Monokine induced synthesis of serum amyloid A protein by hepatocytes, *Nature (London),* 285, 498, 1980.
58. **Ramadori, G., Sipe, J. D., Dinarello, C. A., Mizel, S. B., and Colten, H. R.,** Pretranslational modulation of acute phase hepatic protein synthesis by murine recombinant interleukin 1 (IL-1) and purified human IL-1, *J. Exp. Med.*, 162, 930, 1985.

59. **Magielska-Zero, D., Bereta, J., Czuba-Pelech, B., Pajdak, W., Gauldie, J., and Koj, A.,** Inhibitory effect of human recombinant interferon gamma on synthesis of acute phase proteins in human hepatoma HepG2 cells stimulated by leukocyte cytokines, TNFα and IFN-β2/BSF-2/IL-6, *Biochem. Int.*, 17, 17, 1988.
60. **Mackiewicz, A., Ganapathi, M. K., Schultz, D., Brabenec, A., Weinstein, J., Kelley, M. F., and Kushner, I.,** Transforming growth factor β regulates production of acute phase proteins, *Proc. Natl. Acad. Sci. U.S.A.*, 87, 1491, 1990.
61. **Bereta, J., Szuba, K., Fiers, W., Gauldie, J., and Koj, A.,** Transforming growth factor-β and epidermal growth factor modulate basal and interleukin-6 induced amino acid uptake and acute phase protein synthesis in cultured rat hepatocytes, *FEBS Lett.*, 266, 48, 1990.
62. **Thompson, D., Harrison, S. P., Evans, S. W., and Whicher, J. F.,** Insulin modulation of acute phase protein production in a human hepatoma cell line, *Cytokine,* 3, 619, 1991.
63. **Campos, S. P. and Baumann, H.,** Insulin is a prominent modulator of cytokine-stimulated expression of acute phase plasma protein genes, *Mol. Cell. Biol.*, 12, 1789, 1992.
64. **Pullicino, E. A., Carli, F., Poole, S., Rafferty, B., Malik, S. T. A., and Elia, M.,** The relationship between the circulating concentrations of interleukin 6 (IL-6), tumor necrosis factor (TNF) and the acute phase response to elective surgery and accidental injury, *Lymphokine Res.*, 9, 231, 1990.
65. **Baumann, H. and Wong, G. G.,** Hepatocyte-stimulating factor III shares structural and functional identity with leukemia-inhibitory factor, *J. Immunol.*, 143, 1163, 1989.
66. **Metcalf, D., Nicola, N. A., and Gearing, D. P.,** Effects of injected leukemia inhibitory factor on hematopoietic and other tissue in mice, *Blood,* 76, 50, 1990.
67. **Hilton, D. J. and Gough, N. M.,** Leukemia inhibitory factor: a biological perspective, *J. Cell. Biochem.*, 46, 21, 1991.
68. **Arai, K., Lee, F., Kiyajima, A., Miyatake, S., Arai, N., and Yokota, T.,** Cytokines: coordinators of immune and inflammatory responses, *Annu. Rev. Biochem.*, 59, 783, 1990.
69. **Besedovsky, H., de Rey, A., Sorkin, E., and Dinarello, C. A.,** Immunoregulatory feedback between IL-1 and glucocorticoid hormones, *Science,* 233, 652, 1986.
70. **Woloski, B. M. R. N. J., Smith, E. M., Meyer, W. J., III, Fuller, G. M., and Blalock, J. F.,** Corticotropin-releasing activity of monokines, *Science,* 230, 1035, 1985.
71. **Ritchie, D. G. and Fuller, G. M.,** An in vitro bioassay for leukocytic endogenous mediator(s) using cultured rat hepatocytes, *Inflammation,* 5, 275, 1981.
72. **Sanders, K. D. and Fuller, G. M.,** Kupffer cell regulation of fibrinogen synthesis in hepatocytes, *Thromb. Res.*, 32, 133, 1983.
73. **Otto, J. M., Grenett, H. E., and Fuller, G. M.,** The coordinated regulation of fibrinogen gene transcription by hepatocyte-stimulating factor and dexamethasone, *J. Cell Biol.*, 105, 1067, 1987.
74. **Koj, A., Gauldie, J., Regoeczi, E., Sauder, D. N., and Sweeney, G. D.,** The acute phase response of cultured rat hepatocytes. System characterization and the effect of human cytokines, *Biochem. J.*, 224, 505, 1984.
75. **Baumann, H., Jahreis, G. P., and Gaines, K. C.,** Synthesis and regulation of acute phase plasma proteins in primary cultures of mouse hepatocytes, *J. Cell Biol.*, 97, 866, 1983.
76. **Prowse, K. R. and Baumann, H.,** Interleukin-1 and interleukin-6 stimulate acute phase protein production in primary mouse hepatocytes, *J. Leuk. Biol.*, 45, 55, 1989.
77. **Tatsuta, E., Sipe, J. D., Shirahama, T., Skinner, M., and Cohen, A. S.,** Different regulatory mechanisms for serum amyloid A and serum amyloid P synthesis by cultured mouse hepatocytes, *J. Biol. Chem.*, 258, 5414, 1983.
78. **Castell, J. V., Gomez-Lechon, M. J., David, M., Andus, T., Geiger, T., Trullenque, R., Fabra, R., and Heinrich, P. C.,** Interleukin 6 is the major regulator of the acute phase protein synthesis in adult human hepatocytes, *FEBS Lett.*, 242, 237, 1989.
79. **Clayton, D. E. and Darnell, J. E., Jr.,** Changes in liver-specific compared to common gene transcription during primary culture of mouse hepatocytes, *Mol. Cell. Biol.*, 3, 1552, 1983.
80. **Guillouzo, A., Delers, F., Clement, B., Bernard, N., and Engler, R.,** Long term production of acute phase proteins by adult rat hepatocytes cocultured with another liver cell type in serum-free medium, *Biochem. Biophys. Res. Commun.*, 120, 311, 1984.
81. **Rippe, R. A., Brenner, D. A., and Leffert, H. L.,** DNA-mediated gene transfer into adult rat hepatocyte in primary culture, *Mol. Cell. Biol.*, 10, 689, 1990.
82. **Salas-Prato, M., Tanguay, J.-F., Lefebvre, Y., Wojciechowicz, D., Liem, H. H., Barnes, D. W., Ouellette, G., and Muller-Eberhand, U.,** Attachment and multiplication, morphology and protein production of human fetal primary liver cells cultured in hormonally defined media, *In Vitro,* 24, 230, 1988.
83. **Leffert, H. L. and Paul, D.,** Studies on primary cultures of differentiated fetal liver cells, *J. Cell Biol.*, 52, 559, 1972.

84. **Panduro, A., Shalaby, F., and Sharfritz, D.,** Changing patterns of transcriptional and post-transcriptional control of liver-specific gene expression during rat development, *Genes Dev.*, 1, 1172, 1987.
85. **Isom, H. C., Tevethia, M. J., and Taylor, J. M.,** Properties of simian virus (SV40)-transformed hepatocytes, *Ann. N.Y. Acad. Sci.*, 349, 391, 1980.
86. **Liao, W. S. L., Ma, K. T., Woodworth, C. P., Mengel, L., and Isom, H. C.,** Stimulation of the acute phase response in simian virus 40-hepatocyte cell lines, *Mol. Cell. Biol.*, 9, 2779, 1989.
87. **Baumann, H., Onorato, V., Gauldie, J., and Jahreis, G. P.,** Distinct sets of acute phase plasma proteins are stimulated by separate human hepatocyte-stimulating factors and monokines in rat hepatoma cells, *J. Biol. Chem.*, 262, 9756, 1987.
88. **Baumann, H., Prowse, K. R., Marinkovic, S., Won, K.-A., and Jahreis, G. P.,** Stimulation of hepatic acute phase response by cytokines and glucocorticoids, *Ann. N.Y. Acad. Sci.*, 557, 280, 1989.
89. **Geiger, T., Andus, T., Klapproth, J., Northoff, H., and Heinrich, P. C.,** Induction of α_1-acid glycoprotein by recombinant human interleukin-1 in rat hepatoma cells, *J. Biol. Chem.*, 263, 7141, 1988.
90. **Fey, G. H., Hattori, M., Morthemann, W., Abraham, L. J., Baumann, M., Braciak, T. A., Fletcher, R. G., Gauldie, J., Lee, F., and Reymond, M. F.,** Regulation of rat liver acute phase genes by interleukin-6 and production of hepatocyte stimulating factors by rat hepatoma cells, *Ann. N.Y. Acad. Sci.*, 557, 317, 1989.
91. **Baumann, H., Jahreis, G. P., Sauder, D. N., and Koj, A.,** Human keratinocytes and monocytes release factors which regulate the synthesis of major acute phase plasma proteins in hepatic cells from man, rat and mouse, *J. Biol. Chem.*, 259, 7331, 1984.
92. **Darlington, G. J., Wilson, D. R., and Lachman, L. B.,** Monocyte conditioned medium, interleukin 1 and tumor necrosis factor stimulate the acute phase response in human hepatoma cells in vitro, *J. Cell Biol.*, 103, 787, 1986.
93. **Goldman, N. D. and Liu, T. Y.,** Biosynthesis of human CRP in cultured hepatoma cells is induced by a monocyte factor(s) other than interleukin 1, *J. Biol. Chem.*, 262, 2363, 1987.
94. **Perlmutter, D. H., Colten, H. R., Adams, S. P., May, T., Sehgal, B. P., and Fallons, R. J.,** A cytokine-selective defect in interleukin-1β-mediated acute phase gene expression in a subclone of the human hepatoma cell line (HEPG2), *J. Biol. Chem.*, 264, 7669, 1989.
95. **Ganapathi, M. K., Schultz, D., Machiewicz, A., Samols, D., Hu, S.-I., Brabenec, A., Macintyre, S. S., and Kushner, I.,** Heterogenous nature of the acute phase response. Differential regulation of human serum amyloid A, C-reactive protein and other acute phase proteins by cytokines in Hep3B cells, *J. Immunol.*, 141, 564, 1988.
96. **Arcone, R., Gualandir, G., and Ciliberto, G.,** Identification of sequences responsible for acute phase induction of human C-reactive protein, *Nucleic Acids Res.*, 16, 3195, 1988.
97. **Baumann, H. and Gauldie, J.,** Regulation of hepatic acute phase plasma protein genes by hepatocyte stimulating factors and other mediators of inflammation, *Mol. Biol. Med.*, 7, 147, 1990.
98. **Baumann, H., Morella, K. K., Jahreis, G. P., and Marinkovic, S.,** Distinct regulation of the interleukin-1 and interleukin-6 response elements of the rat haptoglobin gene in rat and human hepatoma cells, *Mol. Cell. Biol.*, 10, 5967, 1990.
99. **Gauldie, J., Richards, C., Harnish, D., Lansdorp, P., and Baumann, H.,** Interferon-β_2/BSF-2 shares identity with monocyte derived hepatocyte stimulating factor (HSF) and regulates the major acute phase protein response in liver cells, *Proc. Natl. Acad. Sci. U.S.A.*, 84, 7251, 1987.
100. **Dinarello, C. A.,** The biology of interleukin 1 and comparison to tumor necrosis factor, *Immunol. Lett.*, 16, 227, 1987.
101. **May, L. T., Ghrayeb, J., Santhanam, U., Tatter, S. B., Sthoeger, Z., Helfgott, D. C., Chiorazzi, N., Grieninger, G., and Seghal, P. B.,** Synthesis and secretion of multiple forms of β2-interferon/B-cell differentiation factor 2/hepatocyte-stimulating factor by human fibroblasts and monocytes, *J. Biol. Chem.*, 263, 7760, 1988.
102. **Woloski, B. M. R. N. J. and Fuller, G. M.,** Identification and partial characterization of hepatocyte stimulating factor from leukemic cell lines: comparison with interleukin 1, *Proc. Natl. Acad. Sci. U.S.A.*, 82, 1443, 1985.
103. **Fey, G. and Gaulide, J.,** The acute phase response of the liver in inflammation, in *Progress in Liver Disease*, Vol. 9, Popper, H. and Schaffner, F., Eds., W. B. Saunders, Philadelphia, 1990, 89.
104. **Rokita, H., Bereta, J., Koj, A., Gordon, A. H., and Gauldie, J.,** Epidermal growth factor and transforming growth factor-beta differently modulate the acute phase response elicited by interleukin-6 in cultured liver cells from man, rat and mouse, *Comp. Biochem. Physiol.*, 95, 41, 1990.
105. **Baumann, H., Jahreis, G. P., Morella, K. K., Won, K.-A., Pruitt, S. C., Jones, V. E., and Prowse, K. R.,** Transcriptional regulation through cytokine- and glucocorticoid-response elements of rat acute phase plasma protein genes by C/EBP and junB, *J. Biol. Chem.*, 266, 20390, 1991.

106. **Brasier, A. R., Ron, D., Tate, J. E., and Habener, J. F.,** A family of constitutive C-EBP-like DNA binding proteins attenuate the IL-1α induced, NFkB mediated *trans*-activation of the angiotensinogen gene acute phase response element, *EMBO J.*, 9, 3933, 1990.
107. **Brasier, A. R., Ron, D., Tate, J. E., and Habener, J. F.,** Synergistic enhansons located within an acute phase responsive enhancer modulate glucocorticoid induction of angiotensinogen gene transcription, *Mol. Endocrinol.*, 74, 1921, 1990.
108. **Prowse, K. R. and Baumann, H.,** Hepatocyte-stimulating factor, β_2-interferon, and interleukin-1 enhance expression of the rat α_1-acid glycoprotein gene via a distal upstream regulatory region, *Mol. Cell. Biol.*, 8, 42, 1988.
109. **Baumann, H., Won, K.-A., and Jahreis, G. P.,** Human hepatocyte-stimulating factor-III and interleukin-6 are structurally and immunologically distinct but regulate the production of the same acute phase plasma proteins, *J. Biol. Chem.*, 264, 8046, 1989.
110. **Baumann, H. and Schendel, P.,** Interleukin-11 regulates the hepatic expression of the same plasma protein genes as interleukin-6, *J. Biol. Chem.*, 266, 20424, 1991.
111. **Baumann, H., Marinkovic-Pajovic, S., Won, K.-A., Jones, V. E., Campos, S. P., Jahreis, G. P., and Morella, K. K.,** The action of IL-6 and LIF on liver cells, *Ciba Found. Symp.*, in press.
112. **Sipe, J. D., Vogel, S. N., Douches, S., and Neta, R.,** Tumor necrosis factor/cachectin is a less potent inducer of serum amyloid A synthesis than interleukin 1, *Lymphokine Res.*, 6, 93, 1987.
113. **Oliviero, S., Morrone, G., and Cortese, R.,** The human haptoglobin gene: transcriptional regulation during development and acute phase induction, *EMBO J.*, 6, 1905, 1987.
114. **Baumann, M., Baumann, H., and Fey, G. H.,** Molecular cloning, characterization and functional expression of the rat liver interleukin 6 receptor, *J. Biol. Chem.*, 265, 19853, 1990.
115. **Baumann, H. and Maquat, L. E.,** Localization of DNA sequences involved in dexamethasone-dependent expression of the rat α_1-acid glycoprotein gene, *Mol. Cell. Biol.*, 6, 2551, 1986.
116. **Poli, V. and Cortese, R.,** Interleukin 6 induces a liver-specific nuclear protein that binds to the promoter of acute phase genes, *Proc. Natl. Acad. Sci. U.S.A.*, 86, 8202, 1989.
117. **Hattori, M., Abraham, L. J., Northemann, W., and Fey, G. H.,** Acute phase reaction induces a specific complex between hepatic nuclear proteins and the interleukin 6 response element of the rat α_2-macroglobulin gene, *Proc. Natl. Acad. Sci. U.S.A.*, 87, 2364, 1990.
118. **Huber, P., Laurent, M., and Dalman, J.,** Human β-fibrinogen gene expression. Upstream sequences involved in its tissue specific expression and its dexamethasone and interleukin 6 stimulation, *J. Biol. Chem.*, 265, 5695, 1990.
119. **Won, K.-A. and Baumann, H.,** The cytokine response element of the rat α_1-acid glycoprotein gene is a complex of several interacting regulatory sequences, *Mol. Cell. Biol.*, 10, 3965, 1990.
119a. **Ron, D., Brasier, A. R., Wright, K. A., Tate, J. E., and Habener, J. F.,** An inducible 50-kilodalton NFkB-like protein and a constitutive protein both bind the acute phase response element of the angiotensinogen gene, *Mol. Cell. Biol.*, 10, 1023, 1990.
120. **Wilson, D. R., Juan, T. S.-C., Wilde, M. D., Fey, G. H., and Darlington, G. J.,** A 58-base-pair region of the human C3 gene confers synergistic inducibility by interleukin-1 and interleukin-6, *Mol. Cell. Biol.*, 10, 6181, 1990.
121. **Woo, P., Sipe, J., Dinarello, C. A., and Colten, H. R.,** Structure of a human serum amyloid A gene and modulation of its expression in transfected L cells, *J. Biol. Chem.*, 262, 15790, 1987.
122. **Perlmutter, D. H., Goldberger, G., Dinarello, C. A., Mizel, S. B., and Colten, H. R.,** Regulation of class III major histocompatibility complex gene products by interleukin-1, *Science*, 232, 850, 1986.
123. **Reinke, R. and Feigelson, P.,** Rat α_1-acid glycoprotein gene sequence and regulation by glucocorticoids in transfected L-cells, *J. Biol. Chem.*, 260, 4397, 1985.
124. **Edbrooke, M. R., Burt, D. W., Cheshire, J. K., and Woo, P.,** Identification of cis-acting sequences responsible for phorbol ester induction of human serum amyloid A gene expression via a nuclear factor kappa B-like transcription factor, *Mol. Cell. Biol.*, 9, 1908, 1989.
125. **Knowles, B. B., Howe, C. C., and Aden, D. P.,** Human hepatocellular carcinoma cell lines secrete the major plasma proteins and hepatitis B surface antigen, *Science*, 209, 497, 1980.
126. **Guertin, M., LaRue, H., Bernier, D., Wrange, O., Chevrette, M., Gingras, M.-C., and Belanger, L.,** Enhancer and promoter elements directing activation and glucocorticoid repression of the α_1-feto-protein gene in hepatocytes, *Mol. Cell. Biol.*, 8, 1398, 1988.
127. **Marinkovic, S. and Baumann, H.,** Structure, hormonal regulation, and identification of the interleukin-6 and dexamethasone-responsive element of the rat haptoglobin gene, *Mol. Cell. Biol.*, 10, 1573, 1990.
128. **Baumann, H., Jahreis, G. P., and Morella, K. K.,** Interaction of cytokine- and glucocorticoid-response elements of the acute phase plasma protein genes, *J. Biol. Chem.*, 265, 22275, 1990.
129. **Nonaka, M. and Huang, Z.-M.,** Interleukin-1-mediated enhancement of mouse factor B gene expression via NFkB-like hepatoma nuclear factor, *Mol. Cell. Biol.*, 10, 6283, 1990.

130. **Oliviero, S. and Cortese, R.,** The human haptoglobin gene promoter: interleukin-6-responsive elements interact with a DNA-binding protein induced by interleukin-6, *EMBO J.*, 8, 1145, 1989.
131. **Ganter, U., Arcone, R., Toniatti, C., Morone, G., and Ciliberto, G.,** Dual control of C-reactive protein gene expression by interleukin 1 and interleukin 6, *EMBO J.*, 8, 3773, 1989.
132. **Li, S.-P., Liu, T.-Y., and Goldman, N. D.,** Cis-acting elements responsible for interleukin-6 inducible C-reactive protein gene expression, *J. Biol. Chem.*, 265, 4136, 1990.
133. **Klein, E. S., Reinke, R., Feigelson, P., and Ringold, G. M.,** Glucocorticoid regulated expression from the 5′-flanking region of the rat alpha-1-acid glycoprotein gene, *J. Biol. Chem.*, 262, 520, 1987.
134. **Poli, V., Mancini, F. P., and Cortese, R.,** IL-6 DBP, a nuclear protein involved in interleukin-6 signal transduction, defines a new family of leucine zipper proteins related to C/EBP, *Cell*, 63, 643, 1990.
135. **Li, X. and Liao, W. S.-L.,** Expression of rat serum amyloid A1 gene involves both C/EBP-like and NFkB-like transcription factor, *J. Biol. Chem.*, 266, 15192, 1991.
136. **Chang, C.-J., Chen, T.-T., Lei, H.-Y., Chen, D.-S., and Lee, S.-C.,** Molecular cloning of a transcription factor, AGP/EBP, that belongs to members of the C/EBP family, *Mol. Cell. Biol.*, 10, 6642, 1990.
137. **Descombes, P., Chojkier, M., Lichtsteiner, S., Falvey, E., and Schibler, U.,** LAP, a novel member of the C/EBP gene family, encodes a liver-enriched transcriptional activator protein, *Genes Dev.*, 4, 1541, 1990.
138. **Cao, Z., Umek, R., and McKnight, S. L.,** Regulated expression of three C/EBP isoforms during adipose conversion of 3T3-L1 cells, *Genes Dev.*, 5, 1538, 1991.
139. **Williams, S. C., Cantwell, C. A., and Johnson, P. F.,** A family of C/EBP-related proteins capable of forming covalently linked leucine zipper dimers in vivo, *Genes Dev.*, 5, 1553, 1991.
140. **Friedman, A. D., Landschulz, W. H., and McKnight, S. L.,** C/EBP activates the promoter of the serum albumin gene in cultured hepatoma cells, *Genes Dev.*, 3, 1314, 1989.
141. **Li, X., Huang, J. H., Reinhoff, H. Y., Jr., and Liao, W. S.-L.,** Two adjacent C/EBP-binding sequences that participate in the cell-specific expression of the mouse serum amyloid A3 gene, *Mol. Cell. Biol.*, 10, 6624, 1990.
142. **Akira, S., Isshiki, H., Sugita, T., Tanabe, O., Kinoshita, S., Nishio, Y., Nakajima, T., Hirano, T., and Kishimoto, T.,** A nuclear factor for IL-6 expression (NF-IL6) is a member of a C/EBP family, *EMBO J.*, 9, 1897, 1990.
143. **Williams, P. M., Ratajczak, T., Lee, S. C., and Ringold, G. M.,** AGP/EBP (LAP) expressed in rat hepatoma cells interacts with multiple promoter sites and is necessary for maximal glucocorticoid induction of the rat alpha-1 acid glycoprotein gene, *Mol. Cell. Biol.*, 11, 4959, 1991.
144. **Won, K.-A. and Baumann, H.,** NF-AB: a liver specific and cytokine inducible nuclear factor that interacts with the IL-1 response element of the rat α_1-acid glycoprotein gene, *Mol. Cell. Biol.*, 11, 3001, 1991.
145. **Palmiter, R. D. and Brinster, R. L.,** Germ-line transformation of mice, *Annu. Rev. Genet.*, 20, 465, 1986.
146. **Dewey, M., Rheaume, C., Berger, F. G., and Baumann, H.,** Inducible and tissue-specific expression of rat α_1-acid glycoprotein in transgenic mice, *J. Immunol.*, 144, 4392, 1990.
147. **Dente, L., Ruther, U., Tripodi, M., Wagner, E. F., and Cortese, R.,** Expression of human α_1-acid glycoprotein genes in cultured cells and in transgenic mice, *Genes Dev.*, 2, 259, 1988.
148. **Syners, L., Fontaine, V., and Content, J.,** Modulation of interleukin-6-receptors in human cells, *Ann. N.Y. Acad. Sci.*, 557, 388, 1989.
149. **Pruitt, S. C.,** Expression vectors permitting cDNA cloning and enrichment for specific sequences by hybridization/selection, *Gene*, 66, 121, 1988.

D. The Cytokines and Hormones Implicated in Acute Phase Protein Regulation

Chapter 15

BIOLOGICAL PERSPECTIVES OF CYTOKINE AND HORMONE NETWORKS

Aleksander Koj, Jack Gauldie, and Heinz Baumann

TABLE OF CONTENTS

0-8493-6913-4/93/$0.00 + $.50

I. HISTORIC BACKGROUND: A LONG WAY FROM LEUKOCYTIC ENDOGENOUS MEDIATOR TO CYTOKINE NETWORK

It has long been recognized that the liver acute phase response must be elicited by a blood-borne messenger produced at the site of injury by damaged cells or activated leukocytes/macrophages[1] but the nature of this "leukocyte endogenous mediator"[2] or "hepatocyte stimulating factor"[3] remained elusive until recent years (for references, see References 1 and 4 to 12). As we know today, the inherent difficulty with isolating the responsible substance was not necessarily related to its chemical instability or scarcity in the tissues, but, rather, was caused by a multiplicity of messengers representing a novel group of signaling molecules with overlapping biological activities. Discovery of interleukin-1 (IL-1)[13,14] led initially to controversy over whether this is indeed the universal and unique regulator of the liver acute phase response[5,15] until Gauldie et al.[16] identified interferon-β_2/B-cell differentiation factor 2 (BSF-2) as the main hepatocyte-stimulating factor, renamed formally interleukin-6.[17,18] However, the existence of other hepatocyte-stimulating factors had already been postulated, leading eventually to discovery of the hepatic function of leukemia inhibitory factor (LIF),[19,20] interleukin-11 (IL-11),[21] and oncostatin M.[22] The list is still open and likely to expand in the near future, although results obtained in tissue culture may not necessarily reflect the *in vivo* importance of the regulatory molecules.

The name "interleukin", implying a communication signal between leukocytes, is clearly misleading and is being gradually replaced by "cytokine", which emphasizes cell origin and various cellular targets.[5] As defined by Nathan and Sporn,[23] "cytokine is a soluble (glyco)protein, non-immunoglobulin in nature, released by living cells of the host, which acts non-enzymically in picomolar to nanomolar concentrations to regulate host cell function. Cytokines make up the fourth major class of soluble intercellular signalling molecules, alongside neurotransmitters, endocrine hormones and autacoids". It is tempting to distinguish a subclass of cytokines involved in regulation of the liver acute phase response — "acute phase cytokines"[8] — but the demarcation line is blurred not only because of overlapping biological activities, but also by the fact that some cytokines exhibit proinflammatory and other antiinflammatory properties. IL-1, tumor necrosis factor-α (TNFα), and IL-8 enhance inflammation, and in addition may be regarded as early cytokines (see Section III), while IL-6 restrains the development of tissue injury and promotes healing by inhibiting the synthesis of TNF and IL-1 and by stimulating production of acute phase proteins, such as antiproteases, with antiinflammatory properties.[5,12,24]

It is clear today that the majority of cytokines and peptide growth factors are multifunctional and exhibit pleiotropic effects.[25-29] This redundancy of function may appear wasteful, but it emphasizes the importance of the acute phase response as the primary defense mechanism of old evolutionary origin.[4,5,24] As reviewed by Beck and Habicht,[30] at least two major cytokines, IL-1 and TNFα, are detected in invertebrates. We are only now beginning to realize that cytokines represent a rich and highly complex signaling intercellular language permitting better integration of multicellular organisms[23,26] and providing a way to maintain homeostasis disturbed by injury or a hostile environment. For these reasons, the complexity of the cytokine network should not be perplexing, but its presentation is not easy. Although several authors proposed schemes illustrating the cytokine network in higher animals,[8,27] no satisfactory picture has yet emerged, and we are waiting for a convincing and unifying theory. Indeed, the concept may be as complex as the presentation of the biochemical pathways of intermediary metabolism.

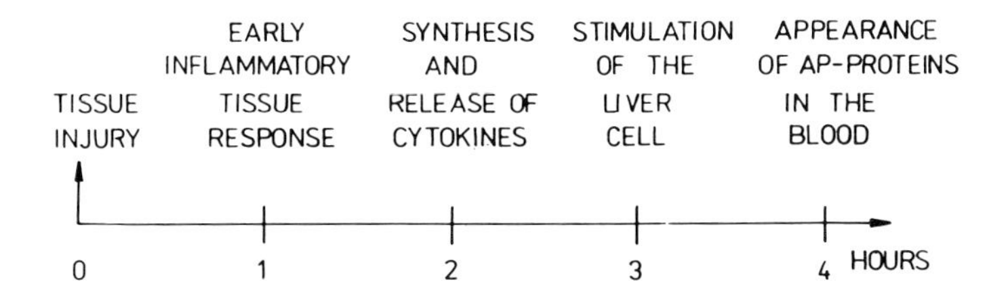

FIGURE 1. Sequence of events during the acute phase response *in vivo*. Approximate time scale is based on experiments with rats.

II. SEQUENCE OF EVENTS DURING THE ACUTE PHASE RESPONSE

Animal organisms respond to an injury with a characteristic set of events, but the time scale is species specific, and in man is certainly longer than in small experimental animals. Figure 1 shows a simplified response for the well-studied rat model.[1,31] Direct administration of IL-6 to the rat shortens the delay time and, in such cases, increases in liver mRNAs coding for acute phase proteins are detected already after 1 h,[31] as are changes in the expression of IL-6 receptor in the liver.[32] Thus, it should be remembered that changes at the cellular level begin earlier than they are manifested in the blood or body fluids, and the delay time shown in Figure 1 is only an approximation.

A. THE PRIMARY SIGNALS FOR THE ACUTE PHASE RESPONSE

Surprisingly little is known about the primary signals in the initiation of the acute phase (AP) response, although bacterial endotoxin (LPS) is generally regarded as a typical triggering factor for the synthesis of AP cytokines.[33-35] A similar function has been ascribed to viruses and some polynucleotides.[36-39] However, there are well-known cases of tissue injury, such as ischemic necrosis during heart infarct or necrosis reperfusion injury in transplantation, when signals other than LPS or viruses trigger the AP response. Among candidates to function as ''alarm molecules'' are free radicals, autacoids (prostaglandins and other products of the activation of the cyclooxygenase pathway), and modified proteins recognized as foreign materials. The situation is complex, since free radicals can directly affect cytokine-producing cells or modifying proteins in the intercellular fluids. Proteins modified by free radicals or leukocyte or lysosomal proteinases will then be taken up by ''scavenger-type'' or specific receptors present on macrophages, fibroblasts, or endothelial cells, and by an unknown mechanism switch on the synthesis of AP-initiating cytokines. Components of the complement cascade represent a typical model of proteins modified during the early stages of the inflammatory reaction,[40] but their role in promoting cytokine production is poorly understood (see also Chapter 11 of this volume).

It has been demonstrated that advanced glucosylation end products (i.e., proteins subjected to nonenzymatic reactions with glucose and found abundantly in diabetes) stimulate the synthesis of TNFα and IL-1.[41] The plasma membrane receptor for these products was found not only in macrophages, but also in fibroblasts and endothelial cells (for references, see Reference 42).

Proteinase-inhibitor complexes or proteolytically modified inhibitors represent other candidates for the initiating primary signals of the AP response. Perlmutter and co-workers[43] reported that α_1-antitrypsin-elastase complexes mediate increased expression of the inhibitor in human Hep G2 cells mediated by a serpin-enzyme complex receptor,[44] and a similar effect was achieved with synthetic peptides corresponding to the C-terminal portion of the serpin molecule adjacent to the cleavage site.[45] On the other hand, Kurdowska and Travis[46] found

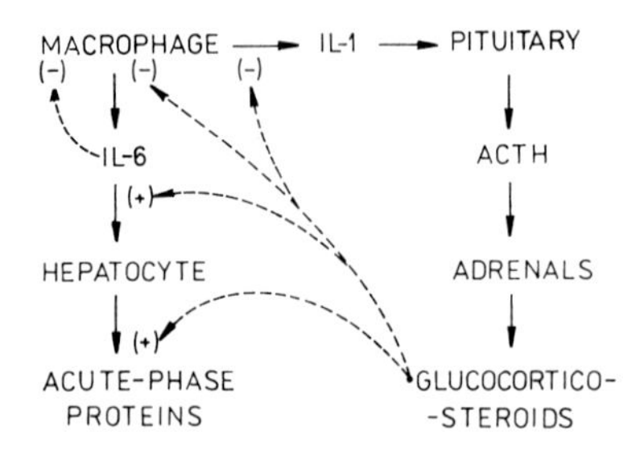

FIGURE 2. Some interactions between cytokines and glucocorticosteroids during the acute phase response *in vivo*. (+), positive effect; (−), inhibitory effect. For further details, see text.

that Hep G2 cells could be stimulated to produce AP proteins only indirectly by conditioned media from human lung fibroblasts that were exposed to modified serpins or serpin-proteinase complexes. In this case, fibroblasts were first stimulated by modified serpins to produce IL-6.[47]

A somewhat similar situation concerns fibrinogen degradation products: although Princen et al.[48] reported that some peptides isolated from plasmin digests of this protein stimulate fibrinogen synthesis in rats, other experiments *in vitro* suggest an indirect pathway of regulation of fibrinogen biosynthesis involving one or more cytokines.[49,50] This has since been confirmed by Grenett and Fuller,[51] who showed that fibrin fragment E released during fibrinolysis directly stimulates macrophage and endothelial cells to produce IL-6, leading to fibrinogen synthesis to replenish plasma levels.

We can only speculate that there is a common mechanism in stimulating the synthesis of AP cytokines by various modified proteins taken up by different cells. Perhaps this is related to a signal generated by the endocytosis of loaded receptors. In this respect, the observations of Krutmann et al.[52] may be relevant. They reported that cross-linking of Fc receptors on monocytes stimulates IL-6 production. It should be added that abnormal proteins trigger the activation of heat-shock genes, but the mechanism seems to be different.[53]

The promoter region of the human IL-6 gene contains several regulatory elements, and many factors can induce expression of this model cytokine.[10,54,55] As emphasized by Sehgal,[10] IL-6 gene expression is enhanced in almost every type of tissue and cell in response to noxious stimuli *in vitro*, but *in vivo* fibroblasts, monocytes/macrophages, epithelial, endothelial, and endometrial cells appear to be the main source of this cytokine. Activation of at least one of three major signal transduction pathways (diacylglycerol, cAMP, and calcium-activated pathways) turn on the synthesis of IL-6. With so many inducers of IL-6, the AP response may perpetuate for a long time (as happens in certain forms of chronic inflammation); hence, some inhibitory mechanisms are likely to have evolved.

B. TERMINATION OF THE ACUTE PHASE RESPONSE

It is likely that complex mechanisms are responsible for the relatively fast diminution of the AP response once the eliciting stimulus disappears, but the most important appears to be the action of glucocorticoids.

As shown by Woloski and Jamieson,[56] local inflammation leads to a four- to tenfold increase in the plasma corticotropin level and prompt release of glucocorticoids from adrenal cortex.[32] Also, recombinant IL-1[57,58] and IL-6[59] release adrenocorticotrophic hormone (ACTH) and other hormones from pituitary cells, and IL-6 increases ACTH plasma levels in freely moving rats.[60] In turn, production of IL-1[58] and other AP cytokines[33,61,62] is strongly inhibited by glucocorticoids; thus, a feedback mechanism can be postulated (Figure 2). Ray et al.[61] demonstrated that ligand-activated glucocorticoid receptor represses the IL-6 gene by occlusion not only of the inducible IL-6 enhancer region, but also of the basal promoter elements. Lee and co-workers[63] found that glucocorticoids inhibit transcription of the IL-1 gene and decrease the stability of IL-1 mRNA. In the case of TNF, glucocorticoids are able to inhibit cytokine biosynthesis if administered prior to LPS;[33] moreover, dexamethasone-elicited suppression of TNF synthesis is prevented by interferon-γ.[64]

Glucocorticoids enhance expression of the IL-6 receptor *in vitro* (see Section III) and *in vivo*,[32] so that reduction of circulating steroids in the later stages of the AP response will decrease the expression of IL-6 receptors on hepatocytes, leading to the cessation of receptor-mediated signal transduction, but quantitative data are not available in this respect. Another possibility is suggested by Nesbitt and Fuller,[65] who postulate the existence in hepatocytes of specific short-lived proteins engaged in the degradation of induced mRNAs coding for AP proteins.

Thus, cessation of the AP response may have more to do with removal of the initiating events than the presence of multiple active inhibitors. Finally, it was demonstrated that IL-6 itself inhibits TNF production in cultured monocytes[66] and IL-1 mRNA induction by LPS in rat *in vivo*.[35] Thus, an additional loop of feedback regulation may exist, helping to terminate the AP response (Figure 2).

III. CYTOKINE NETWORK OR CYTOKINE CASCADE

According to Balkwill and Burke,[27] cytokines interact in a network in a triple fashion: by inducing each other, by modulating specific cell-surface receptors, and by additive or antagonistic effects on functions of target cells. IL-1 and TNF belong to ''early cytokines'', since they appear first after stimulation with LPS, and in turn can induce the expression of IL-6 in tissue culture and *in vivo* (for references, see References 35, 67, and 68). Thus, interactions between cytokines can be presented not only in the form of ''a network'', but also as ''a cascade'' initiated by IL-1 and TNF, while IL-6 is one of the secondary messengers. However, not all stimuli work equally well in inducing IL-6 gene expression in every tissue, e.g., IL-1 and TNF may stimulate IL-6 synthesis in fibroblasts and endothelial cells *in vitro*, but exogenous IL-1 does not appear to stimulate IL-6 production in monocytes.[10] Moreover, even if IL-6 is synthesized by some hepatomas *in vitro*,[69] it functions *in vivo* as an exocrine hormone in the induction of the liver AP response.[70]

Cellular receptors for cytokines show a high degree of specificity, but they share common structural features and belong to hematopoietin receptor superfamilies.[28,54] Liver IL-6 receptor (probably identical to leukocytic receptor) has been cloned and is upregulated by glucocor ticoids, IL-1, and IL-6[71,73] (also see Chapter 20 of this volume). On the other hand, Nesbitt and Fuller[103] observed that inflammatory cytokines, individually or together, downregulated both mRNA and cell-surface expression of IL-6-R in primary rat hepatocytes *in vitro*, while Geisterfer et al.[32] found upregulation of IL-6-R in the liver following administration of IL-6 to rats *in vivo*. Interdependence of receptors for different cytokines, or receptor ''cross-talking'', constitutes an important feature of the cytokine network and regulation of the AP response.

The interaction of cytokines with target cells may be impaired or blocked by specific proteins competing for the receptors, such as IL-1 receptor antagonist (IL-1ra).[74] As shown by Bevan and Raynes,[75] IL-1ra affects the expression of AP proteins in human hepatoma cells. A functional relationship between IL-1ra and natural inhibitors of IL-1 and TNF isolated from the urine of febrile patients[76,77] remains to be defined.

As indicated by tissue culture studies, IL-6 is the principal cytokine regulating the expression of the majority of the AP protein genes in all species, but its function can be partly replaced by LIF, IL-11, or oncostatin M,[20-22,78] while IL-1 and TNFα regulate a

TABLE 1
Acute Phase Protein Genes Regulated by Inflammatory Cytokines

Type 1 (induced by IL-1/TNF and IL-6/IL-11/LIF/OM)	Type 2 (induced by IL-6/IL-11/LIF/OM only)
C-reactive protein (human)	Fibrinogen
α_1-Acid glycoprotein	Haptoglobin (human)
SAA	α_1-Proteinase inhibitor
SAP	Cysteine proteinase inhibitor (rat)
C3	α_2-Macroglobulin (rat)
Factor B	α_1-Antichymotrypsin
Haptoglobin (rat)	Ceruloplasmin
Hemopexin (rat)	C1 esterase inhibitor

different set of genes in man, rat, and mouse liver cells.[9,12,46,79] Taking into account experiments with rat H-35 and human Hep G2 hepatoma cells, Baumann and Gauldie[9,12,79] distinguished two classes of AP proteins: type 1, regulated by IL-1, IL-6, and dexamethasone (hemopexin, complement C3, haptoglobin, α_1-acid glycoprotein, and serum amyloid A), and type 2, regulated by IL-6 and dexamethasone (α_1-antitrypsin, α_1-antichymotrypsin, thiostatin or cysteine proteinase inhibitor, fibrinogen, and α_2-macroglobulin). There is evidence of synergistic stimulation by IL-1 and IL-6 such as seen with α_1-acid glycoprotein, and inhibitory activity such as seen with IL-1 inhibition of IL-6 stimulation of fibrinogen[15] (Table 1). Also, in human hepatoma lines Hep G2 and Hep 3B, combinations of cytokines and glucocorticoids are required to achieve maximum stimulation of such proteins as serum amyloid A and C-reactive protein,[80] while induction of fibrinogen by IL-6 was inhibited by IL-1, TNFα and transforming growth factor-β (TGFβ).[81]

Interferon-γ (IFNγ) is also regarded as an inflammatory cytokine, and when tested with human Hep G2 cells at concentrations above 5 ng/ml, it depressed the synthesis of haptoglobin and α_1-antichymotrypsin, but enhanced the production of α_2-macroglobulin[82] and especially of C1 inactivator.[83] The effects of IL-6 and IFNγ were opposite in the case of α_1-antichymotrypsin and haptoglobin, but additive for α_2-macroglobulin and C1 inactivator. Thus, it appears that the final outcome of the liver AP response depends on the relative concentrations of several cytokines and the sequence in which they interact with the receptors on the hepatocyte.

IV. MODULATION OF ACUTE PHASE PROTEIN SYNTHESIS BY HORMONES AND CELLULAR GROWTH FACTORS

In vivo, cytokines act in a complex milieu containing several hormones and growth factors, the importance of which has not yet been fully elucidated. Due to profound species differences,[46,84] generalizations are of limited value, but undoubtedly glucocorticoids play a pivotal role (for a review of earlier studies, see Reference 5).

As shown in Figure 2, adrenal hormones inhibit the synthesis of cytokines but enhance the production of AP proteins. The latter effect is achieved in two separate ways: by increased expression of hepatic receptors for IL-6 and by a direct effect on the transcription of AP protein genes. The promoters of these genes in many cases contain glucocorticoid-responsive elements (GREs) which act cooperatively with IL-6 and IL-1 regulatory elements and may require specific nuclear factors.[9,85,86] For optimal expression of many AP proteins, a combination of dexamethasone, IL-6, and IL-1 is needed, especially in rat liver cells.[9,12]

Compared to glucocorticoids, other hormones play less important roles during the liver AP response: thyroxine is reported to act cooperatively with dexamethasone in stimulating the synthesis of α_2-macroglobulin in cultured rat hepatocytes,[87] and androgens or estrogens may influence the response pattern of some proteins, as indicated by the sexual dimorphism of α_1-proteinase inhibitor found in mice[88] or a protein belonging to the pentraxin family found in female hamster.[89]

Until recently, little attention was paid to insulin as a potential regulator of the AP response. Magielska-Zero et al.[90] observed that although insulin enhanced the synthesis of AP proteins in cultured adult rat hepatocytes, it decreased the basal production of cysteine proteinase inhibitor (thiostatin) and fibrinogen in Morris hepatoma cells. Thompson and coworkers[91] reported inhibition of fibrinogen synthesis by insulin in human Hep G2 cells, and Campos and Baumann[92] found that insulin is a rapid, nonspecific, dose-dependent inhibitor of cytokine and glucocorticoid stimulation of AP gene expression in H-35 and Hep G2 cells. It appears that insulin acts independently from the transcriptional factor C/EPBβ that has been proposed as a critical regulator of AP protein genes.[93,94] It should be remembered, however, that insulin plays a double role *in vivo* — as a classical metabolic hormone and as a cellular growth factor — and regulation of gene expression by insulin is highly complex.[95]

The effects of TGFβ and epidermal growth factor (EGF) on the AP response of cultured liver cells is species dependent.[84] In human liver cells, TGFβ enhances the synthesis of α_1-antichymotrypsin and inhibits the synthesis of fibrinogen and albumin.[96,97] In mouse, TGFβ increases the synthesis of fibrinogen, C3 complement, haptoglobin, α_1-proteinase inhibitor, and contrapsin, while EGF shows a slight inhibitory activity,[98] but in cultured rat hepatocytes, some stimulation by EGF and inhibition by TGFβ is observed.[99]

Liver regeneration is a highly complex process triggered by hepatocyte growth factors (HGFs) which also appear to be potent mitogens for some epidermal and epithelial cells, but which inhibit proliferation of hepatomas (for references, see References 100 and 101). We observed that human recombinant HGF stimulates the synthesis of α_2-macroglobulin in primary cultures of rat hepatocytes.[102] Its effects on the synthesis of AP proteins are broader in Hep G2 cells, where HGF inhibits the synthesis of albumin, but increases the basal production of fibrinogen, α_1-antichymotrypsin, and α_1-proteinase inhibitor (Figure 3). However, in cells stimulated by high concentrations of IL-6, HGF decreases the maximum response of haptoglobin, fibrinogen, and α_1-antichymotrypsin. A similar phenomenon was observed with insulin (Figure 3). On the other hand, TGFβ increased the production of α_1-antichymotrypsin and α_1-proteinase inhibitor in Hep G2 cells, even at high concentrations of IL-6. These results emphasize again the existence of intricate networks of cytokines and cellular growth factors in the regulation of the liver AP response.

V. CONCLUSIONS AND FUTURE PERSPECTIVES

The growing complexity of cytokine and hormonal regulation of the liver AP response provides a serious challenge to this area of research. Contradictory reports on the action of individual cytokines may arise from variable concentrations of other effectors of the AP reaction in the tested system, as well as from the diversified responses of cells, depending on their growth and differentiation stage, even in established lines and clones which are subjected to genetic variability on prolonged culture. Moreover, it is apparent that the results obtained in tissue culture experiments cannot be extrapolated to the whole organism. With these precautions in mind, we list a few areas expected to be the focus of investigations in the near future:

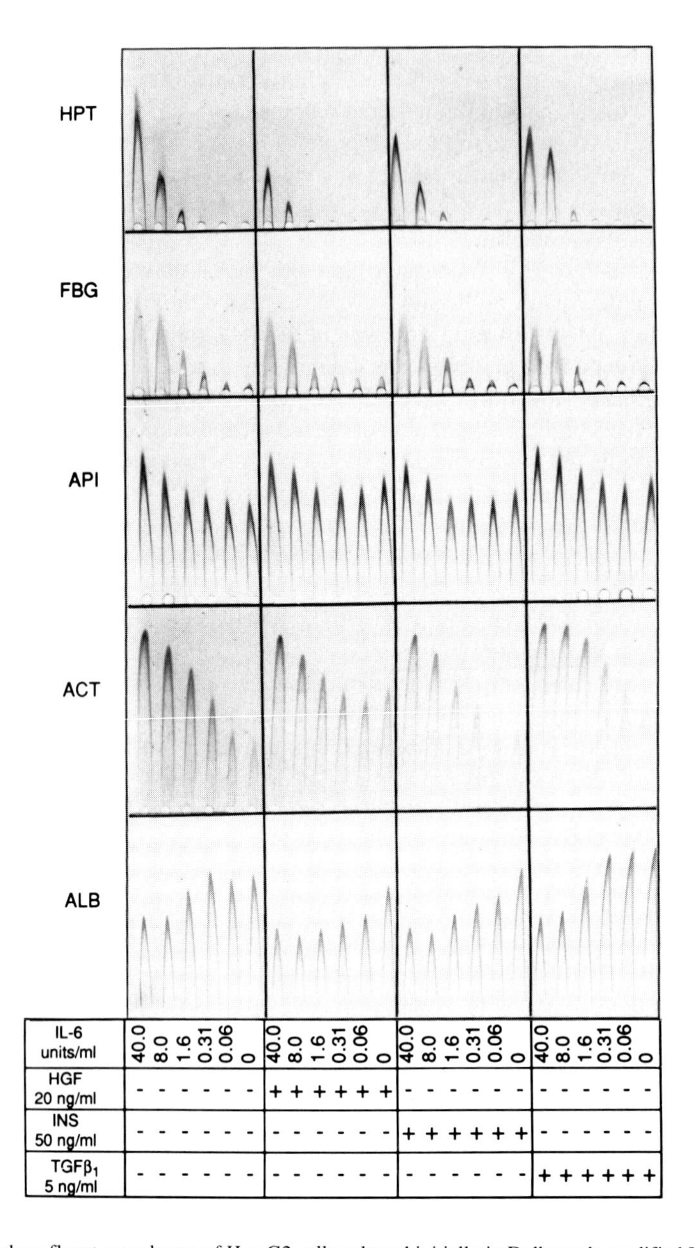

FIGURE 3. Subconfluent monolayers of Hep G2 cells cultured initially in Dulbecco's modified Eagle's medium (DMEM) containing 10% fetal calf serum were rinsed with serum-free DMEM and then cultured for 24 h in serum-free DMEM containing 1 μM dexamethasone with indicated amounts of IL-6, HGF, insulin (INS), and TGFβ_1. The media were directly analyzed by electroimmunoassay with monospecific antisera to human plasma proteins. (From Koj, A. and Baumann, H., unpublished observations.)

1. Elucidating the nature and mechanism of action of the primary signals triggering the AP response and release of different cytokines
2. Completing the list of AP cytokines and their biological activities, taking into account species specificity
3. Understanding the interaction between cytokines and specific receptors, including the action of inhibitors and receptor antagonists
4. Elucidating the nature of the transduction signal elicited in target cells by different cytokines

5. Better understanding the regulation of expression of AP protein genes by *trans*- and *cis*-acting elements
6. Elucidating the molecular mechanisms involved in termination of the AP response, especially in relation to chronic inflammation and neoplastic diseases

Almost all the subjects listed above bear medical implications, and undoubtedly clinical research will expand in the area of the AP response, taking advantage of basic studies in biochemistry, immunology, and molecular biology.

ACKNOWLEDGMENTS

This research has been supported in part by a grant from the Polish State Committee for Scientific Research (KBN 2238/4/91) awarded to A.K.

REFERENCES

1. **Koj, A.,** The acute phase reactants: their synthesis, turnover and biological significance, in *Structure and Function of Plasma Proteins,* Vol. 1, Allison, A. C., Ed., Plenum Press, New York, 1974, 73.
2. **Kampschmidt, R. F., Upchurch, H. F., and Worthington, M. L., III,** Further comparisons of endogenous pyrogens and leukocytic endogenous mediators, *Infect. Immun.,* 41, 6, 1983.
3. **Ritchie, D. G. and Fuller, G. M.,** Hepatocyte stimulating factor: a monocyte-derived acute phase regulatory protein, *Ann. N.Y. Acad. Sci.,* 408, 498, 1983.
4. **Kushner, I.,** The phenomenon of the acute phase response, *Ann. N.Y. Acad. Sci.,* 389, 39, 1982.
5. **Koj, A.,** Hepatocyte stimulating factor and its relationship to interleukin 1, in *The Acute Phase Response to Injury and Infection,* Gordon, A. H. and Koj, A., Eds., Elsevier, New York, 1985, 145.
6. **Fey, G. H. and Fuller, G. M.,** Regulation of acute phase gene expression by inflammatory mediators, *Mol. Biol. Med.,* 4, 323, 1987.
7. **Fey, G. H. and Gauldie, J.,** The acute phase response of the liver in inflammation, in *Progress in Liver Diseases,* Vol. 9, Popper, H. and Schaffner, F., Eds., W. B. Saunders, Philadelphia, 1990, 89.
8. **Koj, A.,** The role of interleukin-6 as the hepatocyte stimulating factor in the network of inflammatory cytokines, *Ann. N.Y. Acad. Sci.,* 557, 1, 1989.
9. **Baumann, H. and Gauldie, J.,** Regulation of hepatic acute phase plasma protein genes by hepatocyte stimulating factors and other mediators of inflammation, *Mol. Biol. Med.,* 7, 147, 1990.
10. **Sehgal, P. B.,** Interleukin-6: a regulator of plasma protein gene expression in hepatic and non-hepatic tissues, *Mol. Biol. Med.,* 7, 117, 1990.
11. **Heinrich, P. C., Castell, J. V., and Andus, T.,** Interleukin-6 and the acute phase response, *Biochem. J.,* 265, 621, 1990.
12. **Gauldie, J. and Baumann, H.,** Cytokines and acute phase protein expression, in *Cytokines and Inflammation,* Kimball, E. S., Ed., CRC Press, Boca Raton, FL, 1991, 275.
13. **Oppenheim, J. J. and Gery, I.,** Interleukin-1 is more than an interleukin, *Immunol. Today,* 3, 113, 1982.
14. **Dinarello, C.,** Interleukin 1, *Rev. Infect. Dis.,* 6, 51, 1984.
15. **Koj, A., Kurdowska, A., Magielska-Zero, D., Rokita, H., Sipe, J. D., Dayer, J. M., Demczuk, S., and Gauldie, J.,** Limited effects of recombinant human and murine interleukin 1 and tumor necrosis factor on production of acute phase proteins by cultured rat hepatocytes, *Biochem. Int.,* 14, 553, 1987.
16. **Gauldie, J., Richards, C., Harnish, D., Lansdorp, P., and Baumann, H.,** Interferon-β2 shares identity with monocyte-derived hepatocyte-stimulating factor and regulates the major acute phase protein response in liver cells, *Proc. Natl. Acad. Sci. U.S.A.,* 84, 7251, 1987.
17. **Andus, T., Geiger, T., Hirano, T., Northoff, H., Ganter, V., Bauer, J., Kishimoto, T., and Heinrich, P. C.,** Recombinant B cell stimulatory factor (BSF-2/IFN-β2) regulates β-fibrinogen and albumin mRNA levels in Fao-9 cells, *FEBS Lett.,* 221, 18, 1987.
18. **Sehgal, P. B., Grieninger, G., and Tosato, G.,** Regulation of the acute phase and immune responses: interleukin-6, *Ann. N.Y. Acad. Sci.,* 557, 1989.
19. **Baumann, H. and Wong, G. G.,** Hepatocyte stimulating factor III shares structural and functional identity with leukemia inhibitory factor, *J. Immunol.,* 143, 1163, 1989.

20. **Baumann, H., Won, K. A., and Jahreis, G. P.,** Human hepatocyte stimulating factor III and interleukin 6 are structurally and immunologically distinct but regulate the production of the same acute phase plasma proteins, *J. Biol. Chem.*, 264, 8046, 1989.
21. **Baumann, H. and Schendel, P.,** Interleukin-11 regulates the hepatic expression of the same plasma protein genes as interleukin-6, *J. Biol. Chem.*, 266, 20424, 1991.
22. **Richards, C. D., Brown, T. J., Shoyab, M., Baumann, H., and Gauldie, J.,** Recombinant oncostatin M stimulates the production of acute phase proteins in HepG2 cells and rat primary hepatocytes *in vitro*, *J. Immunol.*, 148, 1731, 1992.
23. **Nathan, C. and Sporn, M.,** Cytokines in context, *J. Cell Biol.*, 113, 981, 1991.
24. **Gauldie, J.,** Acute phase response, in *Encyclopedia of Human Biology*, Vol. 1, Dulbecco, R., Ed., Academic Press, San Diego, 1991, 25.
25. **O'Garra, A., Umland, S., De France, T., and Christiansen, J.,** "B-cell factors" are pleiotropic, *Immunol. Today*, 9, 45, 1988.
26. **Sporn, M. B. and Roberts, A. B.,** Peptide growth factors are multifunctional, *Nature (London)*, 332, 217, 1988.
27. **Balkwill, F. R. and Burke, F.,** The cytokine network, *Immunol. Today*, 10, 299, 1989.
28. **Cosman, D., Lyman, S. D., Idzerda, R. L., Beckmann, N. P., Park, L. S., Goodwin, R. G., and March, C. J.,** A new cytokine receptor superfamily, *TIBS*, 15, 265, 1990.
29. **Hilton, D. J.,** LIF: lots of interesting functions, *TIBS*, 17, 72, 1992.
30. **Beck, G. and Habicht, G. S.,** Primitive cytokines: harbingers of vertebrate defense, *Immunol. Today*, 12, 180, 1991.
31. **Geiger, T., Andus, T., Klapproth, J., Hirano, T., Kishimoto, T., and Heinrich, P. C.,** Induction of rat acute phase proteins by interleukin 6 *in vivo*, *Eur. J. Immunol.*, 18, 717, 1988.
32. **Geisterfer, M., Richards, C., Baumann, M., Fey, G., Gwynne, D., and Gauldie, J.,** Regulation of Il-6 and IL-6 receptor in acute inflammation in vivo, *Cytokine*, 5, 1, 1993.
33. **Beutler, B. and Cerami, A.,** The biology of cachectin/TNF — a primary mediator of the host response, *Annu. Rev. Immunol.*, 7, 625, 1989.
34. **Han, J., Brown, T., and Beutler, B.,** Endotoxin-responsive sequences control cachectin/tumor necrosis factor biosynthesis at the translational level, *J. Exp. Med.*, 171, 465, 1990.
35. **Ulrich, T. R., Guo, K., Remick, D., Del Castillo, J., and Yin, S.,** Endotoxin-induced cytokine gene expression *in vivo*. III. IL-6 mRNA and serum protein expression and the *in vivo* hematologic effects of IL-6, *J. Immunol.*, 146, 2316, 1991.
36. **Van Damme, J., Cayphas, S., Opdenakker, G., Billiau, A., and Van Snick, J.,** Interleukin-1 and poly(rI)·poly(rC) induce production of a hybridoma growth factor by human fibroblasts, *Eur. J. Immunol.*, 17, 1, 1987.
37. **Ray, A., Tatter, S. B., May, L, T., and Sehgal, P. B.,** Activation of the human β-2-interferon/interleukin-6 promoter by cytokines, viruses and second messenger agonists, *Proc. Natl. Acad. Sci. U.S.A.*, 85, 6701, 1988.
38. **Sehgal, P. B., Helfgott, D. C., Santhanam, V., Tatter, S. B., Clarick, R. H., Ghrayeb, J., and May, L. T.,** Regulation of the acute phase and immune responses in viral diseases, *J. Exp. Med.*, 167, 1951, 1988.
39. **Campbell, I. L.,** Cytokines in viral diseases, *Curr. Opin. Immunol.*, 3, 486, 1991.
40. **Muller-Eberhard, H. J. and Schreiber, R. D.,** Molecular biology and chemistry of the alternative pathway of complement, *Adv. Immunol.*, 29, 1, 1980.
41. **Vlassara, H., Brownlee, M., Manogue, K. R., Dinarello, C. A., and Pasagian, A.,** Cachectin/TNF and IL-1 induced by glucose-modified proteins: role in normal tissue remodelling, *Science*, 240, 1546, 1988.
42. **Yang, Z., Makita, Z., Horii, Y., Brunelle, S., Cerami, A., Sehajpal, P., Suthanthiran, M., and Vlassara, H.,** Two novel rat liver membrane proteins that bind advanced glycosylation endproducts: relationship to macrophage receptor for glucose-modified proteins, *J. Exp. Med.*, 174, 515, 1991.
43. **Perlmutter, D. H., Glover, G. I., Rivetna, M., Schasteen, C. S., and Fallon, R. J.,** Identification of a serpin-enzyme complex receptor on human hepatoma cells and human monocytes, *Proc. Natl. Acad. Sci. U.S.A.*, 87, 3753, 1990.
44. **Perlmutter, D. H., Joslin, G., Nelson, P., and Schasteen, C.,** Endocytosis and degradation of α1-antitrypsin-protease complexes is mediated by the serpin-enzyme complex (SEC) receptor, *J. Biol. Chem.*, 265, 16713, 1990.
45. **Joslin, G., Fallon, R. J., Bullock, J., Adams, S. P., and Perlmutter, D. H.,** The SEC receptor recognizes a pentapeptide neodomain of α-1-protease complexes, *J. Biol. Chem.*, 266, 11282, 1991.
46. **Kurdowska, A. S. and Travis, J.,** Acute phase protein stimulation by α-1-antichymotrypsin-cathepsin G complexes. Evidence for the involvement of interleukin-6, *J. Biol. Chem.*, 265, 21023, 1991.

47. **Koj, A., Rokita, H., Kordula, T., Kurdowska, A., and Travis, J.,** Role of cytokines and growth factors in the induced synthesis of proteinase inhibitors belonging to acute phase proteins, *Biomed. Biochim. Acta,* 50, 421, 1991.
48. **Princen, H. M. G., Moshage, H. J., Emeis, J. J., De Haard, H. J. W., Nieuwenhuizen, W., and Yap, S. H.,** Fibrinogen fragments X, Y, D and E increase levels of plasma fibrinogen and liver mRNAs coding for fibrinogen polypeptides in rats, *Thromb. Haemost.,* 53, 212, 1985.
49. **Ritchie, D. G., Levy, B. A., Adams, M. A., and Fuller, G. M.,** Regulation of fibrinogen synthesis by plasmin-derived fragment of fibrinogen and fibrin: an indirect pathway, *Proc. Natl. Acad. Sci. U.S.A.,* 79, 1530, 1982.
50. **Moshage, H. J., Princen, H. M. G., Van Pelt, J., Roelofs, H. M. J., Nieuwenhuizen, W., and Yap, S. H.,** Differential effects of endotoxin and fibrinogen degradation products (FDPS) on liver synthesis of fibrinogen and albumin: evidence for the involvement of a novel cytokine in the stimulation of fibrinogen synthesis induced by FDPA, *Int. J. Biochem.,* 22, 1393, 1990.
51. **Grenett, H. E. and Fuller, G. M.,** Regulating IL-6 mRNA expression in endothelial cells during wound healing, *Cytokine,* 3, 453, 1991.
52. **Krutmann, J., Kirnbauer, R., Kock, A., Schwarz, T., Schopf, E., May, L. T., Sehgal, P. B., and Luger, T. A.,** Cross-linking Fc receptors on monocytes triggers IL-6 production, *J. Immunol.,* 145, 1337, 1990.
53. **Ananthan, J., Goldberg, A. L., and Voellmy, R.,** Abnormal proteins serve as eukaryotic stress signals and trigger the activation of heat shock genes, *Science,* 232, 522, 1986.
54. **Hirano, T., Akira, S., Taga, T., and Kishimoto, T.,** Biological and clinical aspects of interleukin 6, *Immunol. Today,* 11, 443, 1990.
55. **Krueger, J., Ray, A., Tamm, I., and Sehgal, P. B.,** Expression and function of interleukin-6 in epithelial cells, *J. Cell. Biochem.,* 45, 1, 1991.
56. **Woloski, B. M. R. N. J. and Jamieson, J. C.,** Rat corticotropin, insulin and thyroid hormone levels during the acute phase response to inflammation, *Comp. Biochem. Physiol.,* 86A, 15, 1987.
57. **Bernton, E. W., Beach, J. E., Holaday, J. W., Smallridge, R. C., and Fein, H. G.,** Release of multiple hormones by a direct action of interleukin-1 on pituitary cells, *Science,* 238, 519, 1987.
58. **Besedovsky, H., Del Rey, A., Sorkin, E., and Dinarello, C.,** Immunoregulatory feedback between interleukin-1 and glucocorticoid hormones, *Science,* 233, 652, 1986.
59. **Spangelo, B. L., Judd, A. M., Isakson, P. C., and MacLeod, R. M.,** Interleukin-6 stimulates anterior pituitary hormone release *in vitro, Endocrinology,* 125, 575, 1989.
60. **Naitoh, Y., Fukata, J., Tominaga, T., Nakai, Y., Tamai, S., Mori, K., and Imura, H.,** IL-6 stimulates the secretion of adrenocorticotropic hormone in conscious, freely-moving rats, *Biochem. Biophy. Res. Commun.,* 155, 1459, 1988.
61. **Ray, A., LaForge, K. S., and Sehgal, P. B.,** On the mechanism for efficient repression of the interleukin-6 promoter by glucocorticoids: enhancer, TATA box, and RNA start site (Inr motif) occlusion, *Mol. Cell. Biol.,* 10, 5736, 1990.
62. **Parant, M., Le Contel, C., Parant, F., and Chedid, L.,** Influence of endogenous glucocorticoid on endotoxin-induced production of circulating TNFα, *Lymphokine Cytokine Res.,* 10, 265, 1991.
63. **Lee, S. W., Tsou, A. P., Chan, H., Thomas, J., Petrie, K., Eugui, E. M., and Allison, A. C.,** Glucocorticoids selectively inhibit the transcription of the interleukin 1β gene and decrease the stability of interleukin 1β mRNA, *Proc. Natl. Acad. Sci. U.S.A.,* 85, 1204, 1988.
64. **Luedke, C. E. and Cerami, A.,** Interferon-γ overcomes glucocorticoid suppression of cachectin/tumor necrosis factor biosynthesis by murine macrophages, *J. Clin. Invest.,* 86, 1234, 1990.
65. **Nesbitt, J. E. and Fuller, G. M.,** Transcription and translation are required for fibrinogen mRNA degradation in hepatocytes, *Biochim. Biophys. Acta,* 1089, 88, 1991.
66. **Aderka, D., Le, J., and Vilcek, J.,** IL-6 inhibits lipopolysaccharide-induced tumor necrosis factor production in cultured human monocytes, U937 cells, and in mice, *J. Immunol.,* 143, 3517, 1989.
67. **Jablons, D. M., Mule, J. J., McIntosh, J. E., Sehgal, P. B., May, L. M., Huang, C. M., Rosenberg, S. A., and Lotze, M. T.,** IL-6/IFN-β-2 as a circulating hormone. Induction by cytokine administration in humans, *J. Immunol.,* 143, 1542, 1989.
68. **Baigrie, R. J., Lamont, P. M., Dallman, M., and Morris, P. J.,** The release of interleukin-1β (IL-1) precedes that of interleukin 6 (IL-6) in patients undergoing major surgery, *Lymphokine Cytokine Res.,* 10, 253, 1991.
69. **Northemann, W., Hattori, M., Baffet, G., Braciak, T. A. S., Fletcher, R. G., Abrahams, L. J., Gauldie, J., Baumann, M., and Fey, G.,** Production of interleukin 6 by hepatoma cells, *Mol. Biol. Med.,* 7, 273, 1990.
70. **Gauldie, J., Northemann, W., and Fey, G. H.,** IL-6 functions as an exocrine hormone in inflammation. Hepatocytes undergoing acute phase response require exogenous IL-6, *J. Immunol.,* 144, 3804, 1990.

71. **Bauer, J., Lengyel, G., Bauer, T. M., Acs, G., and Gerok, W.,** Regulation of interleukin 6 receptor expression in human monocytes and hepatocytes, *FEBS Lett.*, 249, 27, 1989.
72. **Baumann, M., Baumann, H., and Fey, G. H.,** Molecular cloning, characterization and functional expression of the rat liver interleukin 6 receptor, *J. Biol. Chem.*, 265, 19583, 1990.
73. **Schooltink, H., Stoyan, T., Lenz, D., Schmitz, H., Hirano, T., Kishimoto, T., Heinrich, P. C., and Rose-John, S.,** Structural and functional studies on the human hepatic interleukin-6 receptor. Molecular cloning and overexpression in Hep G2 cells, *Biochem. J.*, 277, 659, 1991.
74. **Arend, W. P.,** Interleukin 1 receptor antagonist. A new member of the interleukin 1 family, *J. Clin. Invest.*, 88, 1445, 1991.
75. **Bevan, S. and Raynes, J. G.,** IL-1 receptor antagonist regulation of acute phase protein synthesis in human hepatoma cells, *J. Immunol.*, 147, 2574, 1991.
76. **Mazzei, G. J., Seckinger, P. L., Dayer, J. M., and Shaw, A. R.,** Purification and characterization of a 26-kDa competitive inhibitor of interleukin 1, *Eur. J. Immunol.*, 20, 683, 1990.
77. **Seckinger, P., Isaaz, S., and Dayer, J. M.,** Purification and biologic characterization of a specific tumor necrosis factor α inhibitor, *J. Biol. Chem.*, 264, 1966, 1989.
78. **Kordula, T., Rokita, H., Koj, A., Fiers, W., Gauldie, J., and Baumann, H.,** Effects of interleukin-6 and leukemia inhibitory factor on the acute phase response and DNA synthesis in cultured rat hepatocytes, *Lymphokine Cytokine Res.*, 10, 23, 1991.
79. **Baumann, H., Richards, C., and Gauldie, J.,** Interaction among hepatocyte stimulating factors, interleukin 1, and glucocorticoids for regulation of acute phase plasma proteins in human hepatoma (Hep G2) cells, *J. Immunol.*, 139, 4122, 1987.
80. **Ganapathi, M. K., Rzewnicki, D., Samols, D., Jiang, S. L., and Kushner, I.,** Effect of combinations of cytokines and hormones on synthesis of serum amyloid A and C-reactive protein in Hep 3B cells, *J. Immunol.*, 147, 1261, 1991.
81. **Mackiewicz, A., Speroff, T., Ganapathi, M. K., and Kushner, I.,** Effect of cytokine combinations on acute phase protein production by two human hepatoma cell lines, *J. Immunol.*, 146, 3032, 1991.
82. **Magielska-Zero, D., Bereta, J., Czuba-Pelech, B., Pajdak, W., Gauldie, J., and Koj, A.,** Inhibitory effect of human recombinant interferon gamma on synthesis of acute phase proteins in human hepatoma Hep G2 cells stimulated by leukocyte cytokines, TNFα anFN-β2/BSF-2/IL-6, *Biochem. Int.*, 17, 17, 1988.
83. **Zuraw, B. L. and Lotz, M.,** Regulation of the hepatic synthesis of C1 inhibitor by the hepatocyte stimulating factors interleukin 6 and interferon γ, *J. Biol. Chem.*, 265, 12664, 1990.
84. **Rokita, H., Bereta, J., Koj, A., Gordon, A. H., and Gauldie, J.,** Epidermal growth factor and transforming growth factor-β differently modulate the acute phase response elicited by interleukin-6 in cultured liver cells from man, rat and mouse, *Comp. Biochem. Physiol.*, 95A, 41, 1990.
85. **Baumann, H., Jahreis, G. P., and Morella, K. K.,** Interaction of cytokine- and glucocorticoid-response elements of acute phase plasma protein genes, *J. Biol. Chem.*, 265, 22275, 1990.
86. **Di Lorenzo, D., Williams, P., and Ringold, G.,** Identification of two distinct nuclear factors with DNA binding activity within the glucocorticoid regulatory region of the rat alpha-1-acid glycoprotein promoter, *Biochem. Biophys. Res. Commun.*, 176, 1326, 1991.
87. **Bauer, J., Tran-Thi, T. A., Northoff, H., Hirsch, F., Schlayer, H. J., Gerok, W., and Heinrich, P. C.,** The acute phase induction of α2-macroglobulin in rat hepatocyte primary cultures: action of hepatocyte-stimulating factor, triiodothyronine and dexamethasone, *Eur. J. Cell Biol.*, 40, 86, 1986.
88. **Kueppers, F. and Mills, J.,** Trypsin inhibition by mouse serum: sexual dimorphism controlled by testosterone, *Science*, 219, 182, 1983.
89. **Coe, J. E. and Ross, M. J.,** Hamster female protein. A divergent acute phase protein in male and female Syrian hamsters, *J. Exp. Med.*, 157, 1421, 1983.
90. **Magielska-Zero, D., Guzdek, A., Bereta, J., Kurdowska, A., Cieszka, K., and Koj, A.,** Effects of dexamethasone and insulin on the acute phase response of Morris hepatoma cells and of rat hepatocytes in culture, *Acta Biochim. Pol.*, 35, 287, 1988.
91. **Thompson, D., Harrison, S. P., Evans, S. W., and Whicher, J. T.,** Insulin modulation of acute phase protein production in a human hepatoma cell line, *Cytokine*, 3, 619, 1991.
92. **Campos, S. P. and Baumann, H.,** Insulin is a prominent modulator of the cytokine-stimulated expression of acute phase plasma protein genes, *Mol. Cell. Biol.*, 12, 1789, 1992.
93. **Akira, S., Isshiki, H., Sugita, T., Tanabe, O., Kinoshita, S., Nishio, Y., Nakajima, T., Hirano, T., and Kishimoto, T.,** A nuclear factor for IL-6 expression (NF-IL6) is a member of a C/EBP family, *EMBO J.*, 9, 1897, 1990.
94. **Poli, V., Mancini, F. P., and Cortese, R.,** IL-6 DBP, a nuclear protein involved in interleukin-6 signal transduction, defines a new family of leucine zipper proteins related to C/EBP, *Cell*, 63, 643, 1990.
95. **O'Brien, R. M. and Granner, D. K.,** Regulation of gene expression by insulin, *Biochem. J.*, 278, 609, 1991.

96. **Mackiewicz, A., Ganapathi, M. K., Schultz, D., Brabenec, A., Weinstein, J., Kelley, M. F., and Kushner, I.,** Transforming growth factor β_1 regulates production of acute phase proteins, *Proc. Natl. Acad. Sci. U.S.A.*, 87, 1491, 1990.
97. **Morrone, G., Cortese, R., and Sorrentino, V.,** Post-transcriptional control of negative acute phase genes by transforming growth factor beta, *EMBO J.*, 8, 3767, 1989.
98. **Rokita, H. and Szuba, K.,** Regulation of acute phase reaction by transforming growth factor β in cultured murine hepatocytes, *Acta Biochim. Pol.*, 38, 241, 1991.
99. **Bereta, J., Szuba, K., Fiers, W., Gauldie, J., and Koj, A.,** Transforming growth factor-β and epidermal growth factor modulate basal and interleukin-6-induced amino acid uptake and acute phase protein synthesis in cultured rat hepatocytes, *FEBS Lett.*, 266, 48, 1990.
100. **Matsumoto, K. and Nakamura, T.,** Hepatocyte growth factor: molecular structure and implications for a central role in liver regeneration, *J. Gastroenterol. Hepatol.*, 6, 509, 1991.
101. **Shiota, G., Rhoads, D. B., Wang, T. C., Nakamura, T., and Schmidt, E. V.,** Hepatocyte growth factor inhibits growth of hepatocellular carcinoma cells, *Proc. Natl. Acad. Sci. U.S.A.*, 89, 373, 1992.
102. **Pierzchalski, P., Nakamura, T., Takehara, T., and Koj, A.,** Modulation of acute phase protein synthesis in cultured rat hepatocytes by human recombinant hepatocyte growth factor, *Growth Factors*, 7, 161, 1992.
103. **Nesbitt, J. E. and Fuller, G. M.,** Differential regulation of interleukin-6 receptor and gp130 gene expression in rat hepatocytes, *Mol. Biol. Cell.*, 3, 103, 1992.

Chapter 16

INTERLEUKIN-6

Pravin B. Sehgal

TABLE OF CONTENTS

0-8493-6913-4/93/$0.00 + $.50

I. INTRODUCTION

Interleukin-6 (IL-6) was originally cloned in 1980 as the cDNA copy of a poly(I) · poly(C)-inducible 1.3-kb mRNA isolated from human fibroblasts (''interferon-β_2'').[1,2] The recognition in 1986–1987 that the inflammation-associated cytokines tumor necrosis factor (TNF) and interleukin-1 (IL-1), the growth factors platelet-derived growth factor (PDGF) and epidermal growth factor (EGF), and serum, viral, and bacterial infection, including bacterial products such as endotoxin, induced IL-6 production suggested that this cytokine participated in the host response to a broad range of tissue injury.[3-9] The recognition in 1986–1987 that ''interferon-β_2'' was the same as B-cell differentiation factor 2 (BSF-2),[10,11] and subsequently that this cytokine was also the ''hepatocyte-stimulating factor'' (HSF),[12] pointed to the role of IL-6 in mediating many of the systemic physiological, biochemical, and immunological responses of the host to tissue injury, as exemplified by the acute phase plasma protein response. Indeed, IL-6 appears to be a key systemic or long-distance alarm signal that is indicative of tissue damage somewhere in the body.[13] Additionally, in 1986–1987, it was recognized that proteins independently characterized as hybridoma/plasmacytoma growth factor (HPGF),[14-16] monocyte-granulocyte inducer type 2 (MGI-2),[17] T-cell activating factor (TAF),[18] monocyte-derived growth factor for human B cells,[19] and cytolytic T-cell differentiation factor (CDF)[20] all turned out to be IL-6. The consensus term interleukin-6 was adopted in December 1988 by the various investigators who had independently discovered this cytokine.[13]

Since 1987–1988, there has been a veritable explosion of interest in all facets of IL-6 research. Today, IL-6 is the focus of a broad spectrum of research activity that encompasses genetic, molecular, biochemical, cellular, immunological, and clinical studies. Furthermore, in addition to IL-6,[13,21] cytokines such as leukemia inhibitory factor (LIF),[22,23] interleukin-11 (IL-11),[24] and oncostatin M (OM),[25] when added to hepatocyte or hepatoma cell lines in culture, can also elicit many of the same alterations in plasma protein synthesis that are characteristic of the acute phase response. The available data support the thesis that IL-6 is the primary systemic mediator of the hepatic acute phase response *in vivo*. Elevated levels of IL-6 are readily detected in the peripheral circulation of the infected or injured host,[26-30] and neutralizing antibodies to IL-6 can largely block the acute phase plasma protein response in appropriate animal models and in myeloma patients.[31-35]

In addition to the group of cytokines that can each directly elicit many of the acute phase alterations in hepatic plasma protein synthesis (e.g., IL-6, LIF, IL-11, and OM), a second group of cytokines are essentially implicated in ''modulating'' this hepatic response. The latter include IL-1, TNF, and transforming growth factor-β (TGFβ)[36-39] (also reviewed elsewhere in this volume). These modulating cytokines, together with the glucocorticoids, can contribute to the overall magnitude and to the particular spectrum of plasma proteins whose hepatic synthesis is enhanced or inhibited, at least as judged by experiments in hepatocyte or hepatoma cell lines in culture. The contribution of these modulating cytokines to the acute phase plasma protein response *in vivo* remains to be adequately understood. Indeed, while in the turpentine abscess model an anti-IL-1 receptor antibody can block IL-6 induction and the consequent acute phase plasma protein response,[40] in the *Listeria*-infected mouse, the anti-IL-1 receptor antibody blocks neither the IL-6 response nor the acute phase plasma protein response.[41] It is very likely that extrapolation of conclusions from cell culture to *in vivo* situations or even between different *in vivo* situations will be difficult (compare References 42 and 43).

This chapter, in addition to presenting a brief overview of the IL-6 field, focuses on recent data from this laboratory on the influence of glucocorticoids on IL-6 gene expression, and discusses unexpected insights into IL-6 structure that raise new questions about how

IL-6 actually exists and functions in the human body. These new data raise questions of general relevance to the structure and function of other cytokines *in vivo*. The interactions of IL-6 with its receptor and the transcriptional and posttranscriptional events triggered by IL-6 that ultimately lead to modulation of the synthesis of particular plasma proteins are discussed in detail in other chapters of this volume.

II. THE IL-6 GENE AND ITS EXPRESSION

The IL-6 gene, located on chromosome 7 (at p21) in the human genome[44-46] and on chromosome 5 in the mouse genome,[47] consists of five exons that have the same overall exon-intron structure as the genes for human and rodent granulocyte colony-stimulating factor (G-CSF) and chicken myelomonocytic differentiation factor.[48-50] In addition to the major inducible RNA start site at + 1, there is a second minor RNA start site at approximately − 21.[5,51] The 5′ flanking regions of the human and murine IL-6 genes are highly conserved (>95% identity).[49] At least three polyadenylation sites have been mapped — two close together giving rise to 1.3-kb transcripts and a downstream site giving rise to transcripts approximately 2.5 kb in length.[5,6,52] The human IL-6 gene consists of at least three independently segregating allele systems as determined by restriction polymorphism analyses — three MspI alleles, two BglI alleles, and at least four BstNI alleles.[46] While the MspI and BglI alleles appear to be due to point mutations, the BstNI alleles, which are also detected using several other restriction enzymes, are the result of insertions/deletions in an AT-rich region immediately 3′ to the upstream polyadenylation sites (that give rise to the 1.3-kb mRNA).[46,53] Furthermore, the fourth intron contains an Alu repetitive DNA element immediately upstream of the fifth exon.[46] Thus, the 3′ half of the gene is subjected to considerable variability. Nevertheless, to date none of these polymorphisms appear to alter the coding sequence for the IL-6 protein.

A. INDUCIBLE EXPRESSION

IL-6 gene transcription is readily induced in a variety of different normal tissues in response to RNA and DNA virus infection, bacterial products such as endotoxin, serum, inflammation-associated cytokines such as IL-1, TNF, PDGF, and interferons.[10,13,21,29] Based on extensive studies carried out in this and other laboratories, Figure 1 summarizes the transcription regulatory elements present in the 5′ flanking region of the human IL-6 gene.[54-61] An interesting principle to emerge is that the IL-6 and c-*fos* promoters are strikingly similar in their overall function. The c-*fos* serum response enhancer (SRE) element exhibits strong nucleotide sequence similarity to that of the multiple cytokine (IL-1, TNF, and serum) and second messenger (cAMP, phorbol ester)-responsive enhancer (MRE) region in IL-6 (− 173 to − 145).[51,55,62] Indeed, in functional cross-competition assays in intact cells, the IL-6 and c-*fos* regulatory DNA elements cross-compete with each other in a reciprocal manner; in gel-shift competition assays using induced HeLa cell nuclear extracts, oligonucleotides corresponding to regions within the IL-6 MRE compete with those from within the c-*fos* SRE region, and overexpression of c-*fos* using constitutive expression plasmids represses both the IL-6 and the c-*fos* promoters.[55] Overexpression of c-*jun* alone or of c-*jun* in combination with c-*fos* has little overall effect on IL-6 transcription.[111]

The complex IL-6 MRE region consists of two partially overlapping DNA elements, each of which, when attached *in single copy* to the heterologous herpesvirus thymidine kinase/chloramphenicol acetyltransferase (TK/CAT) reporter construct, is responsive, albeit to differing extents, to all of the inducers tested.[55,56] MRE I, − 173 to − 151, contains the typical GACGTCA cAMP/phorbol ester-responsive (CRE/TRE) motif. MRE II, − 158 to − 145, contains an imperfect dyad repeat which bears little resemblance to a typical CRE/TRE motif but is nevertheless strongly inducible by both phorbol ester and cAMP, and by

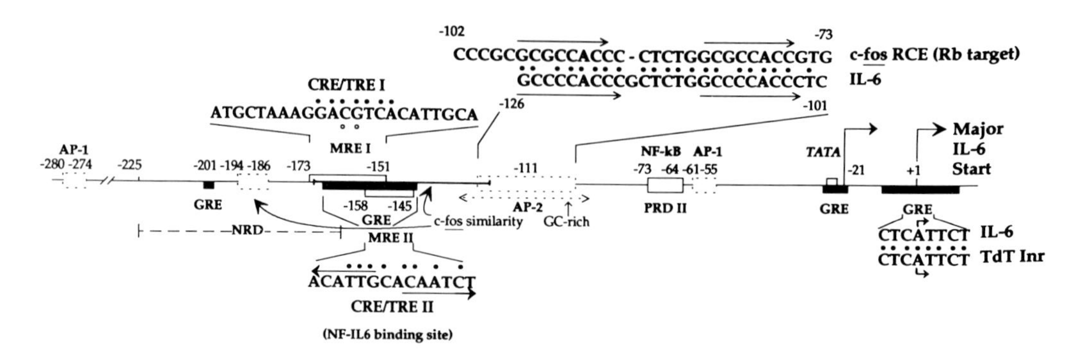

FIGURE 1. Schematic representation of positive and negative transcription regulatory elements in the 5′ flanking region of the IL-6 gene. Solid lines (either boxes or arrows) indicate DNA regulatory elements that have already been functionally implicated in IL-6 gene expression, while those marked by broken lines or boxes are based on DNA sequence analyses. The inducible transcription start sites were derived by S1 nuclease mapping (ratio of major +1 to minor −21 was 99:1).[51] The presence of a negative regulatory domain (NRD) between −225 and −165 was inferred from results published earlier.[55] The typical GACGTCA CRE/TRE motif in MRE I and the nucleotides in the novel CRE/TRE in MRE II which match with nucleotides in the CRE identified in bovine cytochrome $P450_{17\alpha}$ promoter are highlighted by solid circles. The mutation of the CG residues (open circles) to GT reduces the responsiveness of MRE I to TPA and forskolin;[51] similarly, point mutations in MRE II reduce inducibility by these agents.[111] PRDII refers to the NF-κB-like domain in the interferon-β promoter. Other investigators[60] as well as this laboratory have confirmed that point mutations at this location within the context of the complete IL-6 promoter also reduce IL-6 inducibility. Inr refers to the initiator RNA start site motif as functionally characterized in the terminal deoxynucleotidyltransferase (TdT) gene. PCE is the Rb-repressible DNA target in the c-*fos* promoter. (Adapted from Ray, A., La Forge, K. S., and Sehgal, P. B., *Mol. Cell. Biol.*, 10, 5736, 1990. With permission.)

IL-1 and TNF. The imperfect dyad repeat in MRE II is the site of binding of an IL-1- or IL-6-activated member of the C/EBP family of transcription factors designated C/EBPβ (also called NF-IL-6 and IL-6DBP).[58,59] The NF-κB site in the IL-6 promoter also appears to contribute to the activation of this gene in some cell types.[57,60,61] The major RNA start site in the IL-6 promoter corresponds to the ''initiator'' (Inr) motif which has been recently described to be similar to the TATA box in its ability to direct accurate transcription in the absence of a TATA box.[56] The IL-6 DNA sequence from −126 to −101 not only contains a direct repeat with strong similarity to the c-*fos* basal transcription element, but also contains a 21/26 nucleotide match with the Rb-repressible RCE (''Rb control element'') target motif recently identified in the c-*fos* gene.[56] Indeed, in functional assays, we have observed that the overexpression of wt Rb or wt p53 in HeLa cells strongly represses IL-6 promoter constructs.[63] Transforming mutants of p53 generally have a reduced ability to repress the IL-6 promoter.[63]

The marked ''superinduction'' of IL-6 production in the presence of inhibitors of protein synthesis, a consequence primarily of the increased stability of IL-6 mRNA under these conditions,[62] can be viewed as a remarkable adaptation to the ''alarm signal'' function of IL-6. Cells inflicted with injury secrete enhanced levels of IL-6 as their macromolecular synthesis becomes compromised.

B. REPRESSION OF THE IL-6 PROMOTER BY GLUCOCORTICOIDS AND ESTROGENS

We have shown that glucocorticoids and estradiol-17β strongly repress IL-6 expression in a variety of cell types.[64,65] In a series of detailed studies, we observed that the glucocorticoid receptor (GR) footprinted across the entire MRE region, the major TATA box, and the major RNA start site (the Inr element).[56] In functional assays, IL-6/CAT reporter plasmids driven by any of these elements from within the IL-6 promoter were repressed by dexamethasone

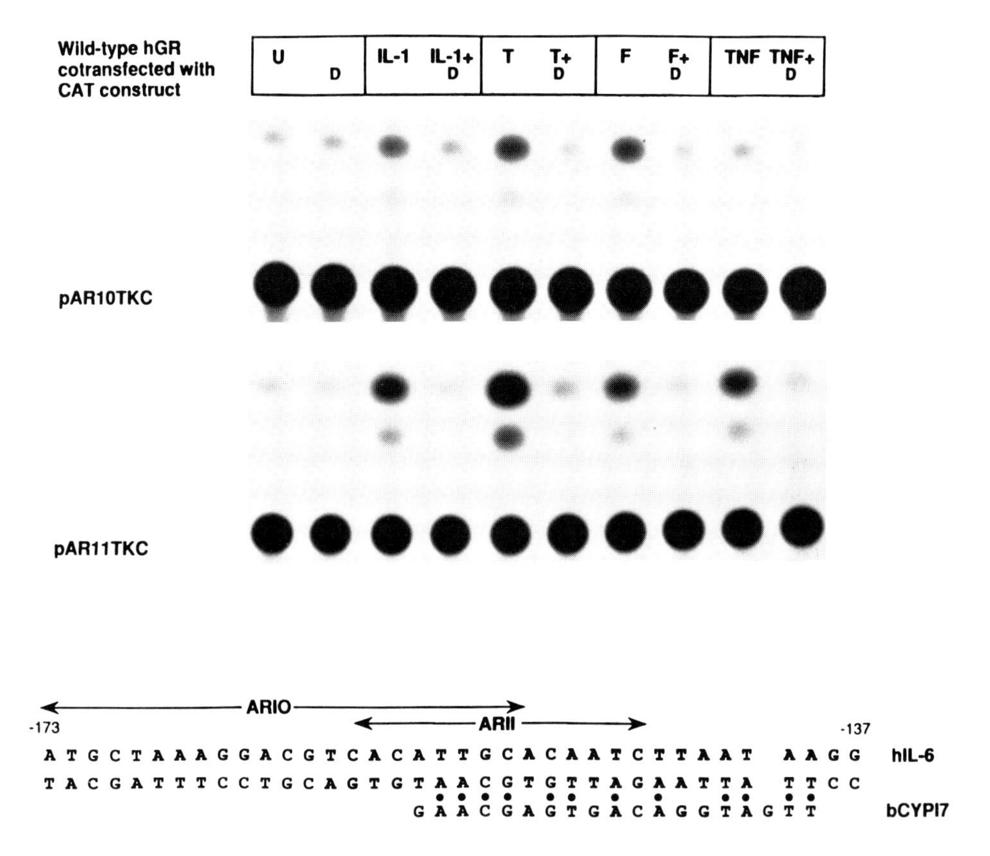

FIGURE 2. Repression of IL-6 enhancer elements by glucocorticoids. Figure illustrates the repression of IL-1-, phorbol ester-, forskolin-, and TNF-induced MRE I/TK/CAT and MRE II/TK/CAT gene expression by dexamethasone (Dex) in HeLa transfected with a wild-type GR constitutive expression plasmid. IL-6 promoter oligonucleotides AR10 (−173 to −151) and AR11 (−158 to −145) were linked to pTK$_{105}$CAT. HeLa cells in 100-mm Falcon dishes were transfected with a mixture of control pSV2neo (3 μg), RSVhGRα (5 μg), and either pAR10TKC (2.5 μg) or pAR11TKC (2.5 μg). The cells were then left untreated (U) or treated with IL-1α (5 ng/ml), phorbol ester (T, 100 ng/ml), a combination of forskolin (50 μ*M*) and isobutylmethylxanthine (IBMX, 0.5 m*M*)(F), or TNF (100 ng/ml) in the presence or absence of Dex (D, 1 μ*M*). The sequences of AR10 and AR11 are presented in the lower part of the figure. The similarity of the CRE/TRE II motif in AR11 to an atypical CRE identified in the bovine cytochrome P450$_{17\alpha}$ gene is also shown. (Adapted from Ray, A., La Forge, K. S., and Sehgal, P. B., *Mol. Cell. Biol.*, 10, 5736, 1990. With permission.)

(Dex) in HeLa cells irrespective of the inducer used, provided that these cells had been transfected with constitutive expression vectors producing wt human GR (Figure 2), giving rise to the hypothesis that IL-6 promoter occlusion was part of the mechanism by which wt GR repressed IL-6 expression.[56] It is now clear that transcriptional repression by GR in other experimental systems also involves direct interactions between GR and c-*jun* (and other c-*jun* family members), and that whether one observes Dex-responsive transcriptional activation or repression can be determined by the relative levels of c-*fos* and c-*jun* in a cell type-specific and target promoter-specific manner.[66]

In a series of experiments designed to investigate the effect of artificially engineered mutations in the DNA binding domain (DBD) of GR on the ability of this molecule to mediate repression of IL-6 gene transcription, we made the unexpected discovery that a class of mutations involving the first Zn finger (deletion of the first finger, point mutation in the Zn catenation site of the first finger or in the steroid specificity domain at the base of the first finger) converted GR from a Dex-responsive *repressor* to a Dex-responsive *activator* that could enhance basal and IL-1-induced IL-6 promoter function in HeLa cells.[67] In a manner consistent with the previous characterization of these GR mutants, none of the first Zn-finger mutants activated the murine mammary tumor virus-long terminal repeat (MTV-LTR)/CAT construct or the c-*fos*/CAT reporter genes; additionally, mutations in the second Zn finger or deletion of the entire DBD were completely inactive irrespective of the promoter tested. The first Zn-finger GR mutants that aberrantly activated IL-6 transcription in HeLa cells failed to bind to the target IL-6 promoter in the conventional DNA binding-immunoprecipitation assay using extracts from HeLa cells appropriately transfected with the GR mutant expression plasmids. These data suggest the novel hypothesis that these GR mutants might activate the IL-6 promoter without directly binding to the target promoter. These data also raise the possibility that naturally occurring point mutations or deletions of the first Zn finger of GR (which is coded for by a separate exon) or of other steroid receptors (such as the estrogen receptor which also ordinarily represses IL-6 expression) may unleash aberrant transcriptional activity leading to a dysregulated overexpression of IL-6 and other cellular genes.

C. A VIEW OF IL-6/HORMONAL INTERACTIONS

The inhibition of IL-6 expression by glucocorticoids, although a phenomenon of great therapeutic value, is but one facet of IL-6/hormonal interactions during the acute phase response that affects the function of the hypothalamus-pituitary-adrenal-gonadal axis. IL-6 and other cytokines (e.g., IL-1) produced in response to infections, tissue injury, or even psychological stress are directly or indirectly a stimulus for the secretion of corticotropin-releasing factor by the hypothalamus, which in turn leads to the enhanced secretion of adrenocorticotrophic hormone (ACTH) by the anterior pituitary and the subsequent increase in the levels of circulating corticosteroids.[68,69] Additionally, IL-6, which can itself be produced by the folliculostellate cells of the anterior pituitary,[70] has been reported to directly stimulate the release of the anterior pituitary hormones ACTH, prolactin, growth hormone, and luteinizing hormone.[71] The administration of IL-6-inducing cytokines such as TNF or of bacterial endotoxin to human volunteers leads to the appearance of circulating IL-6 and to elevations in plasma ACTH and cortisol levels.[72,73]

Elevated levels of circulating glucocorticoids during the acute phase response synergize with IL-6 in inducing the increased hepatic synthesis and secretion of plasma proteins such as fibrinogen, various antiproteinases, complement factors, and scavenger proteins (e.g., haptoglobin, hemopexin, and C-reactive protein)[13] (also see other chapters in this volume). In addition, the role of IL-6 in the activation of B- and T-cell function also contributes to the ability of the host to combat infection and tissue damage.[13]

However, this acute phase reaction is self-limiting in that glucocorticoids strongly inhibit IL-6 gene expression in different tissues. Repression of the IL-6 gene is a component in the well-known antiinflammatory effect of glucocorticoids. The downregulation of IL-6 gene expression by estradiol-17β in estrogen-sensitive tissues such as endometrial stromal cells[65] probably represents an additional feedback regulatory loop affecting circulating IL-6 levels in women.

III. IL-6 PROTEIN SPECIES

Two levels of complexity have emerged with respect to the nature of natural IL-6. The first level, which was not unexpected, relates to the various cell type-dependent posttranslational modifications of the IL-6 polypeptide per se. These alterations include differential amino-terminal processing, differential *N*- and *O*-glycosylation, and phosphorylation. The second level, which was unexpected, relates to the discovery that IL-6 exists in human plasma primarily in the form of high molecular mass complexes that appear to include other plasma proteins in addition to IL-6. The antigenic and biological properties of IL-6 present in such complexes appear to be markedly different from those of essentially monomeric IL-6 preparations.

A. MONOMERIC IL-6 PROTEINS

The amino acid sequence deduced from the nucleotide sequence of human IL-6 cDNA contains 212 residues in the open reading frame.[5,6,10] By immunoprecipitation analyses under completely reducing and denaturing conditions using polyclonal rabbit antisera to *Escherichia coli*-derived IL-6, this cytokine is secreted from different cell types as proteins with a molecular mass of 19 to 30 kDa.[19,74-77] Particular B- or T-cell lines (sfBJAB, TCL-Na1) appear to secrete largely the nonglycosylated 19- to 21-kDa IL-6 protein.[75,77] Other B-cell lines (e.g., KH) can also secrete IL-6 exclusively as an *N*- and *O*-glycosylated 30-kDa protein.[75] Cell types such as fibroblasts, endothelial cells, monocytes, keratinocytes, and others can secrete a mixture of several 23- to 25-kDa exclusively *O*-glycosylated, 23- to 25-kDa exclusively *N*-glycosylated, and 28- to 30-kDa *N*- and *O*-glycosylated IL-6 molecules.[74,76,78] The amino terminus of the 23- to 25-kDa species of fibroblast-derived IL-6 is at Ala_{28}, and that of the 28- to 30-kDa species at Val_{30};[75] Pro_{29} has been identified as the N terminus of the nonglycosylated IL-6 secreted by the TCL-Na1 cell line.[10,77] Thus, mature secreted human IL-6 contains 184 to 186 residues in its polypeptide chain.

Of the several potential *O*-glycosylation sites in human IL-6, Thr_{166} (alternatively numbered as Thr_{138} in the mature protein, which is considered to start at Ala_{28} in the open reading frame) has been directly identified as an *O*-glycosylation site.[79] Of the two potential *N*-glycosylation sites (at Asn_{73} and Asn_{172}), the *N*-linkage of carbohydrate to Asn_{73} (alternatively numbered as Asn_{45} in the mature protein) has been directly identified.[79] Cell type-dependent variations in the proportion of di-, tri-, and tetraantennary carbohydrate side chains add to the complexity of this molecule.[76,79] Fibroblast-, endothelial cell-, monocyte-, and endometrial stromal cell-secreted IL-6 is phosphorylated at serine residues in a cell type- and induction protocol-dependent manner.[80,111] For example, whereas the 23- to 25- and 28- to 30-kDa IL-6 species secreted by fibroblasts are both phosphorylated, only the 23- to 25-kDa species secreted by monocytes appear to be phosphorylated; monocytic 28- to 30-kDa IL-6 species exhibit relatively little phosphorylation.[80] Furthermore, the inclusion of cycloheximide in the induction regimen inhibits phosphorylation of IL-6 proteins secreted by fibroblasts.[81] A major site of phosphorylation has been identified to be Ser_{82} (alternatively numbered as Ser_{54} in the mature protein).[81]

The mature murine IL-6 protein, which is derived from an open reading frame of 211 residues,[82,83] consists of 187 residues which lack the Asn-X-Ser/Thr motifs required for *N*-linked carbohydrate linkage; nevertheless, murine IL-6 as secreted by different cell types consists of a heterogeneous set of differentially *O*-glycosylated proteins of mass 21 to 28 kDa.[83] Murine IL-6, like human IL-6, contains four conserved Cys residues that form two intramolecular S-S bridges that appear to be important for optimal biological activity.

The amino-terminal 28 residues of mature human IL-6, which are highly divergent from those of mature murine IL-6,[15,82,83] appear to be dispensable for hybridoma growth factor and hepatocyte-stimulating factor activity.[84,85] The carboxy-terminal four residues of IL-6 appear to be crucial for both of these activities.[84-86]

B. HIGHER-ORDER STRUCTURE OF CELL CULTURE-DERIVED IL-6

Prior literature reported HSF activity in culture medium from endotoxin-induced human monocytes in fractions that eluted between 30 to 70 kDa from gel filtration columns.[87,88] Inasmuch as monocyte-derived HSF activity is largely due to IL-6, these observations pointed to a complex higher-order structure of natural human IL-6.

Under nondenaturing conditions, IL-6 secreted by fibroblasts (in serum-free culture medium) behaves as a high molecular weight multimeric aggregate.[89] The association between the 23- to 25-kDa exclusively *O*-glycosylated IL-6 and the 28- to 30-kDa *N*- and *O*-glycosylated IL-6 species was suggested by the binding and specific elution of the 23- to 25-kDa IL-6 from wheat germ or lentil lectin columns together with the 28- to 30-kDa *N*-glycosylated species even though the 23- to 25-kDa species lacked the requisite sugars for binding to these lectin columns. When electrophoresced through nondenaturing NP40-polyacrylamide gels, natural fibroblast-derived human IL-6 appears to be largely a trimeric 85-kDa complex containing both the 23- to 25-kDa *O*-glycosylated and the 28- to 30-kDa *N*- and *O*-glycosylated species. Additional homo- and heteromeric complexes of 45 to 120 kDa are present in fibroblast-derived IL-6; their proportion depends on the relative amount of the 23- to 25-kDa IL-6. IL-6 complexes in preparations enriched in the 28- to 30-kDa monomeric IL-6 species are essentially all of the 85-kDa size, whereas preparations enriched in the 23- to 25-kDa IL-6 contain significant amounts of the 45- to 70-kDa complexes in addition to the apparently trimeric 85-kDa complex as judged by NP40-polyacrylamide gel electrophoresis (PAGE). The formation of these multimeric forms of natural IL-6 is not dependent on intermolecular disulfide bridges because these aggregates can be observed even under reducing conditions. (This is in contrast to the artifactual formation of reducing agent-sensitive dimers and tetramers in rIL-6 preparations).

Sephadex G-200 gel filtration chromatography of fibroblast-derived IL-6 reveals that this cytokine elutes in a broad range of sizes from 20 to 70 kDa.[89] The high molecular mass fractions (>60 kDa by gel filtration) are enriched in the 28- to 30-kDa species. Although all of the IL-6-containing fractions exhibit hepatocyte-stimulating factor activity as monitored by the increase in α_1-antichymotrypsin secretion by Hep 3B cells, the high molecular mass fractions (55 to 70 kDa) are approximately tenfold less active than the lower molecular mass fractions (25 to 45 kDa) in B9 hybridoma growth factor activity.[89]

A distinction between the high and low molecular mass fractions of fibroblast IL-6 eluted off a G-200 column is also observed in immunoassays.[90] Using various pairs of anti-IL-6 monoclonal antibodies as probes for IL-6 structure, we have observed that the 4IL6/5IL6 enzyme-linked immunosorbent assay (ELISA) preferentially detects the high molecular mass IL-6, while the IG61/5IL6 ELISA preferentially detects the lower molecular mass IL-6. Although the latter ELISA is highly sensitive to rIL-6 (sensitivity, 1 to 5 pg/ml), it fails to adequately react with fibroblast-derived IL-6 >60 kDa eluted off gel filtration columns. It is noteworthy that only the trapping mAb ("4IL6" compared to "IG61") is different in these two ELISAs; the biotinylated reporter mAb ("5IL6") is the same. This unexpected preferential reactivity of the 4IL6/5IL6 ELISA with the high molecular mass IL-6 is not confined to fibroblast cell culture-derived preparations, but also holds in a rather dramatic fashion for endogenous IL-6 present in human plasma.

C. IL-6 IN HUMAN PLASMA AND SERUM

Early studies evaluating the concentration and nature of IL-6 in *human* body fluids, particularly serum, in health and disease using different bioassays presented a dilemma.[13,26-29,72,73,91] When the hybridoma growth factor assay (in *murine* cell lines such as B9 or MH60.BSF-2) was used, sera from normal individuals had little detectable activity, whereas sera (or other body fluids) from individuals in rejection following renal transplantation, those with rheumatoid arthritis or various bacterial infections, and even volunteers injected with TNF or endotoxin had detectable hybridoma growth factor activity in the range from barely detectable to a maximum of a few hundred biological units per milliliter of serum. Given the conversion factor that approximately 15 pg/ml of natural human IL-6 can correspond to 1 unit of hybridoma growth factor activity (e.g., in B9 cells),[90] these data suggested that the concentration of IL-6 ranged from a few pg/ml to a maximum of approximately 1 ng/ml. Nevertheless, when human sera were assayed using the hepatocyte-stimulating factor assay in *human* cells (stimulation of α_1-antichymotrypsin secretion by human Hep 3B cells; the stimulation verified to be essentially completely blocked by polyclonal anti-rIL-6 antiserum), we estimated circulating IL-6 concentrations in the range of 2.5 to 120 ng/ml in patients with sepsis and in volunteers given endotoxin or TNF.[28,72,73] In at least one instance (volunteers given TNF), the same serum samples that contained a few hundred units of HGF were simultaneously estimated to contain >100 ng/ml HSF activity.[72] In a second study (patients with active psoriasis), plasma samples that appeared to contain IL-6 HSF activity (1 to 10 ng/ml) did not contain any detectable B9 HGF activity (despite the use of the conventional heat inactivation step).[92]

Furthermore, two different laboratories reported that gel filtration of serum followed by assays of the eluted fractions using the *murine* B9 cells to detect *human* IL-6 showed that all of the detectable IL-6 had a molecular mass of approximately 20 to 25 kDa.[26,93] However, when we purified IL-6 from sera of volunteers administered endotoxin or TNF using a polyclonal anti-rIL-6 immunoaffinity column and characterized it by Western blots under "nominally denaturing and reducing conditions", all of the IL-6 appeared to be of mass 43 to 45 kDa.[72,73] Under completely denaturing conditions, the monomeric IL-6 species purified from serum, cerebrospinal fluid, or amniotic fluid of patients with bacterial infection consisted of proteins in the 23- to 30-kDa range.[28,94]

Since these early studies, numerous other investigators have used the hybridoma growth factor assay as carried out in *murine* cell lines and have reported elevations of *human* IL-6 levels in sera of patients with various diseases;[13,29,30,91] invariably, the concentrations reported are in the range from barely detectable to a few hundred units per milliliter. Also, various investigators have since developed mAb-based ELISAs for IL-6 using various recombinant or cell culture-derived IL-6 preparations for raising the mAbs and for calibration of the assays. These ELISAs, which are generally highly sensitive when calibrated using rIL-6 or cell culture-derived IL-6 (sensitivity down to 1 to 5 pg/ml), have as a class revealed that IL-6 is present in sera from normal individuals at a concentration of 10 to 75 pg/ml, whereas individuals with various disease states (infections, neoplasia, and autoimmune diseases) have elevations in IL-6 levels that are usually a few hundred picograms per milliliter, but can be a maximum of 1 to 2 ng/ml (see Reference 95 for an example).

Using the murine B9 hybridoma growth factor assay and the highly sensitive (for rIL-6) IG61/5IL6 ELISA, we essentially confirmed the above observations in this laboratory. For example, using these two assays, we found that plasma from normal individuals had little IL-6 and that from patients with psoriasis had IL-6 concentrations that ranged from undetectable to a few hundred picograms per milliliter at the most. As additional confirmation of the prior IL-6 literature, Figure 3A illustrates B9 and IG61/5IL6 ELISA data obtained when a 0.8-ml sample of serum from a bone marrow transplant patient with severe intercurrent infection (patient succumbed the next day) was fractionated through a Sephadex G-200 column.[96] The highest level of B9 HGF activity in the unfractionated serum from this patient was approximately 300 units/ml on the day prior to death. Figure 3A shows that all of the B9-active IL-6 was found in a low molecular mass (15- to 20-kDa) peak that coincided with the bulk of the IG61/5IL6-detectable IL-6 antigen. The highest concentration of IL-6 antigen detected using this ELISA in any fraction was approximately 200 pg/ml.

Despite our ability to confirm the *experimental observations* that have gained currency in the IL-6 field using particular assays and techniques, our answer to the question, "What is the nature of IL-6 in human plasma or serum?", is unorthodox. Our new data indicate that all of the previous IL-6 literature, in part even some from this laboratory, has presented a largely incomplete picture.

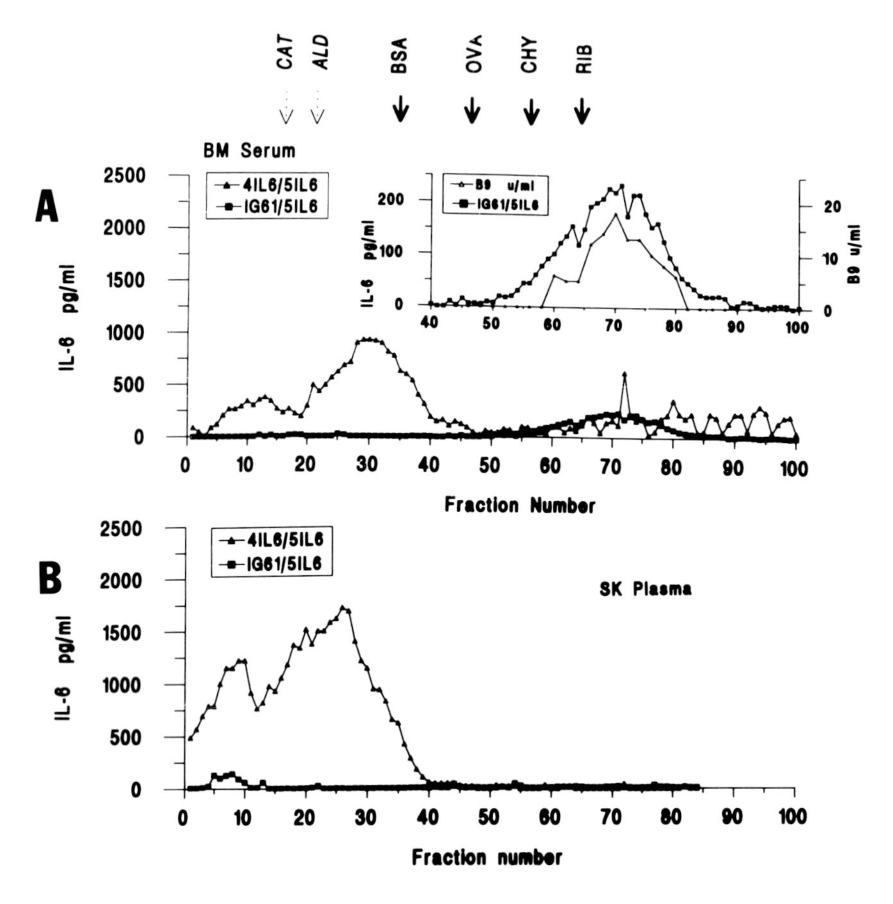

FIGURE 3. Sephadex G-200 fractionation of IL-6 in human serum (A) and in fresh heparinized plasma (B). (A) Serum (0.8 ml) from a patient 21 d after a bone marrow transplant; patient had severe intercurrent infection. (B) Fresh heparinized plasma (3 ml) from a patient with epidermolysis bullosa. In each instance, the serum or plasma was applied to a 300-ml bed volume G-200 column, and all of the eluted fractions (approximately 2.5 ml each) were assayed for IL-6 using the IG61/5IL6 (squares) and 4IL6/5IL6 (triangles) ELISAs, each calibrated using the *E. coli*-derived 88/514 standard and the B9 HGF bioassay (also calibrated using the 88/514 standard). Inset in A shows the B9 activity observed; no B9 activity was observed in any of the eluted fractions in B even after "heat-inactivation". (Adapted from May, L. T., Santhanam, U., Viguet, H., Kenney, J., Ida, N., Allison, A. C., and Sehgal, P. B., *J. Biol. Chem.*, 267, 19698, 1992. With permission.)

1. IL-6 in Complexes with Other Plasma Proteins

The use of the 4IL6/5IL6 ELISA (which preferentially reacts with the high molecular mass fibroblast-derived IL-6 aggregates)[90] to assay for IL-6 in serum fractions eluting off the Sephadex G-200 column led to a novel observation (Figure 3).[96] While this ELISA was able to detect the low molecular mass B9-active IL-6, the bulk of the IL-6 antigen was in two peaks of 100 to 150 and 400 to 500 kDa (Figure 3A). Eluted fractions across these two peaks were completely devoid of B9 HGF activity (as a control, the addition of exogenous fibroblast-derived or *E. coli*-derived IL-6 to these fractions did not inhibit the B9 activity of the exogenous cytokine). Nevertheless, the high molecular mass peak fractions were active in the hepatocyte assay in Hep 3B cells. The 100- to 150- and 400- to 500-kDa high molecular mass complexes are exclusive to endogenous IL-6 because mixing fibroblast-derived ^{35}S-labeled IL-6 with serum followed by gel filtration does not generate these complexes.

G-200 column fractionation of fresh plasma (with or without addition of heparin or EDTA) from patients with active cutaneous disease (epidermolysis bullosa or psoriasis) or even from disease-free individuals (e.g., this author) revealed that virtually all of the IL-6 reactive in the 4IL6/5IL6 ELISA and present at concentrations in the nanogram per milliliter range in various fractions (and reactive to a much lesser extent in the IG61/5IL6 ELISA) was reproducible in the two high molecular mass peaks[96] (Figure 3B). None of the fractions had any detectable B9 HGF activity; this activity could be artifactually generated by preparing serum from plasma of individuals with disease.

When we used the 4IL6/5IL6 ELISA and a natural fibroblast-derived preparation as the calibrating standard to assay for IL-6 in unfractionated heparinized plasma from normal individuals, the concentrations of IL-6 were, surprisingly, estimated to be in the range of 2 to 10 ng/ml.[96] Somewhat lower (approximately twofold) estimates were obtained using an *E. coli*-derived IL-6 preparation (preparation 88/514) as the calibrating standard.

Did the two high molecular mass peaks reactive in the 4IL6/5IL6 ELISA contain the IL-6 protein or did they represent accidental cross-reaction with other materials? IL-6 present in pools of fractions corresponding to the 100- to 150- and 400- to 500-kDa regions in the elutions illustrated in Figure 3 and from additional plasma elutions were purified by immunoaffinity chromatography through a 5IL6-mAb column. Western-blot analyses of the immunoaffinity column eluate under completely reducing and denaturing conditions using a polyclonal anti-rIL-6 rabbit antiserum revealed that the 100- to 150-kDa complexes contained very high concentrations of both the 23- to 25- and 28- to 30-kDa monomeric IL-6 species, whereas the 400- to 500-kDa complexes contained exclusively the 28- to 30-kDa IL-6 monomer.[96] Most strikingly, the immunoaffinity column eluates, including those from the author's 100- to 150-kDa plasma complexes, were now active in the B9 HGF to an extent commensurate with the Western-blottable IL-6 antigen.[96]

Polyacrylamide gel electrophoresis and Coomassie blue staining of the 5IL6-immunoaffinity column eluate of the 100- to 150-kDa pool from Figure 3A revealed several protein bands in the size range from 14 to 80 kDa.[96] These bands were transferred to PVDF membranes and the amino terminus of each visible band was sequenced. The 14- and 23-kDa bands were IL-6 proteins starting at Ala_{28}; a 28-kDa band was also IL-6, starting at Val_{30}; the other "IL-6-associated" proteins were fragments of C-reactive protein, complement C3, complement C4, and serum albumin.[96] The intensity of the serum albumin band precludes an assessment of whether the soluble IL-6 receptor (fragment of p80) was also part of this complex. Recent reports document the presence of high concentrations (10 to 30 ng/ml) of the soluble IL-6 receptor and the gp130 molecule in the peripheral circulation in man.[97] It is noteworthy that thus far we have not detected α_2-macroglobulin in these complexes.

By direct amino acid sequencing,[96] we estimate that the concentration of IL-6 protein present in the original 0.8-ml aliquot of serum loaded onto the G-200 column in Figure 3A must have been at least 5 μg/ml. Most of this IL-6 was not visible in the conventional B9 HGF or the so-called "highly sensitive" IG61/5IL6 ELISA. We believe that most, if not all, IL-6 ELISAs in use today that have been selected and "optimized" using recombinant IL-6 preparations are likely to be of the IG61/5IL6 type and not of the 4IL6/5IL6 variety. Thus, the numerical validity of large bodies of the ever-burgeoning IL-6 clinical literature needs to be carefully reexamined.

The new data raise novel questions about IL-6 biology. What is the spectrum of biological activities that are truly attributable to IL-6 as it exists in human blood and acts on the parenchymal cells lining the sinusoids of the liver? What is the nature of IL-6 in the extravascular tissue fluid (it is this that bathes many of the individual target cells of this cytokine in other organs)? What is the detailed biochemical structure and stoichiometry of the various components in the 100- to 150- and 400- to 500-kDa IL-6 complexes ("inflammasome I and II" in our current laboratory parlance)? In essence, we are beginning to ask a layer of questions that had hitherto not been asked about cytokines: are there mechanisms that control the bioavailability of circulating cytokines? At a more practical level, the answer to the question of how does one reliably assay for IL-6 in human body fluids remains tentative and open to further reexamination. With IL-6 entering early clinical testing, reliable answers to these questions are of critical importance.

IV. IL-6 IN EXPERIMENTAL ANIMAL MODELS

The exogenous administration of rIL-6 to experimental animals, particularly when IL-6 is administered together with glucocorticoids, can result in dramatic alterations in the levels of acute phase plasma proteins in blood.[98] The administered rIL-6 is cleared rapidly from the circulation. Typically, approximately 80% of rIL-6 administered intravenously as a bolus is localized in the liver within 20 min after injection.[98] High levels of endogenously synthesized IL-6 are readily detected in the peripheral circulation in a variety of experimental models of bacterial and viral infection, sterile inflammation, cytokine administration, and even solid-tumor implantation.[13,91,98] In each instance, whereas the appearance of TNF or IL-1 may or may not include a systemic component, depending on the severity of the disease process in the animal model, the IL-6 response always includes a systemic component. Furthermore, while in some models the prior or intercurrent induction of TNF or IL-1 can contribute to the subsequent increase in circulating IL-6 levels (e.g., in the endotoxin- or *E. coli*-treated baboon or the turpentine abscess-bearing mouse).[40-42] in others, IL-6 induction can occur independent of TNF or IL-1 induction (e.g., murine models of systemic *Listeria* infection, adenovirus type 5 pneumonia, and solid-tumor implantation).[41,43,99-101] Elevations of circulating IL-6 levels in mice, rats, or baboons in these experimental models are correlated with the subsequent alterations characteristic of the acute phase plasma protein response.

To what extent can the "acute phase" host response in these various models be ascribed to endogenously produced IL-6? A traditional approach to this question is to flood the experimental animal with neutralizing antibody to the cytokine under consideration and observe whether such passive immunization blocks the acute phase phenomenon or affects the progress of the disease process. Using this approach, Fletcher-Starnes et al.[31] reported that the anti-MuIL-6 monoclonal antibody MP5-20F3 blocked the adverse effects of endotoxin challenge in the mouse. Neta and colleagues have reported that mAb MP5-20F3 blocks the increase in fibrinogen and other acute phase plasma proteins in a murine model of radiation injury.[34] Gauldie and colleagues concur that a polyclonal anti-rat IL-6 antiserum blocks the acute phase response in the endotoxin-treated rat.[33]

The interpretation that antibodies such as MP5-20F3 affect biological phenomena solely because of "neutralization" of circulating IL-6 does not appear to be fully justified by additional recent data. Animals treated with a neutralizing mAb to IL-6 (e.g., baboons injected with 5IL6-H17 or mice injected with MP5-20F3) and an inducer of IL-6 (e.g., endotoxin, adenovirus type 2, turpentine abscess, TNF, or IL-1) can have a ten- to 100-fold increase in the levels of circulating IL-6 as judged using either appropriate ELISAs or the B9 hybridoma growth factor assay.[111] The increase in plasma/serum levels of biologically active IL-6 as assayed *ex vivo* in cell culture assays following appropriate dilution of serum/plasma even in the presence of "neutralizing" mAb is dramatic. Nevertheless, despite the marked increase in *ex vivo*-measurable B9 HGF activity in murine plasma/serum, mAbs such as MP5-20F3 at least partially block the *in vivo* alterations in acute phase plasma protein synthesis in the same mice. The possibility that although there is a large amount of IL-6 in plasma, it is held in a form that is not "bioavailable" *in vivo,* i.e., in some sort of a depot or inaccessible form, may help explain this paradox. These data bring us back to a consideration of "IL-6 complexes" in the peripheral circulation — this time in experimental animals (mouse, baboon) and from a different direction than that described in earlier sections. While the new data render interpretation of "IL-6 blockade" experiments in animal models complex, they point to interesting unexplored facets of IL-6 structure and function as they relate to how this cytokine actually exists *in vivo*.

A. IL-6 IN THE PARANEOPLASTIC SYNDROME

IL-6 is an invariant element in the host-tumor interaction.[13,101-103] Mice bearing a variety of different solid tumor implants exhibit a weight-losing syndrome that is accompanied by increases in the positive acute phase plasma proteins and a decrease in albumin. In such models, IL-6, but not TNF or IL-1, is detected in the peripheral circulation.[101,103] Circulating levels of IL-6 correlate directly with the tumor size and extent of cachexia. It had been reported that anti-TNF antibodies blocked tumor-induced cachexia, but the blocking effect was only partial.[104] It has now been demonstrated that the anti-MuIL-6 mAb MP5-20F3 largely blocks tumor-induced weight loss and markedly inhibits the serum amyloid P response in mice bearing the C-26.IVX tumor.[32] These data are the first demonstration that endogenous IL-6 is a contributor to the development of cancer cachexia. The dilemma created by these data arises from the observation that administration of exogenous IL-6 to mice or other experimental animals does not elicit a weight-losing syndrome.[98] Perhaps the continuous presence of endogenous IL-6 in the tumor-bearing host elicits biological effects different from those of exogenous IL-6.

V. IL-6 FUNCTIONS: COMMENTS

It is now clear that circulating IL-6 levels are elevated in a wide variety of human diseases associated with tissue injury, neoplasia, and autoimmune diseases. Recombinant or natural IL-6 affects a wide variety of cellular functions in cell culture experiments.[13,30,91] Which of these numerous effects of IL-6 observed in cell culture are of relevance or importance in human health and disease? The question is posed from the point of view that among all of the myriad effects of IL-6, there must exist a relative rank order that allows for simplified working hypotheses about IL-6 function in given physiological or pathological situations. This question is also posed from the perspective that investigators outside the IL-6 field should be able to grasp some of the key principles involved without becoming enmeshed in confusing and complex descriptions of IL-6 effects and its interactions with other cytokines in cell culture, descriptions shorn of a sense of *in vivo* importance.

In this laboratory, we have found it convenient to think about IL-6 as a long-distance alarm signal that alerts the hepatocytes to the presence of tissue damage. We view the role of IL-1 and TNF primarily as local or paracrine and that of IL-6 as both local and systemic. From this perspective, the hepatic effects of IL-6 take on major significance. The alterations in plasma protein synthesis elicited by IL-6 can be viewed as increasing the ability of the host to limit tissue damage — a concept explored in considerable detail in other chapters of this volume. Within this context, it is rather efficient that the same cytokine that elicits these "nonspecific" damage-limiting responses, i.e., IL-6, can also be the one that enhances the function of the specific immune responses — immunoglobulin synthesis and T-cell activation — that then contribute to host recovery. The stimulation of thrombocytopoiesis and participation in enhancing hemopoietic progenitor cell growth by IL-6 can be viewed as part of this overall host-response picture. The interactions of glucocorticoids with IL-6 — the stimulation of ACTH secretion by IL-6 (and therefore the subsequent cortisol release) and the synergistic enhancement of IL-6 effects on the hepatocyte by glucocorticoids, but the inhibition by glucocorticoids of further IL-6 synthesis as a part of a feedback loop — point to an exquisitely regulated system geared to function as part of the host defense against injury.

VI. THERAPEUTIC USE OF IL-6 OR ANTI-IL-6 ANTIBODIES

IL-6 is being evaluated in phase I trials in patients with various solid tumors (e.g., breast cancer, colon cancer, lung cancer, etc.) who had failed other modalities of therapy.[99] In preclinical work, IL-6 inhibited tumor growth and even preformed metastases in animals implanted with a variety of different solid tumors.[105,106] The most dramatic results were observed using the metastatic melanoma model where IL-6 was used in combination with other agents such as cyclophosphamide.[107] Various laboratories are also exploring the use of IL-6 as a therapeutic agent that might promote thrombocytopoiesis in primary or secondary thrombocytopenic situations.[108,109]

Because IL-6, primarily paracrine IL-6, appears to contribute to the enhanced proliferation of certain neoplastic cells such as myelomas and plasma cell dyscrasias, neutralizing anti-IL-6 mAb are being used in phase I trials in patients with multiple myeloma and Castleman's disease.[35,110] In myeloma patients, anti-IL-6 mAb markedly decreases the number of mitotic cells in the circulation, elicits a dramatic return of C-reactive protein (CRP) levels to close to baseline, and blocks the pyrexia seen in these patients. Indeed, the extent of reduction in CRP levels by anti-IL-6 mAb administration correlates with whether the mAb administration will inhibit the markers of active myeloma. In other words, a decrease in CRP levels in a patient administered anti-IL-6 mAb appears to be a good indicator of whether adequate amounts of mAb were administered and whether a response can be expected in a particular patient. Unfortunately, available data also indicate that the response to murine anti-HuIL-6 mAb in such myeloma patients is transient — 6 to 8 weeks after the beginning of mAb administration, the myeloma markers return to pretreatment levels. Humanized anti-IL-6 mAb is presently under initial therapeutic evaluation.

ACKNOWLEDGMENTS

Research in the author's laboratory is supported in part by Research Grant AI-16262 from the National Institutes of Health, and a contract from The National Foundation for Cancer Research.

REFERENCES

1. **Weissenbach, J., Chernajovsky, Y., Zeevi, M., Shulman, L., Soreq, H., Nir, U., Wallach, D., Perricaudet, M., Tiollais, P., and Revel, M.,** Two interferon mRNAs in human fibroblasts: in vivo translation and *Escherichia coli* cloning studies, *Proc. Natl. Acad. Sci. U.S.A.*, 77, 7152, 1980.
2. **Sehgal, P. B. and Sagar, A. D.,** Heterogeneity of poly(I) · poly(C)-induced human fibroblast interferon mRNA species, *Nature (London)*, 287, 95, 1980.
3. **Content, J., DeWit, L., Poupart, P., Opdenakker, G., VanDamme, J., and Billiau, A.,** Induction of 26 kDa-protein mRNA in human cells treated with an interleukin-1-related, leukocyte-derived factor, *Eur. J. Biochem.*, 152, 253, 1985.
4. **Kohase, M., Henriksen-DeStefano, D., May, L. T., Vilček, J., and Sehgal, P. B.,** Induction of β_2-interferon by tumor necrosis factor: a homeostatic mechanism in the control of cell proliferation, *Cell*, 45, 659, 1986.
5. **Zilberstein, A., Ruggieri, R., Korn, J. H., and Revel, M.,** Structure and expression of cDNA and genes for human interferon-beta-2, a distinct species inducible by growth stimulatory cytokines, *EMBO J.*, 5, 2529, 1986.
6. **May, L. T., Helfgott, D. C., and Sehgal, P. B.,** Anti-β-interferon antibodies inhibit the increased expression of HLA-B7 mRNA in tumor necrosis factor-treated human fibroblasts: structural studies of the β_2 interferon involved, *Proc. Natl. Acad. Sci. U.S.A.*, 83, 8957, 1986.

7. **Kohase, M., May, L. T., Tamm, I., Vilček, J., and Sehgal, P. B.,** A cytokine network in human diploid fibroblasts: interactions of β-interferons, tumor necrosis factor, platelet-derived growth factor, and interleukin-1, *Mol. Cell. Biol.*, 7, 273, 1987.
8. **Helfgott, D. C., May, L. T., Sthoeger, Z., Tamm, I., and Sehgal, P. B.,** Bacterial lipopolysaccharide (endotoxin) enhances expression and secretion of β_2-interferon by human fibroblasts, *J. Exp. Med.*, 166, 1300, 1987.
9. **Sehgal, P. B., Helfgott, D. C., Santhanam, U., Tatter, S. B., Clarick, R. H., Ghrayeb, J., and May, L. T.,** Regulation of the acute phase and immune responses in viral disease: enhanced expression of the "β_2-interferon/hepatocyte stimulating factor/interleukin-6" gene in virus-infected human fibroblasts, *J. Exp. Med.*, 167, 1951, 1988.
10. **Hirano, T., Yasukawa, K., Harada, H., Taga, T., Watanabe, Y., Matsuda, T., Kashiwamura, S., Nakajima, K., Koyama, K., Iwamatsu, A., Tsunasawa, S., Sakiyama, F., Matsui, H., Takahara, Y., Taniguchi, T., and Kishimoto, T.,** Complementary DNA for a novel human interleukin (BSF-2) that induces B lymphocytes to produce immunoglobulin, *Nature (London)*, 324, 73, 1986.
11. **Sehgal, P. B., May, L. T., Tamm, I., and Vilček, J.,** Human β_2 interferon and B-cell differentiation factor BSF-2 are identical, *Science*, 235, 731, 1987.
12. **Gauldie, J., Richards, C., Harnish, D., Lansdorp, P., and Baumann, H.,** Interferon β_2/B-cell stimulatory factor type 2 shares identity with monocyte-derived hepatocyte-stimulating factor and regulates the major acute phase protein response in the liver, *Proc. Natl. Acad. Sci. U.S.A.*, 84, 7251, 1987.
13. **Sehgal, P. B., Grieninger, G., and Tosato, G., Eds.,** Regulation of the acute phase and immune responses: interleukin-6, *Ann. N.Y. Acad. Sci.*, 557, 1, 1989.
14. **Nordan, R. P. and Potter, M.,** A macrophage-derived factor required by plasmacytomas for survival and proliferation in vitro, *Science*, 233, 566, 1986.
15. **Van Snick, J., Cayphas, S., Vink, A., Uttenhove, C., Coulie, P. G., Rubira, M. R., and Simpson, R. J.,** Purification and NH_2-terminal amino acid sequence of a T-cell derived lymphokine with growth factor activity for B-cell hybridomas, *Proc. Natl. Acad. Sci. U.S.A.*, 83, 9679, 1986.
16. **Aarden, L. A., Lansdorp, P. M,. and DeGroot, E. R.,** A growth factor for B cell hybridomas produced by human monocytes, *Lymphokines*, 10, 175, 1985.
17. **Shabo, Y., Lotem, J., Rubinstein, M., Revel, M., Clark, S. C., Wolf, S. F., Kamen, R., and Sachs, L.,** The myeloid blood cell differentiation-inducing protein MGI-2A is interleukin-6, *Blood*, 72, 2070, 1988.
18. **Garman, R., Jacobs, K., Clark, S. C., and Raulet, D.,** B-cell-stimulatory factor 2 ($beta_2$ interferon) functions as a second signal for interleukin 2 production by mature murine T cells, *Proc. Natl. Acad. Sci. U.S.A.*, 84, 7629, 1987.
19. **Tosato, G., Seamon, K. B., Goldman, N. D., Sehgal, P. B., May, L. T., Washington, G. C., Jones, K. D., and Pike, S. E.,** Identification of a monocyte-derived human B cell growth factor as interferon-β_2 (BSF-2, IL-6), *Science*, 239, 502, 1988.
20. **Ming, J. E., Cernetti, C., Steinman, R. L., and Granelli-Piperno, A.,** Development of cytolytic T lymphocytes in thymus culture requires a cytokine sharing homology with IL-6, *Ann. N.Y. Acad. Sci.*, 557, 396, 1989.
21. **Sehgal, P. B.,** Interleukin-6: a regulator of plasma protein gene expression in hepatic and non-hepatic tissues, *Mol. Biol. Med.*, 7, 117, 1990.
22. **Baumann, H. and Wong, G. G.,** Hepatocyte-stimulating factor II shares structural and functional identity with leukemia inhibitory factor, *J. Immunol.*, 143, 1163, 1989.
23. **Baumann, H., Marinkovic-Pajovic, S., Won, K. A., Jones, V. E., Campos, S. P., Jahreis, G. P., and Morella, K. K.,** The actions of IL-6 and LIF on liver cells, in *Polyfunctional Cytokines: IL-6 and LIF*, Vol. 167, Metcalf, D., Ed., CIBA Found. Symp., Wiley, Chichester, 1992, 100.
24. **Baumann, H. and Schendel, P.,** Interleukin-11 regulates the hepatic expression of the same plasma protein genes as interleukin-6, *J. Biol. Chem.*, 266, 20424, 1991.
25. **Gauldie, J.,** Effect of oncostatin M on protein synthesis in liver cells. Discussion, in *Polyfunctional Cytokines: IL-6 and LIF*, Vol. 167, Metcalf, D., Ed., Wiley, Chichester, 1992, 116.
26. **Van Oers, M. H. J., Van Der Heyden, A. A. P. A. M., and Aarden, L. A.,** Interleukin-6 in serum and urine of renal transplant recipients, *Clin. Exp. Immunol.*, 71, 314, 1988.
27. **Nijsten, M. W., DeGroot, E. R., Tenduis, H. J., Klesen, H. J., Hack, C. E., and Aarden, L. A.,** Serum levels of interleukin-6 and acute phase responses, *Lancet*, 2, 921, 1987.
28. **Helfgott, D. C., Tatter, S. B., Santhanam, U., Clarick, R. H., Bhardwaj, N., May, L. T., and Sehgal, P. B.,** Multiple forms of IFN-β_2/IL-6 in serum and body fluids during acute bacterial infection, *J. Immunol.*, 142, 948, 1989.
29. **Sehgal, P. B.,** Interleukin-6 in infection and cancer, *Proc. Soc. Exp. Biol. Med.*, 195, 183, 1990.
30. **Hirano, T.,** Interleukin-6 and its relation to inflammation and disease, *Clin. Immunol. Immunopathol.*, 62, S60, 1992.

31. **Fletcher-Starnes, H., Jr., Pearce, M. K., Tewari, A., Yim, J. H., Zou, J. C., and Abrams, J. S.,** Anti-IL-6 monoclonal antibodies protect against lethal *Escherichia coli* infection and lethal tumor necrosis factor-α challenge in mice, *J. Immunol.,* 145, 4185, 1990.
32. **Strassmann, G., Fong, M., Kenney, J., and Jacob, C. O.,** Evidence for the involvement of IL-6 in experimental cancer cachexia, *J. Clin. Invest.,* 89, 1681, 1992.
33. **Gauldie, J., Geisterfer, M., Richards, C., and Gwyne, D.,** IL-6 regulation of the hepatic acute phase response, in *IL-6: Physiopathology and Clinical Applications,* Vol. 88, Revel, M., Ed., Ares-Serono Symp., Raven Press, New York, 1992, 151.
34. **Neta, R., Perlstein, R., Vogel, S., Whitnall, M., and Abrams, J.,** The contribution of IL-6 to the in vivo radioprotection and endocrine response to IL-1 and TNF, in *IL-6: Physiopathology and Clinical Applications,* Vol. 88, Revel, M., Ed., Ares-Serono Symp., Raven Press, New York, 1992, 151.
35. **Klein, B., Zhang, X. G., Jourdan, M., Portier, M., and Bataille, R.,** Interleukin-6 is a major myeloma cell growth factor in vitro and in vivo especially in patients with terminal disease, *Curr. Topics Microbiol. Immunol.,* 166, 23, 1990.
36. **Baumann, H., Prowse, K. R., Marinkovic, S., Won, K. A., and Jahreis, G. P.,** Stimulation of hepatic acute phase response by cytokines and glucocorticoids, *Ann. N.Y. Acad. Sci.,* 557, 280, 1989.
37. **Baumann, H. and Gauldie, J.,** Regulation of hepatic acute phase plasma protein genes by hepatocyte stimulating factors and other mediators of inflammation, *Mol. Biol. Med.,* 7, 147, 1990.
38. **Mackiewicz, A., Ganapathi, M. K., Shultz, D., Brabenec, A., Weinstein, J., Kelley, M. F., and Kushner, I.,** Transforming growth factor β_1 regulates production of acute phase proteins, *Proc. Natl. Acad. Sci. U.S.A.,* 87, 1491, 1990.
39. **Kushner, I. and Mackiewicz, A.,** Acute phase response: an overview, in *Acute Phase Proteins: Molecular Biology, Biochemistry, Clinical Applications,* Mackiewicz, A., Kushner, I., and Baumann, H., Eds., CRC Press, Boca Raton, FL, 1993, chap. 1.
40. **Gershenwald, J. E., Fong, Y., Fahey, T. J., III, Calvano, S. E., Chizzonite, R., Kilian, P. L., Lowry, S. F., and Moldawer, L. L.,** Interleukin-1 receptor blockade attenuates the host inflammatory response, *Proc. Natl. Acad. Sci. U.S.A.,* 87, 4966, 1990.
41. **Havell, E. A., Moldawer, L. L., Helfgott, D., Lilian, P. L., and Sehgal, P. B.,** Type I IL-1 receptor blockade exacerbates murine listeriosis, *J. Immunol.,* 148, 1486, 1992.
42. **Fong, Y., Tracey, K. J., Moldawer, L. L., Hesse, D. G., Manogue, K. B., Kenney, J. S., Lee, A. T., Kuo, G. C., Allison, A. C., Lowry, S. F., and Cerami, A.,** Antibodies to cachectin/tumor necrosis factor reduce interleukin-1β and interleukin-6 appearance during lethal bacteremia, *J. Exp. Med.,* 170, 1627, 1989.
43. **Havell, E. A. and Sehgal, P. B.,** Tumor necrosis factor-independent IL-6 production during murine listeriosis, *J. Immunol.,* 146, 756, 1991.
44. **Sehgal, P. B., Zilberstein, A., Ruggieri, M. R., May, L. T., Ferguson-Smith, A., Slate, D. L., Revel, M., and Ruddle, F. H.,** Human chromosome 7 carries the β_2 interferon gene, *Proc. Natl. Acad. Sci. U.S.A.,* 83, 5219, 1986.
45. **Ferguson-Smith, A. C., Chen, Y. F., Newman, M. S., May, L. T., Sehgal, P. B., and Ruddle, F. H.,** Regional localization of the "β_2-interferon/B-cell stimulatory factor 2/hepatocyte stimulating factor" gene to human chromosome 7p15-p21, *Genomics,* 2, 203, 1988.
46. **Bowcock, A. M., Kidd, J. R., Lathrop, M., Daneshvar, L., May, L. T., Ray, A., Sehgal, P. B., Kidd, K. K., and Cavalli-Sforza, L. L.,** The human "beta-2 interferon/hepatocyte stimulating factor/ interleukin-6" gene: DNA polymorphism studies and localization to chromosome 7p21, *Genomics,* 3, 8, 1988.
47. **Mock, B. A., Nordan, R. P., Justice, M. J., Kozak, C., Jenkins, N. A., Copeland, N. G., Clark, S. C., Wong, G. G., and Rudikoff, S.,** The murine *Il-6* gene maps to the proximal region of chromosome 5, *J. Immunol.,* 142, 1372, 1989.
48. **Yasukawa, K., Hirano, T., Watanabe, Y., Muratani, K., Matsuda, T., and Kishimoto, T.,** Structure and expression of human B cell stimulatory factor 2 (BSF-2/IL-6) gene, *EMBO J.,* 6, 2939, 1987.
49. **Tanabe, O., Akira, S., Kamiya, T., Wong, G. G., Hirano, T., and Kishimoto, T.,** Genomic structure of the murine IL-6 gene: high degree conservation of potential regulatory sequences between mouse and human, *J. Immunol.,* 41, 3875, 1988.
50. **Leutz, A., Damm, K., Sterneck, E., Kowenz, E., Ness, S., Frank, R., Gausepohl, H., Pan, Y. C., Smart, J., Hayman, H., and Graf, T.,** Molecular cloning of the chicken myelomonocytic growth factor (cMGF) reveals a relationship to interleukin-6 and granulocyte colony stimulating factor, *EMBO J.,* 8, 175, 1989.
51. **Ray, A., Tatter, S. B., Santhanam, U., Helfgott, D. C., May, L. T., and Sehgal, P. B.,** Regulation of expression of interleukin-6: molecular and clinical studies, *Ann. N.Y. Acad. Sci.,* 557, 353, 1989.
52. **Northemann, W., Braciak, T. A., Hattori, M., and Fey, G. H.,** Rat interleukin-6: mRNA and gene structure and expression in macrophage cultures, *Ann. N.Y. Acad. Sci.,* 557, 536, 1989.

53. **Bowcock, A. M., Ray, A., Ehrlich, H. A., and Sehgal, P. B.,** Rapid detection and sequencing of alleles in the 3′ flanking region of the interleukin-6 gene, *Nucleic Acids Res.*, 17, 6855, 1989.
54. **Ray, A., Tatter, S. B., May, L. T., and Sehgal, P. B.,** Activation of the "β_2-interferon/hepatocyte-stimulating factor/interleukin-6" promoter by cytokines, viruses, and second-messenger agonists, *Proc. Natl. Acad. Sci. U.S.A.*, 85, 6701, 1988.
55. **Ray, A., Sassone-Corsi, P., and Sehgal, P. B.,** A multiple cytokine- and second messenger-responsive element in the enhancer of the human interleukin-6 gene: similarities with c-*fos* gene regulation, *Mol. Cell. Biol.*, 9, 5537, 1989.
56. **Ray, A., LaForge, K. S., and Sehgal, P. B.,** On the mechanism for efficient repression of the interleukin-6 promoter by glucocorticoids: enhancer, TATA box, and RNA start site (Inr motif) occlusion, *Mol. Cell. Biol.*, 10, 5736, 1990.
57. **Shimizu, H., Mitomo, K., Watanabe, T., Okamoto, S., and Yamamoto, K. I.,** Involvement of an NF-κB-like transcription factor in the activation of the interleukin-6 gene by inflammatory lymphokines, *Mol. Cell. Biol.*, 10, 561, 1990.
58. **Isshiki, H., Akira, S., Tanabe, O., Nakajima, T., Shimamoto, T., Hirano, T., and Kishimoto, T.,** Constitutive and interleukin-1 (IL-1)-inducible factors interact with the IL-1-responsive element in the IL-6 gene, *Mol. Cell. Biol.*, 10, 2757, 1990.
59. **Akira, S., Isshiki, H., Sugita, T., Tanabe, O., Kinoshita, S., Nishio, Y., Nakajima, T., Hirano, T., and Kishimoto, T.,** A nuclear factor for IL-6 expression (NF-IL6) is a member of a C/EBP family, *EMBO J.*, 9, 1897, 1990.
60. **Liebermann, T. A. and Baltimore, D.,** Activation of interleukin-6 gene expression through the NF-κB transcription factor, *Mol. Cell. Biol.*, 10, 2327, 1990.
61. **Zhang, Y., Lin, J. X., and Vilcek, J.,** Interleukin-6 induction by tumor necrosis factor and interleukin-1 in human fibroblasts involves activation of a nuclear factor binding to a κB-like sequence, *Mol. Cell. Biol.*, 10, 3818, 1990.
62. **Walther, Z., May, L. T., and Sehgal, P. B.,** Transcriptional regulation of the interferon-β_2/B cell differentiation factor BSF-2/hepatocyte stimulating factor gene in human fibroblasts by other cytokines, *J. Immunol.*, 140, 974, 1988.
63. **Santhanam, U., Ray, A., and Sehgal, P. B.,** Repression of the interleukin 6 gene promoter by p53 and the retinoblastoma susceptibility gene product, *Proc. Natl. Acad. Sci. U.S.A.*, 88, 7605, 1991.
64. **Kohase, M., Henriksen-DeStefano, D., Sehgal, P. B., and Vilček, J.,** Dexamethasone inhibits feedback regulation of the mitogenic activity of tumor necrosis factor, interleukin-1 and epidermal growth factor in human fibroblasts, *J. Cell. Physiol.*, 132, 271, 1987.
65. **Tabibzadeh, S. S., Santhanam, U., Sehgal, P. B., and May, L. T.,** Cytokine-induced production of interferon-β_2/interleukin-6 by freshly explanted human endometrial stromal cells: modulation by estradiol-17β, *J. Immunol.*, 142, 3134, 1989.
66. **Diamond, M. I., Miner, J. N., Yoshinaga, S. K., and Yamamoto, K. R.,** Transcription factor interactions: selectors of positive or negative regulation from a single DNA element, *Science,* 249, 1266, 1990.
67. **Ray, A., LaForge, K. S., and Sehgal, P. B.,** Repressor to activator switch by mutations in the first Zn finger of the glucocorticoid receptor: is direct DNA binding necessary?, *Proc. Natl. Acad. Sci. U.S.A.*, 88, 7086, 1991.
68. **Woloski, B. M. R. N. J., Smith, E. M., Meyer, W. J., III, Fuller, G. M., and Blalock, J. E.,** Corticotropin-releasing activity of monokines, *Science,* 230, 1035, 1985.
69. **Naitoh, Y., Fukata, J., Tominaga, T., Tamai, S., Mori, K., and Imura, H.,** Interleukin-6 stimulates the secretion of adrenocorticotrophic hormone in conscious, freely-moving rats, *Biochem. Biophys. Res. Commun.*, 155, 1459, 1988.
70. **Vankelecom, H., Carmeliet, P., Van Damme, J., Billiau, A., and Denef, D.,** Production of interleukin-6 by folliculo-stellate cells of the anterior pituitary gland in a histiocytic cell aggregate culture system, *Neuroendocrinology,* 49, 102, 1989.
71. **Spangelo, B. L., Judd, A. M., Isakson, P. C., and MacLeod, R. M.,** Interleukin-6 stimulates anterior pituitary hormone release in vitro, *Endocrinology,* 125, 575, 1989.
72. **Jablons, D. M., Mulé, J. J., McIntosh, J. K., Sehgal, P. B., May, L. T., Huang, C. M., Rosenberg, S. A., and Lotze, M. T.,** Interleukin-6/interferon-β_2 as a circulating hormone: induction by cytokine administration in man, *J. Immunol.*, 142, 1542, 1989.
73. **Fong, Y., Moldawer, L. L., Marano, M., Tatter, S. B., Clarick, R. M., Santhanam, U., Sherris, D., May, L. T., Sehgal, P. B., and Lowry, S. F.,** Endotoxemia elicits increased circulating β_2-IFN/IL-6 in man, *J. Immunol.*, 142, 2321, 1989.
74. **May, L. T., Ghrayeb, J., Santhanam, U., Tatter, S. B., Sthoeger, S. B., Helfgott, D. C., Chiorazzi, N., Grieninger, G., and Sehgal, P. B.,** Synthesis and secretion of multiple forms of β_2-interferon/B-cell differentiation factor 2/hepatocyte stimulatory factor by human fibroblasts and monocytes, *J. Biol. Chem.*, 263, 7760, 1988.

75. **May, L. T., Shaw, J. E., Khanna, A. K., Zabriskie, J. B., and Sehgal, P. B.,** Marked cell-type-specific differences in glycosylation of human interleukin-6, *Cytokine,* 3, 204, 1991.
76. **Santhanam, U., Ghrayeb, J., Sehgal, P. B., and May, L. T.,** Post-translational modifications of human interleukin-6, *Arch. Biochem. Biophys.,* 274, 161, 1989.
77. **Hirano, T., Taga, T., Nakano, N., Yasukawa, K., Kashiwamura, S., Shimizu, K., Nakajima, K., Pyun, K. H., and Kishimoto, T.,** Purification to homogeneity and characterization of human B cell differentiation factor (BCDF or BSFp-2), *Proc. Natl. Acad. Sci. U.S.A.,* 82, 5490, 1985.
78. **May, L. T., Santhanam, U., Tatter, S. B., Ghrayeb, J., and Sehgal, P. B.,** Multiple forms of human interleukin-6: phosphoglycoproteins secreted by many different tissues, *Ann. N.Y. Acad. Sci.,* 557, 114, 1989.
79. **Heinrich, P. C., Dufhues, G., Graeve, L., Kruttgen, A., Lenz, D., Mackiewicz, A., Lutticken, C., Schoolnick, H., Stoyan, T., van Dam, M., Zohlnhofer, D., and Rose-John, S.,** Studies on the structure and function of IL-6 and its hepatic receptor, in *IL-6: Physiopathology and Clinical Applications,* Vol. 88, Revel, M., Ed., Ares-Serono Symp., Raven Press, New York, 1992, 63.
80. **May, L. T., Santhanam, U., Tatter, S. B., Bhardwaj, N., Ghrayeb, J., and Sehgal, P. B.,** Phosphorylation of secreted forms of human β_2-interferon/hepatocyte stimulating factor/interleukin-6, *Biochem. Biophys. Res. Commun.,* 152, 1144, 1988.
81. **May, L. T. and Sehgal, P. B.,** Phosphorylation of interleukin-6 at serine[54]: an early event in the secretory pathway in human fibroblasts, *Biochem. Biophys. Res. Commun.,* 185, 524, 1992.
82. **Van Snick, J., Cayphas, S., Szikora, J.-P., Renauld, J.-C., Van Roost, E., Boon, T., and Simpson, R. J.,** cDNA cloning of murine interleukin-HP1: homology with human interleukin-6, *Eur. J. Immunol.,* 18, 193, 1988.
83. **Fuller, G. M. and Grenett, H. E.,** The structure and function of the mouse hepatocyte stimulating factor, *Ann. N.Y. Acad. Sci.,* 557, 31, 1989.
84. **Brakenhoff, J. P., Hart, M., and Aarden, L. A.,** Analysis of human IL-6 mutants expressed in *Escherichia coli*: biologic activities are not affected by deletion of amino acids 1-28, *J. Immunol.,* 143, 1175, 1989.
85. **Snouwaert, J. N., Kariya, K., and Fowlkes, D. M.,** Effects of site-specific mutations on biologic activities of recombinant human IL-6, *J. Immunol.,* 146, 585, 1991.
86. **Krüttgen, A., Rose-John, S., Moller, C., and Heinrich, P.,** Structure-function analysis of human interleukin-6 — evidence for the involvement of the carboxy-terminus in function, *FEBS Lett.,* 262, 323, 1990.
87. **Barry, M. R., Woloski, B. M., and Fuller, G. M.,** Identification and partial characterization of hepatocyte-stimulating factor from leukemia cell lines: comparison with interleukin-1, *Proc. Natl. Acad. Sci. U.S.A.,* 82, 1443, 1985.
88. **Fuller, G. M., Otto, J. M., Woloski, B. M., McGary, C. T., and Adams, M. A.,** The effects of hepatocyte stimulating factor in fibrinogen biosynthesis in hepatocyte monolayers, *J. Cell Biol.,* 101, 1481, 1985.
89. **May, L. T., Santhanam, U., and Sehgal, P. B.,** On the multimeric nature of natural human interleukin-6, *J. Biol. Chem.,* 266, 9950, 1991.
90. **Kenney, J., Masada, M., Santhanam, U., May, L. T., Viguet, H., Ida, N., Sakurai, S., Allison, A. C., and Sehgal, P. B.,** Marked immunological heterogeneity of the biochemically distinct species of cell culture-derived human interleukin-6, in preparation.
91. **Van Snick, J.,** Interleukin-6: an overview, *Annu. Rev. Immunol.,* 8, 253, 1990.
92. **Grossman, R. M., Krueger, J., Yourish, D., Granelli-Piperno, A., Murphy, D. P., May, L. T., Kupper, T. S., Sehgal, P. B., and Gottlieb, A. B.,** Interleukin-6 is expressed in high levels in psoriatic skin and stimulates proliferation of cultured human keratinocytes, *Proc. Natl. Acad. Sci. U.S.A.,* 86, 6367, 1989.
93. **Urbanski, A., Schwarz, T., Neuner, P., Krutmann, J., Kirnbauer, R., Kock, A., and Luger, T. A.,** Ultraviolet light induces circulating interleukin-6 in humans, *J. Invest. Dermatol.,* 94, 808, 1990.
94. **Romero, R., Avila, C., Santhanam, U., and Sehgal, P. B.,** Amniotic fluid interleukin-6 in preterm labor: association infection, *J. Clin. Invest.,* 85, 1392, 1990.
95. **Honda, M., Kitamura, K., Mizutani, Y., Oishi, M., Arai, M., Okura, T., Igarahi, K., Yasukawa, K., Hirano, T., Kishimoto, T., Mitsuyasu, R., Chermann, J. C., and Tokunaga, T.,** Quantitative analysis of serum IL-6 and its correlation with increased levels of serum IL-2R in HIV-induced diseases, *J. Immunol.,* 145, 4059, 1990.
96. **May, L. T., Santhanam, U., Viguet, H., Kenney, J., Ida, N., Allison, A. C., and Sehgal, P. B.,** Human IL-6 exists as a high molecular mass complex in plasma and serum, in preparation.
97. **Taga, T., Murakami, M., Hibi, M., Narazaki, M., Saito, M., Yawata, H., Yasukawa, K., Hamaguchi, M., and Kishimoto, T.,** IL-6 receptor and signal transduction through gp130, in *IL-6: Physiopathology and Clinical Applications,* Vol. 88, Revel, M., Ed., Ares-Serono Symp., Raven Press, New York, 1992, 43.

98. **Heinrich, P. C., Castell, J. V., and Andus, T.,** Interleukin-6 and the acute phase response, *Biochem. J.,* 265, 621, 1990.
99. **Revel, M., Ed.,** *IL-6: Physiopathology and Clinical Applications,* Vol. 88, Ares-Serono Symp., Raven Press, New York, 1992.
100. **Ginsberg, H., Moldawer, L. L., Sehgal, P. B., Redington, M., Kilian, P. L., Chanock, R. M., and Prince, G. A.,** A mouse model for investigating the molecular pathogenesis of adenovirus pneumonia, *Proc. Natl. Acad. Sci. U.S.A.,* 88, 1651, 1991.
101. **McIntosh, J. K., Jablons, D. J., Mulé, J. J., Nordan, R. P., Rudikoff, S., Lotze, M. T., and Rosenberg, S. A.,** In vivo induction of IL-6 by administration of exogenous cytokines and detection of de novo serum levels of IL-6 in tumor-bearing mice, *J. Immunol.,* 143, 162, 1989.
102. **Tabibzadeh, S. S., Poubouridis, D., May, L. T., and Sehgal, P. B.,** Interleukin-6 immunoreactivity in human tumors, *Am. J. Pathol.,* 135, 427, 1989.
103. **Gelin, J., Moldawer, L. L., Lonroth, C., deMan, P., Svanborg-Eden, C., Lowry, S. F., and Lundholm, K. G.,** Appearance of hybridoma growth factor/interleukin-6 in the serum of mice bearing a methylcholanthrene-induced sarcoma, *Biochem. Biophys. Res. Commun.,* 157, 575, 1988.
104. **Gelin, J., Moldawer, L. L., Lonnroth, C., Sherry, B., Chizzonite, R., and Lundholm, K.,** Role of endogenous tumor necrosis factor-α and interleukin-1 for experimental tumor growth and the development of cancer cachexia, *Cancer Res.,* 41, 415, 1991.
105. **Mulé, J. J., McIntosh, J. K., Jablons, D. J., and Rosenberg, S. A.,** Anti-tumor activity of recombinant interleukin-6 in mice, *J. Exp. Med.,* 171, 629, 1990.
106. **Revel, M.,** Antitumor potentials of interleukin-6, *Interferons Cytokines,* 19, 5, 1991.
107. **Mulé, J. J., Marcus, S. G., Jablons, D. M., McIntosh, J. K., Yang, J. C., Weber, J. S., and Rosenberg, S. A.,** Interleukin-6: induction in tumor bearing mice and patients, an in vivo anti-tumor effects in mice, *Physiopathology and Clinical Applications,* Revel, M., Ed., Ares-Serono Symp., Raven Press, New York, 1992, 233.
108. **Ishibashi, T., Kimura, H., Shikama, Y., Uchida, T., Kariyone, S., Hirano, T., Kishimoto, T., Takatsuki, F., and Akiyama, Y.,** Interleukin-6 is a potent thrombopoietic factor in vivo in mice, *Blood,* 74, 1241, 1989.
109. **Hill, R. J., Warren, K., and Levin, J.,** Stimulation of thrombopoiesis in mice by human recombinant interleukin 6, *J. Clin. Invest.,* 85, 1242, 1990.
110. **Klein, B. and Battaile, R.,** Interleukin-6 is the major growth factor for human myeloma cells in vitro and in vivo and serum CRP level, reflecting IL-6 activity in vivo, is a strong prognosis factor in myeloma, in *IL-6: Physiopathology and Clinical Applications,* Revel, M., Ed., Ares-Serono Symp., Raven Press, New York, in press.
111. **May, L. T., Vignet, H., Kenney, J. S., Ida, N., Allison, A. C., and Sehgal, P. B.,** High levels of "complexed" interleukin-6 in human blood. *J. Biol. Chem.,* 267, 19698, 1992.

Chapter 17

INTERLEUKIN-11: MOLECULAR BIOLOGY, BIOLOGICAL ACTIVITIES, AND POSSIBLE SIGNALING PATHWAYS

Yu-Chung Yang

TABLE OF CONTENTS

0-8493-6913-4/93/$0.00 + $.50

I. INTRODUCTION

The hematopoietic microenvironment consists of many different cell types, including endothelial cells, fibroblasts, adventitial reticular cells, macrophages, and adipocytes. It appears that the optimum environment for the maintenance, proliferation, and differentiation of stem cells both *in vivo* and *in vitro* is dependent on a complex interaction between hematopoietic growth factors, various cell types, and matrix proteins of the hematopoietic microenvironment.[1] We have studied the growth factor production of one of the stromal cell lines, PU-34, which was originally derived from primate bone marrow cells infected with a retrovirus expressing simian virus-40 (SV40) large-T antigen.[2] This immortalized line has been shown to produce several known cytokines and support hematopoiesis in long-term bone marrow cultures. Using this primate stromal cell line, we isolated a novel cytokine based on its ability to stimulate the proliferation of an IL-6-dependent mouse plasmacytoma cell line, T1165.[3] The sequence of this cDNA has no sequence homology with other cytokines and was named interleukin (IL)-11. The human IL-11 genomic sequence was subsequently determined and the chromosomal location mapped.[4] Like most of the hematopoietic growth factors studied so far, IL-11 exerts various biological activities in different *in vitro*[3,5-14,26-28] and *in vivo*[10,15,31-34] biological assay systems. Since most of the biological activities of IL-11 examined so far overlapped with those of IL-6, it is possible that these two cytokines may share common receptor(s) or utilize similar signal transduction pathways. Biochemical characterization of the IL-11 receptor(s) and IL-11-mediated signaling has been carried out using IL-11-responsive cell lines. This review focuses on the molecular biology, *in vitro* and *in vivo* biological activities of IL-11, and the possible signal transduction pathways mediated by this cytokine.

II. MATERIALS AND METHODS

A. SCREENING OF PU-34 cDNA EXPRESSION LIBRARY

PU-34 cells were stimulated for 24 h with IL-1α at a concentration of 2 units/ml and poly(A) + RNA was prepared from these cells by standard methods. A cDNA expression library was constructed using pXM vector by the procedures described previously.[3] The cDNA clones encoding the T1165 mitogenic activity were isolated by screening the COS cell supernatants derived from cDNA-transfected cells using [^{3}H]thymidine incorporation in T1165 cells. Sib selection was utilized until a single clone was isolated.

B. SCREENING OF HUMAN GENOMIC LIBRARY AND *IN SITU* CHROMOSOMAL MAPPING

The human IL-11 genomic clones were isolated by screening human genomic libraries with IL-11 cDNA. Positive clones were analyzed by restriction enzyme mapping and Southern analysis. Restricted fragments were then subcloned into plasmid vectors for sequence analysis. *In situ* chromosomal mapping of the IL-11 gene was performed according to procedures published previously.[4]

C. BIOLOGICAL ASSAYS

The mouse megakaryocyte assay was performed as described.[3,5] The human megakaryocyte assay in serum-free conditions was described in publications by Bruno et al.[6] and Teramura et al.[7] The blast cell colony assay was carried out using bone marrow or spleen cells from 5-fluorouracil-treated mice.[8,9] The murine plaque-forming assay was performed according to protocols published previously.[3,10] Induction of acute phase protein synthesis by IL-11 was examined by Baumann and Schendel.[11] The adipogenesis inhibitory activity of IL-11 was demonstrated by procedures described previously.[12-14]

D. *IN VIVO* STUDIES

In vivo effects of IL-11 in normal and myelosuppressed mice were examined as described previously.[10,15] Basically, mice were injected twice a day (once intraperitoneally, once subcutaneously) with 1 μg per injection of IL-11 for various periods of time. Peripheral blood, spleen, and bone marrow cells were then utilized in various biological assays to evaluate the effects of IL-11 in different cellular compartments. Myelosuppression in mice was the result of the injection of 200 mg/kg body weight of cyclophosphamide for 24 h. *In vivo* studies by other investigators are detailed in the respective publications.[31-34]

E. RECEPTOR BINDING ASSAY

For IL-11 receptor binding, cells (attached cells were first detached with buffer containing EDTA) were washed once with Dulbecco's modified Eagle's medium (DMEM) and incubated with various concentrations of [^{125}I]IL-11 in the absence or presence of a 1000-fold excess of unlabeled IL-11 or other cytokines at 4°C for 3 h with rotation. The specific binding and Scatchard plot were performed according to published formulas.[14,16]

F. BIOCHEMICAL CHARACTERIZATION OF IL-11 RECEPTOR

The molecular weight of the IL-11 receptor was estimated by cross-linking experiments. Radiolabeled ligand and cells, plus or minus a large excess of unlabeled ligand, were incubated in the presence of the cross-linking agent DSS, and the cross-linked products analyzed on sodium dodecyl sulfate-polyacrylamide gel (SDS-PAGE) to obtain the molecular weights of the IL-11 receptor/ligand complexes.[14]

G. PROTEIN TYROSINE PHOSPHORYLATION

Protein tyrosine phosphorylation stimulated by IL-11 in IL-11-responsive cell lines was analyzed by immunoblotting using anti-phosphotyrosine monoclonal antibodies, as described.[14]

III. RESULTS

A. IDENTIFICATION AND MOLECULAR BIOLOGY OF IL-11

1. Characterization of the Cell Line PU-34 and Bioactivities of PU-34-Conditioned Medium

PU-34 cells were derived by infecting primate bone marrow cells with a retrovirus expressing SV40 large-T antigen. Immunofluorescence studies demonstrated the presence of vimentin and fibronectin and the absence of cytokeratin, desmin, laminin, collagen type IV, and factor VIII-related antigen. These studies demonstrated that PU-34 expresses a fibroblastic phenotype consistent with a mesenchymal origin. The ability of these cells to support long-term hematopoiesis was demonstrated using *in vitro* progenitor assays.[2] Plating of nonadherent cells from these cultures yielded predominantly myeloid colonies and some erythroid and mixed colonies.[2]

Upon induction with IL-1, PU-34 cells produce granulocyte/macrophage colony-stimulating factor (GM-CSF), IL-6, G-CSF, IL-7, and leukemia inhibitory factor (LIF).[2] In addition to these known cytokines, IL-1-induced PU-34-conditioned medium contained a novel mitogenic activity on an IL-6-dependent mouse plasmacytoma cell line, T1165.

2. Molecular Characteristics of IL-11 cDNA and Genomic Sequences

The cDNA encoding the T1165 mitogenic activity was isolated from IL-1-induced PU-34-cells by an expression cloning method, and the cDNA was named IL-11.[3] The IL-11 cDNA contains a long open reading frame of a 199-amino acid polypeptide with a calculated molecular weight of 23 kDa. The amino terminus contains several hydrophobic residues

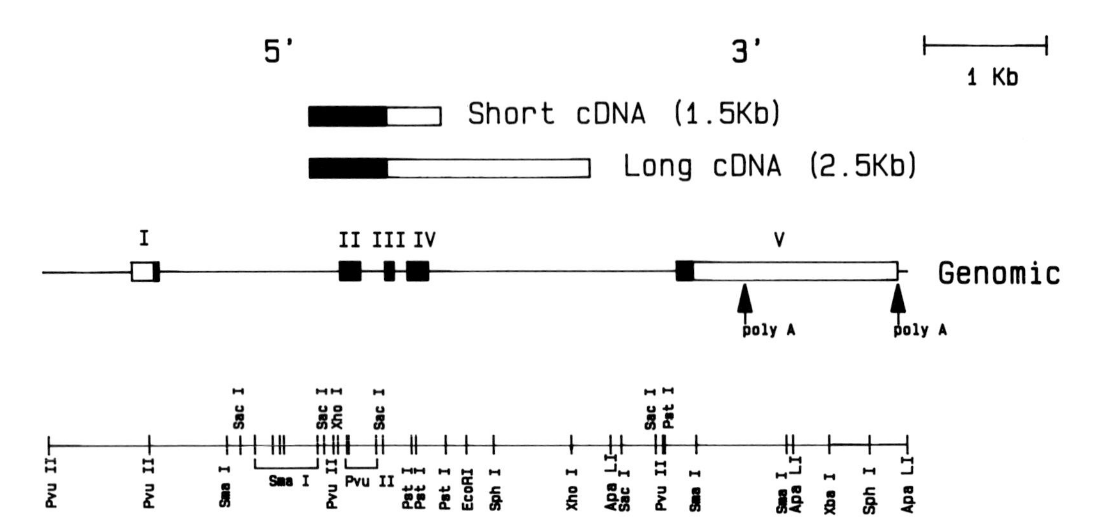

FIGURE 1. Exon/intron organization and restriction map of the human IL-11 gene. The coding and noncoding regions are indicated by solid and open boxes, respectively.

typical of a secretory molecule. Unlike other cytokines, the IL-11 cDNA does not contain cysteine residues or potential glycosylation sites. Several repeats of the ATTTA sequence found in the 3′ noncoding regions of other growth factors and oncogenes are also found in the 3′ flanking region of IL-11 cDNA.[17] The 7-kb IL-11 genomic sequence has been determined, and the gene consists of five exons and four introns (Figure 1). Several transcriptional control elements were identified in the 5′ flanking region of the IL-11 gene.[4] The presence of recognition sequences for AP-1 may account for the TPA inducibility of IL-11 gene expression in human fetal lung fibroblasts and a human stromal cell line (see below). Interestingly, a 30-base pair DNA segment repeated four times was identified in the fourth intron of the IL-11 gene.[4] In addition, several copies of the human Alu repetitive sequence have been found in the 3′ noncoding region of the IL-11 gene.[4] The significance of these repetitive sequences in controlling IL-11 gene expression remains unclear. A DNA sequence (ACATGGCAAAACCC) that has 71% similarity with the IL-1-responsive element of the IL-6 gene[18] was found in the 3′ flanking region of the human IL-11 gene.[4] This element has been shown to be essential for the IL-1-inducible expression of the IL-6 gene in certain cell types.[18] Reporter constructs containing potential transcriptional control sequences present in the 5′, 3′, or introns will be utilized to dissect the *cis*- and *trans*-elements that are essential for IL-11 expression in various cell types under different induction conditions. Based on the activities of human IL-11 on mouse and rat cells, we believe that the factor can act across different species and that the sequence may be well conserved during evolution.

3. Chromosomal Location of the Human IL-11 Gene

In situ hybridization of metaphase chromosomes with IL-11 cDNA has localized IL-11 on the long arm of human chromosome 19 at 19q13.313.4.[4] Interestingly, several zinc-finger genes[19] and genes involved in lipoprotein metabolism, including apolipoproteins, E, CI, and CII,[20] have all been mapped to the nearby region. Furthermore, a chromosome 19q-specific repetitive sequence has been found to repeat four times in the intron of the apolipoprotein CII sequence.[21] This repetitive sequence, however, does not share sequence similarity with the 30-bp repeats in the fourth intron of the IL-11 gene. Interestingly, this region of chromosome 19 has been shown to be associated with recurring translocation in neoplastic cells in certain patients with chronic lymphocytic leukemia.[22] Moreover, *axl,* a transforming gene encoding a novel receptor tyrosine kinase isolated from primary human myeloid leukemia cells, has been mapped to the 19q13.2 region.[23]

4. Expression of IL-11 in Different Cell Lines

IL-11 transcripts have been detected in several cell lines, including primate[3] and human[12] stromal fibroblasts, human fetal lung fibroblasts, and human trophoblasts.[3] The sizes of IL-11 transcripts in stromal and fetal lung fibroblasts are 2.5 and 1.5 kb,[3,12] while only the 2.5-kb transcript can be detected in the trophoblast cell line.[3] The cDNAs corresponding to both transcripts have been isolated and shown to encode the same functional IL-11 protein.[3,12] The expression of IL-11 in these cells appears to be constitutive, based on Northern analysis and transient transfection experiments using reporter plasmids containing the 5′ flanking region of the IL-11 gene.[44] The expression of IL-11 transcripts can be further induced by reagents such as IL-1, TPA, and calcium ionophore.[3,12]

B. BIOLOGICAL ACTIVITIES OF IL-11

1. Growth Promotion of Plasmacytoma and Hybridoma Cell Lines

IL-11 was originally identified based on its mitogenic activity on an IL-6-dependent mouse plasmacytoma cell line, T1165. Subsequent studies have shown that IL-11 can also stimulate the proliferation of an IL-6-dependent mouse hybridoma cell line, B9.[45] An IL-11-dependent plasmacytoma cell line, T10, and an IL-11-dependent hybridoma cell line, B9TY1, were subsequently established by culturing T1165 or B9 cells in the presence of high concentrations of IL-11. The ability of IL-11 to support the growth of these cells suggests that this cytokine may be involved in the establishment and maintenance of plasmacytoma *in vivo* and may play an important role in tumorigenesis and lymphopoiesis. IL-11, however, does not appear to be essential for the growth of cells of myeloma origin.[24]

2. Hematopoietic Colony-Stimulating Activity

Analysis of the effects of IL-11 in a variety of hematopoietic culture systems revealed striking effects on megakaryocyte development in both human and mouse systems.[3,5-7] IL-11 alone cannot support the growth of megakaryocyte colonies, but does stimulate increases in the number, size, and ploidy values of megakaryocyte colonies in combination with IL-3. Interestingly, using highly enriched CD34+DR+ human bone marrow cells, IL-11 was able to synergize with IL-3 in supporting BFU-MK, but not CFU-MK, formation in a serum-free culture condition.[6] Furthermore, Teramura et al.[7] have shown that IL-11 in combination with IL-3 can promote a shift in ploidy toward higher values. The direct effect of IL-11 in the increases of ploidy values has also been demonstrated with bone marrow-derived CD41+ cells.[7] IL-11 therefore is not a megakaryocyte colony-stimulating factor by itself, but, rather, acts like a megakaryocyte potentiator. These results suggest that IL-11 may play an important role in megakaryocytopoiesis and possibly in the *in vivo* production of platelets.

3. Blast Cell Growth Factor Activity

In steady-state bone marrow, the hematopoietic stem cells are thought to be in a quiescent state of the cell cycle called Go. The blast cell colony assay is designed to measure the proliferation of progenitors after they exit from the Go state. Using this assay, Ogawa and co-workers have identified IL-1, IL-6, granulocyte colony-stimulating factor (G-CSF), and c-kit ligand as blast cell growth factors that can shorten the Go period of the progenitor cells to respond to intermediate or late-acting factors such as IL-3 and GM-CSF.[25] Like these growth factors, IL-11 has also been shown to be a synergistic factor for IL-3-dependent proliferation of primitive progenitors, and that part of the synergism is to shorten the Go period of the early hematopoietic progenitors.[8,9] It has also been shown that IL-11 in combination with other cytokines such as c-kit ligand, IL-3, and IL-6 supports the survival and expansion of primitive murine progenitor cells.[26] In the human system, IL-11 has been shown to increase the tritiated thymidine suicide rate of fetal, but not adult, CFU-MIX, CFU-GM, and BFU-E.[27]

4. Biological Effects with Lymphoid Cells

Like IL-6, IL-11 is effective in enhancing the generation of sheep red blood cell (SRBC)-specific plaque-forming cells (PFC) in the *in vitro* mouse spleen cell culture system.[10] Kinetic studies using either normal spleen cells or *in vivo* SRBC-primed spleen cells showed that IL-11, unlike IL-6, has to be added at an early stage of cell culture in order to significantly augment Ag-specific PFC. Cell-depletion studies revealed that L3T4 (CD4)+ T-cells, but not Lyt-2 (CD8)+ T-cells, are required in the IL-11-stimulated augmentation of SRBC-specific antibody responses.[10] In the human system, IL-11 has been shown to act directly on T (not B)-cells, particularly the CD4+CD45RA− T-cell subset, to enhance immunoglobulin secretion.[28]

5. Induction of Acute Phase Protein Synthesis

Studies with liver cells have indicated that several cytokines implicated in the proliferation and differentiation control of hematopoietic cells can also modulate gene expression in hepatocytes.[29] Stimulation of rat hepatoma cells with IL-11 resulted in the synthesis of several major acute phase plasma proteins, including fibrinogen, thiostatin, α_1-antitrypsin, hemopexin, and haptoglobin.[11] The changes were comparable to those mediated by IL-6 or LIF. Like IL-6, IL-11 synergized with IL-1 in stimulating the synthesis of type 1 acute phase proteins. The response of IL-11 plus dexamethasone in acute phase protein synthesis is more similar to LIF plus dexamethasone than IL-6 plus dexamethasone.[11]

6. Adipogenesis Inhibitory Activity of IL-11

Kawashima et al.[12,13] have recently isolated a cDNA encoding a novel adipogenesis inhibitory factor (AGIF) from a human bone marrow-derived stromal cell line that inhibits the process of adipogenesis and the associated lipoprotein lipase (LPL) activity in mouse 3T3-L1 preadipocytes. Analysis of the sequence of this cDNA revealed the identity of this factor with IL-11. Many other cytokines with the ability to inhibit adipogenesis have been reported previously. These include interferon-γ IL-1, TNF, transforming growth factor-β (TGFβ), LIF, and IL-6. Many of these factors can be expressed in the bone marrow stromal cell lines, suggesting that they may play an important role in stromal cell-associated hematopoiesis through their regulatory action on adipocyte differentiation in the bone marrow microenvironment. The mechanisms of inhibition, however, vary among different cytokines. It has been shown that TNF affects LPL activity by decreasing LPL gene transcription and by another posttranscriptional mechanism,[30] whereas IL-11,[14] like IL-1,[30] affects LPL activity through a posttranscriptional mechanism. The fact that an adipocyte cell line can respond to a stromal fibroblast-derived growth factor may provide a useful system to study cell-cell interactions in the hematopoietic microenvironment.

7. *In Vivo* Effects of IL-11

Administration of IL-11 to normal C3H/HeJ mice resulted in an increase in the number of spleen SRBC-specific PFC as well as serum SRBC-specific antibody titer in both the primary and secondary immune responses.[10] IL-11 administration to normal mice also increased the absolute numbers of bone marrow-derived CFU-GM and stimulated marrow and splenic hematopoietic progenitor cells to a higher cell-cycling rate.[15] IL-11 has also been shown to stimulate megakaryocyte maturation and increase in peripheral platelet number in normal mice[31] and nonhuman primates.[32]

In mice immunosuppressed by cyclophosphamide treatment, IL-11 administration significantly augmented the number of spleen SRBC-specific PFC as well as serum SRBC-

TABLE 1
IL-11 Receptor Distribution and Protein Tyrosine Phosphorylation Stimulated by IL-11 in Various IL-11-Responsive Cell Lines

Cell line	Receptor distribution (sites/cell)	Tyrosine phosphorylated proteins (kDa)
3T3-L1	5140	151, 97, 47, and 44
C3H10t1/2	2760	151, 97, 47, and 44
H35	502	97 and 44
T10	138	ND[a]
B9TY1	75	97

[a] Not determined.

specific antibody titer when compared with cyclophosphamide-treated mice without IL-11 treatment.[10] IL-11 also decreased the time required to regain normal levels of leukocyte and platelet counts in peripheral blood. Furthermore, IL-11 accelerated the reconstitution to a normal range of myeloid progenitors from the bone marrow and spleen of these myelosuppressed mice.[15]

The *in vivo* effects of IL-11 on the recovery of peripheral blood cell counts and proliferation of progenitor and stem cells have also been examined using a mouse bone marrow and spleen cell transplantation model.[33] The data suggested that IL-11 can accelerate the recovery of peripheral blood leukocytes, mainly neutrophils, and platelets in transplant mice. Similar results have also been obtained by transplantation of IL-11-retrovirus-infected bone marrow cells into lethally irradiated mice.[34]

C. SIGNAL TRANSDUCTION MEDIATED BY IL-11

1. Expression of IL-11 Receptor(s) on Different Cell Lines

Scatchard plot analysis was performed on various IL-11-responsive cell lines, including T10 (an IL-6- or IL-11-dependent mouse plasmacytoma cell line), B9TY1 (an IL-6- or IL-11-dependent mouse B-cell hybridoma), 3T3-L1 (a mouse preadipocyte cell line), and H-35 (a rat hepatoma cell line). The numbers for IL-11 receptor(s) on different cell lines are summarized in Table 1. Due to the high receptor number present on the mouse preadipocyte cell line 3T3-L1, we have also examined the IL-11 receptor distribution on C3H10t1/2 (a mouse cell line with the potential to differentiate into preadipocytes, myoblasts, and chondroblasts) cells at different stages of adipocyte differentiation. The results showed that the IL-11 receptor number decreases as these cells become more mature adipocytes (data not shown). Competitive binding studies have shown that IL-6 and IL-11 did not compete for binding, suggesting the utilization of different ligand-binding proteins for the two cytokines (data not shown).

2. Biochemical Characterization of IL-11 Receptor on 3T3-L1 and C3H10t1/2 Cells

The molecular weight of IL-11 was determined by affinity cross-linking [^{125}I]IL-11 to 3T3-L1 and C3H10t1/2 cells with the homobifunctional cross-linker, DSS (Figure 2). Addition of a 1000-fold excess of unlabeled IL-11 caused the disappearance of the 174-kDa band, suggesting that the band is a complex of IL-11 and its corresponding receptor. Subtraction of [^{125}I]IL-11 from the complex resulted in a molecular weight of IL-11 receptor on these cells of 151 kDa.[14] Like the other cytokine receptors studied to date, the binding of IL-11 to its receptor was followed by rapid internalization (data not shown).

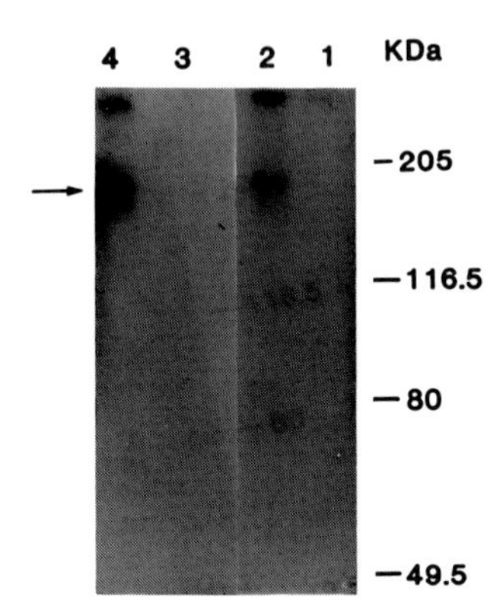

FIGURE 2. Affinity cross-linking of [^{125}I]IL-11 to 3T3-L1 and C3H10t1/2 cells. [^{125}I]IL-11 was incubated with C3H10t1/2 (lanes 1 and 2) or 3T3-L1 cells (lanes 3 and 4) in the presence (lanes 1 and 3) or absence (lanes 2 and 4) of a 1000-fold excess of unlabeled IL-11. The sizes of the molecular mass standards are given in kilodaltons.

3. Signal Transduction Pathways Mediated by IL-11

The possible signaling pathways mediated by IL-11 have been examined by immunoblotting with anti-phosphotyrosine antibody.[14] Studies on 3T3-L1, C3H10t1/2, H-35, and B9TY1 cells are summarized in Table 1. Stimulation of 3T3-L1 and C3H10t1/2 cells with IL-11 resulted in the appearance of tyrosine-phosphorylated proteins with molecular weights of 151, 97, 47, and 44 kDa. In H-35 cells, IL-11 induced the expression of 97- and 44-kDa tyrosine-phosphorylated proteins. Only 97-kDa tyrosine-phosphorylated protein was detected in B9TY1 cells following IL-11 induction. In general, IL-11-induced protein tyrosine phosphorylation is transient and concentration dependent. The specificity of IL-11-induced protein tyrosine phosphorylation has been demonstrated by comparison with those induced by IL-6 (data not shown). Despite the overlapping biological activities shared between IL-6 and IL-11, the patterns of protein tyrosine phosphorylation stimulated by these two cytokines differ in certain cell lines (data not shown).

IV. DISCUSSION

IL-11, like many other cytokines identified to date, is produced by more than one cell type and can act on many different target cells. The cloning of the human IL-11 gene has allowed us to begin to study the regulation of IL-11 gene expression in different cell types and the role of IL-11 in the hematopoietic microenvironment. The availability of the purified recombinant IL-11 protein has also enabled us to test *in vitro* biological activities, evaluate its biological effects *in vivo,* and study the signal transduction mechanisms utilized by this cytokine.

Like most of the cytokines, IL-11 is most likely to exert its biological activities through specific ligand-receptor interactions on various target cells. Although most of the interleukin or CSF receptors do not possess tyrosine kinase activity, it appears that some of these receptor signal transduction pathways, including those for IL-2,[35] IL-3,[36] IL-6,[37] IL-7,[38] GM-CSF,[39] and LIF,[40] do utilize tyrosine kinase activities, as evidenced by the appearance of tyrosine-phosphorylated proteins upon stimulation with these interleukins or CSFs. To understand the molecular mechanisms involved in the multiple effector functions of IL-11, initial biochemical characterizations of IL-11 receptor and protein tyrosine phosphorylation have been performed on several IL-11-responsive cell lines. Cross-linking studies demonstrated that the IL-11 receptor consists of a single 151-kDa polypeptide chain. Different-sized tyrosine-phosphorylated proteins are induced by IL-11 in various IL-11-responsive cell lines. Interestingly, a 97-kDa tyrosine-phosphorylated protein has been shown to be present in all the cell types examined (Table 1). A protein molecule of similar molecular weight has also been found to be a major substrate for tyrosine kinases activated following stimulation with cytokines such as IL-3, GM-CSF, and Epo.[41] The identity and physiological role of the 97-kDa protein requires further investigation.

There is growing evidence that many cytokines may utilize common protein components to bind ligands or transduce signals across the cell membrane.[42,43] In addition, Lord et al.[40] have recently shown the common signal transduction pathways utilized by IL-6 and LIF, two cytokines with redundant biological activities. Based on the overlapping biological activities observed between IL-6 and IL-11, we speculate that these two cytokines may utilize common signal transduction pathways. Preliminary studies on protein tyrosine phosphorylation, activation of primary response genes, and production of second messengers have indicated convergent and divergent signaling pathways utilized by IL-6 and IL-11.[44]

The molecular cloning of cDNA(s) encoding IL-11 receptor and the biochemical characterization of the receptor complexes are essential for understanding the ligand-receptor interactions. In addition, studies of IL-11-mediated signal transduction pathways compared to those of other cytokines such as IL-6, LIF, and oncostatin M[43] will help to explain some of the biological activities shared among these cytokines. Since IL-11 has been shown to have synergistic effects with several cytokines in various bioassays, understanding the signal transduction pathways utilized by IL-11 will also be beneficial in the future utilization of this cytokine in combination therapy.

ACKNOWLEDGMENTS

The author thanks her collaborators and colleagues for their contributions to the work summarized in this review.

REFERENCES

1. **Dexter, T. M., Allen, T. D., and Lajtha, L. G.,** Conditions controlling the proliferation of hematopoietic stem cells in vitro, *J. Cell. Physiol.*, 91, 335, 1977.
2. **Paul, S. R., Yang, Y.-C., Donahue, R. E., Goldring, S., and Williams, D. A.,** Stromal cell-associated hematopoiesis: immortalization and characterization of a primate bone marrow-derived stromal cell line, *Blood*, 77, 1723, 1991.
3. **Paul, S. R., Bennet, F., Calvetti, J. A., Kelleher, K., Wood, C. R., O'Hara, R. M., Jr., Leary, A. C., Sibley, B., Clark, S. C., Williams, D. A., and Yang, Y.-C.,** Molecular cloning of a cDNA encoding interleukin 11, a stromal cell-derived lymphopoietic and hematopoietic cytokine, *Proc. Natl. Acad. Sci. U.S.A.*, 87, 7512, 1990.
4. **McKinley, D., Wu, Q., Yang-Feng, T., and Yang, Y. C.,** Genomic sequence and chromosomal location of human interleukin (IL)-11 gene, *Genomics*, 13, 814, 1992.
5. **Larson, D., Leary, A., Hahn-Cordes, L., Fitzgerald, M., Giannoti, J., Sibley, B., Bennett, F., Calvetti, J., Turner, K., and Clark, S. C.,** Interleukin-11 (IL-11) synergizes with IL-3 in promoting human and murine megakaryocyte colony formation in vitro, *Blood*, 76, 464a, 1990.
6. **Bruno, E., Briddell, R. A., Cooper, R. J., and Hoffman, R.,** Effects of recombinant interleukin 11 on human megakaryocyte progenitor cells, *Exp. Hematol.*, 19, 378, 1991.
7. **Teramura, M., Kobayashi, S., Hoshino, S., Oshimi, K., and Mizoguchi, H.,** Interleukin-11 enhances human megakaryocytopoiesis in vitro, *Blood*, 79, 327, 1992.
8. **Musashi, M., Yang, Y.-C., Paul, S. R., Clark, S. C., Sudo, T., and Ogawa, M.,** Direct and synergistic effects of interleukin 11 on murine hemopoiesis in culture, *Proc. Natl. Acad. Sci. U.S.A.*, 88, 765, 1991.
9. **Musashi, M., Clark, S. C., Sude, T., Urdal, D. L., and Ogawa, M.,** Synergistic interactions between interleukin-11 and interleukin-4 in support of proliferation of primitive hematopoietic progenitors of mice, *Blood*, 78, 1448, 1991.

10. **Yin, T., Schendel, P., and Yang, Y.-C.,** Enhancement of in vitro and in vivo antigen-specific antibody responses by interleukin 11, *J. Exp. Med.*, 175, 211, 1992.
11. **Baumann, H. and Schendel, P.,** Interleukin-11 regulates the hepatic expression of the same plasma protein genes as interleukin-6, *J. Biol. Chem.*, 266, 20424, 1991.
12. **Kawashima, I., Ohsumi, J., Mita-Honjo, K., Shimoda-Takano, K., Ishikawa, H., Sakakibara, S., Miyadai, K., and Takiguchi, Y.,** Molecular cloning of cDNA encoding adipogenesis inhibitory factor and identity with interleukin-11, *FEBS Lett.*, 283, 199, 1991.
13. **Ohsumi, J., Miyadai, K., Kawashima, I., Ishikawa-Ohsumi, H., Sakakibara, S., Mita-Honjo, K., and Takiguchi, Y.,** Adipogenesis inhibitory factor: a novel inhibitory regulator of adipose conversion in bone marrow, *FEBS Lett.*, 288, 13, 1991.
14. **Yin, T., Miyazawa, K., and Yang, Y.-C.,** Characterization of interleukin-11 receptor and protein tyrosine phosphorylation induced by interleukin-11 in mouse 3T3-L1 cells, *J. Biol. Chem.*, 267, 8347, 1992.
15. **Hangoc, G., Yin, T., Cooper, S., Schendel, P., Broxmeyer, H. E., and Yang, Y.-C.,** In vivo effects of recombinant interleukin-11 on myelopoiesis in mice, *Blood*, 81, 965, 1993.
16. **Scatchard, G.,** The attraction of proteins for small molecules and ions, *Ann. N.Y. Acad. Sci.*, 51, 660, 1949.
17. **Shaw, G. and Kamen, R.,** A conserved AU sequence from the 3′ untranslated region of GM-CSF mRNA mediates selective mRNA degradation, *Cell*, 46, 659, 1986.
18. **Isshiki, H., Akira, S., Tanabe, O., Nakajima, T., Shimatomo, T., Hirano, T., and Kishimoto, T.,** Constitutive and interleukin-1 (IL-1)-inducible factors interact with the IL-1-responsive element in the IL-6 gene, *Mol. Cell. Biol.*, 10, 2757, 1990.
19. **Hromas, R., Collins, S. J., Hickstein, D., Raskind, W., Deaven, L. L., O'Hara, P., Hagen, F. S., and Kaushansky, K.,** A retinoic acid-responsive human zinc finger gene, MZF-1, preferentially expressed in myeloid cells, *J. Biol. Chem.*, 266, 14183, 1991.
20. **Jackson, C. J., Bruns, G. A. P., and Breslow, J. L.,** Isolation and sequence of a human apolipoprotein CII cDNA clone and its use to isolate and map to human chromosome 19 the gene for apolipoprotein CII, *Proc. Natl. Acad. Sci. U.S.A.*, 81, 2945, 1984.
21. **Das, H. K., Jackson, C. L., Miller, D. A., Leff, T., and Breslow, J. L.,** The human apolipoprotein C-II gene sequence contains a novel chromosome 19-specific minisatellite in its third intron, *J. Biol. Chem.*, 262, 4787, 1987.
22. **McKeithan, T. W., Rowley, J. D., Shows, T. B., and Diaz, M. O.,** Cloning of the chromosomal translocation breakpoint junction of the t(14;19) in chronic lymphocytic leukemia, *Proc. Natl. Acad. Sci. U.S.A.*, 84, 9257, 1987.
23. **O'Bryan, J. P., Frye, R. A., Cogswell, P. C. Neubauer, A., Kitch, B., Prokop, C., Espinosa, R., III, Le Beau, M. M., Earp, H. S., and Liu, E. T.,** axl, a transforming gene isolated from primary human myeloid leukemia cells, encodes a novel receptor tyrosine kinase, *Mol. Cell. Biol.*, 11, 5016, 1991.
24. **Paul, S. R., Barut, B. A., Bennett, F., Cochran, M. A., and Anderson, K. C.,** Lack of a role for interleukin 11 in the growth of multiple myeloma, *Leukemia Res.*, 16, 247, 1992.
25. **Ogawa, H., Hirayama, F., and Leary, A. G.,** Studies in culture of cell-cycle dormant hemopoietic progenitors in man and mouse, *J. Cell. Biochem.*, 16C, 52, 1992.
26. **Donaldson, D., Neben, S., and Turner, K.,** Interleukin-11 in combination with other hematopoietic growth factors supports the survival and expansion of primitive murine progenitor cells in vitro, *J. Cell. Biochem.*, 16C, 73, 1992.
27. **Schibler, K. R., Yang, Y.-C., and Christensen, R. D.,** Effect of interleukin-11 on cycling status and clonogenic maturation of fetal and adult hematopoietic progenitors, *Blood*, 80, 900, 1992.
28. **Anderson, K. C., Morimoto, C., Paul, S. R., Chauhan, D., Williams, D., Cochran, M., and Barut, B.,** Interleukin-11 promotes accessory cell-dependent B-cell differentiation in humans, *Blood*, 80, 2797, 1992.
29. **Fey, G. and Gauldie, J.,** The acute phase response of the liver in inflammation, *Prog. Liver Dis.*, 9, 89, 1990.
30. **Zechner, R., Newman, T. C., Sherry, B., Cerami, A., and Breslow, J. L.,** Recombinant human cachectin/tumor necrosis factor but not interleukin-1 downregulates lipoprotein lipase gene expression at the transcriptional level in mouse 3T3-L1 adipocytes, *Mol. Cell. Biol.*, 8, 2394, 1988.
31. **Neben, T., Loebelenz, J., Hayes, L., McCarthy, K., Stoudemire, J. B., Schaub, R. G., and Goldman, S.,** Recombinant human interleukin-11 (rhIL-11) stimulates megakaryocytopoiesis and increases peripheral platelets in normal and splenectomized mice, *Blood*, 81, 901, 1993.
32. **Bree, A., Schlerman, F., Timony, G., McCarthy, K., and Stoudimire, J.,** Pharmacokinetics and thrombopoietic effects of recombinant human interleukin-11 (rhIL-11) in nonhuman primates and rodents, *Blood*, 78, 132a, 1991.
33. **Du, X. X., Neben, T., Goldman, S., and Williams, D. A.,** Effects of recombinant human interleukin 11 on hematopoietic reconstitution in transplant mice: acceleration of recovery of peripheral blood neutrophils and platelets, *Blood*, 81, 27, 1993.

34. **Wood, C. R., Paul, S. R., Goldman, S., Muench, M., Palmer, R., Pedneault, G., Morris, G. E., Bree, M., Hayes, L., Hoysradt, J., McCarthy, K., Neben, T., Bennett, F., Kaufman, R. J., Schaub, R., Moore, M. A. S., and Clark, S. C.,** IL 11 expression in donor bone marrow cells improves hematological reconstitution in lethally irradiated recipient mice, *J. Cell. Biochem.*, 16C, 93, 1992.
35. **Saltzman, E. M., Lukowskyj, S. M., and Casnellie, J. E.,** The 75,000-dalton interleukin-2 receptor transmits a signal for the activation of a tyrosine protein kinase, *J. Biol. Chem.*, 264, 19979, 1989.
36. **Koyasu, S., Tojo, A., Miyajima, A., Akiyama, T., Kasuga, M., Urabe, A., Schreurs, J., Arai, K., Takaku, F., and Yahara, I.,** Interleukin 3-specific tyrosine phosphorylation of a membrane glycoprotein of Mr 150,000 in multi-factor-dependent myeloid cell lines, *EMBO J.*, 6, 3979, 1987.
37. **Nakajima, K. and Wall, R.,** Interleukin-6 signals activating junB and TIS11 gene transcription in a B-cell hybridoma, *Mol. Cell. Biol.*, 11, 1409, 1991.
38. **Uckun, F. M., Dibirdik, I., Smith, R., Tuel-Ahlgren, L., Chandan-Langlie, M., Schieven, G. L., Waddick, K. G., Hanson, M., and Ledbetter, J. A.,** Interleukin 7 receptor ligation stimulates tyrosine phosphorylation, inositol phospholipid turnover and clonal proliferation of human B-cell precursors, *Proc. Natl. Acad. Sci. U.S.A.*, 88, 3589, 1991.
39. **Cannistra, S. A., Groshek, P., Garlick, R., Miller, J., and Griffin, J. D.,** Regulation of surface expression of the granulocyte/macrophage colony-stimulating factor receptor in normal human myeloid cells, *Proc. Natl. Acad. Sci. U.S.A.*, 87, 93, 1990.
40. **Lord, K. A., Abdollahi, A., Thomas, S. M., Demarco, M., Brugge, J. S., Hoffman-Liebermann, B., and Liebermann, D. A.,** Leukemia inhibitory factor and interleukin-6 trigger the same immediate early response including tyrosine phosphorylation, upon induction of myeloid leukemia differentiation, *Mol. Cell. Biol.*, 11, 4371, 1991.
41. **Showers, M. O., Moreau, J.-F., and D'Andrea, A. D.,** The role of the erythropoietin receptor in multistage Friend virus-induced erythroleukemia, *J. Cell. Biochem.*, 16C, 58, 1992.
42. **Kitamura, T., Sato, N., Arai, K., and Miyajima, A.,** Expression cloning of the human IL-3 receptor cDNA reveals a shared beta subunit for the human IL-3 and GM-CSF receptors, *Cell*, 66, 1165, 1991.
43. **Gearing, D. P., Comeau, M. R., Friend, D. J., Gimpel, S. D., Thut, C. J., McGourty, J., Brasher, K. K., King, J. A., Gillis, S., Mosley, B., Ziegler, S. F., and Cosman, D.,** The IL-6 signal transducer, gp130: an oncostatin M receptor and affinity converter for the LIF receptor, *Science*, 255, 1434, 1992.
44. **Yang, Y.-C., et al.,** unpublished results.
45. **Yin, T. et al.,** unpublished results.

Chapter 18

THE ROLE OF ONCOSTATIN M IN THE ACUTE PHASE RESPONSE

Carl D. Richards and Mohammed Shoyab

TABLE OF CONTENTS

0-8493-6913-4/93/$0.00 + $.50

I. ACUTE PHASE MEDIATORS

Tissue damage due to various events such as trauma, infection, or invasion by tumor cells can induce a marked and characteristic increase in the serum levels of acute phase proteins. These proteins are produced predominantly by the liver and represent a heterogenous array of functions and characteristics. In searching for specific mechanisms by which this occurs, investigators have identified soluble mediators (cytokines) released by various tissues that act specifically on hepatocytes to induce the production of typical acute phase proteins. These soluble products are referred to collectively here as *acute phase mediators*. The study of the control of the liver acute phase response by cytokines has, in the last 5 to 6 years, provided a better understanding of how this relatively ubiquitous, nonspecific, but important physiological response to tissue damage occurs. This understanding has occurred, in part, by merging previously separate lines of investigation and investigators who are joined by the realization of the redundancy and pleiotrophic nature of most cytokines.

Historically, the first acute phase mediator definitively implicated in acute phase responses was interleukin-1 (IL-1).[1] IL-1 caused the synthesis of some acute phase proteins *in vitro* and induced typical acute phase responses *in vivo* in rodents.[2] It became clear that other factors were implicated and that IL-1 action *in vivo* appeared to be indirect. In 1987, it was shown that IL-6 (previously studied by at least four separate groups focused on different biological activities) was the primary product from lipopolysaccharide (LPS)-stimulated monocytes that induced liver cells *in vitro*.[3,4] In 1989, Baumann and Wong showed that leukemia inhibitory factor (LIF) possessed similar activity,[5] and more recently, IL-11, a bone marrow stromal cell product, was also shown to possess hepatocyte-inducing activity.[6] These studies have formed new collaborations between individuals in various disciplines and those interested in liver acute phase responses. We have now shown that another cytokine, oncostatin M, has potent activity in inducing this response *in vitro*[7] and that it may be the most potent cytokine in this regard (as yet characterized).

II. ONCOSTATIN M IS AN ACUTE PHASE MEDIATOR

Oncostatin M was originally identified and characterized as a growth regulator for certain tumor and nontumor-derived cell lines. Oncostatin M (OM) is expressed in activated human T-lymphocytes[8] and monocytes,[9] and is secreted as a M_r 28,000 single-chain polypeptide with a unique primary structure[10] and specific cell-surface receptors.[11] OM inhibits the growth of human malignant melanoma and carcinoma cell lines[12] as well as cultured bovine aortic endothelial cells,[13] but stimulates the growth of certain cultured fibroblasts.[10,12] Other bioactivities include stimulation of plasminogen activator activity in cultured bovine aortic endothelial cells,[13] upregulation of expression of low density lipoprotein receptors in cultured human hepatoma cells,[14] and induction of IL-6 expression in cultured human endothelial cells.[15] In addition, OM has been implicated as a growth factor for Kaposi's sarcoma cells.[16,17]

Genomic and cDNA clones for OM have been published.[9] Structure-based comparisons of OM and LIF sequences predict similar functional properties of these two cytokines.[18,19] OM binds to specific surface receptors that are present on a number of cell types.[14] Because of its activity in IL-6 induction, we became interested in examining the role of OM in the hepatic acute phase response. We tested OM in human and rat hepatoma cells as well as primary rat hepatocytes, and found potent activity *in vitro*.

OM induced Hep G2 cells and primary rat hepatocytes to produce acute phase proteins in a dose-dependent fashion (Figure 1). Elevated levels of haptoglobin, α_1-antichymotrypsin, α_1-acid glycoprotein, α_1-protease inhibitor, and ceruloplasmin were evident in OM-stimulated Hep G2 cells, and induction of α_2-macroglobulin, α_1-cysteine protease inhibitor (thiostatin), and α_1-acid glycoprotein (α_1-AGP) were found in OM-stimulated rat hepatocytes

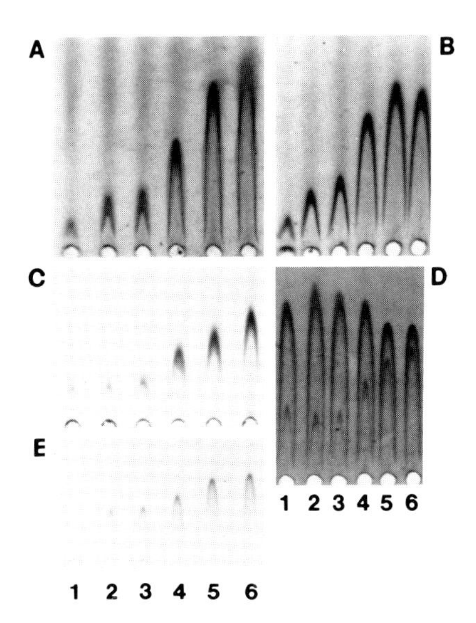

FIGURE 1. OM induction of acute phase protein in hepatocytes. Hep G2 human hepatoma cells (B. Knowles, Wistar Institute) were cultured and passaged in Dulbecco's modified Eagle's medium (DMEM) supplemented with 10% FCS. Cells were split and plated in 24-well cluster plates and allowed to grow to confluency before assay. Monolayers were then treated for 24 h in 250 μl of medium (with and without 1 μ*M* dexamethasone) with the indicated cytokines. Production of α_1-ACH, fibrinogen, haptoglobin, and α_1-AGP by Hep G2 cells was analyzed by rocket electrophoresis as previously described.[28,29] Amounts of protein synthesized (micrograms of protein per 24 h per 10^6 cells) were compared to background output by unstimulated cells. Primary rat hepatocytes were isolated from collagenase-digested liver as previously described[30] and cultured on collagen-coated plates in William's E medium supplemented with 10% FCS and 1 μ*M* dexamethasone. The assay of acute phase protein synthesis by rocket electrophoresis was performed as previously published[31] and described above. OM was used at concentrations of 0 (lane 1), 0.1 (lane 2), 1 (lane 3), 10 (lane 4), 50 (lane 5), and 100 (lane 6) ng/ml. Hep G2 cells were analyzed for haptoglobin (A), α_1-antichymotrypsin (B), and α_1-acid glycoprotein (C). Rat hepatocytes were analyzed for α_2-macroglobulin (lower peaks in D), albumin (upper peaks in D), and α_1-cysteine protease inhibitor (E).

concomitant with a decrease in albumin production. When compared to Hep G2 cells, rat hepatocytes were less sensitive than the human cells to OM (greater concentration needed for the half-maximal response). This may reflect species specificity displayed by rat hepatocyte OM receptors (human recombinant OM was used). When compared to other cytokines such as IL-6 and LIF, OM induced the greatest maximal output of haptoglobin, α_1-antichymotrypsin, and α_1-acid glycoprotein in Hep G2 cells (Table 1). Combinations of OM with IL-1 or glucocorticoid resulted in a synergistic output of α_1-AGP or haptoglobin, indicating that the OM interaction with IL-1 and glucocorticoid is similar to that shown by IL-6 or LIF.[7] We have furthermore shown that OM is capable of stimulating expression of CAT genes downstream of putative IL-6 response elements upon transfection of CAT constructs into Hep G2 cells.[7] OM again showed the greatest activity when compared to IL-6 and LIF. Thus, OM acts in a similar fashion toward acute phase mediators such as IL-6 and LIF on hepatocytes, and although interacting with apparently unique receptors, may initiate gene expression through (at least in part) similar mechanisms.

OM has been shown to interact at the high-affinity LIF receptor,[20] which is composed of the LIF-R "α-chain" associated with the IL-6 signal transducer gp130.[21] OM has been reported to bind with low affinity to gp130 and may interact with a higher affinity with a putative OM-R "α-chain" associated with gp130. The association of gp130 with signaling for IL-6, LIF, and OM may thus provide a rationale for the similar effects on gene regulation in hepatocytes occurring in response to these cytokines.

TABLE 1
Comparative Potency of Acute-Phase Mediators

	Fold stimulation		
	OM	IL-6	LIF
Hep G2			
α_1-ACH	10	7.5	6
HP	20	16	8
Rat Heps			
α_2-M	6	10	6
α_1-CPI	3	5	3

Note: Amounts of haptoglobin (Hp) and α_1-antichymotrypsin (α_1-ACH) produced by Hep G2 cells under stimulation by maximal doses of OM, IL-6, and LIF were calculated and are expressed as fold stimulation over control cells. Similarly, α_2-macroglobulin (α_2-M) and α_1-cysteine protease inhibitor (α_1-CPI) production from rat primary hepatocytes (Rat Heps) was analyzed.

III. ONCOSTATIN M ENHANCES IL-6 EXPRESSION

Although Brown et al.[15] have shown that the production of IL-6 by endothelial cells increases markedly in response to OM, we have not seen such responses in fibroblast cultures to OM alone. However, OM clearly has a potentiating effect on IL-6 production in combination with IL-1 (Figure 2). At concentrations of 1, 5, and 10 ng/ml, OM enhanced IL-6 production in the presence of 2 ng/ml of IL-1α. This suggests that OM could enhance cytokine-mediated IL-6 production by connective tissue cells (stroma) at sites of inflammation or tissue damage.

IV. ONCOSTATIN M ENHANCES ANTIPROTEASE EXPRESSION

A striking feature of the acute phase proteins induced by OM, IL-6, and LIF in hepatocytes is that many are antiproteinase in action.[22-24] A major antiprotease of connective tissue cells is termed TIMP (tissue inhibitor of metalloproteinases) because of the potent inhibitory activity of its metalloproteinase enzyme function. TIMP thus plays a role in the extracellular matrix (ECM) metabolism of normal physiological processes as well as in the pathological breakdown of ECM in chronic inflammation and tumor metastasis.[25-27] We have examined TIMP-1 expression in Hep G2 cells as well as fibroblast cultures and found that OM potently enhances TIMP-1 mRNA levels and protein production (Figure 3).[36] Both cell types show marked enhancement due to OM stimulation, even though basal levels of TIMP-1 mRNA are much greater in fibroblasts (Figure 3). Also, OM was far more potent in enhancing TIMP-1 than IL-6 or LIF.[36] This suggests that OM is a potent regulator of antiproteinase expression not only in hepatocytes, but also in connective tissue cells from extrahepatic sites. The induction of tissue site antiproteases by OM, IL-6, and LIF may well prove to be important in the modulation of local responses at sites of inflammation.

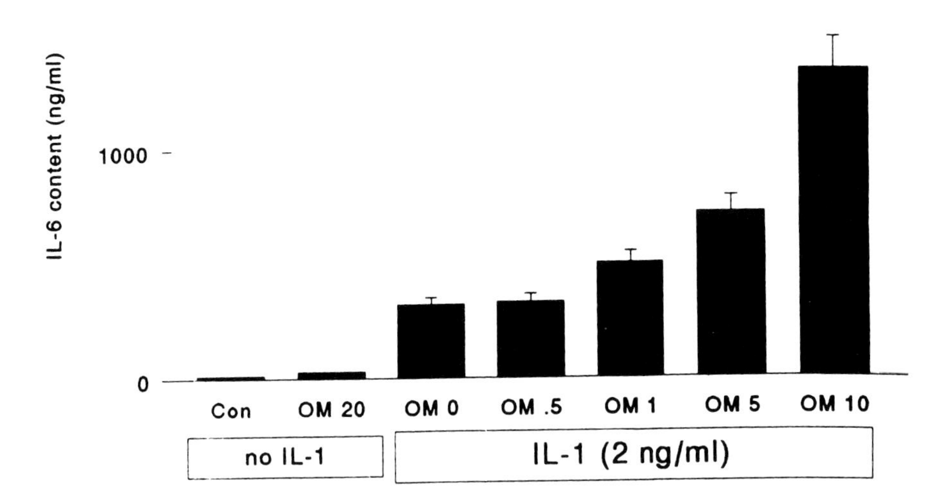

FIGURE 2. OM enhances IL-1-induced IL-6 production. OM (at 0, 0.5, 1, 5, 10, and 20 ng/ml) and IL-1 (2 ng/ml) were added as indicated to fibroblast cultures for 18 h. Supernatants were then decanted and stored for analysis. The B9 hybridoma proliferation assay was used to assay for IL-6, performed according to previously published protocols,[32] and analyzed using colorimetric development.[33] Error bars represent standard error.

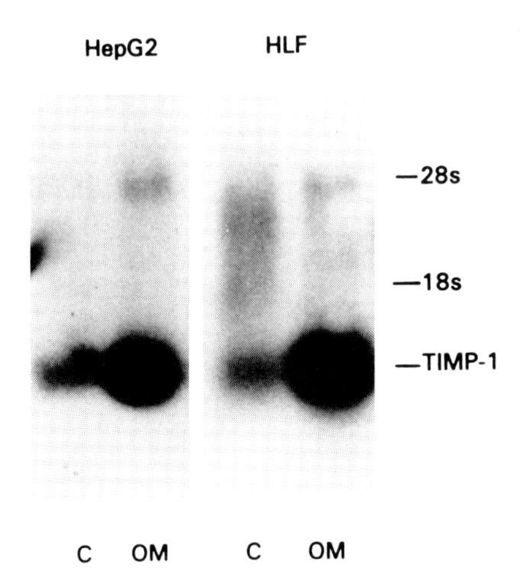

FIGURE 3. OM induces TIMP-1 expression in Hep G2 cells and fibroblasts. Total RNA was prepared from 18-h OM-stimulated Hep G2 cells and human lung fibroblasts (HLF), and isolated using the method of Chomczynski and Sacchi.[34] Northern blots were prepared by standard techniques and probed with cDNA probe to human TIMP-1 (courtesy of Dr. A. J. P. Docherty) labeled by the random primer technique. Although equal amounts of RNA were loaded, the signal for TIMP-1 (in parallel experiments done simultaneously) was four to five times greater in fibroblasts than in Hep G2 cells. Thus, the exposure for fibroblasts shown here is fivefold less than for Hep G2 cells.

V. CONCLUSIONS

OM may contribute to acute phase responses on three levels: (1) direct action on hepatocytes, (2) induction of IL-6 expression from endothelial cells and enhancement of IL-1-induced IL-6 expression in stromal cells, and (3) induction of antiprotease in extrahepatic sites of local inflammation. Whether OM expression is altered in inflammatory responses at the local or systemic level is not presently known; however, future analysis will provide data to evaluate this aspect of OM biology. Interestingly, OM has been found to be present in human serum in association with a binding protein. OM-binding protein complex does not exhibit biological activity.[35] The origin and function of serum OM remains to be elucidated.

ACKNOWLEDGMENTS

The authors thank Donna Green for excellent technical assistance and D. Labonte for preparation of the manuscript. This work was supported by operating grants from the Arthritis Society and MRC of Canada.

REFERENCES

1. **Bornstein, D. L.,** Leukocytic pyrogen: a major mediator of the acute phase reaction, *Ann. N.Y. Acad. Sci.*, 389, 323, 1982.
2. **Kampschmidt, R. F. and Mesecher, M.,** Interleukin-1 from P388D: effects upon neutrophils, plasma iron, and fibrinogen in rats, mice, and rabbits, *Proc. Soc. Exp. Biol. Med.*, 179, 197, 1985.
3. **Gauldie, J., Richards, C., Harnish, D., Lansdorp, P., and Baumann, H.,** Interferon beta2/BSF-2 shares identity with monocyte derived hepatocyte stimulating factor (HSF) and regulates the major acute phase protein response in liver cells, *Proc. Natl. Acad. Sci. U.S.A.*, 84, 7251, 1987.
4. **Baumann, H., Richards, C., and Gauldie, J.,** Interaction between hepatocyte-stimulating factors, interleukin-1 and glucocorticoids for regulation of acute phase plasma proteins in human hepatoma (HepG2) cells, *J. Immunol.*, 139, 4122, 1987.
5. **Baumann, H. and Wong, G. G.,** Hepatocyte-stimulating factor III shares structural and function identity with leukemia inhibitory factor, *J. Immunol.*, 143, 1163, 1989.
6. **Baumann, H. and Schendel, P.,** Interleukin-11 regulates the hepatic expression of the same plasma protein genes as interleukin-6, *J. Biol. Chem.*, 266, 1, 1991.
7. **Richards, C. D., Brown, T. J., Shoyab, M., Baumann, H., and Gauldie, J.,** Recombinant oncostatin M stimulates the production of acute phase proteins in HepG2 cells and rat primary hepatocytes in vitro, *J. Immunol.*, 148, 1731, 1992.
8. **Brown, T. J., Lioubin, M. N., and Marquardt, H.,** Purification and characterization of cytostatic lymphokines produced by activated human T lymphocytes, *J. Immunol.*, 139, 2977, 1987.
9. **Malik, N., Kallestad, J. C., Gunderson, N. L., Austin, S. D., Neubauer, M. G., Ochs, V., Marquardt, H., Zarling, J. M., Shoyab, M., Wei, C.-M., Linsley, P. S., and Rose, T. M.,** Molecular cloning, sequence analysis, and functional expression of a novel growth regulator, oncostatin M, *Mol. Cell. Biol.*, 9, 2847, 1989.
10. **Zarling, J. M., Shoyab, M., Marquardt, H., Hanson, M. B., Lioubin, M. N., and Todaro, G. J.,** Oncostatin M: a growth regulator produced by differentiated histiocyte lymphoma cells, *Proc. Natl. Acad. Sci. U.S.A.*, 83, 9739, 1986.
11. **Linsley, P. S., Bolton-Hanson, M., Horn, D., Malik, N., Kallestad, J. C., Ochs, V., Zarling, J. M., and Shoyab, M.,** Identification and characterization of cellular receptors for the growth regulator, oncostatin M, *J. Biol. Chem.*, 264, 4282, 1989.
12. **Horn, D., Fitzpatrick, W. C., Gompper, P. T., Ochs, V., Bolton-Hansen, M., Zarling, J., Todaro, G. J., and Linsley, P. S.,** Regulation of cell growth by recombinant oncostatin-M, *Growth Factors*, 2, 157, 1990.
13. **Brown, T. J., Rowe, J. M., Shoyab, M., and Gladstone, P.,** Oncostatin M: a novel regulator of endothelial cell properties, *Mol. Biol. Cardiovasc. Syst.*, 131, 195, 1990.
14. **Grove, R. I., Mazzucco, C. E., Radka, S. F., Shoyab, M., and Keiner, P. A.,** Oncostatin M upregulates low density lipoprotein receptors in HepG2 cells by a novel mechanism, *J. Biol. Chem.*, 266, 18194, 1992.
15. **Brown, T. J., Rowe, J. M., and Shoyab, M.,** Regulation of interleukin-6 expression by oncostatin M, *J. Immunol.*, 147, 2175, 1991.

16. **Miles, S. A., Martinez-Maza, O., Rezai, A., Magpantay, L., Kishimoto, T., Nakamura, S., Radka, S. F., and Linsley, P. S.**, Oncostatin M as a potent mitogen for AIDS-Kaposi's sarcoma-derived cells, *Science*, 255, 1432, 1992.
17. **Nair, B. C., DeVico, A. L., Nakamura, S., Copeland, T. D., Chen, Y., Patel, A., O'Neil, T., Oroszlan, S., Gallo, R. C., and Sarngadharan, M. G.**, Identification of a major growth factor for AIDS-Kaposi's sarcoma cells as oncostatin M, *Science*, 255, 1430, 1992.
18. **Rose, T. M. and Bruce, A. G.**, Oncostatin M is a member of the cytokine family which includes LIF, GM-CSF and IL-6, *Proc. Natl. Acad. Sci. U.S.A.*, 88, 8641, 1991.
19. **Bazan, F.**, Neuropeptide cytokines in the hematopoietic fold, *Neuron*, 7, 197, 1991.
20. **Gearing, D. P. and Bruce, A. G.**, Oncostatin M binds the high-affinity leukemia inhibitory factor receptor, *New Biol.*, 4, 61, 1991.
21. **Gearing, D. P., Comeau, M. R., Friend, D. J., Gimpel, S. D., Thut, C. J., McGourty, J., Brasher, K. K., King, J. A., Gillis, S., Mosley, B., Ziegler, S. F., and Cosman, D.**, The IL-6 signal transducer, gp130: an oncostatin M receptor and affinity converter for the LIF receptor, *Science*, 255, 1434, 1992.
22. **Gauldie, J., Richards, C. D., Harnish, D., and Baumann, H.**, Interferon beta2 is identical to monocytic HSF and regulates the full acute phase protein response in liver cells, in *Monokines and Other Non-Lymphocytic Cytokines*, Powanda, M. C., Ed., Alan R. Liss, New York, 1988, 15.
23. **Gauldie, J., Richards, C., Northemann, W., Fey, G., and Baumann, H.**, IFNbeta2/BSF2/IL-6 is the monocyte-derived HSF that regulates receptor-specific acute phase gene regulation in hepatocytes, *Ann. N.Y. Acad. Sci.*, 557, 46, 1989.
24. **Richards, C. D., Gauldie, J., and Baumann, H.**, Cytokine control of acute phase protein expression, *Eur. Cyt. Net.*, 2, 89, 1991.
25. **Khokha, R. and Denhardt, D. T.**, Matrix metalloproteinases and tissue inhibitor of metalloproteinases: a review of their role in tumorigenesis and tissue invasion, *Invasion Metast.*, 9, 391, 1989.
26. **Ponton, A., Coulombe, B., and Skup, D.**, Decreased expression of tissue inhibitor of metalloproteinases in metastatic tumor cells leading to increased levels of collagenase activity, *Cancer Res.*, 51, 2138, 1992.
27. **Mignatti, P., Robbins, E., and Rifkin, D. N.**, Tumor invasion through the human amniotic membrane: requirement for a proteinase cascade, *Cell*, 47, 487, 1986.
28. **Gauldie, J., Richards, C., Harnish, D., Lansdorp, P., and Baumann, H.**, Interferon-beta2/B-cell stimulatory factor type 2 shares identity with monocyte hepatocyte-stimulating factor and regulates the major acute phase protein response in liver cells, *Proc. Natl. Acad. Sci. U.S.A.*, 84, 7251, 1987.
29. **Baumann, H., Richards, C., and Gauldie, J.**, Interaction between hepatocyte-stimulating factors, interleukin-1 and glucocorticoids for regulation of acute phase proteins in human hepatoma (Hep-G2) cells, *J. Immunol.*, 139, 4122, 1987.
30. **Sweeney, G. D., Garfield, R. E., Jones, K. G., and Lathan, A. N.**, Studies using sedimentation velocity on heterogeneity of size and function of hepatocytes from mature male rats, *J. Lab. Clin. Med.*, 91, 432, 1978.
31. **Koj, A., Gauldie, J., Regoeczi, E., Sauder, D. N., and Sweeney, G. D.**, The acute phase response of cultured rat hepatocytes. System characterisation and the effect of human cytokines, *Biochem. J.*, 224, 505, 1984.
32. **Aarden, L. A., De Groot, E. R., Schaap, O. L., and Lansdorp, P. M.**, Production of hybridoma growth factor by human monocytes, *Eur. J. Immunol.*, 17, 1411, 1987.
33. **Mosmann, T.**, Rapid colorimetric assay for cellular growth and survival: application to proliferation and cytotoxicity assays, *J. Immunol. Methods*, 65, 55, 1983.
34. **Chomczynski, P. and Sacchi, N.**, Single-step method of RNA isolation by acid guanidinium thiocyanate-phenol-chloroform extraction, *Anal. Biochem.*, 162, 156, 1987.
35. **Shoyab, M. et al.**, unpublished results.
36. **Richards, C. D., Shoyab, M., Brown, T. J., and Gauldie, J.**, Selective regulation of metalloproteinase inhibitor (TIMP-1) by oncostatin M in fibroblasts in culture, *J. Immunol.*, in press.

Chapter 19

TUMOR NECROSIS FACTOR

Peter G. Brouckaert and Claude Libert

TABLE OF CONTENTS

0-8493-6913-4/93/$0.00 + $.50

I. INTRODUCTION

Tumor necrosis factor (TNF) was originally described as a factor present in the serum of induced mice and capable of causing hemorrhagic necrosis in methylcholanthrene-induced sarcomas in Balb/c mice.[1] Since its cloning, however, it has become clear that this factor has many more activities. It plays a role in the induction of cachexia, pathogenesis of septic shock, and reaction of the body to an infection or injury.[2]

In this chapter, we briefly review its structure and function, and discuss the bidirectional relationship between TNF and the liver.

II. STRUCTURE AND FUNCTION OF TNF

A. STRUCTURE OF TNF

The genes for both human TNF[3] and murine TNF[4] were cloned independently by several research groups. The gene for TNF is a single-copy gene and is located close to the HLA-B locus in the MHC cluster on the short arm of human chromosome 6. The murine gene is similarly located on murine chromosome 17. The gene for lymphotoxin (LT), which is both structurally and functionally closely related to TNF, is located about 1200 bp 5′ to the TNF gene. Both genes are about 3000 bp long and consist of four exons and three introns.

The mature protein is a trimer of 52 kDa. It is composed of three identical subunits of 17 kDa each, consisting of 157 (human) or 156 (murine) amino acids. Human TNF is not glycosylated, but murine TNF contains one *N*-glycosylation site. The mature TNF is derived from a precursor protein containing an unusually long presequence of 76 (human) or 79 (murine) amino acids that is highly conserved. This precursor protein is a transmembrane protein of 26 kDa with an intracellular domain and a membrane anchor, and can exert at least part of the activities of TNF.[5] The mature protein is cleaved by a mechanism that possibly involves a serine protease. Its three-dimensional structure has been determined by X-ray diffraction. The shape of the trimer is a triangular pyramid reminiscent of the arrangement of many viral capsids. The highest structural homology is with the satellite tobacco necrosis virus capsid. Mutational analysis is being performed to determine the active site. It has been proposed that the receptor-binding domain is in the lower half of the pyramid, in the groove between the two subunits.[6]

The three-dimensional structure of LT is similar to that of TNF. There is also about 30% homology at the amino acid level. An important difference, however, is that it does not contain the long presequence, but, instead, a classical signal sequence of 34 amino acids. Hence, it is a secreted protein, not a membrane protein. LT is often called TNFβ.

TNF is produced mainly by macrophages activated by bacterial products, viruses, parasites, and probably other stimuli. CD2-LFA-3 interactions may form a physiological induction signal. This signal can also be produced by other cells such as lymphocytes, LAK cells, natural killer (NK) cells, neutrophils, astrocytes, endothelial cells, smooth muscle cells, and several tumor cells. In some cell types, the gene may have been developmentally silenced.[7] Its induction is under both transcriptional and posttranscriptional regulation.[8] Further details and references can be found in two recent reviews by Fiers[9] and Vilcek and Lee.[10]

B. THE RECEPTORS FOR TNF

The receptors for TNF have been identified recently. In both humans and mice, two distinct receptor types can be found: a 55-kDa (TNF-R55) and 75-kDa type (TNF-R75).[11-13] Most interestingly, in the mouse, human TNF (hTNF) does not bind to the TNF-R75 receptor.[14] Receptors for TNF have been detected on almost every cell type, with some exceptions, such as erythrocytes and unstimulated T-cells. The extracellular domains of both receptors are quite homologous, but the intracellular domains are completely different.[15] The extracellular domain is also related to that of the nerve growth factor receptor, to the CDw40 and OX40 antigens, and to the transcriptionally active open reading frame T2 from the Shope fibroma virus. Triggering of the TNF receptors can be mimicked by antibodies directed against these receptors.[16-18] This indicates that the receptors are triggered by clustering.

Both receptors can shed their extracellular domain and give rise to soluble TNF-binding proteins. The receptors can be found in human serum and urine, and can inhibit TNF effects on target cells. Although lymphotoxin binds to both cellular receptors, it binds only very slightly or not at all to the soluble binding proteins.[19]

C. CELLULAR EFFECTS OF TNF

Binding to the receptor results in nucleus-dependent and nucleus-independent effects. Examples of nucleus-independent effects are the cytotoxic effect toward malignant cells, the release of arachidonic acid, the respiratory burst, and degranulation of neutrophils.[9]

Other effects of TNF are the result of transcriptional activation. These include transcription factors such as c-*fos* and c-*jun*, cytokines such as IL-1, IL-6, IL-8, macrophage colony-stimulating factor (M-CSF), and granulocyte/macrophage colony-stimulating factor (GM-CSF), cell-adhesion molecules such as ELAM-1 and ICAM-1, MHC molecules, collagenase, and viruses such as HIV-1.[10]

Although the signaling pathways are largely unknown, there is some evidence that a G-protein is involved and that protein phosphorylation occurs. The intracellular domain of the receptor gives no hints about possible signaling pathways.[9]

The cellular effects of TNF eventually lead to activation, differentiation, or death of the target cells. Besides its cytotoxic action on transformed (but not on normal) cells, TNF exerts important effects on other cell types. Within the immune system, TNF acts on macrophages where it induces IL-1, TNF, and cytotoxicity,[20] and on neutrophils where it leads to a respiratory burst and degranulation, leading to the release of elastase, lysozyme, and other enzymes.[21,22] Another important target cell of TNF is undoubtedly the endothelial cell. In these cells, TNF induces the synthesis and release of platelet-activating factor (PAF), procoagulant activity, IL-1, IL-6, GM-CSF, ELAM-1 and ICAM-1, HLA-A and -B, plasminogen activator inhibitor, urokinase-type plasminogen activator, and prostacyclin, while it inhibits activated protein C-protein S complex.[23-27] This leads to an increase in lymphocyte and neutrophil adhesion and eventually to an altered permeability of the vessel wall. Recently, it has been shown that TNF induces EDRF. This may be important as a mechanism for TNF-induced hypotension.[28]

Other target cells include hepatocytes (see further), chondrocytes,[29] astrocytes,[30] fibroblasts,[31] and osteoclasts.[32]

D. PHYSIOLOGICAL, PATHOLOGICAL, AND THERAPEUTIC SIGNIFICANCE

These pleiotropic effects of TNF are the basis for its involvement in pathophysiological situations. The main physiological role of TNF will have to be found in antibacterial resistance[33] and resistance against parasites.[34,35] Whether it has still other physiological roles such as tissue remodeling and a role in development requires further investigation. TNF is involved in various pathological situations. It is a central mediator in the systemic inflammatory response syndrome (SIRS, formerly called sepsis),[36] and is involved in inflammation,[37] malaria,[38] anemia,[37] graft-vs.-host disease,[39] severe infectious purpura,[40] AIDS,[41] pulmonary fibrosis,[42] arthritis,[43] and autoimmune diseases.[44] It is likely that many more diseases will join this group.

TNF is, however, also under investigation as a treatment for cancer. Although its shock-inducing properties hamper the administration of therapeutic amounts of TNF in the general circulation, it has been shown to be highly effective in locoregional treatments such as the isolation-perfusion treatment for metastatic melanoma confined to one limb.[45]

E. THE SELECTIVE SPECIES-SPECIFICITY OF TNF

Since most of the rTNF that is used in research is of human origin and since most of the experiments, especially those *in vivo,* are done in animals (in particular, rats and mice), it seems worthwhile to draw attention to the phenomenon of selective species-specificity. Furthermore, the molecular basis of this phenomenon is such that comparison of the effects of human and murine TNF could result in valuable information about the function of the two receptor types for TNF.

Indeed, recent research has shown that human TNF does not interact with murine TNF-R75.[14] Triggering of this receptor type does not seem to be necessary to obtain most TNF effects *in vitro* and in particular the cytotoxic effect toward transformed cells, since both human and murine TNF can exert these effects. However, some *in vitro* effects, especially those on murine T-cells, can only be exerted by murine TNF, and hence are mediated by TNF-R75 triggering. Examples of such effects are the proliferation of murine thymocytes,[46] induction of genes in PC-60 cells,[47] and proliferation of CT6 cells.[48] It is, however, possible that a more extensive comparison of the effects of human and murine TNF on a variety of murine cells will reveal still other functions for which TNF-R75 triggering is necessary or involved. It may be that such differences will be found not only in the nature, but also in the vigor of these effects.

The most striking differences between human and murine TNF were observed *in vivo:* while human TNF is not lethal in healthy mice when administered alone, murine TNF is lethal at doses of about 10 to 20 μg per mouse (depending on the mouse strain used).[49] When other sensitizing agents such as GalN,[49] IL-1,[50] or the glucocorticoid antagonist RU38486[51] are coadministered, both human and murine TNF become equally lethal. This implies that triggering of both receptor subtypes is necessary to obtain lethality, but that triggering of TNF-R75 is no longer needed when sensitizing agents are present. A similar situation was observed when we investigated the antitumor activity of TNF against murine B16BL6 melanoma *in vivo*.[52] Murine TNF was active while human TNF was not, and the addition of a sensitizing agent, i.c. interferon-γ (IFNγ), to the treatment abolished the species specificity. For other effects such as hypothermia and IL-6 induction, the difference was to be found in the sustained character of the effect after murine TNF vs. the transient character after human TNF administration rather than in the amplitude of the effect.[51]

These observations make it necessary to use the species' own cytokine when one wants to investigate the possible involvement of the cytokine in physiological and pathological situations.

III. TNF AND OTHER INDUCERS OF ACUTE PHASE PROTEINS

Besides its direct activities on hepatocytes, which are discussed later in this chapter, TNF can also influence liver function and acute phase protein synthesis through its capacity to augment the levels of other inducers of these proteins. Thus, TNF can induce IL-1[24] and glucocorticoid hormones.[53] In this chapter, we focus on its capacity to induce IL-6.

The induction of IL-6 by TNF has been documented *in vitro* as well as in animal models and human patients. *In vitro,* TNF can induce IL-6 in a variety of cells such as those of fibroblastic[54] and endothelial[55] origin.

In mice, we demonstrated an induction of IL-6 by both human and murine TNF[51] as well as by lipopolysaccharide (LPS) and IL-1.[56] Although TNF induces lower peak levels (at the plateau of the dose response) than LPS or IL-1,[56] we observed peak levels of 10 to 100 ng/ml 2 to 3 h after administration of a bolus injection of TNF.[51] Interestingly, the pattern of IL-6 induction differs after the administration of human recombinant TNF (hrTNF) vs. murine recombinant TNF (mrTNF): after hrTNF, only a transient induction of IL-6 could be observed, whereas after mrTNF, IL-6 levels remained high for over 12 h.[51] Since human TNF is not lethal in mice while murine TNF is lethal,[49] and since we had previously observed that the difference between the IL-6 levels after lethal vs. nonlethal doses of LPS was to be found in the sustained pattern of this induction rather than in the peak levels,[57] we investigated whether IL-6 would be causally involved in the lethal shock induced by TNF. Using two monoclonal antibodies, one directed against the mIL-6R and one against mIL-6, we observed a protection when we challenged mice with a dose of mTNF or LPS causing just 100% lethality.[58] These results are in agreement with the observations of Starnes et al.[59] No protection could be observed, however, when slightly higher doses of TNF or LPS were administered.[58] Neither could we observe protection when we administered lethal combinations of human TNF and IL-1 or the glucocorticoid receptor antagonist RU38486, two combinations that induce the same sustained pattern of IL-6 induction previously observed after mTNF.[51] Also, the lethality observed after a combination of TNF and galactosamine could not be inhibited by the administration of these monoclonals.[58] The fact that these antibodies no longer protected against the lethality caused by these challenges was not due to an insufficient neutralizing capacity, since these antibodies were able to protect against challenges inducing much higher amounts of IL-6.[58]

Although the induction of IL-6 by TNF seems to be only marginally involved in the pathogenesis of endotoxic shock, it might be quite important for the induction of the acute phase response. This induction of IL-6 by TNF could be inhibited by both indomethacin and glucocorticoids.[60]

The induction of IL-6 by TNF is not restricted to animals. Indeed, in human cancer patients to whom TNF was administered as a 24-h continuous infusion during a phase I clinical trial, peak levels of circulating IL-6 were detected 3 to 6 h after the start of the TNF infusion.[61] These peak values correlated with the dose of TNF administered. Addition of IFNγ to the treatment resulted in higher levels of IL-6 than were obtained with a comparable dose of TNF alone. No IL-6 was observed during the infusion with IFNγ alone.

These observations show that the induction of IL-6 by TNF may be responsible for part of the observed *in vivo* effects of TNF.

IV. TNF AS AN INDUCER OF ACUTE PHASE PROTEINS

Injection of TNF in rats caused a significant elevation of the serum concentrations of the major acute phase proteins fibrinogen (Fb), haptoglobulin (Hp), α_2-macroglobulin (α_2-M), and α_1-acid glycoprotein (α-AGP), and a reduction of the serum albumin levels.[62] Since these early studies, however, a number of *in vitro* studies have challenged the view that TNF would be an important inducer of acute phase proteins by its direct effects on the liver. Using rat primary hepatocytes, neither Andus et al.[63] nor Baumann et al.[64] observed any effect of TNF on the levels of α_2-M, α_1-proteinase inhibitor (α_1-PI) or albumin protein, or mRNA. In contrast, addition of TNF to the rat hepatoma line H-35 caused an induction of the proteins C3, Hp, and α_1-AGP.[64] Also, the well-known human cell line Hep G2 developed a response to TNF, since induction of C3, factor B, and α_1-antichymotrypsin (α_1-ACT), and a reduction of albumin and transferrin were observed.[65] The question arises as to what extent the observed TNF effects are direct effects. Indeed, it was recently observed that the augmented production of serum amyloid A (SAA) and C-reactive protein (CRP) seen in human primary hepatocyte cultures after stimulation with interleukin-1 (IL-1) could be blocked by antibodies against interleukin-6 (IL-6).[66] It could well be that the same holds for the effects seen after stimulation with TNF. Indeed, we observed that TNF induced low but significant concentrations of IL-6 in the supernatant of all four hepatomas we tested:

Hep G2, BWTG-3, Fa-32, and FU-5.[112] Hence, experiments in which these cell lines are stimulated with TNF in the presence of antibodies against IL-6 will be necessary to discriminate between direct and indirect effects.

In most cases, IL-6 is the best stimulator of synthesis of acute phase proteins (APPs), especially when corticoids are present.[67] The set of APPs induced by IL-6 and IL-1 is overlapping but not identical. Furthermore, depending on the APP studied, IL-1 can be either synergistic or antagonistic for the IL-6-driven inductions.[68,69] It is worth mentioning that the IL-6-inducible genes contain a consensus sequence CTGGGA/T upstream of the start codon.[70,71] The human TNF-R55 TNF receptor also contains this sequence at − 163 bp, and expression of the protein as well as the mRNA of TNF-R55 is upregulated by IL-6. This stimulation by IL-6 also is inhibited by IL-1, as is the case for some APPs.[72]

When animals are injected with IL-6, a significant but less impressive APR is observed than in animals injected with TNF or IL-1.[73] This argues against a predominant role for IL-6 in the induction of the APR. An explanation of this paradox might be found in the difference between the duration of the exposure to IL-6 (which has a very short half-life of about 5 min in circulation) after an injection of IL-6 vs. an induction of IL-6 by TNF or IL-1, which can provide significant levels of IL-6 for several hours. When the *in vivo* effects of TNF and IL-1 regarding their potency to induce an APR are compared, it is observed that IL-1 seems to be the more potent inducer of APPs.[74] Perhaps IL-1 is a stronger inducer of IL-6 than is TNF.[75] Next to the importance of exposition time, synergistic activities could also account for these *in vivo* observations. So, a synergism between TNF and IL-1 was found for the induction of fibrinogen and IL-6.[51,74] Synergies and/or antagonisms are of considerable pathophysiological importance, since TNF as well as IL-1 and IL-6 are present in infection or injury. In experimental models, TNF seems to appear in the circulation before IL-1 and IL-6, but the importance of the contribution of TNF to the induction of the other two cytokines, has not been completely elucidated yet.

V. THE LIVER AS A TARGET ORGAN FOR TNF

The effects of TNF on the liver are seldom direct, but are nevertheless worthwhile discussing. Injection of TNF in mice or rats caused an increase in liver weight and enhanced DNA, RNA, and protein synthesis,[76] while infusion of TNF for several days leads to hypertrophy of the liver.[77] Single injections of TNF also result in enhanced levels of serum transaminases, indicating that some liver damage has occurred. Also, during clinical trials involving cases of chronic administration of TNF, toxicity to the liver was a dose-limiting side-effect. These observations argue for the importance of the liver as a target organ for TNF.

Several observations have been made regarding individual systems or enzymes in the liver. For example, the expression of most members of the P450 family of membrane-bound isozymes, comprising an important part of classical detoxification systems, is downregulated. TNF shares this activity with other molecules such as IL-1, LPS, and IFN.[78] *In vitro* studies revealed that only IL-1 is capable of downregulating the transcripts of cytochrome P450 molecules and that other factors — TNF, IFN, and LPS — work indirectly.[78]

Also, the dramatically enhanced transport of amino acids to and uptake in the liver, which coincides with the enhanced breakdown of muscle proteins seen in several diseases and especially during shock, might be a phenomenon in which TNF is involved. *In vivo,* both TNF and IL-1 can upregulate amino acid uptake by the liver.[79] This effect might, however, be indirect since only IL-6 is capable of doing so *in vitro*.[80] In such conditions, the metabolism of amino acids is also upregulated, such as the transformation of amino acids to histidine and putrescin, essentially by histidine decarboxylase and ornithine decarboxylase. Both enzymes are activated by TNF.[81] The amino acids are used for the synthesis of the proteins whose synthesis is upregulated and as a substrate for gluconeogenesis. The gluconeogenesis might, however, become inhibited by suppression of the induction of phosphoenolpyruvate carboxykinase (PEPCK), a corticoid-induced key enzyme in the pathway of gluconeogenesis. LPS, TNF, IL-1, and IL-6 inhibit the glucocorticoid-induced upregulation of PEPCK.[82]

Also typical for hepatocytes in a shock-suffering animal is the enhanced activity of fatty acid synthetase, hydroxymethylglutamylCoA reductase, and the TNF-activated acetylCoA carboxylase. The acute changes in hepatic lipid metabolism, however, are not due to changes in gene expression, but to the increase of citrate levels, citrate being an allosteric activator of rate-limiting enzymes in fatty acid synthesis. Taken together, this leads to higher cholesterol and fatty acid and lower ketone concentrations.[83]

VI. PROTECTION AGAINST THE TOXIC EFFECTS OF TNF PROVIDED BY THE LIVER

Although not many functions of APPs are known, several researchers consider these proteins to be part of a large negative feedback loop during sepsis and shock. Since most aspects of shock are mediated by the activities and interplay of IL-1, TNF, and other cytokines, a study of the regulatory connections between these cytokines and APPs is required. The effect of cytokines on the induction of APPs is well known, and more is now known about the effects of APPs on the induction, release, and activities of cytokines.

To study the involvement and protective properties of the liver against TNF-induced shock, we studied the lethal activities of TNF or LPS in normal or galactosamine (GalN)-sensitized mice. In normal mice, murine TNF (mTNF) is lethal at a dose of 10 to 15 μg per mouse. When GalN is coinjected, the lethal dose of TNF is reduced to about 0.1 μg per mouse.[84] An even more spectacular synergy is observed between LPS and GalN.[85] GalN is a sugar derivative that acts specifically on hepatocytes. Injection or infusion of GalN causes a dramatic depletion of UTP, leading to a rapid cessation of transcription and translation.[86,87] Changes in lipid composition, membrane potential,[88] and RNA methylation[89] are commonly observed in GalN-treated hepatocytes. Animals that receive a combination of LPS and GalN suffer from neutrophilia, lymphopenia, and hypothermia,[90] and die 6 h after the injection with enormous levels of transaminases in the serum, indicating fulminant hepatitis.[91] Animals challenged with a combination of LPS and GalN can be protected with anti-TNF antibodies.[92] LPS can be replaced by the sulfopeptidoleukotriene D4 (LTD4).[93] The fact that TNF becomes much more toxic when the liver is eliminated was also indirectly observed when hepatectomized rats were found to be considerably more vulnerable to serum obtained from LPS-injected rats than were sham controls.[94] These findings can be explained by suggesting a role for the liver in TNF clearance. However, the kinetics of TNF injected in GalN-treated mice do not differ from those in control mice.[60]

It is interesting to be able to eliminate the protective capacity of the liver, but it is still more useful to know how to restore or superinduce it and to know if any correlation with APPs can be made. We use two models to confer protection against the lethal effects of TNF. In the first approach, mice can be made tolerant to the lethal TNF effects for a period of about 10 d by pretreating them daily with TNF for 5 d. This tolerance cannot be induced with IL-1, and a connection with the liver has not yet been made.[95] In the second approach, mice can be fully desensitized against lethal LPS, LPS + GalN, TNF, or TNF + GalN by pretreatment with TNF or IL-1.[96,97] LPS, IFN, and IL-4 can also confer this protection, although to a much lesser extent. This is a particular case of the long-known "nonspecific resistance to infection". IL-1-induced desensitization requires 4 h to be induced and lasts for about 24 h. In a controlled experiment, we demonstrated that the liver is involved in the induced protection.[97]

Consequently, we tried to identify the systems or proteins involved in this protection. A possible involvement of glutathione transferases or P450 cytochromes was excluded, since these are downregulated by IFN and IL-1, respectively. APPs, however, are strongly induced by IL-1 in C57Bl mice.[74] Since IL-6 is the most potent inducer of APPs, we tried to induce desensitization with IL-6, murine or human, in various doses and according to several regimens, in both the presence and absence of glucocorticoids, but failed to observe any protective effect.[97] This result, together with our unpublished observation that IL-1-induced protection was also observed in adrenalectomized mice, does not suggest an involvement of APPs. In another study, protection against the same lethal events was induced by hydrazine sulfate, but not in hypophysectomized mice.[98] Recently, a Japanese group found that IL-1-induced desensitization to bacterial infections and LPS could be explained by a faster clearance of LPS in the desensitized mice,[99] but we found no changed kinetics of TNF after injection in IL-1-vs. PBS-treated mice.[97] Obviously, considerable differences exist between TNF and LPS as a challenge, as some globins (e.g., α_1-AGP and α_1-PI), inducible by IL-1, are known to suppress the release of LPS-induced TNF from leukocytes.[100] Indeed, α_1-PI protects mice against a lethal challenge of LPS + GalN because of suppressed TNF release.[101] These findings easily explain the often elevated serum TNF levels in alcoholic cirrhosis.[102]

From a theoretical point of view, at least some APPs should be able to protect against toxic events induced by LPS or TNF. α_1-PI not only can suppress the release of TNF, but also exhibits some IL-1 inhibitory activity, at least *in vitro*.[103] IL-1 is believed to be a mediator in the toxic events seen after LPS administration, and recent observations with a recombinant IL-1 receptor antagonist confirmed this belief.[104,105] α_1-PI and α_1-ACT can neutralize important enzymes such as elastase and collagenase, whose overproduction is likely to be involved in the tissue damage which is part of this toxicity.[106] However, the overproduction of radicals, another mediating event in the pathogenesis of the induced shock,[107] can easily destroy the α_1-PI.[106] Another APP, α_2-M, which is the major APP in the rat, can bind in a covalent and Zn-dependent way some important cytokines such as TNF, IL-1, IL-2, IL-6, platelet-derived growth factor (PDGF), TGFβ, and others. This binding does not lead to a loss of biological activities.[108] In some studies, the TNF-induced phospholipid platelet activating factor (PAF) was considered to be an important mediator of TNF-induced phenomena. CRP can form complexes with PAF and remove it from the circulation.[106] Furthermore, the concanavalin A (ConA) unreactive form of α_1-AGP was reported to induce, in macrophages, an IL-1 inhibitory activity.[109] Finally, the complex of α_1-ACT with one of its substrates, cathepsin, is able to induce IL-6, and this may serve as a positive feedback loop for the ongoing APR.[110]

Although the aforementioned observations would provide ample mechanisms for a protective role of APPs against the shock-inducing properties of TNF and LPS, the results of attempts to use them as protective agents have been disappointing. Perhaps this reflects our lack of knowledge about the activities of the different APPs in the various species. It could be that in addition to their known activities, these proteins also exert other activities that may be harmful during an endotoxic reaction. Further evidence for such a possible harmful role comes from the fact that antibodies against IL-6 could block the lethality induced by TNF or LPS.[58,59]

VII. INFLUENCE OF ACUTE PHASE PROTEINS ON TNF RELEASE

One of the potential beneficial effects deserves particular attention. As mentioned above, mature TNF is derived from a transmembrane precursor protein which is cleaved off to form the circulating protein. Some APPs might act as negative feedback regulators by preventing this cleavage. Scuderi et al. showed that the release of TNF is blocked by inhibitors of serine proteases such as TAME[111] and by some APPs.[100] Regarding the latter, α_1-PI, α_1-AGP, and α_2-M inhibited TNF release, while haptoglobin was inactive at concentrations up to 5 mg/ml. α_1-PI was the most effective inhibitor when the proteins were compared on a weight basis, but α_2-M was more effective on a molar basis. Niehörster et al.[101] showed that *in vivo*, cotreatment with α_1-PI prevented the appearance of circulating TNF after a challenge with the combination of endotoxin and galactosamine. Furthermore, they showed that α_1-PI could protect mice against hepatitis induced by a combination of GalN and LPS, but not against hepatitis induced by a combination of GalN and TNF. We observed that α_1-PI could inhibit lethality in mice challenged with LPS, but not with mTNF.[112] Together with the *in vitro* experiments of Scuderi et al. showing that the addition of α_1-PI influenced neither TNF mRNA levels nor membrane-bound TNF levels, these *in vivo* experiments support the thesis that the action of the protease inhibitor is restricted to the release of TNF and does not influence its activity. In addition to the existence of a negative feedback loop, these results show that although the transmembrane form of TNF may exert some of the activities of TNF and although it may be that TNF induces TNF, these activities may not be important in TNF-induced lethality.

VIII. SUMMARY

TNF is a cytokine that has an important role in inflammatory reactions and in the response against infection. Besides its action on other organs and cell types such as neutrophils and endothelial cells, it has a pronounced effect on liver function, through both its direct actions on hepatocytes and its induction of other cytokines and hormones whose target organ is the liver. Hence, it is involved in the induction of APPs and in the metabolic derangements that accompany infection. The relationship between TNF and the liver is bidirectional, since the liver can also produce substances that protect against the lethal activities of TNF. In addition to this interference with the action of TNF, by producing protease inhibitors which inhibit the release of TNF, the liver is also involved in a negative feedback loop regarding the production of TNF.

REFERENCES

1. **Carswell, E., Old, L., Kassel, R., Green, S., Fiore, N., and Williamson, B.,** An endotoxin induced serum factor that causes necrosis of tumors, *Proc. Natl. Acad. Sci. U.S.A.*, 72, 3666, 1975.
2. **Old, L.,** Tumor necrosis factor, *Sci. Am.*, 258(5), 41, 1988.
3. **Pennica, D., Nedwin, G., Hayflick, J., Seeburg, P., Derynck, R., Palladino, M., Kohr, W., Aggarwal, B., and Goeddel, D.,** Human tumour necrosis factor: precursor structure, expression and homology to lymphotoxin, *Nature (London)*, 312, 724, 1984.
4. **Fransen, L., Muller, R., Marmenout, A., Tavernier, J., Van der Heyden, J., Kawashima, E., Chollet, A., Tizard, R., Van Heuverswyn, H., Van Vliet, A., Ruysschaert, M., and Fiers, W.,** Molecular cloning of mouse tumour necrosis factor cDNA and its eukaryotic expression, *Nucleic Acids Res.*, 13, 4417, 1985.

5. **Perez, C., Albert, I., DeFay, K., Zachariades, N., Gooding, L., and Kriegler, M.,** A non-secretable cell surface mutant of tumor necrosis factor (TNF) kills by cell-to-cell contact, *Cell,* 63, 251, 1990.
6. **Van Ostade, X., Tavernier, J., Prangé, T,. and Fiers, W.,** Localization of the active site of human tumour necrosis factor (hTNF) by mutational analysis, *EMBO J.,* 10, 827, 1991.
7. **Beutler, B. and Brown, T.,** A CAT reporter construct allows ultrasensitive estimation of TNF synthesis, and suggests that the TNF gene has been silenced in non-macrophage cell lines, *J. Clin. Invest.,* 87, 1336, 1991.
8. **Han, J., Brown, T., and Beutler, B.,** Endotoxin-responsive sequences control cachectin/TNF biosynthesis at the translational level, *J. Exp. Med.,* 171, 465, 1990.
9. **Fiers, W.,** Tumor necrosis factor. Characterization at the molecular, cellular and in vivo level, *FEBS Lett.,* 285, 199, 1991.
10. **Vilcek, J. and Lee, T.,** Tumor necrosis factor. New insights into the molecular mechanisms of its multiple actions, *J. Biol. Chem.,* 266, 7313, 1991.
11. **Loetscher, H., Pan, Y., Lahm, H., Gentz, R., Brockhaus, M., Tabuchi, H., and Lesslauer, W.,** Molecular cloning and expression of the human 55 kDa tumor necrosis factor receptor, *Cell,* 61, 351, 1990.
12. **Schall, T. J., Lewis, M., Koller, K. J., Lee, A., Rice, G. C., Wong, G. H. W., Gatanaga, T., Granger, G. A., Lentz, R., Raab, H., Kohr, W. J., and Goeddel, D.,** Molecular cloning and expression of a receptor for human tumor necrosis factor, *Cell,* 61, 361, 1990.
13. **Smith, C. A., Davis, T., Anderson, D., Solam, L., Beckmann, M. P., Jerzy, R., Dower, S. K., Cosman, D., and Goodwin, R. G.,** A receptor for tumor necrosis factor defines an unusual family of cellular and viral proteins, *Science,* 248, 1019, 1990.
14. **Lewis, M., Tartaglia, L., Lee, A., Bennett, G., Rice, G., Wong, G., Chen, E., and Goeddel, D.,** Cloning and expression of cDNAs for two distinct murine tumor necrosis factors demonstrate one receptor is species specific, *Proc. Natl. Acad. Sci. U.S.A.,* 88, 2830, 1991.
15. **Dembic, Z., Loetscher, H., Gubler, U., Pan, Y., Lahm, H., Gentz, R., Brockhaus, M., and Lesslauer, W.,** Two human TNF receptors have similar extracellular, but distinct intracellular, domain sequences, *Cytokine,* 2, 231, 1990.
16. **Engelmann, H., Holtmann, H., Brakebusch, C., Avni, Y. S., Sarov, I., Nophar, Y., Hadas, E., Leitner, O., and Wallach, D.,** Antibodies to a soluble form of tumor necrosis factor (TNF) receptor have TNF-like activity, *J. Biol. Chem.,* 265, 14497, 1990.
17. **Tartaglia, L. A., Weber, R. F., Figari, I. S., Reynolds, C., Palladino, M. A., and Goeddel, D. V.,** The two distinct receptors for tumor necrosis factor mediate distinct cellular responses, *Proc. Natl. Acad. Sci. U.S.A.,* 88, 9292, 1991.
18. **Shalaby, M. R., Sundan, A., Loetscher, H., Brockhaus, M., Lesslauer, W., and Espevik, T.,** Binding and regulation of cellular functions by monoclonal antibodies against tumor necrosis factor receptors, *J. Exp. Med.,* 172, 1517, 1991.
19. **Loetscher, H., Steinmetz, M., and Lesslauer, W.,** Tumor necrosis factor: receptors and inhibitors, *Cancer Cells,* 3, 221, 1991.
20. **Philip, R. and Epstein, L.,** Tumour necrosis factor as immunomodulator and mediator of monocyte cytotoxicity induced by itself, gamma-interferon and interleukin-1, *Nature (London),* 323, 86, 1986.
21. **Klebanoff, S., Vadas, M., Harlan, J., Sparks, L., Gamble, J., Agosti, J., and Waltersdorph, A.,** Stimulation of neutrophils by human tumour necrosis factor, *J. Immunol.,* 136, 4220, 1986.
22. **Shalaby, M., Aggarwal, B., Rinderknecht, E., Svedersky, L., Finkle, B., and Palladino, M.,** Activation of human polymorphonuclear neutrophil functions by interferon-gamma and tumour necrosis factors, *J. Immunol.,* 135, 2069, 1985.
23. **Collins, T., Lapierre, L., Fiers, W., Strominger, J., and Pober, J.,** Recombinant human tumor necrosis factor increases mRNA levels and surface expression of HLA-A, B antigens in vascular endothelial cells and dermal fibroblasts in vitro, *Proc. Natl. Acad. Sci. U.S.A.,* 83, 446, 1986.
24. **Libby, P., Ordovas, J., Auger, K., Robbins, A., Birinyi, L., and Dinarello, C.,** Endotoxin and tumor necrosis factor induce interleukin-1 gene expression in adult human vascular endothelial cells, *Am. J. Pathol.,* 124, 179, 1986.
25. **Nawroth, P. and Stern, D.,** Modulation of endothelial cell hemostatic properties by tumor necrosis factor, *J. Exp. Med.,* 163, 740, 1986.
26. **Pober, J.,** Effects of tumor necrosis factor and related cytokines on vascular endothelial cells, in *Tumour Necrosis Factor and Related Cytotoxins,* John Wiley & Sons, Chichester, 1987, 173.
27. **Fiers, W., Beyaert, R., Brouckaert, P., Everaerdt, B., Grooten, J., Haegeman, G., Libert, C., Suffys, P., Takahashi, N., Tavernier, J., Van Bladel, S., Vanhaesebroeck, B., Van Ostade, X., and Van Roy, F.,** Tumour necrosis factor and interleukin-6: structure and mechanism of action of the molecular, cellular and in vivo level, in *Vectors as Tools for the Study of Normal and Abnormal Growth and Differentiation,* Lother, H., Dernick, R., and Ostertag, W., Eds., Springer-Verlag, Berlin, 1989, 229.

28. **Kilbourn, R., Gross, S., Jubran, A., Adams, J., Griffith, O., Levi, R., and Lodato, R.,** N^G-Methyl-L-arginine inhibits tumor necrosis factor-induced hypotension: implications for the involvement of nitric oxide, *Proc. Natl. Acad. Sci. U.S.A.,* 87, 3629, 1990.
29. **Saklatvala, J.,** Tumor necrosis factor alpha stimulates resorption and inhibits synthesis of proteoglycan in cartilage, *Nature (London),* 322, 547, 1986.
30. **Barna, B., Estes, M., Jacobs, B., Hudson, S., and Ransohoff, R.,** Human astrocytes proliferate in response to tumor necrosis factor alpha, *J. Neuroimmunol.,* 30, 239, 1990.
31. **Vilcek, J., Palombella, V., Henriksen-DeStefano, D., Swenson, C., Feinman, R., Hirai, M., and Tsujimoto, M.,** Fibroblast growth enhancing activity of tumor necrosis factor and its relationship to other polypeptide growth factors, *J. Exp. Med.,* 163, 632, 1986.
32. **Bertolini, D., Nedwin, G., Bringman, R., Smith, D., and Mundy, G.,** Stimulation of bone resorption and inhibition of bone formation in vitro by human tumour necrosis factors, *Nature (London),* 319, 516, 1986.
33. **Havell, E.,** Evidence that tumor necrosis factor has an important role in antibacterial resistance, *J. Immunol.,* 143, 2894, 1989.
34. **Liew, F., Parkinson, C., Millott, S., Severn, A., and Carrier, M.,** Tumour necrosis factor (TNF-alpha) in leishmaniasis. I. TNF-alpha mediates host protection against cutaneous leishmaniasis, *Immunology,* 69, 570, 1990.
35. **Butcher, G. and Clark, I.,** The inhibition of Plasmodium falciparum growth in vitro by sera from mice infected with malaria or treated with TNF, *Parasitology,* 101, 321, 1990.
36. **Cerami, A. and Beutler, B.,** The role of cachectin/TNF in endotoxic shock and cachexia, *Immunol. Today,* 9, 28, 1988.
37. **Tracey, K., Wei, H., Manogue, K., Fong, Y., Hesse, D., Nguyen, H., Kuo, G., Beutler, B., Cotran, R., Cerami, A., and Lowry, S.,** Cachectin/tumor necrosis factor induces cachexia, anemia and inflammation, *J. Exp. Med.,* 167, 1211, 1988.
38. **Grau, G., Taylor, T., Molyneux, M., Wirima, J., Vassalli, P., Homel, M., and Lambert, P.,** Tumor necrosis factor and disease severity in children with falciparum malaria, *N. Engl. J. Med.,* 320, 1586, 1989.
39. **Piguet, P., Grau, G., Allet, B., and Vassalli, P.,** Tumor necrosis factor/cachectin is an effector of skin and gut lesions of the acute phase of graft-vs-host disease, *J. Exp. Med.,* 166, 1280, 1987.
40. **Girardin, E., Grau, G., Dayer, J., Roux-Lombard, P., The J5 study group, and Lambert, P.,** Tumor necrosis factor and interleukin-1 in the serum of children with severe infectious purpura, *N. Engl. J. Med.,* 319, 397, 1988.
41. **Tracey, K. and Cerami, A.,** The role of cachectin/tumor necrosis factor in AIDS, *Cancer Cells,* 1, 62, 1989.
42. **Piguet, P., Collart, M., Grau, G., Sappino, A., and Vassalli, P.,** Requirement of tumour necrosis factor for development of silica-induced pulmonary fibrosis, *Nature (London),* 344, 245, 1990.
43. **Yocum, D., Esparza, L., Durby, S., Benjamin, J., Volz, R., and Scuderi, P.,** Characteristics of tumor necrosis factor production in rheumatoid artritis, *Cell. Immunol.,* 122, 131, 1989.
44. **Jacob, C. and McDevitt, H.,** Tumour necrosis factor-alpha in murine autoimmune lupus nephritis, *Nature (London),* 331, 356, 1988.
45. **Lienard, D., Lejeune, F., Delmotte, J., Renard, N., and Ewalenko, P.,** High-dose recombinant tumor necrosis factor alpha in combination with interferon-gamma and melphalan in isolation perfusion of the limbs for melanoma and sarcoma, *J. Clin. Oncol.,* 10, 1, 1992.
46. **Ranges, G., Zlotnik, A., Espevik, T., Dinarello, C., Cerami, A., and Palladino, M.,** Tumor necrosis factor alpha/cachectin is a growth factor for thymocytes, *J. Exp. Med.,* 167, 1472, 1988.
47. **Plaetinck, G., Declercq, W., Tavernier, J., Nabholz, M., and Fiers, W.,** Recombinant tumor necrosis factor can induce interleukin 2 receptor expression and cytolytic activity in a rat × mouse T cell hybrid, *Eur. J. Immunol.,* 17, 1835, 1987.
48. **Ranges, G., Bombara, R., Aiyer, R., Rice, G., and Palladino, M.,** Tumor necrosis factor as a proliferative signal for an IL-2-dependent T cell line: strict species specificity of action, *J. Immunol.,* 142, 1203, 1989.
49. **Brouckaert, P., Libert, C., Everaerdt, B., and Fiers, W.,** Selective species specificity of tumor necrosis factor for toxicity in the mouse, *Lymphokine Cytokine Res.,* 11, 193, 1992.
50. **Everaerdt, B., Brouckaert, P., Shaw, A., and Fiers, W.,** Four different interleukin-1 species sensitize to the lethal action of tumour necrosis factor, *Biochem. Biophys. Res. Commun.,* 163, 378, 1989.
51. **Brouckaert, P., Everaerdt, B., and Fiers, W.,** The glucocorticoid antagonist RU38486 mimics interleukin-1 in its sensitization to the lethal and interleukin-6-inducing properties of tumor necrosis factor, *Eur. J. Immunol.,* 22, 887, 1992.
52. **Brouckaert, P. G. G., Leroux-Roels, G. G., Guisez, Y., Tavernier, J., and Fiers, W.,** In vivo antitumour activity of recombinant human and murine TNF, alone and in combination with murine IFN-gamma, on a syngeneic murine melanoma, *Int. J. Cancer,* 38, 763, 1986.
53. **Rothwell, N.,** The endocrine significance of cytokines, *J. Endocrinol.,* 128, 171, 1991.

54. **Defilippi, P., Poupart, P., Tavernier, J., Fiers, W., and Content, J.**, Induction and regulation of mRNA encoding 26-kDa protein in human cell lines treated with recombinant human tumor necrosis factor, *Proc. Natl. Acad. Sci. U.S.A.*, 84, 4557, 1987.
55. **Jirik, F., Podor, T. J., Hirano, T., Kishimoto, T., Loskutoff, D., Carson, D., and Lotz, M.**, Bacterial lipopolysaccharide and inflammatory mediators augment IL-6 secretion by human endothelial cells, *J. Immunol.*, 142, 144, 1989.
56. **Libert, C., Brouckaert, P., Shaw, A., and Fiers, W.**, Induction of interleukin-6 by human and murine recombinant interleukin-1 in mice, *Eur. J. Immunol.*, 20, 691, 1990.
57. **Brouckaert, P. G., Libert, C., Everaerdt, B., Takahashi, N., and Fiers, W.**, A role for interleukin-1 in the *in vivo* actions of tumor necrosis factor, *Lymphokine Res.*, 8, 269, 1989.
58. **Libert, C., Vink, A., Coulie, P., Brouckaert, P. G., Everaerdt, B., Van Snick, J., and Fiers, W.**, Limited involvement of interleukin-6 in the pathogenesis of lethal sepsis as revealed by the effect of monoclonal antibodies against itnerleukin-6 or its receptor in various murine models, *Eur. J. Immunol.*, 22, 2625, 1992.
59. **Starnes, H. F., Pearce, M. K., Tewari, A., Yim, J. H., Zou, J. C., and Abrams, J. S.**, Anti-IL-6 monoclonal antibodies protect against lethal *E. coli* infection and lethal tumor necrosis factor-alpha challenge in mice, *J. Immunol.*, 145, 4185, 1990.
60. **Libert, C., Brouckaert, P., and Fiers, W.**, The influence of modulating substances on tumor necrosis factor and interleukin-6 levels after injection of murine tumor necrosis factor or lipopolysaccharide in mice, *J. Immunother.*, 10, 227, 1991.
61. **Brouckaert, P., Spriggs, D. R., Demetri, G., Kufe, D. W., and Fiers, W.**, Circulating interleukin-6 during a continuous infusion of tumor necrosis factor and interferon-gamma, *J. Exp. Med.*, 169, 2257, 1989.
62. **Gresser, I., Delers, F., Tran Quangs, N., Marion, S., Engler, R., Maury, C., Soria, C., Soria, J., Fiers, W., and Tavernier, J.**, Tumor necrosis factor induces acute phase proteins in rats, *J. Biol. Regul. Homeost. Agents*, 1, 173, 1987.
63. **Andus, T., Heinrich, P., Bauer, J., Tran-Thi, T., Decker, K., Mannel, D., and Northoff, H.**, Discrimination of hepatocyte-stimulating activity from human recombinant tumor necrosis factor, *Eur. J. Immunol.*, 17, 1193, 1987.
64. **Baumann, H., Prowse, K., Marinkovic, S., Won, K., and Jahreis, G.**, Stimulation of hepatic acute phase response by cytokines and glucocorticoids, *Ann. N.Y. Acad. Sci.*, 557, 280, 1989.
65. **Perlmutter, D., May, L., and Sehgal, P.**, Interferon beta 2/interleukin 6 modulates synthesis of alpha-1-antitrypsin in human mononuclear phagocytes and in human hepatoma cells, *J. Clin. Invest.*, 84, 138, 1989.
66. **Yap, S., Moshage, H., Hazenberg, B., Roelofs, H., Bijzet, J., Limburg, P., Aarden, L., and Van Rijswijk, M.**, Tumor necrosis factor (TNF) inhibits interleukin (IL)-1 and/or IL-6 stimulated synthesis of C-reactive protein (CRP) and serum amyloid A (SAA) in primary cultures of human hepatocytes, *Biochim. Biophys. Acta*, 1091, 405, 1991.
67. **Baumann, H. and Gauldie, J.**, Regulation of hepatic acute phase plasma protein genes by hepatocyte stimulating factors and other mediators of inflammation, *Mol. Biol. Med.*, 7, 147, 1990.
68. **Ramadori, G., Van Damme, J., Rieder, H., and Meyer zum Buschenfelde, K.**, Interleukin-6, the third mediator of acute phase reaction, modulates hepatic protein synthesis in human and mouse. Comparison with interleukin-1 beta and tumor necrosis factor-alpha, *Eur. J. Immunol.*, 18, 1259, 1988.
69. **Richie, D. and Zuckerman, S.**, Restoration of the LPS responsive phenotype in C3H/HeJ macrophage hybrids: LPS regulation of hepatocyte stimulating factor production, *Immunology*, 61, 429, 1987.
70. **Tsuchiya, Y., Hattori, M., Hayashida, K., Ishibashi, H., Okubo, H., and Sakaki, Y.**, Sequence analysis of the putative regulatory region of rat alpha-2-macroglobulin, *Gene*, 57, 73, 1987.
71. **Heinrich, P., Castell, J., and Andus, T.**, Interleukin-6 and the acute phase response, *Biochem. J.*, 265, 621, 1990.
72. **Van Bladel, S., Libert, C., and Fiers, W.**, Interleukin-6 enhances the expression of tumor necrosis factor receptors on hepatoma cells and hepatocytes, *Cytokine*, 3, 149, 1991.
73. **Neta, R., Vogel, S., Sipe, J., Wong, G., and Nordan, R.**, Comparison of in vivo effects of human recombinant IL 1 and human recombinant IL 6 in mice, *Lymphokine Res.*, 7, 403, 1988.
74. **Mortensen, R., Shapiro, J., Lin, B., Douches, S., and Neta, R.**, Interaction of recombinant IL 1 and recombinant tumor necrosis factor in the induction of mouse acute phase proteins, *J. Immunol.*, 140, 2260, 1988.
75. **Van Damme, J. and Van Snick, J.**, Induction of hybridoma growth factor (HGF), identical to IL-6, in human fibroblasts by IL-1: use of HGF activity in specific and sensitive biological assays for IL-1 and IL-6, *Dev. Biol. Stand.*, 69, 31, 1988.
76. **Moldawer, L., Andersson, C., Gelin, J., and Lundholm, K.**, Regulation of food intake and hepatic protein synthesis by recombinant derived cytokines, *Am. J. Physiol.*, 254, G450, 1988.

77. **Feingold, K., Barker, M., Jones, A., and Grunfeld, C.**, Localization of tumor necrosis factor-stimulated DNA synthesis in the liver, *Hepatology*, 13, 773, 1991.
78. **Bertini, R., Bianchi, M., Villa, P., and Ghezzi, P.**, Depression of liver drug metabolism and increase in plasma fibrinogen by interleukin 1 and tumor necrosis factor: a comparison with lymphotoxin and interferon, *Int. J. Immunopharmacol.*, 10, 525, 1988.
79. **Argiles, J., Lopez-Soriano, F., Wiggins, F., and Williamson, D.**, Comparative effects of tumour necrosis factor-alpha (cachectin), interleukin-1-beta and tumour growth on amino acid metabolism in the rat in vivo: absorption and tissue uptake of α-amino[1-^{14}C]isobutyrate, *Biochem. J.*, 261, 357, 1989.
80. **Bereta, J., Kurdowska, A., Koj, A., Hirano, T., Kishimoto, T., Content, J., Fiers, W., Sehgal, P., Van Damme, J., and Gauldie, J.**, Different preparations of natural and recombinant human interleukin-6 (IFN-β2, BSF-2) similarly stimulate acute phase protein synthesis and uptake of α-aminoisobutyric acid by cultured rat hepatocytes, *Int. J. Biochem.*, 21, 361, 1989.
81. **Endo, Y.**, Induction of histidine and ornithine decarboxylase activities in mouse tissues by recombinant interleukin-1 and tumor necrosis factor, *Biochem. Pharmacol.*, 38, 1287, 1989.
82. **Stith, R., McCallum, R., and Hill, M.**, Effect of interleukin-6/interferon β-2 on glucocorticoid action in rat hepatoma cells, *J. Steroid Biochem.*, 34, 479, 1989.
83. **Grunfeld, C. and Feingold, K.**, The metabolic effects of TNF and other cytokines, *Biotherapy*, 3, 143, 1991.
84. **Brouckaert, P., Everaerdt, B., Libert, C., Takahashi, N., and Fiers, W.**, Species specificity and involvement of other cytokines in endotoxic shock action of recombinant tumor necrosis factor in mice, *Agents Actions*, 26, 196, 1989.
85. **Galanos, C., Freundenberg, M., and Reutter, W.**, Galactosamine-induced sensitization to the lethal effects of endotoxin, *Proc. Natl. Acad. Sci. U.S.A.*, 76, 5939, 1979.
86. **Decker, K. and Keppler, D.**, Galactosamine hepatitis: key role of the nucleotide deficiency period in the pathogenesis of cell injury and cell death, *Rev. Physiol. Biochem. Pharmacol.*, 71, 77, 1974.
87. **Freundenberg, M., Keppler, D., and Galanos, C.**, Requirement for lipopolysaccharide-responsive macrophages in galactosamine-induced sensitization to endotoxin, *Infect. Immun.*, 51, 891, 1986.
88. **Petkova, D., Momchilova, A., Markovska, T., and Koumanov, K.**, D-Galactosamine induced changes in rat liver plasma membrane. Lipid composition and some enzyme activities, *Int. J. Biochem.*, 19, 289, 1987.
89. **Cawson, G., Sesno, J., Milam, K., and Wang, Y.**, The hepatocyte protein synthesis defect induced by galactosamine involves hypomethylation of ribosomal RNA, *Hepatology*, 11, 428, 1990.
90. **Tiegs, G., Niehorster, M., and Wendel, A.**, Leukocyte alterations do not account for hepatitis induced by endotoxin or TNFα in galactosamine-sensitized mice, *Biochem. Pharmacol.*, 40, 1317, 1990.
91. **Tiegs, G. and Wendel, A.**, Leukotriene-mediated liver injury, *Biochem. Pharmacol.*, 37, 2569, 1988.
92. **Hishinuma, I., Nagakawa, J., Hirota, K., Miyamoto, K., Tsukidate, K., Yamanaka, T., Katayama, K., and Yamatsu, I.**, Involvement of tumor necrosis factor-α in development of hepatic injury in galactosamine-sensitized mice, *Hepatology*, 12, 1187, 1990.
93. **Wendel, A., Tiegs, G., and Werner, C.**, Evidence for the involvement of a reperfusion injury in galactosamine/endotoxin-induced hepatitis in mice, *Biochem. Pharmacol.*, 36, 2637, 1987.
94. **Fukushima, H., Ikeuchi, J., Tohkin, M., Matsubara, T., and Harada, M.**, Lethal shock in partially hepatectomized rats administered tumor necrosis serum, *Circ. Shock*, 26, 1, 1988.
95. **Takahashi, N., Brouckaert, P., and Fiers, W.**, Induction of tolerance allows separation of lethal and antitumor activities of tumor necrosis factor in mice, *Cancer Res.*, 51, 2366, 1991.
96. **Wallach, D., Holtmann, H., Engelmann, H., and Nophar, Y.**, Sensitization and desensitization to lethal effects of tumor necrosis factor and IL-1, *J. Immunol.*, 140, 2994, 1988.
97. **Libert, C., Van Bladel, S., Brouckaert, P., Shaw, A., and Fiers, W.**, Involvement of the liver, but not of IL-6, in IL-1-induced desensitization to the lethal effects of tumor necrosis factor, *J. Immunol.*, 146, 2625, 1991.
98. **Silverstein, R., Turly, B., Christoffersen, C., Johnson, D., and Morrison, D.**, Hydrazine sulfate products D-galactosamine-sensitized mice against endotoxin and tumor necrosis factor/cachectin lethality: evidence of a role for the pituitary, *J. Exp. Med.*, 173, 357, 1991.
99. **Morikage, T., Mizushima, Y., and Yano, S.**, Prevention of bacterial infection and LPS-induced lethality by interleukin-1α in mice, *Tohoku J. Exp. Med.*, 163, 47, 1991.
100. **Scuderi, P., Dorr, R. T., Liddil, J., Finley, P. R., Meltzer, P., Raitano, A. B., and Rybski, J.**, Alpha-globulins suppress human leukocyte tumor necrosis factor secretion, *Eur. J. Immunol.*, 19, 939, 1989.
101. **Niehörster, M., Tiegs, G., Schade, U., and Wendel, A.**, In vivo evidence for protease-catalysed mechanism providing bioactive tumor necrosis factor-α, *Biochem. Pharmacol.*, 40, 1601, 1990.
102. **Deviere, J., Content, J., Denys, C., Vandenbussche, P., Schandene, L., Wybran, J., and Dupont, E.**, Excessive in vitro bacterial lipopolysaccharide-induced production of monokines in cirrhosis, *Hepatology*, 11, 628, 1990.

103. **Liao, Z., Grimshaw, R., and Rosenstreich, D.,** Identification of a specific interleukin 1 inhibitor in the urine of febrile patients, *J. Exp. Med.,* 159, 126, 1984.
104. **Alexander, H., Doherty, G., Buresh, C., Venzon, D., and Norton, J.,** A recombinant human receptor antagonist to interleukin 1 improves survival after lethal endotoxemia in mice, *J. Exp. Med.,* 173, 1029, 1991.
105. **Ohlsson, K., Bjork, P., Bergenfeldt, M., Hageman, R., and Thompson, R.,** Interleukin-1 receptor antagonist reduces mortality from endotoxin shock, *Nature (London),* 348, 550, 1990.
106. **Fleck, A.,** Acute phase response: implications for nutrition and recovery, *Nutrition,* 4, 109, 1988.
107. **Tiegs, G., Werner, C., Wolter, M., and Wendel, A.,** Involvement of reactive oxygen species in endotoxin-induced hepatitis in galactosamine-sensitized mice, in *Proceedings of the 4th Biennial General Meeting of the Society for Free Radical Research,* Hayaishi, E., Niki, E., Kondo, M., and Yoshikawa, T., Eds., Elsevier, Amsterdam, 1988, 1379.
108. **James, K.,** Interactions between cytokines and α2-macroglobulin, *Immunol. Today,* 11, 163, 1990.
109. **Bories, P., Guenounou, M., Feger, J., Kodari, E., Agneray, J., and Durand, G.,** Human α1-acid glycoprotein-exposed macrophages release interleukin 1 inhibitory activity, *Biochem. Biophys. Res. Commun.,* 147, 710, 1987.
110. **Kurdowska, A. and Travies, J.,** Acute phase protein stimulation by α1-antichymotrypsin-cathepsin G complexes/evidence for the involvement of interleukin-6, *J. Biol. Chem.,* 265, 21023, 1990.
111. **Scuderi, P.,** Suppression of human leukocyte tumor necrosis factor secretion by the serine protease inhibitor p-toluenesulfonyl-L-arginine methyl ester (TAME), *J. Immunol.,* 143, 168, 1989.
112. **Libert, C.,** unpublished results.

Chapter 20

INTERLEUKIN-6 RECEPTOR

Stefan Rose-John and Peter C. Heinrich

TABLE OF CONTENTS

0-8493-6913-4/93/$0.00 + $.50

I. INTRODUCTION

Interleukin-6 (IL-6) is synthesized and secreted by many different cells after appropriate stimulation (Table 1). As shown in Figure 1, IL-6 is a multifunctional cytokine acting on many different cells. It is involved in (1) the induction of immunoglobulin production in activated B-cell,[1,2] (2) the induction of proliferation of hybridoma/plasmacytoma/myeloma cells,[3-6] (3) the induction of IL-2 production, cell growth, and cytotoxic T-cell differentiation of T-cells,[7-9] (4) the stimulation of multipotent colony formation in hematopoietic stem cells,[10] (5) the regulation of acute phase proteins,[11,12] (6) growth inhibition and induction of differentiation into macrophages of myeloid leukemic cell lines,[13] and (7) the induction of neural differentiation.[14]

IL-6 confers its signal to target cells by binding to IL-6-specific cell-surface receptors. A cDNA coding for an IL-6 receptor molecule has been cloned from the human natural killer-like cell line YT.[15] The IL-6 receptor cDNA encodes a protein consisting of 468 amino acids, including a signal peptide of 19 amino acids. The 90-amino acid-long N-terminal part of the extracellular domain shows homology to the immunoglobulin superfamily. The cytoplasmic part of 82 amino acids lacks a tyrosine kinase domain, unlike other growth factor receptors. The deletion of the cytoplasmic and transmembrane domains of the 80-kDa subunit of the IL-6 receptor showed that neither was required for IL-6 signaling.[16] This finding led to the discovery of the second subunit needed for the signal transduction of IL-6, a protein of a molecular weight of 130 kDa (gp130).[16]

Taga et al.[16] clearly demonstrated an IL-6-triggered aggregation of the two subunits of the IL-6 receptor. The cDNA coding for the gp130 has been cloned[17] and predicts a signal peptide of 22 amino acids, an extracellular region of 597 amino acids, a membrane-spanning region of 22 amino acids, and a cytoplasmic domain of 277 amino acids. Interestingly, the extracellular domain of gp130 contains six fibronectin type III modules. It should be noted that expression studies of gp130 revealed that the gp130 mRNA is not only present in IL-6-responsive cells, but in all cells tested so far.[17] Figure 2 schematically shows how IL-6 and the two receptor subunits are believed to interact.[16,17] IL-6 first binds to the 80-kDa receptor subunit, possibly resulting in a conformational change of the receptor. Subsequently, the complex of IL-6/gp80 interacts with the signal transducing subunit gp130, leading to the generation of a signal.

In this chapter, we present data on the structural and functional analysis of IL-6 and its hepatic receptor. Furthermore, we discuss experiments on the synthesis and biological role of a soluble form of the 80-kDa IL-6 receptor subunit.

II. MATERIALS AND METHODS

A. CHEMICALS AND ENZYMES

Enzymes were purchased from Boehringer Mannheim (Mannheim, Germany). Radiochemicals were obtained from Amersham Int. (Amersham, U.K.). Geneticin (G418 sulfate) was purchased from GIBCO (Eggenstein, Germany) and FCS from Seromed (Berlin, Germany). Vent DNA Polymerase was from Biolabs (Schwalbach, Germany), and α-2,3- and α-2,6-sialyltransferase were obtained from Dr. H. Conradt (GBF, Braunschweig, Germany).

B. CELL CULTURES

Hep G2, NIH/3T3, and COS-7 cells were grown in Dulbecco's modified Eagle's medium (DMEM) supplemented with 10% fetal calf serum (FCS) at 37°C in a humid atmosphere. B9 cells were obtained from Dr. L. Aarden (Amsterdam) and grown as described.[18]

TABLE 1
IL-6-Producing Cells

Cells	Major stimulator
Monocytes/macrophages	LPS
Fibroblasts	IL-1
Endothelial cells	IL-1
Chondrocytes	IL-1
Endometrial stromal cells	IL-1
Smooth muscle cells	IL-1
Astrocytes	IL-1

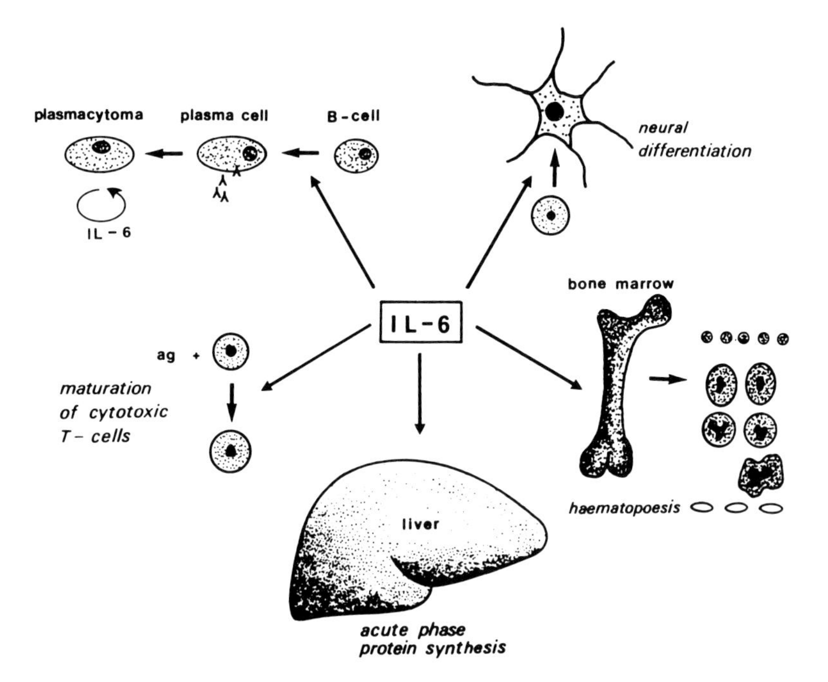

FIGURE 1. Pleiotropic actions of IL-6. (From Heinrich, P.C., Dufhues, G., Flohe, S., Horn, F., Krause, E., Krüttgen, A., Legres, L., Leuz, D., Lütticken, C., Schooltink, H., Stoyan, T., Conradt, H., and Rose-John, S., *Molecular Aspects of Inflammation,* Springer-Verlag, Heidelberg, 1991. With permission.)

C. MOLECULAR BIOLOGY METHODS

All cloning, subcloning, sequencing, and DNA amplification procedures, construction of deletion mutants, and site-directed mutagenesis were carried out using standard methods.[19]

D. TRANSFECTION OF CELLS

Cells were transfected using the calcium phosphate method, as described.[20] Stably transfected cell clones were selected in the presence of 100 or 160 μg/ml G 418 for Hep G2 or NIH/3T3 cells, respectively.

E. NORTHERN-BLOT ANALYSIS

Total RNA was prepared using the phenol extraction method as previously described;[20] 5 μg of RNA was used for Northern-blot analysis. Hybridization was performed with random primed cDNA fragments.

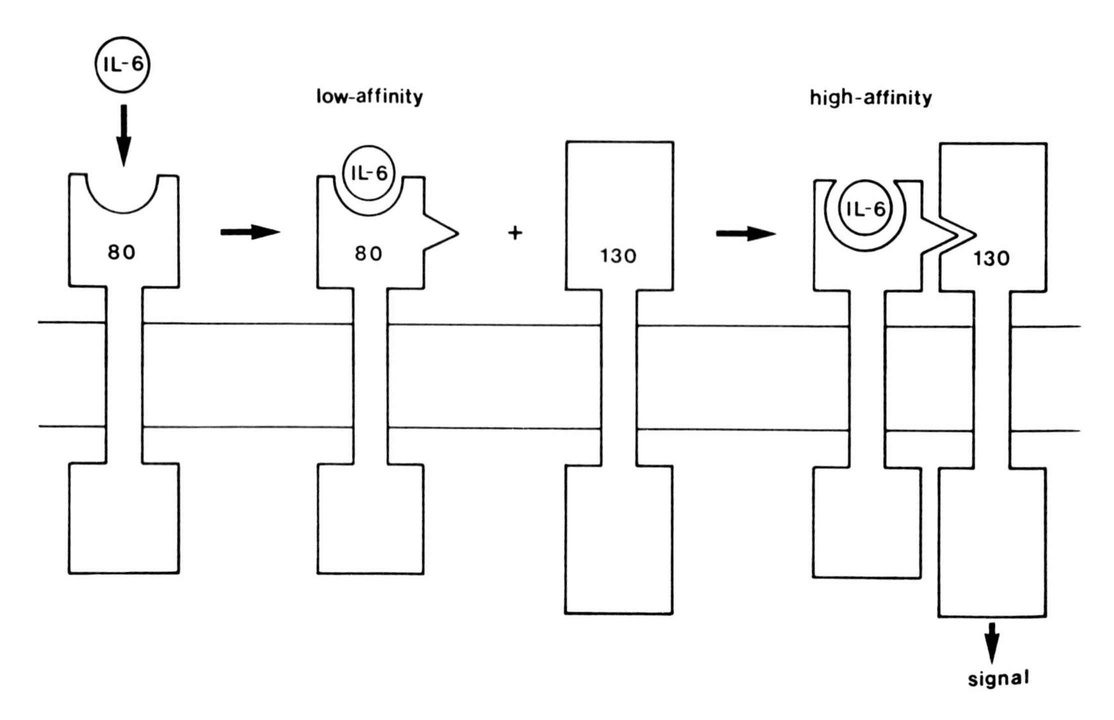

FIGURE 2. IL-6-induced aggregation of the two IL-6 receptor subunits.

F. IODINATION OF IL-6

Recombinant human IL-6 (rhIL-6) was iodinated according to the procedure of Markwell,[21] with modifications as previously described.[20] A specific activity of 250 kBq/μg was obtained. The biological activity of the iodinated IL-6 essentially was unchanged.

G. BINDING AND INTERNALIZATION STUDIES

Binding studies were performed in 24 multiwell dishes using [^{125}I]rhIL-6 at different concentrations (1 to 2000 p*M*). Nonspecific binding was determined by incubating the cells with [^{125}I]rhIL-6 and a 100-fold excess of nonlabeled ligand. For internalization studies, cells were preloaded with [^{125}I]rhIL-6 at 4°C for 2 h, followed by a temperature shift to 37°C. IL-6 was considered to be internalized when it could not be removed from the cells by a wash with 0.5 *M* NaCl, pH 1.0, for 3 min.

H. AFFINITY CROSS-LINKING

[^{125}I]rhIL-6 was cross-linked to Hep G2 cells using the homobifunctional cross-linker BSS (Pierce, Rockford, IL), as described.[20] Membrane proteins were separated by sodium dodecyl sulfate-polyacrylamide gel electrophoresis (SDS-PAGE) and cross-linked complexes were visualized by autoradiography.

I. PROTEIN DETERMINATION

α_1-Antichymotrypsin and haptoglobin secreted by Hep G2 cells and accumulated in the culture medium were measured by an electroimmunoassay as described.[22,23]

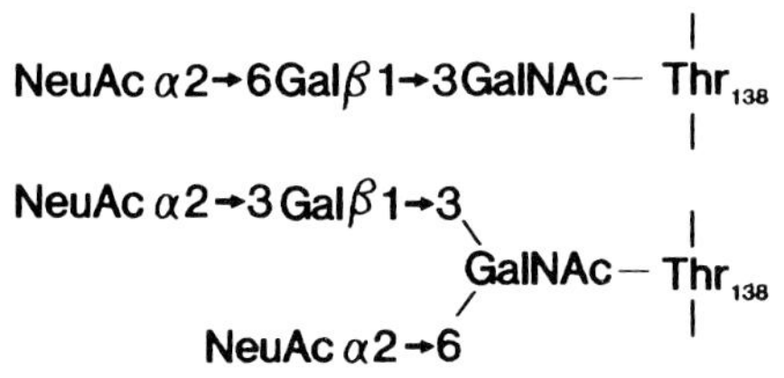

NeuAc α2→6Galβ1→3GalNAc— Thr_{138}

NeuAc α2→3 Galβ1→3
GalNAc— Thr_{138}
NeuAc α2→6

FIGURE 3. Predicted structures of the *O*-linked carbohydrate side chains derived from a component analysis. The *O*-carbohydrate-carrying peptide was subjected to methanolysis, reacetylation, and trimethylsilylation. Monosaccharide derivatives were analyzed by gas liquid chromatography.

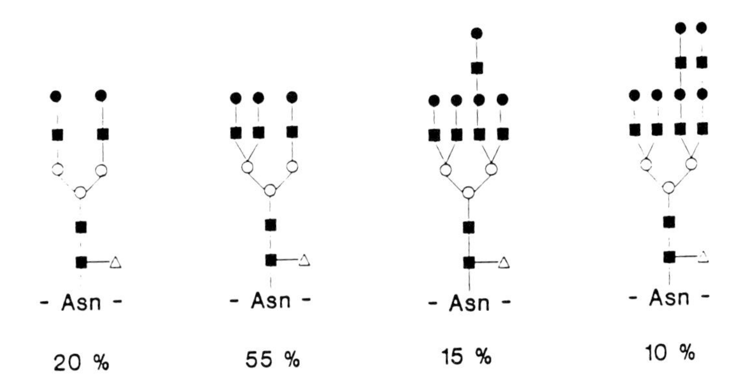

FIGURE 4. Oligosaccharide structures of the *N*-linked carbohydrate side chain of human IL-6 from mouse fibroblasts transfected with human IL-6 cDNA.[29] ●, GlcNAc; ○, man; ■, gal; △, fuc.

III. RESULTS AND DISCUSSION

A. INTERLEUKIN-6

1. Glycosylation of IL-6

IL-6 is synthesized as a precursor with an extra-amino-terminal extension of 28 amino acids. The mature protein consists of 184 amino acids and contains two sequential disulfide bridges.[24] IL-6 secreted by mammalian cells is *N*- and *O*-glycosylated.[25-27] We have determined the *N*- and *O*-glycosylation sites of native IL-6. After the removal of all peripheral sialic acid residues, resialylation with *O*- and *N*-glycane-specific sialyltransferases and radioactively labeled CMP-sialic acid, digestion with trypsin, separation of the tryptic peptides by HPLC chromatography, and identification of the radioactively labeled peptides by Edman degradation, Asp_{45} and Thr_{138} were identified as the only *N*- and *O*-glycosylation sites, respectively. The structure of the *O*-linked oligosaccharide side chain is Gal-β1-3GalNAc, carrying one or two sialic acid residues[28] (Figure 3). For the N-linked oligosaccharide side chain, a high degree of heterogeneity was found (Figure 4). Comparison of the biological activities of glycosylated, desialylated, and deglycosylated IL-6 using an IL-6-specific proliferation assay (B9 cells) and, in addition, the γ-fibrinogen induction assay with Hep G2 cells showed glycosylated IL-6 to be three to four times as active as the deglycosylated cytokine (Table 2). Interestingly, desialylated IL-6 is as active as the glycosylated cytokine in the B9 proliferation assay, but only 60% of the biological activity of native IL-6 was found in the γ-fibrinogen induction assay, very likely due to the presence of asialoglycoprotein receptors on liver cells. These receptors have been shown to rapidly internalize glycoproteins carrying peripheral galactose residues.[29] In addition, we have shown that *N*- and *O*-glycosylation of IL-6 prolongs the plasma clearance.[30]

TABLE 2
Effect of Glycosylation on the Specific Biological Activity of Human IL-6

	Biol. activity (%)		Half-life
	B9	Hep G2	(min)
Glycosylated IL-6	100	100	5
Desialylated IL-6	98	63	n.d.
Deglycosylated IL-6	33	26	3

2. Structure/Functional Relationship of IL-6

No experimental data on the tertiary structure of human IL-6 are presently available. From the amino acid sequence of human IL-6, a secondary structure of the molecule had been predicted: 58% α-helix, 14% β-structure, and 28% turn and coil.[31,32] It should be noted that the C terminus of human IL-6 exhibits an α-helical structure (see below). A content of 67% α-helix was determined from the circular dichroism spectrum of rhIL-6,[31] verifying the high content predicted.

Once the primary structure of human IL-6 had been deduced from the cDNA sequence, the question was asked, which parts of the IL-6 molecule are indispensable for its biological function. Brakenhoff et al.[33] showed that 28 amino acids can be removed from the N terminus without affecting the biological activity of IL-6. Further removal of amino acids 29 and 30 resulted in an approximately 50-fold decrease of activity, whereas the deletion of amino acids 31 to 34 completely abolished the activity. The authors concluded from their study that the amino acids starting from 29 are important for the function of IL-6.

Experiments from our laboratory have shown that in contrast to the N terminus, the far C-terminal end of IL-6 is of particular importance for its biological function. Stepwise truncation of the C-terminal amino acids of human IL-6 resulted in a stepwise loss of biological activity (Figure 5). Removal of $methionine_{184}$ led to an 80% decrease of biological activity. No further change was detected after deletion of $glutamine_{183}$, whereas biological activity was completely abrogated when $methionine_{184}$, $glutamine_{183}$, and $arginine_{182}$ were removed from the C-terminal end of IL-6.[31,32,34]

When point mutations were introduced into the full-length IL-6 molecule, evidence for the importance of a positive charge (Arg_{182}) and an α-helical structure of the C terminus for biological activity of human IL-6 was obtained.[35]

3. IL-6 Chimeras

We asked the question, which domain(s) in human IL-6 is (are) important for the interaction of human IL-6 with the human IL-6 receptor. To answer this question, various regions of the human IL-6 were replaced by murine sequences. The reason for the construction of chimeras of mouse and human IL-6 was based on the observation that human IL-6 is recognized by the human as well as by the mouse IL-6 receptor, whereas murine IL-6 exerts its action only on the mouse IL-6 receptor. The mouse/human chimeras constructed are shown in Figure 6. All chimeras were active in the murine B9 proliferation assay. Murine IL-6, however, is inactive in the human fibrinogen induction assay with Hep G2 cells and also in a binding assay, where iodinated human IL-6 and soluble human IL-6-R were used to test the chimeras for their capability to compete with iodinated human IL-6 for binding to the soluble human receptor (third column). When 39 amino acids from the N terminus of human IL-6 were replaced by the murine sequence, the chimeric molecule was active in all three assays. When we substituted amino acids 40 to 95 in human IL-6 with the corre-sponding murine sequence, we found that the activity in the human assay systems (fibrinogen

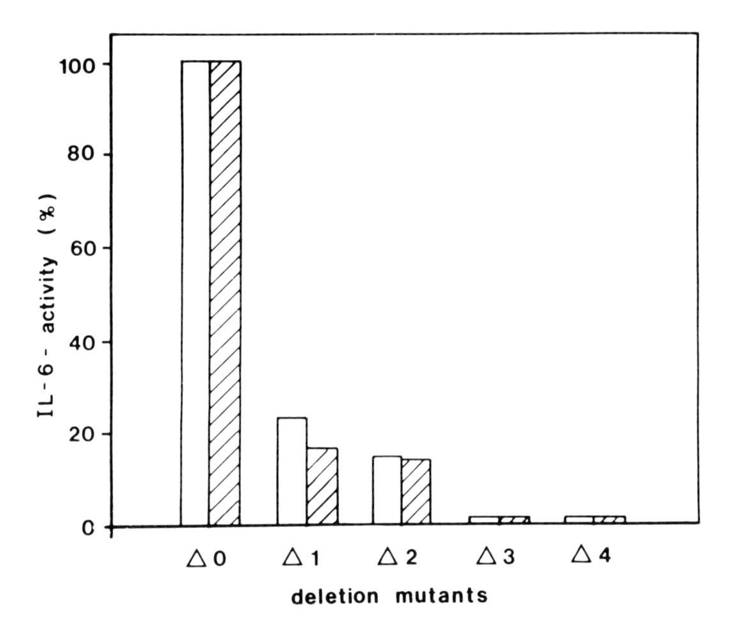

FIGURE 5. Comparison of the biological activity of IL-6 mutants as measured in the B9 and Hep G2 cell assay. Values, normalized for the different amounts of IL-6 synthesized, were compared using the dilution leading to half-maximal stimulation in the respective test. The biological activity exerted by full-length IL-6 (△, 0) was set to 100%. Open bars, B9 cell proliferation assay; hatched bars, Hep G2 cell fibrinogen assay.

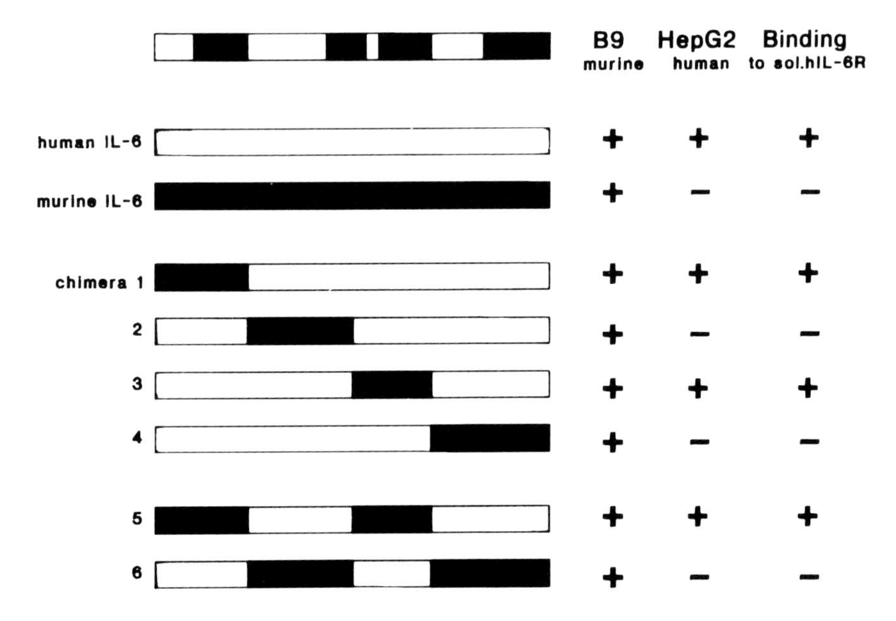

FIGURE 6. Chimeras of mouse and human IL-6. Chimeric IL-6 cDNAs were expressed in *E. coli*, refolded, purified, and tested in three different assay systems (murine B9 cell proliferation, human γ-fibrinogen induction in Hep G2 cells, and binding assay with a soluble human IL-6-R).

induction and binding assay) was lost, indicating that this domain in the human IL-6 molecule must be involved in the recognition of human IL-6 by the human receptor. When amino acids 96 to 134 were replaced in human IL-6 by the corresponding mouse sequence, no effect on the induction of fibrinogen in Hep G2 cells as well as in the binding assay was

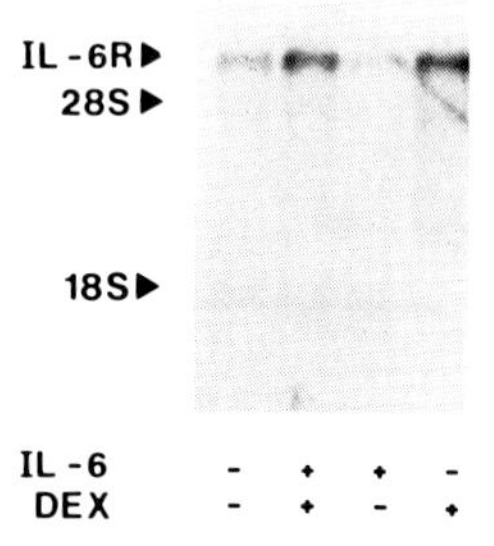

FIGURE 7. Dose dependence of gp80 mRNA expression in Hep G2 cells treated with IL-6 and dexamethasone. Hep G2 cells were treated for 18 h with 10^{-6} *M* dexamethasone and 100 U/ml of IL-6 as indicated in the figure. Northern-blot analysis was hybridized with a gp80 cDNA probe labeled by random priming.

found. However, replacement of the C-terminal part of human IL-6 by amino acids 135 to 184 of murine IL-6 led to the complete loss of biological activity in both human assays, indicating that this region is also important for the interaction of human IL-6 with its receptor. We conclude that the second and fourth domains of IL-6 are responsible for species specificity and therefore also for receptor recognition. This interpretation is further corroborated by chimeras in which double exchanges have been performed. A molecule carrying the second and fourth domains of the murine protein behaves like murine IL-6 (chimera 6), whereas the reverse exchange results in a molecule of human character (chimera 5).

4. Plasma Clearance, Carrier Proteins, and Target Cells for IL-6

When [^{35}S]methionine- or ^{125}I-labeled human IL-6 was intravenously injected into rats, a very rapid biphasic disappearance from the circulation was observed. A rapid initial elimination corresponding to a half-life of about 3 min and a second, slower decrease corresponding to a half-life of about 55 min have been estimated.[36] Twenty minutes after intravenous injection, about 80% of the [^{125}I]IL-6 had disappeared from the circulation and was found in the liver. Autoradiography showed that [^{125}I]IL-6 was exclusively localized on the surface of parenchymal cells, suggesting the existence of IL-6 receptors on the hepatocytes.

B. INTERLEUKIN-6 RECEPTOR

In view of the fact that the biological responses of various target cells of IL-6 differ, we asked whether different IL-6 receptors are expressed on various cells and tissues. The cDNA cloning of the IL-6 receptor from human[37] and rat[38] liver revealed that liver cells and leukocytes[15] express the same type of IL-6 receptor, indicating that different biological responses are obtained via an identical receptor.

1. Regulation of the Hepatic IL-6 Receptor Subunits

It has been shown that acute phase protein synthesis in hepatocytes requires IL-6 and dexamethasone.[11,12,39,40] It was therefore of interest to examine whether Hep G2 cells can be stimulated by IL-6 and dexamethasone via changes in the hepatic IL-6 receptor mRNA. We could clearly show that only dexamethasone induced gp80 mRNA synthesis in Hep G2 cells, whereas IL-6 had no effect (Figure 7).[37,41] In contrast, gp130 mRNA synthesis could be stimulated by IL-6 and particularly by IL-6/dexamethasone treatment of Hep G2 cells (Figure 8).[42]

The stimulation of gp80 mRNA synthesis by dexamethasone was also found at the level of the functional 80-kDa IL-6 receptor. We treated Hep G2 cells with 10^{-7} *M* dexamethasone for the times indicated in Figure 9. Cells were then incubated with iodinated human IL-6, cross-linked with a noncleavable bifunctional cross-linker, immunoprecipitated with an IL-6

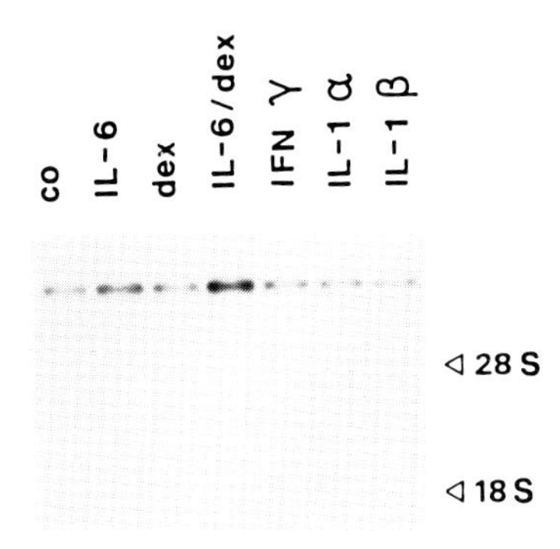

FIGURE 8. Gp130 mRNA expression in Hep G2 cells treated with various cytokines and dexamethasone. Hep G2 cells were treated for 18 h with 100 U/ml rhIL-6, 10^{-6} *M* dexamethasone, 100 U/ml IL-1α, 100 U/ml IL-1β, or 100 U/ml IFNγ. Total RNA was isolated and subjected to Northern-blot analysis. RNAs were probed with a gp130 cDNA labeled by random priming. (From Schooltink, H., Schmitz-Van de Leur, H., Heinrich, P. C., and Rose-John, S., *FEBS Lett.*, 297, 263, 1992. With permission.)

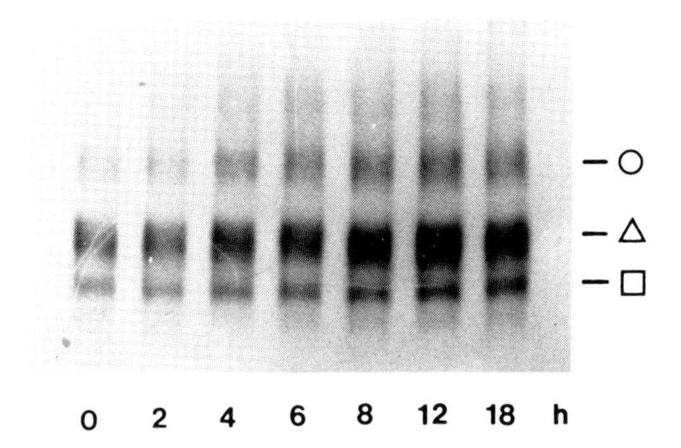

FIGURE 9. Time course of IL-6-binding protein induction by dexamethasone. Affinity cross-linking of [^{125}I] rhIL-6 to Hep G2 cells incubated with 10^{-7} *M* dexamethasone for the times indicated in the figure. Autoradiography of the cross-linked IL-6-binding proteins after separation by SDS/PAGE.

antiserum, and subjected to SDS-PAGE and autoradiography. We found that dexamethasone also led to an increase in the synthesis of functional IL-6 receptor protein. The cross-linking resulted in the detection of three IL-6-containing complexes with apparent molecular masses of 100 (□), 120 (△), and 200 kDa (○). We now have evidence that the 100-kDa band corresponds to gp80 + IL-6, the 120-kDa band to gp130 + IL-6, and the 200-kDa band to gp80 + gp 130 + IL-6.[59] Since the gp80, but not the gp130, mRNA is upregulated by dexamethasone, we conclude that the gp80 levels are rate limiting for the formation of the ternary receptor complex (compare Figure 2).

The upregulation of the three IL-6-containing complexes is even more pronounced when, in addition to dexamethasone, PMA is added to the culture medium.[59]

2. Binding and Internalization of IL-6

The number of IL-6 receptors found on different cells is generally low (between several hundred and several thousand per cell). Two types of binding sites with dissociation constants of 10 to 30 p*M* and 700 p*M* for high- and low-affinity binding, respectively, have been identified in several instances.[15,20] We transfected Hep G2 cells with a cDNA coding for the 80-kDa subunit of the IL-6 receptor under the control of an inducible promoter. Upon induction of the metallothionein promoter by $ZnCl_2$, 10 to 20,000 low-affinity binding sites were induced (Figure 10). This experiment clearly indicates that the 80-kDa receptor subunit

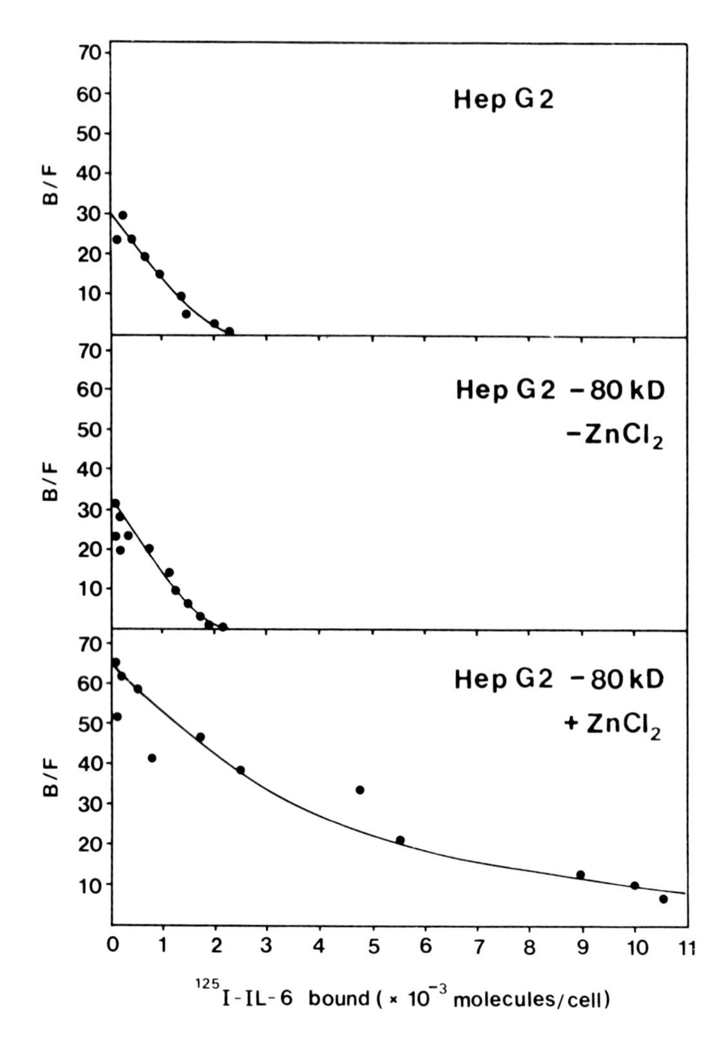

FIGURE 10. Involvement of the 80-kDa IL-6 receptor subunit in the formation of low-affinity IL-6-binding sites. Hep G2 cells were transfected with human IL-6 receptor cDNA under the transcriptional control of a mouse metallothionein promoter. In the transfected cells, IL-6 receptor expression can be upregulated by $ZnCl_2$. Normal Hep G2 cells (A), transfected Hep G2 cells (B), and transfected Hep G2 cells treated with $ZnCl_2$ (C) were incubated with [^{125}I]IL-6 at different concentrations at 4°C. Specific binding was measured and the data transformed using a Scatchard analysis.

is responsible for the formation of low-affinity IL-6-binding sites. When gp130 was transfected into cells with very low levels of gp130, high-affinity binding sites for IL-6 could be created.[17]

To learn more about the fate of IL-6 and its receptor after their interaction, Hep G2 cells as well as Hep G2-80kDa cells were preincubated with [^{125}I]IL-6 at 4°C. After 2 h, the temperature was shifted to 37°C, and surface-bound and internalized IL-6 was measured. Figure 11 shows that iodinated IL-6 was internalized within 60 min (filled triangles); concomitantly, surface receptors disappeared (filled circles). In an experiment which was similarly carried out, we have found that 80% of the internalized iodinated IL-6 is degraded. To find out whether the internalized receptor is also degraded or recycled, Hep G2 cells were incubated with saturating amounts of unlabeled IL-6 at 37°C for 2 h, leading to a downregulation of the receptors. After removal of the ligand, binding studies were carried out at different times. Figure 12 shows that more than 8 h are required to restore the IL-6

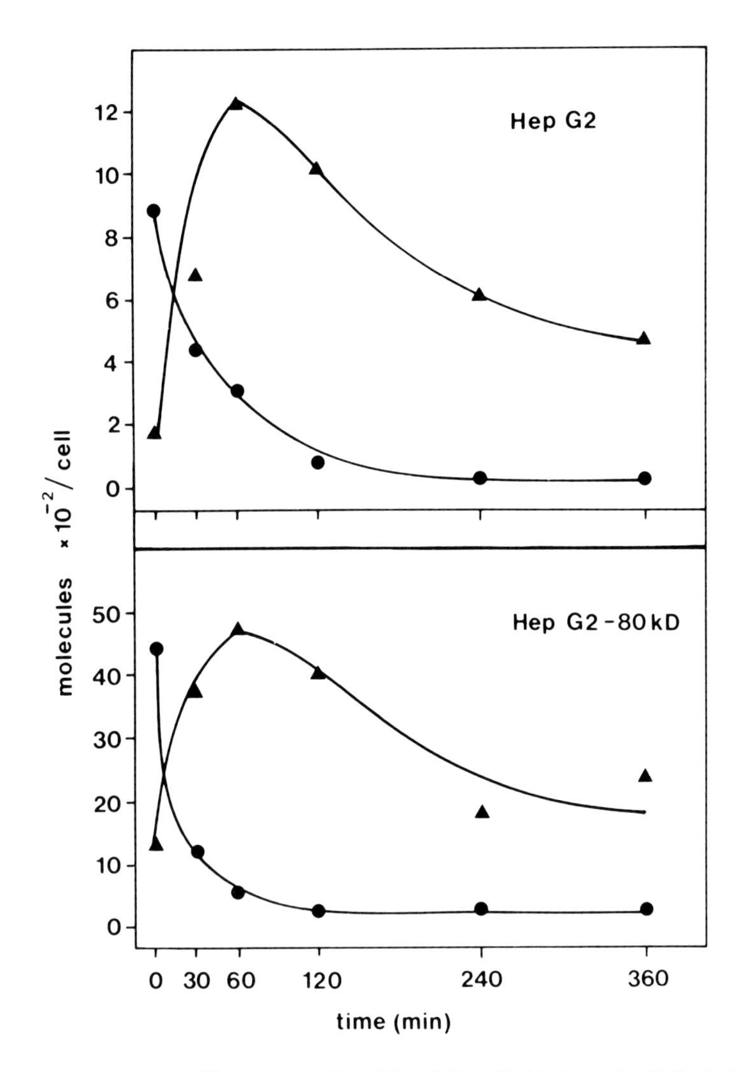

FIGURE 11. Internalization of [^{125}I]rhIL-6 by Hep G2 and Hep G2-80-kDa cells. Cells (2.5 × 10^6) were preincubated with 1 n*M* [^{125}I]rhIL-6 at 4°C for 2 h. The temperature was shifted to 37°C, and surface-bound [^{125}I]rhIL-6 (circles) and internalized [^{125}I]rhIL-6 (triangles) were determined as described in Section II. Upper panel, normal Hep G2 cells; lower panel, Hep G2-80-kDa cells.

receptors on the surface of these cells. This observation and the fact that cycloheximide prevented reappearance of the receptors is evidence for an IL-6 receptor *de novo* synthesis rather than a recycling.

C. THE SOLUBLE IL-6 RECEPTOR

1. Generation of a Soluble Form of the IL-6 Receptor

When the biosynthesis of the 80-kDa IL-6 receptor subunit was studied in a pulse-chase experiment, we found that the radioactivity of the membrane-bound receptor disappeared with time after pulse. The decrease of immunoprecipitable radioactivity in cells was paralleled by a corresponding time-dependent increase of a soluble protein which could be immunoprecipitated by the gp80 antiserum (Figure 13A). We conclude from the electrophoretic mobility of the soluble IL-6 receptor form that it is generated by the action of a specific proteinase. The generation of this soluble IL-6 receptor form is drastically accelerated when the pulse-chase experiment is carried out in the presence of phorbol ester (Figure 13B).

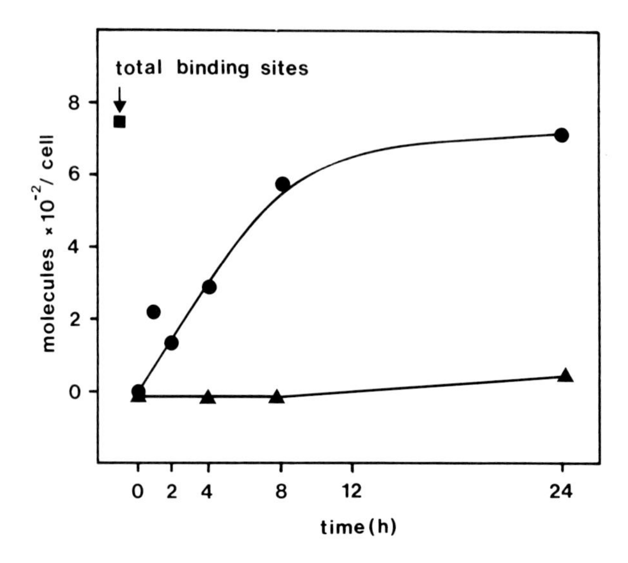

FIGURE 12. Reappearance of IL-6-binding sites at the cell surface after downregulation of the IL-6-R in Hep G2 cells. For the downregulation of the IL-6-R, 2.5 × 10^6 Hep G2 cells were incubated with 1 n*M* of unlabeled rhIL-6 at 37°C for 2 h. After removal of the ligand by three subsequent washes with binding medium, cells were further incubated at 37°C without (circles) or with 10 μg/ml of cycloheximide (triangles). At the times indicated in the figure, the number of binding sites per cell was determined as described in the legend to Figure 1. The square represents the number of total binding sites before the start of the experiment.

Inducibility of processes by the action of phorbol esters is an indication for the involvement of proteinase kinase C. A soluble form of the human IL-6 receptor has been detected in urine of normal individuals.[43] Since no transcripts coding for a soluble form of the IL-6 receptor were found,[15,37] we propose that the naturally occurring soluble IL-6 receptor is generated by limited proteolysis (shedding).

2. Function of the Soluble IL-6 Receptor

In order to find a physiological function of the soluble IL-6 receptor, we established a model for an inflamed liver. On the basis of our findings that high IL-6 concentrations lead to a downregulation of receptors (see Figure 11), we reasoned that liver cells permanently exposed to high IL-6 levels should be completely desensitized, i.e., they should not be able to synthesize and secrete acute phase proteins. This has indeed been observed in Hep G2 cells transfected with IL-6 cDNA. Figure 14 shows that such cells do not synthesize α_1-antichymotrypsin or haptoglobin mRNA as well as the respective proteins (lane 3). The addition of the soluble IL-6 receptor to the transfected Hep G2 cells completely restored the responsiveness. Interestingly, the Hep G2-IL-6 cells are not desensitized towards cytokines such as leukemia inhibitory factor, transforming growth factor-β, and interferon-γ (lanes 5 to 7). It is remarkable that the soluble IL-6 receptor/IL-6 complex acts as an agonist on liver cells by inducing acute phase protein synthesis. Under conditions of downregulated IL-6 receptors by high IL-6 levels, soluble IL-6 receptor could play an important physiological role in modulating the biological responses to IL-6. In fact, very recent work in our laboratory has shown that in 9 out of 40 sera of patients with systemic lupus erythematodes, soluble IL-6 receptor levels were elevated.

D. SUPERFAMILIES OF CYTOKINES AND CYTOKINE RECEPTORS

Bazan[44] discovered the existence of an α-helical cytokine family comprising growth hormone, prolactin, myelomonocytic growth factor, erythropoietin, granulocyte colony-

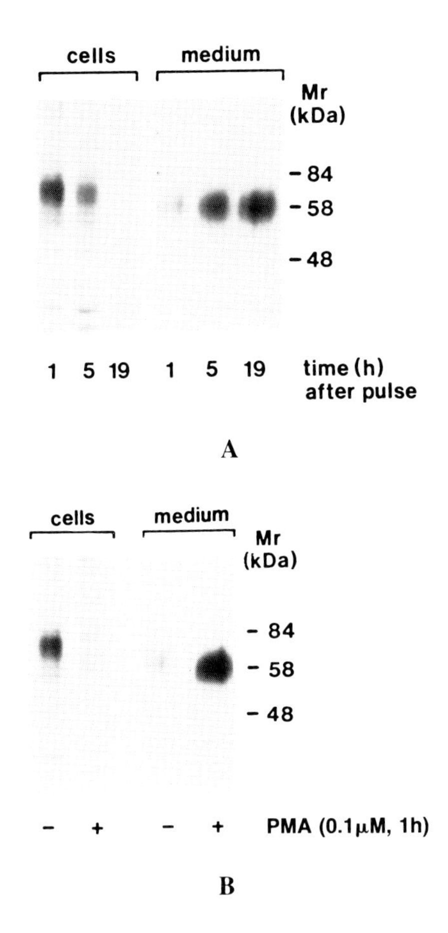

FIGURE 13. Generation of the soluble IL-6 receptor by membrane shedding. COS-7 cells were transfected with a cDNA coding for the 80-kDa subunit of the IL-6 receptor. Cells were labeled for 2 h with [^{35}S]methionine. After 2 h, [^{35}S]methionine was replaced by unlabeled methionine. At the times indicated, proteins from cells and supernatants were immunoprecipitated with a specific IL-6 receptor antiserum. (A) Time course; (B) PMA stimulation of the appearance of soluble IL-6 receptor in the supernatants as shown by SDS-PAGE and fluorography of the immunoprecipitated protein.

stimulating factor, and IL-6. Although there is no detectable homology in the amino acid sequence of the members of this family, secondary structure elements seem to appear at comparable positions within these molecules. Furthermore, similar gene organizations are found for the different members of this cytokine family. Since the crystal structure of growth hormone has been determined and shows a characteristic bundle of four antiparallel α-helices,[45] and since prolactin is a homolog of growth hormone, the author assumes that prolactin is similarly folded. Circular dichroism spectra suggested that erythropoietin, granulocyte colony-stimulating factor, and IL-6 are characterized by a high α-helical content comparable to that seen in the crystal structure of growth hormone. On the basis of these observations, Bazan[44] proposed the model for the different members of the helical cytokine family shown in Figure 15.

Molecular cloning of the cDNAs coding for the receptors of IL-2 (β and γ subunit), IL-3, IL-4, IL-5, IL-6 (80- and 130-kDa subunits), IL-7, IL-9, growth hormone, prolactin, erythropoietin, granulocyte colony-stimulating factor, granulocyte/macrophage colony-stimulating factor (both subunits), leukemia inhibitory factor, and the comparison of the respective amino acid sequences led to the recognition of an absolute conservation of four cysteine

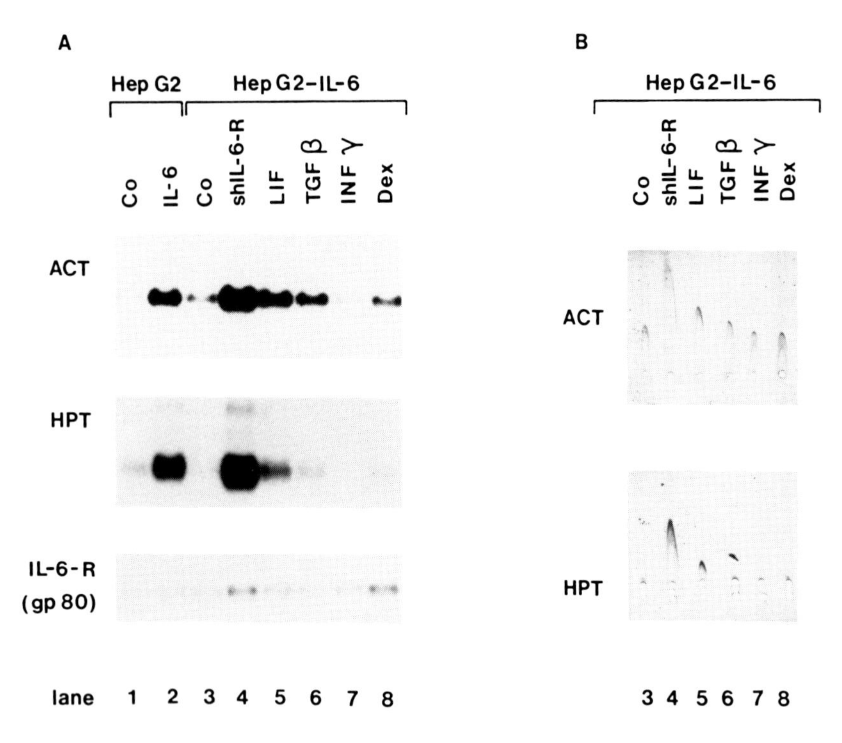

FIGURE 14. Expression of α_1-antichymotrypsin and haptoglobin by transfected Hep G2 cells stimulated with different cytokines, dexamethasone, or shIL-6-R. Transfected Hep G2 cells (Hep G2-IL-6) were incubated with soluble IL-6-R, 10 U/ml LIF, 4 ng/ml TGFβ, 100 U/ml IFNγ, and 10^{-7} *M* dexamethasone for 21 h. As controls, normal Hep G2 cells were incubated in the absence or presence of 100 U/ml IL-6. (A) Northern analysis and (B) protein determination of secreted α_1-antichymotrypsin and haptoglobin.

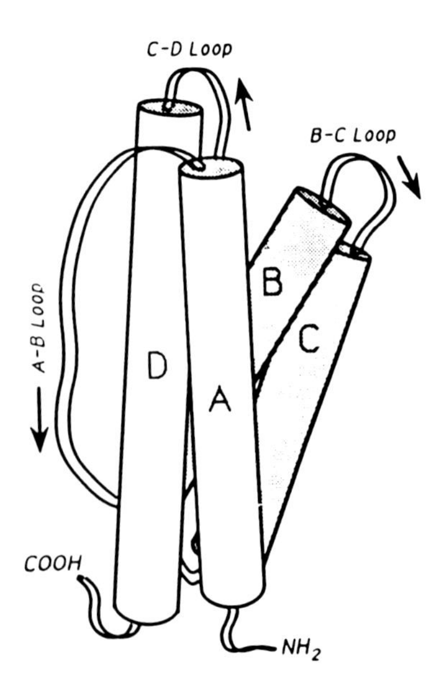

FIGURE 15. Proposed model for the secondary and tertiary structure of cytokines of the helical cytokine family. (From Bazan, F., *Immunol. Today*, 11, 350, 1990. With permission.)

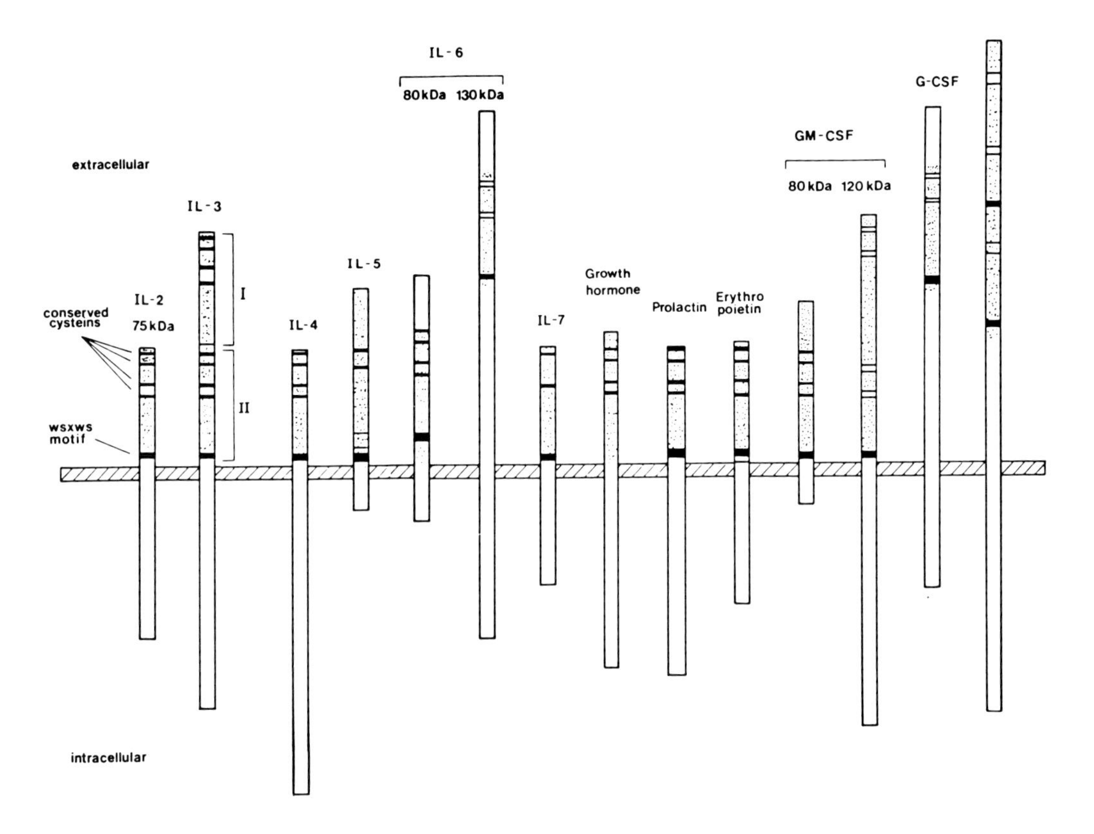

FIGURE 16. Schematic representation of the structures of the presently known members of the cytokine receptor superfamily. Horizontal bars represent conserved cysteine residues. The black boxes represent the conserved Trp-Ser-X-Trp-Ser (WSXWS) motif. The stippled areas define the stretch of homology between the different receptors of the superfamily. G-CSF, granulocyte colony-stimulating factor; GM-CSF, granulocyte/macrophage colony-stimulating factor; LIF, leukemia inhibitory factor.

residues and a Trp-Ser-X-Trp-Ser (WSXWS) motif. However, the overall sequence homology between the cytokine receptors mentioned is only about 20%.[46,47] In addition, Bazan proposes that seven consensus β-strands form an antiparallel β-sandwich with a topology analogous to an immunoglobulin constant domain. The receptors listed above form a new family of cytokine receptors also designated as a hematopoietic or hemopoietin receptor superfamily (Figure 16).

The structural features of the members of this new cytokine superfamily discriminate them from other receptor families such as the growth factor receptor tyrosine kinases,[48] the immunoglobulin superfamily,[49] the receptors related to the β-adrenergic receptors,[50] and a newly emerging group that includes nerve growth factor receptor and tumor necrosis factor receptor.[51]

Very recently, the molecular structure of the human growth hormone/receptor complex (a receptor homodimer binds one growth hormone molecule) has been solved by X-ray analysis at a resolution of 2.8 Å.[52] In general, the model predicted by Bazan has been confirmed by this study, except that the spatial order of the 7β-strands resembles the D2 domain of CD4 rather than the immunoglobulin constant domain.[52] The fact that IL-6 as well as both of its receptor subunits belong to superfamilies points to an evolutionary emergence from single-ancestor cytokine and cytokine receptor genes, respectively.

E. SIGNAL TRANSDUCTION

Not much is presently known about the signal transduction of IL-6. Truncations and amino acid substitutions were introduced into the cytoplasmic region of the IL-6 signal-transducing subunit gp130, identifying a stretch of the first 61 amino acids from the membrane as critical for signal transduction.[53] Within this region, two short segments show significant homology to other cytokine receptor family members.

The fastest effect 5 min after the action of IL-6 on its receptor reported so far is the tyrosine phosphorylation of a 160-kDa protein in M1 myeloblastic leukemia cells and murine B-cell hybridoma and plasmacytoma cells.[54,55] Subsequently, transcriptional activation of several genes (*jun*B, *jun*D, and c-*jun*, tis 11, ICAM-1, MyD 88, and MyD 116) has been observed 15 to 60 min after IL-6 addition.[54,55] This rapid transcriptional activation, particularly of nuclear factors, may be a prerequisite for the stimulation of acute phase protein genes which occurs around 18 h after IL-6 administration. There exist two classes of liver acute phase protein genes, characterized by different *cis*-acting IL-6 response elements and *trans*-acting nuclear factors. The transcriptional factors involved in the regulation of acute phase proteins of both categories are rapidly activated after the interaction of IL-6 with its hepatic receptor.[56-58]

IV. PERSPECTIVES

The ultimate aim should be to unravel the chain of events from the interaction of IL-6 with its hepatic receptor to the transcriptional activation of acute phase protein genes in the nucleus. For the receptor as well as the ternary complex, an X-ray analysis of the various structures needs to be performed. Most obscure is the pathway of the IL-6 signal to the nucleus. It can be anticipated that several kinases involved in IL-6 signaling will be identified and characterized. Finally, the molecular mechanisms by which acute phase protein genes are switched on have to be elucidated. Different combinations of transcription factors might represent the molecular basis for the pleiotropic spectrum of action of IL-6 in different biological systems.

REFERENCES

1. **Hirano, T., Taga, T., Nakano, N., Yasukawa, K., Kashiwamura, S., Shimizu, K., Nakajima, K., Pyun, K. H., and Kishimoto, T.,** Purification to homogeneity and characterization of human B cell differentiation factor (BCDF or BSFp-2), *Proc. Natl. Acad. Sci. U.S.A.*, 82, 5490, 1985.
2. **Hirano, T., Yasukawa, K., Harada, H., Taga, T., Watanabe, Y., Matsuda, T., Kashiwamura, S., Nakajima, K., Koyama, K., Iwamatu, A., Tsunasawa, S., Sakiyama, F., Matsui, H., Takahara, Y., Taniguchi, T., and Kishimoto, T.,** Complementary DNA for a novel human interleukin (BSF-2) that induces B lymphocytes to produce immunoglobulin, *Nature (London)*, 324, 73, 1986.
3. **Van Snick, J., Cayphas, S., Vink, A., Uyttenhove, C., Coulie, P. G., Rubira, M. R., and Simpson, R. J.,** Purification and NH-2-terminal amino acid sequence of T-cell-derived lymphokines with growth factor activity for B-cell hybridomas, *Proc. Natl. Acad. Sci. U.S.A.*, 83, 9679, 1986.
4. **Van Snick, J., Vink, A., Cayphas, S., and Uyttenhove, C.,** Interleukin-HP1, a T cell-derived hybridoma growth factor that supports the in vitro growth of murine plasmacytomas, *J. Exp. Med.*, 165, 641, 1987.
5. **Nordan, R. P. and Potter, M.,** A macrophage-derived factor required by plasmacytomas for survival and proliferation in vitro, *Science*, 233, 566, 1986.
6. **Kawanao, M., Hirano, T., Matsuda, T., Taga, T., Horii, Y., Iwato, K., Asaoku, H., Tang, B., Tanabe, O., Tanaka, H., Kuramoto, A., and Kishimoto, T.,** Autocrine generation and requirement of BSF-2/IL-6 for human multiple myelomas, *Nature (London)*, 332, 83, 1988.

7. **Garman, R. D., Jacobs, K. A., Clark, S. C., and Raulet, D. H.**, B-cell-stimulatory factor 2 (β_2-interferon) functions as a second signal for interleukin-2 production by mature murine T cells, *Proc. Natl. Acad. Sci. U.S.A.*, 84, 7629, 1987.
8. **Lotz, M., Jirik, F., Kabouridis, R., Tsoukas, C., Hirano, T., Kishimoto, T., and Carson, D. A.**, BSF-2/IL-6 is a costimulant for human thymocytes and T-lymphocytes, *J. Exp. Med.*, 167, 1253, 1988.
9. **Okada, M., Kitahara, M., Kishimoto, S., Matsuda, T., Hirano, T., and Kishimoto, T.**, BSF-2/IL-6 functions as killer helper factor in the in vitro induction of cytotoxic T cells, *J. Immunol.*, 141, 1543, 1988.
10. **Ikebuchi, K., Wong, G. C., Clark, S. C., Ihle, J. N., Hirai, Y., and Ogawa, M.**, Interleukin-6 enhancement of interleukin-3-dependent proliferation of multipotential hemopoietic progenitors, *Proc. Natl. Acad. Sci. U.S.A.*, 84, 9035, 1987.
11. **Andus, T., Geiger, T., Hirano, T., Northoff, H., Ganter, U., Bauer, J., Kishimoto, T., and Heinrich, P. C.**, Recombinant human B cell stimulatory factor 2 (BSF-2/IFNβ2) regulates β-fibrinogen and albumin mRNA levels in Fao-9 cells, *FEBS Lett.*, 221, 18, 1987.
12. **Gauldie, J., Richards, C., Harnish, D., Landsdorp, P., and Baumann, H.**, Interferon β_2-/B-cell stimulatory factor type 2 shares identity with monocyte-derived hepatocyte-stimulating factor and regulates the major acute phase protein response in liver cells, *Proc. Natl. Acad. Sci. U.S.A.*, 84, 7251, 1987.
13. **Miyaura, C., Onozaki, K., Akiyama, Y., Taniyama, T., Hirano, T., Kishimoto, T., and Suda, T.**, Recombinant human interleukin-6 (B-cell stimulatory factor 2) is a potent inducer of differentiation of mouse myeloid leukemia cells (M1), *FEBS Lett.*, 234, 17, 1988.
14. **Satoh, T., Nakamura, S., Taga, T., Matsuda, T., Hirano, T., Kishimoto, T., and Kaziro, Y.**, Induction of neural differentiation in PC12 cells by B cell stimulatory factor 2/interleukin-6, *Mol. Cell. Biol.*, 8, 3546, 1988.
15. **Yamasaki, K., Taga, T., Hirata, Y., Yawata, H., Kawanishi, Y., Seed, B., Taniguchi, T., Hirano, T., and Kishimoto, T.**, Cloning and expression of the human interleukin-6 (BSF-2/IFNβ2) receptor, *Science*, 241, 825, 1988.
16. **Taga, T., Hibi, M., Hirata, Y., Yamasaki, K., Yasukawa, K., Matsuda, T., Hirano, T., and Kishimoto, T.**, Interleukin-6 triggers the association of its receptor with a possible signal transducer, gp130, *Cell*, 58, 573, 1989.
17. **Hibi, M., Murakami, M., Saito, M., Hirano, T., Taga, T., and Kishimoto, T.**, Molecular cloning and expression of an IL-6 signal transducer, gp 130, *Cell*, 63, 1149, 1990.
18. **Aarden, L. A., De Groot, E. R., Schaap, O. L., and Lansdorp, P. M.**, Production of hybridoma growth factor by human monocytes, *Eur. J. Immunol.*, 17, 1411, 1987.
19. **Sambrook, J., Fritsch, E. F., and Maniatis, T.**, *Molecular Cloning: A Laboratory Manual*, Cold Spring Harbor Laboratory, Cold Spring Harbor, NY, 1989.
20. **Rose-John, S., Hipp, E., Lenz, D., Legrès, L. G., Korr, H., Hirano, T., Kishimoto, T., and Heinrich, P. C.**, Structural and functional studies on the human interleukin-6-receptor, *J. Biol. Chem.*, 266, 3841, 1991.
21. **Markwell, M. A.**, A new solid-state reagent to iodinate proteins, *Anal. Biochem.*, 125, 427, 1982.
22. **Laurell, C. B.**, Electroimmunoassay, *Scand. J. Clin. Lab. Invest. Suppl.*, 124, 21, 1972.
23. **Mackiewicz, A., Ganapathi, M. K., Schultz, D., and Kushner, I.**, Monokines regulate acute phase protein glycosylation, *J. Exp. Med.*, 166, 253, 1987.
24. **Clogston, C. L., Boonie, T. C., Crandall, B. C., Mendiaz, E. A., and Lu, H. S.**, Disulfide structures of human interleukin-6 are similar to those of human granulocyte-colony stimulating factor, *Arch. Biochem. Biophys.*, 272, 144, 1989.
25. **May, L. T., Ghrayeb, J., Santhanam, U., Tatter, S. B., Stoeger, Z., Helfgott, D. C., Chiorazzi, N., Grieninger, G., and Sehgal, P. B.**, Synthesis and secretion of multiple forms of β2-interferon/B-cell differentiation factor 2/hepatocyte-stimulation factor by human fibroblasts and monocytes, *J. Biol. Chem.*, 263, 7760, 1988.
26. **Gross, V., Andus, T., Castell, J., Vom Berg, D., Heinrich, P. C., and Gerok, W.**, O- and N-glycosylation lead to different molecular weight forms of human monocyte interleukin-6, *FEBS Lett.*, 247, 323, 1989.
27. **Schiel, X., Rose-John, S., Dufhues, G., Schooltink, H., and Heinrich, P. C.**, Microheterogeneity of human interleukin-6 synthesized by transfected NIH/3T3 cells: comparison with human monocytes, fibroblasts, and endothelial cells, *Eur. J. Immunol.*, 20, 883, 1990.
28. **Dufhues, G., Conradt, H., Rose-John, S., Müllberg, J., Lenz, D., Grabenhorst, E., Nimtz, M., Schaper, K., Rollwage, K., Getzlaff, R., and Heinrich, P. C.**, Human interleukin-6 of transfected NIH/3T3 cells: structures of N- and O-linked carbohydrate chains and their location within the polypeptide, *Eur. J. Biochem.*, in press.
29. **Ashwell, G. and Harford, J.**, Carbohydrate-specific receptors of the liver, *Annu. Rev. Biochem.*, 51, 531, 1982.
30. **Castell, J. V., Klapproth, J., Gross, V., Walter, E., Andus, T., Snyers, L., Content, J., and Heinrich, P. C.**, Fate of IL-6 in the rat: involvement of skin in its catabolism, *Eur. J. Biochem.*, 189, 113, 1990.

31. **Krüttgen, A., Rose-John, S., Möller, C., Wroblowski, B., Wollmer, A., Müllberg, J., Hirano, T., Kishimoto, T., and Heinrich, P. C.**, Structure-function analysis of human interleukin-6. Evidence for the involvement of the carboxy-terminus in function, *FEBS Lett.*, 262, 323, 1990.
32. **Brakenhoff, J. P. J., Hart, M., de Groot, E. R., Di Padova, F., and Aarden, L. A.**, Structure-function analysis of human IL-6. Epitope mapping of neutralizing monoclonal antibodies with amino- and carboxy-terminal deletion mutants, *J. Immunol.*, 145, 561, 1990.
33. **Brakenhoff, J. P. J., Hart, M., and Aarden, L. A.**, Analysis of human IL-6 mutants expressed in *Escherichia coli*. Biologic activities are not affected by deletion of amino acids 1-28, *J. Immunol.*, 143, 1175, 1989.
34. **Krüttgen, A., Rose-John, S., Dufhues, G., Bender, S., Lütticken, C., Freyer, P., and Heinrich, P. C.**, The three carboxy-terminal amino acids of human interleukin-6 are essential for its biological activity, *FEBS Lett.*, 273, 95, 1990.
35. **Lütticken, C., Krüttgen, A., Möller, C., Heinrich, P. C., and Rose-John, S.**, Evidence for the importance of a positive charge and an α-helical structure of the C-terminus for biological activity of human IL-6, *FEBS Lett.*, 282, 265, 1991.
36. **Castell, J. V., Geiger, T., Gross, V., Andus, T., Walter, E., Hirano, T., Kishimoto, T., and Heinrich, P. C.**, Plasma clearance, organ distribution and target cells of interleukin-6/hepatocyte stimulating factor in the rat, *Eur. J. Biochem.*, 177, 357, 1988.
37. **Schooltink, H., Stoyan, T., Lenz, D., Schmitz, H., Hirano, T., Kishimoto, T., Heinrich, P. C., and Rose-John, S.**, Structural and functional studies on the human hepatic IL-6-receptor: molecular cloning and overexpression in HepG2 cells, *Biochem. J.*, 277, 659, 1991.
38. **Baumann, M., Baumann, H., and Fey, G. H.**, Molecular cloning, characterization and functional expression of the rat liver interleukin-6-receptor, *J. Biol. Chem.*, 265, 19853, 1990.
39. **Castell, J. V., Gomez-Lechon, M. J., David, M., Andus, T., Geiger, T., Trullenque, R., Fabra, R., Gerok, W., and Heinrich, P. C.**, Interleukin-6 is the major regulator of acute phase protein synthesis in adult human hepatocytes, *FEBS Lett.*, 242, 237, 1989.
40. **Castell, J. V., Gomez-Lechon, M. J., David, M., Fabra, R., Trullenque, R., and Heinrich, P. C.**, Acute phase response of human hepatocytes: regulation of acute phase protein synthesis by interleukin-6, *Hepatology*, 12, 1179, 1990.
41. **Rose-John, S., Schooltink, H., Lenz, D., Hipp, E., Dufhues, G., Schmitz, H., Schiel, X., Hirano, T., Kishimoto, T., and Heinrich, P. C.**, Studies on the structure and regulation of the human hepatic interleukin-6-receptor, *Eur. J. Biochem.*, 190, 79, 1990.
42. **Schooltink, H., Schmitz-Van de Leur, H., Taga, T., Kishimoto, T., Heinrich, P. C., and Rose-John, S.**, Up-regulation of the interleukin-6-signal transducing protein (gp130) by interleukin-6 and dexamethasone in HepG2 cells, *FEBS Lett.*, 297, 263, 1992.
43. **Novick, D., Engelmann, H., Wallach, D., and Rubinstein, M.**, Soluble cytokine receptors are present in normal human urine, *J. Exp. Med.*, 170, 1409, 1989.
44. **Bazan, F.**, Haemopoietic receptors and helical cytokines, *Immunol. Today*, 11, 350, 1990.
45. **Abdel-Meguid, S. S., Shieh, H. S., Smith, W. W., Dayringer, H. E., Violand, B. N., and Bentle, L. A.**, Three-dimensional structure of a genetically engineered variant of porcine growth hormone, *Proc. Natl. Acad. Sci. U.S.A.*, 84, 6434, 1987.
46. **Bazan, J. F.**, A novel family of growth factors: a common binding domain in the growth hormone, prolactin, erythropoietin and IL-6 receptors, and the p75 IL-2 receptor β-chain, *Biochem. Biophys. Res. Commun.*, 164, 788, 1989.
47. **Bazan, J. F.**, Structural design and molecular evolution of a cytokine receptor superfamily, *Proc. Natl. Acad. Sci. U.S.A.*, 87, 6934, 1990.
48. **Yarden, Y. and Ullrich, A.**, Growth factor receptor tyrosine kinases, *Annu. Rev. Biochem.*, 57, 443, 1988.
49. **Williams, A. F. and Barclay, A. N.**, The immunoglobulin superfamily-domains for cell surface recognition, *Annu. Rev. Immunol.*, 6, 381, 1988.
50. **O'Dowd, B. F., Lefkowitz, R. J., and Caron, M. G.**, Structure of the adrenergic and related receptors, *Annu. Rev. Neurosci.*, 12, 67, 1989.
51. **Schall, T. J., Lewis, M., Koller, K. J., Lee, A., Rice, G. C., Wong, H. W., Gatanaga, T., Granger, G. A., Lentz, R., Raab, H., Kohr, W. J., and Goeddel, D. V.**, Molecular cloning and expression of a receptor for human tumor necrosis factor, *Cell*, 61, 361, 1990.
52. **De Vos, A. M., Ultsch, M., and Kossiakoff, A.**, Human growth hormone and extracellular domain of its receptor: crystal structure of the complex, *Science*, 255, 306, 1992.
53. **Murakami, M., Narazaki, M., Hibi, M., Yawata, H., Yasukawa, K., Hamaguchi, M., Taga, T., and Kishimoto, T.**, Critical cytoplasmic region of the interleukin 6 signal transducer gp130 is conserved in the cytokine receptor family, *Proc. Natl. Acad. Sci. U.S.A.*, 88, 11349, 1991.

54. **Lord, K. A., Abdollahi, A., Thomas, S. M., DeMarco, M., Brugge, J. S., Hoffman-Liebermann, B., and Liebermann, D. A.,** Leukemia inhibitory factor and interleukin-6 trigger the same immediate early response, including tyrosine phosphorylation, upon induction of myeloid leukemia differentiation, *Mol. Cell. Biol.,* 11, 4371, 1991.
55. **Nakajima, K. and Wall, R.,** Interleukin-6 signals activating junB and TIS11 gene transcription in a B-cell hybridoma, *Mol. Cell. Biol.,* 11, 1409, 1991.
56. **Isshiki, H., Akira, S., Sugita, T., Nishio, Y., Hashimoto, S., Pawlowski, T., Suematsu, S., and Kishimoto, T.,** Reciprocal expression of NF-IL6 and C/EBP in hepatocytes: possible involvement of NF-IL6 in acute phase protein gene expression, *New Biol.,* 3, 63, 1991.
57. **Krause, E., Wegenka, U., Möller, C., Horn, F., and Heinrich, P. C.,** Gene expression of the high molecular weight proteinase inhibitor α_2-macroglobulin, *Biol. Chem. Hoppe-Seyler,* in press.
58. **Horn, F. and Heinrich, P. C.,** Regulation of the α_2-macroglobulin gene, in *Acute-Phase Proteins: Molecular Biology, Biochemistry, Clinical Applications,* Mackiewicz, A., Kushner, I., and Baumann, H., Eds., CRC Press, Boca Raton, FL, 1993, chap. 25.
59. **Pietzko, D., Zohlnhöfer, D., Graeve, L., Fleischer, D., Stoyan, T., Schooltink, H., Rose-John, S., and Heinrich, P. C.,** The hepatic interleukin-6 receptor: Studies on its structure and regulation by 12-myristate 13-acetate-dexamethasone, *J. Biol. Chem.,* 268, 4250, 1993.

E. Regulation of Acute Phase Protein Gene Expression

Chapter 21

REGULATION OF C-REACTIVE PROTEIN, HAPTOGLOBIN, AND HEMOPEXIN GENE EXPRESSION

Dipak P. Ramji, Riccardo Cortese, and Gennaro Ciliberto

TABLE OF CONTENTS

0-8493-6913-4/93/$0.00 + $.50

I. INTRODUCTION

Among the several functions performed by the liver, an important one is to act as a large sensory organ for all hormonal and biochemical changes that occur during homeostasis, and reach the liver parenchyma through its abundant portal and hepatic circulation. The major changes in liver physiology which occur in response to extracellular stimuli are well known under the name of acute phase response (APR). This is defined as the total fluctuation in the serum concentration of several plasma proteins, which are predominantly synthesized in the liver, in response to infection, inflammation, stress, trauma, and other pathological conditions.[1-7] The nature of this variation allows the distinction of AP proteins into two major groups: the so-called positive AP reactants (e.g., hemopexin, C-reactive protein, haptoglobin, and α_1-acid glycoprotein) whose plasma concentration increases during inflammation, and the negative AP reactants, including albumin and transferrin, whose levels decrease (Table 1). Although the level of decrease is generally two- to threefold, the extent of increase varies among different proteins, and ranges from two- to threefold, for α_1-antitrypsin up to 1000-fold in the case of C-reactive protein (CRP).

Monocytes play a crucial role in the regulation of APR by migrating from peripheral blood to the sites of tissue damage, and secreting a large number of factors which mediate inflammation. However, other cell types such as fibroblasts, endothelial cells, and T-lymphocytes also contribute to the production of the whole spectrum of molecules involved in the inflammatory response.[1-7]

The study of APR has recently attracted substantial interest not only for its medical implications, but also because it provides an excellent system with which to elucidate the molecular mechanisms involved in the modulation of gene expression.

A. REGULATION OF ACUTE PHASE GENE EXPRESSION IS MAINLY TRANSCRIPTIONAL

During the past few years, several aspects of APR have been analyzed *in vivo* using living organisms.[8] However, more detailed and precise information has only been obtained with the implementation of *in vitro* studies using both primary hepatocytes and established hepatoma cell lines such as Hep G2 and Hep 3B.[9-14] The major conclusion from the initial studies using these simplified systems was that the observed changes in the levels of both positive and negative AP reactants can be correlated with variations in mRNA levels or, more precisely, with changes in the rate of transcription of the corresponding genes. In particular, upregulation of positive AP genes occurs exclusively at the transcriptional level, as shown by nuclear run-on experiments with probes for several AP genes.[15]

TABLE 1
Acute Phase Plasma Proteins

Increase by 50%
Ceruloplasmin
Complement C3
Increase Two- to Fourfold
Angiotensinogen
α_1-Acid Glycoprotein
α_1-Chymotrypsin
α_1-Antitrypsin
Haptoglobin
Hemopexin
Increase Several Hundredfold
C-reactive protein
Serum amyloid A
Decrease
Albumin
α_2-HS glycoprotein
Prealbumin
Transferrin

A major breakthrough in this field was the demonstration by Darlington et al.[9] that conditioned medium from peripheral monocytes (MoCM), stimulated with lipopolysaccharides (LPS), can induce in Hep 3B cells the whole spectrum of changes in plasma protein production typical of the APR. This constituted the basis for the subsequent fractionation of this and other cell-conditioned media in order to identify the molecular nature of hepatocyte-stimulating factors (HSFs). Since then, several hormones and cytokines have been shown to play an important role in the regulation of APR. These include interleukin-1 (IL-1),[16-19] tumor necrosis factor (TNF),[9,20] interleukin-6 (IL-6),[5,6,8,13,14] leukemia inhibitory factor (LIF),[21] glucocorticoids,[12,22-24] interleukin-11 (IL-11),[25] and transforming growth factor-β (TGFβ).[26] Among these, IL-1 and IL-6 are the principal mediators, and their effect on the pattern of gene expression in hepatocytes allows a further classification of AP genes into two classes.[12] Maximal induction of genes in class I requires IL-1 plus IL-6 and a combination of these two plus glucocorticoids, and includes C-reactive protein, complement C3, haptoglobin, hemopexin, and serum amyloid A. Class II genes, including α_1-antitrypsin, fibrinogen, and α_2-macroglobulin (α_2-M), are regulated only by IL-6, but maximal expression is achieved with a combination of IL-6 plus glucocorticoids.

The initial indications by run-on experiments that regulation of AP gene expression occurs at the transcriptional level have been further substantiated by the isolation of the promoter and enhancer regions of several AP genes, and analysis of their activity in transfection experiments using both human and murine hepatoma cells. Subsequent genetic analysis of AP regulatory regions has allowed the identification of relatively small *cis*-acting sequences, often defined as APREs (acute phase response elements), which mediate transcriptional activation in response to specific inflammatory mediators.

B. THREE PROTOTYPE CLASS I GENES: HEMOPEXIN (HPX), HAPTOGLOBIN (HP), AND C-REACTIVE PROTEIN (CRP)

In the last few years, our laboratories have been investigating the molecular mechanisms responsible for the regulation of Class I genes, with particular emphasis on the three prototype genes belonging to this class: Hpx, Hp, and CRP.

Before starting with a thorough description of the features that characterize the regulation of expression of this set of genes, it is important to include a brief historical presentation of the structure and function of these three proteins, and of interesting characteristics of the corresponding genes.

1. Hemopexin

Hpx is a heme-binding plasma glycoprotein composed of a single 439-amino acid-long polypeptide chain.[27,28] It is exclusively synthesized in the liver and secreted into the bloodstream, where it plays an important role in heme disposal.[27] The human Hpx gene spans approximately 12 kb and is interrupted by nine introns.[29] The levels of both the Hpx mRNA and protein are low in the fetus and newborn, and do not approach those observed in the adults until after the first year.[30] Thus, regulation of Hpx gene expression is controlled at three distinct levels: (1) developmental control responsible for its increase after birth, (2) tissue-specific control for its selective expression in hepatocytes, and (3) activation by inflammatory stimuli during APR. While maximal expression of Hpx during APR requires IL-1, IL-6, and glucocorticoids, only activation by IL-6 has been investigated in detail. In hepatoma Hep 3B cells, IL-6 causes a sixfold increase in Hpx mRNA.[31] This induction does not require *de novo* protein synthesis, since it can also be observed in the presence of cycloheximide.

2. Haptoglobin

Hp is a tetrachain ($\alpha_2\beta_2$) glycoprotein that is synthesized in the liver and secreted into the plasma, where it plays a role in binding hemoglobin after hemolysis, and thus prevents kidney damage and loss of iron through urinary excretion.[32,33] Hp is transcribed at very low levels in fetal liver and increases about 47-fold in the adult.[34] Thus, similar to the Hpx gene, Hp expression is regulated at three different levels: (1) developmental control for lack of expression in fetal liver, (2) tissue-specific control for selective expression in hepatocytes, and (3) modulation of expression during APR. The human Hp locus is located on chromosome 16 and shows an unusual polymorphism usually involving duplication and, rarely, triplication of part of the coding regions.[35,36] Analysis of the genomic clones for Hp has resulted in the identification of a related gene, called Hpr, which is adjacent to the Hp gene.[35,37] The two genes are clustered and span a region of at least 21 kb.[38] In addition, the number of introns (four) and the position of the intron-exon junctions in the two genes are identical except for the first intron, which is 1.3 and 9.5 kb long in the Hp and Hpr gene, respectively.[38,39] Although the sequences of the two genes are highly homologous, the colinearity is interrupted by the presence of an Alu sequence in the 5′ flanking region and a retrovirus-like sequence in the first intron of the Hpr gene.[38,39] However, the Hpr gene is not expressed in human liver despite having no features characteristic of pseudogenes.[38] This is surprising, since the first 183 bp upstream from the putative CAP site in both genes are identical, and in transient expression studies, the Hpr promoter, extracted from its genomic context, is more active than the Hp counterpart.[34] In addition, the Alu repeated elements in the Hpr promoter have been shown to have a strong *cis*-acting enhancer effect on liver-specific transcription.[40] It is thus likely that nonexpression of the Hpr gene in its normal genetic context is probably due to a particular chromatin configuration, possibly produced by the insertion sequence in the first intron.

Similar to Hpx, maximal expression of human Hp gene also requires a combination of hormones and cytokines. However, only the IL-6 response has been studied in detail. In human hepatoma Hep 3B and Hep G2 cells, IL-6 rapidly induces a severalfold increase in Hp mRNA, and this is not blocked by inhibitors of protein synthesis.[13,34]

The organization and regulation of the Hp gene shows species-specific differences. For example, rats and New World monkeys contain a single-copy Hp gene,[24,41] whereas at least two are present in humans and Old World monkeys.[34,41] In addition, while the principal AP mediators for stimulation of Hp production in human liver cells are IL-6 and LIF[13,21] (and this action is synergistically enhanced by glucocorticoids[12,42]), IL-1, TNFα, IL-6, LIF, and glucocorticoids, alone or in various combinations, stimulate expression of the Hp gene in hepatoma cells from rats, mice, and rabbits.[43-45]

3. C-Reactive Protein

CRP is present in the blood of normal individuals in trace amounts (below 1 mg/l), but its concentration rises to more than 300 mg/l soon after the onset of inflammation, and then returns to a low basal level after the inflammatory stimulus is removed. It thus represents a sensitive, but nonspecific marker, to monitor the evolution of several diseases such as cancer, bacterial meningitis, heart infarction, and rheumatic disease.[46]

CRP and the serum amyloid protein (SAP) are members of the pentraxin family; they circulate in blood as pentamers or decamers, respectively, of identical 21-kDa subunits, which are held together by noncovalent bonds to form one (CRP) or two (SAP) rings.[46] The exact function of CRP is currently not clear, but several opsonic properties have been described such as cell- or platelet-mediated cytotoxicity, monocyte binding, and interaction with the complement system.[47,48]

CRP and SAP are probably derived by duplication of a common ancestor gene, since they show 54% identity in their amino acid sequences and 59% identity in their nucleotide sequence. Furthermore, both genes have been mapped to the same region on human chromosome 1 (between bands 1q12 and q23).[49,50] However, the behavior of these two genes during APR is species specific. For example, SAP is highly induced in mouse but not in man, while CRP is a prototype AP reactant in man and rabbits.[46] CRP, therefore, represents an excellent model to investigate the regulation of gene expression in higher eukaryotes due to its strict liver-specific expression, high degree of inducibility in man, and species-specific participation in APR.

The CRP gene has been cloned by several laboratories and consists of a transcriptional unit of 2263 bp from the CAP to the polyadenylation site.[51,52] There is a small and unique intron of 278 bp in the gene which is characterized by the presence of a 38-base-long purine-pyrimidine stretch with the potential of forming a Z-DNA structure.[51] The most remarkable feature of the CRP transcript, however, is the presence of a long 3′ untranslated region of 1.2 kb which has been suggested to serve as a target for endoribonucleases responsible for the rapid degradation of the transcripts once the inflammatory stimulus has been removed.[53]

In preliminary studies, the regulation of human CRP expression has also been analyzed in transgenic mice. The human CRP gene, containing 16 kb of 5′ flanking and 10 kb of 3′ noncoding regions, was introduced into fertilized mouse eggs and two distinct transgenic families were analyzed.[54] The CRP gene was exclusively transcribed in the liver and its expression was strictly dependent on experimental inflammation. Furthermore, nuclear run-on experiments showed that transcription was only detectable in the presence of inflammatory stimuli. The kinetics of induction of both RNA and proteins were extremely rapid, with RNA being first detectable after 2 h in the liver and the protein appearing after 6 h in the serum. The maximum level of CRP (300 mg/l) was comparable to that observed in human diseases.

DNase-I-hypersensitive sites (DHSs) reflect changes in the conformational structure of the chromatin in response to the interaction of regulatory *trans*-acting factors with their target sequences.[55] In order to identify putative regions involved in the control of CRP expression, the DHSs within and adjacent to the CRP gene were mapped in both uninduced and induced transgenic mice.[53] In uninduced liver, a constitutive and tissue-specific DHS was identified in the 3′ flanking sequence downstream from the polyadenylation site. This region is probably involved in developmental control of CRP expression, a conclusion supported by the observation that transgenic mice obtained with CRP constructs carrying a deletion of this DNA region do not express the transgene.[53] The inflammatory stimulus LPS induces the appearance of three closely spaced, liver-specific DHSs. Two of these map around the CAP site and −250 bp, and the third one is located approximately 600 bp upstream.[53] More recent work with transgenic mice has resulted in a better definition of the sequences involved in both liver-specific and inducible expression.[56] Several segments of the original construct were tested either alone or in various combinations. The conclusion of this study is that correct regulation requires a cooperation of signals located in the immediate 5′ and 3′ flanking sequence of the gene (which includes the DHSs previously mapped), and a 2-kb region containing the CRP pseudogene,[57] which is located 6 to 8 kb downstream from the CRP gene.[56] The role of this last region is to ensure a consistently low background level of expression in noninduced mice and a high degree of inducibility.

The absolute levels of expression of the CRP gene are low in the available human hepatoma cell lines. Despite this, initial studies with both Hep 3B and Hep G2 cells have shown that, similar to Hpx and Hp, CRP mRNA is also induced by the addition of MoCM and IL-6.

C. THE GOAL OF THIS REVIEW

In this chapter, we review our experimental approach to the characterization of the promoter structure of these three genes. We explain how the coordinated regulation of their expression by IL-6 is achieved through the activation of a common transcription factor, IL-6DBP. Initial characterization of this crucial protein in IL-6 signal transduction will also be presented. In addition, IL-6DBP properties of belonging to a large class of dimeric transcriptional activators involved in several aspects of cell proliferation and differentiation, and the possible role of heterodimeric interactions in regulation of AP gene transcription, are discussed.

II. MATERIALS AND METHODS

The majority of the techniques used in the studies reported in this review have been previously described. The reader is thus referred to the materials and methods section in the following references: Hpx,[30,31] Hp,[34,58] CRP,[59-62] and cloning and characterization of IL-6DBP.[63] New techniques and modifications of previously published ones are presented below.

A. RECOMBINANT PLASMIDS

The construction of eukaryotic expression plasmids for C/EBP, IL-6DBP, and truncated IL-6DBP, containing only the DNA binding domain, in the vector PhD[64] has been described by Poli et al.[63] The IL-6DBP/C/EBP chimera was prepared by first cloning the SmaI/ClaI fragment from C/EBP (positions 733 to 1900 bp) into the corresponding sites of PhD, followed by in-frame insertion of the N-terminal SmaI fragment (1 to 599) from IL-6DBP-exp.[63] For the insertion of the tag sequence, two complementary oligonucleotides flanked by NcoI overhangs (5′ CATGCCGGGGCCCTACACCGACATCGAAATGAATCGTCTT-GGAAAGTAA 3′ and 5′ CATGTTACTTTCCAAGACGATTCATTTCGATGTCGG-TGTAGGGCCCCGG 3′) were annealed. These were then phosphorylated with polynucleotide kinase and ATP, and ligated to the unique NcoI site in the chimera (position 922).

B. CELL CULTURE, TRANSIENT TRANSFECTIONS, AND CAT ASSAYS WITH RECOMBINANT IL-6DBP DERIVATIVES

Monolayers of human hepatoma cell line Hep 3B were maintained in Dulbecco's modified Eagle's medium supplemented with 10% fetal calf serum. Human IL-6 was obtained from recombinant vaccinia virus as described previously[31] and used at a concentration of 1000 U/ml. DNA transfections were performed according to the calcium phosphate method.[65] For CAT assays, the precipitate contained 12 μg of reporter plasmid (4xαCRP-CAT[61] or pBLD9-CAT[63]), 5 μg of recombinant plasmid, and 1 μg of RSV-luciferase[66] as an internal control for transfection efficiency. The precipitate was then divided into two 6-cm dishes and left for 12 h. After washing with phosphate-buffered saline (PBS), the cells were fed with new culture medium, one of which was treated with IL-6 for 36 h. Cells were harvested 48 h after transfection and analyzed for both CAT[67] and luciferase activity.[66] The CAT activities were normalized to the luciferase values, and each transfection was repeated at least three times.

C. WESTERN-BLOT AND IMMUNOFLUORESCENCE ANALYSIS

Western-blot analysis on whole-cell extracts from Hep 3B cells were performed using standard procedures.[68] Polyclonal antibodies against bacterially expressed IL-6DBP were used at a 1:1000 dilution, and immunoreactive products were detected using goat anti-rabbit antibodies coupled to alkaline phosphatase.

For the transfected protein, the DNA precipitate added to 15-cm plates contained 50 μg of IL-6DBP, containing the peptide tag sequence, and 2 μg of RSV-luciferase (internal standard). Whole-cell extracts were prepared 48 h after transfection, and equal amounts of protein, based on the luciferase value, were size fractionated on 12.5% SDS-polyacrylamide gels.[69] The proteins were then transferred onto nitrocellulose and probed with a 1:100 dilution of anti-tag antibody[70] (a generous gift from T. E. Kreis, E. M. B. L., Heidelberg). Antigen-antibody complexes were then detected by use of alkaline phosphatase-coupled goat anti-mice antibodies.

Immunofluorescence staining was performed on cells grown on coverslips and transfected with 2.5 μg of recombinant plasmid. Cells were processed 36 h after transfection by a brief washing with PBS followed by fixing in 5% acetic acid in ethanol for 5 min at −20°C. Fixed cells were washed twice for 5 min each with Solution A (PBS, 0.2% gelatine), and then blocked with Solution B (PBS containing 3% bovine serum albumin, 20 m*M* $MgCl_2$, and 0.3% Tween 20). Probing with primary (anti-tag) and secondary (rhodamine-conjugated anti-mice) antibodies was performed in PBS for 1 h in a humid chamber. The slides were washed twice with solution A, containing 1% Triton X100, after incubation with each antibody. Finally, the slides were mounted using paraffin and analyzed by microscopy.

III. RESULTS

A. *CIS*-ACTING ELEMENTS INVOLVED IN INDUCTION OF HPX, HP, AND CRP GENE TRANSCRIPTION DURING THE APR

During the last few years, substantial progress has been made in identifying the *cis*-acting elements required for transcriptional activation of Hpx, Hp, and CRP genes during the APR. The general strategy consisted of analyzing the activity of deletion or substitution mutants in the promoter region of the genes, linked to the coding sequence of bacterial chloramphenicol acetyltransferase (CAT), in transient transfections using hepatoma cell lines such as Hep 3B or Hep G2. For induction experiments, the cells were treated with either MoCM or recombinant IL-6. Individual IL-6 response elements (REs) were also tested for their ability to confer IL-6 responsiveness to heterologous promoters, such as the early SV40 promoter.

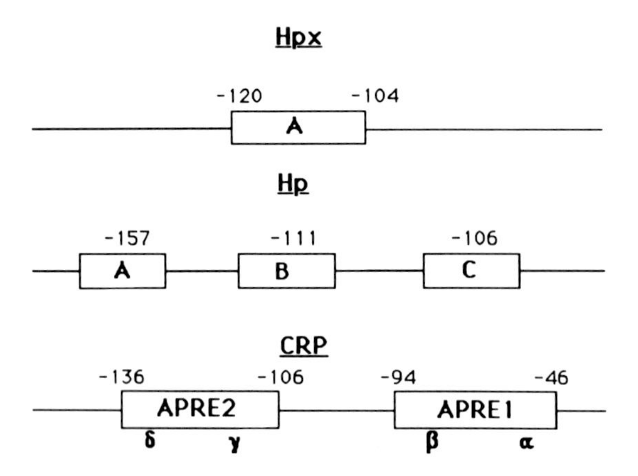

FIGURE 1. Schematic representation of the APREs in the promoter region of Hpx, Hp, and CRP genes. The Hpx IL-6 RE occurs between positions −120 and −104, and is designated the Hpx A-site. For the Hp gene, the precise positions of the three important elements (A, B, and C) are not known. The 3′ boundaries of the 6-bp mutations which affect IL-6 inducibility are thus shown. The two APREs of the CRP gene can be further divided into two subregions (α and β for APRE1, γ and δ for APRE2) on the basis of DNA-protein interaction and site-directed mutagenesis.

A schematic representation of the location of the APREs identified in the promoter of the three genes is shown in Figure 1. For Hpx, a minimal element, between positions −130 and −86, is sufficient for both cell-specific expression and IL-6 induction.[30,31] On the basis of DNA-protein interaction studies, the sequence between position −120 and −104, designated the Hpx A-site, is most important, since its mutation abolishes IL-6 induction.[31] In addition, multimers of this site can confer IL-6 inducibility to a surrogate promoter.[31] For Hp, a more complex situation is found. Mutagenesis experiments resulted in the identification of three distinct regions, A, B, and C, from the most distal to the most proximal, respectively, which contribute to the full activity of the promoter.[58] Of these, only A and C are involved in the IL-6 response, because their multimerization confers IL-6 responsiveness to a heterologous promoter.[58]

Regulation of the rat Hp gene has also been analyzed in detail and displays interesting similarities, but also differences, with the human counterpart. Marinkovic and Baumann[24] initially localized an IL-6 and glucocorticoid-inducible region in the proximal promoter region of the rat gene. This region, however, does not contain any classical glucocorticoid responsive elements (GREs), and thus the exact mechanism by which the hormone induces expression is currently unclear. Interestingly, although the 5′ flanking sequence of human and rat Hp are considerably different, the arrangement of functional modules is surprisingly similar (Figure 2). However, only the A and B elements in the rat gene are functional, and this probably accounts for the observation that the magnitude of the IL-6 response for the human promoter is substantially greater than the rat counterpart. In addition, the human and rat A elements differ by a single nucleotide substitution, and the inducibility of the rat promoter by IL-6 can be improved severalfold by substituting that base with the corresponding human nucleotide.[24]

Although these initial studies with Hep G2 cells indicated the presence of responsive elements for IL-6 and glucocorticoids, no evidence for an IL-1 regulatory sequence could be identified.[24] In contrast to Hep G2, the promoter-proximal 165-bp region displayed a severalfold enhanced response to a combination of IL-1, IL-6, and dexamethasone in Reuber H-35 rat hepatoma cells, which respond to the inflammatory cytokines by increasing the production of most AP proteins, including Hp, to levels observed in an AP liver.[43] Basically,

ELEMENT A

Rat	AGTATGAAGCAAG
Human	...G.........

ELEMENT B

Rat	CTGG-AACAGTCA
Human	.... ..A..AT.

ELEMENT C

Rat	AAATGGAAGAAAAGGA
Human	G...T-.C....T...

FIGURE 2. Sequence comparison of elements A, B, and C from the promoter regions of rat and human Hp genes. The alignment is based on Marinkovic and Baumann.[24] For the human elements, only differences with the rat sequences are given.

a short sequence from −165 to −147, designated the A element (Figure 2), was essential for all hormonal regulation because two copies of it linked to a heterologous promoter responded to all three hormones, albeit to a lesser extent compared to the natural promoter.[71] Interestingly, the additional presence of upstream sequences selectively abolished the response to IL-1 but not to IL-6.

The differential hormonal response displayed by the same DNA sequence in different cell lines (Hep G2 and H-35) requires a certain cautiousness in the evaluation of results obtained with established hepatoma cell lines. For example, malignant transformation and/or *in vitro* cell cloning might sometimes result in partial or total loss in cytokine signal transduction pathways, and thus mask the functional involvement of important regulatory sequences. A further example of differential response to cytokines in various cell lines is described below for CRP gene regulation.

In the case of the CRP promoter, two distinct and redundant APREs are involved in IL-6-induced transcriptional activation.[59-61] The downstream APRE1 maps between nucleotides −94 and −46, while the upstream APRE2 resides between −137 and −106, although its precise boundaries have not been determined. Interestingly, these two elements function independently of each other, since removal of the downstream APRE does not affect inducibility.[53,60] Each of the CRP APREs has a complex structure which cannot be further dissected without losing IL-6 inducibility, and this property is probably due to the cooperative interaction of adjacently bound transcription factors. Indeed, the generation of linker scanning mutants in APRE1 defines two subregions which are important for CRP expression, α and β.[61] Contrary to what has been observed for Hp and Hpx, none of these, when multimerized, generates an IL-6-responsive promoter. Similarly, two important regions, γ and δ, have been defined in APRE2 on the basis of DNA-protein interactions (see Sections III.B and III.C).

Regulation of CRP expression shows some interesting differences in two different hepatoma cell lines, NPLC/PRF/5 and Hep 3B. While IL-6 is sufficient for maximal induction of CRP synthesis in NPLC/PRF/5, a cooperative effect of both IL-1 and IL-6 is necessary in Hep 3B cells.[72] In order to better define the role of these two hormones in expression of the CRP gene and precisely map the corresponding *cis*-acting elements, transient expression experiments were performed on the same series of promoter deletion mutants that were used initially for mapping the APREs. These studies were performed in Hep 3B cells that were treated with 10% MoCM as a control, IL-6 alone, IL-1 alone, or a combination of these two cytokines. As shown in Table 2, the stimulation of CAT expression observed by IL-6

TABLE 2
Cooperative Induction of CRP Expression by IL-1 and IL-6

Template	% of CAT activity given by cytokines compared to MoCM[a]			Cooperativity
	IL-1	IL-6	IL-1 + IL-6	
5′ Δ-786[b]	7.5	27.8	78.6	Yes
5′ Δ-355[b]	4.6	25.3	90.0	Yes
5′ Δ-265[b]	4.0	16.3	90.0	Yes
5′ Δ-219[b]	6.0	27.8	72.7	Yes
5′ Δ-121[b]	4.3	14.5	68.6	Yes
5′ Δ-94[b]	6.8	15.9	42.6	Yes
5′ Δ-46[b]	NE	NE	NE	No
3′ Δ-87[c]	1.0	73.0	75.0	No
3′ Δ-137[c]	NE	NE	NE	No
3′ Δ-121/-46[d]	4.8	67.2	63.2	No
3′ Δ-94/-46[d]	11.6	67.7	67.1	No
3′ Δ-121/-90[d]	0	41.2	33.6	No

[a] The cells transfected with each recombinant plasmid were stimulated with 10% MoCM, IL-6 (500 U/ml), IL-1 (1000 U/ml), and a combination of IL-1 and IL-6. The CAT activity obtained with 10% MoCM is represented as 100%, and the values obtained with the monokines are reported as a percentage of the activity with MoCM. NE, absence of expression (below 2 to 3% CAT conversion).

[b] These represent the 5′ deletion mutants on the −2500 to +15 region of the CRP promoter linked to the CAT gene. The endpoint of the 5′ deletion is indicated.

[c] These two mutants contain 3′ deletion of the CRP 5′ flanking region from −786 to +15, linked to the SV40 early promoter and CAT gene. The 3′ endpoint of the deletion is shown.

[d] The CRP promoter fragment indicated is fused to the SV40 early promoter region and CAT gene.

with the 5′ deletion mutants is only 15 to 25% compared to that obtained with MoCM. While IL-1 is only marginally active (4 to 7.5% activity compared to MoCM), a combination of both IL-1 and IL-6 acts in a highly cooperative manner. In constructs carrying both APREs (e.g., 5′ Δ-786, 5′ Δ-355, and 5′ Δ-265), the induction of CAT expression is identical to that observed with MoCM (80 to 100%). However, this cooperative effect is lost with the 3′ deletion mutants where the CRP TATA box and downstream sequences have been replaced by the heterologous SV40 promoter. In this case, IL-6 is the only effective cytokine and its activity parallels that of MoCM. In accordance with this finding, the two APREs cloned separately in front of the SV40 promoter (3′ Δ-121/−46, 3′ Δ-94/-46) are both induced by IL-6. Thus, IL-6 acts at the level of the two APREs, whereas the target sequence for IL-1 is located between −42 and +15.

To further investigate the mechanism for IL-6/IL-1 cooperativity, the CRP promoter regions from −219 to +15, −94 to +15, and −786 to −84, linked to the SV40 promoter, were fused to the bacterial neomycin gene. Analysis of the neomycin phosphotransferase activity also indicated that cooperativity was maintained by the two 5′ deletion mutants, but lost in the 3′ deletion construct.[60] Surprisingly, however, S1 mapping experiments performed with RNA from transfected cells showed that the level of correctly initiated mRNA from all three constructs was similar in the presence of MoCM, IL-6, or IL-1 plus IL-6.[60] This discrepancy between mRNA and protein levels suggests that IL-6 is the transcriptional activator of CRP expression, while IL-1 is the translational modulator. We believe that CRP transcripts are formed only in the presence of IL-6, but can be efficiently translated when IL-1 is present. The translational modulation by IL-1 is lost when the first 15 bp of the CRP transcription unit are replaced by that of the heterologous SV40 transcript. This is the first example of IL-1 acting as a translational modulator, and is in contrast with other AP genes, such as serum amyloid A, angiotensinogen, and complement C3, where IL-1 acts at the transcriptional level.[17-19] We do not currently know whether IL-1 is also active in inducing CRP production *in vivo,* since IL-6 alone is able to induce secretion of CRP in primary cultures of human hepatocytes.[42]

Li et al.[73] have also analyzed the *cis*-acting sequences necessary for IL-6-induced expression of the CRP gene in the human hepatoma cell line PLC/PRF/5. Two IL-6-inducible elements were identified. While the first element overlaps with APRE1 (−86 to −60), the second element was identified between positions −234 and −200. The reasons for this discrepancy are unclear, but in this case could be due to the different cell lines used. In addition, these workers also defined two constitutive enhancers (positions −885/−835 and −835/−904) and two negatively acting regulatory regions between −144 and −125, and −107 and −88. Both negative regions contain similarities in a short sequence motif (GCTCTGACA at −104/−96 and GTTCTGAAA at −138/−130), but the possible proteins interacting with all these sequences have not yet been identified.

B. IL-6-INDUCIBLE *TRANS*-ACTING FACTORS INTERACT WITH THE HPX, HP, AND CRP PROMOTER

The interaction of transcription factors with IL-6 REs was studied by both gel retardation and DNase-I footprinting analysis. The gel retardation studies were more informative, since no qualitative or quantitative differences in the footprint pattern could be observed with extracts from uninduced or IL-6-treated cells.[31,58,61]

In the case of CRP β- and γ-sites, a single retarded complex can be observed.[61] The formation of this complex is identical using extracts from uninduced or induced cells, and is specific to hepatocytes since it cannot be observed in HeLa cells. The protein interacting with this site has been designated H-APF-1 (hepatocyte acute phase factor 1), and is described later in detail (Section III.C).

With the Hpx A-site, Hp A-, B-, and C-regions, and CRP α- and δ-sites, several DNA-protein complexes can be observed in gel retardation assays with nuclear extracts from Hep 3B cells that are either untreated or stimulated with IL-6 (Figure 3; other data not shown). All these complexes generate an identical footprint and methylation interference pattern on the promoter regions. In the case of the Hp B-site, no IL-6-dependent changes in the formation of complexes can be observed.[58] However, with the other sites, a specific DNA-protein complex (indicated by the arrow in Figure 3) is reproducibly induced in extracts from IL-6-treated Hep 3B cells. The protein responsible for this complex was thus designated IL-6DBP (interleukin-6-dependent DNA binding protein).

Using gel retardation assays, several experiments were performed to analyze the properties of IL-6DBP.[31,58,61] The results (not shown) can be summarized as follows.

1. The induction of IL-6DBP binding activity is extremely rapid; it is already detectable 1 h after IL-6 treatment and maximal levels are achieved within 3 h.
2. The induction of IL-6DBP is not universally associated with IL-6 treatment, since it cannot be observed in CESS cells which secrete IgG in response to IL-6.

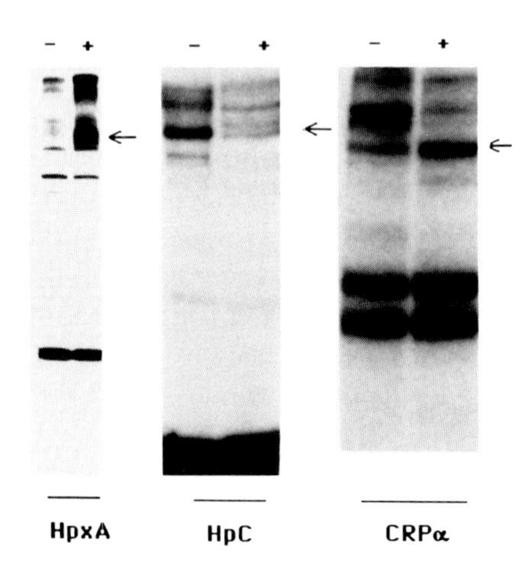

FIGURE 3. Gel retardation analysis with IL-6 REs from the Hpx, Hp, and CRP genes. Oligonucleotides containing the IL-6 RE from the promoter region of Hpx (HpxA), Hp (HpC), and CRP (CRPα) were incubated with nuclear extracts from untreated (–) or IL-6-stimulated (+) Hep 3B cells. The position of the IL-6-induced DNA-protein complex is indicated by an arrow. Similar results were obtained with the Hp A- and CRP δ-sites (data not shown).

3. In agreement with the characteristics of the induction of Hpx, Hp, and CRP gene transcription, activation of IL-6DBP also does not require *de novo* protein synthesis. This suggests that the activity of IL-6DBP is modulated by a posttranslational mechanism.

Mixing of either nuclear or whole-cell extracts from induced or uninduced cells in different ratios does not affect IL-6DBP activity. In addition, IL-6DBP is resistant to heat treatment (up to 95°C) and can be efficiently renatured after denaturation with guanidium hydrochloride. Neither treatment results in the induction of IL-6DBP in extracts from uninduced cells, thus suggesting that IL-6DBP activity is probably not regulated through its association with negative cofactors.

C. H-APF-1 IS IDENTICAL TO HEPATOCYTE NUCLEAR FACTOR 1 (LF-B1/HNF-1)

Sequence comparison between the CRP β- and γ-sites shows some homology with the consensus binding site of LF-B1/HNF-1[74] (hereafter referred to as LF-B1; Figure 4A). In gel retardation assays with hepatoma Hep 3B cells, a DNA-protein complex is formed with both the β- and γ-sites, and this complex can be specifically competed with the LF-B1 binding sites from the albumin and α_1-antitrypsin promoter (Figure 4B; other data not shown). By cross-competition experiments, it can be shown that LF-B1 interacts with the CRP sites with about ten- to 20-fold less affinity compared to the E-site from the albumin promoter. This is consistent with the observation that the CRP β- and γ-sites deviate substantially from the consensus LF-B1 binding sequence (Figure 4A). For example, the two invariant T-residues at positions 4 and 5 are substituted by two As in the β-site, while there is an insertion of an A residue in the γ-site between the two conserved Gs at positions 2 and 3, respectively.

Deletions or base-pair substitutions in either of these two LF-B1 sites abolish IL-6-inducible expression. However, two important lines of evidence indicate that LF-B1 is not the intranuclear mediator of IL-6-induced *trans*-activation: (1) in gel retardation assays, no qualitative or quantitative difference can be observed in DNA-protein interactions with both the β- and γ-sites using nuclear extracts from uninduced and IL-6-treated cells[61] (other data

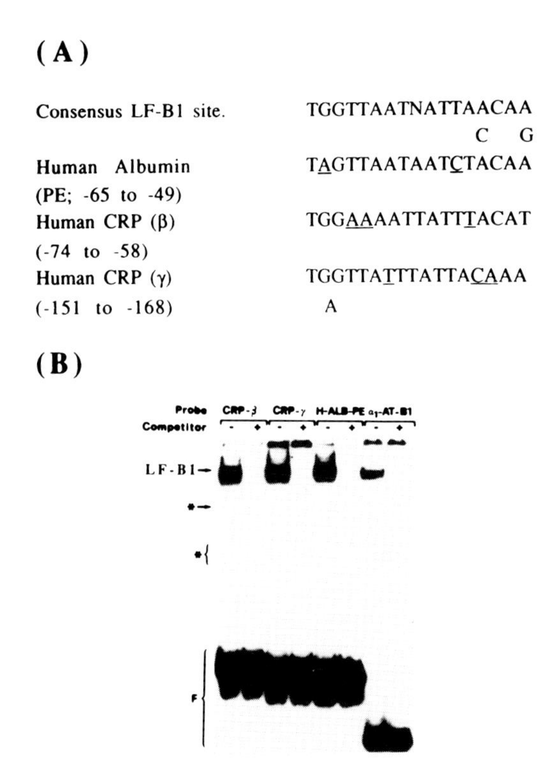

FIGURE 4. The β- and γ-sites from the CRP promoter interact with LFB1. (A) β- and γ-sites compared with the consensus LF-B1 binding site. The consensus for LF-B1 and the human albumin (PE) sequence are from Frain et al.[74] The nucleotides which diverge from the consensus are underlined. The presence of an A residue in the CRPγ site, between the two conserved Gs at positions 2 and 3 of the LF-B1 consensus, is shown below the CRP sequence. (B) Gel mobility shift analysis between nuclear extracts from Hep 3B cells and oligonucleotides containing the different canonical LF-B1 binding sites (−): β- and γ-sites from CRP (CRPβ and CRPγ), E-site from the proximal promoter of the albumin gene (H-ALB-PE),[74-76] and the B1 site from the α_1-antitrypsin promoter (α_1-AT-B1).[74] Competitions (+) were performed with 100-fold molar excess of the corresponding cold oligonucleotides. The LF-B1/DNA complex (LF-B1) and free probe (F) are shown. Asterisks indicate nonspecific DNA binding.

not shown) and (2) synthetic promoters containing multimerized copies of different LF-B1 binding sites (four copies of the β-site or seven copies of the LF-B1 site in the albumin promoter) are not transcriptionally activated by IL-6.[62]

An expression vector coding for full-length LF-B1 is capable of *trans*-activating transcription from the wild-type CRP promoter, but not from mutants that have lost the ability to interact with the protein.[62] LF-B1 has recently been shown to cooperate with other *trans*-acting factors to stimulate transcription from the albumin promoter.[75,76] Thus, the possibility that the interaction of two LF-B1 molecules to the CRP β- and γ-sites could also act synergistically to stimulate transcription was investigated. Three CRP promoter constructs were used for this analysis: Δ-94CRP-CAT, Δ-219CRP-CAT Δ-βα, and Δ-219 CRP-CAT. The first two contain only one site (β and γ, respectively), while in the third, the two sites are separated by their natural distance of about 80 bp. These three constructs were cotransfected into Hep 3B cells with increasing amounts of expression plasmid coding for LF-B1, and CAT activities were determined. The results from the transfection of LF-B1 expression vectors at concentrations of 0.1, 0.25, 0.5, and 1 μg are shown in Figure 5. At any given point, the activation of Δ-219 CRP-CAT is substantially higher than the sum of the activities of the other two constructs. In addition, although the three constructs are highly responsive, even at a low concentration of cotransfected *trans*-activator (i.e., 0.1 μg), only the Δ-219

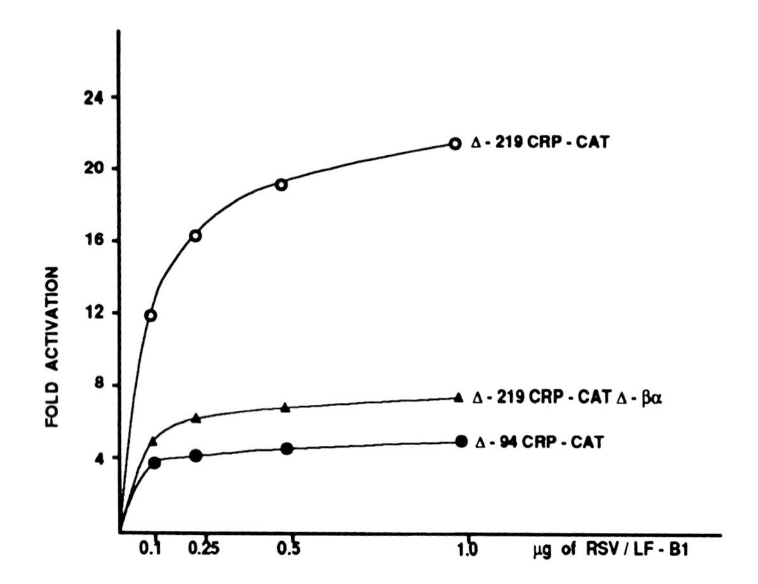

FIGURE 5. Synergistic *trans*-activation by LF-B1 interacting with two distinct sites in the CRP promoter. Constructs Δ-219 CRP-CAT (carrying two LF-B1 sites), Δ-219CRP-CAT Δ-βα, and Δ-94CRP-CAT (containing a single LF-B1 site) were transfected into Hep 3B cells with increasing amounts of RSV/LF-B1 plasmid as indicated. For each point, the ratio between the CAT activities obtained in the presence or absence of RSV/LF-B1 was calculated and reported as fold activation. The results shown are the mean of three separate experiments with different plasmid preparations.

CRP-CAT does not attain a plateau level at a high concentration of *trans*-activator. This result strongly suggests that two LF-B1 molecules bound simultaneously to sites distant from each other in the CRP promoter act synergistically to activate gene expression.

D. IL-6 REs FROM THE HPX, HP, AND CRP GENES ALL INTERACT WITH IL-6DBP

As shown in Section III.B, Hp, Hpx, and CRP IL-6 REs interact with a group of proteins, one of which (IL-6DBP) displays IL-6-dependent DNA binding properties. Alignment of the IL-6 REs from the three genes reveals interesting similarities and allows the derivation of a consensus sequence (Figure 6A), suggesting that they bind the same transcription factor(s). To investigate this possibility, binding competition experiments were performed. As shown in Figure 6B, the IL-6-induced complexes observed with the CRPα site can be specifically competed with a 100-fold excess of oligonucleotides representing the Hpx A-site and the haptoglobin A- and C-sites. Similar results were obtained when the experiment was performed with labeled oligonucleotides representing the Hpx and Hp IL-6 RE[31] (other data not shown). In addition, IL-6DBP binding sites can also be identified in the regulatory regions of other AP protein genes (e.g., α_1-acid glycoprotein) and several cytokine genes, such as TNF, IL-8, granulocyte colony-stimulating factor G-CSF, and, most importantly, IL-6 itself.[77] Furthermore, IL-6DBP interacts with its own promoter.[78] This suggests the existence of an autocrine amplification loop in which IL-6-dependent activation of IL-6DBP activity results in the induction of expression of both the cytokine and its related transcription factor.

E. A FAMILY OF FACTORS RELATED TO C/EBP INTERACT WITH IL-6 REs

The consensus sequence of IL-6 REs shows strong similarities with the recognition site of C/EBP,[79] a positive regulator of gene transcription in hepatocytes and adipocytes.[79-85]

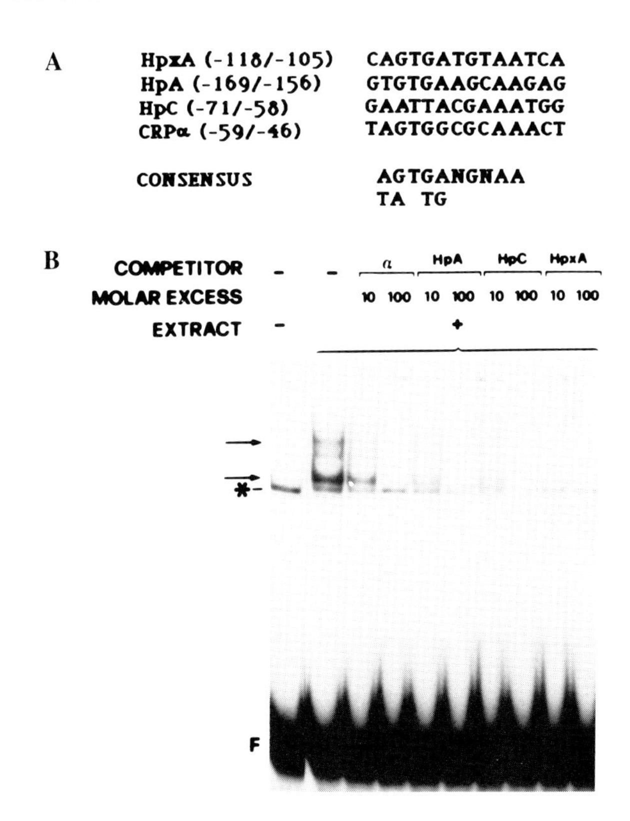

FIGURE 6. Interaction of a common IL-6-inducible transcription factor with the IL-6 RE from the Hpx, Hp, and CRP genes. (A) Sequences of the IL-6 RE from the Hpx (site A), Hp (sites A and C), and CRP (α-site); (B) Gel-shift competition experiment with radiolabeled CRP α-oligonucleotide using nuclear extracts from uninduced (−) or IL-6-stimulated (+) Hep 3B cells. Competitions were performed with a ten- or 100-fold molar excess of unlabeled oligonucleotides as indicated. The position of two IL-6-inducible complexes are shown by arrows. F and asterisks represent free probe and nonspecific DNA binding, respectively.

C/EBP has a bipartite DNA binding domain consisting of a dimerization interface, termed the leucine zipper, and a DNA contact surface containing clusters of basic amino acids.[79,86-87] Indeed, in gel retardation assays, recombinant C/EBP interacts with IL-6 REs from the Hpx, Hp, and CRP genes, and this can be specifically competed with an excess of HpxA oligonucleotide (Figure 7A). In addition, the majority of DNA-protein complexes between hepatoma nuclear extracts and HpxA can be competed with an excess of peptide containing the leucine zipper of C/EBP, but not from the unrelated GCN4.[63] This suggests that the DNA binding domain of IL-6DBP contains a leucine-zipper motif related to C/EBP.

The cDNA clone for IL-6DBP was isolated and characterized in detail. The recombinant protein specifically interacted with IL-6 REs from the Hpx, Hp, and CRP genes,[63] and generated an identical *o*-phenanthroline footprint with the Hpx promoter as IL-6DBP from Hep 3B cells (Figure 7B). The cDNA codes for a 297-amino acid-long polypeptide with a predicted molecular weight of 36 kDa. The same gene has been independently isolated from rat, mice, and humans by several other laboratories, and designated LAP,[88] AGP/EBP,[89] CRP2,[90] C/EBPβ,[85] and NF-IL-6.[77]

The carboxy-terminal DNA binding domain of IL-6DBP contains a heptad repeat of leucine residues (leucine zipper) preceded by clusters of basic amino acids. This is consistent with the observations that recombinant IL-6DBP binds DNA as a dimer.[63] Interestingly, this C-terminal region of IL-6DBP is 96% identical to the corresponding region of C/EBP.[63]

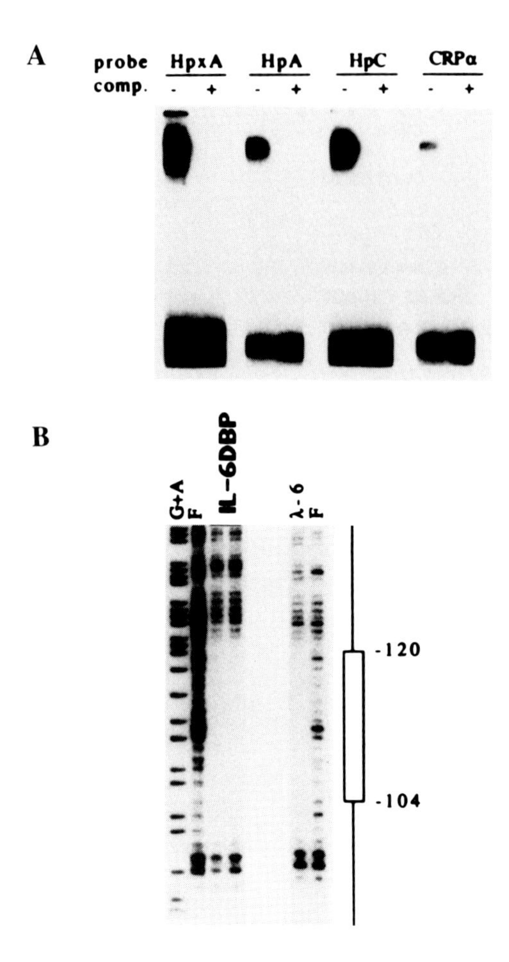

FIGURE 7. Analysis of DNA-protein interactions between recombinant C/EBP and IL-6DBP with IL-6 REs. (A) Extracts from bacterially expressed C/EBP were used in gel retardation analysis with different radiolabeled probes (HpxA, HpA, HpC, and CRPα), carrying the respective IL-6 REs. Either none (−) or 10 pmol of cold HpxA double-stranded oligonucleotide (+) was used as a specific competitor. (B) *In situ* footprinting was performed on the DNA-protein complex between the Hpx promoter fragment and either recombinant IL-6DBP (λ6) or IL-6DBP from IL-6-treated Hep 3B cells (IL-6DBP). The G + A sequence ladder of the promoter fragment is also indicated. F, free probe.

Upstream from this region, the two proteins show only 26% identity. The high degree of homology in the C-terminal DNA-binding domain suggested that IL-6DBP and C/EBP can form heterodimers with each other. To investigate this possibility, gel retardation experiments were performed with C/EBP and truncated IL-6DBP (containing only the DNA-binding domain) synthesized *in vitro*. While both C/EBP and truncated IL-6DBP generated a slower and faster migrating complex, respectively, a new retarded complex of intermediate mobility could be observed when the two proteins were either cotranslated or mixed after translation, indicating the formation of a heterodimer between the two polypeptide species (Figure 8). Similar results were obtained using truncated C/EBP and full-length IL-6DBP.[63]

Although subunits of C/EBP and truncated IL-6DBP exchange rapidly,[63,79] a comparable exchange with full-length IL-DBP can only be observed when the temperature of mixing is increased.[63] This difference in the exchange kinetics between truncated and full-length IL-6DBP suggests that its N-terminal portion affects dimer formation or stability, probably through a modification of the exposure of the leucine zipper. It is thus possible that the role

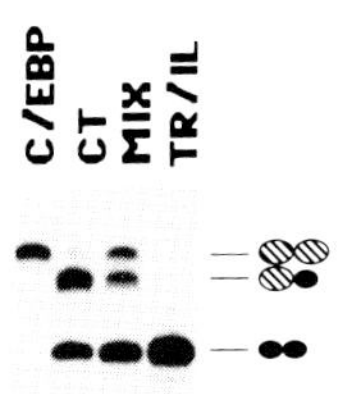

FIGURE 8. Heterodimeric interactions *in vitro* between IL-6DBP and C/EBP. Full-length C/EBP or truncated IL-6DBP (TR/IL) were either cotranslated (CT) or mixed (MIX) after translation for 15 min at 20°C. The proteins were then used in a gel retardation assay with the HpxA oligonucleotide as a probe. The composition of the different complexes is shown on the right side of the figure: large or small hatched circles represent full-length C/EBP and truncated IL-6DBP, respectively.

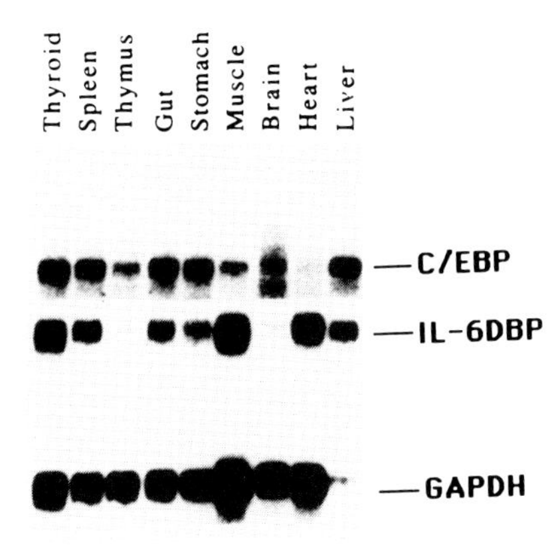

FIGURE 9. Northern-blot analysis. Poly A^+ RNA (4 μg) from the rat tissues indicated was size fractionated by agarose gel electrophoresis and transferred onto a nylon membrane. The same filter was subsequently hybridized with radiolabeled fragments from IL-6DBP and C/EBP, which do not cross-react, and with a labeled glyceraldehyde phosphate dehydrogenase (GAPDH) cDNA used as an internal control.

of exchange of IL-6DBP subunits might depend on changes in the structural conformation of the protein. If IL-6 influences IL-6DBP folding, then this change might have the consequence of shifting the equilibrium between homodimers and heterodimers with other members of the C/EBP family. If homodimers and heterodimers have different binding affinities for different promoters (through positive or negative interactions with nearby transcription factors), this would result in redirecting transcription factors from some promoters to others. This would explain the partial transcriptional inactivation of promoters, such as that of the albumin gene during the APR, and the induction of other promoters such as those described in this study.

F. EXPRESSION ANALYSIS OF IL-6DBP

The tissue distribution of both IL-6DBP and C/EBP was investigated by Northern-blot analysis on RNA extracted from various rat tissues (Figure 9). After normalization to constitutive glyceraldehyde phosphate dehydrogenase mRNA, both C/EBP and IL-6DBP mRNAs are found to be relatively more abundant in the liver. The levels of IL-6DBP are also high in thyroid, skeletal muscle, and heart. In contrast, C/EBP levels are low in muscle and heart. In all the other tissues analyzed, C/EBP and IL-6DBP mRNA levels are comparable. Coexpression of IL-6DBP and C/EBP in several tissues suggests the existence of heterodimers *in vivo,* in conjunction with homodimers. The *in vivo* mRNA levels for IL-6DBP and C/EBP undergo an inverse regulation during inflammation: IL-6DBP mRNA increases, but C/EBP mRNA is strongly downregulated.[77,91]

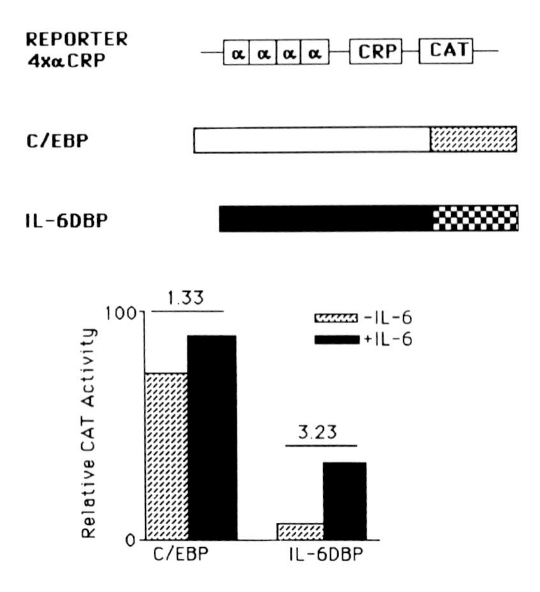

FIGURE 10. Transcriptional activity of IL-6DBP is induced by IL-6. DNA transfections and CAT assays were performed as described in Section II. A schematic representation of the constructs used (IL-6DBP, C/EBP, and reporter 4xαCRP-CAT) is shown. The histograms represent CAT activities obtained from Hep 3B cells that are either untreated or stimulated with IL-6. Numbers over the histograms indicate the induction ratios.

The tissue distribution of IL-6DBP mRNA, however, does not reflect the actual protein levels. For example Descombes et al.[88] showed that in rats, LAP (IL-6DBP) transcripts are most abundant in lung, about five- to tenfold less abundant in liver, spleen, and kidney, and sparser in brain and testes. In contrast, the protein is most abundant in liver nuclei, five- to tenfold lower in lung, and almost undetectable in spleen, brain, testis, and kidney. Analogous results have been obtained by Williams et al.[90] Unfortunately, similar experiments have not yet been performed following experimental inflammation. It is therefore not possible to say if, in conditions which induce IL-6DBP or downregulate C/EBP transcription, the translational control is removed or maintained.

G. THE *TRANS*-ACTIVATION POTENTIAL OF IL-6DBP IS INDUCED BY IL-6

In order to demonstrate that IL-6DBP acts as a transcriptional activator, a series of cotransfection experiments were performed using hepatoma Hep 3B cells. In this system, expression plasmids bearing IL-6DBP or C/EBP cDNAs, under the control of strong eukaryotic promoters, are tested for their ability to *trans*-activate a IL-6 RE-CAT reporter gene in the presence or absence of IL-6. The reporter plasmid used, 4xαCRP-CAT, contains four copies of the IL-6DBP site from the APRE1 of the CRP gene linked to its own TATA box and transcriptional initiation site.[61] The results are shown in Figure 10. In contrast to C/EBP, which *trans*-activates in an IL-6-independent manner, activation by IL-6DBP is induced on average three- to fourfold by IL-6. In some cases, up to an eightfold induction has been observed (data not shown). Similar results were obtained using a different reporter gene, pBLD9-CAT, containing nine copies of the albumin D-site, that interacts with both C/EBP and IL-6DBP, fused to the thymidine kinase promoter.[63]

H. IL-6DBP AND C/EBP FUNCTIONALLY INTERACT *IN VIVO*

As IL-6DBP and C/EBP form heterodimers *in vitro*, it is likely that they also interact *in vivo*. To explore this possibility, the activity of the 4xαCRP-CAT reporter gene in cells cotransfected with equal amounts of both IL-6DBP and C/EBP was investigated (Figure

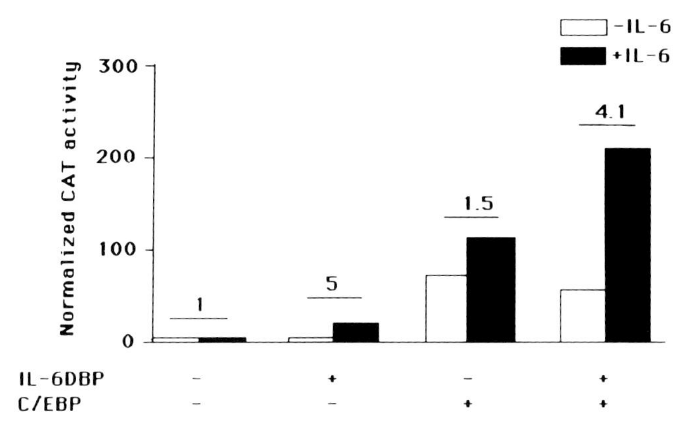

FIGURE 11. Cooperative interaction between C/EBP and IL-6DBP in the presence of IL-6. The reporter plasmid, 4xαCRP-CAT (see Figure 10) was cotransfected into Hep 3B cells with 1.25 μg of expression plasmids for IL-6DBP, C/EBP, or both. Histograms represent the CAT activities obtained in untreated (untr.) or IL-6-treated (IL-6) Hep 3B cells. The induction ratios obtained are shown above the histograms.

11). The amount of expression plasmid used (1.25 μg) is in the range of the linear increase in the activity of target gene transcription. In the absence of IL-6, the activity of the reporter gene is partially, but reproducibly, reduced compared to transfection of C/EBP alone. In IL-6-treated cells, the activation observed is substantially higher than the sum of values obtained when C/EBP and IL-6DBP are separately transfected. The fourfold induction observed is comparable to that obtained with IL-6DBP. The simplest explanation of these results is that C/EBP:IL-6DBP heterodimers are less active than C/EBP in constitutive *trans*-activation (absence of IL-6), but more active than the corresponding homodimers in the presence of IL-6. This is an example of an IL-6-independent activator, C/EBP, being recruited into an IL-6-dependent pathway by formation of heterodimers with IL-6DBP.

I. THE AMINO TERMINUS OF IL-6DBP CONTAINS SUFFICIENT INFORMATION FOR BOTH CONSTITUTIVE AND IL-6-INDUCED *TRANS*-ACTIVATION

To identify regions in IL-6DBP that are essential for transcriptional activation, domain-swap experiments were performed with the C/EBP DNA binding domain (see Figure 10). A truncated form of IL-6DBP (IL-6DBP/BD), coding only for the DNA binding domain, fails to *trans*-activate despite being expressed at a level comparable to the full-length protein, and also efficiently translocated to the nucleus (Figure 12A; other data not shown). This suggests that the transcriptional activation domain of IL-6DBP resides in the amino terminus of the protein, but does not allow separation of the information for constitutive and IL-6-induced activity. The same conclusion is supported by the observation that the N-terminal portion of IL-6DBP can impart the same level of constitutive and IL-6-induced activity of the reporter gene when fused to the heterologous C/EBP DNA binding domain (Figure 12A).

We are currently mapping regions in the N terminus of IL-6DBP that are responsible for IL-6-induced transcriptional activation. To help in this investigation, a sequence encoding a peptide tag from the C terminus of the vesicular stomatitis virus G-protein[70] was inserted into all the recombinant IL-6DBP derivatives. A Western-blot analysis of cells transfected with IL-6DBP containing this tag sequence is shown in Figure 12B. Apart from providing a cleaner detection system, the presence of the tag allows an easy discrimination between transfected and endogenous protein. It is thus possible to accurately measure the expression of manipulated proteins (e.g., bearing mutations or deletions) in transfected cells, and also to identify whether they are efficiently translocated to the nucleus. The insertion of this

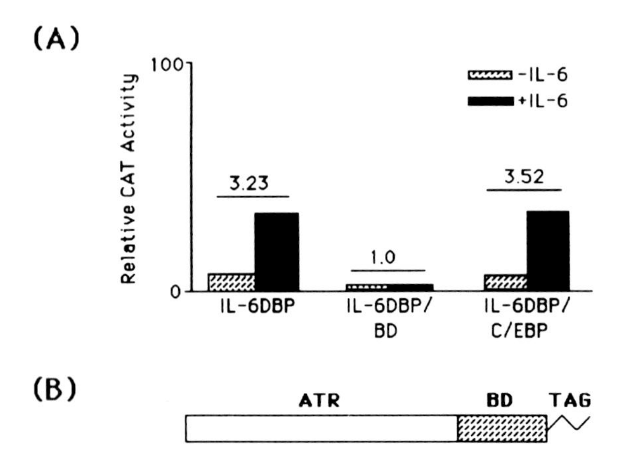

FIGURE 12. Activities and expression analysis of manipulated IL-6DBP derivatives. (A) IL-6DBP/C/EBP contains the N terminus of IL-6DBP fused in-frame with the DNA binding domain of C/EBP, while IL-6DBP/BD represents the DNA binding domain of IL-6DBP. The CAT activities obtained from Hep 3B cells that are either untreated or stimulated with IL-6 are shown as histograms, with induction ratios indicated above them. (B) Use of peptide tag in detection of recombinant IL-6DBP. Hep 3B cells were transfected with the IL-6DBP/C/EBP chimera containing the tag sequence in the DNA binding domain. The cells were then either untreated (−) or treated with (+) IL-6 and analyzed for Western blot using the anti-tag antibody. Two species of 36 and 33 kDa can be detected. ATR represents the N terminus of IL-6DBP, while BD is the DNA binding domain of C/EBP.

peptide does not interfere with several biochemical properties of the protein such as DNA binding, dimerization, nuclear translocation, and *trans*-activation (data not shown).

J. THE ACTIVITY OF IL-6DBP IS INDUCED BY A POSTTRANSLATIONAL MECHANISM

In order to understand the molecular mechanisms responsible for IL-6DBP activation, polyclonal antibodies against the protein were used in Western-blot analysis of cell extracts from Hep 3B cells that are either untreated or stimulated with IL-6. As shown in Figure 13A, IL-6DBP is expressed constitutively in Hep 3B cells and its levels are not increased by IL-6 treatment. Recently, Metz and Ziff[92] have shown that in rat pheochromocytoma PC12 cells, cAMP stimulates IL-6DBP to *trans*-locate to the nucleus and activate c-*fos* transcription. However, no change in nuclear translocation of IL-6DBP can be observed in Hep 3B cells in either the presence or absence of IL-6 (Figure 13B). This suggests that IL-6 induces the activity of preexisting IL-6DBP by a posttranslational mechanism. This modification is subtle, since no significant alterations in the electrophoretic mobility of IL-6DBP can be observed by both one- and two-dimensional gel electrophoresis (data not shown).

IV. DISCUSSION

A. IL-6-RESPONSIVE TRANSCRIPTION FACTORS: IL-6DBP AND IL-6 RE-BP

Studies described in Section III of this chapter clearly demonstrate that IL-6DBP plays a key role in the coordinate and inducible expression of Hpx, Hp, and CRP. However, IL-6DBP is not the only factor involved in IL-6-induced activation of AP gene transcription. For example, in Hep 3B stably expressing exogenous copies of NF-IL-6, the average induction of Hp production in response to IL-6 is substantially lower compared to that of nontransfected cells, although the absolute production and secretion of Hp is much higher.[93]

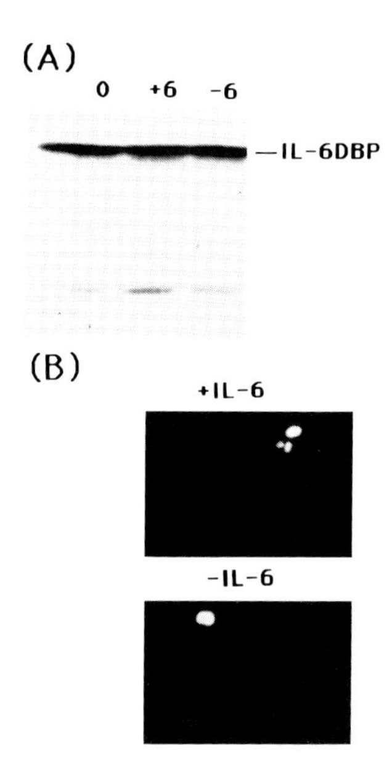

FIGURE 13. Western-blot and indirect immunofluorescence analysis of IL-6DBP. (A) Hep 3B cells were either untreated (−) or treated (+) with IL-6 for the indicated times. Equal amounts of total cell extracts were then size fractionated by SDS-PAGE and analyzed for Western blot using polyclonal antibodies against IL-6DBP. (B) Indirect immunofluorescence performed on Hep 3B cells transfected with recombinant IL-6DBP, and either untreated or treated with IL-6.

This suggests that other factors, in conjunction with IL-6DBP, are responsible for full activation. In addition, transfection of NF-IL-6 in the same cells does not exert any positive effect on IL-6-dependent induction of the fibrinogen gene, a type II AP reactant.[93]

The existence of other transcription factors that are activated by IL-6 is also suggested by studies on another type II gene, α_2-macroglobulin (α_2-M). Although IL-6 alone is capable of inducing transcription of the α_2-M gene, this effect is amplified by glucocorticoids.[5,94] Using a series of 5′ deletion mutants, Kunz et al.[94] identified two IL-6 REs (from −852 to −777 and −336 to −165). These regions contain a core sequence motif (CTGGGAA) that is also present in the promoter regions of several other AP genes, including Hp and CRP (Figure 14). A direct role for this consensus in IL-6-induced transcription was provided by the observation that two copies of the element from positions −176 to −159 of α_2-M promoter (ATCCTT**CTGGGAA**TTCTG) can confer a 20-fold stimulation of a heterologous SV40 promoter in response to IL-6.[5,94]

Hattori et al.[101] have performed a detailed analysis of transcription factors interacting with this type of IL-6 RE (referred to as type II IL-6 RE to distinguish it from the type I IL-6 RE which forms the binding site for IL-6DBP). Maximal induction of transcription by IL-6 was observed when the 5′ flanking sequences of α_2-M, from positions −209 to −43, directed expression from the gene's own TATA box and transcriptional initiation site. Removal of the CTGGGA motif (−164 to −159) abolished 60 to 70% of the IL-6 induction in rat hepatoma FAO1 cells. Using a fragment of α_2-M promoter (−209 to −43) in gel retardation assays, a specific DNA-protein complex (complex I) was observed with nuclear extracts from normal rat liver. In contrast, a faster migrating complex (complex II) was assembled with nuclear proteins from either AP rat livers or Hep 3B cells treated with IL-6. Furthermore, complex II could be specifically competed with oligonucleotides containing the conserved IL-6 RE from other AP genes (e.g., rat α_1-acid glycoprotein, murine serum amyloid A3 [SAA3], and γ-fibrinogen). The protein responsible for the formation of this

GENE	LOCATION	SEQUENCE	REFERENCES.
rα2M	-168	CTGGGAA	5, 94
rAGP	-5199	CTGGGAA	95
hHp	-117	CTGGAAA	34,58
hCRP	-76	TTGGAAA	53, 59-61
mSAA	-79	GTGGGAT	96
rα-fibrin.	-129	CTGGGAT	97
rβ-fibrin.	-150	CTGGGAA	97
rγ-fibrin.	-151	CTGGGAA	97
rTTR	-322	CTGGGAA	98
rT-kin.	-111	CTGGGTA	99
hα1-AT	-116	CTGGGAT	100
CONSENSUS		CTGGGAA	

FIGURE 14. The presence of the core consensus sequence (CTGGGAA) in the promoter regions of several AP genes. r, rat; h, human; m, mouse; α_2-M, α_2-macroglobulin; AGP, α_1-acid glycoprotein; Hp, haptoglobin; CRP, C-reactive protein; SAA, serum amyloid A; fibrin, fibrinogene; TTR, transthyretin; T-kin, T-kininogen; α_1-AT, α_1-antitrypsin.

complex, designated IL-6 RE-BP, is thus a specific second nuclear target for the IL-6 response of AP genes containing the conserved CTGGGAA motif. This protein is different from IL-6DBP, since it is much larger (46 kDa compared to 36 kDa for IL-6DBP) and is heat sensitive.[102] In addition, the IL-6-induced formation of complex II requires ongoing protein synthesis.[102] Thus, IL-6 probably activates simultaneously multiple pathways, resulting in the cytokine-specific changes of at least two independent *trans*-acting factors, IL-6DBP and IL-6 RE-BP. In addition to factors specific for APR, other transcription factors might also be induced by IL-6. For example, IL-6 induces *jun*B transcription in a B-cell hybridoma.[103]

Interestingly, the conserved CTGGGAA motif is also present in the promoter region of the Hp and CRP genes (Figure 14). In the Hp gene, this region corresponds to the B-site (Section III.A), while in the CRP gene, it overlaps the LF-B1 binding site (Sections III.B and III.C). In both cases, circumstantial evidence has been provided for the role of this region in IL-6-induced activation. For example, site-directed mutagenesis of the Hp B-site drastically affects IL-6 inducibility.[58] In addition, of the two mutants in the CRP CTGGGAA motif which diverge from the consensus sequence, one affects IL-6 inducibility.[61] In this case, the mutation introduces four base changes, the first of which substitutes one of the constant A-residues of the consensus into a G. For several reasons, however, a convincing role for this motif in IL-6-dependent regulation of Hp and CRP gene expression cannot currently be provided. (1) The Hp B-site and CRP β-site (binds LF-B1) cannot confer IL-6 inducibility to a heterologous promoter.[58,61] (2) No qualitative or quantitative changes in the DNA-protein interactions can be observed with both these sites using extracts from uninduced and IL-6-induced cells.[58,61] (3) A CRP promoter mutant that precedes the core consensus motif severely impairs inducibility without touching the consensus sequence.[61] Nevertheless, we cannot exclude the possibility that IL-6 RE-BP may play an as yet unidentified, but important role in modulating IL-6-dependent expression of both Hp and CRP genes.

B. IL-6DBP BELONGS TO A FAMILY OF C/EBP-RELATED FACTORS: MODULATION OF AP GENE TRANSCRIPTION BY HETERODIMERIC INTERACTIONS

IL-6DBP belongs to the C/EBP class of transcription factors. These proteins are positive regulators of gene transcription and bind DNA as dimers. Their DNA binding domain consists of a dimerization interface, termed the leucine zipper, and a DNA contact surface containing clusters of basic amino acid residues. Several members of this family have recently been cloned and show substantial homology in their DNA binding domain (Figure 15). These include: C/EBP (C/EBPα),[79-87] IL-6DBP (NF-IL-6,[77] LAP,[88] AGP/EBP,[89] CRP2,[90] and C/EBPβ[85]), and Ig/EBP-1[104] (C/EBPγ), CRP1,[90] and C/EBPδ.[85]) With the exception of Ig/EBP-1, which is expressed ubiquitously,[104] the other members show a restricted, but partially overlapping, expression pattern in different tissues.[85,88,90] The different members are capable of forming heterodimers in all intrafamilial combinations.[85,88,90,104] Thus, heterodimeric interactions between the different members may potentially play an important role in the modulation of AP gene transcription. Indeed, functional interaction between IL-6DBP and C/EBP has been demonstrated (Section III.H). In the presence of equimolar amounts of IL-6DBPand C/EBP, the latter, which normally activates transcription in an IL-6-independent manner, is recruited into heterodimers whose activity is regulated by IL-6.

On the basis of the above observations, a possible mechanism for modulation of gene transcription could be by AP-mediated changes in the levels of expression of different C/EBP members, and their ability to form heterodimers with IL-6DBP. Indeed, Akira et al.[77] have shown that NF-IL-6 mRNA is present in trace amounts in normal mice, but can be induced severalfold by treatment with LPS or IL-6. In contrast, the levels of C/EBP mRNA decrease considerably after IL-6 stimulation.[91] Similarly, D. Ron[105] has recently shown that in 3T3-L1 fibroblasts, TNF induces a decrease in C/EBP protein levels, by affecting its rate of synthesis, and a reciprocal increase in IL-6DBP. In addition, AP-induced alterations in other members of the C/EBP family were also observed using radiolabeled IL-6DBP DNA-binding domain as a probe.

Interestingly IL-6DBP, CRP1, C/EBP, and C/EBPδ, all contain a conserved cysteine residue at or near the C terminus, immediately following the leucine zipper (Figure 15). This allows the formation of efficient covalently linked dimers *in vitro*.[90] Although no gross differences between covalent and noncovalent dimers in binding to DNA were detected,[90] it is conceivable that linked dimers are necessary for stabilizing subunit interactions *in vivo*, and thus maintain a homeostatic population of dimers within a cell. In this framework, changes in the redox potential of the cell could provide a potential regulatory mechanism for modulating transcription.

It must be emphasized that the presence of a binding site for IL-6DBP does not necessarily imply that the gene is induced by IL-6. For example, IL-6DBP interacts with the promoter of the albumin gene, which is a negative AP reactant. This suggests that sequences adjacent to the recognition site or the context of the site with respect to binding sites for other transcription factors might determine whether a particular gene is regulated by IL-6. In addition, as all the C/EBP-related factors interact with identical recognition sequences, the partners used in heterodimeric interaction may also play a role in determining IL-6 inducibility.

C. UNIQUE PROPERTIES OF IL-6DBP: REGULATION BY NUCLEAR TRANSLOCATION AND GENERATION OF INHIBITOR PROTEIN

IL-6DBP also interacts with the serum response element of the c-*fos* promoter. Metz and Ziff[92] have recently shown that the activity of IL-6DBP is regulated by cAMP in rat pheochromocytoma PC12 cells. Following forskolin treatment, IL-6DBP becomes

```
C/EBP    MESADFYEAEPRPPMSSHLQSPPHAPSNARLWLSPGRGPRAAPSPTCRPG    50
CRP1     --------------------------------------------------
IL6-DBP  MHRLLAWDAACLPPPPAAFRPMEVANFYYEP-DCLAYGAKAARAAPRAPA    49
C/EBPδ   MSAALFSLDS---PVRGTPWPTEPAAFYEPGRVDKPGRGPEPGELGELGS    47
C/EBPγ   --------------------------------------------------

C/EBP    AAGRICEHETSIDISAYIDP-----------AAFNDEFLADLFQHSRQQE    89
CRP1     ----MCEHEASIDLSAYIES-------------GEEQLLSDLFAMKPTPE    33
IL6-DBP  AEPAIGEHERAIDFSPYLEPLAPAAADFAAPAPAHHDFLSDLFAD---DY    96
C/EBPδ   TTPAMYDDESAIDFSAYIDSMAAVPTLELC----HDELFADLFNSNHK--    91
C/EBPγ   --------------------------------------------------

C/EBP    KAKAAAGPA--------GGGGDFDYPGAPAGPGGAVMSAGAHGPPPGYGC   131
CRP1     ARSLKGPGTPSFPHYLPADPRPFAYPSHTFGPDRKALGPGIYSNPGSYDP    83
IL6-DBP  GAKPS---------------KKPSDYGYVSLGRAGAKAAPPACFPPP---   128
C/EBPδ   ----------------------------AAGAGGLELLQGGPTRPPGVGS   113
C/EBPγ   --------------------------------------------------

C/EBP    AAAGYLDGRLEPLYERVGAPALRPLVIKQEPREEDEAKQLALAGLFPYQP   181
CRP1     RA----------------------VAVKEEPRGPEGNRGTGRGSYNP---   108
IL6-DBP  ----------------------PPAALKAEPGFEPADCKRADDAPAMAAG   156
C/EBPδ   VARG---------------------PLKREPDWGDGDAPG----------   132
C/EBPγ   -----------------------HELAGCEAWVGSGRAAPRGAARSGGRL    27

C/EBP    PPPPPPPHPHASPAHLAAPHLQFQIAHCGQTTM-------------HLQP   218
CRP1     --------------------LQYQVAHCGQTAV-----------------   121
IL6-DBP  FPFALR------------AYLGYQATPSGSSGS-------------LSTS   181
C/EBPδ   ------------------SLLPAQVAVCAQTVV------------SLAAA   152
C/EBPγ   GAGGPAGAQVH-------VKIFWQLSVASSDHIALISTYSVLAKERAQMS    70
```

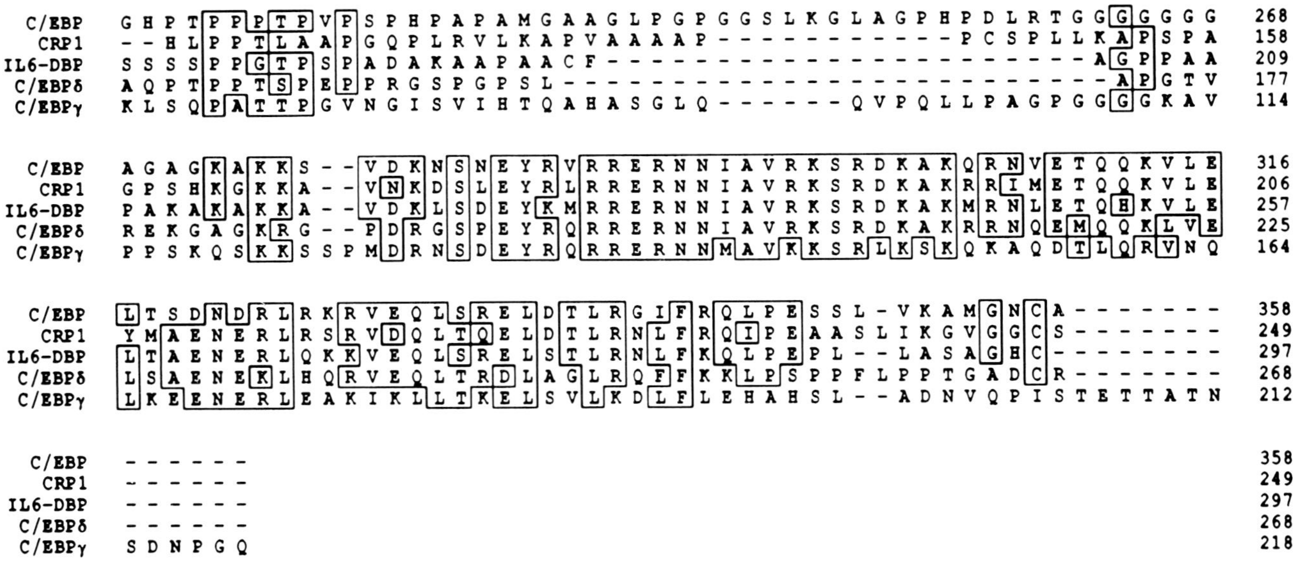

```
C/EBP    G H P T P P P T P V P S P H P A P A M G A A G L P G P G G S L K G L A G P H P D L R T G G G G G G G   268
CRP1     - - H L P P T L A A P G Q P L R V L K A P V A A A A P - - - - - - - - - - - - P C S P L L K A P S P A   158
IL6-DBP  S S S S P P G T P S P A D A K A A P A A C F - - - - - - - - - - - - - - - - - - - - - - A G P P A A   209
C/EBPδ   A Q P T P P T S P E P P R G S P G P S L - - - - - - - - - - - - - - - - - - - - - - - - - A P G T V   177
C/EBPγ   K L S Q P A T T P G V N G I S V I H T Q A H A S G L Q - - - - - - - Q V P Q L L P A G P G G G G K A V   114

C/EBP    A G A G K A K K S - - V D K N S N E Y R V R R E R N N I A V R K S R D K A K Q R N V E T Q Q K V L E   316
CRP1     G P S H K G K K A - - V N K D S L E Y R L R R E R N N I A V R K S R D K A K R R I M E T Q Q K V L E   206
IL6-DBP  P A K A K A K K A - - V D K L S D E Y K M R R E R N N I A V R K S R D K A K M R N L E T Q H K V L E   257
C/EBPδ   R E K G A G K R G - - P D R G S P E Y R Q R R E R N N I A V R K S R D K A K R R N Q E M Q Q K L V E   225
C/EBPγ   P P S K Q S K K S S P M D R N S D E Y R Q R R E R N N M A V K K S R L K S K Q K A Q D T L Q R V N Q   164

C/EBP    L T S D N D R L R K R V E Q L S R E L D T L R G I F R Q L P E S S L - V K A M G N C A - - - - - - - -   358
CRP1     Y M A E N E R L R S R V D Q L T Q E L D T L R N L F R Q I P E A A S L I K G V G G C S - - - - - - - -   249
IL6-DBP  L T A E N E R L Q K K V E Q L S R E L S T L R N L F K Q L P E P L - - L A S A G H C - - - - - - - - -   297
C/EBPδ   L S A E N E K L H Q R V E Q L T R D L A G L R Q F F K K L P S P P F L P P T G A D C R - - - - - - - -   268
C/EBPγ   L K E E N E R L E A K I K L L T K E L S V L K D L F L E H A H S L - - A D N V Q P I S T E T T A T N   212

C/EBP    - - - - - -     358
CRP1     - - - - - -     249
IL6-DBP  - - - - - -     297
C/EBPδ   - - - - - -     268
C/EBPγ   S D N P G Q     218
```

FIGURE 15. Comparison of the amino acid sequences of the different C/EBP-related factors. Identical residues are boxed.

phosphorylated and *trans*-locates to the nucleus, where it induces c-*fos* transcription. However, this is not likely to be a general mechanism for IL-6-induced activation of IL-6DBP in all IL-6-responsive cell types, since the protein is localized in the nucleus in Hep 3B cells, in both the presence and absence of IL-6 (Figure 13B).

IL-6DBP (also called LAP[88]) contains three in-frame AUGs that are differentially recognized by the translational machinery due to a leaky ribosome scanning mechanism.[106] The first two are close to each other, and their utilization leads to the formation of transcriptionally active proteins — LAP* (the longer) and LAP (the shorter). The third AUG occurs upstream of the DNA binding domain, and its use generates the transcriptional inhibitor, LIP, that lacks the activation domain and has a higher DNA binding activity for its cognate DNA sequences. Transcriptional activation does not reflect the absolute amount of LAP, but, rather, the LAP/LIP ratio.[106] For example, a modest increase in the LAP/LIP ratio results in a significantly higher activation of the target genes. The LAP/LIP ratio increases about fivefold during terminal rat liver differentiation, and is thus likely to play a role in the modulation of liver-specific gene expression in intact animals.[106] However, this mechanism is probably not involved in IL-6-induced activation of IL-6DBP, at least in Hep 3B cells, since deletion of this region has no effect on the *trans*-activation potential of IL-6DBP in transient expression studies.[107] In addition, no alterations in the LAP/LIP ratio by IL-6 can be detected.[108]

D. IL-6 INDUCES IL-6DBP ACTIVITY VIA A POSTTRANSLATIONAL MECHANISM

Several lines of evidence suggest that direct posttranslational modification of IL-6DBP is probably involved in IL-6-induced transcriptional activation. For example, the induction of Hpx, Hp, and CRP transcription and the appearance of IL-6DBP activity in Hep 3B cells does not require *de novo* protein synthesis (Sections III.A and III.B). In addition, IL-6DBP is constitutively expressed in Hep 3B cells and its levels are not increased by IL-6 treatment (Figure 13A). Phosphorylation is the most likely modification, since phosphatase treatment abolishes the DNA-protein complex formation between IL-6DBP and its target sequences (see the discussion sections in References 61 and 91). Indeed, changes in the phosphorylation status of several transcription factors have recently been shown to be involved in modulating their activity.[109-114] This includes cAMP-mediated phosphorylation of IL-6DBP in PC12 cells and its subsequent translocation to the nucleus.[92]

The N terminus of IL-6DBP contains sufficient information for both constitutive and IL-6-induced *trans*-activation (Figure 12A). Within this domain, distinct regions necessary for IL-6 induction, basal *trans*-activation, and modulation of the activation potential have been mapped.[107] The IL-6-responsive signal is located in a 60-amino acid stretch in the amino terminus of IL-6DBP.

E. CONCLUSIONS AND FUTURE PERSPECTIVES

The three members of class I AP genes analyzed by our laboratory all contain in their promoter region at least one IL-6 APRE which is able to interact with C/EBP-related transcription factors. All these factors are characterized by a leucine zipper/basic amino acid DNA binding domain through which they can form heterodimers *in vivo* and *in vitro*. Among these proteins, IL-6DBP emerges as a clear signal transducer of IL-6 response. However, we cannot exclude the possibility that modulation of heterodimeric interactions between the different members, with potentially distinct activation properties, could influence the regulation of AP gene transcription. Given the number of members belonging to the family, which is likely to expand in the future, it is vital to establish an *in vitro* transcription system that mimics aspects of the AP reaction. This would allow direct evaluation of the role of individual members in transcriptional activation. Alternatively, cell lines lacking the different members can be used to directly assess their role *in vivo* by transfection studies.

Despite the complexity of the system, we have been able to demonstrate that IL-6DBP plays a key role in IL-6-dependent activation of AP gene transcription. The N terminus of IL-6DBP contains sufficient information for both basal and IL-6-induced gene transcription. The *trans*-activation potential of IL-6DBP is induced by IL-6 via a posttranslational mechanism. Identification of this mechanism is a necessary step in understanding the IL-6 signal transduction pathway.

REFERENCES

1. **Fey, G. H. and Fuller, G. M.,** Regulation of acute phase gene expression by inflammatory mediators, *Mol. Biol. Med.,* 4, 323, 1987.
2. **Ciliberto, G.,** Transcriptional regulation of acute phase response genes with emphasis on the human C-reactive protein gene, in *Acute Phase Proteins and the Acute Phase Response,* Pepys, M., Ed., Springer-Verlag, New York, 1989, 29.
3. **Baumann, H.,** Hepatic acute phase reaction *in vivo* and *in vitro, In Vitro Cell. Dev. Biol.,* 25, 115, 1989.
4. **Fey, G. H. and Gauldie, J.,** The acute phase response of the liver in flammation, in *Progress in Liver Diseases,* Vol. 9, Popper, H. and Schaffner, F., Eds., W. B. Saunders, Philadelphia, 1990, 89.
3. **Heinrich, P. C., Castell, J. V., and Andus, T.,** Interleukin-6 and the acute phase response, *Biochem. J.,* 265, 621, 1990.
6. **Sehgal, P. B.,** Interleukin-6: a regulator of plasma protein gene expression in hepatic and non-hepatic tissues, *Mol. Biol. Med.,* 7, 117, 1990.
7. **Baumann, H. and Gauldie, J.,** Regulation of hepatic acute phase plasma protein genes by hepatocyte stimulating factors and other mediators of inflammation, *Mol. Biol. Med.,* 7, 147, 1990.
8. **Geiger, T., Andus, T., Klapproth, J., Hirano, T., Kishimoto, T., and Heinrich, P. C.,** Induction of rat acute phase proteins by interleukin 6, *Eur. J. Immunol.,* 18, 717, 1988.
9. **Darlington, G. J., Wilson, D. R., and Lachman, L. B.,** Monocyte-conditioned medium, interleukin-1, and tumor necrosis factor stimulate the acute phase response in human hepatoma cells *in vitro, J. Cell Biol.,* 103, 787, 1986.
10. **Gauldie, J., Richards, C., Harnish, D., Lansdorp, P., and Baumann, H.,** Interferon-β_2 shares identity with monocyte-derived hepatocyte-stimulating factor and regulates the major acute phase protein response in liver cells, *Proc. Natl. Acad. Sci. U.S.A.,* 84, 7251, 1987.
11. **Baumann, H., Onorato, V., Gauldie, J., and Jahreis, G. P.,** Distinct sets of acute plasma proteins are stimulated by separate human hepatocyte-stimulating factors and monokines in rat hepatoma cells, *J. Biol. Chem.,* 262, 9756, 1987.
12. **Baumann, H., Richards, C., and Gauldie, J.,** Interaction among hepatocyte-stimulating factors, interleukin-1 and glucocorticoids for regulation of acute phase plasma proteins in human hepatoma (HepG2) cells, *J. Immunol.,* 138, 4122, 1987.
13. **Morrone, G., Ciliberto, G., Oliviero, S., Arcone, R., Dente, L., Content, J., and Cortese, R.,** Recombinant interleukin 6 regulates the transcriptional activation of a set of human acute phase genes, *J. Biol. Chem.,* 263, 12554, 1988.
14. **Gauldie, J., Richards, C., Northemann, W., Fey, G., and Baumann, H.,** IFNβ2/BSF2/IL-6 is the monocyte-derived HSF that regulates receptor-specific acute phase gene regulation in hepatocytes, *Ann. N.Y. Acad. Sci.,* 557, 46, 1989.
15. **Schreiber, G., Howlett, G., Nagashima, M., Millership, A., Martin, H., Urban, J., and Kotler, L.,** The acute phase response of plasma protein synthesis during experimental inflammation, *J. Biol. Chem.,* 257, 10271, 1982.
16. **Ramadori, G., Sipe, J. D., Dinarello, C. A., Mizel, S. B., and Colten, H. R.,** Pretranslational modulation of acute phase hepatic protein synthesis by murine recombinant interleukin 1 (IL-1) and purified human IL-1, *J. Exp. Med.,* 162, 930, 1985.
17. **Woo, P., Sipe, J., Dinarello, C. A., and Colten, H. R.,** Structure of a human serum amyloid A gene and modulation of its expression in transfected L cells, *J. Biol. Chem.,* 262, 15790, 1987.
18. **Brasier, A. R., Ron, D,. Tate, J. E., and Habener, J. F.,** A family of constitutive C/EBP-like DNA binding proteins attenuate the IL-1α induced, NFκB mediated trans-activation of the angiotensinogen gene acute phase response element, *EMBO J.,* 9, 3933, 1990.

19. **Wilson, D. R., Juan, T. S.-C., Wilde, M. D., Fey, G. H., and Darlington, G. J.,** A 58-base-pair region of the human C3 gene confers synergistic inducibility by interleukin-1 and interleukin-6, *Mol. Cell. Biol.*, 10, 6181, 1990.
20. **Perlmutter, D. H., Dinarello, C. A., Punsal, P. J., and Colten, H. R.,** Cathetin/tumor necrosis factor regulates hepatic acute phase gene expression, *J. Clin. Invest.*, 78, 1349, 1986.
21. **Baumann, H. and Wong, G. G.,** Hepatocyte-stimulating factor III shares structural and functional identity with leukemia inhibitory factor, *J. Immunol.*, 143, 1163, 1989.
22. **Brasier, A. R., Ron, D., Tate, J. E., and Habener, J. F.,** Synergistic enhansons located within an acute phase responsive enhancer modulate glucocorticoid induction of angiotensinogen gene transcription, *Mol. Endocrinol.*, 4, 1921, 1990.
23. **Ron, D., Brasier, A. R., Wright, K. A., and Habener, J. F.,** The permissive role of glucocorticoids on interleukin-1 stimulation of angiotensinogen gene transcription is mediated by an interaction between inducible enhancers, *Mol. Cell. Biol.*, 10, 4389, 1990.
24. **Marinkovic, S. and Baumann, H.,** Structure, hormonal regulation, and identification of interleukin-6- and dexamethasone-responsive element of the rat haptoglobin gene, *Mol. Cell. Biol.*, 10, 1573, 1990.
25. **Baumann, H. and Schendel, P.,** Interleukin-11 regulates the hepatic expression of the same plasma protein genes as interleukin-6, *J. Biol. Chem.*, 266, 20424, 1991.
26. **Morrone, G., Cortese, R., and Sorrentino, V.,** Posttranscriptional control of negative acute phase genes by transforming growth factor beta, *EMBO J.*, 8, 3767, 1989.
27. **Muller-Eberhard, U. and Liem, H. H.,** Hemopexin, in *Structure and Function of Plasma Proteins*, Vol. 1, Allison, A. C., Ed., Plenum Press, London, 1974, 35.
28. **Altruda, F., Poli, V., Restagno, G., Argos, P., Cortese, R., and Silengo, L.,** The primary structure of human hemopexin deduced from cDNA sequence: evidence for internal, repeating homology, *Nucleic Acids Res.*, 13, 3841, 1985.
29. **Altruda, F., Poli, V., Restagno, G., and Silengo, L.,** Structure of the human hemopexin gene and evidence for intron-mediated evolution, *J. Mol. Evol.*, 27, 102, 1988.
30. **Poli, V., Silengo, L., Altruda, F., and Cortese, R.,** The analysis of the human hemopexin promoter defines a new class of liver-specific genes, *Nucleic Acids Res.*, 17, 935, 1989.
31. **Poli, V. and Cortese, R.,** Interleukin 6 induces a liver-specific nuclear protein that binds to the promoter of acute phase genes, *Proc. Natl. Acad. Sci. U.S.A.*, 86, 8202, 1989.
32. **Bowman, B. H. and Kurosky, A.,** Haptoglobin: the evolutionary product of duplication, unequal crossing over and mutation, *Adv. Hum. Genet.*, 12, 189, 1982.
33. **Putnam, F. W.,** Haptoglobin, in *The Plasma Proteins, Structure, Function and Genetic Control*, Vol. 2, Putnam, F. W., Eds., Academic Press, New York, 1975, 1.
34. **Oliviero, S., Morrone, G., and Cortese, R.,** The human haptoglobin gene: transcriptional regulation during development and acute phase induction, *EMBO J.*, 6, 1905, 1987.
35. **Maeda, N., Yang, F., Barnett, D. R., Bowman, B. H., and Smithies, O.,** Duplication within the haptoglobin Hp^2 gene, *Nature (London)*, 309, 131, 1984.
36. **Oliviero, S., DeMarchi, M., Carbonara, A. O., Bernin, L. F., Bensi, G., and Raugei, G.,** Molecular evidence of triplication in the haptoglobin Johnson variant gene, *Hum. Genet.*, 71, 49, 1985.
37. **Oliviero, S., DeMarchi, M., Bensi, G., Raugei, G., and Carbonara, A. O.,** A new restriction fragment length polymorphism in the haptoglobin gene region, *Hum. Genet.*, 71, 66, 1985.
38. **Bensi, G., Raugei, G., Klefenz, H., and Cortese, R.,** Structure and expression of the human haptoglobin locus, *EMBO J.*, 4, 119, 1985.
39. **Maeda, N.,** Nucleotide sequence of the haptoglobin-related gene pair, *J. Biol. Chem.*, 260, 6698, 1985.
40. **Oliviero, S. and Monaci, P.,** RNA polymerase III promoter elements enhance transcription of RNA polymerase II genes, *Nucleic Acids Res.*, 16, 1285, 1988.
41. **McEvoy, S. M. and Maeda, N.,** Complex events in the evolution of the haptoglobin gene cluster in primates, *J. Biol. Chem.*, 263, 15740, 1988.
42. **Castell, J. V., Gomez-Lechon, M. J., David, T., Hirano, T., Kishimoto, T., and Heinrich, P. C.,** Recombinant human interleukin-6 (IL-6/BSF/2/HSF) regulates the synthesis of acute phase proteins in human hepatocytes, *FEBS Lett.*, 232, 347, 1988.
43. **Baumann, H., Prowse, K. R., Marinkovic, S., Won, K.-A., and Jahreis, G. P.,** Stimulation of hepatic acute phase response by cytokines and glucocorticoids, *Ann. N.Y. Acad. Sci.*, 5, 280, 1989.
44. **Mackiewicz, A., Ganapathi, M. K., Schultz, D., Samols, D., Reese, J., and Kushner, I.,** Regulation of rabbit acute phase protein biosynthesis by monokines, *Biochem. J.*, 253, 851, 1988.
45. **Prowse, K. R. and Baumann, H.,** Interleukin-1 and interleukin-6 stimulates acute phase protein production in primary mouse hepatocytes, *J. Leuk. Biol.*, 44, 55, 1989.
46. **Pepys, M. B. and Baltz, M. L.,** Acute phase proteins with special reference to C-reactive protein and related proteins (pentraxins) and serum amyloid A protein, *Adv. Immunol.*, 34, 141, 1983.

47. **Zanetakis, E., Burnett, R., Buruel, J., and Ballou, S.,** Specific binding of C-reactive protein to human monocytes *in vitro, Arthritis Rheum.*, 30, 567, 1987.
48. **Bout, D., Joseph, M., Pontet, M., Vorng, H., Deslee, K., and Capron, A.,** Rat resistance to schistosomiasis: platelet-mediated cytotoxicity induced by C-reactive protein, *Science*, 231, 153, 1986.
49. **Mantzouranis, E. C., Dowton, S. B., Whitehead, A. S., Edge, M. D., Bruns, G. A. P., and Colten, H. R.,** Human serum amyloid P component. cDNA isolation, complete sequence of pre-serum amyloid P component, and localization of the gene to chromosome 1, *J. Biol. Chem.*, 260, 7752, 1985.
50. **Floyd-Smith, G., Whitehead, A. S., Colten, H. R., and Francke, U.,** The human C-reactive protein gene (CRP) and serum amyloid P component gene (APCS) are located on the proximal long arm of chromosome 1, *Immunogenetics*, 24, 171, 1986.
51. **Woo, P., Korenberg, J. R., and Whitehead, A. S.,** Characterization of genomic and complementary DNA sequence of human C-reactive protein, and comparison with the complementary DNA sequence of serum amyloid P component, *J. Biol. Chem.*, 260, 13384, 1985.
52. **Lei, K.-J., Liu, T., Zon, G., Soravia, E., Liu, T.-Y., and Goldman, N. D.,** Genomic DNA sequence for human C-reactive protein, *J. Biol. Chem.*, 260, 13377, 1985.
53. **Toniatti, C., Arcone, R., Majello, B., Ganter, U., Arpaia, G., and Ciliberto, G.,** Regulation of the human C-reactive protein gene, a major marker of inflammation and cancer, *Mol. Biol. Med.*, 7, 199, 1990.
54. **Ciliberto, G., Arcone, R., Wagner, E. F., and Ruther, U.,** Inducible tissue-specific expression of human C-reactive protein in transgenic mice, *EMBO J.*, 6, 4017, 1987.
55. **Gross, D. and Garrard, W. T.,** Nuclease hypersensitive sites in chromatin, *Annu. Rev. Biochem.*, 57, 159, 1988.
56. **Murphy, C., Ciliberto, G., and Ruther, G.,** unpublished results, 1992.
57. **Ciliberto, G., Pizza, M. G., Arcone, R., and Dente, L.,** DNA sequence of a pseudogene for human C-reactive protein, *Nucleic Acids Res.*, 15, 5895, 1987.
58. **Oliviero, S. and Cortese, R.,** The human haptoglobin gene promoter: interleukin-6-responsive elements interact with a DNA-binding protein induced by interleukin-6, *EMBO J.*, 8, 1145, 1989.
59. **Arcone, R., Gualandi, G., and Ciliberto, G.,** Identification of sequences responsible for acute phase induction of human C-reactive protein, *Nucleic Acids Res.*, 16, 3195, 1988.
60. **Ganter, U., Arcone, R., Toniatti, C., Morrone, G., and Ciliberto, G.,** Dual control of C-reactive protein gene expression by interleukin-1 and interleukin-6, *EMBO J.*, 8, 3773, 1989.
61. **Majello, B., Arcone, R., Toniatti, C., and Ciliberto, G.,** Constitutive and IL-6-induced nuclear factors that interact with the human C-reactive protein promoter, *EMBO J.*, 9, 457, 1990.
62. **Toniatti, C., Demartis, A., Monaci, P., Nicosia, A., and Ciliberto, G.,** Synergistic trans-activation of the human C-reactive protein promoter by transcription factor HNF-1 binding at two distinct sites, *EMBO J.*, 9, 4467, 1990.
63. **Poli, V., Mancini, F. P., and Cortese, R.,** IL-6DBP, a nuclear protein involved in interleukin-6 signal transduction, defines a new family of leucine zipper proteins related to C/EBP, *Cell*, 63, 643, 1990.
64. **Muller, G., Ruppert, S., Schmid, E., and Schutz, G.,** Functional analysis of alternatively spliced tyrosinase gene transcripts, *EMBO J.*, 7, 2723, 1988.
65. **Graham, F. L. and van der Eb, A. J.,** A new technique for the assay of infectivity of human adenovirus 5 DNA, *Virology*, 52, 456, 1973.
66. **de Wet, J. R., Wood, K. V., DeLuca, M., Helinski, D. R., and Subramani, S.,** Firefly luciferase gene: structure and expression in mammalian cells, *Mol. Cell. Biol.*, 7, 725, 1987.
67. **Gorman, C. M., Merlino, G. T., Willingham, M. C., Pastan, I., and Howard, B. H.,** The Rous sarcoma virus long terminal repeat is a strong promoter when introduced into a variety of eukaryotic cells by DNA-mediated transfection, *Proc. Natl. Acad. Sci. U.S.A.*, 79, 6777, 1982.
68. **Harlow, E. and Lane, D.,** *Antibodies: A Laboratory Manual*, Cold Spring Harbor Laboratory, Cold Spring Harbor, NY, 1988, 471.
69. **Laemmli, U. D.,** Cleavage of structural proteins during the assembly of the head of bacteriophage T4, *Nature (London)*, 227, 680, 1970.
70. **Kries, T. E.,** Microinjected antibodies against the cytoplasmic domain of vesicular stomatitis virus glycoprotein block its transport to the cell surface, *EMBO J.*, 5, 931, 1986.
71. **Baumann, H., Morella, K. K., Jahreis, G. P., and Marinkovic, S.,** Distinct regulation of the interleukin-1 and interleukin-6 response elements of the rat haptoglobin gene in rat and human hepatoma cells, *Mol. Cell. Biol.*, 10, 5967, 1990.
72. **Ganapathi, M. K., Schultz, D., Mackiewicz, A., Samols, D., Hu, S.-I., Brabenec, A., Macintyre, S. S., and Kushner, I.,** Differential regulation of human serum amyloid A, C-reactive protein, and other acute phase proteins by cytokines in Hep3B cells, *J. Immunol.*, 141, 564, 1988.
73. **Li, S.-P., Liu, T.-Y., and Goldman, N. D.,** Cis-acting elements responsible for interleukin-6 inducible C-reactive protein gene expression, *J. Biol. Chem.*, 265, 4136, 1990.

74. **Frain, M., Swart, G., Monaci, P., Nicosia, A., Stampfli, S., Frank, R., and Cortese, R.,** The liver-specific transcription factor LF-B1 contains a highly diverged homeobox DNA binding domain, *Cell,* 59, 145, 1989.
75. **Tronche, F., Rollier, A., Bach, I., Weiss, M. C., and Yaniv, M.,** The rat albumin promoter: cooperation with upstream elements is required when binding of APF/HNF1 to the proximal element is partially impaired by mutation or bacterial methylation, *Mol. Cell. Biol.,* 9, 4759, 1989.
76. **Lichtsteiner, S. and Schibler, U.,** A glycosylated liver-specific transcription factor stimulates transcription of the albumin gene, *Cell,* 57, 1179, 1989.
77. **Akira, S., Isshiki, H., Sugita, T., Tanabe, O., Kinoshita, S., Nishio, Y., Nakajima, T., Hirano, T., and Kishimoto, T.,** A nuclear factor for IL-6 expression (NF-IL6) is a member of a C/EBP family, *EMBO J.,* 9, 1897, 1990.
78. **Kishimoto, T.,** personal communication, 1991.
79. **Landschulz, W. H., Johnson, P. F., and McKnight, S. L.,** The DNA binding domain of the rat liver nuclear protein C/EBP is bipartite, *Science,* 243, 1681, 1989.
80. **Friedman, A. D., Landschulz, W. H., and McKnight, S. L.,** CCAAT/enhancer binding protein activates the promoter of the serum albumin gene in cultured hepatoma cells, *Genes Dev.,* 3, 1341, 1989.
81. **Friedman, A. D. and McKnight, S. L.,** Identification of two polypeptide segments of CCAAT/enhancer-binding protein required for transcriptional activation of the serum albumin gene, *Genes Dev.,* 4, 1416, 1990.
82. **Christy, R. J., Yang, V. W., Ntambi, J. M., Geiman, D. E., Landschulz, W. H., Friedman, A. D., Nakabeppu, Y., Kelly, T. J., and Lane, M. D.,** Differentiation-induced gene expression in 3T3-L1 preadipocytes: CCAAT/enhancer binding protein interacts with and activates the promoters of two adipocyte-specific genes, *Genes Dev.,* 3, 1323, 1989.
83. **Christy, R. J., Kaestner, K. H., Geiman, D. E., and Lane, M. D.,** CCAAT/enhancer binding protein gene promoter: binding of nuclear factors during differentiation of 3T3-L1 preadipocytes, *Proc. Natl. Acad. Sci. U.S.A.,* 88, 2593, 1991.
84. **Umek, R. M., Friedman, A. D., and McKnight, S. L.,** CCAAT/enhancer binding protein: a component of a differentiation switch, *Science,* 251, 288, 1991.
85. **Cao, Z., Umek, R. M., and McKnight, S. L.,** Regulated expression of three C/EBP isoforms during adipose conversion of 3T3-L1 cells, *Genes Dev.,* 5, 1538, 1991.
86. **Landschulz, H. W., Johnson, P. F., and McKnight, S. L.,** The leucine zipper: a hypothetical structure common to a new class of DNA binding proteins, *Science,* 240, 1759, 1988.
87. **Vinson, C. R., Sigler, P. B., and McKnight, S. L.,** Scissors-grip model for DNA recognition by a family of leucine zipper proteins, *Science,* 246, 911, 1989.
88. **Descombes, P., Chojkier, M., Lichtsteiner, S., Falvey, E., and Schibler, U.,** LAP, a novel member of the C/EBP family, encodes a liver-enriched transcriptional activator protein, *Genes Dev.,* 4, 1541, 1990.
89. **Chang, C.-J., Chen, T.-T., Lei, H.-Y., Chen, D.-S., and Lee, S.-C.,** Molecular cloning of a transcription factor, AGP/EBP, that belongs to members of the C/EBP family, *Mol. Cell. Biol.,* 10, 6642, 1990.
90. **Williams, S. C., Cantwell, C. A., and Johnson, P. F.,** A family of C/EBP-related proteins capable of forming covalently linked leucine zipper dimers *in vitro, Genes Dev.,* 5, 1553, 1991.
91. **Isshiki, H., Akira, S., Sugita, T., Nishio, Y., Hashimoto, S., Pawlowski, T., Suematsu, S., and Kishimoto, T.,** Reciprocal expression of NF-IL6 and C/EBP in hepatocytes: possible involvement of NF-IL6 in acute phase gene expression, *New Biol.,* 3, 63, 1991.
92. **Metz, R. and Ziff, E.,** cAMP stimulates the C/EBP-related transcription factor rNFIL-6 to trans-locate to the nucleus and induce c-fos transcription, *Genes Dev.,* 5, 1754, 1991.
93. **Natsuka, S., Isshiki, H., Akira, S., and Kishimoto, T.,** Augmentation of haptoglobin production in Hep3B cell line by a nuclear factor NF-IL6, *FEBS Lett.,* 291, 58, 1991.
94. **Kunz, D., Zimmermann, R., Heisig, M., and Heinrich, P. C.,** Identification of the promoter sequences involved in the interleukin-6 dependent expression of the rat α_2-macroglobulin gene, *Nucleic Acids Res.,* 17, 1121, 1989.
95. **Won, K.-A. and Baumann, H.,** The cytokine response element of the rat α1-acid glycoprotein gene is a complex of several interacting regulatory sequences, *Mol. Cell. Biol.,* 10, 3965, 1990.
96. **Lowell, C. A., Potter, D. A., Stearman, R. S., and Morrow, J. F.,** Structure of the murine serum amyloid A gene family, *J. Biol. Chem.,* 261, 8442, 1986.
97. **Fowlkes, D. M., Mullis, N. T., Comeau, C. M., and Crabtree, G. R.,** Potential basis for regulation of the coordinately expressed fibrinogen genes: homology in the 5′ flanking regions, *Proc. Natl. Acad. Sci. U.S.A.,* 81, 2313, 1984.
98. **Fung, W. P., Thomas, T., Dickson, P. W., Alred, A. R., Milland, J., Dziadek, M., Power, B., Hudson, P., and Schreiber, G.,** Structure and expression of the rat transthyretin (prealbumin) gene, *J. Biol. Chem.,* 263, 480, 1988.

99. **Fung, W. P. and Schreibler, G.,** Structure and expression of the genes for major acute phase α_1-protein (thiostatin) and kininogen in the rat, *J. Biol. Chem.*, 262, 9298, 1987.
100. **Ciliberto, G., Dente, L., and Cortese, R.,** Cell-specific expression of a transfected human α1-antitrypsin gene, *Cell,* 41, 531, 1985.
101. **Hattori, M., Abraham, L. J., Northemann, W., and Fey, G. H.,** Acute phase reaction induces a specific complex between hepatic nuclear proteins and the interleukin 6 response element of the rat α_2-macroglobulin gene, *Proc. Natl. Acad. Sci. U.S.A.*, 87, 2364, 1990.
102. **Fey, G. H.,** personal communication, 1991.
103. **Nakajima, K. and Wall, R.,** Interleukin-6 signals activating *jun*B and TIS11 gene transcription in a B-cell hybridoma, *Mol. Cell. Biol.*, 11, 1409, 1991.
104. **Roman, C., Platero, J. S., Shuman, J. D., and Calame, K.,** Ig/EBP-1: a ubiquitously expressed immunoglobulin enhancer binding protein that is similar to C/EBP and heterodimerizes with C/EBP, *Genes Dev.*, 4, 1404, 1990.
105. **Ron, D.,** personal communication, 1991.
106. **Descombes, P. and Schibler, U.,** A liver-enriched transcriptional activator protein, LAP, and a transcriptional inhibitory protein, LIP, are translated from the same mRNA, *Cell,* 67, 569, 1991.
107. **Ramji, D. P., Tronche, F., Gallinari, P., Vitelli, A., Ciliberto, G., and Cortese, R.,** in preparation.
108. **Ramji, D. P.,** unpublished results, 1991.
109. **Gonzalez, G. A. and Montminy, M. R.,** Cyclic AMP stimulates somatostatin gene transcription by phosphorylation of CREB at serine 133, *Cell,* 59, 675, 1989.
110. **Boyle, W. J., Smeal, T., Defize, L. H. K., Angel, P., Woodgett, J. R., Karin, M., and Hunter, T.,** Activation of protein kinase C decreases phosphorylation of c-jun at sites that negatively regulate its DNA-binding activity, *Cell,* 64, 573, 1991.
111. **Binetruy, B., Smeal, T, and Karin, M.,** Ha-ras augments c-jun activity and stimulates phosphorylation of its activation domain, *Nature (London),* 351, 122, 1991.
112. **Sorger, P. K. and Pelham, H. R. B.,** Yeast heat shock factor is an essential DNA-binding protein that exhibits temperature-dependent phosphorylation, *Cell,* 54, 855, 1988.
113. **Manak, J. R. and Prywes, R.,** Mutation of serum response factor phosphorylation sites and the mechanism by which its DNA-binding activity is increased by casein kinase II, *Mol. Cell. Biol.*, 11, 3652, 1991.
114. **Roberts, S. B., Segil, N., and Heintz, N.,** Differential phosphorylation of the transcription factor Oct1 during the cell cycle, *Science,* 253, 1022, 1991.

Chapter 22

SERUM AMYLOID A GENE REGULATION

Patricia Woo, Mark R. Edbrooke, Jonathan Betts, Glenda Watson, and Phillippa Francis

TABLE OF CONTENTS

0-8493-6913-4/93/$0.00 + $.50

I. INTRODUCTION

Serum amyloid A (SAA) is a group of closely related proteins coded for by a multigene family in many species, e.g., mice,[1] Syrian hamster,[2] mink,[3] and dog[4] as well as humans.[5] Most of these proteins are apolipoproteins carried by the high density lipoprotein (HDL) fraction of plasma and are acute phase reactants. They respond to conditioned medium from stimulated monocytes or from recombinant cytokines such as interleukin 1 (IL-1), interleukin 6 (IL-6), and tumor necrosis factor-α (TNFα).

The murine SAA gene family contains three expressed genes and a pseudogene mapped to a 79-kb region of mouse chromosome 7. All three genes are acute phase, but show differential patterns of expression, both temporarily and fractionally following various inflammatory stimuli.[6] In addition, only proteins encoded by the SAA1 and SAA2 genes circulate in the HDL fraction of plasma during an acute phase response. The murine SAA1 and SAA2 genes show similar patterns of expression and are regulated at the transcriptional level. A cytokine-responsive element between −185 and −118, and a liver-specific element between −118 and −63 have been described in a mouse SAA3 gene.[7] Posttranscriptional regulation may occur as well because of the significant time lapse between peak transcription rate (>300-fold at 3 h) and peak murine SAA mRNA (at 9 to 12 h).[6] More recent studies on the SAA3 gene expressed in the mouse liver-derived line BNL also suggest a posttranscriptional mechanism for induction by monocyte-conditioned medium.[8] In this study, Rienhoff and Groudine suggest that there is attenuation of an mRNA-degrading activity.

II. HUMAN SERUM AMYLOID A GENE FAMILY

Our recent studies have revealed that there are at least four genetic loci for the human SAA gene family (Figure 1). The loci coding for SAA1 and SAA2 proteins are polymorphic. These two genes are highly homologous and code for acute phase proteins in the HDL fraction of the plasma, like the murine SAA1 and 2 genes. The SAA3 gene has been described by Sack and Talbot[9] and Kluve-Beckerman et al.[10] Sack and Talbot show that the SAA3-derived protein sequence has strong homology in the N-terminal region with the rabbit SAA collagenase inducer described by Brinkerhoff et al.[11] Although the sequences of the Kluve-Beckerman and Sack SAA3 genes are broadly similar, there is a distinct difference in that there is a stop codon in exon 3 in the gene described by Kluve-Beckerman, indicating that this gene is not expressed (Figure 2). We have also confirmed this by our limited population study, where we could not detect sequences similar to that described by Sack and Talbot. To date, neither cDNA nor protein corresponding to the SAA3 gene has been described. Therefore, it is very probable that SAA3 is not expressed.

The SAA4 locus is 10 kb downstream from the SAA2 locus, and our results have shown that this encodes a novel SAA protein with considerable sequence dissimilarity with SAA1 and 2 in the N-terminal region, giving a distinctly different predicted tertiary structure compared to other human SAA proteins. Therefore, it may have a different function. Another striking dissimilarity is an insertion of eight amino acids in the beginning of exon 4,[12] similar to the insertions in the dog, mink, and cat SAA genes (Figure 3). However, the amino acid sequence of SAA4 is very different from the SAAs of these species and is more homologous to mouse and rabbit. There are amino acid substitutions in the derived amino acid sequence within the highly conserved region in exon 3. This is the first SAA gene of any species that has been described that has dissimilarities in the conserved region. The cDNA sequence of this new SAA gene has also been cloned by Whitehead et al., and a small level of protein has been detected in HDL which does not increase during an acute phase response.[13] The mRNA for this protein did not increase in liver cells in response to monocyte-conditioned

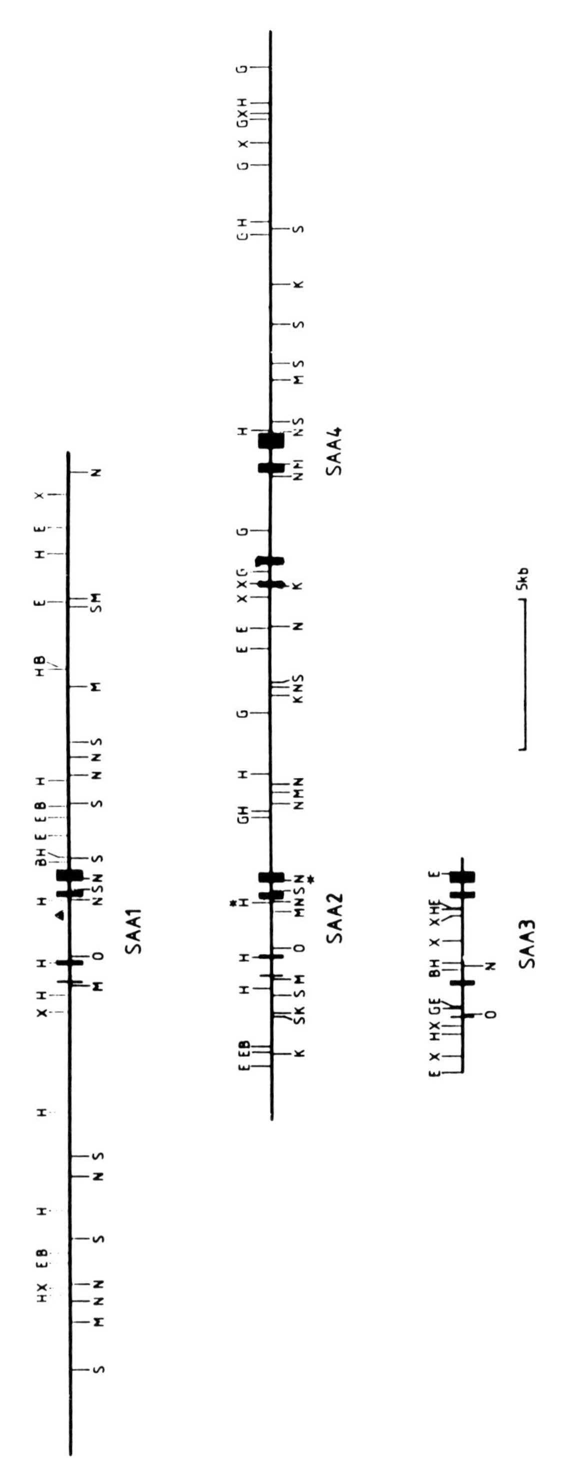

FIGURE 1. Restriction maps of the four human SAA loci. At present, only SAA2 and SAA4 have been oriented in relation to each other.

```
                                                                   ↓
GGG ACT AAA GAC ATG TGG AAA GCC TAC TCT GAC ATG AAA GAA GCC AAT TAC AAA AAA TTC
Gly Thr Lys Asp Met Trp Lys Ala Tyr Ser Asp Met Lys Glu Ala Asn Tyr Lys Lys Phe

AGA CAA ATA CTT CCA TGC TTG GGG GAA CTA |TGA| TGC TGT ACA AAG GGG GCT TGG GGC TGT
Arg Gln Ile Leu Pro Cys Leu Gly Glu Leu |End| Cys Cys Thr Lys Gly Ala Trp Gly Cys

CTG GGC TAC AGA AGT GAT CAG
Leu Gly Tyr Arg Ser Asp Gln
```

FIGURE 2. Nucleotide and amino acid sequence of exon 3 of SAA3.[10] Arrow indicates nucleotide absent from the Sack sequence.

medium. Thus, a new SAA protein which apparently is not acute phase, with a distinctive amino acid structure, has been found in humans.

III. REGULATION OF HUMAN SAA2 BY CYTOKINES

Studies of murine and human primary hepatocyte cultures have shown that SAA protein synthesis is increased predominantly by IL-6, but also by IL-1.[14,15] Studies of hepatoma cell lines have yielded variable results, but all three cytokines (IL-1, IL-6, and TNFα) have been shown to increase protein synthesis. Both these tissue culture models have their drawbacks in that primary hepatocyte cultures are already stimulated, even though they have been allowed to ''settle'' before the experiments. As far as hepatoma cell lines are concerned, the response is individualistic, depending on the cell lines. The HuH7 cell line is probably most similar to human liver cells in terms of differentiation characteristics, and in this system all three cytokines induce SAA protein synthesis.[16]

IV. TRANSCRIPTIONAL REGULATION OF HUMAN SAA2 BY CYTOKINES

Our previous work has shown that transcription of the human SAA2β gene is enhanced by the cytokines IL-1, TNFα, IL-6, and interferon-γ (IFNγ) in mouse L-cells that have been persistently transfected with a genomic clone containing the SAA2β gene.[17] Our subsequent work has revealed that there is a phorbol ester-responsive element in the 5′ flanking region of the human SAA2β gene which is capable of binding an NF-κB-like transcription factor.[18] We have also shown that this is the IL-1-responsive enhancer element in transfection studies using the two hepatoma cell lines Hep G2 and HuH7, in common with findings by Osborn et al., who showed that IL-1-activated factors binding to the κB elements in the LTR of HIV.[19] There are two NF-κB binding sites in the SAA 5′ flanking region: a proximal site acting as an IL-1-responsive enhancer element and a distal NF-κB site acting as part of a negative transcriptional control element (Figure 4A and B).[20] Results from DNase footprint experiments demonstrated the presence of a constitutive nuclear factor binding to the margin of the distal NF-κB site and a C/EBP recognition site (Figure 5A). There is no constitutive expression of the reporter gene (CAT) in transfection experiments, indicating that the SAA promoter is normally silent or repressed. On induction with IL-1, DNase footprint experi-

```
               1                                                    50
Human SAA4     MRLFTGIVFC SLVMGVTSES WRSFFKEALQ GVGDMGRAYW DIMISNHQNS
Mink SAA1      MKLFTGLIFC SLVLGVSSQ. WYSFIGEAAQ GAWDMYRAYS DMIEAKYKNS
Mink SAA2      MKLFTGLIFC SLVLGVSSQ. WYSFIGEAVQ GAWDMYRAYS DMREANYKNS
Dog SAA                           Q. WYSFVGEAAQ GAWDMLRAYS DMREANYKNS
Cat SAA                           E. WYSFLGEAAQ GAWDMWRAYS DMREANYIGA
Bovine SAA                           .WSFFGEAYE GAKDMWRAYS DMREANYKGA
Horse SAA                            LLSFLGEAAR GTWDMLRAYN DMREANYIGA
Hamster SAA1   MKPFVAIIFC FLVLGVDSQR WFQFMKEAGQ GTRDMWRAYT DMREANWKNS
Hamster SAA2   MKPFLSIIFC FLVLGVDSQR WFQFMKEAGQ GTRDMWRAYT DMREANWKNS
Hamster SAA3   MKPFLAIIFC FLILGVDSQR WFQFMKEAGQ GSTDMWRAYS DMREANWKNS
Mouse SAA1     MKLLTSLVFC SLLLGVCHGG FFSFVHEAFQ GAGDMWRAYT DMKEANWKNS
Mouse SAA2     MKLLTSLVFC SLLLGVCHGG FFSFIGEAFQ GAGDMWRAYT DMKEAGWKDG
Human SAA1     MKLLTGLVFC SLVLGVSSRS FFSFLGEAFD GARDMWRAYS DMREANYIGS
Macaque SAA                       RS WFSFLGEAYD GARDMWRAYS DMKEANYKNS
Rabbit SAA     MKLLSGLLLC SLVLGVSGQG WFSFIGEAVR GAGDMWRAYS DMREANYINA
Human SAA3     MKLSTGIIFC SLVLGVSSQG WLTFLKAAGQ GAKDMWRAYS DMKEANYKKS
Duck SAA                      DNPFTR GGRFVLDAAG GAWDMLRAYR DMREANHIGA

               51                                                  100
Human SAA4     NRYLYARGNY DAAQRGPGGV WAAKLISRSR VYLQGLIDYY L.FGNSSTVL
Mink SAA1      DKYFHARGNY DAAQRGPGGA WAAKVISDAR ERSQRITD.L IKYGDSGHGV
Mink SAA2      DKYFHARGNY DAAQRGPGGA WAAKVISDAR ERSQRVTD.L FKYGDSGHGV
Dog SAA        DKYFHARGNY DAAQRGPGGA WAAKVISDAR ENSQRITD.L LRFGDSGHGA
Cat SAA        DKYFHARGNY DAAQRGPGGA WAAKVISDAR ENSQRVTD.F FRHGNSGHGA
Bovine SAA     DKYFHARGNY DAAQRGPGGA WAAKVISDAR ENIQRFTDPL FKGTTSGQGQ
Horse SAA      DKYFHARGNY DAAKRGPGGA WAAKVISDAR ENFQRFTDR. FSFGGSGRAB
Hamster SAA1   DKYFHARGNY DAAQRGPGGA WAAKVISDAR EGFKRIT... ......GRGI
Hamster SAA2   DKYFHARGNY DAAQRGPGGA WAAKVISDAR EGFKRMR... ......GRGI
Hamster SAA3   DKYFHARGNY DAAKRGPGGA WAAKVISDAR EGIQRFT... ......GRGA
Mouse SAA1     DKYFHARGNY DAAQRGPGGV WAAEKISDGR EAFQEFF... ......GRGH
Mouse SAA2     DKYFHARGNY DAAQRGPGGV WAAEKISDAR ESFQEFF... ......GRGH
Human SAA1     DKYFHARGNY DAAKRGPGGV WAAEAISDAR ENIQRFF... ......GHGA
Macaque SAA    DKYFHARGNY DAAQRGPGGV WAAEVISDAR ENIQKLL... ......GHGA
Rabbit SAA     DKYFHARGNY DAAQRGPGGV WAAKVISDVR EDLQRLM... ......GHGA
Human SAA3     DKYFHARGNY DAVQRGPGGV WATEVISDAR ENVQRLT... ......GDHA
Duck SAA       DKYFHARGNY DAARRGPGGA WAARVISDAR ENWQ...... ..GGVSGRGA

               101                          131
Human SAA4     EDSKSNEKAE EWGRSGKDPD RFRPDGLPKK Y
Mink SAA1      EDSKADQAAN EWGRSGKDPN HFRPPGLPDK Y
Mink SAA2      EDSKADQAAN EWGRSGKDPN HFRPSGLPDK Y
Dog SAA        EDSKADQAAN EWG....... .......... .
Cat SAA        EDSKADQ... EWG....... .......... .
Bovine SAA     EDSRADQAA. .......... .......... .
Horse SAA      Z......... .......... .......... .
Hamster SAA1   EDSRADQFAN EWGRSGKDPN FFRPPGLPSK Y
Hamster SAA2   EDSRADQFAN EWGRSGKDPN FFRPPGLPSK Y
Hamster SAA3   ADSRADQFAN KWGRSGKDPN HFRPAGLPSK Y
Mouse SAA1     EDTIADQEAN RHGRSGKDPN YYRPPGLPDK Y
Mouse SAA2     EDTMADQEAN RHGRSGKDPN YYRPPGLPAK Y
Human SAA1     EDSLADQAAN EWGRSGKDPN HFRPAGLPEK Y
Macaque SAA    EDT....... .......... .......... .
Rabbit SAA     EDSMADQAAN EWGRSGKDPN HFRPKGLPDK Y
Human SAA3     EDSLAGQATN KWGQSGKDPN HFRPAGLPEK Y
Duck SAA       EDTRADQEAN AWGRNGGDPN RYRPPGLP.. .
```

FIGURE 3. Comparison of SAA protein sequences of different species. The sequences within the box denote the conserved region.

ments showed that this negative constitutive nuclear factor is displaced by a NF-κB-like factor, and transfection experiments show a corresponding six- to sevenfold increase in CAT expression (Figure 5B).

Similar analysis of the SAA2 promoter by DNA mobility shift assays, DNA footprint analysis, and transfection studies using SAA promoter-CAT constructs shows that IL-6 acts by increasing the binding of a constitutive nuclear factor, NF-IL6 (Figure 6). This factor was first described by Akira et al.[21] as the nuclear factor mediating transcription of the

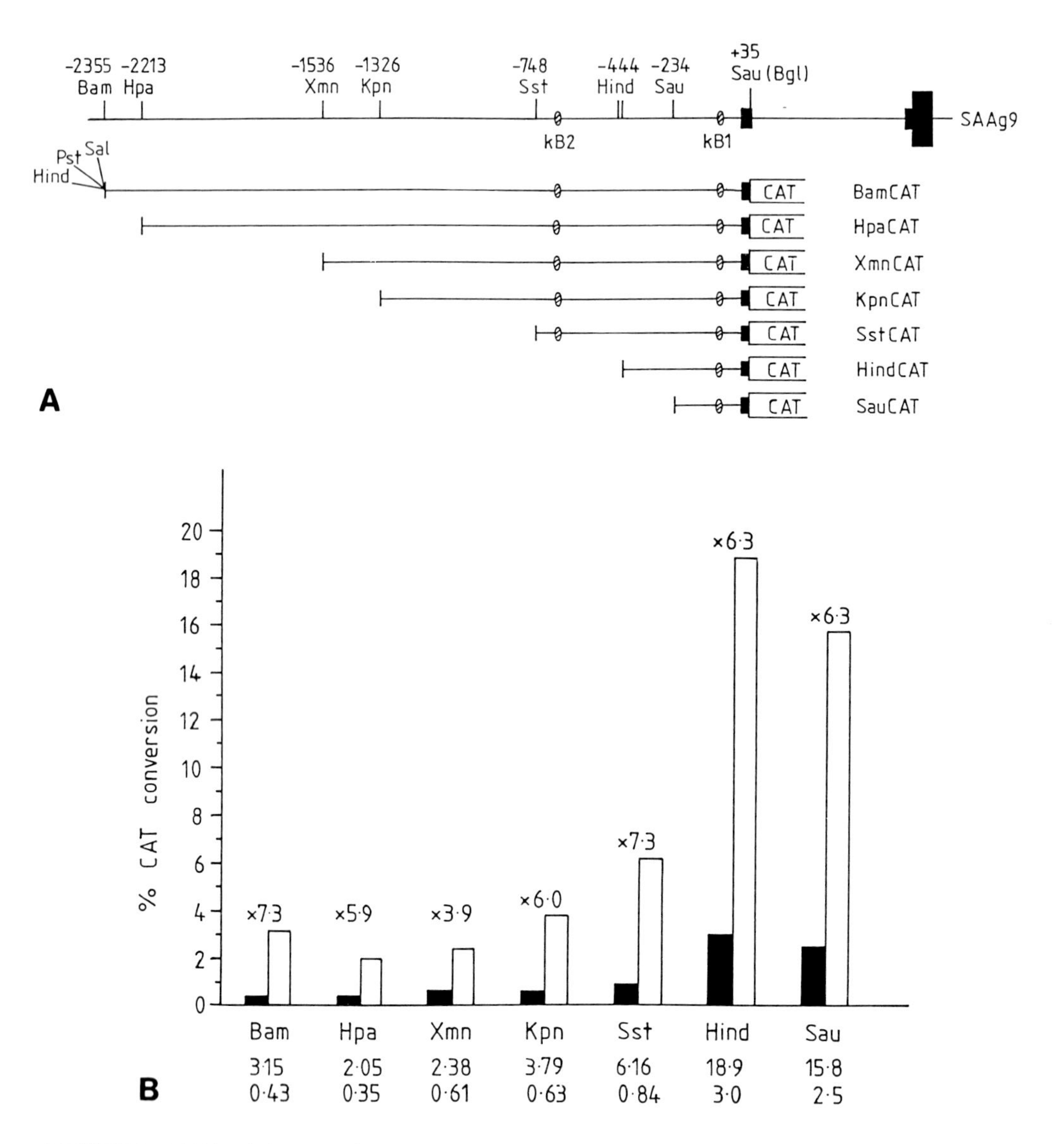

FIGURE 4. (A) Restriction map of the promoter of a genomic clone of the SAA2β gene (SAAg9). Also shown are the different-length clones linked to the reporter gene CAT. Numbers relate to the distance, 5′(−) or 3′(+), to the transcriptional start site. κB1, proximal NF-κB recognition site; κB2, distal NF-κB recognition site; ■, exons 1 and 2 of the SAA2β gene. (B) Histograms showing the expression of SAA-CAT DNA constructs under the influence of IL-1 in transfection experiments in Hep G2 cells. ■, no IL-1β; □, with IL-1β (1 U/ml). Figures at the bottom denote induced and uninduced percent CAT conversion. Figures above the columns indicate fold increase of induced over uninduced percent CAT conversion.

IL-6 gene in glioblastoma cells in the presence of IL-1. *In vitro* mutagenesis studies in our laboratory show that the intact proximal NF-κB recognition sequence (previously described as the IL-1 and phorbol ester response element in human SAA) is also necessary in the full expression of the SAA-CAT constructs when the transfected Hep G2 cells are stimulated by IL-6. Since there is always a low level of NF-κB binding to this enhancer site, it is possible that the maximum IL-6 response is produced when NF-IL-6 interacts with subunits of NF-κB constitutively bound to this enhancer site. Interestingly, we have shown an increase in NF-IL-6 binding to the SAA promoter when IL-1 is the stimulus, although it has a negligible effect compared to NF-κB in transfection studies.

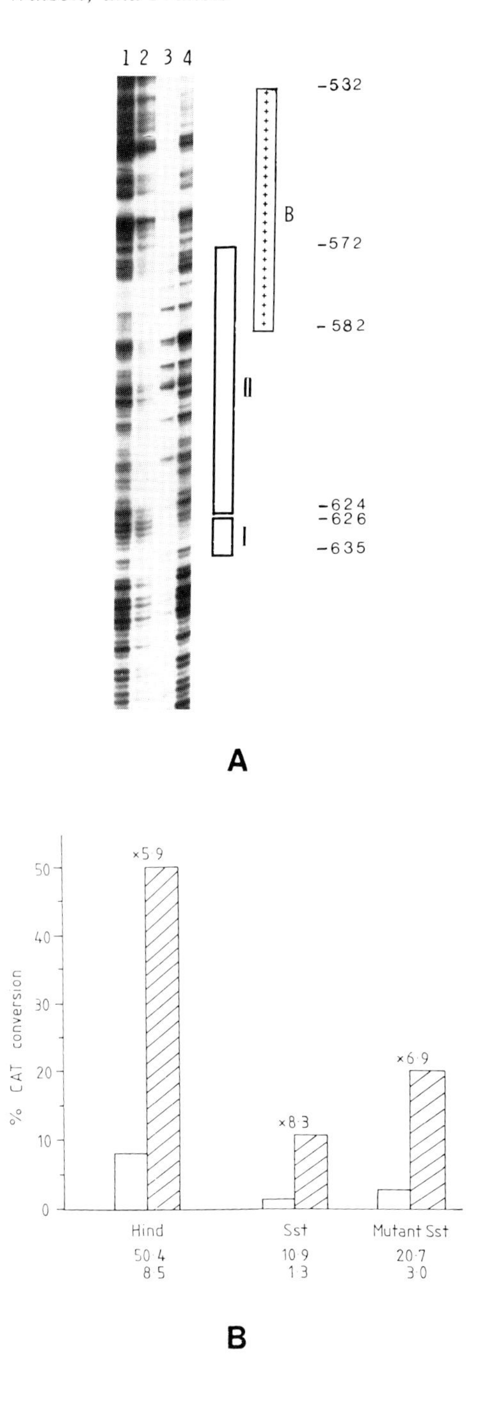

FIGURE 5. (A) Binding of nuclear proteins to region near the distal SAAκB2 DNase footprint analysis using ^{32}P-labeled DNA fragment Sst/Pst, and constitutive and IL-1-stimulated Hep G2 nuclear extracts. Lane 1, fragment alone; lane 2, fragment + 200 μg of constitutive nuclear protein; lane 3, fragment + 10 μg of IL-1-induced nuclear protein; lane 4, fragment + 50 μg of IL-1-induced nuclear protein. □I, DNAase footprint of distal SAA NF-κB binding sequence (κB2); □II, DNAase footprint of constitutive factor binding sequence; ⊞B, region of homology between SAA1 and SAA2. The rest of the sequence is unique to SAA2. (B) Transient transfection assays show the role of the SAA κB2 sequence in transcriptional regression. HindCAT, SstCAT, and SstCAT DNA with the SAA κB2. Sequences mutated by *in vitro* mutagenesis were transfected in parallel into Hep G2 cells. □, constitutive CAT expression; ▨, IL-1-induced CAT expression. Numbers over the histograms denote fold CAT increases over constitutive levels. Numbers below histograms show the exact CAT levels.

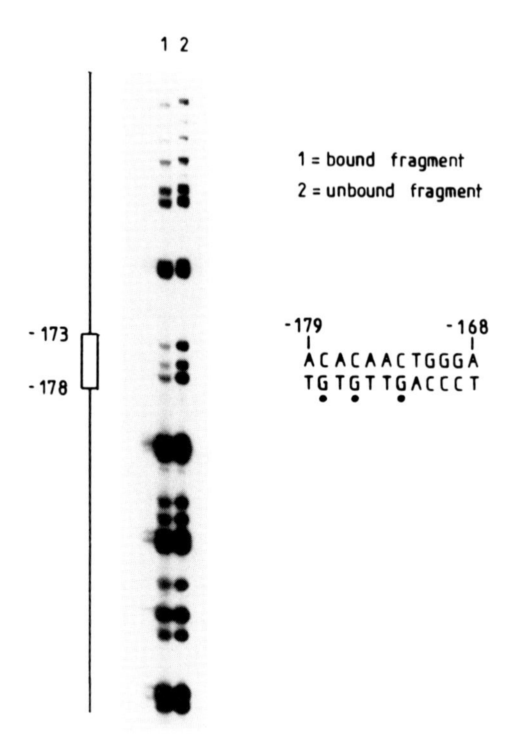

FIGURE 6. Methylation interference DNA footprint of the fragment Sau-Sau (+35 to −234 bp) and IL-6-stimulated nuclear protein extract from Hep G2 cells. □, NF-IL-6 footprint denoted by the dots below the sequence −179 to −168.

During acute and chronic inflammation, multiple cytokines usually are expressed *in vivo,* and therefore we have investigated the response of the SAA2 gene to a combination of IL-1, IL-6, and TNFα. The synergistic action of IL-1 and IL-6 produces maximum transcription from the SAA-CAT construct containing 440 bp of promoter sequence (Figure 7). Thus far, transcriptional studies on acute phase genes by many laboratories have highlighted the fundamental importance of two classes of transcription factors. Members of the NF-κB family, in addition to regulating human SAA, are required for rat SAA,[22] rat angiotensinogen,[23] and mouse factor B gene expression.[24] Members of the C/EBP family, including NF-IL-6, DBP,[25] and AGP/EBP,[26] are more widely involved in regulating acute phase genes. Our experimental results summarized below show that both families of transcription factors are involved in the synergistic and transcriptional induction of the human SAA2 gene by IL-1 and IL-6.[27]

Northern-blot analysis of Hep G2 cell mRNA shows a greatly increased quantity of SAA mRNA in response to induction by IL-1 and IL-6 (Figure 8), confirming that the synergism occurs at a pretranslational level. Transfection experiments using constructs containing mutant or wild-type NF-κB sites in the hepatoma cell line Hep G2 show that the NF-κB recognition site is necessary in the synergistic control of SAA transcription by IL-1 and IL-6. Although IL-1 also induces increased NF-IL-6 binding to the SAA promoter, *in vitro* mutagenesis of this particular recognition element did not abolish the synergistic response to a combination of IL-1 and IL-6 (Figure 9). The synergism between these cytokines does not involve increased binding by NF-κB-like factors to the NF-κB element. Further experiments to detect any increase in the mRNA of NF-κB DNA binding subunits (p100 and p105) compared with the effect of IL-1 were negative. Furthermore, IL-6 does not induce

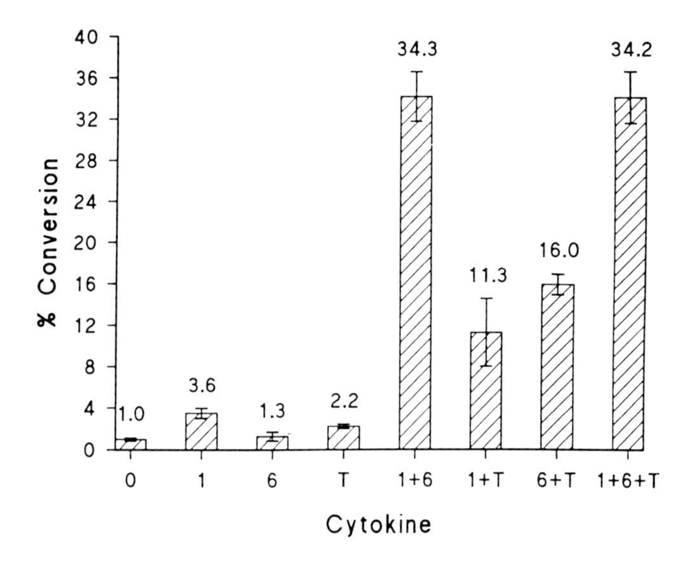

FIGURE 7. Expression of CAT from 440-bp SAA-CAT construct in transfection experiments into Hep G2 cells. 0, constitutive CAT expression; 1, IL-1-induced CAT expression; 6, IL-6-induced CAT expression; T, TNFα-induced CAT expression. Figures above the histogram denote fold increase over constitutive CAT expression.

any increase in NF-κB mRNA. These results suggest that the synergistic actions of IL-1 and IL-6 is probably at the nuclear protein level, involving interactions of the DNA-binding nuclear proteins and subunits that are responsible for activating the SAA promoter.

V. POST-TRANSCRIPTIONAL REGULATION OF SAA

In view of the murine studies indicating that SAA3 is regulated post-transcriptionally and that there is a significant time lapse between the peak of the transcriptional rate and that of mRNA production of murine SAA1 and 2, we have examined the mRNA half-life of human SAA2 in our tissue culture model, consisting of the SAA2β gene permanently transfected in mouse L-cells. Results from [^{35}S]UTP pulse-labeling experiments show that the mRNA half-life induced by IL-1 is approximately 3 h and that the constitutive SAA mRNA half-life is less than 30 min. Therefore, IL-1 increases mRNA stability in this system; however, the results do not explain why the peak of mRNA production is, in fact, closer to 16 to 24 h.

VI. SUMMARY

IL-1 and IL-6 appear to act mainly transcriptionally on both human and mouse SAA genes, although mRNA stabilization can be demonstrated. Our work on the human SAA gene promoter has shown multiple interactions of positive and negative factors. The SAA gene is normally repressed by a negative factor binding to the promoter near a distal NF-κB recognition site. Induction of transcription by IL-1 is mediated by NF-κB-like factors, but also by increased binding by factors of the NF-IL-6 family. IL-6, however, mediates its transcriptional control mainly by the transcription factor NF-IL-6. Cooperation between these two families of factors is a probable mechanism in the synergistic response to IL-1 and IL-6. Further experiments using purified factors are necessary to identify the exact nature of the nuclear protein interactions.

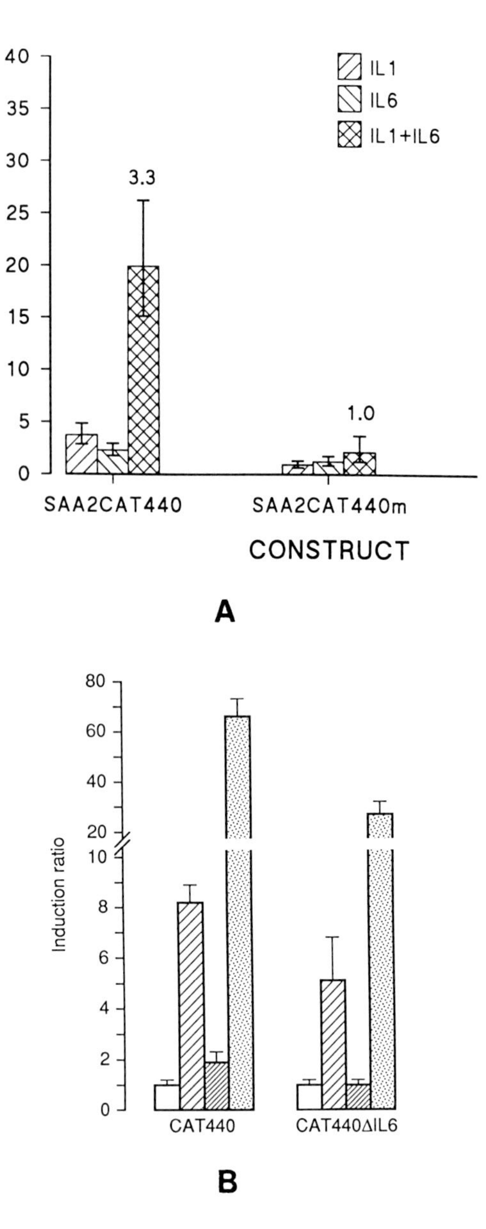

FIGURE 8. (A) Comparison of CAT expression by SAA-CAT construct containing 400 bp of SAA 5′ flanking sequences and a similar-length construct with the NF-κB binding site mutated (*GGG*ACTTTCC to *CTC*ACTTTCC). Figures above histograms denote fold increase over constitutive CAT expression. Synergism is observed when IL-1 and IL-6 are used in combination, and abolished by mutation of the NF-κB site. (B) Comparison of CAT expression by SAA440 CAT and the same-length DNA construct with the NF-IL-6 binding site mutated in three positions. □, constitutive CAT expression; ▨, IL-6-induced CAT expression; ▨, IL-1-induced CAT expression; ▒, IL-1- and IL-6-induced CAT expression.

ACKNOWLEDGMENTS

We thank J. Cheshire, D. Faulkes, and S. Coade for their technical assistance, and G. Brown for her secretarial assistance.

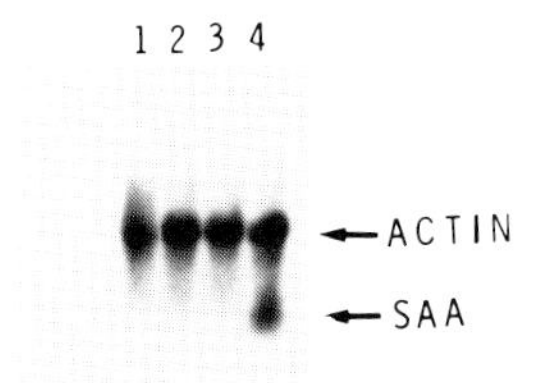

FIGURE 9. Northern-blot analysis showing the presence of SAA mRNA in the total RNA from Hep G2 cells stimulated by a combination of IL-1 and IL-6. SAA mRNA is only detectable when Hep G2 cells are stimulated by IL-1 or IL-6 alone. Control RNA is actin mRNA.

REFERENCES

1. **Lowell, C. A., Potter, D. A., Stearman, R. S., and Morrow, J. F.,** Structure of the murine serum amyloid A gene family in gene conversion, *J. Biol. Chem.*, 261, 8442, 1986.
2. **Webb, C. F., Tucker, P. W., and Dowton, S. B.,** Expression and analyses of serum amyloid A in the Syrian hamster, *Biochemistry*, 28, 4785, 1989.
3. **Marhaug, G., Husby, G., and Dowton, S. B.,** Mink serum amyloid A protein: expression and primary structure based on cDNA sequence, *J. Biol. Chem.*, 265, 10049, 1990.
4. **Sellar, G. C., deBeer, M. C., Lelias, J. M., Snyder, P. W., Glickman, L. T., Felsburg, P. J., and Whitehead, A. S.,** Dog serum amyloid A protein, *J. Biol. Chem.*, 266, 3505, 1991.
5. **Betts, J. C., Edbrooke, M. R., Thakker, R. V., and Woo, P.,** The human acute phase serum amyloid A gene family: structure, evolution and expression in hepatoma cells, *Scand. J. Immunol.*, 34, 471, 1991.
6. **Lowell, C. A., Stearman, R. S., and Morrow, J. F.,** Transcriptional regulation of serum amyloid A gene expression, *J. Biol. Chem.*, 261, 8453, 1986.
7. **Huang, J. H., Rienhoff, H. Y., Jr., and Liao, W. S. L.,** Regulation of mouse serum amyloid A gene expression in transfected hepatoma cells, *Mol. Cell. Biol.*, 10, 3619, 1990.
8. **Rienhoff, H. Y., Jr. and Groudine, M.,** Regulation of amyloid A gene expression in cultured cells, *Mol. Cell. Biol.*, 8, 3710, 1988.
9. **Sack, G. H., Jr. and Talbot, C. C., Jr.,** The human serum amyloid A (SAA)-encoding gene GSAA1, nucleotide sequence and possible autocrine collagenase inducer function, *Gene*, 84, 509, 1989.
10. **Kluve-Beckerman, B., Drumm, M. L., and Benson, M. D.,** Non-expression of the human serum amyloid A three (SAA3) gene, *DNA Cell Biol.*, 10, 651, 1991.
11. **Brinkerhoff, C. E., Mitchell, T. I., Karmilowicz, M. J., Kluve-Beckerman, B., and Benson, M. D.,** Autocrine induction of collagenase by serum amyloid A-like, and β2-microglobulin-like proteins, *Science*, 243, 655, 1989.
12. **Watson, G., Coade, S., and Woo, P.,** Structure and comparative analysis of a novel human serum amyloid A gene — SAA4, *Scand. J. Immunol.*, 36, 703, 1992.
13. **Whitehead, A. S., deBeer, M. C., Steel, D. M., Rits, M., Lelias, J. M., Lane, W., and deBeer, F. C.,** Identification of novel members of the serum amyloid A protein superfamily as constitutive apolipoproteins of high density lipoprotein, *J. Biol. Chem.*, 267, 3862, 1992.
14. **Ramadori, G., Sipe, J. D., Dinarello, C. A., Mizel, S. B., and Colten, H. R.,** Pretranslational modulation of acute phase hepatic protein synthesis by murine recombinant interleukin 1 (IL-1) and purified human IL-1, *J. Exp. Med.*, 162, 930, 1985.
15. **Moshage, H. J., Roedofs, H. M. J., van Pelt, J. F., Hazenburg, B. P. C., van Leeuwen, M. A., Limburg, P. C., Aarden, L. A., and Yap, S. H.,** The effect of interleukin-1, interleukin-6 and their interrelationship on the synthesis of serum amyloid A and C-reactive protein in primary cultures of adult human hepatocytes, *Biochem. Biophys. Res. Commun.*, 155, 112, 1988.
16. **Raynes, J. G., Eagling, S., and McAdam, K. P. W. J.,** Acute phase protein synthesis in human hepatoma cells: differential regulation of SAA and haptoglobin by IL1 and IL6, *Clin. Exp. Immunol.*, 83, 488, 1991.
17. **Woo, P., Sipe, J., Dinarello, C., and Colten, H. R.,** Structure of a human serum amyloid A gene and modulation of its expression in transfected L cells, *J. Biol. Chem.*, 262, 15790, 1987.
18. **Edbrooke, M. R., Burt, D. W., Cheshire, J. K., and Woo, P.,** Identification of *cis*-acting sequences responsible for phorbol ester induction of human serum amyloid A gene expression via a nuclear factor κB-like transcription factor, *Mol. Cell. Biol.*, 9, 1908, 1989.

19. **Osborn, L., Kunkel, S., and Nabel, G. J.,** TNF and IL1 stimulate the HIV enhancer by activation of NFκB, *Proc. Natl. Acad. Sci. U.S.A.,* 86, 2336, 1989.
20. **Edbrooke, M. R., Foldi, J., Cheshire, J. K., Li, F., Faulkes, D. J., and Woo, P.,** Constitutive and NF-κB-like proteins in the regulation of the serum amyloid A gene by interleukin 1, *Cytokine,* 3, 380, 1991.
21. **Akira, S., Isshiki, H., Sugita, T., Tanabe, O., Kinoshita, S., Nishio, Y., Nakajima, T., Hirano, T., and Kishimoto, T.,** A nuclear factor for IL-6 gene expression (NF-IL6) is a member of a C/EBP gene family, *EMBO J.,* 9, 1897, 1990.
22. **Li, X. X. and Liao, W. S.,** The expression of rat serum amyloid A gene involves both C/EBP-like and NF kappa B-like transcription factors, *J. Biol. Chem.,* 266, 15192, 1991.
23. **Ron, D., Brasier, A. R., Wright, K. A., Tate, J. E., and Habener, J. F.,** An inducible 50kD NFκB-like protein and a constitutive protein both bind the APRE of the angiotensinogen gene, *Mol. Cell. Biol.,* 10, 1023, 1990.
24. **Nonaka, M. and Huang, Z. M.,** IL-1-mediated enhancement of mouse factor B gene expression via NFκB-like hepatoma nuclear factor, *Mol. Cell. Biol.,* 10, 6283, 1990.
25. **Poli, V., Mancini, F. P., and Cortese, R.,** IL6DBP, a nuclear protein involved in IL6 signal transduction, defines a new family of leucine zipper proteins related to C/EBP, *Cell,* 63, 643, 1990.
26. **Chang, C. J., Chen, T. T., Lei, H. Y., Chen, D. S., and Lee, S. C.,** Molecular cloning of a transcription factor, AGP/EBP, that belongs to members of the C/EBP family, *Mol. Cell. Biol.,* 10, 6642, 1990.
27. **Betts, J. C., Akira, S., Kishimoto, T., and Woo, P.,** The role of NFκB and NF-IL6 transactivating factors in the synergistic activation of human serum amyloid A gene expression by interleukin 1 and interleukin 6, submitted.

Chapter 23

REGULATION OF α_1-ACID GLYCOPROTEIN GENES AND RELATIONSHIP TO OTHER TYPE 1 ACUTE PHASE PLASMA PROTEINS

Heinz Baumann, Karen R. Prowse, Kwang-Ai Won, and Sanja Marinkovic-Pajovic

TABLE OF CONTENTS

0-8493-6913-4/93/$0.00 + $.50

I. INTRODUCTION

α_1-Acid glycoprotein (AGP), also known as orosomucoid, is a plasma protein that has caught the attention of various scientists: biochemists, because of its high degree of carbohydrate heterogeneity; physiologists, because of its elusive function as a major acute phase protein; molecular biologists, because of its multihormone-responsive and tissue- and developmental-specific expression pattern; and geneticists, because of species-specific gene duplication events.[1] The acute phase-mediated change in AGP gene activity in liver cells was found in several species to be of impressive magnitude and proved to be evolutionarily conserved in vertebrates. The elucidation of the molecular events involved in AGP gene expression was aimed at defining regulatory mechanisms that were also applicable to other acute phase plasma protein genes. Indeed, several independent studies on AGP genes have uncovered regulatory features that were general to acute phase protein genes (e.g., cytokine regulatory elements and *trans*-acting factors), but also that were unique (e.g., mechanism of glucocorticoid regulation).

The hepatic acute phase reaction is considered to be the culmination of the effects of inflammatory mediators and hormones whose circulating concentrations have been influenced by systemic tissue injury. Since the liver response is a stochastic process that involves temporally coordinated changes in the expression of multiple genes and activities of numerous gene products, the molecular characterization of AGP gene regulation has to ve viewed in the context of coregulated genes. The information derived from studies on AGP genes and other gene systems has contributed to the present understanding of the regulatory pathways in liver cells. The course of and approach to the analysis of the AGP genes has also been influenced and redirected by findings made independently with other genes. The goal of this chapter is to review the principal features of AGP gene regulation, and to compare them with those of coregulated acute phase plasma protein genes. For detailed descriptions of other genes, the reader is referred to appropriate chapters in this volume.

II. REGULATED EXPRESSION OF AGP GENE

The plasma concentration of AGP in mammals rises from two- to more than 100-fold during the acute phase.[2,3] The substantial variability in the magnitude of stimulation is mainly due to the quite variable basal level of AGP ($<$0.02 to 0.8 mg/ml) in "normal" organisms.[4,5] The peak acute-phase level in various mammals has been measured to be 1 to 2.5 mg/ml at 24 to 48 h postinjury.[2-6] The acute phase-mediated increase in the plasma concentration is a reflection of similar changes of the AGP mRNA in the liver.[7,8] The AGP mRNA copy number per cell in unstimulated liver ranges from 20 to 500 and increases to approximately 7000 to 10,000 within 12 to 24 h of the acute phase.[7,9] It has been shown that for the rat, the increased transcription rate of the AGP gene[10] primarily accounts for the change in mRNA concentration. However, based on quantitative comparison of transcription rates and mRNA accumulation, posttranscriptional mechanisms such as transcript processing and mRNA stabilization also appear to contribute to the observed acute phase changes.[8,11,12] Although a significant regulation by the acute phase of the level of AGP mRNA translation, protein processing, and secretion has not been observed, in view of the known effects of the acute phase on protein glycosylation,[13-16] a controlling action at the posttranslational level is conceivable.

Taken together, *in vivo* analyses have indicated that (1) in adult liver, the AGP gene is transcriptionally activated by the acute phase, (2) AGP mRNA accumulation, AGP synthesis, and the increase of AGP in plasma follow similar kinetics,[7,8] (3) the magnitude of change is consistently above twofold, and (4) AGP belongs to the category of major acute phase plasma proteins because of the relatively high amounts produced by an acute phase liver (0.6 to 0.9 μg/h × 10^6 rat hepatocytes).[3,17]

The prominent acute phase response of the AGP gene served as one of the systems for identifying the inflammatory mediators that interact with the liver to elicit changes in plasma protein gene expression. The goal was to define the regulatory pattern, as established *in vivo,* and to determine the contributions of specific components. Hepatic tissue culture systems which have retained the ability to regulate the AGP gene *ex vivo* were developed and permitted direct testing of potential factors.[18] Many of the initial assays have been carried out on primary cultures of rat,[9,19,20] mouse,[21,22] rabbit,[23] and human[24,25] hepatocytes. However, with identification of rat (HTC,[9,26,27] Fao,[28] and H-35[29,30]) and human (Hep G2[31] and Hep 3B[32]) hepatoma cell lines, each of which regulates expression of the AGP gene similarly to normal liver, the characterization of hormone functions and cell response was greatly facilitated. The hepatoma cell lines subsequently proved to be crucial assets for elucidation of the genetic elements controlling AGP gene expression.[18,33,34]

III. HORMONES WHICH STIMULATE AGP GENE EXPRESSION

Interleukin-1 (IL-1), tumor necrosis factor-α (TNFα), IL-6, and glucocorticoids have been recognized as the principal components which stimulate AGP production.[18] Each of these factors contributes to activation of the AGP gene. The qualitative action of these factors is remarkably similar in rodent and human cells, although there are also noteworthy cell line-specific differences.

IL-1 and TNFα appear to be equally effective, and the maximal response to IL-1 and TNFα is generally greater than that to IL-6 (in rodent liver[28-30,36] but not in human liver[24,32-34,37]). There is a strong synergism between IL-1 (or TNFα) and IL-6, and the action of any cytokine combination is prominently enhanced by glucocorticoids (e.g., dexamethasone).[28,30,33,37,38] The maximal expression of AGP genes in almost all cell systems is achieved by the combination of IL-1, IL-6, and glucocorticoids.[18,33,37-39] Although oncostatin M,[40] leukemia inhibitory factor (LIF),[41] and IL-11[42] exert IL-6-like action on liver cells, these, when combined with IL-1 and dexamethasone, were only equally (IL-11 and oncostatin M) or less effective (LIF) than IL-6. Moreover, addition of these cytokines to the mixture of IL-1, IL-6, and dexamethasone failed to further enhance the AGP gene response.

Glucocorticoids alone are stimulatory on the AGP gene in rodent liver and primary hepatocyte cultures,[7,9,10,19,31] but produce nondetectable or far less stimulation in hepatoma cells.[17,29,33,38,43] An exception is HTC cells, which respond to dexamethasone (but not to any cytokines) by a prominent activation of AGP gene transcription[44,45] and accumulation of AGP mRNA.[9,12,26] As in primary liver cells, the stimulation is detectable in HTC cells within 30 min.[9,26,44,45] The rapid and sustained increase in transcription rates induced by steroids is dependent on *de novo* protein synthesis.[9,12,35] The rapid loss of AGP gene inducibility in cycloheximide-treated HTC cells suggested a short half-life (<2 h) of the transcription-controlling component that was needed in addition to the activated glucocorticoid receptor.[46]

Measurement of AGP gene transcription rates in H-35 cells indicated that this gene was activated only by combinations of cytokines and dexamethasone.[47] The activation process was delayed by 4 to 4.5 h and was prevented in cycloheximide-treated cells. It is speculated that the different response kinetics of normal rat liver, HTC cells, and H-35 cells is due to different pool sizes of the labile transcription factors.[46,47] Furthermore, the AGP gene regulation in H-35 cells seems to have been converted into an indirect process; a cytokine-inducible factor must first be synthesized in order to activate the AGP gene.

The hormone specificity of AGP gene regulation is qualitatively similar to that of several other major acute phase plasma proteins.[30,43] These proteins, termed type 1 acute phase proteins, include complement C3[32,48] (see Chapter 24 of this volume), hemopexin, serum amyloid A[49] (see Chapter 22 of this volume), haptoglobin (in rat[50]), and C-reactive protein (in human[51]) (see Chapters 22 and 28 of this volume). The common feature among these proteins is that a combination of IL-1 and IL-6 is necessary for achieving maximal expression. Although not all type 1 genes are simultaneously stimulated,[52] it seems likely that these genes nevertheless depend on common regulatory components. Since the hepatic response pattern and cytokine specificity have been conserved during mammalian evolution, comparable AGP gene regulatory systems are expected in the various species.

The molecular cloning and functional analysis of rat, mouse, and human AGP genes were aimed at identifying the *cis*- and *trans*-acting elements which mediate the transcriptional stimulation by IL-1, IL-6, LIF, IL-11, oncostatin M, and glucocorticoid. Moreover, the same molecular resources have been utilized in an attempt to delineate the mechanism responsible for liver-specific expression of the AGP genes.

IV. STRUCTURE OF AGP GENE

The molecular biology of AGP genes started with the cloning of the AGP cDNA from rat liver,[53] HTC cells,[9,27] human liver,[54,55] and mouse liver.[56-58] The amino acid sequences derived from these cDNA confirmed the structural homologies of AGP from the different species and were in agreement with the then already established protein sequence.[56] The results from mRNA translation[60,61] and cDNA analysis[54,56,57] also indicated that human and mouse expressed more than one AGP form, suggesting either allelic forms or two or more structural genes per haploid genome. Characterization of the genomic AGP gene organization and cloning of structural genes confirmed a single gene in the rat[63,64] and the presence of triplicated AGP genes (AGP-A, AGP-B, and AGP-B′) in human,[65] triplicate genes in *Mus domesticus* (AGP-1, AGP-2, and AGP-3),[66,67] and eight genes in *M. caroli,* a wild mouse species from Southeast Asia.[58] Since transcripts from only one or two of the multiple AGP genes were represented in the liver mRNA, it was concluded that extra copies are either inactive genes or have extremely low activity.[58,65-67]

In humans, the three AGP genes are clustered within 18 kb on chromosome 9q31-qter.[68,69] The transcriptionally active and acute phase-regulated AGP-A gene is located at the 5′ end of the cluster.[54,65]

In *M. domesticus*, the two active AGP-1 and -2 genes (*Orm*-1 and *Orm*-2) are colocalized to chromosome 4 near the *Lps* locus.[60] However, AGP-1 and -2 genes are not as close to each other as the human AGP genes, since none of the isolated genomic λ-phage clones contained sequences of both active genes.[66,67] AGP-1 and AGP-2 have been defined here as the genes encoding the proteins originally numbered "1" and "2" in genetic studies.[60] However, Cooper and Papaconstantinou[56] numbered their *M. domesticus* cDNA and genes independently of the genetic nomenclature. Sequence comparison indicated that the AGP-1 of these authors is identical to AGP-2.[57]

Although AGP-1 and AGP-2 are coordinately stimulated by the acute phase, the relative level of expression differs between the two genes. At the peak of the acute phase, AGP-1 is two to ten times higher than AGP-2,[60] but under chronic inflammatory conditions, AGP-2 exceeds AGP-1 by a factor of 1.5.[70] Based on cDNA and genomic sequence analysis, AGP-3 appears to represent an inactive gene.

Characterization of the acute phase response in various mouse species revealed that there is a substantial variation in the structure, copy number, and expression of AGP genes.[61] An extreme case was found in *M. caroli.* This mouse species carried a cluster of eight AGP genes within a 45-kb region.[56] The genes are equally spaced and separated by 2- to 3-kb intergenic sequences. Only the most 5′ gene (AGP-1) and the most 3′ gene (AGP-8) are

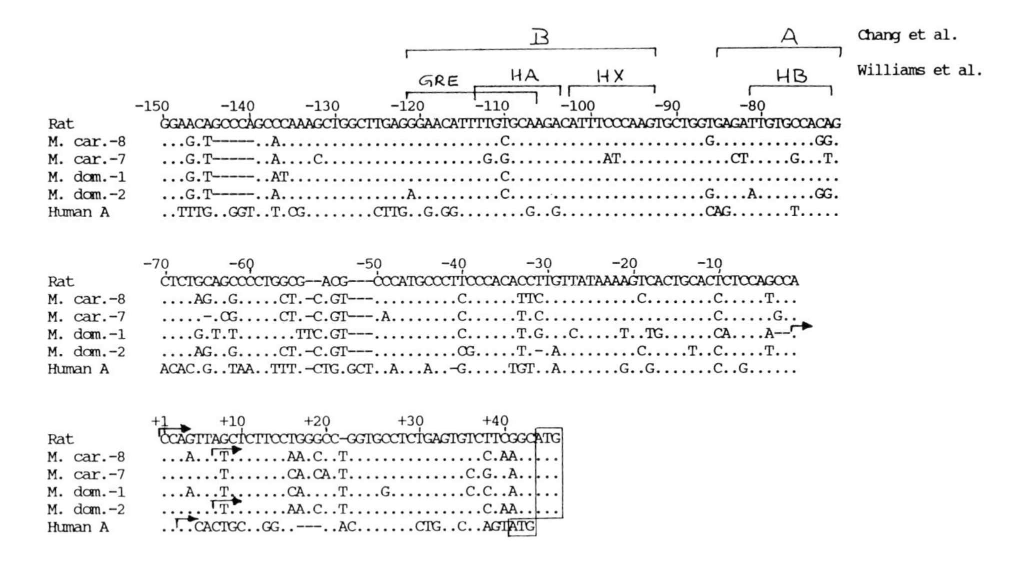

FIGURE 1. Comparison of the 5′ flanking regions of AGP genes. The nucleotide sequence of the rat AGP gene promoter from position −150 to +46[63] is compared to the AGP-8 and AGP-7 of *M. caroli*,[58] AGP-1,[74] and AGP-2[66] of *M. domesticus* and AGP-A of human.[54] The positions where proteins interact with the glucocorticoid regulatory regions have been established by nuclease protection assay and are indicated at the top. The glucocorticoid receptor binding site (~GRE) and the three AGP/EBP binding sites (HA, HX, and HB) in the rat gene are described by Williams et al.[73] and the AGP/EBP binding sites (A and B) in mouse AGP-1 gene by Chang et al.[74]

expressed in liver. The acute phase stimulates the two genes by a factor of 10. AGP-1 and -8 of *M. caroli* are the structural homologs of AGP-1 and -2 of *M. domesticus*.

In contrast to mice, the rat has only one AGP gene copy per haploid genome located on chromosome 5.[9] Although rat AGP and mouse AGP-2 are similar in that both contain six *N*-glycosylation sites,[53,57] the overall sequence of the rat AGP is equally diverged from either AGP-1 and AGP-2 of *M. domesticus* or AGP-1 and AGP-8 of *M. caroli*.[86]

All mammalian AGP genes cloned thus far exhibit an identical overall organization of the transcribed region: six protein-coding exons within a 3.2- to 3.5-kb transcription unit. The exon-intron arrangements have been maintained in the inactive human and mouse AGP genes[58,65] (for more details regarding AGP structure, see Chapter 6).

The immediate promoter proximal regions of the rodent genes show a high degree of sequence similarity, but differ substantially from the human AGP-A gene (Figure 1). The phylogenetic comparison of the active AGP genes indicates that only a few segments have been evolutionarily conserved. Although it is tempting to speculate that these sequences are involved in controlling AGP gene activity, the functional significance of these gene regions has yet to be documented.

V. *CIS*- AND *TRANS*-ACTING REGULATORY ELEMENTS

A. GLUCOCORTICOID RESPONSE ELEMENT OF THE RAT AGP GENE

The prominent stimulation of the AGP gene in HTC cells and rat liver has prompted a search for the molecular regulatory elements.[35,63,71] The structural gene, including several hundred bp 5′ flanking regions, was stably transformed into mouse fibroblasts (L-cells). A prominent dexamethasone stimulation was detected which is characteristic of the regulation in hepatic cells.[17,63] It was concluded that the glucocorticoid response element (GRE) must be contained in the selected gene sequence and that expression of the rat AGP gene was independent of liver-specific components. In analogous experiments in nonhepatic cells, such as HeLa cells. Dente et al.[34] found that the human AGP-A gene was also active.

The GRE in the rat AGP gene has been localized by testing various 5′ flanking regions which have been integrated into CAT gene reporter plasmids and transfected into L-cells[71] or rat hepatoma cells.[35] The region at position −120 to −64 relative to the transcription start site (Figure 1) was required for achieving maximal dexamethasone response (~50-fold). The glucocorticoid receptor binding site was localized to region −121 to −107.[46] This receptor binding site alone, when integrated in a reporter CAT gene construct, was able to mediate only a fewfold stimulation by glucocorticoid.[46,52,58] The sequence from −107 to −64 was required in addition to the glucocorticoid receptor binding site to restore full stimulation. The same sequence was found to be the target of the cycloheximide-sensitive component necessary for glucocorticoid regulation.[46]

Using this 5′ flanking region as a binding substrate for nuclear proteins from HTC cells, two sites of protein interaction aside from the one for the glucocorticoid receptor were detected. These sites, HA at −113 to −104 and HB at −81 to −72 (Figure 1), were recognized by the protein ANF-2 which was present in cells with or without dexamethasone treatment.[72] The significance of that interaction was demonstrated by creating a point mutation in these sites and observing a loss of glucocorticoid responsiveness.[73]

Independent of these studies, Chang et al.[74] cloned a binding protein, AGP/EBP, that interacts with the mouse AGP-1 gene homolog of the ANF-2 binding site. The primary structure of AGP/EBP proved that this protein is the mouse homolog of rat IL-6DBP[75] or LAP[76] and identical to C/EBPβ.[77] Based on biochemical, physical, and immunological criteria, the ANF-2 of HTC cells seems to be identical to AGP/EBP.[73] Purified AGP/EBP bound to the previously detected HA and HB sites and, in addition, to a third site, HX at position −102 to −93. The interaction at the latter site, however, does not seem to be relevant for glucocorticoid regulation, since a mutation within that sequence was inconsequential.[73] AGP/EBP, which has been found to be necessary for glucocorticoid stimulation, was noted to be a stable factor within HTC cells and, therefore, did not appear to represent the labile factor predicted from the cycloheximide treatment experiments.[73] One explanation could be that the combined action of AGP/EBP and glucocorticoid receptor is dependent upon a third (cycloheximide-sensitive) factor. However, experiments involving cotransfection with an expression vector for AGP/EBP and other C/EBP isoforms do not seem to support that possibility.

Transfection of the AGP promoter (−139 to +20)-CAT construct together with expression plasmids for mouse C/EBPα,[52] C/EBPβ, or C/EBPδ[77] into Hep G2 cells yielded a >100-fold *trans*-activation in the absence of dexamethasone.[52,78] When Hep G2 cells were provided with an extra glucocorticoid receptor[47] and the amount of episomally derived C/EBP isoforms were below the maximal level, a strong cooperativity between C/EBPs and glucocorticoid receptor was detectable at the level of reporter gene expression.[52] This cooperativity was dependent upon the physical linkage of the glucocorticoid receptor binding site and the C/EBP binding site HB.[52]

The functional consequence of the overlapping HA (AGP/EBP or C/EBP binding site) and GRE (Figure 1) remains an unsolved issue. Two possible explanations are (1) during dexamethasone stimulation, the glucocorticoid receptor and AGP/EBP form one complex or (2) the binding of the two factors is mutually exclusive.[73] In either case, the activation of the AGP gene is seen as a function of the glucocorticoid receptor, which is in agreement with the characteristic dose-dependent stimulation of the AGP gene by dexamethasone.[9,12] If the glucocorticoid receptor is critical, the question arises as to why C/EBP transactivates Hep G2 cells even in the absence of steroid treatment. Alternative explanations come to mind: (1) at high concentrations of C/EBP, all three binding sites (HA, HX, and HB; Figure 1) are occupied[89] and that triggers transcription or (2) an inhibitor for C/EBP, whose activity is reduced during dexamethasone treatment, is out-titrated by excess C/EBP. Since all the resources are in place, answers to these questions will be forthcoming.

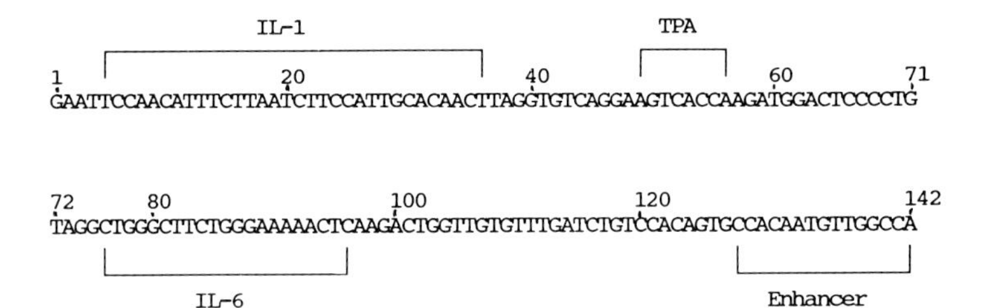

FIGURE 2. Nucleotide sequence of the cytokine response element of the rat AGP genes. The 142-bp distal regulatory element (DRE) of the rat AGP gene (located at −5300) has been functionally subdivided into regions responsive to IL-1, phorbol ester (TPA), and IL-6. The region marked "Enhancer", when combined with the response elements, increases the magnitude of response. The brackets indicate the positions where binding of proteins to the regulatory regions were detected by nuclease protection assays.[81]

B. GRE IN MOUSE AGP GENES

AGP genes in mouse and rat liver are regulated with the same hormone specificity, except that the response to glucocorticoids is not as prominent in the mouse as in the rat.[21,22,31,79] The active AGP-1 and AGP-2 genes from *M. domesticus* have been cloned.[66,67] The nucleotide sequence of the promoters shows a high degree of similarity with the rat (Figure 2). In particular, the sequence of AGP-1, within the critical glucocorticoid response region −130 to −70, is identical to the rat gene except that there is a cytidine in place of a thymidine at position −110. This single base substitution modifies the consensus glucocorticoid receptor site recognition[46] and renders it nonfunctional as a GRE.[58] However, the sequence modification does not seem to affect the binding of AGP/EBP (or C/EBPs) to this region "B", as demonstrated by nuclease protection assay.[74] Since the immediate adjacent 3′ flanking sequence was identical to the rat, the protein-DNA interaction analysis confirmed that the AGP/EBP binding site "A" at position −85 to −71 is congruent with site HB in the rat (Figure 1).[73,74,90] The "A" site in the AGP-2 gene differs from AGP-1 by three nucleotides. In particular, the two adjacent changes at positions −73 and −72 produce a negative influence on AGP/EBP binding, since site-directed mutation at the homologous site in HB of the rat reduced glucocorticoid stimulation and, by inference, AGP/EBP (ANF-2) binding.[73] At present, no glucocorticoid regulation of the *M. domesticus* AGP gene promoter has been documented. However, since the three AGP/EBP binding sites are intact (at least as shown for AGP-1), the structural prerequisite for *trans*-activation by C/EBP isoforms exists.[90] It is also tempting to speculate that the lower magnitude of acute phase stimulation of AGP-2 relative to AGP-1 is in part due to the unfavorable sequence within AGP/EBP binding site "A".

Since the lineages leading to *M. domesticus* and the rat diverged roughly 30 million years ago,[80] evolution of slightly different regulatory regions obviously has occurred. To place these two end points into an evolutionary perspective, the AGP genes of *M. caroli* have been characterized.[58] This mouse species diverged from *M. domesticus* approximately 5 million years ago and is exceptional in that it carries eight AGP genes (AGP-1 through -8) per haploid genome. Based on the protein coding regions, *M. caroli* AGP-1 and AGP-8 represent the homologs of AGP-1 and -2 of *M. domesticus*. *M. caroli* AGP-1 and AGP-8 differ from the two *M. domesticus* genes in that both are equally active.[58] Since at present the promoter for AGP-1 remains to be cloned, the promoter sequences of AGP gene 7 and 8 at the 3′ end of the AGP gene cluster are included in the sequence comparison (Figure 1). AGP-7 serves as a representative for the six inactive AGP genes of *M. caroli*. The promoter sequence of AGP-8 is highly similar to AGP-2 of *M. domesticus* and contains the characteristic nucleotide changes at −110 (impaired glucocorticoid receptor recognition) and at −72 and −73 (reduced *trans*-activation through "HB" or "A").[52,58] These changes together probably explain the failure of AGP-8 promoter to respond to dexamethasone.[58]

By testing additional 5′ flanking regions of the *M. caroli* AGP gene 8 for hormone responsiveness, a GRE was localized to an 83-bp fragment at position −825 to −742.[58] This GRE does not share sequence similarities with the promoter proximal rat GRE and functions independently of AGP/EBP (or other C/EBPs).[52] The analogous region of the *M. caroli* AGP gene 7 and the rat AGP gene did not function as GREs. This finding indicates that a different organization of regulatory sequences was established during rodent evolution which has ensured the glucocorticoid responsiveness of AGP genes.

The studies on glucocorticoid regulation of rat and mouse AGP genes thus far have suggested a surprising plasticity in the molecular mechanisms involved. However, since the information is based on only two examples, AGP gene regulation in additional species including human, in order to identify the regulatory elements which have been conserved during evolution.

C. CYTOKINE RESPONSE ELEMENTS

Far more prominent than the glucocorticoid response is the regulation of AGP genes by the combination of IL-1 (or TNFα) and IL-6. The isolated genes of rat,[17] mouse,[58] and human[34] have been tested for the presence of relevant *cis*-acting elements by transfection into cytokine regulatable hepatoma cells. Only in the case of the rat AGP gene was a discrete cytokine response element detected that was separate from the GRE. Based primarily on the functional assays in Hep G2 cells, the IL-1/IL-6 response element (RE) has been localized to position −5300 to −5150 and termed the distal regulatory element (DRE).[33] Fine mapping of the functionally important sequences within the 142-bp DRE was accomplished by deletion, recombination, and mutations (Figure 2).[81] The IL-1 RE was confined to position 1 to 36, IL-6 RE to 76 to 97, and a phorbol ester-sensitive NF-κB-like sequence to 45 to 55. Each element functions on its own; however, strong cooperativity is achieved when these elements are arranged in a complex.

The DRE elements are recognized by nuclear proteins from rat liver and rat and human hepatoma cells. The pattern of interaction as revealed by gel mobility shift or by nuclease protection assay was highly complex and was not qualitatively or quantitatively modified by either the acute phase *in vivo* (liver) or IL-1, IL-6, and dexamethasone treatment (hepatoma cells).[82] Based on Southwestern-blot analysis (i.e., separation of nuclear proteins from hepatic cells under denaturing conditions on polyacrylamide gel, renaturation and transfer onto nitrocellulose, and probing for specific DNA binding activities by incubation with labeled DRE fragments), an active protein for the IL-1 RE, but not IL-6 RE, was observed.[82] This activity, originally termed NF-AB and subsequently identified as C/EBPβ (≡ NF-IL-6, IL-6DBP, and AGP/EBP),[78] is strongly stimulated in hepatic cells by IL-1, IL-6, and dexamethasone (Figure 3). The kinetics of C/EBPβ stimulation is correlated with the transcriptional activation of the endogenous AGP gene.[82] Interaction of C/EBPβ is restricted to region 10 to 36 of IL-1 RE, which, as such, fails to respond to IL-1.[82] In cotransfection experiments, all C/EBP isoforms (C/EBPα, β, and δ) have been found to *trans*-activate through sequence 10 to 36.[32,78] It appears that IL-1 regulation of the AGP element might be brought about indirectly by first elevating C/EBPβ, which binds to sequence 10 to 36 and then cooperates with an additional unknown factor that is bound to position 3 to 10. Indeed, a single base substitution at position 28 (T to G) abolished IL-1 regulation and drastically reduced C/EBP binding and *trans*-activation. Only when C/EBP isoforms are present in excess, such as in cells transfected with high-expression plasmids, is a *trans*-activation achieved without IL-1 stimulation.[52]

In rat hepatic cells, AGP and haptoglobin genes are coordinately regulated by IL-1 and IL-6.[30] Structural and functional analysis of the haptoglobin promoter sequence indicated a

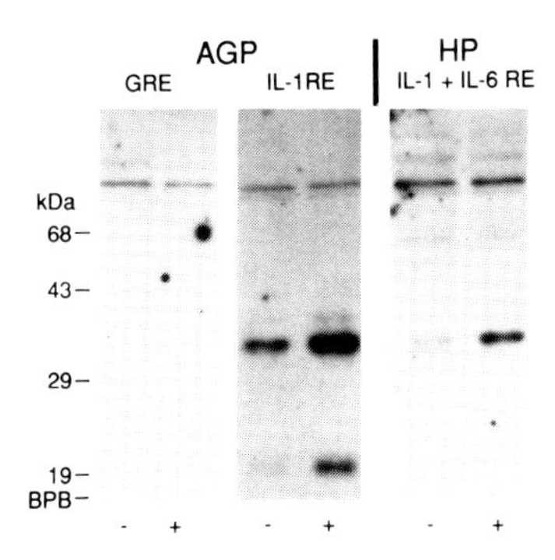

FIGURE 3. Detection of hormone-inducible C/EBPβ (or AGP/EBP) from H-35 cells by interaction with the IL-1 response element of rat AGP genes. Nuclear proteins were extracted from untreated H-35 cells (−) or cells treated for 16 h with IL-1β (10 U/ml), IL-6 (100 U/ml), and dexamethasone (1 μm (+) and 80 μg) of each preparation in three paired lanes and were separated by gel electrophoresis. The proteins were renatured and electrotransferred into nitrocellulose. The DNA binding activities were probed with ^{32}P-end-labeled rat AGP promoter fragment (−120 to −42) (GRE), IL-1 response region 1 to 46 in the DRE of rat AGP gene (IL-1 RE), or IL-1/IL-6 response element of the rat haptoglobin (HP) promoter (−165 to −56). The specific activity of the probes, radioactive concentrations, binding reactions, and washing procedures were identical for all three blots. The autoradiogram after 24 h of exposure is shown (for details, see Reference 82).

strong cytokine response element at position −165 to −56.[50,83] This haptoglobin sequence is also recognized by C/EBPβ, although the binding is slightly less avid than with the IL-1 RE of the AGP gene (Figure 3). The GRE of the AGP gene is known to be a target of C/EBPβ,[73,74] but when used for proteins in Southwestern blots under conditions optimal for IL-1 RE, no interaction was detected (Figure 3). These results suggest that the AGP promoter proximal sequences are far less favorable C/EBPβ binding sites than the IL-1 REs of either the AGP DRE or haptoglobin promoter.

Based on the functional data, one would predict that a rat AGP gene with a deleted DRE sequence loses the regulation by cytokines and is, therefore, less stimulated in liver. However, the stimulation observed is still accomplished via the promoter proximal GRE. The functional relevance of the DRE has been verified in transgenic mice. Two types of transgenic animals were generated, one receiving the entire structural gene from −4700 to +4800 (missing DRE) and one from −5300 to +5400 (with DRE).[79] All transgenic lines expressed the rat gene in the liver and responded to dexamethasone treatment by a twofold increase of transgene activity. Only in mice with the DRE-containing rat gene was the endotoxin-induced acute phase able to elevate transgene expression twofold above that observed with the glucocorticoid treatment. The precise cytokine-specific regulation of the transgene was then verified in primary cultures of transgenic hepatocytes.

The combined results indicate that the rat AGP gene has two major regulatory sites, DRE and GRE. Both are targets of the same C/EBP-related transcription factor. In each case, additional factors establish the hormone-specific response pattern. Obviously, there is no operating independence between the two elements. It can be argued that one of the limiting factors is the concentration of C/EBP isoforms. The cytokine-induced production of C/EBP isoforms is, consequently, an important aspect in explaining the massive transcriptional activation of the AGP gene during the acute phase.

VI. CYTOKINE EFFECT IN HTC CELLS

A variety of hepatoma cells have been tested for their responsiveness to cytokines. Several cell lines failed to show any change in the expression of plasma protein genes when

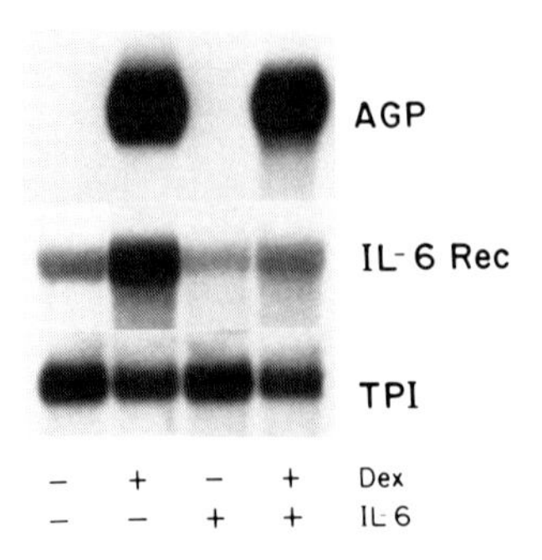

FIGURE 4. Effect of dexamethasone and IL-6 on HTC cells. Confluent monolayers of HTC cells were treatd for 4 h with 1 μ*M* dexamethasone, 100 U/ml of IL-6, or both. Polyadenylated mRNA were prepared, and 15 μg were analyzed by Northern blotting for the mRNA encoding AGP, IL-6 receptor,[85] and the internal marker triose-phosphate isomerase (TPI).

treated with IL-1, IL-6, or both. The failure to respond has been associated with lack of receptor function, ineffective signal transduction pathways, or irreversibly inactivated plasma protein genes.[18] HTC cells are an interesting example of cells whose AGP gene is strongly stimulated by dexamethasone, but not by IL-1 and/or IL-6.[17] HTC cells express the liver-characteristic 5.1-kb IL-6 mRNA receptor (Figure 4) and display functional IL-6 binding activity at its surface equivalent to H-35 cells.[87] Dexamethasone treatment increased the IL-6 receptor mRNA concentration three- to fivefold, while IL-6 treatment produced a twofold reduction (Figure 4). Although no IL-6-mediated modulation of AGP gene expression was observed, IL-6 nevertheless triggers an intracellular reaction that leads to an activation of the *jun*B gene with the same kinetics and magnitude as in H-35 cells.[52] The "partial" IL-6 response in HTC cells suggests that IL-6 signaling does not involve a common pathway for all IL-6-responsive genes, but that separate components with a different genetic target specificity are activated.[91] It seems that HTC cells are deficient in the component (or components) required for cytokine stimulation of the AGP gene. The hypothetical component is *not* C/EBPβ (or AGP/EBP), since C/EBPβ is present in HTC cells[73,78] and functions as IL-1 RE binding activity when tested by Southwestern blotting.[78] The deficient IL-6 response of HTC cells may prove to be a unique system in which to test by complementation those prospective components which restore full cytokine regulation of the AGP gene.

VII. OTHER ISSUES CONCERNING AGP GENE REGULATION

A. TISSUE-SPECIFIC EXPRESSION

The characterization of human AGP gene was originally approached in an attempt to define the genetic elements governing liver-specific expression.[54,65] A survey of adult human and rodent tissues has indicated that liver is the major site of AGP expression. The only other significant expression of AGP occurs in the decidua of the placenta during early fetal development.[84]

When human AGP-A gene constructs, containing as little as 1.2 kb 5′ and 2 kb 3′ flanking region, were introduced into transgenic mice, a liver-specific, acute phase-inducible expression was observed that was comparable to the endogenous mouse AGP genes.[34] A similar result was obtained with the rat AGP gene, except that a minor ($^{1}/_{100}$ of liver) but consistent expression of the transgene, but not endogenous mouse AGP gene, was observed in the heart.[79] Analysis of the placenta of transgenic mice showed that both rat and mouse AGP mRNA were detectable at about fivefold higher levels than in the maternal liver.[88]

As a prelude for studies aimed at delineating the liver-specific elements, the same human or rat AGP gene constructs that had been used for making transgenic animals were also introduced into nonhepatic tissue culture cells, HeLa and L-cells, respectively.[17,34] In each case, the transgene was expressed, suggesting that the activity of the AGP gene was not strictly dependent upon liver-specific factors. The relative high level of human AGP gene activity in HeLa cells even led to the conclusion that tissue culture analysis of gene sequences for tissue-specific expression yields unreliable information.[34] No further advancements have been made in deciphering the elements underlying tissue-specific expression. With the identification of the transcription factors needed for hormone-specific regulation, potential candidates for tissue-specific expression might be recognized. One significant role in establishing the basal activity of AGP gene in liver could be ascribed to C/EBP isoforms.[52] However, supporting evidence for that proposition remains to be established.

B. CONTROL OF AGP EXPRESSION THROUGH POSTTRANSCRIPTIONAL EVENTS

Transcriptional activation of the AGP gene is a major contributor to the increase in AGP expression in acute phase liver,[8,10,11] dexamethasone-treated HTC cells,[44,45] and cytokine plus dexamethasone-treated H-35 cells.[52] However, the magnitude of transcription rate changes has been reported in several instances to be less than the change in AGP mRNA accumulation.[8,11,26] This observation has been interpreted to mean that cytokine and glucocorticoid treatment also affects posttranscriptional events. Since the half-life of cytoplasmic AGP mRNA does not appear to be affected by hormonal treatment, the controlling influence has been proposed to occur at the level of nuclear transcript processing and export.[26] Although there is a prominent and fairly rapid shortening of the polyadenylate tail of mature AGP mRNA (as well as other mRNAs) in both liver and hepatoma cells, this process does not seem to contribute to the change in mRNA concentration.[11,44]

In order to gain an experimental system for studying potential posttranscriptional regulation of AGP mRNA, the entire structural rat AGP gene has been placed under control of the highly active and hormone-independent immediate early promoter of human cytomegalovirus. This gene construct was transiently introduced into Hep G2 cells.[45] Treatment of these cells with IL-1, IL-6, and dexamethasone caused a three- to fivefold increase of the transcription rate of the endogenous human AGP gene and a 20- to 30-fold increase of human AGP mRNA. However, the steady-state concentration of the mature rat AGP mRNA, derived from the transgene construct, was unaffected by the same treatments. Although this experimental system has not provided supportive evidence for posttranscriptional AGP regulation, alternative approaches have to be pursued (e.g., use of different AGP gene constructs and cell targets, and analysis of nuclear precursor DNAs) in order to document the presence and define the mode of the proposed regulatory system.

VIII. CONCLUSION

The analysis of AGP gene regulation has been rewarding on one hand and frustrating on the other. Nevertheless, this gene remains a fascinating subject for further studies, since it will hold secrets that, when uncovered, will open new perspectives to show how acute phase protein genes are regulated, what determines tissue and developmental specificity, and why the regulation has been so tightly conserved during vertebrate evolution.

ACKNOWLEDGMENTS

Research in the authors' laboratory was supported by NIH grant CA26122 and DK33886. We thank Karen K. Morella and Gerald P. Jahreis for technical assistance, Dr. Susana P. Campos for help in manuscript preparation, and Marcia Held for secretarial support.

REFERENCES

1. **Baumann, P., Eap, C. B., Muller, W. E., and Tillement, J.-P., Eds.,** Alpha$_1$-acid glycoprotein. Genetics, biochemistry, physiological functions and pharmacology, *Prog. Clin. Biol.*, 300, 1, 1989.
2. **Schreiber, G., Hawlett, G., Nagashima, M., Millership, A., Martin, H., Urban, J., and Kotler, L.,** The acute phase response of plasma protein synthesis during experimental inflammation, *J. Biol. Chem.*, 257, 10271, 1982.
3. **Koj, A.,** Definition and classification of acute phase proteins, in *The Acute Phase Response to Injury and Infection*, Vol. 5, Gordon, A. H. and Koj, A., Eds., Elsevier, Amsterdam, 1985, 139.

3a. **Kushner, I. and Mackiewicz, A.,** Acute phase proteins as disease markers, *Dis. Markers*, 5, 1, 1987.

4. **Schmid, K.,** α_1-Acid glycoprotein, in *The Plasma Proteins, Structure Function and Genetics*, Vol. 1, Putnam, F. W., Ed., Academic Press, New York, 1975, 183.
5. **Shibata, K., Okubo, H., Ishibashi, H., and Tsuda, K.,** Rat α_1-acid glycoprotein. Purification and immunological estimation of its serum concentration, *Biochim. Biophys. Acta*, 495, 37, 1977.
6. **Jeanloz, R. W.,** α_1-Acid glycoprotein, in *Glycoproteins*, Gottschalk, A., Ed., Elsevier/North-Holland, New York, 1966, 362.
7. **Ricca, G. A., Hamilton, R. W., McLean, J. W., Conn, A., Kallinyak, J. E., and Taylor, J. M.,** Rat α_1-acid glycoprotein messenger RNA, *J. Biol. Chem.*, 256, 10362, 1981.
8. **Birch, H. E. and Schreiber, G.,** Transcriptional regulation of plasma protein synthesis during inflammation, *J. Biol. Chem.*, 261, 8077, 1986.
9. **Baumann, H., Firestone, G. L., Burgess, T. L., Gross, K. W., Yamamoto, K. R., and Held, W. A.,** Dexamethasone regulation of alpha$_1$-acid glycoprotein and other acute phase reactants in rat liver and hepatoma cells, *J. Biol. Chem.*, 258, 563, 1983.
10. **Kulkarni, A. B., Reinke, R., and Feigelson, P.,** Acute phase mediators and glucocorticoids elevate α_1-acid glycoprotein gene transcription, *J. Biol. Chem.*, 260, 15386, 1985.
11. **Shiels, B. R., Northemann, W., Gehring, M. R., and Fey, G. H.,** Modified nuclear processing of α_1-acid glycoprotein RNA during inflammation, *J. Biol. Chem.*, 262, 12826, 1987.
12. **Vannice, J. L., Ringold, G. M., McLean, J. W., and Taylor, J. M.,** Induction of the acute phase reactants α_1-acid glycoprotein by glucocorticoids in rat hepatoma cells, *DNA*, 2, 205, 1983.
13. **Nicollet, I., Lebreton, J. P., Fontaine, M., and Hiron, M.,** Evidence for alpha-1-acid glycoprotein populations of different pI values after concanavalin A affinity chromatography: study of their evolution during inflammation in man, *Biochim. Biophys. Acta*, 668, 235, 1981.
14. **Koj, A., Dubin, A., Kasperczak, H., Bereta, J., and Gordon, A. H.,** Changes in the blood level and affinity to concanavalin A of rat plasma glycoproteins during acute inflammation and hepatoma growth, *Biochem. J.*, 206, 545, 1982.
15. **Mackiewicz, A., Ganapathi, M. K., Schulz, D., and Kushner, I.,** Monokines regulate glycosylation of acute phase proteins, *J. Exp. Med.*, 166, 253, 1987.
16. **Pos, O., van Dijk, W., Ladiges, N., Linthorst, C., Sala, M., van Tiel, D., and Boers, W.,** Glycosylation of four acute phase glycoproteins secreted by rat liver cells in vivo and in vitro. Effects of inflammation and dexamethasone, *Eur. J. Cell Biol.*, 46, 121, 1988.
17. **Baumann, H. and Prowse, K. R.,** Genetic element involved in regulated expression of rodent α_1-acid glycoprotein genes, in *Genetics, Biochemistry, Physiological Functions and Pharmacology of Alpha 1-Acid Glycoprotein*, Alan R. Liss, New York, 1989, 99.
18. **Baumann, H.,** Hepatic acute phase reaction in vitro and in vivo, *In Vitro Cell. Dev. Biol.*, 25, 115, 1989.
19. **Baumann, H. and Held, W. A.,** Biosynthesis and hormone-regulated expression of secretory glycoproteins in rat liver and hepatoma. Effect of glucocorticoids and inflammation, *J. Biol. Chem.*, 256, 10145, 1981.
20. **Koj, A., Gauldie, J., Regoeci, E., Sander, D. N., and Sweeney, G. P.,** The acute phase response of cultured rat hepatocytes, *Biochem. J.*, 224, 505, 1984.
21. **Baumann, H., Jahreis, G. P., and Gaines, K. C.,** Synthesis and regulation of acute phase plasma proteins in primary cultures of mouse hepatocytes, *J. Cell Biol.*, 97, 866, 1983.
22. **Prowse, K. R. and Baumann, H.,** Interleukin-1 and interleukin-6 stimulate acute phase protein production in primary mouse hepatocytes, *J. Leuk. Biol.*, 45, 55, 1989.
23. **Mackiewicz, A., Ganapathi, M. K., Schultz, D., Samols, D., Reese, J., and Kushner, I.,** Regulation of rabbit acute phase protein synthesis by monokines, *Biochem. J.*, 253, 851, 1988.
24. **Castell, J. V., Gomez-Lechon, M. J., David, M., Andus, T., Geiger, T., Trullenque, R., Fabra, R., and Heinrich, P. C.,** Interleukin 6 is the major regulator of the acute phase protein synthesis in adult human hepatocytes, *FEBS Lett.*, 242, 237, 1989.
25. **Castell, J. V., Gomez-Lechon, M. J., David, M., Hirano, T., Kirshimoto, T., and Heinrich, P. C.,** Recombinant human interleukin-6 (IL-6/BSF-2/HSF) regulates the synthesis of acute phase proteins in human hepatocytes, *FEBS Lett.*, 232, 347, 1988.

26. **Vannice, J. L., Taylor, J. M., and Ringold, G. M.,** Glucocorticoid-mediated induction of α_1-acid glycoprotein evidence for hormone-regulated RNA processing, *Proc. Natl. Acad. Sci. U.S.A.*, 81, 4241, 1984.
27. **Feinberg, R. F., Sun, L. H. K., Ordahl, C. P., and Frankel, F. R.,** Identification of glucocorticoid-induced genes in rat hepatoma cells by isolation of cloned cDNA sequences, *Proc. Natl. Acad. Sci. U.S.A.*, 80, 5042, 1983.
28. **Geiger, T., Andus, T., Klapproth, J., Northoff, H., and Heinrich, P. C.,** Induction of α_1-acid glycoprotein by recombinant human interleukin-1 in rat hepatoma cells, *J. Biol. Chem.*, 263, 7141, 1988.
29. **Baumann, H., Onorato, V., Gauldie, J., and Jahreis, G. P.,** Distinct sets of acute phase plasma proteins are stimulated by separate human hepatocyte-stimulating factors and monokines in rat hepatoma cells, *J. Biol. Chem.*, 262, 9756, 1987.
30. **Baumann, H., Prowse, K. R., Marinkovic, S., Won, K.-A., and Jahreis, G. P.,** Stimulation of hepatic acute phase response by cytokines and glucocorticoids, *Ann. N.Y. Acad. Sci.*, 557, 280, 1989.
31. **Baumann, H., Jahreis, G. P., Sauder, D. N., and Koj, A.,** Human keratinocytes and monocytes release factors which regulate the synthesis of major acute phase plasma proteins in hepatic cells from man, rat and mouse, *J. Biol. Chem.*, 259, 7331, 1984.
32. **Darlington, G., Wilson, D. R., and Lachman, L. B.,** Monocyte-conditioned medium, interleukin-1 and tumor necrosis factor stimulate the acute phase response in human hepatoma cells in vitro, *J. Cell Biol.*, 103, 787, 1986.
33. **Prowse, K. R. and Baumann, H.,** Hepatocyte-stimulating factor, β2-interferon and interleukin-1 enhance expression of the rat α_1-acid glycoprotein gene via a distal upstream regulatory region, *Mol. Cell. Biol.*, 8, 42, 1988.
34. **Dente, L., Ruther, U., Tripoli, M., Wagner, E. F., and Cortese, R.,** Expression of human α_1-acid glycoprotein genes in cultured cells and in transgenic mice, *Genes Dev.*, 2, 259, 1988.
35. **Klein, E. S., Reinke, R., Feigelson, P., and Ringold, G. M.,** Glucocorticoid regulated expression from the 5′-flanking region of the rat alpha-1-acid glycoprotein gene, *J. Biol. Chem.*, 262, 520, 1987.
36. **Andus, T., Geiger, T., Hirano, T., Kishimoto, T., and Heinrich, P. C.,** Action of recombinant human interleukin 6, interleukin 1β and tumor necrosis factor α on the mRNA induction of acute phase proteins, *Eur. J. Immunol.*, 18, 739, 1988.
37. **Gauldie, J., Richards, C., Harnish, D., Lansdorp, P., and Baumann, H.,** Interferon β2/B-cell stimulatory factor type 2 shares identity with monocyte-derived hepatocyte-stimulating factor and regulates the major acute phase response in liver cells, *Proc. Natl. Acad. Sci. U.S.A.*, 84, 7251, 1987.
38. **Baumann, H., Richards, C., and Gauldie, J.,** Interaction among hepatocyte-stimulating factors, interleukin-1 and glucocorticoids for regulation of acute phase plasma proteins in human hepatoma (HepG2) cells, *J. Immunol.*, 139, 4122, 1987.
39. **Fey, G. and Gauldie, J.,** The acute phase response of the liver in inflammation, in *Progress in Liver Disease*, Vol. 9, Popper, H. and Schaffner, F., Eds., W. B. Saunders, Philadelphia, 1990, 89.
40. **Richards, C. D., Brown, T. J., Shoyab, M., Baumann, H., and Gauldie, J.,** Oncostatin M stimulates the production of acute phase proteins in HepG2 cells and rat primary hepatocytes, *J. Immunol.*, 148, 1731, 1992.
41. **Baumann, H. and Wong, G. G.,** Hepatocyte-stimulating factor-III shares structural and functional identity with leukemia-inhibitory factor, *J. Immunol.*, 143, 1163, 1989.
42. **Baumann, H. and Schendel, P.,** Interleukin-11 regulates the hepatic expression of the same plasma protein genes as interleukin-6, *J. Biol. Chem.*, 266, 10424, 1991.
43. **Baumann, H. and Gauldie, J.,** Regulation of hepatic acute phase plasma protein genes by hepatocyte stimulating factors and other mediators of inflammation, *Mol. Biol. Med.*, 7, 147, 1990.
44. **Carter, K. C., Bryan, S., Gadson, P., and Papaconstantinou, J. C.,** Deadenylation of α_1-acid glycoprotein mRNA in cultured hepatic cells during stimulation by dexamethasone, *J. Biol. Chem.*, 264, 4112, 1989.
45. **Baumann, H.,** Transcriptional control of the rat α_1-acid glycoprotein gene, *J. Biol. Chem.*, 265, 19420, 1990.
46. **Klein, E. S., DiLorenzo, D., Posseckert, G., Beato, M., and Ringold, G. M.,** Sequences downstream of the glucocorticoid regulatory element mediate cycloheximide inhibition of steroid induced expression from the rat alpha 1-acid glycoprotein promoter: evidence for a labile transcription factor, *Mol. Endocrinol.*, 2, 1343, 1988.
47. **Baumann, H., Jahreis, G. P., and Morella, K. K.,** Interaction of cytokine- and glucocorticoid-response elements of acute phase plasma protein genes. Importance of glucocorticoid receptor level and cell type for regulation of the elements from rat α_1-acid glycoprotein and β-fibrinogene genes, *J. Biol. Chem.*, 265, 22275, 1990.
48. **Wilson, D. R., Juan, T. S.-C., Wilde, M. D., Fey, G. H., and Darlington, G. J.,** A 58 base pair region of the human C3 gene confers synergistic inducibility by interleukin 1 and interleukin 6, *Mol. Cell. Biol.*, 10, 6181, 1990.

49. **Huang, J. H., Rienhoff, H. Y., and Liao, W. S. L.,** Regulation of mouse serum amyloid A gene expression in transfected hepatoma cells, *Mol. Cell. Biol.,* 10, 3619, 1990.
50. **Marinkovic, S. and Baumann, H.,** Structure, hormonal regulation, and identification of the interleukin-6 and dexamethasone-responsive element of the rat haptoglobin gene, *Mol. Cell. Biol.,* 10, 1573, 1990.
51. **Ganter, U., Arcone, R., Toniatti, C., Morone, G., and Ciliberto, G.,** Dual control of C-reactive protein gene expression by interleukin 1 and interleukin 6, *EMBO J.,* 8, 3773, 1989.
52. **Baumann, H., Jahreis, G. P., Morella, K. K., Won, K.-A., Pruitt, S. C., Jones, V. E., and Prowse, K. R.,** Transcription regulation through cytokine- and glucocorticoid-response elements of rat acute phase plasma protein genes by C/EBP and JunB, *J. Biol. Chem.,* 266, 10390, 1991.
53. **Ricca, G. A. and Taylor, J. M.,** Nucleotide sequence of rat α_1-acid glycoprotein messenger RNA, *J. Biol. Chem.,* 256, 11199, 1981.
54. **Dente, L., Ciliberto, G., and Cortese, R.,** Structure of the human α_1-acid glycoprotein gene: sequence homology with other human acute phase protein genes, *Nucleic Acids Res.,* 13, 3941, 1985.
55. **Board, G., Jones, I. M., and Bentley, A. K.,** Molecular cloning and nucleotide sequence of AAG cDNA, *Gene,* 44, 127, 1986.
56. **Cooper, R. and Papaconstantinou, J.,** Evidence for the existence of multiple α_1-acid glycoprotein genes in the mouse, *J. Biol. Chem.,* 261, 1849, 1986.
57. **Lee, S.-C., Chang, C.-J., Lee, Y.-M., Lei, H.-Y., Lai, M.-Y., and Chen, D. S.,** Molecular cloning of cDNAs corresponding to two genes of α_1-acid glycoprotein and characterization of two alleles of AGP-1 in the mouse, *DNA,* 8, 245, 1989.
58. **Prowse, K. R. and Baumann, H.,** Molecular characterization and acute-phase expression of the multiple Mus caroli α_1-acid glycoprotein (AGP) genes: difference in glucocorticoid stimulation and regulatory elements between the rat and mouse AGP genes, *J. Biol. Chem.,* 265, 10201, 1990.
59. **Oschmid, K., Kaufmann, H., Isemura, S., Bauer, F., Emura, J., Motoyama, T., Ishiguro, M., and Nanno, S.,** Structure of α_1-acid glycoprotein. The complete amino acid substitutions and homology with the immunoglobulins, *Biochemistry,* 12, 2711, 1973.
60. **Baumann, H., Held, W. A., and Berger, F. G.,** The acute phase response of mouse liver. Genetic analysis of the major acute phase reactants, *J. Biol. Chem.,* 259, 566, 1984.
61. **Baumann, H. and Berger, F. G.,** Genetics and evolution of the acute phase proteins in mice, *Mol. Gen. Genet.,* 201, 505, 1985.
62. **Szpirer, C., Riviere, M., Szpirer, J., Geneti, M., Dreze, P., Islam, M. Q., and Levan, G.,** Assignment of 12 loci to rat chromosome 5: evidence that this chromosome is homologous to mouse chromosome 4 and to human chromosomes 9 and 1 (1parm), *Genomics,* 6, 679, 1990.
63. **Reinke, R. and Feigelson, P.,** Rat α_1-acid glycoprotein gene sequence and regulation by glucocorticoids in transfected L-cells, *J. Biol. Chem.,* 260, 4397, 1985.
64. **Liao, Y.-C., Taylor, J. M., Vannice, J. L., Clawson, G. A., and Smuckler, E. A.,** Structure of the rat α_1-acid glycoprotein gene, *Mol. Cell. Biol.,* 5, 3634, 1985.
65. **Dente, L., Pizza, M. G., Metspalu, A., and Cortese, R.,** Structure and expression of the genes coding for human α_1-acid glycoprotein, *EMBO J.,* 6, 2289, 1987.
66. **Cooper, R., Eckley, D. M., and Papaconstantinou, J.,** Nucleotide sequence of the mouse α_1-acid glycoprotein gene 1, *Biochemistry,* 26, 5244, 1987.
67. **Chang, C.-J., Lin, J.-H., Yao, T.-P., and Lee, S.-C.,** Regulation of expression of mouse α_1-acid glycoprotein (AGP) genes, in *Structure and Function of Nucleic Acids and Proteins,* Wu. F. Y. and Wu, C. W., Eds., Raven Press, New York, 1990, 273.
68. **Eiberg, H., Mohr, J., and Nielsen, L. S.,** δ-Amino levulinase dehydrogenase: synteny with ABO, AK1, ORM and assignment to chromosome 9, *Clin. Genet.,* 23, 150, 1983.
69. **Cox, D. and Francke, U.,** Direct assignment of orosomucoid to chromosome 9 and α_2HS-glycoprotein to chromosome 3 using human fetal liver $\times$ rat hepatoma hybrids, *Hum. Genet.,* 70, 109, 1985.
70. **Glibetic, M. D. and Baumann, H.,** Influence of chronic inflammation on the level of mRNAs for acute phase reactants in the mouse liver, *J. Immunol.,* 137, 1616, 1986.
71. **Baumann, H. and Maquat, L.,** Localization of DNA sequences involved in dexamethasone-dependent expression of the rat alpha 1-acid glycoprotein gene, *Mol. Cell. Biol.,* 6, 2551, 1986.
72. **DiLorenzo, D., Williams, P. M., and Ringold, G. M.,** Identification of two distinct nuclear factors with DNA binding activity within the glucocorticoid regulatory region of the rat alpha-1-acid glycoprotein gene promoter, *Biochem. Biophys. Res. Commun.,* 176, 1326, 1991.
73. **Williams, P. M., Ratajczak, T., Lee, S. C., and Ringold, G. M.,** AGP/EBP (LAP) expressed in rat hepatoma cells interacts with multiple promoter sites and is necessary for maximal glucocorticoid induction of the rat alpha-1 acid glycoprotein gene, *Mol. Cell. Biol.,* 11, 4959, 1991.
74. **Chang, C.-J., Chen, T.-T., Lei, H.-Y., Chen, D.-S., and Lee, S.-C.,** Molecular cloning of a transcription factor, AGP/EBP, that belongs to members of the C/EBP family, *Mol. Cell. Biol.,* 10, 6642, 1990.

75. **Poli, V., Manaini, F. P., and Cortese, R.**, IL-6 DBP, a nuclear protein involved in interleukin-6 signal transduction defines a new family of leucine zipper proteins related to C/EBP, *Cell*, 63, 643, 1990.
76. **Descombes, P., Chojkier, M., Lichsteiner, S., Falvey, E., and Schibler, U.**, LAP, a novel member of the C/EBP gene family, encodes a liver-enriched transcriptional activator protein, *Genes Dev.*, 4, 1541, 1990.
77. **Cao, Z., Umek, R., and McKnight, S. L.**, Regulated expression of three C/EBP isoforms during adipose conversion of 3T3-L1 cell, *Genes Dev.*, 5, 1538, 1991.
78. **Baumann, H., Morella, K. K., Campos, S. P., Cao, Z., and Jahreis, G. P.**, Role of C/EBP isoforms in the cytokine regulation of acute phase plasma protein genes, *J. Biol. Chem.*, 267, 19744, 1992.
79. **Dewey, M. J., Rheaume, C., Berger, F. G., and Baumann, H.**, Inducible and tissue specific expression of rat α_1-acid glycoprotein in transgenic mice, *J. Immunol.*, 144, 4392, 1990.
80. **Bonhomme, F., Catalan, J., Britton-Davidson, J., Chapman, V., Moriwaki, K., Nevo, E., and Thaler, L.**, Biochemical diversity and evolution of the genus Mus, *Biochem. Genet.*, 22, 275, 1984.
81. **Won, K.-A. and Baumann, H.**, The cytokine response element of the rat α_1-acid glycoprotein gene is a complex of several interacting regulatory sequences, *Mol. Cell. Biol.*, 10, 3965, 1990.
82. **Won, K.-A. and Baumann, H.**, NF-AB: a liver specific and cytokine inducible nuclear factor that interacts with the IL-1 response element of the rat α_1-acid glycoprotein gene, *Mol. Cell. Biol.*, 11, 3001, 1991.
83. **Baumann, H., Morella, K. K., Jahreis, G. P., and Marinkovic, S.**, Distinct regulation of the interleukin-1 and interleukin-6 response elements of the rat haptoglobin gene in rat and human hepatoma cells, *Mol. Cell. Biol.*, 10, 5967, 1990.
84. **Thomas, T., Fletcher, S., Yeoh, G. C. T., and Schreiber, G.**, The expression of α_1-acid glycoprotein mRNA during rat development, *J. Biol. Chem.*, 264, 5784, 1989.
85. **Baumann, M., Baumann, H., and Fey, G. H.**, Molecular cloning, characterization and functional expression of the rat liver interleukin-6 receptor, *J. Biol. Chem.*, 265, 19853, 1990.
86. **Baumann, H. and Berger, F. G.**, unpublished results.
87. **Baumann, H.**, unpublished results.
88. **Dewey, M. J. et al.**, unpublished results.
89. **Ratajczak, T., Williams, P. M., DiLorenzo, D., and Ringold, G. M.**, Multiple elements within the glucocorticoid regulatory unit of the rat α_1-acid glycoprotein gene are recognition sites for C/EBP. *J. Biol. Chem.*, 267, 11111, 1992.
90. **Alam, T. and Papaconstantinou, J.**, Interaction of acute phase-inducible and liver-enriched nuclear factors with the promoter of the mouse α_1-acid glycoprotein gene-1. *Biochemistry*, 31, 1928, 1992.
91. **Baumann, H., Morella, K. K., and Campos, S. P.**, Interleukin-6 signal communication to the α_1-acid glycoprotein gene, but not *junB* gene, is impaired in HTC cells. *J. Biol. Chem.*, 268, in press.

Chapter 24

TRANSCRIPTIONAL REGULATION OF THE HUMAN C3 GENE

Gretchen J. Darlington, Deborah R. Wilson, and Todd S.-C. Juan

TABLE OF CONTENTS

0-8493-6913-4/93/$0.00 + $.50

I. INTRODUCTION

The human acute phase response is characterized by the elevation of a number of serum proteins. C-reactive protein and protein serum amyloid A are dramatically increased in serum in response to inflammation, whereas other proteins such as the third component of complement (C3) increase only 50% in serum levels. In spite of this modest increase in the level of the protein in the serum of the whole organism, we have found that the transcriptional regulation of the C3 gene is dramatic and complex.

C3 is critical to the function of both the classical and alternative pathways of complement activation. The vast majority of serum C3 is synthesized by the liver (approximately 90%[1]). The expression of C3 differs from that of the bulk of the acute phase genes, however, in that it has a fairly broad tissue distribution. C3 synthesis occurs in a wide range of cell types, including fibroblasts,[2-4] monocytes,[5-7] macrophages,[5,8,9] astrocytes,[10] and endothelial[11,12] and some epithelial cells.[13-15]

The human C3 gene is 41 to 42 kb in length with 41 exons.[16,17] Sequence analysis of the 5′ flanking region has revealed a substantial number of elements with homology to known regulatory sequences, including those for NF-κB, AP-1, estrogen and glucocorticoid receptors, and interferon-γ (IFNγ)- and interleukin-6 (IL-6)-responsive factors.[16,17] There are also two consensus sequences for the C/EBP family of transcription factors (bZIP) within the C3 promoter.[18] C3 falls into a subset of acute phase reactants including mouse serum amyloid A (SAA),[19,20] whose hepatic synthesis is increased primarily in response to the cytokine IL-1. Since C3 synthesis is influenced by a number of cytokines/hormones in addition to IL-1 (e.g., tumor necrosis factor [TNF],[21] IL-6,[18,21,22] estradiol,[14] glucocorticoids,[11] and IFNγ,[11,23]) it is reasonable to expect a complex array of regulatory elements. We have used transfection analysis of deletional and/or site-directed mutants of the promoter and 5′ flanking sequence of the C3 gene to functionally identify *cis*-acting elements involved in its transcriptional regulation. Gel retardation and footprint analyses have provided information about the proteins that bind at some of these *cis*-regulatory sites.

II. MATERIALS AND METHODS

A. MATERIALS

Unless otherwise indicated, restriction and DNA modification enzymes were purchased from BRL (Gaithersburg, MD) or New England Biolabs (Beverly, MA). Recombinant rat C/EBP was kindly provided by Steven L. McKnight. Recombinant IL-1β and IL-6 were the gifts of Biogen (Geneva, Switzerland) and Genetics Institute (Cambridge, MA), respectively. Expression vectors for LAP/IL-6DBP and for DBP were kindly supplied by Dr. Ueli Schibler. A double-stranded NF-κB oligomer was kindly provided by Warren S. L. Liao and has the sequence CGGGGACTTTCCG. A double-stranded oligonucleotide, CATGGATCTGGGGCAGCCCCAAAAC, containing the NF-κB-like element of the C3 gene was synthesized for gel retardation analysis. Nine oligonucleotides were prepared for site-directed mutagenesis in this study:

- GGAAATGGTATTGAGAAATGATATCCAGCCCCAAAAGG (C3MutERV)
- GGAAATGGTATTGTTCGACTGGGGCAGC (C3MutSal)
- GAAATCTGGGGCCGGCAGAAAAGGGGAGAG (C3MutNae)
- GACTGAAAAGCTTATCGCGAGGTATTGAGAAA (C3MutNru)
- GAAATCTGGGGCAGAGACAAAAGGGGAGAG (C3MutBsA)
- GGTATTGAGAAATGAGGGGCAGCCCCAAAAGG (C3MutCT)
- ATGGTATTGAGAAATCTTAGGCAGCCCCAAAAGG (C3MutGG1)

- ATGGTATTGAGAAATCTGGTCCAGCCCCAAAAG (C3MutGG2)
- GCAGCCCCAAAAGGATATCGGCCATGGGGAGG (C3MutERV-2)

These oligonucleotides were synthesized by the Nucleic Acids Core in the Institute for Molecular Genetics at Baylor College of Medicine using an automated DNA synthesizer.

B. CELL CULTURE AND TRANSFECTION ANALYSIS

Hep 3B2 cells were maintained as monolayer cultures in 3 MEM:1 MAB (MEM, Eagles's minimal essential medium; MAB, MAB 87/3; GIBCO, Gaithersburg, MD), 2% fetal bovine serum (FBS: Hazleton, Lenexa, KS), and 8% horse serum (Hazleton) and passaged weekly by trypsinization. Cells used for Northern analysis were processed as previously described.[21] In experiments where cycloheximide (Sigma Chemical Co., St. Louis, MO) was employed, 5 μg/ml of the inhibitor was added to the culture medium 15 min before the addition of cytokines. Cells used for preparation of nuclear extracts were treated as reported.[18] Cell extracts for luciferase assays were prepared either by the freeze-thaw procedure of DeWet et al.[24] with subsequent processing as previously described[18] or by Triton X-100 lysis,[25] in which the cells were harvested into 300 μl of glycyl-glycine, pH 7.8, 1 m*M* dithiothreitol, 1% Triton X-100. Five to 30 μl of cell extract were assayed for luciferase activity in 100 m*M* KPB (potassium phosphate buffer), pH 7.9, 15 m*M* $MgSO_4$, 5 m*M* ATP, using a Monolight 2010 luminometer (Analytical Luminescence Laboratories, San Diego, CA). The luminescence obtained from a 30-μl buffer sample was subtracted from each measurement prior to quantitation. Luciferase activities are expressed as relative luminescence units per milligram or per microgram cellular protein.

C. PLASMID CONSTRUCTION

The construction of the C3-promoted luciferase plasmids has been reported.[18] Where necessary, restriction fragment ends were blunted with the large fragment of DNA Pol I before ligation. The expression vector CMV-hC/EBPα was made by inserting the NruI to XhoI fragment of the human C/EBP gene[69] into the unique NotI site of pCMVϕ (prepared by digesting pCMVβ[26] with NotI and resealing to delete the β-galactosidase sequence). The rat α-fibrinogen clone 4α5 was obtained from Gerald Crabtree. The clone was digested with SphI and DdeI, and the fragment comprising the region from −177 to −70 bp subcloned between the SphI and HincII sites of pUC19 to generate pUCαfibC/EBP. A second fragment was subcloned from 4α5 (AluI to PstI, −83 to −33 bp) to generate pUCαfibHNF I. A portion of the human albumin promoter from −190 to −90 (Tth111I to Fnu4HI)[27] was subcloned into HincII-digested pUC19 to generate pUC101.

D. GEL RETARDATION AND FOOTPRINT ANALYSES

Nuclear extracts were prepared and DNase I footprinting and gel retardation analyses were performed as previously described.[18] The 20-μl binding reactions for DNase I footprinting contained 15,000 cpm DNA, 2.5 μg poly(dI-dC):poly(dI-dC) (Pharmacia), 50 μg of Hep 3B2 nuclear extract, 30 m*M* Tris, pH 8.0, 5 m*M* HEPES, pH 7.9, 0.66 m*M* EDTA, 7.5 m*M* $MgCl_2$, 60 m*M* KCl, 1.2 m*M* dithiothreitol, and 14% glycerol. Reactions were incubated for 30 min at 30°C. Following binding, the footprinting reactions were treated with DNase I (BRL) for 3 min at 0°C. DNase I concentrations were empirically determined; two different concentrations were used with each extract concentration in order to ensure an appropriate level of digestion. Electrophoresis was through an 8% polyacrylamide-8.3 *M* urea gel with Maxam-Gilbert sequencing reactions used as markers. For gel retardation analyses, binding reactions contained 10 μg of nuclear extract with 3.0 μg of poly(dI-dC):poly(dI-dC) and 10,000 cpm of labeled fragment. After incubation at 30°C for 15 min, the binding reactions were analyzed by electrophoresis in 6% high-ionic-strength polyacrylamide gels.[28] Heterologous fragments used for competitors were gel purified from restriction digestions of the human albumin[27] and rat α-fibrinogen[29] promoters. The DNA concentration was determined by comparing the relative staining intensity of electrophoretically separated aliquots to markers of known concentration. A 100-fold molar excess of cold competitors were preincubated with extracts for 10 min before probe addition. Probes used in gel retardation analysis were the HindIII to NcoI fragment of pT81lucH/N127[18] and the double-stranded NF-κB-like oligonucleotide representing the C3 promoter region from −103 to −85 bp.

E. NORTHERN-BLOT ANALYSIS

Northern-blot analysis was performed by standard procedures.[30] Lanes contained 20 μg of total cellular RNA. Probes used in hybridization were inserts isolated from the human albumin cDNA clone, F47,[31] the human C3 cDNA clone, pHC3.11,[32] the human α-fibrinogen cDNA clone, p115.6,[33] and the human γ-fibrinogen cDNA clone, p253.[34]

F. NUCLEAR RUN-ON ANALYSES

Transcriptional analysis by nuclear run-on was done according to the method previously published.[35] Hybridization was carried out using 5×10^6 to 10^7 cpms per reaction. The following plasmids containing human cDNA segments were immobilized on filters for transcriptional analysis: albumin, F47 (Alb), C3, pHC3.11 (C3), α_1-acid glycoprotein, pAGP-1 (AGP),[36] and α-fibrinogen, p115.6 (Fib).

G. SITE-DIRECTED MUTAGENESIS

Site-directed mutants were produced by the procedure of Kunkel,[37] with some modifications.[18] Two constructs, C3R4Pst4 or C3R4Pst4MutERV, were used for generating single-stranded DNA templates. C3R4Pst4 was derived from pTZ19 (Pharmacia, Piscataway, NJ) and contained nearly 500 base pairs of the C3 promoter/enhancer region. C3R4Pst4mutERV is identical to C3R4Pst4 except that it contains a 6-bp mutation of CTGGGG found in the wild-type C3 promoter to an EcoRV site, GATATC. Single-stranded DNA templates were recovered from the supernatants of transformed *Escherichia coli* infected with the helper phage M13K07 (Pharmacia, Piscataway, NJ). Primer extension was then performed by hybridizing the single-stranded DNA templates with one of the single-stranded mutant oligonucleotides and elongating with T4 DNA polymerase as described.[18] Mutant clones C3R4Pst4MutERV, C3R4Pst4MutSal, C3R4Pst4MutNru, C3R4Pst4MutNae, C3R4Pst4-MutBsA, and C3R4Pst4MutERV2 were done by using wild-type C3R4Pst4 as template; mutations were identified by restriction digestion using enzymes that recognized the mutant sequences we substituted. Mutants C3R4Pst4MutCT, C3R4Pst4MutGG1, and C3R4-Pst4MutGG2 were generated by using C3R4Pst4MutERV as template. Loss of the EcoRV site was used to identify clones containing substitutions within this region. All mutant constructs were inserted into the unique NcoI site of plasmid C3luc199ΔN/N (which has a deletion of 82 bp of the wild-type promoter) in order to create luciferase reporter constructs with mutant C3 promoters. Each one of these reporter constructs was further analyzed by diagnostic restriction digestion as well as by DNA sequencing to verify the presence of the mutation.

III. RESULTS

A. *IN VITRO* MODEL SYSTEM

We characterized an *in vitro* model system of the acute phase response and used it to study the molecular basis for the acute phase response. The human hepatoma cell line Hep

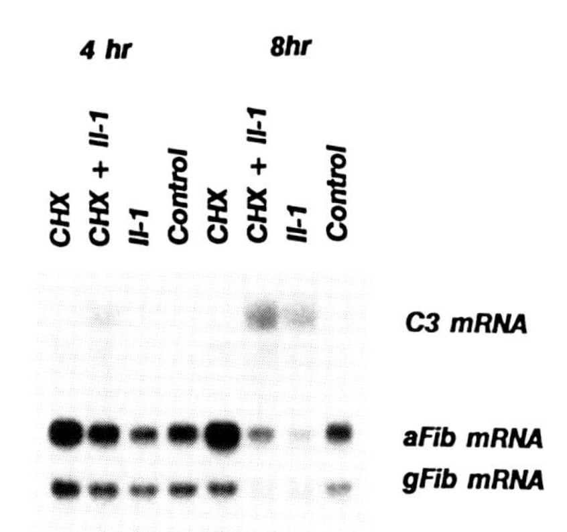

FIGURE 1. An IL-1-induced increase in C3 mRNA occurs in Hep 3B2 in the absence of protein synthesis, as determined by Northern-blot analysis. Hep 3B2 cells receiving cycloheximide (CHX) were pretreated for 15 min prior to addition of IL-1. Cells were incubated with IL-1 for either 4 h (left side) or 8 h (right side). Control cultures received only fresh medium. The Northern blot was probed for C3 and α- and γ-fibrinogen mRNAs. (From Darlington, G. J., Wilson, D. R., Revel, M., and Kelly, J. H., *Ann. N.Y. Acad. Sci.*, 557, 310, 1989. With permission.)

3B2, originally isolated by Knowles et al.,[38] expresses many hepato-specific gene products.[39] Hep 3B2 cells increase the level of C-reactive protein *in vitro* in response to the presence of conditioned monocyte medium.[21,40,41] Several other acute phase proteins, including C3, α_1-acid glycoprotein, fibrinogen, and haptoglobin, are also induced by monocyte-conditioned medium.[21] Analysis of cytokine factors, including IL-1, IL-6, and TNF, showed that these peptide hormones were able to regulate the expression of the acute phase response genes, and the C3 gene in particular, when used individually to treat Hep 3B2 cells.[21] This *in vitro* cell model system provides the opportunity to carry out gene transfer studies, dissect the molecular structures and features of the C3 gene, and determine the molecular basis for its regulation.

B. C3 TRANSCRIPTS ARE ELEVATED BY IL-1 AND IL-6

In conjunction with the increase in protein expression, messenger RNA levels for C3 are also elevated by treatment with interleukin-1. As Figure 1 illustrates, the level of C3 message is extremely low in untreated Hep 3B2 cells. Following 8 h of IL-1 stimulation, the level of C3 message increases dramatically. Shown in the same figure for comparison are the levels of α- and γ-fibrinogen. These two genes respond negatively to IL-1. Although they are positive acute phase reactants in that the levels of α- and γ-fibrinogen increase in serum during inflammation, IL-1 stimulation downregulates α- and γ-fibrinogen message levels in Hep 3B2 cells.

IL-6 also causes an increase in C3 mRNA. Figure 2B illustrates the elevation of C3 transcripts in response to IL-6. After 4 h (lane C) or 8 h (lane G) of treatment with IL-6, C3 message is increased above the level in untreated cells (lane D and H, respectively). In contrast to their response to IL-1, both fibrinogen genes are induced by IL-6 (Figure 2A, lanes C and G) compared to controls (lanes D and H). Thus, the alterations in C3 protein expressed by cytokine-treated Hep 3B2 cells reflect elevations in C3 message levels.

C. TRANSCRIPTIONAL REGULATION OF C3 BY CYTOKINES

The response of C3 to monocyte-conditioned medium, IL-1, or IL-6 is regulated, at least in part, at the transcriptional level. Nuclear run-on analyses have shown that both cytokines stimulate the transcription of the C3 gene. The transcriptional stimulation resulting from IL-1 (Figure 3A) is greater than that due to IL-6 (Figure 3B) over a 24-h period. TNF

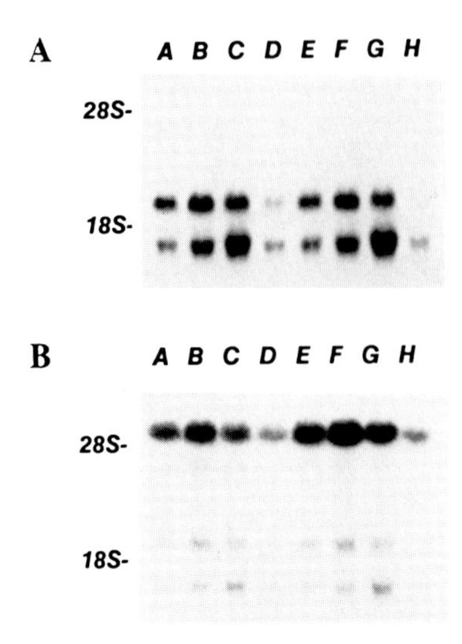

FIGURE 2. IL-6-induced changes in C3 and α- and γ-fibrinogen mRNA occur in Hep 3B2 in the absence of protein synthesis, as determined by Northern-blot analysis. RNA was prepared from Hep 3B2 cells treated as follows: cycloheximide (lanes A and E); cycloheximide pretreatment for 15 min, then addition of IL-6 for 4 h (lane B) or 8 h (lane F); IL-6 alone for 4 h (lane C) or 8 h (lane G); fresh medium (control) for 4 h (lane D) or 8 h (lane H). (A) α (upper band)- and γ (lower band)-fibrinogen mRNAs; (B) C3 mRNA.

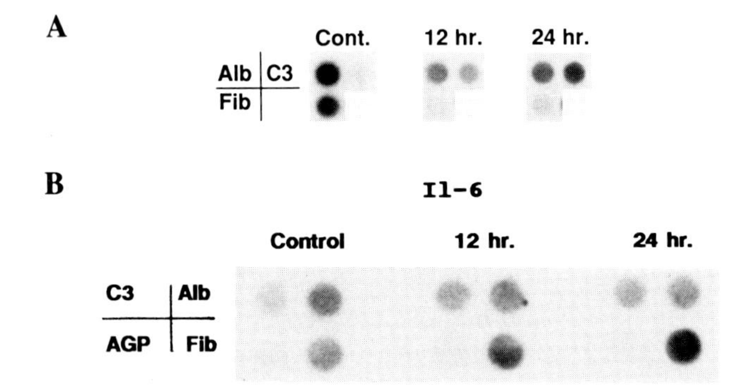

FIGURE 3. IL-1 and IL-6 increase C3 transcription, as determined by nuclear run-on analysis. Nuclei from Hep 3B2 cells were harvested after 12 or 24 h of IL-1 (A) or IL-6 (B) stimulation. Immobilized cDNAs represent albumin (Alb), C3, α-fibrinogen (Fib), and α_1-acid glycoprotein (AGP). (Figure 3A from Darlington, G. J., Wilson, D. R., Revel, M., and Kelly, J. H., *Ann. N.Y. Acad. Sci.*, 557, 310, 1989. With permission.)

also elevates C3 expression, but we have not examined its action on transcription of the endogenous gene.

D. CYTOKINE INDUCTION DOES NOT REQUIRE NEW PROTEIN SYNTHESIS

It is interesting to note that the elevation of mRNA for C3 and the downregulation of α- and γ-fibrinogen transcription by IL-1 is insensitive to cycloheximide (Figure 1), as is the increase of all three genes in response to IL-6 (Figures 2A and B, lanes B and F). The fact that no new protein synthesis is required for the response to IL-1 and IL-6 suggests that modification and/or nuclear translocation of transcription factor(s) which are preexisting in the cells is responsible for the acute phase reaction. Carrying this observation further, cycloheximide insensitivity suggests that the acute phase response can be quickly mobilized, presumably by the loss or alteration of activity of a labile repressor molecule or posttranslational modification of the factor itself. Indeed, cycloheximide alone elevates C3 mRNA (Figure 2B, lanes A and E), and studies of transcription by nuclear run-on analysis have shown that cycloheximide increased transcription directly.[70] Thus, the inducing effect of cycloheximide is due to a rapid response at the transcriptional level.

E. STRUCTURE OF THE HUMAN C3 PROMOTER

We have used transient transfection analysis in Hep 3B2[18] to identify a region of the C3 promoter that functions to mediate the transcriptional response to IL-1. This region is also modestly responsive to IL-6 and controls a synergistic response to simultaneous treatment with IL-1 plus IL-6.

Portions of the human C3 promoter were linked to luciferase as a reporter gene. Deletion analysis of the promoter within 320 bp of the transcription start site indicated that IL-1 and IL-6 responses could both be localized to a 58-bp element lying between −71 and −127 (Table 1). Construct C3luc127, containing 127 bp of 5′ flanking sequence, was induced approximately 20-fold by IL-1, fourfold by IL-6, and, synergistically, 47-fold by the combination of IL-1 plus IL-6. To confirm that this *cis* element was truly responsible for IL-1- and IL-6-inducible expression, a single copy of this promoter fragment (−127 to −71 bp: construct pT81H/N127) was coupled to a heterologous promoter from the herpes simplex virus thymidine kinase (TK) gene. The *cis* element from the C3 gene conferred IL-1- and IL-6-responsiveness, as well as the synergistic response to IL-1 plus IL-6, to the TK promoter and conclusively demonstrated the functional importance of this domain.

Within the region of functional response to cytokines lie several sequences with homology to factor-binding sites defined in other genes (Figure 4). Two C/EBP-like consensus sequences (−110 to −102 and −123 to −115), an IL-6-responsive element (−101 to −96) overlapping an NF-κB-like sequence (−99 to −89), can be found. Analysis of an additional 5′ deletion mutant, C3luc114, has further delineated the functional domain to be within a 45-bp region from −114 to −71 (data not shown). Thus, the distal C/EBP-like consensus sequence positioned at −115 to −123 can be deleted without substantially altering the basic IL-1 response of C3. Transfection experiments also revealed that an additional IL-1-responsive element(s) exists between −156 and −237 bp (Table 1). This upstream element(s) added a further 20-fold increase in the IL-1 response and a 40-fold increase in the IL-1 plus IL-6 response.

We used the construct C3luc237 to examine the kinetics of induction by IL-1 or IL-6 alone and in combination. The IL-1 response was quite rapid, achieving a 20-fold increase in reporter gene activity within 5 h of stimulation with only a 30% additional increase by 24 h. Costimulation with IL-1 and IL-6 gave a level of induction at 5 h equivalent to that achieved with IL-1 alone. The IL-6 response was much slower; a twofold increase in IL-6-induced activity was seen at 13.5 h with the combined IL-1 plus IL-6 treatment, resulting in a 150% increase over IL-1 stimulation alone. The synergistic response was greater at 24 h, reaching 175% of the IL-1-induced activity. These results are consistent with our nuclear run-on analyses, which indicated that IL-1 was faster than IL-6 in inducing an increase in C3 gene transcription.

F. FOOTPRINT ANALYSIS REVEALS A C/EBP BINDING SITE

Analysis of the sequence of the 58-bp, cytokine-responsive *cis* element by DNase I footprinting revealed a protected region of some 30 bp (Figure 5). The sequence of the protected region showed strong similarity to a sequence described by Akira et al. in the promoter of the IL-6 gene.[42] Like C3, the IL-6 gene *cis* element is IL-1 responsive. We

TABLE 1
Induction of C3 Transcription by IL-1 and IL-6

5′ end	Fraction unstimulated C3luc199 activity[b]	Fold induction[a] IL-1	IL-6	IL-1 + 6
−237	2.6 ± 0.7	45.5 ± 10.7	3.6 ± 0.4	84.9 ± 20.7
−199	1.0	55.5 ± 14.5	6.4 ± 0.8	77.4 ± 18.3
−156	0.8 ± 0.1	14.5 ± 1.2	2.8 ± 0.3	32.3 ± 5.3
−127	0.4 ± 0.1	21.0 ± 1.3	3.9 ± 0.3	46.9 ± 4.1
−75	0.2 ± 0.0	1.3 ± 0.1	1.8 ± 0.4	1.8 ± 0.6
Controls				
pXP1	0.01 ± 0.01	1.2 ± 0.1	0.9 ± 0.2	1.0 ± 0.1
pSV2AL	14.2 ± 7.0	2.6 ± 0.9	1.4 ± 0.8	1.9 ± 0.6
Heterologous Constructs[c]				
pT81luc	1.0	0.5 ± 0.1	1.1 ± 0	0.4 ± 0
pT81lucH/N127	2.4 ± 0.8	7.3 ± 2.3	1.8 ± 0.3	12.9 ± 1.9

Note: Values presented for the 5′ deletion mutants are the averages of at least three experiments ± SEM. Values for the remaining constructs are the averages for two experiments.

[a] Induction values represent the increase in luciferase expression promoted by a given construct following cytokine treatment relative to the level of expression of that construct in unstimulated cells.

[b] These values represent the level of the luciferase expression promoted by a given construct in unstimulated cells relative to that promoted by C3luc199 (also in unstimulated cells), thus allowing for a comparison of baseline values obtained with increasing promoter length. C3luc199 was selected as the reference point, as it is the minimum length tested that gives the full response to cytokine treatment.

[c] Values for the unstimulated heterologous constructs are expressed as the fraction of unstimulated pT81luc activity.

Data from Wilson, D. R., Juan, T. S.-C., Wilde, M. D., Fey, G. H., and Darlington, G. J., *Mol. Cell. Biol.*, 10, 6181, 1990. With permission.

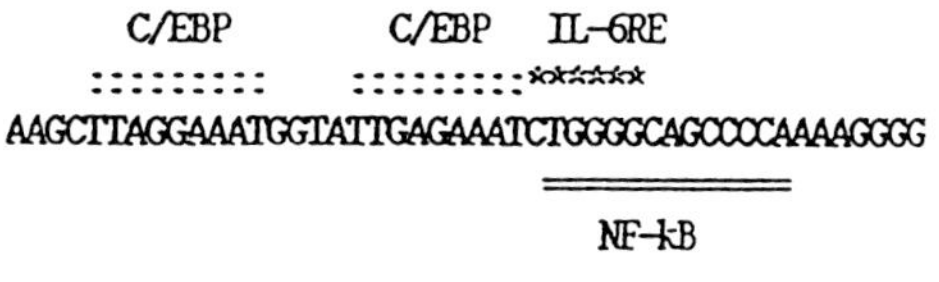

C/EBP C/EBP IL-6RE
AAGCTTAGGAAATGGTATTGAGAAATCTGGGGCAGCCCCAAAAGGGG
NF-kB

FIGURE 4. Consensus sequences contained within the C3 cytokine-responsive element.

have further demonstrated that the protected region of the C3 promoter has, on one strand, a DNase I footprint identical to, and on the other strand, one nearly identical to those generated from the binding of recombinant rat C/EBPα, confirming that the sequence of the functional *cis* element was capable of binding C/EBPα. C/EBPα is one of a family of bZIP DNA binding proteins that contains a region called the leucine zipper (which promotes dimerization) plus DNA binding and *trans*-activation domains. The various family members share considerable sequence homology throughout the zipper and DNA binding regions.

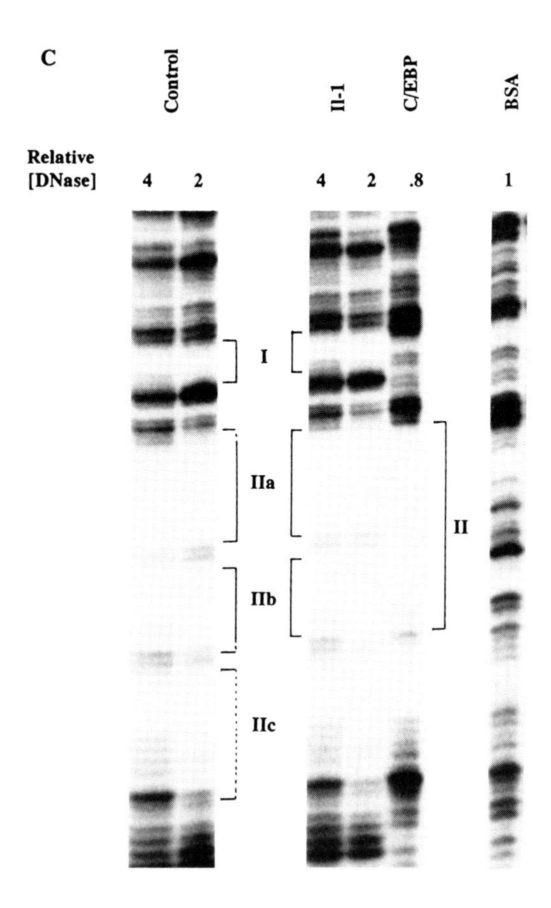

FIGURE 5. DNase I footprint analysis of the anti-sense strand of the proximal C3 promoter. Footprinting was performed using a C3 fragment extending from −1 to −156 bp. Extract or protein sources are indicated above the lanes: control, nuclear extracts prepared from unstimulated Hep 3B2 cells; IL-1, nuclear extracts prepared from IL-1-stimulated Hep 3B2 cells; BSA, bovine serum albumin; C/EBP, rat recombinant C/EBP. The relative DNase concentration used to digest each binding reaction is indicated by the number immediately above the lane. Protected regions are indicated by brackets; those occurring within the 58-bp functionally defined regulatory region are denoted by Roman numerals. Dashed lines indicate areas of variable protection, depending on the relative amount of DNase I used in the digestion. (From Wilson, D. R., Juan, T. S.-C., Wilde, M. D. Fey, G. H., and Darlington, G. J., *Mol. Cell. Biol.*, 10, 6181, 1990. With permission.)

Although DNase I footprint analysis identified C/EBPα as a binding factor, there were no qualitative differences in the footprint pattern generated using nuclear extracts from control or IL-1-stimulated cells.

G. ADDITIONAL PROTEIN-DNA COMPLEXES APPEAR IN RESPONSE TO CYTOKINE STIMULATION

In contrast to footprint analysis, gel-shift studies show that new DNA-protein complexes appear in the nuclear extracts derived from cells treated with IL-1 or IL-6. This observation suggests that cytokine stimulation may result in an accumulation of proteins binding to the bZIP-DNA complex or a qualitative change in the proteins that bind to this *cis* element. Furthermore, the alterations seen in the banding pattern in response to cytokine stimulation were qualitatively equivalent regardless of whether treatment was with IL-1 or IL-6. For C3, there were several differences between the banding patterns produced with extracts from control and IL-1- or IL-6-stimulated cells (Figure 6). Control lanes contained at least two

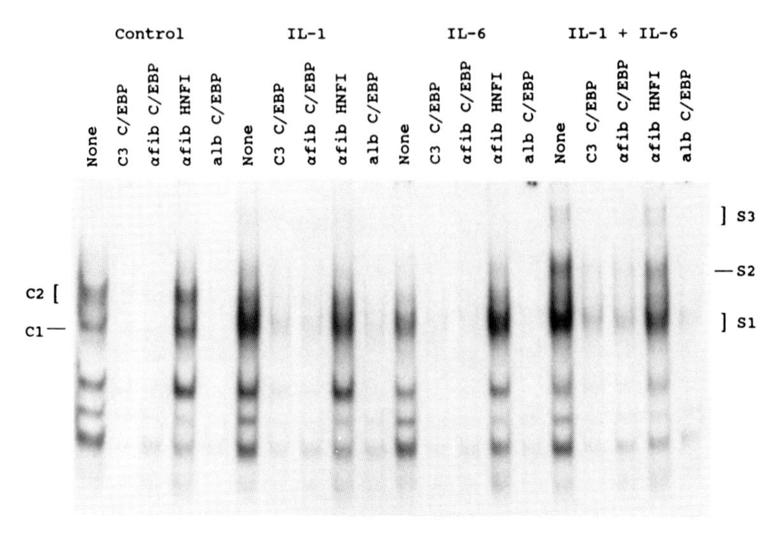

FIGURE 6. Gel retardation analysis of the cytokine-responsive region of the C3 promoter. The extracts used are indicated horizontally above the lanes: control, nuclear extracts prepared from unstimulated Hep 3B2 cells; IL-1, nuclear extracts prepared from IL-1-stimulated Hep 3B2 cells; IL-6, nuclear extracts prepared from IL-6-stimulated Hep 3B2 cells; IL-1 + IL-6, nuclear extracts prepared from Hep 3B2 cells simultaneously stimulated with IL-1 and IL-6. Competitors used are indicated vertically above the lanes: C3 C/EBP, the BamHI-NcoI fragment of pT81lucH/N127 (−127 to −74 bp of the human C3 promoter); αfib C/EBP, the SphI-XbaI fragment of pUCαfibC/EBP (−177 to −70 bp of the rat α-fibrinogen promoter); αfibHNF I, the SphI-XbaI fragment of pUCαfibHNF I (−83 to −33 of the rat α-fibrinogen gene); alb C/EBP, the EcoRI-HindIII fragment of pluc101 (−190 to −90 bp of the human albumin promoter). Complexes specific to or highly enriched in extracts from control (C) cells are bracketed on the left; bands specific to or highly enriched in extracts from stimulated (S) cells are bracketed on the right. The probe represents the human C3 promoter region from the HindIII to NcoI sites (−127 to −74 bp).

closely migrating complexes (bracketed as C2) and, to a lesser extent, an additional complex (C1) that were not seen or were greatly reduced in lanes with extracts from stimulated cells. Lanes containing extracts from stimulated cells (either IL-1 or IL-6) had two cytokine-specific bands (bracketed as S1) not present in control lanes. All of these binding activities could be competed by fragments isolated from the human albumin and rat α-fibrinogen genes that each contain a C/EBP binding consensus, but not by the HNF-1 site of the α-fibrinogen gene.

Additional binding activities were produced on the C3 promoter using extracts made from cells stimulated simultaneously with IL-1 and IL-6. These extracts produced one very retarded (S2) and two extremely retarded complexes (bracketed as S3) that were practically unique (they can be seen faintly in the lanes containing IL-1 extracts). Formation of these complexes could also be competed by fragments containing a C/EBP binding consensus but not by an HNF-1 binding site, suggesting that one or more members of the C/EBP family of factors was involved in their formation. These highly retarded complexes may reflect protein-protein interactions responsible for the synergistic induction of this promoter.

All the complexes seen with nuclear extracts prepared from cytokine-stimulated cells could be reproduced with extracts made from cells stimulated with cytokines in the presence of cycloheximide (data not shown). This result suggests that protein(s) involved in the formation of the cytokine-specific bands is preexisting in the cell and requires modification/activation/translocation for binding, a result consistent with the observation that increased transcription of the C3 gene in response to cytokine stimulation occurs in the absence of new protein synthesis.

H. SITE-DIRECTED MUTAGENESIS OF ONE C/EBP-LIKE BINDING SITE ABOLISHES THE IL-1 RESPONSE

We tested the C/EBP-like binding site for a functional role in the regulation of the C3 gene by preparation and transfection of a mutant in which the AGAAAT of the downstream C/EBP-like site was changed to TGTCGA (C3MutSal, Table 2); the GTCGA of this mutation plus the adjacent C-residue constitute a Sal I site. Results obtained with this mutant were equivalent to those obtained with the minimal promoter, C3luc76, i.e., no response to IL-1 or IL-6 stimulation. This result shows that the region identified by footprint analysis as a binding site for nuclear protein is critical for promoter activity. In contrast, the upstream C/EBP-like consensus element can be mutated (C3MutNru) and still retain the basic IL-1 responsiveness of the promoter.

I. AN NF-κB-LIKE SEQUENCE IS CRITICAL TO THE IL-1 RESPONSE

An NF-κB-like consensus sequence has been found to be important in the regulation of several of the IL-1-responsive acute phase genes and other IL-1-responsive genes. These include the acute phase reactants SAA,[43] complement factor B,[44] and angiotensinogen,[45] and those for IL-6,[46-48] the immunoglobulin κ-light chain,[49] the IL-2 receptor α-chain,[50] IL-8,[51] and the IL-1-responsive human immunodeficiency virus enhancer.[52] Several of these NF-κB-like sites are linked to (SAA), or overlap with (angiotensinogen), C/EBP-like consensus sequences. The proximity of these sites has been found to be significant: for SAA, both NF-κB-like and C/EBP-like sequences are required for the full cytokine-induced expression;[71] for angiotensinogen, binding to a C/EBP-like site attenuates the IL-1 inducibility by NF-κB.[45] The functionally defined, 45-bp region of the C3 promoter also contains an NF-κB-like consensus sequence adjacent to a C/EBP-like binding site. Site-directed mutagenesis of this sequence (C3MutNae and C3MutBsA, Table 2) significantly disrupted the IL-1 responsiveness of the C3 promoter.

A third sequence, CTGGGA, has been identified by site-directed mutagenesis as being important in IL-6 regulation of the rat α_2-macroglobulin gene.[53] This sequence, termed the IL-6 response element (IL-6 RE), has also been found within the IL-6-responsive regions of the rat β-fibrinogen[54] and AGP[55] genes. The human C3 gene has a similar sequence, CTGGGG, that overlaps with both the C/EBP-like and NF-κB consensus sequences. We have made three dinucleonucleotide substitutions within this sequence; each substantially affects the cytokine responsiveness of the promoter. Due to the complexity of this region, we cannot as yet determine whether a separate factor binds to this sequence to mediate IL-6 inducibility or whether the IL-6 response is mediated through the C/EBP-like and/or NF-κB elements.

J. A NUCLEAR PROTEIN BINDS TO THE NF-κB-LIKE ELEMENT OF THE CYTOKINE-RESPONSIVE *CIS* ELEMENT

An oligonucleotide corresponding to the IL-6 RE/NF-κB-like element binds one major complex in gel-shift analysis using Hep 3B2 nuclear extracts (Figure 7). The appearance of this complex is constitutive, with no qualitative or major quantitative differences in binding activity between extracts from unstimulated or cytokine-stimulated cells. An oligonucleotide corresponding to an authentic NF-κB binding site from the immunoglobulin enhancer competes with this fragment for complex formation, suggesting that NF-κB or an NFκB-like factor is involved in its formation (data not shown).

K. bZIP FACTORS *TRANS*-ACTIVATE THE C3 GENE

A number of members of the bZIP family have been cloned.[42,56-63] Of these, C/EBPα and C/EBPβ (NF-IL-6, IL-6DBP) have been tested for their ability to bind to key regulatory regions or *trans*-activate reporter gene constructs for several of the acute phase genes. Both

TABLE 2
Transfection Analysis of C3 Site-Directed Mutants

Name	Sequence	Baseline activity[a]	IL-1 induced	Fold induction
	*−127 ... *−75			
Wildtype	AAGCTTAGGAAATGGTATTGAGAAATCTGGGGCAGCCCCAAAAGGGGAGAGGCCATGG	54.2	3779.7	69.8
MutNru	AAGCTTATCGCGAGGTATTGAGAAATCTGGGGCAGCCCCAAAAGGGGAGAGGCCATGG	88.2	3805.7	43.2
MutSal	AAGCTTAGGAAATGGTATTGTGTCGACTGGGGCAGCCCCAAAAGGGGAGAGGCCATGG	11.3	41.4	3.7
MutERV-1	AAGCTTAGGAAATGGTATTGAGAAATGATATCCAGCCCCAAAAGGGGAGAGGCCATGG	7.0	12.0	1.7
MutCT	AAGCTTAGGAAATGGTATTGAGAAATGAGGGGCAGCCCCAAAAGGGGAGAGGCCATGG	6.8	19.4	2.9
Mut GG1	AAGCTTAGGAAATGGTATTGAGAAATCTTAGGCAGCCCCAAAAGGGGAGAGGCCATGG	7.3	48.5	6.7
Mut GG2	AAGCTTAGGAAATGGTATTGAGAAATCTGGTCCAGCCCCAAAAGGGGAGAGGCCATGG	24.7	25.9	1.1
MutNae	AAGCTTAGGAAATGGTATTGAGAAATCTGGGGCCGGCAGAAAAGGGGAGAGGCCATGG	11.1	30.5	2.7
MutBsA	AAGCTTAGGAAATGGTATTGAGAAATCTGGGGCAGAGACAAAAGGGGAGAGGCCATGG	28.7	146.1	5.1
MutERV-2	AAGCTTAGGAAATGGTATTGAGAAATCTGGGGCAGCCCCAAAAGGATATCGGCCATGG	30.2	6634.0	219.5

Note: Doubly underlined nucleotide sequences represent the sites of mutation.

[a] Luciferase activities are expressed as relative luminometer units per microgram of cell extract. Values represent the average of two experiments.

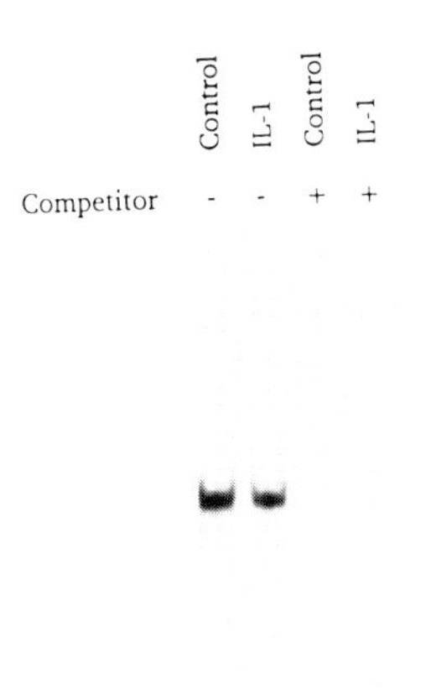

FIGURE 7. Gel retardation analysis of an oligonucleotide comprising the NF-κB-like consensus sequence of the cytokine-response region of the C3 promoter. Extracts used are as follows: control, nuclear extracts prepared from unstimulated Hep 3B2 cells; IL-1, nuclear extracts prepared from IL-1-stimulated Hep 3B2 cells. Competitor lanes contained a 100-fold molar excess of the unlabeled double-stranded oligonucleotide.

serve as *trans*-activators for CRP-promoted gene constructs.[57] Cells cotransfected with C/EBPβ (IL-6DBP) as the *trans*-activator showed a further increase in CRP-promoted reporter gene expression in response to treatment with IL-6, suggesting that the IL-6 response was mediated through binding of this factor, C/EBPβ may also participate in IL-1-induced acute phase regulation. Cotransfection experiments showed human C/EBPβ (NF-IL-6) to be capable of *trans*-activating reporter gene constructs for the IL-6 gene with a further increase in activity in response to either IL-1 or IL-6.[42,64] We have tested C/EBPα, C/EBPβ, and other bZIP factors for a potential involvement in C3 gene regulation. Purified protein corresponding to two of these genes (α and γ) interacted with the 58-bp regulatory region at a region which includes the C/EBP binding consensus.[18,72] Cotransfection experiments using expression vectors for three of the family members showed tham to be capable of *trans*-activating a C3-promoted reporter gene construct in Hep 3B2, although the magnitude of the response and the cytokine inducibility differed among the various vectors (Figure 8). C/EBPα stimulated the basal level ten- to 30-fold, depending on the ratio of *trans*-activator to reporter. NF-IL-6 (C/EBPβ) was also effective in *trans*-activation (up to a 150-fold increase), which increased further (two- to threefold) when IL-1 was added to the culture medium. DBP and LAP/IL-6DBP (both kindly provided by Dr. U. Schiber) were also tested. DBP required higher ratios (4:1) of expression vector to reporter construct to be effective as a *trans*-activator. LAP/IL-6DBP (rat C/EBPβ) increased basal and stimulated activity nearly 50-fold with an additional 50% increase with IL-1. There is apparently considerable redundancy in the factors able to interact with the C3 promoter. Clarification of the role of each of these factors *in vivo* will be an important step in determining the molecular basis for the acute phase response.

IV. DISCUSSION

Sequence analysis suggests that the C3 promoter is complex, with an array of potentially important consensus sequences. We have tested three of these sequences (two C/EBP-like and one NF-κB-like) for function by site-directed mutagenesis; two of them are critical for the IL-1 inducibility of the promoter; the third (the upstream C/EBP-like site) modulates the

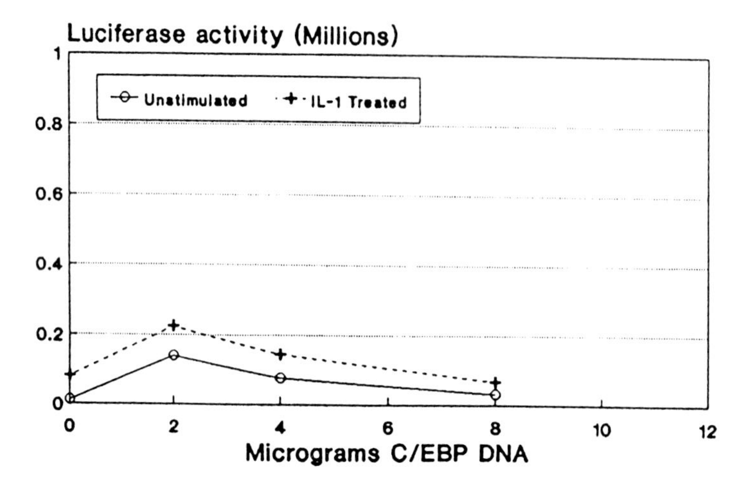

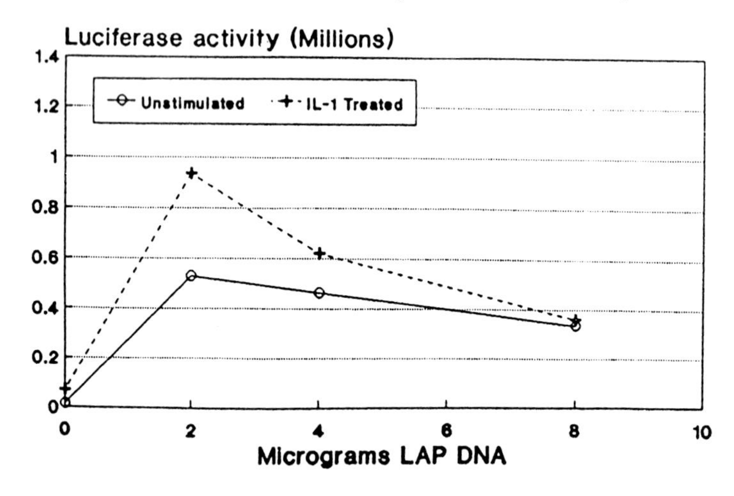

FIGURE 8. *Trans*-activation of C3luc114 in Hep 3B2 by bZIP family members. Hep 3B2 cells were cotransfected with 2 μg of C3luc114 plus 8 μg of CMV-promoted plasmid. The ratio of putative *trans*-activator (all CMV promoted) to reporter construct was varied from 1:1 to 4:1 by addition of pCMVβ.[26] Following transfection, cells received either fresh medium (unstimulated) or fresh medium + IL-1 (IL-1 treated). The *trans*-activators tested were C/EBPα[56] (C/EBP), C/EBPβ (NF-IL-6,[42] LAP[58]), and DBP.[61]

magnitude of the IL-1 response. Our transient transfection analyses in Hep 3B2 indicate that elements upstream of these sites further modulate both basal and induced levels of expression. Additionally, Vik et al.[16] located two overlapping, IFNγ-responsive consensus sequences at nucleotide positions −127 to −113 and −116 to −100. These potential IFNγ-responsive elements also overlap the C/EBP-like consensus sequences. We have preliminary evidence from transient transfection of C3luc320 into a subclonal derivative of the human endothelial cell line SK Hep[65] that IFNγ acts via this promoter fragment to decrease C3 transcription in these cells. This construct has a reasonable level of expression in the endothelial cell line, which can be modestly (approximately twofold) increased in response to IL-1. Both the baseline and IL-1-responsive activities, however, are dramatically reduced (to 10 to 20%) by treatment with IFNγ. Additional transfection experiments are currently planned to further delineate the responsive element.

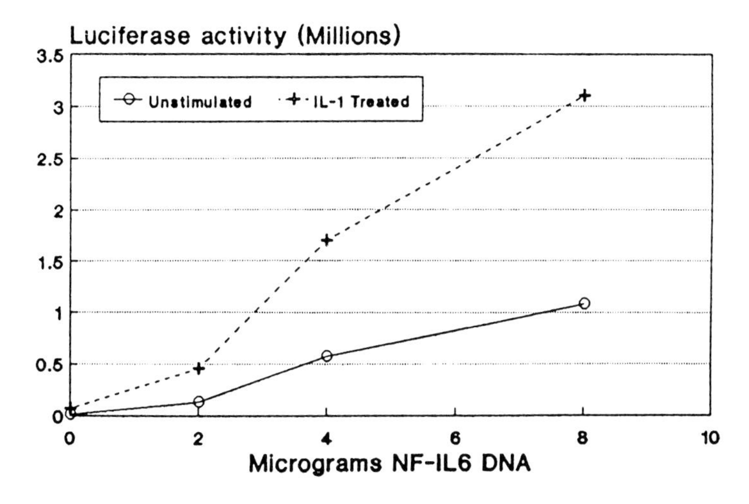

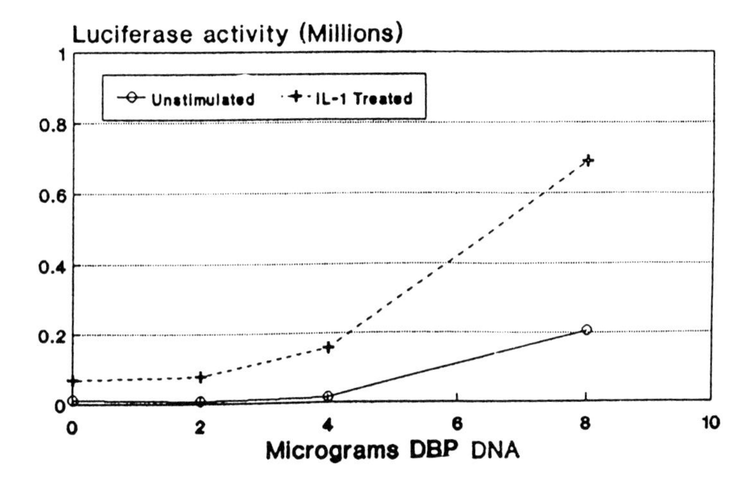

FIGURE 8 (continued).

Fong et al.[17] reported that the C3 promoter contained an HNF-1 consensus at −166. The sequence that they identified, however, matches only $^5/_6$ of the HNF-1 hexanucleotide core and does not show the palindromic nature of the HNF-1 sites identified in albumin,[66] α_1-antitrypsin,[67,68] and fibrinogen.[68] We have used cotransfection of C3luc199 with an expression vector for rat HNF-1 into Hep 3B2 to determine that C3luc199 is not *trans*-activated by HNF-1. The absence of a strongly tissue-specific regulatory element such as HNF-1 in the promoter of this gene would be consistent with its relatively broad tissue distribution.

Both C/EBP-like and NF-κB-like binding sites are required for cytokine induction of the C3 promoter. Site-directed mutagenesis of either of these sequences abolished the ability of the promoter to respond to IL-1 stimulation. Each of these elements binds one or more nuclear proteins, as determined by gel retardation analysis, although no IL-1-inducible complexes were apparent using an oligonucleotide corresponding to the NF-κB-like region. Gel retardation of the wild-type, cytokine-responsive region showed not only the formation of several new complexes using extract from cytokine-stimulated cells, but also the loss of multiple complexes found using extract from unstimulated cells. This alteration in complex formation could represent the displacement of one or more factors with little *trans*-activating ability with a factor(s) having a much greater ability to *trans*-activate. This second factor(s) is preexisting in the cells, as the change in electrophoretic patterns occurs using extracts from cells treated with cytokines in the presence of cycloheximide. Identification of this factor and elucidation of the mechanism of its activation by cycloheximide and/or cytokines is being pursued in our laboratory.

REFERENCES

1. **Alper, C. A., Johnson, A. M., Birtch, A. G., and Moore, F. D.,** Human C3: evidence for the liver as the primary site of synthesis, *Science,* 163, 286, 1969.
2. **Senger, D. R. and Hynes, R. O.,** C3 component of complement secreted by established cell lines, *Cell,* 15, 375, 1978.
3. **Katz, Y. and Strunk, R. C.,** IL-1 and tumor necrosis factor: similarities and differences in stimulation of expression of alternative pathway of complement and IFN-beta$_2$/IL-6 genes in human fibroblasts, *J. Immunol.,* 142, 3862, 1989.
4. **Guiguet, M., Exilie Frigere, M.-F., Dethieux, M.-C., Bidan, Y., and Mack, G.,** Biosynthesis of the third component of complement in rat liver epithelial cell lines and its stimulation by effector molecules from cultured human mononuclear cells, *In Vitro Cell. Dev. Biol.,* 23, 821, 1987.
5. **Cole, F. S., Beatty, D., Davis, A. E., and Colten, H. R.,** Complement biosynthesis by human breast milk macrophages and blood monocytes, *Immunology,* 46, 429, 1982.
6. **Einstein, L. P., Hansen, P. J., Ballow, M., et al.,** Biosynthesis of the third component of complement (C3) *in vitro* by monocytes from normal and homozygous C3-deficient humans, *J. Clin. Invest.,* 60, 963, 1977.
7. **Whaley, K.,** Biosynthesis of the complement components and the regulatory proteins of the alternative complement pathway by human peripheral blood monocytes, *J. Exp. Med.,* 151, 501, 1980.
8. **Cole, F. S., Matthews, W. Y., Marino, J. T., Gash, D. Y., and Colten, H. R.,** Control of complement synthesis and secretion in bronchoalveolar and peritoneal macrophages, *J. Immunol.,* 125, 1120, 1980.
9. **Falus, A. and Meretey, K.,** Effect of histamine on the gene expression and biosynthesis of complement components C2, factor B and C3 in mouse peritoneal macrophages, *Immunology,* 60, 547, 1987.
10. **Rus, H. G., Kim, L. M., Niculescu, F. I., and Shin, M. L.,** Induction of C3 expression in astrocytes is regulated by cytokines and Newcastle disease virus, *J. Immunol.,* 148, 928, 1992.
11. **Dauchel, H., Julen, N., Lemercier, C., et al.,** Expression of complement alternative pathway proteins by endothelial cells. Differential regulation by interleukin 1 and glucocorticoids, *Eur. J. Immunol.,* 20(8), 1669, 1990.
12. **Brooimans, R. A., Van Der Ark, A. A., Buurman, W. A., Van Es, L. A., and Daha, M. R.,** Differential regulation of complement factor H and C3 production in human umbilical vein endothelial cells by IFN-gamma and IL-1, *J. Immunol.,* 144(10), 3835, 1990.
13. **Strunk, R. C., Eidlen, D. M., and Mason, R. J.,** Pulmonary alveolar type II epithelial cells synthesize and secrete proteins of the classical and alternative complement pathways, *J. Clin. Invest.,* 81, 1419, 1988.
14. **Sundstrom, S. A., Komm, B. S., Ponce-de-Leon, H., Yi, Z., Teuscher, C., and Lyttle, C. R.,** Estrogen regulation of tissue-specific expression of complement C3, *J. Biol. Chem.,* 264, 16941, 1989.
15. **Brooimans, R. A., Stegmann, A. P., Van Dorp, W. T., et al.,** Interleukin 2 mediates stimulation of complement C3 biosynthesis in human proximal tubular epithelial cells, *J. Clin. Invest.,* 88(2), 379, 1991.
16. **Vik, D. P., Amiguet, P., Moffat, G. J., et al.,** Structural features of the human C3 gene: intron/exon organization, transcriptional start site, and promoter region sequence, *Biochemistry,* 30(4), 1080, 1991.
17. **Fong, K. Y., Botto, M., Walport, M. J., and So, A. K.,** Genomic organization of human complement component C3, *Genomics,* 7, 579, 1990.
18. **Wilson, D. R., Juan, T. S.-C., Wilde, M. D., Fey, G. H., and Darlington, G. J.,** A 58-base-pair region of the human C3 gene confers synergistic inducibility by interleukin-1 and interleukin-6, *Mol. Cell. Biol.,* 10, 6181, 1990.
19. **Neta, R., Vogel, S. N., Sipe, J. D., Wong, G. G., and Nordan, R. P.,** Comparison of in vivo effects of human recombinant IL 1 and human recombinant IL 6 in mice, *Lymphokine Res.,* 7, 403, 1988.

20. **Huang, J. H., Rienhoff, H. Y., Jr., and Liao, W. S. L.**, Regulation of mouse serum amyloid A gene expression in transfected hepatoma cells, *Mol. Cell. Biol.*, 10, 3619, 1990.
21. **Darlington, G. J., Wilson, D. R., and Lachman, L. B.**, Monocyte-conditioned medium, interleukin-1, and tumor necrosis factor stimulate the acute phase response in human hepatoma cells in vitro, *J. Cell Biol.*, 103(3), 787, 1986.
22. **Darlington, G. J., Wilson, D. R., Revel, M., and Kelly, J. H.**, Response of liver genes to acute phase mediators, *Ann. N.Y. Acad. Sci.*, 557, 310, 1989.
23. **Lappin, D. F., Birnie, G. D., and Whaley, K.**, Modulation by interferons of the expression of monocyte complement genes, *Biochemistry*, 268, 387, 1990.
24. **De Wet, J. R., Wood, K. V., DeLuca, M., Helinski, D. R., and Subramani, S.**, Firefly luciferase gene: structure and expression in mammalian cells, *Mol. Cell. Biol.*, 7, 725, 1987.
25. **Brasier, A. R., Tate, J. E., and Habener, J. F.**, Optimized use of the firefly luciferase assay as a reporter gene in mammalian cell lines, *BioTechniques*, 7, 1116, 1989.
26. **MacGregor, G. R. and Caskey, C. T.**, Construction of plasmids that express *E. coli* B-galactosidase in mammalian cells, *Nucleic Acids Res.*, 17, 2365, 1989.
27. **Parker Ponder, K., Dunbar, R. P., Wilson, D. R., Darlington, G. J., and Woo, S. L. C.**, Evaluation of relative promoter strength in primary hepatocytes, *Human Gene Ther.*, 2, 41, 1991.
28. **Chodosh, L. A.**, Mobility shift DNA-binding assay using gel electrophoresis, in *Current Protocols in Molecular Biology*, Ausubel, F. M., Brent, R., Kingston, R. E., et al., Eds., Greene Publishing Associates and Wiley-Interscience, New York, 1989, 12.2.1.
29. **Fowlkes, D., Mullis, N., Comeau, C., and Crabtree, G.**, Potential basis for regulation of the coordinately expressed fibrinogen genes: homology in the 5′ flanking regions, *Proc. Natl. Acad. Sci. U.S.A.*, 81, 2313, 1984.
30. **Maniatis, T., Fritsch, E. F., and Sambrook, J.**, *Molecular Cloning: A Laboratory Manual*, Cold Spring Harbor Laboratory, Cold Spring Harbor, NY, 1982.
31. **Lawn, R. M., Adelman, J., Bock, S. C., et al.**, The sequence of human serum albumin cDNA and its expression in *E. coli*, *Nucleic Acids Res.*, 9, 6103, 1981.
32. **de Bruijn, M. H. and Fey, G. H.**, Human complement component C3: cDNA coding sequence and derived primary structure, *Proc. Natl. Acad. Sci. U.S.A.*, 82, 708, 1985.
33. **Kant, J. A., Lord, S. T., and Crabtree, G. R.**, Partial mRNA sequences for human Aα, Bβ, and γ fibrinogen chains: evolutionary and functional implications, *Proc. Natl. Acad. Sci. U.S.A.*, 80, 3953, 1983.
34. **Bolyard, M. G. and Lord, S. T.**, High level expression of a functional human fibrinogen gamma chain in *Escherichia coli*, *Gene*, 66, 183, 1989.
35. **Kelly, J. H. and Darlington, G. J.**, Modulation of the liver specific phenotype in the human hepatoblastoma line Hep G2, *In Vitro Cell. Dev. Biol.*, 25, 217, 1989.
36. **Dente, L., Ciliberto, G., and Cortese, R.**, Structure of the human alpha 1-acid glycoprotein gene: sequence homology with other human acute phase protein genes, *Nucleic Acids Res.*, 13(11), 3941, 1985.
37. **Kunkel, T. A.**, Rapid and efficient site-specific mutagenesis without phenotypic selection, *Proc. Natl. Acad. Sci. U.S.A.*, 82, 488, 1985.
38. **Knowles, B. B., Howe, C. C., and Aden, D. P.**, Human hepatocellular carcinoma cell lines secrete the major plasma proteins and hepatitis B surface antigen, *Science*, 209, 497, 1980.
39. **Darlington, G. J.**, Liver cell lines, *Methods Enzymol.*, 151, 19, 1987.
40. **Goldman, N. D. and Liu, T. Y.**, Biosynthesis of human C-reactive protein in cultured hepatoma cells is induced by a monocyte factor(s) other than interleukin-1, *J. Biol. Chem.*, 262, 2363, 1987.
41. **Ganapathi, M. K., Schultz, D., Mackiewicz, A., et al.**, Heterologous nature of the acute phase response. Differential regulation of human serum amyloid A, C-reactive protein, and other acute phase proteins by cytokines in Hep 3B cells, *J. Immunol.*, 141, 564, 1988.
42. **Akira, S., Isshiki, H., Sugita, T., et al.**, A nuclear factor for IL-6 expression (NF-IL6) is a member of a C/EBP family, *EMBO J.*, 9, 1897, 1990.
43. **Li, X. and Liao, W.**, Expression of rat serum amyloid A1 gene involves both C/EBP-like and NF-κB-like transcription factors, *J. Biol. Chem.*, 266(23), 15192, 1991.
44. **Nonaka, M. and Huang, Z. M.**, Interleukin-1-mediated enhancement of mouse factor B gene expression via NF-κB-like hepatoma nuclear factor, *Mol. Cell. Biol.*, 10(12), 6283, 1990.
45. **Braiser, A. R., Ron, D., Tate, J. E., and Habener, J. F.**, A family of constitutive C/EBP-like DNA binding proteins attenuate the IL-1α induced, NF-κB mediated transactivation of the angiotensinogen gene acute phase response element, *EMBO J.*, 9, 3933, 1990.
46. **Zhang, Y., Lin, J. X., and Vilcek, J.**, Interleukin-6 induction by tumor necrosis factor and interleukin-1 in human fibroblasts involves activation of a nuclear factor binding to a κB-like sequence, *Mol. Cell. Biol.*, 10(7), 3818, 1990.

47. **Libermann, T. A. and Baltimore, D.,** Activation of interleukin-6 gene expression through the NF-κB transcription factor, *Mol. Cell. Biol.*, 10(5), 2327, 1990.
48. **Shimizu, H., Mitomo, K., Watanabe, T., Okamoto, S., and Yamamoto, K. I.,** Involvement of a NF-κB like transcription factor in the activation of the interleukin-6 gene by inflammatory lymphokines, *Mol. Cell. Biol.*, 10(2), 561, 1990.
49. **Shirakawa, F., Chedid, M., Suttles, J., Pollok, B. A., and Mizel, S. B.,** Interleukin 1 and cyclic AMP induce κ immunoglobulin light-chain expression via activation of an NF-κB like DNA-binding protein, *Mol. Cell. Biol.*, 9(3), 959, 1989.
50. **Lowenthal, J. W., Ballard, D. W., Bohnlein, E., and Greene, W. C.,** Tumor necrosis factor α induces proteins that bind specifically to κB-like enhancer elements and regulate interleukin 2 receptor α-chain gene expression in primary human T lymphocytes, *Proc. Natl. Acad. Sci. U.S.A.*, 86, 2331, 1989.
51. **Mukaida, N., Mahe, Y., and Matsushima, K.,** Cooperative interaction of nuclear factor-κB- and cis-regulatory enhancer binding protein-like factor binding elements in activating the interleukin-8 gene by pro-inflammatory cytokines, *J. Biol. Chem.*, 265(34), 21128, 1990.
52. **Osborn, L., Kunkel, S., and Nabel, G. J.,** Tumor necrosis factor α and interleukin 1 stimulate the human immunodeficiency virus enhancer by activation of the nuclear factor κB, *Proc. Natl. Acad. Sci. U.S.A.*, 86, 2336, 1989.
53. **Hattori, M., Abraham, L. J., Northemann, W., and Fey, G. H.,** Acute phase reaction induces a specific complex between hepatic nuclear proteins and the interleukin 6 response element of the rat alpha$_2$-macroglobulin gene, *Proc. Natl. Acad. Sci. U.S.A.*, 87, 2364, 1990.
54. **Baumann, H., Jahreis, G. P., and Morella, K. K.,** Interaction of cytokine- and glucocorticoid-response elements of acute phase plasma protein genes, *J. Biol. Chem.*, 265(36), 9060, 1990.
55. **Won, K.-A. and Baumann, H.,** The cytokine response element of the rat alpha1-acid glycoprotein gene is a complex of several interacting regulatory sequences, *Mol. Cell. Biol.*, 10, 3965, 1990.
56. **Landschulz, W. H., Johnson, P. F., Adashi, E. Y., Graves, B. J., and McKnight, S. L.,** Isolation of a recombinant copy of the gene encoding C/EBP, *Genes Dev.*, 2, 786, 1988.
57. **Poli, V., Mancini, F. P., and Cortese, R.,** IL-6DBP, a nuclear protein involved in interleukin-6 signal transduction, defines a new family of leucine zipper proteins related to C/EBP, *Cell*, 63, 643, 1990.
58. **Descombes, P., Chojkier, M., Lichtsteiner, S., Falvey, E., and Schibler, U.,** LAP, a novel member of the C/EBP gene family, encodes a liver-enriched transcriptional activator protein, *Genes Dev.*, 4, 1541, 1990.
59. **Williams, S. C., Cantwell, C. A., and Johnson, P. F.,** A family of C/EBP-related proteins capable of forming covalently linked leucine zipper dimers in vitro, *Genes Dev.*, 5, 1553, 1991.
60. **Chang, C. J., Chen, T. T., Lei, H. Y., Chen, D. S., and Lee, S. C.,** Molecular cloning of a transcription factor, AGP/EBP, that belongs to members of the C/EBP family, *Mol. Cell. Biol.*, 10(12), 6642, 1990.
61. **Mueller, C. R., Maire, P., and Schibler, U.,** DBP, a liver-enriched transcriptional activator, is expressed late in ontogeny and its tissue specificity is determined posttranscriptionally, *Cell*, 61, 279, 1990.
62. **Roman, C., Platero, J. J., Shuman, J., and Calame, K.,** Ig/EBP-1: a ubiquitously expressed immunoglobulin enhancer-binding protein that is similar to C/EBP and heterodimerizes with C/EBP, *Genes Dev.*, 4, 1404, 1990.
63. **Cao, Z., Umek, R. M., and McKnight, S. L.,** Regulated expression of three C/EBP isoforms during adipose conversion of 3T3-L1 cells, *Genes Dev.*, 5, 1538, 1991.
64. **Isshiki, H., Akira, S., Sugita, T., et al.,** Reciprocal expression of NF-IL6 and C/EBP in hepatocytes: possible involvement of NF-IL6 in acute phase protein gene expression, *New Biol.*, 3, 63, 1991.
65. **Heffelfinger, S. C., Hawkins, H. K., Barrish, J., Taylor, L., and Darlington, G. J.,** SK-HEP-1: a human cell line of endothelial origin, *In Vitro Cell. Dev. Biol.*, 28A(2), 136, 1992.
66. **Frain, M., Hardon, E., Ciliberto, G., and Sala-Trepat, J. M.,** Binding of a liver-specific factor to the human albumin gene promoter and enhancer, *Mol. Cell. Biol.*, 10(3), 991, 1990.
67. **Hardon, E. M., Frain, M., Paonessa, G., and Cortese, R.,** Two distinct factors interact with the promoter regions of several liver-specific genes, *EMBO J.*, 7(6), 1711, 1988.
68. **Courtois, G., Morgan, J. G., Campbell, L. A., Fourel, G., and Crabtree, G. R.,** Interaction of a liver-specific nuclear factor with the fibrinogen and alpha 1-antitrypsin promoters, *Science*, 238, 688, 1987.
69. **Wilson, D. R. et al.,** in preparation.
70. **Darlington, G. J.,** unpublished observations.
71. **Li, X. and Liao, W. S. L.,** Cooperative effects of C/EBP-like and NF-κB-like binding sites or rat serum amyloid A1 gene expression in liver cells, *Nucleic Acids Res.*, 18, 4765, 1992.
72. **Roman, C.,** unpublished observations.

Chapter 25

REGULATION OF THE α_2-MACROGLOBULIN GENE

Friedemann Horn, Ursula M. Wegenka, and Peter C. Heinrich

TABLE OF CONTENTS

0-8493-6913-4/93/$0.00 + $.50

I. INTRODUCTION

α_2-Macroglobulin (α_2-M), together with the complement factors C3, C4, C5, rat α_1-M, rat α_1-inhibitor 3, and human pregnancy zone protein, belongs to the thiol ester protein family (for reviews, see References 1 and 2). This family is characterized by the presence of an internal β-cysteinyl γ-glutamyl thiolester. α_2-M is an extracellular high molecular weight glycoprotein (720 kDa) and consists of four identical subunits.[1,2] It inhibits proteinases of all four major classes. Proteinase binding by α_2-M occurs by the "trap" mechanism in which proteolytic cleavage of a particularly exposed peptide stretch near the middle of the 180-kDa subunit, the so-called bait region, results in a conformational change of the α_2-M tetramer, thereby trapping the proteinase.[3] The α_2-M-proteinase complexes are then rapidly cleared from the circulation by high-affinity receptors on various cell types.[4] In man, α_2-M is constitutively present in plasma at concentrations of about 2 mg/ml.[5] While the hepatocytes of the liver are believed to be the main producers of plasma α_2-M,[6] other sites of synthesis have been described, including fibroblasts,[7] monocytes/macrophages,[8] and astrocytes in brain.[9] Rat α_2-M — originally known as α-macrofetoprotein — is synthesized during fetal development.[10,11] In the adult rat, plasma levels are very low (10 to 100 μg/ml), but increase dramatically during an acute phase response (up to 3 to 8 mg/ml).[12,13] Under these conditions, in addition to α_2-M, the hepatocytes of the liver synthesize and secrete further proteins, which collectively are called acute phase proteins. According to their regulation by inflammatory mediators, the acute phase proteins have been divided into two classes.[14] Class I proteins (α_1-acid glycoprotein, C-reactive protein, complement C3, haptoglobin, hemopexin, and serum amyloid A) are regulated by IL-1 or by combinations of IL-1 and IL-6, whereas class II proteins (α_1-antitrypsin, α_1-antichymotrypsin, α_2-M, cysteine protease inhibitor [also known as thiostatin or T-kininogen], and fibrinogen) are inducible by IL-6, leukemia inhibitory factor, and glucocorticoids.

Since the synthesis of acute phase proteins is mainly regulated at the transcriptional level, the 5′ flanking regions of many acute phase proteins have been studied extensively. In the course of these studies, two types of cytokine-responsive enhancer elements have been found. In the promoter regions of haptoglobin,[15] hemopexin,[16] and C-reactive protein,[17] DNA sequences designated as type I element have been identified. This element, with the consensus sequence $T^T/_G$NNGNAA$^T/_G$, binds the transcripton factor NF-IL-6[18] or IL-6DBP,[19] which is a member of the C/EBP family of leucine zipper proteins. Type II elements contain the hexanucleotide core CTGGGA and were originally proposed by Fowlkes et al.,[20] when they compared the 5′ flanking regions of the three rat fibrinogen genes. Enhancer elements containing this hexanucleotide sequence are required for IL-6 induction in genes for α_1-acid glycoprotein,[21] haptoglobin,[15] and β-fibrinogen.[22] The transcription factor binding to this type II element has not yet been purified or cloned.

TABLE 1
Oligonucleotides Used in this Work

Oligonucleotide	Sequence
Core	5'-gATCCTTCTGGGAATTCcta GAAGACCCTTAAGgatctag-5'
Palindrome	5'-gatcttccgggaa aaggcccttctag-5'
Core tandem	5'-ATCCTTCTGGGAATTCTGATCCTTCTGGGAATTCTG acgtTAGGAAGACCCTTAAGACTAGGAAGACCCTTAAGACgatc-5'
Core-like	5'-GAGAAAAAGTGAGCAGTAACTGGAAAGTCCCTTAgat acgtCTCTTTTTCACTCGTCATTGACCTTTCAGGGAATctagatc
Short core-like	5'-gatcc*AACTGGAAAGTCCTa gTTGACCTTTCAGGAtctag-5'
Rat APRE	5'-agcttCAGTAACTGGAAAGTCCTTAATCCTTCTGGGAATTCTGt aGTCATTGACCTTTCAGGAATTAGGAAGACCCTTAAGACagatc-5'
Human APRE	5'-GCTGTACGGTAAAAGTGAGCTCTTACGGGAATGGGAAT acgTCGACATGCCATTTTCACTCGAGAATGCCCTTACCCTTAGatc-5'
AGP C	5'-gatCTGGGCTTCTGGGAAAAACTCAAG ACCCGAAGACCCTTTTTGAGTTCctag-5'
CRPα	5'-agcttCATAGTGGCGCAAACTCCCTTACTGa aGTATCACCGCGTTTGAGGGAATGACtctag-5'

Note: The sequences of the double-stranded oligonucleotides used in gel retardation assays are shown. The motifs matching the APRF binding consensus (see Table 2) are underlined. The sequences given in capital letters correspond to rat α_2-M (core, core tandem, core-like, short core-like, and rat APRE), human α_2-M (human APRE), rat α_1-acid glycoprotein (AGP C), and C-reactive protein (CRPα) promoter sequences, respectively. The asterisk indicates the T → C exchange in the APRF binding site of the short core-like oligonucleotide.

In this chapter, we show that the rat α_2-M promoter is regulated by a type II element binding a nuclear factor, which is rapidly activated by a posttranslational mechanism after the action of IL-6 or leukemia inhibitory factor.

II. MATERIALS AND METHODS

A. MATERIALS

Escherichia coli lipopolysaccharide (LPS) was from Sigma (Munich, Germany). Recombinant human interleukin-6 (rhIL-6) was a gift from Drs. T. Kishimoto and T. Hirano (Osaka, Japan).

B. PLASMID CONSTRUCTIONS AND SYNTHETIC OLIGONUCLEOTIDES

The rat and human α_2-M-promoter/CAT plasmids were constructed by subcloning the α_2-M promoter fragments into the vectors pUC-CAT1 and pUC-CAT2, as described.[23] The sequences of the double-stranded oligonucleotides used in gel retardation assays are given in Table 1.

C. ANIMALS AND CELL CULTURE

Male Sprague Dawley rats of 200 to 300 g body weight were injected intraperitoneally with 10 mg of LPS per kilogram of body weight or 60,000 BSF-2 units of rhIL-6 per kilogram of body weight. After the times indicated, the animals were killed by asphyxiation,

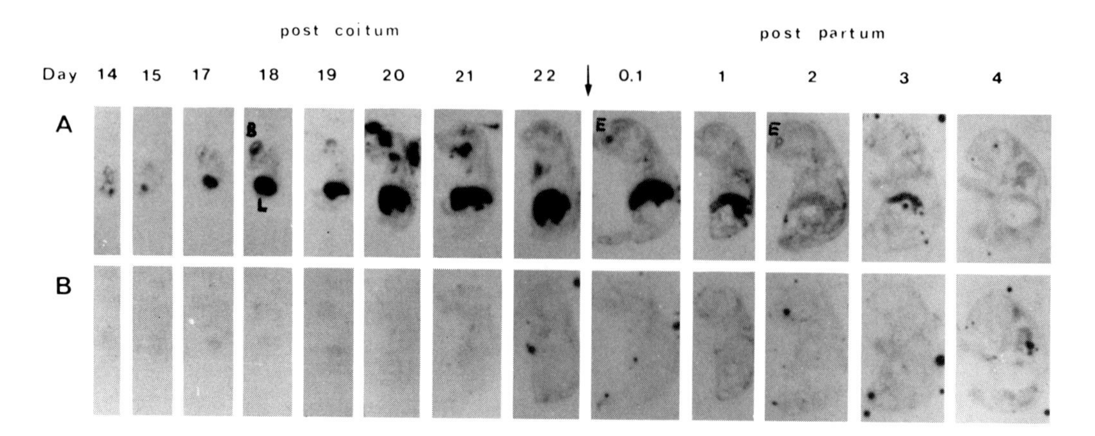

FIGURE 1. *In situ* hybridization of α_2-M cDNA to whole-body sections of fetal and newborn rats. ^{32}P-labeled α_2-M cDNA (A) or ^{32}P-labeled pBR322 vector DNA as control (B) was hybridized to saggital whole-body sections from fetal and newborn rats at different stages of development. Autoradiography was for 5 d. The arrow separates the sections of fetal from newborn animals. B, brain; E, eye; L, liver. (From Kodelja et al., *EMBO J.*, 5, 3151, 1986. With permission.)

and the livers removed immediately. Hep G2 cells were grown in a 1:1 mixture of DME and Ham's F12 media containing 10% fetal calf serum.

D. GEL RETARDATION ASSAYS

Nuclear extracts from Hep G2 cells or rat livers were prepared according to Dignam et al.[24] Oligonucleotides or DNA fragments were labeled by filling in 5′ DNA overhangs with the Klenow fragment of DNA polymerase using [α^{32}P]dATP. One to 5 μg of protein of nuclear extracts were incubated for 10 min with about 10 fmol (5000 cpm) of the probes in gel-shift incubation buffer [10 m*M* Hepes, pH 7.8, 50 m*M* KCl, 1 m*M* EDTA, 5 m*M* $MgCl_2$, 10% glycerol, 5 m*M* dithiothreitol, 0.7 m*M* phenylmethyl sulfonylfluoride, 1 mg/ml bovine serum albumin, 0.1 to 0.2 mg/ml poly(dIdC)] at room temperature. The DNA-protein complexes formed were separated by electrophoresis on a 4% polyacrylamide gel as described.[25] The gels were dried and exposed overnight to an X-ray film.

E. TRANSIENT TRANSFECTIONS

Transient transfectons of Hep G2 cells were carried out by the calcium phosphate coprecipitaton technique.[26] β-Galactosidase[27] and CAT assays[28] of the cell extracts were carried out as described.

III. RESULTS

A. α_2-MACROGLOBULIN REGULATION IN THE RAT

1. *In Vivo* Studies

a. *α_2-Macroglobulin Expression During Pregnancy and Fetal Development*

While α_2-M in serum of normal adult rats is barely detectable, its serum levels increase dramatically during pregnancy and fetal development.[10,11] By *in situ* hybridization of whole-body sections of rats of various developmental stages, the sites of α_2-M mRNA synthesis have been localized.[29] In the decidua, the α_2-M message is first detected at day 8 of gestation, with high levels observed from days 10 to 21 of gestation.[30] In late gestation, liver is the most prominent site of α_2-M mRNA synthesis. Figure 1 shows that α_2-M mRNA can be detected as early as 14 d after fertilization. Its concentrations increase during gestation and sharply decrease after birth. Lower levels of α_2-M mRNA could also be detected in brain,

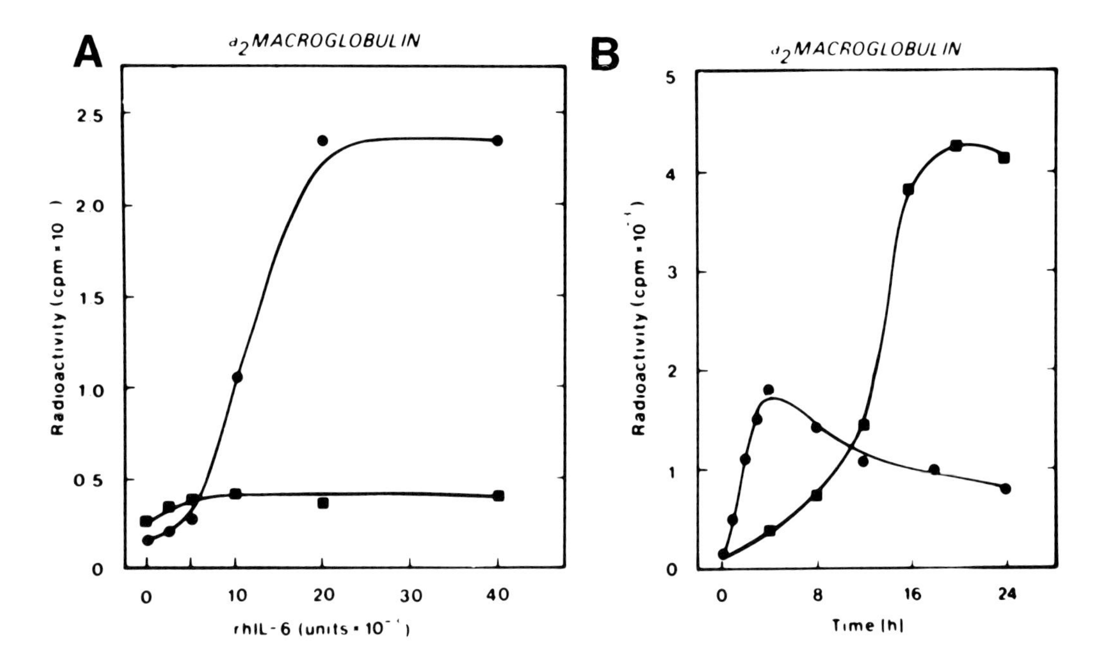

FIGURE 2. Induction of α_2-M mRNA by rhIL-6 and turpentine in rat liver. (A) Dependence of α_2-M mRNA induction on the dose of injected rhIL-6. Male (●) or female (■) Wistar rats of 180 to 200 g body weight were injected with the amounts of rhIL-6 (in BSF-2 units) indicated. Four hours after rhIL-6 injection, the animals were killed, livers excised, and total RNA extracted and analyzed by dot-blot hybridization. (B) Comparison of the time courses of α_2-M induction by rhIL-6 or turpentine. Male Wistar rats of 180 to 200 g body weight were injected either i.p. with 40,000 BSF-2 units of rhIL-6 (●) or i.m. with 0.8 ml of turpentine (■). At the times indicated, the animals were killed, the livers excised, and total RNA extracted and analyzed as above. (From Geiger et al., *Eur. J. Immunol.*, 18,717, 1988. With permission.)

eye, and spinal cord.[29] In the pregnant rat, placenta and uterus were found to be the major organs producing α_2-M mRNA at days 12 to 15.[31]

b. α_2-Macroglobulin Expression During Acute Inflammation

α_2-M serum levels are markedly elevated following an experimentally induced acute inflammatory reaction. Commonly used stimuli to generate a local inflammation are turpentine oil or lipopolysaccharide. After intramuscular injection of turpentine into adult rats, α_2-M serum levels increase up to 2 mg/ml within 24 h.[13,32] This increase in α_2-M serum levels is prevented by adrenalectomy[33,34] or hypophysectomy,[35] reflecting the requirement of glucocorticoids for the induction of α_2-M synthesis. The injection of the synthetic glucocorticoid dexamethasone in addition to turpentine administration to hypophysectomized rats resulted in an α_2-M induction comparable to the one where only turpentine had been injected into normal rats.[35] Administration of dexamethasone alone did not lead to a stimulation of α_2-M synthesis, indicating the action of additional mediators produced after turpentine injection.

In 1987, interleukin-6 (IL-6) was identified as the major mediator for α_2-M induction.[36,37] When recombinant human IL-6 was intraperitoneally injected into adult male rats, α_2-M mRNA levels increased in a dose- and time-dependent manner (Figure 2).[38,39] Whereas maximal α_2-M mRNA induction was achieved 4 h after IL-6-administration (Figure 2B, circles), the maximal response to intramuscular injection of turpentine required about 18 h (Figure 2B, squares).

When α_2-M gene transcription rates were measured by nuclear run-on experiments, the relative changes in transcription activities were far smaller than the changes measured for mRNA levels. Whereas α_2-M mRNA levels increased between 60- and 200-fold after turpentine administration, changes in transcriptional rates were only three- to eightfold.[40-42]

2. α_2-Macroglobulin Synthesis in Cultured Cells

Analogous to the *in vivo* studies on the regulation of rat α_2-M synthesis, rat hepatocytes in primary culture have been shown to be able to synthesize α_2-M.[43] Subsequently, supernatants from LPS-stimulated rat Kupffer cells[35,44-46] or polymorphonuclear leukocytes[47] were found to contain a "factor" which induces α_2-M synthesis in rat hepatocytes. This factor present in conditioned media of Kupffer cells was identified as IL-6.[36,37,48] In rat hepatocytes in primary culture, α_2-M synthesis was stimulated by IL-6 in a dose-dependent manner, whereas interleukin-1β (IL-1β) and tumor necrosis factor-α showed no effect (Figure 3A).[37] When the induction of α_2-M synthesis was examined at increasing concentrations of IL-1β, α_2-M synthesis was strongly impaired, particularly at high IL-1β concentrations (Figure 3B). Recently, leukemia inhibitory factor[49] and interleukin-11[50] have been identified as inducers of α_2-M synthesis in rat hepatocytes.

As observed *in vivo*, α_2-M induction in hepatocytes in primary culture also requires glucocorticoids.[46]

3. Isolation of the Rat α_2-Macroglobulin Gene and Analysis of its Promoter

a. Organization of the Rat α_2-Macroglobulin Gene

Starting with a cDNA probe representing the 3′ end of the α_2-M mRNA, the whole gene was isolated by chromosome walking.[51-53] The rat α_2-M gene is approximately 50 kb long and contains 36 exons.[53] A highly homologous gene structure has recently been found for human α_2-M.[67]

b. Localization of an Acute Phase Response Element (APRE) in the 5′ Flanking Sequence of the Rat α_2-Macroglobulin Gene

In order to localize the APRE(s) of the rat α_2-M gene, we fused 1321 bp of the 5′ flanking region and various 5′ deletions of it to the bacterial reporter gene chloramphenicol acetyltransferase (CAT). These constructs were transiently transfected into Hep G2 cells and promoter activity was determined by measuring CAT activity in cell extracts. In this assay, IL-6 induced α_2-M promoter activity 20- to 30-fold.[23] As shown in Figure 4, deletion from −215 to −165 resulted in a dramatic decrease in IL-6-induced promoter activity, indicating the presence of an APRE in this region.

In our previous study,[23] we reported the existence of a further regulatory element between −852 and −777 in the rat α_2-M promoter. This, however, turned out to be an artifact due to a change of vectors used for these constructs (data not shown).

c. The APRE Contains Two Type II-Like Elements and Confers Interleukin-6 Responsiveness to Control Promoters

Inspection of the APRE sequence defined above revealed two motifs, CTGGAA and CTGGGA (Figure 5), with homology to the type II IL-6 response elements found in the rat α_1-acid glycoprotein,[21] α-, β-, and γ-fibrinogen,[20] and human haptoglobin[15] genes (see Section I). In the α_2-M APRE, the two hexanucleotides are referred to as "core-like" and "core" motifs, respectively.[54] To test whether the APRE element is sufficient to confer IL-6 inducibility to a control promoter normally unresponsive to IL-6, we used a construct containing the enhancerless SV40 early promoter fused to the CAT gene. When the whole α_2-M APRE region (−159 to −196) was inserted into this construct, a 51-fold induction of CAT activity by IL-6 was observed after transient transfection into Hep G2 cells (Figure 6). The fragment −159 to −176 containing only the core motif resulted in only a 7.8-fold IL-6 effect. With two of these fragments in tandem, however, a 49-fold induction was

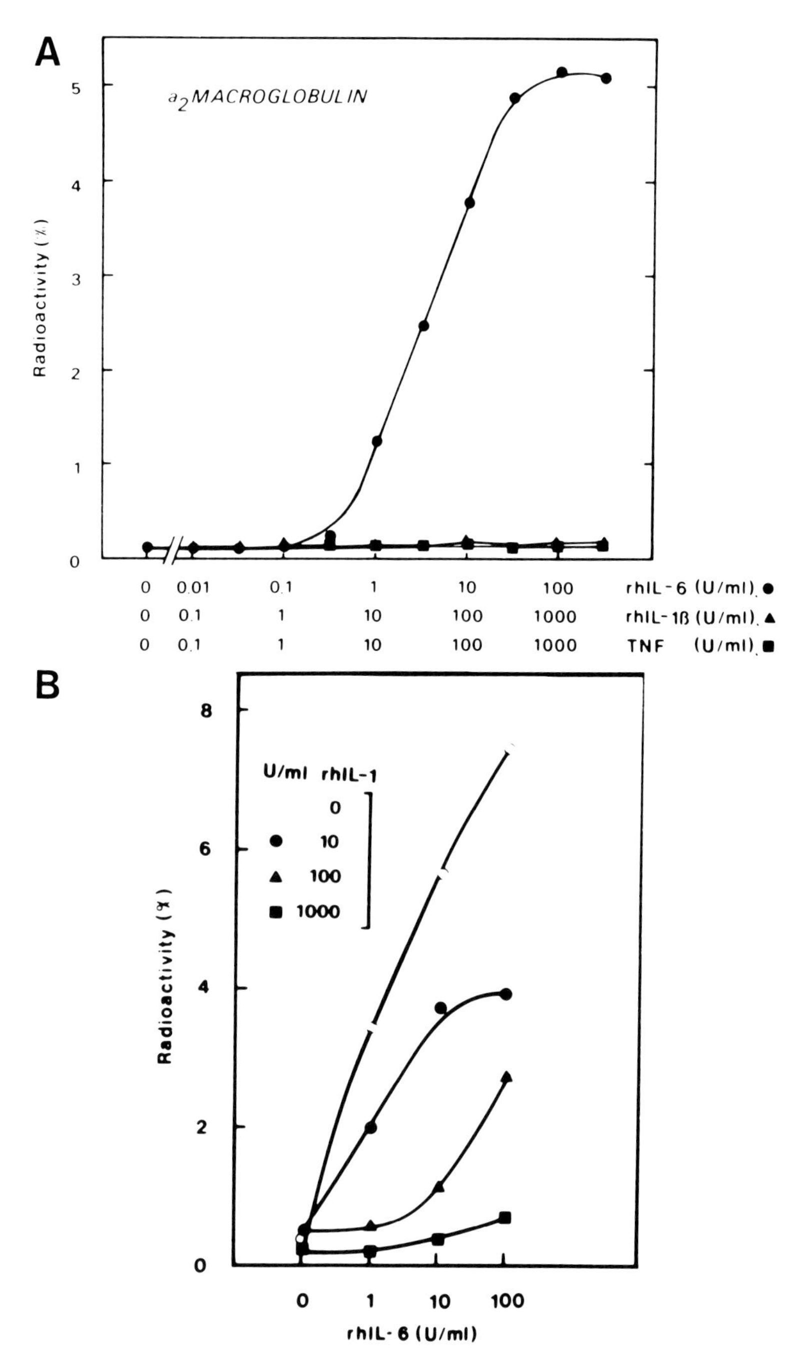

FIGURE 3. Regulation of α_2-M protein synthesis by rhIL-6, rhIL-1β, and rhTNFα in rat hepatocyte primary cultures. (A) Dose response. Rat hepatocyte primary cultures (2.5×10^5 cells per well) were incubated in the presence of 1 μM dexamethasone and rhIL-6, rhIL-1β, and rhTNFα at the concentrations indicated. After 12 h, the cells were labeled with [^{35}S]methionine for 3 h, and α_2-M was immunoprecipitated from the hepatocyte media and subjected to SDS-PAGE and fluorography. The radioactivity of the immunoprecipitated proteins is given as the percent of total TCA-precipitable material (10^5 cpm). Data are means of three different experiments. For reasons of better clarity, SEMs have been omitted. The SEM usually varied by $<15\%$. (B) rhIL-1β inhibits the stimulatory effect of rhIL-6. Rat hepatocyte primary cultures were incubated with 1 μM dexamethasone and combinations of rhIL-6 and rhIL-1β as indicated. Labeling, immunoprecipitation, and quantitation of α_2-M protein was as above, except that incubation with the cytokines was done for 18 h and labeling for 2 h. (From Andus et al., *Eur. J. Biochem.*, 173, 287, 1988. With permission.)

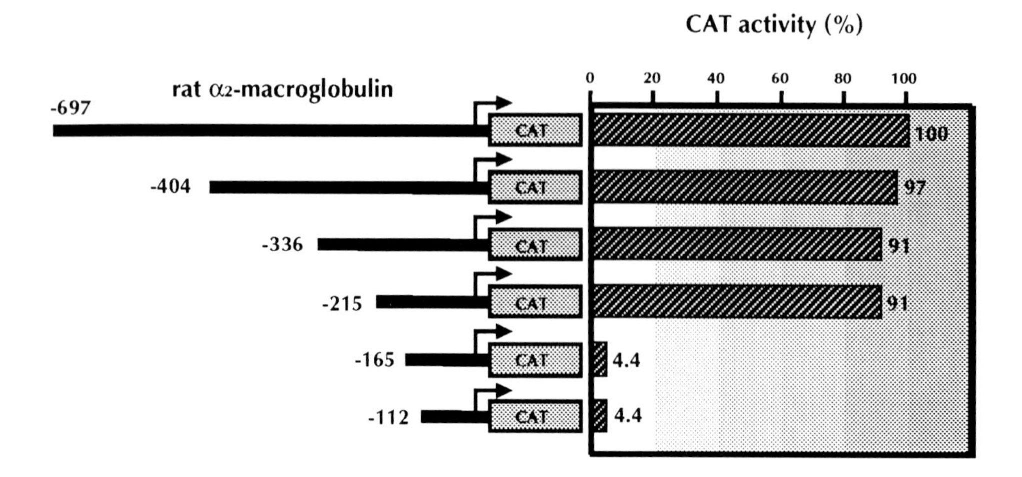

FIGURE 4. Mapping of the acute phase response element (APRE) of the rat α_2-M gene by transient transfection studies. Plasmid constructs containing the indicated part of the 5′ flanking region of the rat α_2-M gene fused to the bacterial reporter gene CAT were transiently transfected into Hep G2 cells using the calcium phosphate coprecipitation method. Twenty micrograms per dish of the α_2-M promoter/CAT plasmids were cotransfected with 1 μg of pCH110 carrying the gene for β-galactosidase; 24 h after transfection, the cells were stimulated with 100 BSF-2 units of rhIL-6 per milliliter of medium and harvested 24 h later. The CAT activity measured in the cell extracts was normalized for transfection efficiency by determining the β-galactosidase activity. The results are given as the percent of value obtained with the longest construct.

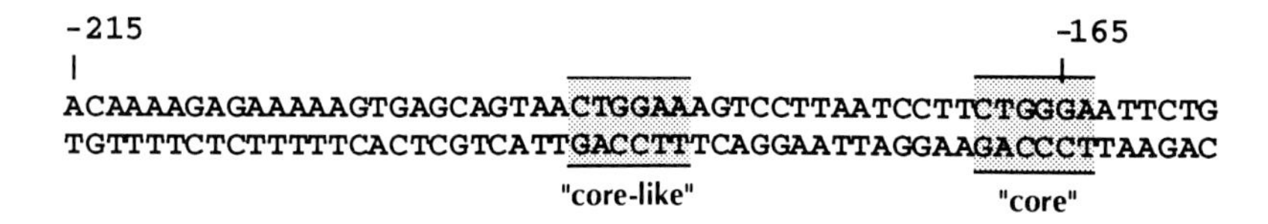

FIGURE 5. Sequence of the rat α_2-M APRE. The sequence from −215 to −158 of the α_2-M gene is shown. This region contains the APRE as defined in Figure 4. The two hexanucleotide motifs (core and core-like) homologous to the type II IL-6 response elements of other acute phase genes are boxed.

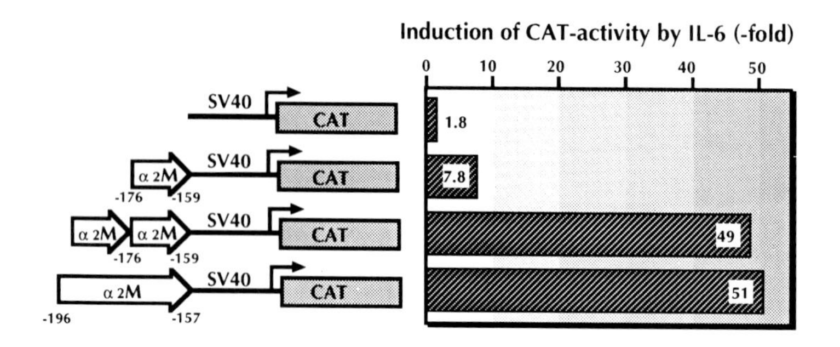

FIGURE 6. The rat α_2-M APRE confers IL-6-responsiveness to a control promoter. Constructs containing the rat α_2-M APRE (−196 to −157 of the α_2-M gene) or one or two copies of half this element (−176 to −157) fused to a SV40 early promoter/CAT vector were transiently transfected into Hep G2 cells as described in the legend of Figure 3. Cells were stimulated with 100 BSF-2 units of rhIL-6 per milliliter of medium and the induction of promoter activity was determined by measuring the CAT activity in extracts.

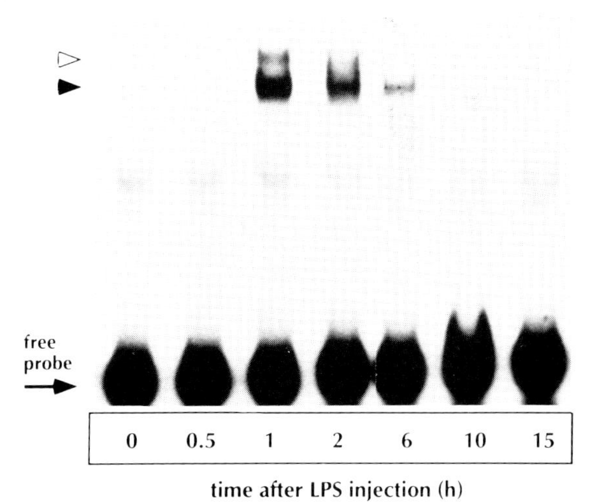

FIGURE 7. Gel retardation assay of liver nuclear extracts from LPS-treated rats. Rats were injected intraperitoneally with 10 mg/kg LPS. At the times indicated, the animals were killed, livers excised, and nuclear extracts prepared. One microgram of nuclear protein was incubated with a ^{32}P-labeled core tandem probe (see Table 1). The DNA-protein complexes formed were separated by electrophoresis on a native 4% polyacrylamide gel. The gel was dried and exposed overnight to an X-ray film. The filled arrowhead shows the position of the APRF complex. The open arrowhead indicates the complex in which both APRF binding sites of the probe are located (see text).

obtained, indicating that the two type II elements act synergistically to confer full IL-6-responsiveness.

4. Transcription Factors Regulating α_2-Macroglobulin Expression

a. In Rat Liver, Lipopolysaccharide Rapidly Induces a Nuclear Factor ("Acute Phase Response Factor", APRF) Binding to the APRE In Vitro

Bacterial lipopolysaccharides (LPSs) are very potent agents for evoking an acute phase response in man and in animal models. We isolated nuclear extracts from rats after i.p. injection of LPS and analyzed the DNA-binding proteins by gel retardation assays. As a probe we used a ^{32}P-labeled oligonucleotide comprising the tandem of two α_2-M APRE core motifs which conferred strong IL-6 inducibility to a control promoter (see Figure 6). After incubation of nuclear extracts with this probe, the protein-DNA complexes formed were separated by native gel electrophoresis. With liver nuclear extracts from control rats and from rats 30 min after LPS administration, only a very weak, retarded band was observed (Figure 7). One hour after LPS injection, however, the signal increased dramatically, then decreased gradually between 2 and 6 h, and returned to basal levels 10 h after treatment. To reflect the LPS responsiveness and the specific binding to the APRE, we propose to call the factor forming this complex the acute phase response factor (APRF).

b. APRF Cooperatively Binds to Two Binding Sites in the Rat α_2-Macroglobulin APRE

In gel retardation assays using liver nuclear extracts from rats treated with LPS for 1 h, an additional retarded band with lower mobility than the major APRF band appeared (Figure 7). Since a core tandem was used as a probe for this assay, this band could represent probe molecules with both core motifs occupied by APRF. This view is supported by the observation that a core monomer probe gave rise to only one retarded band (Figure 8). In addition, when low concentrations of nuclear extracts were used in gel retardation assays with the core tandem probe, only the faster migrating complex was formed. Upon increasing the extract concentrations, the low-mobility band also appeared, and became the predominant one at

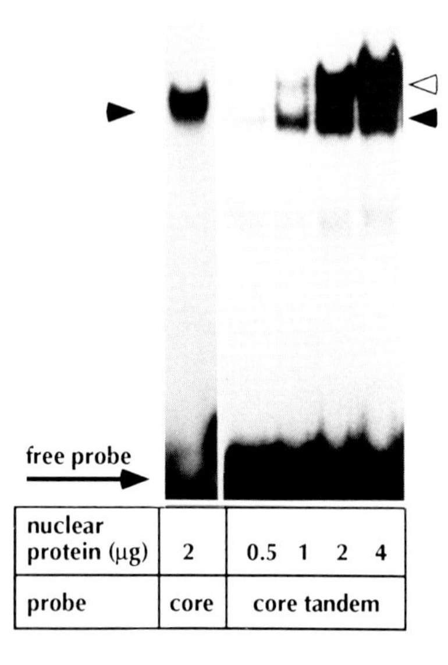

FIGURE 8. APRF binds cooperatively to adjacent binding sites. The indicated amounts of liver nuclear extracts from rats treated with LPS for 1 h were incubated with ^{32}P-labeled core or core tandem oligonucleotides. The DNA-protein complexes were separated in a gel retardation assay as described in Section II.

the highest extract amounts used. Together with the observation that both retarded bands can be competed by an excess of unlabeled core monomer oligonucleotide (see Section III.A.4.d), these findings prove that the low-mobility band is in fact formed by two APRF molecules bound to the probe. From the formation of this complex even in the presence of a large excess of free probe, one must conclude that binding of APRF to two adjacent binding sites occurs cooperatively. This characteristic corresponds well to the synergistic action of two "core" motifs in transient transfection studies as shown above.

When probes containing the entire rat α_2-M APRE were used in gel retardation assays, a doublet of retarded bands indistinguishable from that observed with the core tandem probe was obtained (Figure 9A). This indicates the existence of two binding sites for APRF in the α_2-M APRE to which the factor also binds in a cooperative manner. As shown in Figure 5, a second type II-like motif (the core-like motif) is found 5′ of the core. To prove the binding of APRF to this motif, a labeled oligonucleotide containing the core-like sequence was used in a gel retardation assay. With nuclear extracts from LPS-treated rats, a retarded band appeared at the same position as observed for the core probe (Figure 9B). In addition, an excess of unlabeled core-like oligonucleotide competed for binding of APRF to the core motif in competition gel retardation assays (results not shown). These findings strongly indicate that the same factor, APRF, binds to both the core-like and core sites. From Figure 9B, it is also evident that the APRF band formed with the core-like probe is much weaker than the one observed with the core probe. We conclude that the core-like motif represents a lower-affinity binding site for APRF.

When extracts from untreated rats were incubated with the core-like probe, a specific protein-DNA complex was formed which showed a higher mobility than the APRF band (Figure 9B). The intensity of this band decreased upon treatment of the rats with LPS. To compare the binding specificity of the factor forming this complex to APRF, we used a shorter core-like probe which contains the hexanucleotide CTGGAAA but introduced a T→C exchange at position −193 (see Table 1). In accordance with the consensus binding sequence defined below (Section III.A.4.c), APRF did not bind to the short core-like oligonucleotide (Figure 9B). However, the constitutive complex was formed with this probe. This finding clearly demonstrates that the constitutive factor binding to the core-like motif is different from APRF.

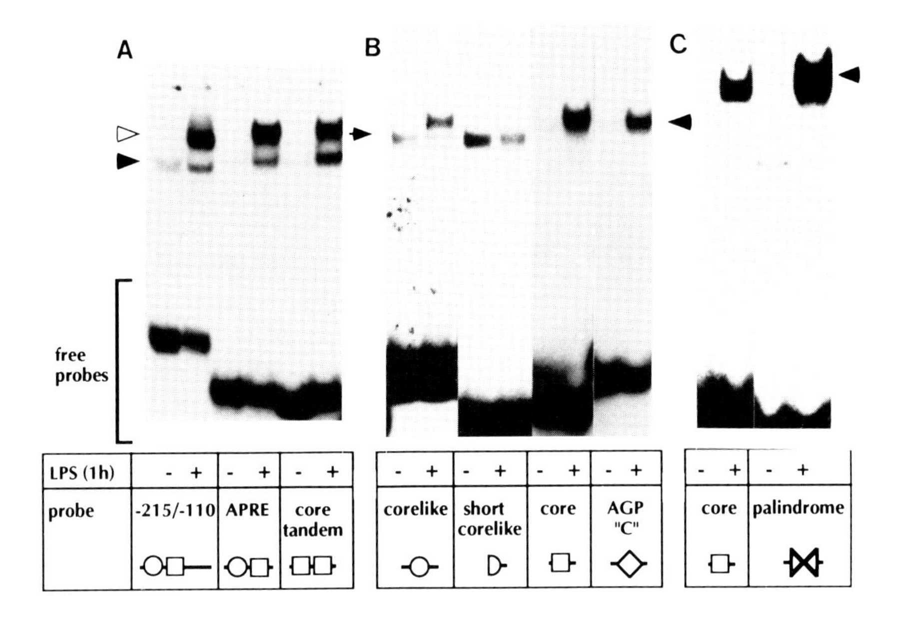

FIGURE 9. Binding of APRF to different binding sites in gel retardation assays. Two micrograms of liver nuclear extracts from rats treated with LPS for 1 h were incubated with different ^{32}P-labeled oligonucleotides. DNA-protein complexes were separated as described above. The APRF binding sites present on the probes were indicated by symbols: ○, α_2-M core-like motif; □, α_2-M core motif; ⊠, palindromic sequence shown in Table 2; ◇, sequence in the α_1-acid glycoprotein (AGP) C element homologous to the α_2-M core. See Table 1 for the sequence of the oligonucleotides used. Filled and open arrowheads indicate complexes with one and two binding sites occupied by APRF, respectively. The small arrow marks the position of the complex formed by a constitutive protein with core-like probes. (A) APRF binds cooperatively to two binding sites in the rat α_2-M APRE. Binding of APRF to the core tandem probe was compared to its binding to either labeled APRE oligonucleotide or a fragment of the α_2-M promoter (Rsa I/Hinf I) containing the APRE. (B) Binding of APRF to the α_2-M core-like, core, and AGP C motifs; (C) comparison of APRF binding to the palindrome (see also Table 2) and α_2-M core probes.

c. APRF Binds to the Acute Phase Response Elements of Other Acute Phase Genes

As mentioned before, the core and core-like motifs in the rat α_2-M promoter belong to the type II acute phase response elements also found in other acute phase genes such as the fibrinogen, haptoglobin, or α_1-acid glycoprotein genes. As one example for such elements, we studied the rat α_1-acid glycoprotein gene C-site, which is part of the distal regulatory element (DRE) located about 5 kb upstream of the transcriptional start site of this gene.[21] The C-site has been shown to be important for the IL-6 responsiveness of the DRE and contains a sequence identical to the rat α_2-M core motif (Table 2). In fact, with this oligonucleotide, a band of the same mobility and intensity as with the α_2-M core probe was found (Figure 9B). We conclude that APRF binds to the α_1-acid glycoprotein gene C-site with an affinity comparable to that of the α_2-M core.

Similarly, a fragment of the human haptoglobin promoter containing the B element was analyzed. Mutation of this element was reported to impede IL-6 induction of the haptoglobin promoter.[15] In gel retardation assays, this fragment was also shown to bind APRF (results not shown).

When the sequences of the APRF binding sites in rat α_2-M, human α_2-M (see below), rat α_1-acid glycoprotein, and human haptoglobin are compared, a conserved sequence of palindromic structure becomes evident (Table 2). Half of the site of this palindrome (GGAA) is conserved in all but one binding site, whereas the other half (TTCC) shows a higher degree of degeneracy. In fact, APRF bound to an oligonucleotide containing the perfect palindromic sequence with substantially higher affinity than to the wild-type core motif (Figure 9C).

TABLE 2
Binding Sites for APRF in Different Promoters

Sequence	Promoter
T T C T G G G A A	Rat α_2-M core, −172[53]
T A A C T G G A A	Rat α_2-M core-like, −193[53]
T T A C G G G A A	Human α_2-M, −226[53]
G T A C G G T A A	Human α_2-M, −246[53]
T T C T G G G A A	Rat α_1-acid glycoprotein (''C'')[21]
T T A C T G G A A	Human haptoglobin (''B'')[15]
T T C C N G G A A	Palindromic sequence

Note: The sequences of the APRF binding sites shown in Table 1. The homologies to the perfect palindromic binding consensus are highlighted by shaded boxes.

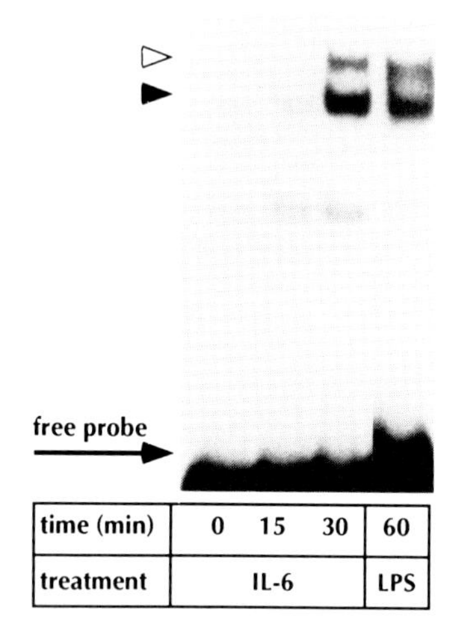

FIGURE 10. Induction of APRF by rhIL-6 in rat liver. Rats were injected intraperitoneally with either rhIL-6 (60,000 units/kg) or LPS (10 mg/kg). At the times indicated, the livers were removed and nuclear extracts prepared. Two micrograms of each extract were subjected to a gel retardation assay using the core tandem oligonucleotide as ^{32}P-labeled probe.

d. APRF Activity is Induced by Interleukin-6 in both Rat Liver and Human Hepatoma Cells

IL-6 is known to be the major mediator of acute phase protein synthesis in the rat *in vivo*[38,39] and in rat and human hepatocytes in primary culture.[37,55,56] We tested whether IL-6 is able to induce APRF activity in rat liver and in human hepatoma (Hep G2) cells. IL-6 was injected intraperitoneally into rats, and nuclear extracts were prepared from the livers. As early as 15 min after IL-6 administration, APRF was detectable (Figure 10). After 30 min, about the same level of APRF activity was reached as was found 60 min after LPS injection. Thus, activation of APRF by IL-6 occurs to the same extent as, but more rapidly than, after LPS treatment. Also, in human hepatoma (Hep G2) cells, IL-6 rapidly induced a factor binding to the rat α_2-M core motif (Figure 11). Maximal induction was seen

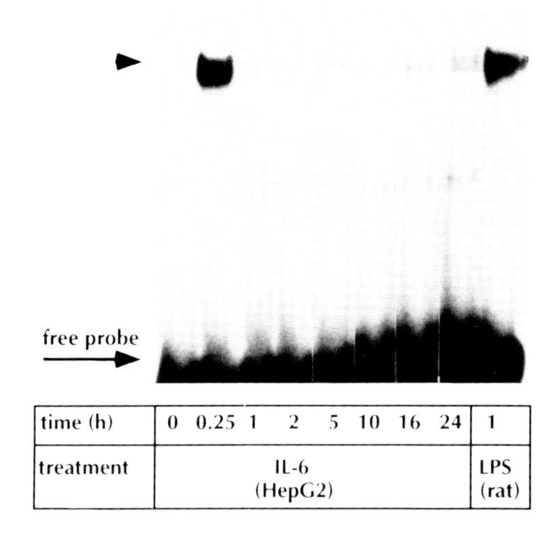

FIGURE 11. Induction of APRF by rhIL-6 in Hep G2 cells. Hep G2 cells were treated with rhIL-6 (100 BSF-2 units/ml) and harvested at the times indicated. Nuclear extracts were prepared, and 5 μg of nuclear proteins were subjected to a gel retardation assay using the core oligonucleotide as a probe.

15 min after addition of IL-6 to the culture medium. After 1 h, the signal had returned to levels only slightly higher than in the control cells and remained low during the following 24 h. During that period, the highest levels were found 5 to 24 h after IL-6 treatment, but signal intensities comparable to those after a 15-min treatment with IL-6 were never reached.

In gel retardation assays with Hep G2 extracts, the APRF band showed the same mobility as when rat liver nuclear extracts were used (Figure 11). To study whether human APRF also has binding specificity identical to the factor from rat liver, we used the same labeled oligonucleotides in a gel retardation assay as in the experiments discussed above for rat liver nuclear extracts. As observed for rat APRF, the human factor binds cooperatively to two binding sites in the rat α_2-M APRE and also binds the C-site of the rat α_1-acid glycoprotein gene (Figure 12). Leukemia inhibitory factor has been shown to induce in hepatocytes the same set of acute phase proteins as IL-6.[49] To study whether this effect is due to an activation of the same transcription factor, we performed a gel retardation assay with nuclear extracts from Hep G2 cells treated for 15 min with leukemia inhibitory factor. As shown in Figure 13, leukemia inhibitory factor rapidly induced a factor bindingto the rat α_2-M core site. The complex formed by this factor has the same mobility in native gels as APRF, indicating that it is probably identical or very similar to APRF. Thus, the signal transduction pathways of IL-6 and leukemia inhibitory factor may converge to activate APRF.

IL-6 is known to activate a transcription factor of the C/EBP family, NF-IL-6[18] or IL-6DBP,[19] which binds to the type I IL-6 response elements of acute phase genes (see Section I). Although the consensus binding sequence for NF-IL-6 ($T^T/_GNNGNAA^T/_G$) differs from the one defined for APRF (see above), we wanted to prove that APRF is different from NF-IL-6 and does not bind to its binding sites. For this purpose, we used an oligonucleotide containing a NF-IL-6 binding site from the promoter of the human C-reactive protein gene in a competition gel retardation assay. Whereas the APRE and core oligonucleotides competed efficiently for binding of APRF to the rat α_2-M APRE, the C-reactive protein oligonucleotide did not (Figure 14). In addition, bacterially expressed NF-IL-6 strongly bound to the C-reactive protein probe, but did not interact with the rat α_2-M APRE (results not shown). These findings clearly show that APRF and NF-IL-6 are different factors.

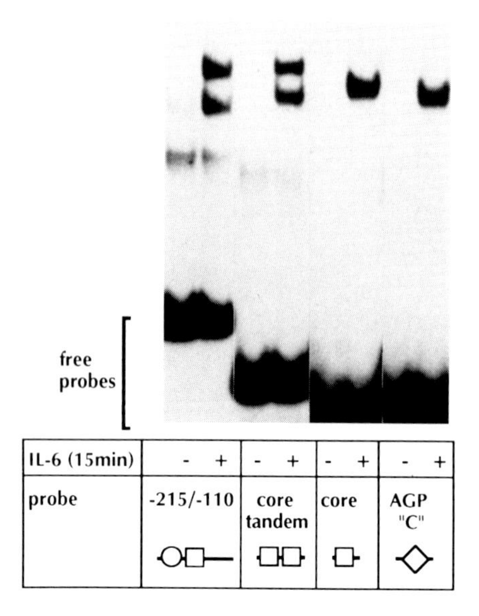

FIGURE 12. Binding of human APRF to different binding sites. A gel retardation assay was performed using 5 μg of nuclear extracts from Hep G2 cells treated with rhIL-6 (100 BSF-2 units/ml) for 15 min and different ^{32}P-labeled oligonucleotides. (See Table 1 for the sequence of the oligonucleotides used.) The symbols indicate the different APRF binding sites present on the probes (see legend of Figure 9).

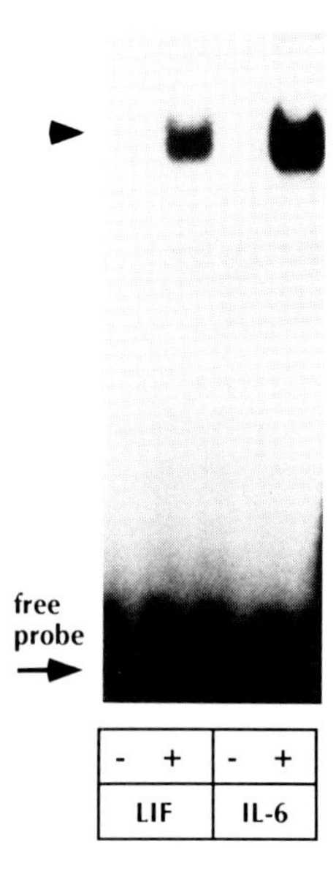

FIGURE 13. Activation of APRF by leukemia inhibitory factor. Hep G2 cells were treated with either leukemia inhibitory factor (10 units/ml) or rhIL-6 (100 BSF-2 units/ml) for 15 min. After preparation of nuclear extracts, APRF activation was tested in a gel retardation assay using a labeled core probe.

e. *The Rapid APRF Induction by Interleukin-6 is Due to a Posttranslational Modification of a Preexisting Molecule*

To investigate whether the induction of APRF activity by IL-6 is due to *de novo* synthesis of APRF protein or to the activation of a preexisting, inactive form of APRF, protein synthesis in Hep G2 cells was inhibited by cycloheximide. As shown in Figure 15 by gel retardation

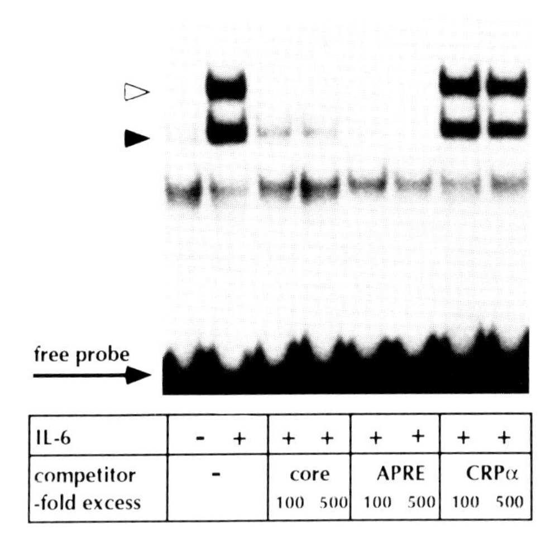

FIGURE 14. Competition gel retardation assays. Five micrograms of nuclear extracts from Hep G2 cells treated with rhIL-6 (100 BSF-2 units/ml) for 15 min were incubated with the ^{32}P-labeled APRE oligonucleotide and the indicated excess of unlabeled oligonucleotides. The protein-DNA complexes formed were separated as described in Section II.

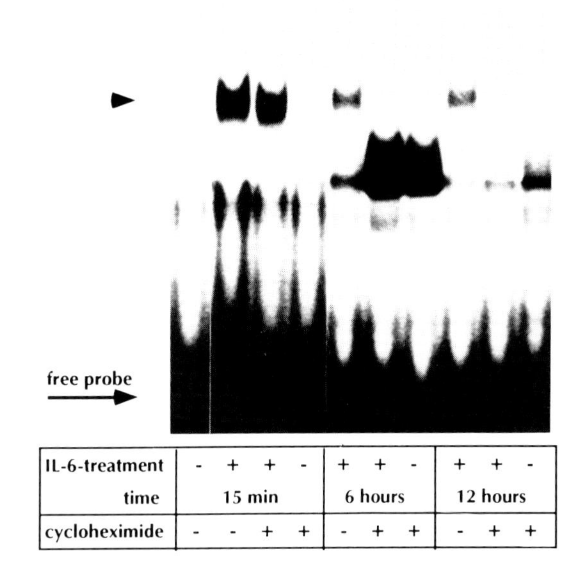

FIGURE 15. Ongoing protein synthesis is not required for the rapid activation of APRF. Hep G2 cells were preincubated with 10 μg/ml cycloheximide for 1 h. Then rhIL-6 at 100 BSF-2 units/ml was added to the medium and the incubation was continued for the indicated times. After cell harvesting and preparation of nuclear extracts, 5 μg of nuclear proteins were subjected to a gel retardation assay using a ^{32}P-labeled core oligonucleotide.

assay, pretreatment with cycloheximide prior to the addition of IL-6 did not prevent the appearance of active APRF in nuclear extracts from Hep G2 cells stimulated for 15 min with IL-6. Therefore, ongoing protein synthesis is not required for the rapid activation of APRF. In contrast, the low APRF levels observed after long-term stimulation with IL-6 were sensitive to cycloheximide treatment. The signal obtained after a 6-h IL-6 stimulation was reduced, and that after 12 h was completely abolished by concomitant treatment with cycloheximide (Figure 15).

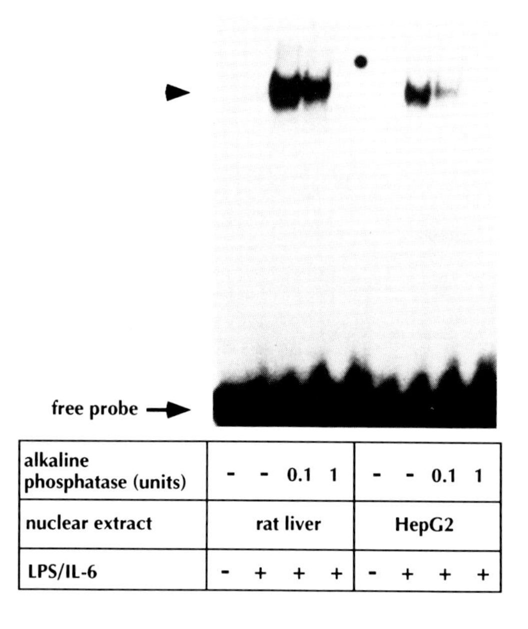

FIGURE 16. Phosphorylation of APRF is required for DNA binding. Two micrograms of liver nuclear extract from rats treated with LPS for 1 h or 5 μg of nuclear proteins from Hep G2 cells stimulated with rhIL-6 for 15 min were preincubated with the indicated amount of calf intestine alkaline phosphatase for 15 min at 37°C. DNA-protein complexes were analyzed in a gel retardation assay with an α_2-M core probe.

Since phosphorylation has been found to play an important role in the regulation of transcription factor activity,[57] we examined whether IL-6-activated APRF contains phosphate groups required for its activity. Preincubation of the nuclear extracts from either LPS-stimulated rat livers or IL-6-treated Hep G2 cells with alkaline phosphatase completely abolished the ability of APRF to bind to its binding sites (Figure 16). Phosphorylation of APRF might therefore be a key event in IL-6 signal transduction.

B. α_2-MACROGLOBULIN REGULATION IN MAN

Human hepatocytes in primary culture have been shown to synthesize and secrete α_2-M.[55,56,58] Unlike rat α_2-M, synthesis of the human protein is not regulated by IL-6 in these cells.[55] In the human hepatoma cell line Hep G2, a slight increase (~twofold) of α_2-M mRNA has been observed after addition of IL-6 (results not shown). In order to elucidate the mechanisms underlying the difference in IL-6 responsiveness of the rat and human α_2-M genes, we decided to compare both promoter regions with respect to regulation by IL-6. In collaboration with Drs. Y. Sakaki (Nagasaki, Japan) and P. Marynen (Leuven, Belgium), we isolated 4.8 kb of the 5′ flanking region of the human α_2-M gene and fused it to the CAT reporter gene. In Hep G2 cells transfected with this construct, CAT activity could be induced by IL-6 to an extent comparable to that observed for the rat promoter (Figure 17). After construction of 5′ deletion mutants, we found that most of the IL-6 inducibility was lost upon deletion of the region from −261 to −232. The region from −248 to −211 conferred IL-6 responsiveness to the SV40 early promoter (Figure 17, lower part). Two motifs showing homology to APRF binding sites exist in that region (see Table 2). To prove the existence of such APRF binding sites, gel retardation assays with a labeled human APRE oligonucleotide were performed. As observed for the rat APRE, two retarded bands appeared when nuclear extracts isolated from Hep G2 cells 15 min after IL-6 treatment were used (Figure 18). This finding demonstrates cooperative binding of APRF to two sites in the human APRE.

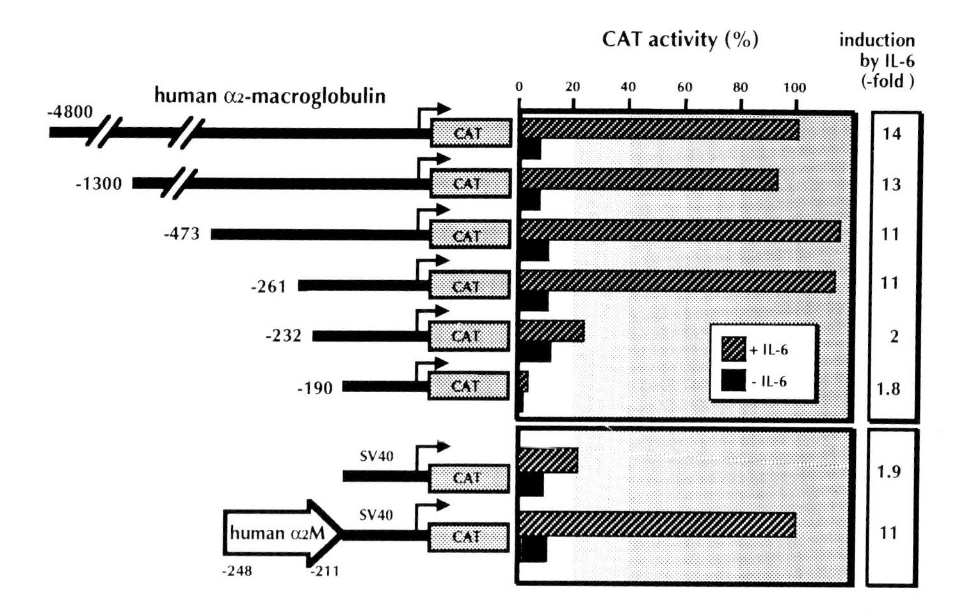

FIGURE 17. Mapping of the APRE of the human α_2-M gene by transient transfection studies. Upper part: plasmid constructs containing the indicated part of the 5′ flanking region of the human α_2-M gene fused to the bacterial reporter gene CAT were transiently transfected into Hep G2 cells as described in the legend of Figure 4. Cells were stimulated with 100 BSF-2 units of rhIL-6 per milliliter of medium and the induction of promoter activity was determined by measuring the CAT activity in extracts. The CAT activity normalized to β-galactosidase activity is given as the percent of activity measured with the longest construct. Lower part: the human α_2-M APRE confers IL-6-responsiveness to a control promoter. Constructs containing the human α_2-M APRE (−248 to −211 of the α_2-M gene) fused to a SV40 early promoter/CAT vector were transiently transfected into Hep G2 cells. Promoter activities were determined as above.

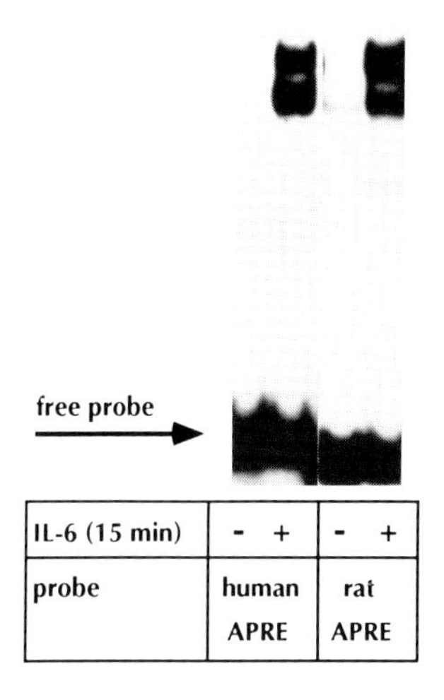

FIGURE 18. APRF binds to two binding sites in the human α_2-M APRE. Five micrograms of nuclear extracts from Hep G2 cells treated without or with rhIL-6 (100 BSF-2 units/ml) for 15 min were incubated with ^{32}P-labeled oligonucleotides comprising either the human or rat APRE sequence. DNA-protein complexes were analyzed by gel retardation assay

IV. DISCUSSION

A. RAPID ACTIVATION OF APRF BY INTERLEUKIN-6

The acute phase response element in the rat α_2-M promoter has been identified independently by three different laboratories.[23,54,59] It contains two sequence motifs, the core and core-like motifs homologous to the so-called type II IL-6 response elements defined in other acute phase genes. The factor(s) binding to these type II elements are not well characterized, and in the case of the α_1-acid glycoprotein C element[21] and the haptoglobin B-site,[15] no IL-6-induced change of the pattern of DNA-protein complexes was observed. First attempts to characterize the nuclear factors binding to the rat α_2-M APRE were made by Ito et al.[54] and Hattori et al.[59] In DNase I protection assays using rat liver nuclear extracts, two footprints in the APRE region appeared covering the core-like and core motifs, respectively. These footprints were observed regardless of whether extracts from control or inflamed rats were used. In gel retardation assays, both laboratories detected a specific DNA-protein complex formed upon incubation of rat liver nuclear proteins with a labeled DNA fragment containing the APRE region.[54,59] The mobility of this complex was reported to increase slightly during an acute phase response, whereas the intensity of the observed band remained unchanged. Similar DNA-protein complexes were found when nuclear extracts from either untreated or IL-6-treated human (Hep 3B) or rat (FAO) hepatoma cells were used.[59] More recently, Hocke et al. reported that this mobility shift was induced maximally 18 h after IL-6 treatment of Hep 3B cells, and that it is inhibited by cycloheximide and therefore requires ongoing protein synthesis.[60] Southwestern-blot experiments suggested that the factor has a molecular weight of 46 kDa and is present in livers from both control and inflamed rats.[60] The authors conclude from their data that probably the same factor binds to the APRE in normal and acute phase liver nuclei, but that during an acute phase reaction, the factor becomes posttranslationally modified, resulting in a different mobility in the gel retardation assay.

In contrast, we now have shown evidence that IL-6 rapidly (within minutes) and dramatically induces the binding of a factor which we call acute phase response factor, APRF, to two binding sites in the rat α_2-M APRE. The activation of APRF is observed in both rat liver and human hepatoma (Hep G2) cells and occurs at the posttranslational level, as shown by the finding that ongoing protein synthesis is not required. The activation is completely reversed 30 to 60 min after IL-6 treatment, but active APRF reappears to a much smaller extent after long-term treatment with IL-6 (5 to 24 h). Interestingly, this second activation peak is sensitive to cycloheximide and therefore depends on ongoing protein synthesis.

In contrast to the findings by Hocke et al.,[60] we showed that APRF does not bind to the α_2-M APRE prior to its activation. We observed the formation of a specific DNA-protein complex with the core-like motif in liver nuclear extracts from control rats. However, we were able to demonstrate that the factor forming this complex is different from APRF, since it recognizes a binding site in a mutated core-like oligonucleotide which has lost its ability to bind APRF.

Comparison of the identified binding sites for APRF led us to propose a palindromic structure of the binding sequence, TTCCNGGAA. As palindromic binding sites are a well-known characteristic of transcription factors binding to DNA as dimers, e.g., the steroid hormone receptor, leucine zipper factor, and helix-loop-helix factor families, APRF might also form homo- or heterodimers.

Since APRF binds in the same IL-6-inducible manner to other type II elements, it may be a general factor regulating many IL-6-induced genes. Direct proof for the *trans*-activating potential of APRF will require the use of purified APRF protein in *in vitro* transcription assays and eventually the cloning of its cDNA. Circumstantial evidence for its role in α_2-M gene regulation originates from the observation that a single point mutation of the APRE core motif abolishing APRF binding also destroys the IL-6 responsiveness of the element in transient transfection studies.[68] It is intriguing to speculate that due to the biphasic activation kinetics, APRF may be implicated in the *trans*-activation of two different sets of genes. On the other hand, APRF is likely to play an important role in the slow, but long-lasting induction of various acute phase genes. On the other hand, the rapid induction of high levels of active APRF by IL-6 may suggest that it also is involved in the induction of immediate early genes after IL-6. The genes for c-*jun*, *jun*B, *jun*D, TIS11, and others have been shown to respond within 30 min to IL-6 in different cell types.[61,62] A possible involvement of APRF in the regulation of these genes merits future investigation.

As shown above, the rapid activation of APRF occurs at the posttranslational level. Several mechanisms for this activation seem possible: (1) modification of APRF, e.g., by phosphorylation or dephosphorylaton, (2) release of active APRF from an inhibitor protein analogous to the situation described for NF-κB/I-κB,[63] and (3) the assembly of a multimeric transcription factor complex triggered by IL-6, as has been reported for the activation of the transcription factor ISGF3 by interferon-α.[64] We have presented evidence that phosphorylation of APRF is required for its activation. In fact, protein kinases are likely to be involved in IL-6 signal transduction. A rapid phosphorylation of a 160-kDa protein at tyrosine residues after IL-6 action has been documented.[61,62] In addition, the induction of immediate early genes by IL-6 can be blocked by specific inhibitors of both tyrosine and serine/threonine protein kinases.[61] The activation of APRF is the most rapid nuclear event, after binding of IL-6 to its receptor, so far reported. It therefore represents an excellent tool for elucidation of the still unknown signal transduction pathway for IL-6.

Leukemia inhibitory factor induces the same set of acute phase proteins in liver hepatocytes as IL-6.[49] The regulatory element responsible for the activation of the rat α_2-M promoter by leukemia inhibitory factor has been shown to reside in the APRE.[60] We have demonstrated that leukemia inhibitory factor activates APRF with the same rapid kinetics as IL-6 (Figure 13). Therefore, the two signal transduction pathways of IL-6 and leukemia inhibitory factor are likely to converge at a yet unknown step and induce α_2-M expression by activation of the same transcription factor.

B. THE DIFFERENT REGULATION OF HUMAN AND RAT α_2-MACROGLOBULIN

Whereas α_2-M concentrations in rat serum are dramatically elevated during an acute phase reaction, α_2-M serum levels in man are constitutively high. This difference is also found in cultured cells. Rat hepatocytes in primary culture synthesize α_2-M only marginally, but produce large amounts after IL-6 stimulation. In contrast, constitutive high α_2-M protein and mRNA levels are found in primary human lung fibroblasts and hepatocytes, and in Hep G2 cells. To understand this different regulation, we compared the IL-6 responsiveness of the human and rat α_2-M promoters by transient transfection studies in Hep G2 cells. Unexpectedly, the human α_2-M promoter responded about as well as the rat promoter to IL-6. The acute phase response element for the human promoter was localized and shown to be organized similar to the rat α_2-M APRE: it contains two binding sites for APRF arranged in a spacing similar to that in the rat APRE, and the APRF binds to the two sites in a cooperative manner. Therefore, we have no evidence from our data that the different regulation of human and rat α_2-M synthesis can be accounted for by differently regulated promoters. We cannot exclude, however, the possibility that additional regulatory elements enhancing the basal promoter activity are located either upstream of -4.8 kb or downstream of the transcriptional start site.

Another possible explanation for this paradox could be that a major part of the regulation of α_2-M synthesis occurs at the posttranscriptional level. In fact, the discrepancy between measurements of the transcription rates of the rat α_2-M gene (three- to eightfold) and the increase in mRNA levels observed after IL-6 (60- to 200-fold) may reflect such a mechanism. Thus, in the rat, IL-6 might stabilize the α_2-M mRNA, whereas in man, α_2-M mRNA could be stable even in the absence of IL-6. Determination of the stabilities of rat vs. human α_2-M mRNA and of the transcription rates of the human gene will be required to clarify the molecular basis of the different regulation of human and rat α_2-M.

C. THE PERMISSIVE ROLE OF GLUCOCORTICOIDS

The induction of α_2-M synthesis in rat by interleukin-6 *in vivo* shows an absolute requirement for glucocorticoids (see Section III.A.1). The molecular basis for this requirement is presently not understood. Several mechanisms may be involved. (1) Glucocorticoids may be permissive for the expression of either IL-6 receptor subunits or factors involved in IL-6 signal transduction (e.g., protein kinases). There is evidence from our laboratory that the synthetic glucocorticoid dexamethasone upregulates the mRNA for the IL-6-binding subunit of the IL-6 receptor in Hep G2 cells.[65] (2) Alternatively, glucocorticoids could directly activate the α_2-M promoter. Such a mechanism has been demonstrated for the stimulation of α_1-acid glycoprotein expression by glucocorticoids, where a proximal glucocorticoid response element (GRE) in the promoter directs this effect.[66] In fact, at -37 and -62 bp upstream of the transcription start site in the rat α_2-M promoter, two half-sites of the palindromic GRE consensus are present. Binding studies with bacterially expressed DNA-binding fragments of the glucocorticoid receptor revealed only a weak interaction of receptor monomers with these sites. (3) In an indirect mechanism, glucocorticoids could induce a factor required for transcriptional or posttranscriptional regulation of α_2-M expression, e.g., a factor interacting with APRF, aiding its *trans*-activating effect, or a factor stabilizing the α_2-M mRNA.

Experimental attempts to answer these questions are hampered by the fact that in contrast to the *in vivo* situation, in most liver- or hepatoma-derived cell culture systems, the permissive effect of glucocorticoids for acute phase protein synthesis is not observed. After transfection of the hepatoma cell lines Hep G2, Hep 3B, and FAO with α_2-M promoter/luciferase[60] or α_2-M promoter/CAT constructs,[68] IL-6 induced the reporter gene expression even in the absence of glucocorticoids, and dexamethasone enhanced this induction only less than two-fold. An 18-bp fragment of the α_2-M promoter containing the APRE core motif was shown to confer this glucocorticoid effect.[60] Since no direct interaction of the glucocorticoid receptor with this sequence or with factors binding to it was observed, this effect might be mediated by the activation of an intermediate gene.[60] Whether the small effect of glucocorticoids on the α_2-M promoter observed in hepatoma cell lines reflects the glucocorticoid requirement *in vivo*, however, or whether another mechanism is involved remains to be elucidated.

Our observation that IL-6 rapidly triggers binding of the nuclear factor APRF to the α_2-M APRE and other type II IL-6 response elements by a posttranslational mechanism opens new avenues for studying the permissive effect of glucocorticoids on α_2-M expression. It will be very interesting to examine whether, in the absence of glucocorticoids (e.g., in hypophysectomized rats), APRF activation by IL-6 is prevented. In that case, glucocorticoids would be required for the synthesis of factors involved in IL-6 signaling, i.e., the IL-6 receptor, protein kinases, or even APRF itself. An APRF activation by IL-6 even in the absence of glucocorticoids, however, would imply that regulation at the transcriptional or posttranscriptional level is involved.

REFERENCES

1. **Travis, J. and Salvesen, G. S.,** Human plasma proteinase inhibitors, *Annu. Rev. Biochem.,* 52, 655, 1983.
2. **Sottrup-Jensen, L.,** Alpha-macroglobulins: structure, shape, and mechanism of proteinase complex formation, *J. Biol. Chem.,* 264, 11539, 1989.
3. **Barrett, A. J. and Starkey, P. M.,** The interaction of α_2-macroglobulin with proteinases. Characteristics and specificity of the reaction, and a hypothesis concerning its molecular mechanism, *Biochem. J.,* 133, 709, 1973.
4. **Ohlsen, K.,** Elimination of ^{125}I-rhIL-6-trypsin α_2-macroglobulin complexes from blood by reticuloendothelial cells in dog, *Acta Physiol. Scand.,* 81, 269, 1971.
5. **Ganrot, P. O. and Schersten, B.,** Serum α_2-macroglobulin concentration and its variation with age and sex, *Clin. Chim. Acta,* 15, 113, 1967.
6. **Koj, A.,** Comparison of synthesis and secretion of plasma albumin, fibrinogen and α_2-macroglobulin by slices of Morris hepatomas and rat liver, *Br. J. Exp. Pathol.,* 61, 332, 1980.
7. **Mosher, D. F., Saksela, O., and Vaheri, A.,** Synthesis and secretion of alpha-2-macroglobulin by cultured adherent lung cells: comparison with cell strains derived from other tissues, *J. Clin. Invest.,* 60, 1036, 1977.
8. **Hovi, T., Mosher, D. F., and Vaheri, A.,** Cultured human monocytes synthesize and secrete α_2-macroglobulin, *J. Exp. Med.,* 145, 1580, 1977.
9. **Bauer, J., Gebicke-Haerter, P.-J., Ganter, U., Richter, I., and Gerok, W.,** Astrocytes synthesize and secrete α_2-macroglobulin: differences between the regulation of α_2-macroglobulin in synthesis in rat liver and brain, *Adv. Exp. Med. Biol.,* 240, 199, 1988.
10. **Weimer, H., Humbelbaugh, C., and Roberts, D. M.,** The α_2-AP globulin of maternal and neonatal rat serums, *Am. J. Physiol.,* 213, 418, 1967.
11. **Panrucker, D. E., Lai, P. C. W., and Lorscheider, F. L.,** Distribution of acute phase α_2-macroglobulin in rat fetomaternal compartments, *Am. J. Physiol.,* 245, E138, 1983.
12. **Okubo, H., Miyanaga, O., Nagano, M., Ishibashi, H., Kudo, J., Ikuta, T., and Shibata, K.,** Purification and immunological determination of α_2-macroglobulin in serum from injured rats, *Biochim. Biophys. Acta,* 668, 257, 1981.
13. **Koj, A., Dubin, A., Kasperczyk, H., Bereta, J., and Gordon, A. H.,** Changes in the blood level and affinity to concanavalin A of rat plasma glycoproteins during acute inflammation and hepatoma growth, *Biochem. J.,* 206, 545, 1982.
14. **Baumann, H., Prowse, K. R., Marinkovic, S., Won, K.-A., and Jahreis, G. P.,** Stimulation of hepatic acute phase response by cytokines and glucocorticoids, *Ann. N.Y. Acad. Sci.,* 557, 280, 1989.
15. **Oliviero, S. and Cortese, R.,** The human haptoglobin gene promoter: interleukin-6-responsive elements interact with a DNA-binding protein induced by interleukin-6, *EMBO J.,* 8, 1145, 1989.
16. **Poli, V. and Cortese, R.,** Interleukin 6 induces a liver-specific nuclear protein that binds to the promoter of acute phase genes, *Proc. Natl. Acad. Sci. U.S.A.,* 86, 8202, 1989.
17. **Ganter, U., Arcone, R., Toniatti, C., Morrone, G., and Ciliberto, G.,** Dual control of C-reactive protein gene expression by interleukin-1 and interleukin-6, *EMBO J.,* 8, 3773, 1989.
18. **Akira, S., Isshiki, H., Sugita, T., Tanabe, O., Kinoshita, S., Nishio, Y., Nakajima, T., Hirano, T., and Kishimoto, T.,** A nuclear factor for IL-6 expression (NF-IL6) is a member of a C/EBP family, *EMBO J.,* 9, 1897, 1990.
19. **Poli, V., Mancini, F. P., and Cortese, R.,** IL-6DBP, a nuclear protein involved in interleukin-6 signal transduction, defines a new family of leucine zipper proteins related to C/EBP, *Cell,* 63, 643, 1990.
20. **Fowlkes, D. M., Mullis, N. T., Comeau, C. M., and Crabtree, G. R.,** Potential basis for regulation of the coordinately expressed fibrinogen genes: homology in the 5′ flanking regions, *Proc. Natl. Acad. Sci. U.S.A.,* 81, 2313, 1984.
21. **Won, K.-A. and Baumann, H.,** The cytokine response element of the rat α1-acid glycoprotein gene is a complex of several interacting regulatory sequences, *Mol. Cell. Biol.,* 10, 3965, 1990.
22. **Huber, P., Laurent, M., and Dalmon, J.,** Human β-fibrinogen gene expression. Upstream sequences involved in its tissue specific expression and its dexamethasone and interleukin-6 stimulation, *J. Biol. Chem.,* 265, 5695, 1990.

23. **Kunz, D., Zimmermann, R., Heisig, M., and Heinrich, P. C.,** Identification of the promoter sequences involved in the interleukin-6 dependent expression of the α_2-macroglobulin gene, *Nucleic Acids Res.,* 17, 1121, 1989.
24. **Dignam, J. D., Lebovitz, R. M., and Roeder, R. G.,** Accurate transcription initiation by RNA polymerase II in a soluble extract from isolated mammalian nuclei, *Nucleic Acids Res.,* 11, 1475, 1983.
25. **Sawadogo, M., Van Dyke, M. W., Gregor, P. D., and Roeder, R. G.,** Multiple forms of the human gene-specific transcription factor USF. I. Complete purification and identification of USF from HeLa cell nuclei, *J. Biol. Chem.,* 263, 11985, 1988.
26. **Graham, F. L. and Van der Eb, A. J.,** A new technique for the assay of infectivity of human adenovirus 5 DNA, *Virology,* 52, 456, 1973.
27. **Nielsen, D. A., Chou, J., MacKrell, A. J., Casadaban, M. J., and Steiner, D. F.,** Expression of a preproinsulin-β-galactosidase gene fusion in mammalian cells, *Proc. Natl. Acad. Sci. U.S.A.,* 80, 5198, 1983.
28. **Gorman, C. M., Moffat, L. F., and Howard, B. H.,** Recombinant genomes which express chloramphenicol acetyltransferase in mammalian cells, *Mol. Cell. Biol.,* 2, 1044, 1982.
29. **Kodelja, V., Heisig, M., Northemann, W., Heinrich, P. C., and Zimmermann, W.,** α_2-Macroglobulin gene expression during rat development studied by in situ hybridization, *EMBO J.,* 5, 3151, 1986.
30. **Fletcher, S., Thomas, T., Schreiber, G., Heinrich, P. C., and Yeoh, G. C. T.,** The development of rat α_2-macroglobulin: studies in vivo and in cultured fetal rat hepatocytes, *Eur. J. Biochem.,* 171, 703, 1988.
31. **Hayashida, K., Tsuchiya, Y., Kurokawa, S., Hattori, M., Ishibashi, H., Okubo, H., and Sakaki, Y.,** Expression of rat α_2-macroglobulin gene during pregnancy, *J. Biochem.,* 100, 989, 1986.
32. **Sarcione, E. J. and Bogden, A. E.,** Hepatic synthesis of alpha$_2$ (acute phase)-globulin of rat plasma, *Science,* 153, 547, 1966.
33. **Heim, W. G. and Ellenson, S. R.,** Involvement of the adrenal cortex in the appearance of rat slow alpha$_2$-globulin, *Nature (London),* 208, 1330, 1965.
34. **Sobocinski, P. Z., Canterburry, W. J., Knutsen, G. L., and Hauer, E. C.,** Effect of adrenalectomy on cadmium- and turpentine-induced hepatic synthesis of metallothionein and α_2-macrofetoprotein in rat, *Inflammation,* 5, 153, 1981.
35. **Bauer, J., Birmelin, M., Northoff, G.-H., Northemann, W., Tran-Thi, T.-A., Ueberberg, H., Decker, K., and Heinrich, P. C.,** Induction of rat α_2-macroglobulin in vivo and in hepatocyte primary cultures: synergistic action of glucocorticoids and a Kupffer cell derived factor, *FEBS Lett.,* 177, 89, 1984.
36. **Gauldie, J., Richards, C., Harnish, D., Landsdorp, P., and Baumann, H.,** Interferon β_2/B-cell stimulatory factor type 2 shares identity with monocyte-derived hepatocyte-stimulating factor and regulates the major acute phase protein response in liver cells, *Proc. Natl. Acad. Sci. U.S.A.,* 84, 7251, 1987.
37. **Andus, T., Geiger, T., Hirano, T., Kishimoto, T., Tran-Thi, T.-A., Decker, K., and Heinrich, P. C.,** Regulation of synthesis and secretion of major rat acute phase proteins by recombinant human interleukin-6 (BSF-2/IL-6) in hepatocyte primary cultures, *Eur. J. Biochem.,* 173, 287, 1988.
38. **Geiger, T., Andus, T., Klapproth, J., Hirano, T., Kishimoto, T., and Heinrich, P. C.,** Induction of rat acute phase proteins by interleukin-6 in vivo, *Eur. J. Immunol.,* 18, 717, 1988.
39. **Marinkovic, S., Jahreis, G. P., Wong, G. G., and Baumann, H.,** IL-6 modulates the synthesis of a specific set of acute phase plasma proteins in vivo, *J. Immunol.,* 142, 808, 1989.
40. **Northemann, W., Heisig, M., Kunz, D., and Heinrich, P. C.,** Molecular cloning of cDNA sequences for rat α_2-macroglobulin and measurements of its transcription during experimental inflammation, *J. Biol. Chem.,* 260, 6200, 1985.
41. **Birch, H. E. and Schreiber, G.,** Transcriptional regulation of plasma protein synthesis during inflammation, *J. Biol. Chem.,* 261, 8077, 1986.
42. **Gehring, M. R., Shiels, B. R., Northemann, W., de Bruijn, M. H. L., Kan, C.-C., Chain, A. C., Noonan, D. J., and Fey, G. H.,** Sequence of rat liver α_2-macroglobulin and acute phase control of its messenger RNA, *J. Biol. Chem.,* 262, 446, 1987.
43. **Gross, V., Andus, T., Tran-Thi, T.-A., Bauer, J., Decker, K., and Heinrich, P. C.,** Induction of acute phase proteins by dexamethasone in rat hepatocyte primary cultures, *Exp. Cell. Res.,* 151, 46, 1984.
44. **Bauer, J., Tran-Thi, T.-A., Northoff, H., Hirsch, F., Schlayer, H.-J., Gerok, W., and Heinrich, P. C.,** The acute-phase induction of α_2-macroglobulin in rat hepatocyte primary cultures: action of a hepatocyte-stimulating factor, triiodothyronine and dexamethasone, *Eur. J. Cell Biol.,* 40, 86, 1986.
45. **Hirata, Y., Ishibashi, H., Kimura, H., Hayashida, K., Nagano, M., and Okubo, H.,** α_2-Macroglobulin secretion enhanced in rat hepatocytes by partially characterized factor from Kupffer cells, *Inflammation,* 9, 201, 1985.
46. **Kurokawa, S., Ishibashi, H., Hayashida, K., Tsuchiya, Y., Hirata, Y., Sakaki, Y., Okubo, H., and Niho, Y.,** Kupffer cell stimulation of alpha$_2$-macroglobulin synthesis in rat hepatocytes and the role of glycocorticoid, *Cell Struct. Function,* 12, 35, 1987.

47. **Hirata, Y., Kurokawa, S., Ishibashi, H., Hayashida, K., Kimura, H., Nagano, M., and Okubo, H.,** Polymorphonuclear leukocytes and the induction of α_2-macroglobulin synthesis, *J. Clin. Lab. Immunol.*, 21, 125, 1986.
48. **Andus, T., Geiger, T., Hirano, T., Northoff, H., Ganter, U., Bauer, J., Kishimoto, T., and Heinrich, P. C.,** Recombinant human B cell stimulatory factor-2 (BSF2/IFN-β_2) regulates β-fibrinogen and albumin mRNA levels in Fao-9 cells, *FEBS Lett.*, 221, 18, 1987.
49. **Baumann, H. and Wong, G. G.,** Hepatocyte-stimulating factor III shares structural and functional identity with leukemia-inhibitory factor, *J. Immunol.*, 143, 1163, 1989.
50. **Baumann, H. and Schendel, P.,** Interleukin-11 regulates the hepatic expression of the same plasma protein genes as interleukin-6, *J. Biol. Chem.*, 266, 20424, 1991.
51. **Northemann, W., Heisig, M., Kunz, D., Hanson, R. W., and Heinrich, P. C.,** Regulation of rat α_2-macroglobulin gene activity, in *Protides of the Biological Fluids*, Peters, H., Ed., Pergamon Press, New York, 1985, 165.
52. **Northemann, W., Shiels, B. R., Braciak, T. A., Hanson, R. W., Heinrich, P. C., and Fey, G. H.,** Structure and acute phase regulation of the rat α_2-macroglobulin gene, *Biochemistry*, 27, 9194, 1988.
53. **Hattori, M., Kusakabe, S., Ohgusu, H., Tsuchiya, Y., Ito, T., and Sakaki, Y.,** Structure of the rat α_2-macroglobulin-coding gene, *Gene*, 77, 333, 1989.
54. **Ito, T., Tanahasi, H., Misumi, Y., and Sakaki, Y.,** Nuclear factors interacting with an interleukin-6 responsive element of rat α_2-macroglobulin gene, *Nucleic Acids Res.*, 17, 9425, 1989.
55. **Castell, J. V., Gomez-Lechon, M., David, M., Hirano, T., Kishimoto, T., and Heinrich, P. C.,** Recombinant human interleukin-6 (IL-6/BSF-2/HSF) regulates the synthesis of acute phase proteins in human hepatocytes, *FEBS Lett.*, 232, 347, 1988.
56. **Castell, J. V., Gomez-Lechon, M. J., David, M., Andus, T., Geiger, T., Trullenque, R., Fabra, R., Gerok, W., and Heinrich, P. C.,** Interleukin-6 is the major regulator of acute phase protein synthesis in adult human hepatocytes, *FEBS Lett.*, 242, 237, 1989.
57. **Bohmann, D.,** Transcription factor phosphorylation: a link between signal transduction and the regulation of gene expression, *Cancer Cells*, 2, 337, 1990.
58. **Munck Petersen, C., Christiansen, B. S., Heickendorff, L., and Ingerslev, J.,** Synthesis and secretion of α_2-macroglobulin by human hepatocytes in culture, *Eur. J. Clin. Invest.*, 18, 543, 1988.
59. **Hattori, M., Abraham, L. J., Northemann, W., and Fey, G. H.,** Acute phase reaction induces a specific complex between hepatic nuclear proteins and the interleukin-6 responsive element of the rat α_2-macroglobulin gene, *Proc. Natl. Acad. Sci. U.S.A.*, 87, 2364, 1990.
60. **Hocke, G., Baffet, G., Cui, M.-Z., Brechner, T., Barry, D., Goel, A., Fletcher, R., Abney, C., Hattori, M., and Fey, H.,** Transcriptional control of liver acute phase genes by interleukin-6 and leukemia inhibitory factor, in *Molecular Aspects of Inflammation*, Sies, H., Flohe, L., and Zimmer, G., Eds., Springer-Verlag, Berlin, 1991, 147.
61. **Nakayima, K. and Wall, R.,** Interleukin-6 signals activating *junB* and TIS11 gene transcription in a B-cell hybridoma, *Mol. Cell. Biol.*, 11, 1409, 1990.
62. **Lord, K. A., Abdollahi, A., Thomas, S. M., DeMarco, M., Brugge, J. S., Hoffman-Liebermann, B., and Liebermann, D. A.,** Leukemia inhibitory factor and interleukin-6 trigger the same immediate early response, including tyrosine phosphorylation, upon induction of myeloid leukemia differentiation, *Mol. Cell. Biol.*, 11, 4371, 1990.
63. **Lenardo, M. J. and Baltimore, D.,** NF-κB: a pleiotropic mediator of inducible and tissue-specific gene control, *Cell*, 58, 227, 1989.
64. **Kessler, D. S., Veals, S. A., Fu, X.-Y., and Levy, D. E.,** Interferon-α regulates nuclear translocation and DNA-binding affinity of ISGF3, a multimeric transcriptional activator, *Genes Dev.*, 4, 1753, 1990.
65. **Rose-John, S., Schooltink, H., Lenz, D., Hipp, E., Dufhues, G., Schmitz, H., Schiel, X., Hirano, T., Kishimoto, T., and Heinrich, P. C.,** Studies on the structure and regulation of the human hepatic interleukin-6-receptor, *Eur. J. Biochem.*, 190, 79, 1990.
66. **Baumann, H. and Maquat, L. E.,** Localization of DNA sequences involved in dexamethasone-dependent expression of the rat α_1-acid glycoprotein gene, *Mol. Cell. Biol.*, 6, 2551, 1986.
67. **Marynen, P.,** personal communication.
68. **Horn, F., Wegenka, U. M., and Heinrich, P. C.,** unpublished results.

Chapter 26

REGULATION OF THE RAT α_2-MACROGLOBULIN GENE BY INTERLEUKIN-6 AND LEUKEMIA INHIBITORY FACTOR

Gertrud M. Hocke, Mei-Zhen Cui, Jürgen A. Ripperger, and Georg H. Fey

TABLE OF CONTENTS

0-8493-6913-4/93/$0.00 + $.50

I. INTRODUCTION

The acute phase genes represent a convenient system to study cytokine-controlled gene expression. Rat acute phase genes are particularly attractive because they offer the advantage that the inflammatory response can be provoked and controlled in living animals as well as in cultured primary hepatocytes and hepatoma cell lines. In addition, rat liver is available in sufficient supply to study the regulatory pathways in biochemical detail. This becomes important when investigators need to study cytokine-induced posttranslational alterations of nuclear transcription factors or intermediate cytoplasmic steps of the cytokine signal cascades. Certainly, it is possible to clone cDNAs coding for human nuclear factors controlling acute phase gene transcription and human intermediate components of the signal pathways and to use these clones and derived recombinant proteins to study their functions. Indeed, one of the first known cytokine response factors, the nuclear factor NF-IL-6 mediating the control of the interleukin-6 (IL-6) gene by interleukin-1 (IL-1) and IL-6, was first cloned from a human cDNA library.[1] It is further feasible to express the derived recombinant human proteins in hepatoma cells or primary hepatocyte cultures and to study their cytokine-induced modifications. However, it is ultimately necessary to verify that the modifications observed in cultured hepatic cells are the same as those occurring in the intact liver because cultured cells may differ in their patterns of modifications from intact livers. This last verification is difficult to achieve with human and easy with rat liver biopsy material. Therefore, we have chosen to work with rat acute phase genes.

The rat acute phase genes have recently been divided into two classes according to the cytokines that are their main inducers (Table 1).[2,3] Class 1 genes respond to IL-1 alone or combinations of IL-1 + IL-6, IL-1 + LIF (leukemia inhibitory factor), or IL-1 + IL-6 + glucocorticoids. Class 2 genes respond to IL-6 or LIF alone, combinations of IL-6 + glucocorticoids, or IL-6 + LIF + glucocorticoids. This distinction has its correlate on the molecular level. The two classes of genes carry different types of *cis*-acting DNA control elements, which mediate their cytokine control. Class 1 genes carry type I IL-6 response elements (IL-6 REs) that are binding sites for the nuclear transcription factor NF-IL-6; class 2 genes carry type II IL-6 REs that are binding sites for a factor different from NF-IL-6, which has so far not been cloned and sequenced. Other classifications of the acute phase proteins have been given in the literature, based on either the relative increase in their plasma concentrations or their functions.[4-6] In this chapter, emphasis is placed on the mechanism of control of acute phase genes rather than the function of the corresponding proteins. Therefore, we use the classification given in Table 1.

An informative two-dimensional gel electrophoretic analysis of the major rat acute phase proteins has been published.[7,8] We have chosen to study the rat α_2-macroglobulin (α_2-M) gene as our main experimental paradigm because it is one of the genes with the greatest relative increases in both plasma protein and liver mRNA concentrations during the acute phase response. Normal plasma concentrations of this protein in healthy adult male rats range from 10 to 40 μg/ml (Figure 1).[7] Acute phase concentrations reach approximately 2 to 3 mg/ml after 2 to 3 d, depending on the intensity and type of the inflammatory stimulus (Figure 1). This represents an increase of approximately 100-fold, accompanied by corresponding changes in hepatic mRNA concentrations (Figure 2).[9-11] The increase in mRNA concentrations was preceded by a corresponding increase in transcription rates. The rates reached maximum values usually 8 to 24 h after a single injection of an inflammatory stimulus (Figure 3).[10,12,13] Increases in transcription rates were usually in the range of five- to tenfold. These results clearly demonstrated that the overall acute phase induction of the α_2-M gene involved transcriptional control as a major component. After a one-time injection of an inflammatory agent, the acute phase induction of this gene peaked after 2 to 3 d. Subsequently, mRNA and protein levels declined and reached baseline values after about 7

TABLE 1
Two Classes of Rat Liver Acute Phase Genes

	Genes	Inducers
	Class 1	
Human	C-reactive protein (CRP)	IL-1
	Serum amyloid A (SAA)	IL-1 + IL-6
	Complement C3	IL-1 + LIF
	Hemopexin	IL-1 + IL-6 + glucocorticoids
	Haptoglobin and others	
Rat	SAA	
	Complement C3	
	Hemopexin	
	Haptoglobin	
	α_1-Acid glycoprotein	
	Class 2	
Rat	α_1-Antitrypsin	IL-6
	α_1-Antichymotrypsin	LIF
	Thiostatin (= α_1-cysteine proteinase inhibitor)	IL-6 + LIF
	α_2-Macroglobulin (α_2-M)	IL-6 + LIF + glucocorticoids
	γ-Fibrinogen	

Note: Classification as proposed by Baumann and Gauldie.[2]

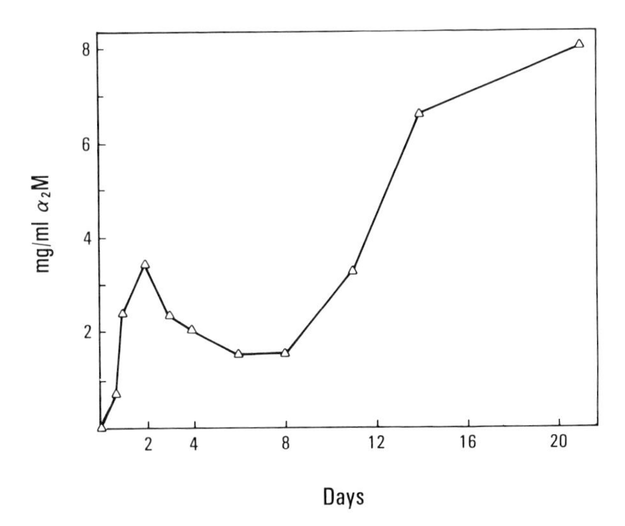

FIGURE 1. Plasma concentrations of rat α_2-M during acute and chronic inflammation. Plasma concentrations (ordinate; in milligrams per milliliter) were determined by quantitative immunoelectrophoresis and quantitative two-dimensional gel electrophoresis, both calibrated with purified rat α_2-M.[7]

to 10 d (Figure 1).[7] When an agent was used that caused a chronic inflammatory condition, such as complete Freund's adjuvant (causing an adjuvant-induced polyarthritis in rats), a biphasic response curve was observed. The first phase was an acute phase response, reaching a maximum again after 2 to 3 d and declining thereafter. The second phase was a secondary increase to very high plasma protein levels, reaching 8 to 10 mg/ml after 2 to 3 weeks

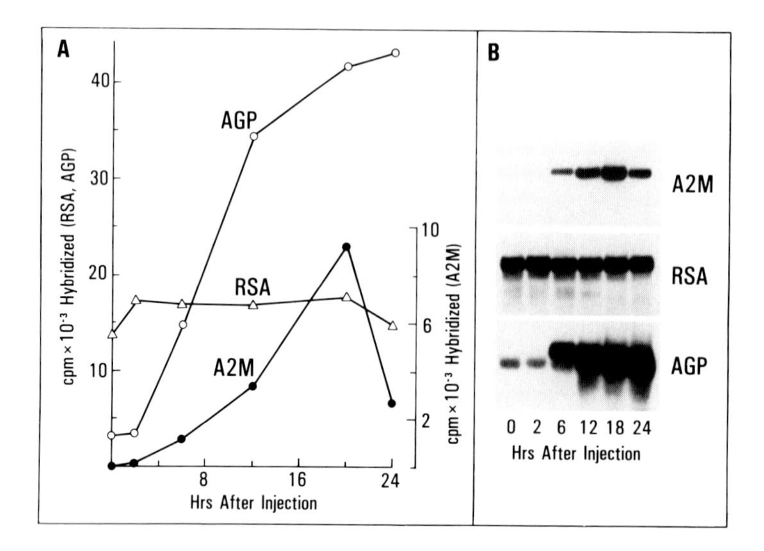

FIGURE 2. Acute phase induction of hepatic α_2-M mRNA concentrations. mRNA concentrations were determined by quantitative Northern-blot hybridization with a radiolabeled α_2-M cDNA probe.[9]

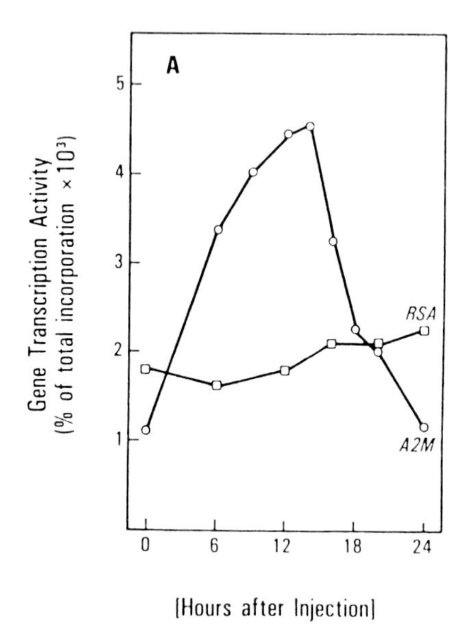

FIGURE 3. Transcription rates of the α_2-M gene during an acute phase response. Transcription rates were measured by nuclear run-on experiments. (From Northemann, W., Heisig, M., Kunz, D., and Heinrich, P. C., *Eur J. Biochem.*, 137, 257, 1985. With permission.)

(Figure 1).[7] Other treatments have been described that induced α_2-M plasma concentrations to similarly elevated levels, such as a combined administration of mediators of inflammation plus glucocorticoids plus catecholamines.[14] Glucocorticoids play an important role in inflammatory processes. On the one hand, they have antiinflammatory effects on chronic inflammatory processes such as rheumatoid arthritis, and are therefore used extensively as antiinflammatory drugs. This effect is due at least in part to glucocorticoids preventing the upregulation of key genes, such as collagenase, by mediators of inflammation.[15] On the other hand, glucocorticoids have the opposite effect on inflammatory processes controlled

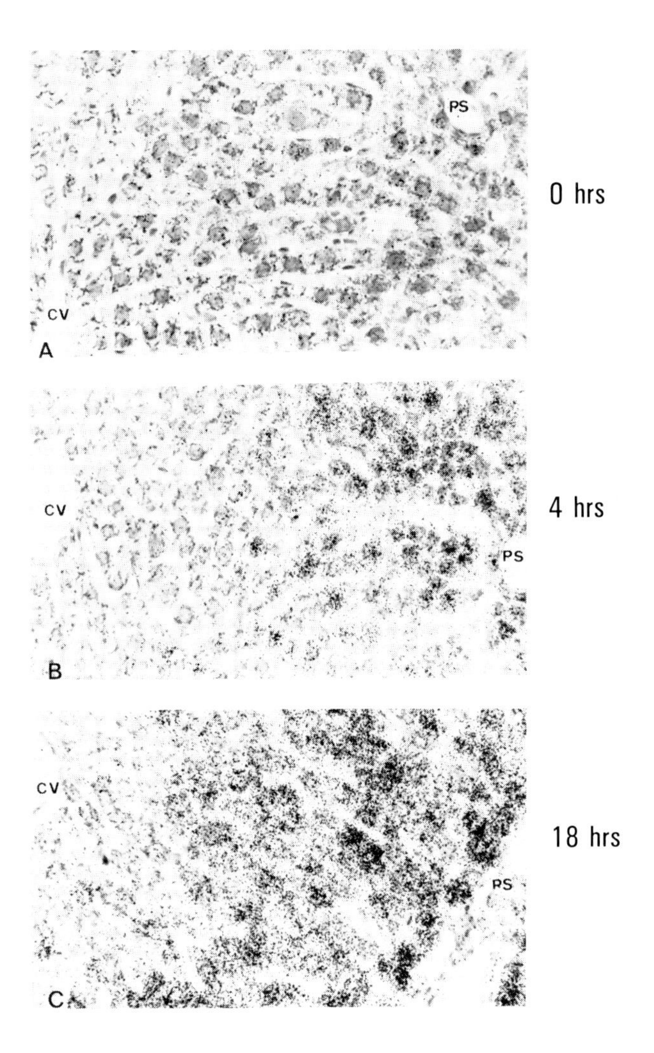

FIGURE 4. Progressive recruitment of hepatocytes into synthesis of α_2-M mRNA during an acute phase reaction. Liver sections were produced from rats at various times after triggering an experimental acute phase response (0, 4, and 18 h; panels A, B, and C, respectively). *In situ* hybridization was performed with radiolabeled α_2-M cDNA. PS, portal space (portal vein); CV, centrilobular vein.[21]

by the liver. Here, they play a proinflammatory role, often referred to as the "permissive effect of glucocorticoids". In certain tissue culture models, the presence of glucocorticoids was required before an effect of cytokines such as IL-6 could be observed.[16,17] For several years, it was not clear whether this result was a peculiarity of those culture systems or reflected an essential requirement of liver cells for glucocorticoids in order to become sensitive to the effect of inflammatory cytokines such as IL-6 and LIF. The results presented below demonstrated that glucocorticoids are not absolutely required for the transcriptional induction of acute phase genes by cytokines, but play an important synergistic role.

During a systemic acute phase response resulting from infection or tissue injury in the periphery, these cytokines are locally produced at sites of tissue damage by skin keratinocytes, blood vessel endothelial cells, and macrophages/monocytes. They disseminate over the body with the bloodstream, and a major target is the liver. Hepatocytes carry IL-6 and LIF receptors on their cell surface.[18-20] After binding of the ligand to these receptors, transcriptional activation of the acute phase genes follows. Figure 4 shows the progressive recruitment of hepatocytes into synthesis of α_2-M mRNA during an experimentally induced acute phase response in rat liver.[21] Prior to the arrival of the stimulus, only background levels of α_2-M mRNA were detected, uniformly distributed over the liver section. Four hours after induction of the response, hepatocytes surrounding the portal vein were the first to show increased levels of α_2-M mRNA. At 18 h, a far greater number of hepatocytes produced a positive hybridization signal, in a pattern representing a radial, centrifugal spread starting from the portal vein. At 24 h, most hepatocytes across the hepatic lobe were positive (data not shown). This result demonstrated a progressive recruitment starting from the portal vein and finally reaching the centrilobular vein. Several interpretations are possible. One is that a diffusible, hormone-like signal entered through the portal vein and diffused through the pericellular space. A second interpretation is that a signal emanating from the portal vein reached the first layer of surrounding hepatocytes, and that these in turn produced a secondary signal, either a diffusible substance or a cell-cell contact signal, that reached the next surrounding layer of cells, and so on, thus producing a centrifugal wave. The possibility that the signal was neuronally mediated could be excluded because a similar pattern of radial diffusion was observed with perfused livers severed from neuronal contact. The generally adopted view is that this radial spread is brought about by the diffusion of inflammatory cytokines. These arrive with the blood through the portal vein and spread in an unidirectional flow across the lobe toward the centrilobular vein.[21] At normal times, no cytokines or only negligible concentrations are found in the liver.[22-24] During a systemic acute phase response, these agents are produced at extrahepatic sites and act on the liver. This picture is to be distinguished from a localized inflammation of the liver itself. In that case, Kupffer cells, the resident tissue macrophages of the liver, are capable of synthesizing cytokines, including IL-6. It is not clear whether, under these circumstances, liver hepatocytes also produce some of these cytokines. This may be the case because certain rat and human hepatoma-derived cell lines can be stimulated by treatment with lipopolysaccharides (LPSs) and IL-1 to produce IL-6 and LIF.[23-26] Additionally, after partial hepatectomy, regenerating rat liver cells, but not their resting counterparts, produce LIF.[24]

II. MATERIALS AND METHODS

Materials and methods for each experiment are described in detail in the quoted original articles and the figure legends.

III. RESULTS

A. IDENTIFICATION AND CHARACTERIZATION OF AN IL-6 RESPONSE ELEMENT OF THE RAT α_2-M GENE

The rat α_2-M gene was cloned, and the transcription start sites and promoter upstream regions were identified and sequenced.[27-30] Control elements that mediate inducibility by hormones are often found within 1 to 2 kilobase (kb) pairs 5′ of the transcription start sites of a broad variety of genes. Therefore, our initial search for IL-6-responsive control elements of the rat α_2-M gene was focussed on this region. A series of plasmid constructs were prepared that carried various subsets of these sequences driving the expression of a firefly luciferase reporter gene. Two constructs were used that carried either the sequences from −2.2 kb to +17 bp or from −1151 to +17 bp of the α_2-M gene (Figure 5). They were cotransfected with a neomycin (G 418) resistance plasmid into FAO rat hepatoma cells. This cell line was chosen because extensive pilot studies performed in our laboratory had shown that it responded most closely to rat liver cells regarding the regulation of the α_2-M gene by a variety of stimuli, including IL-6 and dexamethasone (a synthetic model glucocorticoid).[23] A series of stably transfected, clonally derived neomycin-resistant cell lines were

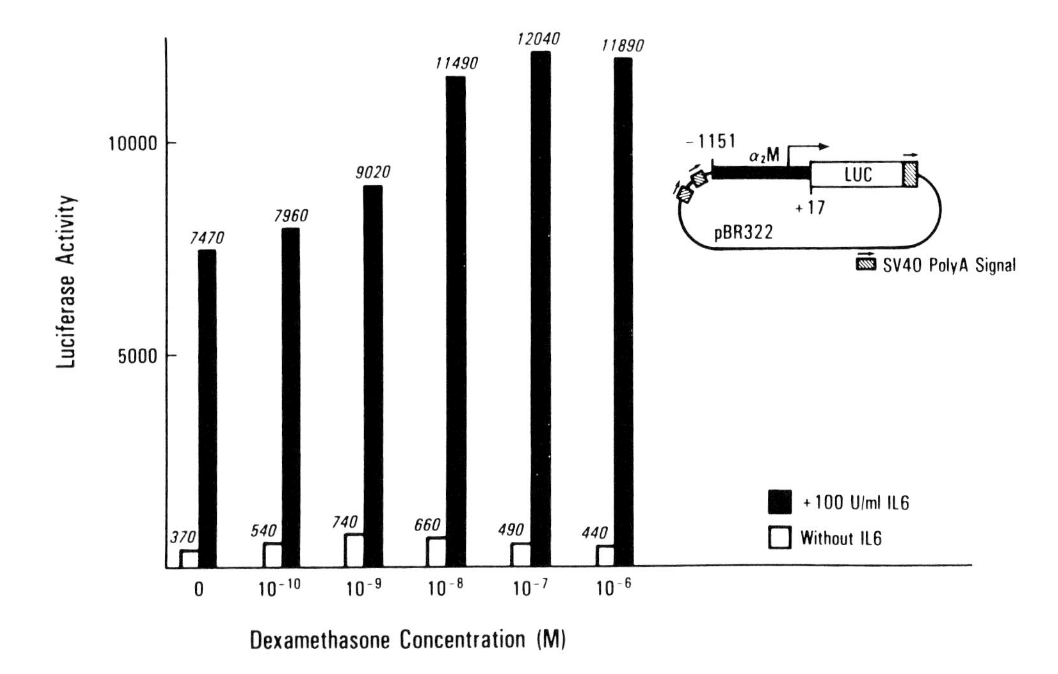

FIGURE 5. IL-6 is the main transcriptional inducer of the α_2-M gene promoter. The plasmid construct pα_2-M HB·Luc (upper right corner) was based on the vector pBR322 and carried protein-coding sequences for firefly luciferase and polyadenylation signals from SV40. Expression of luciferase was controlled by the fragment −1151 to +17 bp from the promoter region of the rat α_2-M gene.[27,32] The plasmid was cotransfected with a neomycin-resistance (G418) plasmid into FAO rat hepatoma cells, and stably transfected, neomycin-resistant cell lines were established by single-cell subcloning. They were treated with a saturating concentration of IL-6 alone (100 BSF-2 units/ml), increasing concentrations of dexamethasone alone (open bars), or combinations of the same constant dose of IL-6 plus increasing concentrations of dexamethasone (black bars). Luciferase activities are given for a standard assay containing 100 μg of cellular protein extract as described.[23,25,32]

established by multiple rounds of single-cell subcloning.[31,32] One of these lines carrying the −1151 to +17 bp construct was designated FAO/HB3 and further characterized. When these cells were treated with physiological concentrations of glucocorticoids alone (10^{-9} to 10^{-6} *M*), they expressed baseline promoter activities (Figure 5). Treatment with IL-6 alone showed a 20-fold increase in promoter activity, and with IL-6 plus glucocorticoids, up to a 31-fold induction. When the IL-6 concentration was kept constant at a saturating level and the glucocorticoid concentration was raised progressively, the relative induction showed a clear dose dependence on the dexamethasone concentration (Figure 5). Thus, the promoter was inducible by IL-6 alone, but not significantly by glucocorticoids alone, and was synergistically induced by both agents. Synergism was defined as a superadditive effect. The result showed that an IL-6 RE was contained in the −1151 to +17 bp region of this gene. Similar results were also obtained with a construct carrying the sequences from −2.2 kb to +17 bp. Therefore, it was concluded that the region between −1151 bp and −2.2 kb carried no additional important IL-6 REs. Subsequent studies with transient transfection showed that this interpretation was correct. Maximum transcriptional activation of the 1151 bp construct was achieved between 4 and 6 h after addition of IL-6 alone (Figure 6A). When IL-6 and glucocorticoids were added simultaneously, a first peak of activation was observed at 4 to 6 h and a second peak at around 24 h (Figure 6A). Recent studies showed that the rat liver IL-6 receptor gene can be upregulated by glucocorticoids.[20,33] Therefore, this secondary induction was probably due to an upregulation of the IL-6 receptor gene by glucocorticoids, which occurred with slower kinetics than the primary response of the α_2-M gene.

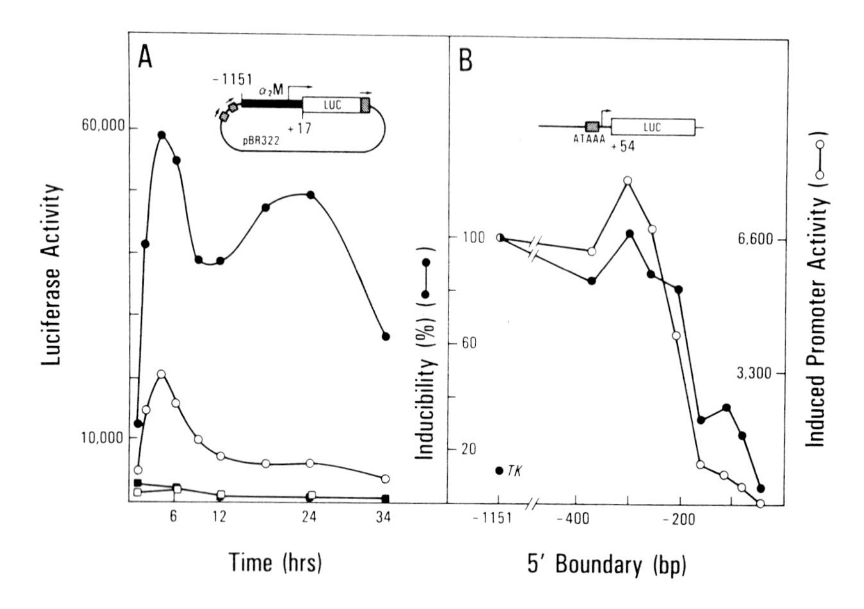

FIGURE 6. Kinetics of cytokine-mediated transcriptional induction and localization of an IL-6 response element by 5′ deletion mutagenesis. (A) The stably transfected cell line FAO/HB3, carrying the plasmid construct shown in Figure 5 stably integrated in its chromosomes, was treated with medium alone (mock treatment; open squares), 1 μ*M* dexamethasone alone (black squares), IL-6 alone (open circles), or IL-6 plus dexamethasone (closed circles). Cellular extracts were prepared at various times (hours; abscissa) after addition of the treatment, and luciferase activities are given in light units per 200μg of cellular protein extract in a standard luciferase assay. (B) A series of progressive deletion constructs were prepared, starting from a construct similar to that shown in A and differing only by the 3′ end of the α_2-M promoter sequences, which extended to +54 bp. The 3′ boundary was kept constant in all constructs, and the 5′ boundary was progressively reduced, as shown on the abscissa. These constructs were transfected with the calcium phosphate technique into FAO rat hepatoma cells, and the cells subsequently treated with either dexamethasone alone or IL-6 plus dexamethasone. Promoter activities were measured as luciferase units in transient assays and are given either as a percent of the maximum activity (inducibility, %, black dots) or as absolute induced activity (open symbols).[32]

For a more precise mapping of the response element within this −1151 to +17 bp region, mutagenesis and transient transfection studies were performed. A series of 5′ deletion constructs were produced and transiently transfected into FAO rat hepatoma cells. The cells were then treated with either IL-6 alone or a combination of IL-6 plus dexamethasone, and induced promoter (luciferase) activities were measured for each construct. As a result, promoter activities were approximately the same for all constructs with 5′ boundaries from −1151 bp to −220 bp (Figure 6B). A significant drop in inducibility occurred when sequences located between −210 bp and −150 bp were removed (Figure 6B). Therefore, an important IL-6 RE must have been contained in this region. This element was mapped more precisely by a series of internal deletions and linker scan mutations (Figure 7). Deletion or substitution of the sequence block around −160 bp (mutant m205, Figure 7) had the single most important effect, reducing inducibility by IL-6 to 40% of the value measured for the wild-type construct. Therefore, we concluded that the sequence altered in mutant m205, CTGGGA, was an essential part of an IL-6 RE. However, since this mutant still retained a residual inducibility of 40%, it was also clear that these were not the only sequences in the promoter region of the α_2-M gene that contributed to IL-6 inducibility. Other investigators have mapped an IL-6 RE within a few hundred base pairs in the same region by a similar approach.[28,33,34] By comparing the relevant sequence of the α_2-M gene with the promoter upstream regions of other acute phase genes, it was found that the sequence identified by linker scan analysis was the most strongly conserved region between a number

A.

```
       -210
       |            |             |              |          |
       GTACAAAAGAGAAAAAGTGAGCAGTAACTGGAAAGTCCTTAA
              M201        M202        M203       M204
                                                    -130
              |             |             |         |
       TCCTTCTGGGAATTCTGGCTAACGGGTCAGGAATTAACCTTG
              M205        M206                M214
```

B.

	IL6 (−)	IL6 (+)	Fold	(%)
-1151 (Wild-Type)	100	1800	18.0	100
Δ-160/-115	64	970	15.2	84
Δ-209/-160	76	511	6.7	37
Δ-115/-80	44	871	19.8	110
Δ-160/-80	80	1100	13.8	77
Δ-209/-80	69	214	3.1	17
TK	465	1098	2.4	13

C.

	IL6 (−)	IL6 (+)	Fold	(%)
-1151(Wild-Type)	100	2476	24.3	100
M201(TGGATCCT)	74	1873	25.4	105
M202(TGGATCCT)	70	1457	20.8	86
M203(TGGATCCT)	78	1521	19.5	80
M204(TAGGATCCCA)	66	1768	26.8	110
M205(GATATC)	56	562	10.0	40
M206(GATATC)	66	2306	34.9	144
M214(TCCT)	58	1171	20.2	83
TK	433	1431	3.3	14

FIGURE 7. Mapping the IL-6 RE by internal deletion and linker scan mutagenesis. A series of plasmids based on those described in Figures 5 and 6 were constructed carrying either internal deletions of the portions preceded in B by the symbol Δ, or 6- to 10-bp substitutions, shown by underscores in A. The substituted sequences are shown in parentheses in C. These constructs were transfected into FAO rat hepatoma cells and the cells were either mock treated (−) or treated with IL-6 (+), and induced promoter activity was evaluated as luciferase activity in transient expression assays. The relative induction over the mock-treated control level is given for each construct in the column "Fold", and as a fraction of the relative induction obtained with the wild-type construct (last column to the right, "%"). TK, enhancerless construct carrying only the minimal promoter of the herpes simplex virus thymidine kinase gene.[32]

Rat α_2M	TAATCCT	TCTGGGAA	TTCTGGC
Rat α_1 AGP	CTGGGCT	TCTGGGAA	AAACTCA
Rat T_1-Kininogen	TTTGTTC	TCTGAGAA	GAGGGCA
Rat T_2-Kininogen	TTTATTC	TCTGGGAA	GAGGGCA
Rat γ-Fibrinogen	TGCAAAA	TCTGGGAA	TCCCTCG

FIGURE 8. IL-6 RE is conserved in several rat liver acute phase genes. The region of the IL-6 RE core element of the α_2-M gene was compared by computer analysis with the 5′ flanking regions of other known IL-6-inducible rat liver acute phase genes. Conserved sequences encompassing the central hexanucleotide of the core element are boxed. This sequence was later shown to be not only conserved in the α_1-AGP gene, but also to serve as a functional element. It conferred IL-6 responsiveness to heterologous promoters in mutagenesis and transfection studies.[38,62,63,74]

of IL-6-controlled rat liver acute phase genes (Figure 8).[35] Direct proof that this sequence functioned as an IL-6 RE was provided by demonstrating that it conferred IL-6 inducibility to heterologous minimal promoters, which by themselves were not responsive to IL-6 (see Tables 2 and 3).[36,37] The loss in inducibility resulting from deletion of the region −150 to −210 bp was greater than the loss produced by mutation m205 (Figures 6B and 7). Detailed analysis revealed that the region −210 to −150 contained a second element of similar sequence located 20 bp upstream of the first. This element was functionally weaker than the first, probably because of the altered sequence, and was revealed by linker scan mutant m203 (Figure 7B) as well as protein-DNA binding studies. Thus, the IL-6 RE region from −210 to −150 bp consists of two subelements, called the IL-6 RE core and core homology

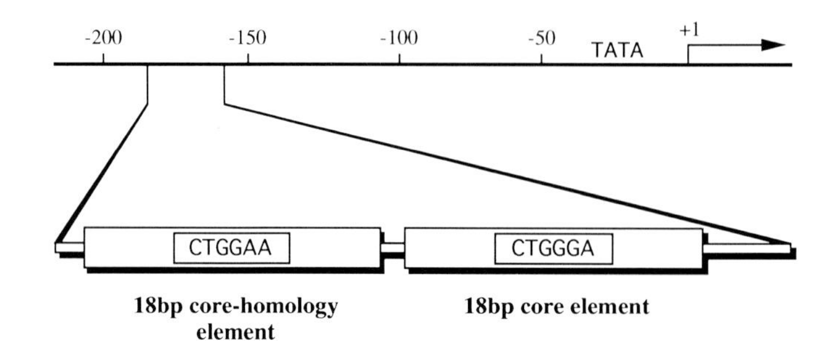

FIGURE 9. The IL-6 RE of the α_2-M gene consists of two subelements. The two subelements were defined by both functional studies (mutagenesis and transfection; linker scan mutants m203 and m205, Figure 7) and protein DNA-binding studies. In DNase 1 footprinting experiments, two protected windows were found, each about 18 to 20 bp in length, that corresponded in position to these two subelements.[32] Methylation interference and gel-mobility shift experiments confirmed this conclusion, as well as similar findings reported by other laboratories.[32-34,39] The central hexanucleotide sequences are boxed. The functional element requires more than the central hexanucleotides. The exact requirements have not yet been identified at the one-nucleotide resolution level, but the central hexanucleotide plus approximately six nucleotides on both sides from the α_2-M gene sequence produced a functional element.

elements, that are located 20 bp apart (Figure 9). Each subelement is approximately 18 to 20 bp long, as defined by protein-binding studies. Both contain a central hexanucleotide that is essential for their function (linker scan analysis) and for protein binding. This hexanucleotide sequence is CTGGGA for the core element and CTGGAA for the core homology element (Figure 9). This central part of the sequence is the most strongly conserved portion that was also found in the control regions of other IL-6-induced rat liver acute phase genes (Figure 8). The conserved element found in the α_1-AGP gene is not only fortuitously conserved in sequence, but also functions as an IL-6 RE in that gene.[38] However, the central hexanucleotides alone are not sufficient to produce the function; the protein-binding site and the minimal functional element are approximately 18 bp in length as shown in DNase I footprinting, methylation interference, and gel-mobility shift experiments.[32] Methylation interference revealed that two of the G-residues of the central hexanucleotides of both the core and core homology elements were directly involved in protein-DNA contacts (Figure 10).[32] Binding at these sites was sequence specific, as demonstrated by competition gel-mobility shift experiments.[32,36,39] Quantitative binding studies further revealed that the core homology site had a weaker affinity for the protein than the core site, and that binding at both sites occurred cooperatively.[39] A synthetic double-stranded oligonucleotide containing two copies of the 18-bp core element in the proper 20-bp distance was used as a target for protein binding. Using this target, binding at the second site occurred with 80-fold greater affinity when the first site was occupied than when it was empty.[39] The currently favored hypothesis is that the same protein binds at each of the two subelements in a cooperative manner. If this interpretation is correct, this protein must contain at least three functional domains: one for DNA binding, one for cooperative interactions with itself, and one for transcriptional activation.[39]

To prove that the 18-bp core element was sufficient to generate IL-6 responsiveness, a series of reporter constructs were produced based on the firefly luciferase reporter gene (Figure 11). This series comprised four groups, which used the minimal enhancerless promoter of the herpes virus thymidine kinase (TK) gene, the minimal promoter of the simian virus-40 early gene (SV40e), the minimal promoter of the rat α_2-M gene itself, and the minimal promoters with added IL-6 REs, respectively. These constructs were transfected into Hep G2 human hepatoma cells. The cells were then treated with IL-6, and induced promoter activities were measured using the luciferase reporter assay (Table 2). Multiple

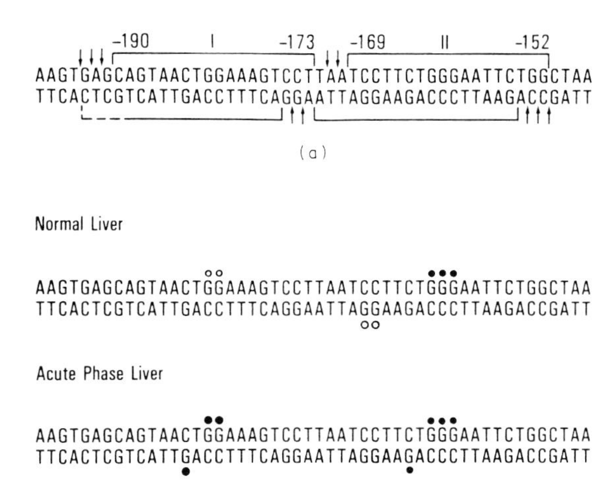

FIGURE 10. Definition of the protein-binding regions of the IL-6 RE by DNase I footprinting and methylation interference. (a) Windows of protection against DNase I digestion on the upper and lower DNA strands by bound nuclear proteins. Arrows indicate sites that have become hypersensitive to DNase I as a result of protein binding. Windows I and II correspond to the core homology and core regions, respectively. (b) Nucleotides in protein-DNA contact as revealed by methylation interference, using either nuclear protein extracts from normal or acute phase rat livers. Open circles, weak contacts/interference; black dots, strong contacts/interference.[32,39]

copies of the 18-bp element conferred a copy number-dependent inducibility to the two heterologous promoters (TK and SV40e) and the autologous minimal promoter of the α_2-M gene. Four copies of the element produced a 9.3-fold activation of the SV40e promoter and a 187.3-fold activation of the α_2-M gene's own minimal promoter by IL-6. When the central hexanucleotide sequence was destroyed, the constructs lost their inducibility (Table 2). Therefore, the 18-bp element possessed an intrinsic IL-6 responsiveness that functioned irrespective of the minimal promoter with which it was combined. However, the minimal promoter of the α_2-M gene functioned better in combination with this element than did those of the two heterologous genes. It is conceivable that the autologous minimal promoter assembled a protein complex that interacted more favorably with the factor bound at the IL-6 RE than those assembled over the heterologous minimal promoters. The precise explanation of this difference is currently unknown. These studies clearly demonstrated that the 18-bp core element was sufficient to generate IL-6 responsiveness.

B. SYNERGISM OF IL-6 AND GLUCOCORTICOIDS IS MEDIATED BY THE IL-6 RESPONSE ELEMENT

The same 18-bp element that mediated the response to IL-6 was also sufficient to confer a synergistic response to glucocorticoids and IL-6 (Table 3).[36] The synergistic effect was defined as the ratio of the induction achieved with both hormones over that achieved with IL-6 alone. Four copies of the 18-bp core element produced a 4.5-fold synergistic effect in Hep 3B human hepatoma cells in combination with the SV40 minimal promoter and a threefold effect in combination with the autologous minimal promoter of the α_2-M gene. Mutation of the central hexanucleotide of the element abolished the effect, and no significant synergism was produced by the minimal promoters alone (Table 3). This 18-bp core element conferred the synergistic effect to IL-6 and glucocorticoids, although it contained no consensus glucocorticoid response element (GRE; binding site for the glucocorticoid receptor, GR). However, the GR occasionally binds at DNA sequences that show only weak conservation with its consensus binding site. Therefore, we investigated whether the GR could

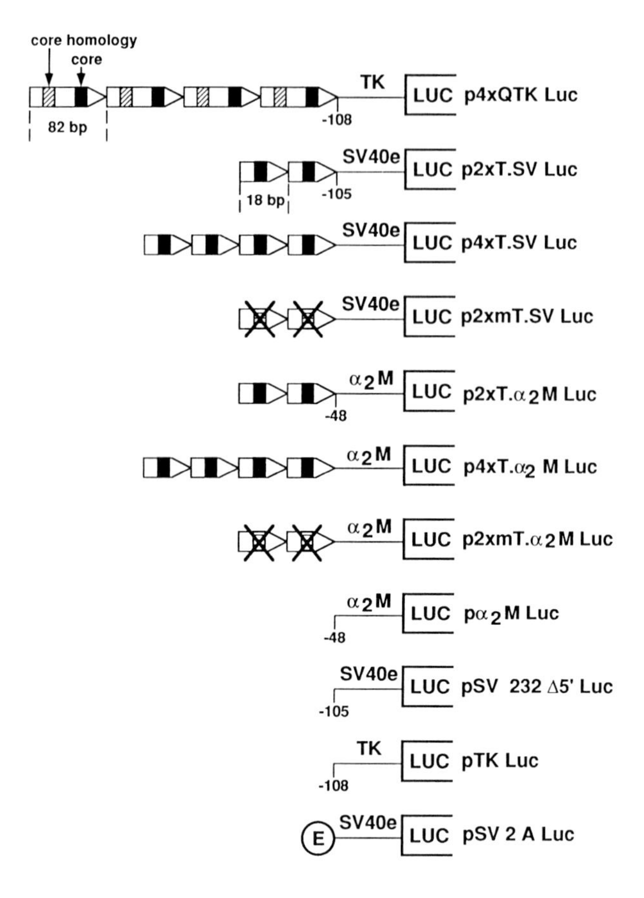

FIGURE 11. Constructs used to demonstrate transcription control function of the IL-6 RE. Luc: coding sequences for firefly luciferase. TK, SV40e, α_2-M: minimal promoters (without enhancers) of the thymidine kinase, SV40 early, and α_2-M genes. Black box, shaded box: central hexanucleotides of the core and core homology regions, respectively. Crossed-out box: mutated hexanucleotide, as in mutant m205 (Figure 6). E: enhancer of the SV40 early region promoter.[36]

bind at the 18-bp IL-6 RE sequence. A fragment of recombinant rat GR containing the DNA binding domain was expressed in *Escherichia coli* and purified.[40] It was used for DNA-binding studies with synthetic oligonucleotide probes containing either multiple copies of the 18-bp sequence or an authentic GRE, a fragment from the long terminal repeat (LTR) of the mouse mammary tumor virus (MMTV). DNA binding was observed with the authentic GRE probe, but no binding was observed with the 18-bp element. In addition, a polyclonal antibody against the purified recombinant rat GR[41] was used in gel-mobility shift experiments. No specific supershifts were obtained when the antiserum was added to a reaction mixture containing the IL-6 RE and nuclear extracts from acute phase rat livers or IL-6-treated hepatoma cells. Thus, until now, all experiments aimed at showing a direct interaction between the GR fragment and the IL-6 RE have produced negative results. On the other hand, we know that the synergistic response between glucocorticoids and IL-6 involved the GR because it was prevented by treatment with the inhibitor RU486, a specific antagonist of the GR.[42] The preliminary conclusion was that the effect of glucocorticoids on the IL-6 response was indirect, i.e., glucocorticoids probably activated an intermediary gene, and the product of this gene rather than the GR itself interacted with the IL-6 signal cascade. However, this conclusion is preliminary because in our experiments, only a fragment of the GR was used. There is still a possibility that the GR could have engaged in a direct protein-

TABLE 2
Multiple Copies of the 18-bp Core Element Are Sufficient to Confer IL-6 and LIF Responsiveness

Construct transfected	Luciferase activity after treatment with			Relative induction[b]	
	Medium alone	LIF[a]	IL-6	LIF	IL-6
p4xQ·TKLuc	519,232	1,536,514	2,858,572	3.0	5.5
p2xT·SVLuc	79,100	218,564	523,602	2.8	6.6
p2xmT·SVLuc	92,477	90,064	114,684	1.0	1.2
p4xT·SVLuc	261,866	1,083,518	2,434,943	4.1	9.3
p2xT·α_2MLuc	7,234	55,477	322,938	7.7	44.6
p2xmT·α_2MLuc	5,064	5,285	7,616	1.0	1.5
p4xT·α_2MLuc	7,743	293,685	1,450,274	37.9	187.3
pα_2MLuc	9,207	9,796	14,499	1.1	1.6
pSV232Δ5′Luc	103,247	111,534	122,361	1.1	1.2
pTKLuc	238,884	259,950	293,822	1.1	1.2
pSV2ALuc	19,777,802	18,964,833	23,483,519	1.0	1.2

[a] Light units are given for a standard assay with 200 μg of protein extract per assay.
[b] Relative inductions were defined as the ratios of the luciferase activities measured after treatment with the cytokine divided by those of the corresponding cultures transfected with the same constructs, but mock treated with medium alone (value given in first column). Values are averages of two separately transfected dishes. Each experiment was performed three times. The source of LIF was the supernatant of a CHO cell line secreting 625,000 units/ml of recombinant human LIF into the culture medium (kind gift from Genetics Institute, Cambridge, MA). This supernatant was used at a final dilution of 1:1000. Recombinant human IL-6 from Amgen was purchased from R&D, Inc.

TABLE 3
The Synergistic Effect of IL-6 and Glucocorticoids Is Mediated by the 18-bp IL-6 Response Element (IL-6 RE)

	Promoter (luciferase) activity[a]				Relative increase[b]			
	Untreated	Dex	IL-6	IL-6 + Dex	Dex	IL-6	IL-6 + Dex	Synergistic effect[c]
2xT·SVLuc	703	1,054	620	3,074	1.5	0.9	4.9	4.9
2xmT·SVLuc	2,488	1,402	1,631	1,773	0.6	0.6	0.7	1.1
4xT·SVLuc	1,671	3,161	3,562	15,963	1.9	2.1	9.6	4.5
SV232Δ5′Luc	222	160	242	131	0.7	1.1	0.6	<1
2xT·AMLuc	2,950	2,991	9,810	16,725	1.0	3.4	5.8	1.7
4xT·AMLuc	409	696	6,354	19,231	1.7	15.5	47.0	3.0
AMLuc	1,383	1,294	1,289	1,808	0.9	0.9	1.3	1.4

Note: Plasmid constructs were those described in Figure 11 and Table 2.[36] Constructs were transfected into Hep 3B human hepatoma cells. Cells were subsequently treated with medium alone (mock, untreated), dexamethasone (Dex) alone (2 μm), IL-6 alone, or a combination of IL-6 plus Dex.

[a] Promoter activities were luciferase units in a standard assay with 100 μg of cellular protein extract per assay.
[b] Relative increase was defined as value from treated culture divided by value from corresponding mock-treated (untreated) culture.
[c] Synergistic effect was defined as the ratio of the activities obtained after treatment with IL-6 plus Dex divided by the activity of the corresponding culture treated with IL-6 alone.

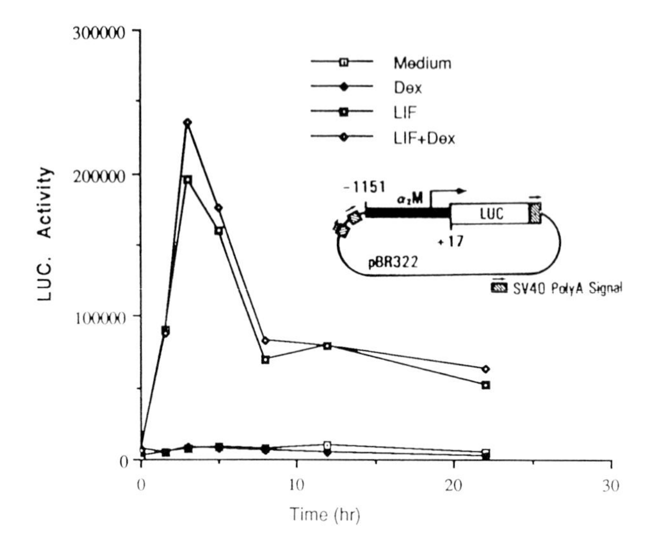

FIGURE 12. The 1151-bp 5′ flanking region of the rat α_2-M gene contains a LIF RE. The stably transfected cell line FAO/HB3 carried the construct shown in the insert as described in Figure 5. These cells were treated with medium alone (dotted squares), dexamethasone alone (closed diamonds), LIF alone (open squares), or LIF plus dexamethasone (open diamonds). Cellular extracts were prepared at various times (hours, abscissa) after addition of the treatment and evaluated for induced promoter activity by measuring luciferase reporter activities (ordinate). Activities are given for 200 μg of cellular protein extract in a standard luciferase assay.

protein interaction with the IL-6 response factor bound at the IL-6 RE in an intact cell, and that this interaction was not detected in our experiments. This would be expected if the interaction between these two proteins involved a domain of the GR that was missing in the GR fragment used in our experiments. This possibility must be considered because a direct protein-protein interaction between the GR and the factor NF-IL-6 bound to its cognate recognition sequence has recently been observed.[77]

Synergism between glucocorticoids and inflammatory cytokines is a long-recognized important feature of inflammatory reactions. The synergism between IL-6 and glucocorticoids is mediated through the 18-bp IL-6 RE sequence, but until now a direct protein-protein interaction between the GR and an IL-6 response factor bound at this element has not been documented.

C. THE LIF RESPONSE ELEMENT OVERLAPS THE IL-6 RESPONSE ELEMENT AND IS PROBABLY IDENTICAL

The cytokine LIF is produced by many of the same cell types that also produce IL-6, and shows striking similarities with IL-6 in its actions on a variety of target cell types and target genes.[43,44] In particular, LIF induces a set of acute phase genes in rat hepatocytes similar to those induced by IL-6.[45,46] Moreover, the LIF receptor has recently been cloned and sequenced, and shown to belong to the same protein superfamily as the IL-6 receptor.[47] In view of these strong similarities in the physiological effects of both cytokines, it has been speculated that they may trigger similar or identical intracellular signal cascades. Therefore, we attempted, to map a LIF RE in the promoter region of the rat α_2-M gene by taking advantage of the existing series of reporter constructs described above. First, we examined whether the cell line FAO/HB3, carrying a stably integrated luciferase construct with the 1151 bp 5′ flanking portion of the gene, responded to LIF. Treatment of this cell line with LIF indeed resulted in an induction of luciferase (promoter) activity, indicating that a LIF RE was contained within the −1151 to +17 bp region of the gene (Figure 12). Next,

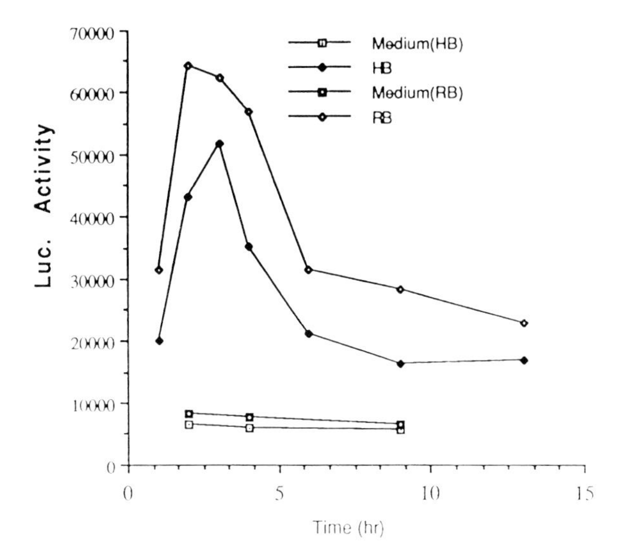

FIGURE 13. LIF RE of the rat α_2-M gene is contained in the 220-bp region 5′ of the promoter. FAO rat hepatoma cells were transiently transfected with the constructs HB, carrying 1151 bp of 5′ flanking sequences (described in Figures 5 and 12), and RB, a deletion variant that carried only 220 bp of 5′ flanking sequences. Cells were then mock treated with medium alone (open and dotted squares) or LIF (open and closed diamonds for constructs HB and RB, respectively), and luciferase reporter activities were measured as described for Figure 12.

transient transfection experiments were performed with the FAO cell line and two plasmid constructs, HB and RB, carrying the regions from −1151 to +17 and −220 to +17 of the gene, respectively. Treatment of cells with LIF after transfection with these constructs led to a very similar activation of the luciferase reporter (Figure 13). Therefore, the location of a LIF RE was confined to the region −220 to +17 bp. This region was further dissected by testing the hypothesis that the LIF RE may be overlapping or identical with the IL-6 RE. The constructs shown in Figure 11 were transfected into Hep G2 and Hep 3B human hepatoma cells, and the cells treated with LIF. For comparison, they were also treated with IL-6 in parallel experiments. Luciferase activities were then determined as a measure of induced promoter activity (Table 2, Figure 14). The result was that the same 18-bp IL-6 RE core element that mediated IL-6 responsiveness also conferred LIF responsiveness to heterologous promoters and the autologous minimal promoter of the α_2-M gene. Four copies of the element produced a 187.3-fold induction of the autologous promoter by IL-6 and a 37.9-fold induction by LIF. Similarly, four copies produced a 9.3-fold induction of the SV40 minimal promoter by IL-6 and a 4.1-fold induction by LIF. Mutation of the central hexanucleotide abolished the inductions by both IL-6 and LIF (Table 2).[32,36,37] Thus, a LIF RE was contained in the 18-bp sequence that also functioned as an IL-6 RE. The essential requirement for the central hexanucleotide for inducibility by both cytokines allows the conclusion that both elements must have overlapped and are possibly identical. We cannot yet make a definitive statement about identity because saturation mutagenesis of each nucleotide of the 18-bp sequence has not yet been performed. Therefore, we concluded that the signal cascades for IL-6 and LIF in Hep G2 cells must converge and utilize this common endpoint.

D. ADDITIVITY OF THE IL-6 AND LIF SIGNALS

Although the IL-6 and LIF receptors belong to the same superfamily of cytokine receptors, they are distinct molecules. Therefore, the two signal cascades for IL-6 and LIF must differ at least in their initial steps. However, both cascades converge at a common endpoint, the IL-6 RE that also functions as a LIF RE, as demonstrated above. Strict identity

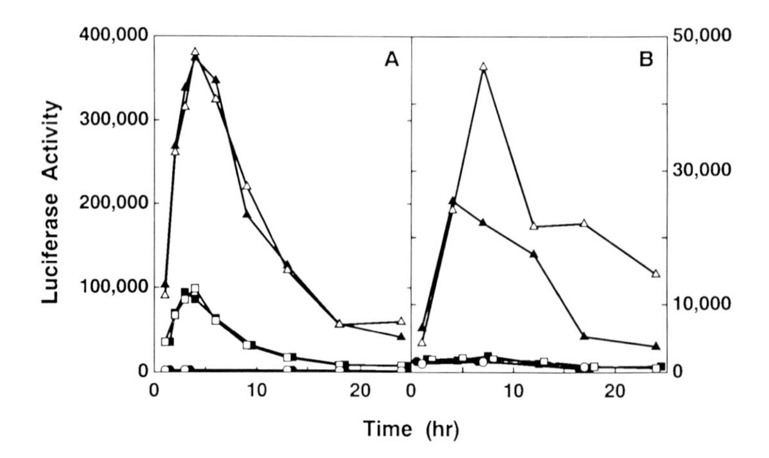

FIGURE 14. Hep G2 human hepatoma cells respond to both IL-6 and LIF; Hep 3B cells respond only to IL-6, and not to LIF. (A) Hep G2 cells were transfected with the construct p4xT.α_2M Luc carrying four tandem copies of the 18-bp core element (Figure 11), and treated with medium alone (black dots), medium plus dexamethasone (open circles), LIF alone (black squares), LIF plus dexamethasone (open squares), IL-6 alone (black triangles), or IL-6 plus dexamethasone (open triangles). (B) Hep 3B cells were treated similarly. At various times after addition of the treatment (hours, abscissa), cellular protein extracts were prepared and assayed for luciferase activity. Activities are given for standard assays with 200 μg of cellular protein extracts per assay.

of the IL-6 RE with the LIF RE sequence has not yet been shown, and therefore it is not known whether both cascades end in the same nuclear protein binding at this sequence, or in two different factors interacting with this element. If both end in the same protein, it may be possible to saturate the common portion of the cascade by treating a cell with saturating doses of one of the cytokines, and thus render it temporarily unresponsive to the other. We have attempted this type of experiment. Dose-response curves for IL-6 and LIF were first separately established for Hep G2 cells transiently transfected with the reporter construct p4xT.α_2M Luc carrying four tandem copies of the 18-bp element (Figure 15A and B). Saturation was observed after treatment for 4h, a previously determined optimal length of treatment for each cytokine. Six hundred units/ml of LIF and 200 BSF-2 units/ml of IL-6 were found to be saturating doses for these cytokines, respectively. Hep G2 cells were then transfected with the same construct and treated with a saturating dose of LIF for 4 h. The cultures then received a second treatment, with either another saturating dose of LIF or a saturating dose of IL-6. A control culture was carried along in parallel that had received only the first treatment. After the second 4-h interval, cellular extracts were prepared and luciferase activities measured. The first treatment with LIF alone produced a 21-fold increase, the cultures treated with LIF plus IL-6 showed a 187-fold increase, and the cultures treated with two consecutive doses of LIF showed a 20-fold increase (Figure 15C). Thus, after saturating the LIF cascade, a full additional response to IL-6 was still obtained, but no further increase in response to a second treatment with LIF, confirming that the LIF cascade had indeed been saturated. A corresponding experiment was also performed with the reverse order of addition. In this case, the transfected cells were first treated with saturating amounts of IL-6 for 4 h, and then with saturating amounts of LIF or a second dose of IL-6 for 4 h. The increase obtained with IL-6 alone was 106-fold, with LIF alone 21-fold, and the combination showed a 121-fold increase. In the control culture after a second dose of IL-6, the total increase was 168-fold. Thus, the two cascades also generated approximately additive signals in this order of addition. The simultaneous addition of saturating doses of both cytokines again created a cumulative total effect of 126-fold. Other authors have performed similar experiments with M1 murine myeloid leukemia cells and obtained comparable re-

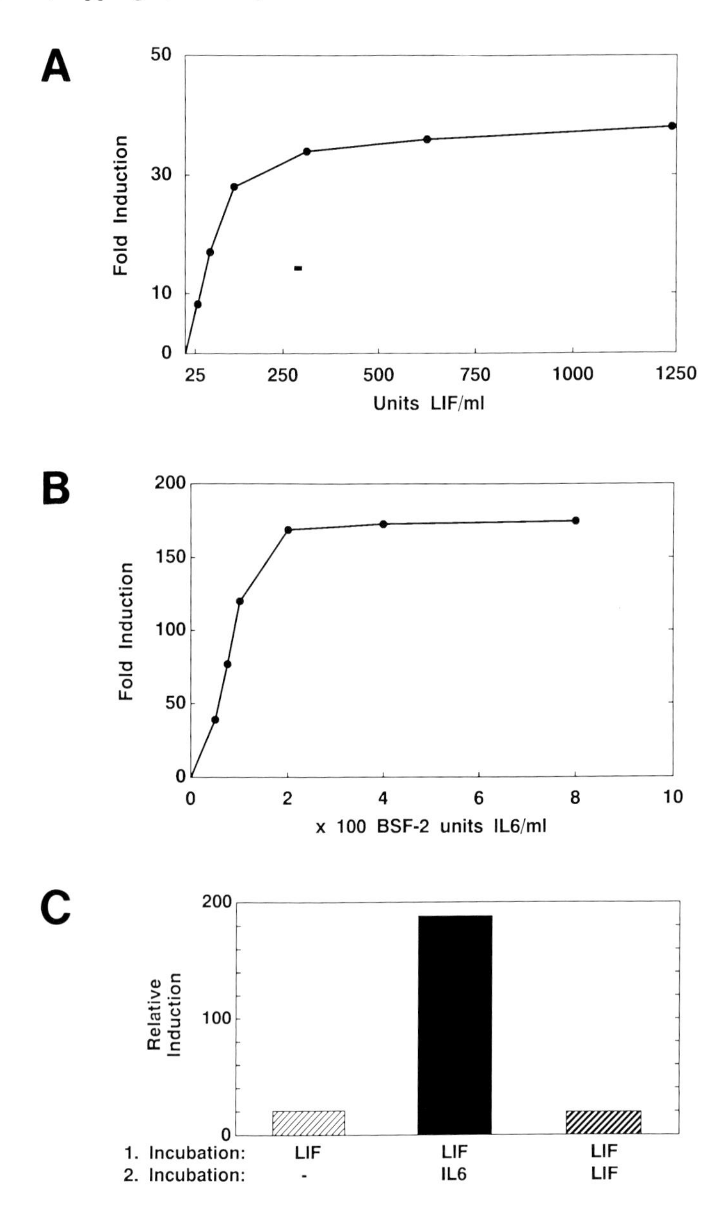

FIGURE 15. The signals for IL-6 and LIF in Hep G2 cells are additive. Dose-response curves were first determined for LIF (A) and IL-6 (B) after transient transfection of the construct p4xTα_2M·Luc into Hep G2 cells. The concentration of LIF generating half maximum induction in FAO/HB3cells had previously been defined as 50 units/ml. The source of recombinant human LIF was a Chinese hamster ovary (CHO) cell line that secreted 625,000 units/ml of LIF (with this definition of LIF units) into the culture medium (kind gift from Genetics Institute, Cambridge, MA). Recombinant human IL-6 was a kind gift of Drs. T. Hirano and T. Kishimoto, Osaka University, Osaka, Japan. BSF-2 units (B-cell stimulatory factor 2 units) are standard IL-6 units as defined by the Osaka group and measured in an antibody production assay. (C) Transfected Hep G2 cells were treated with 625 units/ml of LIF (1:1000 dilution of the supernatant of the CHO cell line from Genetics Institute, that secreted recombinant human LIF) for a first 4-h incubation period. Then one batch received only a mock treatment for a second 4-h period, one batch was given an additional saturating dose (200 BSF-2 units/ml) of IL-6, and a third batch received a second treatment with another saturating dose of LIF. Relative inductions were evaluated as multiples of the activity obtained after mock treatment without cytokines.[54]

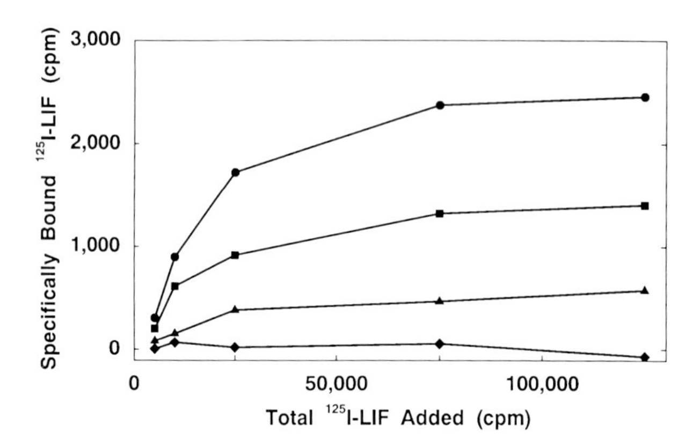

FIGURE 16. Hep 3B human hepatoma cells lack high-affinity binding sites for LIF. Recombinant purified human LIF (Amgen) was purchased from R&D, Inc. and radioiodinated using a Bolton & Hunter kit (Amersham). Ligand-binding studies with Hep G2 (circles), FAO (squares), HeLa (triangles), and Hep 3B cells (diamonds) were performed as described.[18,54] Total ligand bound and ligand bound in the presence of a 200-fold molar excess of nonradioactive LIF competitor were measured separately, and specifically bound ligand was calculated by subtracting the non-specifically bound counts (in the presence of the cold competitor) from the total counts.

sults.[48,49] We concluded that Hep G2 human hepatoma cells were able to handle both signal cascades in an additive fashion. This finding is compatible with two interpretations: either both cascades end in a different protein binding at the IL-6 RE/LIF RE, or both end in the same protein, which is present in nonlimiting amounts. For reasons detailed below (Section III.F), we believe both cascades terminate at the same nuclear factor that binds at this common 18-bp element. If this interpretation is correct, then the saturation of the two cascades observed in Figure 15 must be due to a limiting component different from this nuclear factor. One obvious candidate for a limiting component is the number of high-affinity binding sites per cell for each of these cytokines.

E. CELL-TYPE DISTRIBUTION AND REGULATION OF THE LIF RECEPTOR

Construct p4xT.α_2M Luc, carrying four tandem copies of the 18-bp core element, was transfected in parallel into Hep G2 and Hep 3B human hepatoma cells, and the cells then treated with IL-6 or LIF. Hep G2 cells responded to both IL-6 and LIF, whereas Hep 3B cells responded only to IL-6, but not to LIF (Figure 14). As shown above, both cascades terminate in a common portion which was present in Hep 3B cells, since these cells responded to IL-6. Therefore, the absence of a LIF signal in Hep 3B cells showed that they differed in the part of the LIF cascade that is not shared with the IL-6 cascade, i.e., in the receptor proximal part and possibly the receptor itself. Therefore, the density of high-affinity LIF binding sites was determined for a variety of cell lines by ligand-binding studies with radioiodinated LIF. FAO rat hepatoma cells carried elevated numbers of binding sites, Hep G2 cells intermediate numbers, HeLa cells low numbers, and Hep 3B cells no detectable high-affinity sites (Figure 16). The high-affinity LIF binding site consists of two polypeptide chains — the LIF ligand-binding chain, also referred to as the LIF receptor, and a second associated chain.[47] The absence of high-affinity LIF binding sites on Hep 3B cells explains the observed absence of a functional LIF cascade (Figure 14), and could be due to the absence of the LIF receptor, the associated chain, or both, or to their failure to associate upon arrival of the ligand. We were surprised to find LIF receptors on HeLa cells, an epidermal carcinoma cell line which had not previously been reported as a targert for LIF. Moreover, HeLa cells had been reported to carry very low to vanishing levels of IL-6 receptors.[18,19] Therefore, it became interesting to determine whether HeLa cells carried only the LIF receptor or a complete functional LIF cascade. In the latter case, one would have gained an argument in favor of the proposal that the intracellular portion of the cascade is fairly ubiquitous, and that the cell-type specificity of the IL-6 and LIF responsiveness is determined primarily by the cell-type distribution of high-affinity IL-6 and LIF binding sites. Therefore, the construct p4xT.α_2M Luc was transfected into HeLa cells, and the cells then treated with IL-6 or LIF. A weak, but definite response to LIF was obtained, but no response to IL-6, confirming that HeLa cells carried a complete, functional LIF cascade. The weak response was probably due to the low number of high-affinity binding sites on these cells.[78] Therefore, we surmise that the intracellular portion of the IL-6 and LIF signal cascade is probably present in a large number of different cell types and represents a fairly ubiquitous signaling mechanism.

LIF is a known growth factor for murine embryonal stem cells and established stem cell lines.[50-53] After withdrawal of LIF, the cells stop proliferating and start to differentiate. We were curious to learn whether the density of the high-affinity LIF binding sites was altered during this induction of differentiation. Ligand binding studies with radioiodinated purified recombinant LIF were performed with the murine embryonal stem cell line ES1 at various stages of proliferation (in the presence of LIF) and after induction of differentiation by withdrawal of LIF. Significant densities of high-affinity LIF binding sites were found on rapidly proliferating cells, but after induction of differentiation, the density was reduced approximately tenfold to almost undetectable levels.[54] The observed result was interpreted to indicate two things: (1) the receptor density was not a constant fixture of these cells, but, rather, a variable that changed with the differentiation status of these cells and (2) the fact that the receptor was downregulated could mean that once the cells had engaged on the pathway to differentiation beyond a certain point of no return, this process was essentially irreversible. Beyond that point, there was no more use for the LIF receptor as a growth factor receptor, and it was switched off.

Taken together, these results with human hepatoma, HeLa, and murine ES cell lines indicate that the density of high-affinity binding sites for these cytokines is a highly regulated property. It is subject to more variability than the intracellular trunk of the signal cascade. This makes it tempting to speculate that the receptors should be suitably critical targets for pharmacological attempts to interfere with these signal cascades.

F. A CHARACTERISTIC PROTEIN-DNA COMPLEX WITH THE IL-6 RE IS INDUCED BY IL-6 AND LIF IN THEIR RESPECTIVE TARGET CELL TYPES

A DNA fragment comprising both the IL-6 RE core and core homology elements or a double-stranded oligonucleotide with two tandem copies of the 18-bp core element were used for gel-mobility shift experiments, showing a characteristic protein-DNA complex with several nuclear extracts. This complex, called complex II (Figure 17), was obtained with nuclear extracts from acute phase rat livers, but not with extracts from untreated control rat livers. A complex II of indistinguishable mobility was also obtained with nuclear extracts from IL-6-treated Hep G2 and Hep 3B human hepatoma cells and LIF-treated Hep G2 cells, but not with extracts from LIF-treated Hep 3B cells (Figure 17, tracks 3 to 8). Therefore, the ability to form complex II was a cytokine-induced property of these cells. Complex II was shown to require specifically the presence of the central hexanucleotide of the core element by competition gel-mobility shift experiments. A similar complex II was also obtained with nuclear extracts from IL-6- and LIF-treated FAO rat hepatoma cells (Figure 17, tracks 9 and 10). However, for this cell line, a basal level of complex II was observed even in extracts from untreated cells. This was probably due to the fact that untreated FAO cells produced, secreted, and reinternalized trace amounts of endogenous IL-6 sufficient to trigger this signal cascade and induce the ability to form this complex.[23,25] Induction of complex

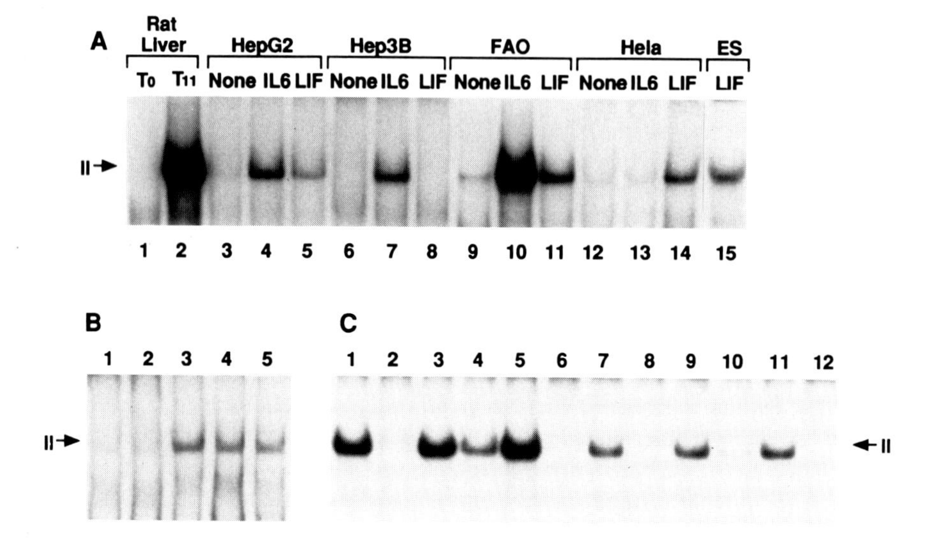

FIGURE 17. A characteristic complex II between the IL-6 RE/LIF RE and nuclear DNA-binding proteins is generated with extracts from various cytokine-treated cells. A synthetic double-stranded oligonucleotide (TB2 probe) containing two tandem copies of the 18-bp IL-6 RE core element was end-labeled with $[^{32}P]\gamma ATP$ and polynucleotide kinase. It was combined with nuclear protein extracts from a variety of cell types, and the protein-DNA complexes were analyzed by gel-mobility shift experiments as described.[32,36,39,54] (A) Extracts from control rat livers and livers of rats undergoing an experimentally induced acute phase response 11 h after triggering the response (T_0, T_{11} extracts; tracks 1, 2); extracts from untreated, IL-6-, and LIF-treated Hep G2 cells (tracks 3 to 5); untreated, IL-6- and LIF-treated Hep 3B cells (tracks 6 to 8); FAO (tracks 9 to 11); HeLa (tracks 12 to 14); and ES1 murine embryonal stem cells rapidly proliferating in the presence of LIF (track 15). The position of the characteristic cytokine-induced complex II is indicated by an arrow; the sequence specificity of this complex for IL-6 RE had previously been established[32,36,39] and is reconfirmed in C. (B) Time course of induction of complex II by LIF in Hep G2 cells. Cells were treated with a saturating dose of recombinant human LIF (625 units/ml; CHO culture supernatant at 1:1000 dilution) for 0, 1.5, 3.5, 8, and 17 h (tracks 1 to 5, respectively). Nuclear protein extracts were then prepared and analyzed for DNA binding as in A. The appearance of the induced complex after 3.5 h correlated well with the kinetics of transcriptional induction of reporter constructs by LIF (Figures 12 to 14). (C) Sequence specificity of complex II induced by IL-6 (tracks 1 to 6) and LIF (tracks 7 to 12) in Hep G2 cells. Gel-mobility shift competition experiment. The cells were treated for 4 h with a saturating dose of IL-6 or LIF, and nuclear protein extracts prepared. Constant amounts of extracts were then allowed to react with the TB2 probe in the absence (tracks 1, 7) or presence of a 100-fold molar excess of unlabeled competitor oligonucleotides. Tracks 2, 8: competitor oligonucleotide TB1, representing the core site; tracks 3, 9: competitor oligonucleotide mTB1 with a mutated core site that abolished its activity (same mutation as that in linker scan mutant m205, Figure 7); tracks 4, 10: competitor oligonucleotide TB3, representing the core homology site; tracks 5, 11: competitor oligonucleotide mTB3 with a mutated core homology site that abolished its function (same mutation as in linker scan mutant m203, Figure 7); tracks 6, 12: competitor oligonucleotide CA1, a 42-bp oligonucleotide that contained one core element and one core homology element, as in their natural configuration in the α_2-M gene.[36,39]

II in rat and human hepatoma cells by IL-6 was prevented by pretreatment of these cells with cycloheximide, and thus required intermediate protein synthesis.[36] It is not clear whether this intermediate synthesis was required to produce more of the DNA binding protein itself or a labile enzymatic activity that altered a preexisting protein, allowing it to form complex II, or both. It has been reported by other authors that a phosphorylation event was required for nuclear proteins to gain the capacity to form complex II, and that this ability was abolished by phosphatase treatment of nuclear extracts.[34] However, these experiments were performed with only partially purified nuclear extracts, and therefore do not provide definitive proof for the contention that the essential phosphorylation/dephosphorylation events occurred on the DNA binding protein itself rather than some other protein that influenced the ability to form complex II. Complex II was also seen in nuclear extracts from HeLa cells after treatment with LIF, but not after treatment with IL-6 (Figure 17, tracks 12 to 14). Finally, it was formed with nuclear extracts from ES1 murine embryonal stem cells grown in the presence of LIF, and it disappeared after withdrawal of LIF and induction of differentiation of these cells (data not shown).[54] Therefore, the presence or absence of this complex showed a complete correlation with the presence or absence of a functional IL-6 or LIF signal cascade in all of these cell types, as determined by the aforementioned functional assays and ligand binding studies. Moreover, the kinetics of induction of this complex in Hep G2 cells by IL-6 and LIF showed a tight temporal correlation with the kinetics of transcriptional activation of transfected promoter constructs by these cytokines (Figures 6, 12 to 14, and 17B). Therefore, we propose that this complex is not only a secondary consequence or a passive indicator of a successful IL-6/LIF signal cascade, but that it probably occurs *in vivo* and participates as an active element in mediating this transcriptional activation. Our data so far do not provide definitive proof for a functional role for complex II and its protein constituents as transcription factors. This must await cloning of the factor(s) involved in complex II, their availability as purified recombinant proteins, and their test in transcription assays. In addition, it is also not yet clear that complex II from rat and human liver cells has the same molecular composition, and that this composition is the same as that of complex II from HeLa cells and ES1 murine embryonal stem cells. However, we have performed two-dimensional gel-mobility shift experiments with complex II from rat livers, human Hep G2 cells, and ES1 cells.[36] In these studies, a regular first-dimension gel mobility shift experiment was first carried out. Subsequently, complex II was UV irradiated in the gel to cross-link the protein covalently to the DNA, and the covalent complex was then analyzed on a second-dimension SDS polyacrylamide gel. In this type of experiment, the major protein component contained in complex II had an apparent molecular weight close to 200 kDa. When only a monomeric DNA target was used (i.e., only one copy of the 18-bp IL-6 RE core element), a corresponding, sequence-specific, cytokine-induced complex III was observed (data not shown). The protein contained in complex III had an apparent molecular weight of 102 kDa in this type of two-dimensional analysis.[36] Therefore, our currently preferred hypothesis is that complex II consists of two copies of this 102-kDa protein, one each bound at each of the two subelements of the dimeric DNA probe. If this interpretation is correct, an equivalent protein with an approximate molecular weight of 100 kDa was induced in all of these cell types, including primary rat liver, rat and human hepatoma cell lines, and murine ES1 embryonal stem cells. We therefore presume that this protein constitutes the endpoint of a fairly universal signal cascade that responds to both IL-6 and LIF in a large variety of their target cell types. Our current efforts are directed at further purifying, cloning, and sequencing this protein.

IV. DISCUSSION

The studies described above have identified a novel IL-6 and LIF RE used in a subset of rat liver acute phase genes. However, this element is not the only *cis* element that mediates IL-6 and LIF responsiveness of liver acute phase genes. As shown in Table 1, the acute phase genes can be divided into at least two categories, according to their principal inducers. The prototype human acute phase genes, the C-reactive protein (CRP), hemopexin (Hx), and haptoglobin (Hp) genes, belong to class 1. The prototype rat genes, including the α_2-M gene, are members of class 2 (Table 1).[2,39] The IL-6 RE of class 1 genes has previously been mapped by other authors.[39,55-65] The consensus sequence (Figure 18) differs entirely from the sequence of the element contained in class 2 genes.[65] Therefore, these elements were called type I and type II elements. Type I elements are binding sites for the nuclear factor NF-IL-6.[1] This factor belongs to the C/EBP family. The prototype factor of this family is C/EBP, a key determinant of adipogenesis and an important regulator of many liver

Type I	Hpx A	-117 to -108	A G T G A T G T A A
	Hp A	-167 to -158	T G T G A A G C A A
	Hp C	-60 to -68	A T T T C G T A A
	CRPα	-56 to -47	A G T G G C G C A A
		Consensus	a g T $^{T}_{G}$ N N G Y A A $^{T}_{G}$
Type II	α_2M (Rat)		T C T G G G A A
	γFbg		
	α_1AGP		
	T_2KG (α_1CPI)		

FIGURE 18. Two types of IL-6 REs. Type I REs as defined by the α-site of the human C-reactive protein gene (CRPα), the A-site of the hemopexin gene (HpxA), and the A- and C-sites of the haptoglobin gene.[55-64] The consensus sequence as defined by Poli and co-workers[65] represents a binding site for the nuclear factor NF-IL-6.[1] Nucleotides in contact with the bound factor are marked by dots. The type II element is, as defined here, for the major class 2 rat acute phase genes. Only the conserved central hexanucleotide and a few surrounding nucleotides are shown. This is a binding site for an unknown factor that is distinct from NF-IL-6, NF-κB, IL-6 DBP/LAP, and DBP in its binding sequence, molecular weight, and immunological reactivities (see text). The factor has not yet been cloned, nor has its consensus binding sequence been determined.

genes.[66] This factor is one of the first-known members of the family of leucine zipper proteins. A characteristic of these proteins is an interactive domain, the leucine zipper, that allows them to form dimers with other factors of the same family. Examples are the factors *jun* and *fos* that form a heterodimer through their leucine zipper, called AP1. This important complex mediates the response of many genes to phorbol esters. Similarly, C/EBP can form homodimers with itself or heterodimers with other members of the family. This ability to form heterodimers is one of the characteristic properties of factors of this family. It potentiates the regulatory possibilities of these factors. The rat equivalent of human NF-IL-6 is called IL-6DBP (IL-6 DNA-binding protein)[67] or LAP (liver activator protein).[68] Another member of this family in rats is the factor DBP,[69] and additional members have been described in mice that participate in the regulation of immunoglobulin genes and others.[70] Interestingly, NF-IL-6/IL-6DBP/LAP also has been reported to have the ability to form homodimers with itself and heterodimers with C/EBP, and possibly with other members of this family. In addition, NF-IL-6 was recently found to engage in protein-protein interactions with the factor NF-κB, a factor involved in the regulation of many inflammation-controlled genes,[71] and with the glucocorticoid receptor.

The factor binding at the type II element of the α_2-M gene is clearly different from NF-IL-6. First, its DNA binding sequence differs entirely from the consensus binding site for NF-IL-6 (Figure 18). Second, we have shown in direct binding studies with recombinant LAP and DBP that neither was able to bind at the type II IL-6 RE of the α_2-M gene.[39] In addition, no complex II was assembled with acute phase rat liver extracts over a consensus binding site for DBP and LAP, the D-site of the mouse albumin gene.[68,69] Antisera against recombinant LAP and DBP failed to react with complex II. Finally, the molecular weight of the protein involved in complexes III and II was approximately 102 kDa,[36] and was thus significantly larger than that of NF-IL-6, LAP, and DBP. Similarly, we have accumulated a substantial body of circumstantial evidence showing that the factor forming complex II is also different from NF-κB.[32,39] Final statements about the identity of this factor must await its molecular cloning and the elucidation of its sequence. However, a convincing amount of available evidence clearly suggests that it is an unknown factor.

The IL-6 signal cascade splits into at least two divergent branches inside a liver cell: one branch activates class 1 acute phase genes through NF-IL-6/LAP and type I response

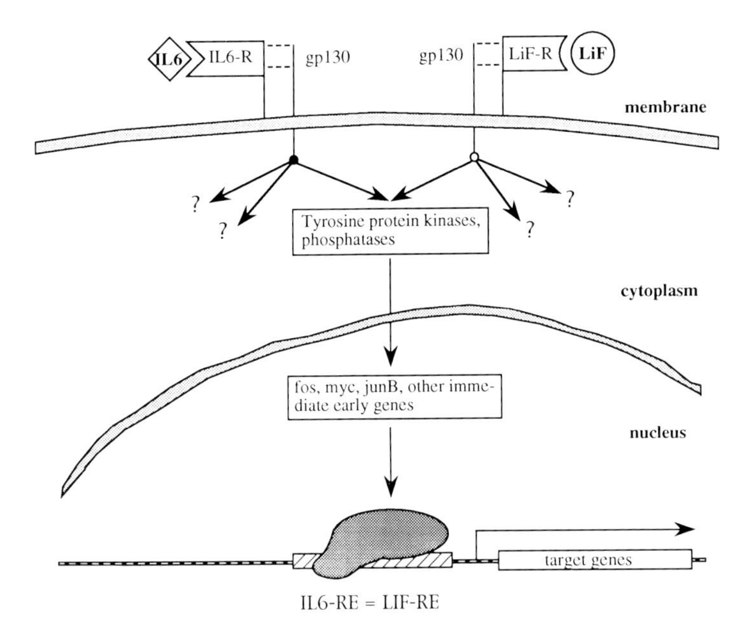

FIGURE 19. The IL-6 and LIF signal pathways converge at the gp130 signal transducer chain. Both the IL-6 and LIF ligand-binding chains can couple to the gp130 signal transducer chain after loading with their respective ligands.The common pathway includes transfer of the signal from gp130 to an as-yet unidentified tyrosine-protein kinase, and possibly additional cytoplasmic steps, proposed to include protein kinases and phosphatases.[48,49,72-74] Immediate early nuclear events include the transcriptional activation of *myc, fos,* and *jun*B.[48,49,73,74] These events and the posttranslational activation of the factor NF-IL-6 occur independently of *de novo* protein synthesis, whereas the subsequent alteration of the nuclear factor that forms complex II with the type II IL-6 RE of class 2 genes requires ongoing protein synthesis.[36] Arrows ending in question marks symbolize additional hypothetical signal pathways specific for each cytokine.

elements; a second branch activates class 2 genes through type II elements and the corresponding novel factor(s). We refer to this phenomenon of multiple parallel signal pathways for one cytokine as ''signal divergence''. Signal divergence is a typical property of many cytokines and has been recognized for other key cytokines in hematopoiesis. It will be interesting to determine whether the factor binding at type II elements is also a member of the C/EBP family, whether it can heterodimerize with other members of this family, and whether it can interact with NF-κB and the GR, analogous to NF-IL-6.

In contrast to signal divergence is the signal convergence of the IL-6 and LIF cascades within one liver cell (Figure 19). This phenomenon was puzzling for a while, and has recently been clarified by the discovery that the signal transducer chain of the LIF receptor is the same gp130 protein as the transducer for the IL-6 receptor. According to this simple scheme, the point of convergence of both cascades is the gp130 chain. This protein is apparently ubiquitous and was found on many different cell types, whereas the distribution of the IL-6 and LIF ligand-binding chains shows far greater cell-type specificity. Therefore, in order to regulate the presence or absence of high-affinity binding sites, it is sufficient to regulate the gene coding for the ligand-binding chain. The next step in signal transduction is believed to be the activation of a tyrosin-protein kinase (TPK) that associates with the cytoplasmic portion of gp130.[72] The molecular identity of this TPK is still unknown and forms the subject of intense ongoing research in a number of laboratories. Only vague knowledge about the further cytoplasmic events of the cascade is available. It has been shown that protein kinases and phosphatases are activated, but their identities are unknown.[48,72-74] A little more is known about the immediate early responses to IL-6 and LIF in the nuclei of hepatoma and leukemia cells. Among the earliest-recognized events are an activation of transcription of the *fos, myc,* and *jun*B genes that occurs within minutes after binding of the ligand to the receptor and in the absence of *de novo* protein synthesis.[48,49,73,74] We suspect that the changes in nuclear proteins leading to the activation of NF-IL-6 and the factor binding at type II elements that forms complex II are mediated through this activation of *fos, myc,* and *jun*B. However, no direct evidence is yet available to support this hypothesis. Activation of NF-IL-6 and the factor(s) forming complex II are also likely to proceed along different pathways because activation of NF-IL-6 does not require intermediate protein synthesis, whereas activation of complex II does require it and was prevented by cyclohex-imide.[36,60,67]

We believe that the rat liver system will facilitate the future analysis of the IL-6 and LIF signal pathways because it allows us to take the "reverse approach" instead of the classical "forward approach". The classical approach starts with the receptor and works forward from the cell surface to the nucleus, by attempting to define the next substrates for each step of the cascade. The problem with this approach is that at each step, there are often multiple next substrates, and thus the pathways branch and fan open like a Christmas tree. It becomes a labor-intensive task to follow all these potential paths in order to identify the few that are actually relevant. In the reverse approach, one starts with the endpoint, e.g., the cytokine-induced modifications of NF-IL-6 or the factor building complex II over type II IL-6 REs. Then one works backwards toward the cell surface and identifies the sequential enzymatic activities responsible for this modification. In this manner, one should theoretically be able to avoid much of the complexity of the forward approach and focus more narrowly on the actually relevant pathway. Rat liver is available in sufficient supplies to determine biochemical details of the individual steps of the signal cascade, and therefore it is reasonable to anticipate that much of the knowledge will first be gained in the rat liver system, and subsequently transferred to human liver cells, lymphoid and hematopoietic cells, embryonal stem cells, myeloid leukemia cells, and others that respond to IL-6 and LIF.

One intriguing question is the following: if the IL-6 and LIF pathways converge at the gp130 transducer, what is the benefit for the cell of carrying separate receptors for both cytokines? Why would one cytokine not have been sufficient to induce the acute phase genes? We do not know the answer to this question. We can only argue in hindsight that since the cell did evolve with both receptors rather than one, there must have been a selective advantage to it. Therefore, in addition to common effects that both cytokines elicit through the common trunk of their pathways mediated by gp130, we believe it is likely that each cytokine sends additional signals to the cell not shared with the other that justify the presence of different receptors. How these cytokine-specific "second signals" are communicated to the cell is unknown. This is symbolically represented in Figure 19 by arrows ending in question marks.

IL-6 and LIF are members of a group of cytokines that also includes IL-11 and oncostatin M (OM), and possibily still other unknown members. Both IL-11 and OM have recently been shown to activate a spectrum of liver acute phase genes similar to that activated by IL-6 and LIF, and to act in a qualitatively similar fashion as IL-6 and LIF. However, IL-6 is the most potent member of this group, with the highest specific activity in inducing the transcription of acute phase genes.[75] It is now known that OM also uses the high-affinity LIF binding site as its receptor, and it is suspected that IL-11 may also use gp130 to transduce its signal, although the IL-11 receptor and its signal transducer have not yet been positively identified. However, it was shown that the same *cis* elements that mediate transcriptional activation of certain liver acute phase genes by IL-6 and LIF also serve as REs for IL-11 and OM.[75,76] The same question comes to mind that was raised above: why do liver cells need the ability to respond to four different cytokines if they all couple to the same signal transducer chain and provoke essentially the same intracellular response? Why was one receptor not sufficient? The answer must be sought along lines similar to those described above. We expect that the rat liver system will be a significant asset in determining the biochemical basis of all these signal pathways.

Ultimately, we are interested in elucidating the mechanism of terminal differentiation of myeloid leukemia cells by IL-6 and LIF. However, we believe it to be advantageous to first determine the elementary steps in liver cells, and then transfer the knowledge to myeloid leukemia cells, rather than attempt their direct elucidation in myeloid leukemia cells. We expect this expedition to resemble mountain climbing, where the ''direttissima'' is not always the guaranteed route to success.

ACKNOWLEDGMENTS

We thank Drs. Taga and Kishimoto for sharing unpubished data about NF-IL-6 and Drs. Schibler, Yamamoto, Miner, and Groner for reagents and valuable discussion. The generous gift of a cell line secreting recombinant human LIF from Genetics Institute, Cambridge, MA, is gratefully acknowledged. This work was supported by grants AI 23351 and AI 22166 from the National Institutes of Health to G.H.F. and a fellowship from Deutsche Forschungsgemeinschaft to G.M.H. J.R. was the recipient of partial support from the Boehringer Ingelheim Foundation and a graduate fellowship from the state of Bavaria. We thank Helga Vieten for expert assistance with the production of the manuscript.

REFERENCES

1. **Akira, S., Isshiki, H., Sugita, T., Tanabe, O., Kinoshita, S., Nishio, Y., Nakajima, T., Hirano, T., and Kishimoto, T.,** A nuclear factor for IL6 expression (NF-IL6) is a member of the C/EBP family, *EMBO J.*, 9, 1897, 1990.
2. **Baumann, H. and Gauldie, J.,** Regulation of hepatic acute phase plasma protein genes by hepatocyte stimulating factors and other mediators of inflammation, *Mol. Biol. Med.*, 7, 147, 1990.
3. **Hocke, G., Baffet, G., Cui, M.-Z., Brechner, T., Barry, D., Goel, A., Fletcher, R., Abney, C., Hattori, M., and Fey, G. H.,** Transcriptional control of liver acute phase genes by interleukin-6 and leukemia inhibitory factor, in *Molecular Aspects of Inflammation, 42nd Mosbach Colloquium,* Sies, H., Flohe, L., and Zimmer, G., Eds., Springer-Verlag, Berlin, 1991, 149.
4. **Gordon, A. H. and Koj, A.,** The acute phase response to injury and infection, in *Research Monographs in Cell and Tissue Physiology,* Vol. 10, Elsevier/North-Holland, Amsterdam, 1985.
5. **Fey, G. H. and Fuller, G. M.,** Regulation of acute phase gene expression by inflammatory mediators, *Mol. Biol. Med.*, 4, 323, 1987.
6. **Fey, G. H. and Gauldie, J.,** The acute phase response of the liver in inflammation, in *Progress in Liver Disease,* Vol. 9, Popper, H. and Schaffner, F., Eds., W. B. Saunders, Philadelphia, 1990, 89.
7. **Lonberg-Holm, K., Reed, D. L., Roberts, R. C., Herbert, R. R., Hillmann, M. C., and Kutney, R. M.,** Three high molecular weight proteinase inhibitors of rat plasma, *J. Biol. Chem.*, 262, 438, 1986.
8. **Eggertsen, G., Hudson, G., Shiels, B., Reed, D., and Fey, G. H.,** Sequence of rat α_1-macroglobulin, a broad-range proteinase inhibitor from the α-macroglobulin-complement family, *Mol. Biol. Med.*, 8, 287, 1991.
9. **Gehring, M. R., Shiels, B. R., Northemann, W., de Bruijn, M. H. L., Kan, C.-C., Chain, A. C., Noonan, D., and Fey, G. H.,** Sequence of rat liver α_2-macroglobulin and acute phase control of its messenger RNA, *J. Biol. Chem.*, 262, 446, 1987.
10. **Northemann, W., Heisig, M., Kunz, D., and Heinrich, P. C.,** Molecular cloning of cDNA for α_2-macroglobulin and measurements of its transcription during experimental inflammation, *Eur. J. Biochem.*, 137, 257, 1985.
11. **Hayashida, K., Okubo, H., Noguchi, M., Yashida, H., Kanagawa, K., Matsuo, H., and Sakaki, Y.,** Molecular cloning of a cDNA complementary to rat α_2-macroglobulin mRNA, *J. Biol. Chem.*, 260, 14224, 1985.

12. **Birch, H. and Schreiber, G.,** Transcriptional regulation of plasma protein synthesis during inflammation, *J. Biol. Chem.*, 261, 8077, 1986.
13. **Schreiber, G.,** Synthesis, processing and secretion of plasma proteins by the liver (and other organs) and their regulation, in *The Plasma Proteins*, Vol. 5, Putnam, F. W., Ed., Academic Press, New York, 1987, chap. 5.
14. **Van Gool, J., Boers, W., Sala, M., and Ladiges, N. C. J. J.,** Glucocorticoids and catecholamines as mediators of acute phase proteins, especially rat α-macrofetoprotein, *Biochem. J.*, 220, 125, 1984.
15. **Yang-Yen, H.-F., Chambard, J.-C., Sun, Y.-L., Smeal, T., Schmidt, T. J., Drouin, J., and Karin, M.,** Transcriptional interference between c-jun and the glucocorticoid receptor: mutual inhibition of DNA binding due to direct protein-protein interaction, *Cell*, 67, 1205, 1990.
16. **Andus, T., Geiger, T., Hirano, T., and Heinrich, P. C.,** Action of recombinant human interleukin 6, interleukin 1β, and tumor necrosis factor α on the mRNA induction of acute phase proteins, *Eur. J. Immunol.*, 18, 739, 1988.
17. **Gross, V., Andus, T., Tran-Thi, T. A., and Heinrich, P. C.,** Induction of acute phase proteins by dexamethasone in primary rat hepatocyte cultures, *Exp. Cell Res.*, 151, 46, 1984.
18. **Taga, T., Kawanishi, Y., Hardt, R. R., Hirano, T., and Kishimoto, T.,** Receptors for B-cell stimulatory factor 2, *J. Exp. Med.*, 166, 967, 1987.
19. **Yamasaki, K., Taga, T., Hirata, Y. U., Yawata, H., Kawanishi, Y., Seed, B., Taniguchi, T., Hirano, T., and Kishimoto, T.,** Cloning and expression of the human interleukin 6 (BSF2/IFNβ2) receptor, *Science*, 241, 825, 1988.
20. **Baumann, M., Baumann, H., and Fey, G. H.,** Molecular cloning, characterization and functional expression of the rat liver interleukin 6 receptor, *J. Biol. Chem.*, 265, 19853, 1990.
21. **Bernuau, D., Legres, L., Lamri, Y., Giuily, N., Abraham, L. J., Fey, G. H., and Feldmann, G.,** Heterogeneous lobular distribution of hepatocytes expressing acute phase genes during the acute inflammatory reaction, *J. Exp. Med.*, 170, 349, 1989.
22. **Gauldie, J., Northemann, W., and Fey, G. H.,** IL6 functions as an exocrine hormone in inflammation: hepatocytes undergoing acute phase responses require exogenous IL6, *J. Immunol.*, 144, 3804, 1990.
23. **Northemann, W., Hattori, M., Baffet, G., Braciak, T. A., Fletcher, R. G., Abraham, L. J., Gauldie, J., Baumann, M., and Fey, G. H.,** Production of interleukin 6 by hepatoma cells, *Mol. Biol. Med.*, 7, 273, 1990.
24. **Baffet, G., Fletcher, R., Cui, M.-Z., Northemann, W., and Fey, G. H.,** Structure of the gene coding for rat leukemia inhibitory factor and its expression in hepatoma cells and macrophages, submitted.
25. **Baffet, G., Braciak, T. A., Fletcher, R. G., Gauldie, J., Fey, G. H., and Northemann, W.,** Autocrine activity of interleukin 6 secreted by hepatocarcinoma cell lines, *Mol. Biol. Med.*, 8, 141, 1991.
26. **Lotz, M., Zuraw, B. L., Carson, D. A., and Jirik, F. R.,** Hepatocytes produce interleukin 6, *Ann. N.Y. Acad. Sci.*, 557, 509, 1989.
27. **Northemann, W., Shiels, B. R., Braciak, T. A., Hanson, R. W., Heinrich, P. C., and Fey, G. H.,** Structure and acute phase regulation of the rat α_2-macroglobulin gene, *Biochemistry*, 27, 9194, 1988.
28. **Kunz, D. R., Zimmermann, R., Heisig, M., and Heinrich, P. C.,** Identification of the promoter sequences involved in the interleukin 6 dependent expression of the rat α_2 macroglobulin gene, *Nucleic Acids Res.*, 17, 1121, 1989.
29. **Tsuchiya, Y., Hattori, M., Hayashida, K., Ishibashi, H., Okubo, H., and Sakaki, Y.,** Sequence analysis of the putative regulatory region of the rat α_2 macroglobulin gene, *Gene*, 57, 73, 1987.
30. **Hattori, M., Kusakabe, S.-I., Ohgusu, H., Tsuchiya, Y., Ito, T., and Sakaki, Y.,** Structure of the rat α_2 macroglobulin-coding gene, *Gene*, 77, 333, 1989.
31. **Hattori, M., Abraham, L. J., and Fey, G. H.,** Identification of an interleukin 6 response element in the rat α_2 macroglobulin gene, *Ann. N.Y. Acad. Sci.*, 557, 499, 1989.
32. **Hattori, M., Abraham, L. J., Northemann, W., and Fey, G. H.,** Acute phase reaction induces a specific complex between hepatic nuclear proteins and the interleukin 6 response element of the rat α_2 macroglobulin gene, *Proc. Natl. Acad. Sci. U.S.A.*, 87, 2364, 1990.
33. **Heinrich, P. C., Castell, J. V., and Andus, T.,** Interleukin 6 and the acute phase response, *Biochem. J.*, 265, 621, 1990.
34. **Ito, T., Tanahashi, H., Misumi, Y., and Sakaki, Y.,** Nuclear factors interacting with an interleukin 6 responsive element of the rat α_2 macroglobulin gene, *Nucleic Acids Res.*, 17, 9425, 1989.
35. **Fey, G. H., Hattori, M., Northemann, W., Abraham, L. J., Baumann, M., Braciak, T. A., Fletcher, R. G., Gauldie, J., Lee, F., and Reymond, M. F.,** Regulation of rat liver acute phase genes by interleukin 6 and production of hepatocyte stimulating factors by rat hepatoma cells, *Ann. N.Y. Acad. Sci.*, 557, 317, 1989.
36. **Hocke, G., Barry, D., and Fey, G. H.,** Synergistic action of interleukin 6 and glucocorticoids is mediated by the interleukin 6 response element of the rat α_2 macroglobulin gene, *Mol. Cell. Biol.*, 12, 2282, 1992.

37. **Hocke, G., Cui, M.-Z., Baffet, G., Fletcher, R., Barry, D., and Fey, G. H.,** Regulation of liver acute phase genes by interleukin 6 and leukemia inhibitory factor, *Cell. Mol. Aspects Cirrhosis,* 216, 49, 1992.
38. **Won, K.-A. and Baumann, H.,** The cytokine response element of the rat α_1 acid glycoprotein gene is a complex of several interacting regulatory sequences, *Mol. Cell. Biol.,* 10, 3965, 1990.
39. **Brechner, T., Hocke, G., Goel, A., and Fey, G. H.,** The interleukin 6 response factor of the rat α_2 macroglobulin gene binds cooperatively at two adjacent sites in the promoter upstream region, *Mol. Biol. Med.,* 8, 267, 1991.
40. **Diamond, M. I., Miner, J. N., Yoshinaga, S. K., and Yamamoto, K. R.,** Transcription factor interactions: selectors of positive or negative regulation from a single DNA element, *Science,* 249, 1266, 1990.
41. **Hoeck, W., Rusconi, S., and Groner, B.,** Down-regulation and phosphorylation of glucocorticoid receptors in cultured cells, *J. Biol. Chem.,* 264, 14396, 1989.
42. **Beaulieu, E. E.,** Contragestion and other clinical applications of RU 486, an antiprogesterone at the receptor, *Science,* 245, 1351, 1989.
43. **Metcalf, D.,** The molecular control of cell division, differentiation commitment and maturation in hemopoietic cells, *Nature (London),* 339, 27, 1989.
44. **Hilton, D. J.,** LIF: lots of interesting functions, *TIBS,* 17, 72, 1992.
45. **Baumann, H. and Wong, G.,** Hepatocyte stimulating factor III shares structural and functional identity with leukemia inhibitory factor, *J. Immunol.,* 143, 1163, 1989.
46. **Baumann, H., Won, K.-A., and Jahreis, G. P.,** Human hepatocyte-stimulating factor III and interleukin 6 are structurally and immunologically distinct but regulate the production of the same acute phase plasma proteins, *J. Biol. Chem.,* 264, 8046, 1989.
47. **Gearing, D. P., Thut, C. J., VandenBos, T., Gimpel, S. D., Delaney, P. B., King, J., Price, V., Cosman, D., and Beckmann, M. P.,** Leukemia inhibitory factor receptor is structurally related to the IL6 signal transducer, gp 130, *EMBO J.,* 10, 2839, 1991.
48. **Lord, K. A., Abdollahi, A., Thomas, S. M., De-Marco, M., Brugge, J. S., Hoffmann-Liebermann, B., and Liebermann, D. A.,** Leukemia inhibitory factor and interleukin 6 trigger the same immediate early response, including tyrosine phosphorylation, upon induction of myeloid leukemia differentiation, *Mol. Cell. Biol.,* 11, 4371, 1991.
49. **Hoffmann-Liebermann, B. and Liebermann, D.,** Interleukin 6 and leukemia inhibitory factor-induced terminal differentiation of myeloid leukemia cells is blocked at an intermediate stage by constitutive *c-myc, Mol. Cell. Biol.,* 11, 2375, 1991.
50. **Smith, A. G., Health, J. K., Donaldson, D. D., Wong, G. G., Moreau, J., Stahl, M., and Rogers, D.,** Inhibition of pluripotential embryonic stem cell differentiation by purified polypeptides, *Nature (London),* 336, 688, 1988.
51. **Williams, R. L., Hilton, D. J., Pease, S., Willson, T. A., Stewart, C. L., Gearing, D. P., Wagner, E. F., Metcalf, D., Nicola, N. A., and Gough, N. M.,** Myeloid leukemia inhibitory factor maintains the development potential of embryonic stem cells, *Nature (London),* 336, 684, 1988.
52. **Pease, S., Braghetta, P., Gearing, D., Grail, D., and Williams, R. L.,** Isolation of embryonic stem (ES) cells in media supplemented with recombinant leukemia inhibitory factor (LIF), *Dev. Biol.,* 141, 344, 1990.
53. **Cooper, H. M., Tamura, R. N., and Quaranta, V.,** The major laminin receptor of mouse embryonic stem cells is a novel form of the $\alpha_6\beta_1$ integrin, *J. Cell. Biol.,* 115, 843, 1991.
54. **Hocke, G., Cui, M.-Z., Barry, D., and Fey, G. H.,** Receptors and nuclear response proteins for leukemia inhibitory factor (LIF) in murine embryonal stem cells, *Mol. Endocrinol.,* submitted.
55. **Arcone, R., Gualandi, G., and Ciliberto, G.,** Identification of sequences responsible for acute phase induction of human C-reactive protein, *Nucleic Acids Res.,* 16, 3195, 1988.
56. **Toniatti, C., Arcone, R., Majello, B., Ganter, U., Arpaia, G., and Ciliberto, G.,** Regulation of the human C-reactive protein gene, a major marker of inflammation and cancer, *Mol. Biol. Med.,* 7, 199, 1990.
57. **Li, S.-P., Liu, T.-Y., and Goldman, N. D.,** Cis-acting elements responsible for interleukin 6 inducible C-reactive protein gene expression, *J. Biol. Chem.,* 265, 4136, 1990.
58. **Oliviero, S., Morrone, G., and Cortese, R.,** The human haptoglobin gene: transcriptional regulation during development and acute phase induction, *EMBO J.,* 6, 1905, 1987.
59. **Oliviero, S. and Cortese, R.,** The human haptoglobin gene promoter: interleukin 6 responsive elements interact with a DNA-binding protein induced by interleukin 6, *EMBO J.,* 8, 1145, 1989.
60. **Poli, V. and Cortese, R.,** Interleukin 6 induces a liver-specific nuclear protein that binds to the promoter of acute phase genes, *Proc. Natl. Acad. Sci. U.S.A.,* 86, 8202, 1989.
61. **Wilson, D. R., Juan, T. S. C., Wilde, M. D., Fey, G. H., and Darlington, G. J.,** A 58 base-pair region of the human C3 gene confers synergistic inducibility by interleukin 1 and interleukin 6, *Mol. Cell. Biol.,* 10, 6181, 1990.

62. **Prowse, K. R. and Baumann, H.,** Hepatocyte stimulating factor, β2-interferon, and interleukin 1 enhance expression of the rat α_1 acid glycoprotein gene via a distal upstream regulatory region, *Mol. Cell. Biol.,* 8, 42, 1988.
63. **Baumann, H., Jahreis, G. P., and Morella, K. K.,** Interaction of cytokine and glucocorticoid-response elements of acute phase plasma protein genes, *J. Biol. Chem.,* 265, 22275, 1990.
64. **Marinkovic, S. and Baumann, H.,** Structure, hormonal regulation and identification of the interleukin 6 and dexamethasone responsive elements of the rat haptoglobin gene, *Mol. Cell. Biol.,* 10, 1573, 1990.
65. **Poli, V., Silengo, L., Altruda, F., and Cortese, R.,** The analysis of the human hemopexin promoter defines a new class of liver-specific genes, *Nucleic Acids Res.,* 17, 9351, 1989.
66. **Landschulz, W. H., Johnson, P. F., Adashi, E. Y., Graves, B. J., and McKnight, S. L.,** Isolation of a recombinant copy of the gene encoding C/EBP, *Genes Dev.,* 2, 786, 1988.
67. **Poli, V., Mancini, F. P., and Cortese, R.,** IL6-DBP, a nuclear protein involved in interleukin 6 signal transduction, defines a new family of leucine zipper proteins related to C/EBP, *Cell,* 63, 643, 1990.
68. **Descombes, P., Chojkier, M., Lichtsteiner, S., Falvey, E., and Schibler, U.,** LAP, a novel member of the C/EBP family, encodes a liver-enriched transcriptional activator protein, *Genes Dev.,* 4, 1541, 1990.
69. **Mueller, C. R., Maire, P., and Schibler, U.,** DBP, a liver-enriched transcriptional activator, is expressed late in ontogeny and its tissue-specificity is determined post-transcriptionally, *Cell,* 61, 279, 1990.
70. **Roman, C., Platero, J. S., Shuman, J., and Calame, K.,** Ig/EGP-1: a ubiquitously expressed immunoglobulin enhancer binding protein that is similar to C/EBP and heterodimerizes with C/EBP, *Genes Dev.,* 4, 1404, 1990.
71. **Lenardo, M. J. and Baltimore, D.,** NF-κB: a pleiotropic mediator of inducible and tissue-specific gene control, *Cell,* 58, 227, 1989.
72. **Hirano, T.,** Interleukin 6 (IL6) and its receptor: their role in plasma cell neoplasia, *Int. J. Cell Cloning,* 9, 166, 1991.
73. **Nakajima, K. and Wall, R.,** Interleukin 6 signals activating junB and TIS11 gene transcription in a B cell hybridoma, *Mol. Cell. Biol.,* 11, 1409, 1991.
74. **Baumann, H. J., Jahreis, G. P., Morella, K. K., Won, K. A., Pruitt, S. C., Jones, V. E., and Prowse, K. R.,** Transcriptional regulation through cytokine and glucocorticoid response elements of rat acute phase plasma protein genes by C/EBP and junB, *J. Biol. Chem.,* 266, 20390, 1991.
75. **Baumann, H. and Schendel, P.,** Interleukin 11 regulates the hepatic expression of the same plasma protein genes as interleukin 6, *J. Biol. Chem.,* 266, 20424, 1991.
76. **Richards, C. D., Brown, T. J., Shoyab, M., Baumann, H., and Gauldie, J.,** Recombinant oncostatin M stimulates the production of acute phase proteins in HepG2 cells and primary rat hepatocytes in vitro, *J. Immunol.,* 148, 1731, 1992.
77. **Taga, T., Kishimoto, T., et al.,** personal communication.
78. **Cui, M.-Z. and Fey, G.,** unpublished data.

Chapter 27

REGULATION OF THE RAT THIOSTATIN GENE

Timothy J. Cole and Gerhard Schreiber

TABLE OF CONTENTS

0-8493-6913-4/93/$0.00 + $.50

I. INTRODUCTION

The acute phase response to injury and infection is a carefully orchestrated series of systemic and local changes intended to restore homeostasis.[1,2] Systemic changes include fever and an increase in the concentration of leukocytes in the blood. Locally, there is an activation of the complement and clotting cascades. In the liver during the acute phase response, the pattern of rates of expression of the plasma protein genes is dramatically altered.[3,4] Expression in the liver of those genes whose products are important during the acute phase response is induced, and some genes whose protein products are not required are downregulated, indicating a shift in priorities for gene expression.

Some plasma protein genes, such as the albumin-, transthyretin-, and retinol-binding protein genes, encode negative acute phase proteins.[4] The transthyretin mRNA level decreases in the liver to 25% of normal (see Chapter 2 of this volume) 36 h after induction of an inflammation, and at this time, transcription of the gene is almost undetectable.[5,6] Other genes, such as those for human C-reactive protein, rat α_2-macroglobulin, and rat fibrinogen β-chain (called positive acute phase proteins), are highly activated. Their mRNA levels increase after induction of the acute phase response, due mainly to induction at the level of transcription.[6]

Thiostatin is a prominent positive acute phase protein in the rat[4,7] (see Chapter 13 of this volume). A strong increase in the concentration of protein(s) in the α_1-globulin region of plasma was first detected in the rat after trauma by Darcy.[8] The physiochemical properties of one particular α_1-globulin, thiostatin, were first described in detail by Urban et al.[7] as the most prominent acute phase protein (then called major acute phase α_1-protein or MAP). The concentration of thiostatin in plasma increases up to 16-fold 48 h after experimentally induced inflammation, due to a 20-fold increase in its synthesis rate by the liver.[7,9] Except for the guinea pig,[10] a thiostatin homolog has not been detected in the plasma of other mammals. The function of thiostatin in the rat remained unclear until it was shown to be immunochemically identical to a cysteine proteinase inhibitor isolated from rat plasma.[11,12] This cysteine proteinase inhibitor occurred in the plasma of Wistar rats as two electrophoretically distinct forms and was also a positive acute phase protein.[12] The cDNAs for thiostatin mRNAs were cloned for Buffalo,[13,14] Wistar,[15] and Sprague-Dawley[16] rats. Two slightly different thiostatin cDNAs were obtained. Analysis of the deduced amino acid sequences indicated similarity to human α_2-thiol proteinase inhibitor[17] and low molecular weight (LMW) kininogens[18] (see Chapter 13 of this volume). Kininogens, precursors for the vasoactive kinins, have been shown to be very similar in amino acid sequence to the plasma cysteine proteinase inhibitors.[19] Also cloned was a rat LMW kininogen,[15] which was approximately 90% homologous in nucleotide sequence to that of thiostatin. The thiostatins were also named T-kininogens because of their similarity to the previously characterized human and bovine kininogens. Thiostatin/T-kininogen contains the peptide Ile-Ser-bradykinin (called T-kinin), which is released *in vitro* by trypsin and cathepsin D.[20] A T-kininogenase has been purified from the submandibular gland,[21] and on induction of an acute inflammation, limited release of T-kinin has been detected *in vivo*.[22] It is still unclear whether thiostatin/T-kininogen functions as a physiological source of vasoactive kinin *in vivo*. Thiostatin/T-kininogen has been shown to be a strong inhibitor of cysteine proteinases, such as the lysosomal cathepsins and papain.[23] In this chapter, the name thiostatin is used, as it describes the more likely function *in vivo* of this protein as a potent cysteine proteinase inhibitor. This review is restricted to a summary of the structure of the thiostatin genes and their regulation during the acute phase response. For a description of the synthesis and function of thiostatin in the rat during the acute phase response, see Chapter 13 of this volume.

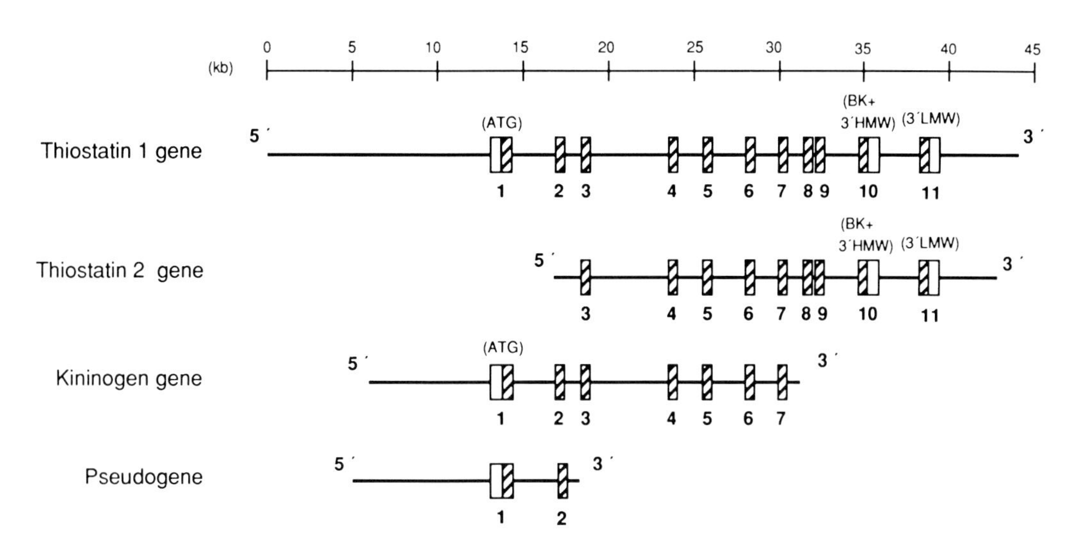

FIGURE 1. Organization of the rat thiostatin and kininogen genes. Four types of genes called thiostatin 1, thiostatin 2, kininogen, and a pseudogene are represented in a 5′-to-3′ arrangement. Coding exons are represented by striped boxes and noncoding parts by open boxes. Introns and flanking regions are represented by thick lines. ATG, codon initiating translation; BK, bradykinin sequence; HMW, high molecular weight domain; LMW, low molecular weight domain. (Modified from Fung, W.-P. and Schreiber, G., *J. Biol. Chem.*, 262, 9298, 1987. With permission.)

II. MATERIALS AND METHODS

All materials and methods are as described in the quoted original references.

III. RESULTS AND DISCUSSION

A. THE RAT THIOSTATIN GENES

Thiostatin cDNAs were independently cloned from liver cDNA libraries prepared from Buffalo,[13,14] Wistar,[15] and Sprague-Dawley[16] rats. There are two distinct thiostatin mRNAs which are 96% identical at the nucleotide level. Both mRNAs encode a protein of 412 amino acids and, depending on the rat strain, these proteins differ in up to 27 amino acids spread throughout their length. These proteins are referred to as thiostatin 1 and thiostatin 2, where thiostatin 1 represents the protein encoded by the cDNA described by Cole et al.[14] Both proteins can be detected in the plasma of rats and their plasma concentrations increase dramatically during the acute phase response.[24]

1. Structural Organization

Thiostatin and related genes have been isolated by a number of groups.[25-27] Two different thiostatin genes, one kininogen gene, and one structurally related pseudogene were obtained by Fung and Schreiber.[25] Their structures are shown schematically in Figure 1. Only the thiostatin 1 gene was isolated in full, and covers 24 kb. It is composed of 11 exons, has a genomic organization similar to the previously isolated bovine and human kininogen genes,[18] and also encodes the so-called HMW- and LMW-specific sequences on separate adjacent exons (10 and 11). Kitagawa et al.[26] isolated part of the two thiostatin genes and kininogen gene corresponding to the regions around exons 10 and 11 (see Figure 1). Recently, the entire gene for a rat T-kininogen (thiostatin) was isolated and characterized,[27] and it showed an organization identical to that of the thiostatin 1 gene.[25] Exon 10 of both thiostatin genes

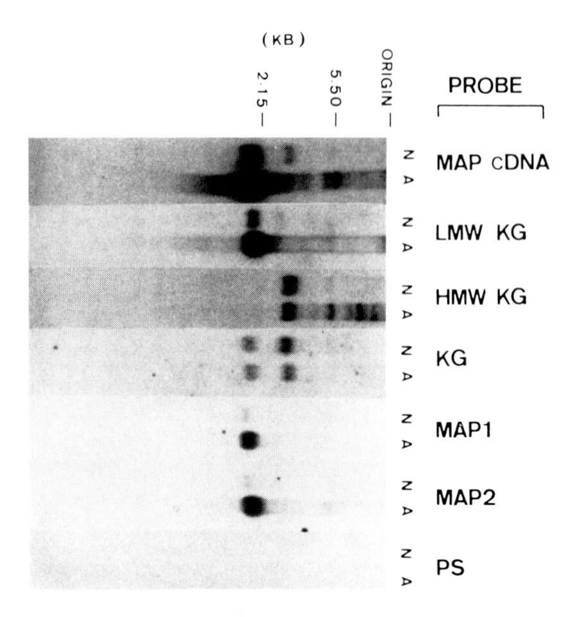

FIGURE 2. Analysis of rat thiostatin and kininogen RNA by Northern-blot hybridization. Polyadenylated RNA, 3 μg per lane, prepared from normal rat liver (N) and from liver 24 h after inducing an acute inflammation (A), were separated by electrophoresis and transferred onto nitrocellulose membranes. Probes used in hybridizations were thiostatin 1 cDNA, DNA fragments encoding the carboxyl termini of LMW and HMW kininogen, and synthetic oligonucleotides with sequences specific for the kininogen (KG), thiostatin 1 (MAP 1), thiostatin 2 (MAP-2), and pseudo (PS) genes. The 28S and 18S ribosomal RNA bands in a total RNA sample (not shown), which are 5.50 and 2.15 kb, respectively, were used as molecular weight markers. (From Fung, W.-P. and Schreiber, G., *J. Biol. Chem.*, 262, 9298, 1987. With permission.)

encodes the unique T-kinin sequence[20] as well as sequences equivalent to the light-chain region of HMW kininogen. It is not yet known whether these four related genes are linked together on the same chromosome in the rat. The human kininogen gene has recently been assigned to chromosome 3.[28]

2. Relationship to the Kininogen Genes

Analysis of cloned parts of the thiostatin and rat kininogen genes indicates that their genomic organization is very similar (Figure 1). Nucleotide sequence analysis of the thiostatin and kininogen genes shows a high similarity of approximately 90%, and it is clear that these genes share an evolutionary origin (discussed below). Functionally, the thiostatin and kininogen genes differ in two important respects. Expression of the thiostatin genes is induced during the acute phase response, whereas the kininogen gene is not.[25] The kininogen gene gives rise to two mRNAs encoding LMW and HMW kininogen through the differential splicing of its two 3′ terminal exons (10 and 11),[18] while only LMW thiostatin mRNA has been detected for both thiostatin genes, even though these genes encode similar HMW-specific sequences in exon 10. This is discussed further in the next section.

B. THE EXPRESSION PATTERN OF THIOSTATINS IN RAT LIVER

Through the use of gene-specific probes, expression of the thiostatin and kininogen genes in liver before and after induction of the acute phase response has been analyzed by Northern blots.[25] This is shown in Figure 2. Using a full-length thiostatin 1 cDNA as a probe, mRNAs of approximately 3.0 and 1.6 kb were detected and found to correspond to both thiostatin and kininogen mRNAs. Only the level of the 1.6 kb mRNA was induced by acute inflammation. An oligonucleotide probe specific for the kininogen gene detected both

LMW and HMW kininogen mRNAs (KG, Figure 2), and their levels of expression were unaffected by induction of the acute phase response. Only the specific thiostatin 1 and 2 mRNAs of 1.6 kb are inducible, but no HMW mRNA form of either thiostatin is detectable. This was also confirmed by Kitagawa et al.[26] using S1 nuclease assays. The thiostatin genes only express a LMW mRNA form.This indicates differences in pre-mRNA splicing between the highly similar thiostatin and kininogen genes. The molecular basis of this has been studied in detail and results from several mutational changes in the HMW-specifying regions of both the thiostatin 1 and 2 genes.[25,26] The nucleotide sequences around exons 10 and 11 for the thiostatin 1 and 2 genes are shown in Figure 3. These mutations to the thiostatin genes include an A-to-C substitution at nucleotide 4 in the putative polyadenylation/processing signal site (AATCAA, double underlined in Figure 3) for the HMW mRNA. There is a single nucleotide deletion in the HMW-specifying region of exon 10 for the thiostatin 2 gene, resulting in a shift in reading frame and premature termination of translation. Kitagawa et al.,[26] by comparison to the rat kininogen gene, showed that the HMW-specifying region of exon 10 for both thiostatin genes also contains the insertion of two type 2 Alu-equivalent sequences.

The mode of mRNA splicing for the thiostatin and kininogen genes is depicted in Figure 4. The kininogen gene is transcribed into both HMW and LMW kininogen mRNA, while the thiostatin genes only give rise to a LMW thiostatin mRNA. Kakizuka et al.[29,30] have suggested an RNA splicing model which proposes that stretches of repeated sequences exist in exon 10 of the kininogen gene that are complementary to the U1 snRNA of the spliceosome. This would allow formation of a stable, nonfunctional spliceosome and allow time for the pre-mRNA to undergo polyadenylation at the end of the HMW-encoding exon sequence and form a HMW-kininogen mRNA. The HMW-specific sequences of exon 10 are therefore not always spliced out. In the case of the thiostatin genes, mutations (described above) could disrupt this proposed interaction with the U1 snRNA and thereby prevent formation of a stable spliceosome. Therefore, a functional processive spliceosome arises and efficiently splices the pre-mRNA into LMW-thiostatin mRNA.

C. REGULATION OF THE THIOSTATIN GENES DURING THE ACUTE PHASE RESPONSE

1. At the Level of Transcription

The increase in the synthesis rates of thiostatins in liver during the acute phase response is paralleled by an increase in their mRNA levels. This is shown in Figure 5 for both thiostatin genes using specific oligonucleotide probes. Thiostatin mRNA levels increase six- to eightfold 36 h after turpentine-induced inflammation. This was also demonstrated by S1 nuclease analysis after lipopolysaccharide-induced inflammation.[31]

Changes in mRNA levels can result from regulation at the level of transcription, mRNA stability, or mRNA degradation. The transcription rate of the thiostatin genes during the acute phase response was analyzed and found to also increase dramatically[6] (Figure 6; α_1-major acute phase protein is the old name for thiostatin). The changes in the transcription rate paralleled the increase in mRNA levels, indicating that the thiostatin genes are primarily regulated at the level of transcription.

Thiostatin mRNA also has a relatively long half-life compared to other acute phase protein genes.[4] Its upper threshold value has been calculated as being approximately 35 h. Not all genes for the acute phase proteins are solely transcriptionally regulated. The transcriptional activities for the rat α_2-macroglobulin and α_1-acid glycoprotein genes after induction are low (Figure 6) and do not account for their changes in mRNA level. In addition, the α_1-acid glycoprotein gene has been shown to be regulated at the level of mRNA stability.[32]

```
AAGCTTGTGAGAAGGCCCTATGGTGGCACAGCTGAGAACGTTTAGCCATAAAGACCTTTAAGGATAGTAACAGCAATGGTTGCAACAGAAATGATATCAAAGTGCATACCCCCGGGATTA   120
                         A              A                                        A

CGTGCTCAATCACCCTGTTTTCTAAATGAAGAAAAACTCTTAGCCTTAGAGGACTATTAAGAATAATCAGGACAGGGCTGAGGCTGTAACTCAGTGAAGGGGTACCTGCTCAACACTTAC   240
                                                            T  G

AAAGTTCAATCCCTTGGACCACAAACAAAGCCTACACATTTCAAGAATAATCAAGGGAGAAGTATTGTTGATTCAGTCTTCTAGGAAGGTAATATAGCATTTGAAATCTGGGAAAGATTC   360
                                     G                                     A T                                       C

                                                                                                         Met Ile Ser Arg
ACGCTGTTGCAATACAGAATAGCATCACCATGTTCCTAGTCTCTTCTTCCTTCACAGACAGCCTGATGCTAACACGATGCATTCCTCTCAATATTTCTGTGTAG ATG ATT TCT AGG   476
                    - -

                                                                                HMW -->
                                        Gln
Pro Pro Gly Phe Ser Pro Phe Arg Leu Val Arg Val Gln Glu Thr Lys Glu Gly Thr Thr Val Ser Pro Pro Tyr Ile Ala Arg Val Gln
CCT CCA GGA TTT TCA CCT TTC CGG CTG GTG CGA GTA CAA GAA ACT AAA GAA GGA ACG ACT GTA AGT CCA CCC TAC ATT GCC AGG GTG CAA   566
                                         A                               A

                            Asp Lys Asp Pro Ser Met Ala Met Ala Gly Phe Met Gln Ser Lys ***
Glu Glu Arg Asp Pro Gly Asn Glu Gln Gly Pro Ile His Gly His Gly Trp Phe His Ala Lys Gln Ile Lys Asp Lys Asn His Gln Gly
GAA GAG AGG GAT CCA GGA AAT GAA CAA GGA CCC ATC CAT GGC CAT GGC TGG TTT CAT GCA AAG CAA ATA AAG GAT AAG AAT CAC CAA GGT   656
                              -

His Lys His Gly His Asp His Gly Cys Trp Pro Pro Lys Arg His Gly Leu Ser Asn Gly His Gln Lys Pro His Gly Leu Gly His Gly
CAT AAG CAT GGG CAT GAC CAT GGC TGT TGG CCC CCA AAG CGG CAT GGT CTT AGC AAT GGA CAC CAG AAA CCA CAT GGC CTT GGT CAT GGA   746

His Lys His Lys Leu Asp Asp Leu Lys Gln Gln Arg Asp Asp Gly Tyr Asn Tyr Arg His Pro Met Gly His Gly His Gly His Gly Leu
CAT AAA CAT AAA CTT GAT GAC CTT AAA CAG CAA AGA GAT GAT GGC TAT AAT TAT AGA CAT CCA ATG GGA CAT GGT CAT GGT CAT GGT CTT   836
                                          G                       C                                           T

Gly His Gly Tyr Gly Gln Gly His Asp                                          His Gly Glu Gly Lys His Thr Asn Lys Asp Lys
GGT CAT GGT TAT GGC CAG GGT CAT GAT ---------------------------------------- CAC GGT GAG GGC-AAA CAC ACA AAT AAA GAC AAA   937
                  T   C       C  T  CACAGTCATGGTCTTGGTCATGGTTATGGCCAGGTCAGGT                G

Asn Asn Val Lys His Thr Asp Gln Arg Ala Glu Ser Leu Thr Ser Ser Ser Glu Asp Ser Thr Thr Ser Thr Gln Ile Gln Gly Arg Thr
AAC AAT GTC AAG CAC ACT GAC CAG AGA GCA GAG TCT TTG ACA AGC TCT TCA GAA GAC AGT ACT ACA TCT ACA CAG ATA CAG GGG AGG ACA   1027
                                            C                   T

Glu Gly Phe Thr Leu Asn Pro Pro Leu Ala Gln Pro Ala Val Ile Ser His Gly Phe Gln Asp Ser Gly Phe Thr Glu Gly Val Ile Ala
GAG GGC TTC ACC TTG AAC CCT CCC CTA GCT CAG CCA GCT GTT ATC TCT CAT GGT TTT CAG GAC TCA GGT TTC ACT GAA GGT GTG ATA GCT   1117
                                                 A               G

Thr Thr Ser Pro Tyr Asp Thr Glu Thr His Asp Val Leu Ile Pro Asp Ile His Val Gln Pro Asp Ser Leu Ser Phe Lys Leu Ile Ser
ACC ACA TCA CCA TAT GAC ACG GAG ACC CAT GAT GTT TTG ATC CCT GAT ATC CAT GTA CAA CCA GAT AGC CTT TCA TTT AAG CTG ATA TCT   1207
                                 A           A

Asp Phe Pro Asp Phe Pro Lys Leu Ser Ser Val Pro Met Glu Phe Ile ***
GAT TTT CCA GAC TTC CCA AAA CTC TCC TCG GTG CCC ATG GAA TTC ATT TAG TAGGAAGGATCCAACCACAGAAACAGCAGAATTTTCTGATTTTGATCTCCTT   1310
                        C     G      G  CA              GC   GG  G
```

FIGURE 3. Nucleotide sequences and the deduced amino acid sequences of the 3′ ends of the rat thiostatin 1 and 2 genes near exons 10 and 11. The nucleotide sequence of the thiostatin 1 gene is shown in full and only those sequences which are different are shown for the thiostatin 2 gene. Hyphens are introduced to optimize the alignment of sequences. The amino acid sequence deduced from the thiostatin 1 gene is shown above the nucleotide sequence. Only differences in amino acid sequence are shown for the thiostatin 2 gene. The beginning of the section coding for the sequence homologous to the carboxyl terminus of HMW kininogen is indicated as HMW and shown by an arrow. The bradykinin moiety is enclosed by a box. The sequences homologous to consensus sequences for RNA splicing sites are underlined. The site of a polyadenylation signal in exon 11 is indicated by a box drawn with a broken line. A putative polyadenylation signal (AATCAA) in exon 10, containing a one-nucleotide change, is double underlined. (From Fung, W.-P. and Schreiber, G., *J. Biol. Chem.*, 262, 9298, 1987. With permission.)

The utilization of transcription start sites or RNA cap sites for the thiostatin and kininogen genes have been well studied. Fung and Schreiber[25] detected, by primer extension analysis, one major transcriptional start site utilized by both the thiostatin 1 (MAP1) and kininogen (KG) genes. This is shown in Figure 7 and comprised at least five nucleotides. The thiostatin-related pseudogene (PS) was transcriptionally inactive. The position of this transcriptional start site is shown in Figure 8 (marked 2), where nucleotide sequences of the 5′portions of the thiostatin 1, thiostatin 2, and kininogen genes are compared. Kageyama et al.[33] subsequently detected two additional transcriptional start sites positioned close to the aforementioned site. These two transcriptional start sites are marked 1 and 3 in Figure 8. Of the three

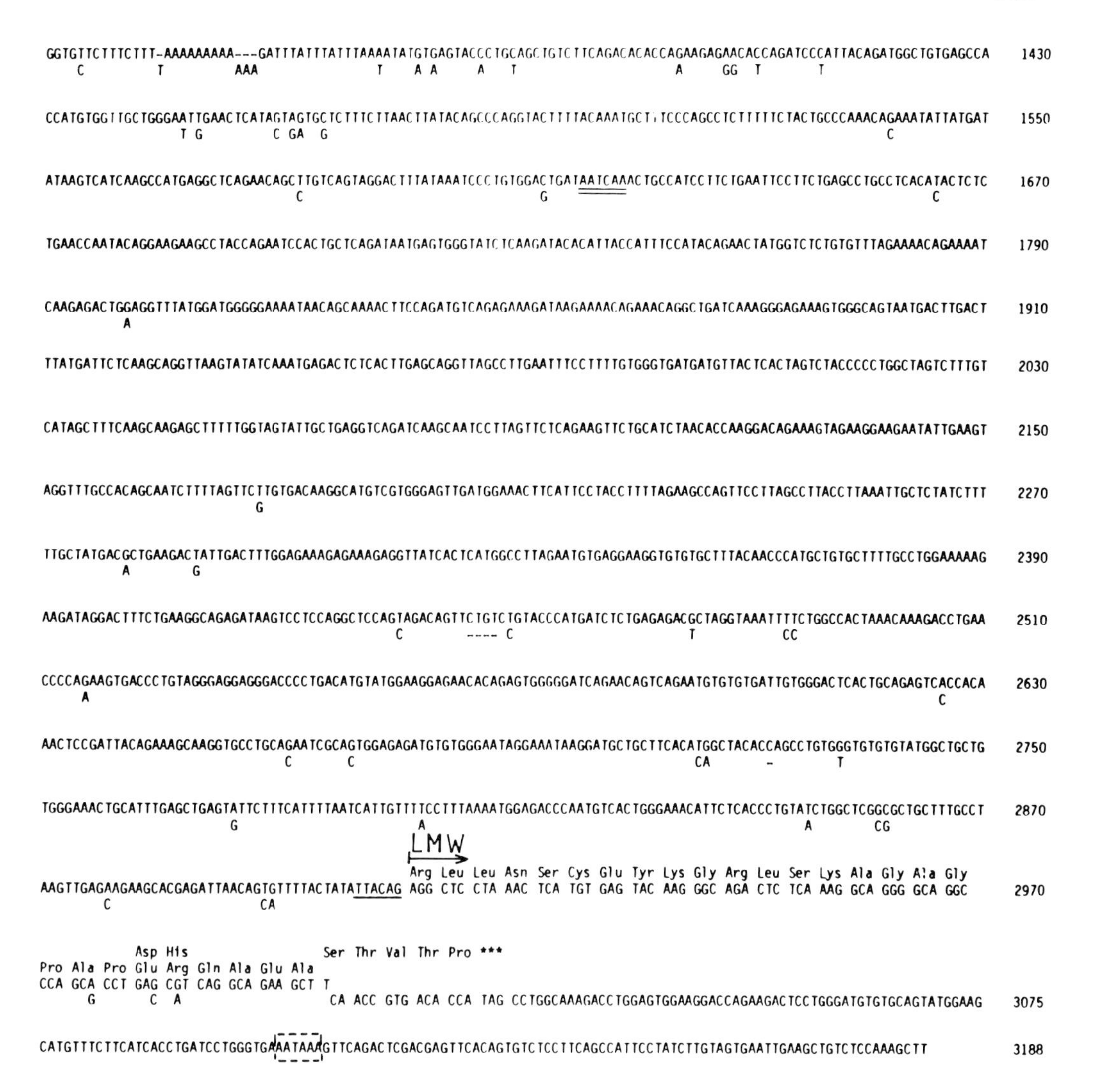

FIGURE 3 (continued).

transcription start sites, the two situated most 3′ (start sites 2 and 3 in Figure 8) are preferentially used by the thiostatin genes during the acute phase response.[33] It is interesting that the 5′ flanking regions of both the thiostatin and kininogen gene promoters do not contain the normal TATA or CAAT box elements usually associated with cell-specific promoters.[34]

2. By Interleukin-6 and Interleukin-1

It is now known that cytokines are released from activated macrophages and monocytes following inflammation and are responsible for induction of acute phase protein synthesis.[35] These cytokines include interleukin-6 (IL-6), interleukin-1 (IL-1), and tumor necrosis factor-α (TNFα). IL-6, a 26-kDa protein also involved in B-cell differentiation and T-cell activation, is now recognized as responsible for the induction of the majority of acute phase proteins, including thiostatin in the rat.[35] IL-6 modulates the expression of the thiostatin genes in a dose- and time-dependent way *in vivo*[36] and in hepatocyte cultures *in vitro*.[37] In these primary hepatocyte cultures, treatment with IL-6 stimulates thiostatin synthesis eightfold.[37] Treatment

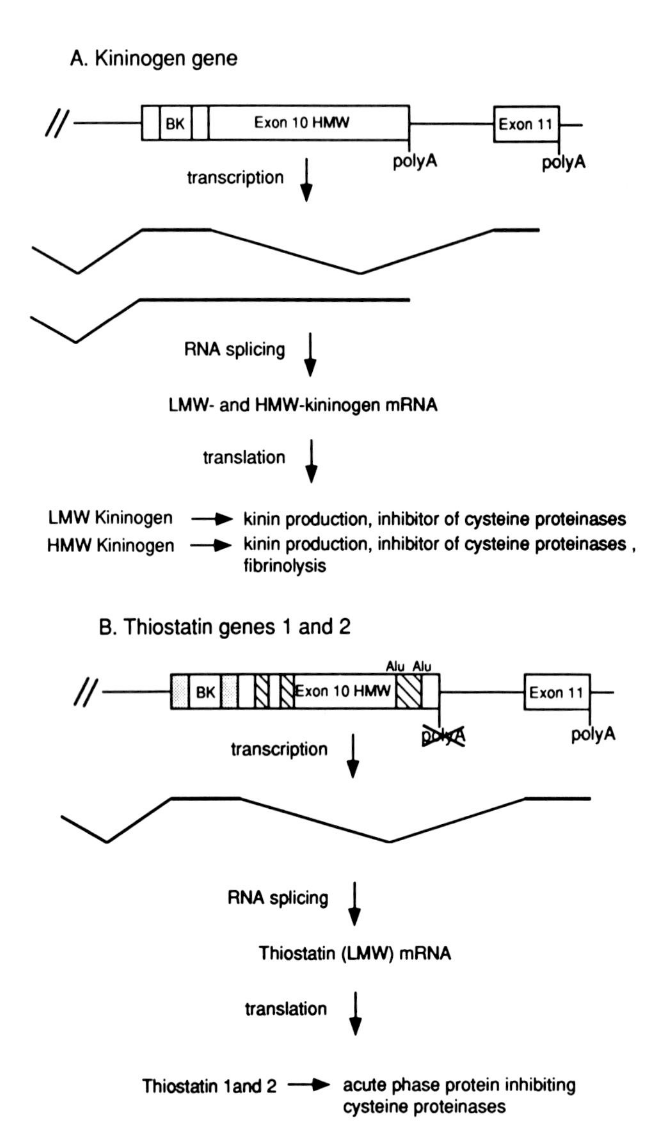

FIGURE 4. Schematic diagram of RNA processing for the rat kininogen (A) and thiostatin (B) genes at the 3′ exons 10 and 11. Exons are boxed. BK indicates the position of the bradykinin moiety. Striped boxes within exon 10 of the thiostatin genes represent insertions. Alu marks the position of alu type 2 repeats. Filled boxes for the thiostatin exon 10 represent changes to kallikrein recognition sequences.

with IL-1 or TNFα shows no effect, but when IL-6 and IL-1 were added together, thiostatin synthesis was severely inhibited. Similar results were obtained using a rat hepatoma H-35 cell line.[38] Thiostatin expression was found to be downregulated *in vivo* by treatment with IL-1.[39]

IL-6 exerts its effect by first binding to a specific cell-surface IL-6 receptor found on most cells. How this signal is transmitted into the nucleus is unclear, but it is now known for a number of acute phase protein genes that IL-6 induces binding of a *trans*-acting factor called IL-6 DNA binding protein (IL-6DBP) to specific response elements upstream of their promoters.[40] The genes in the human for hemopexin, haptoglobin, and C-reactive protein contain *cis*-acting IL-6 response elements (IL-6 REs) upstream of their promoters which are necessary and sufficient for induction of transcription by IL-6.[40-42] The 5′ flanking regions

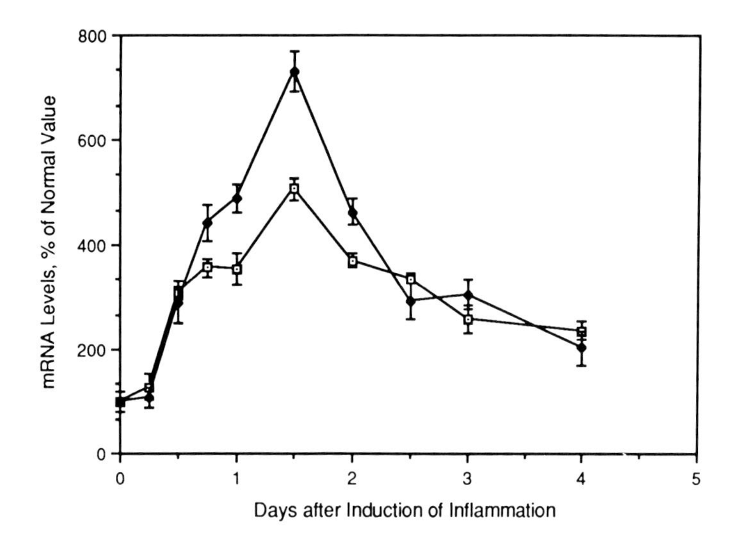

FIGURE 5. Level of thiostatin 1 and 2 mRNA in rat liver during turpentine-induced acute inflammation measured at different lengths of time (shown in days). Cytoplasmic extracts were prepared from livers at various times after induction of inflammation, spotted onto nitrocellulose membranes, and hybridized with ^{32}P-labeled thiostatin 1- and 2-specific oligonucleotides.[25] Messenger RNA levels were calculated as radioactivity bound to filters as a percentage of the value for normal liver. Each point is the mean ± SE (standard error; indicated by bars for each time point) for eight rats. Thiostatin 1 mRNA levels are denoted by open boxes, and thiostatin 2 mRNA levels by closed circles.

of the thiostatin genes and the kininogen gene are compared in Figure 8. They are very similar up to 1 kb upstream of the transcriptional start sites. Transfection experiments using deletion constructs of the 5′ region of the thiostatin 1 gene showed that a 321-bp fragment was sufficient to confer IL-6 responsiveness, even when placed in front of the basic heterologous promoter.[43] Analysis of this fragment revealed the presence of an IL-6-like RE (CTGGAAT) about 230 bp upstream of transcription start site 2 (Figure 8). This sequence is similar to the IL-6 RE consensus, CTGGGAA.[35] The highly homologous 5′ region upstream of the kininogen gene promoter also contains this IL-6 RE, except for a single nucleotide change (see Figure 8), and it is not known whether this substitution blocks induction of gene expression by IL-6 or if other factors are involved in the nonresponsiveness to IL-6 of this gene.

3. By Glucocorticoids

The regulation of thiostatin expression is complicated further by the effect of steroid hormones, such as glucocorticoids. Dexamethasone was shown to be required for the maximal induction of thiostatin by IL-6 in hepatoma cells[44] and has been shown to act synergistically with IL-6 to induce thiostatin expression in transfection experiments.[43] The expression of the endogenous thiostatin gene in a rat hepatoma cell line was also induced five- and twofold by dexamethasone and estradiol, respectively.[45] Transfection experiments using 5′ gene deletion constructs of the thiostatin 1 gene have identified a number of steroid-responsive regions upstream of the thiostatin 1 transcription start sites.[43,45] This region contained a number of consensus sequences to the previously described glucocorticoid (GRE) and estrogen (ERE) response elements.[46,47] Some of these elements are boxed and marked in Figure 8. Anderson and Lingrel[45] described a hormone-responsive region between −167 and −100 nucleotides which contained a closely positioned GRE and ERE (boxed as 1/2GRE and 1/2ERE, Figure 8). Chen et al.[43] have located a second, partially palindromic GRE further upstream (from nucleotide −313, Figure 8) which was located in a 321-bp fragment between

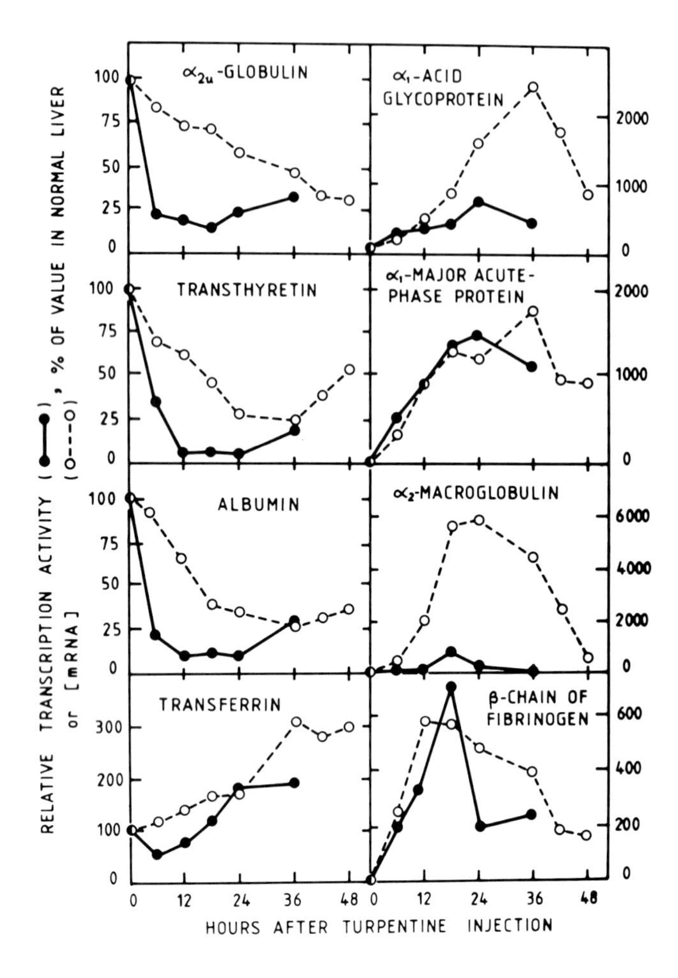

FIGURE 6. Relative transcription activities and mRNA levels for acute phase proteins in liver during experimental inflammation. (From Birch, H. E. and Schreiber, G., *J. Biol. Chem.*, 261, 8077, 1986. With permission.)

−427 and −106. This fragment synergistically conferred both IL-6 (see above) and dexamethasone responsiveness in transfection experiments. Furthermore, it has been reported that the administration of dexamethasone after induction of an acute inflammation will downregulate the expression of both thiostatin genes in the liver by 24%.[48] Thiostatin expression in the liver has also recently been shown to be induced during aging, and this has been suggested as being related to changes in hormonal control.[49]

D. EVOLUTION OF THE THIOSTATIN GENES

From the percent of nucleotide divergences between the rat thiostatin and kininogen genes, one can deduce a scheme for the evolution and divergence of the thiostatin genes[25,26] (Figure 9). The apparent absence of a thiostatin homolog in the mouse[26] and the lower percent of site divergences not under selective pressure between mouse kininogen and rat thiostatin (13 to 14%), compared to other genes between rat and mouse (18%), indicates that the first duplication for the kininogen and thiostatin genes occurred after the divergence of rat and mouse (approximately 15 million years ago). This was followed by the introduction of a variety of mutations to produce the thiostatin gene. These mutations included a 6-bp deletion in the sequences at the N terminus of the bradykinin moiety, thus altering the recognition site for kallikreins. The region of the primordial thiostatin gene encoding the

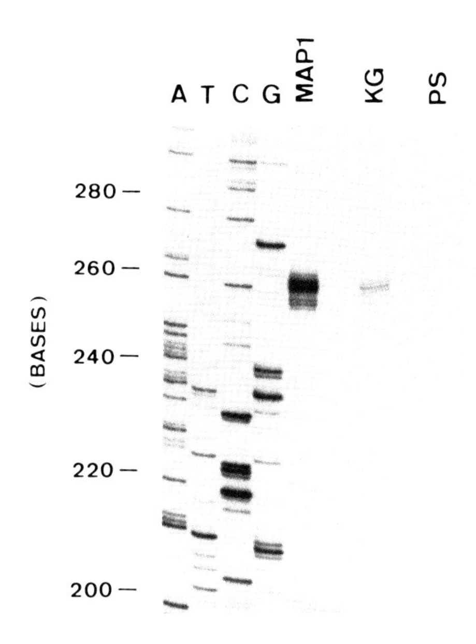

FIGURE 7. Determination of the initiation sites of transcription of the thiostatin 1 gene (MAP1), kininogen gene (KG), and pseudogene (PS). Synthetic oligonucleotides with sequences complementary to the sequences in the translated regions of exon 1 of thiostatin 1 (MAP1), kininogen (KG), and pseudogene (PS) genes were used as primers to anneal with rat liver polyadenylated RNA in primer extension analysis. Nucleotide sequencing of a DNA fragment of known sequence was used as a molecular weight marker. (From Fung, W.-P. and Schreiber, G., *J. Biol. Chem.*, 262, 9298, 1987. With permission.)

HMW domain then underwent further changes, including a nucleotide substitution at the polyadenylation signal sequence and the insertion of type 2 Alu-equivalent sequences. The third important change was the generation of an inducible promoter, bringing expression under the positive control of the acute phase response. This allowed synthesis of a protein in large amounts with antiproteinase activity, but without possibly detrimental hypotensive function brought about by the release of large amounts of kinin. A second gene duplication occurred, followed by additional structural mutations, to produce the two thiostatin genes as they occur today. The presence of a nonfunctional thiostatin PS[25] indicates that a third gene duplication may have occurred. Thus, through a process of gene duplication and gene conversion, which in this case has involved both a loss and a gain of function, the rat has increased its gene diversity and created the thiostatin/kininogen multigene family.

IV. SUMMARY AND PERSPECTIVES

Thiostatin plays an important role in the acute phase response to trauma in the rat as a potent proteinase inhibitor and perhaps as a source of vasoactive kinin. Two very similar thiostatins are detected in plasma, and are expressed in rat liver from two separate, highly similar genes. Thiostatin is one of the major positive acute phase proteins synthesized by rat liver, and expression of both genes is induced during the acute phase response at the level of transcription. Thiostatins are very similar in amino acid sequence to rat LMW kininogen. This is further emphasized in their homologous gene structures, and suggests that the thiostatin and kininogen genes evolved from a common ancestor.

The two thiostatin genes and other acute phase protein genes are under the control of complex regulatory pathways during the acute phase response which involve the cytokines IL-6 and IL-1, and possibly steroid hormones. The most important direct inducer of thiostatin

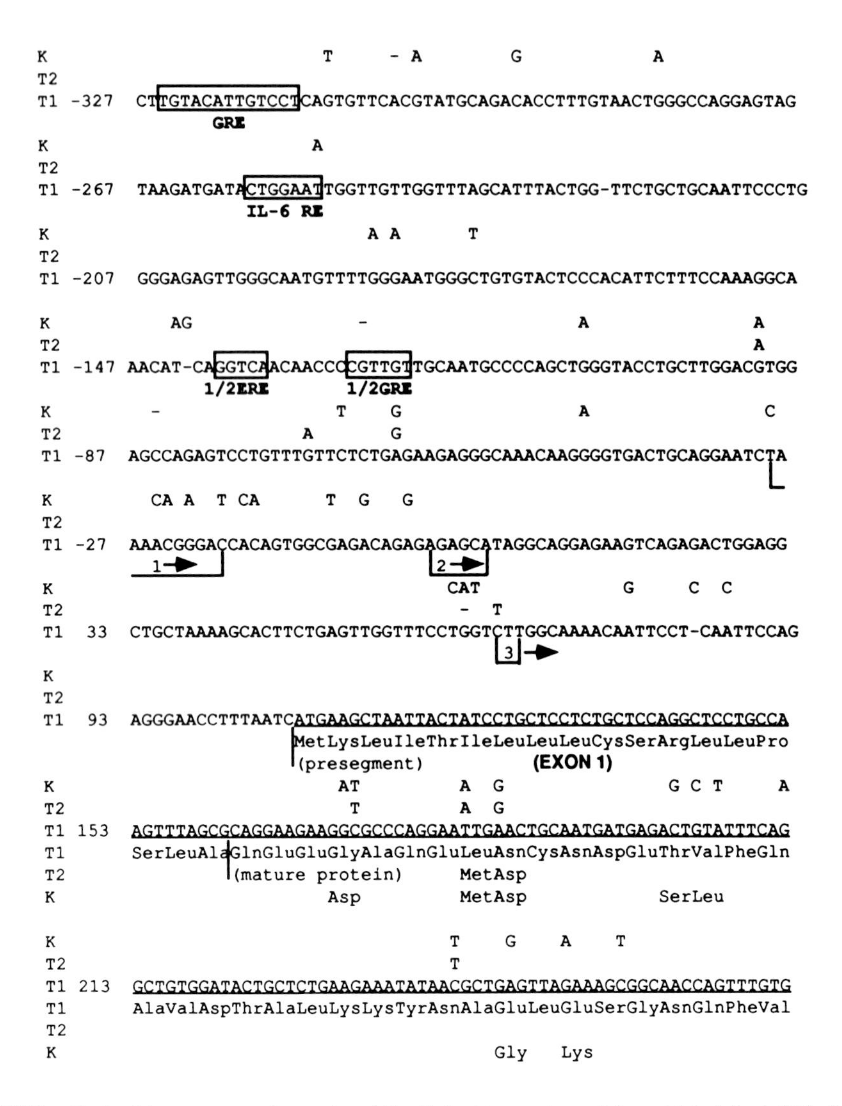

FIGURE 8. Nucleotide sequences of exon 1 and the 5′ flanking regions of the rat thiostatin 1 (T1), thiostatin 2 (T2), and kininogen (K) genes. The nucleotide sequence of the thiostatin 1 gene is shown in full. Nucleotide differences for the T2 and K gene are displayed above this sequence. Hyphens have been introduced in aligning the sequences of the three genes. The deduced amino acid sequence for exon 1 of the thiostatin 1 gene is displayed below the nucleotide sequence. Only different amino acids are displayed for the T2 and K genes. The initiation sites of transcription are indicated by arrows and marked 1, 2, and 3. Putative half- or full-consensus sites for glucocorticoid response elements (GRE), estrogen response elements (ERE), and IL-6 response elements (IL-6 RE), are boxed and labeled. (Modified from Fung, W.-P. and Schreiber, G., *J. Biol. Chem.*, 262, 9298, 1987. With permission.)

gene expression in the liver is IL-6, and this is mediated by IL-6-responsive DNA binding proteins which interact with specific sequences upstream of the thiostatin gene promoter. Many aspects of the mechanisms involved in the regulation of the thiostatin genes in the liver during the acute phase response remain unclear. Further insights will be gained by continuing the analysis on the *trans*-acting protein factors which bind to thiostatin gene regulatory sequences and also by further study on the signal transduction pathway initiated by IL-6 on binding to its cell-surface receptor.

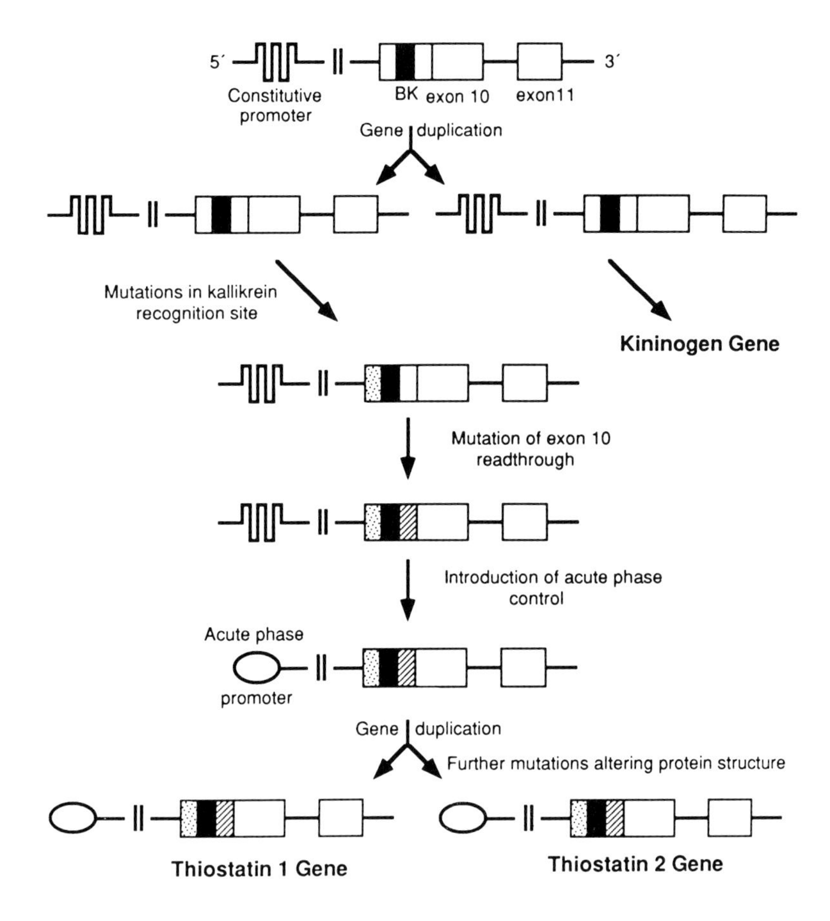

FIGURE 9. Proposed evolutionary scheme of the thiostatin/kininogen gene family in the rat. The gene promoter and exons 10 and 11 are depicted diagrammatically. The filled box represents the bradykinin moiety (BK) encoded in exon 10. Mutations in exon 10 affecting the kallikrein recognition sequence and readthrough of exon 10 are represented by dotted and striped regions, respectively. Two gene duplications are depicted. The first generates the kininogen gene and a primitive thiostatin gene. The thiostatin gene then undergoes a series of mutations, and a second gene duplication then produces the two functional thiostatin genes existing today. (Modified from Schreiber, G., Tsykin, A., Aldred, A. R., Thomas, T., Fung, W.-P., Dickson, P. W., Cole, T., Birch, H., De Jong, F. A., and Milland, J., *Ann. N.Y. Acad. Sci.*, 557, 61, 1989. With permission.)

ACKNOWLEDGMENTS

We thank M. Achen for critical discussion and J. Withers for photographic assistance. This work was supported by grants from the National Health and Medical Research Council of Australia and the Australian Research Council.

REFERENCES

1. **Koj, A.,** *The Acute Phase Response to Injury and Infection,* Vol. 10, Gordon A. H. and Koj, A., Eds., Elsevier/North-Holland, Amsterdam, 1985.
2. **Hurley, J. V.,** *Acute Inflammation,* 2nd ed., Churchill Livingstone, Edinburgh, 1983.
3. **Schreiber, G.,** Synthesis, processing, and secretion of plasma proteins by the liver and other organs and their regulation, in *The Plasma Proteins,* Vol. 5, Putnam, F. W., Ed., Academic Press, New York, 1987, chap. 5.

4. **Schreiber, G., Tsykin, A., Aldred, A. R., Thomas, T., Fung, W.-P., Dickson, P. W., Cole, T., Birch, H., De Jong, F. A., and Milland, J.,** The acute phase response in the rodent, *Ann. N.Y. Acad. Sci.*, 557, 61, 1989.
5. **Dickson, P. W., Howlett, G. J., and Schreiber, G.,** Rat transthyretin (prealbumin); molecular cloning, nucleotide sequence and gene expression in liver and brain, *J. Biol. Chem.*, 260, 8214, 1985.
6. **Birch, H. E. and Schreiber, G.,** Transcriptional regulation of plasma protein synthesis during inflammation, *J. Biol. Chem.*, 261, 8077, 1986.
7. **Urban, J., Chan, D., and Schreiber, G.,** A rat serum glycoprotein whose synthesis rate increases greatly during inflammation, *J. Biol. Chem.*, 254, 10565, 1979.
8. **Darcy, D. A.,** Immunological demonstration of a substance in rat blood associated with tissue growth, *Br. J. Cancer,* 11, 137, 1957.
9. **Schreiber, G., Howlett, G., Nagashima, M., Millership, A., Martin, H., Urban, J., and Kotler, L.,** The acute phase response of plasma protein synthesis during experimental inflammation, *J. Biol. Chem.*, 257, 10271, 1982.
10. **Yoshida, K., Saito, A., and Sinohara, H.,** Isolation and partial characterization of a novel form of low molecular weight kininogen from guinea pig plasma, *Biochem. Int.*, 19, 1339, 1989.
11. **Urban, J., Nagashima, M., and Schreiber, G.,** A hepatic precursor form of the major acute phase α_1 protein in the rat, *Biochem. Int.*, 4, 75, 1982.
12. **Esnard, F. and Gauthier, F.,** Rat α_1-cysteine proteinase inhibitor: an acute phase reactant identical with α_1 acute phase globulin, *J. Biol. Chem.*, 258, 12443, 1983.
13. **Cole, T., Inglis, A., Nagashima, M., and Schreiber, G.,** Major acute phase alpha (1) protein in the rat: structure, molecular cloning, and regulation of mRNA levels, *Biochem. Biophys. Res. Commun.*, 126, 719, 1985.
14. **Cole, T., Inglis, A. J., Roxburgh, C. M., Howlett, G. J., and Schreiber, G.,** Major acute phase α_1-protein of the rat is homologous to bovine kininogen and contains the sequence for bradykinin: its synthesis is regulated at the mRNA level, *FEBS Lett.* 182, 57, 1985.
15. **Furuto-Kato, S., Matsumoto, A., Kitamura, N., and Nakanishi, S.,** Primary structures of the mRNAs encoding the rat precursors for bradykinin and T-kinin, *J. Biol. Chem.*, 260, 12054, 1985.
16. **Anderson, K. P. and Heath, E. C.,** The relationship between rat major acute phase protein and the kininogens, *J. Biol. Chem.*, 260, 12065, 1985.
17. **Ohkubo, I., Kurachi, K.,Takasawa, T., Shiokawa, H., and Sasaki, M.,** Isolation of a human α_2-thiol proteinase inhibitor and its identity with low molecular weight kininogen, *Biochemistry,* 23, 5691, 1984.
18. **Kitamura, N., Kitagawa, H., Fukushima, D., Takagaki, Y., Miyata, T., and Nakanishi, S.,** Structural organization of the human kininogen gene and a model for its evolution, *J. Biol. Chem.*, 260, 8610, 1985.
19. **Müller-Esterl, W., Iwanaga, S., and Nakanishi, S.,** Kininogens revisited, *Trends Biochem. Sci.*, 11, 336, 1986.
20. **Okamoto, H. and Greenbaum, L. M.,** Kininogen substrates for trypsin and cathepsin D in human, rabbit, and rat plasmas, *Life Sci.*, 32, 2007, 1983.
21. **Barlas, A., Gao, X., and Greenbaum, L. M.,** Isolation of a thiol-activated T-kininogenase from the rat submandibular gland, *FEBS Lett.*, 218, 266, 1987.
22. **Barlas, A., Okamoto, H., and Greenbaum, L. M.,** Release of T-kinin and bradykinin in carrageenin-induced inflammation in the rat, *FEBS Lett.*, 190, 268, 1985.
23. **Moreau, T., Gutman, N., El Moujahed, A., Esnard, F., and Gauthier, F.,** Relationship between the cysteine-proteinase-inhibitory function of rat T-kininogen and the release of immunoreactive kinin upon trypsin treatment, *Eur. J. Biochem.*, 159, 341, 1986.
24. **Cole, T. and Schreiber, G.,** Synthesis of thiostatins (major acute phase α_1 proteins) in different strains of *Rattus Norvegicus, Comp. Biochem. Physiol.*, 93B, 813, 1989.
25. **Fung, W.-P. and Schreiber, G.,** Structure and expression of the genes for major acute phase α_1-protein (thiostatin) and kininogen in the rat, *J. Biol. Chem.*, 262, 9298, 1987.
26. **Kitagawa, H., Kitamura, N., Nayashida, H., Miyata, T., and Nakanishi, S.,** Differing expression patterns and evolution of the rat kininogen gene family, *J. Biol. Chem.*, 262, 2190, 1987.
27. **Anderson, K. P., Croyle, M. L., and Lingrel, J. B.,** Primary structure of a gene encoding rat T-kininogen, *Gene,* 81, 119, 1989.
28. **Fong, D., Smith, D. L., and Hsieh, W.-T.,** The human kininogen gene (KNG) mapped to chromosome 3q26-qter by analysis of somatic cell hybrids using the polymerase chain reaction, *Hum. Genet.*, 87, 189, 1991.
29. **Kakizuka, A., Kitamura, N., and Nakanishi, S.,** Localization of DNA sequences governing alternative mRNA production of rat kininogen genes, *J. Biol. Chem.*, 263, 3884, 1988.
30. **Kakizuka, A., Ingi, T., Murai, T., and Nakanishi, S.,** A set of U1 snRNA-complementary sequences involved in governing alternative RNA splicing of the kininogen genes, *J. Biol. Chem.*, 265, 10102, 1990.

31. **Kageyama, R., Kitamura, N., Ohkubo, H., and Nakanishi, S.,** Differential expression of the multiple forms of rat prekininogen mRNAs after acute inflammation, *J. Biol. Chem.*, 260, 12060, 1985.
32. **Shiels, B. R., Northemann, W., Gehring, M. R., and Fey, G. H.,** Modified nuclear processing of α_1-acid glycoprotein RNA during inflammation, *J. Biol. Chem.*, 262, 12826, 1987.
33. **Kageyama, R., Kitamura, N., Ohkubo, H., and Nakanishi, S.,** Differing utilization of homologous transcription initiation sites of rat K and T kininogen genes under inflammation condition, *J. Biol. Chem.*, 262, 2345, 1987.
34. **Breathnach, R. and Chambon, P.,** Organization and expression of eukaryotic split genes coding for proteins, *Annu. Rev. Biochem.*, 50, 349, 1981.
35. **Heinrich, P. C., Castell, J. V., and Andus, T.,** Interleukin-6 and the acute phase response, *Biochem. J.*, 265, 621, 1990.
36. **Geiger, T., Andus, T., Klapproth, J., Hirano, T., Kishimoto, T., and Heinrich, P. C.,** Induction of rat acute phase proteins by interleukin 6 *in vivo*, *Eur. J. Immunol.*, 18, 717, 1988.
37. **Andus, T., Geiger, T., Hirano, T., Kishimoto, T., Tran-Thi, T.-A., Decker, K., and Heinrich, P.,** Regulation of synthesis and secretion of major acute phase proteins by recombinant human interleukin-6 (BSF-2/IL-6) in hepatocyte primary cultures, *Eur. J. Biochem.*, 173, 287, 1988.
38. **Baumann, H., Onorato, V., Gauldie, J., and Jahreis, G. P.,** Distinct sets of acute phase plasma proteins are stimulated by separate human hepatocyte-stimulating factors and monokines in rat hepatoma cells, *J. Biol. Chem.*, 262, 9756, 1987.
39. **De Jong, F. A., Birch, H. E., and Schreiber, G.** Effects of recombinant interleukin-1 on mRNA levels in rat liver, *Inflammation,* 12, 613, 1988.
40. **Poli, V., Mancini, F. P., and Cortese, R.,** IL-6DBP, a nuclear protein involved in interleukin-6 signal transduction, defines a new family of leucine zipper proteins related to C/EBP, *Cell,* 63, 643, 1990.
41. **Poli, V. and Cortese, R.,** Interleukin 6 induces a liver-specific nuclear protein that binds to the promoter of acute phase genes, *Proc. Natl. Acad. Sci. U.S.A.*, 86, 8202, 1989.
42. **Majello, B., Arcone, R., Toniatti, C., and Ciliberto, G.,** Constitutive and IL-6-induced nuclear factors that interact with the human C-reactive protein promoter, *EMBO J.*, 9, 457, 1990.
43. **Chen, H.-M., Considine, K. B., and Liao, W. S. L.,** Interleukin-6 responsiveness and cell-specific expression of the rat kininogen gene, *J. Biol. Chem.*, 266, 2946, 1991.
44. **Baumann, H., Richards, C., and Gauldie, J.,** Interaction among hepatocyte-stimulating factors, interleukin 1, and glucocorticoids for regulation of acute phase plasma proteins in human hepatoma (HepG2) cells, *J. Immunol.*, 139, 4122, 1987.
45. **Anderson, K. P. and Lingrel, J. B.,** Glucocorticoid and estrogen regulation of a rat T-kininogen gene, *Nucleic Acids Res.*, 17, 2835, 1989.
46. **Strähle, U., Kloch, G, and Schütz, G.,** A DNA sequence of 15 base pairs is sufficient to mediate both glucocorticoid and progesterone induction of gene expression, *Proc. Natl. Acad. Sci. U.S.A.*, 84, 7871, 1987.
47. **Walker, P., Germond, J.-E., Brown-Luedi, M., Givel, F., and Wahli, W.,** Sequence homologies in the region preceding the transcription initiation site of the liver estrogen-responsive vitellogenin and apo-VLDLII genes, *Nucleic Acids Res.*, 12, 8611, 1984.
48. **Howard, E. F., Thompson, Y. G., Lapp, C. A., and Greenbaum, L. M.,** Reduction of T-kininogen messenger RNA levels by dexamethasone in the adjuvant-treated rat, *Life Sci.*, 46, 411, 1990.
49. **Sierra, F., Fey, G. H., and Guigoz, Y.,** T-kininogen gene expression is induced during aging, *Mol. Cell. Biol.*, 9, 5610, 1989.

Chapter 28

CYTOKINE REGULATION OF THE MOUSE SAA GENE FAMILY

Jean D. Sipe, Hanna Rokita, and Frederick C. de Beer

TABLE OF CONTENTS

0-8493-6913-4/93/$0.00 + $.50

I. INTRODUCTION

A. MOUSE SAA GENE FAMILY

1. Structure and Number of Members

Inbred strains of mice have been powerful tools in the analysis of serum amyloid A (SAA) synthesis and catabolism in normal host defense and in dysfunctions such as amyloidosis.[1-40] Six inbred strains (C57BL, C3H, BALB/c, DBA/2, CBA, and A) constituted about 70% of the strains used in 1600 wide-ranging research studies reviewed by Festing;[41] with the possible exception of DBA/2, these strains are also most frequently employed to study SAA.

The SAA gene family in mice has been assigned to a 79-kb fragment of chromosome 7.[29,42,43] Cloning, sequencing, and mapping studies of the SAA gene family in BALB/c mice originally identified three active genes and a pseudogene that lacks exons 1, 2, and a 25-bp segment in exon 3, with the net result of an in-frame stop codon.[6,14-16] The products of genes designated SAA1 and SAA2 are secreted from liver and, in close association with the HDL3 subclass of high-density serum lipoproteins (HDLs), are rapidly transported from the plasma compartment to peripheral tissues, where they are catabolized.[17-19,45] Computer modeling predicts a pI of 10.35 for the protein product of gene SAA3, which is transcribed in liver concomitantly with SAA1 and SAA2, but with different kinetics.[6,46] A protein corresponding to the SAA3 gene product has been detected as a minor constituent of HDL fractions from the plasma of the novel CE/J and SAM mouse strains.[77] Meek et al.[47] have described a protein in several mouse tissue that reacts with antibodies to the cloned product of the mouse SAA3 gene.

The active SAA genes are similar to other apolipoproteins in structure: >3000 bp in length, consisting of four exons, and specifying primary translation products with 18-amino acid signal peptide sequences (Figure 1). The pseudogene in BALB/c mice has been designated SAA4.[16] Recently, during a search for the SAA3 gene product among HDL3 proteins in BALB/c mice, a protein with pI 8.1 was identified as a previously undescribed member of the SAA gene family, and was designated SAA5.[46] Unlike apo-SAA1, 2, and 3, apo-SAA5 has amino acid substitutions at positions 33, 36, 42, and 44 in the so-called constant or conserved region between amino acids 33 and 44, a region for which the SAA1 to 4 genes are identical at the nucleotide level.[6,15,16,48-50] Because of differences in structure between SAA5 and the SAA1 to 3 gene products, and because maximal concentrations of apo-SAA5 in plasma appear almost 10 h later than apo-SAA1 and apo-SAA2, it has been suggested that the function of apo-SAA5 may differ from that of apo-SAA1 and 2.[46]

While the complete structural details of the SAA gene superfamily are as yet incomplete, there is evidence for three subfamilies: (1) the apo-SAA1 and 2 subfamily, in which the major members are 96% identical and minor members have been identified, (2) the apo-SAA3 subfamily, which, in mice, exhibits pronounced extrahepatic expression, and (3) the apo-SAA5 subfamily, which contains amino acid substitutions in the region corresponding to the "constant region" of apo-SAA1/2 and 3, and which is expressed later than the former subfamilies.

2. Structural Variation among Inbred and Wild-Derived Mice

Taylor and Rowe surveyed 48 inbred strains and substrains for restriction fragment length polymorphisms (RFLPs) using the restriction endonucleases EcoRI, HindIII, BamHI, and PvuII.[29] An apo-SAA-deficient mutant strain was undetectable by these mapping studies; 33 strains exhibited common restriction fragment patterns with all four enzymes. BALB/c was designated the prototype for this group, called type A. The A/J, CBA/J, C3HeB/FeJ, and C57BL strains, all of which have been used for studies of apo-SAA gene structure and

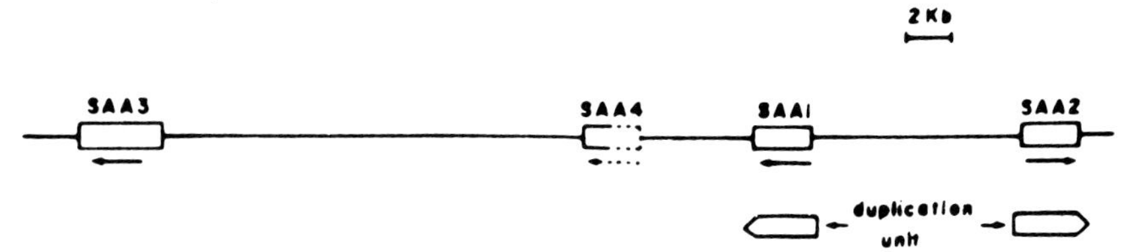

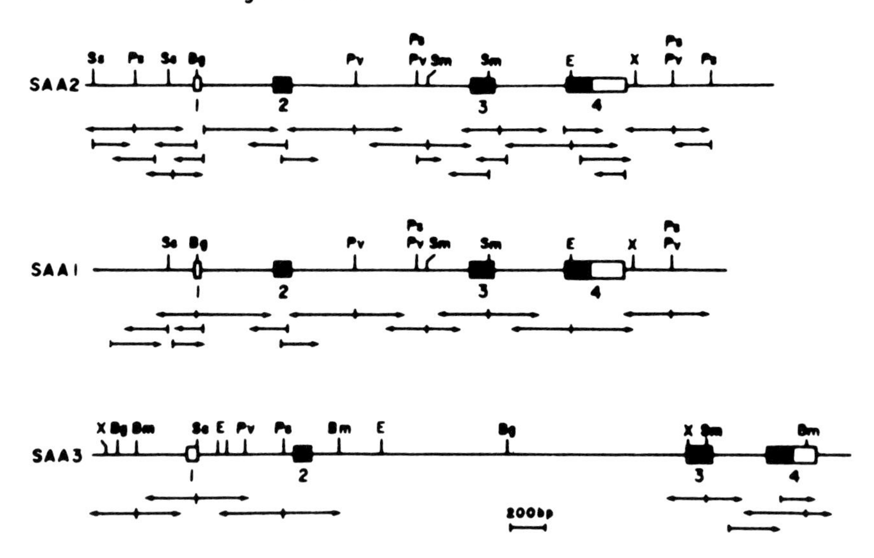

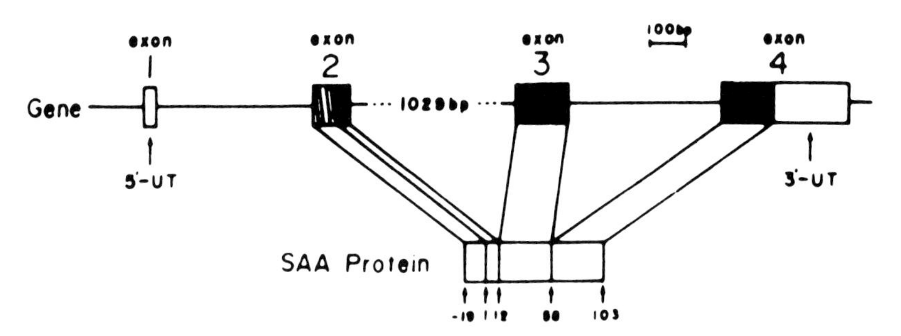

FIGURE 1. Organization and structure of the mouse SAA genes. (A) Schematic representation of the 79-kb region of mouse chromosome 7 containing four SAA genes. The transcriptional orientation of the SAA1, SAA2, and SAA3 genes is indicated by arrows; a possible transcriptional orientation of the SAA4 gene deduced from sequence analysis is indicated by a dashed arrow. Open, large arrows denote a region of homology between the SAA1 and SAA2 genes. (B) Exon/intron structure of the SAA genes and sequence strategy. Exons are shown as boxes and introns as lines; open boxes denote untranslated regions. Restriction enzymes: Bg, BglII; Bm, BamHI; E, EcoRI; Pv, PvuII; Ps, PstI; Sm, SmaI; Ss, SstI; X, XbaI. (C) Relation of the genomic structure (SAA2 gene) to SAA protein structure. The exons are shown as boxes; open, hatched, and closed boxes denote untranslated, signal peptide, and mature protein regions, respectively. (From Yamamoto, K.-I., Goto, N., Kosaka, J., Shiroo, M., Yeul, Y. D., and Migita, S., *J. Immunol.*, 139, 1683, 1987. With permission.)

expression, fall within the type A haplotype group. Eight of 15 strains exhibited a common set of RFLP variants different from type A; these strains were designated haplotype type B. Type B strains include ABP/J, BDP/J, I/LnJ, P/J, PRO/1ReJ, SJL/J, 129/J, and 129/Sv. The prototype strain for the type B haplotype group is 129/J. The restriction endonuclease digests of DNA from type B mice (SJL/J) contain SAA gene fragments that differ from those in corresponding (type A) BALB/c digests in the lengths of PvuII, BamHI, and EcoRI fragments.

Baumann and co-workers described changes in translatable liver apo-SAA mRNAs in ten inbred strains (A/J, AKR/J, BALB/cByJ, C3H/HeJ, C57BL1/6ByJ, C57BL/6J, C57L/J, DBA/2J, DE/Cv, and SWR/J) and wild-derived *Mus spretus* mice by two-dimensional gel electrophoresis.[43,44] RNA from nine of the inbred strains falling within the type A haplotype grouping of Taylor and Rowe[29] directed the synthesis of proteins designated phenotype SAA-B. RNA from DE/Cv mice (not examined in the Taylor and Rowe study) directed the synthesis of variant, acidic SAA(s) designated SAA-A. RNA from (C3H/HeJ × DE/Cv)F1 hybrids directed the synthesis of the so-called phenotypes SAA-A and SAA-B, and RNA from *M. spretus*, a stock of random bred wild-derived mice, was translated into electrophoretically distinct SAA phenotypes, designated SAA-C and SAA-D. A still different pattern of murine SAA gene expression is evident from the *in vitro* studies of Prowse and Baumann,[51] in which hepatocytes isolated from *M. caroli* mice produced two electrophoretic classes of SAA regulated differently by cytokines and dexamethasone. Synthesis of one class, designated SAA-B, was relatively unchanged, while synthesis of the other, SAA-A, was synergistically stimulated in the presence of both interleukin-1 (IL-1) and IL-6.

We have attempted to correlate the phenotypes identified by two-dimensional polyacrylamide gel electrophoresis in Baumann's laboratory[43,44,51] with the apo-SAA isoform patterns revealed by isoelectric focusing (see e.g., Reference 46). We compared the isoelectric focusing patterns of the apo-SAA isoform populations of type A and B and *M. spretus* mice (Figure 2). It appears that pI 6.45 (apo-SAA1) and pI 6.3 (apo-SAA2) comigrate as the single spot SAA-B on two-dimensional gel electrophoresis.[43] Since *M. spretus* does not produce apo-SAA1 and apo-SAA2, but, rather pI 6.15 and pI 5.9 apo-isoforms (Figure 2) as well as apo-SAA5,[46,78] the so-called phenotypes SAA-C and SAA-D must correspond to the latter three apo-SAA species.

Taken together, these studies indicate that the generic term SAA is unsuitable for future studies of SAA regulation by cytokines; since SAA is not a single entity, there is a need to specifically identify the isoform being measured. Because there appear to be differences in the kinetics of induction of the apo-SAA1/2, apo-SAA3, and apo-SAA5 subfamilies,[6,46] it is necessary when working with uncharacterized mouse strains to define the SAA gene family further in terms of gene structure and gene products.

B. REGULATION BY CYTOKINES

Initiation and termination of the acute phase SAA response is known to involve cytokines;[3-5,10,11,13] however, the range of stimulatory factors appears to be broader than proinflammatory cytokines, and serum and other elements such as phorbol esters have been implicated.[23,38,49] The importance of lymphoid cells to the acute phase SAA response in mice was established by the adoptive transfer studies of Rosenstreich and McAdam using the LPS-responsive and -nonresponsive strain pair C3H/HeN and C3H/HeJ.[2] The *in vivo* role of a circulating mediator was described by Sipe and co-workers;[3] the direct effect of cytokines on hepatocytes was demonstrated by the *in vitro* studies of Selinger and co-workers.[4] The importance of the cytokine IL-1 in SAA gene regulation was established by the studies of McAdam and Dinarello,[11] in which endogenous pyrogen (EP), now known to be subserved by IL-1, was shown to modulate *in vivo* SAA expression, and by the studies

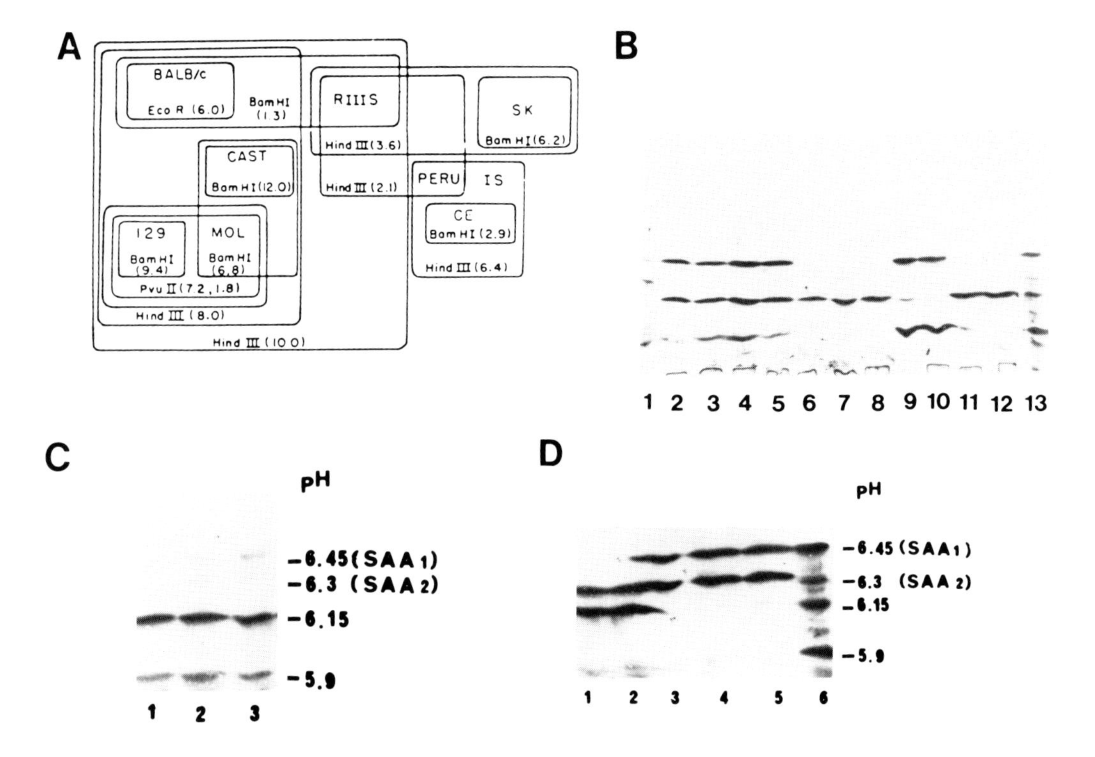

FIGURE 2. (A) Distribution of SAA gene restriction fragment variants among inbred strains. (From Taylor, B. A. and Rowe, L., *Mol. Gen. Genet.*, 195, 491, 1984. With permission.) (B) Immunochemical staining with anti-AA antiserum of apo-SAA isoforms from (lane 1) BALB/c, (lanes 2 to 5) CAST, 6-8-PERU, 9-ABP, 10-129/J, 11-IS/CAM, 12-CE/J, and 13-PN/CR; immunochemical staining of apo-SAA isoforms with rabbit anti-mouse AA. Lanes 1 to 3, apo-SAA isoforms with pIs 6.15 and 5.9 present in the plasma of *Mus spretus* mice 20 h after intraperitoneal injection of 100 μg of LPS; (D) Immunochemical staining of apo-SAA isoforms with rabbit anti-mouse AA. Lanes 1 and 2, apo-SAA isoforms with pIs 6.3 and 6.15 present in the plasma of *M. musculus czech* mice 20 h after i.p. injection of 100 μg of LPS; lanes 3 to 5, apo-SAA isoforms with pIs 6.45 and 6.3 present in the plasma of BALB/c mice (haplotype A) 20 h after i.p. injection with 100 μg of LPS; lane 6, apo-SAA isoforms present in 50 μg of electrofocused HDL from LPS-injected SJL/J mice (haplotype B).

of Sztein et al.[52] using purified natural IL-1 and Ramadori and co-workers using recombinant murine IL-1.[5]

The kinetics of induction of SAA proteins and stimulation of constitutive proteins such as fibrinogen (Fg) are markedly different: SAA concentrations reach maximal levels after 16 to 36 h and then promptly return to threshold, while maximum concentrations of the constitutive protein Fg are reached after 36 to 48 h and are sustained for several days.[53] A naturally occurring IL-1 receptor antagonist[54,55] has recently been used by us to suggest a mechanism by which the multiple effects of IL-1 can regulate temporal differences in plasma SAA and Fg profiles during the acute phase response.[57] Furthermore, our recent *in vitro* studies have defined IL-1 and IL-6 ratios and concentrations under which SAA expression can be stimulated to the order of magnitude that would be needed to maintain *in vivo* concentrations of apo-SAA of more than 100 μg/ml in the face of rapid clearance; the $T_{1/2}$ of apo-SAA in rodents is less than 2 h.[21,45] The quantity of apo-SAA proteins produced by earlier studies involving primary mouse hepatocyte cultures, nanograms per 1 million cells over a 24-h period, was insufficient to account for the apo-SAA concentrations of almost 1 mg/ml at the peak of an acute phase response.[1,23,24] Our recent studies show that the particular combination, ratio, and quantity of cytokines can dramatically regulate the quantity of SAA gene expression over three orders of magnitude.[56,57]

The studies of Lowell and co-workers showed the coordinate transcription of the SAA1 and 2 genes in mouse liver and the concomitant transcription of SAA3, but with different kinetics.[6] Extrahepatic SAA gene expression was demonstrated in macrophages and at numerous peripheral sites in mice following inflammatory stimulation and administration of IL-1;[34-36] the majority of extrahepatic expression in mice is from the SAA3 gene. Maximal SAA5 expression occurs about 10 h later than SAA1/2 and 3 in mice stimulated with lipopolysaccharide (LPS), a cell-wall constituent of Gram-negative bacteria.[46] The *in vitro* studies of Rienhoff and Groudine[22] further support differential regulation of SAA3. While the SAA3 gene is transcribed, the SAA1 and SAA2 genes are inactive in BALB/c-derived mouse BNL hepatoma cells. Huang and co-workers[58] have described an IL-1 response element in the mouse SAA3 gene using an SAA3-CAT hybrid gene that is specifically expressed in liver cells. They observed that SAA3 was responsive to IL-1, but not IL-6, although IL-6 together with IL-1 was synergistic. Studies involving human and rat as well as mouse SAA gene expression have implicated both C/EBP- and NF-κB-related proteins as important regulatory factors that contribute to tissue specificity and the rapid and high rates of changes in SAA transcription.[58-61]

C. REGULATION BY NONCYTOKINE FACTORS

SAA expression can be induced by many acute phase stimuli, including LPS, trauma, infections, and malignancy, all of which are thought to act, in large part, through cytokine signals. Until recently, experimental SAA production required the use of an animal model; these *in vivo* studies are inherently technically difficult and susceptible to artifacts from fighting, environmental endotoxins, etc. Furthermore, it is difficult to establish that a substance is a direct stimulus through *in vivo* studies. For example, dexamethasone maximally stimulates *in vivo* SAA synthesis in mice as early as 6 h,[38] but this is probably an indirect effect, since SAA genes are not described as having glucocorticoid response elements like other acute phase proteins such as Fg and α_1-acid glycoprotein.[62] Apart from these technical considerations, there is evidence that elements other than cytokines, such as serum and phorbol esters, can act directly on hepatocytes to stimulate SAA gene expression. For example, Tatsuta and co-workers[23] showed that serum stimulates SAA synthesis by hepatocytes from CBA and C3H mice in primary culture. Recently, in passive immunization experiments, we observed stimulation of SAA by control serum, suggesting an *in vivo* role for serum in addition to its *in vitro* effects.[78] Phorbol esters (PMA) have been used *in vivo* and *in vitro* to stimulate expression of murine SAA genes or a human SAA gene transfected into murine cells.[38,60,61] Bacterial products such as LPS, endotoxin-associated protein (EAP),[63] and muramyl dipeptide[64] stimulate SAA production. While the *in vivo* effects of LPS are thought to be mediated primarily by secretory products of activated mononuclear phagocytes, it is possible that LPS acts directly to effect SAA gene expression, since the effects of LPS cannot be blocked by passive immunization with anti-IL-1 and anti-IL-6.[79]

D. AMYLOIDOSIS

For more than 70 years, the mouse model has been employed to investigate the pathogenesis of inflammation-associated or AA amyloidosis.[1,9,12,19,25-28,30,31,36,39,40] In CBA/J mice, apo-SAA2, but not apo-SAA1, is converted into AA fibrils.[9,30,31] SJL/J mice have been employed in studies of SAA expression and amyloidosis.[16,27,28] Because of an isolated report of spontaneous amyloidosis in SJL/J mice in which the fibril protein could not be sequenced because of a blocked amino terminus[28] and because of the apparent lack of apo-SAA2 production,[16] SJL/J mice were thought to be resistant to AA amyloidosis. However, our recent study using the AEF model demonstrates that SJL/J mice develop amyloidosis of the AA type in a course that is temporally and quantitatively similar to the formation of AA fibrils from apo-SAA2 in CBA/J mice.[27]

SJL/J and some other strains of mice are prone to age-associated spontaneous deposition of apo-AII amyloid fibrils.[33] Inflammation, cytokines, and the acute phase response are not implicated in this process. Therefore, the SJL/J strain is of particular interest to the study of amyloidosis because of its unique capability to deposit two kinds of fibrils.

In addition to the structure of the apo-SAA fibril precursor, other factors are thought to be important in AA fibrillogenesis, including amyloid-enhancing factor (AEF) and mononuclear phagocyte function; impaired proteolysis and impaired clearance of SAA and AA leads to the growth of AA amyloid deposits.[65]

E. SUMMARY AND FUTURE PERSPECTIVES

The interspecies SAA nomenclature is sometimes confusing because the total number of members of the SAA gene family has not yet been determined for any species. In addition to mouse, direct and derived amino acid sequence data are available for cat, cow, dog, duck, hamster, horse, human, mink, rabbit, rat, and sheep. Nearly all genes share a constant region in exon 3, and most of the structural differences reside in exon 4. Two separate branches of SAA proteins have been described;[66] cat, cow, dog, horse, and mink have an insert spanning positions 72 to 81 that is not present in the mouse genes characterized to date.

Apo-SAA is thought to alter HDL charge as a result of its association with HDL particles. For example, in human, HDL3 particles during the acute phase response are larger than, and have reduced electrophoretic mobility compared with, HDL3 particles isolated from individuals during homeostasis.[67] In Syrian hamsters, lipoprotein metabolism is disrupted during the acute phase response, with decreased amounts of cholesterol in HDLs and retarded electrophoretic mobility of HDLs that coincides with an increased content of SAA in HDL.[68] It seems likely that the apo-SAA function in mice is similar to that in human and hamsters.

II. MATERIALS AND METHODS

A. ANALYSIS OF SAA PROTEINS

1. Enzyme-Linked Immunosorbant Assay (ELISA)

The measurement of SAA, like that of other lipophilic plasma proteins, has proven difficult. Many of the technical problems stem from conformational changes in the SAA proteins; such conformational changes have been observed when SAA is analyzed by immunodiffusion or charge shift electrophoresis.[69] Changes in apo-SAA conformation are induced by heating, acid, alkali, guanidine hydrochloride, and extraction with organic solvents (reviewed in Reference 70). Inaccurate measurements of SAA by immunoassays have frequently resulted from the tendency of SAA to interact nonspecifically with the surfaces of laboratory vessels and with itself and other molecules of similar structure. These difficulties have been circumvented by the use of monoclonal antibodies that recognize different conformationally stable epitopes on SAA proteins[71] or by the use of a direct binding ELISA for SAA in which noncovalent interactions of SAA with other plasma constituents are disrupted, and a fraction of SAA is bound to polyvinylchloride surfaces according to its concentration and relative affinity for the surface relative to that of other constituents.[72]

2. Urea-SDS-Polyacrylamide Gel Electrophoresis (PAGE)

In SAA-rich HDL fractions from BALB/c and other type A mice, although the actual molecular weights calculated from the derived amino acid sequences are 11,750 and 11,650, respectively,[6] apo-SAA2 can be separated from apo-SAA1 by one-dimensional electrophoresis on urea-SDS gels, probably due to differential unfolding. This feature has been utilized in studies of amyloidogenesis; most frequently, 11.4% acrylamide and 6.4 *M* urea have been employed.[9,30] However, urea-SDS-PAGE cannot be used for the separation of apo-SAA

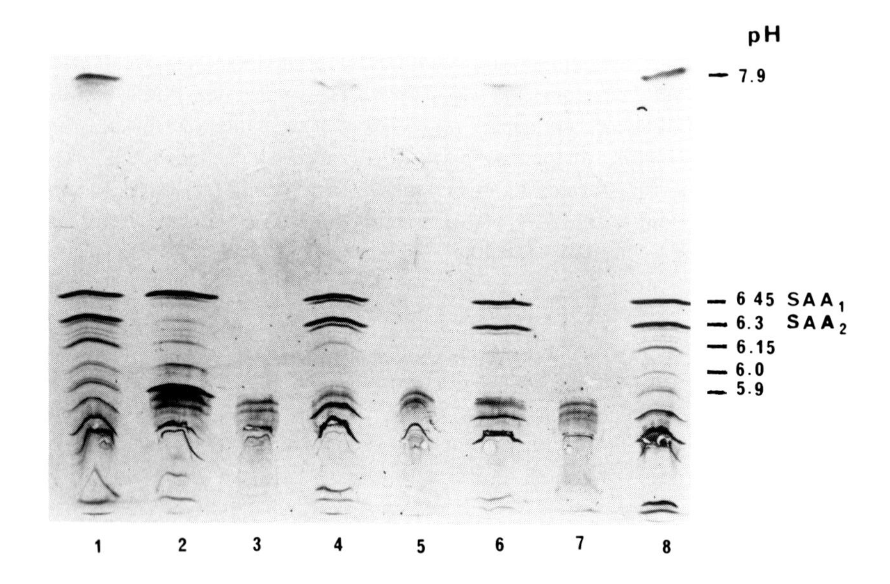

FIGURE 3. Isoelectric focusing analysis of mouse HDL. Coomassie Blue staining of 200 μg per lane of electrofocused HDL.[46,74] Lanes 1 and 8, SAA-rich HDL standards from BALB/c mice >24 h after administration of LPS; lane 2, HDL from SJL/J mice 20 h after LPS; lane 3, HDL from control SJL/J mice; lane 4, HDL from C3Heb/FeJ mice 20 h after LPS; lane 5, HDL from control C3Heb/FeJ mice; lane 6, HDL from BALB/c mice 20 h after LPS; lane 7, HDL from control BALB/c mice.

isoforms from type B mice. For example, Yamamoto and co-workers used 6 *M* urea, 13% SDS-PAGE slab gels to separate the pre-apo-SAA1 and pre-apo-SAA2 products of the cell-free translation of acute phase liver RNA from BALB/c, SJL/J, and Swiss mice, and noted the apparent absence of apo-SAA2 in the SJL/J strain.[16] It cannot be predicted whether the pre-form of the novel SJL/J apo-SAA isoform would have been resolved from pre-apo-SAA1 under the conditions used; a more definitive analysis by isoelectric focusing was required to identify the pI 5.9 isoform (Figure 3).[74] Moreover, use of the urea-SDS PAGE technique precludes identification of apo-SAA5, as its migration in this system has not been defined.

3. Isoelectric Focusing (IEF)

The technique of IEF was useful for resolution of the three subfamilies of murine apo-SAA isoforms and generally proves useful in resolving the full spectrum of apo-SAA isoforms. HDL was isolated from the plasma of inbred and wild-derived mice by sequential ultracentrifugation, and delipidated HDL was electrofocused on ultrathin (0.3 mm) polyacrylamide gels containing 7 *M* urea. To identify basic isoforms, an ampholine ratio of 20% (v/v) pH 3 to 10, 40% (v/v) pH 4 to 6.5, and 40% (v/v) pH 7 to 9 was used. Acidic apo-SAA isoforms were exposed using an ampholine ratio of 20% (v/v) pH 3 to 10 and 80% (v/v) pH 4 to 6.5. Immunochemical analysis of apo-SAA isoforms was performed after pressure blotting of the IEF gels onto nitrocellulose membranes (Figures 2 and 3).[46,74]

B. NORTHERN HYBRIDIZATION OF SAA ISOTYPES

When dealing with a specific strain, one should provide evidence from gene or protein structure that the hybridization probes in use will hybridize with SAA mRNA transcribed in that strain. Difficulties can result from the use of oligonucleotide probes based on type A structure for hybridization with mRNA from type B mice. For example, SJL/J are identical to type A mice in exon 2 of the SAA1/2 and 3 genes, but our studies have identified differences between SJL/J and type A mice in exon 4 in the 3′ untranslated region.[74]

1. Oligonucleotide Probes

Although the haplotype A SAA1 and 2 genes and gene products are very similar in structure,[6] it is possible to distinguish between them with oligonucleotide probes, such as the set of three 18-mers which was designed by Yamamoto and colleagues[16,73] to distinguish between SAA1/2 and 3 transcripts in the regions of exon 2 encoding residues 4 to 9 or the set selected by Meek and Benditt[34] for exon 4 18-bp segments in the 3′ UT 14 bases beyond the termination codons. However, these probes are valid for use only in BALB/c and other type A strains. Their use with dissimilar strains would require structural analysis of the structure of the SAA genes and proteins if the information is not available. A 17-mer exon 3 probe corresponding to residues 37 to 42 common to the mouse SAA1/2 and 3 genes was used by us to measure total murine SAA mRNA in type A and B mice.[57,74]

2. cDNA Probes

The first murine cDNA, isolated and characterized by Stearman and co-workers, was a partial clone of the SAA3 gene in BALB/c mice containing exons 2, 3, and 4.[15] Full-length clones encoding the SAA1 and SAA2 genes in BALB/c mice were described by Yamamoto and Migita.[14] The partial cDNA clone of SAA3 was used by Taylor and Rowe to characterize polymorphisms in murine SAA gene structure.[29] Rienhoff and Groudine described an SAA3-specific probe, pAA2, that was constructed from a 1-kb XbaI-Xba-I fragment of SAA3 containing exons 3 and 4.[22] Human cDNA clones containing exon 3 have been used for Northern analysis of murine mRNA.[5,35,36,73] Lowell and co-workers[6] constructed an SAA3-specific probe from the AvaI/PstI fragment from the 3′ UTS of the SAA3 cDNA plasmid cloned into the PstI and SmaI sites of the M13mp11 vector.

3. Genomic Fragment Probes

In order to distinguish between SAA3 and SAA1/2 mRNAs, clones of BALB/c genomic fragments have been used. Lowell et al.[6] cloned the AvaII/XhoI fragment from the 3′ untranslated region of the SAA2 gene into the SmaI site of the M13 vector mp8.

III. RESULTS

A. SAA PROTEINS

When HDL isolated from EDTA anticoagulated plasma obtained 20 h after intraperitoneal injection of 100 μg of LPS (W) from *S. typhosa* was analyzed by urea-SDS-PAGE, we observed that the apo-SAA of type A mice resolved into two bands, apo-SAA1 and apo-SAA2, while in type B mice, all apo-SAA isoforms migrated as a single band. Our data in Figures 2 and 3 show that the single band in SJL/J mice (type B) can be resolved into several isoforms and that BALB/c mice (type A) produce, in addition to apo-SAA1 and apo-SAA2, several quantitatively minor novel apo-SAA isoforms.

Taylor and Rowe diagrammed the distribution of SAA gene restriction fragment variants among inbred strains.[29] Figure 2 from their paper is reproduced in Figure 2A. They classified strains according to restriction fragment variants distinguishing individual strains. They state: "Strains enclosed by the same box share the particular fragment designated on the base of the box. Nonintersecting boxes suggest mutational divergence of genetically isolated haplotypes, while intersecting boxes suggest that recombination may have occurred during haplotype evolution. . . . Nonrandom associations among variants are evident." Based on Taylor and Rowe's RFLP analysis[29] and the biosynthetic studies from Baumann's laboratory,[43,44,51] we examined non-type A,B inbred strains and wild-derived mice, including CE/J, MOL, and CAST in comparison with BALB/c (type A) and ABP, 129/J, and PN/CR (type B) (Figure 2B to D). We also examined *M. spretus,* the random bred wild-derived mice studied by Baumann and co-workers (Figure 2C). The fact that *M. spretus* produces apo-SAA1 or 2 only in trace amounts and the novel pI 5.0 and pI 6.15 isoforms as dominant products is consistent with the findings of Baumann et al. that *M. spretus* produces two novel SAA phenotypes called SAA-C and SAA-D. It also illustrates how essential it is to define the subfamilies to which apo-SAA species under study belong, perhaps by a combination of IEF and amino-terminal sequence analysis.

There is a close correlation between Taylor and Rowe's RFLP patterns and our IEF patterns (Figure 2A to D). For example, PERU, IS, and CE are phenotypically identical and share common RFLPs distinct from haplotype A and haplotype B mice. There is a consistency among members of type A and type B groupings, i.e., the apo-SAA pattern of A/J mice is the same as that of BALB/c mice, and that of PN/CR is the same as that of SJL/J mice. We have examined at least five individuals of 5-member strains of each haplotype group, and found the IEF patterns to be identical. In some strains such as CAST/ei, three SAA isoforms are predominant (apo-SAA1, apo-SAA2, and the pI 6.15 isoform), whereas in others such as *M. musculus czech,* a unique pattern is present, with apo-SAA2 and the pI 6.15 isoforms as the quantitatively major products of the SAA gene family.

Complete amino acid sequence analysis of the pI 5.9 apo-SAA isoform of SJL/J mice was performed. The pI 5.9 variant is identical to the apo-SAA2 isoform from BALB/c mice except to the substitutions of aspartic acid for alanine at position 101.[74] We confirmed by IEF the observation by Baumann's laboratory of the presence of novel pI 5.9 and pI 6.15 isoforms corresponding to SAA-C and SAA-D in acute phase plasma from *M. spretus,* but, in addition, identified apo-SAA1 and apo-SAA2 as minor constituents.

We analyzed the inheritance of apo-SAA isoforms in type B mice using the F1 hybrid strain CSJLF1/J (BALB/c × SJL/J) and in part of the set of recombinant inbred (RI) strains CXJ derived from the type A strain BALB/cKe and type B strain SJL/J. While there is codominant expression of the three major apo-SAA isoforms of the parent strains in F1 hybrids, only one of six RI substrains after 20 generations of brother-sister matings retained the type B phenotype. Taylor and Rowe's RFLP analysis revealed that all of the SAA-specific DNA restriction fragment variants segregate as a single haplotype and that only two of ten have retained the type B haplotype.

IEF has been useful for the identification of apo-SAA isoforms, including the products of previously undiscovered genes. The IEF technique has served to illustrate the complexities of murine SAA gene structure and expression, and the need for consistent terminology in SAA studies.

B. SAA mRNA

We compared SAA expression in type B (SJL/J) mice with that of type A (BALB/c) mice using seven different oligonucleotide probes corresponding to the three known active genes in the BALB/c family.[6] The 18-mer exon 2 probes were designed to distinguish between SAA1/2 and 3 transcripts in the region encoding residues 4 to 9.[14] The 17-mer exon 3 probe corresponds to residues 37 to 42 common to the mouse SAA1/2 and 3 genes. The set of exon 4 probes was designed to distinguish between SAA1/2 and 3 transcripts in the 18-base region in the 3′ untranslated region starting 14 bases beyond the termination codon.[34] On the basis of hybridization with the exon 2 and exon 3 probes, BALB/c, SJL/J, and the F1 hybrids exhibit equivalent transcription of SAA genes following administration of LPS (Figure 4). However, Northern-blot hybridization analysis (Figure 4) revealed that exon 4 of the type B pI 5.9 isoform differs from apo-SAA2 mRNA in the 3′ untranslated region, since the oligonucleotide probe specific for exon 4 of BALB/c apo-SAA2 mRNA failed to hybridize with SJL/J mRNA, although the mRNA for the major pI 5.9 isoform appears to be abundant by hybridization with the oligonucleotide probe specific for exon 2 of BALB/c mRNA (Figure 4). Furthermore, SJL/J mice appear to differ in the 3′ untranslated region

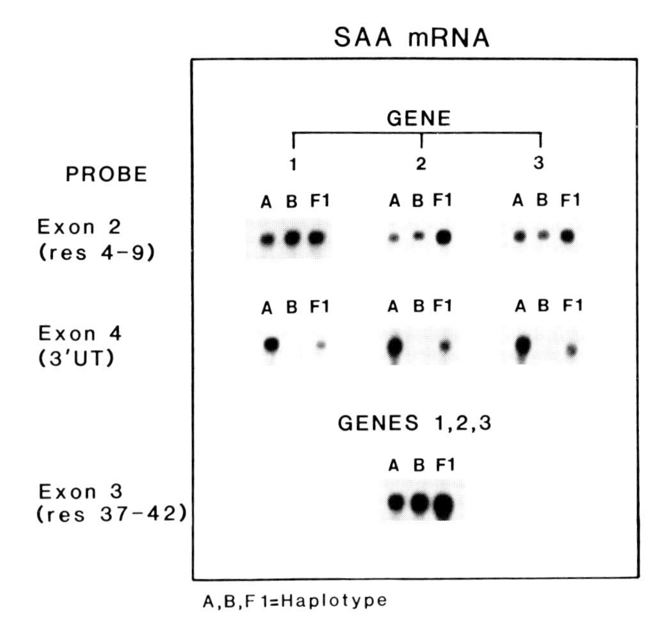

FIGURE 4. Gene-specific hybridization analysis of SAA mRNA in BALB/c (A), SJL/J (B), and CSJLF1/J (F1) mice. Mice were injected intraperitoneally with 100 μg of LPS type W from *Salmonella typhosa*. Livers were removed 16 h later, and polyadenylated RNA was extracted, denatured, size fractionated by electrophoresis (2 μg per lane), and transferred to nylon filters. Hybridization of identical blots was carried out with oligonucleotide probes corresponding to BALB/c SAA genes 1, 2, and 3.[6] The set of 18-mer exon 2 probes corresponds to the region of exon 2 encoding residues 4 to 9.[14,15] A single 17-mer probe corresponds to the portion of exon 3 encoding residues 37 to 42, identical in SAA1, SAA2, and SAA3 gene transcripts, and the set of probes designed for the 3′ untranslated region[34] corresponds to 18 bases located 14 bases beyond the termination codons. (From de Beer, M. C., de Beer, F. C., Beach, C. M., Carreras, I., and Sipe, J. D., *Biochem. J.*, 283, 673, 1992. With permission.)

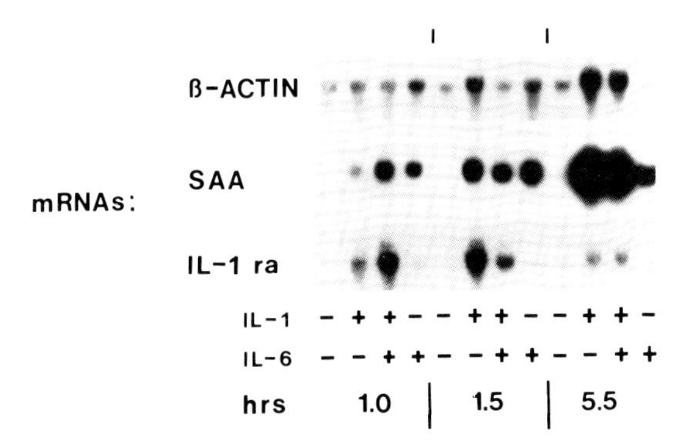

FIGURE 5. Kinetics of SAA and IL-1ra mRNA expression in livers of mice after administration of IL-1 and IL-6 alone and in combination. Northern blots were prepared from mouse liver RNA enriched in polyadenylated mRNA, as described for Figure 4, 3 to 5 μg per lane, using the exon 3 SAA oligonucleotide probe, and IL-1ra and actin-specific probes.[35,55,74]

of apo-SAA1 as well, suggesting that all of the SJL/J SAA genes differ from BALB/c in exon 4.

We have obtained preliminary evidence for the role of IL-1ra in regulating the magnitude and kinetics of the acute phase SAA response in mice. The time course of IL-1ra expression in liver of CD2F-1 mice following administration of IL-1and IL-6 alone and in combination was compared with the induction of SAA mRNA (Figure 5). Kinetic analysis of liver mRNA by Northern-blot hybridization of SAA and IL-1ra mRNA in CD2F-1 mice following administration of IL-1 showed early synergy (1 h) between IL-1 and IL-6 in the stimulation of SAA gene expression along with the appearance of IL-1ra mRNA in response to IL-1 (at 1.0 and 1.5 h), but not IL-6.

IV. DISCUSSION

The mouse model offers many advantages for studying SAA structure and function. In particular, it is suited to studies of the genomic structure of SAA subfamilies, the differential regulation of the three subfamilies by cytokines, and the role of SAA structure in amyloidosis. Our results establish the production of apo-SAA isoforms other than, and in addition to, apo-SAA1 and 2 following administration of LPS to a variety of inbred and wild-derived mice. The existence of previously unrecognized species of SAA has been confirmed by immunoblotting and amino acid sequence analysis in addition to their properties of LPS inducibility and association with HDL (Figures 2 and 3, References 46 and 74). Future studies of murine SAA gene expression will encompass apo-SAA isoforms derived from three subfamilies of genes (apo-SAA1/2, apo-SAA3, and apo-SAA5), and we propose that a thorough analysis of protein structure as well as molecular biologic approaches is necessary in view of evidence of the differential regulation of the three subfamilies.

Our studies indicate that the use of urea-SDS-PAGE for separation of apo-SAA isoforms should be avoided, since the novel pI 5.9 isoform in SJL/J mice is expressed in quantity comparable to apo-SAA2, is intact, and yet seemingly larger than apo-SAA2 upon analysis by urea-SDS-PAGE because it comigrates with apo-SAA1 with a completely unfolded structure. The studies of Creighton on urea-induced unfolding of proteins[75,76] suggest that apo-SAA1 and the pI 5.9 isoform undergo identical "melting" in 6.4 *M* urea. This serves as further proof of structural differences between the pI 5.9 isoform and apo-SAA2. We conclude that the resolution of apo-SAA2 from apo-SAA1 by 6.4 *M* urea SDS-PAGE results from incomplete unfolding of apo-SAA2, thus allowing it to migrate further into the gel than the completely unfolded apo-SAA1.

Just as urea-SDS-PAGE studies of SAA proteins can be misleading, so can the use of oligonucleotides for Northern hybridization analysis in the absence of knowledge of gene structure. Evidence for differences in the gene structure of type B and type A mice is provided by hybridization analysis with exon 2- and exon 4-specific probes derived from the structure of the BALB/c genes.[74] It appears that isoelectrically identical apo-SAA1 molecules in BALB/c and SJL/J mice are translated from mRNAs that differ in exon 4 and that the gene for SAA3 in SJL/J mice differs from BALB/c in exon 4 (Figure 4).

The cytokine regulation of SAA gene expression in mice appears to be similar to that in other species, and the availability of genetically uniform animals may be expected to continue to be experimentally powerful. *In vitro,* IL-1 and IL-6 independently regulate production of human SAA proteins to a limited extent, but together they synergistically stimulate expression of the SAA gene family to a 100-fold or greater extent than they do acting alone.[57] IL-1ra blocks SAA expression stimulated by IL-1 alone and reduces the synergistic expression stimulated by IL-1 and IL-6 together to the level achieved with IL-6 alone. Our *in vivo* studies are consistent with a close relationship between the regulation of mouse and human SAA gene families. The results of our study establish that IL-1 induces IL-1ra expression, which, in turn, reduces IL-1, but not IL-6, stimulation of SAA production. We conclude that, *in vivo,* IL-1, through secondary stimulation of IL-6 and stimulation of the IL-1ra, controls both the magnitude and duration of SAA production. Future work may be expected to define the combinations of IL-1- and IL-6-specific nuclear transcription factors (e.g., NF-κB, C/EBP) that result in the prompt, exponential increases in SAA expression and the signal transduction pathways through which IL-1ra downregulates SAA gene expression.

In the past few years, our understanding of the SAA superfamily has grown to the extent that it is recognized as complex in structure and in the regulation of expression. Future studies must take into consideration structural features of the particular SAA under study.

REFERENCES

1. **McAdam, K. P. W. J. and Sipe, J. D.,** Murine model for human secondary amyloidosis: genetic variability of the acute phase serum protein SAA response to endotoxins and casein, *J. Exp. Med.,* 144, 1121, 1976.
2. **Rosenstreich, D. L. and McAdam, K. P. W. J.,** Lymphoid cells in endotoxin induced production of the amyloid-related serum protein SAA, *Infect. Immunol.,* 23, 181, 1979.
3. **Sipe, J. D., Vogel, S. N., Ryan, J. L., McAdam, K. P. W. J., and Rosenstreich, D. L.,** Detection of a macrophage-derived mediator for induction of the acute phase SAA response in mice by endotoxin, *J. Exp. Med.,* 150, 597, 1979.
4. **Selinger, M. J., McAdam, K. P. W. J., Kaplan, M. M., Sipe, J. D., Rosenstreich, D. L., and Vogel, S. N.,** Monokine-induced synthesis of serum amyloid A protein by hepatocytes, *Nature (London),* 285, 498, 1980.
5. **Ramadori, G., Sipe, J. D., Dinarello, C. A., Mizel, S. B., and Colten, H. R.,** Pretranslational modulation of acute phase hepatic protein synthesis by recombinant generated mouse interleukin-1 and purified human IL-1, *J. Exp. Med.,* 162, 930, 1985.
6. **Lowell, C. A., Potter, D. A., Stearman, R. S., and Morrow, J. F.,** Structure of the murine serum amyloid A gene family, *J. Biol. Chem.,* 261, 8442, 1986.
7. **Benson, M. D., Scheinberg, M. A., Shirahama, T., Cathcart, E. S., and Skinner, M.,** Kinetics of serum amyloid protein A in casein-induced murine amyloidosis, *J. Clin. Invest.,* 59, 412, 1977.
8. **Benson, M. D. and Kleiner, E.,** Synthesis and secretion of serum amyloid protein A (SAA) by hepatocytes in mice treated with casein, *J. Immunol.,* 124, 495, 1980.
9. **Dwulet, F. E. and Benson, M. D.,** Primary structure of amyloid fibril protein AA in azocasein-induced amyloidosis of CBA/J mice, *J. Lab. Clin. Med.,* 110, 322, 1987.
10. **Sipe, J. D., Vogel, S. N., Sztein, M. B., Skinner, M., and Cohen, S. A.,** The role of interleukin 1 in acute phase SAA and SAP synthesis, *Ann. N.Y. Acad. Sci.,* 389, 137, 1982.
11. **McAdam, K. P. W. J. and Dinarello, C. A.,** Induction of serum amyloid A synthesis by human leukocyte pyrogen, in *Bacterial Endotoxin and Host Response,* Agarwal, M. K., Ed., Elsevier/North-Holland, Amsterdam, 1980, 167.
12. **Sipe, J. D., McAdam, K. P. W. J., and Uchino, F.,** Biochemical evidence for the biphasic development of experimental amyloidosis, *Lab. Invest.,* 38, 110, 1978.
13. **Vogel, S. N. and Sipe, J. D.,** The role of macrophages in the acute phase serum amyloid A (SAA) response to endotoxin, *Surv. Immunol. Res.,* 1, 235, 1982.
14. **Yamamoto, K.-I. and Migita, S.,** Complete primary structures of two major murine serum amyloid A proteins deduced from cDNA sequences, *Proc. Natl. Acad. Sci. U.S.A.,* 82, 2915, 1985.
15. **Stearman, R. S., Lowell, C. A., Peltzman, C. G., and Morrow, J. F.,** The sequence and structure of a new serum amyloid A gene, *Nucleic Acids Res.,* 14, 797, 1986.
16. **Yamamoto, K.-I., Shiroo, M., and Migita, S.,** Diverse gene expression for isotypes of murine serum amyloid A protein during acute phase reaction, *Science,* 232, 227, 1986.
17. **Hoffman, J. S. and Benditt, E. P.,** Changes in high density lipoprotein content following endotoxin administration in the mouse. Formation of serum amyloid protein-rich subfractions, *J. Biol. Chem.,* 257, 10510, 1982.
18. **Benditt, E. P., Hoffman, J. S., Erikson, N., and Walsh, K. A.,** SAA, an apoprotein of HDL: its structure and function, *Ann. N.Y. Acad. Sci.,* 389, 193, 1982.
19. **Baltz, M. L., Rowe, I. F., Caspi, D., Turnell, W. G., and Pepys, M. B.,** Is the serum amyloid A protein in acute phase high density lipoprotein the precursor of tissue AA amyloid deposits?, *Clin. Exp. Immunol.,* 66, 701, 1986.
20. **Meek, R. L., Eriksen, N., and Benditt, E. P.,** Mouse SAA3: detection in mouse tissues with specific antibody, in *VI International Symposium on Amyloidosis,* Natvig, J., Ed., Kluwer, Dordrecht, 1991.

21. **Hoffman, J. S. and Benditt, E. P.,** Plasma clearance kinetics of the amyloid-related high density lipoprotein apoprotein, serum amyloid protein (ApoSAA) in the mouse, *J. Clin. Invest.*, 71, 926, 1983.
22. **Rienhoff, H. Y. and Groudine, M.,** Regulation of amyloid A gene expression in cultured cells, *Mol. Cell. Biol.*, 8, 3710, 1988.
23. **Tatsuta, E., Sipe, J. D., Shirahama, T., Skinner, M., and Cohen, A. S.,** Different regulatory mechanisms for serum amyloid A and serum amyloid P synthesis by cultured mouse hepatocytes, *J. Biol. Chem.*, 258, 5414, 1983.
24. **Benson, M. D.,** In vitro synthesis of the acute phase reactant SAA by hepatocytes, *Ann. N.Y. Acad. Sci.*, 389, 116, 1982.
25. **Hoffman, J. S., Ericsson, L. H., Eriksen, N., Walsh, K. A., and Benditt, E. P.,** Murine tissue amyloid protein AA; NH2-terminal sequence identity with only one of two serum amyloid protein (apoSAA) gene products, *J. Exp. Med.*, 159, 641, 1984.
26. **Shiroo, M., Kawahara, E., Nakanishi, I., and Migita, S.,** Specific deposition of serum amyloid A protein 2 in the mouse, *Scand. J. Immunol.*, 26, 709, 1987.
27. **Rokita, H., Shirahama, T., Cohen, A. S., and Sipe, J. D.,** Serum amyloid A gene expression and AA amyloid formation in A/J and SJL/J mice, *Br. J. Exp. Pathol.*, 70, 327, 1987.
28. **Scheinberg, M. A., Cathcart, E. S., Eastcott, J. W., Skinner, M., Benson, M. D., Shirahama, T., and Bennett, M.,** The SJL/J mouse: a new model for spontaneous age-associated amyloidosis. I. Morphologic and immunochemical aspects, *Lab. Invest.*, 35, 47, 1976.
29. **Taylor, B. A. and Rowe, L.,** Genes for serum amyloid A proteins map to chromosome 7 in the mouse, *Mol. Gen. Genet.*, 195, 491, 1984.
30. **Meek, R. L., Hoffman, J. S., and Benditt, E. P.,** Amyloidogenesis: one serum amyloid A isotype is selectively removed from the circulation, *J. Exp. Med.*, 163, 499, 1986.
31. **Hebert, L. and Gervais, F.,** Apo-SAA1/apo-SAA2 isotype ratios during casein and AEF induced secondary amyloidosis in A/J and C57BL/6J mouse strains, *Scand. J. Immunol.*, 31, 167, 1990.
32. **Morrow, J. F., Stearman, R. S., Peltzman, C. G., and Potter, D. A.,** Induction of hepatic synthesis of serum amyloid A protein and actin, *Proc. Natl. Acad. Sci. U.S.A.*, 78, 4718, 1981.
33. **Warden, C., Bee, L., Lusis, A. J., Lerner, C., Chai, C. K., Gorevic, P. D., Qian, L. P., and Munoz, P. C.,** The low leukocyte (LLC) mouse, a second model of ''senescence-accelerated'' amyloidosis associated with apolipoprotein AIIpro5-gln; prevalence of this substitution among inbred strains of mice, in *Amyloid and Amyloidosis 1990,* Natvig, J. B., Foree, O., Husby, G., Husebekk, A., Skogen, B., Sletten, K., and Westermark, P., Eds., Kluwer, Dordrecht, 1991, 397.
34. **Meek, R. L. and Benditt, E. P.,** Amyloid A gene family expression in different mouse tissues, *J. Exp. Med.*, 164, 2006, 1986.
35. **Ramadori, G., Sipe, J. D., and Colten, H. R.,** Expression and regulation of the murine serum amyloid A (SAA) gene in extrahepatic sites, *J. Immunol.*, 135, 3645, 1985.
36. **Rokita, H., Shirahama, T., Cohen, A. S., Meek, R. L., Benditt, E. P., and Sipe, J. D.,** Differential expression of the amyloid SAA 3 gene in liver and peritoneal macrophages of mice undergoing dissimilar inflammatory episodes, *J. Immunol.*, 139, 3849, 1987.
37. **Neta, R., Vogel, S. N., Sipe, J. D., Wong, G. G., and Nordan, R.,** Comparison of in vivo effects of human recombinant IL 1 and human recombinant IL 6 in mice, *Lymphokine Res.*, 7, 403, 1988.
38. **Ghezzi, P. and Sipe, J. D.,** Dexamethasone modulation of LPS, IL-1 and TNF stimulated serum amyloid A synthesis in mice, *Lymphokine Res.*, 7, 157, 1988.
39. **Sipe, J., Rokita, H., Shirahama, T., and Cohen, A. S.,** Deposition of amyloid A fibrils in spleen is accompanied by decreased hepatic and splenic and increased macrophage serum amyloid A expression, in *Amyloid and Amyloidosis,* Isobe, T., Araki, S., Uchino, F., Kito, S., and Tsubura, E., Eds., Plenum Press, New York, 1988, 81.
40. **Wohlgethan, J. R. and Cathcart, E. S.,** Amyloid resistance in A/J mice. Studies with a transfer model, *Lab. Invest.*, 42, 663, 1980.
41. **Festing, M. F. W.,** *Inbred Strains in Biomedical Research,* Oxford University Press, Oxford, 1979.
42. **Yamamoto, K.-I., Goto, N., Kosaka, J., Shiroo, M., Yeul, Y. D., and Migita, S.,** Structural diversity of murine serum amyloid A genes. Evolutionary implications, *J. Immunol.*, 139, 1683, 1987.
43. **Baumann, H., Held, W. A., and Berger, F. G.,** The acute phase response of mouse liver. Genetic analysis of the major acute phase reactants, *J. Biol. Chem.*, 259, 566, 1984.
44. **Baumann, H. and Berger, F. G.,** Genetics and evolution of the acute phase proteins in mice, *Mol. Gen. Genet.*, 201, 505, 1985.
45. **Bausserman, L. L.,** SAA kinetics in animals, in *Amyloidosis,* Marrink, J. and van Rijswijk, M. H., Eds., Martinus Nijhoff, Boston, 1986, 337.
46. **de Beer, M. C., Beach, C. M., Shedlofsky, S. I., and de Beer, F. C.,** Identification of a novel serum amyloid A protein in BALB/c mice, *Biochem. J.*, 280, 45, 1991.

47. **Meek, R. M., Eriksen, N., and Benditt, E. P.,** Mouse SAA3: detection in mouse tissues with specific antibody, in *Amyloid and Amyloidosis 1990,* Natvig, J. B. et al., Eds., Kluwer, Dordrecht, 1991, 75.
48. **Sipe, J. D., Colten, H. R., Goldberger, G., Edge, M. D., Tack, B. F., Cohen, A. S., and Whitehead, A. S.,** Human serum amyloid A (SAA): biosynthesis and post synthetic processing of pre-SAA and structural variants defined by complementary DNA, *Biochemistry,* 24, 2931, 1985.
49. **Woo, P., Sipe, J. D., Dinarello, C. A., and Colten, H. R.,** Structure of a human serum amyloid A gene and modulation of its expression in transfected L cells, *J. Biol. Chem.,* 262, 15790, 1988.
50. **Kluve-Beckerman, B., Long, G. L., and Benson, M. D.,** DNA sequence evidence for polymorphic forms of human serum amyloid A, *Biochem. Genet.,* 24, 795, 1986.
51. **Prowse, K. R. and Baumann, H.,** Interleukin-1 and interleukin-6 stimulate acute phase protein production in primary mouse hepatocytes, *J. Leuk. Biol.,* 45, 55, 1989.
52. **Sztein, M. B., Vogel, S. N., Sipe, J. D., Murphy, P. A., Mizel, S. B., Oppenheim, J. J., and Rosenstreich, D. L.,** The role of macrophages in the acute phase response: SAA inducer is closely related to lymphocyte activating factor and endogenous pyrogen, *Cell. Immunol.,* 63, 164, 1981.
53. **Gitlin, J. D. and Colten, H. R.,** Molecular biology of the acute phase plasma proteins, *Lymphokines,* 14, 123, 1987.
54. **Arend, W.,** Interleukin 1 receptor antagonist, *J. Clin. Invest.,* 88, 1445, 1991.
55. **Zahedi, K., Seldin, M. F., Rits, M., Ezekowitz, R. A. B., and Whitehead, A. S.,** Mouse IL-1 receptor antagonist protein. Molecular characterization, gene mapping, and expression of mRNA in vitro and in vivo, *J. Immunol.,* 146, 4228, 1991.
56. **Sipe, J. D., Bartle, L. M., and Loose, L. D.,** Modification of proinflammatory cytokine production by the antirheumatic agents tenidap and naproxen: a possible correlate with clinical acute phase response, *J. Immunol.,* 148, 480, 1992.
57. **Sipe, J. D., Rokita, H., Bartle, L. M., Loose, L. D., and Neta, R.,** The IL-1 receptor antagonist simultaneously inhibits SAA and stimulates fibrinogen synthesis in vivo and in vitro: a proposed mechanism of action, *Cytokine,* 3, 497, 1991.
58. **Huang, J. H., Rienhoff, H. Y., and Liao, W. S. L.,** Regulation of mouse serum amyloid A gene expression in transfected hepatoma cells, *Mol. Cell. Biol.,* 10, 3619, 1990.
59. **Li, X., Huang, J. H., Rienhoff, H. Y., and Liao, W. S.-L.,** Two adjacent C/EBP-binding sequences that participate in the cell-specific expressions of the mouse serum amyloid A3 gene, *Mol. Cell. Biol.,* 12, 6624, 1990.
60. **Woo, P., Betts, J., and Edbrooke, M.,** The human serum amyloid A genes and their regulation by inflammatory cytokines, in *Amyloid and Amyloidosis 1990,* Nativig, J. B. et al., Eds., Kluwer, Dordrecht, 1991, 13.
61. **Edbrooke, M. R., Burt, D. W., Cheshire, J. K., and Woo, P.,** Identification of cis-acting sequences responsible for phorbol ester induction of human serum amyloid A gene expression via a nuclear factor kB-like transcription factor, *Mol. Cell. Biol.,* 9, 1908, 1989.
62. **Baumann, H. and Gauldie, J.,** Regulation of hepatic acute phase plasma protein genes by hepatocyte stimulating factors and other mediators of inflammation, *Mol. Biol. Med.,* 7, 147, 1990.
63. **Sipe, J. D., Johns, M. A., Ghezzi, P., and Knapschaefer, G.,** Modulation of serum amyloid A gene expression by cytokines and bacterial cell wall components, in *Eicosamoids, Apolipoprotein Particles and Atherosclerosis,* Malmendier, C. L. and Alaupovic, P., Eds. Plenum Press, New York, 1988, 193.
64. **McAdam, K. P. W. J., Foss, N. T., Garcia, C., DeLellis, R., Chedid, L., Rees, R. J. W., and Wolff, S. M.,** Amyloidosis and the serum amyloid A protein response to muramyl dipeptide analogs and different mycobacterial species, *Infect. Immun.,* 39, 1147, 1983.
65. **Sipe, J. D.,** Amyloidosis, *Annu. Rev. Biochem.,* 61, 947, 1992.
66. **Syvverson, P. V., Sletten, K., and Husby, G.,** Evolutionary aspects of protein SAA, in *Amyloid and Amyloidosis 1990,* Natvig, J. B. et al., Eds., Kluwer, Dordrecht, 1991, 111.
67. **Strachan, A., de Beer, F. C., Coetzee, G. A., Hoppe, H. C., Jeenah, M. S., and van der Westhuyzen, D. R.,** Characteristics of apo-SAA-containing HDL3 in humans, *Colloquium Prot. Biol. Fluids,* 34, 359, 1986.
68. **Hayes, K. C., Lim, M., Pronczuk, A., and Sipe, J. D.,** Disruption in lipoprotein metabolism during the acute phase response in hamsters, *FASEB J.,* 5, A1287, 1991.
69. **Linke, R. P.,** Amphipathic properties of the low molecular weight component of serum amyloid-A protein shown by charge-shift electrophoresis, *Biochim. Biophys. Acta,* 668, 388, 1981.
70. **Sipe, J. D., Gonnerman, W. A., Loose, L. D., Knapschaefer, G., Xie, W.-J., and Franzblau, C.,** Direct binding enzyme-linked immunosorbant assay (ELISA) for serum amyloid A (SAA), *J. Immunol. Methods,* 125, 125, 1989.
71. **Hazenberg, B. P. C., Limburg, P. C., Bijzet, J., and van Rijswijk, M. H.,** Monoclonal antibody based ELISA for human SAA, in *Amyloid and Amyloidosis 1990,* Natvig, J. B. et al., Eds., Kluwer, Dordrecht, 1991, 898.

72. **Sipe, J. D., de Beer, F. C., Pepys, M., Husebekk, A., Skogen, B., Kisilevsky, R., Selkoe, D., Buxbaum, J., Linke, R. P., and Gertz, M. A.,** Report of special session on bioassays and standardization of amyloid proteins and precursors, in *Amyloid and Amyloidosis 1990,* Natvig, J. B. et al., Kluwer, Dordrecht, 1991, 745.
73. **Zahedi, K., Gonnerman, W. A., de Beer, F. C., de Beer, M. C., Steel, D. M., Sipe, J. D., and Whitehead, A. S.,** Major acute phase protein synthesis during chronic inflammation in amyloid susceptible and resistant mouse strains, *Inflammation,* 15, 1, 1991.
74. **de Beer, M. C., de Beer, F. C., Beach, C. M., Carreras, I., and Sipe, J. D.,** Mouse serum amyloid A protein: complete amino acid sequence and mRNA analysis of a new isoform, *Biochem. J.,* 283, 673, 1992.
75. **Creighton, T. E.,** Electrophoretic analysis of the unfolding of proteins by urea, *J. Mol. Biol.,* 137, 61, 1979.
76. **Creighton, T. E.,** Kinetic study of protein unfolding and refolding using area gradient electrophoresis, *J. Mol. Biol.,* 137, 61, 1980.
77. **de Beer, F. C.,** unpublished observations.
78. **Sipe, J. D., Rokita, H., and de Beer, F. C.,** unpublished observations.
79. **Otterness, I.,** personal communication.

F. Signal Transduction of Cytokines in Hepatocytes

Chapter 29

SIGNAL TRANSDUCTION MECHANISMS REGULATING CYTOKINE-MEDIATED INDUCTION OF ACUTE PHASE PROTEINS

Mahrukh K. Ganapathi

TABLE OF CONTENTS

0-8493-6913-4/93/$0.00 + $.50

I. INTRODUCTION

The acute phase response represents one of the organism's first line of defense following inflammatory stimuli. This response, unlike the immune response, is nonspecific in nature and is characterized by changes in a wide variety of systemic and metabolic processes, including alterations in the plasma concentrations of a group of proteins of hepatic origin referred to as the acute phase proteins[1,2] (see Chapter 1 of this volume). It was originally thought that acute phase protein synthesis was regulated synchronously by a single cascade of events initiated by a defined mediator, which was considered to be interleukin (IL)-1. However, extensive studies carried out during the past decade have revealed significant heterogeneity in the molecular mechanisms regulating the synthesis of acute phase proteins.[3]

The key mediators of acute phase protein synthesis in hepatocytes are cytokines. Several cytokines have been implicated in the acute phase response.[4-11] IL-6 seems to be the principal cytokine involved in the acute phase response, since it regulates the synthesis of almost all acute phase proteins in different species. IL-1 also regulates the synthesis of a large, although not as extensive, spectrum of acute phase proteins. More recently, leukemia inhibitory factor (LIF), IL-11, and oncostatin M, all of which share biological properties with IL-6, have been shown to induce a set of acute phase proteins similar to that induced by IL-6. In addition, other cytokines, such as tumor necrosis factor-α (TNFα), interferon-γ, and transforming growth factor-β have been shown to influence the synthesis of a limited set of acute phase proteins. *In vitro* studies have demonstrated that these cytokines can act individually or interact with each other or with other hormones, such as corticosteroids, and lead to the production of specific subsets of acute phase proteins.[8-17] These data, therefore, suggest that the pattern of acute phase proteins produced *in vivo* following different inflammatory stimuli or pathophysiologic conditions may vary, depending on the type and number of extracellular signals that the hepatocyte is exposed to and the mechanisms by which the various signal transduction pathways integrate these signals. Thus, a comprehensive understanding of these complex regulatory processes requires not only a knowledge of the different cytokines involved in regulating acute phase protein synthesis, but also an understanding of how individual cytokines transduce their signals and the mechanisms by which the various signal transduction pathways interact and led to altered synthesis of acute phase proteins. This chapter focuses on the signaling mechanisms by which various cytokines, in particular IL-6, transduce their signals and the mechanisms that are involved in the regulation of acute phase protein synthesis.

II. SIGNAL TRANSDUCTION MECHANISMS INVOLVED IN THE PLEIOTROPIC ACTIONS OF CYTOKINES

Cytokines are multifunctional signals that regulate diverse and, in most cases, redundant biological functions.[18-21] It is generally thought that each individual cytokine transduces its message via one or more clearly defined pathways, which can act independently or converge at some step downstream to regulate the pleiotropic functions of cytokines. In the broadest sense, signal transduction cascades encompass a coordinated series of pretranscriptional events that occur sequentially between the plasma membrane and the nucleus.

A. CYTOKINE RECEPTORS AND RECEPTOR-ASSOCIATED PLASMA MEMBRANE PROTEINS: STRUCTURE-FUNCTION RELATIONSHIP

The initial event in cytokine-mediated signal transduction cascades involves interaction of the cytokine with specific cell-surface receptors. The receptors for several cytokines belong to a distinct receptor family designated "cytokine receptor family", which includes the receptors for IL-6 (both the 80-kDa binding subunit and gp130, the signal transducing subunit), LIF, oncostatin M, IL-2 (β-chain), IL-3, IL-4, IL-7, granulocyte/macrophage colony-stimulating factor (GM-CSF), granulocyte colony-stimulating factor (G-CSF), erythropoetin, prolactin, and growth hormone.[19,22-28] Members of this family share significant amino acid homology in the extracellular domain; consensus regions including fibronectin type III modules, four cysteine residues in the N-terminal region, and a common amino acid motif, WSXWS, near the transmembrane region are found. The cytoplasmic regions of these receptors are more diverse, although some similarities within this region are observed between certain members of this family. For example a serine-rich region is present in the middle of the cytoplasmic domain of the receptors for IL-2 (β-chain), IL-4, G-CSF, and gp130. The cytoplasmic regions of the IL-6 and GM-CSF receptor are too short and, by themselves, incapable of activating signal transduction. Unlike the receptors for several growth factors, e.g., epidermal growth factor, platelet-derived growth factor, insulin, insulin-like growth factor, and colony-stimulating factor 1, which posess intrinsic tyrosine kinase activity, the receptors belonging to the cytokine receptor family do not encode sequences found in the catalytic domains of protein kinases. Thus, these findings suggest that the intracellular postreceptor signaling mechanisms likely differ and do not involve activation of receptor-associated tyrosine kinases as a first step in the signaling pathway.

Several cytokine receptors function as multicomponent systems, consisting primarily of a binding component(s) and a signal transducing component(s). The binding subunit often represents the low-affinity receptor, which can interact with the signal transducing subunit, following ligand binding, to form the high-affinity receptor complex capable of initiating signal transduction. In the case of the IL-2 receptor, both subunits, α and β, individually bind IL-2 with low and intermediate affinities, respectively, whereas the αβ complex binds IL-2 with high affinity.

IL-6-stimulated signal transduction requires at least a two-component receptor system consisting of a binding (M_r, 80 kDa) and signal transducing subunit (M_r, 130 kDa).[29] Following interaction of IL-6 with the 80-kDa binding subunit (low-affinity binding), the IL-6-receptor complex associates with another membrane glycoprotein, gp130, to form a trimolecular high-affinity complex, (IL-6- IL-6 receptor-gp130), which is capable of initiating signal transduction. The interaction between gp130 and the 80-kDa binding subunit occurs within the extracellular regions of these molecules. Thus, soluble IL-6 receptors, consisting only of the extracellular region and devoid of the transmembrane and cytoplasmic regions of the 80-kDa subunit, are capable of stimulating signal transduction. This phenomenon may be physiologically advantageous, since cells lacking IL-6 receptors can be made to transduce the IL-6 signal.

Structure-function analysis of gp130 has demonstrated that a 61-amino acid region proximal to the transmembrane domain is critical for IL-6-signal transduction.[30,31] This region contains two conserved amino acid segments (P-X-P preceded by a cluster of hydrophobic amino acids, and a cluster of hydrophobic amino acids ending with one or two positively charged amino acids) that are also present in the cytoplasmic region of other members of the cytokine receptor family, including KH97, a GM-CSF receptor-associated molecule. The cytoplasmic region of gp130 contains consensus sequences present in the nucleotide binding domain of protein kinases and in GTP binding sites similar to those found in the ras proteins.[30] However, these regions are not essential for signal transduction. Consensus sequences found in the catalytic domains of different protein kinases are absent in the cytoplasmic region of gp130.

Recent studies carried out for the receptor systems of LIF and oncostatin M, two cytokines that share many biological properties with IL-6, including induction of acute phase proteins, demonstrate several common features with the IL-6 receptor system.[26,27] The receptor systems for these three cytokines employ gp130 as one of the subunits. The LIF receptor system is similar to that of IL-6, consisting of a binding subunit (M_r ~190 kDa), which is related to gp130, and a signal transducing subunit, gp130. In contrast, oncostatin M binds to gp130 with low affinity and requires the LIF receptor binding subunit for intermediate-affinity binding. The redundancies observed in the biological activities of IL-6, LIF, and oncostatin M are likely due to activation of similar signal transduction pathways by gp130, which is shared by the receptor systems of these three cytokines. The receptor systems of GM-CSF, IL-3, and IL-5 also employ a common signal transducing, high-affinity converting subunit, KH97.[28]

IL-1 is another cytokine that is important for regulating the synthesis of several acute phase proteins. Two receptors for IL-1, type I (mr, 80 kDa) and type II (mr, 60 kDa), have been cloned.[32,33] Both these receptors are similar in their overall structure, but totally different from those belonging to the cytokine receptor family. The extracellular regions of these two receptors are similar and consist of three immunoglobulin-like domains that are involved in ligand binding (IL-α, IL-1β, and IL-1 receptor antagonist). The IL-6 receptor also contains one immunoglobulin-like domain which is not involved in binding IL-6. The major difference between the type I and type II receptors is in the cytoplasmic domain, which consists of 215 amino acids in the type I receptor and only 29 amino acids in the type II receptor, which suggests that the two receptors signal via different mechanisms. While the type I receptor is potentially capable of stimulating signal transduction, the type II receptor must associate with other membrane proteins and lead to signal transduction. Two receptors of TNFα (mr of 55 and 75 kDa) have also been cloned and shown to possess primary structures that are different from those found in the receptors belonging to the cytokine receptor family.[34-36] These two receptors exhibit significant homology in the extracellular regions, but differ considerably in their intracellular cytoplasmic domains, thus suggesting that they may be capable of triggering different signal transduction pathways.

B. POSTRECEPTOR SIGNALING MECHANISMS

Signaling mechanisms that occur downstream of the receptor are poorly understood. It is now clearly evident that regulation of a given biological function, especially more complicated processes such as cell proliferation or differentiation, involves activation of a complex network consisting of interactions between diverse signal transduction pathways, which ultimately converge at some distal step downstream in the signal transduction cascade. These coordinated series of cytosolic events would then regulate the activities of nuclear transcription factors responsible for modulating the transcription of specific response genes mediating the cellular effects of extracellular signals. Studies carried out to date have identified certain key events that are essential in the signal transduction cascades; however, information about the ordered sequence of events within the network is still forthcoming.

Cytokines, especially those involved in regulating the synthesis of acute phase proteins, have been shown to activate novel signal transduction pathways. Several studies have demonstrated that IL-6-stimulated signal transduction mechanisms are not mediated by the classical second-messenger pathways involving cAMP, Ca^{2+}, phosphatidylinositol turnover, or activation of protein kinase A or protein kinase C (PKC).[29,37-40] However, in one study, IL-6 was shown to lead to a transient, small increase in the concentration of inositol 1,4,5-trisphosphate and Ca^{2+} in rat mesangial cells, wheres in another study, IL-6 led to a moderate decrease in the concentration of inositol phosphates in pituitary cells, thereby antagonizing TRH-stimulated prolactin secretion.[41,42] The defined second-messenger signal transduction pathways have been variably shown to mediate the pleiotropic functions of IL-1, TNFα, and interferon-γ.[43] Since two receptors for both IL-1 and TNFα have been identified, it is possible that one type of receptor may be capable of signaling via defined signal transduction pathways. Similarly, the interferon-γ receptor system, which consists of a binding component and at least one signaling component, may display variability in the signaling component in different cell types, some of which may signal by activating known second-messenger systems.

Activation of G-proteins, which leads to increased GTP binding and hydrolysis of GTP, is another common intermediate step involved in several signal transduction pathways. Indeed, it has been demonstrated that this mechanism is employed in both IL-1 and TNFα stimulated signal transduction pathways.[43,44] However, IL-6-stimulated signal transduction does not seem to employ this mechanism, since the GTP binding site present in gp 130 is not essential for IL-6-activated signal transduction.[30]

The receptors for most cytokines do not encode amino acid sequences found in the catalytic domains of protein kinases, and therefore activation of receptor kinases does not represent a first step in their signal transduction cascade. Nevertheless, studies of postreceptor signaling mechanisms for several cytokines have revealed that protein phosphorylation-dephosphorylation plays a central role in cytokine-mediated signaling cascades. Thus, cytokine receptors may be capable of interacting with other kinases which trigger the phosphorylation-dephosphorylation cascade. Indeed, it has been demonstrated that gp130 can be tyrosine phosphorylated, which is prevented in receptors mutated in the critical region within the cytoplasmic domain of gp130.[30] This mutation also leads to abrogation of IL-6 signal transduction. In addition, IL-1 and interferon-γ receptors are rapidly phosphorylated on serine-threonine residues following ligand binding.[45,46] However, it is not clear whether these phosphorylations are a consequence of activation of signal transduction pathways or essential for signal transduction.

Recent studies of IL-6-mediated signal transduction pathways have identified key events. Two independent studies, examining IL-6-stimulated activation of B-cell hybridoma and plasmacytoma proliferation and IL-6- and LIF-stimulated terminal differentiation of myeloid leukemia cells, demonstrated that activation of novel protein kinase signal transduction cascades were required for IL-6- and LIF-stimulated signal transduction.[39,40] Two key events in this signal transduction pathway were identified and shown to involve activation of a tyrosine kinase that led to rapid phosphorylation of a 160-kDa cellular protein, followed by activation of a novel H7-sensitive protein kinase. The latter protein kinase does not correspond to any known protein kinase inhibited by H7, including PKC, cAMP-, or -cGMP-dependent protein kinase, Ca^{2+}/calmodulin-dependent protein kinase, Raf-1, microtubule-associated protein kinase, or casein kinase II. Both these events are required for transcriptional activation of several primary response genes thought to mediate the effects of IL-6 and LIF.

Another study examining the signal transduction mechanisms involved in IL-6- as well as interferon- and TGFβ-induced growth arrest (G_0/G_1 phase block) of the M1 myeloblastic cell line demonstrated two critical events in the signaling cascade.[47] These events involved dephosphorylation of the retinoblastoma (RB) gene product and inhibition of expression of the c-*myc* gene, due to activation of parallel signal transduction pathways, which were regulated independently of each other. If was further demonstrated that dephosphorylation of RB was not simply an indirect consequence of the cytokine-induced changes in cell-cycle distribution. Thus, these findings suggest that cytokines could increase the activity of a cell cycle-specific phosphatase(s) or decrease the activity of phase-specific kinase(s), which leads to an underphosphorylated state of RB. Indeed, activation of protein phosphatases has been implicated in the signal transduction mechanisms involved in the proliferative effects of IL-4 in leukemic cell lines[48] and in IL-6-mediated induction of acute phase proteins.[49]

The signal transduction pathways involved in the actions of IL-1, TNFα, and interferon-γ also involve modulation of the activities of protein kinases and/or protein phosphatases. These cytokines increase the phosphorylation of certain cellular proteins within minutes of

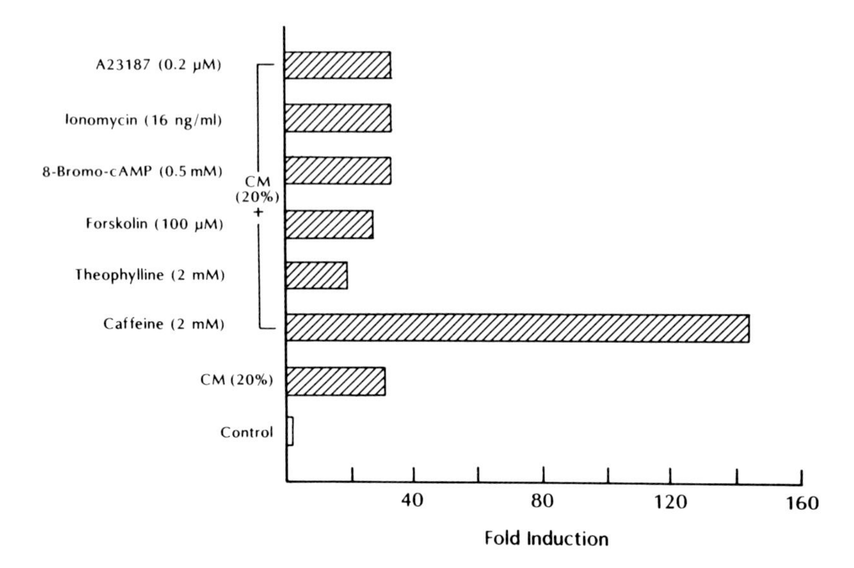

FIGURE 1. Effect of different pharmacologic agents on the induction of CRP by monocyte-conditioned medium in Hep 3B cells. Hep 3B cells were incubated in serum-free RPMI-1640 containing 1 μM dexamethasone, 0.02 U/ml insulin, and 20% conditioned medium (CM) in the absence or presence of various agents for 18 to 24 h at 37°C and labeled with L-[^{35}S]methionine for 4 h. Control cells were incubated in the absence of conditioned medium. Radiolabeled CRP secreted into the culture medium was analyzed by immunoprecipitation, followed by fractionation of the immunoprecipitates on SDS-polyacrylamide gels. Radioactivity in total proteins was estimated by trichloroacetic acid precipitation, and radioactivity in the acute phase protein bands was determined following digestion with hydrogen peroxide. Fold induction corresponds to the ratio of radioactivity (cpm) in treated cultures to that in control cultures. Values correspond to the average of two or more independent experiments. (From Ganapathi, M. K., Mackiewicz, A., Samols, D., Brabenec, A., Kushner, I., Schultz, D., and Hu, S.-H., *Biochem. J.*, 269, 41, 1990. With permission.)

stimulation. Two proteins that have been identified to be phosphorylated following IL-1 and TNFα stimulation are the epidermal growth factor receptor and the small heat-shock protein, hsp27,[50,51] the latter being phosphorylated by a novel protein kinase.[51] The spectrum of proteins phosphorylated by IL-1 and TNFα is very similar and involves phosphorylation primarily on serine/threonine residues, although tyrosine phosphorylation is also observed.[52,54] Enhanced phosphorylation is thought to result from activation of multiple protein kinases, some of which are likely to represent novel enzymes, distinct from the well-characterized kinases.

III. SIGNAL TRANSDUCTION MECHANISMS INVOLVED IN INDUCTION OF ACUTE PHASE PROTEINS

Studies of the signal transduction mechanisms involved in the regulation of acute phase proteins have been minimally undertaken. These studies have revealed significant heterogeneity in the intracellular signaling mechanisms. It has also been demonstrated that cytokine-mediated induction of acute phase proteins does not involve activation of well-defined signal transduction pathways; rather, novel signaling mechanisms are employed. The second messengers cAMP and inositol 1,4,5-trisphosphate/Ca^{2+} do not mediate cytokine-induced changes in the synthesis of several acute phase proteins, including C-reactive protein (CRP).[37,38] In one study,[38] it was shown that several agents that mimic the effect of increased levels of cAMP or Ca^{2+} did not influence the synthesis of CRP in either the absence or presence of cytokines (a crude preparation of lipopolysaccharide-stimulated conditioned medium from monocytes) in the human hepatoma cell line, Hep 3B (Figure 1). Monocyte-conditioned

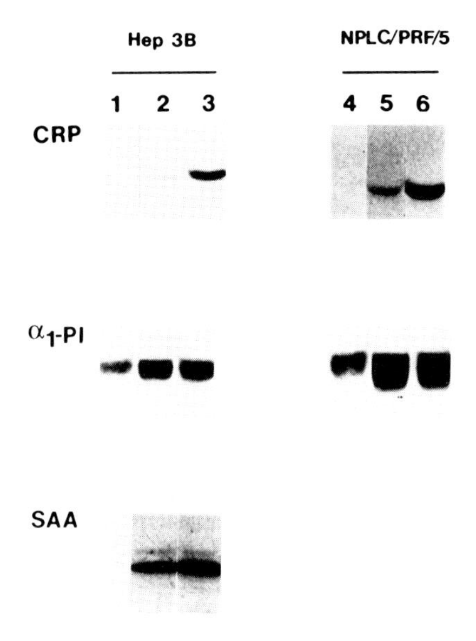

FIGURE 2. Effect of caffeine on induction of acute phase proteins by IL-6 plus IL-1α in Hep 3B cells and by IL-6 in NPLC/PRF/5 cells. Hep 3B cells (lanes 1 to 3) were incubated with IL-6 (25 ng/ml) plus IL-1α (0.04 ng/ml) in the absence (lane 2) or presence of caffeine (lane 3). Control cells were incubated in the absence of cytokines (lane 1). NPLC/PRF/5 cells (lanes 4 to 6) were treated with IL-6 (25 ng/ml) alone in the absence (lane 5) or presence of caffeine (lane 6). Control NPLC/PRF/5 cells were incubated in the absence of cytokines (lane 4). Cells were radiolabeled with L-[^{35}S]methionine, and labeled CRP, SAA, and α_1-PI synthesized and secreted into the culture medium were determined by immunoprecipitation, SDS-PAGE, and autoradiography, as described in the legend to Figure 1. (From Ganapathi, M. K., Mackiewicz, A., Samols, D., Brabenec, A., Kushner, I., Schultz, D., and Hu, S.-H., *Biochem. J.*, 269, 41, 1990. With permission.)

medium also did not lead to changes in the concentration of cAMP, although this cell line was capable of exhibiting a 200- to 300-fold increase in the level of cAMP in response to forskolin, a potent activator of adenylate cyclase. Interestingly, it was observed that the methylxanthine caffeine, but not theophylline (both of which have been shown to lead to increased levels of cAMP and Ca^{2+} [55,56]), led to significant potentiation (three- to fourfold) of the induction of CRP synthesis and mRNA levels by conditioned medium or the defined cytokines IL-6 alone or in combination with IL-1α in the NPLC/PRF/5 and Hep 3B cell lines, respectively (Figure 2). The total increase in the synthesis of CRP observed in the presence of caffeine and cytokines ranged between 40- and 180-fold above that seen in control cultures. This degree of induction approximates responses seen *in vivo* in many inflammatory states, suggesting that caffeine might be mimicking or modifying one or more signal transduction pathway(s) involved in the induction of CRP. The exact mechanism by which caffeine potentiates the induction of CRP is not known, but is independent of increases in the levels of cAMP or Ca^{2+}.

In contrast to the potentiating effect of caffeine on the induction of CRP by cytokines, caffeine minimally augmented the induction of several other APP such as serum amyloid A (SAA), α_1-protease inhibitor (α_1-PI), and α_1-antichymotrypsin by cytokines. These observations suggest that the intracellular mechanisms regulating the synthesis of CRP differ from those regulating some other APP, and it is possible that caffeine might activate a unique intracellular step involved in the induction of CRP, but not other acute phase proteins.

A. ROLE OF PROTEIN KINASE C

Several studies have evaluated the role of PKC in the induction of acute phase proteins by cytokines. Evans et al.[57] observed that 12-*O*-tetradecanoylphorbol-13-acetate (TPA) and

1-oleoyl-2-acetyl-sn-glycerol (OAG), two activators of PKC, led to increased mRNA levels of several acute phase proteins, including fibrinogen, in a rat and human hepatoma cell line, which led them to conclude that the induction of acute phase proteins by cytokines was mediated by PKC. TPA was also shown to mimic the inducing effect of several cytokines on SAA mRNA levels in transfected mouse L-cells,[58] suggesting that activation of PKC mediates the induction of SAA by cytokines. However, it is not clear from these studies whether the effect of cytokines is mediated by PKC or whether cytokines and TPA lead to the induction of acute phase proteins via distinct mechanisms. Subsequent studies carried out by Baumann et al.[59] and Kurdowska et al.[60] addressed this question. It was shown that although TPA led to a moderate induction of several acute phase proteins in rat and human hepatoma cell lines and in rat primary hepatocyte cultures, the induction of acute phase proteins by IL-6 and/or IL-1 was PKC independent. However, activation of PKC could modulate the effect of cytokines on the induction of acute phase proteins; the effect of IL-6 was upmodulated by PKC, whereas the effect of IL-1 was downmodulated due to the downregulation of IL-1 receptors by TPA.[59]

In our studies examining the role of PKC in the induction of the acute phase proteins CRP, SAA, fibrinogen, and α_1-PI in the human hepatoma Hep 3B and NPLC/PRF/5 cell lines, different results were observed. In these two cell lines, induction of CRP and SAA requires different extracellular signals, suggesting that the cell lines may differ in the intracellular signal transduction mechanisms regulating the synthesis of CRP and SAA. In NPLC/PRF/5 cells, IL-6 alone is capable of inducing the synthesis of these two acute phase proteins as well as fibrinogen and α_1-PI. In contrast, in the Hep 3B cell line, IL-6 alone is capable of inducing fibrinogen, α_1-PI, and some other acute phase proteins, whereas induction of CRP and SAA requires cooperative interaction between IL-6 and IL-1.

The phorbol esters, TPA and 4-phorbol-12,13-didecanoate (PDD), had no effect on the synthesis of CRP, SAA, fibrinogen, and α_1-PI, but inhibited in a concentration-dependent manner (10 to 80 ng/ml) the induction of CRP, SAA, and fibrinogen by IL-6 plus IL-1α, and of fibrinogen by IL-6 alone in Hep 3B cells (Figure 3A). In contrast, induction of CRP and fibrinogen by IL-6 in NPLC/PFR/5 cells was minimally affected by TPA (Figure 3B), and induction of α_1-PI by IL-6 alone or in combination with IL-1α was not affected by TPA in either cell line (Figure 3). The synthetic diacylglycerol analog, OAG, exerted a minimal inhibitory effect on the induction of these three acute phase proteins in Hep 3B cells.

The inhibitory effect of TPA on the induction of acute phase proteins by cytokines was not due to downregulation of the IL-1 receptor, since induction of fibrinogen by IL-6 alone was also inhibited by TPA, nor was it due to downregulation of PKC. The latter conclusion was based on results of the effect of TPA on the levels of PKC isoforms present in the hepatoma cell lines. Three isoforms of PKC (PKCα, β_{II}, and ζ) that were detected in the two cell lines were differentially downregulated by various treatments with TPA (Figure 4). (1) PCKα was downregulated by TPA at a concentration of 40 ng/ml, but not 10 ng/ml, although both concentrations inhibit the induction of CRP, SAA, and fibrinogen by cytokines. (2) PKCα was not affected following treatment of Hep 3B cells with 40 ng/ml TPA for only the first hour of an 18- to 20-h incubation with medium alone or IL-6 plus IL-1α, even though this treatment leads to inhibition of the induction of CRP, SAA, and fibrinogen by cytokines. (3) PKCβ_{II} and PKCζ were not affected by TPA at concentrations of 10 and 40 ng/ml or following brief exposure to 40 mg/ml TPA. Thus, only one TPA treatment led to downregulation of one of the PKC isoforms, PKCα, indicating that downregulation of PKC by TPA cannot be responsible for the inhibitory effect of TPA on the induction of these acute phase proteins by cytokines. In addition, the cytokines IL-6 and IL-1α, acting alone or in combination with each other, did not influence the amount of these three PKC isoforms in Hep 3B cells.

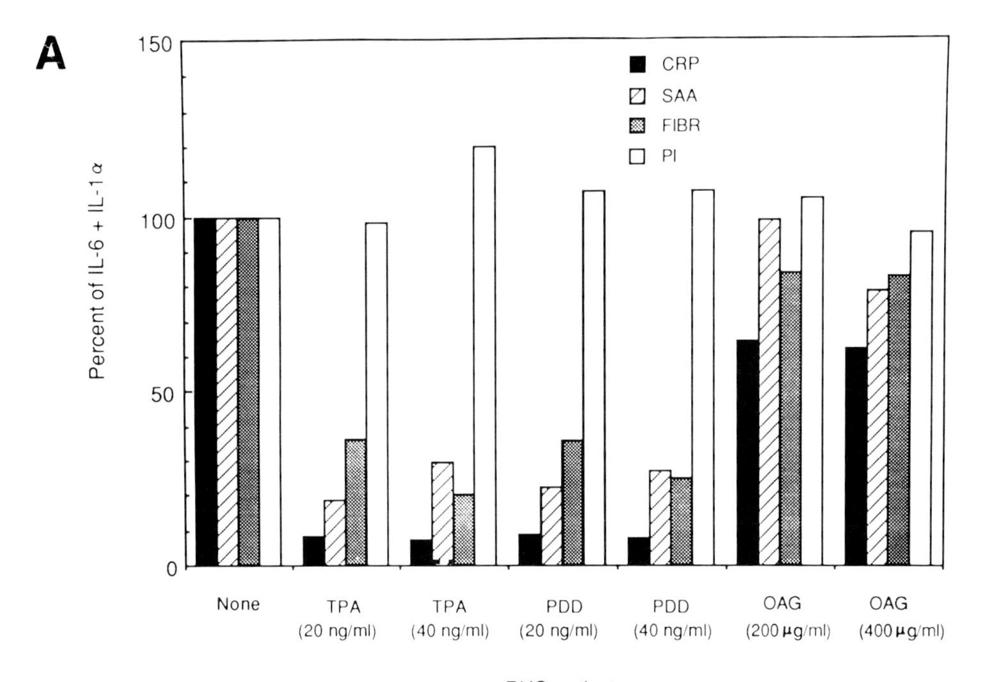

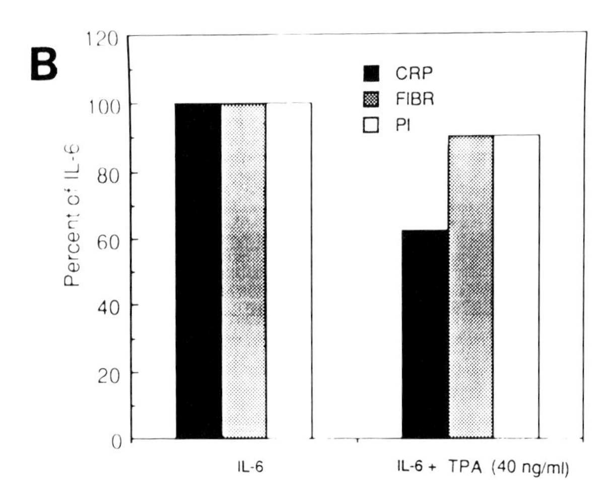

FIGURE 3. Effect of PKC activators on induction of acute phase proteins by IL-6 plus IL-1α in Hep 3B cells (A) and by IL-6 in NPLD/PRF/5 cells (B). Hep 3B cells were incubated with IL-6 (20 ng/ml) plus IL-1α (0.04 ng/ml) in the absence or presence of various concentrations of PKC activators. NPLC/PRF/5 cells were incubated in the absence or presence of IL-6 (20 ng/ml) or IL-6 (20 ng/ml) plus TPA (40 ng/ml). L-[^{35}S]methionine-labeled CRP, SAA, fibrinogen (FIBR), and α_1-PI secreted into the culture medium were determined by immunoprecipitation, SDS-PAGE, and autoradiography, as described in the legend to Figure 1. Radioactivity in the acute phase protein bands was determined following digestion with hydrogen peroxide, and the amount present in cells treated with cytokines was normalized to 100%. The ordinate corresponds to the percent ratio of the radioactivity in cells treated with cytokines plus PKC activators to that present in cells treated with cytokines alone.

When the induction of CRP, SAA, fibrinogen, and α_1-PI by IL-6 plus IL-1α was observed following pretreatment of Hep 3B cells with TPA (10 and 40 ng/ml) for 24 h, almost complete inhibition of the induction of CRP was obtained, whereas induction of SAA and fibrinogen was only partially (13 to 45%) blocked following pretreatment with 40 ng/ml TPA and not affected following pretreatment with 10 ng/ml TPA. TPA was further capable of inhibiting

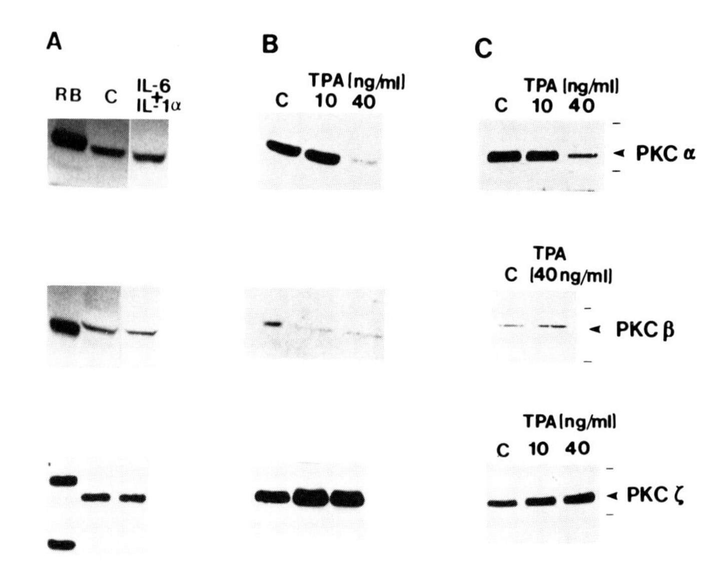

FIGURE 4. Effect of TPA and IL-6 plus IL-1α on expression of PKC isoforms in Hep 3B (Panels A and B) and NPLC/PRF/5 (Panel C) cells. Hep 3B and NPLC/PRF/5 cells were incubated in serum-free RPMI-1640 containing 1 μ*M* dexamethasone and 0.02 U/ml of insulin for 24 h in the absence or presence of IL-6 (29 ng/ml) plus IL-1α (0.67 ng/ml) or different concentrations of TPA. RB corresponds to rat brain extract and C to control cultures incubated in serum-free RPMI-1640 containing 1 μ*M* dexamethasone and 0.02 U/ml of insulin. Aliquots of cell extracts prepared in 20 m*M* Tris-HCL, pH 7.4, 1% NP-40, were subjected to immunoblot analyses. PKC isoforms were detected by the ECL 1 m MPMSF, 200 μg/ml leupehin, 25 μg/ml aprohnin and 20 μg/ml pepstatin (enhanced chemiluminescence) Western-blotting detection system (Amersham). Equal amounts of protein were loaded in each lane for a given experiment, except in B (Western blot of PKCζ). In this Western blot, the protein concentration loaded in lane C (control cultures) is much less than in the other two lanes; protein concentrations in C, TPA (10 ng/ml), and TPA (40 ng/ml) were 15, 37,and 31 μg per lane, respectively. Thus, the intensity of the band in control cultures is underrepresented.

the induction of SAA and fibrinogen by IL-6 plus IL-1α in cells preexposed to TPA. These data are in agreement with the above conclusion that activation of PKC by TPA leads to inhibition of induction of CRP, SAA, and fibrinogen by cytokines, since TPA is not readily metabolized and is therefore available for continuous activation of a PKC isoform(s) that is not downregulated by TPA. The difference in the effect of TPA pretreatment on the induction of CRP by IL-6 plus IL-1α and on the induction of SAA and fibrinogen suggests that induction of CRP by IL-6 plus IL-1α may be more sensitive to the inhibitory effect of TPA. The differential effect of TPA on the downregulation of various PKC isoforms indicates that results of classical "PKC downregulation" experiments, in which the effects on individual isoforms are not determined, should be interpreted with caution.

The reason for the discrepancies in the role of PKC in the regulation of acute phase protein induction in different studies is not clear, but could be due to differences in cell model systems (rat hepatocytes, rat hepatoma cell lines, transfected mouse L-cells, human hepatoma cell lines Hep G2, Hep 3B, and NPLC/PRF/5 cells). Since most of these cell lines are transformed, subtle differences in some step in the intracellular signaling cascade could led to altered cellular responses. The *in vivo* consequence of this finding may be that activation of PKC could lead to up- or downmodulation of the synthesis of different acute phase proteins under different pathophysiological conditions, due to interaction between different signal transduction pathways. Since different PKC isoforms may exert differential effects on the synthesis of various acute phase proteins, the net outcome may depend on the type of isoform present in a given cell line and its ability to interact with the cytokine-mediated signal transduction pathways. However, it seems clear that in hepatocytes, IL-6- or IL-1-mediated induction of acute phase proteins is PKC independent. PKC involvement in the IL-1-mediated induction of transfected SAA genes in fibroblasts may be due to activation of a signal transduction pathway different from that activated by IL-1 in hepatocytes, possibly via distinct IL-1 receptors. Indeed, PKC has been shown to play a variable role in IL-1-stimulated signal transduction pathways in different systems.

The mechanism by which PKC influences the signal transduction pathways for the induction of acute phase proteins is not known at the present time. However, the above data indicate that a coordinated cascade of protein phosphorylation-dephosphorylation reactions is important in regulating acute phase protein synthesis. The inhibitory effect of PKC on the induction of CRP, SAA, and fibrinogen by cylokines in Hep 3B cells suggests that one of the steps in the signal transduction pathway involves dephosphorylation of a key protein(s), which could lead to activation of transcription or a posttranscriptional event regulating the synthesis of these acute phase proteins. If, indeed, a dephosphorylation event is important, this would suggest that one possible mechanism by which IL-6 or IL-6 plus IL-1α transduce their signal and lead to induction of some acute phase proteins involves activation of a protein phosphatase(s).

B. ROLE OF PROTEIN PHOSPHATASES

Until recently, protein phosphatases were not considered to be key players in signal transduction cascades and were thought to be important only for reversing the effect of protein kinases. However, this viewpoint has changed and protein phosphatases are considered to play central and specific roles in cellular physiology.[61-63] The role of protein phosphatases has been evaluated in the signal transduction mechanisms regulating the synthesis of acute phase proteins by IL-6 in the absence or presence of cytokines, and these studies have revealed that the serine protein phosphatase 1 and/or 2A may be important in regulating the induction of some of the acute phase proteins.[49]

Okadaic acid (OA), an inhibitor of protein phosphatases 1 and 2A, has served as a valuable tool for evaluating the role of protein phosphatases in several cellular functions. Examination of the effect of OA on the induction of acute phase proteins by the cytokines IL-6 and IL-1 in Hep 3B and NPLC/PRF/5 cells revealed that [49] (1) in Hep 3B cells, OA inhibited the induction of CRP, SAA, and fibrinogen by IL-6 plus IL-1α, and of fibrinogen by IL-6 alone in a concentration-dependent manner (5 to 20 n*M*) (Figure 5A), and (2) in NPLC/PRF/5 cells, OA inhibited the induction of CRP, fibrinogen, and α_1-PI by IL-6, albeit at a higher concentration (20 to 80 n*M*) (Figure 5B). The induction of CRP by IL-6 plus IL-1α was most sensitive to the inhibitory effect of OA; a concentration of 10 n*M* OA in Hep 3B cells and 40 n*M* in NPLC/PRF/5 cells led to almost complete inhibition of the induction of CRP, whereas induction of other acute phase proteins was inhibited by only 40 to 60%. In Hep 3B cells, concentrations of OA above 20 n*M* were toxic and OA below this concentration had no significant effect on the induction of α_1-PI by IL-6 plus IL-1α or IL-6 alone.

These results strongly suggest that induction of at least some acute phase proteins by IL-6 alone or in combination with IL-1 is mediated by activation of protein phosphatase 1 and/or 2A, thus supporting the above hypothesis that dephosphorylation of a key protein(s) may represent an important mechanism regulating the synthesis of acute phase proteins. Both OA and TPA, acting via distinct mechanisms (inhibition of protein phosphatase 1 and 2A and activation of PKC, respectively) would lead to increased phosphorylation of several cellular proteins, some of which could lead to repression of gene transcription or inhibition of a posttranscriptional event involved in the regulation of acute phase protein synthesis. The role of protein phosphatases has also been suggested for the signal transduction pathways

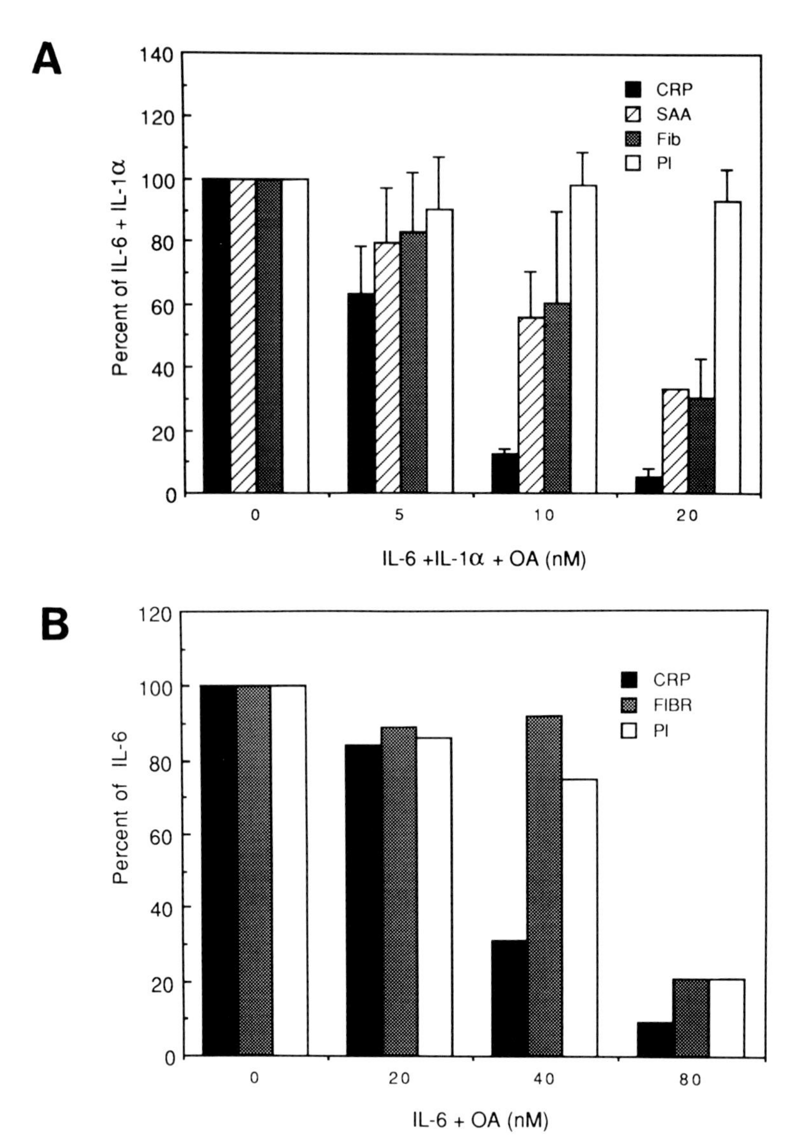

FIGURE 5. Dose-response curve for the effect of OA on induction of acute phase proteins by IL-6 plus IL-1α in Hep 3B cells and by IL-6 in NPLC/PRF/5 cells. Hep 3B cells (A) and NPLC/PRF/5 cells (B) were incubated with IL-6 (26 ng/ml) plus IL-1α (0.67 ng/ml) or IL-6 (29 ng/ml) alone, respectively, in the absence or presence of various concentrations of OA for 18 to 24 h and labeled with L-[^{35}S]methionine for 4 h. Radiolabeled CRP, SAA, fibrinogen (Fib), and α_1-PI secreted into the culture medium were analyzed by immunoprecipitation, followed by fractionation of the immunoprecipitate on SDS-polyacrylamide gels, fluorography, and autoradiography of the dried gels, as described in the legend to Figure 1. The relative intensities of the autoradiographic signals were determined by densitometric scanning. The value for IL-6 plus IL-1α or for IL-6 alone were normalized to 100%. The ordinate represents the percent ratio of the induction of acute phase proteins by cytokines in the presence of OA to that in the absence of OA. Values correspond to the mean ± SD of three independent experiments carried out in duplicate (A) or to the mean of two or more experiments (B). (From Ganapathi, M. K., *Biochem. J.*, 289, 645, 1992. With permission.)

involved in the inhibitory effect of IL-6 on the proliferation of hematopoeitic cells, since dephosphorylation of the RB gene product was observed.

The differential dose response of OA in inhibiting the synthesis of acute phase proteins by cytokines in the Hep 3B and NPLC/PRF/5 cell lines and the differential effect of OA on the induction of α_1-PI by cytokines in the two cell lines suggest that (1) the two cell lines may contain different amounts of protein phosphatase 1 and/or 2A and (2) induction of different acute phase proteins may require activation of different phosphatases. This could also explain why TPA, which almost completely abolished the induction of CRP, SAA, and fibrinogen by IL-6 plus IL-1 or IL-6 alone in Hep 3B cells, minimally inhibited the induction of CRP, fibrinogen, and α_1-PI in the NPLC/PRF/5 cell line.

IV. FUTURE DIRECTIONS

Our understanding of the signal transduction mechanisms regulating acute phase proteins is only beginning to be unraveled. The data obtained so far indicate that regulation of acute phase protein synthesis involves novel signal transduction mechanisms which are highly heterogeneous. The potential for differential interaction between various signal transduction pathways stimulated by either a single or multiple extracellular signal(s) allows for fine tuning of the cellular responses that can be achieved *in vivo*.

The complexity of the intracellular signal transduction network necessitates that initial studies be directed at identifying critical events in the signal transduction cascades stimulated by a single cytokine. These studies should be facilitated in mutant or transformed cell lines defective in specific signal transduction pathways. In our studies examining the signal transduction mechanisms involved in the induction of acute phase proteins, we have taken advantage of two hepatoma cell lines,Hep 3B and NPLC/PRF/5, which display different characteristics for the induction of CRP and SAA. In the Hep 3B cell line, IL-6 plus IL-1 is required for the induction of CRP and SAA, whereas in the NPLC/PRF/5 cell line, IL-6 alone induces these two acute phase proteins.[13] This suggests that the signal transduction mechanisms in these two cell lines differ. It is possible that Hep 3B cells are defective in one of the IL-6-stimulated signal transduction pathways that is essential for the induction of CRP and SAA, but not fibrinogen or α_1-PI, since fibrinogen and α_1-PI are induced by IL-6 alone in both cell lines. Alternatively, it is possible that the signal transduction pathways activated by IL-1 in Hep 3B cells (which in combination with IL-6 leads to the induction of CRP and SAA) is constitutively activated in NPLC/PRF/5 cells. In either case, it would imply that the induction of CRP and SAA requires activation of more than one parallel signal transduction pathway, which could act independently or converge at some step downstream.

It has been unequivocally established that coordinated phosphorylation-dephosphorylation reactions are involved in the pleiotropic functions of cytokines, including IL-6, LIF, and IL-1. Many of the protein kinases and phosphatases that catalyze these reactions have been shown to be novel. Thus, future efforts should be directed toward identifying and charcterizing these novel protein kinases and phosphatases and their physiologic substrates. Certainly, other covalent posttranslational events, distinct from phosphorylation-dephosphorylation reactions, could play a role in signal transduction; therefore, these need to be examined. Since several nuclear transcriptional factors, e.g., NF-κB[64] and NF-IL-6[65,66] have been identified in regulating the transcription of acute phase proteins, it is necessary to determine how signal transduction pathways affect nuclear events and regulate the activities of transcriptional factors.

ACKNOWLEDGMENTS

This work was supported by the Ohio Board of Regents, Research Initiation Grant BBR WM72428, Case Western University; Cuyahoga County Hospital Foundation Grant IMED-203; and Cancer Center Grant P30 CA 43703.

REFERENCES

1. **Kushner, I.,** The phenomenon of the acute phase response, *Ann. N.Y. Acad. Sci.,* 389, 39, 1982.
2. **Fey, G. H. and Gauldie, J.,** The acute phase response of the liver in inflammation, *Prog. Liver Dis.,* 9, 89, 1990.
3. **Kushner, I., Ganapathi, M., and Schultz, D.,** The acute phase response is mediated by heterogeneous mechanisms, *Ann. N.Y. Acad. Sci.,* 557, 19, 1989.
4. **Gauldie, J., Richards, C., Harnish, D., Lansdorp, P., and Baumann, H.,** Interferon β_2/B-cell stimulatory factor type 2 shares identity with monocyte-derived hepatocyte-stimulating factor and regulates the major acute phase protein response in liver cells, *Proc. Natl. Acad. Sci. U.S.A.,* 84, 7251, 1987.
5. **Ramadori, G., Sipe, J. D., Dinarello, C. A., Mizel, S. B., and Colten, H. R.,** Pretranslational modulation of acute phase hepatic protein synthesis by murine recombinant interleukin 1 (IL-1) and purified human IL-1, *J. Exp. Med.,* 162, 930, 1985.
6. **Perlmutter, D. H., Dinarello, C. A., Punsal, P. I., and Colten, H. R.,** Cachectin/tumor necrosis factor regulates hepatic acute phase gene expression, *J. Clin. Invest.,* 78, 1349, 1986.
7. **Magielska-Zero, D., Bereta, J., Czuba-Pelech, B., Pajdak, W., Gauldie, J., and Koj, A.,** Inhibitory effect of human recombinant interferon gamma on synthesis of acute phase proteins in human hepatoma Hep G2 cells stimulated by leukocyte cytokines, TNFα and IFN-β_2/BSF-2/IL-6, *Biochem. Int.,* 17, 17, 1988.
8. **Mackiewicz, A., Ganapathi, M. K., Schultz, D., Brabenec, A., Weinstein, J., Kelley, M. F., and Kushner, I.,** Transforming growth factor β1 regulates synthesis of acute phase proteins, *Proc. Natl. Acad. Sci. U.S.A.,* 87, 1491, 1990.
9. **Baumann, H. and Wong, G. G.,** Hepatocyte-stimulating factor III shares structural and functional identity with leukemia-inhibitory factor, *J. Immunol.,* 143, 1163, 1989.
10. **Baumann, H. and Schendel, P.,** Interleukin-11 regulates the hepatic expression of the same plasma protein genes as interleukin-6, *J. Biol. Chem.,* 266, 20424, 1991.
11. **Richards, C. D., Brown, T. J., Shoyab, M., Baumann, H., and Gauldie, J.,** Recombinant oncostatin M stimulates the production of acute phase proteins in Hep G2 cells and rat primary hepatocytes in vitro, *J. Immunol.,* 148, 1731, 1992.
12. **Rokita, H., Bereta, J., Koj, A., Gordon, A. H., and Gauldie, J.,** Epidermal growth factor and transforming growth factor-β differently modulate the acute phase response elicited by interleukin-6 in cultured liver cells from man, rat and mouse, *Comp. Biochem. Physiol.,* 95, 41, 1990.
13. **Ganapathi, M. K., May, L. T., Schultz, D., Brabenec, A., Weinstein, J., Sehgal, P. B., and Kushner, I.,** Role of interleukin-6 in regulating synthesis of C-reactive protein and serum amyloid A in human hepatoma cell lines, *Biochem. Biophys. Res. Commun.,* 157, 271, 1988.
14. **Koj, A.,** The role of interleukin-6 as the hepatocyte stimulating factor in the network of inflammatory cytokines, *Ann. N.Y. Acad. Sci.,* 557, 1, 1989.
15. **Baumann, H., Prowse, K. R., Marinkovic, S., Won, K.-A., and Jahreis, G. P.,** Stimulation of hepatic acute phase response by cytokines and glucocorticoids, *Ann. N.Y. Acad. Sci.,* 557, 280, 1989.
16. **Mackiewicz, A., Speroff, T., Ganapathi, M. K., and Kushner, I.,** Effect of cytokine combinations on acute phase protein production in two human hepatoma cell lines, *J. Immunol.,* 146, 3032, 1991.
17. **Ganapathi, M. K., Rzewnicki, D., Samols, D., Jiang, S.-L., and Kushner, I.,** Effect of combinations of cytokines and hormones on synthesis of serum amyloid A and C-reactive protein in Hep 3B cells, *J. Immunol.,* 147, 1261, 1991.
18. **Balkwill, F. R. and Burke, F.,** The cytokine network, *Immunol. Today,* 10, 299, 1989.
19. **Smith, K. A.,** Cytokine in the nineties, *Eur. Cytokine Net.,* 1, 7, 1990.
20. **Sporn, M. B. and Roberts, A. B.,** Peptide growth factors are multifunctional, *Nature (London),* 332, 217, 1988.
21. **Mizel, S. B.,** The interleukins, *FASEB J.,* 3, 2379, 1989.
22. **Bazan, J. F.,** A novel family of growth factor receptors: a common binding domain in the growth hormone, prolactin, erythropoietin and IL-6 receptors, and the p75 IL-2 receptor β chain, *Biochem. Biophys. Res. Commun.,* 164, 788, 1989.
23. **Akira, S., Hirano, T., Taga, T., and Kishimoto, T.,** Biology of multifunctional cytokines: IL 6 and relatted molecules (IL1 and TNF), *FASEB J.,* 4, 2860, 1990.
24. **Bazan, J. F.,** Structural design and molecular evolution of a cytokine receptor superfamily, *Proc. Natl. Acad. Sci. U.S.A.,* 87, 6934, 1990.
25. **Cosman, D., Lyman, S. D., Idzerda, R. L., Beckmann, M. P., Park, L. S., Goodwin, R. G., and March, C. J.,** *Trends Biochem. Sci.,* 15, 265, 1990.
26. **Gearing, D. P., Thut, C. J., VandenBos, T., Gimpel, S. D., Delaney, P. B., King, J., Price, V., Cosman, D., and Beckmann, M. P.,** Leukemia inhibitory factor receptor is structurally related to the IL-6 signal transducer, gp 130, *EMBO J.,* 10, 2839, 1991.

27. **Gearing, D. P., Comeau, M. R., Friend, D. J., Gimpel, S. D., Thut, C. J., McGourty, J., Basher, K. K., King, J. A., Gillis, S., Mosley, B., Ziegler, S. F., and Cosman, D.,** The IL-6 signal transducer, gp130: an oncostatin M receptor and affinity converter for the LIF receptor, *Science,* 255, 1434, 1992.
28. **Nicola, N. A. and Metcalf, D.,** Subunit promiscuity among hemopoietic growth factor receptors, *Cell,* 67, 1, 1991.
29. **Taga, T., Hibi, M., Hirata, Y., Yamasaki, K., Yasukawa, K., Matsuda, T., Hirano, T., and Kisimoto, T.,** Interleukin-6 triggers the association of its receptor with a possible signal transucer, gp130, *Cell,* 58, 573, 1989.
30. **Murakami, T., Narazaki, M., Hibi, M., Yawata, H., Yasukawa, K., Hamaguchi, M., Taga, T., and Kishimoto, T.,** Critical cytoplasmic region of the interleukin 6 signal transducer gp130 is conserved in the cytokine receptor family, *Proc. Natl. Acad. Sci. U.S.A.,* 88, 11349, 1991.
31. **Hibi, M., Murakami, M., Saito, M., Hirano, T., Taga, T., and Kishimoto, T.,** Molecular cloning and expression of an IL-6 signal transducer, gp130, *Cell,* 63, 1149, 1990.
32. **Sims, J. E., March, C. J., Cosman, D., Widner, M. B., MacDonald, H. R., McMahan, C. J., Grubin, C. E., Wignall, J. M., Jackson, J. L., Call, S. M., Friend, D., Alpert, A. R., Gillis, S., Urdall, D. L., and Dower, S. K.,** cDNA expression cloning of the IL-1 receptor, a member of the immunoglobulin superfamily, *Science,* 241, 585, 1988.
33. **McMahan, C. J., Slack, J. L., Mosley, B., Cosman, D., Lupton, S. D., Brunton, L. L., Grubin, C. E., Wignall, J. M., Jenkins, N. A., Brannan, C. I., Copeland, N. G., Hueber, K., Croce, C. M., Cannizzarro, L. A., Benjamin, D., Dower, S. K., Spriggs, M. K., and Sims, J. E.,** A novel IL-1 receptor, cloned from B cells by mammalian expression, is expressed in many cell types, *EMBO J.,* 10, 2821, 1991.
34. **Loetscher, H., Pan, Y.-C., Lahm, H.-W., Gentz, R., Brockhaus, M., Tabuchi, H., and Lessluaer, W.,** Molecular cloning and expression of the human 55 kd tumor necrosis factor receptor, *Cell,* 61, 351, 1990.
35. **Schall, T. J., Lewis, M., Koller, K. J., Lee, A., Rice, G. C., Wong, G. H., Gatanaga, T., Granger, G. A., Lentz, R., Raab, H., Kohr, W. J., and Goeddel, D. V.,** Molecular cloning and expression of a receptor for tumor necrosis factor, *Cell,* 61, 361, 1990.
36. **Smith, C. A., Davis, T., Anderson, D., Solam, L., Beckmann, M. P., Jerzy, R., Dower, S. K., Cosman, D., and Goodwin, R. G.,** A receptor for tumor necrosis factor defines an unusual family of cellular and viral proteins, *Science,* 248, 1019, 1990.
37. **Heinrich, P. C., Castell, J. V., and Andus, T.,** Interleukin-6 and the acute phase response, *Biochem. J.,* 265, 621, 1990.
38. **Ganapathi, M. K., Mackiewicz, A., Samols, D., Brabenec, A., Kushner, I., Schultz, D., and Hu, S. I.,** Induction of C-reactive protein by cytokines in human hepatoma cell lines is potentiated by caffeine, *Biochem. J.,* 269, 41, 1990.
39. **Nakajima, K. and Wall, R.,** Interleukin-6 signals activating junB and TIS11 gene transcription in a B-cell hybridoma, *Mol. Cell. Biol.,* 11, 1409, 1991.
40. **Lord, K. A., Abdollahi, A., Thomas, S. M., DeMarco, M., Brugge, J. S., Hoffmann-Lieberman, B., and Lieberman, D. A.,** Leukemia inhibitory factor and interleukin-6 trigger the same intermediate early response, including tyrosine phosphorylation, upon induction of myeloid leukemia differentiation, *Mol. Cell. Biol.,* 11, 4371, 1991.
41. **Fukunaga, M., Fujiwara, Y., Fujibayashi, M., Ochi, S., Yokoyama, K., Ando, A., Hirano, T., Ueda, N., and Kamada, T.,** Signal transduction mechanisms of interleukin 6 in cultured rat mesangial cells, *FEBS Lett.,* 285, 265, 1991.
42. **Florio, T., Landolfi, E., Grimaldi, M., Meucci, O., Ventra, C., Scorziello, A., Marino, A., and Schettini, G.,** Interleukin-6 mediates neural-immune interactions: study on prolactin release and intracellular transducing mechanisms, *Pharm. Res.,* 22, 54, 1990.
43. **O'Neill, L. A. J., Bird, T., Gearing, A. J. H., and Saklatvala, J.,** Interleukin-1 signal transduction, *Immunol. Today,* 11, 392, 1990.
44. **Yanaga, F., Abe, M., Koga, T., and Hirata, M.,** Signal transduction by tumor necrosis factor α is mediated through a guaninine nucleotide-binding protein in osteoblast-like cell line, MC3T3-E1, *J. Biol. Chem.,* 267, 5114, 1992.
45. **Gallis, B., Prickett, K. S., Jackson, J., Slack, J., Schooley, K., Sims, J. E., and Dower, S. K.,** IL-1 induces rapid phosphorylation of the IL-1 receptor, *J. Immunol.,* 143, 3235, 1989.
46. **Hershey, G. K., McCourt, D. W., and Schreiber, R. D.,** Ligand-induced phosphorylation of the human interferon-gamma receptor. Dependence on the presence of functionally active receptor, *J. Biol. Chem.,* 265, 17868, 1990.
47. **Resnitzky, D., Tiefenbrun, N., Berissi, H., and Kimchi, A.,** Interferons and interleukin 6 suppress phosphorylation of the retinoblastoma protein in growth-sensitive hematopoietic cells, *Proc. Natl. Acad. Sci. U.S.A.,* 89, 402, 1992.

48. **Mire-Sluis, A. R. and Thorpe, R.,** Interleukin-4 proliferative signal transduction involves the activation of a tyrosine-specific phosphatase and the dephosphorylation of an 80-kDa protein, *J. Biol. Chem.,* 266, 18113, 1991.
49. **Ganapathi, M. K.,** Okadaic acid, an inhibitor of protein phosphatases 1 and 2a, inhibits induction of acute phase proteins by interleukin 6 alone or in combination with interleukin 1 in human hepatoma cell lines, *Biochem. J.,* 284, 645, 1992.
50. **Saklatvala, J., Kaur, P., and Guesdon, F.,** Phosphorylation of the small heat-shock protein is regulated by interleukin 1, tumour necrosis factor, growth factors, bradykinin and ATP, *Biochem. J.,* 277, 635, 1991.
51. **Guesdon, F. and Saklatvala, J.,** Identification of a cytoplasmic protein kinase regulated by IL-1 that phosphorylates the small heat shock protein, hsp27, *J. Immunol.,* 147, 3402, 1991.
52. **Kaur, P. and Saklatvala, J.,** Interleukin 1 and tumour necrosis factor increase phosphorylation of fibroblast proteins, *FEBS Lett.,* 241, 6, 1988.
53. **Guy, G. R., Chua, S. K., Wong, N. S., Ng, S. B., and Tan, Y. H.,** Interleukin 1 and tumor necrosis factor activate common multiple protein kinases in human fibroblasts, *J. Biol. Chem.,* 266, 14343, 1991.
54. **Guy, G. R., Cao, X., Chua, P., and Tan, Y. H.,** Okadaic acid mimics multiple changes in early protein phosphorylation and gene expression induced by tumor necrosis factor or interleukin 1, *J. Biol. Chem.,* 267, 1846, 1992.
55. **Rall, T. W.,** Evolution of the mechanism of action of methylxanthines: from calcium mobilizers to antagonists of adenosine receptors, *Pharmacologist,* 24, 277, 1982.
56. **Daniel, E. E.,** Cellular calcium mobilization, *J. Cardiovasc. Pharmacol.,* 6, S622, 1984.
57. **Evans, E., Courtois, G. M., Kilian, P. L., Fuller, G. M., and Crabtree, G. R.,** Induction of fibrinogen and a subset of acute phase response genes involves a novel monokine which is mimicked by phorbol esters, *J. Biol. Chem.,* 262, 10850, 1987.
58. **Edbrooke, M. R. and Woo, P.,** Regulation of human SAA gene expressin by cytokines, in *Acute Phase Proteins in the Acute Phase Response,* Pepys, M. B., Ed., Springer-Verlag, London, 1989, 21.
59. **Baumann, H., Isseroff, H., Latimer, J. J., and Jahreis, G. P.,** Phorbol ester modulates interleukin 6- and interleukin 1-regulated expression of acute phase plasma proteins in hepatoma cells, *J. Biol. Chem.,* 263, 17390, 1988.
60. **Kurdowska, A., Bereta, J., and Koj, A.,** Comparison of the action of interleukin-6, phorbol myristate acetate and glucagon on the acute phase protein production and amino acid uptake by cultured rat hepatocytes, *Ann. N.Y. Acad. Sci.,* 557, 506, 1989.
61. **Cohen, P.,** The structure and regulation of protein phosphatases, *Annu. Rev. Biochem.,* 58, 453, 1989.
62. **Hunter, T.,** Protein-tyrosine phosphatases: the other side of the coin, *Cell,* 58, 1013, 1989.
63. **Cohen, P.,** The structure and regulation of protein phosphatases, *Adv. Second Messenger Phosphoprotein Res.,* 24, 230, 1990.
64. **Edbrooke, M. R., Burt, D. W., Cheshire, J. K., and Woo, P.,** Identification of cis-acting sequences responsible for phorbol ester induction of human serum amyloid A gene expression via a nuclear factor κB-like transcription factor, *Mol. Cell. Biol.,* 9, 1908, 1989.
65. **Poli, V., Mancini, F. P., and Cortese, R.,** IL-6DBP, a nuclear protein involved in the interleukin-6 signal transduction, defines a new family of leucine zipper proteins related to C/EBP, *Cell,* 63, 643, 1990.
66. **Akira, S., Isshiki, H., Sugita, T., Tanabe, O., Kinoshita, S., Nishio, Y., Nakajima, T., Hirano, T., and Kishimoto, T.,** A nuclear factor for IL-6 expression (NF-IL6) is a member of a C/EBP family, *EMBO J.,* 9, 1897, 1990.

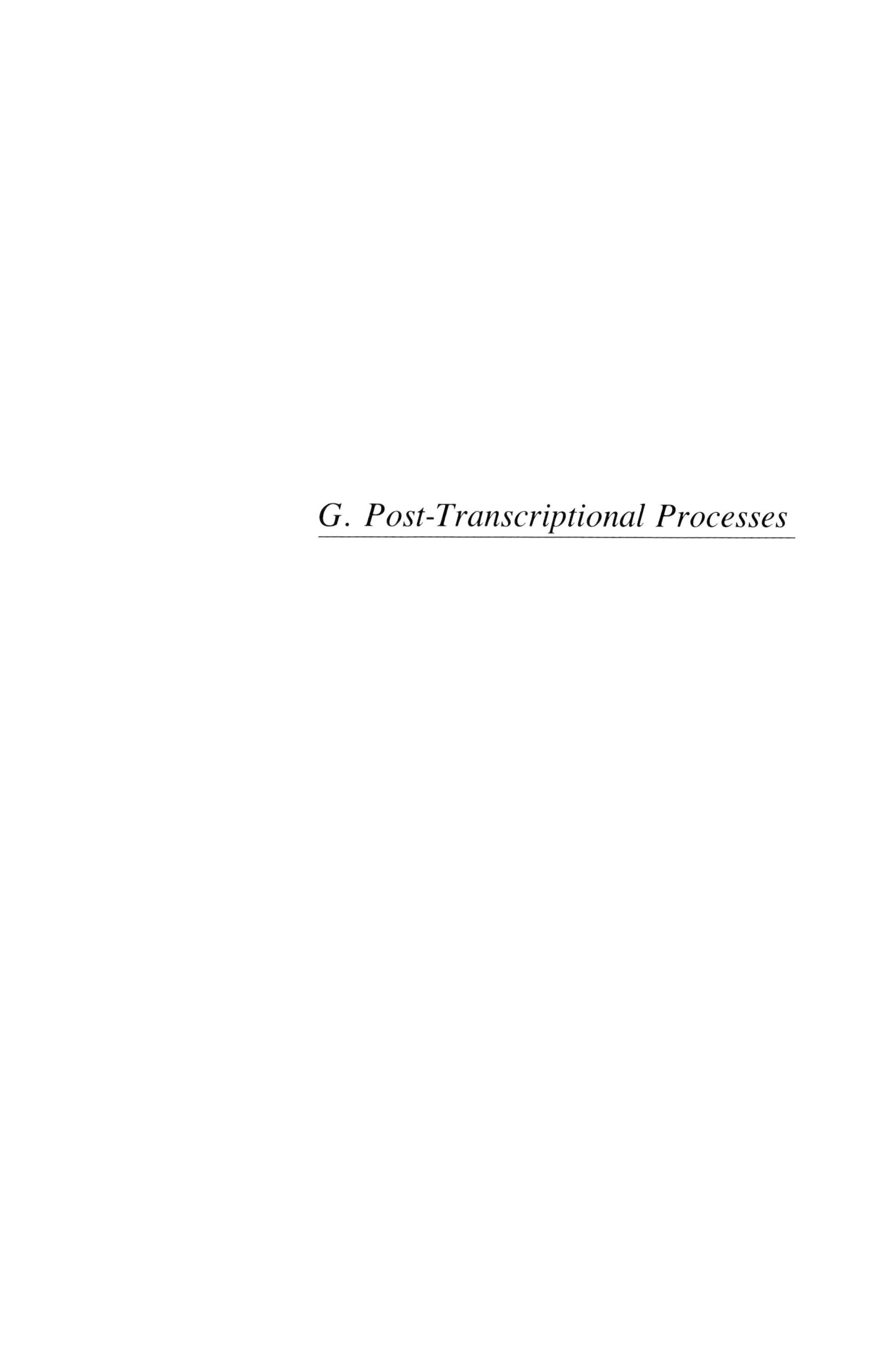

G. Post-Transcriptional Processes

Chapter 30

INTRACELLULAR MATURATION OF ACUTE PHASE PROTEINS

Erik Fries and E. Mathilda Sjöberg

TABLE OF CONTENTS

0-8493-6913-4/93/$0.00 + $.50

I. INTRODUCTION

The mere assembly of amino acids into a polypeptide is not sufficient for the formation of an active protein. The newly synthesized polypeptide must also acquire a specific conformation and in many cases selected amino acid residues must undergo chemical modification. For secretory proteins, this maturation occurs during their transport from the site of synthesis to the surface of the cell. To understand the mechanisms of this process, it is necessary to recognize that the secretory pathway consists of a series of compartments and that each compartment contains a set of modifying enzymes. The first section of this chapter therefore deals with the compartmentation of the secretory pathway. In the remainder, we use acute phase proteins as examples to describe various modifications. The number of modifications that secretory proteins can undergo is large and only some of these have been found on acute phase proteins; readers who are interested in other modifications should consult References 1 and 2.

II. INTRACELLULAR TRANSPORT

A. ENDOPLASMIC RETICULUM

During the assembly of a secretory protein, the first 15 to 30 amino acid residues — the signal sequence — serve to direct the synthetic machinery to the endoplasmic reticulum (ER). Further elongation results in the progressive translocation of the polypeptide into the lumen of the ER and the removal of the signal sequence.[3] The secretory proteins are then transported by vesicles to the Golgi complex (GC) via a tubulo-vesicular system, usually referred to as the intermediate compartment (Figure 1).[4,5]

Early electron microscopic studies showed that the ER consists of two domains, one with and one without attached ribosomes: the rough and the smooth ER. The proteins of the rough ER seem to form an insoluble network[6] which might exclude other proteins and thereby create the two domains of the ER.[7] Interestingly, immunoelectron microscopy has indicated that certain secretory proteins are unevenly distributed within the ER,[8,9] thus suggesting a further compartmentation.

B. GOLGI COMPLEX

A number of observations have led to the notion that the GC is divided into three functionally different compartments, *cis*-,medial, and *trans*-Golgi (Figure 1), and that secretory proteins are carried between these compartments by vesicles.[10] One of the first observations supporting this idea was the finding that some Golgi enzymes could be partially separated upon density gradient centrifugation.[11] Subsequent immunolocalization of these enzymes provided more direct evidence for a spatial separation.[12-14] In contrast to this model, Roth et al.[15] found that two enzymes that were located strictly in the *trans*-Golgi in one cell type of the intestine occurred in the whole Golgi stack except the first *cis* cisterna in another cell type of the same tissue. These results indicate that the compartmentation within the GC may not be essential for the secretory process. Some Golgi enzymes compete for the same substrate, and therefore differential distribution of these enzymes within the Golgi stack could be a reason why different cells may process the same protein differently.[16] The *trans*-Golgi cisternae are connected with an extensive tubular reticulum[14] often referred to as the *trans*-Golgi network (Figure 1).[17] Proteins destined for the cell surface pass through this membrane system[18] before they are packaged into transport vesicles.[19]

C. RECYCLING BETWEEN ER AND GC

The lumen of the ER has a very high concentration of protein — 30 to 100 mg/ml.[20] Most of the proteins in this fluid are permanent residents of the ER.[20] The cell must therefore

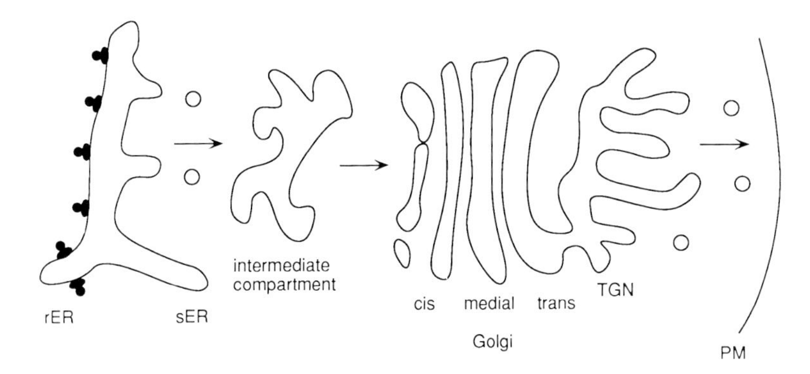

FIGURE 1. Schematic representation of the compartments of the secretory pathway. rER and sER, rough and smooth endoplasmic reticulum, respectively; TGN, *trans*-Golgi network; PM, plasma membrane.

have an efficient system for selectively transferring the secretory proteins to the GC. It appears that the secretory proteins are passively transported along the secretory pathway by liquid bulk-flow;[21] soluble ER-specific proteins seem to be salvaged from this flow in the intermediate compartment by binding reversibly to a receptor which recognizes their specific C-terminal sequence (for a review, see Reference 22). The idea that transport of secretory proteins occurs by liquid bulk-flow implies that large amounts of membrane material must also be transported from the ER to the GC. Based on the observation that the transfer of membrane lipid from the cell surface to the interior is relatively small, Wieland et al.[21] suggested that the major flow of lipid back to the ER is from the GC. The question of how the secretory proteins could be excluded from this transport is still not resolved.

Recent experiments with the drug brefeldin A have substantiated the idea of recycling of membrane from the GC to the ER. When cells are treated with this drug, protein secretion stops,[23] the Golgi stack disassembles,[24] and *cis*- and medial Golgi enzymes appear in the ER.[25] Consequently, glycosylated ER proteins become processed by Golgi enzymes;[25] for unknown reasons, this is not seen with hepatocytes.[26] One explanation for these effects is that in the presence of brefeldin A, the selectivity in the transport from the GC to the ER is lost due to the formation of direct connections.[27,28]

D. RETENTION OF SECRETORY PROTEINS IN ER

Studies during the last 10 years have revealed that all cells possess a system for stopping incompletely folded or otherwise defective secretory proteins from leaving the cell (reviewed in Reference 22). The key component in this system is a soluble ER protein called the immunoglobulin heavy-chain binding protein, or BiP. This protein was discovered when it was observed to coprecipitate with the heavy chain of IgG in cells producing only this subunit.[29] Later, BiP was found to bind to other incomplete or aberrant proteins.[30] BiP apparently binds to polypeptide structures which are normally hidden.[31] Proteins that fail to leave the ER, such as the PiZ α_1-antitrypsin variant, are eventually degraded in a nonlysosomal compartment through a process that is poorly understood.[32,33]

It has been recognized for some time that different secretory proteins are transported intracellularly at different rates.[34,35] Thus, for example, pulse-labeled albumin and α_1-protease inhibitor are secreted with half-times of 30 to 45 min, whereas transferrin and fibrinogen require 110 to 140 min.[34,36,37] This difference is due mainly to different rates of transport from the ER to the GC,[36,38] which may be caused by binding to resident ER proteins.[10,34] The fact that some proteins are transiently retained in the ER implies that their concentration in this organelle may be relatively high. Indeed, the amount of transferrin in the ER is

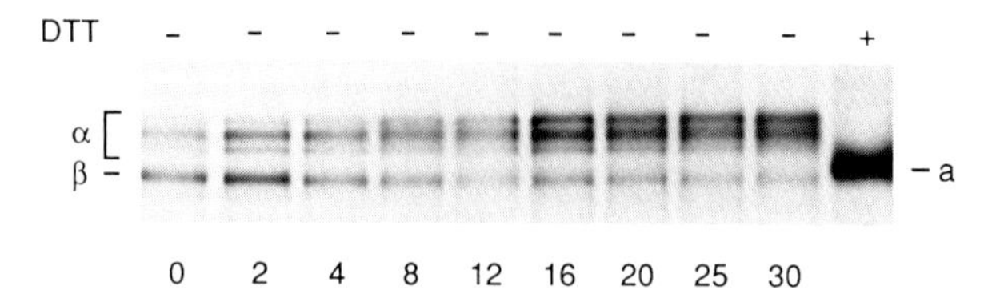

FIGURE 2. Detection of slowly formed disulfide bonds in α_1-I_3. Rat hepatocytes were pulse labeled with [^{35}S]methionine and chased for the number of minutes indicated. The cells were then solubilized and antibodies against α_1-I_3 added. The immune complexes were analyzed by SDS-PAGE in the absence or presence of the reducing agent dithiothreitol (DTT), as indicated. The sample to the far right was prepared from cells labeled for 60 min. Note the difference in mobility between a and β, and the shift with time from β to α. (From Sjöberg, M., Esnard, F., and Fries, E., *Eur. J. Biochem.*, 197, 61, 1991. With permission.)

similar to that of albumin, although its rate of synthesis is several times lower than that of albumin.[34,38]

III. INTRACELLULAR MODIFICATIONS

A. DISULFIDE FORMATION

Nascent polypeptides appear to cross the ER membrane in an essentially unfolded state.[39] In most cases, the subsequent folding is completed as soon as the polypeptide has emerged in the lumen of the ER.[40] In the simplest situation, the different domains fold as the corresponding sequences appear and the cysteine residues rapidly form disulfide bonds.[41,42] In their mature, stable conformation, however, many proteins contain disulfide bonds that are not sequential. Their formation is catalyzed by the enzyme protein disulfide isomerase (PDI), which resides in the lumen of the ER.[43] This is, in fact, one of the most abundant proteins in the ER.[44]

As stated above, the disulfides of many proteins are formed cotranslationally. However, in some proteins, these bonds are formed more slowly and their appearance can be monitored relatively easily in pulse-chase experiments.[45] In the presence of sodium dodecyl sulfate (SDS), a protein with free cysteine residues is less compact than the same protein in which these residues form disulfide bonds. The reduced form of a protein therefore (usually) migrates more slowly upon polyacrylamide gel electrophoresis (PAGE) in the presence of SDS than its oxidized counterpart. Thus, Lodish and Kong[46] found that pulse-labeled transferrin ran as a broad band upon SDS-PAGE under nonreducing conditions, whereas after 1 h of chase, a distinct, fast-migrating band had formed.

The mature form of α_1-inhibitor 3 (α_1-I_3) is atypical in that it has a lower electrophoretic mobility under nonreducing than under reducing conditions (Figure 2; compare bands α and a).[47] Electron microscopic analysis indicates that α_1-I_3 has the shape of a ring.[48] Presumably, this disulfide-stabilized structure binds little SDS, accounting for its low electrophoretic mobility. In contrast, pulse-labeled α_1-I_3 runs faster under nonreducing than under reducing conditions (Figure 2; compare bands β and a). Upon chase, the low-mobility form (band α) appears with a half-time of 4 to 12 min, and the transition appears to proceed via two or three intermediates.[49] Analysis of α_1-I_3 by limited proteolytic digestion suggests that it consists of five domains, three of which are linked to their neighbors by disulfide bonds.[50] The stepwise decrease in the mobility of α_1-I_3 seen in the pulse-chase experiment may reflect the formation of these interdomain bonds.

B. OLIGOMERIZATION

Some secretory proteins consist of different polypeptides coded for by more than one gene.[45] These genes may be expressed at different levels. For example, in the human hepatoma cell line Hep G2, the rate of synthesis of the Bβ-chain of fibrinogen is lower than those of the Aα- and γ-chains. Fibrinogen assembly commences on nascent Bβ-chains, and the Aα- and γ-chains are taken from pools of protein made earlier. The redundant Aα- and γ-chains are apparently degraded intracellularly.[51] In rat hepatocytes, on the other hand, the synthesis of the Aα-chain is the rate-limiting step.[52]

C. FORMATION OF INTERNAL THIOL ESTER

Most of the proteins belonging to the α-macroglobulin family have an internal thiol ester whose function is to form a link to other proteins.[53] When heated in a denatured form, these thiol ester-containing proteins are autolytically cleaved at the peptide bond adjacent to the glutamyl residue of the thiol ester.[54] In a study of the biosynthesis of α_1-I_3 in rat hepatocytes, we found that the ability to undergo autolytic cleavage upon heating appeared only 10 to 20 min after the synthesis of the protein[49]. Thus, the thiol ester was formed after the protein had acquired its disulfide bonds (cf., Figure 2). This sequence of events makes sense in view of the fact that the thiol ester, to avoid premature hydrolysis by water, must be buried in a hydrophobic region.[55,56] Upon interaction with a proteinase, α_1-I_3 presumably undergoes a conformational change leading to the exposure of the thiol ester.[57]

D. GLYCOSYLATION

1. Biosynthesis of *N*-Linked Oligosaccharides

Most secretory proteins are glycosylated, and the oligosaccharide is usually linked to the amide nitrogen of an asparagine residue. As discussed elsewhere in this volume, these *N*-linked carbohydrates appear in many different forms on secretory proteins. Despite this variety, they are all synthesized from a common precursor: $Glc_3Man_9(GlcNAc)_2$. This oligosaccharide structure is synthesized on a lipid in the ER membrane and is then transferred cotranslationally to the polypeptide chain. As soon as the oligosaccharide tree has been transferred, it is modified. First, the Glc residues are excised and then one Man residue (on the average) is removed. Thus, secretory proteins isolated from the ER typically have eight Man residues.[36] When the proteins reach the GC, the number of Man residues is reduced to five and a GlcNac residue is added; this modification is followed by the removal of two Man and the addition of one or several GlcNac. The great variation of the *N*-linked oligosaccharides seen on secreted proteins is due to the fact that the removal of Man residues may be incomplete and/or the number of added GlcNAc residues may be variable.[58] At present it is not clear whether the reduction to five Man occurs in the *cis*- or medial Golgi.[59,60] However, the enzyme carrying out the subsequent step has been located to the medial Golgi.[12] Finally, as the glycoprotein reaches the *trans*-Golgi, Gal and sialic acid residues are added (for reviews, see References 61 and 16). The enzymes mediating these late modifications seem to be segregated within the *trans*-Golgi/*trans*-Golgi network.[62-64]

The carbohydrate processing and intracellular transport of secretory proteins can to some extent be monitored by pulse-chase experiments followed by SDS/PAGE. For many proteins, the increase in size due to the addition of galactose and sialic acid residues in the *trans*-Golgi results in an easily detectable decrease in electrophoretic mobility.[65-67] Alternatively, the immunoprecipitated protein is treated with the enzyme endoglycosidase H before electrophoresis. This enzyme will remove all early (high-mannose) forms and thus produce a clear shift in the mobility of the protein;[65-67] resistance to the enzyme marks the arrival of the pulse-labled protein in the medial Golgi.[12]

2. Inhibitors of *N*-Linked Glycosylation

The formation of *N*-linked carbohydrates can be inhibited by the addition of the antibiotic tunicamycin.[68] Use of this compound has made it possible to study the role of *N*-linked carbohydrates in secretion. When Hep G2 cells or hepatocytes are treated with tunicamycin, the secretion of some, but not all, proteins is impaired.[66,67,69,70] In another type of cell, it has been shown that a protein's sensitivity to tunicamycin treatment is correlated to its binding to BiP,[30] suggesting an effect of the glycan moiety on the conformation of the polypeptide. A number of compounds have been discovered which inhibit specific hydrolases of the *N*-glycosylation pathway.[68] Thus, for example, deoxynojirimycin inhibits the ER enzymes which remove the glucoses. When Hep G2 cells are incubated with this compound, the secretion of some proteins is inhibited.[69,71] The mechanism of this differential retention of proteins with incompletely processed oligosaccharides is unknown. Deoxymannojirimycin, on the other hand, inhibits the Golgi enzymes that reduce the number of mannose residues down to five; in the presence of this compound, the transport rates are unaffected.[72,73]

3. *O*-Linked Oligosaccharides

Oligosaccharides may also be linked to proteins via an oxygen atom of a serine or threonine residue. However, this is a relatively rare modification.[74,75] It is possible that the scarcity of examples is a reflection of the fact that *O*-linked oligosaccharides are relatively difficult to detect. In contrast to *N*-linked glycosylation, the potential sites for *O*-linked glycosylation cannot be predicted from amino acid sequence.[75] Furthermore, since *O*-linked sugars are often small, their (enzymatic)[76] removal may not lead to a noticeable shift in the electrophoretic mobility of a protein. A lectin which binds oligopeptides with *O*-linked sugars should be useful in the detection of this modification.[77]

E. SULFATION

In 1984, Hille et al.[78] reported that after the injection of [^{35}S]sulfate into rats, labeled proteins could be detected in all tissues investigated, including blood plasma. Further analysis showed that the sulfate groups on proteins in tissues other than blood were linked mainly to carbohydrates, whereas those in plasma were linked mainly to tyrosine residues. Based on this and the earlier observation that certain secretory proteins contain sulfated tyrosines, they proposed that sulfate residues might have a function in protein secretion. However, reagents that could inhibit this modification showed that tyrosine sulfation was irrelevant to secretion.[79] Tyrosine sulfation has been shown to take place in the *trans*-Golgi, apparently after sialylation.[80,81]

When rat hepatocytes are incubated with [^{35}S]sulfate, the secretory proteins that incorporate most label are α_1-I_3,[49] bikunin,[82] and presumably pre-α-inhibitor.[82] The sulfate on α_1-I_3 is apparently attached to a tyrosine residue,[49] whereas those of bikunin and pre-α-inhibitor are on chrondroitin sulfate chains. Hortin et al. have investigated the sulfated proteins that are secreted by Hep G2 cells.[83] They found that many, but not all, of the proteins against which they had antibodies were labeled by [^{35}S]sulfate. Thus, transferrin, ceruloplasmin, α_1-antitrypsin, α_1-acid glycoprotein, α_2-HS, prothrombin, fibrinogen, C2, C4, and C5 were labeled, but albumin and apolipoprotein A were not. Of these proteins, C4 had the highest relative molar content — 3 mol/mol protein — whereas the other proteins had a ratio <0.4. α_2-HS, which was the most abundant sulfated protein, was analyzed in more detail and found to have its sulfates attached to *N*-linked oligosaccharides; these sulfate groups apparently substitute for sialic acid residues.

The sulfate-bearing tyrosine residues have now been identified on a number of proteins.[84] It has been noted that these tyrosines are situated in segments containing many acidic and few basic amino acids.[85] Furthermore, for some of these proteins, it has been shown that the sulfate groups are important for the interaction with other proteins.[86-88]

F. PROTEOLYTIC CLEAVAGE

Many secretory proteins are synthesized as precursors which must be proteolytically cleaved to become biologically active. One example is haptoglobin, which occurs in plasma

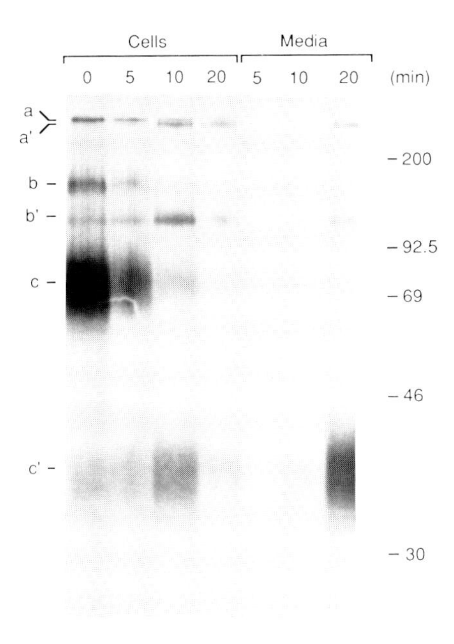

FIGURE 3. Late intracellular processing of bikunin, a chondroitin sulfate-containing protein. Rat hepatocytes were labeled with [^{35}S]sulfate for 7 min and then chased for the times indicated. Saponin extracts of cells and media were incubated with antibodies against bikunin, and the immunocomplexes analyzed by SDS-PAGE. Band c is the precursor of bikunin, which also contains α_1-microglobulin. The precursor undergoes proteolytic cleavage, giving rise to free bikunin (c′). Bands a and b are oligomeric proteins in which the precursor is linked to other polypeptides via the chondroitin sulfate chain of bikunin; a′ and b′ that form upon chase are presumably inter-x-inhibitor and pre-x-inhibitor, respectively. (From Sjöberg, E. M. and Fries, E., *Arch. Biophys. Biochem.*, 275, 217, 1992. With permission).

as an $\alpha_2\beta_2$ tetramer of 90 kDa. The α- and β-chains (of 9 and 35 kDa, respectively) are formed from a 42-kDa precursor which is cleaved in the ER.[89] With haptoglobin the only known exception, intracellular proteolytic processing of secretory proteins occurs late during intracellular transport — in the *trans*-Golgi network and/or secretory vesicles.[90-92] The enzyme carrying out this processing specifically cleaves at dibasic sequences, such as those found in, e.g., proalbumin and pro-C4.[93-94]

Bikunin is synthesized as a precursor in which it is linked to α_1-microglobulin by two arginine residues.[95,96] In plasma, most bikunin occurs as a subunit of the three-chain protein inter-α-inhibitor and of the two-chain protein pre-α-inhibitor.[97] In both these proteins, all polypeptides are linked by bikunin's chondroitin sulfate chain.[98,99] To study the late steps in the biosynthesis of bikunin, we did pulse-chase experiments with rat hepatocytes labeled with [^{35}S]sulfate. This labeling revealed a short-lived intermediate of the α_1-microglobulin/bikunin precursor with a full-size chondroitin sulfate chain (Figure 3, band c). Chasing for 5 to 10 min showed that this precursor was cleaved to form a 40-kDa band — free bikunin (band c′). Thus, cleavage occurs after sulfation, indicating a spatial separation of the correspoding enzymes. In blood, α_1-microglobulin occurs in complex with an unidentified brown-colored ligand.[100] This binding capacity of α_1-microglobulin could pose a hazard to the cell during intracellular transport and the binding might therefore be blocked in the precursor.

IV. CONCLUSIONS

Acute phase proteins, like all other secretory proteins, use their time in the cell to mature and become ready for the real life outside. For those modifications that occur in the ER, there exists a control system that allows only proteins that have been properly modified to proceed to the GC. In contrast, later modifications are not critical for secretion, and secretory proteins may therefore be released in a variety of different forms.[101] The widespread use of recombinant DNA techniques for the production of secretory proteins in heterologous cells underscores the importance of intracellular processing because the enzymes mediating the processing differ between cells.[101] Great efforts are therefore presently being made to define the postranslational modifications of secretory proteins and to determine the capacity of different cells to perform these modifications.

REFERENCES

1. **Wold, F.,** In vivo chemical modification of proteins. Post-translational modification, *Annu. Rev. Biochem.*, 50, 783, 1981.
2. **Freedman, R. B.,** Post-translational modification and folding of secretory proteins, *Biochem. Soc. Proc.*, 17, 331, 1989.
3. **Walter, P. and Lingappa, V. R.,** Protein translocation across the endoplasmic reticulum membrane, *Annu. Rev. Cell Biol.*, 2, 499, 1986.
4. **Saraste, J. and Kuismanen, E.,** Pre- and post-Golgi vacuoles operate in the transport of Semliki forest virus membrane glycoproteins to the cell surface, *Cell*, 38, 535, 1984.
5. **Schweitzer, A., Fransen, J. A. M., Matter, K., Kreis, T. E., Ginser, L., and Hauri, H.-P.,** Identification of an intermediate compartment involved in protein transport from endoplasmic reticulum to Golgi apparatus, *Eur. J. Cell Biol.*, 53, 185, 1990.
6. **Crimaudo, C., Hortsch, M., Gausepohl, H., and Meyer, D. I.,** Human ribophorins I and II: the primary structure and membrane topology of two highly conserved rough endoplasmic reticulum-specific glycoproteins, *EMBO J.*, 6, 75, 1987.
7. **Rose, J. K. and Doms, R. W.,** Regulation of protein export from the endoplasmic reticulum, *Annu. Rev. Cell Biol.*, 4, 257, 1988.
8. **Yokota, S. and Fahimi, H. D.,** Immunocytochemical localization of albumin in the secretory appratus of rat liver parenchymal cells, *Proc. Natl. Acad. Sci. U.S.A.*, 78, 4970, 1981.
9. **Vertel, B. M., Velasco, A., La France, S., Walters, L., and Kaczman-Daniel, K.,** Precursors of chondroitin sulfate proteoglycan are segregated within a subcompartment of the chondrocyte endoplasmic reticulum, *J. Cell Biol.*, 109, 1827, 1989.
10. **Pfeffer, S. R. and Rothman, J. E.,** Biosynthetic protein transport and sorting by the endoplasmic reticulum and Golgi, *Annu. Rev. Biochem.*, 56, 829, 1987.
11. **Dunphy, W. G., Fries, E., Urbani, L. J., and Rothman, J. E.,** Early and late functions associated with the Golgi apparatus reside in distinct compartments, *Proc. Natl. Acad. Sci. U.S.A.*, 78, 7453, 1981.
12. **Dunphy, W. G., Brands, R., and Rothman, J. E.,** Attachment of terminal N-acetylglucosamine to asparagine-linked oligosaccharides occurs in central cisternae of the Golgi stack, *Cell*, 40, 463, 1985.
13. **Roth, J. and Berger, E. G.,** Immunocytochemical localization of galactosyltransferase in HeLa cells: codistribution with thiamine pyrophosphatase in *trans*-Golgi cisternae, *J. Cell Biol.*, 92, 223, 1982.
14. **Roth, J., Taatjes, D. J., Lucocq, J. M., Weinstein, J., and Paulson, J. C.,** Demonstration on an extensive trans-tubular network continuous with the Golgi apparatus stack that may function in glycosylation, *Cell*, 43, 287, 1985.
15. **Roth, J., Taatjes, D. J., Weinstein, J., Paulson, J. C., Greenwell, P., and Watkins, W. M.,** Differential subcompartmentation of terminal glycosylation in the Golgi apparatus of intestinal absorptive and goblet cells, *J. Biol. Chem.*, 261, 14307, 1986.
16. **Roth, J.,** Subcellular organization of glycosylation in mammalian cells, *Biochim. Biophys. Acta*, 906, 405, 1987.
17. **Griffiths, G. and Simons, K.,** The *trans* Golgi network: sorting at the exit site of the Golgi complex, *Science*, 234, 438, 1986.
18. **Taatjes, D. J. and Roth, J.,** The trans-tubular network of the hepatocyte Golgi apparatus is part of the secretory pathway, *Eur. J. Cell Biol.*, 42, 344, 1986.
19. **Strous, G. J. A. M., Willemsen, R., van Kerkhof, P., Slot, J. W., Geuze, H. J., and Lodish, H. F.,** Vesicular stomatitis virus glycoprotein, albumin, and transferrin are transported to the cell surface via the same Golgi vesicles, *J. Cell Biol.*, 97, 1815, 1983.

20. **Koch, G. L. E., Smith, M. J., Macer, D. R. J., Booth, C., and Wooding, F. B. P.,** Structure and assembly of the endoplasmic reticulum, *Biochem. Soc. Proc.*, 17, 328, 1989.
21. **Wieland, F. T., Gleason, M. L., Serafini, T. A., and Rothman, J. E.,** The rate of bulk flow from the endoplasmic reticulum to the cell surface, *Cell,* 50, 289, 1987.
22. **Pelham, H. R. B.,** Control of protein exit from the endoplasmic reticulum, *Annu. Rev. Cell Biol.*, 5, 1, 1989.
23. **Misumi, Y., Misumi, Y., Miki, K., Takatsuki, A., Tamura, G., and Ikehara, Y.,** Novel blockade by brefeldin A of intracellular transport of secretory proteins in cultured rat hepatocytes, *J. Biol. Chem.*, 261, 11398, 1986.
24. **Fujiwara, T., Oda, K., Yokota, S., Takatsuki, A., and Ikehara, Y.,** Brefeldin A causes disassembly of the Golgi complex and accumulation of secretory proteins in the endoplasmic reticulum, *J. Biol. Chem.*, 263, 18545, 1988.
25. **Doms, R. W., Russ, G., and Yewdell, J. W.,** Brefeldin A redistributes resident and itinerant Golgi proteins to the endoplasmic reticulum, *J. Cell Biol.*, 109, 61, 1989.
26. **Oda, K., Fujiwara, T., and Ikehara, Y.,** Brefeldin A arrests the intracellular transport of viral envelope proteins in primary cultured rat hepatocytes and HepG2 cells, *Biochem. J.*, 265, 161, 1990.
27. **Lippincott-Schwartz, J., Donaldson, J. G., Schweizer, A., Berger, E. G., Hauri, H.-P., Yuan, L. C., and Lausner, R. D.,** Microtubule-dependent retrograde transport of proteins into the ER in the presence of brefeldin A suggests an ER recycling pathway, *Cell,* 60, 821, 1990.
28. **Pelham, H. R. B.,** Multiple targets for brefeldin A, *Cell,* 67, 449, 1991.
29. **Haas, I. G. and Wabl, M.,** Immunoglobulin heavy chain binding protein, *Nature (London),* 306, 387, 1983.
30. **Dorner, A. J., Bole, D. G., and Kaufman, R. J.,** The relationship of N-linked glycosylation and heavy chain-binding protein association with the secretion of glycoproteins, *J. Cell Biol.*, 105, 2665, 1987.
31. **Flynn, G. C., Pohl, J., Flocco, M. T., and Rothman, J. E.,** Peptide-binding specificity of the molecular chaperone BiP, *Nature (London),* 353, 726, 1991.
32. **Le, A., Graham, K. S., and Sifers, R. N.,** Intracellular degradation of the transport-impaired human PiZ α_1-antitrypsin variant, *J. Biol. Chem.*, 265, 14001, 1990.
33. **Bonifacino, J. S. and Lippincott-Schwartz, J.,** Degradation of proteins within the endoplasmic reticulum, *Curr. Opinions Cell Biol.*, 3, 592, 1991.
34. **Morgan, E. H. and Peters, T., Jr.,** Intracellular aspects of transferrin synthesis and secretion in the rat, *J. Biol. Chem.*, 246, 3508, 1971.
35. **Ledford, B. E. and Davis, D. F.,** Kinetics of serum protein secretion by cultured hepatoma cells. Evidence for multiple secretory pathways, *J. Biol. Chem.*, 258, 3304, 1983.
36. **Lodish, H. F., Kong, N., Snider, M., and Strous, G. J. A. M.,** Hepatoma secretory proteins migrate from rough endoplasmic reticulum to Golgi at characteristic rates, *Nature (London),* 304, 80, 1983.
37. **Parent, J. B., Bauer, H. C., and Olden, K.,** Three secretory rates in human hepatoma cells, *Biochim. Biophys. Acta,* 846, 44, 1985.
38. **Fries, E., Gustafsson, L., and Petersson, P. A.,** Four secretory proteins synthesized by hepatocytes are transported from endoplasmic reticulum to Golgi complex at different rates, *EMBO J.*, 3, 147, 1984.
39. **Eilers, M. and Schatz, G.,** Protein unfolding and the energetics of protein translocation across biological membranes, *Cell,* 52, 481, 1988.
40. **Tsou, C.-L.,** Folding of the nascent peptide chain into a biologically active protein, *Am. Chem. Soc.*, 27, 1809, 1988.
41. **Bergman, L. W. and Kuehl, W. M.,** Formation of an intrachain disulfide bond on nascent immunoglobulin light chains, *J. Biol. Chem.*, 254, 8869, 1979.
42. **Peters, T., Jr. and Davidson, L. K.,** The biosynthesis of rat serum albumin, *J. Biol. Chem.*, 257, 8847, 1982.
43. **Freedman, R. B.,** Native disulphide bond formation in protein biosynthesis: evidence for the role of protein disulphide isomerase, *Trends Biochem. Sci.*, 9, 438, 1984.
44. **Ohba, H., Harano, T., and Omura, T.,** Presence of two different types of protein-disulfide isomerase on cytoplasmic and luminal surfaces of endoplasmic reticulum of rat liver cells, *Biochem. Biophys. Res. Commun.*, 77, 830, 1977.
45. **Hurtley, S. M. and Helenius, A.,** Protein oligomerization in the endoplasmic reticulum, *Annu. Rev. Cell Biol.*, 5, 277, 1989.
46. **Lodish, H. F. and Kong, N.,** Cyclosporin A inhibits an initial step in folding of transferrin within the endoplasmic reticulum, *J. Biol. Chem.*, 266, 14835, 1991.
47. **Saito, A. and Sinohara, H.,** Rat plasma murinoglobulin: isolation, characterization and comparison with rat α-1- and α-2-macroglobulins, *J. Biochem. (Tokyo),* 98, 501, 1985.

48. **Ikai, A., Nishigai, M., Saito, A., Sinohara, H., Muto, Y., and Arata, Y.,** Electron microscopic demonstration of a common structural motif in human complement factor C3 and rat α1-inhibitor 3 (murinoglobulin), *FEBS Lett.*, 260, 291, 1990.
49. **Sjöberg, M., Esnard, F., and Fries, E.,** Intracellular modifications of rat α_1-inhibitor$_3$. Formation of disulphides, internal thiolester and sulphation, *Eur. J. Biochem.*, 197, 61, 1991.
50. **Rubenstein, D. S., Enghild, J. J., and Pizzo, S. V.,** Limited proteolysis of the α-macroglobulin rat α_1-inhibitor-3. Implications for a domain structure, *J. Biol. Chem.*, 266, 11252, 1991.
51. **Yu, S., Sher, B., Kudryk, B., and Redman, C. M.,** Intracellular assembly of human fibrinogen. Order of assembly of fibrinogen chains, *J. Biol. Chem.*, 259, 10574, 1988.
52. **Hirose, S., Oda, K., and Ikehara, Y.,** Biosynthesis, assembly and secretion of fibrinogen in cultured rat hepatocytes, *Biochem. J.*, 251, 373, 1988.
53. **Sottrup-Jensen, L.,** α-Macroglobulins: structure, shape, and mechanism of proteinase complex formation, *J. Biol. Chem.*, 264, 11539, 1989.
54. **Howard, J. B., Vermeulen, M., and Swenson, R. P.,** The temperature-sensitive bond in human α_2-macroglobulin is the alkylamine-reactive site, *J. Biol. Chem.*, 255, 3820, 1980.
55. **Van Leuven, F.,** Human α_2-macroglobulin: structure and function, *Trends Biochem. Sci.*, 7, 185, 1982.
56. **Law, S. K. A. and Dodds, A. W.,** C3, C4, and C5: the thioester site, *Biochem. Soc. Trans.*, 18, 1155, 1990.
57. **Enghild, J. J., Salvesen, G., Thogersen, I. B., and Pizzo, S. V.,** Proteinase binding and inhibition by the monomeric macroglobulin rat α1-inhibitor-3, *J. Biol. Chem.*, 264, 11428, 1989.
58. **Schachter, H., Narasimhan, S., Gleeson, P., and Vella, G.,** Control of branching during biosynthesis of asparagine-linked oligosaccharides, *Can. J. Biochem. Cell Biol.*, 61, 1049, 1983.
59. **Dunphy, W. G. and Rothman, J. E.,** Compartmental organization of the Golgi stack, *Cell*, 42, 13, 1985.
60. **Balch, W. E. and Keller, D. S.,** ATP-coupled transport of vesicular stomatitis virus G protein. Functional boundaries of secretory compartments, *J. Biol. Chem.*, 261, 14690, 1986.
61. **Kornfeld, R. and Kornfeld, S.,** Assembly of asparagine-linked oligosaccharides, *Annu. Rev. Biochem.*, 54, 631, 1985.
62. **Berger, E. G. and Hesford, F. J.,** Localization of galactosyl- and sialyltransferase by immunofluorescence: evidence for different sites, *Proc. Natl. Acad. Sci. U.S.A.*, 82, 4736, 1985.
63. **Bergeron, J. J. M., Paiement, J., Khan, M. N., and Smith, C. E.,** Terminal glycosylation in rat hepatic Golgi fractions: heterogeneous locations for sialic acid and galactose acceptors and their transferases, *Biochim. Biophys. Acta*, 821, 393, 1985.
64. **Chege, N. W. and Pfeffer, S. R.,** Compartmentation of the Golgi complex: brefeldin A distinguishes *trans*-Golgi cisternae from the *trans*-Golgi network, *J. Cell Biol.*, 111, 893, 1990.
65. **Katz, N. R., Giffhorn, S., Goldfarb, V., and Muller-Eberhard, U.,** Partial characterization of intra- and extracellular forms of the glycoprotein hemopexin in liver cell culture and cell-free translation, *Biochem. Biophys. Res. Commun.*, 131, 593, 1985.
66. **Carlson, J. and Stenflo, J.,** The biosynthesis of rat α_1-antitrypsin, *J. Biol. Chem.*, 257, 12987, 1982.
67. **Bauer, J., Kurdowska, A., Tran-Thi, T.-A., Budek, W., Koj, A., Decker, K., and Heinrich, P. C.,** Biosynthesis and secretion of α_1 acute phase globulin in primary cultures of rat hepatocytes, *Eur. J. Biochem.*, 146, 347, 1985.
68. **Elbein, A. D.,** Inhibitors of the biosynthesis and processing of N-linked oligosaccharide chains, *Annu. Rev. Biochem.*, 56, 497, 1987.
69. **Lodish, H. F. and Kong, N.,** Glucose removal from N-linked oligosaccharides is required for efficient maturation of certain secretory glycoproteins from the rough endoplasmic reticulum to the Golgi complex, *J. Cell Biol.*, 98, 1720, 1984.
70. **Katz, N. R., Goldfarb, V., Liem, H., and Muller-Eberhard, U.,** Synthesis and secretion of hemopexin in primary cultures of rat hepatocytes. Demonstration of an intracellular precursor of hemopexin, *Eur. J. Biochem.*, 146, 155, 1985.
71. **Gross, V., Andus, T., Tran-Thi, T.-A., Schwartz, R. T., Decker, K., and Heinrich, P. C.,** 1-Deoxynojirimycin impairs oligosaccharide processing of α_1-proteinase inhibitor and inhibits its secretion in primary cultures of rat hepatocytes, *J. Biol. Chem.*, 258, 12203, 1983.
72. **Gross, V., Steube, K., Tran-Thi, T.-A., McDowell, W., Schwartz, R. T., Decker, K., Gerok, W., and Heinrich, P. C.,** Secretion of high-mannose-type α_1-proteinase inhibitor and α_1-acid glycoprotein by primary rat hepatocytes in the presence of the mannosidase I inhibitor 1-deoxymannojirimycin, *Eur. J. Biochem.*, 150, 41, 1985.
73. **Persson, R., Schnell, C. R., Borg, L. A. H., and Fries, E.,** Accumulation of Golgi-processed secretory proteins in an organelle of high density upon reduction of ATP concentration in rat hepatocytes, *J. Biol. Chem.*, 267, 2760, 1992.
74. **Baenziger, J. U.,** The oligosaccharides of plasma glycoproteins: synthesis, structure and function, in *The Plasma Proteins*, Vol. 4, 2nd ed., Putnam, F. W., Ed., Academic Press, Orlando, 1984, chap. 5.

75. **Wilson, I. B. H., Gavel, Y., and von Heijne, G.,** Amino acid distributions around o-linked glycosylation sites, *Biochem. J.*, 275, 529, 1991.
76. **Umemoto, J., Bhavanandan, V. P., and Davidson, E. A.,** Purification and properties of an endo-α-N-acetyl-D-galactoseaminidase from *Diplococcus pneumoniae*, *J. Biol. Chem.*, 252, 8609, 1977.
77. **Hortin, G. L.,** Isolation of glycopeptides containing O-linked oligosaccharides by lectin affinity chromatography on jacalin-agarose, *Anal. Biochem.*, 191, 262, 1990.
78. **Hille, A., Rosa, P., and Huttner, W. B.,** Tyrosine sulfation: a post-translational modification of proteins destined for secretion?, *FEBS Lett.*, 177, 129, 1984.
79. **Hortin, G. L. and Graham, J. P.,** Sulfation of tyrosine residues does not influence secretion of α_2-antiplasmin or C4, *Biochem. Biophys. Res. Commun.*, 151, 417, 1988.
80. **Baeuerle, P. A. and Huttner, W.,** Tyrosine sulfation is a *trans*-Golgi-specific protein modification, *J. Cell Biol.*, 105, 2655, 1987.
81. **Marcks von Würtemberg, M. and Fries, E.,** Secretion of $^{35}SO_4$-labelled proteins from isolated rat hepatocytes, *Biochemistry*, 28, 4088, 1989.
82. **Sjöberg, E. M. and Fries, E.,** Biosynthesis of bikunin (urinary trypsin inhibitor) in rat hepatocytes, *Arch. Biophys. Biochem.*, 295, 217, 1992.
83. **Hortin, G., Green, E. D., Baenziger, J. U., and Strauss, A. W.,** Sulphation of proteins secreted by a human hepatoma-derived cell line. Sulphation of N-linked oligosaccharides on α_2HS-glycoprotein, *Biochem. J.*, 235, 407, 1986.
84. **Huttner, W. B.,** Tyrosine sulfation and the secretory pathway, *Annu. Rev. Physiol.*, 50, 363, 1988.
85. **Hortin, G., Folz, R., Gordon, J. I., and Strauss, A. W.,** Characterization of sites of tyrosine sulfation in proteins and criteria for predicting their occurrence, *Biochem. Biophys. Res. Commun.*, 141, 326, 1986.
86. **Suiko, M. and Liu, M.-C.,** Change in binding affinities of 3Y1 secreted fibronectin upon desulfation of tyrosine-O-sulfate, *Biochem. Biophys. Commun.*, 154, 1094, 1988.
87. **Hortin, G. L., Farries, T. C., Graham, J. P., and Atkinson, J. P.,** Sulfation of tyrosine residues increases activity of the fourth component of complement, *Proc. Natl. Acad. Sci. U.S.A.*, 86, 1338, 1989.
88. **Leyte, A., van Schijndel, H. B., Niers, C., Huttner, W. B., Verbeet, M. Ph., Mertens, K., and van Mourik, J. A.,** Sulfation of Tyr^{1680} of human blood coagulation factor VIII is essential for the interaction of factor VIII with bon Willebrand factor, *J. Biol. Chem.*, 266, 740, 1991.
89. **Hanley, J. M., Haugen, T. H., and Heath, E. C.,** Biosynthesis and processing of rat haptoglobin, *J. Biol. Chem.*, 258, 7858, 1983.
90. **Ikehara, Y., Oda, K., and Kato, K.,** Conversion of proalbumin into serum albumin in the secretory vesicles of rat liver, *Biochem. Biophys. Res. Commun.*, 72, 319, 1976.
91. **Misumi, Y., Takami, N., and Ikehara, Y.,** Biosynthesis and processing of pro-C3, a precursor of the third component of complement in rat hepatocytes: effect of secretion-blocking agents, *FEBS Lett.*, 175, 63, 1984.
92. **Bennet, M. K., Wandinger-Ness, A., and Simons, K.,** Release of putative exocytotic transport vesicles from perforated MDCK cells, *EMBO J.*, 7, 4075, 1988.
93. **Brenna, S. O. and Peach, R. J.,** Calcium-dependent KEX2-like protease found in hepatic secretory vesicles converts proalbumin to albumin, *FEBS Lett.*, 229, 167, 1988.
94. **Hosaka, M., Nagahama, M., Kim, W.-S., Watanabe, T., Hatsuzawa, K., Ikemizu, J., Murakami, K., and Nakayama, K.,** Arg-X-Lys/Arg-Arg motif as a signal for precursor cleavage by furin within the constitutive secretory pathway, *J. Biol. Chem.*, 266, 12127, 1991.
95. **Kaumeyer, J. F., Polazzi, J. O., and Kotick, M. P.,** The mRNA for a proteinase inhibitor related to the HI-30 domain of inter-α-trypsin inhibitor also encodes α1-microglobulin (protein HC), *Nucleic Acids Res.*, 14, 7839, 1986.
96. **Lindqvist, A., Bratt, T., Altieri, M., Kastern, W., and Åkerström, B.,** Rat α_1-microglobulin: coexpression in liver with the light chain of inter-α-inhibitor, *Biochim. Biophys. Acta*, 1130, 63, 1992.
97. **Enghild, J. J., Thorgersen, I. B., Pizzo, S. V., and Salvesen, G.,** Analysis of inter-α-trypsin inhibitor and a novel trypsin inhibitor, pre-α-trypsin inhibitor, from human plasma. Polypeptide stoichiometry and assembly by glycan, *J. Biol. Chem.*, 264, 15975, 1989.
98. **Jessen, T. E., Faarvang, K. L., and Ploug, M.,** Carbohydrate as covalent crosslink in human inter-α-trypsin inhibitor: a novel plasma protein structure, *FEBS Lett.*, 230, 195, 1988.
99. **Enghild, J. J., Salvesen, G., Hefta, S. A., Thorgersen, I. B., Rutherfurd, S., and Pizzo, S. V.,** Chondroitin 4-sulfate covalently cross-links the chains of the human blood protein pre-α-inhibitor, *J. Biol. Chem.*, 266, 747, 1991.
100. **Åkerström, B. and Lögdberg, L.,** An intriguing member of the lipocalin family: α_1-microglobulin, *Trends Biochem. Sci.*, 15, 240, 1990.
101. **Parekh, R. B.,** Mammalian cell gene expression: protein glycosylation, *Curr. Opinion Biotechnol.*, 2, 730, 1991.

Chapter 31

CONTROL OF GLYCOSYLATION ALTERATIONS OF ACUTE PHASE GLYCOPROTEINS

Willem van Dijk and Andrzej Mackiewicz

TABLE OF CONTENTS

0-8493-6913-4/93/$0.00 + $.50

I. INTRODUCTION

During the acute phase response, the plasma concentrations of acute phase proteins (APP) are subject to changes (see Chapter 1 of this volume). This is a result of the liver response to cytokines that are released by monocytes and other cells following an inflammatory stimulus.[1] Most of the APP are glycoproteins containing two or more carbohydrate side chains. Specific alterations of the carbohydrate moieties of APP often accompany the quantitative changes in APP concentration during the acute phase response.[2-18] Several groups have demonstrated the usefulness of the determination of these alterations of a number of APP for the diagnosis, differentiation, and monitoring of patients with various disorders.[2-18]

Glycoproteins synthesized in the liver in general bear oligosaccharides which are conjugated to the asparagine of the polypeptide chain through *N*-glycosidic linkage of *N*-acetylglucosamine (GlcNAc). These oligosaccharides may have from two to four branches (biantennary, triantennary, and tetraantennary structures) arising from the $\alpha1\rightarrow3$- and $\alpha1\rightarrow6$-linked mannose (Man) residues of the core: Man$\alpha1\rightarrow$ 3(Man$\alpha1\rightarrow6$)Man$\beta1\rightarrow$ 4GlcNAc$\beta1\rightarrow$4GlcNAc. These branches generally consist of Gal$\beta1\rightarrow$4GlcNAc units which can, in turn, bear sialic acid, fucose, or other sugars in a number of different configurations. Moreover, in these antennary structures, a bisecting GlcNAc in an $\alpha1\rightarrow4$ linkage on the β-linked Man residue of the core may be present.[19] Variations in the structure of the oligosaccharides present at a given glycosylation site have been referred to as microheterogeneity. Two types of microheterogeneity have been distinguished: major microheterogeneity, which reflects differences in the number of branches on the antennary structures, and minor microheterogeneity, referring to variations in sialic acid or fucose content.[20]

For most APP, a major microheterogeneity is observed, resulting from the fact that the sites for *N*-glycosylation can be occupied by different complex-type glycans. However, preferential occupation of a particular site by a certain type of chain can occur. As a result, different glycoforms of an APP can be identified in serum using various methods based on interactions with lectins such as agarose affinity chromatography[21,22] or crossed-affinity immunoelectrophoresis (CAIE).[2-18] Lectins are (glyco)proteins which possess binding sites for defined oligosaccharides (for a review, see Reference 23). The most commonly used lectin is concanavalin A (Con A), which binds the unsubstituted groups of α-linked 2-*O*-substituted mannose (Man) residues at carbons 3, 4, and 6, with at least two interacting Man molecules being required for the binding.[24] As a result, this lectin binds with bi-, but not with tri- or tetraantennary structures. In the case of multiheteroglycan proteins, the degree of reactivity with Con A depends on the number of biantennary structures present on the molecule.[21,22] Thus, human α_1-acid glycoprotein (AGP) can be separated into Con A nonreactive (AGP-A), weakly reactive (AGP-B), and reactive (AGP-C) glycoforms. Of the five *N*-linked glycans present per molecule, none, one, and, two are of the biantennary type, respectively.[21] One of the types of minor microheterogeneity, variation in the content of fucose, can be studied using other lectins such as *Lens culinaris* agglutinin (LCA) and *Aleuria aurantia* lectin (AAL). AAL reacts with $\alpha1\rightarrow6$- and/or $\alpha1\rightarrow3$-linked fucose residues present on *N*-linked glycans.[23] LCA reacts with an $\alpha1\rightarrow6$-linked fucose residue present on the asparagine-linked GlcNAc residue of biantennary glycans only.[23]

In vitro studies have shown that the inflammation-related changes in glycosylation can have implications for the functioning of APP. Thus, the various glycoforms of AGP were found to differ in their ability to (1) inhibit the CD3-stimulated proliferation of T-cells[25] and (2) induce an IL-1 inhibitory activity by macrophages.[26]

The rise in serum concentration of APP during inflammation results mainly from cytokine-induced increases of their synthesis by the liver, as is extensively documented in other chapters of this volume. It could be expected, therefore, that the accompanying changes in glycosylation also occur in the liver. Indeed, liver damage has been reported to be accompanied by a change in the glycosylation of AGP and α_1-protease inhibitor (PI) in serum.[21] Furthermore, changes have been noticed in components of the glycoprotein biosynthetic pathway in the liver during inflammation.[28-31] These findings appear to support the view that changes in the glycosylation of serum glycoproteins take place in the liver during inflammation. However, a few groups have reported extrahepatic secretion of APP. Leukocytes and alveolar macrophages are able to synthesize PI,[32-34] and AGP can be secreted by human lymphocytes, granulocytes, and monocytes.[35] As inflammation is associated with the proliferation of leukocytes, part of the increased APP concentration in serum as well as the observed changes in the glycosylation of APP could originate from such cells. Finally, glycosidases released in the inflamed tissue or by the liver[36] might have modified the oligosaccharide chains of APPs during circulation in the inflamed plasma.

In this chapter, evidence is presented that the inflammation-induced alterations in glycosylation of the acute phase glycoproteins AGP and PI result from cytokine-induced changes in the posttranslational glycosylation process of the biosynthesis of the glycoproteins in the liver. Moreover, the profile of glycosylation changes is controlled by a network consisting of cytokines, cytokine receptors, and glucocorticoids. The evidence relies on studies with sera of humans and of mice transgenic for human AGP, in combination with *in vitro* studies employig the human hepatoma cell lines Hep 3B and Hep G2, and primary monolayer cultures of human hepatocytes. Crude cytokine preparations obtained from monocytes isolated from patients with active rheumatic disorders as well as defined, purified, or genetically engineered cytokines were used. CAIE with lectins was employed to determine the microheterogeneity in sera and secretion media from cultured cells, and electroimmunoassay to assay the rate of synthesis of APP secreted by these cells.

II. MATERIALS AND METHODS

A. PATIENT SERA

Pooled human serum was obtained from apparently healthy individuals.

Serum samples of previously healthy burn patients were taken within hours following the injury and at regular intervals during the next 2 months; the samples were stored at −70°C. The patients had burns covering 32 to 50% of the total body surface area. Initial therapy was directed toward repletion of fluid losses, mainly with saline infusions. The patients displayed acute phase responses, reflected by prolonged temperature elevation, tachycardia, and sharp increases in C-reactive protein levels.[13]

Serum samples from patients with severe rheumatoid arthritis (RA) and severe systemic lupus erythematosus (SLE), as well as RA and SLE patients with intercurrent infection, were analyzed. RA and SLE activity was evaluated at the time of blood sampling. RA activity was assessed by multivariate analysis based on morning stiffness, pain scale, grip strength, articular index, hemoglobin, erythrocyte sedimentation rate, and C-reactive protein concentration, while SLE activity was determined based on clinical criteria.[6-8]

B. MOUSE SERA

Mouse sera were obtained from transgenic mice for human AGP (see below).

C. DETECTION OF GENETIC VARIANTS OF AGP

Variants of human AGP were determined according to Eap and Baumann.[37] After desialylation with neuraminidase (type V, Sigma Chemical Co., St. Louis, MO), the samples were subjected to isoelectric focusing (IEF), followed by immunoblotting using rabbit anti-(human AGP) antiserum and goat anti-rabbit IgG conjugated with alkaline phosphatase (Sigma) for detection. Each sample was analyzed in duplicate after serial dilutions to avoid an excess of antigen. The type of variant obtained was identified by running standard sera.[31] Relative intensities of the three main variants were determined by laser densitometry; each track was determined on two positions.

D. PREPARATION OF CONDITIONED MEDIUM (CM) FROM HUMAN PERIPHERAL BLOOD MONOCYTES

Monocytes were isolated from the peripheral blood (50 ml) of seven healthy donors, eight active RA patients, and five active SLE patients by gradient centrifugation and adherence to plastic. Adherent cells were washed and incubated for 24 h in serum-free MEM Eagle's medium (Biomed, Lublin, Poland) in the absence [CM-LPS(−)] and presence [CM-LPS(+)] of 20 μg/ml of lipopolysaccharide (LPS) from *Escherichia coli* 055:B5 (Sigma).[38,39]

E. INDUCTION OF APP IN HEP 3B AND HEP G2 HUMAN HEPATOMA CELL LINES

Hep 3B and Hep G2 cells were cultured as described.[40,41] Cells were exposed to CM prepared as described above and to cytokines and cytokine combinations. The cytokines used included recombinant human (rh) interleukin (IL)-6 (specific activity, 2.5×10^6 U/ml; generous gift of Dr. G. Wong, Genetics Institute, Cambridge, MA), rhIL-1α (specific activity, 2×10^1 U/mg; generous gift of Dr. P. Lomedico, Hoffman La-Roche Inc., Nutley, NJ), rh tumor necrosis factor-α (TNFα) (specific activity, 2×10^7 U/mg; Genzyme, Boston, MA), rh leukemia inhibitory factor (LIF), which represented conditioned medium of Chinese hamster ovary (CHO) cells transfected with pC10-6R (1×10^5 U/ml;[42] generous gift of Dr. H. Baumann, Buffalo, NY), platelet-derived purified transforming growth factor-β_1 (TGFβ) (R&D Systems Inc., Minneapolis, MN), and rh interferon-γ (INFγ) (Bioferon, Germany). IL-6 was used in doses of 5, 10, 50, 100, and 1000 U/ml; IL-1 in doses of 50, 100, 200, and 500 U/ml; TNFα in doses of 100, 200, and 500 U/ml; TGFβ in doses of 5, 10, and 20 ng/ml; LIF in doses of 10, 50, 100, and 200 U/ml; and INFγ in doses 12.5, 25, 50, and 100 ng/ml. Cytokines were added to the culture separately and in different combinations. Incubation was carried out in the presence and absence of 1 μ*M* dexamethasone (Dex). Moreover, IL-6 and soluble human IL-6 receptor (shIL-6-R) (generous gift of Drs. S. Rose-John and P. C. Heinrich, Aachen, Germany) in combination were added to the Hep G2 cells. Cells were incubated with these preparations for an additional 48 h, with replacement of the medium every 24 h. Analyses were carried out in media collected over the final 24 h.

F. INDUCTION OF APP GLYCOSYLATION CHANGES IN Hep G2 CELLS STABLY TRANSFECTED WITH IL-6 cDNA

Hep G2 cells were stably transfected with an expression vector (pBMGNeo)[43] containing the mouse metallothionein I promoter driving the transcription of human IL-6 cDNA.[44] These cells (Hep G2-IL-6) constitutively secreted 2 μg per 10^6 cells per 24 h and 20 μg per 10^6 cells per 24 h of rhIL-6 upon induction with $ZnCl_2$. Hep G2-IL-6 cells were maintained in culture, and induced with cytokines and shIL-6-R as nontransfected Hep G2 cells.[45]

G. INDUCTION OF APP GLYCOSYLATION CHANGES IN HUMAN HEPATOCYTES IN PRIMARY MONOLAYER CULTURE

Postmortem human liver tissue was obtained from aged kidney donors. Isolation of the hepatocytes by a two-step perfusion technique and the preparation of a cell culture substrate were performed as described by Moshage et al.[46] Cells were seeded at a density of 175×10^3 hepatocytes per cm^2. After isolation and attachment (day 1), cells were cultured in the presence of 50 n*M* Dex (days 1 and 2).[47] From days 3 to 5, the hepatocytes were exposed to recombinant cytokines (IL-1, 100 U/ml; IL-6, 2000 U/ml; TNF, 30 U/ml) in the presence of 1 m*M* Dex with replacement of the medium (including cytokines and Dex) every 24 h, as described previously.[47] Analyses were carried out in tenfold-concentrated secretion media on days 4 and 5, i.e., 24 and 48 h after addition of the cytokines. Prior to concentration, the media were dialyzed for 24 h against 50 m*M* NH_4HCO_3/0.1 m*M* phenylmethylsulfonylfluoride/0.02% NaN_3. The resultant solutions were lyophilized and stored frozen until used in CAIE.

H. MICE TRANSGENIC FOR HUMAN AGP

The generation of transgenic mice in which intact copies of the human AGP-A gene were expressed has been described by Dente et al.[48] The transgenic mice carried the human gene stably integrated into their genome and could transmit it to their progeny. Although human AGP was secreted into the serum, full expression was reached by injection of LPS (100 μg) 24 h before serum samples were collected.

I. QUANTITATION OF APP

The amounts of PI and AGP present in human and mouse sera or secreted into the culture medium by human hepatocytes and hepatoma cells were quantitated by electroimmunoassay.[49] Human PI and AGP were determined with goat anti-human or rabbit anti-human specific antibodies (Atlantic Antibodies, Scarborough, ME, or Dakopats, Copenhagen). A human serum calibrator kit (Atlantic Antibodies, Scarborough, ME) was used as a standard for PI and AGP determination. It has been previously shown[50,51] that increased accumulation of plasma proteins in the culture medium occurred parallel to the increase of newly synthesized [^{35}S]methionine-labeled proteins and the increase in its intracellular mRNA concentration. Accordingly, the terms "accumulation" and "synthesis" are used interchangeably throughout this chapter.

J. DETERMINATION OF MICROHETEROGENEITY OF APP

The microheterogeneity of PI and AGP in sera and culture media was studied by means of agarose CAIE with free Con A as a ligand, according to the method of Bøg-Hansen.[52-54] Briefly, 50 μ*M* of Con A (EY Labs, San Mateo, CA, for human hepatocytes and type IV, Sigma Chemical Co., for hepatoma cells) was included in the first-dimension gel. Electrophoresis was carried out for 1.5 h at 10 V/cm, and the gel was transferred onto the second-dimension plate. Gel containing the aforementioned antibodies and 50 mg/ml of α-methyl-D-mannoside was cast. Electrophoresis in the second dimension was carried out for 16 to 18 h at 1.5 V/cm. The concentrations of antibodies and the amounts of sample added to the well were adjusted in different experiments to yield precipiation arcs of approximately comparable intensity and height for the different samples studied. After completing the electrophoresis and staining the gel, the area covered by the PI and AGP was determined by planimetry and the relative amounts of the different microheterogeneous forms were expressed as percentages of the total. Moreover, a reactivity coefficient (RC) was calculated

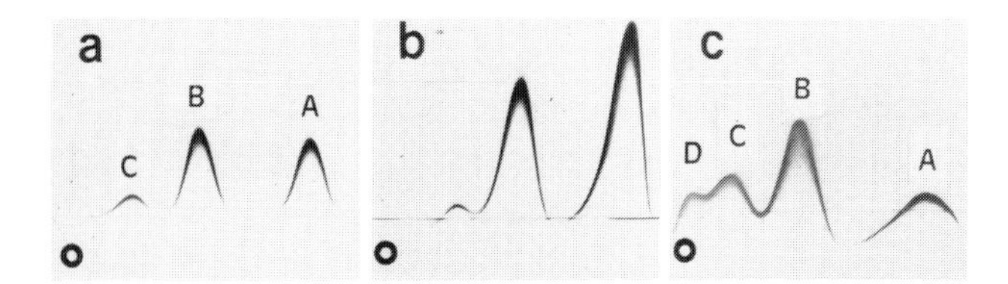

FIGURE 1. CAIE with Con A as a ligand of AGP in sera of (a) a healthy individual, (b) an active rheumatoid arthritis patient (type II glycosylation changes), and (c) a patient with systemic lupus erythematosus with intercurrent infection (type I glycosylation changes). Glycoform A, nonreactive with Con A; B, weakly reactive; C, reactive with Con A; D, strongly reactive with Con A.

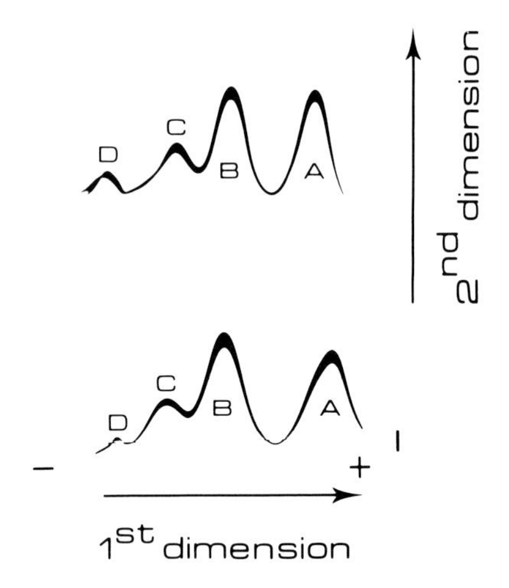

FIGURE 2. Con A reactivity of human AGP in serum of a burn patient (top) and a transgenic mouse in which only the A-gene of human AGP is expressed (bottom). The Con A reactivity was analyzed by CAIE using Con A as an affino-component in the first-dimension gel. The serum was obtained from burn patient 1 3 d after injury; A, AGP nonreactive with Con A; B to D, reactive forms of AGP. (From Van Dijk, W., Pos, O., Van der Stelt, M.E., Moshage, H. J., Yap, S.-H., Dente, L., Baumann, P., and Eap, C. B., *Biochem. J.*, 276, 343, 1991. With permission.)

according to the formula: sum of Con A-reactive variants/Con A-nonreactive variant. Statistical analyses were carried out using the Mann-Whitney test.

III. RESULTS

A. DIFFERENT TYPES OF CHANGES OF THE GLYCOSYLATION PATTERNS OF AGP IN ACUTE (TYPE I) AND CHRONIC INFLAMMATION (TYPE II)

CAIE with Con A revealed four heterogenous variants of AGP: variant A — nonrective with Con A, variant B — weakly reactive with Con A, variant C — reactive with Con A, and variant D — strongly reactive with Con A. Three AGP variants (A, B, and C) were observed in the sera of healthy individuals, as well as in the sera of patients suffering from rheumatic disorders (Figure 1). A fourth AGP variant was found in the sera of infected RA and SLE patients (Figure 1), severely burned patients (Figure 2), and LPS-treated transgenic mice in which only one of the human AGP genes was expressed (Figure 2).

Changes in glycosylation occurring in acute inflammation are well illustrated by the variation in RC values of AGP in the sera of severely burned patients during the period

TABLE 1
Con A Reactivity of Human AGP and Relative Occurrence of the Genetic Variants ORM1 S, ORM1 F1, and ORM2 A in Sera and Secretion Media of Primary Cultures of Human Hepatocytes and Hep 3B Cells Under Various Conditions

Source of AGP	Condition	Conc. of AGP (%)	Con A reactivity (RC)	Occurrence of genetic variants (%) AGP-A gene product ORM1 F1	ORM1 S	AGP-B/B′ gene product ORM2 A
Patient 1	Day 1	100	0.7	31.2 ± 0.3	40.3 ± 0.4	28.5 ± 0.6
	Day 3	125	2.2	30.1 ± 1.9	47.6 ± 1.3	22.3 ± 1.0
	Day 12	320	1.8	33.1 ± 0.1	42.8 ± 0.6	24.1 ± 0.6
	Day 60	154	0.9	35.6 ± 1.6	40.4 ± 0.0	24.0 ± 1.6
Patient 2	Day 1	100	2.1	40.7 ± 1.2	35.9 ± 0.9	23.4 ± 0.7
	Day 5	311	2.7	42.4 ± 1.1	37.7 ± 0.6	19.9 ± 0.6
	Day 22	307	1.0	42.0 ± 2.3	33.3 ± 2.3	24.7 ± 0.7
	Day 50	215	0.9	41.1 ± 2.4	32.7 ± 0.9	26.2 ± 1.7
Patient 3	Day 1	100	0.4	36.0 ± 0.4	30.5 ± 0.9	33.5 ± 0.6
	Day 4	208	1.7	44.0 ± 1.4	35.4 ± 1.0	20.6 ± 0.8
	Day 14	526	1.1	39.1 ± 0.9	36.4 ± 0.9	24.5 ± 0.7
Med. HH	Control	100	1.0	46.0 ± 1.7	33.9 ± 2.7	20.1 ± 1.3
	IL-1α	145	2.8	41.6 ± 1.8	37.3 ± 0.7	21.1 ± 1.7
	IL-6	182	3.8	45.0 ± 3.0	35.9 ± 2.6	19.1 ± 0.5
	CM	191	2.7	43.1 ± 2.1	37.8 ± 2.1	19.1 ± 1.3
Med. Hep 3B	Control	100	0	88[a]	n.d.	12[a]
	CM	450	0	87[a]	n.d.	13[a]

Note: IEF patterns of human AGP in sera of burn patients at various days after injury and in secretion media of control and cytokine-treated primary cultures of human hepatocytes were analyzed by laser densitometry. Each value is the mean of four determinations. The concentrations of AGP are given relative to the values at day 1 (sera of burn patients) or to control media of isolated human hepatocytes (HH) and Hep 3B cells. The reactivity of AGP with Con A (RC) is expressed as the ratio of the sum of the areas of Con A-reactive forms to the area of the nonreactive form. n.d., not detectable.

[a] Means of duplicate measurements.

From Van Dijk, W., Pos, O., Van der Stelt, M. E., Moshage, H. J., Yap, S.-H., Dente, L., Baumann, P., and Eap, C. B., *Biochem. J.*, 276, 343, 1991. With permission.

after injury (Figure 2, Table 1).[13] The reactivity with Con A (RC) as well as the serum level of AGP was strongly increased in the first days following injury. During the second and third week, the RC values declined to those of healthy persons, but the serum level of AGP was still elevated in most of the patients. The same type of change (APP glycosylation change, type I) was apparent in the sera of RA and SLE patients with intercurrent infection (Figure 1, Table 2).[6,7]

Decreased RC values of AGP (APP glycosylation change, type II) were observed in the sera of patients with severe RA (Table 2).

In patients with severe SLE, no glycosylation alterations were seen despite on increased concentration of AGP (Table 2).

B. INFLAMMATION-INDUCED CHANGES IN Con A REACTIVITY OF AGP ARE NOT DEPENDENT ON CHANGES IN ITS GENETIC EXPRESSION

Three genetic variants of AGP, ORM1 F1 and S (products of the A-gene of AGP[55]) and ORM2 (product of the B- or B′-gene of AGP[55] were detected by IEf in the sera of burn

TABLE 2
α_1-Acid Glycoprotein-Con A Reactivity (RC) of Severe and Infected Rheumatoid Arthritis (RA) and Systemic Lupus Erythematosus (SLE) Patients

Clinical diagnosis	n	AGP-RC[a]
Healthy donors	44	1.35 ± 0.2
Severe RA	40	0.70 ± 0.15[b]
Severe SLE	25	1.36 ± 0.4
RA + infection	15	1.87 ± 0.23[b,c]
SLE + infection	20	2.81 ± 0.76[b,d]

Note: n, number of patients.

[a] Mean ± SD.
[b] Significant difference when compared with healthy donors.
[c] Significant difference when compared with noninfected RA.
[d] Significant difference when compared with noninfected SLE.

patients (Table 1). Small changes in relative occurrence of the products of A-gene and the B- or B′- gene of AGP were found during the substantial rise and fall of the total AGP level in the sera of all three patients during inflammation. These small changes in the relative occurrence of the genetic variants cannot account for the large changes in the RC vaues of the AGP. This was further substantiated in studies with sera of transgenic mice in which only the A-gene of AGP was expressed. Up to four glycosylation variants of AGP could be distinguished in these mouse sera (Figure 2), although only one human variant of AGP (ORM1 F1) was present.[55]

C. HUMAN HEPATOCYTES ARE ABLE TO SECRETE THE GENETIC AND THE GLYCOSYLATION VARIANTS OF AGP OCCURRING IN NORMAL AND INFLAMED SERA

Human hepatocytes in primary monolayer culture secreted both of the ORM1 variants and the ORM2 variant of AGP, the latter representing about 20% of the total AGP (Table 1). This value is comparable to the relative occurrence of the variants observed in human sera. "Acute" inflammatory conditions were simulated *in vitro* by incubating hepatocytes with IL-1α, IL-6, or a mixture of crude cytokines (CM). The cytokines induced secretion of the various genetic variants of AGP to the same extent, since the relative occurrence of the ORM1 and ORM2 variants was not affected, despite the fact that the total amount of secreted AGP was increased under the same conditions (Table 1).

CAIE with Con A revealed three glycosylation variants of AGP (A, B, and C) in the secretion media of primary monolayer cultures of human hepatocytes. The pattern remained the same throughout the various days of culture (Figure 3a and b) and was comparable with the pattern of normal human serum (Figure 1a). Incubation of the hepatocytes with cytokines resulted in an increased reactivity of AGP with Con A (Table 1). The *in vitro* "acute" inflammatory conditions stimulated the secretion of AGP-B, -C, and -D, and even a new variant, AGP-E (Figure 3). The inducing effects of the cytokines were only found in the presence of Dex.[47]

D. CHANGES IN GLYCOSYLATION AND SECRETION ARE INDEPENDENTLY REGULATED

As described above, the change in the glycosylation of AGP observed in the sera of severely burned individuals lasts for a much shorter period than the increase of the total

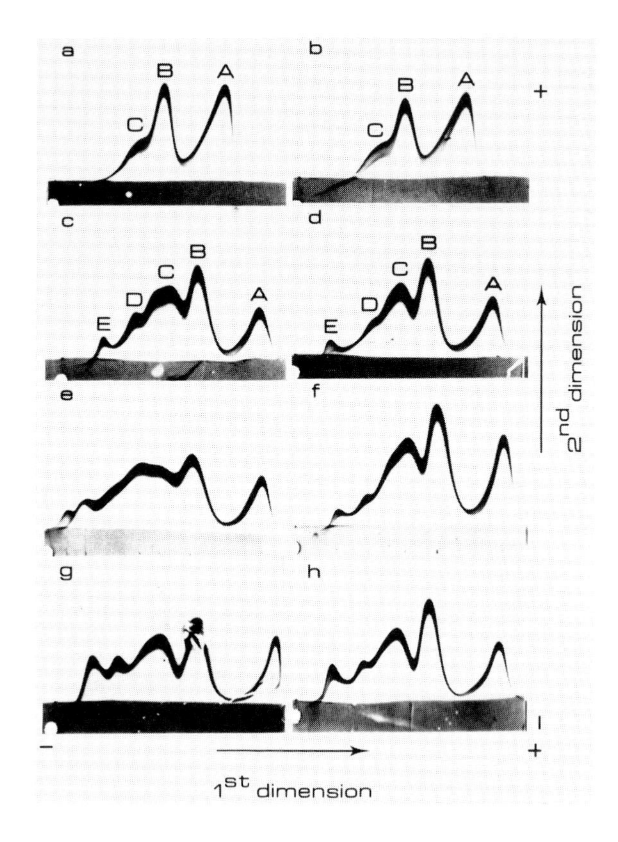

FIGURE 3. CAIE with Con A as a ligand of AGP secreted by isolated human hepatocytes cultured for 24 h (left) or 48 h (right) in the absence (a, b) or presence of cytokines (c to h). The same amount of tenfold-concentrated medium was applied to each gel. (a, b) control; (c, d) IL-1; (e, f) IL-6; (g, h) IL-1 + IL-6. Glycoform A, nonreactive with Con A; B, weakly reactive with Con A; C to E, strongly reactive with Con A.

amount of AGP. This was also found for another APP, PI, in the same sera.[13] Moreover, type II found in severe RA and type I of AGP glycosylation changes seen in RA with intercurrent infection were accompanied by an increase of serum AGP levels. Finally, the increased concentration of AGP was not accompanied by an altered pattern of glycosylation in severe SLE patients. These findings suggested that changes in the glycosylation and concentration of APP in patient sera are regulated independently. The occurrence of different mechanisms was supported by *in vitro* studies using the human hepatoma cell lines Hep G2 and Hep 3B incubated with a variety of inflammatory mediators. PI was used in these studies as a model acute phase glycoprotein, since AGP secreted by both cell lines did not react with Con A under the various conditions (Table 1).

Changes in the synthesis and secretion of PI induced by cytokines were consistently parallel in both cell lines, with one exception: Hep 3B cells were not responsive to LIF (Table 3). INFγ was studied in Hep G2 cells only. IL-6, LIF, and TGFβ caused a dose-dependent increase, while INFγ caused a decrease of the synthesis of PI. IL-1, TNFα, and Dex did not affect PI synthesis.

CAIE with Con A revealed three microheterogeneous forms of PI: a variant nonreactive with Con A, a variant weakly reactive with Con A, and a variant reactive with Con A. In both cell lines, the three glycoforms were always present in the culture medium (Figure 4).

Incubation of Hep 3B cells with IL-6, TGFβ, and Dex caused a dose-dependent increase of the reactivity of PI with Con A (type I glycosylation changes). Other preparations tested had no effect on the PI glycosylation pattern (Table 3).

TABLE 3
Regulation by Cytokines of Synthesis and ConA Reactivity (RC) of α_1-Protease Inhibitor in Hep 3B and Hep G2 Cells

	Synthesis (%)		RC	
Inducer	**Hep 3B**	**Hep G2**	**Hep 3B**	**Hep G2**
Control	100[a]	100	1.0	1.0
IL-6 (50 U/ml)	170	146	1.7	0.35
LIF (100 U/ml)	100	151	1.0	0.68
INFγ (100 ng/ml)	n.d.	70	n.d.	0.78
TGFβ (10 ng/ml)	150	130	1.7	3.6
TNF (500 U/ml)	100	100	0.7	0.6
IL-1 (200 U/ml)	100	100	0.92	0.95

Note: All experiments were carried out in the presence of 1 μ*M* dex; n.d., not detectable.

[a] Mean of ten experiments carried out in duplicate.

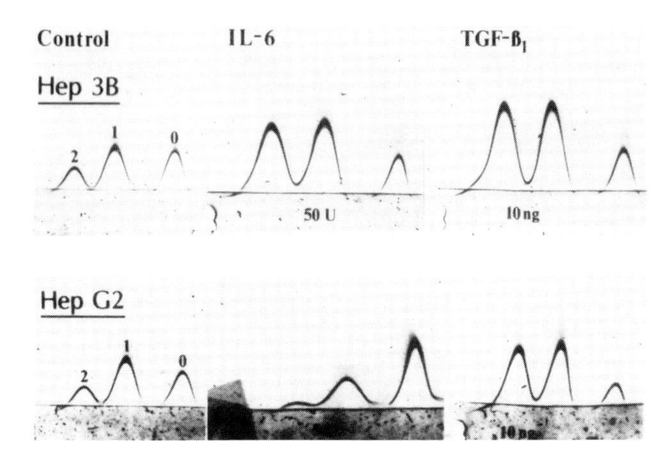

FIGURE 4. CAIE with Con A as a ligand of PI secreted by Hep 3B and Hep G2 cells upon induction with IL-6 and TGFβ. Glycoform 0, nonreactive with Con A; 1, weakly reactive with Con A; 2, reactive with Con A.

In contrast to Hep 3B cells, exposure of Hep G2 cells to IL-6, LIF, and high doses of TNFα led to a dose-dependent decrease of the reactivity of PI with Con A (Table 3). The same effect was displayed by INFγ (type II). However, TGFβ and Dex caused an increase of PI-ConA reactivity (type I). IL-1 did not significantly affect the distribution of Con A-PI glycoforms.

E. DIFFERENT CAPABILITIES OF PERIPHERAL BLOOD MONOCYTES OF RA AND SLE PATIENTS TO INDUCE CHANGES IN APP GLYCOSYLATION

All CM preparations affected the reactivity with Con A of PI secreted by Hep G2 cells (Table 4). CM-LPS(+) obtained from healthy donors and SLE patients had stronger effects on PI-Con A reactivity than CM-LPS(−) prepared from the same cells. The degree of alteration caused by CM-LPS(−) from SLE patients was the same as that caused by the corresponding CM from healthy donors. In contrast, CM-LPS(−) obtained from RA patients affected PI-Con A reactivity to the same degree as CM-LPS(+), which was similar to that of CM-LPS(+) from SLE and healthy donors. The inducing capabilities (for the glycosy-

TABLE 4
Capabilities of Conditioned Media (CM) from Healthy Individuals (HI), Rheumatoid Arthritis, and Systemic Lupus Erythematosus (SLE) Patients to Induce α_1-Protease Inhibitor Changes in Hep G2 Cells

Diagnosis	n	CM-LPS(−)	CM-LPS(+)
HI	7	0.50 ± 0.05[a]	0.25 ± 0.01[b]
RA	8	0.30 ± 0.05[b]	0.25 ± 0.005[b]
SLE	5	0.52 ± 0.05	0.3 ± 0.005[b]

Note: n, number of patients.

[a] Mean ± SD; results expressed as the ratio of an RC to the RC of control cultures (with no CM).
[b] Significant difference ($p < 0.001$) when compared with CM-LPS(−) of HI.

TABLE 5
Effect of Dex on Cytokine-Induced Glycosylation Changes of PI Secreted by Hep G2 Cells

	RC	
Cytokine	Dex (−)	Dex (+)
Control	1.0	1.29 ± 0.05[a,b]
LIF (100 U/ml)	0.44 ± 0.03	0.53 ± 0.03[b]
INFγ (25 ng/ml)	0.58 ± 0.02	1.18 ± 0.06[b]
INFγ (100 ng/ml)	1.0 ± 0.06	0.73 ± 0.04[b]
IL-6 (10 U/ml)	0.43 ± 0.04	0.24 ± 0.02[b]
IL-6 (100 U/ml)	0.29 ± 0.02	0.23 ± 0.02[b]
TGFβ (2 ng/ml)	1.33 ± 0.05	3.3 ± 0.1[b]

[a] Mean ± SD of six experiments carried out in duplicate.
[b] $p < 0.05$ compared to experiments without Dex.

lation of APP) of nonactivated monocytes from healthy donors [CM-LPS(−)], which served here as a control, may be due to the partial activation of these cells by adherence to the plastic and/or to possible endotoxin contamination in laboratory materials.[57]

F. NETWORK OF CYTOKINES, CYTOKINE RECEPTORS, AND GLUCOCORTICOIDS CONTROLLING TYPE I AND TYPE II GLYCOSYLATION CHANGES OF APP

1. Effect of Binary Combinations of LIF, INFγ, IL-6, TGFβ, IL-1, TNFα, and Dex on PI-Con A Reactivity

The combination of LIF with either INFγ or IL-6 had an additive effect, resulting in a further decrease of PI-Con A reactivity over the effect of each cytokine used alone (type II). Addition of TGFβ on Dex to LIF caused a reduction of the effect of LIF on PI-Con A reactivity. This effect was also additive (Tables 5 and 6, Figures 5 and 6).

The combination of INFγ with IL-6 and IL-6 with TNFα had an additive effect, further decreasing PI-Con A reactivity. In contrast, TGFβ and Dex diminished the effect of INFγ in an additive manner.

The combination of IL-6 with TGFβ had an additive effect that reduced the effects of each of the cytokines used alone. Surprisingly, the addition of Dex to IL-6 did not reduce the effect of IL-6, but increased it.

TABLE 6
Effect of Binary Cytokine Combinations on Glycosylation Changes of PI Secreted by Hep G2 Cells

Cytokines	RC	
	Dex (−)	Dex (+)
LIF(100 U/ml) + INFγ(100 ng/ml)	0.37 ± 0.03[a]	0.45 ± 0.03[b]
LIF(100 U/ml) + IL-6(100 U/ml)	0.25 ± 0.02	0.33 ± 0.02[b]
LIF(100 U/ml) + TGFβ(2 ng/ml)	0.56 ± 0.04	0.96 ± 0.05[b]
IL-6(100 U/ml) + INFγ(100 U/ml)	0.22 ± 0.02	0.18 ± 0.01[b]
INFγ(100 U/ml) + TGFβ(2 ng/ml)	1.10 ± 0.03	1.57 ± 0.05[b]
IL-6(100 U/ml) + TGFβ(2 ng/ml)	0.37 ± 0.02	1.22 ± 0.05[b]
IL-6(100 U/ml) + TNFα(500 U/ml)	0.22 ± 0.01	n.d.
TGFβ(2 ng/ml) + IL-1(200 U/ml)	1.02 ± 0.02	n.d.
TGFβ(2 ng/ml) + TNFα(500 U/ml)	1.01 ± 0.02	n.d.

Note: n.d., not done.

[a] Mean ± SD of three experiments carried out in duplicate.
[b] $p < 0.05$ when compared to experiments without Dex.

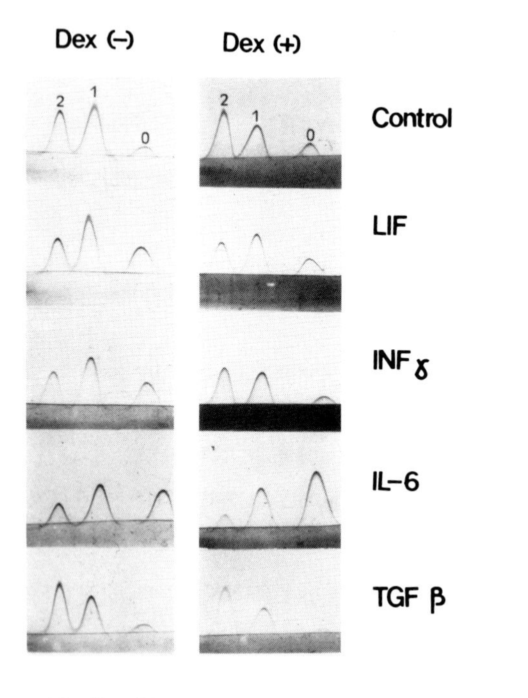

FIGURE 5. CAIE of PI secreted by Hep G2 cells upon induction with cytokines in the absence and presence of Dex. PI glycoforms as in Figure 4.

The combination of TGFβ and Dex had a synergistic effect, resulting in a further increase of PI-Con A reactivity (type I). In contrast, the addition of IL-1 and TNF to TGFβ diminished its effect.

In experiments in which increasing amounts of INFγ were added at a constant concentration of Dex, the final effect resulted from the ratio of the two mediators used (Table 5, Figure 5).

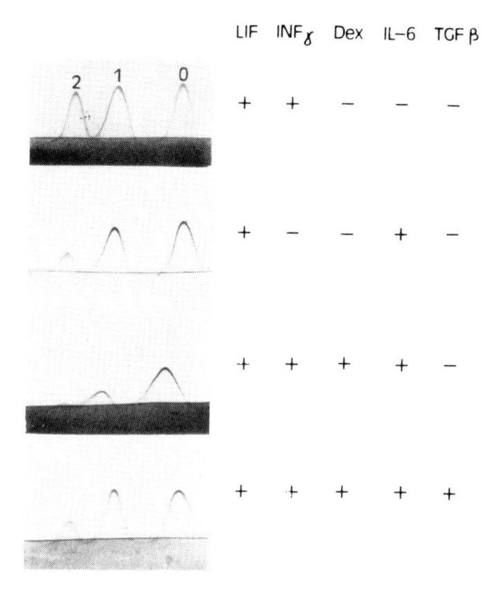

FIGURE 6. CAIE of PI secreted by Hep G2 cells upon induction with combinations of cytokines.

TABLE 7
Effect of Complex Cytokine Combinations on Glycosylation Changes of PI Secreted by Hep G2 Cells

Cytokines	RC	
	Dex (−)	Dex (+)
LIF + IL-6 + INFγ (25 ng/ml)[a]	0.21 ± 0.02[b]	0.13 ± 0.01[c]
LIF + IL-6 + INFγ (100 ng/ml)	0.14 ± 0.01	0.14 ± 0.01
LIF + IL-6 + TGFβ	0.35 ± 0.02	0.35 ± 0.02
LIF + IL-6 + TGFβ + INFγ (25 ng/ml)	0.31 ± 0.02	0.24 ± 0.01[c]
LIF + IL-6 + TGFβ + INFγ (100 ng/ml)	0.22 ± 0.01	0.23 ± 0.01

[a] LIF was used at a dose of 100 U/ml, IL-6 at 100 U/ml, and TGFβ at 2 ng/ml.
[b] Mean ± SD of three experiments carried out in duplicate.
[c] $p < 0.05$ when compared with experiments without Dex.

2. Effect of Complex Combinations of LIF, INFγ, IL-6, TGFβ, and Dex on PI-Con A Reactivity

The addition of Dex to binary combinations of LIF and INFγ, LIF and IL-6, LIF and TGFβ, INFγ and TGFβ, and IL-6 and TGFβ increased the Con A reactivity of PI secreted by Hep G2 cells compared to the Con A reactivity of PI which resulted from the effect of binary combinations of these cytokines (Table 7). In contrast, the addition of Dex to INFγ combined with IL-6 further decreased the Con A reactivity of secreted PI.

The addition of INFγ to a combination of LIF and IL-6 further decreased PI-Con A reactivity. In contrast, TGFβ added to such a combination increased PI-Con A reactivity. In turn, TGFβ combined with a mixture of LIF, IL-6, and INFγ reduced their ability to decrease PI-Con A reactivity (Figure 6, Table 7).

The addition of Dex to combinations of LIF, IL-6, and INFγ with or without TGFβ further decreased PI-Con A reactivity. However, Dex had no effect when INFγ was used in a dose of 100 ng/ml. The addition of Dex to a combination of LIF, IL-6, and TGFβ resulted in no change of PI-Con A reactivity.

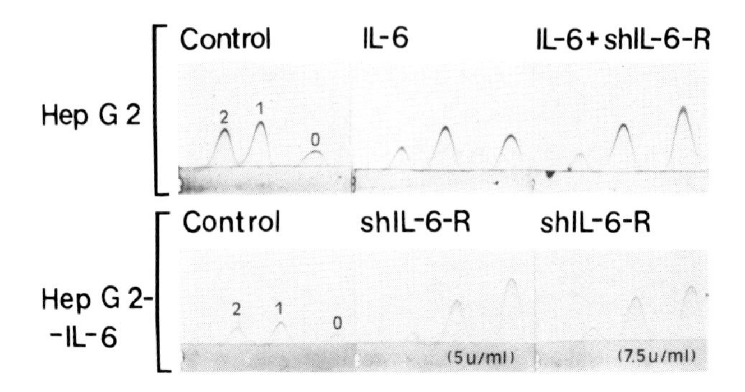

FIGURE 7. CAIE of PI secreted by Hep G2 cells (top) upon induction with IL-6 and shIL-6-R and Hep G2-IL-6-R (bottom) upon induction with shIL-6-R.

3. Effect of shIL-6-R on IL-6-Dependent Glycosylation Changes of APP

A constant amount of shIL-6-R added to increasing concentrations of IL-6 or added in increasing doses to constant amounts of IL-6 very significantly augmented, in a dose-dependent manner, the ability of IL-6 to decrease the Con A reactivity of PI secreted by Hep G2 cells (Figure 7).

Hep G2-IL-6 cells showed similar levels of mRNA and secreted comparable quantities of PI and other APP as nontransfected Hep G2 cells in the absence of IL-6.[45] The Con A reactivity of PI was also comparable to that of PI from intact, nonstimulated Hep G2 cells (Figure 7). Treatment of Hep G2-IL-6 cells with $ZnCl_2$ or addition of exogenous IL-6 did not change the PI-Con A reactivity. Exposure of these cells to LIF, INFγ, or TGFα affected PI-Con A reactivity. The addition of shIL-6 to Hep G2-IL-6 cells resulted in a dose-dependent (with a maximum at 5 U/ml) significant decrease on Con A reactivity of PI secreted by these cells (Figure 7).

IV. DISCUSSION

During acute and chronic inflammatory processes, APP undergo qualitative changes.[10,16,58] In sera of patients with acute inflammation, an increase of biantennary vs. more complex glycans on the polypeptide backbone, referred to as type I glycosylation alterations, was found. In these studies, ''acute inflammation'' was represented by burn patients and patients with RA and SLE who experienced intercurrent infection. On the other hand, in some chronic inflammatory states, a relative decrease of biantennary vs. more complex glycans, as shown here in severe RA patients, was observed (type II). Finally, there are some disorders, such as SLE, where the glycosylation profile of APP is not altered.[6] As demonstrated in the course of RA, the same patient may display various types of glycosylation alterations, depending on the clinical status. A flare-up of the disease caused type II changes, while an intercurrent infection evoked type I alterations.[7,8]

It has been suggested that changes of the APP glycosylation pattern observed in sera may result from selective clearance of particular microheterogeneous forms from the circulation through a lectin-like receptor system on the hepatocyte membrane. This receptor system binds terminal galactose residues, and thus may eliminate asialo-glycoproteins, with a preference for Con A nonreactive variants.[59] However, recent studies of patients with various types of hepatitis[24] and patients with RA[17] have demonstrated parallel increases of asialo-AGP and AGP-Con A nonreactive forms. Thus, these data, taken together with data on the AGP carbohydrate moiety structure,[21] rule out the hypothesis that ''selective elimination'' is a major mechanism responsible for changes of APP reactivity with Con A.

Another possible mechanism included variations in the expressin of the genetic variants of particular APPs. Using AGP as a model, it has been shown that minor changes in the relative occurrence of the products of three AGP genes found in the sera of burn patients could not account for the major alterations of AGP glycosylation. Moreover, this finding was supported in a transgenic mouse model in which only one human AGP gene was expressed, while multiple AGP glycoforms were found in serum.[55,56] These results seem to exclude the above hypothesis as well.

Others have postulated that increased synthesis of acute phase glycoproteins in the liver is not compensated for by an increase of the synthesis or activity of glycosylating enzymes.[3] The data presented here, which demonstrate that an increased serum concentration of APP might be accompanied by type I, type II, or no glycosylation changes, suggest that this is not a major mechanism. However, recent studies in primary culture of mouse hepatocytes with transgenic rat AGP have shown that AGP-Con A reactivity was associated with the level of rat AGP synthesis.[60,61] The question remains, however, whether this was a true effect or an artifact of the transgenic mouse model.

Human hepatocytes, human hepatoma cell lines Hep G2 and Hep 3B, and Hep G2 cells transfected with IL-6 cDNA were used as a model to investigate the regulation of the major microheterogeneity of AGP and PI occurring in the sera of patients suffering from acute or chronic inflammation. The validity of these cell systems for the study of inflammation-related events has been established in studies of the effects of cytokines, cytokine receptors, and glucocorticoids on the regulatory mechanisms of APP gene expression.[1,41,45,47,62-69]

Two different types of microheterogeneity of APP were revealed upon incubation of isolated human hepatocytes, and Hep 3B and Hep G2 cells with conditioned medium from activated peripheral blood monocytes (crude cytokine preparation). The changes in the glycosylation of the APP observed in the human hepatocytes and the Hep 3B cells resembled the type I alterations seen in the sera of patients with acute inflammation (increased synthesis and reactivity with Con A). The Hep G2 cells displayed the type II changes observed in the sera of patients with some chronic inflammatory states (increased synthesis of APP and decreased reactivity with Con A). The contribution of modifications in the protein moiety to these changes could be excluded for AGP (Table 1). These results strongly support the theory that changes in the microheterogeneity of APP in sera result from cytokine-induced variations in the posttranslational glycosylation process in the parenchymal cells of the liver. The same conclusion was reached for a number of other human[40] and rat APP.[54,70] Moreover, the mechanism that regulates the glycosylation changes of APP is at least partly independent from those governing alterations of their synthesis. This uncoupling of the cytokine regulation of the production and glycosylation of human AGP has also been reported by Hiron and co-workers.[70] It cannot be excluded, however, that the increased protein production will also affect the microheterogeneity of APP, as was observed in isolated hepatocytes of mice transgenic for rat AGP.[61]

A number of cytokines have been shown to be involved in the regulation of APP glycosylation *in vitro*.[47,65,71-75] Due to the different effects on the major microheterogeneity of APPs in human hepatoma cell lines, cytokines could be divided into at least four classes: (1) those responsible for controlling both type I and type II alterations (IL-6), (2,3) those which induce type I (TGFβ) and type II (LIF, INFγ, TNFα) alterations, and (4) those which have no direct effect, but modulate the activity of other cytokines (IL-1). Moreover, glucocorticoids were found to be directly and indirectly involved in these processes. In Hep G2 cells, they directly induced type I changes and, when combined with other cytokines, modulated their effect. Interestingly, each of the cytokines was distinctly affected by Dex. Dex had no influence on LIF activity since it acted only additively, which resulted in a diminishing LIF effect as both factors acted oppositely. When Dex was combined with low doses of INFγ, both factors acted additively having an effect similar to the combination of LIF and Dex. However, when INFγ was used in high doses, which were ineffective in changing PI glycosylation in Hep G2 cells, the addition of Dex led to changes analogous to those caused by INFγ but not by Dex when they were used separately. Another type of modulation of cytokine activity by Dex was seen in the case of IL-6. The effect of IL-6 on PI glycosylation in both low and high doses was enhanced by Dex despite the fact that both factors had opposite effects when used separately. Finally, Dex combined with TGFβ had a synergistic effect in increasing the number of binantennary vs. triantennary glycans of PI.

It is becoming clear that the biological effects of cytokines, including regulation of APP gene expression, result from the cooperation of cytokines and glucocorticoids rather than the effect of a single factor.[74] Accordingly, reasoning that various types of glycosylation alterations of APP found in patient sera might be controlled by a specific set of interacting factors, the effects of binary and complex combinations of a number of cytokines and glucocorticoids were evaluated. In Hep G2 cells, binary and complex combinations of LIF, INFγ, and IL-6, three cytokines which similarly affected the PI glycosylation profile, had an additive effect in increasing the relative number of tri- vs. biantennary glycans. However, a more detailed analysis has shown that the combination of LIF and IL-6 was less effective than the simple addition of their separate effects. Moreover, high doses of INFγ, which were ineffective when used separately, in combination with LIF and/or IL-6 further enhanced their stimulatory effect. TGFβ, in both binary and complex combinations with other cytokines, also had an additive effect. However, as TGFβ acted oppositely to the aforementioned cytokines by decreasing the relative amount of triantennary vs. biantennary glycans of PI, it diminished their effect. Dex acted additively synergistically, or had no effect when added to binary and complex combinations of cytokines. The type of interaction depended on the composition and the amount of interacting cytokines.

Cytokine activity can be modulated at different levels: (1) regulation of synthesis and secretion, (2) regulation by soluble inhibitors or agonists, and (3) regulation of cell-surface receptor expression.[75] As an example of differences in the synthesis and secretion of cytokines involved in controlling the glycosylation of APP, we conducted an experiment in which conditioned media prepared from blood monocytes isolated from patients with severe RA and severe SLE differed in their ability to induce alterations of the glycosylation pattern of PI secreted by Hep G2 cells. *Ex vivo,* stimulation of SLE monocytes with lipopolysaccharide resulted in induction of the synthesis and secretion of cytokines which were able to induce PI glycosylation changes *in vitro.*[38,39] Regulation of cytokine activity by soluble inhibitors and agonists, particularly soluble cytokine receptors, has recently attracted considerable interest.[75] Soluble receptors are truncated forms of membrane-bound receptors that lack their transmembrane and cytoplastic domains.[76] They are generally believed to compete for the ligand with their membrane counterparts, which finally leads to the decrease of cytokine biological activity. Data presented here indicate that shIL-6-R displays an unique activity, resulting in enhancement of the effects of IL-6 on hepatocytes.[45,77] Finally, modulation of the effects of cytokines on hepatocytes by regulation of the expression of membrane receptors could be demonstrated using Hep G2-IL-6 cells. Chronic exposure of hepatocytes to IL-6 rendered them unresponsive (homologous desensitization) to the IL-6, most probably preventing overstimulation of the hepatocytes.[45] The underlying mechanism involved is probably related to downregulation of the gp80 subunit of the IL-6 receptor[78]-IL-6 binding protein. This is supported by the fact that shIL-6-R functionally substituted gp80, reconstituting the responsiveness of IL-6-desensitized hepatocytes. Another mechanism involved in regulation of the expression of membrane receptors may be attributed to the effect of glucocorticoids. Upregulation of hepatic IL-6 and INFγ receptors on human monocytes has been recently reported.[79,80] The enhancing effect of Dex (which, when applied alone, induced type I alterations) on the ability of IL-6 supports the hypothesis that the effect of Dex is related to the increase of a number of IL-6 receptors on the hepatocyte.

The enhancing effect of shIL-6-R on IL-6 activity as well as its ability to reconstitute the responsiveness to IL-6 of hepatocytes chronically exposed to the cytokine indicate that shIL-6-R may be the active constituent of the network controlling glycosylation of APP. Moreover, shIL-6-R was found in biological fluids.[81]

Recent studies of Bierhuizen et al.[21] have indicated that the AGP-Con A nonreactive variant is composed only of tri- and tetraantennary units, while the weakly reactive variant has one biantennary unit and the reactive variant, two biantennary units. Assuming that the number of biantennary units is the sole determinant of Con A binding and employing our findings in AGP as a standard, one can infer the heteroglycan structures of the forms of PI secreted by the two cell lines we studied.[73] Since PI possesses three complex-type heteroglycans which are *N*-linked to a polypeptyde chain and composed of bi- and triantennary oligosaccharides,[82] our results suggest that for both Hep 3B and Hep G2 cells, the nonreactive variant of PI has three triantennary structures, the weakly reactive variant one biantennary and two triantennary units, and the reactive variant one triantennary and two biantennary structures. These conclusions are in agreement with the findings of Vaughan et al.,[22] based on affinity chromatography studies, which indicate that the degree of Con A binding of serum PI was directly related to the number of biantennary units. Therefore, the cytokine-induced variations in Con A rectivity patterns apparently reflect changes in the biantennary glycan content of the APP.

Posttranslational modification of oligosaccharide side chains of glycoproteins is a multistep enzymatic process. A series of highly specific glycosidases and glycosyltransferases sequentially process an oligosaccharide precursor to yield various types of *N*-linked glycans.[19] The branches that occur on complex-type oligosaccharides are initiated by the incorporation of an N-acetylglucosamine (GlcNAc) residue. This reaction is catalyzed by GlcNAc transferases (GnT). GnT III catalyzes the formation of biantennary structures with bisecting GlcNAc, while GnT IV and GnT V catalyze the formation of tri- and tetraantenary units. Control of the level of relative activity of different GlcNAc transferases is one of the regulating mechanisms of branching during the synthesis of complex-type oligosaccharides.[83]

Recently, IL-6 has been demonstrated to increase GnT IV and GnT V activity, and decrease GnT III activity in a human myeloma cell line.[84] Changes in GlcNAc transferase activities were accompanied by an increase of tri- and tetraantenary structures of glycoproteins on the membranes of these cells. In addition, it should be noted that reduced GnT III activity might also contribute to the formation of more branched structures, since GnT III and GnT V compete for the same substrate.[85] It has also been reported that cytokines and Dex regulate the activity and gene expression of other enzymes involved in oligosaccharide processing.[30,36,62,86,87] Moreover, discrepancies were found between the effect of Dex (stimulating) and conditioned medium derived from the keratocarcinoma cell line Colo-16, which contains LIF (ineffective), in the regulation of sialyltransferase mRNA in the rat hepatoma cell line H-35.[87] All these facts, taken together, indicate that the changes of PI-Con A reactivity may result from the altered relative activity (gene expression?) of glycosylating enzymes. Inflammatory mediators may either up- or downregulate expression of the same and/or different GnT genes, which results in the altered structure of oligosaccharide chains.

However, other possible mechanisms such as altered rate of intracellular transport[88] or altered route of transit cannot be excluded.

The results presented have described only the effect of cytokines, cytokine receptors, and glucocorticoids on the major microheterogeneity of certain APP, i.e., changes in the degree of branching. Minor microheterogeneity, connected with differences in sialic acid or fucose content, has also been reported to be associated with inflammation.[14,17,89] Very recently, Van Dijk and co-workers[89] have obtained evidence that expression of the sialylated Lewis X structures, NeuAcα2→3Galβ1→4(Fucα1→3)GlcNAc-, on human AGP increases markedly during various acute inflammatory processes. This structure has been shown to be the ligand present on leukocytes for E-selectin, the endothelial leukocyte adhesion molecule-1 expressed on the activated endothelial cells in inflamed tissues.[90] Studies on the effect of these changes in the glycosylation of AGP on the interaction of leukocytes and activated endothelial cells lining inflamed tissues are in progress.

ACKNOWLEDGMENTS

This research was supported by The Netherlands Organization for Scientific Research Grant 523-063, KBN Grant 40791, and II Maria Curie-Sklodowska Fond No. MZ-HHS-92-104.

REFERENCES

1. **Baumann, H. and Gauldie, J.,** Regulation of hepatic acute phase plasma protein genes by hepatocyte stimulating factors and other mediators of inflammation, *Mol. Biol. Med.*, 7, 147, 1990.
2. **Nicollet, I., Lebreton, J. P., Fontaine, M., and Hiron, M.,** Evidence for alpha-1-acid glycoprotein populations of different pI values after concanavalin A affinity chromatography: study of their evolution during inflammation in man, *Biochim. Biophys. Acta,* 668, 235, 1981.
3. **Raynes, J.,** Variations in the relative proportions of microheterogeneous forms of plasma glycoproteins in pregnancy and disease, *Biomedicine,* 36, 77, 1982.
4. **Hansen, J. E., Jensen, S. P., Nørgaard-Pedersen, B., and Bøg-Hansen, T. C.,** Electrophoretic analysis of the glycan microheterogeneity of orosomucoid in cancer and inflammation, *Electrophoresis,* 7, 180, 1986.
5. **Hansen, J. E., Iversen, J., Lihme, A., and Bøg-Hansen, T. C.,** Acute phase reaction, heterogeneity, and microheterogeneity of serum proteins as non specific tumor markers in lung cancer, *Cancer,* 60, 1630, 1987.
6. **Mackiewicz, A., Marcinkowska-Pieta, R., Ballou, S., Mackiewicz, S., and Kushner, I.,** Microheterogeneity of alpha-1-acid glycoprotein in the detection of intercurrent infection in systemic lupus erythematosus, *Arthritis Rheum.*, 30, 513, 1987.
7. **Pawlowski, T., Mackiewicz, S., and Mackiewicz, A.,** Microheterogeneity of alpha-1-acid glycoprotein in the detection of intercurrent infection in rheumatoid arthritis, *Arthritis Rheum.*, 32, 347, 1989.
8. **Mackiewicz, A., Pawlowski, T., Mackiewicz-Pawlowska, A., Wiktorowicz, K., and Mackiewicz, S.,** Microheterogeneity forms of alpha-1-acid glycoprotein as indicators of rheumatoid arthritis activity, *Clin. Chim. Acta,* 163, 185, 1987.
9. **Mackiewicz, A., Khan, M. A., Reynolds, T. L., Van der Linden, S., and Kushner, I.,** Serum IgA, α1-acid glycoprotein microheterogeneity and acute phase proteins in ankylosing spondylitis, *Ann. Rheum. Dis.*, 48, 99, 1989.
10. **Breborowicz, J. and Mackiewicz, A.,** Affinity electrophoresis for diagnosis of cancer and inflammatory conditions, *Electrophoresis,* 10, 568, 1989.
11. **Jezequel, M., Seta, N. S., Corbic, M. M., Feger, J. M., and Durand, G. M.,** Modifications of concanavalin A patterns of α1-acid glycoprotein and α2-HS glycoprotein in alcoholic liver disease, *Clin. Chim. Acta,* 176, 49, 1988.
12. **Hachulla, E., Laine, A., and Hayem, A.,** α-1-antichymotrypsin microheterogeneity in crossed immunoaffinoelectrophoresis with free concanavalin A: a useful diagnostic tool in inflammatory syndrome, *Clin. Chem.*, 34, 911, 1988.
13. **Pos, O., Van der Stelt, M. E., Wolbink, G.-J., Nijsten, M. W. N., Van der Tempel, G. L., and Van Dijk, W.,** Changes in the serum concentration and the glycosylation of human α_1-acid glycoprotein and α_1-protease inhibitor in severely burned persons: relation to interleukin-6 levels, *Clin. Exp. Immunol.*, 82, 579, 1990.
14. **Serbource-Goguel Seta, N., Durand, G., Corbic, M., Agneray, J., and Feger, J.,** Alterations in relative proportions of microheterogeneous forms of human α1-acid glycoprotein in liver disease, *J. Hepatol.*, 2, 245, 1986.

15. **Serbource-Goguel, N., Corbic, M., Erlinger, S., Durand, G., Agneray, J., and Feger, J.,** Measurement of serum α_1-acid glycoprotein and α_1-antitrypsin desialylation in liver diseae, *Hepatology*, 3, 356, 1983.
16. **Mackiewicz, A., Pawlowski, T., Gorny, A., and Kushner, I.,** Glycoforms of α_1-acid glycoprotein in the management of rheumatic patients, in *Affinity Electrophoresis: Principles and Application*, Bręborowicz, J. and Mackiewicz, A., Eds., CRC Press, Boca Raton, FL, 1992, 229.
17. **Pawlowski, T. and Mackiewicz, A.,** Minor microheterogeneity of α1-acid glycoprotein in rheumatoid arthritis, *Prog. Clin. Biol. Res.*, 300, 223, 1989.
18. **Bręborowicz, J., Gorny, A., Drews, K., and Mackiewicz, A.,** Glycoforms of alpha$_1$-acid glycoprotein in cancer, in *Affinity Electrophoresis: Principles and Application*, Bręborowicz, J. and Mackiewicz, A., Eds., CRC Press, Boca Raton, FL, 1992, 191.
19. **Baenziger, J. U.,** The oligosaccharides of plasma proteins: synthesis, structure and function, in *The Plasma Proteins*, Vol. 4, Putman, F., Ed., Academic Press, New York, 1984, 271.
20. **Hatton, M. W. C., Marz, L., and Regoeczi, E.,** On the significance of heterogeneity of plasma glycoproteins possessing N-glycans of the complex type: a perspective, *Trends Biochem. Sci.*, 92, 287, 1983.
21. **Bierhuizen, M., De Wit, M., Govers, C., Ferwerda, W., Koeleman, C., Pos, O., and Van Dijk, W.,** Glycosylation of three molecular forms of human α_1-acid glycoprotein having different interactions with concanavalin A. Variations in the occurrence of di-, tri-, and tetraantenary glycans and the degree of sialylation, *Eur. J. Biochem.*, 175, 387, 1988.
22. **Vaughan, L., Lorier, M., and Carrell, W.,** α1-Antitrypsin microheterogeneity. Isolation and physiological significance of isoforms, *Biochim. Biophys. Acta*, 701, 339, 1982.
23. **Debray, H. and Montreuil, J.,** Specificity of lectins toward oligosaccharide sequences belonging to N- and O-glycosylproteins, in *Affinity Electrophoresis: Principles and Application*, Bręborowicz, J. and Mackiewicz, A., Eds., CRC Press, Boca Raton, FL, 1992, 23.
24. **Narasimhan, S., Freed, J. C., and Schachter, H.,** The effect of a "bisecting" N-acetylglycosaminyl group on the binding of biantennary, complex oligosaccharides to concanavalin A, Phaseolus vulgaris erythroagglutinin (E-PHA), and Ricinus communis agglutinin (RCA-120) immobilized on agarose, *Carbohydr. Res.*, 149, 65, 1986.
25. **Pos, O., Oostendorp, R. A. J., Van der Stelt, M. E., Scheper, R. J., and Van Dijk, W.,** Con A-nonreactive human α_1-acid glycoprotein (AGP) is more effective in modulation of lymphocyte proliferation than Con A-reactive AGP serum variants, *Inflammation*, 14, 133, 1990.
26. **Bories, P. N., Feger, J., Benbernou, N., Agneray, J., and Durand, G.,** Prevalence of tri- and tetrantennary glycans of human α_1-acid glycoprotein in release of macrophage inhibitor of interleukin-1 activity, *Inflammation*, 14, 315, 1990.
27. **Monnet, D., Durand, D., Biou, D., Feger, J., and Durand, G.,** D-Galactosamine-induced liver injury: a rat model to study the heterogeneity of the oligosaccharide chains of α_1-acid glycoprotein, *J. Clin. Chem. Clin. Biochem.*, 23, 249, 1985.
28. **Lombart, C., Sturgess, J., and Schachter, H.,** The effect of turpentine-induced inflammation on rat liver glycosyl transferases and Golgi complex ultrastructure, *Biochim. Biophys. Acta*, 629, 1, 1980.
29. **Kaplan, H. A., Woloski, B. M. R. N. J., Hellman, M., and Jamieson, J. C.,** Studies on the effect of inflammation on rat liver and serum sialyltransferase. Evidence that inflammation causes release of Gal1$\rightarrow$4GlcNAcα2$\rightarrow$6-sialyltransferase from liver, *J. Biol. Chem.*, 258, 11505, 1983.
30. **Van Dijk, W., Boers, W., Sala, M., Lasthuis, A. H. M., and Mookerjea, S.,** Activity and secretion of sialyltransferase in primary monolayer cultures of rat hepatocytes cultured with and without dexamethasone, *Biochem. Cell. Biol.*, 64, 79, 1986.
31. **Jamieson, J. C., Kaplan, H. A., Woloski, B. M. R. N. J., Hellman, M., and Ham, K.,** Glycoprotein biosynthesis during the acute phase response to inflammation, *Can. J. Biochem. Cell Biol.*, 61, 1041, 1983.
32. **Budek, W., Bunning, P., and Heinrich, P. C.,** Rat lung tissue is a site of α_1-proteinase inhibitor synthesis: evidence by cell-free translation, *Biochem. Biophys. Res. Commun.*, 122, 394, 1984.
33. **Van Furth, R., Kramps, J. A., and Diesselhof-Den Dulk, M. M. C.,** Synthesis of α_1-antitrypsin by human monocytes, *Clin. Exp. Immunol.*, 51, 551, 1983.
34. **Gahmberg, C. G. and Andersson, L. C.,** Leukocyte surface origin of human α_1-acid glycoprotein (orosomucoid), *J. Exp. Med.*, 148, 507, 1978.
35. **Rogers, J., Kalsheker, N., Wallis, S., Speer, A., Coutelle, C. H., Woods, D., and Humphries, S. E.,** The isolation of a clone for human α_1-antitrypsin and the detection of α_1-antitrypsin in mRNA from liver and leukocytes, *Biochem. Biophys. Res. Commun.*, 116, 375, 1983.
36. **Woloski, B. M. R. N. J., Gospodarek, E., and Jamieson, J. C.,** Studies on monokines as mediators of the acute phase response. Effects on sialyltransferase, α_1-acid glycoprotein and β-*N*-acetylhexosaminidase, *Biochem. Biophys. Res. Commun.*, 130, 30, 1985.
37. **Eap, C. B. and Baumann, P.,** Isoelectric focusing of α_1-acid glycoprotein in immobilized pH gradients with 8 M urea: detection of its desialylated variants using an alkaline phosphatase-linked secondary antibody system, *Electrophoresis*, 9, 650, 1988.

38. **Mackiewicz, A., Sobieska, M., Kapcińska, M., and Pawlowski, T.,** Capabilities of peripheral blood monocytes to induce changes in glycosylation of plasma proteins *in vitro,* in *Affinity Electrophoresis: Principles and Application,* Bręborowicz, J. and Mackiewicz, A., Eds., CRC Press, Boca Raton, FL, 1992, 155.
39. **Mackiewicz, A., Sobieska, M., Kapcińska, M., Mackiewicz, S. H., Wiktorowicz, K., and Pawlowski, T.,** Different capabilities of monocytes from patients with systemic lupus erythematosus and rheumatoid arthritis to induce glycosylation alterations of acute phase proteins in vitro, *Ann. Rheum. Dis.,* 51, 67, 1992.
40. **Mackiewicz, A., Ganapathi, M. K., Schultz, D., and Kushner, I.,** Monokines regulate glycosylation of acute phase proteins, *J. Exp. Med.,* 166, 253, 1987.
41. **Mackiewicz, A., Schultz, D., Mathison, J., Ganapathi, M. K., and Kushner, I.,** Effect of cytokines on glycosylation of acute phase proteins in human hepatoma cell lines, *Clin. Exp. Immunol.,* 75, 70, 1989.
42. **Baumann, H. and Wong, G. G.,** Hepatocyte-stimulating factor III shares structural and functional identity with leukemia-inhibitory factor, *J. Immunol.,* 143, 1163, 1989.
43. **Karasuyama, H. and Melchers, F.,** Establishment of mouse cell lines which constitutively secrete large quantities of interleukin 2, 3, 4 or 5, using modified cDNA expression vectors, *Eur. J. Immunol.,* 18, 97, 1988.
44. **Schiel, X., Rose-John, S., Dufhues, G., Schooltink, H., and Heinrich, P. C.,** Microheterogeneity of human interleukin-6 synthesized by transfected NIH/3T3 cells: comparison with human monocytes, fibroblasts, and endothelial cells, *Eur. J. Immunol.,* 20, 883, 1990.
45. **Mackiewicz, A., Schooltink, H., Heinrich, P. C., and Rose-John, S.,** Soluble human interleukin-6-receptor upregulates synthesis of acute phase proteins, *J. Immunol.,* 149, 2021, 1992.
46. **Moshage, H. I., Rijntjes, P. J. M., Hafkenscheid, J. C. M., Roelofs, H. M. J., and Yap, S. H.,** Primary culture of human hepatocytes on homologous extracellular matrix: influence of monocytic products on albumin synthesis, *J. Hepatol.,* 7, 34, 1988.
47. **Pos, O., Moshage, H. J., Yap, S. H., Schnieders, J. P. M., Aarden, L. A., Van Gool, J., Boers, W., Brugman, A. M., and Van Dijk, W.,** Effects of monocytic products, recombinant interleukin-1, and recombinant interleukin-6 on glycosylation of α_1-acid glycoprotein: studies with primary human hepatocyte cultures and rats, *Inflammation,* 13, 415, 1989.
48. **Dente, L., Uhlrich, R., Tripodi, M., Wagner, E. F., and Cortese, R.,** Expression of human α_1-acid glycoprotein genes in cultured cells and transgenic mice, *Genes Dev.,* 2, 259, 1988.
49. **Laurell, C. B.,** Electroimmunoassay, *Scand. J. Clin. Lab. Invest. Suppl.,* 124, 21, 1972.
50. **Ganapathi, M. K., Schultz, D., Mackiewicz, A., Samols, D., Hu, S .-I., Brabenec, A., Macintyre, S. S., and Kushner, I.,** Heterogeneous nature of the acute phase response: differential regulation of human serum amyloid A, C-reactive protein and other acute phase proteins by cytokines in Hep 3B cells, *J. Immunol.,* 141, 564, 1988.
51. **Baumann, H., Hill, R. E., Sauder, D. N., and Jahreis, G. P.,** Regulation of major acute phase plasma proteins by hepatocyte-stimulating factors of human squamous carcinoma cells, *J. Cell Biol.,* 102, 370, 1986.
52. **Bøg-Hansen, T. C.,** Crossed immuno-affinoelectrophoresis: an analytical method to predict the result of affinity chromatography, *Anal. Biochem.,* 56, 480, 1973.
53. **Mackiewicz, A. and Mackiewicz, S.,** Determination of lectin-sugar dissociation constants by agarose affinity electrophoresis, *Anal. Biochem.,* 156, 481, 1986.
54. **Pos, O., Van Dijk, W., Ladiges, N., Linthorst, C., Sala, M., Van Tiel, D., and Boers, W.,** Glycosylation of four acute phase glycoproteins secreted by rat liver cells in vivo and in vitro: effects of inflammation and dexamethasone, *Eur. J. Cell. Biol.,* 46, 121, 1988.
55. **Van Dijk, W., Pos, O., Van der Stelt, M. E., Moshage, H. J., Yap, S.-H., Dente, L., Baumann, P., and Eap, C. B.,** Inflammation-induced changes in expression and glycosylation of genetic variants of α_1-acid glycoprotein. Studies with human sera, primary cultures of human hepatocytes and transgenic mice, *Biochem. J.,* 276, 343, 1991.
56. **Tomei, L., Eap, C. B., Baumann, P., and Dente, L.,** Use of transgenic mice for the characterization of human α_1-acid glycoprotein (orosomucoid) variants, *Hum. Genet.,* 84, 89, 1989.
57. **Dinarello, C. A.,** Interleukin 1, *Rev. Infect. Dis.,* 6, 51, 1984.
58. **Fassbender, K., Zimmerli, W., Kissling, R., Sobieska, M., Aeschlimann, A., Kellner, M., and Muller, W.,** Glycosylation of α_1-acid glycoprotein in relation to duration of disease in acute and chronic infection and inflammation, *Clin. Chim. Acta,* 203, 315, 1991.
59. **Ashwell, G. and Harford, J.,** Carbohydrate-specific receptors of the liver, *Annu. Rev. Biochem.,* 51, 531, 1982.
60. **Dewey, M. J., Rheaume, C., Berger, F. G., and Baumann, H.,** Inducible and tissue-specific expression of rat α_1-acid glycoprotein in transgenic mice, *J. Immunol.,* 144, 4392, 1990.

61. **Mackiewicz, A., Dewey, M. J., Berger, F., and Baumann, H.,** Acute phase mediated changes in glycosylation of rat α_1-acid glycoprotein in transgenic mice, *Glycobiology,* 7, 265, 1991.
62. **Woloski, B. M., Fuller, G. M., Jamieson, J. C., and Gospodarek, E.,** Studies on the effect of the hepatocyte-stimulating factor on galactose-β-4-N-acetylglucosamine α2→6-sialyltransferase in cultured hepatocytes, *Biochim. Biophys. Acta,* 885, 185, 1986.
63. **Baumann, H., Hill, R. E., Sauder, D. N., and Jahreis, G. P.,** Regulation of major acute phase plasma proteins by hepatocyte stimulating factors of human squamous carcinoma cells, *J. Cell Biol.,* 102, 370, 1986.
64. **Castell, J. V., Gomez-Lechon, M. J., David, M., Hirano, T., Kishimoto, T., and Heinrich, P. C.,** Recombinant human interleukin-6 (IL-6/BSF-2/HSF) regulates the synthesis of acute phase proteins in human hepatocytes, *FEBS Lett.,* 232, 347, 1988.
65. **Mackiewicz, A. and Kushner, I.,** Affinity electrophoresis for studies of mechanisms regulating glycosylation of plasma proteins, *Electrophoresis,* 10, 830, 1990.
66. **Baumann, H., Won, K.-A., and Jahreis, G. P.,** Human hepatocyte-stimulating factor-III and interleukin-6 are structurally and immunologically distinct but regulate the production of the same acute phase plasma proteins, *J. Biol. Chem.,* 264, 8046, 1989.
67. **Castell, J. V., Gomez-Lechon, M. J., David, M., Andus, T., Geiger, T., Trullenque, R., Fabra, R., and Heinrich, P. C.,** Interleukin-6 is the major regulator of acute protein synthesis in adult hepatocytes, *FEBS Lett.,* 242, 237, 1989.
68. **Moshage, H. J., Roelofs, H. M. J., van Pelt, J. F., Hazenberg, B. P. C., van Leeuwen, M. A., Limburg, P. C., Aarden, L. A., and Yap, S. H.,** The effect of interleukin-1, interleukin-6 and their relationship on the synthesis of serum amyloid A and C-reactive protein in primary cultures of adult human hepatocytes, *Biochem. Biophys. Res. Commun.,* 155, 112, 1988.
69. **Mackiewicz, A., Ganapathi, M. K., Schultz, D., Samols, D., Reese, J., and Kushner, I.** Regulation of rabbit acute phase protein biosynthesis by monokines, *Biochem. J.,* 253, 851, 1988.
70. **Hiron, M., Daveau, M., and Lebreton, J.-P.,** Microheterogeneity of α_1-acid glycoprotein and α_2-HS in cultured rat and human hepatocytes and in cultures of human hepatoma cells: role of cytokines in the uncoupling of changes in secretion and in Con A reactivities of acute phase glycoprotein, in *Affinity Electrophoresis: Principles and Application,* Bręborowicz, J. and Mackiewicz, A., Eds., CRC Press, Boca Raton, FL, 1992, 163.
71. **Mackiewicz, A., Pos, O., Van der Stelt, M. E., Yap, S.-H., Kapcinska, M., Laciak, M., Dewey, M. J., Berger, F. G., Baumann, H., Kushner, I., and Van Dijk, W.,** Regulation of glycosylation of acute phase proteins by cytokines *in vitro,* in *Affinity Electrophoresis: Principles and Application,* Bręborowicz, J. and Mackiewicz, A., Eds., CRC Press, Boca Raton, FL, 1992, 135.
72. **Mackiewicz, A. and Kushner, I.,** Interferon β2/BSF/IL-6 affects glycosylation of acute phase proteins in human hepatoma cell lines, *Scand. J. Immunol.,* 29, 257, 1989.
73. **Mackiewicz, A. and Kushner, I.,** Transforming growth factor β1 influences glycosylation of α_1-proteinase inhibitor in human hepatoma cell lines, *Inflammation,* 14, 485, 1990.
74. **Mackiewicz, A., Speroff, T., Ganapathi, M. K., and Kushner, I.,** Effects of cytokine combinations on acute phase protein production in two human hepatoma cell lines, *J. Immunol.,* 146, 3032, 1991.
75. **Fernandez-Botran, R.,** Soluble cytokine receptors: their role in immunoregulation, *FASEB J.,* 5, 2567, 1991.
76. **Taga, T., Hibi, M., Hirata, Y., Yamasaki, K., Yasukawa, K., Matsuda, T., Hirano, T., and Kishimoto, T.,** Interleukin-6 triggers the association of its receptor with a possible signal transducer, gp130, *Cell,* 58, 573, 1989.
77. **Mackiewicz, A., Rose-John, S., Schooltink, H., Laciak, M., Gorny, A., and Heinrich, P. C.,** Soluble human interleukin-6-receptor modulates interleukin-6-dependent N-glycosylation of α1-protease inhibitor secreted by Hep G2 cells, *FEBS Lett.,* 306, 257, 1992.
78. **Zohlnhofer, D., Graeve, L., Rose-John, S., Schooltink, H., Dittrich, E., and Heinrich, P. C.,** The hepatic interleukin-6-receptor: internalization and downregulation by interleukin-6, *Eur. J. Immunol.,* in press.
79. **Rose-John, S., Schooltink, H., Lenz, D., Hipp, G., Dufhues, G., Schmitz, H., Schiel, X., Hirano, T., Kishimoto, T., and Heinrich, P. C.,** Studies on the structure and regulation of the human hepatic interleukin-6 receptor, *Eur. J. Biochem.,* 190, 79, 1990.
80. **Strickland, R. W., Wahl, L. M., and Finbloom, D. S.,** Corticosteroids enhance the binding of recombinant interferon γ to cultured human monocytes, *J. Immunol.,* 1986, 137, 1577.
81. **Novick, D., Engelmann, H., Wallach, D., and Rubinstein, M.,** Soluble cytokine receptors are present in normal human urine, *J. Exp. Med.,* 170, 1409, 1989.
82. **Mega, T., Lujan, E., and Yoshida, A.,** Studies on the oligosaccharide chains of human α1-protease inhibitor. Structure of oligosaccharides, *J. Biol. Chem.,* 255, 4057, 1980.

83. **Schachter, H., Narasimhan, S., Gleeson, P., and Vella, G.,** Control of branching during the biosynthesis of asparagine-linked oligosaccharides, *Can. J. Biochem. Cell Biol.,* 16, 1049, 1983.
84. **Nakao, H., Nishikawa, A., Karasuno, T., Nishiura, T., Iida, M., Kanayama, Y., Yonezawa, T., Tarui, S., and Tanigushi, N.,** Modulcation of N-acteyl-glucosaminyltransferase III, IV, and V activates an alteration of the surface oligosaccharide structure of a myeloma cell line by interleukin 6, *Biochem. Biophys. Res. Commun.,* 172, 1260, 1990.
85. **Brockhausen, I., Hull, E., Hindsgaul, O., Schachter, H., Shah, R. N., Michnick, S. W., and Carver, J. P.,** Control of glycoprotein synthesis. Detection and characterization of a novel branching enzyme from hen oviduct, UDP-N-acetylglucosamine:GLCNAcβ1-6(GlcNAcβ1-2)Manα-R (GlcNAc to Man). Beta-4-N-acetylglucosaminyltransferase VI, *J. Biol. Chem.,* 264, 11272, 1989.
86. **Sarkar, M. and Mookerjea, S.,** Effect of dexamethasone on the synthesis of dolichol-linked saccharides and glycoproteins in hepatocytes prepared from control and inflamed rats, *Biochem. J.,* 227, 675, 1985.
87. **Wang, X. C., O'Hanlon, T. P., and Lau, J. T. Y.,** Regulation of β-galactoside α2,6-sialyltransferase gene expression by dexamethasone, *J. Biol. Chem.,* 264, 1854, 1989.
88. **Drechou, A., Rouzeau, J.-D., Feger, J., and Durand, G.,** Variations in the rate of secretion of different glycosylated forms of rat α1-acid glycoprotein, *Biochem. J.,* 263, 961, 1989.
89. **De Graaf, T. W., Van der Stelt, M. E., Anbergen, W. G., and Van Dijk, W.,** Inflammation induced increase in sialyl-Lewis^x bearing glycans on α1-acid glycoprotein in human sera, *J. Exp. Med.,* 177, 657, 1993.
90. **Springer, T. A. and Lasky, L. A.,** Sticky sugars for selectins, *Nature (London),* 349, 196, 1991.

Chapter 32

POST-TRANSLATIONAL REGULATION OF C-REACTIVE PROTEIN SECRETION

Stephen S. Macintyre and Patricia A. Kalonick

TABLE OF CONTENTS

0-8493-6913-4/93/$0.00 + $.50

I. INTRODUCTION

Recent intensive investigation has led to an increasing appreciation of the complexity of the processes which take place within the endoplasmic reticulum (ER), the site for the synthesis, assembly, and initial modification of both integral membrane and secretory proteins. Despite much progress in our understanding of these processes, little is known of the post-translational mechanisms which may regulate the secretion of plasma proteins by the liver. This chapter reviews the recent advances in our understanding of the assembly and sorting of proteins within the ER and presents data which indicate that the intracellular trafficking of newly synthesized C-reactive protein (CRP) is governed by a novel post-translational mechanism which is regulated during the acute phase response.

Proteins which are cotranslationally inserted into the lumen of the ER during their synthesis have varied destinations, including the plasma membrane, the extracellular space, lysosomes, elements of the Golgi apparatus, and the ER itself. Thus, the need for specific delivery of a diversity of proteins to multiple locations represents a formidable task in protein trafficking. The mechanism by which newly synthesized lysosomal enzymes are specifically targeted to lysosomes has been well characterized.[1] Considerable progress has also been made in eluciding the role of the carboxyl terminal KDEL,[2] or homologous,[3-5] sequence in the continuous retrieval of soluble ER resident proteins from downstream compartments to their proper location within the lumen of the ER.[2,6,7] More recently, several reports have begun to identify sequence and/or structural motifs of certain transmembrane proteins which allow for their specific localization to the membranes of the ER and Golgi elements.[8-12]

In a process referred to as quality control,[13-15] proteins destined for exit from the ER appear to require a degree of proper folding and/or assembly, possibly facilitated by molecular chaperones such as BiP (grp78),[16] in order to be released from the ER. Proteins which do not meet these criteria are subject to degradation within the ER.[17-20] In the case of the T-cell antigen receptor, the processes of ER retention, assembly of subunits, and degradation of improperly assembled complexes appear to be tightly coupled.[21,22]

While much evidence indicates that the process of quality control within the ER can prevent secretion or accumulation of abnormal proteins, the extent to which ER retention may also serve to post-translationally regulate the intracellular trafficking of normally assembled integral membrane and secretory proteins is not clear. For example, although it is widely recognized that different secretory proteins exit the ER at varying rates,[23-27] the mechanisms underlying this observation are not understood. A currently prevailing hypothesis suggests that in the absence of specific targeting signals, secretory proteins are transported to the cell surface by a default pathway of rapid bulk flow of vesicular contents.[28] However, there is also evidence that "positive" sorting signals may be required to facilitate the exit of α_1-antitrypsin and proalbumin from the ER.[29-31]

CRP has been a particularly useful protein in illuminating the mechanisms regulating the intracellular transport of secretory proteins. In the remainder of this chapter, we review evidence which demonstrates that the secretion of CRP becomes more efficient during the acute phase response, and present new data on the molecular mechanisms which regulate the intracellular sorting of CRP under differing physiologic conditions.

II. MATERIALS AND METHODS

A. ANIMALS AND CELL CULTURES

Primary hepatocyte cultures were prepared from male New Zealand white rabbits (obtained from Howard Gutman, Madison, OH) by an *in situ* collagenase (Type I, sigma Chemical Co., St. Louis, MO) perfusion technique, as described previously.[32] Acute tissue injury was induced in some rabbits by the intramuscular injection of 1 ml of turpentine in each thigh 18 to 24 h prior to cell preparation. Blood obtained from the marginal ear vein at the time of sacrifice was used for serum CRP determinations by radial immunodiffusion, as described.[33]

In the pulse-chase studies, cells were allowed to acclimate to culture conditions for 20 to 22 h. Incubation conditions for the pulse-chase studies were as described before.[34] Medium from a replicate dish which had received no medium change other than the rinse following cell attachment was used to determine rates of extracellular accumulation of both CRP and albumin employing radioimmunoassays (RIAs), as described below.

B. SUBCELLULAR FRACTIONATION

For the preparation of subcellular fractions from cultured cells, medium from a minimum of ten culture dishes per time sampling was removed and the dishes were rinsed in ice-cold homogenization buffer consisting of 0.25 *M* sucrose, 20 m*M* Hepes, and 10 m*M* KCl. Cells were scraped from the dishes in a total of 6 ml of homogenization buffer and were homogenized in a glass Dounce-type homogenizer with 15 passes of a tight glass pestle, followed by ten passes of a Teflon pestle. Microsomal subfractions corresponding to rough, smooth, and Golgi were prepared by Carey and Hirschberg's modification of the technique of Fleischer and Kervina, and all operations were at 4°C.[35,36] A 1-ml aliquot of the lysate was removed, and both the aliquot and the remaining lysate were centrifuged at 11,000 × *g* for 10 min. The 1-ml portion was used to estimate homogenization-induced leakage of pulse-labeled proteins, as described below.

For use in the initial microsomal binding assays (see below), subcellular fractions were prepared from whole rabbit liver by the same procedure except that homogenization was performed as described,[35] employing motor-driven graded Teflon pestles with clearances of 0.026 and 0.012 in. in a Potter-Elvehjem homogenizer. In subsequent binding assays employing purified rough microsomes, subcellular fractionation was performed as described previously.[37] In those cases where unfractionated microsomes were studied, the initial 11,000 × *g* supernate was centrifuged at 110,000 × *g* for 60 min to produce a total microsome pellet.

C. ESTIMATION OF CELL BREAKAGE AND LEAKAGE AND ADSORPTION OF PULSE-LABELED PROTEINS

Homogenization-induced leakage of pulse-labeled proteins and adsorption of leaked proteins to microsomal fractions were estimated employing a strategy described previously.[38] Adsorption was determined in initial control experiments by including trace amounts of ^{125}I-labeled CRP or rabbit albumin (prepared as described below) in the homogenization buffer added to unlabeled hepatocytes prior to homogenization. The homogenate was centrifuged as usual at 11,000 × *g* for 10 min and the resulting postmitochondrial supernate was centrifuged at 143,000 × *g* for 60 min. The distribution of added radioactivity in the initial pellet, the microsomal pellet, and the soluble supernate was determined by counting in a Nuclear Chicago model 1085 γ-counter. Homogenization-induced cell breakage and leakage of pulse-labeled proteins (10 min, [^{35}S]methionine) were estimated using the 1-ml aliquot of lysate referred to in the section above. The 11,000 × *g* pellet from this sample was suspended in 5 ml of lysis buffer (10 m*M* Tris, 0.15 *M* NaCl, 1% Triton X-100, and 0.5% sodium desoxycholate ([DOC]), while the supernate was centrifuged at 143,000 × *g* to produce a microsomal pellet and a soluble supernate. The microsomal pellet was suspended in 5 ml of lysis buffer, and Triton X-100 and DOC were added to the supernate to final concentrations of 1.0 and 0.5%, respectively. Labeled CRP and albumin were specifically immunoprecipitated (see below) from each of these three samples, using 10% of the available sample for albumin and 80% for CRP. The immunoprecipitates were subjected to sodium dodecyl sulfate-polyacrylamide gel electrophoresis (SDS-PAGE) and specific protein bands quantitated as described previously.[34] The proportion of labeled protein in the initial pellet, as a percentage of the total of the three fractions, was considered to reflect the presence of whole cells, since adsorption was found to be negligible. Leakage of labeled proteins was estimated from the radioactivity in the soluble supernate as a percentage of the radioactivity present in the microsomal pellet plus that present in the soluble supernate. In the pulse-chase studies, leakage of individual proteins was determined from the percentage of radioactivity present in the 11,000 $\times$ *g* supernate, which remained soluble following the 143,000 $\times$ *g* centrifugation, and no corrections were made for the adsorption of leaked proteins. In a typical preparation, ten culture dishes (5 $\times$ 10^7 cells) yielded approximately 1.5 mg of protein in total microsomes, 400 μg in rough microsomes, 180 μg in smooth microsomes, and 150 μg in Golgi fractions.

D. ENZYME AND IMMUNOASSAYS

The efficacy of the fractionation procedure was assessed by determination of specific marker enzyme activities. Glucose-6-phosphatase activity was determined as described previously,[39] except that the A_{660} of samples was determined 3 h after the addition of the 1-amino-2-naphthol-4-sulfonic acid reagent. Galactosyl transferase activity was measured as described previously,[40] except that uridine diphospho-D-[U-^{14}C]galactose (Amersham Corp., Arlington Heights, IL) was employed at a specific activity adjusted to 4.3 mCi/mmol. Sample values were obtained by subtracting the values obtained with exogenous substrate (Trypsin inhibitor type III-O, Sigma Chemical Co., St. Louis, MO) from those determined without substrate. For both enzyme assays, determinations were made on at least two different volumes of sample membranes. Distribution of marker enzyme-specific activities within the homogenate as compared to the three subcellular fractions indicated that the Golgi fraction was enriched 17-fold in galactosyl transferase activity and the rough microsome fraction 3.4-fold in glucose-6-phosphatase activity, values in reasonable agreement with those reported previously (27- and 3.4-fold, respectively) for fractions prepared from murine whole liver.[35]

CRP and albumin contained in cell lysates or subcellular fractions were quantitated by radioimmunoassay as described previously.[33,34] Membrane pellets were suspended in 5 ml of lysis buffer, sonicated (model 185E Sonifier; Heat Systems-Ultrasonics, Plainville, NY) with the small probe at 30 W for 30 s on ice, and the 12,000 $\times$ *g*, 15-min supernate was used in the RIAs. Radiolabeled CRP and albumin present in these same supernates were specifically immunoprecipitated as described previously.[34] Transferrin was precipitated employing 10 μg of carrier rabbit transferrin and 50 μl of goat anti-rabbit transferrin (both from Cappel, Cooper Biomedical, Malvern, PA). The volumes of supernate used for immunoprecipitation were 2 ml for CRP, 1 ml for albumin, and 0.5 ml for transferrin. This strategy allowed for adequate radioactivity to be recovered for each protein within fractions from the two chase times and was based upon individual differences in methionine composition, relative rate of synthesis, and transit time. Portions of the initial cell homogenates were adjusted to 1% Triton X-100 and 0.5% DOC to determine total intracellular immunoprecipitable proteins. These samples combined with immunoprecipitates from culture medium were used to determine the efficacy of the 75-min chase incubation. Immunoprecipitates were washed with lysis buffer and subjected to SDS-PAGE on 12.5% gels. Following autoradiography, the gels were rehydrated in water and stained bands corresponding to added carrier proteins were excised, dissolved in 30% H_2O_2, and the radioactivity determined as described.[34]

To investigate the possible association of newly synthesized CRP with BiP (grp78), pulse-labeled cell lysates were incubated with rat monoclonal anti-human BiP (generous gift of Dr. David Bole) under ATP-depleting conditions. Following a 30-min incubation with [^{35}S]methionine as described above, four dishes of cells were rinsed with Hanks' buffered saline and scraped in 3 ml of 0.15 *M* NaCl, 20 m*M* Hepes, pH 7.4, 1 m*M* $MgCl_2$, 5 m*M* glucose, 10 U/ml hexokinase (Calbiochem, LaJolla, CA), 1% NP40 (Pierce Chemicals, Rockford, IL), and 1 m*M* PMSF (Eastman Kodak, Rochester, NY) or in the same buffer containing 1.5 m*M* $CaCl_2$. After 15 min at 4°C, the lysate was centrifuged at 12,000 × *g* and 1 ml of the supernate was used for immunoprecipitation of CRP, as described above. One half milliliter of the supernate was incubated with 100 μl of anti-BiP and 50 μl of Protein A-Sepharose (Pharmacia Fine Chemicals, Piscataway, NJ) made up as a 50% suspension in 0.15 *M* NaCl, 20 m*M* Hepes, pH 7.4, and 0.1% BSA, and the suspension incubated for 1 h at 4°C with rotation. The Sepharose was pelleted by microcentrifugation (Model 59A, Fisher Scientific, Fair Lawn, NJ) and washed twice with 1 ml of 0.4 *M* NaCl, 20 m*M* Hepes, pH 7.4, 1% NP40, 0.1% SDS, and 0.5% DOC, and once with 1 ml of 20 m*M* Hepes and 0.15 *M* NaCl, pH 7.4, prior to boiling in SDS-PAGE sample buffer and analysis by autoradiography of 12.5% gels, as described.[34]

Nitrocellulose blots used for ligand probing were prepared by transferring samples separated on 10% SDS gels to nitrocellulose in 25 m*M* Tris, 190 m*M* glycine, and 20% methanol buffer,[41] employing a Genie electroblotter (Idea Scientific Co., Minneapolis, MN), and the blot was blocked in 1% gelatin overnight. Following three washes in 0.15 *M* NaCl, 20 m*M* Hepes, and 0.1% Chaps, pH 7.4, the blots were probed with [^{125}I]CRP (prepared as below) at a concentration of 0.5 to 1 μg/ml (3 to 5 × 10^6 cpm/ml) in the same buffer, containing 1.5 m*M* $CaCl_2$, and 1.0% BSA. Following incubation at 4°C for 2 h, blots were washed in three changes of 0.15 *M* NaCl, 20 m*M* Hepes, 0.1% Chaps, 1.5 m*M* $CaCl_2$, and 0.1% BSA, pH 7.4, dried, and subjected to autoradiography. Western blots probed with anti-BiP (1:5000 dilution of hybridoma culture medium) were then incubated with a 1:3000 dilution of alkaline phosphatase-conjugated goat anti-rat IgG (Pierce Chemicals, Rockford, IL) and developed with BCIP/NBT (Bio-Rad, Richmond, CA) as per the manufacturer's instructions.

E. PROTEIN PURIFICATION AND MODIFICATION

The purification of rabbit CRP from acute phase rabbit serum and of CRP subunits from purified CRP was as described previously.[33] CRP was radioiodinated as described,[33] except that for equilibrium and kinetic binding studies, [^{125}I]CRP obtained after G200 chromatography was concentrated by affinity chromatography on a 0.25-ml column of phosphocholine-agarose (Pierce Chemicals, Rockford, IL). Specific radioactivities ranged from 6 to 9 × 10^6 cpm/μg and CRP concentrations from 9 to 15 μg/ml. [^{125}I]CRP was stored frozen as aliquots, used within 3 to 4 weeks of preparation, and microfuged for 45 min prior to use in binding assays. Estimates of CRP contained within isolated rough microsomes were made by RIA determinations of microsomes lysed in buffer containing 1.0% Triton X-100, 0.5% DOC, and 10 m*M* sodium citrate in order to dissociate CRP bound by calcium-dependent interactions.

CRP (500 to 750 μg/ml) was biotinylated employing a 30-fold molar excess of freshly prepared NHS-LC-biotin (Pierce Chemcials, Rockford, IL) in 0.15 *M* NaCl and 20 m*M* Hepes, pH 7.4. Following incubation at 20°C for 60 min, glycine was added to a concentration of 100 m*M* and the sample dialyzed exhaustively against 0.15 *M* NaCl and 20 m*M* Hepes, pH 7.4. An aliquot of the resulting sample was radioiodinated as described above and subjected to gel filtration on Sephacryl S200 (Pharmacia Fine Chemicals, Piscataway, NJ) as well as affinity chromatography of phosphocholine-agarose (Pierce Chemicals, Rockford, IL). Greater than 93% of the radioactivity was present as native, functional CRP as judged by pentameric molecular size and ability to bind to phosphocholine. Greater than 70% of this material bound to Streptavidin-agarose (Pierce Chemicals, Rockford, IL) and resisted washing with 0.1% SDS as well as 1 m*M* EDTA, indicating the functional availability of biotin moieties. As a control, BSA was biotinylated in exactly the same manner.

F. PREPARATION OF SAMPLES FOR ELECTRON MICROSCOPY

Rough microsomes were prepared and permeabilized as described above in 0.05% DOC at a concentration of 2 mg of protein per milliliter. Aliquots were incubated for 1 h at 4°C (final concentration of 1 mg of protein per milliliter) in buffer containing 0.15 *M* NaCl, 20 m*M* Hepes, 1.5 m*M* $CaCl_2$, 1 m*M* $MgCl_2$, 1.0% BSA, pH 7.4, and biotinylated CRP (2 μg/ml). Control incubations included biotinylated BSA in place of CRP, biotinylated CRP incubated in the absence of DOC, and biotinylated CRP incubated with DOC in the presence of a 15-fold excess of native CRP. To remove unbound CRP, samples were layered over discontinuous sucrose gradients (Beckman SW41 tubes) containing 3.0 ml of 1.9 *M* sucrose and 7.0 ml of 20% sucrose, 0.15 *M* NaCl, 20 m*M* Hepes, 1.5 m*M* $CaCl_2$, and 1 m*M* $MgCl_2$, pH 7.4, and then centrifuged at 38,000 rpm for 1 h at 4°C. The tops of the gradients were carefully aspirated, rinsed with 0.15 *M* NaCl and 20 m*M* Hepes, pH 7.4, and the rough microsomes present at the 1.9 *M* sucrose interface were collected and dialyzed for 1 h at 4°C against 0.15 *M* NaCl, 20 m*M* Hepes, 1.5 m*M* $CaCl_2$, and 1 m*M* $MgCl_2$, pH 7.4. Dialyzed samples were then incubated for 1 h at 4°C, under permeabilizing conditions, in buffer contaning 0.15 *M* NaCl, 20 m*M* Hepes, 1.5 m*M* $CaCl_2$, 1 m*M* $MgCl_2$, 1.0% BSA, 0.05% DOC, and peroxidase-conjugated Streptavidin at a final dilution of 1:40 of that supplied (Zymed Laboratories, San Francisco, CA). Free enzyme-conjugated Streptavidin was removed by centrifugation through discontinuous sucrose gradients and harvested as before. Material at the 1.9 *M* sucrose interface was collected and diluted to 5 ml with 0.25 *M* sucrose, pH 7.0, and glutaraldehyde was added to a final concentration of 1.0%. Following incubation for 1 h at 4°C, samples were layered over discontinuous sucrose gradients contaning 2 ml of 2 *M* sucrose and 4 ml of 20% sucrose, and then spun at 38,000 rpm for 1 h at 4°C (SW41 rotor). Material at the 2 *M* sucrose interface was adjusted to a volume of 2 ml and a concentration of 0.25 *M* sucrose, and 0.5 ml of 1% 3,3′-diaminobenzidine (Sigma Chemical Co., St. Louis, MO) in 50 m*M* Tris, pH 7.4, was added. Following the addition of 10 μl of 1.0% H_2O_2, samples were incubated for 15 min at 20°C and the reaction terminated by the addition of 9 ml of ice-cold 0.25 *M* sucrose. Samples were cleared of debris by centrifugation at 1500 × *g* for 5 min, the supernates layered over 2 ml of 20% sucrose, and the samples spun at 38,000 rpm for 1 h at 4°C (SW41). Pellets were cut into strips, post-fixed with OsO_4 (1% in H_2O), and dehydrated through graded ethanol. Following imbedding in Spurr, thin sections were cut and examined, without further staining, in a Jeol CX 100 II electron microscope.

G. MICROSOMAL BINDING ASSAY

Microsomal subfractions were prepared, permeabilized, and passed over Sepharose 2B as described above. Incubation buffer included 0.15 *M* NaCl, 20 m*M* Hepes, 1.5 m*M* $CaCl_2$, 1 m*M* $MgCl_2$, 1.0% BSA, and 0.035% DOC (except as indicated in Figure 2), pH 7.4. In typical competitive binding assays, each sample's volume was 250 μl and contained 10 to 15 μg of microsomal protein and 50 to 100 ng of [^{125}I]rabbit CRP. Following incubation for 3 h at 4°C, duplicate 100-μl aliquots were layered over 200 μl of 15% sucrose, 0.15 *M* NaCl, 20 m*M* Hepes, 1.5 m*M* $CaCl_2$, and 1 m*M* $MgCl_2$, pH 7.4, in 0.4-ml polypropylene centrifugation tubes. Following microcentrifugation for 45 min at 11,000 × g (Model 59A,

Fisher Scientific, Fair Lawn, NJ), 10-μl aliquots of the supernatant were removed to determine free radioactivity, the tubes frozen in a solution of dry ice and isopropanol, and the tips cut off and counted to determine bound radioactivity. In control experiments, less than 0.005% of [^{125}I]CRP was recovered in the counted tips when microsomes were omitted from the incubation. Specific binding (usually about 85% of total binding) was defined as total binding less the radioactivity bound in the presence of at least a 50-fold excess of unlabeled CRP as competitor. Additional proteins tested for competition included rabbit albumin and transferrin (Cappel, Cooper Biomedical, West Chester, PA), histones (calf thymus from U.S. Biochemical Corporation, Cleveland, OH, and H2A from Sigma Chemical Co., St. Louis, MO), as well as human CRP prepared from malignant ascites fluid using the same procedures as for rabbit CRP.

In the formal equilibrium binding studies, incubations included paired samples containing [^{125}I]CRP ranging from 26 ng to 2.6 μg (8.3×10^{-10} to 8.3×10^{-8} *M*) incubated in the presence or absence of a 60-fold excess of competing unlabeled CRP, except in the case of the lowest concentration of labeled CRP, in which a 120-fold excess of unlabeled CRP (representing 25 times the estimated concentration of half-maximal binding) was employed. The data obtained were subjected to Scatchard analysis,[42] and the results further interpreted employing the nonlinear curve-fitting program LIGAND,[43] modified for microcomputers[44] and carried out on a Macintosh II computer (Apple Computer, Cupertino, CA).

In kinetic binding studies, labeled CRP was added to a concentration at least approximating that determined to be half-maximal saturation (from equilibrium binding studies). Paired samples, one containing a 50-fold excess of unlabeled CRP, were further processed at timed intervals after the addition of labeled CRP. Experiments designed to determine dissociation kinetics were performed by preincubating labeled CRP with microsomes for 3 h, then adding a 50-fold excess of unlabeled CRP and separating bound from free radioactivity at timed intervals. Nonspecific binding was determined from an incubation in which both labeled and unlabeled CRP were coincubated with microsomes. Analyses of kinetic binding studies were performed with the aid of the nonlinear curve-fitting program KINETIC.[44]

III. RESULTS

A. CHANGES IN THE KINETICS OF CRP SECRETION DURING THE ACUTE PHASE RESPONSE

The kinetics of CRP and albumin secretion were studied in cultures prepared from animals manifesting varying degrees of the acute phase response to *in vivo* inflammatory stimulus.[34] In the case of cells from a maximally responsive animal, following a 15-min pulse with [^{35}S]methionine, labeled CRP was detected in the culture medium within 30 min, and 50% of pulse-labeled CRP had been secreted by 60 to 90 min (the time at which labeled CRP present in both the intracellular and extracellular immunoprecipitates was equivalent) (Figure 1). In contrast, in the same culture, 50% of labeled albumin was secreted within 30 min, indicating a more rapid minimal transit time for albumin than for CRP. Quantitative analyses of the kinetics of CRP and albumin secretion were carried out in three additional cultures (Figure 2). The data have been expressed as intracellular and extracellular radioactivity as a percentage of total radioactivity (the sum of both), to allow for direct comparison of CRP to albumin and to eliminate differences in slopes due to differences in synthetic rates. The kinetics of albumin secretion were very similar in the four cultures (Figures 1 and 2), with 50% of labeled albumin appearing in culture medium within 20 to 40 min of chase. In contrast, the half-time for secretion of pulse-labeled CRP was increasingly rapid in cultures prepared from progressively more responsive animals. The half-time for CRP secretion in control cells was extraordinarily prolonged, being estimated at 18 h. No evidence

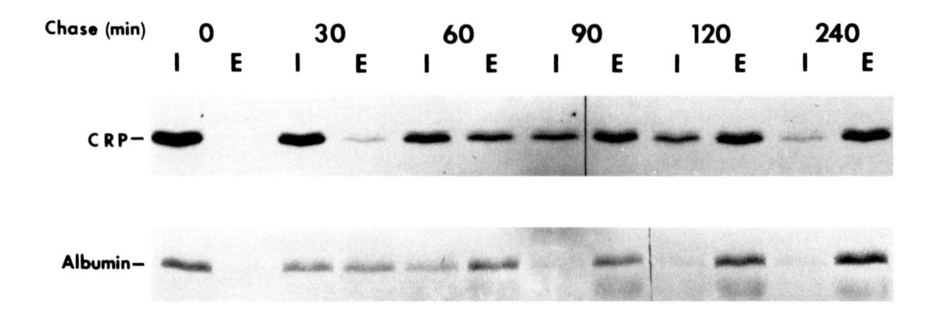

FIGURE 1. Electrophoretic analysis of pulse-chase labeled CRP and albumin. Hepatocytes prepared from a highly responsive animal were pulse labeled for 15 min with [^{35}s]methionine (100 μC/ml) and chased with unlabeled methionine, as described in Section II. At the indicated intervals, intracellular (I) and extracellular (E) CRP and albumin were immunoprecipitated and run on SDS-acrylamide gels prior to autoradiography. (From Macintyre, S., Kushner, I., and Samols, D., *J. Biol. Chem.*, 260, 4169, 1985. With permission.)

of intracellular degradation of CRP or albumin was detected.[34] These findings suggested two possible explanations: (1) secretion of CRP might somehow be actively facilitated during the acute phase respone or (2) CRP might be retained within the normal hepatocyte by a mechanism which is diminished during the acute phase response.

To distinguish between these two possible mechanisms, a fusion gene consisting of the rabbit CRP gene driven by the MMT promoter was transfected into human HeLa and hepatoma cells.[45,46] In these cell lines, the half-time for secretion of rabbit CRP was found to be very rapid (45 min) and did not vary with the rate of synthesis (regulated by manipulation of extracellular zinc).[46] Even at low rates of synthesis, the half-time for CRP secretion was 45 min in the transfected cells, while in rabbit hepatocytes synthesizing CRP at comparable rates, the half-time for secretion was greater than 6 h.[46] We reasoned that the variation in CRP secretion time seen in rabbit hepatocytes was due to differential retention of CRP within these cells, in that it is unlikely that human HeLa cells would be better able than rabbit hepatocytes to express a facilitative transport receptor for rabbit CRP.

B. CRP IS SPECIFICALLY RETAINED WITHIN THE ENDOPLASMIC RETICULUM

To further explore the nature of the intracellular retention of CRP, hepatocytes incubated under pulse-chase conditions were subjected to subcellular fractionation prior to immunoprecipitation of CRP, albumin, and transferrin (included because it has a relatively long transit time) for lystates of the isolated fractions.[47] In cells from a control animal, all radiolabeled proteins were, as expected, localized to the rough (R) and smooth (S) components of the endoplasmic reticulum immediately following the pulse (Figure 3). Quantitation of the data in Figure 3 indicated that 85% of labeled albumin exited from the ER-derived fractions during the 75-min chase, while labeled transferrin decreased by 40% and CRP by only 12%. For comparison, the same studies were carried out in cells from an animal undergoing the acute phase response. While there was little difference in albumin (82% of ER cpm lost during the 75-min chase), or for transferrin (50% increase in chased transferrin from 40 to 59%), a threefold increase in the chase of labeled CRP (from 12 to 35%) was observed (data not shown). These findings corroborated our previous observations of a decrease in transit time for CRP during the course of the acute phase response and are in agreement with the conclusions of most other workers that the rate-limiting step in constitutive secretion of proteins is exit from the ER.[23,25,26,48]

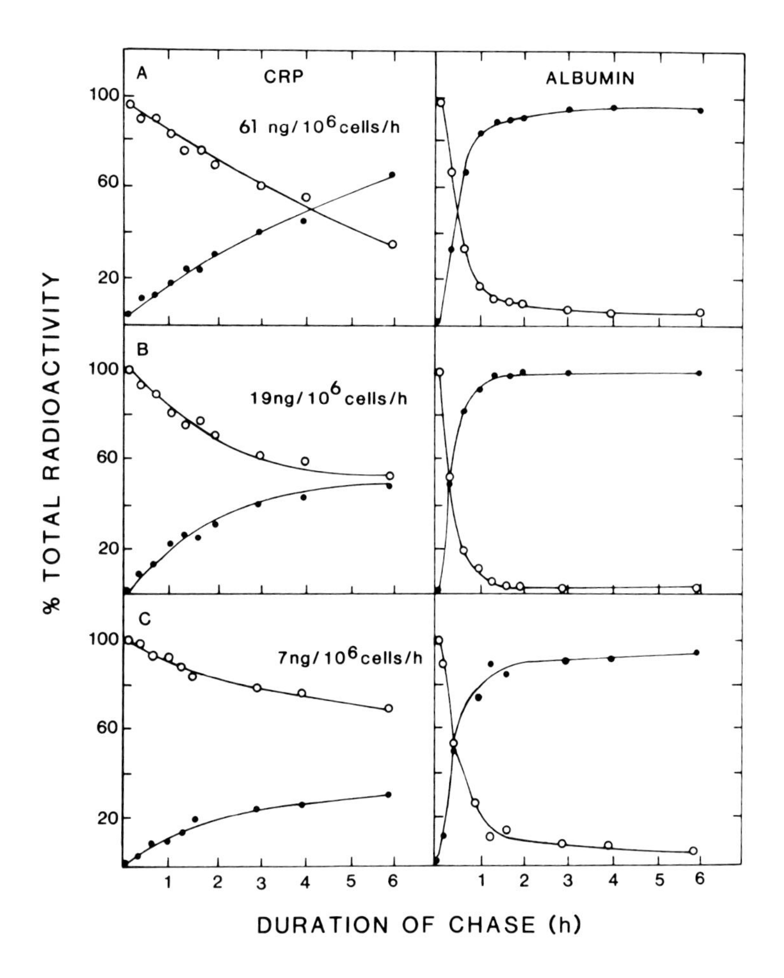

FIGURE 2. Kinetics of pulse-chase labeling of CRP and albumin. Hepatocyte cultures were prepared from rabbits manifesting minimal (C), moderate (B), and marked (A) responses to inflammatory stimulus and were pulsed labeled as in Figure 1. Intracellular (open circles) and extracellular (closed circles) radiolabeled CRP (left panel) and albumin (right panel) were immunoprecipitated at the indicated intervals of chase time. Following SDS-gel electrophoresis, the radioactivity present in sections of gel was determined directly, as described in Section II. Data for each protein are expressed as a percent of the total (intracellular + extracellular) radioactivity in CRP or in albumin at each interval. Rates of *in vitro* CRP synthesis are indicated in the left panel. (From Macintyre, S., Kushner, I., and Samols, D., *J. Biol. Chem.*, 260, 4169, 1985. With permission.)

Determination by radioimmunoassay of the relative concentrations of CRP and albumin within subcellular fractions provided evidence of an intracellular pool which we had inferred previously: CRP content represented 12 to 13% of the albumin content within the ER fractions, while it was only 0.8 to 1.0% in Golgi fractions. The ratio of CRP to albumin in the Golgi was in good agreement with the ratio of the net rates of secretion of the two proteins (1.1%), indicating that upon arrival in the Golgi, the kinetic differential between the two proteins has been overcome.

C. EVIDENCE FOR A SPECIFIC CRP BINDING SITE WITHIN PERMEABILIZED ROUGH MICROSOMES

In an attempt to detect a specific binding site for CRP within the ER, we developed techniques which permitted the determination of binding of labeled CRP to detergent-per-

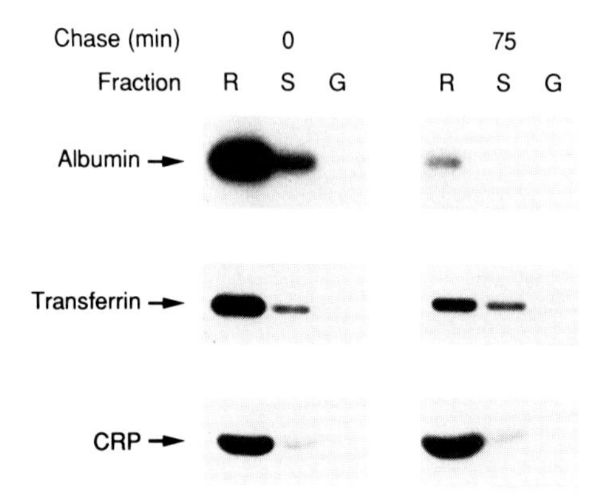

FIGURE 3. Chase of pulse-labeled proteins from subcellular fractions. Cells prepared from an animal manifesting a minimal response to inflammatory stimulus (serum CRP, 16 μg/ml) were pulse labeled with [^{35}S]methionine for 10 min, and subcellular fractions prepared from ten dishes immediately after the pulse and from 10 dishes following 75 min of chase. Radiolabeled albumin, transferrin, and CRP were specifically immunoprecipitated from portions of lysates of fractions and subjected to SDS-PAGE, followed by autoradiography. Panels are from portions of three gels used to analyze the immunoprecipitates of the three proteins present in rough (R), smooth (S), and Golgi (G) fractions. In order to achieve comparable autoradiographic results, the proportions of lysates used for immunoprecipitation of individual proteins (CRP, 40%; albumin, 20%; transferrin, 10%) were designed to account for differences in methionine composition, relative rates of synthesis, and intracellular transit times (see Section II for details). Exposure times were 3 d for transferrin and 5 d for albumin and CRP. (Reproduced from Macintyre, S., *J. Cell. Biol.*, 118, 253, 1992, by copyright permission of the Rockefeller University Press.)

meabilized microsomal subfractions.[47] While little binding of CRP to total microsomes was observed in the absence of detergent, detectable specific calcium-dependent binding of CRP increased with increasing concentrations of the detergent DOC, reaching a maximum value at 0.035% DOC.[47] The observed relationship between DOC concentration and binding activity is in excellent agreement with previous studies demonstrating the relationship between DOC concentration and induced permeability of rough microsomes to macromolecules.[49]

To investigate the distribution of CRP binding within the population of total microsomes, binding assays were performed on detergent-permeabilized subcellular fractions. In two experiments, the mean specific binding of [^{125}I]CRP (nanograms of bound/milligrams of microsomal protein) was found to be 20.5 for rough microsomes, 14.8 for smooth microsomes, and was not detected in Golgi-derived microsomes. Thus, the localization of the binding activity to fractions derived from the ER correlates with the observation that the ER is the compartment in which CRP is regained, as judged by the results of the pulse-chase and steady-state distribution experiments described above. In subsequent studies, purified rough microsomes were used in order to avoid the possibility of heterogeneity in the smooth ER fractions.

The interaction between CRP and permeabilized rough microsomes was examined electron microscopically employing biotinylated CRP as the ligand and peroxidase-conjugated Streptavidin as a second probe. In control incubations (Figure 4B), DOC-permeabilized rough microsomes incubated with biotin-BSA did not accumulate any peroxidase reaction product when compared to untreated rough microsomes (Figure 4A). Examination of rough microsomes not treated with DOC prior to incubation with biotin-CRP and peroxidase-Streptavidin revealed occasional positive microsomes (Figure 4C), possibly due to membrane disruption during isolation. In DOC-permeabilized preparations incubated with biotin-CRP, the majority of microsomes contained reaction product (Figure 4D), occasionally nearly filling the vesicle (large arrows) and/or appearing to preferentially deposit on the inner

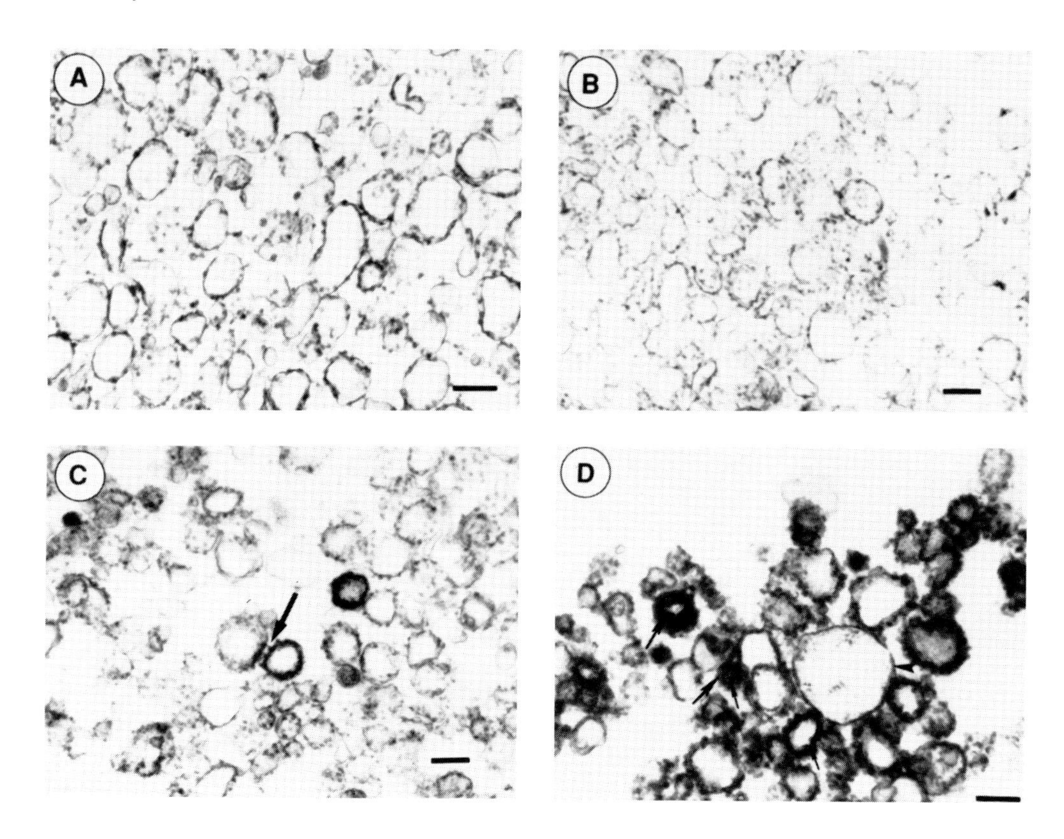

FIGURE 4. Electron microscopic localization of CRP binding to rough microsomes. Rough microsomes were permeabilized with DOC (B, D) and incubated with biotinylated BSA (B) or biotinylated CRP (C, D) prior to incubation with peroxidase-conjugated Streptavidin, followed by peroxidase substrate. The sample in A represents untreated rough microsomes and the sample in C represents rough microsomes not treated with detergent prior to incubation with biotinylated CRP. Bar, 0.5 μm. (Reproduced from Macintyre, S., *J. Cell. Biol.*, 118, 253, 1992, by copyright permission of the Rockefeller University Press.)

surface of the membrane (small arrows), while a contaminating smooth vesicle (arrowhead) was negative. The observed detergent dependence of CRP binding seen in ligand binding studies as well as by electron microscopy (EM) indicates that the detectable binding activity is present on the lumenal face of permeabilized rough microsomes.

D. SPECIFICITY OF BINDING OF CRP TO PERMEABILIZED ROUGH MICROSOMES

The specificity of the interaction between rabbit CRP and permeabilized rough microsomes was investigated in competitive binding studies. As seen in Figure 5A, competition by unlabeled CRP (closed circles) indicated a K_i (concentration at 50% inhibition) of approximately 10^{-8} *M*, with specific binding being greater than 90% of total binding. In contrast, only minimal diminution in binding was observed when rabbit CRP subunits (Figure 5A, open circles) were employed as competitor, suggesting that pentameric structure is critical to the interaction of CRP with the membrane. Rabbit albumin and transferrin (Figure 5B) and human CRP (Figure 5C) did not appear to interact with the CRP binding site to an appreciable degree, as judged by the lack of competition by these proteins for the binding of [^{125}I]CRP to permeabilized rough microsomes. Histones, tested because CRP has been shown to have a calcium-dependent, phosphocholine-inhibitable affinity for histones,[50] also failed to compete with CRP for binding.

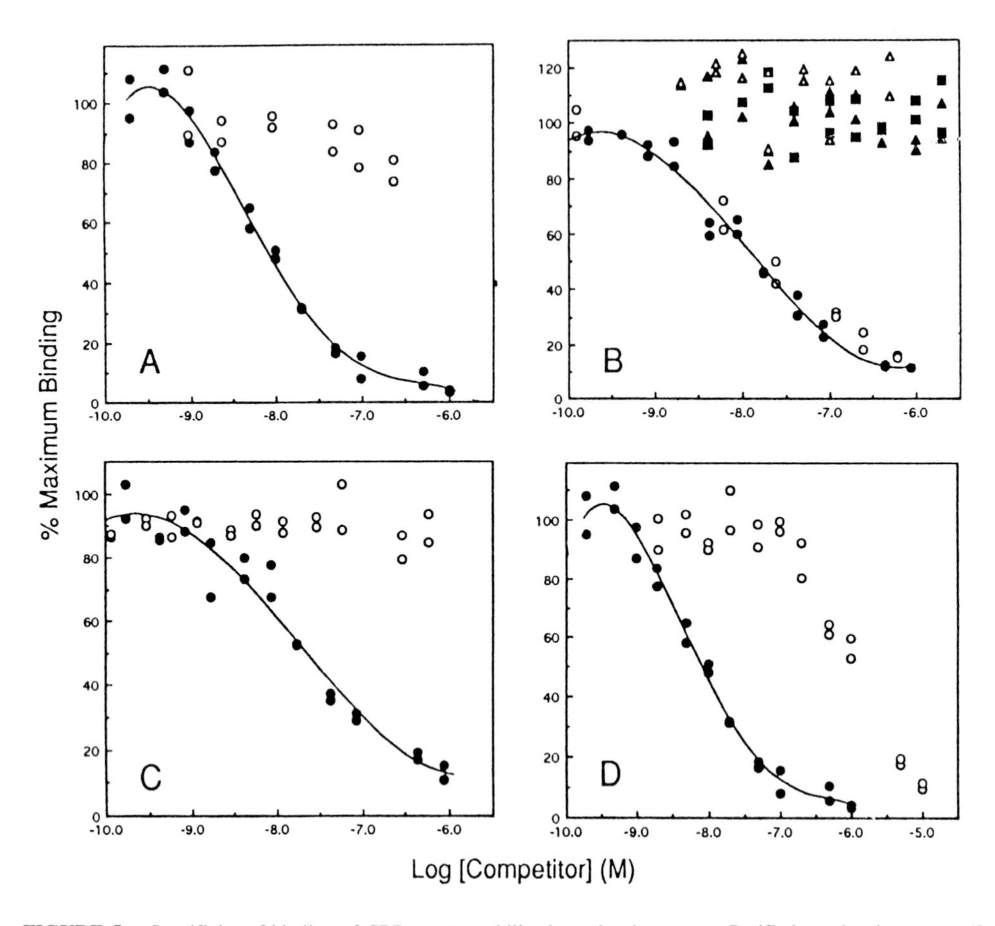

FIGURE 5. Specificity of binding of CRP to permeabilized rough microsomes. Purified rough microsomes (8 to 10 μg) were incubated in the presence of 50 to 100 ng of labeled CRP (5 to 9 × 10^5 cpm) and a variety of competing proteins. Solid circles and solid lines indicate the reference competition for binding obtained with purified unlabeled CRP coincubated at the indicated concentrations. Minimal competition was observed with CRP subunits (A, open circles). No competition was detectable (B) with rabbit transferrin (solid squares), rabbit albumin (solid triangles), or histones (open triangles), while competition by biotinylated rabbit CRP was indistinguishable from native CRP (B, open circles). Human CRP did not compete in the assay (C, open circles) and equivalent inhibition by phosphocholine (D, open circles) required concentrations about 100-fold greater than native CRP. Maximum binding ranged from approximately 6000 to 8000 cpm/250 μl incubation (2000 to 3000 cpm/100 μl centrifuged and counted) and specific binding represented 85 to 95% of total binding, as shown. Reference competitive binding curves for unlabeled CRP (solid circles and solid line) are representative in A and B, and were determined in side-by-side incubations in C and D. (Reproduced from Macintyre, S., *J. Cell. Biol.*, 118, 253, 1992, by copyright permission of the Rockefeller University Press.)

Since CRP has a known calcium-dependent binding capacity for the polar head group of phosphocholine,[51] and binds to disrupted but not intact membranes,[51-54] it is possible that these observations could represent binding of CRP to phosphocholine exposed on the inner surface of permeabilized microsomes. Phosphocholine was found to inhibit the binding of [^{125}I]CRP, but concentrations necessary for competition equivalent to that by CRP were 100-fold greater (Figure 5D), indicating it is unlikely that the binding site is simply phosphocholine itself. The effect of trypsin on the binding of CRP to microsomes also supports this conclusion. CRP binding to permeabilized microsomes which had been pretreated with trypsin, then quenched with PMSF, was found to be only 20% of the binding to permeabilized microsomes which had been preincubated with trypsin which had been premixed with PMSF (data not shown). Possible explanations for the effect of phosphocholine on the binding of CRP to rough microsomes are dealt with in Section IV.

E. KINETIC AND SATURATION BINDING STUDIES

The kinetics of dissociation of CRP from permeabilized rough microsomes isolated from two unstimulated (control) rabbits were determined by preincubating fixed amounts of labeled CRP with microsomes, then adding a 60-fold excess of unlabeled CRP and determining bound and free labeled CRP at timed intervals thereafter. Nonspecific binding was determined from an incubation containing labeled and unlabeled CRP added simultaneously. Analysis (KINETIC)[44] of the dissociation kinetics indicated the presence of two sites (p = 0.005 vs. a one-site fit) with dissociation rate constants (k′) of $3.5 \pm 2.5 \times 10^{-2}$ (± SEM) and $2.6 \pm 0.33 \times 10^{-4}$ min^{-1} (data not shown). Similarly, analysis of association kinetics also suggested the presence of two sites (p = 0.001) with calculated association rate constants (k) of 7.5×10^4 and 7.9×10^6 M^{-1} min^{-1} (data not shown). From these data, calculation of equilibrium dissociation constants (K_d = k′/k) resulted in K_d estimates of 3.3×10^{-10} M for a high-affinity site and 4.7×10^{-7} M for a second, lower-affinity site.

Saturation binding studies were carried out employing increasing amounts of labeled CRP in paired incubations plus and minus at least a 60-fold excess of unlabeled CRP (see Section II). Studies of permeabilized rough microsomes from unstimulated rabbits suggested that saturable binding was approached at concentrations of labeled CRP in excess of 20 n*M* (Figure 6A, inset), and allowed for estimation of a mean K_d (as judged by the concentration of CRP at half saturation) of approximately 3 n*M*. Nonlinear curve-fitting analysis (LIGAND)[43] again indicated that a two-site fit was statistically superior ($p < 0.02$) (Figure 6A). The resulting association constants were $1.14 \times \pm 0.83 \times 10^8$ and $7.29 \times \pm 2.2 \times 10^6$ M^{-1}, yielding K_d values of 8.8×10^{-10} M for the high-affinity site and 1.4×10^{-7} M for the low-affinity site, values in acceptable agreement with those obtained from the kinetic binding studies.

For comparison, saturation binding studies were also performed with permeabilized rough microsomes prepared from two animals stimulated *in vivo* to undergo the acute phase response (Figure 6B). In this case, a single-fit model resulted in an estimated K_d of 1.5×10^{-7} M, corresponding to the lower-affinity site seen with microsomes from unstimulated animals, and the higher affinity was not demonstrable. These data correlate with our previous observations that the half-time for secretion of CRP is markedly longer in normal hepatocytes than in cells from animals undergoing the acute phase response,[34] as a result of retention of CRP within the ER of the normal cell,[47] and indicate that the decreased expression of the high-affinity site could be the mechanism responsible for the observed differences in the kinetics of CRP secretion.

F. PRELIMINARY CHARACTERIZATION OF THE ER BINDING SITE FOR CRP

In an initial approach to the direct identification of the CRP binding site, nitrocellulose blots of electrophoretically resolved microsomal lysates were found to be suitable for the demonstration of a CRP binding band. A blot (Figure 7B) prepared from an SDS gel (Figure 7A) loaded with samples of rough microsomes (lane 1) and Golgi membranes (lane 3) from an unstimulated rabbit, and rough microsomes from a stimulated animal (lane 2) was probed with radioiodinated CRP and demonstrated a 60-kDa CRP binding band present in the lysate of unstimulated rough microsomes, but not in the other samples.

Since evidence suggests that the ER resident protein BiP (grp78) plays a role in determining folding efficiency and the exit of at least some secretory proteins from the ER,[16,55-57] the same blot shown in Figure 7B was subsequently probed with a monoclonal anti-BiP (Figure 7C). In this case, a different band, 80 kDa in size, was identified in equivalent amounts in the rough microsomal samples, but greatly diminished in the Golgi sample, findings which are consistent with the expected size and distribution of BiP and indicate that BiP expression does not change significantly during the acute phase response. Thus,

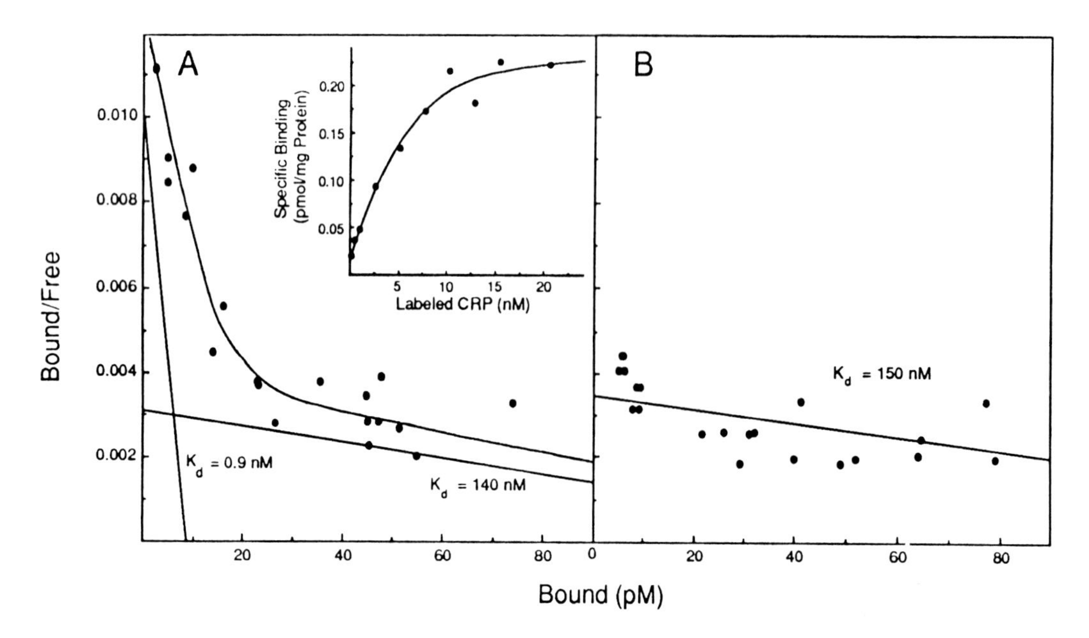

FIGURE 6. Scatchard plots of equilibrium binding studies. Permeabilized rough microsomes from two control rabbits (A) and two animals undergoing the acute phase response (B) were incubated with increasing concentrations of labeled CRP (range of 26 ng to 2.6 μg) in the presence and absence of competing unlabeled CRP. Individual incubations contained 9 to 11 μg of microsomal protein and, in the case of incubations from control microsomes, the mean maximal specific binding in counted aliquots (100 μl) was 5900 cpm, representing 86% of total bound CRP and 1.1% of total labeled CRP. Each plot represents the means of duplicates from two incubations. Inset (A) demonstrates the saturability of specific binding with increasing concentrations of added labeled CRP. (Reproduced from Macintyre, S., *J. Cell. Biol.*, 118, 253, 1992, by copyright permission of the Rockefeller University Press.)

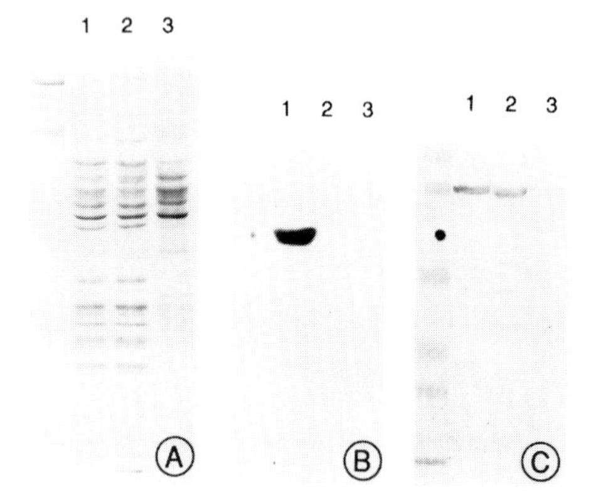

FIGURE 7. Identification of a 60-kDa ER protein capable of binding CRP. Seventy-five micrograms of protein from unstimulated rough microsomes (lane 1), stimulated rough microsomes (lane 2), and unstimulated Golgi membranes (lane 3) were subjected to electrophoresis on a 10% SDS gel. Following transfer to nitrocellulose, the gel was stained with Coomassie Blue (A). The nitrocellulose blot was probed with radiolabeled CRP and an initial autoradiograph of the blot used to align placement of a radioactive dye spot prior to a second autoradiographic exposure (B). The same blot was then incubated with rat anti-human BiP, followed by alkaline phosphatase-conjugated rabbit anti-rat IgG prior to enzymatic color development (C). Molecular weight markers (in kDa) are 97, 66, 45, and 31 in A, and prestained markers of 110, 84, 47, 33, 24, and 16 in C. (Reproduced from Macintyre, S., *J. Cell. Biol.*, 118, 253, 1992, by copyright permission of the Rockefeller University Press.)

the expression of the 60-kDa CRP binding band, a protein distinct from BiP, correlates well with the results of the binding of CRP to permeabilized microsomal subfractions, as judged by both subcellular localization and a decrease in activity in microsomes from stimulated animals.

Preliminary characterization (data not shown) of the 60-kDa CRP binding band described above indicates the activity is a glycoprotein which is sensitive to endoglycosidase H, is degraded by trypsin only if microsomes are trypsinized in the presence of detergent, and the protein does not appear to be an integral membrane protein in that it partitions into the aqueous phase of a Triton X-114 extraction.[58] Of particular interest is the finding that fractionation of microsomal extracts on DEAE cellulose demonstrates the presence of two 60-kDa peaks of binding activity, separable by a difference in net charge, but indistinguishable on SDS gel electrophoresis. The more abundant form (eluting earlier) binds relatively little CRP on ligand-blot assay, while the less abundant form of the 60-kDa glycoprotein binds considerably more CRP. Thus, the characteristics of these two forms, whether they represent two distinct polypeptides of similar size or possibly the same protein with different secondary modifications, correlate with the previously described low- and high-affinity binding sites detected in permeabilized rough microsomes.

IV. DISCUSSION

These studies were undertaken to investigate the mechanisms underlying our previous observation that the half-time for CRP secretion decreases markedly during the acute phase response.[34] The major findings of the studies presented here are:

1. The changes in CRP secretion kinetics observed during the acute phase response can be explained by changes in the degree to which CRP is retained within the ER of the hepatocyte.
2. The specific, detergent-dependent binding of CRP to microsomes is limited to the lumenal face of ER-derived subcellular fractions.
3. The specific binding of CRP is greatly diminished in samples from animals undergoing the acute phase response and appears to involve an ER-localized protein distinct from BiP.

Together, these data suggest that the destination of newly synthesized CRP is regulated by a novel post-translational sorting mechanism.

Since CRP has a known calcium-dependent binding affinity for the polar head group of phosphocholine[51] and adheres preferentially to disrupted but not intact membranes,[52,53] it is possible that phosphocholine exposed by detergent treatment of rough microsomes might be an available ligand for CRP and confound the interpretation of the binding data. However, several lines of evidence indicate that the high-affinity microsomal binding site is not simply exposed phosphocholine. The estimated affinity for CRP is approximately 500-fold greater than that for free phosphocholine.[59] The high-affinity site was not detected in Golgi fractions or rough microsomes from stimulated animals. Furthermore, no specific binding of CRP was found with permeabilized rough microsomes prepared from mice (data not shown), an unusual species in which CRP synthesis is minimal and does not change substantially during the acute phase response.[60] Finally, human CRP, which has the same affinity for phosphocholine as does rabbit CRP,[59,61] did not compete for the binding of rabbit CRP to rough microsomes (Figure 5C).

Nevertheless, we did observe weak inhibition by phosphocholine of the interaction between CRP and rough microsomes. One possible explanation for this finding would be that phosphocholine is a constituent of the rough microsomal binding site for CRP (i.e., the 60-kDa protein could be a phospholipoprotein) and the greater apparent affinity of this site for CRP is due to additional protein structure. Such a phenomenon would be analogous to the observation that the affinity of the cation-dependent mannose-6-phosphate receptor for mannose-6-phosphate expressed in lysosomal enzymes is substantially greater than that for free mannose-6-phosphate.[61] Alternatively, phosphocholine could be exerting an allosteric effect, since it is known that the interaction of phosphocholine with CRP results in a conformational change in CRP.[63] Thus, phosphocholine added to the assay could bind to free CRP and result in a conformational change which lessens the ability of CRP to bind, via another site, to the rough microsomal membrane. Indeed, the observed K_i of about 3 μM (Figure 5D) for phosphocholine in the binding assay is in agreement with what would be expected for the interaction between CRP and free phosphocholine, having a K_d of 5 μM.[61] The lack of inhibition of microsomal binding by human CRP suggests that the effect of phosphocholine on the binding of rabbit CRP to rough microsomes is due to an allosteric effect of phosphocholine on the CRP molecule, although it remains possible that phosphocholine is a constituent of the binding site and that additional protein structure increases the affinity for rabbit CRP, but also sterically interferes with the interaction of human CRP with the phosphocholine moiety of the microsomal binding site.

In addition to a high degree of specificity, the affinity of binding (K_d = 1 nM) detected in permeabilized rough microsomes is considerably greater than the affinities previously reported: 5 μM for phosphocholine,[59] 0.8 μM for chromatin,[64,65] and 0.03 to 0.1 μM for surface receptors present on phagocytic cells.[66-69] The B_{max} determined for the high-affinity site (0.88 pmol CRP/mg of microsomal protein) is within the range of values reported for physiologically significant receptors, including receptors for IL-1 (B_{max} = 0.5 pmol/mg of membrane protein),[70] inositol trisphosphate (B_{max} = 5 pmol/mg of protein),[71] and 5-hydroxytryptamine (B_{max} = 1 pmol/mg of protein).[72] On the basis of these data, the estimated density of the high-affinity site within the ER would be the equivalent of a few thousand cell-surface receptors per cell. A B_{max} of about 0.9 pmol (110 ng) of CRP per milligram of microsomal protein is more than sufficient to account for the amount of CRP contained within rough microsomes isolated from animals synthesizing CRP at low rates (data not shown). Accounting for homogenization-induced leakage, the amount of CRP within control microsomes represents about 7 to 10% of the B_{max}.

The results presented here illustrate a novel mechanism which could effectively reroute the intracellular trafficking of a secretory protein under differing physiologic conditions. What might be the function of such a regulated retention mechanism for CRP? On the basis of previous findings,[34] as well as the pulse-chase data and *in vitro* binding assays reported here, it is apparent that effective retention of CRP within the ER occurs preferentially in hepatocytes synthesizing CRP at relatively low rates. As a result, the cell accumulates a small pool of CRP within the ER. Since the retention (or retrieval) of CRP is calcium dependent, this pool would be rapidly mobilizable in response to transient decreases in local calcium concentration resulting, for example, from signal transduction during the early acute phase response. While there is controversy regarding the effects of calcium ionophores on the fate of ER resident proteins,[73,74] local calcium fluxes within the ER appear to be of great potential physiologic significance.[75] Whether a rapid secretory burst of intracellular CRP might play a role in the early acute phase response is presently unknown.

An alternative explanation for the retention of CRP would be that CRP has a function within the ER of the hepatocyte which is superseded during the acute phase response. In this respect, while the majority of functions ascribed to CRP are related to its role as a major acute phase plasma protein, it is intriguing to note that CRP has been demonstrated to bind to chromatin,[64,65] histones,[50,64] and U1 snRNPs,[76] and further, that it is structurally homologous to nucleoplasmin and contains a nuclear localization signal.[77]

REFERENCES

1. **Kornfeld, S.,** Lysosomal enzyme targeting, *Biochem. Soc. Trans.,* 18, 367, 1990.
2. **Pelham, H. R. B.,** Control of protein exit from the endoplasmic reticulum, *Annu. Rev. Cell Biol.,* 5, 1, 1989.
3. **Andres, D. A., Rhodes, J. D., Meisel, R. L., and Dixon, J. E.,** Characterization of the carboxyl-terminal sequences responsible for protein retention in the endoplasmic reticulum, *J. Biol. Chem.,* 266, 14277, 1991.
4. **Haugejorden, S. M., Srinivasan, M., and Green, M.,** Analysis of the retention signals of two resident luminal endoplasmic reticulum proteins by in vitro mutagenesis, *J. Biol. Chem.,* 266, 6015, 1991.
5. **Robbi, M. and Beaufay, H.,** The COOH terminus of several liver carboxylesterases targets these enzymes to the lumen of the endoplasmic reticulum, *J. Biol. Chem.,* 266, 20498, 1991.
6. **Lewis, M. J. and Pelham, H. R. B.,** A human homologue of the yeat HDEL receptor, *Nature (London),* 348, 162, 1990.
7. **Vaux, D., Tooze, J., and Fuller, S.,** Identification by anti-idiotype antibodies of an intracellular membrane protein that recognizes a mammalian endoplasmic reticulum retention signal, *Nature (London),* 345, 495, 1990.
8. **Jackson, M. R., Nilsson, T., and Peterson, P. A.,** Identification of a consensus motif for retention of transmembrane proteins in the endoplasmic reticulum, *EMBO J.,* 9, 3153, 1990.
9. **Nilsson, T., Jackson, M., and Peterson, P. A.,** Short cytoplasmic sequences serve as retention signals for transmembrane proteins in the endoplasmic reticulum, *Cell,* 58, 707, 1989.
10. **Munro, S.,** Sequences within and adjacent to the transmembrane segment of α-2,6-sialyltransferase specify Golgi retention, *EMBO J.,* 10, 3577, 1991.
11. **Nilsson, T., Lucocq, J. M., Mackay, D., and Warren, G.,** The membrane spanning domain of beta-1,4-galactosyltransferase specifies transgolgi localization, *EMBO J.,* 10, 3567, 1991.
12. **Swift, A. M. and Machamer, C. E.,** A golgi retention signal in a membrane-spanning domain of coronavirus-E1 protein, *J. Cell Biol.,* 115, 19, 1991.
13. **Hurtley, S. M. and Helenius, A.,** Protein oligomerization in the endoplasmic reticulum, *Annu. Rev. Cell Biol.,* 5, 277, 1989.
14. **Klausner, R. D.,** Sorting and traffic in the central vacuolar system, *Cell,* 57, 703, 1989.
15. **Rose, J. K. and Doms, R. W.,** Regulation of protein export from the endoplasmic reticulum, *Annu. Rev. Cell Biol.,* 4, 257, 1988.
16. **Gething, M. J. and Sambrook, J.,** Protein folding in the cell, *Nature (London),* 355, 33, 1992.
17. **de Silva, A. M., Balch, W. E., and Helenius, A.,** Quality control in the endoplasmic reticulum: folding and misfolding of vesicular stomatitis virus G protein in cells and in vitro, *J. Cell Biol.,* 111, 857, 1990.
18. **Stafford, F. J. and Bonifacino, J. S.,** A permeabilized cell system identifies the endoplasmic reticulum as a site of protein degradation, *J. Cell Biol.,* 115, 1225, 1991.
19. **Wikstrom, L. and Lodish, H. F.,** Endoplasmic reticulum degradation of a subunit of the asialoglycoprotein receptor *in vitro* — vesicular transport from endoplasmic reticulum is unnecessary, *J. Biol. Chem.,* 267, 5, 1992.
20. **Klausner, R. D. and Sitia, R.,** Protein degradation in the endoplasmic reticulum, *Cell,* 62, 611, 1990.
21. **Bonifacino, J. S., Cosson, P., Shah, N., and Klausner, R. K.,** Role of potentially charged transmembrane residues in targeting proteins for retention and degradation within the endoplasmic reticulum, *EMBO J.,* 10, 2783, 1991.
22. **Bonifacino, J. S., Suzuki, C. K., and Klausner, R. D.,** A peptide sequence confers retention and rapid degradation in the endoplasmic reticulum, *Science,* 247, 79, 1990.
23. **Fries, E., Gustafsson, L., and Peterson, P. A.,** Four secretory proteins synthesized by hepatocytes are transported from endoplasmic reticulum to Golgi complex at different rates, *EMBO J.,* 3, 147, 1984.
24. **Ledford, B. E. and Davis, D. F.,** Kinetics of serum protein secretion by cultured hepatoma cells. Evidence for multiple secretory pathways, *J. Biol. Chem.,* 258, 3304, 1983.
25. **Lodish, H. F., Kong, N., Snider, M., and Strous, G. J. A.,** Hepatoma secretory proteins migrate from rough endoplasmic reticulum to Golgi at characteristic rates, *Nature (London),* 304, 80, 1982.

26. **Scheele, G. and Tartakoff, A.,** Exit of nonglycosylated secretory proteins from the rough endoplasmic reticulum is asynchronous in the exocrine pancreas, *J. Biol. Chem.,* 260, 926, 1985.
27. **Yeo, K. T., Parent, J. B., Yeo, T. K., and Olden, K.,** Variability in transport rates of secretory glycoproteins through the endoplasmic reticulum and Golgi in human hepatoma cells, *J. Biol. Chem.,* 260, 7896, 1985.
28. **Karrenbauer, A., Jeckel, D., Just, W., Birk, R., Schmidt, R. R., Rothman, J. E., and Wieland, F. T.,** The rate of bulk flow from the Golgi to the plasma membrane, *Cell,* 63, 259, 1990.
29. **McCracken, A. A., Kruse, K. B., and Brown, J. L.,** Molecular basis for defective secretion of the Z variant of human alpha-1-proteinase inhibitor: secretion of variants having altered potential for salt bridge formation between amino acids 290 and 342, *Mol. Cell. Biol.,* 9, 1406, 1989.
30. **Sifers, R. N., Rogers, B. B., Hawkins, H. K., Finegold, M. J., and Woo, S. L. C.,** Elevated synthesis of human α1-antitrypsin hinders the secretion of murine α1-antitrypsin from hepatocytes of transgenic mice, *J. Biol. Chem.,* 264, 15696, 1989.
31. **McCracken, A. A. and Kruse, K. B.,** Intracellular transport of rat serum albumin is altered by a genetically engineered deletion of the propeptide, *J. Biol. Chem.,* 264, 20843, 1989.
32. **Macintyre, S. S., Schultz, D., and Kushner, I.,** Synthesis and secretion of C-reactive protein by rabbit primary hepatocyte cultures, *Biochem. J.,* 210, 707, 1983.
33. **Macintyre, S. S.,** C-reactive protein, *Methods Enzymol.,* 163, 383, 1988.
34. **Macintyre, S. S., Kushner, I., and Samols, D.,** Secretion of C-reactive protein becomes more efficient during the course of the acute phase response, *J. Biol. Chem.,* 260, 4169, 1985.
35. **Carey, D. J. and Hirschberg, C. B.,** Kinetics of glycosylation and intracellular transport of sialoglycoproteins in mouse liver, *J. Biol. Chem.,* 255, 4348, 1980.
36. **Fleischer, S. and Kervina, M.,** Subcellular fractionation of rat liver, *Methods Enzymol.,* 31, 6, 1976.
37. **Sztul, E. S., Howell, K. E., and Palade, G. E.,** Biogenesis of the polymeric IgA receptor in rat hepatocytes. I. Kinetic studies of its intracellular forms, *J. Cell Biol.,* 100, 1248, 1985.
38. **Scheele, G. A., Palade, G. E., and Tartakoff, A. M.,** Cell fractionation studies on the guinea pig pancreas. Redistribution of exocrine proteins during tissue homogenization, *J. Cell Biol.,* 78, 110, 1978.
39. **Aronson, N. N. and Touster, O.,** Isolation of rat liver plasma membrane fragments in isotonic sucrose, *Methods Enzymol.,* 31A, 90, 1974.
40. **Bartles, J. R., Feracci, H. M., Stieger, B., and Hubbard, A. L.,** Biogenesis of the rat hepatocyte plasma membrane *in vivo:* comparison of the pathways taken by apical and basolateral proteins using subcellular fractionation, *J. Cell Biol.,* 105, 1241, 1987.
41. **Towbin, H., Staehelin, T., and Gordon, J.,** Electrophoretic transfer of proteins from polyacrylamide gels to nitrocellulose sheets: procedure and some applications, *Proc. Natl. Acad. Sci. U.S.A.,* 76, 4350, 1979.
42. **Scatchard, G.,** The attraction of proteins for small moelcules and ions, *Ann. N.Y. Acad. Sci.,* 51, 660, 1949.
43. **Munson, P. J. and Robard, D.,** LIGAND: a versatile computerized approach for the characterization of ligand binding systems, *Anal. Biochem.,* 107, 220, 1980.
44. **McPherson, B. A.,** Analysis of radioligand binding experiments: a collection of computer programs for the IBM PC, *J. Pharmacol. Methods,* 14, 213, 1985.
45. **Hu, S., Miller, S. M., and Samols, D.,** Cloning and characterization of the gene for rabbit C-reactive protein, *Biochemistry,* 25, 7834, 1986.
46. **Hu, S. I., Macintyre, S. S., Schultz, D., Kushner, I., and Samols, D.,** Secretion of rabbit C-reactive protein by transfected human cell lines is more rapid than by cultured rabbit hepatocytes, *J. Biol. Chem.,* 263, 1500, 1988.
47. **Macintyre, S.,** Regulated export of a secretory protein from the endoplasmic reticulum of the hepatocyte: a specific binding site retaining C-reactive protein within the ER is downregulated during the acute phase response, *J. Cell Biol.,* 118, 253, 1992.
48. **Williams, D. B., Sweidler, S. J., and Hart, G. W.,** Intracellular transport of membrane glycoproteins: two closely related histocompatibility antigens differ in their rates of transit to the cell surface, *J. Cell Biol.,* 101, 725, 1985.
49. **Kreibich, G., Debey, P., and Sabatini, D. D.,** Selective release of content from microsomal vesicles without membrane disassembly. I. Permeability changes induced by low detergent concentrations, *J. Cell Biol.,* 58, 436, 1973.
50. **Du Clos, T. W., Zlock, L. T., and Marnell, L.,** Definition of a C-reactive protein binding determinant on histones, *J. Biol. Chem.,* 266, 2167, 1991.
51. **Volanakis, J. E. and Kaplan, M. H.,** Specificity of C-reactive protein for choline phosphate residues of pneumococcal C-polysaccharide, *Proc. Soc. Exp. Biol. Med.,* 136, 612, 1971.
52. **Kushner, I. and Kaplan, M. H.,** Studies of acute phase protein. I. An immunohistochemical method for the localization of C-reactive protein in rabbits. Association with necrosis in local inflammatory lesions, *J. Exp. Med.,* 114, 961, 1961.

53. **Mold, C., Rodgers, C. P., Richards, R. L., Alving, C. R., and Gewurz, H.,** Interaction of C-reactive protein with liposomes. III. Membrane requirements of binding, *J. Immunol.,* 126, 856, 1981.
54. **Volanakis, J. E. and Wirtz, K. W. A.,** Interaction of C-reactive protein with artificial phosphatidylcholine bilayers, *Nature (London),* 281, 155, 1979.
55. **Dorner, A. J., Bole, D. G., and Kaufman, R. J.,** The relationship of N-linked glycosylation and heavy chain-binding protein association with the secretion of glycoproteins, *J. Cell Biol.,* 105, 2665, 1987.
56. **Dul, J. L. and Argon, Y.,** A single amino acid substitution in the variable region of the light chain specifically blocks immunoglobulin secretion, *Proc. Natl. Acad. Sci. U.S.A.,* 87, 8135, 1990.
57. **Hendershot, L. M.,** Immunoglobulin heavy chain and binding protein complexes are dissociated *in vivo* by light chain addition, *J. Cell Biol.,* 111, 829, 1990.
58. **Bordier, C.,** Phase separation of integral membrane proteins in Triton X-114 solution, *J. Biol. Chem.,* 256, 1604, 1981.
59. **Anderson, J. K., Stroud, R. M., and Volanakis, J. E.,** Studies on the binding specificity of human C-reactive protein for phosphorylcholine, *Fed. Proc.,* 37, 1495, 1970.
60. **Siboo, R. and Kulisek, E.,** A fluorescent immunoassay for quantification of C-reactive protein, *J. Immunol. Methods,* 23, 59, 1978.
61. **Bach, B. A., Gewurz, H., and Osmand, A. P.,** C-reactive protein in the rabbit: isolation, characterization and binding affinity to phosphocholine, *Immunochemistry,* 14, 215, 1977.
62. **Hoflack, B., Fujimoto, K., and Kornfeld, S.,** The interaction of phosphorylated oligosaccharides and lysosomal enzymes with bovine liver cation-dependent mannose-6-phosphate receptor, *J. Biol. Chem.,* 262, 123, 1987.
63. **Young, N. M. and Williams, R. E.,** Comparison of the secondary structures and binding sites of C-reactive protein and the phosphorylcholine-binding murine myeloma proteins, *J. Immunol.,* 121, 1893, 1978.
64. **Du Clos, T. W., Zlock, L. T., and Rubin, R. L.,** Analysis of the binding of C-reactive protein to histones and chromatin, *J. Immunol.,* 141, 4266, 1988.
65. **Robey, F. A., Jones, K. D., Tanaka, T., and Liu, T.,** Binding of C-reactive protein to chromatin and nucleosome core particles, *J. Biol. Chem.,* 259, 7311, 1984.
66. **Ballou, S. P., Buniel, J., and Macintyre, S. S.,** Specific binding of human C-reactive protein to human monocytes *in vitro, J. Immunol.,* 142, 2708, 1989.
67. **Muller, H. and Fehr, J.,** Binding of C-reactive protein to human polymorphonuclear leukocytes: evidence for association of binding sites with Fc receptors, *J. Immunol.,* 136, 2202, 1986.
68. **Tebo, J. M. and Mortensen, R. F.,** Characterization and isolation of a C-reactive protein receptor from the human monocytic cell line U-937, *J. Immunol.,* 144, 231, 1990.
69. **Zahedi, K., Tebo, J. M., Siripont, J., Klimo, G. F., and Mortensen, R. F.,** Binding of human C-reactive protein to mouse macrophages is mediated by distinct receptors, *J. Immunol.,* 142, 2384, 1989.
70. **Paganelli, K. A., Stern, A. S., and Kilian, P. L.,** Detergent solubilization of the interleukin 1 receptor, *J. Immunol.,* 138, 2249, 1987.
71. **Supattapone, S., Worley, P. F., Baraban, J. M., and Snyder, S. H.,** Solubilization, purification, and characterization of an inositol trisphosphate receptor, *J. Immunol.,* 263, 1530, 1988.
72. **McKernan, R. M., Biggs, C. S., Gillard, N., Quirk, K., and Ragan, C. I.,** Molecular size of the 5-HT3 receptor solubilized from NCB 20 cells, *Biochem. J.,* 269, 623, 1990.
73. **Booth, C. and Koch, G. L. E.,** Perturbation of cellular calcium induces secretion of luminal ER proteins, *Cell,* 59, 729, 1989.
74. **Lodish, H. F. and Kong, N.,** Glucose removal from N-linked oligosaccharides is required for efficient maturation of certain secretory glycoproteins from the rough endoplasmic reticulum to the Golgi complex, *J. Cell Biol.,* 98, 1720, 1984.
75. **Sambrook, J. F.,** The involvement of calcium in transport of secretory proteins from the endoplasmic reticulum, *Cell,* 61, 197, 1990.
76. **Du Clos, T. W.,** C-reactive protein reacts with the U1 small nuclear ribonucleoprotein, *J. Immunol.,* 143, 2553, 1989.
77. **Du Clos, T. W., Mold, C., and Stump, R. F.,** Identification of a polypeptide sequence that mediates nuclear localization of the acute phase protein C-reactive protein, *J. Immunol.,* 145, 3869, 1990.

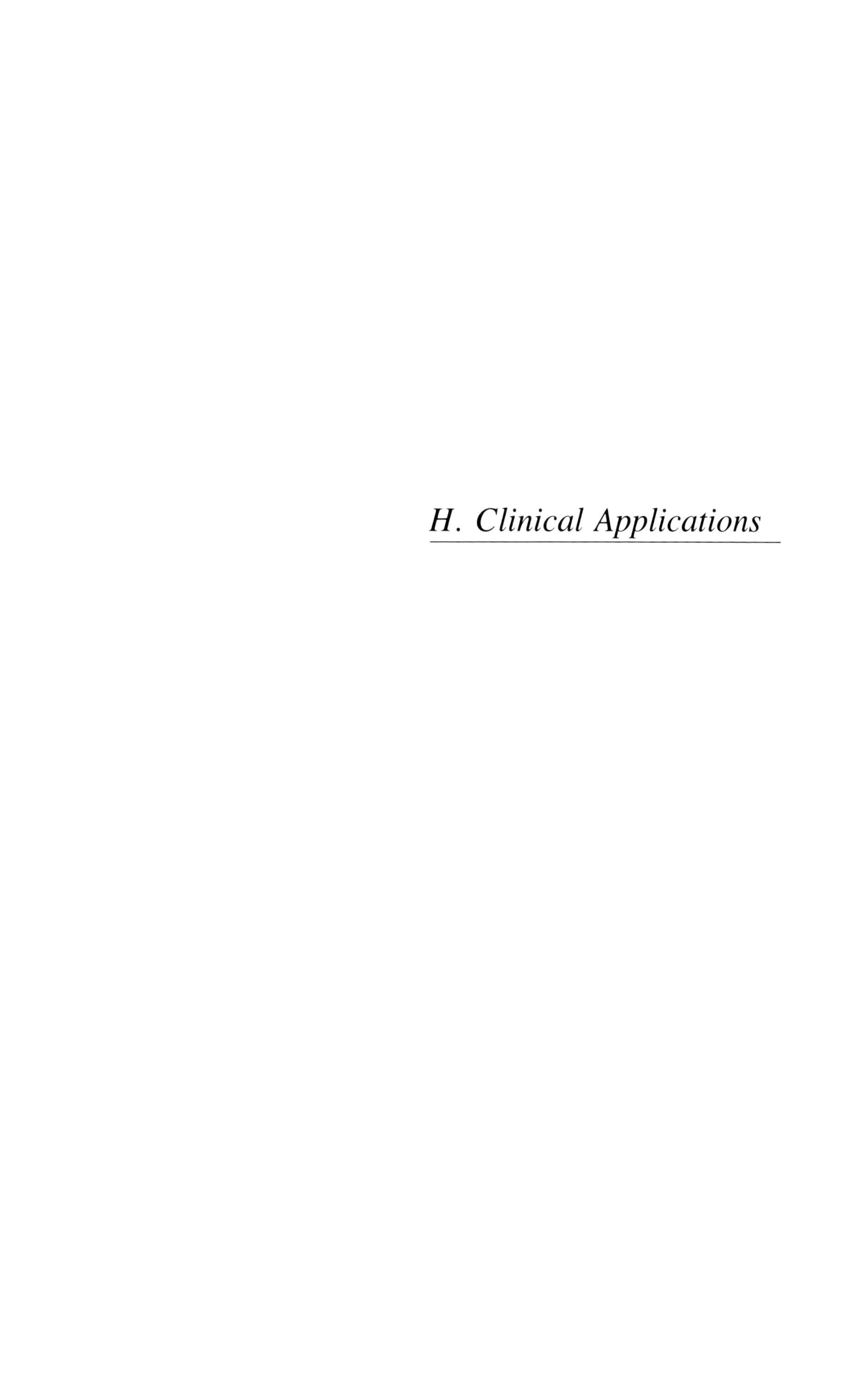

H. Clinical Applications

Chapter 33

CYTOKINE MEASUREMENTS IN DISEASE

David Heney, Rosamonde E. Banks, John T. Whicher, and Stuart W. Evans

TABLE OF CONTENTS

0-8493-6913-4/93/$0.00 + $.50

I. INTRODUCTION

During the last 5 years, the field of cytokine biology has seen enormous expansion, with the cloning and sequencing of many new molecules. We review some aspects of cytokine measurement which, in the future, may be of clinical value in the field of acute phase response assessment. These are situations in which the laboratory measurement of cytokines known to be mediators of the acute phase response may have a role to play in the assessment and management of disease, specifically in inflammation, infection, malignancy, and transplantation. The clinical use of soluble interleukin-2 receptor (sIL-2-R) measurements is included for completeness. A brief overview of the problems associated with the bioassay and immunoassay of cytokines is given before the clinical use of cytokine measurements is discussed.

II. CYTOKINE ASSAYS

The development of a bioassay for any particular cytokine has generally preceded that of the corresponding immunoassay, largely because, almost without exception, cytokines have been discovered and charcterized on the basis of their biological actions. However, the application of such assays, often developed predominantly for use with tissue culture supernatant samples, to the analysis of more complex biological fluids such as blood and urine has been problematic. The establishment of specific immunoassays and bioassays for cytokines has been recently reviewed.[1,2] An overview of the problems associated with the interpretation of such assays is given below. The heterogeneity of cytokine levels which have been reported in any particular clinical situation by different groups may thus, at least in part, be a reflection of these problems, which are only now beginning to be understood and appreciated. It must be emphasized that bioassays and immunoassays can yield completely different information about any cytokine. Both sets of information are essential for obtaining the full picture and thus understanding the role of cytokines in disease.

A. ACCURACY AND SPECIFICITY

The factors affecting specificity and accuracy in cytokine assays are outlined in Table 1. The complex interactions which exist in the cytokine network predict that the likelihood of any cell type responding to only one cytokine is unlikely, making the development of specific bioassays difficult. This is exemplified by the mouse thymocyte assay which, although originally thought to be interleukin-1 (IL-1) specific, is now known to be sensitive to tumor necrosis factor (TNF) and interleukin-6 (IL-6). Additionally, interleukin-2 (IL-2) and interleukin-4 (IL-4) may exert a synergistic influence.[3-5] The use of neutralizing antibodies may allow the relative contributions of the different cytokines to be determined, although these themselves may be contaminated by cytokines.[6] Use of the assay with saturating levels of IL-2 or IL-4 may obviate any synergistic interaction of these cytokines.[7] However, even with these precautions, it is impossible to control for the existence and interference of other undescribed cytokines which may be present in samples. Additionally, the influence of other potentially interfering factors such as steroid hormones is often ignored.

Immunoassays have the advantage of specificity, and the epitope reactivity can be charcterized to some extent if monoclonal antibodies are used. However, it is increasingly obvious that, as for many plasma proteins, cytokines in biological fluids may show molecular heterogeneity. Calibrants and immunogens do not share these charcteristics, as they are almost invariably nonglycosylated recombinant proteins, and it is impossible to duplicate the mixture of various isoforms which may be present in any sample. Monoclonal antisera raised to such antigens may fail to recognize polymorphic forms of the cytokine and reactivity

TABLE 1
Factors To Be Considered in the Interpretation of Cytokine Assays with Biological Samples

Parameter	Bioassay	Immunoassay
Measurement	Biologically active molecules	Immunoreactive molecules
Sensitivity	Can be high, often higher than an immunoassay	Can be reasonably high
Specificity	Poor to high (assay dependent); must be confirmed with neutralizing antibodies	Usually high, but dependent on antibodies; unless epitopes determined, the exact form of cytokine detected in unknown, e.g., fragments, bound form, etc.
Accuracy	May be affected by specific and nonspecific inhibitors, cytokine-binding proteins, and antibodies present in the sample; may be influenced by blood collection procedure (uptake/release by leukocytes, stability)	May be subject to interference from matrix, e.g., complement, binding proteins, rheumatoid factor, heterophilic antibodies; may be influenced by blood collection procedure (uptake/release by leukocytes, stability)
Precision	Relatively poor (CV, 10–100%)	Usually good (CV, 5–10%)
Ease of use	Labor intensive (24 h–4 d)	2–24 h
Standardization	Nationally available standards	Nationally available standards

with precursor molecules or fragments is difficult to assess. Although it is desirable for the matrix of the reference material to be identical to that of the sample, this is often impossible owing to intersample variation. This may complicate the interpretation of results, particularly with samples which contain levels close to background levels of the assay.

Specific and nonspecific inhibitors detected by bioassays have been described for many cytokines. They include the IL-1 receptor antagonist[8] and soluble receptors for cytokines such as IL-2, IL-1, IL-6, interferon-γ (IFNγ), and TNF.[9-12] The influence of these inhibitors on immunoassays is largely unknown, although for TNF it has recently been shown that the soluble receptor interferes in some immunoassays.[13] Several cytokines have been shown to bind to plasma proteins, particularly α_2-macroglobulin, although this does not always affect their biological activity,[14] and circulating autoantibodies to cytokines have been described in normal individuals.[15] In many of these cases, the implications for assays have not yet been satisfactorily determined, although poor recovery of recombinant cytokines in biological fluids subjected to bioassay is often experienced. Figure 1 shows the effect of sample dilution on the inhibitory activity of synovial fluid to which IL-1 has been added.[16] Various strategies such as chloroform extraction[17] and silica adsorption[18] have been used to remove inhibitory activity from bioassays for IL-1.

Immunoassays as sensitive as those needed to measure the low concentrations of cytokines in body fluids are inevitably susceptible to factors known to interfere generally with immunoassays such as rheumatoid factor, heterophilic antibodies, and complement.[19-21] These potential sources of interference must be taken into account in assay design and must be investigated in the validation of the assay. Additionally, binding proteins and inhibitors have the potential to affect the results by masking epitopes.

B. STANDARDIZATION

Standardization of cytokine assays presents the same problem as for other proteins in biological fluids.[22] With the availability of many different preparations of cytokines which vary in potency depending upon the assay used, interlaboratory comparison of results can only be made if similar assays are used and a common reference material has been used such as those available from institutions concerned with international standardization.

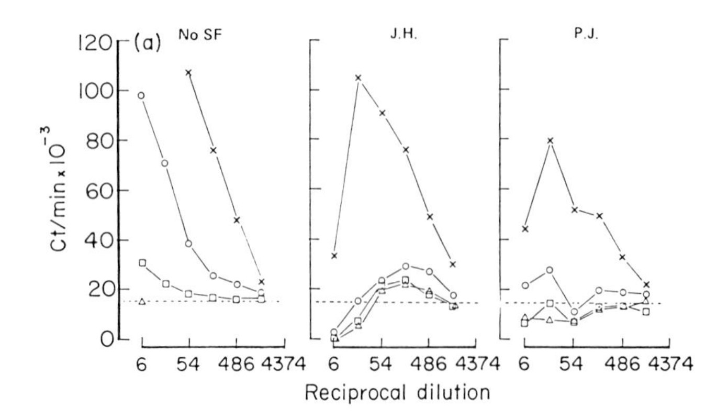

FIGURE 1. Purified Il-1β added to synovial fluid dilutions (J. H. and P. J.) or medium (No SF) to give an equivalent of 1233 pg/ml (×), 123 pg/ml (○), 12 pg/ml (□), or no (△) exogenous IL-1β in the undiluted SF. (From Hopkins, S. J., Humphreys, M., and Jayson, M. I., *Clin. Exp. Immunol.*, 72, 422, 1988. With permission.)

C. SENSITIVITY

Bioassays are generally more sensitive than immunoassays for cytokines. The characteristics of calibration curves vary between different assays, and limits of sensitivity must be set so that only the portion of the curve producing acceptable precision is used. Sensitivity is determined not only by the assay sensitivity, but also by the dilution of the biological sample which may be needed to overcome inhibitory or cytotoxic effects of the sample. Similarly, in immunoassays, the dilution necessary to decrease nonspecific background to acceptable levels may result in a higher minimum detectable level.

D. PRECISION

Precision in bioassays is often poor. Immunoassays have a better signal-to-noise ratio and it is worth expending considerable efforts to achieve high-affinity antisera. Many antisera which provide adequate assays for use with the relatively high levels of cytokines in cell culture supernatants are inadequate for use in body fluids. This can be seen in many commercially available assays, where the range of the standard curve often spans several thousand picograms per milliliter, with many biological samples having concentrations of cytokines which lie in the least precise area at the very bottom of the curve.

III. CYTOKINES IN ACUTE SEPSIS AND SEPTIC SHOCK

There is considerable evidence implicating cytokines in the pathogenesis of acute sepsis. TNF is often considered to play a central role, and this view is supported by a number of animal studies. High levels of TNF have been demonstrated in the plasma of rabbits and baboons within 45 to 100 min after administration of endotoxin,[23,24] recombinant TNF produced circulatory collapse in rodents with clinical and postmortem findings indistinguishable from septic shock,[25] and monoclonal anti-TNF antibodies protected against shock, organ injury, and death following lethal doses of endotoxin.[26] Recent reviews,[27] however, have emphasized the physiological complexity of acute sepsis and have cautioned against ascribing a predominant role to TNF. Cytokines such as IL-1 and IL-6 are also important; administration of low doses of IL-1β induced a shock-like state, the IL-1 receptor antagonist has been used successfully in the treatment of septic shock in animal models, and monoclonal antibodies to IL-6 protected against lethal *Escherichia coli* infection in mice.[28,29]

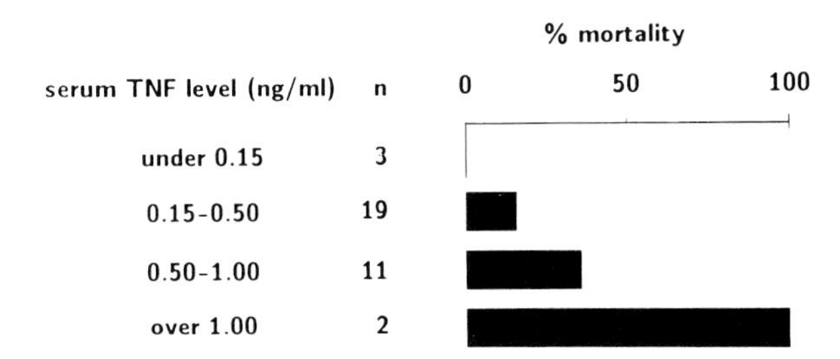

FIGURE 2. Relationship between mortality and serum levels of TNFα on admission in children with severe infectious purpura. (From Girardin, E. P., Grau, G. E., Dayer, J. M., Roux-Lombard, P., and Lambert, P. H., *N. Engl. J. Med.*, 319, 397, 1988. With permission.)

A. TNF

In severe Gram-negative sepsis, TNFα concentrations were increased in most patients, sometimes to a very high level, whereas IL-1 and IFN were increased in fewer than 20% of cases.[30,31] Systemic meningococcal disease in children seems to be one of the strongest stimuli to increased plasma TNFα levels,[31,32] providing a consistent pattern of raised TNFα in severe cases, with levels correlating with mortality (Figure 2). In clinical practice, septicemia often presents with a more protracted course, with repeated endotoxic stimuli, especially in patients with a poor underlying immune system. Treatment may be initiated at varying times and most studies have looked at diverse groups of patients.

Offner et al.[33] found that TNFα levels at the onset of septic shock were raised to some degree in all patients, but increased thereafter and in some cases remained moderately elevated for a prolonged period. Extremely high levels were found in patients who died within 24 h. By contrast, Debets et al.[34] found that only 25% of cases had raised TNFα levels at the time of diagnosis of sepsis, but those with raised TNFα had twice the mortality of patients with normal TNFα levels. Similarly, TNFα was undetectable on day 10 in patients with septic shock who survived, but was still greatly elevated in those patients who subsequently died.[35] De Groot et al.[36] reported raised TNFα levels in only 16% of patients with Gram-negative sepsis. In a study of burn victims who had repeated episodes of bacteremia, TNFα was detected transiently and repeatedly in the circulation, with the frequency of TNFα peaks correlating with both infection and mortality rate.[37]

Most studies measuring TNFα in plasma as a marker of the disease process or its severity agree that there is a relationship between the level or duration of elevated TNFα values and the mortality in septic shock. However, it is difficult to define a value above which prognosis declines markedly, and although TNFα levels may provide a prognostic index for a group as a whole, its ability to provide a prognosis for an individual patient is poor because of a marked interpatient variability. It is clear that too much weight must not be placed on cytokine measurements when simple clinical measurements may have more to offer. Calandra et al.[35] showed that clinical severity, the patient's age, culture of bacteria from blood, urine output, and arterial pH contributed more significantly to prediction of the patient's outcome than serum levels of TNFα.

Leroux-Roels and Offner[38] have suggested that persistently low levels of TNFα may be as harmful, or even more harmful, than high peak values. It is interesting that cancer patients given TNFα therapeutically as a continuous 24-h infusion show a rise to TNFα levels over the first 5 h, but then levels fall to being undetectable by 24 h.[39] This suggests the mobilization of factors which "neutralize" circulating TNFα. The release of soluble TNFα receptors[10] may be one such fctor. An imbalance in the ratio of soluble TNF receptor to cytokine, or an inability to clear TNFα from the circulation, may be more important than the exact level of the cytokine in the development of septic shock. This may partly explain the variable nature of the studies described above.

Other factors contributing to the variation in results include the phasic nature of TNF elevation,[40] and differences in assay technique and reported "normal" values. It should be noted that the central role of TNF is questioned by a number of other studies. For example, the administration of antibodies to endotoxin, although protecting against the lethal effect of subsequent endotoxin challenge, did not lower the increase in TNFα caused by endotoxin admnistration,[41] and antibodies to TNFα administered to mice infected with *Histoplasma capsulatum* were shown to accelerate death,[42] confirming that TNFα also has important protective properties.

B. INTERLEUKIN-6

In studies using human volunteers, an injection of either endotoxin or TNFα was followed by a rise in IL-6 at 30 to 60 min, peaking at 2 h and then declining with a half-life of 30 to 70 minutes.[43] Although IL-6 contributes to many of the immune and inflammatory responses, unlike TNF and IL-1, it does not by itself cause clinical shock in experimental situations.[44]

In acute sepsis, IL-6 appears to be more consistently raised than TNFα, with comparatively higher values. Meningococcal septicemia, again proves to be a good model system, with raised IL-6 levels in 69 out of 79 admission samples.[45] The median serum level of IL-6 was approximately 1000 times higher in patients with septic shock (189 ng/ml) than in patients with either meningitis or bacteremia alone (0.2 ng/ml). Levels of IL-6 were approximately 100 times higher than corresponding TNF levels and also correlated strongly with severity.

Both Gram-negative and Gram-positive infections are capable of inducing a similar rise in IL-6. Increased levels were found in 32 of 37 patients admitted to an intensive care unit with sepsis or septic shock,[46] with the highest values (up to 100 ng/ml) in patients who suffered from septic shock. The degree of elevation correlated with mortality, with 89% of patients with levels above 7.5 ng/ml dying, whereas all of those with levels of <40 pg/ml survived. Calandra et al.[47] examined IL-6 in 70 patients with established septic shock. Raised IL-6 levels were detected in 64% of the patients at study entry, with levels falling rapidly even in the presence of persisting infection and clinical shock, and were largely undetectable by day 2. Although the levels were higher in patients dying of fulminant septic shock than in survivors, no cutoff value predicted the outcome for an individual patient.

1. Predictive Value of Interleukin-6 in Acute Sepsis

Many of the studies described have examined TNFα or IL-6 levels at a time when patients were already critically ill. There are, however, situations in which it is difficult to distinguish patients who will subsequently develop severe sepsis from those who will have a mild infective illness. Even though patients may have a Gram-negative bacteremia at presentation, they are often indistinguishable clinically from those with viremias or less severe bacteremias. Septic shock usually only develops later, and typically after initiation of antibiotic therapy. Evidence is now accumulating that in well-defined situations, IL-6 may be a sensitive predictor of subsequent clinical severity in patients with suspected acute sepsis.

In patients (n = 85) presenting to an emergency department with potential bacteremia, a raised IL-6 was more sensitive than TNFα or IL-1 in predicting the presence of bacteremia.[48] Using a cutoff value for IL-6 of 2 ng/ml, eight of the nine patients exceeding this value had a positive blood cuture, whereas 88% of patients with IL-6 concentrations <2 ng/ml were not bacteremic. Of the bacteremic patients, those with an IL-6 level of <2 ng/ml survived and had a hospital stay of less than 14 d, while those with a value of >2 ng/ml either died or had a hospital stay of greater than 14 d. A clinical scoring system was unable to predict either bacteremia or length of hospitalization.

Raised IL-6 was associated closely with positive bacterial blood cultures in a small group of 20 children with potential bacterial sepsis presenting to an emergency department, and the degree of elevation correlated with clinical severity.[49] The authors concluded that elevated IL-6 in this situation distinguishes children with a subsequent diagnosis of bactermia from those with a similar clinical presentation.

In cancer patients presenting with febrile neutropenia, it is difficult to distinguish patients with potential severe sepsis. Episodes of fever are common in cancer patients, with the majority of patients having ''unexplained fevers'' with negative cultures and no localizing signs of infection. The most important of many associated factors are episodes of neutropenia, which usually occur following pulses of intensive cytotoxic chemotherapy. Gram-negative bacteremias continue to be associated with a significant mortality. In a study examining IL-6 levels in children presenting with febrile neutropenia (n = 47), the median IL-6 value for Gram-negative infections on admission was 1610 pg/ml (range, 896 to 40,000), for Gram-positive infections it was 138 pg/ml (range, 66 to 1045), and for unexplained fevers it was 50 pg/ml (range of 24 to 135, with a single high value of 665 pg/ml).[50] These groups of patients were indistinguishable clinically on admission. Four of the seven patients with Gram-negative bacteremias subsequently developed signs of endotoxic shock. Two of the nine Gram-positive bacteremias subsequently had a severe and toxic clinical course, and the patient with the high IL-6 value in the group with unexplained fever developed signs of cardiovascular compromise. Patients with an IL-6 value of >600 pg/ml on admission included all the Gram-negative bacteremias, as well as the two Gram-positive and the single unexplained fever with a more severe clinical course. The clinical course in other patients was unremarkable.

The above studies were on small groups of patients, and larger studies are needed to confirm these findings. However, IL-6 measurements have the potential for identifying patients who may develop a more severe illness and in whom the early use of therapeutic agents such as anti-endotoxin antibodies may be beneficial.

IV. CYTOKINES IN OTHER INFECTIVE CONDITIONS

A. MENINGITIS

Both IL-1 and TNFα have been implicated in the pathogenesis of bacterial meningitis.[51,52] High levels of IL-1β (mean, 994 pg/ml) and TNFα (mean, 787 pg/ml) in cerebrospinal fluid (CSF) have been reported in the majority of patients with bacterial meningitis at the time of diagnosis.[53] A seprate study has confirmed similar raised values for CSF levels of TNFα.[54] CSF levels of TNFα were not elevated in patients with viral meningitis. No correlation was seen between CSF and plasma TNFα levels, and it was concluded that detectable CSF levels of TNFα could be used as a diagnostic tool to suggest bacterial meningitis, but that the absence of detectable TNFα in an individual patient could not be used to exclude a bacterial etiology of meningitis.

CSF levels of IL-6 have also been reported to be variably raised in bacterial as well as viral meningitis and encephalitis.[55] This highlights the difference in the sensitivity and specificity of TNF and IL-6, and is illustrated in Figure 3A and B. Whereas the CSF level of TNF is only raised in bacterial meningitis as opposed to viral and other causes, it is not sufficiently sensitive in that a significant percentage of patients with bacterial infections have normal values. IL-6 is more sensitive, with all bacterial infections having raised CSF

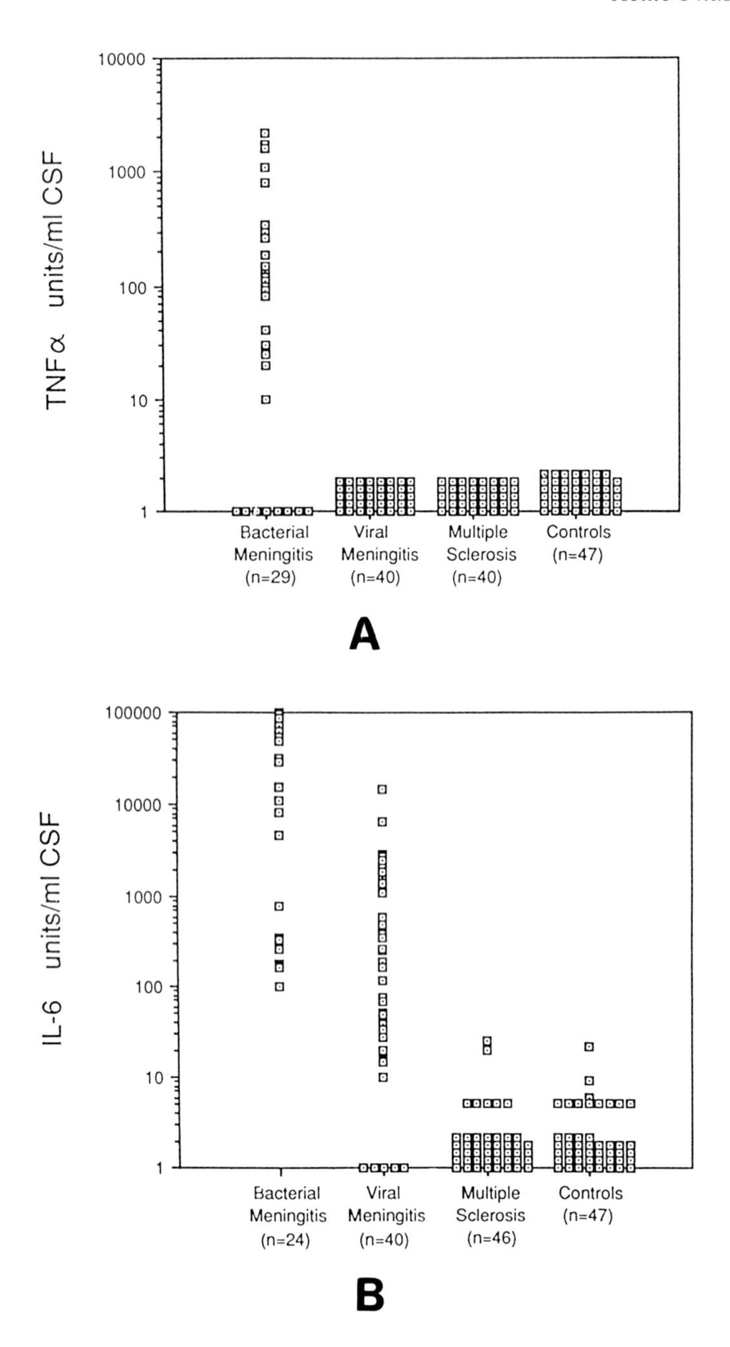

FIGURE 3. TNFα levels (A) and IL-6 levels (B) in CSF of patients with infectious meningitis. (From Frei, K., Nodal, D., and Fontana, A., *Ann. N.Y. Acad. Sci.*, 594, 326, 1990. With permission.)

levels, but is less specific since it is also raised in viral meningitis and other inflammatory conditions. This is a typical pattern also seen in other situations of acute sepsis and suggests that both TNFα and IL-6 should be measured in order to provide a more accurate interpretation.

B. PARASITIC DISEASE

There is a growing body of information documenting cytokine measurements and their role in the pathogenesis of parasitic disorders.[56,57] Malaria is the best studied. TNFα, probably released in response to schizont rupture,[58] was found in the plasma of malaria patients, and was related to the degree of parisitemia, being higher in patients with severe or cerebral malaria.[59] IL-6 was also raised in patients with malaria and correlated with severity.[60] Raised TNFα levels have also been found in leishmaniasis.[61] It is not yet clear if the monitoring of cytokine levels provides any advantage over existing routine measurements.

C. HIV INFECTION

Elevated serum levels of TNFα and IL-1 have been found in patients with AIDS, although usually associated with lymphocytic interstitial penumonitis.[62] Serum IL-6 was elevated in HIV-seropositive asymptomatic carriers, and the levels correlated with the degree of HIV-induced disease progression.[63] A correlation between IL-6 levels and immunoglobulin levels has also been found,[64] and it has been suggested that IL-6 may be an autocrine growth factor for AIDS Kaposi sarcoma cells.[65]

D. NEONATAL INFECTION

Plasma levels of TNFα in neonates are raised in a fashion similar to that of older children and adults, with higher levels in those with bacterial infections, and the highest in those with septic shock.[66] IL-6 was also raised in perinatal infectious complications, but not in noninfectious complications.[67]

V. CYTOKINES IN THE DETECTION AND MONITORING OF INFLAMMATION

The key role of macrophage-derived cytokines in inflammation and the acute phase response has already been discussed. On a theoretical basis, it might be expected that their measurement in serum or other biological fluids would be a more sensitive or rapid indication of inflammatory activity than the secondary effects such as acute phase proteins. Because cytokines are involved in so many aspects of immune and inflammatory responses, it is not surprising that raised levels have been detected in a large number of disorders. In most cases, raised levels as yet do not appear specific enough, and there are only a few situations where correlations with either severity or prognosis have been well defined.

Rheumatoid arthritis (RA) is a disease associated par excellence with an acute phase response, and cytokine measurements have been extensively studied. Plasma IL-1 has been detected in some cases in RA and acute arthritides,[68,69] and in RA there is some correlation with disease activity. Biologically active and immunologically active IL-1 were detected in synovial fluid from patients with RA, osteoarthritis, and other arthritides, and inhibitory activity was present in most cases.[16,70-72]

Levels of TNFα in synovial fluid have been reported as consistently elevated in patients with seropositive RA as opposed to other arthritides,[73] although another study found that synovial fluid TNFα levels were similar in seropositive and seronegative arthritis, but higher in patients with osteoarthritis.[74] TNFα levels in osteoarthritis were related to disease duration.[75]

Elevated levels of IL-6 have been reported in the serum and synovial fluid of patients with various rheumatic diseases.[76-79] Synovial fluid levels of IL-6 were found to be 1000-fold higher and to correlate positively with those in the serum, suggesting localized production in the joint. Synovial fluid levels of IL-6 could also be correlated with the articular index, wheres serum IL-6 levels were positively correlated with CRP and negatively correlated with albumin in RA. Other authors[80] confirm a correlation between synovial fluid IL-6 and clinical parameters. IL-6 plasma levels were also raised in polymyalgia rheumatica and giant cell arteritis.[81]

Urinary IL-6 has been detected in a variety of renal diseases, but was not specific for any individual disease process,[82] and has also been found in urinary tract infections.[83] Raised plasma levels of TNFα and IL-6 have been found in patients with chronic renal failure, but were not related to the degree of uremia.[84]

Fulminant hepatic failure has been associated with raised levels of TNFα, although appearing to be more closely related with associated infections than with disease activity.[85] Raised levels of TNFα, IL-1, and IL-6 were found in patients with chronic alcoholic hepatitis, correlating with disease activity, with TNF levels being a predictor of long-term survival.[86]

IL-1, TNFα, and IL-6 have been measured in inflammatory bowel disease,[87,88] and IL-6 levels were found to correlate with disease activity. In ulcerative colitis, but not in Crohn's disease, IL-6 correlated with other measures of the acute phase response.

In patients with acute pancreatitis, no significant differences were seen between TNFα concentrations in the severe and mild patient groups, although the severe group had a higher median level. Many patients had levels that remained within normal limits.[89] Raised IL-6 values were correlated with severity of illness, CRP response, and mortality.[90]

Cerebrospinal TNFα and IL-6 have been detected in only 20 to 40% of patients with multiple sclerosis.[91,92] CSF levels of TNF have been shown to be related to disease severity, whereas plasma levels of IL-6, about 17 times higher than cerebrospinal levels, did not correlate with disease activity.[93] Cytokine levels have not correlated with other immune and biochemical parameters.

Raised levels of cytokines have been reported in a number of conditions characterized by an acute vasculitis. Raised levels of IFNα, IL-2, and TNFα have been reported in polyarteritis nodosa and Wegener's granulomatosis,[94] and were related to disease activity. IL-6 and TNFα have also been measured in Kawasaki disease and related to disease activity.[95,96]

VI. CYTOKINES IN MALIGNANCY

The therapeutic use of cytokines for the treatment of malignant disease has attracted much attention. Despite this, there are few situations where the measurement of cytokines provides any additional diagnostic or prognostic information. Even if a raised or low level of a particular cytokine is found, its interpretation for individual patients and its role in the pathogenesis of malignancy is far from clear.

TNFα has been reported to be raised in approximately 50% of cancer patients with solid tumors[97] and was raised at presentation in most children with acute lymphoblastic leukemia.[98] IL-6 was raised in some patients with solid tumors,[99] and much interest has been generated by the finding of raised levels in myeloma, although usually only seen in patients with advanced disease.[100]

Low levels of IL-2 have been reported in patients with metastatic solid tumors and have been shown to correlate with reduced survival.[101] sIL-2-R has been measured in a number of malignant disorders, and raised levels have been found in advanced solid tumors, lung cancer, hairy cell leukemia, and in lymphoproliferative disorders in adults and children,[102-106] with higher levels tending to correlate with a poorer prognosis.

Pharmacokinetic data are available for those cytokines used therapeutically,[107-109] but again there are no data to indicate the need to monitor blood levels.

VII. CYTOKINES IN TRANSPLANTATION

Detecting graft rejection has depended largely on measurements of graft function, although some indices of immune activation have previously been studied, e.g., β_2-microglobulin and C-reactive protein. Both show changes in graft rejection, but have only found a limited role in management. In the early posttransplant period, concomitant infection and cyclosporin toxicity are potential problems, and both may be difficult to distinguish from acute rejection. The immune response to infection is in many ways similar to that of rejection, and this has proved to be a major limitation for cytokine measurement in predicting rejection episodes.

A. INTERLEUKIN-2 (IL-2) AND SOLUBLE IL-2 RECEPTOR (sIL-2-R)

Interestingly, high plasma IL-2 levels prior to renal transplantation have been associated with subsequent episodes of rejection.[110,111] This suggests that performing a transplant when the patient's immune system is activated enhances recognition and rejection of the graft. During acute rejection, an increase in both plasma and urinary IL-2 and sIL-2-R was seen.[110,112] Cyclosporin toxicity or acute tubular necrosis did not generate a marked IL-2-mediated response, and the use of these indices was found to provide a good distinction between toxicity and rejection. IL-2 measurements were more specific than sIL-2-R, and urinary IL-2 appeared to be the single-most-sensitive indicator.[110]

Unfortunately, there is still a large overlap with patients who are infected. It is therefore important that cytokine measurement be evaluated with other pathological and biochemical parameters. Authors have also highlighted the necessity for sequential measurements in individual patients in order to overcome the large interpatient variability and use values in a predictive fashion.

The sequential measurement of IL-2 and sIL-2-R in liver allograft recipients has also been used to predict and aid the diagnosis of acute rejection.[113] In liver transplantation, the main differential diagnosis is that of infection, and similar cytokine values have been found, which limits their diagnostic value. In other transplants such as cardiac and bone marrow, IL-2 levels were less informative.

B. TNF AND IL-6

Mononuclear phagocytes are present at an early stage in the tissues of rejecting grafts, and both TNFα and IL-6 were found in the plasma prior to, and in association with, rejection episodes for renal,[114,115] liver,[116,117] cardiac,[118] and bone marrow transplants (BMTs).[119] Again, there remains the problem of values overlapping with those seen in infection and of interpatient variability.

A small subgroup of patients with chronic asymptomatic release of TNFα before admission for BMTs (levels, 412 to 13,160 pg/ml) were protected from subsequent BMT-related complications.[120] The authors suggest the possibility of desensitization by TNFα. The practical value of findings such as these has yet to be established.

VII. THE RELATIONSHIP OF CYTOKINE LEVELS TO ACUTE PHASE PROTEIN MEASUREMENTS

A number of studies have sought to compare changes in the levels of IL-1, IL-6, and TNFα with those of acute phase proteins, although as yet there is insufficient information to assess their relative clinical usefulness.

Several studies[121,122] in patients undergoing surgery have confirmed the absence of a TNFα response (as seen in sterile inflammation in animals) and demonstrated a clear response in IL-6,[123,124] peaking at 4 to 6 h after incision, but showing a relatively poor quantitative relationship to CRP or α_1-antichymotrypsin levels.[121,123] Di Padova et al.,[122] however, showed that mean serum CRP levels did correlate with those of both IL-6 and cortisol. It has been suggested that IL-6 correlates better with duration of surgery and severity of injury than

CRP.[123] In contrast to IL-6, CRP tends to show an all-or-nothing response, and minor surgery may elicit no response. In the case of IL-1β, Di Padova et al.[122] were unable to demonstrate a response, although they detected a transient increase in IL-1 inhibitors.

A relationship between fever and IL-6 response has been described in patients with burns.[125] A more detailed study[126] showed that in extensively burned patients, IL-6 rose earlier than α_1-acid glycoprotein or α_1-antitrypsin and that both proteins remained raised for several days after IL-6 fell to normal (IL-1 and TNF were not detectable in the acute stages). Changes in the glycosylation of these acute phase proteins did, however, parallel the IL-6 levels and were independent of the concentrations of the proteins.

The relationship of CRP levels to those of TNFα have been examined in a variety of conditions.[127] In several diseases (RA, SLE, graft rejection, HIV infection, and sepsis), very high levels of CRP were consistently associated with raised serum levels of TNFα, although the ratio between the two varied. However, in myocardial infarctions, high CRP levels were associated with only modest elevations of TNFα and, indeed, a number of cases had no elevation of TNFα. A rather similar pattern has been seen in bacterial sepsis, possibly reflecting the greater sensitivity of IL-6 than TNF to the stimulus. In patients with graft rejection, TNFα levels were much higher, with only modest or no elevations in CRP.

More information is required before the relative values of different measures can be assessed for clinical use, although IL-6 does appear to be a rapid and sensitive marker of tissue damage and inflammation in many circumstances.

IX. SUMMARY

There is a need for caution when interpreting measurements of cytokines in a large number of varied diseases. Studies to date have looked mainly at circulating levels of cytokines, and it is only now that we are able to initiate studies to examine some of the regulatory components, such as soluble cytokine receptors or cytokine inhibitors. It may be necessary to have this information together with cytokine levels in order to properly interpret results. There is some concern that many studies over-interpret results, and authors often link the finding of cytokines with the pathogenesis of that particular disease. In many instances, they are likely to be secondary and, to some extent, incidental findings. It would be useful if cytokine measurement could identify particular groups of patients or provide a prognostic indicator. At present, indications for their routine use are limited. They need to be used in clearly defined situations where the patient population has been well studied in a longitudinal fashion. In general, IL-6 measurements seem to be potentially more useful clinically than TNF, especially in acute inflammation. In particular, IL-6 may prove useful in identifying groups of patients with potentially severe sepsis.

For all these issues, further studies are needed. The relative values of cytokine levels in comparison to standard clinical and laboratory mesurements also need to be critically examined.

REFERENCES

1. **Meager, A.,** RIA, IRMA, and ELISA assays for cytokines, in *Cytokines, A Practical Approach,* Balkwill, F. R., Ed., IRL Press, Oxford, 1991, 229.
2. **Wadhwa, M., Bird, C., Tinker, A., Mire-Sluis, A., and Thorpe, R.,** Qauntitative biological assays for individual cytokines, in *Cytokines, A Practical Approach,* Balkwill, F. R., Ed., IRL Press, Oxford, 1991, 309.

3. **Falk, W., Krammer, P. H., and Mannel, D. N.,** A new assay for interleukin-1 in the presence of interleukin-2, *J. Immunol. Methods,* 99, 47, 1987.
4. **Uyttenhove, C., Coulie, P. G., and Van Snick, J.,** T cell growth and differentiation induced by interleukin-HP1/IL-6, the murine hybridoma/plasmacytoma growth factor, *J. Exp. Med.,* 167, 1417, 1988.
5. **Gearing, A. J. and Thorpe, R.,** Assay of interleukin-1, in *Interleukin-1, Inflammation and Disease,* Bomford, R. and Henderson, B., Eds., Elsevier, Amsterdam, 1989, 79.
6. **Gearing, A. J., Leung, H., Bird, C. R., and Thorpe, R.,** Presence of the inflammatory cytokines IL-1, TNF, and IL-6 in preparations of monoclonal antibodies, *Hybridoma,* 8, 361, 1989.
7. **Hopkins, S. J. and Humphreys, M.,** Simple, sensitive and specific bioassay of interleukin-1, *J. Immunol. Methods,* 120, 271, 1989.
8. **Hannum, C. H., Wilcox, C. J., Arend, W. P., Joslin, F. G., Dripps, D. J., Heimdal, P. L., Armes, L. G., Sommer, A., Eisenberg, S. P., and Thompson, R. C.,** Interleukin-1 receptor antagonist activity of a human interleukin-1 inhibitor, *Nature (London),* 343, 336, 1990.
9. **Novick, D., Engelmann, H., Wallach, D., and Rubinstein, M.,** Soluble cytokine receptors are present in normal human urine, *J. Exp. Med.,* 170, 1409, 1989.
10. **Seckinger, P., Isaaz, S., and Dayer, J. M.,** Purification and biologic charcterisation of a specific tumor necrosis factor alpha inhibitor, *J. Biol. Chem.,* 264, 11966, 1989.
11. **Fanslow, W. C., Sims, J. E., Sassenfeld, H., Morrissey, P. J., Gillis, S., Dower, S. K., and Widmer, M. B.,** Regulation of allorectivity in vivo by a soluble form of the interleukin-1 receptor, *Science,* 248, 739, 1990.
12. **Engelmann, H., Novick, D., and Wallach, D.,** Tumor necrosis factor-binding proteins purified from human urine. Evidence for immunological cross-reactivity with cell surface tumor necrosis factor receptors, *J. Biol. Chem.,* 265, 1531, 1990.
13. **Engelberts, I., Stephens, S., Francot, G. J. M., Van Der Linden, C. J., and Buurman, W. A.,** Evidence for different effects of soluble TNF-receptors on various TNF measurements in human biological fluids, *Lancet,* 338, 515, 1991.
14. **James, K.,** Interactions between cytokines and α2-macroglobulin, *Immunol. Today,* 11, 163, 1990.
15. **Bendtzen, K., Svenson, M., Jonsson, V., and Hippe, E.,** Autoantibodies to cytokines — friends or foes?, *Immunol. Today,* 11, 167, 1990.
16. **Hopkins, S. J., Humphreys, M., and Jayson, M. I.,** Cytokines in synovial fluid. I. The presence of biologically active and immunoreactive IL-1, *Clin. Exp. Immunol.,* 72, 422, 1988.
17. **Cannon, J. G., van der Meer, J. W., Kwiatkowski, D., Endres, S., Lonnemann, G., Burke, J. F., and Dinarello, C. A.,** Interleukin-1 beta in human plasma: optimization of blood collection, plasma extraction, and radioimmunoassay methods, *Lymphokine Res.,* 7, 457, 1988.
18. **Series, J. J., Shapiro, D., Cameron, I., Smith, J., Fraser, W. D., Garden, O. J., and Shenkin, A.,** Rapid, sensitive detection of plasma IL-1 by extraction and bioassay, *J. Immunol. Methods,* 108, 33, 1988.
19. **Hamilton, R. G.,** Rheumatoid factor interference in immunological methods, *Monogr. Allergy,* 26, 27, 1989.
20. **Weber, T. H., Kapyaho, K. I., and Tanner, P.,** Endogenous interference in immunoassays in clinical chemistry. A review, *Scand. J. Clin. Lab. Invest.,* 50, (Suppl. 201), 77, 1990.
21. **Bormer, O. P.,** Interference of complement with the binding of carcinoembryonic antigen to soild-phase monoclonal antibodies, *J. Immunol. Methods,* 121, 85, 1989.
22. **Whicher, J. T.,** Calibration is the key to immunoassay but the ideal calibrator is unattainable, *Scand. J. Clin. Lab. Invest.,* 51, 21, 1991.
23. **Hesse, D. G., Tracey, K. J., Fong, Y., Manogue, K. R., Cerami, A., Shires, G. T., and Lowry, S. F.,** Cytokine appearance in human endotoxaemia and primate bacteraemia, *Surg. Gynecol. Obstet.,* 166, 147, 1988.
24. **Mathison, J. C., Wolfson, E., and Ulevitch, R. J.,** Participation of tumor necrosis factor in the mediation of gram negative bacterial lipopolysaccharide-induced injury in rabbits, *J. Clin. Invest.,* 81, 1925, 1988.
25. **Tracey, K. J., Beutler, B., Lowry, S. F., Merryweather, J., Wolpe, S., Milsark, I. W., Hariri, R. J., Fahey, T. J., Zentella, A., Albert, J. D., Shires, G. T., and Cerami, A.,** Shock and tissue injury induced by recombinant human cachectin, *Science,* 234, 470, 1986.
26. **Tracey, K. J., Fong, Y., Hesse, D. G., Manogue, K. R., Lee, A. T., Kuo, G. C., Lowry, S. F., and Cerami, A.,** Anti-cachectin/TNF monoclonal antibodies prevent septic shock during lethal bacteraemia, *Nature (London),* 330, 662, 1987.
27. **Bone, R. C.,** The pathogenesis of sepsis, *Ann. Int.Med.,* 115, 457, 1991.
28. **Alexander, H. R., Doherty, G. M., Buresh, C. M., Venzon, D. J., and Norton, J. A.,** A recombinant human receptor antagonist to interleukin 1 improves survival after lethal endotoxaemia in mice, *J. Exp. Med.,* 173, 1029, 1991.

29. **Starnes, H. F., Pearce, M. K., Tewari, A., Yim, J. H., Zou, J. C., and Abrams, J. S.,** Anti-IL-6 monoclonal antibodies protect against lethal Escherichia coli infection and lethal tumor necrosis factor-alpha challenge in mice, *J. Immunol.*, 145, 4185, 1990.
30. **Damas, P., Reuter, A., Gysen, P., Demonty, J., Lamy, M., and Franchimont, P.,** Tumor necrosis factor and interleukin-1 serum levels during severe sepsis in humans, *Crit. Care Med.*, 117, 875, 1989.
31. **Girardin, E. P., Grau, G. E., Dayer, J. M., Roux-Lombard, P., and Lambert, P. H.,** Tumor necrosis factor and interleukin 1 in the serum of children with severe infectious purpura, *N. Engl. J. Med.*, 319, 397, 1988.
32. **Waage, A., Halstensen, A., and Espevik, T.,** Association between tumor necrosis factor in serum and interleukin-1 in serum and fatal outcome in patients with meningococcal disease, *Lancet*, 1, 355, 1987.
33. **Offner, F., Philippe, J., Vogelaers, D., Colardyn, F., Baele, G., Baudrihaye, M., Vermeulin, A., and Leroux-Roels, G.,** Serum tumor necrosis factor levels in patients with infectious disease and septic shock, *J. Lab. Clin. Med.*, 116, 100, 1990.
34. **Debets, J. M. H., Kampmeijer, R., van der Linden, M. P. M. H., Buurman, W. A., and van der Linden, C. J.,** Plasma tumor necrosis factor and mortality in critically ill septic patients, *Crit. Care Med.*,
35. **Calandra, T., Baumgartner, J. D., Grau, G. E., Wu, M. M., Lambert, P. H., Schellekens, J., Verhoef, J., and Glauser, M. P.,** Prognostic values of tumor necrosis factor/cachectin, interleukin-1, interferon-alpha, and interferon-gamma in the serum of patients with septic shock. Swiss-Dutch J5 Immunoglobulin Study Group, *J. Infect. Dis.*, 161, 982, 1990.
36. **De Groot, M. A., Martin, M. A., Densen, P., Pfaller, M. A., and Wenzel, R. P.,** Plasma tumor necrosis factor levels in patients with presumed sepsis. Results in those treated with antilipid A antibody vs placebo, *JAMA*, 262, 249, 1989.
37. **Marano, M., Fong, Y., Moldawer, L. L., Wei, H., Calvano, S. E., Tracey, K. J., Barie, P. S., Manogue, K., Cerami, A., Shires, G. T., and Lowry, S. F.,** Serum cachectin/tumor necrosis factor in critically ill patients with burns correlates with infection and mortality, *Surg. Gynecol. Obstet.*, 170, 32, 1990.
38. **Leroux-Roels, G. and Offner, F.,** Tumor necrosis factor in sepsis, *JAMA*, 263, 1494, 1990.
39. **Spriggs, D. R., Sherman, M. L., and Michie, M.,** Recombinant human tumor necrosis factor administered as a 24 hour IV infusion. A phase I and pharmacological study, *J. Natl. Cancer Inst.*, 80, 1039, 1988.
40. **Fong, Y., Lowry, S. F., and Cerami, A.,** Cachectin/TNF: a macrophage protein that induces cachexia and shock, *J. Parent. Enterol. Nutr.*, 12, 72S, 1988.
41. **Silva, A. A. T., Appelmelk, B. J., Buurman, W. A., Bayston, K. F., and Cohen, J.,** Monoclonal antibody to endotoxin core protects mice from Escherichia coli sepsis by a mechanism independent of tumor necrosis factor and interleukin-6, *J. Infect. Dis.*, 162, 454, 1990.
42. **Smith, J. G., Magee, D. M., Williams, D. M., and Graybill, J. R.,** Tumor necrosis factor-alpha plays a role in host defense against Histoplasma capsulatum, *J. Infect. Dis.*, 162, 1349, 1990.
43. **Jablons, D. M., Mule, J. J., McIntosh, J. K., Sehgal, P. B., May, L. T., Huang, C. M., Rosenberg, S. A., and Lotze, M. T.,** IL-6/IFN-beta-2 as a circulating hormone. Induction by cytokine administration in humans, *J. Immunol.*, 142, 1542, 1989.
44. **Mule, J. J., McIntosh, J. K., Jablons, D. M., and Rosenberg, S. A.,** Antitumour activity of recombinant interleukin 6 in mice, *J. Exp. Med.*, 171, 629, 1990.
45. **Waage, A., Brandtzaeg, P., Halstensen, A., Kierulf, P., and Espevik, T.,** The complex pattern of cytokines in serum from patients with meningococcal septic shock. Association between interleukin-6, interleukin 1, and fatal outcome, *J. Exp. Med.*, 169, 333, 1989.
46. **Hack, C. E., De Groot, E. R., Felt-Bersma, R. J. F., Nuijens, J. H., Strack van Schijndel, R. J. M., Eerenberg-Belmer, A. J. M., Thijs, L. G., and Aarden, L. A.,** Increased plasma levels of interleukin-6 in sepsis, *Blood*, 74, 1704, 1989.
47. **Calandra, T., Gerain, J., Heumann, D., Baumgartner, J.-D., and Glauser, M. P.,** High circulating levels of interleukin-6 in patients with septic shock: evolution during sepsis, prognostic vaue, and interplay with other cytokines, *Am. J. Med.*, 91, 23, 1991.
48. **Moscovitz, H., Mignott, H., Cobb, P., Behrman, A., Scofer, F., Hoover, D., and Kilpatarick, L.,** Interleukin-6 (IL-6) determination in emergency department patients as a predictor of bacteraemia and infectious disease severity, paper presented at Int. Symp. IL-6: Physiopathology and Clinical Potentials, Montreux, October 21, to 23, 1991, P4.
49. **Fleisher, G. R., Saladino, R., Bachman, D., Thompson, C., Erikson, M., Levy, N., and Siber, G.,** Interleukin-6 in children with systemic bacterial infections: correlation with severity of illness and prediction of disease, paper presented at Int. Symp. IL-6: Physiopathology and Clinical Potentials, Montreux, October 21 to 23, 1991, P1.
50. **Heney, D., Lewis, I. J., Evans, S. W., Banks, R., Bailey, C. C., and Whicher, J. T.,** Interleukin-6 and its relationship to C-reactive protein and fever in children with febrile neutropenia, *J. Infect. Dis.*, 165, 886, 1992.

51. **Saez-Llorens, X., Ramilo, O., Mustafa, M. M., Mertsola, J., and McCracken, G. H.,** Molecular pathophysiology of bacterial meningitis: current concepts and therapeutic implications, *J. Pediatr.*, 116, 671, 1990.
52. **Ramilo, O., Saez-Llorens, X., Mertsola, J., Jafari, H., Olsen, K. D., Hansen, E., Yoshinaga, M., Ohkawara, S., Mariuchi, H., and McCracken, G. H.,** Tumour necrosis factor α/cachectin and interleukin-1β initiate meningeal inflammation, *J. Exp. Med.*, 172, 497, 1990.
53. **Mustafa, M. M., Lebel, M. H., Ramilo, O., Olsen, K. D., Reisch, J. S., Beutler, B., and McCracken, G. H.,** Correlation of interleukin-1β and cachectin concentrations of cerebrospinal fluid and outcome from bacterial meningitis, *J. Pediatr.*, 115, 208, 1989.
54. **Nadal, D., Leppert, D., Frei, K., Gallo, P., Lamche, H., and Fontana, A.,** Tumour necrosis factor-α in infectious meningitis, *Arch. Dis. Child.*, 64, 1274, 1989.
55. **Frei, K., Nadal, D., and Fontana, A.,** Intracerebral synthesis of tumour necrosis factor-α and interleukin-6 in infectious meningitis, *Ann. N.Y. Acad. Sci.*, 594, 326, 1990.
56. **Titus, R. G., Sherry, B., and Cerami, A.,** The involvement of TNF, IL-1 and IL-6 in the immune response to protozoan parasites, *Immunol. Today*, 12, A13, 1991.
57. **Stadnyk, A. W. and Gauldie, J.,** The acute phase protein response during parasitic infection, *Immunol. Today*, 12, A7, 1991.
58. **Kwiatkowski, D., Cannon, J. G., Monogue, K. R., Cerami, A., and Dinarello, C. A.,** Tumour necrosis factor production in falciparum malaria and its association with schizont rupture, *Clin. Exp. Immunol.*, 77, 361, 1989.
59. **Shaffer, N., Grau, G. E., Hedberg, K., Davachi, F., Lyamba, B., Hightower, A. W., Breman, J. G., and Nguyen-Dinh, P.,** Tumour necrosis factor and severe malaria, *J. Infect. Dis.*, 163, 96, 1991.
60. **Kern, P., Hemmer, C. J., Van Damme, J., Gruss, H.-J., and Dietrich, M.,** Elevated tumour necrosis factor alpha and interleukin-6 serum levels as markers for complicated Plasmodium falciprum malaria, *Am. J. Med.*, 87, 139, 1989.
61. **Pisa, P., Gennene, M., Soder, O., Ottenhoff, T., Hanson, M., and Kiesslling, R.,** Serum tumor necrosis factor levels and disease dissemination in leprosy and leishmaniasis, *J. Infect. Dis.*, 161, 988, 1990.
62. **Arditi, M., Kabata, W., and Yogev, R.,** Serum tumor necrosis factor alpha, interleukin 1-beta, p24 antigen concentrations and CD4+ cells at various stages of human immunodeficiency virus 1 infection in children, *Pediatr. Infect. Dis. J.*, 10, 450, 1991.
63. **Honda, M., Kitamura, K., Mizutani, Y., Oishi, M., Arai, M., Okura, T., Igarahi, K., Yasukawa, K., Hirano, T., Kishimoto, T., Mitsuyasu, R., Chermann, J.-C., and Tokunaga, T.,** Quantitative analysis of serum IL-6 and its correlation with increased levels of serum IL-2R in HIV-induced diseases, *J. Immunol.*, 145, 4059, 1990.
64. **Birx, D. L., Redfield, R. R., Tencer, K., Fowler, A., Burke, D. S., and Tosato, G.,** Induction of interleukin-6 during human immunodeficiency virus infection, *Blood*, 176, 2303, 1990.
65. **Miles, S. A., Rezai, A. R., Salazar-Gonzalez, J. F., Van der Meyden, M., Stevens, R. H., Logan, D. M., Mitsuyasu, R. T., Taga, T. Hirano, T., Kishimoto, T., and Martinez-Maza, O.,** AIDS Kaposi sarcoma-derived cells produce and repsond to interleukin 6, *Proc. Natl. Acad. Sci. U.S.A.*, 87, 4068, 1990.
66. **Girardin, E. P., Berner, M. E., Grau, G. E., Suter, S., Lacourt, G., and Paunier, L.,** Serum tumor necrosis factor in newborns at risk for infections, *Eur. J. Paediatr.*, 149, 645, 1990.
67. **Miller, L. C., Isa, S., LoPreste, G., Schaller, J. G., and Dinarello, C. A.,** Neonatal interleukin-1 beta, interleukin-6, and tumor necrosis factor: cord blood levels and cellular production, *J. Pediatr.*, 117, 961, 1990.
68. **Eastgate, J. A., Wood, N. G., di Giovine, F. S., Symons, J. A., Grinlinton, F. M., and Duff, G. W.,** Correlation of plasma interleukin 1 levels with disease activity in rheumatoid arthritis, *Lancet*, 2, 706, 1988.
69. **Malawista, S. E., Duff, G. W., Atkins, E., Cheung, H. S., and McCarty, D. J.,** Crystal-induced endogenous pyrogen production. A further look at gouty inflammation, *Arthritis Rheum.*, 28, 1039, 1985.
70. **Fontana, A., Hengartner, H., Weber, E., Fehr, K., Grob, P., and Cohen, G.,** Interleukin 1 activity in the synovial fluid of patients with rheumatoid arthritis, *Rheumatol. Int.*, 2, 49, 1982.
71. **Nouri, A. M., Panayi, G. S., and Goodman, S. M.,** Cytokines and the chronic inflammation of rheumatic disease. I. The presence of interleukin-1 in synovial fluids, *Clin. Exp. Immunol.*, 55, 295, 1984.
72. **Smith, J. B., Bocchieri, M. H., Sherbin Allen, L., Borofsky, M., and Abruzzo, J. L.,** Occurrence of interleukin-1 in human synovial fluid: detection by RIA, bioassay and presence of bioassay-inhibiting factors, *Rheumatol. Int.*, 9, 53, 1989.
73. **Hopkins, S. J. and Meager, A.,** Cytokines in synovial fluid. II. The presence of tumor necrosis factor and interferon, *Clin. Exp. Immunol.*, 73, 88, 1988.
74. **Westacott, C. I., Whicher, J. T., Barnes, I. C., Thompson, D., Swan, A. J., and Dieppe, P. A.,** Synovial fluid concentration of five different cytokines in rheumatic diseases, *Ann. Rheum. Dis.*, 49, 676, 1990.

75. **Di Giovine, F. S., Nuki, G., and Duff, G. W.**, Tumour necrosis factor in synovial exudates, *Ann. Rheum. Dis.*, 47, 768, 1988.
76. **Bhardwaj, N., Santhanam, U., Lau, L. L., Tatter, S. B., Ghrayeb, J., Rivelis, M., Steinman, R. M., Sehgal, P. B., and May, L. T.**, IL-6/IFN-β2 in synovial effusions of patients with rheumatoid arthritis and other arthritides. Identification of several isoforms and studies of cellular sources, *J. Immunol.*, 143, 2153, 1989.
77. **Guerne, P. A., Zuraw, B. L., Vaughan, J. H., Carson, D. A., and Lotz, M.**, Synovium as a source of interleukin 6 in vitro. Contribution to local and systemic manifestations of arthritis, *J. Clin. Invest.*, 83, 585, 1989.
78. **Hirano, T., Matsuda, T., Turner, M., Miyasaka, N., Buchan, G., Tang, B., Sato, K., Shimizu, M., Maini, R., Feldmann, M., and Kishimoto, T.**, Excessive production of interleukin 6/B cell stimulatory factor-2 in rheumatoid arthritis, *Eur. J. Immunol.*, 18, 1797, 1988.
79. **Swaak, A. J., van Rooyen, A., Nieuwenhuis, E., and Aarden, L. A.**, Interleukin-6 (IL-6) in synovial fluid and serum of patients with rheumatic diseases, *Scand. J. Rheumatol.*, 17, 469, 1988.
80. **Miltenburg, A. M., van Laar, J. M., de Kuiper, R., Daha, M. R., and Breedvedl, F. C.**, Interleukin-6 activity in paired samples of synovial fluid. Correlation of synovial fluid interleukin-6 levels with clinical and laboratory parameters of inflammation, *Br. J. Rheumatol.*, 30, 186, 1991.
81. **Dasgupta, B. and Panayi, G. S.**, Interleukin-6 in serum of patients with polymyalgia rheumatica and giant cell arteritis, *Br. J. Rheumatol.*, 29, 456, 1990.
82. **Gordon, C., Richards, N., Howie, A. J., Richardson, K., Michael, J., Adu, D., and Emery, P.**, Urinary IL-6: a marker for mesangial proliferative glomerulonephritis?, *Clin. Exp. Immunol.*, 86, 145, 1991.
83. **Hedges, S., Anderson, P., Llinden-Janson, G., de Man, P., and Svanborg, C.**, Interleukin-6 response to deliberate colonization of the human urinary tract with gram-negative bacteria, *Infect. Immun.*, 59, 421, 1991.
84. **Herbelinn, A., Urena, P., Nguyen, A. T., Zingrall, J., and Descamps-Latscha, B.**, Elevated circulating levels of interleukin-6 in patients with chronic renal failure, *Kidney Int.*, 39, 954, 1991.
85. **De La Mata, M., Meager, A., Rolando, N., Daniels, H. M., Nouri-Aria, K. T., Goka, A. K. J., Eddleston, A. L. W. F., Alexander, G. J. M., and Williams, R.**, Tumour necrosis factor production in fulminant hepatic failure: relation to aetiology and superimposed microbial infection, *Clin. Exp. Immunol.*, 82, 479, 1990.
86. **Khoruts, A., Stahnke, L., McClain, C. J., Logan, G., and Allen, J. I.**, Circulating tumor necrosis factor, interleukin-1 and interleukin-6 concentrations in chronic alcoholic patients, *Hepatology*, 13, 267, 1991.
87. **Satsangi, J., Wolstencroft, R. A., Cason, J., Ainley, C. G., Dumonde, D. C., and Thompson, R. P.**, Interleukin 1 in Crohn's disease, *Clin. Exp. Immunol.*, 67, 594, 1987.
88. **Lobo, A. J., Evans, S. W., Jones, S. C., Banks, R., Axon, A. T. R., and Whicher, J. T.**, Plasma interleukin-6 in inflammatory bowel disease, *Eur. J. Gastroenterol. Hepatol.*, 4, 367, 1992.
89. **Banks, R. E., Evans, S. W., Alexander, D., McMahon, M. J., and Whicher, J. T.**, Is fatal pancreatitis a consequence of excessive leukocyte stimulation? The role of tumor necrosis factor-α, *Cytokine*, 3, 1, 1991.
90. **Leser, H. G., Gross, V., Scheibenbogen, C., Heinisch, A., Salm, R., Lausen, M., Ruckauer, K., Andreesen, R., Farthmann, E. H., and Scholmerich, J.**, Elevation of serum interleukin-6 concentration precedes acute phase response and reflects severity of acute pancreatitis, *Gastroenterology*, 101, 782, 1991.
91. **Maimone, D., Gregory, S., Arnason, B. G., and Reder, A. T.**, Cytokine levels in the cerebrospinal fluid and serum of patients with multiple sclerosis, *J. Neuroimmunol.*, 32, 67, 1991.
92. **Sharief, M. K. and Hentges, R.**, Association between tumor necrosis factor-alpha and disease progression in patients with multiple sclerosis, *N. Engl. J. Med.*, 325, 467, 1991.
93. **Frei, K., Fredrikson, S., Fontana, A., and Link, H.**, Interleukin-6 is elevated in plasma in multiple sclerosis, *J. Neuroimmunol.*, 31, 147, 1991.
94. **Grau, G. E., Roux-Lombard, P., Gysler, C., Lambert, C., Lambert, P. H., Dayer, J. M., and Guillevin, L.**, Serum cytokine changes in systemic vasculitis, *Immunology*, 68, 196, 1989.
95. **Matsubara, T.**, Interleukin 6 activities and tumor necrosis factor-alpha levels in serum of patients with Kawasaki disease, *Arerugi — Jpn. J. Allergol.*, 40, 147, 1991.
96. **Ueno, Y., Takano, N., Kanegane, H., Yokoi, T., Yachie, A., Miyawaki, T., and Taniguchi, N.**, The acute phase nature of interleukin 6: studies in Kawasaki disease and other febrile illnesses, *Clin. Exp. Immunol.*, 76, 337, 1989.
97. **Balkwill, F., Burke, F., Talbot, D.,Tavernier, J., Osborne, R., Naylor, S., Durbin, H., and Fiers, W.**, Evidence for tumour necrosis factor/cachectin production in cancer, *Lancet*, 2, 1229, 1987.
98. **Saarinen, U. M., Koskelo, E.-K., Teppo, A. O. M., and Siimes, M. A.**, Tumour necrosis factor in children with malignancies, *Cancer Res.*, 50, 592, 1990.

99. **Jablons, D. M., McIntosh, J. K., Mule, J. J., Nordan, R. P., Rudikojj, S., and Lotz, M. T.,** Induction of interferon-β_2/interleukin-6 by cytokine administration and detection of circulating interleukin-6 in the tumour bearing state, *Ann. N.Y. Acad. Sci.,* 557, 157, 1989.
100. **Bataille, R., Jourdan, M., Zhang, X-G., and Klein, B.,** Serum levels of interleukin 6, a potent myeloma cell growth factor, as a reflection of disease severity in plasma cell dyscrasias, *J. Clin. Invest.,* 84, 2008, 1989.
101. **Lissoni, P., Barni, S., Rovelli, F., and Tacini, G.,** Lower survival in metastatic cancer patients with reduced interleukin-2 blood concentrations. Preliminary report, *Oncology,* 48, 125, 1991.
102. **Lissoni, P., Barni, S., Rovelli, F., Rescaldani, R., Rizzo, V., Biondi, A., and Tancini, G.,** Correlation of serum interleukin-2 levels, soluble interleukin-2 receptors and T lymphocyte subsets in cancer patients, *Tumori,* 76, 14, 1990.
103. **Marino, P., Cugno, M., Preatoni, A., Cori, P., Rosti, A., Frontini, L., and Circardi, M.,** Increased levels of soluble interleukin-2 receptors in serum of patients with lung cancer, *Br. J. Cancer.,* 61, 434, 1990.
104. **Ambrosetti, A., Semenzato, G., Prior, M., Chilosi, M., Vinante, F., Vincenzi, C., Zanotti, R., Trentin, L., Portuese, A., Menestrina, F., Perona, G., Agostini, C., Todeschini, G., and Pizzolo, G.,** Serum levels of soluble interleukin-2 receptor in hairy cell leukaemia: a reliable marker of neoplastic bulk, *Br. J. Haematol.,* 73, 181, 1989.
105. **Zambello, R., Pizzolo, G., Trentin, L., Agostini, C., Chisesi, T., Vinante, F., Scarselli, E., Zanotti, R., Vespignani, M., De Rossi, G., Pandolfi, F., and Semenzato, G.,** Evaluation of serum levels of soluble interleukin-2 receptor in patients with chronic lymphoproliferative disorders of T-lymphocytes, *Cancer,* 64, 2019, 1989.
106. **Pui, C.-H., Ip, S. H., Kung, P., Dodge, R. K., Berard, C. U., Crist, W. M., and Murphy, S. B.,** High serum interleukin-2 receptor levels are related to advanced disease and a poor outcome in childhood non-Hodgkin's lymphoma, *Blood,* 70, 624, 1987.
107. **Sculier, J. P., Body, J. J., Donnadieu, N., Nejai, S., Gilbert, F., Raymakers, N., and Paesmans, M.,** Pharmacokinetics of repeated i.v. bolus administration of high doses of r-met-Hu interleukin-2 in advanced cancer patients, *Cancer Chemother. Pharmacol.,* 26, 355, 1990.
108. **Chapman, P. B., Lester, T. J., Casper, E. S., Gabrilove, J. L., Wong, G. Y., Kempin, S. J., Gold, P. J., Welt, S., Warren, R. S., Starnes, H. F., Sherwin, S. A., Old, L. J., and Oettgen, H. F.,** Clinical pharmacology of recombinant human tumor necrosis factor in patients with advanced cancer, *J. Clin. Oncol.,* 5, 1942, 1987.
109. **Wills, R. J.,** Clinical pharmacokinetics of interferons, *Clin. Pharmacokinet.,* 19, 390, 1990.
110. **Simpson, M. A., Madras, P. N., Cornaby, A. J., Etienne, T., Dempsey, R. A., Clowes, G. H. A., and Monaco, A. P.,** Sequential determinations of urinary cytology and plasma and urinary lymphokines in the management of renal allograft recipients, *Transplantation,* 47, 218, 1989.
111. **Young-Fadok, T. M., Simpson, M. A., Madras, P. N., Demsey, R. A., O'Connor, K., and Monaco, A. P.,** Predictive value of pretransplant IL-2 levels in kidney transplantation, *Transplant. Proc.,* 23, 1295, 1991.
112. **Colvin, R. B., Preffer, F. I., Fuller, T. C., Brown, C., Ip, S. H., Kung, P. C., and Cosimi, A. B.,** A critical analysis of serum and urine interleukin-2 receptor assays in renal allograft recipients, *Transplantation,* 48, 800, 1989.
113. **Simpson, M. A., Young-Fadok, T. M., Madras, P. N., Freeman, R. B., Dempsey, R. A., Shaffer, D., Lewis, D., Jenkins, R. L., and Monaco, A. P.,** Sequential interleukin 2 and interleukin 2 receptor levels distinguish rejection from cyclosporin toxicity in liver allograft recipients, *Arch. Surg.,* 126, 717, 1991.
114. **McLaughlin, P. J., Aikawa, A. A., Davies, H. M., Bakran, A., Sells, R. A., and Johnson, P. M.,** Tumour necrosis factor in renal transplantation, *Transplant. Proc.,* 23, 1289, 1991.
115. **Yoshimura, N., Oka, T., and Kahan, B.,** Sequential determinations of serum interleukin 6 levels as an immunodiagnostic tool to differentiate rejection from nephrotoxicity in renal allograft recipients, *Transplantation,* 51, 172, 1991.
116. **Tono, T., Mondon, M., Yoshizaka, K., Valdivia, L. A., Nakano, Y., Gotoh, M., Ohzato, H., Doki, Y., Ogata, A., Kishimoto, T., and Mori, T.,** Interleukin 6 levels in bile following liver transplantation, *Transplat. Proc.,* 23, 630, 1991.
117. **Kraus, T., Noronha, I. L., Manner, M., Klar, E., Kuppers, P., and Otto, G.,** Clinical value of cytokine determinations for screening, differentiation, and therapy monitoring of infectious and noninfectious complications after orthotopic transplantation, *Transplat. Proc.,* 23, 1509, 1991.
118. **Chollet-Martin, S., Depoix, J. P., Hvass, U., Pansard, Y., Vissuzaine, C., and Gougerot Pocidalo, M. A.,** Raised plasma levels of tumor necrosis factor in heart allograft rejection, *Transplant. Proc.,* 22, 283, 1990.

119. **Holler, B. E., Kolb, H. J., Moller, A., Kempeni, J., Liensenfeld, S., Pechumer, H., Lehmacher, W., Ruckdeschel, G., Gleixner, B., Riedner, C., Ledderose, G., Brehm, G., Mittermuller, J., and Wilmanns, W.,** Increased serum levels of tumor necrosis factor alpha precede major complications of bone marrow transplantation, *Blood*, 75, 1011, 1990.
120. **Holler, B. E., Hintermeier-Knabe, R., Kolb, H. J., Kempeni, J., Moller, A., Liesenfeld, S., Daum, L., and Wilmanns, W.,** Low incidence of transplant-related complications in patients with chronic release of tumor necrosis factor-alpha before admission to bone marrow transplantation: a clinical correlate of cytokine desensitization?, *Pathobiology*, 59, 171, 1991.
121. **Pullicino, E. A., Carli, F., Poole, S., Rafferty, B., Malik, S. T. A., and Elia, M.,** The relationship between circulating concentrations of interleukin 6 (IL-6), tumour necrosis factor (TNF) and the acute phase response to elective surgery and accidental injury, *Lymphokine Res.*, 9, 231, 1990.
122. **Di Padova, F., Pozzi, C., Tondre, M. J., and Tritapepe, R.,** Selective and early increase of IL-1 inhibitors, IL-6 and cortisol after elective surgery, *Clin. Exp. Immunol.*, 85, 137, 1991.
123. **Cruickshank, A. M., Fraser, W. D., Burns, H. J. G., Van Damme, J., and Shenkin, A.,** Response of serum interleukin-6 in patients undergoing elective surgery of varying severity, *Clin. Sci.*, 79, 161, 1990.
124. **Nishimoto, N., Yoshizaki, K., Tagoh, H., Monden, M., Kishimoto, S., Hirano, T., and Kishimoto, T.,** Elevation of serum interleukin-6 prior to acute phase proteins on the inflammation by surgical operation, *Clin. Immunol. Immunopathol.*, 50, 399, 1989.
125. **Nijsten, M. W. N., De Groot, E. R., Ten Duis, H. J., Klasen, J., Hack, J. C., and Aarden, L. A.,** Serum levels of interleukin-6 and acute phase responses, *Lancet*, 2, 921, 1987.
126. **Pos, O., Van Der Stelt, M. E., Wolbink, G. J., Nijsten, M. W. N., Van Der Tempel, G. L., and Van Dijk, W.,** Changes in the serum concentration and the glycosylation of human α1-acid glycoprotein and α1-protease inhibitor in severely burned persons: relation to interleukin-6 levels, *Clin. Exp. Immunol.*, 82, 579, 1990.
127. **Maury, C. P. J.,** Monitoring the acute phase response: comparison of tumour necrosis factor (cachectin) and C-reactive protein responses in inflammatory and infectious diseases, *J. Clin. Pathol.*, 43, 1078, 1989.

Chapter 34

DIAGNOSTIC AND PROGNOSTIC VALUE OF INTERLEUKIN-6 MEASUREMENTS IN HUMAN DISEASE

Pravin B. Sehgal

TABLE OF CONTENTS

0-8493-6913-4/93/$0.00 + $.50

I. INTRODUCTION

During the last 5 years, it has become abundantly clear that the levels of interleukin-6 (IL-6) in body fluids are increased in a wide variety of disease states.[1-3] These include infections, neoplasia, autoimmune diseases, transplantation, and other injury states such as surgical trauma and thermal and radiation injury. A variety of different assays have been used for IL-6 measurements by various investigators. In general, these assays have proven adequate to point to differences between groups of patients with different diseases and healthy volunteers, and to suggest an overall statistical relationship between IL-6 levels and disease severity or disease outcome. It is an impression that IL-6 levels in body fluids are likely to be more informative than measurements of interleukin-1 (IL-1) or tumor necrosis factor (TNF). However, attempts to formally validate specific IL-6 assays as diagnostic or prognostic indicators that can be applied to individual patients in specific disease have begun only recently. The biochemical complexity of IL-6 as it exists in the human body is a partial reason for the difficulty in the development of ''clinical-quality'' assays for this cytokine and their evaluation in clinical practice.

An important question that has been addressed by several investigators is the relationship between local or circulating IL-6 levels and the observed alterations in acute phase plasma proteins in different diseases (see below). In general, a consensus has emerged that there is indeed an overall correlation between circulating IL-6 levels and the observed increases in indices such as body temperature and levels of C-reactive protein (CRP), α_1-antichymotrypsin (ACT), α_1-antitrypsin (AAT), fibrinogen, and others. Despite the fact that four different cytokines (IL-6, IL-11, leukemia inhibitory factor, and oncostatin M) can directly elicit most of the acute phase alterations of plasma protein synthesis in hepatocyte or hepatoma cells in culture[4-7] and that several others (IL-1, TNF and transforming growth factor-β) modulate this response in cell culutre,[8-10] polyclonal antiserum to human rIL-6 almost completely inhibits the ability of serum, cerebrospinal fluid, synovial fluid, and amniotic fluid from patients with infections or autoimmune diseases to elicit a stimulation of ACT synthesis by Hep 3B hepatoma cells,[11-13] suggesting that IL-6 remains the major cytokine in body fluids that influences acute phase plasma protein synthesis *in vivo*. The observation that most of the IL-6 in human plasma/serum appears to exist in high molecular mass complexes[14] raises an important biolgoical question: is IL-6 in such complexes biologically available to elicit hepatic plasma protein responses? The presence of high levels of complexed IL-6 in the intravascular compartment, even if biologically active, does not necessarily mean that this IL-6 has access to the extravascular tissue and the hepatocellular environment. Because the IL-6-associated proteins appear to include acute phase reactants such as CRP and fragments of complement factors C3 and C4 as well as the soluble IL-6 receptor, does the association with such proteins represent a feedback modulation of IL-6 biological function (see Chapter 16 of this volume)? It seems that, as a practical matter, the greatest difficulties with assays for IL-6 in serum/plasma, e.g., nonlinearity of enzyme-linked immunoassays (ELISAs), may occur in serum samples from individuals with strong acute phase plasma protein reactions. As IL-6 enters clinical evaluation, the reliability of assays for this cytokine in the face of marked elevations of acute phase plasma proteins is a critical question that remains to be validated.

This chapter briefly discusses IL-6 measurements as they correlate with different disease states. A key focus is to assess examples for their ability to provide diagnostic or prognostic information to the practicing clinician. Can we begin to identify disease states in which knowing the IL-6 level in a body fluid in a rapid and timely fashion could alter the management of an individual patient? The technical problems and shortcomings of IL-6 bioassays and ELISAs that are currently in use have already been summarized in Chapter 16 of this volume.

II. IL-6 IN HUMAN DISEASE

One of the first and most dramatic descriptions of the elevation of serum IL-6 levels in patients and its relationship with the increases in body temperature and the subsequent rise and fall of CRP and AAT levels was presented by Nijsten et al.[15] in 1987. Patients with extensive burns exhibited elevations in serum IL-6 levels on the day of admission that were followed the next day (1 d post-burn) by peak levels of CRP and a slower increase in AAT. There was a good overall correlation between the rise in body temperature and the serum IL-6 level. This study has formed the prototype for numerous other descriptions of IL-6 measurements in human disease and their relationship with the observed acute phase plasma protein response. Selected examples of this literature are reviewed in the following sections.

A. EXPERIMENTAL VOLUNTEER STUDIES

A series of early studies in which volunteers were administered endotoxin or TNF, or subjected to injury such as ultraviolet (UV) irradiation established the relationship beteen the IL-6 response and the elicitation of the acute phase reaction in man. Fong et al.[16] observed an increase in circulating IL-6 level within 45 to 60 min of endotoxin administration, peak levels by approximately 2 to 4 h, and a decline by 6 h. Subsequent to this, there was an increase in CRP levels as measured 20 h later. These investigators used the ability of serum from the endotoxin-treated volunteers to stimulate ACT synthesis and secretion by Hep 3B cells and the ability of anti-rIL-6 antiserum to completely neutralize this activity as the bioassay for IL-6. Jablons et al.[17] observed an increase in circulating levels of IL-6 within 30 to 60 min of the start of a bolus TNF infusion into cancer patients. Peak levels were observed by 2 to 4 h and declined thereafter. These investigators used both a plasmacytoma growth assay and the Hep 3B ACT stimulation assay to monitor IL-6 levels. It is instructive that again anti-rIL-6 antibody completely blocked the ability of serum from TNF-infused patients to stimulate ACT synthesis by Hep 3B cells. Jablons et al.[17] also documented the subsequent increase in plasma cortisol levels and in the CRP levels that followed the IL-6 induction. IL-2 caused a more delayed increase in IL-6 levels, but interferon-α administration did not elicit IL-6 induction. The ability of a continuous TNF infusion to enhance circulating IL-6 levels in volunteers was also confirmed by Brouckaert et al.,[18] who, in addition, observed a further increase in circulating IL-6 levels if the TNF-treated patients were also infused with interferon-γ.

A rather dramatic example of experimental IL-6 induction in man was described by Urbanski et al.[19] These investigators exposed volunteers to 20 min of UV irradiation under a suntan lamp (an erythema dose) and then monitored circulating IL-6 and CRP levels and body temperature. There was an increase in circulating IL-6 levels within 1 to 3 h after the UV irradiation, with a peak at approximately 12 h. Subsequent to this, there was an increase in body temperature, with a peak at 12 to 24 h. An elevation in the CRP level was first seen at 24 h, with a peak at 48 h.

In extrapolating these experimental volunteer data to various clinical settings, many investigators have relied heavily on the description by Fong et al.[20] that administration of anti-TNF antibody in the *E. coli*-baboon model blocks the subsequent induction of IL-1 and IL-6. Thus, IL-6 has been considered to be a ''secondary'' cytokine. It should be emphasized that this is not necessarily true in every clinical situation. In appropriate murine models of infection, it can be shown that IL-6 induction can precede the induction of IL-1 or TNF[21-23] and that it is the circulating IL-6 levels that correlate with the increases in acute phase plasma protein levels.

B. INFECTIONS

In one of the first papers of its kind, Houssiau et al.[24] used the 7TD1 hybridoma growth factor assay to demonstrate marked elevations of IL-6 levels in the cerebrospinal fluid (CSF) of patients with acute infection of the central nervous system. The infections included viral meningitis, bacterial meningitis, tuberculous meningitis, and herpes simplex encephalitis. IL-6 levels decreased as the infections resolved. Helfgott et al.[11] used a hepatocyte-stimulating factor assay in Hep 3B cells to document elevations of IL-6 in the CSF, synovial fluid, and serum of patients with various bacterial infections. An anti-rIL-6 antiserum blocked essentially all of the stimulation of ACT synthesis observed in Hep 3B cells exposed to these body fluids. The concentrations of IL-6 in CSF and serum were so high that this cytokine was recovered by immunoaffinity chromatography and demonstrated to consist of multiple monomeric species of molecular mass from 21 to 28 kDa by Western blotting under completely denaturing conditions.

Hack et al.[25] described marked elevations of serum IL-6 levels in the majority of patients with bacterial sepsis. IL-6 levels correlated with increases in circulating levels of complement C3a and C4a. All patients with IL-6 levels below 40 U/ml (in the B9 hybridoma growth assay) survived, whereas 89% of patients with levels exceeding 7900 U/ml died. This study raised the possibility that serum IL-6 measurements could be informative in evaluating the prognosis of a patient in septic shock. This notion has been reinforced by a recent report describing the detection of very high levels of IL-6 in patients in endotoxic shock compared to those in nonendotoxin shock.[26]

The localized production of IL-6 and other cytokines in compartmentalized infections such as meningococcal meningitis, other bacterial meningitides, septic arthritis, and chorioamnionitis has also been clearly documented.[11,13,27,28] Depending upon the circumstances, a spillover of compartmentalized IL-6 into the peripheral circulation has also been observed.[11]

Infection with the human immunodeficiency virus (HIV) and the development of the acquired immunodeficiency syndrome (AIDS) is correlated with increases in circulating IL-6 levels. Breen et al.[29] reported elevations in circulating IL-6 levels in HIV-infected patients. There was an overall correlation between circulating IL-6 levels and the severity of the AIDS syndrome. These elevations correlated with increases in CRP, α_2-macroglobulin, and immunoglobulin levels. These observations have since been confirmed by Honda and colleagues[30] and Brix et al.[31] Central nervous system infection by HIV also leads to elevations in the IL-6 levels in the CSF.[32] No relationship was observed between intrathecal immunoglobulin concentrations and IL-6 levels in the CSF.[24,32] Taken together, these data suggest that HIV infection of man may be a disease state where the measurement of serum IL-6 levels could provide valuable prognostic information.

Kern et al.[33] reported elevations in serum IL-6 and TNF levels in patients with *Plasmodium falciparum* malaria. There was an overall correlation between the levels of the two cytokines and with the extent of organ impairment. Elevations of serum IL-6 levels were observed in patients with acute hepatitis;[34] there was a correlation between IL-6 levels and increase in prothrombin time, suggesting that the more severe the hepatitis, the higher the IL-6 levels.

Romero and colleagues[13,35,36] have observed marked elevations in IL-6 levels in amniotic fluid from women with microbial invasion of the amniotic cavity. These women present in preterm labor with intact membranes or with premature rupture of membranes. An important diagnostic consideration in the management of these patients is the rapid determination of the presence of microbial invasion of the amniotic cavity. Data discussed in a later section show that this clinical situation is likely to be one in which amniotic fluid IL-6 measurements will prove to be of diagnostic and prognostic value directly to the individual patient.[39]

C. NEOPLASIA

The discussion concerning IL-6 in neoplasia can be divided into two categories. On the one hand are neoplastic diseases such as multiple myeloma and other plasma cell dyscrasias, where there is clear evidence that IL-6 (usually paracrine, sometimes autocrine) drives the proliferation of the tumor.[37-41] Generally, higher levels of circulating IL-6 were observed in individuals with more aggressive neoplasms. On the other hand,IL-6 also enters discussions about human neoplasia as the cytokine that is induced and then mediates the host response to the tumor, a response usually characterized by an increase in CRP levels, erythrocyte sedimentation rate, hypoalbuminemia, and sometimes hypergammaglobulinemia.[42-45] Castleman's disease is a well-characterized example where the systemic acute phase plasma protein response that is typical of this disease can be ameliorated by resection of the tumor nodules.[43] In this instance, the tumor secretes copious amounts of IL-6 that then elicits the alterations in hepatic plasma protein synthesis.

In patients with multiple myeloma, the administration of anti-IL-6 monoclonal antibody (mAb) decreases myeloma cell proliferation and serum CRP levels.[41] Indeed, an almost complete decline of CRP levels in such patients is considered to be indicative of adequate anti-IL-6 mAb administration and to be a prognostic indicator of ultimate responsiveness of the patient to the anti-IL-6 mAb therapy.

While the data are not as clear as with multiple myeloma and plasma cell dyscrasias, it has been proposed that IL-6 may also contribute to driving cell proliferation in other situations such as mesangial proliferative glomerulonephritis, renal cell carcinoma, and Kaposi's sarcoma.[45-47] Indeed, IL-6 has been shown to act as a strong proliferative stimulus for some epithelial cells such as mesangial cells and keratinocytes[46,48] and for Kaposi's sarcoma-derived cells[47] in cell culture.

The observation that IL-6 is an almost invariant participant in the host-tumor interaction either because the tumor cells secrete the cytokine or because it is induced in the tumor-associated stroma and inflammatory cells[44,49] leads to a more general question: is there prognostic information in the measurement of serum IL-6 levels? The suggestive answer to this question is in the affirmative. The question is particularly interesting because of the recent demonstration that tumor-induced cachexia can be due to the production of IL-6 in some murine models.[50] Indeed, higher IL-6 levels are observed in more aggressive plasma cell dyscrasias.[39-41] A recent study shows that in patients with metastatic renal carcinoma, pretreatment levels of IL-6 and CRP were higher in those who experienced progressive disease after IL-2 therapy.[45] Thus, serum IL-6 and CRP levels were adverse prognosis factors in patients with metastatic renal cell carcinoma. It was suggested that serum IL-6 level could help in the selection or stratification of patients in IL-2 trials. These data are reminiscent of studies that point to an increase in the erythrocyte sedimentation rate as a predictor of early relapse and survival in early-stage Hodgkin's disease.[51]

It is now clear that serum/plasma IL-6 levels are increased in various neoplastic disease states in a manner that generally correlates with the severity of the disease and the intensity of the host acute phase protein response. Nevertheless, a formal evaluation of whether IL-6 measurements add diagnostic or prognostic information to the other available indices (such as CRP levels) remains open.

D. AUTOIMMUNE DISEASES

Several investigators have described markedly elevated levels of IL-6 in the synovial fluids of patients with rheumatoid arthritis.[12,52-56] There was a correlation between the levels of IL-6 in synovial fluid and the observed elevations in serum IL-6 levels;[52,54,56] the former were approximately 1000-fold higher than the latter. Serum IL-6 levels correlated positively with a local disease activity score. Serum levels of IL-6 was also correlated with those of

CRP, fibrinogen, haptoglobin, complement C3, α_1-acid glycoprotein, and AAT, but not always with the erythrocyte sedimentation rate (ESR).[52,54,56] However, a clear correlation was observed between serum IL-6 levels and both CRP levels and the ESR in patients with psoriatic arthritis and ankylosing spondylitis.[56] Whether measurement of serum IL-6 levels adds any diagnostic or prognostic information to the management of rheumatoid arthritis or the other arthritides is not clear.

Elevated levels of IL-6 have been reported in patients with systemic lupus erythematosus, both in serum and in the CSF of patients with central nervous system involvement.[57,58] Similarly, serum IL-6 levels are elevated in febrile illnesses such as Kawasaki disease.[59] It is noteworthy that in this instance, IL-6 levels correlate positively with body temperature and with increases in CRP, AAT, haptoglobin, and α_1-acid glycoprotein levels.[59] Treatment with aspirin reduced the body temperature, serum IL-6 levels and CRP levels. IL-6 has also been implicated in the pathogenesis of thyroiditis, type I diabetes, and cirrhosis.[60-62] Lower IL-6 levels were observed in bronchoalveolar lavage fluids in patients with sarcoidosis.[63]

IL-6 is involved in the pathophysiology of psoriasis.[48,64,65] While the relationship between elevations in circulating IL-6 levels, the CRP levels, and the ESR in patients with cutaneous psoriasis is not yet clear, patients with psoriatic arthritis exhibit a clear relationship between elevations in IL-6 and CRP levels and the ESR.[56]

E. TRANSPLANTATION

Van Oers and colleagues were the first to draw attention to the increase in IL-6 levels in serum and urine in renal transplant recipients.[66] In contrast to healthy volunteers, both serum and urine of renal transplant recipients contained high levels of IL-6 directly after transplantation and during acute rejection episodes. During rejection episodes, elevations in serum IL-6 levels correlated with increases in serum creatinine levels. Treatment with glucocorticoids not only decreased serum creatinine levels, but also lowered both serum and urine IL-6 levels. An increase in serum and urine IL-6 levels preceeded the recurrence of a rejection episode. On the basis of these studies, it was suggested that serial measurement of IL-6, especially in urine, may be of value in monitoring renal transplant recipients. These investigators also reported the elevation of serum and urine IL-6 levels 2 d after major abdominal surgery.[66]

Other investigators have confirmed the elevation of serum IL-6 levels in renal transplant recipients, particularly during graft rejection.[67-69] The administration of OKT3 antibody in order to blunt graft rejection itself leads to the rapid induction of IL-6 in the peripheral circulation.[68] The increase in serum IL-6 levels can be used to distinguish between graft rejection and cyclosporine nephrotoxicity.[69] In the former, there is a marked increase in serum IL-6, whereas in the latter, there is no detectable IL-6 in the serum (using the MH60.BSF2 hybridoma growth assay). The use of steroids to control graft rejection in this study also lowered circulating IL-6 levels.

Symington and colleagues[70] have evaluated the relationship between serum IL-6 levels and acute graft-vs.-host disease (GVHD) and hepatorenal disease (HRD) after human bone marrow transplantation. Serial serum samples from allogeneic marrow recipients with GVHD of varying severity were evaluated. Serum IL-6 peaks were temporally related to onset of GVHD, onset of the HRD syndrome, or bilateral lung infiltration. Serum IL-6 peaks preceded GVHD onset and may thus serve as an advance warning of impending GVHD. In some patients, high serum IL-6 levels were also correlated with an increase in platelet counts, suggesting that endogenous IL-6 in the bone marrow recipient may contribute to thrombocytopoiesis.

F. TRAUMA AND OTHER CONDITIONS

A variety of different injuries have been reported to lead to elevations in IL-6 levels and the subsequent increase in levels of positive acute phase plasma proteins in the peripheral circulation. The paradigm here is the report by Nijsten et al.[15] discussed earlier. These investigators noted increases in serum IL-6 levels in patients with severe burns, followed subsequently by a rise in body temperature and CRP and AAT levels.

Elevated serum IL-6 levels have been reported in alcoholic hepatitis; the increase parallels disease severity and correlates with the increase in CRP levels.[71,72] As the hepatitis resolves, both the IL-6 and CRP levels decline gradually.

Sturk et al.[73] report a pilot study showing the relationship between serum IL-6 levels to other indices in patients with acute myocardial infarction. IL-6 levels began to increase approximately 14 h after the initial complaints and reached maximal levels after 36 h. No correlation was observed between the size of the infarction or the increase in creatine kinase levels. CRP levels began to increase after 16 h and reached a maximum after 65 h. The increase in IL-6 levels preceeded or was simultaneous with the increase in CRP levels. The maximal IL-6 levels achieved correlated with the maximal CRP levels observed. Taken together, these data indicate that IL-6 is an important endogenous mediator in eliciting the CRP response in patients with acute myocardial infarction.

In a similar vein, other investigators have reported increases in serum IL-6 levels within 1.5 h of the start of elective surgery, with peak levels at 2 to 4 h after the initial incision.[74,75] The increases in IL-6 levels preceded those in CRP and ACT levels.

III. IL-6 IN HUMAN PARTURITION

It is curious that both Biernacki[76] in 1894 and Fahraeus[77] in 1921 were essentially looking for a pregnancy test when they independently discovered the enhanced sedimentation of erythrocytes in columns of blood obtained from pregnant women and soon realized that this phenomenon was more indicative of infections and a variety of other diseases. Romero and colleagues[13,35,36] have initiated several studies to explore the role of IL-6 in human pregnancy and parturition. Because of prior experience with the measurement of TNF and IL-1 in amniotic fluid and the possibility that such cytokine data could be clinically informative, the initial focus has been to ask the following question: is there diagnostic information in an amniotic fluid IL-6 measurement? The clinical situation selected for study was preterm labor. This condition represents a serious clinical diagnostic and therapeutic dilemma. β-Adrenergic tocolytic agents such as ritodrine can result in serious maternal and fetal adverse effects, including maternal death. The presence of microbial invasion in 10 to 20% of these patients, often in a subclinical manner, further complicates the management. It is important to rapidly identify the subset of patients who have microbial invasion of the amniotic cavity as well as the subset that is not likely to respond to tocolysis. The usual Gram stain can detect only 50 to 60% of all cases with a positive amniotic fluid culture. Amniotic fluid IL-6 turns out to be a reporter cytokine *par excellence* in heralding the presence of microbial invasion of the amniotic cavity in this clinical setting.

These investigators[13,35,36,78] used a variety of assays to study IL-6 in amniotic fluid (AF) — the stimulation of ACT synthesis and secretion in Hep 3B cells, the proliferation of B9 hybridoma cells, and two different ELISAs for IL-6 (IG61/IC67 and 4IL6/5IL6). Low levels of AF IL-6 were observed in midtrimester or at term in the absence of labor. Modest increases were observed at term in the presence of labor. However, marked increases were observed in patients presenting in preterm labor who had a positive AF culture. In these patients, the concentrations of AF IL-6 were estimated to be in the 1 to 5-μg/ml range.[13] Patients with a negative AF culture but who were not responsive to tocolysis also exhibited elevations of

AF IL-6 levels. These pilot studies suggested that AF IL-6 measurements may provide diagnostic and prognostic information. While there was an overall statistical correlation, the various IL-6 assays provided numerically different data for many of the AF IL-6 samples. The 4IL6/5IL6 ELISA was selected for further study.

In order to explore the diagnostic and prognostic value of IL-6 measurements, AF samples from 146 consecutive patients presenting in preterm labor were evauated for IL-6 levels. Gram stain, and microbiological studies, and the data tested for predicting clinical outcome.[36,78] All but one patient with positive AF cultures had AF IL-6 levels in excess of 11.2 ng/ml (sensitivity, 93.7% compared to 62.5% for Gram stain). A combination of Gram stain and AF IL-6 >11.2 ng/ml had a sensitivity of 100% and a specificity of 92.3% in the diagnosis of a positive amniotic fluid culture. The ability to make this determination rapidly (perhaps with a dipstick-based ELISA for IL-6) points to the practical value of IL-6 as a diagnostic tool. In all evaluable patients who had AF IL-6 >11.2 ng/ml but a negative AF culture, histological examination of the placenta revealed chorioamnionitis. Thus, AF IL-6 levels appear to rapidly provide key diagnostic information not available in any other fashion. All patients with AF IL-6 >11.2 ng/ml delivered preterm within 48 h. Furthermore, all patients with AF IL-6 levels >6.7 ng/ml also delivered preterm. Of those with AF IL-6 levels <6.7 ng/ml, approximately 85% delivered at term. Thus, AF IL-6 levels also provide key prognostic information. These data provide justification for planning a prospective study to evaluate AF IL-6 measurements as a patient management tool in this clinical setting.

IV. COMMENTS

It is clear that there is likely to be clinically useful information in the numerous descriptions of elevations of IL-6 levels in body fluids in a wide range of different diseases. It is necessary to make the transition to a rigorous evaluation of IL-6 measurements in patient management. A prerequisite that has been as yet incompletely achieved is the validation of reliable "clinical-quality" assays for IL-6 as it exists in human body fluids.

From a biological standpoint, it is amply clear that the appearance of IL-6 in the human body correlates with the observed acute phase plasma protein response. The latter response is invariably preceded or accompanied by elevations in circulating IL-6 levels. The presence of circulating IL-6 in complexes with other proteins that could mask its biolgoical properties suggests that circulating IL-6 may not always be accompanied by an acute phase plasma protein response. These are some of the issues that will have to be evaluated in detail before IL-6 becomes a practical clinical diagnostic tool.

ACKNOWLEDGEMENTS

Research in the author's laboratory was supported by Research Grant AI-16262 from the National Institutes of Health and by a contract from the National Foundation for Cancer Research.

REFERENCES

1. **Sehgal, P. B., Grieninger, G., and Tosato, G., Eds.,** Regulation of the acute phase and immune responses: Interleukin-6, *Ann. N.Y. Acad. Sci.*, 557, 1, 1989.
2. **Van Snick, J.,** Interleukin-6: an overview, *Annu. Rev. Immunol.*, 8, 253, 1990.
3. **Hirano, T.,** Interleukin-6 and its relation to inflammation and disease, *Clin. Immunol. Immunopathol.*, 62 S60, 1992.

4. **Gauldie, J., Richards, C., Harnish, D., Lansdorp, P., and Baumann, H.,** Interferon β_2/B-cell stimulatory factor type 2 shares identity with monocyte-derived hepatocyte-stimulating factor and regulates the major acute phase protein response in the liver, *Proc. Natl. Acad. Sci. U.S.A.*, 84, 7251, 1987.
5. **Baumann, H. and Wong, G. G.,** Hepatocyte-stimulating factor II shares structural and functional identity with leukemia inhibitory factor, *J. Immunol.* 143, 1163, 1989.
6. **Baumann, H. and Schendel, P.,** Interleukin-11 regulates the hepatic expression of the same plasma protein genes as interleukin-6, *J. Biol. Chem.*, 266, 20424, 1991.
7. **Gauldie, J.,** Effect of oncostatin M on protein synthesis in liver cells. Discussion, in *Polyfunctional Cytokines: IL-6 and LIF,* Vol. 167, Metcalf, D., Ed., CIBA Foundation Symp., John Wiley & Sons, Chichester, 1992, 116.
8. **Koj, A., Kurdowska, A., Magielska-Zero, D., Rokita, H., Sipe, J. D., Dayer, J. M., Demczuk, S., and Gauldie, J.,** Limited effects of recombinant human and murine interleukin 1 and tumor necrosis factor on production of acute phase proteins by cultured rat hepatocytes, *Biochem. J.*, 244, 505, 1987.
9. **Mortensen, R. F., Shapiro, J., Lin, B.-F., Douches, S., and Neta, R.,** Interaction of recombinant IL-1 and tumor necrosis factor in the induction of mouse acute phase proteins, *J. Immunol.*, 140, 2260, 1988.
10. **Mackiewicz, A., Ganapathi, M. K., Shultz, D., Brabenec, A., Weinstein, J., Kelley, M. F., and Kushner, I.,** Transforming growth factor β_1 regulates production of acute phase proteins, *Proc. Natl. Acad. Sci. U.S.A.*, 87, 1491, 1990.
11. **Helfgott, D. C., Tatter, S. B., Santhanam, U., Clarick, R. H., Bhardwaj, N., May, L. T., and Sehgal, P. B.,** Multiple forms of IFN-β_2/IL-6 in serum and body fluids during acute bacterial infection, *J. Immunol.*, 142, 948, 1989.
12. **Bhardwaj, N., Santhanam, U., Lau, L. L., Tatter, S. B., Ghrayeb, J., Rivelis, M., Steinman, R. M., Sehgal, P. B., and May, L. T.,** IL-6/IFN-β_2 in synovial effusions of patients with rheumatoid arthritis and other arthritides: identification of several isoforms and studies of cellular sources, *J. Immunol.*, 143, 2153, 1989.
13. **Romero, R., Avila, C., Santhanam, U., and Sehgal, P. B.,** Amniotic fluid interleukin 6 in preterm labor: association with infection, *J. Clin. Invest.*, 85, 1392, 1990.
14. **May, L. T., Viguet, H., Kenney, J. S., Ida, N., Allison, A. C., and Sehgal, P. B.,** High levels of "complexed" interleukin-6 in human blood, *J. Biol. Chem.*, 267, 19698, 1992.
15. **Nijsten, M. W., DeGroot, E. R., Tenduis, H. J., Klesen, H. J., Hack, C. E., and Aarden, L. A.,** Serum levels of interleukin-6 and acute phase responses, *Lancet*, 2, 921, 1987.
16. **Fong, Y., Moldawer, L. L., Marano, M., Wei, H., Tatter, S. B., Clarick, R. H., Santhanam, U., Sherris, D., May, L. T., Sehgal, P. B., and Lowry, S. F.,** Endotoxemia elicits increased circulating β_1-IFN/IL-6 in man, *J. Immunol.*, 142, 2321, 1989.
17. **Jablons, D. M., Mule, J. J., McIntosh, J. K., Sehgal, P. B., May, L. T., Huang, C. M., Rosenberg, S. A., and Lotze, M. T.,** Interleukin-6/interferon-β_1 as a circulating hormone: induction by cytokie admnistration in man, *J. Immunol.*, 142, 1542, 1989.
18. **Brouckaert, P., Spriggs, D. R., Demetri, G., Kufe, D. W., and Fiers, W.,** Circulating interleukin 6 during a continuous infusion of tumor necrosis factor and interferon γ, *J. Exp. Med.*, 169, 2257, 1989.
19. **Urbanski, A., Schwarz, T., Neuner, P., Krutmann, J., Kirnbauer, R., Kock, A., and Luger, T. A.,** Ultraviolet light induces increased circulating interleukin-6 in humans, *J. Invest. Dermatol.*, 94, 808, 1990.
20. **Fong, Y., Tracey, K. J., Moldawer, L. L., Hesse, D. G., Manogue, K. B., Kenney, J. S., Lee, A. T., Kuo, G. C., Allison, A. C., Lowry, S. F., and Cerami, A.,** Antibodies to cachectin/tumor necrosis factor reduce interleukin-1β and interleukin-6 appearance during lethal bacteremia, *J. Exp. Med.*, 170, 1627, 1989.
21. **Havell, E. A. and Sehgal, P. B.,** Tumor necrosis factor-independent IL-6 production during murine listeriosis, *J. Immunol.*, 146, 756, 1991.
22. **Havell, E. A., Moldawer, L. L., Heflgott, D., Kilian, P. L., and Sehgal, P. B.,** Type I IL-1 receptor blockade exacerbates murine listeriosis, *J. Immunol.*, 148, 1486, 1992.
23. **Ginsberg, H., Moldawer, L. L., Sehgal, P. B., Redington, M., Kilian, P. L., Chanock, R. M., and Prince, G. A.,** A mouse model for investigating the molecular pathogenesis of adenovirus pneumonia, *Proc. Natl. Acad. Sci. U.S.A.*, 88, 1651, 1991.
24. **Houssiau, F. A., Bukasa, K., Sindic, C. J. M., Van Damme, J., and Van Snick, J.,** Elevated levels of the 26K human hybridoma growth factur (interleukin 6) in cerebrospinal fluid of patients with acute infection of the central nervous system, *Clin. Exp. Immunol.*, 71, 320, 1988.
25. **Hack, C. E., DeGroot, E. R., Felt-Fersma, R. J. F., Nuijens, J. H., Van Schijndel, R. J. M. S., Erenberg-Belmer, A. J. M., Thijs, L. G., and Aarden, L. A.,** Increased plasma levels of interleukin-6 in sepsis, *Blood*, 74, 1704, 1989.
26. **Yoshimoto, T., Nakanishi, K., Hirose, S., Hiroishi, K., Okamura, H., Takemoto, Y., Kanamaru, A., Hada, T., Tamura, T., Kakishita, E., and Higashino, K.,** High serum IL-6 levels reflect susceptible status of the host to endotoxin and IL-1/tumor necrosis factor, *J. Immunol.*, 148, 3596, 1992.

27. **Waage, A., Halstensen, A., Shalaby, R., Brandtzaeg, P., Kierulf, P., and Espevik, T.,** Local production of tumor necrosis factor α, interleukin 1 and interleukin 6 in meningococcal meningitis: relation to the inflammatory response, *J. Exp. Med.*, 170, 1859, 1989.
28. **Kono, Y., Beagley, K. W., Fujihashi, K., McGhee, J. R., Taga, T., Hirano, T., Kishimoto, T., and Kiyono, H.,** Cytokine regulation of localized inflammation: induction of activated B cells and IL-6-mediated polyclonal IgG and IgA synthesis in inflamed human gingiva, *J. Immunol.*, 146, 1812, 1991.
29. **Breen, E. C., Rezai, A. R., Nakajima, K., Beall, G. N., Mitsuyasu, R. T., Hirano, T., Kishimoto, T., and Martinez-Maza, O.,** Infection with HIV is associated with elevated IL-6 levels and production, *J. Immunol.*, 144, 480, 1990.
30. **Honda, M., Kitamura, K., Mizutani, Y., Oishi, M., Arai, M., Okura, T., Igarahi, K., Yasukawa, K., Hirano, T., Kishimoto, T., Mitsuyasu, R., Chermann, J.-C., and Tokunaga, T.,** Quantitative analysis of serum IL-6 and its correlation with increased levels of serum IL-2R in HIV-induced diseases, *J. Immunol.*, 145, 4059, 1990.
31. **Brix, D. L., Redfield, R. R., Tencer, K., Fowler, A., Burke, D. S., and Tosato, G.,** Induction of interleukin-6 during human immunodeficiency virus infection, *Blood,* 76, 2303, 1990.
32. **Laurenzi, M. A., Siden, A., Persson, M. A. A., Norkrans, G., Hagberg, L., and Chiodi, F.,** Cerebrospinal fluid interleukin-6 activity in HIV infection and inflammatory and noninflammatory diseases of the nervous system, *Clin. Immunol. Immunopathol.*, 57, 233, 1990.
33. **Kern, P., Hemmer, C. J., Van Damme, J., Gruss, H. J., and Dietrich, M.,** Elevated tumor necrosis factor alpha and interleukin-6 serum levels as markers for complicated *Plasmodium falciparum* malaria, *Am. J. Med.*, 87, 139, 1989.
34. **Sun, Y., Tokushige, K., Isono, E., Yamauchi, K., and Obata, H.** Elevated serum interleukin-6 levels in patients with acute hepatitis, *J. Clin. Immunol.*, 12, 197, 1992.
35. **Santhanam, U., Avila, C., Romero, R., Viguet, H., Ida, N., Sakurai, S., and Sehgal, P. B.,** Cytokines in normal and abnormal parturition: elevated amniotic fluid interleukin-6 levels in women with premature rupture of membranes associated with intrauterine infection, *Cytokine,* 3, 155, 1991.
36. **Romero, R., Sepulveda, W., Kenney, J. S., Archer, L. E., Allison, A. C., and Sehgal, P. B.,** Interleukin 6 assay in the detection of microbial invasion of the amniotic cavity, in *Polyfunctional Cytokines: IL-6 and LIF,* Vol. 167, Metcalf, D., Ed., CIBA Foundation Symp., John Wiley & Sons, Chichester, 1992, 205.
37. **Kawano, M., Hirano, T., Matsuda, T., Taga, T., Horii, Y., Iwato, K., Asaoku, H., Tang, B., Tanabe, O., Tanaka, H., Kuramoto, A., and Kishimoto, T.,** Autocrine generation and essential requirement of BSF-2/IL-6 for human multiple myeloma, *Nature (London),* 322, 83, 1988.
38. **Klein, B., Zhang, X. G., Jourdan, M., Content, J., Houssiau, F., Aarden, L., Piechaczyk, M., and Battaile, R.,** Paracrine rather than autocrine regulation of myeloma-cell growth and differentiation by interleukin-6, *Blood,* 73, 517, 1989.
39. **Zhang, X. G., Klein, B., and Bataille, R.,** Interleukin-6 is a potent myeloma-cell growth factor in patients with agressive multiple myeloma, *Blood,* 74, 11, 1989.
40. **Batille, R., Jourdan, M., Zhang, X. G., and Klein, B.,** Serum levels of interleukin 6, a potent myeloma cell growth factor, as a reflection of disease severity in plasma cell dyscrasias, *J. Clin. Invest.*, 84, 2008, 1989.
41. **Klein, B., Wijdenes, J., Zhang, X. G., Jourdan, M., Boiron, J. M., Brochier, J., Liautard, J., Merlin, M., Clement, C., Morel-Fournier, B., Lu, Z. Y., Mannoni, P., Sany, J., and Bataille, R.,** Murine anti-interleukin-6 monoclonal antibody therapy for a patient with plasma cell leukemia, *Blood,* 75, 1198, 1991.
42. **Hirano, T., Taga, T., Yasukawa, K., Nakajima, K., Nakano, N., Takatsuki, F., Shimizu, M., Murashima, A., Tsunasawa, S., Sakiyama, F., and Kishimoto, T.,** Human B cell differentiation factor defined by an anti-peptide antibody and its possible role in autoantibody production, *Proc. Natl. Acad. Sci. U.S.A.*, 81, 228, 1987.
43. **Yoshizaki, K., Matsuda, T., Nishimoto, N., Kuritani, T., Taeho, L., Aozasa, K., Nakahata, T., Kawai, H., Tagoh, H., Komori, T., Kishimoto, S., Hirano, T., and Kishimoto, T.,** Pathogenic significance of interleukin-6 (IL-6/BSF-2) in Castleman's disease, *Blood,* 74, 1360, 1989.
44. **Jablons, D. M., McIntosh, J. K., Mule, J. J., Nordan, R. P., Rudikoff, S., and Lotze, M. T.,** Induction of interferon-β_2≡ interleukin-6 (IL-6) by cytokine administration and detection of circulating interleukin-6 in the tumor-bearing state, *Ann. N.Y. Acad. Sci.*, 557, 157, 1989.
45. **Blay, J. Y., Negrier, S., Combaret, V., Attali, S., Goillot, E., Merrouche, Y., Mercatello, A., Ravault, A., Tourani, J. M., Moskovtchenko, J. F., Philip, T., and Favrot, M.,** Serum level of interleukin-6 as a prognostic factor in metastatic renal cell carcinoma, *Cancer Res.*, 52, 3317, 1992.
46. **Horii, Y., Muraguchi, A., Iwano, M., Matsuda, T., Hirayama, T., Yamada, H., Fujii, Y., Dohi, K., Ishikawa, H., Ohmoto, Y., Yoshizaki, K., Hirano, T., and Kishimoto, T.,** Involvement of IL-6 in mesangial proliferative glomerulonephritis, *J. Immunol.*, 143, 3949, 1989.

47. **Miles, S. A., Rezai, A. R., Salazar-Gonzalez, J. F., Vander Meyden, M., Stevens, R. H., Logan, D. M., Mitsuyasu, R. T., Taga, T., Hirano, T., Kishimoto, T., and Martinez-Maza, O.,** AIDS Kaposi's sarcoma-derived cells produce and respond to interleukin-6, *Proc. Natl. Acad. Sci. U.S.A.*, 87, 4068, 1990.
48. **Grossman, R. M., Krueger, J., Yourish, D., Granelli-Piperno, A., Murphy, D. P., May, L. T., Kupper, T. S., Sehgal, P. B., and Gottlieb, A. B.,** Interleukin-6 is expressed in high levels in psoriatic skin and stimulates proliferation of cultured human keratinocytes, *Proc. Natl. Acad. Sci. U.S.A.*, 86, 6367, 1989.
49. **Tabibzadeh, S. S., Poubourides, D., May, L. T., and Sehgal, P. B.,** Interleukin-6 immunoreactivity in human tumors, *Am. J. Pathol.*, 135, 427, 1989.
50. **Strassman, G., Fong, M., Kenney, J., and Jacob, C. O.,** Evidence for the involvement of IL-6 in experimental cancer cachexia, *J. Clin. Invest.*, 89, 1681, 1992.
51. **Henry-Amar, M., Friedman, S., Hayat, M., Somers, R., Meerwalt, J. H., Carde, P., and Burgers, J. M. V.,** Erythrocyte sedimentation rate predicts early relapse and survival in early-stage Hodgkin's disease, *Ann. Int. Med.*, 114, 361, 1991.
52. **Houssiau, F. A., Devogelaer, J. P., Van Damme, J., and De Deuxchaisnes, C. N.,** Interleukin-6 in synovial fluid and serum of patients with rheumatoid arthritis and other inflammatory arthropathies, *Arthritis Rheum.*, 31, 784, 1988.
53. **Hirano, T., Matsuda, T., Turner, M., Miyasaka, N., Buchan, G., Tang, B., Sato, K., Shimizu, M., Maini, R., Feldmann, M., and Kishimoto, T.,** Excessive production of interleukin-6/B cell stimulatory factor-2 in rheumatoid arthritis, *Eur. J. Immunol.*, 18, 1797, 1988.
54. **Swaak, A. J., Van Rooyen, A., Nieuwenhuis, E., and Aarden, L. A.,** Interleukin-6 (IL-6) in synovial fluid and serum of patients with rheumatic diseases, *Scand. J. Rheumatol.*, 17, 469,1988.
55. **Guerne, P. A., Zuraw, B. L., Vaughn, J. H., Carson, D. A., and Lotz, M.,** Synovium as a source of interleukin 6 in vivo: contribution to local and systemic manifestations of arthritis, *J. Clin. Invest.*, 83, 585, 1989.
56. **Holt, I., Cooper, R. G., and Hopkins, S. J.,** Relationships between local inflammation, interleukin-6 concentration and the acute phase protein response in arthritis patients, *Eur. J. Clin. Invest.*, 21, 479, 1991.
57. **Linker-Israeli, M., Deans, R. J., Wallace, D. J., Prehn, J., Ozeri-Chen, T., and Klinenberg, J. R.,** Elevated levels of endogenous IL-6 in systemic lupus erythematosus: a putative role in pathogenesis, *J. Immunol.*, 147, 117, 1991.
58. **Hirohata, S. and Miyamoto, T.,** Elevated levels of interleukin-6 in cerebospinal fluid from patients with systemic lupus erythematosus and central nervous system involvement, *Arthritis Rheum.*, 33, 644, 1990.
59. **Ueno, Y., Takano, N., Kanegane, H., Yokoi, T., Yachies, A., Miyawaki, T., and Taniguchi, N.,** The acute phase nature of interleukin 6: studies in Kawasaki disease and other febrile illnesses, *Clin. Exp. Immunol.*, 76, 337, 1989.
60. **Bendtzen, K., Buschard, K., Diamant, M., Horn, T., and Svenson, M.,** Possible role of IL-1, TNF-α, and IL-6 in insulin-dependent diabetes mellitus and autoimmune thyroid disease, *Lymphokine Res.*, 8, 335, 1989.
61. **Campbell, I. L., Cutri, A., Wilson, A., and Harrison, L. C.,** Evidence for IL-6 production by and effects on the pancreatic beta-cell, *J. Immunol.*, 143, 1188, 1989.
62. **Deviere, J., Content, J., Denys, C., Vandenbussche, P., Schandene, L., Wybran, J., and Dupont, E.,** High interleukin-6 serum levels and increased production by leucocytes in alcoholic liver cirrhosis — correlation with IgA serum levels and lymphokine production, *Clin. Exp. Immunol.*, 77, 221, 1989.
63. **Jones, K. P., Reynolds, S. P., Capper, S. J., Kalina, S., Edwards, J. H., and Davies, B. H.,** Measurement of interleukin-6 in bronchoalveolar lavage fluid by radioimmunoassay: differences between patients with interstitial lung disease and control subjects, *Clin. Exp. Immunol.*, 83, 30, 1991.
64. **Oxholm, A., Oxholm, P., Staberg, B., and Bendtzen, K.,** Interleukin-6 in the epidermis of patients with psoriasis before and during PUVA treatment, *Acta Dermatol. Venerol.*, 69, 195, 1989.
65. **Neuner, P., Urbanski, A., Trautinger, F., Moller, A., Kirnbauer, R., Kapp, A., Schopf, E., Schwarz, T., and Luger, T. A.,** Increased IL-6 production by monocytes and keratinocytes in patients with psoriasis, *J. Invest. Dermatol.*, 97, 27, 1991.
66. **Van Oers, M. H. J., Van Der Heyden, A. A. P. A. M., and Aarden, L. A.,** Interleukin-6 (IL-6) in serum and urine of renal transplant recipients, *Clin. Exp. Immunol.*, 71, 314, 1988.
67. **Kunzendorf, U., Brockmoller, J., Bickel, U., Jochimsen, F., Walz, G., Roots, I., and Offerman, G.,** Promotion of B cell stimulation in graft recipients through a mechanism distinct from interleukin-6 gene superinduction, *Transplantation*, 51, 1312, 1991.
68. **Bloemna, E., Ten berge, I. J. M., Suracno, J., and Wimink, J. M.,** Kinetics of interleukin-6 during OKT3 treatment in renal allogrft recipients, *Transplantation*, 50, 330, 1990.
69. **Yoshimura, N., Oka, T., and Kahane, B. D.,** Sequential determinations of serum interleukin 6 levels as an immunodiagnostic tool to differentiate rejection from nephrotoxicity in renal allograft recipients, *Transplantation*, 51, 172, 1991.

70. **Symington, F. W., Symington, B. E., Lau, P. Y., Viguet, H., Santhanam, U., and Sehgal, P. B.,** Relationship of serum IL-6 levels to acute graft-versus-host disease and hepatorenal disease after human bone marrow transplantation, *Transplantation*, 54, 457, 1992.
71. **Sheron, N., Bird, G., Goka, J., Alexander, G., and Williams, R.,** Elevated plasma interleukin-6 and increased severity and mortality in alcoholic hepatitis, *Clin. Exp. Immunol.*, 84, 449, 1991.
72. **Hill, D. B., Marsano, L., Cohen, D., Allne, J., Shedlofsky, S., and McClain, C. J.,** Increased plasma interleukin-6 concentrations in alcoholic hepatitis, *J. Lab. Clin. Med.*, 119, 547, 1992.
73. **Sturk, A., Hack, C. E., Aarden, L. A., Brouwer, M., Koster, R. R. W., and Sanders, G. T. B.,** Interleukin-6 release and the acute phase reaction in patients with acute myocardial infarction: a pilot study, *J. Lab. Clin. Med.*, 119, 574, 1992.
74. **Shenkin, A., Fraser, W. D., Series, J., Winstanley, F. P., McCartney, A. C., Burns, H. J. G., and Van damme, J.,** The serum interleukin 6 response to elective surgery, *Lymphokine Res.*, 8, 123, 1989.
75. **Pullicino, E. A., Carli, F., Poole, S., Rafferty, B., Malik, S. T. A., and Elia, M.,** The relationship between the circulating concentrations of interleukin-6 (IL-6), tumor necrosis factor (TNF) and the acute phase response to elective surgery and accidental surgery, *Lymphokine Res.*, 9, 231, 1990.
76. **Biernacki, E.,** W kwestyi wzajemnego stosunku czerwonych cialek osocza we krwi krazacej, *Gaz. Lek*, 14, 274, 1894.
77. **Fahraeus, T.,** The suspension-stability of blood, *Acta Med. Scand.*, 55, 1, 1921.
78. **Romero, R., Sepulveda, W., Kenney, J. S., Archer, L., Allison, A. C., and Sehgal, P. B.,** Amniotic fluid interleukin-6 determinations are of both diagnostic and prognostic value in preterm labor, *Am. Soc. Ob. Gyn.*, submitted.

Chapter 35

THE MEASUREMENT OF ACUTE PHASE PROTEINS AS DISEASE MARKERS

John T. Whicher, Rosamonde E. Banks, Douglas Thompson, and Stuart W. Evans

TABLE OF CONTENTS

0-8493-6913-4/93/$0.00 + $.50

I. INTRODUCTION

The plasma protein changes accompanying inflammation were first recognized by Von den Velden,[1] who, in a report in 1914, described an increase in plasma fibrinogen following experimental inflammation in animals. This finding was put to clinical use by Fahraeus[2] in 1921 following the discovery that red cell aggregation and sedimentation were enhanced by a raised plasma fibrinogen concentration. His studies gave rise to the development of the erythrocyte sedimentation rate (ESR) as a test for inflammation, which, in various forms of development, has persisted until the present time. In 1930, Tillet and Francis,[3] in their classical studies, found that the serum from patients acutely ill with pneumococcal pneumonia precipitated a polysaccharide component of the pneumococcal cell wall, the C-polysaccharide. They named the serum component C-reactive material, and it was found to be present in the acute stage of other diseases, but not during health.[4] The material was identified as a protein, C-reactive protein (CRP), and the term "acute phase" was introduced to describe the serum from patients in whom it was present.[5,6]

Over the following years, an increasing number of plasma proteins were described which increased in concentration in acute phase serum; they became known as the acute phase proteins. They are structurally and functionally diverse, and for a long time the response was thought to be a nonspecific one. However, with an increasing knowledge of the function of these proteins, it has become clear that their synthesis by the liver is mediated by inflammatory cytokines, and they have specific roles to play in inflammation and related events, as described elsewhere in this volume. There are a number of other proteins which increase in concentration in inflammation, whose sites and mechanisms of synthesis are probably very different from those of the acute phase proteins. Thus, molecules such as ferritin, β_2-microglobulin, and alkaline phosphatase, all of which increase in inflammation, are not usually termed acute phase proteins. Some proteins such as Von Willebrand factor (factor VIII) may be released from vascular endothelial cells as a result of the action of cytokines such as interleukin-1 released during inflammation.

II. KINETICS OF THE ACUTE PHASE PROTEIN RESPONSE

The rate of increase in plasma concentrations and incremental change following the inflammatory stimulus varies considerbly between the acute phase proteins. The changes seen in the plasma will also reflect the molecular size, volume of distribution, and catabolic rate of the proteins. A number of acute phase proteins demonstrate increases in catabolic rate during inflammation complicated by various additional pathologies. The most obvious examples are fibrinogen which is consumed in intravascular coagulation, haptoglobin in hemolysis, and α_1-antitrypsin in vasculitis. Such changes give rise to alterations in the normally stereotyped relationship of the acute phase proteins in either acute or chronic inflammation, producing a "disharmonic response".[7]

Proteins showing disharmonic responses due to an increased catabolism in certain circumstances are precluded as markers of inflammation, but may be used to good effect to detect other pathological processes, e.g., the decreased concentrations of complement proteins C3 and C4 are used to detect immune complex disease. A lowered plasma haptoglobin concentration is a sensitive indicator of intravascular hemolysis, and is comparable to the time-consuming and expensive tests of red cell survival. The decreased concentration of α_1-antitrypsin in vasculitis is less constant. If such proteins are used in this way, it is clear that the presence of a marked acute phase response may negate the changes due to consumption, and they should thus be measured in conjunction with another acute phase protein such as CRP.

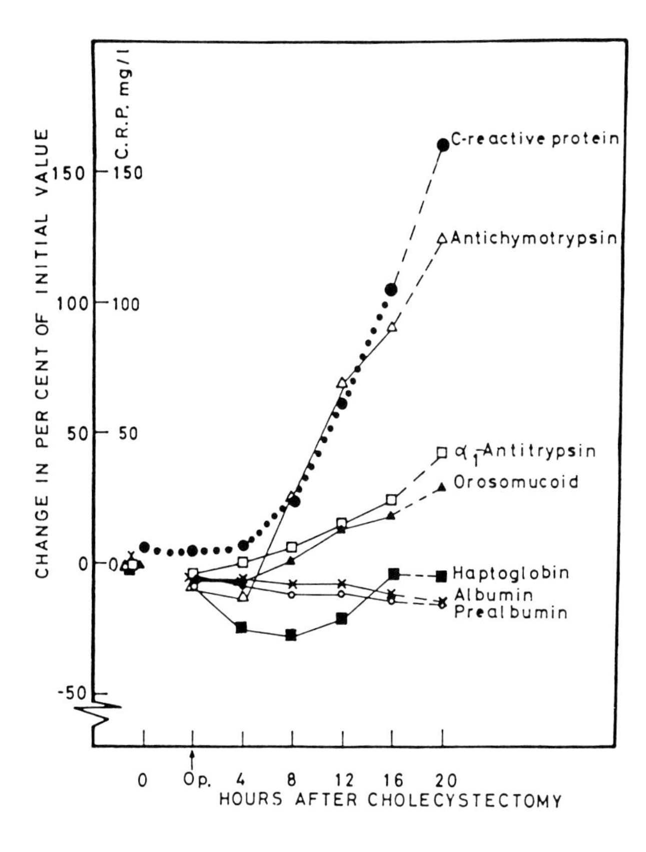

FIGURE 1. Serum albumin, prealbumin, and the acute phase reactants during the first day after cholecystectomy. The results are given as means from blood samples of six patients taken every fourth hour after the operation. CRP is given as milligrams per liter, while the percent deviation from the initial level has been used for the other variables. The broken line between 16 and 20 h denotes that samples were obtained from four of the six patients at 20 h. (From Aronson, K.-F., Ekelund, G., Kindmark, C.O., and Laurell, C.-B., *Scand. J. Clin. Lab. Invest.*, 29 (Suppl. 124), 127, 1972. With permission.)

The rate of increase of acute phase proteins following injury has been studied by a number of workers.[8,9] It is clear that the kinetics of the various acute phase proteins differ, and it is interesting to note that those proteins showing the greatest magnitude of increase also show the most rapid rise (Figure 1). Thus, CRP and SAA rise rapidly to high relative concentrations, whereas fibrinogen and haptoglobin change much more slowly. CRP shows an approximate quantitative relationship to the severity of injury in surgery[9] and to disease activity in rheumatoid arthritis.[10] In acute inflammation, the response is very stereotyped for a given species, but in chronic inflammation it is generally less marked[11] than would be expected for the apparent inflammatory activity of the disease, and the pattern also is somewhat different.[12] In some diseases, such as systemic lupus erythematosus (SLE), scleroderma, polymyositis, and ulcerative colitis, specific unresponsiveness has been suggested to occur.[13,14] The reasons for this are unclear, although defective production of cytokines in SLE and scleroderma has been suggested.[15] Infusion of prostaglandin E_1, which induces a brisk acute phase response in animals and man, fails to elicit this in many patients with scleroderma,[16] although such patients appear to be able to mount a response to infection.[17] Such observations suggest that altered patterns of cytokines or cytokine inhibitor release in different diseases may affect the acute phase response.[18]

TABLE 1
Best-Studied Human Acute Phase Proteins Classified on the Basis of Usual Magnitude of Increase in Plasma Concentration

Group I (about 50% increase)	Group II (about 2 to 4 × increase)	Group III (up to 1000 × increase)
Ceruloplasmin	α_1-Acid glycoprotein	C-reactive protein
Complement components (C3, C4)	α_1-Antitrypsin	Serum amyloid A
	α_1-Antichymotrypsin	
	Haptoglobin	
	Fibrinogen	

From Kushner, I. and Mackiewicz, A., *Dis Markers*, 5, 1, 1987. With permission.

Despite these observations, measurement of the acute phase response may be an aid to detecting the presence of inflammation and monitoring changes in its activity. It is, however, important to appreciate that other factors may affect acute phase protein concentrations in plasma. α_1-Antitrypsin and haptoglobin are notable because of common genetic polymorphisms giving rise to low serum levels. In the case of α_1-antitrypsin, more than 30 allotypes have been described,[19] with three to four variants associated with reduced hepatic synthesis. Concentrations vary from moderate to very low, being lowest in the z-variant; interestingly, these show an acute phase increase in inflammation, but frequently only to subnormal levels. In the case of haptoglobin, deficient synthesis of the α-chain results in low serum concentrations in up to 20% of American negroes.[20] Hormones may influence plasma protein concentrations through alterations in synthesis, volume of distribution, and catabolism.[21]

III. THE SENSITIVITY OF THE ACUTE PHASE PROTEIN RESPONSE TO INFLAMMATION

Sepsis is probably the strongest stimulus to the acute phase response. CRP is elevated up to 200 mg/l in bacterial infection and up to 500 mg/l in severe sepsis. Viral and parasitic infections elicit a response in CRP of up to 100 mg/l, which is more comparable with that of sterile inflammation. It is probable that the potent property of bacterial endotoxins in releasing cytokines such as IL-6 and TNF from macrophages is responsible for this. Sterile inflammation does not induce TNF, but gives rise to a clear-cut acute phase response, as evidenced by surgical injury.[22] Cytokines released by accumulating activated macrophages are thought to be responsible for the acute phase response in conditions such as infarction and trauma.

In many of the clinical conditions in which acute phase protein measurements are used, specificity is relatively poor and rarely better than 70 to 85%. This is because concommitant pathology, inducing inflammation, commonly occurs in disease. For example, in graft rejection, viral infection is particularly likely due to the increased susceptibility of patients as a result of their immunosuppressant therapy. Similarly, when investigating chorioamnionitis, fetal asphyxia and meconium aspiration will also cause a response. Despite these limitations, one of the most useful aspects of acute phase protein measurements is their high sensitivity to the presence of relatively small amounts of inflammation and their use for detecting incremental changes in a disease. Kushner and Mackiewicz[23] have classified the acute phase proteins into three groups on the basis of incremental change (Table 1). Few studies have examined the sensitivity of various acute phase proteins to minimal inflammation and to changes in inflammatory activity and diseases.

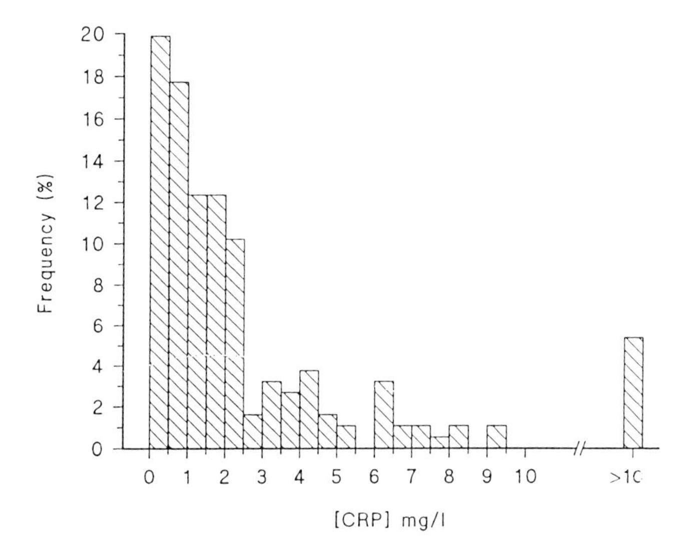

FIGURE 2. Circulating CRP levels in 186 healthy blood donors assayed using an ELISA.[26] Range, 0.09 to 35.3 mg/l; median, 1.525 mg/l.

In a study of the common cold and infuenza, Whicher et al.[24] showed that viral infection provides a significant response in SAA in more than half of the subjects, whereas CRP generally remains within the reference range. Similarly, in experimental inflammation, SAA is usually raised above the 95th percentile of the reference range, whereas CRP remains within the reference range.[25] When CRP is measured with a very sensitive assay, it can be seen to change by an amount comparable to SAA. The population distribution for CRP is markedly skewed (Figure 2). Several studies have found that the majority of CRP values for normal healthy individuals fall below the detection limits of most routinely used assays (range, 0.09 to 35.3 mg/l and median, 1.525 mg/l;[26] range, 0.07 to 29 mg/l and median, 0.8 mg/l;[27] range, 0.068 to 8.2 mg/l and median, 0.58 mg/l[28]), and it is not improbable that this enormously wide reference range (when compared to SAA) may be determined by a genetic predisposition to produce various levels of CRP. Such a notion has considerable support from the observations in mice that serum amyloid P component (SAP, the murine pentraxin) has genetically determined reference ranges in different strains.[29,30] In practice, this means that SAA is a more sensitive measure for detection of inflammation if the baseline level for an individual is not known and a population-based reference value is used.

The sensitivity of acute phase proteins to incremental changes in inflammation has been studied in inflammatory bowel disease.[31] A disease activity index based on clinical and some laboratory parameters was compared with incremental changes in four acute phase proteins. From Figure 3, it is clear that SAA is the most sensitive reflection of increments in disease activity, followed by CRP, α_1-antichymotrypsin, and α_1-acid glycoprotein.

Such information on sensitivity is essential to underpin the informed use of acute phase protein measurements in disease. Their potentially high sensitivity even to subclinical inflammation requires detailed information about the changes in concentration induced by trivial or concomitant pathological states. For example, serum CRP levels may be raised in smokers and after marathon running.[32,33] Several acute phase proteins demonstrate a clear age dependency in serum concentration. SAA has been shown to increase progressively with age in the normal population.[34] Fibrinogen also increases with age and probably forms the basis for the increasing reference range of the Westergren ESR with age.[35,36] Such observations clearly point to the possibility that low-grade inflammation is a frequent association

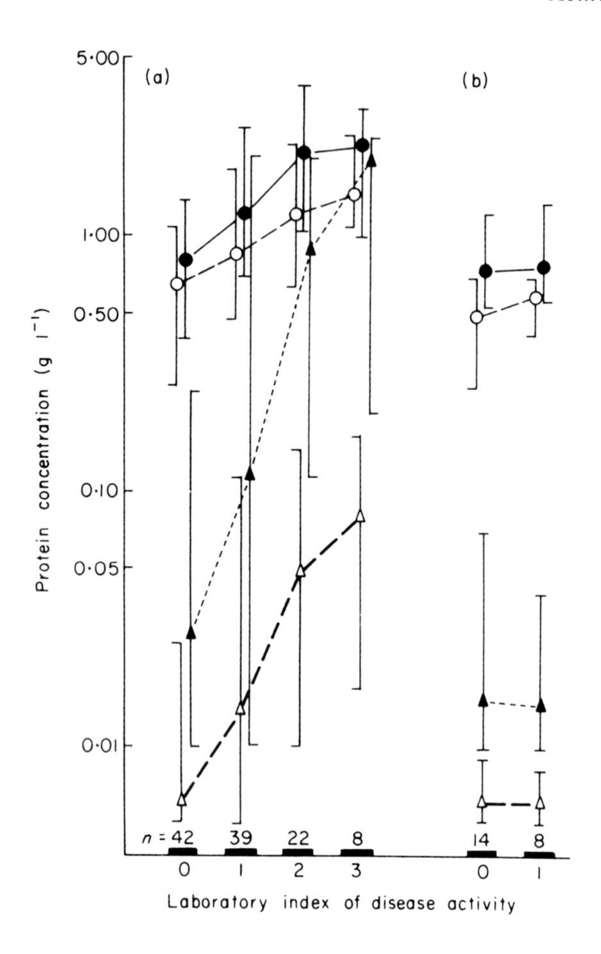

FIGURE 3. Median (grams per liter) and range concentrations of α_1-acid glycoprotein (○), α_1-antichymotrypsin (○), SAA (△), and CRP (△) plotted against laboratory index of disease activity for all patients with (a) inflammatory bowel disease and (b) irritable bowel syndrome. n, number of patients at each index. (From Chambers, R. E., Stross, P., Barry, R. E., and Whicher, J. T., *Eur. J. Clin. Invest.*, 17, 460, 1987. With permission.)

of old age in the absence of clinical symptoms. It is thus of great interest that patients with atherosclerotic vascular disease without recent thrombosis or infarction may have increased concentrations of fibrinogen,[37] with an elevated ESR and viscosity.[38] An association between increased plasma viscosity (partly due to fibrinogen) and the extent of coronary atheroma has been reported,[39] and fibrinogen is a risk factor for stroke and cardiovascular disease[40,41] in healthy adults. These changes may be due to the release of cytokines from macrophages within the intima of the vessel wall.

Normal pregnancy is associated with changes in several acute phase proteins.[42] Fibrinogen, ESR, and viscosity increase with gestational age,[43,44] and α_1-antitrypsin, ceruloplasmin, and α_2-pregnancy-associated glycoprotein (pregnancy zone protein, not an acute phase protein under normal circumstances) also show increases.[21] The effect may not be entirely hormonally controlled, as an acute phase response is seen in allogeneic pregnancy in mice, but is absent or diminished in synergeneic pregnancy.[45] Induction of cytokines in implantation is an area of considerable interest.

IV. MEASUREMENTS OF THE ACUTE PHASE PROTEIN RESPONSE

A. INTEGRATED MEASUREMENTS

The ESR remains the most widely used method for detecting and monitoring the acute phase response. There have been many modifications to the method over the years, from automation to disposable blood tubes. The selected method of the International Committee for Standardisation in Haematology[46] is based on that of Westergren, and is a measure of the rate of sedimentation of erythrocytes in a 200-mm column of blood over 1 h. Sedimentation is a consequence of erythrocyte aggregation caused by loss of the ζ-potential by which red cells repel each other. The relative contributions of plasma proteins to this phenomenon are fibrinogen 55%, α_2-macroglobulin 27%, immunoglobulins 11%, and albumin 7%. It is also influenced by the number, shape, density, and deformability of the erythrocytes. In addition, plasma lipids affect the red cell membrane, and their charges form the basis for the circadian rhythm that is observed in the ESR. The disadvantages of the test as a measure of the acute phase response are its lack of specificity and slowness of response. There is no satisfactory method of adjusting for the number of erythrocytes, and the presence of anemia severely limits its usefulness. Raised plasma immunoglobulins, monoclonal components, and immune complexes may greatly enhance the ESR in the absence of inflammatory disease. The ESR is slow to respond to changes in inflammatory activity, owing to the kinetics of the fibrinogen response upon which it is largely based, and it must be performed within 2 h of venepuncture. The advantage of the ESR is its usefulness as a screening test for disease in general. When anemia is secondary to the primary disease, it increases the apparent sensitivity of the ESR. The test is sensitive to immunoglobulins and immune complexes, which are frequently a feature of the complex hyperproteinemia of chronic disease. The test is simple, cheap, and easy to perform. A number of studies have compared the sensitivity and specificity of the ESR with other measures of the acute phase response, and in most cases it performs reasonably well when compared with specific acute phase protein measurements, as discussed later.

Plasma viscosity, like the ESR, is a measure of large proteins, especially those such as fibrinogen, α_2-macroglobulin, and immunoglobulins with axial asymmetry. The viscosity of plasma has been shown to correlate more closely with fibrinogen concentrations than the Westergren mesurement of ESR. In the presence of anemia, the correlation with ESR falls considerably. The advantages of the viscosity over the ESR are that it is not influenced by red cell characteristics, the reference range is less dependent on age and sex, and it is easier to quality control. It is similarly a rather slow responder and insensitive to changes in inflammatory activity.

B. SPECIFIC PROTEIN MEASUREMENTS

CRP is the most widely used of the acute phase proteins, and hence there is a large database of serum CRP changes in inflammation. It is easily measured using a range of assays such as enzyme immunoassay,[47] radial immunodiffusion,[48] electroimmunoassay,[49] and nephelometry or turbidimetry,[50] with lower limits of sensitivity of 6 to 8 mg/l. However this approximates to the upper part of the reference range, as previously discussed, so that radioimmunoassay (RIA), FIA,[51] or sensitive EIA[52] are desirable. Antisera are readily and cheaply available and a World Health Organization (WHO) standard is available. Because of the wide reference range, significant changes in an individual may occur within this range.

Serum amyloid A (SAA) protein antisera are difficult to raise and no internationally agreed upon reference serum exists. Also, SAA is an apolipoprotein of high density lipoprotein (HDL). This causes difficulties in comparability between different assays owing to the variable influence of the lipid moiety. This is reflected in different reference ranges, e.g., radial immunodiffusion with tissue amyloid calibrant — 1 to 30 mg/l,[53] radioimmunoassay with purified SAA standard — <200 μg/l.[54] Thus, even though serum SAA is a very sensitive acute phase protein, increasing even in response to the common cold,[24] it is not at the moment suitable for routine use.

α_1-Antichymotrypsin has been a much neglected acute phase protein in the past. Like CRP, it is easily measured by a variety of methods.[55] It takes longer to fall to normal following injury than CRP and may be of use, in conjunction with CRP measurements, in the longer-term assessment of inflammation.[56] In the absence of an international reference standard, local reference ranges must be defined.

α_1-Acid glycoprotein measurement has been much used in the past, particularly in inflammatory bowel disease, although it is rather insensitive. It is catabolized in the renal tubule, and the serum concentration is therefore increased in renal disease.[57] Due to its heavy and variable glycosylation, this protein may become more important, since it has been shown that the glycosylation pattern changes during inflammation (see Chapter 36 of this volume).

V. SPECIFIC USES OF ACUTE PHASE PROTEIN MEASUREMENTS

As indicated in the preceeding sections, the choice of acute phase protein is limited and is also dependent on the clinical situation. Although elevated acute phase proteins do signify the presence of inflammation, whether caused by trauma, tumor, infection, infarction, or immunologically mediated damage (although giving no indication of the site of inflammation), the reverse is much less certain. The inflamation of mild chronic tissue damage, localized disease, or certain diseases including ulcerative colitis or SLE is, for unknown reasons, often associated with serum acute phase protein concentrations within the reference range.[58] For these reasons, it is essential to interpret acute phase protein measurements against the full clinical background of the patient.

Acute phase protein measurement may be used to detect organic disease, assess the activity or extent of disease and monitor therapy, and detect infection in circumstances where microbiological diagnosis is slow, difficult, or impossible. In the first 24 h of the onset or resolution of inflammation, a fast-reacting and sensitive test such as the measurement of CRP is essential. After this time, the nature of the protein changes that accompany inflammation becomes more complex. The negative acute phase reactants albumin, transferrin, and prealbumin fall in concentration, and anemia and an immune response frequently develop. In this chronic phase of inflammation, integrated measures such as the ESR may be more useful and certainly complement measurement of specific acute phase protein levels such as CRP.

A. TISSUE NECROSIS

Necrosis rapidly results in a neutrophil and macrophage infiltration to the vascularized periphery of the damaged area. The resorption of necrotic tissue is affected by phagocytosis. Cells in the vicinity of necrotic tissue, such as activated macrophages, fibroblasts, epithelial cells, endothelial cells, and keratinocytes release cytokines, and an acute phase protein response, sometimes including fever and leukocytosis, often accompanies tissue necrosis such as myocardial infarction. Various studies have examined the relationship of acute phase protein increase to the extent of the infarct and the subsequent prognosis.[59-62] As with other forms of inflammation, CRP has been found to be the most sensitive reactant (when compared with orosomucoid, α_1-antitrypsin, and haptoglobin).[63] Increased levels of CRP are detectable in some patients within 1 to 2 h of the onset of chest pain,[64] although the rapid time course suggests that pain does not coincide with the start of tissue damage. Peak concentrations of CRP occur on the third or fourth day, suggesting ongoing macrophage activation (when compared to the 48-h peak of surgical injury).[63,65]

In practical clinical terms, a rise in CRP in the presence of suggested symptoms appears to be a sensitive indicator of myocardial infarction, was present in 49 of 50 patients with acute myocardial infarction and in all of 100 patients with significant Q-wave ECG changes.[66,67] The fall to normal concentrations does not occur for 7 to 10 d, thus potentially providing a marker of damage after the creatine kinase MB isoenzyme has returned to normal (usually after about 3 d). The rise in fibrinogen following infarction may have implications for blood viscosity[68,69] and myocardial perfusion, and interestingly, is predictive of a poor prognosis.[70]

Burns are a potent stimulus to the acute phase response which correlates with the extent of tissue destruction in the acute stage. Later in the course of recovery, raised acute phase proteins may reflect infection or extensive colonization of necrotic tissue by bacteria.[71]

B. MALIGNANCY

Fever and an acute phase protein response are common features of a wide variety of neoplastic processes. The underlying mechanisms may vary from cytokine release from the tumor, as postulated for myelomatosis,[72] to tumor infiltration by activated macrophages. A considerable number of studies have attempted to relate the levels of acute phase proteins to tumor load, metastasis, and prognosis. There is a consensus that raised and increasing concentrations of acute phase proteins generally bode a poor prognosis, although frequently such changes can be correlated with infection. While of interest, there is no evidence to suggest a clinical value for acute phase protein measurements.[73-76]

C. TYPE A AMYLOIDOSIS

The fibrils of A-type or secondary amyloidosis are derived from the proteolytic degradation of SAA. A prolonged increase in plasma SAA such as occurs in juvenile rheumatoid arthritis undoubtedly predispose to AA amyloid fibril formation.[77] There may in addition exist certain allotypes of the protein which are more amyloidogenic. The measurement of SAA is probably no more predictive of amyloid than that of CRP, as the two proteins change in parallel.[78] In established amyloidosis, patients with a more intense acute phase response move rapidly to acute renal failure, and normalization of acute phase proteins should be a therapeutic objective.

D. SEPSIS

Bacterial infection, especially with Gram-negative organisms, is a particularly potent stimulus to the acute phase protein response (see above). Viral and parasitic infections also often elicit a response of more moderate extent.[63,79-81] Thus, local bacterial infection (e.g., of a wound) results in CRP concentrations of between 100 and 200 mg/l (sterile trauma such as major surgery causes similar levels). Septicemia typically elicits CRP responses up to 500 mg/l or higher. Parasitic and viral infections cause moderate to mild responses, usually of less than 100 mg/l, unless there is severe tissue damage.

The presence, type, and extent of infection is, of course, usually investigated microbiologically. However, there are several circumstances in which diagnosis and monitoring may be slow, difficult, or impossible due to inaccessibility of the infected organ or tissue,or because an answer is required before the results of the microbiological tests are available.[82] Suitably raised CRP concentrations give a strong indication of bacterial (usually Gram negative) infection under these circumstances, and should fall in response to appropriate antibiotic treatment.

1. Distinguishing Bacterial from Viral Infection

In childhood, fever is most frequently due to acute viral infections. It may, however, be difficult to distinguish this from bacterial sepsis such as otitis media, tonsillitis, bronchitis, and cystitis, and antibiotics are frequently prescribed unnecessarily. Up to 38% of adenovirus infections and 20% of influenza infections were reported to have CRP concentrations >40 mg/l.[83] In a thorough study, Putto et al.[84] showed that both CRP and ESR could be used to discriminate viral from bacterial infection in children who had been ill for more than 12 h. As would be expected, sensitivity and specificity varied with the levels of the measures that were used for discrimination. Thus, an ESR >30 mm/h had a sensitivity of 91% and a specificity of 89% for detecting bacterial infection, while a CRP concentration >40 mg/l had a sensitivity of 79% and a specificity of 90%.

Similar studies have been carried out comparing viral and bacterial meningitis. Peltola[85] concluded that a CRP concentration above 20 mg/l in children was indicative of bacterial, rather than viral, meningitis. Benjamin et al.[86] concluded that while this was generally true, serum CRP was too nonspecific in unselected cases. As many children with the clinical picture of sepsis undergo lumbar puncture as part of their diagnostic evaluation, in selected cases with negative microscopic evidence of infection in CSF and low-grade pleiocytosis, it might, however, be valuable. De Beer et al.[87] studied a population in which tuberculous meningitis was prevalent and found that it caused raised concentrations of CRP intermediate between those of viral and bacterial infection. They, however, concluded that a CRP concentration 20 mg/l strongly indicated viral meningitis, while levels above 100 mg/l strongly indicated a bacterial origin. Interestingly, measurements of CRP in the CSF only showed a sensitivity of 66%, although a sensitivity of 100% for the presence of bacterial infection was shown in another study.[88]

2. Monitoring High-Risk Groups

There are a number of situations in which bacterial infection is likely to occur and carries a poor prognosis if not recognized rapidly or treated adequately, none more so than in the immunocompromised patients who frequently pose difficult diagnostic problems. Under these circumstances, the greater sensitivity of sequential measurements can be exploited by regular monitoring of a sensitive acute phase reactant such as CRP. It was originally shown in 1979[89] that CRP concentrations increased to above 100 mg/l in neutropenic patients with bacterial infection. More moderate increases are associated with cytotoxic drug therapy and administration of blood products. Many further studies have been performed.[90-96] If CRP concentrations are between 30 and 40 mg/l for 48 h after the onset of fever, infection is unlikely to be the cause. In patients treated with antibiotics, if the CRP concentration does not fall below 100mg/l, an appropriate response has not occurred and therapy must be maintained or changed. Very similar conclusions can be drawn in graft-vs.-host disease.[80,97]

Another risk group for bacterial infection which is difficult to diagnose are mothers and infants following premature rupture of the membranes (PROM). It has been suggested that elevated maternal serum CRP may be useful as an early predictor of clinical chorioamnionitis, although this is disputed.[98-104] However, interpretation is complicated by the use of assays of varying sensitivity, the use of differing discriminatory levels of CRP resulting in differing specificities and sensitivities, few serial samples, and little data on CRP concentrations during normal labor. More recently, a study using a more sensitive assay and sequential samples throughout labor and postpartum showed no significant difference between the serum CRP concentrations of women with PROM and the wide range of CRP values found in the acute phase response normally occurring postpartum, negating the use of CRP as a predictor for chorioamnionitis.[105]

Cord blood CRP concentrations are raised in neonates with subsequently proven bacterial or fungal infections.[106] However, specificity is poor owing to concomitant pathologies such as fetal asphyxia, distress, shock, or muconial aspiration, all of which raise CRP concentrations,[107] so that such measurements are of little clinical value.

3. Anatomically Closed Infections

An enormous number of infectious conditions have been studied in patients of all ages to assess the changes in acute phase proteins and their usefulness. The very first publications on disease-related changes in CRP by Ash[4] in 1933 described elevated levels in two of three children with acute pyelitis. However, there are relatively few circumstances in which such measurements have provided sufficient useful clinical information to become part of the diagnostic armamentarium.

Infection is an important complication of major abdominal surgery. In the absence of such complications, CRP rises from about 6 h after incision, peaks at 48 h, and returns to normal with a halving of concentration about every 48 h thereafter.[9,108] This is true even in patients who have an acute phase response prior to the surgery, provided the positive lesion is resected. Surgical complications such as wound infection, and deep infections such as subclinic abscesses, usually result in a sustained elevation of CRP which may be used to indicate their presence. Such an approach is clearly much more effective if serial measurements are made, or serial samples frozen to be measured retrospectively if a clinical complication occurs. It is important, however, to appreciate that complications such as tissue necrosis and thromboembolism can also cause elevations in acute phase proteins and must be distinguished clinically.

E. RHEUMATOLOGY

The importance of the acute phase proteins in rheumatology is to be expected, since the rheumatic diseases are probably the best example of inflammation as the major pathological feature. In some conditions such as rheumatoid arthritis (RA), a large volume of tissue is affected, while in others (e.g., ankylosing spondylitis) only discrete areas of tissue are affected. This variability is reflected in the usefulness of acute phase protein measurements in the rheumatic diseases.

In patients with RA, an acute phase response is nearly always present, in many cases reflecting disease activity. The question of which of CRP, the ESR, or plasma viscosity is the best assessment of disease activity is still open to debate,[109] although in general, single measurements of the acute phase response may be of limited use prognostically. Studies have shown an association between a persistent acute phase response in patients with RA and a more severe disease progression,[110,111] leading to the suggestion that the effectiveness of treatment could be determined by the measurement of acute phase proteins. McConkey et al.[112] showed that rest, analgesics, nonsteroidal antiinflammatory drugs, and local measures had little or no effect on acute phase protein concentrations. They postulated that such therapies only altered symptomatology and not the underlying inflammation. In contrast, the disease-modifying antitheumatic drugs (DMARDs) such as gold, sulfasalazine, and D-penicillamine caused acute phase protein concentrations to fall.[113-115] However, the relationship between therapy, disease progression, and acute phase reactants is still not clear, although based on existing studies,[116] it appears that in those patients in whom a consistent effective suppression of the acute phase response is achieved, radiological progression is subsequently reduced. More recently, however, it has been reported that serum α_1-antichymotrypsin reflects disease activity, being elevated even in those patients in whom serum CRP was normal despite clinical evidence of disease activity.[117]

Polymyalgia rheumatica is a disease of the elderly characterized by severe morning stiffness associated with diffuse systemic symptoms such as depression and malaise. In most cases, both serum CRP and the ESR are markedly elevated, and there is some controversy as to which of these is the better index of disease activity.[118,119] If untreated, approximately 30% of patients will develop cranial arteritis, with serious risk to eyesight. Measurement of CRP may be used to monitor the effectiveness of treatment since it falls to normal at a rate reflecting clinical improvement,[118] while the ESR may respond more slowly. Neither CRP nor the ESR can be used to predict a relapse, however.

Arthralgia is a common nonspecific symptom which can be caused by a variety of local, systemic, and psychogenic factors. The finding of elevated acute phase proteins may confirm organic disease. Similarly, back pain is usually nonspecific. The presence of an acute phase response can give a strong indication of organic disease such as ankylosing spondylitis or cancer. In ankylosing spondylitis, serum CRP may be elevated before the diagnosis is clinically obvious,[120] although there is currently no consensus about the relationship of CRP or the ESR to disease activity.

Even in SLE and other diseases where an acute phase response is mild or absent, such as systemic sclerosis,[13] measurement of CRP may be useful in distinguishing exacerbations (with minimal acute phase protein response) from infection (when CRP will be significantly increased).[121] It has been proposed that in SLE, a serum CRP concentration exceeding 60 mg/l is strongly indicative of infection,[122] wheres concentrations <30 mg/l make it unlikely that severe infection is present. However, the use of CRP to discriminate between intercurrent infection and disease exacerbation in SLE is not absolute[122,123] and caution should be observed.

Thus, the value of measuring acute phase proteins in rheumatic diseases lies particularly in (1) early detection of synovitis and monitoring of the response to therapy, (2) differential diagnosis of arthralgia, myalgia, and atypical back pain, and (3) detection of intercurrent infection in some connective tissue diseases.

F. GASTROENTEROLOGY

The differentiation between functional and inflammatory bowel disease and assessing the response of inflammatory disease to treatment is a major problem to the gastroenterologist. A number of acute phase proteins have been used for this purpose. Early studies suggested that α_1-acid glycoprotein was the most sensitive marker,[124] but a more recent study found SAA to be a more sensitive marker of inflammatory bowel disease than α_1-acid glycoprotein or CRP.[31] However SAA and α_1-acid glycoprotein lack specificity, whereas CRP showed a very high specificity and only slightly less sensitivity and was therefore recommended, although α_1-antichymotrypsin was very similar in sensitivity and specificity. In patients with irritable bowel syndrome, there were no significant changes in acute phase proteins, whereas the differentiation between Crohn's disease and ulcerative colitis was less clear-cut. SAA and CRP concentrations were significantly lower in ulcerative colitis than in Crohn's disease when mild, but there was no significant difference when ulcerative colitis was severe. This is in agreement with an earlier study[125] which showed that all Crohn's disease patients studied had elevated ESR and CRP, while 50% of ulcerative colitis patients had an increase in these markers. None of the patients with functional bowel syndrome had elevated ESR or CRP. Because a significant proportion of patients with ulcreative colitis have a normal serum CRP, the measurement of this acute phase protein is less useful in the diagnosis and monitoring of this condition than in Crohn's disease. It has been suggested that a high serum CRP concentration (>50 mg/l) in gastrointestinal disease may be indicative of Crohn's disease rather than ulcerative colitis.[126]

More recently, Calvin et al.[56] assessed the relative merits of the acute phase proteins in addition to the ESR in a retrospective study of 170 patients, of whom 130 had gastrointestinal disease. They showed that the combination of serum CRP and α_1-antichymotrypsin measurement provided the best screen for inflammatory bowel disease.

VI. SUMMARY

Acute phase proteins remain unsurpassed as markers of inflammation, although in the future it is conceivable that some cytokines may prove more useful in certain circumstances. It is obvious from the overview here that the particular measurement of the acute phase response chosen must be selected with care and that the limitations described must be borne in mind for any given clinical situation.

REFERENCES

1. **Von den Velden, R.,** Die blutgerinnung nach parenteraler zufuhr von eiweisskorpern, *Dstch. Arch. J. Klin. Med.,* 41, 298, 1914.
2. **Fahraeus, R.,** The suspension stability of the blood, *Acta Med. Scand.,* 55, 1, 1921.
3. **Tillet, W. S. and Francis, T.,** Serological reactions in pneumonia with a non-protein somatic fraction of pneumonococcus, *J. Exp. Med.,* 52, 561, 1930.
4. **Ash, R.,** Non-specific precipitins of fraction C in acute infections, *J. Infect. Dis.,* 53, 89, 1933.
5. **Abernethy, T. J. and Avery, O. T.,** The occurrence during acute infection of a protein not normally present in blood. I. Distribution of the reactive protein in patients' sera and the effect of calcium on the flocculation reaction with the C polysaccharide of pneumococcus, *J. Exp. Med.,* 73, 173, 1941.
6. **MacLeod, C. M. and Avery, O. T.,** The occurrence during acute infections of a protein not normally present in the blood. III. Immunological properties of the C-reactive protein and its differentiation from normal blood proteins, *J. Exp. Med.,* 73, 183, 1941.
7. **Laurell, C.-B.,** The use of electroimmunoassay for determining specific proteins as a supplement to agarose gel electrophoresis, *J. Clin. Pathol.,* 28(Suppl. 6), 22, 1975.
8. **Aronsen, K.-F., Ekelund, G., Kindmark, C.-O., and Laurell, C. B.,** Sequential changes of plasma proteins after surgical trauma, *Scand. J. Clin. Lab. Invest.,* 29(Suppl. 124), 127, 1972.
9. **Colley, C. M., Fleck, A., Goode, A. W., Muller, B. R., and Myers, M. A.,** Early time course of the acute phase protein response in man, *J. Clin. Pathol.,* 36, 203, 1983.
10. **Mallya, R. K., Vergani, D., Tee, D. E. H., Bevis, L., De Beer, F. C., Berry, H., Hamilton, E. D. B., Mace, B. E. W., and Pepys, M. B.,** Correlation in rheumatoid arthritis of concentrations of plasma C3d, serum rheumatoid factor, immune complexes and C-reactive protein with each other and with clinical features of disease activity, *Clin. Exp. Immunol.,* 48, 747, 1982.
11. **Yorston, D., Whicher, J., Chambers, R., Klouda, P., and Easty, D.,** The acute phase response in acute anterior uveitis, *Trans. Opthalmol. Soc. U.K.,* 104, 166, 1985.
12. **Engler, R.,** Observations on plasma proteins, in *Plasma Protein Pathology,* Peeters, H. and Wright, P. H., Eds., Pergamon Press, Oxford, 1979, 13.
13. **Whicher, J. T., Bell, A. M., Martin, M. F. R., Marshall, L. A., and Dieppe, P. A.,** Prostaglandins cause an increase in serum acute phase proteins in man, which is diminished in systemic sclerosis, *Clin. Sci.,* 66, 165, 1984.
14. **Pepys, M. B.,** Serum C-reactive protein, serum amyloid P-component and serum amyloid A protein in autoimmune diseae, *Clin. Immunol. Allergy,* 1, 77, 1981.
15. **Whicher, J. T., Gilbert, A. M., Westacott, C., Hutton, C., and Dieppe, P. A.,** Defective production of leucocytic endogenous mediator (interleukin 1) by peripheral blood leucocytes of patients with systemic sclerosis, systemic lupus erythematosus, rheumatoid arthritis and mixed connective tissue disease, *Clin. Exp. Immunol.,* 65, 80, 1986.

16. **Whicher, J. T., Bell, A. M., Martin, M. F. R., Marshall, L. A., and Dieppe, P. A.,** Prostaglandins cause an increase in serum acute phase proteins in man, which is diminished in systemic sclerosis, *Clin. Sci.*, 66, 165, 1984.
17. **Chellingsworth, M., Scott, M. G. I., Crockson, P. A., and Bacton, P. A.,** "Normal" acute phase response in systemic sclerosis, *Br. Med. J.*, 289, 946, 1984.
18. **Whicher, J. T., Westacott, C. I., and Dieppe, P. A.,** Defective acute phase response in systemic sclerosis: disease mechanisms and consequences, in *Protides of the Biological Fluids, Proceedings of the 34th Colloquium*, Peeters, H., Ed., Pergamon Press, New York, 1986, 605.
19. **Fagerhol, M. K. and Cox, D. W.,** The Pi polymorphism — genetic, biochemical and clinical aspects of human α_1-antitrypsin, *Adv. Hum. Genet.*, 11, 1, 1981.
20. **Putnam, F. W.,** Haptoglobin, in *The Plasma Proteins, Structure, Function and Genetic Control*, Vol. 2, 2nd ed., Putnam, F. W., Ed., Academic Press, New York, 1975, 1.
21. **Laurell, C.-B. and Rannevik, G.,** A comparison of plasma protein changes induced by Danazol, pregnancy and estrogens, *J. Clin. Endocrinol. Metab.*, 49, 719, 1979.
22. **Pullicino, E. A., Carli, F., Poole, S., Rafferty, B., Malik, S. T. A., and Elia, M.,** The relationship between the circulating concentrations of interleukin 6 (IL-6), tumour necrosis factor (TNF) and the acute phase response to elective surgery and accidental injury, *Lymphokine Res.*, 9, 231, 1990.
23. **Kushner, I. and Mackiewicz, A.,** Acute phase proteins as disease markers, *Dis. Markers*, 5, 1, 1987.
24. **Whicher, J. T., Chambers, R. E., Higginson, J., Nashef, L., and Higgins, P. G.,** Acute phase response of serum amyloid A protein and C-reactive protein to the common cold and influenza, *J. Clin. Pathol.*, 38, 312, 1985.
25. **Chambers, R. E., Hutton, C. W., Dieppe, P. A., and Whicher, J. T.,** A comparative study of C reactive protein and serum amyloid A protein in experimental inflammation, *Ann. Rheum. Dis.*, 50, 677, 1991.
26. **Banks, R. E., Thompson, D., and Whicher, J. T.,** unpublished observations, 1991.
27. **Shine, B., De Beer, F. C., and Pepys, M. B.,** Solid phase radioimmunoassays for human C-rective protein, *Clin. Chim. Acta*, 117, 13, 1981.
28. **Claus, D. R., Osmand, A. P., and Gewurz, H.,** Radioimmunoassay of human C-reactive protein and levels in normal sera, *J. Lab. Clin. Med.*, 87, 120, 1976.
29. **Mortensen, R. F., Beisel, K., Zelezik, N. J., and Le, P. T.,** Acute phase reactants of mice. II. Strain dependence of serum amyloid P-component (SAP) levels and response to inflammation, *J. Immunol.*, 130, 885, 1983.
30. **Pepys, M. B., Baltz, M. L., Gomer, K., Davies, A. S., and Doenhoff, M.,** Serum amyloid P component is an acute phase reactant in the mouse, *Nature (London)*, 278, 259, 1979.
31. **Chambers, R. E., Stross, P., Barry, R. E., and Whicher, J. T.,** Serum amyloid A protein compared with C-reactive protein, alpha 1-antichymotrypsin and alpha 1-acid glycoprotein as a monitor of inflammatory bowel disease, *Eur. J. Clin. Invest.*, 17, 460, 1987.
32. **Strachan, A. F., Noakes, T. D., Kotzenberg, G., Nel, A. E., and De Beer, F. C.,** C-reactive protein concentrations durig long distance running, *Br. Med. J.*, 289, 1249, 1984.
33. **Das, I.,** Raised C-reactive protein levels in serum from smokers, *Clin. Chim. Acta*, 153, 9, 1985.
34. **Rosenthal, C. J. and Franklin, E. C.,** Variation with age and disease of an amyloid A related serum component, *J. Clin. Invest.*, 55, 746, 1975.
35. **Meade, T. W., Chakrabarti, R., Haines, A. P., North, W. R. S., and Stirling, Y.,** Characteristics affecting fibrinolytic activity and plasma fibrinogen concentrations, *Br. Med. J.*, 1, 153, 1979.
36. **Milman, N., Graudal, N., and Andersen, H. C.,** Acute phase reactants in the elderly, *Clin. Chim. Acta*, 176, 59, 1988.
37. **Pilgeram, L. O.,** Relation of plasma fibrinogen concentration changes to human arteriosclerosis, *J. Appl. Physiol.*, 16, 660, 1961.
38. **Stuart, J., George, A. J., Davies, A. J., Aukland, A., and Hurlow, R. A.,** Haematological stress syndrome in atherosclerosis, *J. Clin. Pathol.*, 34, 464, 1981.
39. **Lowe, G. D. O., Drummond, M. M., Lorimer, A. R., Hutton, I., Forbes, C. D., Prentice, C. R. M., and Barbenel, J. C.,** Relationship between coronary artery disease and blood viscosity, *Br. Med. J.*, 280, 673, 1980.
40. **Wilhelmsen, L., Svardsudd, K., Korsan-Bengsten, K., Larsson, B., Welin, L., and Tibblin, G.,** Fibrinogen as a risk factor for stroke and myocardial infarction, *N. Engl.J. Med.*, 311, 501, 1984.
41. **Meade, T. W., North, W. R. S., Chakrabarti, R., Stirling, Y., Haines, A. P., Thompson, S. G., and Brozovic, M.,** Haemostatic function and cardiovascular death: early results of a prospective study, *Lancet*, 1, 1050, 1980.
42. **Studd, J.,** The plasma proteins in pregnancy, *Clin. Obstet. Gynaecol.*, 2, 285, 1975.
43. **Gram H. C.** On the causes of the variations of the sedimentation of the corpuscles and the formation of the crusta phlogistica ("size", buffy coat") on the blood, *Arch. Int. Med.*, 28, 312, 1921.

44. **Plass, E. D. and Matthew, C. W.,** Plasma protein fractions in normal pregnancy, labor and puerperium, *Am. J. Obstet. Gynecol.*, 12, 347, 1926.
45. **Waites, G. T., Bell, A. M., and Bell, S. C.,** Acute phase serum proteins in syngeneic and allogeneic mouse pregnancy, *Clin. Exp. Immunol.*, 53, 225, 1983.
46. **International Committee for Standardization in Haematology,** Recommendation for measurement of erythrocyte sedimentation rate of human blood, *Am. J. Clin. Pathol.*, 68, 505, 1977.
47. **Voller, A., Bartlett, A., and Bidwell, D. E.,** Enzyme immunoassays with specific reference to ELISA techniques, *J. Clin. Pathol.*, 31, 507, 1978.
48. **Mancini, G., Vaerman, J. P., Carbonara, A. O., and Heremans, J. F.,** A single radial immunodiffusion method for the immunological quantitation of proteins, in *Protides of the Biological Fluids*, Peeters, H., Ed., Elsevier, Amsterdam, 1964, 370.
49. **Laurell, C.-B.,** Electroimmunoassay, *Scand. J. Clin. Lab. Invest.*, 29 (Suppl. 124), 21, 1972.
50. **Whicher, J. T., Price, C. P., and Spencer, K.,** Immunonephelometric and immunoturbidimetric assays for proteins, *CRC Crit. Rev. Clin. Lab. Sci.*, 18, 213, 1983.
51. **Siboo, R. and Kulisek, E.,** A fluorescent immunoassay for the quantification of C-reactive protein, *J. Immunol. Methods*, 23, 59, 1978.
52. **Salonen, E. M.,** A rapid and sensitive solid-phase enzyme immunoassay for C-reactive protein, *J. Immunol. Methods*, 48, 45, 1982.
53. **Chambers, R. E. and Whicher, J. T.,** Quantitative radial immunodiffusion assay for serum amyloid A protein, *J. Immunol. Methods*, 59, 95, 1983.
54. **Brandwein, S. R., Medsger, T. A., Jr., Skinner, M., Sipe, J. D., Rodnan, G. P., and Cohen, A. S.,** Serum amyloid A protein concentration in progressive systemic sclerosis (scleroderma), *Ann. Rheum. Dis.*, 43, 586, 1984.
55. **Calvin, J. and Price, C. P.,** Measurement of serum $alpha_1$ antichymotrypsin by immunoturbidimetry, *Ann. Clin. Biochem.*, 23, 206, 1986.
56. **Calvin, J., Neale, G., Fotherby, K. J., and Price, C. P.,** The relative merits of acute phase proteins in the recognition of inflammatory conditions, *Ann. Clin. Biochem.*, 25, 60, 1988.
57. **Laurell, C.-B.,** Acute phase proteins — a group of protective proteins, in *Recent Advances in Clinical Biochemistry*, No. 3, Price, C. P. and Alberti, K. G. M. M., Eds., Churchill Livingstone, New York, 1985, 103.
58. **Whicher, J. T. and Dieppe, P. A.,** Acute phase proteins, *Clin. Immunol. Allergy*, 5, 425, 1985.
59. **Shainken-Kestenbaum, R., Winikoff, Y., and Cristal, N.,** Serum amyloid A concentrations during the course of acute ischaemic heart disease, *J. Clin. Pathol.*, 39, 635, 1986.
60. **Lotstrom, G.,** Comparison between reactions of acute phase serum with pneumococcus C-polysaccharide and with pneumococcus type 27, *Br. J. Exp. Pathol.*, 25, 21, 1944.
61. **Losner, S., Volk, B. W., and Wilensky, N. D.,** Fibrinogen concentration in acute myocardial infarction; comparison of clot density determination of fibrinogen with erythrocyte sedimentation rate, *Arch. Int. Med.*, 93, 231, 1954.
62. **Rabinowitz, M. A., Shookhoff, C., and Douglas, A. H.,** The red cell sedimentation time in coronary occlusion, *Am. Heart J.*, 7, 52, 1931.
63. **Voulgari, F., Cummins, P., Gardecki, T. I., Beeching, N. J., Stone, P. C., and Stuart, J.,** Serum levels of acute phase and cardiac proteins after myocardial infarction, surgery and infection, *Br. Heart J.*, 48, 352, 1982.
64. **Kushner, I., Broder, M. L., and Karp, D.,** Control of the acute phase response. Serum C-reactive protein kinetics after acute myocardial infarction, *J. Clin. Invest.*, 61, 235, 1978.
65. **De Beer, F. C., Hind, C. R. K., Fox, K. M., Allan, R. M., Maseri, A., and Pepys, M. B.,** Measurement of serum C-reactive protein concentration in myocardial ischaemia and infarction, *Br. Heart J.*, 47, 239, 1982.
66. **Levinger, E. L., Levy, H., and Elster, S. K.,** Study of C-reactive protein in the sera of patients with acute myocardial infarction, *Ann. Int. Med.*, 46, 68, 1957.
67. **Kozonis, M. C. and Gurevin, I.,** The value of the C-reactive protein determination in coronary artery disease, *Ann. Int. Med.*, 46, 79, 1957.
68. **Ditzel, J., Bang, H. O., and Thorsen, N.,** Myocardial infarction and whole-blood viscosity, *Acta Med. Scand.*, 183, 577, 1968.
69. **Biro, G. P., Beresford-Kroeger, D., and Hendry, P.,** Early deleterious hemorheologic changes following acute experimental coronary occlusion and salutary antihyperviscosity effect of hemodilution with stroma-free hemoglobin, *Am. Heart J.*, 103, 870, 1982.
70. **Haines, A. P., Howart, D., North, W. R. S., Goldenberg, E., Stirling, Y., Meade, T. W., Raftery, E. B., and Miller-Craig, M. W.,** Haemostatic variables and the outcome of myocardial infarction, *Thromb. Haemost.*, 50, 800, 1983.

71. **Kohn, J.,** Occurrence and behaviour of C-reactive protein in burns, in *Protides of Biolgoical Fluids*, Peeters, H., Ed., Elsevier, Amsterdam, 1961, 315.
72. **Bataille, R., Jourdan, M., Zhang, X. G., and Klein, B.,** Serum levels of interleukin-6, a potent myeloma cell growth factor, as a reflection of disease severity in plasma cell dyscrasias, *J. Clin. Invest.*, 84, 2008, 1989.
73. **Weinstein, P. S., Skinner, M., Sipe, J. D., Lokich, J. J., Zamcheck, N., and Cohen, A. S.,** Acute phase proteins or tumour markers: the role of SAA, SAP, CRP and CEA as indicators of metastasis in a broad spectrum of neoplastic diseases, *Scand. J. Immunol.*, 19, 193, 1984.
74. **Cooper, E. H. and Stone, J.,** Acute phase reactant proteins in cancer, *Adv. Cancer Res.*, 30, 1, 1979.
75. **Vaughan Hudson, B., MacLennan, K. A., Bennett, M. H., Easterling, M. J., Vaughan Hudson, G., and Jelliffe, A. M.,** Systemic disturbance in Hodgkin's disease and its relation to histopatholgoy and prognosis (BNLI report no. 30), *Clin. Radiol.*, 38, 257, 1987.
76. **Friedman, S., Henry-Amar, M., Cosset, J.-M., Carde, P., Hayat, M., Dupouy, N., and Tubiana, M.,** Evolution of erythrocyte sedimentation rate as predictor of early relapse in posttherapy early-stage Hodgkin's disease, *J. Clin. Oncol.*, 6, 596, 1988.
77. **De Beer, F. C., Mallya, R. K., Fagan, E. A., Lanham, J. G., Hughes, G. R. V., and Pepys, M. B.,** Serum amyloid-A protein concentration in inflammatory diseases and its relationship to the incidence of reactive systemic amyloidosis, *Lancet*, 2, 231, 1982.
78. **Falck, H. M., Maury, C. P. J., Teppo, A.-M., and Wegelius, O.,** Correlation of persistently high serum amyloid A protein and C-reactive protein concentrations with rapid progression of secondary amyloidosis, *Br. Med. J.*, 286, 1391, 1983.
79. **Gronn, M., Slordahl, S. H., Skrede, S., and Lie, S. O.,** C-reactive protein as an indicator of infection in the immunosuppressed child, *Eur. J. Paediatr.*, 145, 18, 1986.
80. **Walker, S. A., Rogers, T. R., Riches, P. G., White, S., and Hobbs, J. R.,** Value of serum C-reactive protein measurement in the management of bone marrow transplant recipients. I. Early transplant period, *J. Clin. Pathol.*, 37, 1018, 1984.
81. **Cooper, E. H., Forbes, M. A., and Hambling, M. H.,** Serum β2-microglobulin and C reative protein concentrations in viral infections, *J. Clin. Pathol.*, 37, 1140, 1984.
82. **Hanson, L. A., Jodal, U., Sabel, K.-G., and Wadsworth, C.,** The diagnostic value of C-rective protein, *Paediatr. Infect. Dis.*, 2, 87, 1983.
83. **Ruuskanen, O., Putto, A., Sarkkinen, H., Meurman, O., and Irjala, K.,** C-reactive protein in respiratory virus infections, *J. Pediatr.*, 107, 97, 1985.
84. **Putto, A., Ruuskanen, O., Meurman, O., Ekblad, H., Korvenranta, H., Mertsola, J., Peltola, H., Sarkkinen, H., Viljanen, M. K., and Halonen, P.,** C-reactive protein in the evaluation of febrile illness, *Arch. Dis. Child.*, 61, 24, 1986.
85. **Peltola, H. O.,** C-reactive protein for rapid monitoring of infections of the central nervous system, *Lancet*, 1, 980, 1982.
86. **Benjamin, D. R., Opheim, K. E., and Brewer, L.,** Is C-reactive protein useful in the management of children with suspected bacterial meningitis?, *Am. J. Clin. Pathol.*, 81, 779, 1984.
87. **De Beer, F. C., Kirsten, G. F., Gie, R. P., Beyers, N., and Strachan, A. F.,** Value of C-reactive protein measurement in tuberculous, bacterial and viral meningitis, *Arch. Dis. Child.*, 59, 653, 1984.
88. **Corrall, C. J., Pepple, J. M., Moxon, E. R., and Hughes, W. T.,** C-reactive protein in spinal fluid of children with meningitis, *J. Pediatr.*, 99, 365, 1981.
89. **Mackie, P. H., Crockson, R. A., and Stuart, J.,** C-reactive protein for rapid diagnosis of infection of leukaemia, *J. Clin. Pathol.*, 32, 1253, 1979.
90. **Rose, P. E., Johnson, S. A., Meakin, M., Mackie, P. H., and Stuart, J.,** Serial study of C-reactive protein during infection in leukaemia, *J. Clin. Pathol.*, 34, 263, 1981.
91. **Harris, R. I., Stone, P. C., Hudson, A. G., and Stuart, J.,** C reactive protein rapid assay techniques for monitoring resolution of infection in immunosuppressed patients, *J. Clin. Pathol.*, 37, 821, 1984.
92. **Schofield, K. P., Voulgari, F., Gozzard, D. I., Leyland, M. J., Beeching, N. J., and Stuart, J.,** C-reactive protein concentration as a guide to antibiotic therapy in acute leukaemia, *J. Clin. Pathol.*, 35, 866, 1982.
93. **Williams, M., McCallum, J., and Dick, H. M.,** The detection of infection in leukaemia by serial measurement of C-reactive protein, *J. Infect.*, 4, 139, 1982.
94. **Peltola, H., Saarinen, U. M., and Siimes, M. A.,** C-reactive protein in rapid diagnosis and follow-up of bacterial septicaemia in children in leukaemia, *Paediatr. Infect. Dis.*, 2, 370, 1983.
95. **Starke, I. D., De Beer, F. C., Donnelly, J. P., Catovsky, D., Goldman, J. M., Galton, D. A. G., and Pepys, M. B.,** Serum C-reactive protein levels in the management of infection in acute leukaemia, *Eur. J. Cancer and Clin. Oncol.*, 20, 319, 1984.
96. **Gozzard, D. I., French, E. A., Blecher, T. E., and Powell, R. J.,** C-reactive protein levels in neutropenic patients with pyrexia, *Clin. Lab. Haematol.*, 7, 307, 1985.

97. **Rowe, I. F., Worsley, A. M., Donnelly, P., Catovsky, D., Goldman, J. M., Galton, D. A. G., and Pepys, M. B.,** Measurement of serum C reactive protein concentration after bone marrow transplantation for leukaemia, *J. Clin. Pathol.*, 37, 263, 1984.
98. **Hawrylyshun, P., Bernstein, P., Milligan, J. E., Soldin, S., Pollard, A., and Papsin, F. R.,** Premature rupture of membranes: the role of C-reactive protein in the prediction of chorioamnionitis, *Am. J. Obstet. Gynecol.*, 147, 240, 1983.
99. **Romem, Y. and Artal, R.,** C-reactive protein as a predictor for chorioamnionitis in cases of premature rupture of the membranes, *Am. J. Obstetr. Gynecol.*, 150, 546, 1984.
100. **Ismail, M. A., Zinaman, M. J., Lowensohn, R. I., and Moawad, A. H.,** The significance of C-reactive protein levels in women with premature rupture of membranes, *Am. J. Obstet. Gynecol.*, 151, 541, 1985.
101. **Salzer, H. R., Genger, H., Muhar, U., Lischka, A., Schatten, C., and Pollak, A.,** C-reactive protein: an early marker for neonatal bacterial infection due to prolonged rupture of amniotic membranes and/or amnionitis, *Acta Obstet. Gynecol. Scand.*, 66, 365, 1987.
102. **Farb, H. F., Arnesen, M., Geistler, P., and Knox, G. E.,** C-reactive protein as a predictor of infectious morbidity with premature rupture of membranes, *Obstet. Gynecol.*, 62, 49, 1983.
103. **Fisk, N. M., Fysh, J., Child, A. G., Gatenby, P. A., Jefferey, H., and Bradfield, A. H.,** Is C-reactive protein really useful in preterm premature rupture of the membranes?, *Br. J. Obstet. Gynaecol.*, 94, 1159,1987.
104. **Ernest, J. M., Swain, M., Block, S. M., Nelson, L. H., Hatjis, C. G., and Meis, P. J.,** C-reactive protein: a limited test for managing patients with preterm labor or preterm rupture of membranes?, *Am. J. Obstet. Gynecol.*, 156, 499, 1987.
105. **De Villiers, W. J., Louw, J. P., Strachan, A. F., Etsebeth, S. M., Shephard, E. G., and De Beer, F. C.,** C-reactive protein and serum amyloid A protein in pregnancy and labour, *Br. J. Obstet. Gynecol.*, 97, 725, 1990.
106. **Sabel, K. G. and Wadsworth, C.,** C-reactive protein (CRP) in early diagnosis of neonatal septicaemia, *Acta Paediatr. Scand.*, 68, 825, 1979.
107. **Ainbender, E., Cabatu, E. E., Guzman, D. M., and Sweet, A. Y.,** Serum C-reactive protein and problems of newborn infants, *J. Pediatr.*, 101, 438, 1982.
108. **Fischer, C. L., Gill, C., Forrester, M. G., and Nakamura, R.,** Quantitation of "acute phase proteins" postoperatively. Value in detection and monitoring of complications, *Am. J. Clin. Pathol.*, 66, 840, 1976.
109. **McKenna, F.,** Clinical and laboratory assessment of outcome in rheumatoid arthritis, *Br. J. Rheumatol.*, 27 (Suppl. 1), 12, 1988.
110. **McConkey, B., Crockson, R. A., and Crockson, A. P.,** The assessment of rheumatoid arthritis. A study based on measurements of the serum acute phase reactants, *Q. J. Med.*, 41, 115, 1972.
111. **Amos, R. S., Constable, T. J., Crockson, R. A., and McConkey, B.,** Rheumatoid arthritis: relation of serum C-reactive protein and erythrocyte sedimentation rates to radiographic changes, *Br. Med. J.*, 1(6055), 195, 1977.
112. **McConkey, B., Crockson, R. A., Crockson, A. P., and Wilkinson, A. R.,** The effects of some anti-inflammatory drugs on the acute phase proteins in rheumatoid arthritis, *Q. J. Med.*, 42, 785, 1973.
113. **McConkey, B., Davies, P., Crockson, R. A., Crockson, A. P., Butler, M., Constable, T. J., and Amos, R. S.,** Effects of gold, dapsone, and prednisone on serum C-reactive protein and haptoglobin and the erythrocyte sedimentation rate in rheumatoid arthritis, *Ann. Rheum. Dis.*, 38, 141, 1979.
114. **McConkey, B., Amos, R. S., Durham, S., Forster, P. J. G., Hubball, S., and Walsh, L.,** Sulphasalazine in rheumatoid arthritis, *Br. Med. J.*, 280, 442, 1980.
115. **Hind, C. R. K., Winearls, C. G., and Pepys, M. B.,** Correlation of disease activity in systemic vasculitis with serum C-reactive protein measurement. A predictive study of 38 patients, *Eur. J. Clin. Invest.*, 15, 89, 1985.
116. **Situnayake, R. D.,** Can "disease modifying" drugs influence outcome in rheumatoid arthritis?, *Br. J. Rheumatol.*, 27 (Suppl. 1), 55, 1988.
117. **Chard, M. D., Calvin, J., Price, C. P., Cawston, C. E., and Hazleman, B. L.,** Serum α_1-antichymotrypsin concentration as a marker of disease activity in rheumatoid arthritis, *Ann. Rheum. Dis.*, 47, 665, 1988.
118. **Mallya, R. K., Hind, C. R. K., Berry, H., and Pepys, M. B.,** Serum C-reactive protein in polymyalgia rheumatica, *Arthritis Rheum.*, 28, 383, 1985.
119. **Kyle, V., Cawston, T. E., and Hazleman, B. L.,** Erythrocyte sedimentation rate and C reactive protein in the assessment of polymyalgia rheumatica/giant cell arteritis on presentation and during follow-up, *Ann. Rheum. Dis.*, 48, 667, 1989.
120. **Cowling, P., Ebringer, R., Cawdell, D., Ishii, M., and Ebringer, A.,** C-reactive protein, ESR, and klebsiella in ankylosing spondylitis, *Ann. Rheum. Dis.*, 39, 45, 1980.

121. **Becker, G. J., Waldburger, M., Hughes, G. R. V., and Pepys, M. B.,** Value of C-reactive protein measurement in the investigation of fever in systemic lupus erythematosus, *Ann. Rheum. Dis.*, 39, 50, 1980.
122. **Pepys, M. B., Lanham, J. G., and De Beer, F. C.,** C-reactive protein in SLE, *Clin. Rheum. Dis.*, 8, 91, 1982.
123. **Mackiewicz, A., Marcinkowska-Pieta, R., Ballou, S., Mackiewicz, S., and Kushner, I.,** Microheterogeneity of alpha1-acid glycoprotein in the detection of intercurrent infection in systemic lupus erythematosus, *Arthritis Rheum.*, 30, 513, 1987.
124. **Cooke, W. T., Fowler, D. I., Cox, E. V., Gaddie, R., and Meynell, M. J.,** The clinical significance of seromucoids in regional ileitis and ulcerative colitis, *Gastroenterology,* 34, 910, 1985.
125. **Shine, B., Berghouse, L., Lennard Jones, J. E., and Landon, J.,** C-reactive protein as an aid in the differentiation of functional and inflammatory bowel disorders, *Clin. Chim. Acta,* 148, 104, 1985.
126. **Fagan, E. A., Dyck, R. F., Maton, P. N., Hodgson, H. J. F., Chadwick, V. S., Petrie, A., and Pepys, M. B.,** Serum levels of C-reactive protein in Crohn's disease and ulcerative colitis, *Eru. J. Clin. Invest.*, 12, 351, 1982.

Chapter 36

GLYCOFORMS OF α_1-ACID GLYCOPROTEIN AS DISEASE MARKERS

Andrzej Mackiewicz and Aleksander Górny

TABLE OF CONTENTS

0-8493-6913-4/93/$0.00 + $.50

I. INTRODUCTION

Determination of acute phase protein (APP) concentrations in a serum has proven to be very useful in the management of patients with a large variety of pathological conditions. C-reactive protein (CRP) and serum amyloid A (SAA) were found to provide the most valuable information for practitioners. However, their clinical applications have certain limitations, e.g., the elevated serum levels of these proteins are seen in the flare-up of some rheumatic diseases and intercurrent infections in the course of these diseases.[1,2] Such instances create a problem of differential diagnosis which is very often crucial for therapeutic decisions.

Recently, another phenomenon of the acute phase response, changes of the glycosylation profile of APP, has been recognized.[1-14] These changes concern *N*-linked glycans of APP and are referred to as microheterogeneity. Two types of microheterogeneity have been distinguished: major microheterogeneity, which reflects the number of branches on *N*-linked glycans, and minor microheterogeneity, which refers to variation in the sialic acid or fucose content.[15] Two different types of changes of the major microheterogeneity of APP in patient sera have been observed. Type I is characterized by an increase in the relative amount of biantennary over more branched glycans, seen in patients with "acute" inflammatory processes, and type II is characterized by a decrease in the relative amount of biantennary over more branched glycans; this type has been seen in patients with a number of chronic inflammatory states.[9,13] However, there are some exceptions to this classification. In some disorders such as systemic lupus erythematosus (SLE), altered glycosylation of APP is not seen, while an increase of APP concentration is observed.[1] Very recently, van Dijk et al.[16] have demonstrated inflammation-related minor microheterogeneity changes showing that the expression of sialylated Lewis X structures, NeuAcα2$\rightarrow$Galβ1$\rightarrow$4(Fucα1$\rightarrow$3)GlcNAc-, on α_1-acid glycoprotein (AGP) is increased markedly during various acute inflammatory processes.

Agarose crossed-affinity immunoelectrophoresis (CAIE) with lectins as ligands has been widely used for studies of the microheterogeneity of APP in patient sera.[1-14,16-19] Application of the lectin concanavalin A (Con A) allowed determination of a major microheterogeneity.[15] Con A separates glycoforms of plasma proteins bearing various amounts of biantennary glycans.[20] Forms possessing tri- and/or tetraantenary heteroglycans do not react with the lectin in CAIE, forms having on biantennary unit react weakly, forms bearing two biantennary structures react strongly, etc.[21] By assessing the relative amounts of a particular Con A glycoform of the APP studied, it is possible to calculate the Con A reactivity coefficient (RC) for a given glycoprotein. An increase of the RC value corresponds to the type I alterations and a decrease to type II changes.

Originally, Nicollet et al.[3] showed increased reactivity of AGP with Con A in sera from four patients with "acute inflammatory states". Raynes[4] made a similar observation in sera of patients with sepsis and acute pancreatitis, and showed that other APP such as α_1-proteinase inhibitor, α_1-antichymotrypsin, or ceruloplasmin also demonstrated increased Con A reactivity. The same type of changes of glycosylation profile was seen in other acute inflammatory processes such as burns,[22] infected newborn,[23] acute mediastinitis,[24] and intercurrent infection in the course of rheumatoid arthritis (RA) or SLE.[1,2] Moreover, Pos et al.,[22] in sera of burned patients, found different kinetics of increase of AGP-Con A rectivity and α_1-proteinase inhibitor-Con A reactivity. In contrast, in patients having active RA, ankylosing spondylitis (AS), or polymyalgia rheumatica, decreased AGP-Con A reactivity was found.[2,7,6,8,14] In addition, the reactivity of Con A with other APP in sera of RA patients showed the same pattern of changes as AGP.[12] Similarly, Hansen et al.[5,6] observed a decrease of AGP-Con A reactivity in sera of patients with chronic colorectal inflammation, including ulcerative colitis or Crohn's disease. The same type of changes was found in sera of patients with

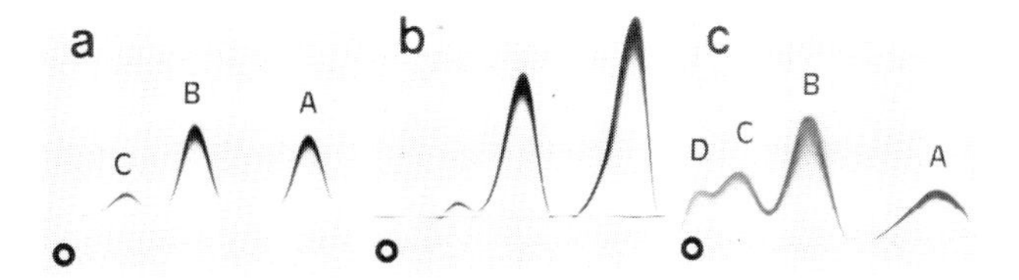

FIGURE 1. Crossed-affinity immunoelectrophoresis with Con A as a ligand of α_1-acid glycoprotein in serum of (a) healthy individual, (b) patient with active rheumatoid arthritis (type II glycosylatin alterations), (c) patient with systemic lupus erythematosus and intercurrent infection (type I). AGP glycoform A, nonreactive with Con A; B, reactive weakly; C, reactive; D, reactive strongly with Con A.

liver damage,[25] pregnancy, or estrogen administration.[26] However, these changes are probably not related to the acute phase response, but, rather, are affected by sex hormones.

In this chapter, we demonstrate some of the potential clinical applications of the determination of the major microheterogeneity (glycoforms) of AGP, a glycoprotein which shows the most marked alterations of glycosylation pattern.

II. DETERMINATION OF AGP GLYCOFORMS

AGP major microheterogeneity was studied by means of CAIE according to Bøg-Hansen,[27] with minor modifications.[28,29] Experiments were performed in 1% agarose (HSA agarose, M_r of -0.13, Litex, Denmark) in 0.02 *M* Tris-barbital-lactate buffer, pH 8.6. The first-dimension gel contained 50 μ*M* of Con A (type V, 53F-7255 Sigma Chemical Co., St. Louis, MO). The second-dimension gel included goat anti-human AGP and 5% α-methyl-mannoside. Electrophoresis in the first dimension was run at 10 V/cm until an albumin bromophenol blue complex migrated 7 cm, and in the second dimension at 1.5 V/cm overnight. The area under the precipitates corresponding to a particular variant was determined by planimetry, and the relative amounts of different forms were expressed as percentages of the total. A reactivity coefficient (RC) of AGP-Con A was calculated for each sample according to the formula: sum of all lectin-reactive variants/lectin-nonreactive variant.

CAIE with Con A revealed four glycoforms of AGP: variant A, nonreactive with Con A; variant B, weakly reactive; variant C, reactive; and variant D, strongly reactive with Con A (Figure 1).

CRP serum levels were determined by electroimmunoassay.[30] CRP-standard serum (Behringwerke, Germany) was used as a calibrator.

The RC and CRP values were compared using the Mann-Whitney test. The sensitivity and specificity of AGP microheterogeneity and CRP determinations for clinical diagnosis were calculated according to Sox and Liang.[31] The relationship between the RC and disease activity grade was determined using Spearman's rank-order correlation coefficient.

III. RHEUMATIC DISEASES

Proper evaluation of a patient with RA, SLE, or AS depends upon having objective and reproducible measures of the disease activity, and the effectiveness of therapy should be judged by its ability to control the activity. An additional clinical problem which appears during the normal course of rheumatic diseases is intercurrent infection. It has been firmly established that there is a higher rate of mortality among patients with RA because of accompanying infections than is seen in the general population.[32] The increased mortality rate among RA patients is not the result of a higher frequency of infections, but, rather, is probably the result of the more severe course of the infections that occur.[33] Thus, the ability to differentiate an intercurrent infection from severe disease in patients with RA is of great clinical value in the management of the disease. Similarly, in SLE, intercurrent infection with or without fever presents a diagnostic problem. Since fever may be a symptom of active SLE, differentiation of an infection from an exacerbation of disease is often difficult.[34]

Since clinical and laboratory findings of polymyositis-dermatomyositis in the early course of disease may overlap those of polymyalgia rheumatica, the ability to differentiate non-characteristic clinical features of polymyositis-dermatomyositis from polymyalgia rheumatica is important in the diagnosis.

Sera from 280 rheumatic patients and healthy individuals were analyzed (75 patients with SLE, 120 patients with RA, 48 patients with AS, 15 patients with polymyalgia rheumatica, 5 patients with polymyalgia with histological proof of giant cell arteritis, and 17 patients with polymyositis-dermatomyositis). At the time of blood sampling, the patient's disease activity was assessed according to clinical criteria. Patients with SLE were divided into three activity groups:[35] I, involvement of one system in the absence of fever (20 patients); II, involvement of one system and presence of the fever, or involvement of more than one system without fever (18 patients); III, fever and invovlement of at least two systems (22 patients). Patients with RA were assigned to one of four activity groups, based on a multivariate analysis[36] involving morning stiffness, pain scale, grip strength, articular index, hemoglobin concentration, and erythrocyte sedimentation rate: I, inactive (ten patients); II, mild active (25 patients); III, moderately active (36 patients); IV, severe (38 patients). AS patients were divided into two groups: 14 patients with inactive and 34 with active disease.[37] From the group of patients with active disease, four patients having peripheral joint involvement (presence of joint fluid) were distinguished and regarded as a separate group. Polymyositis-dermatomyositis patients were classified into one of four diagnostic categories:[38] I, primary idiopathic polymyositis (two patients); II, primary idiopathic dermatomyositis (eight patients); III, polymyositis with malignancy (two patients); IV, overlap syndromes (five patients). In addition, 15 of the 75 SLE patients and 11 of the 120 RA patients experienced 20 and 15 episodes of intercurrent infection, respectively. Infection was diagnosed by positive culture in most cases and by a clinical picture and therapeutic response that was strongly suggestive of infection in a few cases.

The mean values of AGP-RC in sera from healthy individuals and patients are shown in Table 1. Comparison of the RC values with the activity grades of RA, AS, and SLE showed a significant correlation only for RA ($r = 0.75$, $p < 0.001$). A similar correlation of serum CRP values with grades of RA activity gave a coefficient of $r = 0.69$. The results show that AGP glycoforms appear to be valuable biochemical indicators of RA activity; however, similar information may be gained by CRP measurement. Moreover, like CRP, it does not seem likely that the determination of AGP-RC can be used as an independent indicator. In AS patients, a decrease of RC, similar in direction but of less magnitude than that seen in RA, was found. However, no differences between clinically inactive and active patients were seen. Only in patients who had peripheral joint involvement was a significant decrease of RC observed compared to other cases of AS. The results indicate that determination of AGP microheterogeneity may be a useful discriminator of AS, but has no value as an indicator of disease activity. In the sera of SLE patients, in contrst to RA and AS, there were no changes of AGP glycoform proportions despite disease activity. The RC values in patients' sera were similar to those seen in healthy donors.

Like the SLE in patients with polymyositis-dermatomyositis, RC values were similar to those observed in the sera of healthy individuals.

In infected patients with RA and SLE, a significant increase of RC was found. Intercurrent infection in SLE could be diagnosed with a sensitivity of 90% and specificity of

TABLE 1
α_1-Acid Glycoprotein-Con A Reactivity in Sera of Patients with Rheumatic Diseases

Clinical diagnosis	n	AGP-RC	CRP (median)
Healthy individuals	44	1.3 ± 0.20	0
Rheumatoid arthritis I	10	1.20 ± 0.20[a]	4
Rheumatoid arthritis II	25	1.09 ± 0.11[a]	20
Rheumatoid arthritis III	34	0.90 ± 0.17[a]	47
Rheumatoid arthritis IV	38	0.71 ± 0.14[a]	63
Rheumatoid + infection	15	1.87 ± 0.23[a,b]	60
Ankylosing spondylitis — inactive	14	1.08 ± 0.25[a]	4
Ankylosing spondylitis — active	30	1.01 ± 0.19[a]	14
Ankylosing + peripheral joints	4	0.92 ± 0.20[a]	40
SLE I	20	1.41 ± 0.48	0
SLE II	18	1.40 ± 0.38	8
SLE III	22	1.36 ± 0.40	36
SLE + infection	20	2.81 ±0.76[a,c]	40
Polymyalgia rheumatica	15	0.92 ± 0.17[a]	76
Polymyalgia + arteritis	5	0.91 ± 0.12[a]	68
Polymyositis-dermatomyositis	17	1.38 ± 0.52	6

Note: n, number of sera studied; AGP-RC values are mean ± SD.

[a] Significant difference when compared with healthy individuals.
[b] Significant difference when compared with noninfected RA.
[c] Significant difference when compared with noninfected SLE.

92% when the cut-off RC value was 2.0 (Figure 2). In the same patients, high CRP levels (>60 mg/l) for predicting infection had poor sensitivity (39%).

The differentiation of RA patients with infection from those with active RA (grades III and IV) based on RC gave a sensitivity of 82% and specificity of 91%. CRP levels in these two groups of patients were not significantly different. Although measurement of CRP levels is useful in the assessment of RA activity, it is of limited value in the diagnosis of intercurrent infection in patients with RA. The determination of CRP might be helpful in differentiating healthy persons from RA patients experiencing an infection. When the presence of the fourth AGP variant was considered in the analysis, accompanying infections could have been confirmed in the analysis, accompanying infections could have been confirmed in all patients with RA and SLE. An RC above 1.25 was not found in any of the polymyositis rheumatica patients, but was seen in the sera of 13 of 17 patients with polymyositis-dermatomyositis (sensitivity, 76%; specificity, 100%). In 15 of 17 patients with polymyositis-dermatomyositis, AGP variant D was found (sensitivity, 88%; specificity, 100%). These differences permitted differentiation of both diseases. A similar application of CRP determination gave a sensitivity of 60% and specificity of 68%.

IV. CANCER

Quantitative determination of APP was found to be of limited value in the management of cancer patients, but in some cases was helpful in differentiating between benign and malignant neoplasms, e.g., ovarial tumors, and in evaluating the dissemination of cancer, progression and regression of neoplastic diseases, and effectiveness of therapy, or in prognosis. APP determination was found to be most useful in the detection of intercurrent infections in cancer patients undergoing chemotherapy, e.g., in patients with leukemia.[39-45]

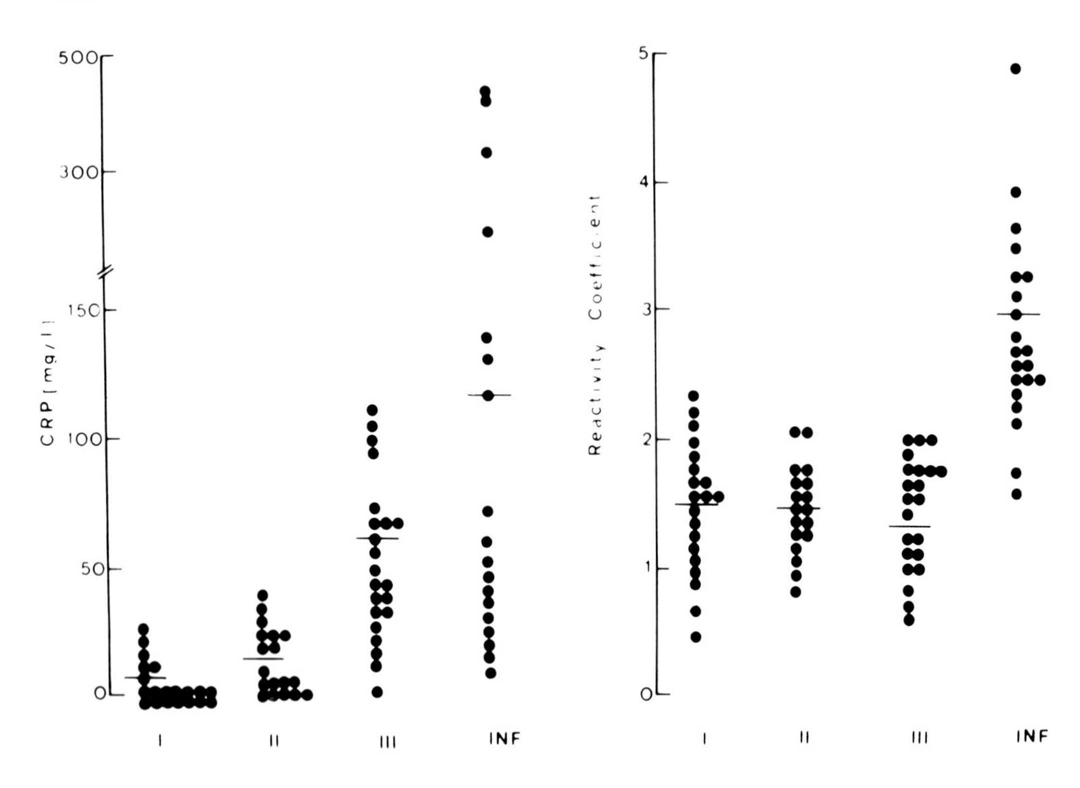

FIGURE 2. α_1-Acid glycoprotein-Con A reactivity coefficient (RC) and C-reactive protein (CRP) in sera of systemic lupus erythematosus patients with various grades of disease activity and intercurrent infection.

Several attempts have been made to apply the determination of AGP glycoforms to the diagnosis and management of cancer. The original reports of Hansen et al.,[46] who differentiated lung cancer from inflammatory lung diseases by AGP glycoform determination, were not confirmed by others.[41] However, recent studies of Hansen et al.[5,6] have demonstrated that the same technique allows differentiation between chronic colorectal inflammation (ulcerative colitis, Crohn's disease) and colorectal cancer. Dobryszycka and Katnik[48] have shown that AGP glycoforms may be helpful in assessing the stage of ovarian carcinoma and evaluating the effectiveness of therapy in stage IV of ovarian cancer. Moreover, they have demonstrated that the determination of haptoglobin microheteogeneity by CAIE with another lectin, wheat germ agglutinin, may provide more precise information for the same purpose.

A. PATIENTS

Serum samples from 265 patients and 44 healthy individuals were studied.

A wide variability was observed in the distribution of AGP glycoforms with no cancer-specific pattern in the sera of patients. Two (A, B) to four (A, B, C, D) AGP glycoforms were seen. In nine patients with cancer (7 of 29 yolk sac tumor, 1 of 16 hepatoma, and 1 of 46 bile duct), "atypical" AGP with γ-mobility in both Con A and control experiments (without a lectin) was seen. However, in defined groups of patients, there were significant changes toward glycoforms being more or less reactive with Con A. The mean values of AGP-RC in sera from healthy individuals and patients are shown in Table 2.

1. Tumors of the Digestive System

Sera from 94 patients with tumors of the digestive system (46 patients with bile duct cancers, 11 with pancreatic cancers, 17 with hepatoma, and 20 with metastatic liver tumors) and from 31 patients with inflammatory disorders of the liver and bile ducts were studied.

TABLE 2
α_1-Acid Glycoprotein Glycoforms in Sera of Cancer Patients

Clinical diagnosis	n	AGP-RC[a]
Healthy adults	44	1.35 ± 0.02
Yolk sac tumor	22	1.75 ± 0.67[b,d]
Hepatoma	16	0.71 ± 0.46[b,c]
Pancreas + bile duct cancers	56	0.91 ± 0.53[b,c]
Metastatic liver tumors	20	1.12 ± 0.44[c,d]
Benign liver disorders	11	0.87 ± 0.34[b,c]
Cholelithiasis	20	1.21 ± 0.70[c,d]
Hydatidiform moles	7	0.53 ± 0.21[b]
Invasive moles	13	0.89 ± 0.22[b,e]
Choriocarcinoma	8	1.21 ± 0.34[e,f]
Myeloblastic leukemia		
Remission	26	1.43 ± 0.45
Relapse	27	1.61 ± 0.54
Remission + infection	7	2.22 ± 0.29[b,g,h]
Relapse + infection	32	1.98 ± 0.58[b,g,h]

Note: n, number of patients studied.

[a] Mean ± SD.
[b] Significant difference when compared with healthy adults.
[c] Significant difference when compared with yolk sac tumor.
[d] Significant difference when compared with hepatoma.
[e] Significant difference when compared with hydatidiform moles.
[f] Significant difference when compared with invasive moles.
[g] Significant difference when compared with remission of leukemia.
[h] Significant difference when compared with relapse of leukemia.

Metastatic tumors originated in lungs in five patients, in stomach in four patients, in large bowel in five patients, in breast in two patients, and in ovaries in four patients. Benign liver disorders included liver cirrhosis (six patients), viral hepatitis (five patients), and bile duct obstruction (cholelithiasis) resulting in periportal fibrosis (20 patients).

In cancers of the digestive system — originating in the liver, pancreas, or bile ducts — there is a decrease in serum AGP-Con A reactivity, while such changes are not seen in metastatic liver tumors. This diversity could be used to differentiate between primary and secondary liver cancer (sensitivity, 80%; specificity, 62%). In patients with liver cirrhosis and patients with viral hepatitis, changes similar to those in hepatoma patients were observed. However, in patients with bile duct obstruction, the AGP glycoform distribution was similar to that seen in healthy individuals. This indicates that differentiation between cancer and liver cirrhosis is impossible, but suggests that this test might be helpful in differentiating between cholelithiasis, hepatoma, and liver cirrhosis.

2. Yolk Sac Tumor

Sera from 29 patients with yolk sac tumor were studied. In 23 cases the cancer originated in testes, in 3 cases in ovaries, and in 3 cases in retroperitoneal tissue. Sera were collected at the time of the first contact with the physician and were α-fetoprotein positive.

A shift of serum AGP glycoforms toward forms more reactive with Con A was found. The RC in the sera of these patients was significantly higher then in patients with cancers of the liver, both primary and secondary, as well as in patients with benign liver disorders. Hence, determination of AGP-RC might be helpful in differentiating between these types of cancers, especially when the AGP serum levels are high.

3. Trophoblastic Diseases

Sera from 28 patients with trophoblastic disease (7 with hydatidiform mole, 13 with invasive moles, and 8 with choriocarcinoma) were studied.

In hydatidiform mole, there was a dramatic decrease of serum AGP-Con A reactivity. Similar changes, but to a lesser extent, were also seen in patients with invasive moles, while in patients with choriocarcinoma, the mean RC vaues were not different from those seen in healthy individuals. Due to the limited number of cases, definitive conclusions about the utility of this test for differentiating between the three trophoblastic diseases cannot be drawn. However, it seems most likely that such differentiation would be possible. Parallel studies of the β-subunit of human chorionic gonadotrpin (β-HCG) in patient sera have demonstrated a significant overlap of results between the groups studies.

4. Acute Myeloblastic Leukemia

Sera from 92 patients with acute myeloblastic leukemia (33 during remission and 59 during relapse) were studied. Intercurrent infections were observed in 7 of 33 patients in remission and in 32 of 59 patients in relapse.

In patients with acute myeloblastic leukemia, the distribution of AGP glycoforms (A, B, and C) was similar to that in healthy individuals. However, intercurrent infections caused a significant shift toward more AGP-Con A-reactive glycoforms. A fourth glycoform (D) was seen in the majority of these patients. Determination of the RC could be used for the detection of intercurrent infections in the course of leukemia (sensitivity, 69% specificity, 78%). Inclusion of glycoform D in the analysis significantly increased the sensitivity (92%) and specificity (100%). However, serum CRP measurement provided better discrimination between these two groups of patients (sensitivity, 100%; specificity,89%).

V. AIDS

Sera from 94 HIV-infected patients were analyzed. Patients were divided into four exclusive groups according to the Centers for Disease Control classification.[49] Group I, acute infection (one patient); group II, asymptomatic infection (21 patients); group III, persistent generalized lymphadenopathy (26 patients); group IV (subgroups a, b, c_1, c_2, and d), other diseases (45 patients). In 17 patients, joints were involved (11 patients had Reiter's syndrome, 4 had psoriatic arthritis, 1 had acute idiopathic oligoarthritis, and 1 had chronic idiopathic arthritis). In sera of patients assigned to groups I, II, III, and IVa and b, the RC was similar to that found in healthy individuals (Table 3). In group IVc_1 and d, especially in patients with *Pneumocystis carini* pneumonia (PCP), a dramatic increase of RC was found. When the cut-off RC value was 2.0, infection could be detected with a specificity of 60% and sensitivity of 96%, while PCP infection with a specificity of 100% and sensitivity of 96%. A similar evaluation of CRP measurement with a 10 mg/l cut-off value gave a sensitivity of 30% and specificity of 85% (for PCP, 20 and 85%, respectively). In contrast, a decrease of RC was observed in patients whose joints were affected, as well as in patients who developed chronic liver diseases such as liver cirrhosis or chronic hepatitis. According to the above classification of glycosylation changes, two types of alterations in the sera of HIV-infected patients were found: type I was observed in patients with secondary infectious diseases and type II in patients who had joint involvement or liver disorders.

TABLE 3
α_1-Acid Glycoprotein Glycoforms in Sera of HIV-Infected Patients

Group	n	RC	CRP Mean ± SD	CRP Median
I	1	1.24	18	18
II	22	1.53 ± 0.68[a]	1.36 ± 3.5	0
III	23	1.4 ± 0.38	4.21 ± 6.9	0
IVa	4	1.45 ± 0.48	0 ± 0	0
+ joints involved	3	0.76 ± 0.23[b,c]	1.5 ± 1.5	0
IVc_1	20	2.21 ± 0.87[b]	9.65 ± 16[g]	4
+ joints involved	7	1.25 ± 0.2[b,d]	3.57 ± 5.1[b]	0
IVc_2	0			
+ joints involved	6	1.27 ± 0.2	23.8 ± 46[b]	5
IVd	3	1.44 ± 0.34	43 ± 73	2

Note: n, number of patients studies.

[a] Mean ± SD.
[b] Significant difference when compared with healthy individuals (in Table 1).
[c] Significant difference when compared with patients (IVa) with no joint involvement.
[d] Significant difference when compared with patients (IVc_1) with no joint involvement.

The high specificity and sensitivity of RC determinations for the detection of secondary infection suggests that it may have potential clinical usefulness. However, when evaluating results, one must consider other factors such as joint involvement or liver diseases.

ACKNOWLEDGMENT

This research has been supported by KBN grants 41121 and 41076.

REFERENCES

1. **Mackiewicz, A., Marcinkowska-Pieta, R., Ballou, S., Mackiewicz, S., and Kushner, I.,** Microheterogeneity of alpha-1-acid glycoprotein in the detection of intercurrent infection in systemic lupus erythematosus, *Arthritis Rheum.*, 30, 513, 1987.
2. **Pawloski, T., Mackiewicz, S., and Mackiewicz, A.,** Microheterogeneity of alpha-1-acid glycoprotein in the detection of intercurrent infection in rheumatoid arthritis, *Arthritis Rheum.*, 32, 347, 1989.
3. **Nicolllet, I., Lebreton, J. P., Fontaine, M., and Hiron, M.,** Evidence for alpha-1-acid glycoprotein populations of different pI values after concanavalin A affinity chromatography: study of their evolution during inflammation in man, *Biochim. Biophys. Acta*, 668, 235, 1981.
4. **Raynes, J.,** Variations in the relative proportions of microheterogeneous forms of plasma glycoproteins in pregnancy and disease, *Biomedicine*, 36, 77, 1982.
5. **Hansen, J. E., Jensen, S. P., Nørgaard-Pedersen, B., and Bøg-Hansen, T. C.,** Electrophoretic analysis of the glycan microheterogeneity of orosomucoid in cancer and inflammation, *Electrophoresis*, 7, 180, 1986.
6. **Hansen, J. E., Iversen, J., Lihme, A., and Bog-Hansen, T. C.,** Acute phase reaction, heterogeneity, and microheterogeneity of serum proteins as non specific tumor markers in lung cancer, *Cancer*, 60, 1630, 1987.

7. **Mackiewicz, A., Pawlowski, T., Mackiewicz-Pawlowska, A., Wiktorowicz, K., and Mackiewicz, S.,** Microheterogeneity forms of alpha-1-acid glycoprotein as indicators of rheumatoid arthritis activity, *Clin. Chim. Acta,* 163, 185, 1987.
8. **Mackiewicz, A., Khan, M. A., Reynolds, T. L., Van der Linden, S., and Kushner, I.,** Serum IgA and glycosylation of α1-acid glycoprotein in ankylosing spondylitis, *Ann. Rheum. Dis.,* 48, 99, 1989.
9. **Breborowicz, J. and Mackiewicz, A.,** Affinity electrophoresis for diagnosis of cancer and inflammatory conditions, *Electrophoresis,* 10, 568, 1989.
10. **Jezequel, M., Seta, N. S., Corbic, M. M., Feger, J. M., and Durand, G. M.,** Modifications of concanavalin A patterns of α1-acid glycoprotein and α2-HS glycorpotein in alcoholic liver disease, *Clin. Chim. Acta,* 176, 49, 1988.
11. **Hachulla, E., Laine, A., and Hayem, A.,** α-1-Antichymotrypsin microheterogeneity in crossed immunoaffinoelectrophoresis with free concanavalin A: a useful diagnostic tool in inflammatory syndrome, *Clin. Chem.,* 34, 911, 1988.
12. **Mackiewicz, A., Pawlowski, T., Wiktorowicz, K., and Mackiewicz, S.,** Microheterogeneity of alpha 1-acid glycoprotein, alpha 1-antichymotrypsin and alpha 1-antitrypsin in rheumatoid arthritis, in *Lectins,* Vol. 5, Bog-Hansen, T. C. and van Driessche, E., Eds., de Gruyter, New York, 1986, 623.
13. **Fassbender, K., Zimmerli, W., Kissling, R., Sobieska, M., Aeschlimann, A., Kellner, M., and Muller, W.,** Glycosylation of α_1-acid glycoprotein in relation to duration of disease in acute and chronic infection and inflammation, *Clin. Chim. Acta,* 203, 315, 1991.
14. **Pawlowski, T., Aeschlimann, A., Kahn, M. F., Vaith, P., Mackiewicz, S. H., and Mueller, W.,** Microheterogeneity of acute phase proteins in the differentiation polymyalgia rheumatica from polymyositis, *J. Rheum.,* 17, 1187, 1990.
15. **Hatton, M. W. C., Marz, L., and Regoeczi, E.,** On the significance of heterogeneity of plasma glycoproteins possessing N-glycans of the complex type: a perspective, *Trends Biochem. Sci.,* 92, 287, 1983.
16. **De Graaf, T. W., Van der Stelt, M. E., Anbergen, W. G., and Van Dijk, W.,** Inflammation induced increase in sialyl-Lewis^x bearing glycans on α1-acid glycoprotein in human sera, *J. Exp. Med.,* in press.
17. **Pepys, M. and Baltz, M. L.,** Acute phase proteins with special reference to C-reactive protein and related proteins (pentaxins) and serum amyloid A protein, *Adv. Immunol.,* 34, 141, 1983.
18. **Mackiewicz, A., Pawlowski, T., Górny, A., and Kushner, I.,** Glycoforms of alpha$_1$-acid glycoprotein in the management of rheumatic patients, in *Affinity Electrophoresis: Principles and Applications,* Bręborowicz, J. and Mackiewicz, A., Eds., CRC Press, Boca Raton, FL, 1992, 229.
19. **Bręborowicz, J., Gorny, A., Drews, K., and Mackiewicz, A.,** Glycoforms of alpha$_1$-acid glycoprotein in cancer, in *Affinity Electrophoresis: Principles and Application,* Bręborowicz, J. and Mackiewicz, A., Eds., CRC Press, Boca Raton, FL, 1992, 191.
20. **Narasimhan, S., Freed, J. C., and Schachter, H.,** The effect of a "bisecting" N-acetylglycosaminyl group on the binding of biantennary, complex oligosaccharides to concanavalin A, Phaseolus vulgaris erythroagglutinin (E-PHA), and Ricinus communis agglutinin (RCA-120) immobilized on agarose, *Carbohydr. Res.,* 149, 65, 1986.
21. **Bierhuizen, M., De Wit, M., Govers, C., Ferwerda, W., Koeleman, C., Pos, O., and Van Dijk, W.,** Glycosylation of three molecular forms of human α1-acid glycoprotein having different interactions with concanavalin A. Variations in the occurrance of di-, tri-, and tetraantenary glycans and the degree of sialylation, *Eur. J. Biochem.,* 175, 387, 1988.
22. **Pos, O., van der Stelt, M. E., Wolbink, G.-J., Nijsten, M. W., van der Tempel, G. L., and van Dijk, W.,** Changes in the serum concentration and Con A reactivity of human α1-acid glycoprotein (AGP) and α1-antitrypsin (AT) in severe burn patients: relationship to IL-6 levels, *Clin. Exp. Immunol.,* 82, 579, 1990.
23. **Seta, N., Lebrun, D., de Cregpy, A., Feger, J., and Durand, G.,** Maternal and newborn serum alpha$_1$-acid glycoprotein at delivery: concentrations and CAIE patterns in the presence of concanavalin A, in *Affinity Electrophoresis: Principles and Application,* Bręborowicz, J. and Mackiewicz, A., Eds., CRC Press, Boca Raton, FL, 1992, 183.
24. **Feger, J., Seta, N., Giai-Brueri, M., and Durand, G.,** Glycoforms of serum alpha$_1$-acid glycoprotein in patients with mediastinal infection, in *Affinity Electrophoresis: Principles and Application,* Bręborowicz, J. and Mackiewicz, A., Eds., CRC Press, Boca Raton, FL, 1992, 257.
25. **Serbource-Goguel Seta, N., Dourand, G., Corbic, M., Agneray, J., and Feger, J.,** Alterations in relative proportions of microheterogeneous forms of human α1-acid glycoprotein in liver disease, *J. Hepatol.,* 2, 245, 1986.
26. **Wells, C., Bøg-Hansen, T. C., Cooper, E. H., and Glass, M. R.,** The use of concanavalin A crossed immuno-affinoelectrophoresis to detect hormone-associated variations in α1-acid glycoprotein, *Clin. Chim. Acta,* 109, 59, 1981.
27. **Bøg-Hansen, T. C.,** Crossed immuno-affinoelectrophoresis: an analytical method to predict the result of affinity chromatography, *Anal. Biochem.,* 56, 480, 1973.

28. **Mackiewicz, A. and Mackiewicz, S.,** Determination of lectin-sugar dissociation constants by agarose affinity electrophoresis, *Anal. Biochem.,* 156, 481, 1986.
29. **Mackiewicz, A. and Bręborowicz, J.,** Determination of dissociation constants of interacting molecules by agarose crossed-affinity immunoelectrophoresis, in *Affinity Electrophoresis: Principles and Application,* Bręborowicz, J. and Mackiewicz, A., Eds., CRC Press, Boca Raton, FL, 1992, 119.
30. **Mackiewicz, A., Wiktorowicz, K., and Mackiewicz, S.,** Comparison of three immunoassays for C-reactive protein determination, *Arch. Immunol. Ther. Exp.,* 33, 693, 1985.
31. **Sox, H. C., Jr. and Liang, M. H.,** The erythrocyte sedimentation rate. Guidelines for rational use, *Ann. Intern. Med.,* 104, 515, 1986.
32. **Baum, J.,** Infection in rheumatoid arthritis, *Arthritis Rheum.,* 14, 135, 1971.
33. **Vanderbroucke, J. P., Kaaks, R., Valkenburg,H. A., Boersma, J. W., Cats, A., Feston, J. J. M., Hartman, A. P., Huber-Bruning, O., Rasker, J. J., and Weber, J.,** Frequency of infections among rheumatoid arthritis patients, before and after disease onset, *Arthritis Rheum.,* 30, 810, 1987.
34. **Rothfield, N.,** Clinical features of systemic lupus erythematosus, in *Textbook of Rheumatology,* 2nd ed., Kelley, W. N., Harris, E. D., Ruddy, S., and Sledge, C. B., Eds., W. B. Saunders, Philadelphia, 1985, 1070.
35. **Rothfield, N. F. and Pace, N.,** Relation of positive LE-cell preparations to activity of lupus erythematosus and corticosteroid therapy, *N. Engl. J. Med.,* 266, 535, 1962.
36. **Mallya, R. K. and Mace, B. E. W.,** The assessment of disease activity in rheumatoic arthritis using multivariate analysis, *Rheumatol. Rehab.,* 20, 14, 1981.
37. **Van der Linden, S., Ferraz, M. B., and Tugwell, P.,** Clinical and functional assessment of ankylosing spondylitis, in *Spine,* Vol. 4, *Ankylosing Spondylitis and Related Spondyloarthropaties,* Khan, M. A., Ed., Hanley & Belfus, Philadelphia, 1990, 583.
38. **Hunder, G. G. and Hazleman, B. L.,** Giant cell arteritis and polymyalgia rheumatica, in *Textbook of Rheumatology,* 2nd ed., Kelley, W. N., Harris, E. D., Ruddy, S., and Sledge, C. B., Eds., W. B. Saunders, Philadelphia, 1985, 1166.
39. **Cooper, E. H. and Stone, J.,** Acute phase proteins in cancer, *Adv. Cancer Res.,* 30, 1, 1979.
40. **Meervaldt, J. H., Haije, W. G., Cooper, E. H., et al.,** Biochemical aids in the monitoring of patients with ovarian cancer, *Gynecol. Oncol.,* 16, 209, 1983.
41. **Raynes, J. G. and Cooper, E. H.,** Comparison of serum amyloid A protein and C-reactive protein concentrations in cancer and non malignant disease, *J. Clin. Pathol.,* 36, 798, 1983.
42. **Barauh, B. D. and Gogol, B. C.,** C-reactive protein in malignant tumors, *Ind. J. Cancer,* 12, 39, 1975.
43. **Trautner, K., Cooper, E. H., Haworth, S., and Ward, A. H.,** An evaluation of serum protein profiles in the long term surveillance of prostatic cancer, *Scand. J. Urol. Nephrol.,* 14, 143, 1980.
44. **Velde, E. R., Berrens, L., Zegers, B. J. M., and Balliieux, R. E.,** Acute phase reactants and complement components as indicators of recurrence in human cervical cancer, *Eur. J. Cancer,* 15, 893, 1971.
45. **Vickers, M.,** Serum haptoglobin. A preoperative detection of metastatic renal carcinoma, *J. Urol.,* 112, 310, 1974.
46. **Hansen, J. E. S., Larsen, V. A., and Bøg-Hansen, T. C.,** The microheterogeneity of α1-acid glycoprotein in inflammatory lung disease, cancer of the lung and normal health, *Clin. Chim. Acta,* 138, 41, 1984.
47. **Bleasby, A. J., Knowles, J. C., and Cooke, N. J.,** Microheterogeneity of α1-acid glycoprotein: lack of discrimination between benign and malignant diseases of the lung, *Clin. Chim. Acta,* 150, 231, 1985.
48. **Dobryszycka, W. and Katnik, I.,** Interaction of haptoglobin with concanavalin A and wheat germ agglutinin. Basic research and clinical applications, in *Affinity Electrophoresis: Principles and Applications,* Bręborowicz, J. and Mackiewicz, A., Eds., CRC Press, Boca Raton, FL, 1992, 211.
49. **U.S. Department of Health and Human Services, Centers for Disease Control,** Classification system for human T-lymphotropic virus type III/lymphadenopathy-associated virus infections, *Ann. Intern. Med.,* 105, 234, 1986.

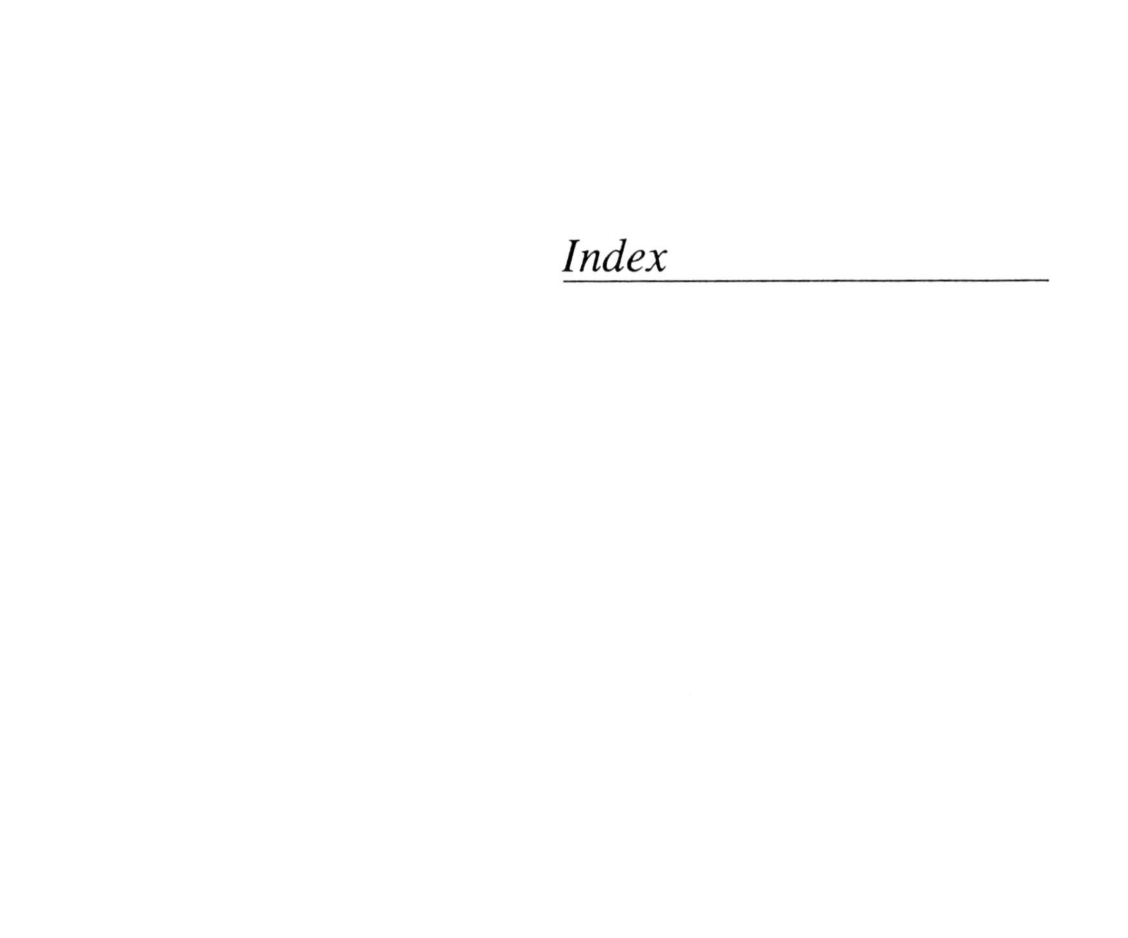

Index

INDEX

A

D

E

F

G

H

I

J

K

L

M

P

R

T

U

V

W

Y